# Strength of Materials

MULTICOLOUR EDITION

# Strength of Materials

## (Mechanics of Solids)

---

(A Textbook for the students of B.E./B.Tech., A.M.I.E.,

U.P.S.C. (Engg. Services) and other Engineering Examinations)

**(SI UNITS)**

R.S. KHURMI

**S. CHAND & COMPANY LTD.**

(AN ISO 9001 : 2000 COMPANY)

**RAM NAGAR, NEW DELHI - 110055**

# S. CHAND & COMPANY LTD.

**(An ISO 9001 : 2000 Company)**

*Head Office* : 7361, RAM NAGAR, NEW DELHI - 110 055

Phones : 23672080-81-82, 9899107446, 9911310888; Fax : 91-11-23677446

Shop at: **schandgroup.com;** E-mail: **schand@vsnl.com**

*Branches* :

- 1st Floor, Heritage, Near Gujarat Vidhyapeeth, Ashram Road, **Ahmedabad**-380 014. Ph. 27541965, 27542369, ahmedabad@schandgroup.com
- No. 6, Ahuja Chambers, 1st Cross, Kumara Krupa Road, **Bangalore**-560 001. Ph : 22268048, 22354008, bangalore@schandgroup.com
- 238-A M.P. Nagar, Zone 1, **Bhopal** - 462 011. Ph : 4274723. bhopal@schandgroup.com
- 152, Anna Salai, **Chennai**-600 002. Ph : 28460026, chennai@schandgroup.com
- S.C.O. 2419-20, First Floor, **Sector- 22**-C (Near Aroma Hotel), **Chandigarh**-160022, Ph-2725443, 2725446, chandigarh@schandgroup.com
- 1st Floor, Bhartia Tower, Badambadi, **Cuttack**-753 009, Ph-2332580; 2332581, cuttack@schandgroup.com
- 1st Floor, 52-A, Rajpur Road, **Dehradun**-248 001. Ph : 2740889, 2740861, dehradun@schandgroup.com
- Pan Bazar, **Guwahati**-781 001. Ph : 2738811, guwahati@schandgroup.com
- Sultan Bazar, **Hyderabad**-500 195. Ph : 24651135, 24744815, hyderabad@schandgroup.com
- Mai Hiran Gate, **Jalandhar** - 144008 . Ph. 2401630, 5000630, jalandhar@schandgroup.com
- A-14 Janta Store Shopping Complex, University Marg, Bapu Nagar, **Jaipur** - 302 015, **Phone : 2719126**, jaipur@schandgroup.com
- 613-7, M.G. Road, Ernakulam, **Kochi**-682 035. Ph : 2378207, cochin@schandgroup.com
- 285/J, Bipin Bihari Ganguli Street, **Kolkata**-700 012. Ph : 22367459, 22373914, kolkata@schandgroup.com
- Mahabeer Market, 25 Gwynne Road, Aminabad, **Lucknow**-226 018. Ph : 2626801, 2284815, lucknow@schandgroup.com
- Blackie House, 103/5, Walchand Hirachand Marg , Opp. G.P.O., **Mumbai**-400 001. Ph : 22690881, 22610885, mumbai@schandgroup.com
- Karnal Bag, Model Mill Chowk, Umrer Road, **Nagpur**-440 032 Ph : 2723901, 2777666 nagpur@schandgroup.com
- 104, Citicentre Ashok, Govind Mitra Road, **Patna**-800 004. Ph : 2300489, 2302100, patna@schandgroup.com
- 291/1, Ganesh Gayatri Complex, 1st Floor, Somwarpeth, Near Jain Mandir, **Pune**-411011. Ph : 64017298, pune@schandgroup.com
- Flat No. 104, Sri Draupadi Smriti Apartment, East of Jaipal Singh Stadium, Neel Ratan Street, Upper Bazar, **Ranchi**-834001. Ph: 2208761, ranchi@schandgroup.com
- Kailash Residency, Plot No. 4B, Bottle House Road, Shankar Nagar, **Raipur.** Ph. 09981200834 raipur@schandgroup.com



Multicolour edition conceptualized by R.K. Gupta, CMD

*First Edition 1968*

*Subsequent Editions and Reprints 1970, 71, 72, 73, 74, 75 (Twice), 76, 77 (Twice), 78 (Twice), 79 (Twice), 80, 81, 82 (Twice), 83, 84 (Twice), 85, 86, 87 (Twice), 88, 89, 90, 91 92, 93, 94, 95, 96, 97, 98, 99, 2000, 2001, 2002, 2003, 2004, 2005, 2006*

***Multicolour Revised Edition 2007,*** *Reprint 2007, 2008 (Thrice)*

*Reprint 2009*

ISBN : 81-219-2822-2

Code : 10 320

PRINTED IN INDIA

*By Rajendra Ravindra Printers (Pvt.) Ltd., 7361, Ram Nagar, New Delhi-110 055 and published by S. Chand & Company Ltd. 7361, Ram Nagar, New Delhi-110 055*

# Preface To The Twenty-Third Edition

It gives a great pleasure in presenting the new multicolour edition of this popular book to innumerable students and academic staff of the Universities in India and abroad. The favourable and warm reception, which the previous editions and reprints of this book have enjoyed all over India and abroad, has been a matter of great satisfaction.

The present edition of this book is in S.I. Units. To make the book really useful at all levels, a number of articles as well as solved and unsolved examples have been added. The mistakes, which had crept in, have been eliminated. Three new chapters of Thick Cylindrical and Spherical Shells, Bending of Curved Bars and Mechanical Properties of Materials have also been added.

Any errors, omissions and suggestions for the improvement of this volume, will be thankfully acknowledged and incorporated in the next edition.

**E-mail :**
khurmieducation@yahoo.com
**Website :**
www.khurmis.com
**Address :**
B-510, New Friends Colony,
New Delhi-110025
**Mobile :** 9810199785

**R.S. KHURMI**
**N. KHURMI**

# Preface To The First Edition

I take an opportunity to present *Strength of Materials* to the students of Degree and Diploma, in general, and A.M.I.E (I) Section 'A' in particular. The object of this book is to present the subject matter in most concise, compact, to the point and lucid manner.

While writing the book, I have always kept in view the examination requirements of the students and various difficulties and troubles, which they face, while studying the subject. I have also, constantly, kept in view the requirements of those intelligent students, who are always keen to increase their knowledge. All along the approach to the subject matter, every care has been taken to deal with each and every topic as well as problem from the fundamentals and in the simplest possible manner, within the mathematical ability of an average student. The subject matter has been amply illustrated by incorporating a good number of solved, unsolved and well graded examples of almost every variety. Most of these examples are taken from the recent examination papers of Indian as well as foreign Universities and professional examining bodies, to make the students, familiar with the types of questions, usually set in their examinations. At the end of each topic, a few exercises have been added, for the students to solve them independently. Answer to these problems have been provided, but it is too much to hope that these are entirely free from errors. At the end of each chapter, *Highlights* have been added, which summarise the main topics discussed in the chapter for quick revision before the examination. In short, it is earnestly hoped that the book will earn the appreciation of the teachers and students alike.

Although every care has been taken to check mistakes and misprints, yet it is difficult to claim perfection. Any errors, omissions and suggestions for the improvement of this volume, brought to my notice, will be thankfully acknowledged and incorporated in the next edition.

**R.S. KHURMI**

# Contents

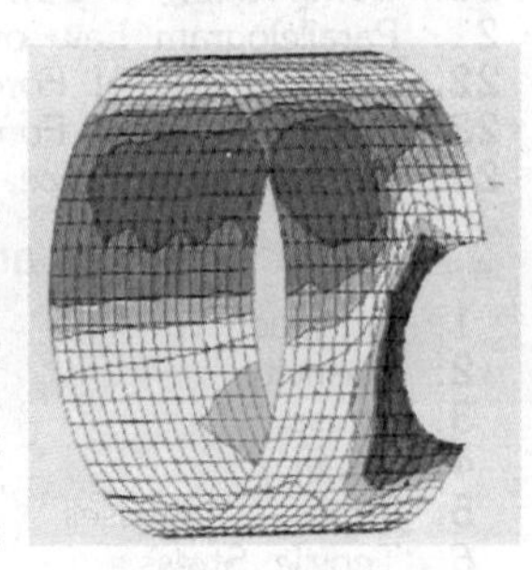

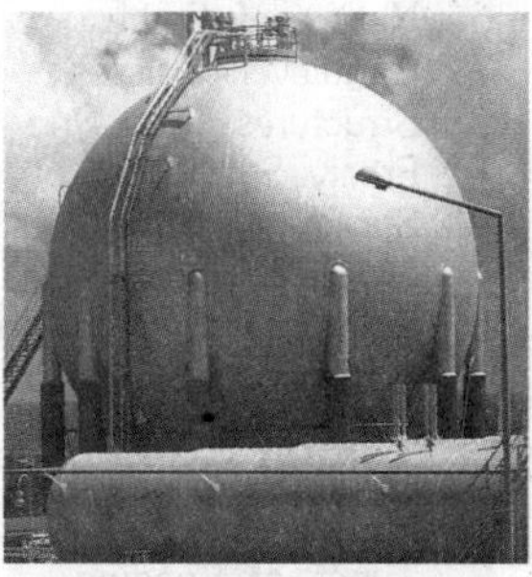

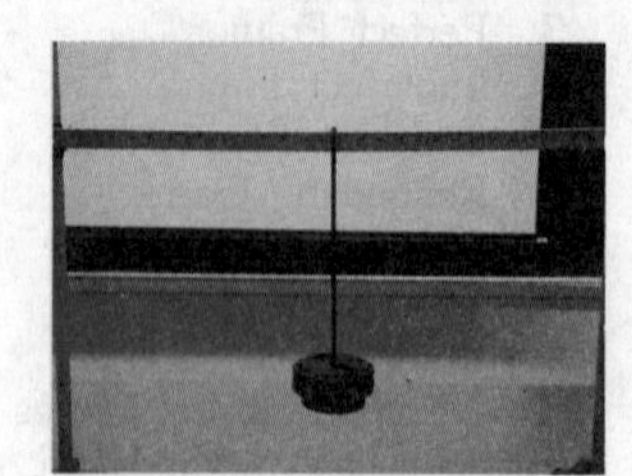

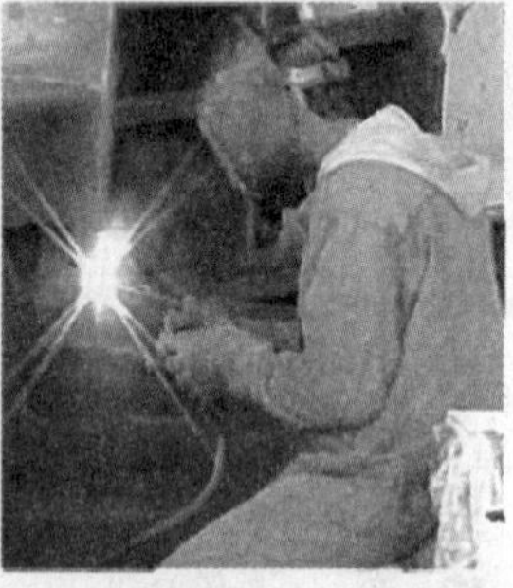

# List of Symbols

| | | |
|---|---|---|
| $A$ | = | Area of cross-section |
| $a$ | = | Rankine's constant |
| $B, b$ | = | Width |
| $C$ | = | Shear modulus of rigidity ($N/mm^2$) |
| $D, d$ | = | Depth |
| | = | Diameter |
| $E$ | = | Young's modulus of elasticity ($N/mm^2$) |
| $e$ | = | Linear strain |
| | = | Eccentricity |
| $G$ | = | Centre of Gravity |
| | = | Centroid of area or lamina |
| $g$ | = | Acceleration due to gravity ($9.81\ m/s^2$) |
| $H, h$ | = | Height (m) |
| $I$ | = | Moment of inertia ($mm^4$) |
| $J$ | = | Polar moment of inertia ($mm^4$) |
| $K$ | = | Bulk modulus of elasticity ($N/mm^2$) |
| $k$ | = | Radius of Gyration |
| $k$ | = | Stiffness of Spring (N/mm) |
| $L, l$ | = | Length (m) |
| $M, m$ | = | Mass (kg) |
| $M$ | = | Bending moment (N-m) |
| $N$ | = | Speed (r.p.m.) |
| $n$ | = | Number |
| $P$ | = | Force (N) |
| $p$ | = | Pressure ($N/mm^2$) |
| $R, r$ | = | Radius |
| $T, t$ | = | Time (s) |
| $T$ | = | Torque (N-m) |
| | = | Twisting Moment |
| $U$ | = | Strain Energy |
| $V$ | = | Volume ($m^3$) |
| $W$ | = | Load or Weight (N) |
| $w$ | = | Load per unit length (N/m) |
| $w$ | = | Specific weight ($kN/m^3$) |
| $x, y, z$ | = | Cartesian co-ordinates |
| $y$ | = | Distance |
| | = | Deflection |
| $Z$ | = | Section modulus |
| $r, \theta$ | = | Polar co-ordinates |
| $\alpha$ | = | Co-efficient of linear expansion (/ °C) |
| $\alpha, \theta, \beta$ | = | Angle (rad) |
| $\mu$ | = | Poisson's ratio or $\left(\frac{1}{m}\right)$ |
| $\eta$ | = | Efficiency |
| $\varepsilon$ | = | Strain |
| $\rho$ | = | Density ($kg/m^3$) |
| $\phi$ | = | Shear strain |
| $i$ | = | Slope |
| $\delta$ | = | Deflection |
| $\Delta$ | = | Deflection |
| $\delta l$ | = | Change in length |
| $\omega$ | = | Angular velocity (rad/s) |
| $\mu$ | = | Co-efficient of friction |
| $\sigma$ | = | Normal stress ($N/mm^2$) |
| $\tau$ | = | Shear stress ($N/mm^2$) |
| $\sigma_c$ | = | Circumferential (or hoop) stress |
| $\sigma_l$ | = | Longitudinal stress |
| $\sigma_r$ | = | Radical stress |
| $\sigma_t$ | = | Tangential stress |
| | = | Tearing stress |
| $\sigma_b$ | = | Bending stress |
| | = | Bearing stress |
| $\sigma_1, \sigma_2, \sigma_3$ | = | Principal streses |
| $\frac{L_e}{k}$ | = | Slenderness ratio |

Chapter 1

# Introduction

## Contents

## 1.1. Definition

In day-to-day work, an engineer comes across certain materials, *i.e.*, steel girders, angle irons, circular bars, cement etc., which are used in his projects. While selecting a suitable material, for his project, an engineer is always interested to know its strength. The strength of a material may be defined as ability, to resist its failure and behaviour, under the action of external forces. It has been observed that, under the action of these forces, the material is first deformed and then its failure takes place. A detailed study of forces and their effects, alongwith some suitable protective measures for the safe working conditions, is known as Strength of Materials. As a matter of fact, such a knowledge is very essential, for an engineer, to enable him, in designing all types of structures and machines.

## 1.2. Fundamental Units

The measurements of physical quantities is one of the most important operations in engineering. Every quantity is measured in terms of some arbitrary, but internationally accepted units, called fundamental units. All the physical quantities, met with in Strength of Materials, are expressed in terms of the following three fundamental quantities :

**1.** Length, **2.** Mass and **3.** Time.

## 1.3. Derived Units

Sometimes, physical quantities are expressed in other units, which are derived from fundamental units, known as derived units, *e.g.*, units of area, velocity, acceleration, pressure, etc.

## 1.4. Systems of Units

Following are only four systems of units, which are commonly used and universally recognised.

**1.** C.G.S. units, **2.** F.P.S. units, **3.** M.K.S. units and **4.** S.I. units.

In this book, we shall use only the S.I. system of units, as the future courses of studies are conducted in this system of units only.

## 1.5. S.I. Units (International System of Units)

The eleventh General Conference* of Weights and Measures has recommended a unified and systematically constituted system of fundamental and derived units for international use. This system of units is now being used in many countries. In India, the Standards of Weights and Measures Act of 1956 (vide which we switched over to M.K.S. units) has been revised to recognise all the S.I. units in industry and commerce.

In this system of units, the †fundamental units are metre (m), kilogram (kg) and second (s) respectively. But there is a slight variation in their derived units. The following derived units will be used in this book :

| | |
|---|---|
| Density (or Mass density) | $kg/m^3$ |
| Force (in Newtons) | $N (= kg.m/s^2)$ |
| Pressure (in Pascals) | $Pa (= N/m^2 = 10^{-6} N/mm^2)$ |
| Stress (in Pascals) | $Pa (= N/m^2 = 10^{-6} N/mm^2)$ |
| Work done (in Joules) | J (= N-m) |
| Power (in Watts) | W (= J/s) |

International metre, kilogram and second are discussed here.

## 1.6. Metre

The international metre may be defined as the shortest distance (at 0°C) between two parallel lines engraved upon the polished surface of the Platinum-Iridium bar, kept at the International Bureau of Weights and Measures at Sevres near Paris.

---

* It is known as General Conference of Weights and Measures (G.C.W.M.). It is an international organisation of which most of the advanced and developing countries (including India) are members. This conference has been ensured the task of prescribing definitions of various units of weights and measures, which are the very basis of science and technology today.

† The other fundamental units are electric current, ampere (A), thermodynamic temperature, kelvin (K) and luminous intensity, candela (cd). These three units will not be used in this book.

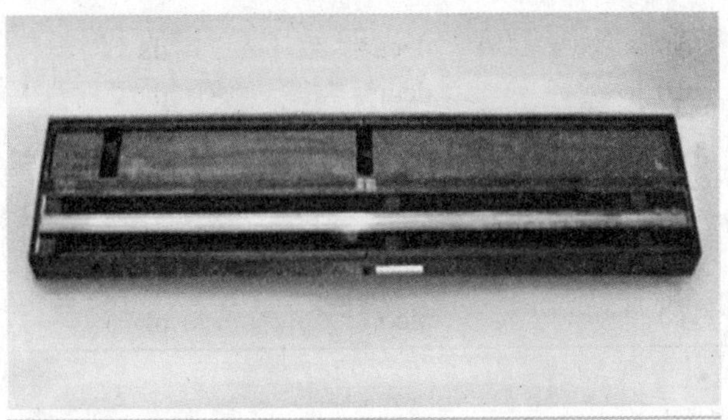

A bar of platinum - iridium metre kept at a temperature of 0º C.

The standard platinum - kilogram is kept at the International Bureau of Weights and Measures at Serves in France.

## 1.7. Kilogram

The international kilogram may be defined as the mass of the Platinum-Iridium cylinder, which is also kept at the International Bureau of Weights and Measures at Sevres near Paris.

## 1.8. Second

The fundamental unit of time for all the four systems is second, which is $1/(24 \times 60 \times 60) = 1/86\ 400$th of the mean solar day. A solar day may be defined as the interval of time between the instants at which the sun crosses the meridian on two consecutive days. This value varies throughout the year. The average of all the solar days, of one year, is called the mean solar day.

## 1.9. Presentation of Units and Their Values

The frequent changes in the present day life are facilitated by an international body known as International Standard Organisation (ISO). The main function of this body is to make recommendations regarding international procedures. The implementation of ISO recommendations in a country is assisted by an organisation appointed for the purpose. In India, Bureau of Indian Standard formerly known as Indian Standards Institution (ISI) has been created for this purpose.

We have already discussed in the previous articles the units of length, mass and time. It is always necessary to express all lengths in metres, all masses in kilograms and all times in seconds. According to convenience, we also use larger multiples or smaller fractions of these units. As a typical example, although metre is the unit of length, yet a smaller length equal to one-thousandth of a metre proves to be more convenient unit especially in the dimensioning of drawings. Such convenient units are formed by using a prefix in front of the basic units to indicate the multiplier. The full list of these prefixes is given in Table 1.1

**TABLE 1.1.**

| *Factor by which the unit is multiplied* | *Standard form* | *Prefix* | *Abbreviation* |
|---|---|---|---|
| 1 000 000 000 000 | $10^{12}$ | Tera | T |
| 1 000 000 000 | $10^{9}$ | giga | G |
| 1 000 000 | $10^{6}$ | mega | M |
| 1 000 | $10^{3}$ | kilo | k |
| 100 | $10^{2}$ | hecto* | h |

| | | | |
|---|---|---|---|
| 10 | $10^{1}$ | deca* | da |
| 0.1 | $10^{-1}$ | deci* | d |
| 0.01 | $10^{-2}$ | centi* | c |
| 0.001 | $10^{-3}$ | milli | m |
| 0.000 001 | $10^{-6}$ | micro | μ |
| 0.000 000 001 | $10^{-9}$ | nano | n |
| 0.000 000 000 001 | $10^{-12}$ | pico | p |

## 1.10. Rules for S.I. Units

The Eleventh General Conference of Weights and Measures recommended only the fundamental and derived units of S.I. system. But it did not elaborate the rules for the usage of these units. Later on, many scientists and engineers held a number of meetings for the style and usage of S.I. units. Some of the decisions of these meetings are :

1. A dash is to be used to separate units, which are multiplied together. For example, a newton-meter is written as N-m. It should not be confused with mN, which stands for millinewton.
2. For numbers having 5 or more digits, the digits should be placed in groups of three separated by spaces (instead of †† commas) counting both to the left and right of the decimal point.
3. In a ††† four digit number, the space is not required unless the four digit number is used in a column of numbers with 5 or more digits.

At the time of revising this book, the author sought the advice of various international authorities regarding the use of units and their values, keeping in view the global reputation of the author as well as his books. It was then decided to †††† present the units and their values as per the recommendations of ISO and ISI. It was decided to use :

| | | | | |
|---|---|---|---|---|
| 4500 | not | 4 500 | or | 4,500 |
| 7 589 000 | not | 7589000 | or | 7,589,000 |
| 0.012 55 | not | 0.01255 | or | .012,55 |
| $30 \times 10^{6}$ | not | $3 \times 10^{7}$ | or | 3,00,00,000 |

The above mentioned figures are meant for numerical values only. Now we shall discuss about the units. We know that the fundamental units in S.I. system for length, mass and time are metre, kilogram and second respectively. While expressing these quantities, we find it time-consuming to write these units such as metres, kilograms and seconds, in full, every time we use them. As a result of this, we find it quite convenient to use the following standard abbreviations, which are internationally recognised. We shall use :

| | |
|---|---|
| m | for metre or metres |
| km | for kilometre or kilometres |
| kg | for kilogram or kilograms |

* The prefixes are generally becoming obsolete probably due to possible confusion. Moreover, it is becoming a conventional practice to use only those powers of ten which confirm to $10^{3n}$ where $n$ is a positive or negative whole number.

† In certain countries, comma is still used as the decimal marker.

††† In certain countries, space is used even in a four digit number.

†††† In some question papers, standard values are not used. The author has tried to avoid such questions in the text of the book, in order to avoid possible confusion. But at certain places, such questions have been included keeping in view the importance of question from the reader's angle.

| | |
|---|---|
| t | for tonne or tonnes |
| s | for second or seconds |
| min | for minute or minutes |
| N | for newton or newtons |
| N-m | for newton × metres (*i.e.*, work done) |
| kN-m | for kilonewton × metres |
| rad | for radian or radians |
| rev | for revolution or revolutions |

## 1.11. Useful Data

The following data summarises the previous memory and formulae, the knowledge of which is very essential at this stage.

## 1.12. Algebra

1. $a^0 = 1$ ; $x^0 = 1$

   (*i.e., Anything raised to the power zero is one.*)

2. $x^m \times x^n = x^{m+n}$

   (*i.e., If the bases are same, in multiplication, the powers are added.*)

3. $\dfrac{x^m}{x_n} = x^{m-n}$

   (*i.e., If the bases are same, in division, the powers are subtracted.*)

4. If $ax^2 + bx + c = 0$

   then $x = \dfrac{-b \pm \sqrt{b^2 - 4ac}}{2a}$

   where $a$ is the coefficient of $x^2$,

   $b$ is the coefficient of $x$ and $c$ is the constant term.

## 1.13. Trigonometry

In a right-angled triangle $ABC$ as shown in Fig. 1.1.

1. $\dfrac{b}{c} = \sin\theta$

2. $\dfrac{c}{a} = \cos\theta$

3. $\dfrac{b}{a} = \dfrac{\sin\theta}{\cos\theta} = \tan\theta$

4. $\dfrac{c}{b} = \dfrac{1}{\sin\theta} = \text{cosec}\,\theta$

5. $\dfrac{c}{a} = \dfrac{1}{\cos\theta} = \sec\theta$

6. $\dfrac{a}{b} = \dfrac{\cos\theta}{\sin\theta} = \dfrac{1}{\tan\theta} = \cot\theta$

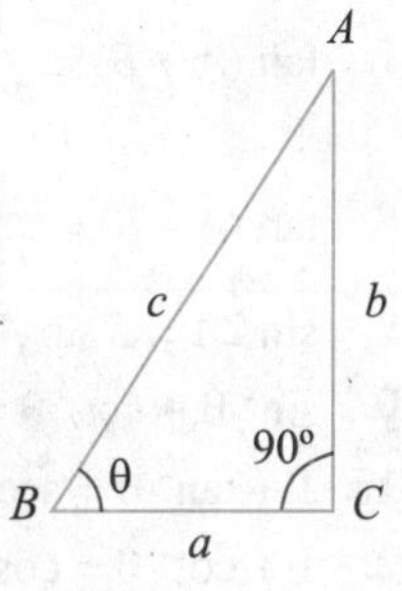


Fig. 1.1

7. The following table shows the values of trigonometrical functions for some typical angles:

| angle | 0° | 30° | 45° | 60° | 90° |
|---|---|---|---|---|---|
| sin | 0 | $\frac{1}{2}$ | $\frac{1}{\sqrt{2}}$ | $\frac{\sqrt{3}}{2}$ | 1 |
| cos | 1 | $\frac{\sqrt{3}}{2}$ | $\frac{1}{\sqrt{2}}$ | $\frac{1}{2}$ | 0 |
| tan | 0 | $\frac{1}{\sqrt{3}}$ | 1 | $\sqrt{3}$ | ∞ |

or in other words, for sin write.

| 0° | 30° | 45° | 60° | 90° |
|---|---|---|---|---|
| $\frac{\sqrt{0}}{2}$ | $\frac{\sqrt{1}}{2}$ | $\frac{\sqrt{2}}{2}$ | $\frac{\sqrt{3}}{2}$ | $\frac{\sqrt{4}}{2}$ |
| 0 | $\frac{1}{2}$ | $\frac{1}{\sqrt{2}}$ | $\frac{\sqrt{3}}{2}$ | 1 |

for cos write the values in reverse order; for tan divide the value of sin by cos for the respective angle.

8. In the first quadrant (*i.e.*, 0° to 90°) all the trigonometrical ratios are positive.
9. In the second quadrant (*i.e.*, 90° to 180°) only sin θ and cosec θ are positive.
10. In the third quadrant (*i.e.*, 180° to 270°) only tan θ and cot θ are positive.
11. In the fourth quadrant (*i.e.*, 270° to 360°) only cos θ and sec θ are positive.
12. In any triangle *ABC*,

$$\frac{a}{\sin A} = \frac{b}{\sin B} = \frac{c}{\sin C}$$

where *a*, *b* and *c* are the lengths of the three sides of a triangle. *A*, *B* and *C* are opposite angles of the sides *a*, *b* and *c* respectively.

13. $\sin(A + B) = \sin A \cos B + \cos A \sin B$.
14. $\sin(A - B) = \sin A \cos B - \cos A \sin B$.
15. $\cos(A + B) = \cos A \cos B - \sin A \sin B$.
16. $\cos(A - B) = \cos A \cos B + \sin A \sin B$.
17. $\tan(A + B) = \frac{\tan A + \tan B}{1 - \tan A . \tan B}$
18. $\tan(A - B) = \frac{\tan A - \tan B}{1 + \tan A . \tan B}$
19. $\sin 2A = 2 \sin A \cos A$.
20. $\sin^2 \theta + \cos^2 \theta = 1$.
21. $1 + \tan^2 \theta = \sec^2 \theta$.
22. $1 + \cot^2 \theta = \text{cosec}^2 \theta$.
23. $\sin^2 A = \frac{1 - \cos 2A}{2}$
24. $\cos^2 A = \frac{1 + \cos 2A}{2}$

**25.** $2 \cos A \sin B = \sin (A + B) - \sin (A - B)$.

**26.** Rules for the change of trigonometrical ratios:

**(A)**

$\sin(-\theta) = -\sin\theta$

$\cos(-\theta) = \cos\theta$

$\tan(-\theta) = -\tan\theta$

$\cot(-\theta) = -\cot\theta$

$\sec(-\theta) = \sec\theta$

$\text{cosec}(-\theta) = -\text{cosec}\,\theta$

**(B)**

$\sin(90° - \theta) = \cos\theta$

$\cos(90° - \theta) = \sin\theta$

$\tan(90° - \theta) = \cot\theta$

$\cot(90° - \theta) = \tan\theta$

$\sec(90° - \theta) = \text{cosec}\,\theta$

$\text{cosec}(90° - \theta) = \sec\theta$

**(C)**

$\sin(90° + \theta) = \cos\theta$

$\cos(90° + \theta) = -\sin\theta$

$\tan(90° + \theta) = -\cot\theta$

$\cot(90° + \theta) = -\tan\theta$

$\sec(90° + \theta) = -\text{cosec}\,\theta$

$\text{cosec}(90° + \theta) = \sec\theta$

**(D)**

$\sin(180° - \theta) = \sin\theta$

$\cos(180° - \theta) = -\cos\theta$

$\tan(180° - \theta) = -\tan\theta$

$\cot(180° - \theta) = -\cot\theta$

$\sec(180° - \theta) = -\sec\theta$

$\text{cosec}(180° - \theta) = \text{cosec}\,\theta$

**(E)**

$\sin(180° + \theta) = -\sin\theta$

$\cos(180° + \theta) = -\cos\theta$

$\tan(180° + \theta) = \tan\theta$

$\cot(180° + \theta) = \cot\theta$

$\sec(180° + \theta) = -\sec\theta$

$\text{cosec}(180° + \theta) = -\text{cosec}\,\theta$

Following are the rules to remember the above 30 formulae :

**Rule 1.** Trigonometrical ratio changes only when the angle is $(90° - \theta)$ or $(90° + \theta)$. In all other cases, trigonometrical ratio remains the same. Following is the law of change:

sin changes into cos and cos changes into sin,
tan changes into cot and cot changes into tan,
sec changes into cosec and cosec changes into sec.

**Rule 2.** Consider the angle $\theta$ to be a small angle and write the proper sign as per formulae 8 to 11 above.

## 1.14. Differential Calculus

1. $\frac{d}{dx}$ is the sign of differentiation.

2. $\frac{d}{dx}(x)^n = nx^{n-1}$ ; $\frac{d}{dx}(x)^8 = 8x^7$, $\frac{d}{dx}(x) = 1$

(*i.e., to differentiate any power of x, write the power before x and subtract one from the power*).

3. $\frac{d}{dx}(C) = 0$ ; $\frac{d}{dx}(7) = 0$

(*i.e., differential coefficient of a constant is zero*).

4. $\frac{d}{dx}(u.\,v) = u.\frac{dv}{dx} + v.\frac{du}{dx}$

[*i.e.*, Differential coefficient of product of any two functions] = [(Ist function × Differential coefficient of second function) + (2nd function × Differential coefficient of first function)]

5. $\frac{d}{dx}\left(\frac{u}{v}\right) = \frac{v.\frac{du}{dx} - u.\frac{dv}{dx}}{v^2}$

[*i.e.*, Differential coefficient of two functions when one is divided by the other] = [(Denominator × Differential coefficient of numerator) – (Numerator × Differential coefficient of denominator) / Square of denominator]

6. Differential coefficient of trigonometrical functions

$\frac{d}{dx}(\sin x) = \cos x$ ; $\frac{d}{dx}(\cos x) = -\sin x$

$\frac{d}{dx}(\tan x) = \sec^2 x$ ; $\frac{d}{dx}(\cot x) = -\text{cosec}^2 x$

$\frac{d}{dx}(\sec x) = \sec x \,.\, \tan x$ ; $\frac{d}{dx}(\text{cosec}\, x) = -\text{cosec}\, x \,.\, \cot x$

**Note.** *The differential coefficient, whose trigonometrical function begins with co, is negative.*

7. If the differential coefficient of a function is zero, the function is either maximum or minimum. *Conversely*, if the maximum or minimum value of a function is required, then differentiate the function and equate it to zero.

## 1.15. Integral Calculus

1. $\int dx$ is the sign of integration.

2. $\int x^n\, dx = \frac{x^{n+1}}{n+1}$ ; $\int x^6\, dx = \frac{x^7}{7}$

(*i.e., to integrate any power of x, add one to the power and divide by the new power*).

3. $\int 7dx = 7x$ ; $\int C\, dx = Cx$

(*i.e., to integrate any constant, multiply the constant by x*).

4. $\int (ax+b)^n \, dx = \dfrac{(ax+b)^{n+1}}{(n+1) \times a}$

(*i.e., to integrate any bracket with power, add one to the power and divide by the new power and also divide by the coefficient of x within the bracket*).

## 1.16. Scalar Quantities

The scalar quantities (or sometimes known as scalars) are those quantities which have magnitude only such as length, mass, time, distance, volume, density, temperature, speed etc.

## 1.17. Vector Quantities

The vector quantities (or sometimes known as vectors) are those quantities which have both magnitude and direction such as force, displacement, velocity, acceleration, momentum etc. Following are the important features of vector quantities :

1. **Representation of a vector.** A vector is represented by a directed line as shown in Fig. 1.2. It may be noted that the length *OA* represents the magnitude of the vector $\overrightarrow{OA}$. The direction of the vector is $\overrightarrow{OA}$ is from *O* (*i.e.*, starting point) to *A* (*i.e.*, end point). It is also known as vector *P*.

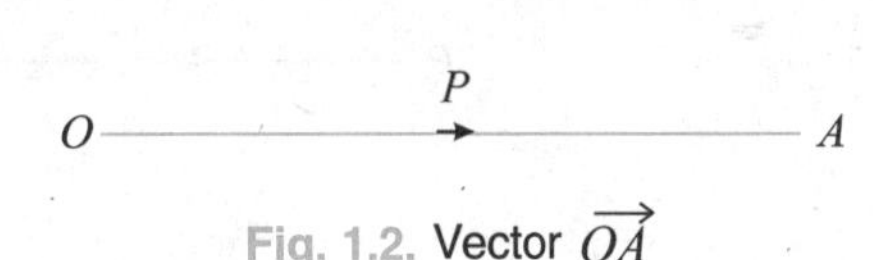


Fig. 1.2. Vector $\overrightarrow{OA}$

2. **Unit vector.** A vector, whose magnitude is unity, is known as unit vector.
3. **Equal vectors.** The vectors, which are parallel to each other and have same direction (*i.e.*, same sense) and equal magnitude are known as equal vectors.
4. **Like vectors.** The vectors, whch are parallel to each other and have same sense but unequal magnitude, are known as like vectors.
5. **Addition of vectors.** Consider two vectors *PQ* and *RS*, which are required to be added as shown in Fig. 1.3 (*a*).

Take a point *A*, and draw line *AB* parallel and equal in magnitude to the vector *PQ* to some convenient scale. Through *B*, draw *BC* parallel and equal to vector *RS* to the same scale. Join *AC* which will give the required sum of vectors *PQ* and *RS* as shown in Fig. 1.3 (*b*).

The velocity of this cyclist is an example of a vector quantity.

This method of adding the two vectors is called the Triangle Law of Addition of Vectors. Similarly, if more than two vectors are to be added, the same may be done first by adding the two vectors, and then by adding the third vector to the resultant of the first two and so on. This method of adding more than two vectors is called Polygon Law of Addition of Vectors.

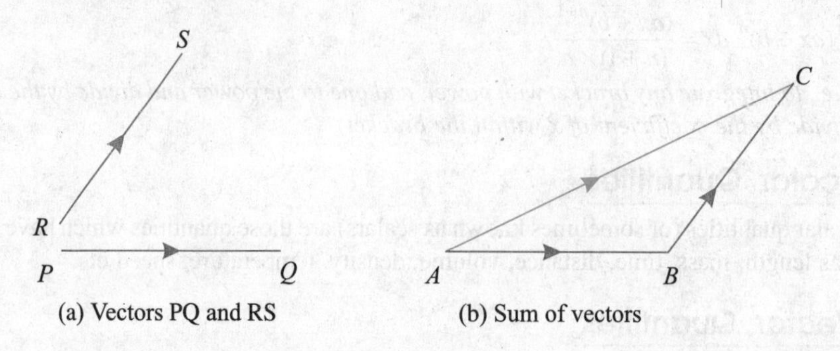


(a) Vectors PQ and RS (b) Sum of vectors

Fig. 1.3

6. **Subtraction of vectors.** Consider two vectors *PQ* and *RS* whose difference is required to be found out as shown in Fig. 1.4 (*a*).

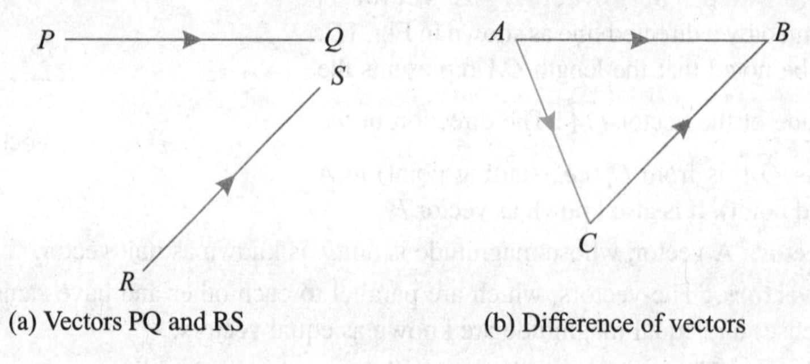


(a) Vectors PQ and RS (b) Difference of vectors

Fig. 1.4

Take a point *A*, and draw line *AB* parallel and equal in magnitude to the vector *PQ* to some convenient scale. Through *B*, draw *BC* parallel and equal to the vector *RS*, but in *opposite direction*, to that of the vector *RS* to the same scale. Join *AC*, which will give the required difference of the vectors *PQ* and *RS* as shown in Fig. 1.4 (*b*).

## 1.18. Force

It is an important factor in the field of Engineering-science, which may be defined as an agent which produces or tends to produce, destroys or tends to destroy motion.

## 1.19. Resultant Force

If a number of forces *P*, *Q*, *R*......... etc., are acting simultaneously on a particle, then a single force, which will produce the same effect as that of all the given forces, is known as a *resultant force*. The forces *P*, *Q*, *R*.... etc., are called component forces. The resultant force of the component forces or the point through which it acts may be found out either mathematically or graphically.

## 1.20. Composition of Forces

It means the process of finding out the resultant force of the given component forces. A resultant force may be found out analytically, graphically or by the following laws :

## 1.21. Parallelogram Law of Forces

It states, "*If two forces acting simultaneously on a particle be represented, in magnitude and direction, by the two adjacent sides of a parallelogram, their resultant may be represented, in magnitude and direction, by the diagonal of the parallelogram passing through the point of their intersection.*"

## 1.22. Triangle Law of Forces

It states, "*If two forces acting simultaneously on a particle be represented in magnitude and direction, by the two sides of a triangle taken in order, their resultant may be represented, in magnitude and direction, by the third side of the triangle taken in opposite order.*"

## 1.23. Polygon Law of Forces

It states, "*If a number of forces acting simultaneously on a particle be represented in magnitude and direction by the sides of a polygon taken in order, their resultant may be represented, in magnitude and direction, by the closing side of the polygon taken in opposite order.*"

## 1.24. Moment of a Force

It is the turning effect, produced by the force, on a body on which it acts. It is mathematically equal to the product of the force and the perpendicular distance between the line of action of the force and the point about which the moment is required.

Chapter 2

# Simple Stresses and Strains

## Contents

## 2.1. Introduction

In our daily life, we see that whenever a load is attached to a thin hanging wire, it elongates and the load moves downwards (sometimes through a negligible distance). The amount, by which the wire elongates, depends upon the amount of load and the nature as well as cross-sectional area of the wire material. It has been experimentally found that the cohesive force, between molecules of the hanging wire, offers resistance against the deformation, and the force of resistance increases with the deformation. It has also been observed that the process of deformation stops when the force of resistance is equal to the external force (*i.e.*, the load attached). Sometimes, the force of resistance, offered by the molecules, is less than the external force. In such a case, the deformation continues until failure takes place.

In the succeeding pages, we shall discuss the effects produced by the application of loads, on the materials. Before entering into the details of the effects, following few terms should be clearly understood at this stage.

## 2.2. Elasticity

We have already discussed in the last article that whenever a force acts on a body, it undergoes some deformation and the molecules offer some resistance to the deformation. It will be interesting to know that when the external force is removed, the force of resistance also vanishes ; and the body springs back to its original position. But it is only possible, if the deformation, caused by the external force, is within a certain limit. Such a limit is called elastic limit. The property of certain materials of returning back to their original position, after removing the external force, is known as elasticity. A body is said to be perfectly elastic, if it returns back completely to its original shape and size, after the removal of external forces. If the body does not return back completely to its original shape and size, after the removal of the external force, it is said to be partially elastic.

It has been observed that if the force, acting on a body, causes its deformation beyond the elastic limit, the body loses, to some extent, its property of elasticity. If the external force, after causing deformation beyond the elastic limit, is completely removed, the body will not return back to its original shape and size. There will be some residual deformation to the body, which will remain permanently.

## 2.3. Stress

Every material is elastic in nature. That is why, whenever some external system of forces acts on a body, it undergoes some deformation. As the body undergoes deformation, its molecules set up some resistance to deformation. This resistance per unit area to deformation, is known as stress. Mathematically stress may be defined as the force per unit area *i.e.*, stress.

$$\sigma = \frac{P}{A}$$

where $P$ = Load or force acting on the body, and

$A$ = Cross-sectional area of the body.

In S.I. system, the unit of stress is pascal (Pa) which is equal to 1 N/m$^2$. In actual practice, we use bigger units of stress *i.e.*, megapascal (MPa) and gigapascal (GPa), which is equal to N/mm$^2$ or kN/mm$^2$ respectively.

## 2.4. Strain

As already mentioned, whenever a single force (or a system of forces) acts on a body, it undergoes some deformation. This deformation per unit length is known as strain. Mathematically strain may be defined as the deformation per unit length. *i.e.*, strain

$$\varepsilon = \frac{\delta l}{l} \quad \text{or} \quad \delta l = \varepsilon . l$$

where $\delta l$ = Change of length of the body, and

$l$ = Original length of the body.

## 2.5. Types of Stresses

Though there are many types of stresses, yet the following two types of stresses are important from the subject point of view :

1. Tensile stress. 2. Compressive stress.

## 2.6. Tensile Stress

When a section is subjected to two equal and opposite pulls and the body tends to increase its length, as shown in Fig. 2.1, the stress induced is called tensile stress. The corresponding strain is called tensile strain. As a result of the tensile stress, the *cross-sectional area of the body gets reduced.

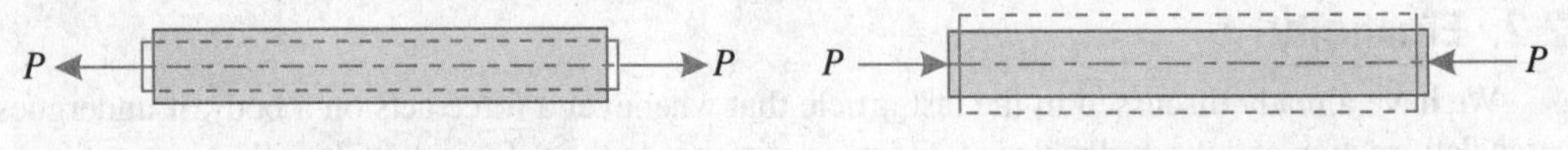


Fig. 2.1. Tensile stress　　Fig. 2.2. Compressive stress

## 2.7. Compressive Stress

When a section is subjected to two equal and opposite pushes and the body tends to shorten its length, as shown in Fig. 2.2, the stress induced is called compressive stress. The corresponding strain is called compressive strain. As a result of the compressive stress, the cross-sectional area of the body gets increased.

## 2.8. Elastic Limit

We have already discussed that whenever some external system of forces acts on a body, it undergoes some deformation. If the external forces, causing deformation, are removed the body springs back to its original position. It has been found that for a given section there is a limiting value of force up to and within which, the deformation entirely disappears on the removal of force. The value of intensity of stress (or simply stress) corresponding to this limiting force is called elastic limit of the material.

Beyond the elastic limit, the material gets into plastic stage and in this stage the deformation does not entirely disappear, on the removal of the force. But as a result of this, there is a residual deformation even after the removal of the force.

## 2.9. Hooke's Law**

It states, "*When a material is loaded, within its elastic limit, the stress is proportional to the strain.*" Mathematically,

$$\frac{\text{Stress}}{\text{Strain}} = E = \text{Constant}$$

It may be noted that Hooke's Law equally holds good for tension as well as compression.

* Since the volume of the body remains constant, therefore an increase in the length will automatically reduce the cross-sectional area of the body. Similarly a decrease in the length will automatically increase the cross-sectional area of the body.

As a matter of fact, there is a relationship between the increase (or decrease) in length of the body and decrease (or increase) in the cross-sectional area of the body. This relation will be discussed in Art. 6.6.

** Named after Robert Hooke, who first established it by experiments in 1678. While making tensile tests on a metallic bar, he took enough precautions, to ensure that the force is applied axially and the bending of the bar is prevented. He assumed that during tension, all the longitudinal fibres of the bar have the same elongation. All the cross-sections of the bar, which were originally plane, remain so even after extension.

## 2.10. Modulus of Elasticity or Young's Modulus (E)

We have already discussed that whenever a material is loaded, within its elastic limit, the stress is proportional to strain. Mathematically stress,

$$\sigma \propto \varepsilon$$

$$= E \times \varepsilon$$

or $$E = \frac{\sigma}{\varepsilon}$$

$\sigma$ = Stress,

$\varepsilon$ = Strain, and

$E$ = A constant of proportionality known as modulus of elasticity or Young's modulus. Numerically, it is that value of tensile stress, which when applied to a uniform bar will increase its length to double the original length if the material of the bar could remain perfectly elastic throughout such an excessive strain.

Young's Modulus Appratus

**TABLE 2.1.**

The value of $E$ (*i.e.*, modulus of elasticity) of materials, in everyday use, are given below :

| S. No. | Material | Modulus of elasticity (E) in GPa i.e. GN/m² or kN/mm² | | |
|---|---|---|---|---|
| 1. | Steel | 200 | to | 220 |
| 2. | Wrought iron | 190 | to | 200 |
| 3. | Cast iron | 100 | to | 160 |
| 4. | Copper | 90 | to | 110 |
| 5. | Brass | 80 | to | 90 |
| 6. | Aluminium | 60 | to | 80 |
| 7. | Timber | 10 | | |

## 2.11. Deformation of a Body Due to Force Acting on it

Consider a body subjected to a tensile stress.

Let $P$ = Load or force acting on the body,

$l$ = Length of the body,

$A$ = Cross-sectional area of the body,

$\sigma$ = Stress induced in the body,

$E$ = Modulus of elasticity for the material of the body,

$\varepsilon$ = Strain, and

$\delta l$ = Deformation of the body.

We know that the stress

$$\sigma = \frac{P}{A} \qquad \text{Strain,} \qquad \varepsilon = \frac{\sigma}{E} = \frac{P}{AE}$$

and deformation,

$$\delta l = \varepsilon . l = \frac{\sigma . l}{E} = \frac{Pl}{AE} \qquad \ldots \left( \because \sigma = \frac{P}{A} \right)$$

**Notes:** **1.** The above formula holds good for compressive stress also.

**2.** For most of the structural materials, the modulus of elasticity for compression is the same as that for tension.

**3.** Sometimes in calculations, the tensile stress and tensile strain are taken as positive, whereas compressive stress and compressive strain as negative.

**EXAMPLE 2.1.** *A steel rod 1 m long and 20 mm × 20 mm in cross-section is subjected to a tensile force of 40 kN. Determine the elongation of the rod, if modulus of elasticity for the rod material is 200 GPa.*

**SOLUTION.** Given : Length ($l$) = 1 m = 1 × $10^3$ mm ; Cross-sectional area ($A$) = 20 × 20 = 400 $mm^2$ ; Tensile force ($P$) = 40 kN = 40 × $10^3$ N and modulus of elasticity ($E$) = 200 GPa = 200 × $10^3$ $N/mm^2$.

We know that elongation of the road,

$$\delta l = \frac{P.l}{A.E} = \frac{(40 \times 10^3) \times (1 \times 10^3)}{400 \times (20 \times 10^3)} = 0.5 \text{ mm} \qquad \textbf{Ans.}$$

**EXAMPLE 2.2.** *A hollow cylinder 2 m long has an outside diameter of 50 mm and inside diameter of 30 mm. If the cylinder is carrying a load of 25 kN, find the stress in the cylinder. Also find the deformation of the cylinder, if the value of modulus of elasticity for the cylinder material is 100 GPa.*

**SOLUTION.** Given : Length ($l$) = 2 m = 2 × $10^3$ mm ; Outside diameter ($D$) = 50 mm ; Inside diameter ($d$) = 30 mm ; Load ($P$) = 25 kN = 25 × $10^3$ N and modulus of elasticity ($E$) = 100 GPa = 100 × $10^3$ $N/mm^2$.

***Stress in the cylinder***

We know that cross-sectional area of the hollow cylinder.

$$A = \frac{\pi}{4} \times (D^2 - d^2) = \frac{\pi}{4} \times [(50)^2 - (30)^2] = 1257 \text{ mm}^2$$

and stress in the cylinder,

$$\sigma = \frac{P}{A} = \frac{25 \times 10^3}{1257} = 19.9 \text{ N/mm}^2 = 19.9 \text{ MPa} \qquad \textbf{Ans.}$$

***Deformation of the cylinder***

We also know that deformation of the cylinder,

$$\delta l = \frac{P.l}{A.E} = \frac{(25 \times 10^3) \times (2 \times 10^3)}{1257 \times (100 \times 10^3)} = 0.4 \text{ mm} \qquad \textbf{Ans.}$$

**EXAMPLE 2.3.** *A load of 5 kN is to be raised with the help of a steel wire. Find the minimum diameter of the steel wire, if the stress is not to exceed 100 MPa.*

**SOLUTION.** Given : Load ($P$) = 5 kN = 5 × $10^3$ N and stress ($\sigma$) = 100 MPa = 100 $N/mm^2$

Let $\qquad d$ = Diameter of the wire in mm.

We know that stress in the steel wire ($\sigma$),

$$100 = \frac{P}{A} = \frac{5 \times 10^3}{\frac{\pi}{4} \times (d)^2} = \frac{6.366 \times 10^3}{d^2}$$

$$\therefore \quad d^2 = \frac{6.366 \times 10^3}{100} = 63.66 \quad \text{or} \quad d = 7.98 \text{ say } 8 \text{ mm} \quad \textbf{Ans.}$$

**EXAMPLE 2.4.** *In an experiment, a steel specimen of 13 mm diameter was found to elongate 0.2 mm in a 200 mm gauge length when it was subjected to a tensile force of 26.8 kN. If the specimen was tested within the elastic range, what is the value of Young's modulus for the steel specimen ?*

**SOLUTION.** Given : Diameter ($d$) = 13 mm ; Elongation ($\delta l$) = 0.2 mm ; Length ($l$) = 200 mm and Force ($P$) = 26.8 kN.

Let $E$ = Value of Young's modulus for the steel specimen.

We know that cross-sectional area of the specimen.

$$A = \frac{\pi}{4} \times (d)^2 = \frac{\pi}{4} \times (13)^2 = 132.73 \text{ mm}^2$$

and elongation of the specimen ($\delta l$)

$$0.2 = \frac{P.l}{A.E} = \frac{26.8 \times 20}{132.73\, E} = \frac{40.38}{E}$$

$$\therefore \quad E = \frac{40.38}{0.2} = 201.9 \text{ kN/mm}^2 = 201.9 \text{ GPa} \quad \textbf{Ans.}$$

**EXAMPLE 2.5.** *A hollow steel tube 3.5 m long has external diameter of 120 mm. In order to determine the internal diameter, the tube was subjected to a tensile load of 400 kN and extension was measured to be 2 mm. If the modulus of elasticity for the tube material is 200 GPa, determine the internal diameter of the tube.*

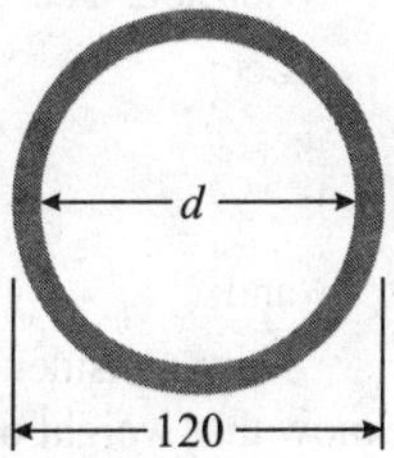

**SOLUTION.** Given : Length ($l$) = 3.5 m = 3.5 × $10^3$ mm ; External diameter ($D$) = 120 mm ; Load ($P$) = 400 kN = 400 × $10^3$ N; Extension ($\delta l$) = 2 mm and modulus of elasticity $E$ = 200 GPa = 200 × $10^3$ N/mm$^2$.

d

120

Fig. 2.3

Let $d$ = Internal diameter of the tube in mm.

We know that area of the tube,

$$A = \frac{\pi}{4} [(120)^2 - d^2] = 0.7854 [(120)^2 - d^2]$$

and extension of the tube ($\delta l$),

$$2 = \frac{P.l}{A.E} = \frac{(400 \times 10^3) \times (3.5 \times 10^3)}{0.7854 [(120)^3 - d^2\ (200 \times 10^3)} = \frac{8913}{14400 - d^2}$$

$$\therefore \quad 28800 - 2d^2 = 8913 \quad \text{or} \quad 2d^2 = 28800 - 8913 = 19887$$

$$\text{or} \quad d^2 = \frac{19887}{2} = 9943.5 \quad \text{or} \quad d = 99.71 \text{ mm} \quad \textbf{Ans.}$$

**EXAMPLE 2.6.** *Two wires, one of steel and the other of copper, are of the same length and are subjected to the same tension. If the diameter of the copper wire is 2 mm, find the diameter of the steel wire, if they are elongated by the same amount. Take E for steel as 200 GPa and that for copper as 100 GPa.*

**SOLUTION.** Given: Diameter of copper wire ($d_C$) = 2 mm ; Modulus of elasticity for steel ($E_S$) = 200 GPa = 200 × $10^3$ N/mm$^2$ and modulus of elasticity for Copper ($E_C$) = 100 GPa = 100 × $10_3$ N/mm$^2$.

Let $d_S$ = Diameter of the steel wire,

$l$ = Lengths of both the wires and

$P$ = Tension applied on both the wires.

We know that area of the copper wire,

$$A_C = \frac{\pi}{4} \times (d_C)^2 = \frac{\pi}{4} \times (2)^2 = 3.142 \text{ mm}^2$$

and area of steel wire, $$A_S = \frac{\pi}{4} \times (d_S)^2 = 0.7854\, d_S^2 \text{ mm}^2$$

We also know that increase in the length of the copper wire

$$\delta l_C = \frac{Pl}{A_C E_C} = \frac{Pl}{3.142 \times (100 \times 10^3)} = \frac{Pl}{314.2 \times 10^3} \quad ...(i)$$

and increase in the length of the steel wire,

$$\delta l_S = \frac{Pl}{A_S E_S} = \frac{Pl}{0.7854\, d_S^2 \times (200 \times 10^3)} = \frac{Pl}{157.1 \times 10^3 \times d_S^2} \quad ...(ii)$$

Since both the wires are elongated by the same amount, therefore equating equations (*i*) and (*ii*).

$$\frac{Pl}{314.2 \times 10^3} = \frac{Pl}{157.1 \times 10^3 \times d_S^2} \quad \text{or} \quad d_S^2 = \frac{314.2}{157.1} = 2$$

$$\therefore \quad d_S = \sqrt{2} = 1.41 \text{ mm} \quad \textbf{Ans.}$$

## 2.12. Deformation of a Body Due to Self Weight

Consider a bar $AB$ hanging freely under its own weight as shown in Fig. 2.4.

Let $l$ = Length of the bar.

$A$ = Cross-sectional area of the bar.

$E$ = Young's modulus for the bar material,

and $w$ = Specific weight of the bar material.

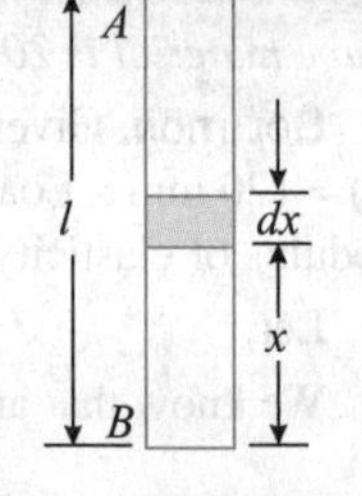

Fig. 2.4

Now consider a small section $dx$ of the bar at a distance $x$ from $B$. We know that weight of the bar for a length of $x$,

$$P = wAx$$

$\therefore$ Elongation of the small section of the bar, due to weight of the bar for a small section of length $x$,

$$= \frac{Pl}{AE} = \frac{(wAx).dx}{AE} = \frac{wx.dx}{E}$$

Total elongation of the bar may be found out by integrating the above equation between zero and $l$. Therefore total elongation,

$$\delta l = \int_0^l \frac{wx.dx}{E}$$

$$= \frac{w}{E} \int_0^l x.dx$$

$$= \frac{w}{E} \left[ \frac{x^2}{2} \right]_0^l$$

or $$\delta l = \frac{wl^2}{2E} = \frac{Wl}{2AE} \quad ...(\because W = wAl = \text{Total weight})$$

**Note.** From the above result, we find that the deformation of the bar, due its own weight, is equal to half of the deformation, if the same body is subjected to a direct load equal to the weight of the body.

**EXAMPLE 2.7.** *A copper alloy wire of 1.5 mm diameter and 30 m long is hanging freely from a tower. What will be its elongation due to self weight? Take specific weight of the copper and its modulus of elasticity as 89.2 kN/m³ and 90 GPa respectively.*

**SOLUTION.** Given: Diameter $(d) = 1.5$ mm ; Length $(l) = 30 \text{ m} = 30 \times 10^3$ mm ; Specific weight $(w) = 89.2 \text{ kN/m}^3 = 89.2 \times 10^{-9} \text{ kN/mm}^3 = 89.2 \times 10^{-6} \text{ N/mm}^3$ and modulus of elasticity $(E) = 90$ GPa $= 90 \times 10^3 \text{ N/mm}^2$.

We know that elongation of the wire due to self weight,

$$\delta l = \frac{wl^2}{2E} = \frac{(89.2 \times 10^{-6}) \times (30 \times 10^3)}{2 \times (90 \times 10^3)} = 0.45 \text{ mm} \quad \textbf{Ans.}$$

**EXAMPLE 2.8.** *An alloy wire of 2 mm² cross-sectional area and 12 N weight hangs freely under its own weight. Find the maximum length of the wire, if its extension is not to exceed 0.6 mm. Take E for the wire material as 150 GPa.*

**SOLUTION.** Given: Cross-sectional area $(A) = 2 \text{ mm}^2$; Weight $(W) = 12$ N ; Extension $(\delta l) = 0.6$ mm and modulus of elasticity $(E) = 150 \text{ GPa} = 150 \times 10^3 \text{ N/mm}^2$.

Let $l$ = Maximum length of the wire,

We know that extension of the wire under its own weight,

$$0.6 = \frac{Wl}{2AE} = \frac{12 \times l}{2 \times 2 \times (150 \times 10^3)} = 0.02 \times 10^{-3}\, l$$

$$l = \frac{0.6}{0.02 \times 10^{-3}} = 30000 \text{ mm} = 30 \text{ m} \quad \textbf{Ans.}$$

**EXAMPLE 2.9.** *A steel wire ABC 16 m long having cross-sectional area of 4 mm² weighs 20 N as shown in Fig. 2.5. If the modulus of elasticity for the wire material is 200 GPa, find the deflections at C and B.*

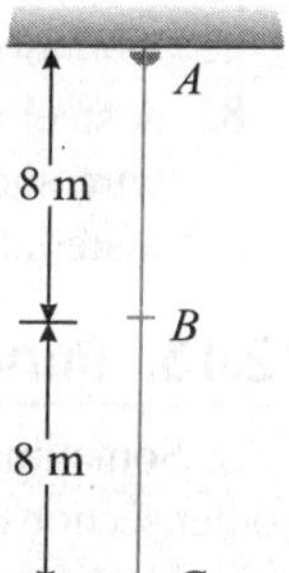

**SOLUTION.** Given: Length $(l) = 16 \text{ m} = 16 \times 10^3$ mm ; Cross-sectional area $(A) = 4 \text{ mm}^2$ ; Weight of the wire $ABC$ $(W) = 20$ N and modulus of elasticity $(E) = 200$ GPa $= 200 \times 10^3 \text{ N/mm}^2$.

Fig. 2.5

***Deflection at C***

We know that deflection of wire at $C$ due to self weight of the wire $AC$,

$$dl_C = \frac{Wl}{2AE} = \frac{20 \times (16 \times 10^3)}{2 \times 4 \times (200 \times 10^3)} = 0.2 \text{ mm} \quad \textbf{Ans.}$$

***Deflection at B***

We know that the deflection at $B$ consists of deflection of wire $AB$ due to self weight plus deflection due to weight of the wire $BC$. We also know that deflection of the wire at $B$ due to self weight of wire $AB$

$$\delta l_1 = \frac{(W/2) \times (l/2)}{2AE} = \frac{10 \times (8 \times 10^3)}{2 \times 4 \times (200 \times 10^3)} = 0.05 \text{ mm} \quad ...(i)$$

and deflection of the wire at $B$ due to weight of the wire $BC$.

$$\delta l_2 = \frac{(W/2)\times(l/2)}{AE} = \frac{10\times(8\times10^3)}{4\times(200\times10^3)} = 0.1 \text{ mm} \quad ...(ii)$$

∴ Total deflection of the wire at $B$.

$$\delta l_B = \delta l_1 + \delta l_2 = 0.05 + 0.1 = 0.15 \text{ mm} \quad \textbf{Ans.}$$

## EXERCISE 2.1

1. A steel bar 2 m long and 150 mm$^2$ in cross-section is subjected to an axial pull of 15 kN. Find the elongation of the bar. Take $E$ = 200 GPa. **[Ans.** 1.0 mm]
2. A straight bar of 500 mm length has its cross-sectional area of 500 mm$^2$. Find the magnitude of the compressive load under which it will decrease its length by 0.2 mm. Take $E$ for the bar material as 200 GPa. **[Ans.** 40 kN]
3. An alloy bar 1 m long and 200 mm$^2$ in cross-section is subjected to a compressive force of 20 kN. If the modulus of elasticity for the alloy is 100 GPa, find the decrease in length of the bar. **[Ans.** 1 mm]
4. A hollow cylinder 4 m long has outside and inside diameters of 75 mm and 60 mm respectively. Find the stress and deformation of the cylinder, when it is carrying an axial tensile load of 50 kN. Take $E$ = 100 GPa. **[Ans.** 31.4 MPa; 1.26 mm]
5. A hollow cast iron column has internal diameter of 200 mm. What should be the external diameter of the column, so that it can carry a load of 1.6 MN without the stress exceeding 90 MPa. **[Ans.** 250 mm]
6. A brass rod 1.5 m long and 20 mm diameter was found to deform 1.9 mm under a tensile load of 40 kN. Calculate the modulus of elasticity of the rod. **[Ans.** 100.5 GPa]
7. A steel wire of 80 m length and 1 mm$^2$ cross-sectional area is freely hanging from a tower. What will be its elongation due to its self weight? Take specific weight of the steel as 78.6 kN/m$^3$ and modulus of elasticity as 200 GPa. **[Ans.** 1.3 mm]
8. A steel wire of 1 mm diameter is freely hanging under its own weight. If the extension of the wire should not exceed 2.5 mm, what should be its maximum length? Take $E$ for the wire material as 200 GPa and its specific weight as 78.5 kN/m$^3$. **[Ans.** 112.87 mm]

## 2.13. Principle of Superposition

Sometimes, a body is subjected to a number of forces acting on its outer edges as well as at some other sections, along the length of the body. In such a case, the forces are split up and their effects are considered on individual sections. The resulting deformation, of the body, is equal to the algebraic sum of the deformations of the individual sections. Such a principle, of finding out the resultant deformation, is called the principle of superposition.

The relation for the resulting deformation may be modified as:

$$\delta l = \frac{Pl}{AE} = \frac{1}{AE}(P_1 l_1 + P_2 l_2 + P_3 l_3 + ...)$$

where $P_1$ = Force acting on section 1,

$l_1$ = Length of section 1,

$P_2, l_2$ = Corresponding values of section 2, and so on.

**EXAMPLE 2.10.** *A steel bar of cross-sectional area 200 mm$^2$ is loaded as shown in Fig. 2.6. Find the change in length of the bar. Take E as 200 GPa.*

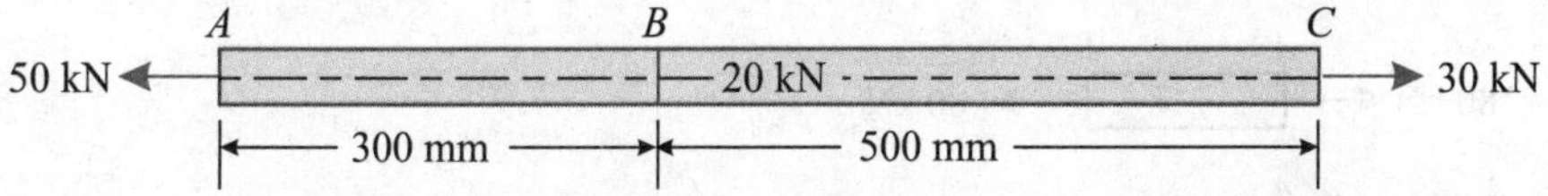

Fig. 2.6

*Find the change in length of the bar. Take E as 200 GPa.*

**SOLUTION.** Given: Cross-sectional area ($A$) = 200 mm$^2$ and modulus of elasticity ($E$) = 200 GPa = $200 \times 10^3$ N/mm$^2$.

For the sake of simplification, the force of 50 kN acting at $A$ may be split up into two forces of 20 kN and 30 kN respectively.

Now it will be seen that part $AB$ of the bar is subjected to a tension of 20 kN and $AC$ is subjected to a tension of 30 kN as shown in *Fig. 2.7.

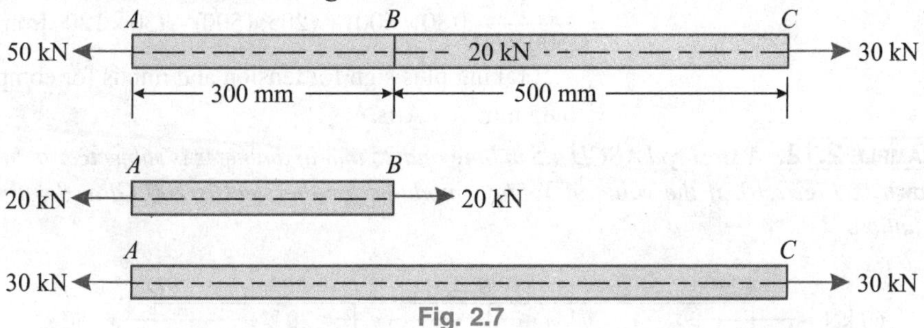


Fig. 2.7

We know that change in length of the bar.

$$\delta l = \frac{1}{AE}(P_1 l_1 + P_2 l_2)$$

$$= \frac{1}{200 \times 200 \times 10^3}\left[[(20 \times 10^3) \times (300)] + [(30 \times 10^3) \times (800)]\right] \text{ mm}$$

$$= 0.75 \text{ mm} \quad \textbf{Ans.}$$

**EXAMPLE 2.11.** *A brass bar, having cross-sectional area of 500 mm$^2$ is subjected to axial forces as shown in Fig. 2.8.*

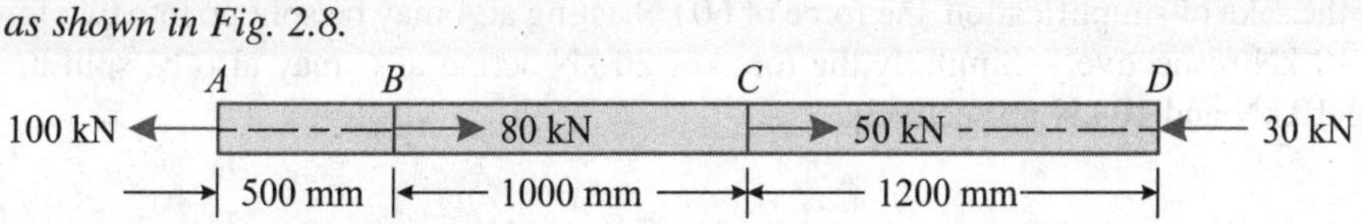

Fig. 2.8

*Find the total elongation of the bar. Take E = 80 GPa.*

**SOLUTION.** Given: Cross-sectional area ($A$) = 500 mm$^2$ and modulus of elasticity ($E$) = 80 GPa = 80 kN/mm$^2$.

For the sake of simplification, the force of 100 kN acting at $A$ may be split up into two forces of 80 kN and 20 kN respectively. Similarly, the force of 50 kN acting at $C$ may also be split up into two forces of 20 kN and 30 kN respectively.

Now it will be seen that the part $AB$ of the bar is subjected to a tensile force of 80 kN, part $AC$ is subjected to a tensile force of 20 kN and the part $CD$ is subjected to a compression force of 30 kN as shown in Fig. 2.9.

* Such a figure is called a free body diagram.

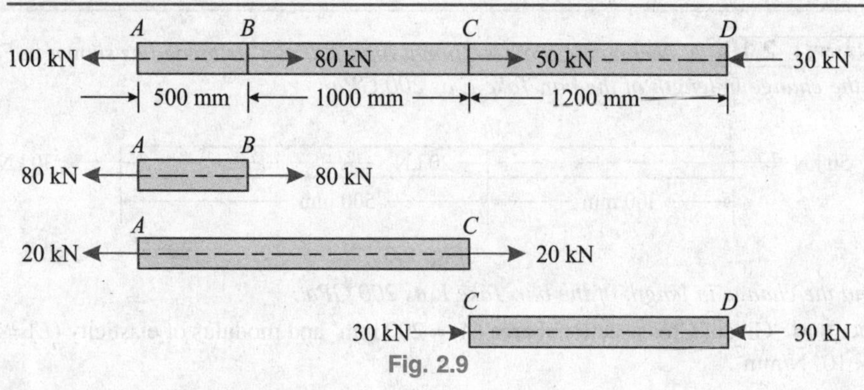


Fig. 2.9

We know that elongation of the bar,

$$\delta l = \frac{1}{AE}\left[P_1 l_1 + P_2 l_2 + P_3 l_3\right]$$

$$= \frac{1}{500 \times 80}\left[(80 \times 500) + (20 \times 1500) - (30 \times 1200)\right] \text{mm}$$

...(Taking plus sign for tension and minus for compression)

$$= 0.85 \text{ mm} \qquad \textbf{Ans.}$$

**EXAMPLE 2.12.** *A steel rod ABCD 4.5 m long and 25 mm in diameter is subjected to the forces as shown in Fig. 2.10. If the value of Young's modulus for the steel is 200 GPa, determine its deformation.*

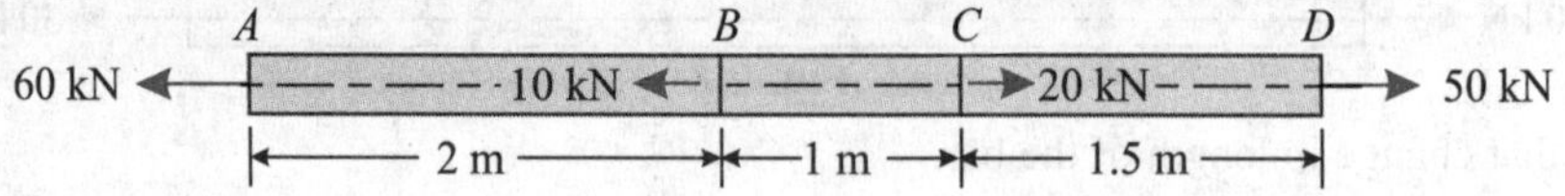

Fig. 2.10

**SOLUTION.** Given: Diameter ($D$) = 25 mm and Young's modulus ($E$) = 200 GPa = 200 kN/mm$^2$.

We know that cross-sectional area of the steel rod.

$$A = \frac{\pi}{4}(D)^2 = \frac{\pi}{4} \times (25)^2 = 491 \text{ mm}^2$$

For the sake of simplification, the force of 60 kN acting at $A$ may be split up into two forces of 50 kN and 10 kN respectively. Similarly the force of 20 kN acting at $C$ may also be split up into two forces of 10 kN and 10 kN respectively.

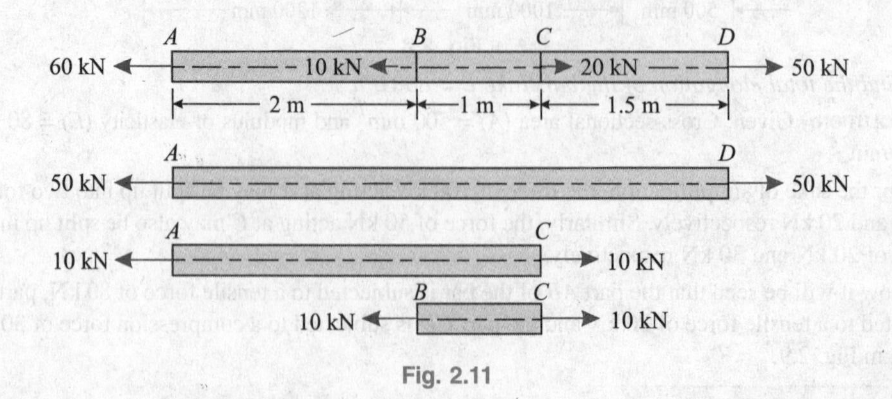


Fig. 2.11

Now it will be seen that the bar *AD* is subjected a tensile force of 50 kN, part *AC* is subjected to a tensile force of 10 kN and the part *BC* is subjected to a tensile force of 10 kN as shown in Fig. 2.11

We know that deformation of the bar,

$$\delta l = \frac{1}{AE}[P_1 l_1 + P_2 l_2 + P_3 l_3]$$

$$= \frac{1}{491 \times 200}\left[[50 \times (4.5 \times 10^3)] + [10 \times (3 \times 10^3)] + [10 \times (1 \times 10^3)\right] \text{mm}$$

$$= \frac{1}{491 \times 200} \times (265 \times 10^3) = 2.70 \text{ mm} \qquad \textbf{Ans.}$$

## EXERCISE 2.2

1. A steel bar *ABC* of 400 mm length and 20 mm diameter is subjected to a point loads as shown in Fig. 2.12.

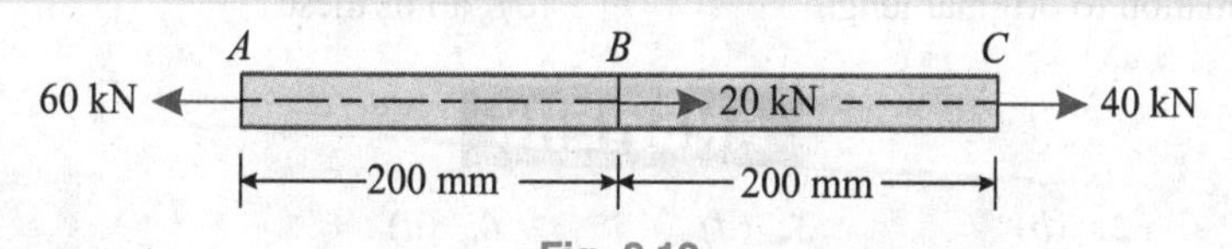


Fig. 2.12

Determine the total change in length of the bar. Take $E = 200$ GPa. **[Ans.** 0.32 mm]

2. A copper rod *ABCD* of 800 mm$^2$ cross-sectional area and 7.5 m long is subjected to forces as shown in Fig. 2.13.

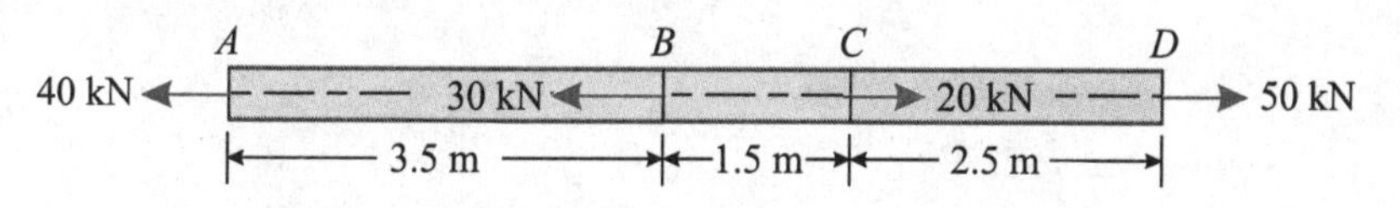


Fig. 2.13

Find the total elongation of the bar. Take *E* for the bar material as 100 GPa. **[Ans.** 4.6 mm]

3. A steel bar of 600 mm$^2$ cross-sectional area is carrying loads as shown in Fig. 2.14.

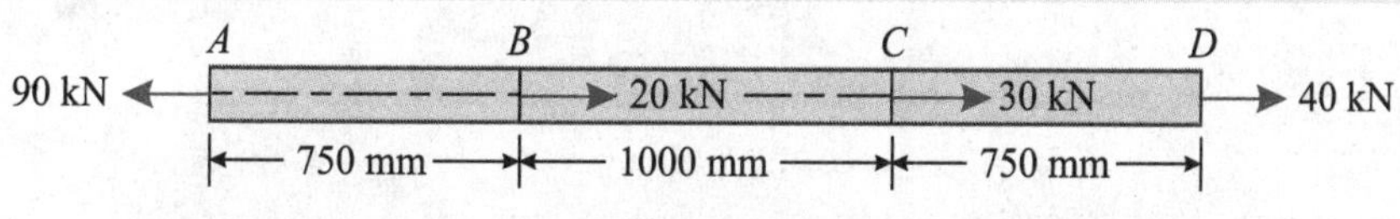


Fig. 2.14

Determine the elongation of the bar. Take *E* for the steel as 200 GPa. **[Ans.** 1.4 mm]

## QUESTIONS

1. Define stress, strain and elasticity. Derive a relation between stress and strain of an elastic body.
2. State clearly the Hooke's law.
3. Derive from fundamental, the relation for the deformation of a body, when it is subjected to: (*a*) a tensile force and (*b*) its own weight.
4. What is principle of the superposition? Explain its uses.

## MULTIPLE CHOICE QUESTIONS

1. If a force acts on a body, it sets up some resistance to the deformation. This resistance is known as
   (*a*) stress (*b*) strain
   (*c*) elasticity (*d*) modulus of elasticity
2. The term deformation per unit length is applied for
   (*a*) stress (*b*) strain
   (*c*) modulus of elasticity (*d*) none of these
3. The term 'Young's modulus' is used
   (*a*) only for young persons (*b*) only for old persons
   (*c*) young and old person (*d*) none of these
4. Modulus of elasticity is the ratio of
   (*a*) stress to strain (*b*) stress to original length
   (*c*) deformation to original length (*d*) all of these

## ANSWERS

**1.** (*a*) **2.** (*b*) **3.** (*d*) **4.** (*a*)

Chapter 3

# Stresses and Strains in Bars of Varying Sections

## Contents

## 3.1. Introduction

In the last chapter, we have discussed the procedure of obtaining stresses and strains in the bars of uniform cross-sectional area. But sometimes we come across bars of varying sections in which we are required to find out the stresses and strains. The procedure for finding out the stresses and strains in same sections is slightly different.

## 3.2. Types of Bars of Varying Sections

Though there are many types of bars of varying sections, in the field of strength of materials yet the following are important from the subject point of view :

1. Bars of different sections
2. Bars of uniformly tapering sections
3. Bars of composite sections.

Now we shall study the procedure for the stresses and strains in the above mentioned bars in the following pages.

## 3.3. Stresses in the Bars of Different Sections

Sometimes a bar is made up of different lengths having different cross-sectional areas as shown in Fig. 3.1.

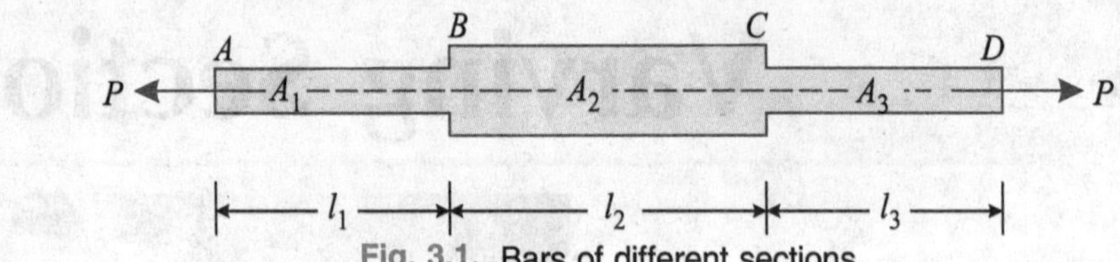


Fig. 3.1. Bars of different sections

In such cases, the stresses, strains and hence changes in lengths for each section is worked out separately as usual. The total changes in length is equal to the sum of the changes of all the individual lengths. It may be noted that each section is subjected to the same external axial pull or push.

Let $P$ = Force acting on the body,

$E$ = Modulus of elasticity for the body,

$l_1$ = Length of section 1,

$A_1$ = Cross-sectional area of section 1,

$l_2, A_2$ = Corresponding values for section 2 and so on.

We know that the change in length of section 1.

$$\delta l_1 = \frac{Pl_1}{A_1 E} \qquad \text{Similarly} \qquad \delta l_2 = \frac{Pl_2}{A_2 E} \quad \text{and so on}$$

∴ Total deformation of the bar,

$$\delta l = \delta l_1 + \delta l_2 + \delta l_3 + \ldots\ldots\ldots$$

$$= \frac{Pl_1}{A_1 E} + \frac{Pl_2}{A_2 E} + \frac{Pl_3}{A_3 E} + \ldots\ldots\ldots$$

$$= \frac{P}{E}\left(\frac{l_1}{A_1} + \frac{l_2}{A_2} + \frac{l_3}{A_3} + \ldots\ldots\ldots\right)$$

**Note.** Sometimes, the modulus of elasticity is different for different sections. In such cases, the total deformation,

$$\delta l = P\left(\frac{l_1}{A_1 E_1} + \frac{l_2}{A_2 E_2} + \frac{l_3}{A_3 E_3} + \ldots\ldots\ldots\right)$$

**Example 3.1.** *An automobile component shown in Fig. 3.2 is subjected to a tensile load of 160 kN.*

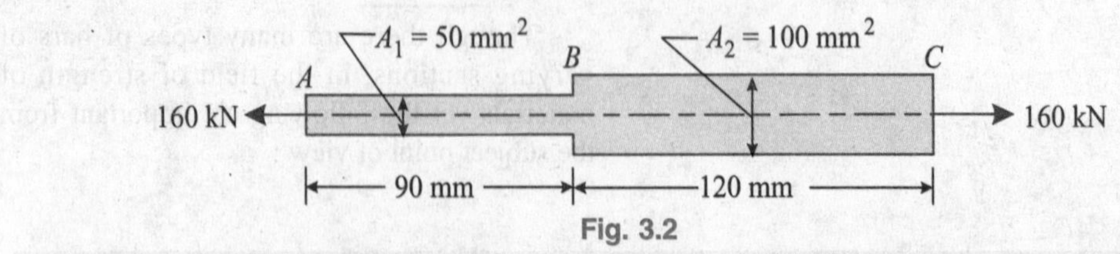


Fig. 3.2

*Determine the total elongation of the component, if its modules of elasticity is 200 GPa.*

**SOLUTION.** Given : Tensile load $(P) = 160$ kN $= 160 \times 10^3$ N ; Length of section 1 $(l_1) = 90$ mm; Length of section 2 $(l_2) = 120$ mm ; Area of section 1 $(A_1) = 50$ mm$^2$ ; Area of section 2 $(A_2) = 100$ mm$^2$ and modulus of elasticity $(E) = 200$ GPa $= 200 \times 10^3$ N/mm$^2$.

We know that total elongation of the component,

$$\delta l = \frac{P}{E}\left(\frac{l_1}{A_1} + \frac{l_2}{A_2}\right) = \frac{160 \times 10^3}{200 \times 10^3}\left(\frac{90}{50} + \frac{120}{100}\right) \text{mm}$$

$$= 0.8 \times 1.8 + 1.2 = 2.4 \text{ mm} \qquad \textbf{Ans.}$$

**EXAMPLE 3.2.** *A member formed by connecting a steel bar to an aluminium bar is shown in Fig. 3.3.*

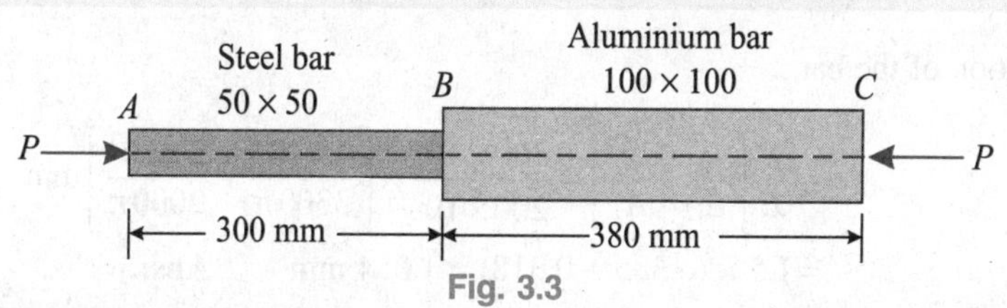


Fig. 3.3

*Assuming that the bars are prevented from buckling sidewise, calculate the magnitude of force P, that will cause the total length of the member to decrease by 0.25 mm. The values of elastic modulus for steel and aluminium are 210 GPa and 70 GPa respectively.*

**SOLUTION.** Given : Decrease in length $(\delta l) = 0.25$ mm ; Modulus of elasticity for steel $(E_S) = 210$ GPa $= 210 \times 10^3$ N/mm$^2$ ; Modulus of elasticity for aluminium $(E_A) = 70$ GPa $= 70 \times 10^3$ N/mm$^2$ ; Area of steel section $(A_S) = 50 \times 50 = 2\,500$ mm$^2$ ; Area of aluminium section $(A_A) = 100 \times 100 = 10000$ mm$^2$ ; Length of steel section $(l_S) = 300$ mm and length of aluminium section $(l_A) = 380$ mm.

Let $\qquad P$ = Magnitude of the force in kN.

We know that decrease in the length of the member $(\delta l)$,

$$0.25 = P\left(\frac{l_S}{A_S\,E_S} + \frac{l_A}{A_A\,E_A}\right)$$

$$= P\left(\frac{300}{2500 \times (210 \times 10^3)} + \frac{380}{10000 \times (70 \times 10^3)}\right)$$

$$= \frac{780\,P}{700 \times 10^6}$$

$$\therefore \qquad P = \frac{0.25 \times (700 \times 10^6)}{780} = 224.4 \times 10^3 \text{ N} = 224.4 \text{ kN} \qquad \textbf{Ans.}$$

**EXAMPLE 3.3.** *A 6 m long hollow bar of circular section has 140 mm diameter for a length of 4 m, while it has 120 mm diameter for a length of 2 m. The bore diameter is 80 mm throughout as shown in Fig. 3.4.*

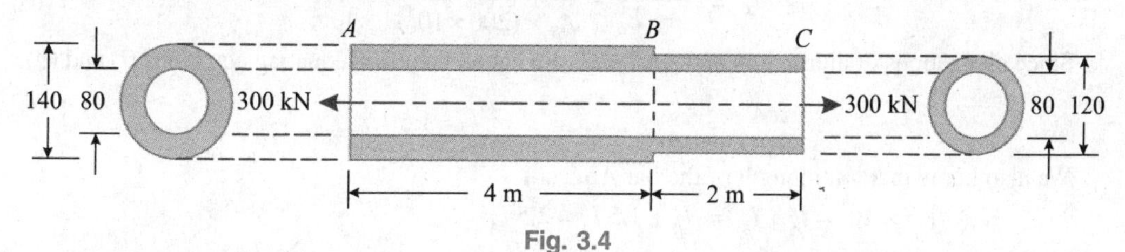


Fig. 3.4

*Find the elongation of the bar, when it is subjected to an axial tensile force of 300 kN. Take modulus of elasticity for the bar material as 200 GPa.*

**SOLUTION.** Given : Total length $(L) = 6 \text{ m} = 6 \times 10^3$ mm ; Diameter of section 1 $(D_1) = 140$ mm; Length of section 1 $(l_1) = 4 \text{ m} = 4 \times 10^3$ mm ; Diameter of section 2 $(D_2) = 120$ mm ; Length of section 2 $(l_2) = 2 \text{ m} = 2 \times 10^3$ mm ; Inner diameter $(d_1) = d_2 = 80$ mm ; Axial tensile force $(P) = 300 \text{ kN} = 300 \times 10$ N and modulus of elasticity $(E) = 200 \text{ GPa} = 200 \times 10^3 \text{ N/mm}^2$.

We know that area of portion $AB$,

$$A_1 = \frac{\pi}{4} \times [D_1^2 - d_1^2] = \frac{\pi}{4} \times [(140)^2 - (80)^2] = 3300 \pi \text{ mm}^2$$

and area of portion $BC$.

$$A_2 = \frac{\pi}{4} \times [D_2^2 - d_2^2] = \frac{\pi}{4} \times [(120)^2 - (80)^2] = 2000 \pi \text{ mm}^2$$

∴ Elongation of the bar,

$$\delta l = \frac{P}{E}\left[\frac{l_1}{A_1} + \frac{l_2}{A_2}\right] = \frac{300 \times 10^3}{200 \times 10^3} \times \left[\frac{4 \times 10^3}{3300\pi} + \frac{2 \times 10^3}{2000\pi}\right] \text{mm}$$

$$= 1.5 \times (0.385 + 0.318) = 1.054 \text{ mm} \qquad \textbf{Ans.}$$

**EXAMPLE 3.4.** *A compound bar ABC 1.5 m long is made up of two parts of aluminium and steel and that cross-sectional area of aluminium bar is twice that of the steel bar. The rod is subjected to an axial tensile load of 200 kN. If the elongations of aluminium and steel parts are equal, find the lengths of the two parts of the compound bar. Take E for steel as 200 GPa and E for aluminium as one-third of E for steel.*

**SOLUTION.** Given: Total length $(L) = 1.5 \text{ m} = 1.5 \times 10^3$ mm ; Cross-sectional area of aluminium bar $(A_A) = 2A_S$ ; Axial tensile load $(P) = 200 \text{ kN} = 200 \times 10^3$ N ; Modulus of elasticity of steel $(E_S) = 200 \text{ GPa} = 200 \times 10^3 \text{ N/mm}^2$ and modulus of elasticity of aluminium $(E_A) = \frac{E_S}{3} = \frac{200 \times 10^3}{3} \text{ N/mm}^2$.

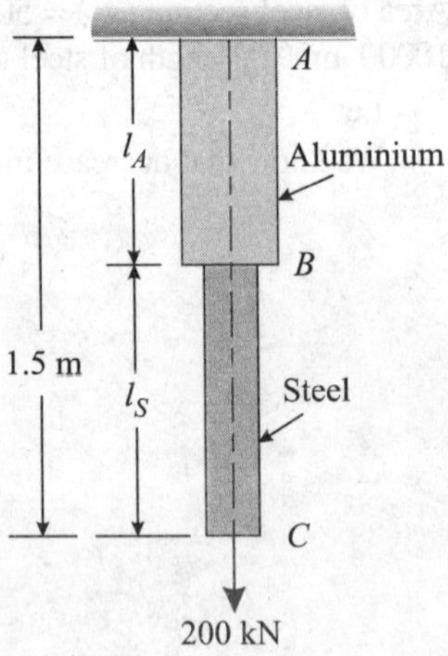

Let, $l_A$ = Length of the aluminium part,

and $l_S$ = Length of the steel part.

We know that elongation of the aluminium part $AB$,

$$\delta l_A = \frac{P.l_A}{A_A.E_A} = \frac{(200 \times 10^3) \times l_A}{2A_S \times \left(\frac{200 \times 10^3}{3}\right)}$$

$$= \frac{1.5\, l_A}{A_S} \qquad ...(i)$$



**Fig. 3.5**

and elongation of the steel part $BC$,

$$\delta l_S = \frac{P.l_S}{A_S.E_S} = \frac{(200 \times 10^3) \times l_S}{A_S \times (200 \times 10^3)} = \frac{l_S}{A_S} \qquad ...(ii)$$

Since elongations of aluminium and steel parts are equal, therefore equating equations (*i*) and (*ii*),

$$\frac{1.5\, l_A}{A_S} = \frac{l_S}{A_S} \quad \text{or} \quad l_S = 1.5\, l_A$$

We also know that total length of the bar $ABC$ $(L)$

$$1.5 \times 10^3 = l_A + l_S = l_A + 1.5\, l_A = 2.5\, l_A$$

$$\therefore \quad l_A = \frac{1.5 \times 10^3}{2.5} = 600 \text{ mm} \quad \textbf{Ans.}$$

and $\quad l_S = (1.5 \times 10^3) - 600 = 900 \text{ mm} \quad$ **Ans.**

**EXAMPLE 3.5.** *An alloy circular bar ABCD 3 m long is subjected to a tensile force of 50 kN as shown in Fig. 3.6.*

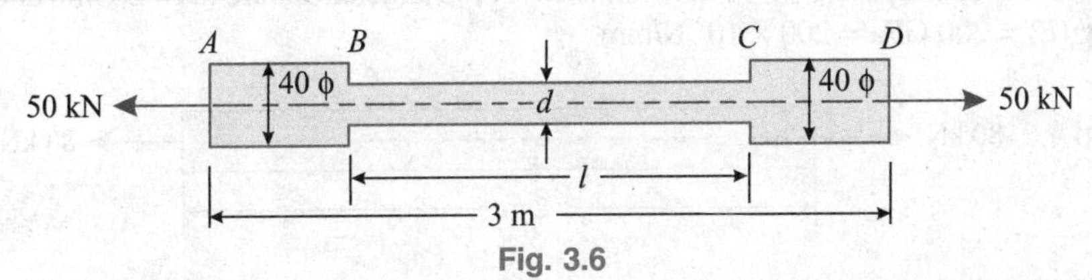


**Fig. 3.6**

*If the stress in the middle portion BC is not to exceed 150 MPa, then what should be its diameter? Also find the length of the middle portion, if the total extension of the bar should not exceed by 3 mm. Take E as 100 GPa.*

**SOLUTION.** Total length of circular bar ($L$) = 3m = $3 \times 10^3$ mm = 3000 mm ; Tensile force ($P$) = 50 kN = $50 \times 10^3$ N ; Maximum stress of portion $BC$ ($\sigma_{BC}$) = 150 MPa = 150 N/mm$^2$ ; Total extension ($\delta l$) = 3 mm and modulus of elasticity ($E$) = 100 GPa = $100 \times 10^3$ N/mm$^2$.

***Diameter of the middle portion BC***

Let $\quad d$ = Diameter of the middle portion in mm.

We know that stress in the middle portion $BC$ ($\sigma_{BC}$),

$$150 = \frac{P}{A} = \frac{50 \times 10^3}{\frac{\pi}{4} \times (d)^2} = \frac{63.66 \times 10^3}{d^2}$$

$$\therefore \quad d^2 = \frac{63.66 \times 10^3}{150} = 424.4 \quad \text{or} \quad d = 20.6 \text{ mm} \quad \textbf{Ans.}$$

length of the middle portion BC

Let $\quad l_{BC}$ = Length of the middle portion in mm.

We know that area of the end portions $AB$ and $CD$,

$$A_1 = \frac{\pi}{4} \times (40)^2 = 1257 \text{ mm}^2$$

and area of the middle portion $BC$,

$$A_2 = \frac{\pi}{4} \times (d)^2 = \frac{\pi}{4} \times (20.6)^2 = 333.3 \text{ mm}^2$$

We also know that total extension of bar ($\delta l$),

$$3 = \frac{P}{E}\left[\frac{l_1}{A_1} + \frac{l_2}{A_2}\right] = \frac{50 \times 10^3}{100 \times 10^3} \times \left[\frac{3000 - l}{1257} + \frac{l}{333.3}\right]$$

$$= 0.5\,[2.387 - 0.0008\,l + 0.003\,l] = 0.5\,[2.387 + 0.0022\,l]$$

$$= 1.194 + 0.0011\,l$$

$$\therefore \quad l = \frac{3 - 1.194}{0.0011} = 1.64 \times 10^3 \text{ mm} = 1.64 \text{ m} \quad \textbf{Ans.}$$

**NOTE.** We have taken total length of the circular bar as $(3000 - l)$ mm.

**EXAMPLE 3.6.** *A steel bar 2 m long and 40 mm in diameter is subjected to an axial pull of 80 kN. Find the length of the 20 mm diameter bore, which should be centrally carried out, so that the total elongation should increase by 20% under the same pull. Take E for the bar material as 200 GPa.*

**SOLUTION.** Given : Length of steel bar ($l$) = 2 m = $2 \times 10^3$ mm = 2000 mm ; Diameter of steel bar ($D$) = 40 mm ; Axial pull ($P$) = 80 kN = $80 \times 10^3$ N ; Diameter of bore ($d$) = 20 mm and modulus of elasticity ($E$) = 200 GPa = $200 \times 10^3$ N/mm$^2$.

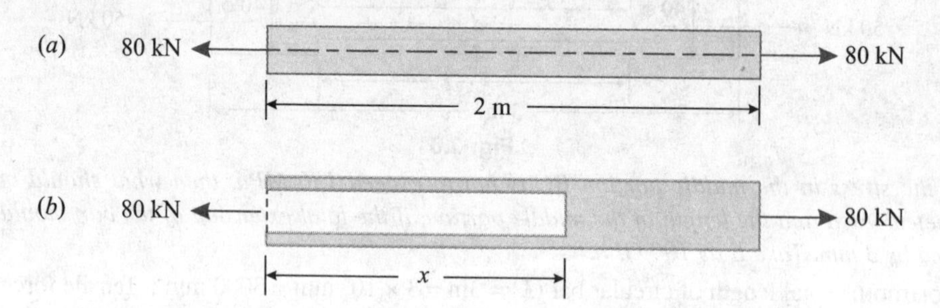


Fig. 3.7

Let $x$ = Length of the bore in mm.

First of all, consider the solid bar before the bore as shown in Fig. 3.7 (*a*). We know that cross-sectional area of the bar,

$$A_1 = \frac{\pi}{4} \times (D)^2 = \frac{\pi}{4} \times (40)^2 = 400\,\pi \text{ mm}^2$$

and elongation of the bar,

$$\delta l = \frac{Pl}{AE} = \frac{(80 \times 10^3) \times (2 \times 10^3)}{400\pi \times (200 \times 10^3)} = 0.64 \text{ mm}$$

Now consider the bar after the bore. Since the elongation of the bar after bore is increased by 20%, therefore total elongation of the bar after bore,

$$= 0.64 + (0.2 \times 0.64) = 0.768 \text{ mm}$$

We also know that cross-sectional area of the bored part

$$A_2 = \frac{\pi}{4}\,[D^2 - d^2] = \frac{\pi}{4}\,[(40)^2 - (20)^2] = 300\,\pi \text{ mm}^2$$

and total elongation of the bar after bore,

$$0.768 = \frac{P}{E}\left[\frac{l_1}{A_1} + \frac{l_2}{A_2}\right] = \frac{80 \times 10^3}{200 \times 10^3}\left[\frac{2000 - x}{400\pi} + \frac{x}{300\pi}\right]$$

$$= 0.4\left[\frac{4x + 3(2000 - x)}{1200\pi}\right]$$

or $$\frac{0.768 \times 1200\pi}{0.4} = 4x + 6\,000 - 3x \quad \text{or} \quad 7\,240 = x + 6\,000$$

$\therefore$ $$x = 7\,240 - 6\,000 = 1240 \text{ mm} = 1.24 \text{ m}$$ **Ans.**

**EXAMPLE 3.7.** *A steel bar ABCD 4 m long is subjected to forces as shown in Fig. 3.8.*

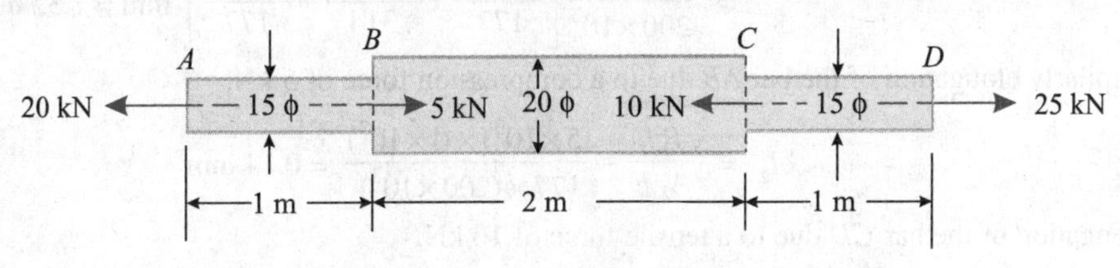


Fig. 3.8

*Find the elongation of the bar. Take E for the steel as 200 GPa.*

**SOLUTION.** Given : Total length of steel bar ($L$) = 4 m = $4 \times 10^3$ mm ; Length of first part ($l_1$) = 1 m = $1 \times 10^3$ mm ; Diameter of first part ($d_1$) = 15 mm ; Length of second part ($l_2$) = 2 m = $2 \times 10^3$ mm ; Diameter of second part ($d_2$) = 20 mm ; Length of third part ($l_3$) = 1 m = $1 \times 10^3$ mm ; Diameter of third part ($d_3$) = 15 mm and modulus of elasticity ($E$) = 200 GPa = $200 \times 10^3$ N/mm$^2$.

We know that area of the first and third parts of the bar,

$$A_1 = A_3 = \frac{\pi}{4} \times (d_1)^2 = \frac{\pi}{4} \times (15)^2 = 177 \text{ mm}^2$$

and area of the middle part of the bar

$$A_2 = \frac{\pi}{4} \times (d_2)^2 = \frac{\pi}{4} \times (20)^2 = 314 \text{ mm}^2$$

For the sake of simplification, the force of 25 kN acting at *D* may be split up into two forces of 15 kN and 10 kN respectively. Similarly the force of 20 kN acting at *A* may also be split up into two forces of 15 kN and 5 kN respectively.

Now it will be seen that the bar *ABCD* is subjected to a tensile force of 15 kN, part *BC* is subjected to a compressive force of 5 kN and the part *CD* is subjected to a tensile force of 10 kN as shown in Fig. 3.9.

We know that elongation of the bar *ABCD* due to a tensile force of 15 kN,

$$\delta l_1 = \frac{P}{E}\left[\frac{l_1}{A_1} + \frac{l_2}{A_2} + \frac{l_3}{A_3}\right]$$

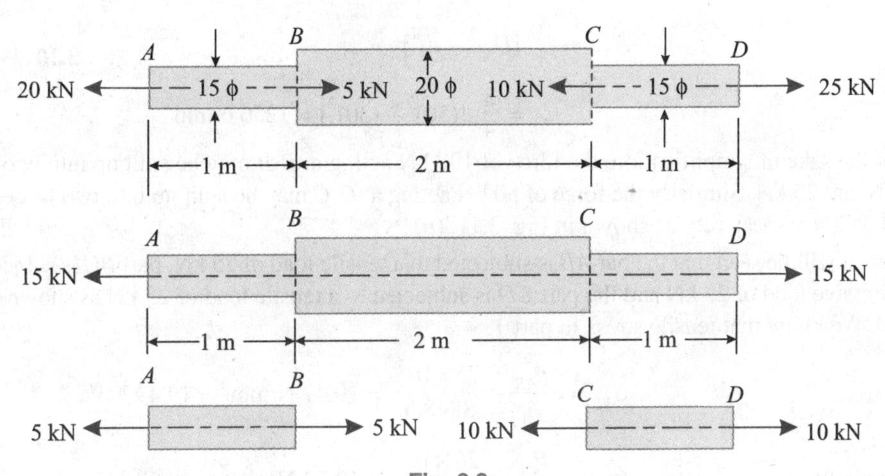


Fig. 3.9

$$= \frac{15\times10^3}{200\times10^3}\left[\frac{1\times10^3}{177}+\frac{2\times10^3}{314}+\frac{1\times10^3}{177}\right] \text{ mm} = 1.32 \text{ mm}$$

Similarly elongation of the bar *AB* due to a compression force of 5 kN,

$$\delta l_2 = \frac{P_1 l_1}{A_2 E} = \frac{(5\times10^3)\times(1\times10^3)}{177\times(200\times10^3)} = 0.14 \text{ mm}$$

and elongation of the bar *CD* due to a tensile force of 10 kN,

$$\delta l_3 = \frac{P_3 l_3}{A_3 E} = \frac{(10\times10^3)\times(1\times10^3)}{177\times(200\times10^3)} = 0.3 \text{ mm}$$

∴ Total elongation of the bar *ABCD*,

$$\delta l = \delta l_1 + \delta l_2 + \delta l_3 = 1.43 + 0.14 + 0.28 = 1.85 \text{ mm} \quad \textbf{Ans.}$$

**EXAMPLE 3.8.** *A circular steel rod ABCD of different cross-sections is loaded as shown in Fig. 3.10.*

*Find the maximum stress induced in the rod and its deformation. Take E = 200 GPa.*

**SOLUTION.** Given : Length of first part *AB* ($l_1$) = 1 m = $1 \times 10^3$ mm ; Diameter of first part *AB* ($D_1$) = 70 mm ; Length of second part *BC* ($l_2$) = 2 m = $2 \times 10^3$ mm ; Diameter of second part *BC* ($D_2$) = 50 mm ; Length of third part *CD* ($l_3$) = 1 m = $1 \times 10^3$ mm ; Diameter of third part *CD* ($D_3$) = 50 mm and internal diameter of hole ($d_3$) = 30 mm.

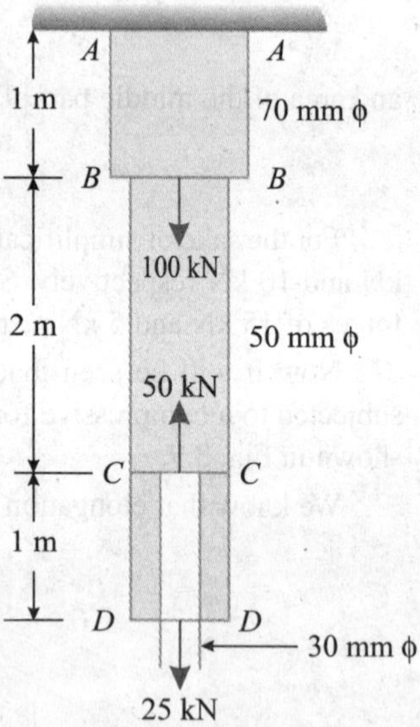

Fig. 3.10

***Maximum stress induced in the rod***

We know that area of the first part (*AB*) of the rod,

$$A_1 = \frac{\pi}{4}(D_1)^2 = \frac{\pi}{4}(70)^2 \text{ mm}^2$$

$$= 3848.5 \text{ mm}^2$$

Similarly area of the second part (*BC*) of the rod,

$$A_2 = \frac{\pi}{4}(D_2)^2 = \frac{\pi}{4}(50)^2 = 1963.5 \text{ mm}^2$$

and area of the third part *CD* of the rod,

$$A_3 = \frac{\pi}{4}[(D_3)^2 - d_3^2]$$

$$= \frac{\pi}{4}[(50)^2 - (30)^2] = 1256.6 \text{ mm}^2$$

For the sake of simplification, the force of 100 kN acting at *B-B* may be split up into two forces of 75 kN and 25 kN. Similarly the force of 50 kN acting at *C-C* may be split up into two forces of 25 kN and 25 kN respectively as shown in Fig. 3.11. (*b*).

Now it will be seen that the bar *AB* is subjected to a tensile load of 75 kN, part *BC* is subjected to a compressive load of 25 kN and the part *CD* is subjected to a tensile load of 25 kN as shown in Fig. 3.11 (*b*). We know that tensile stress in part 1,

$$\sigma_1 = \frac{P_{AB}}{A_1} = \frac{75\times10^3}{3848.5} = 19.49 \text{ N/mm}^2 = 19.49 \text{ MPa}$$

Similarly,

$$\sigma_2 = \frac{P_{BC}}{A_2} = \frac{25\times10^3}{1963.5} = 12.73 \text{ N/mm}^2 = 12.73 \text{ MPa}$$

and $$\sigma_3 = \frac{P_{CD}}{A_3} = \frac{25 \times 10^3}{1256.6} = 19.89 \text{ N/mm}^2 = 19.89 \text{ MPa}$$

From the above three values of the stresses, we find that maximum stress induced in the rod is in *CD* and is equal to 19.89 MPa. **Ans.**

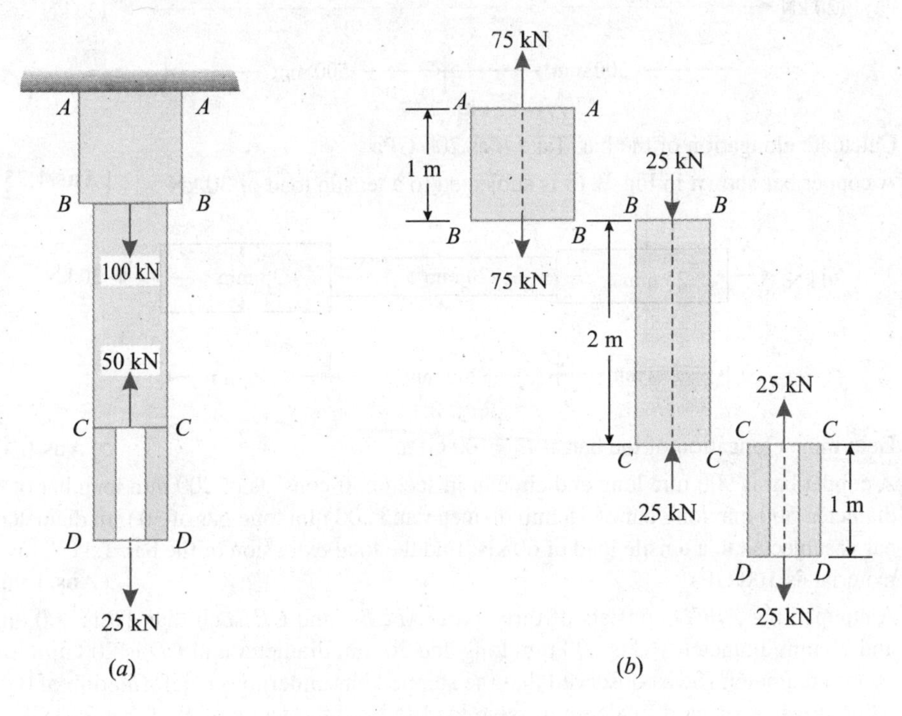


Fig. 3.11

***Deformation of the rod***

We also know that elongation of the part *AB*, due to tensile load of 75 kN,

$$\delta l_1 = \frac{P_1 l_1}{A_1 E} = \frac{(75 \times 10^3) \times (1 \times 10^3)}{3848.5 \times (200 \times 10^3)} = 0.097 \text{ mm}$$

Similarly shortening of the part *BC* due to compressive load of 25 kN.

$$\delta l_2 = \frac{P_2 l_2}{A_2 E} = \frac{(25 \times 10^3) \times (2 \times 10^3)}{1963.5 \times (200 \times 10^3)} = 0.127 \text{ mm}$$

and elongation of the part *CD* due to tensile load of 25 kN.

$$\delta l_3 = \frac{P_3 l_3}{A_3 E} = \frac{(25 \times 10^3) \times (1 \times 10^3)}{1256.6 \times (200 \times 10^3)} = 0.099 \text{ mm}$$

∴ Deformation of the rod,

$$\delta l = \delta l_1 - \delta l_2 + \delta l_3 = 0.097 - 0.127 + 0.099 = 0.069 \text{ mm} \qquad \textbf{Ans.}$$

## EXERCISE 3.1

1. A steel bar shown in Fig. 3.12 is subjected to a tensile force of 120 kN.

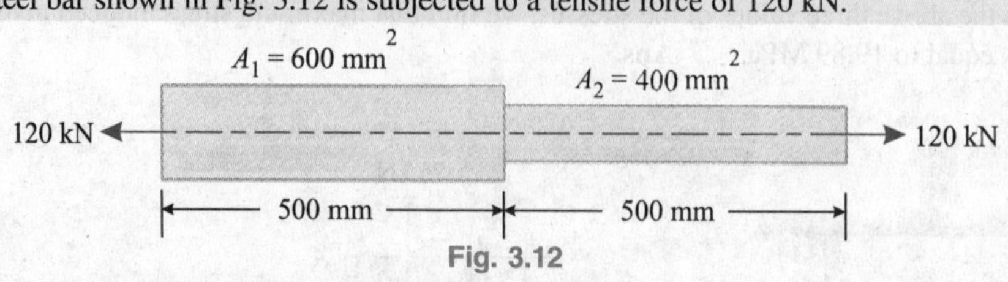


Fig. 3.12

Calculate elongation of the bar. Take $E$ as 200 GPa.

2. A copper bar shown in Fig. 3.13 is subjected to a tensile load of 30 kN. [**Ans.** 1.25 mm]

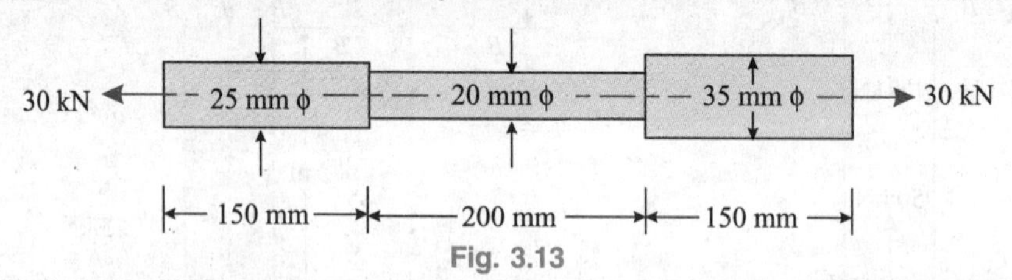


Fig. 3.13

Determine elongation of the bar, if $E$ = 100 GPa. [**Ans.** 0.33 mm]

3. A copper bar is 900 mm long and circular in section. It consists of 200 mm long bar of 40 mm diameter, 500 mm long bar of 15 mm diameter and 200 mm long bar of 30 mm diameter. If the bar is subjected to a tensile load of 60 kN, find the total extension of the bar. Take $E$ for the bar material as 100 GPa. [**Ans.** 1.963 mm]

4. A stepped bar $ABCD$ consists of three parts $AB$, $BC$ and $CD$ such that $AB$ is 300 mm long and 20 mm diameter, BC is 400 mm long and 30 mm diameter and $CD$ is 200 mm long and 40 mm diameter. It was observed that the stepped bar undergoes a deformation of 0.42 mm, when it was subjected to a compressive load $P$. Find the value of $P$, if $E$ = 200 GPa. [**Ans.** 50 kN]

5. A member $ABCD$ is subjected to point load as shown in Fig. 3.14.

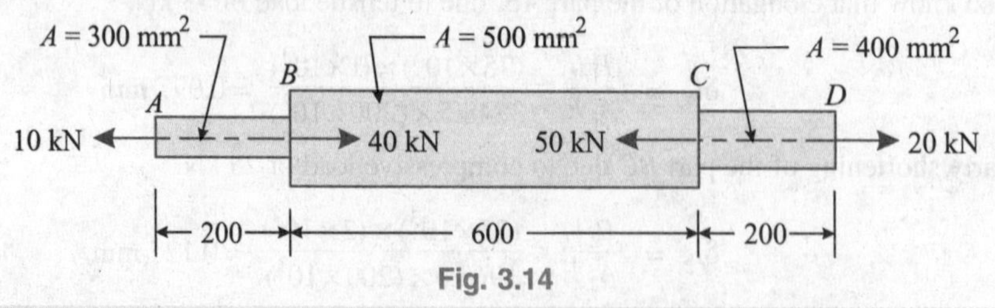


Fig. 3.14

Determine the total change in length of the member. Take $E$ = 200 GPa. [**Ans.** 0.096 mm (decrease)]

6. A steel bar $ABCD$ is subjected to point loads of $P_1$, $P_2$, $P_3$ and $P_4$ as shown in Fig. 3.15.

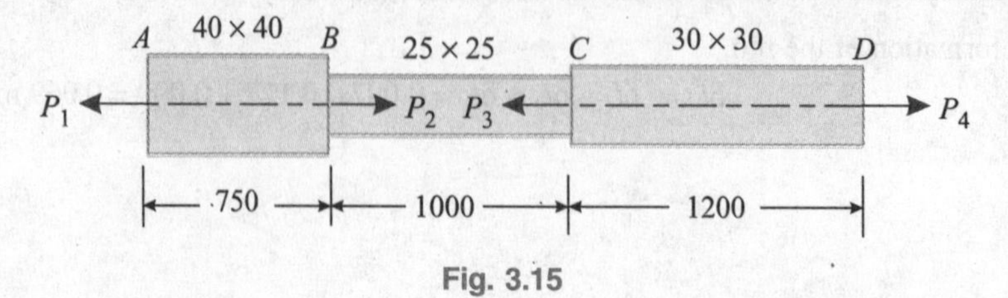


Fig. 3.15

Determine the magnitude of the force $P_3$ necessary for the equilibrium, if $P_1 = 120$ kN, $P_2 = 220$ and $P_4 = 160$ kN. Also determine the net change in the length of the steel bar. Take $E = 200$ GPa.

[**Ans.** 260 kN ; 0.55 mm]

[**Hint.** *AB* will be subjected to 120 kN (tension). *BC* will be subjected to 100 kN (compression) and *CD* will be subjected to 160 kN (tension).

## 3.4. Stresses in the Bars of Uniformly Tapering Sections

In the last article, we have discussed the stresses in the bars of different sections or stepped sections. But now we shall discuss the stresses in the bars of uniformly tapering sections. Following two types of uniformly tapering sections are important from the subject point of view :

1. Bars of uniformly tapering circular sections.
2. Bars of uniformly tapering rectangular sections.

Now we shall discuss the stresses in the bars of both the above mentioned types of uniformly tapering sections.

## 3.5. Stresses in the Bars of Uniformly Tapering Circular Sections

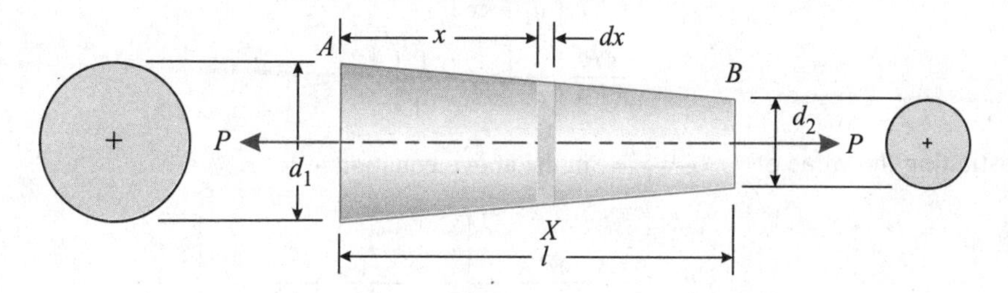


Fig 3.16

Consider a circular bar *AB* of uniformly tapering circular section as shown in Fig. 3.16.

Let $P$ = Pull on the bar.

$l$ = Length of the bar,

$d_1$ = Diameter of the bigger end of the bar, and

$d_2$ = Diameter of the smaller end of the bar.

Now consider a small element of length $dx$ of the bar, at a distance $x$ from the bigger end as shown in Fig. 3.16. We know that diameter of the bar at a distance $x$, from the left end $A$,

$$dx = d_1 - (d_1 - d_2)\,\frac{x}{l} = d_1 - k\,x, \qquad \text{...( where } k = \frac{d_1 - d_2}{l}\text{)}$$

and cross-sectional area of the bar at this section,

$$A_X = \frac{\pi}{4}\,(d_1 - kx)^2$$

∴ Stress,

$$\sigma_X = \frac{P}{\frac{\pi}{4}\,(d_1 - kx)^2} = \frac{4P}{\pi\,(d_1 - kx)^2}$$

and strain,

$$\varepsilon_X = \frac{\text{Stress}}{E} = \frac{\frac{4P}{\pi\,(d_1 - kx)^2}}{E} = \frac{4P}{\pi\,(d_1 - kx)^2\,E}$$

∴ Elongation of the elementary length

$$= \varepsilon_x \cdot dx = \frac{4P \cdot dx}{\pi (d_1 - kx)^2 E}$$

Total extension of the bar may be found out by integrating the above equation between the limit 0 and $l$. Therefore total elongation,

$$\delta l = \int_0^l \frac{4P \cdot dx}{\pi (d_1 - kx)^2 E}$$

$$= \frac{4P}{\pi E} \int_0^l \frac{dx}{(d_1 - kx)^2}$$

$$= \frac{4P}{\pi E} \left[ \frac{(d_1 - kx)^{-1}}{-1 \times -k} \right]_0^l$$

$$= \frac{4P}{\pi E k} \left[ \frac{1}{d_1 - kx} \right]_0^l$$

$$= \frac{4P}{\pi E k} \left[ \frac{1}{d_1 - kl} - \frac{1}{d_1} \right]$$

Substituting the value of $k = \dfrac{d_1 - d_2}{l}$ in the above equation,

$$\delta l = \frac{4P}{\pi E \dfrac{(d_1 - d_2)}{l}} \left[ \frac{1}{d_1 - \dfrac{(d_1 - d_2)\, l}{l}} - \frac{1}{d_1} \right]$$

$$= \frac{4Pl}{\pi E (d_1 - d_2)} \left[ \frac{1}{d_2} - \frac{1}{d_1} \right] = \frac{4Pl}{\pi E (d_1 - d_2)} \left[ \frac{d_1 - d_2}{d_2 d_1} \right]$$

$$\therefore \qquad \delta l = \frac{4Pl}{\pi E d_2 d_1}$$

**Cor.** If the bar had been of uniform diameter $d$ throughout, then

$$\delta l = \frac{4Pl}{\pi E d^2} = \frac{Pl}{\frac{\pi}{4} \times d^2 E} = \frac{Pl}{AE}$$ ...(Same as in Art. 2.12)

**EXAMPLE 3.9.** *A circular alloy bar 2 m long uniformly tapers from 30 mm diameter to 20 mm diameter. Calculate the elongation of the rod under an axial force of 50 kN. Take E for the alloy as 140 GPa.*

**SOLUTION.** Given : Length of bar ($l$) = 2 m = $2 \times 10^3$ mm ; Diameter of section 1 ($d_1$) = 30 mm; Diameter of section 2 ($d_2$) = 20 mm ; Axial force ($P$) = 50 kN = $50 \times 10^3$ N and modulus of elasticity ($E$) = 140 GPa = $140 \times 10^3$ N/mm$^2$.

We know that elongation of the rod,

$$\delta l = \frac{4Pl}{\pi E d_1 d_2} = \frac{4 \times (50 \times 10^3) \times (2 \times 10^3)}{\pi \times (140 \times 10^3) \times 30 \times 20} = 1.52 \text{ mm}$$ **Ans.**

**EXAMPLE 3.10.** *If the tension test bar is found to taper from (D + a) diameter to (D – a) diameter, prove that the error involved in using the mean diameter to calculate Young's modulus is* $\left(\frac{10\,a}{D}\right)^2$ *per cent.*

**SOLUTION.** Given : Larger diameter $(d_1) = (D + a)$ and smaller diameter $(d_2) = (D - a)$.

Let $P$ = Pull on the bar,

$l$ = Length of the bar,

$E_1$ = Young's modulus by the tapering formula,

$E_2$ = Young's modulus by the mean diameter formula and

$\delta l$ = Extension of the bar.

First of all, let us find out the values of Young's modulus for the test bar by the tapering formula and then by the mean diameter formula. We know that extension of the bar by uniformly varying formula.

$$\delta l = \frac{4\,Pl}{\pi E_1 d_1 d_2} = \frac{4\,Pl}{\pi E_1 (D+a)(D-a)} = \frac{4\,Pl}{\pi E_1 (D^2 - a^2)}$$

or
$$E_1 = \frac{4\,Pl}{\pi (D^2 - a^2)\,.\,\delta l} \qquad \ldots(i)$$

and extension of the bar by mean diameter (*D*) formula,

$$\delta l = \frac{Pl}{AE_2} = \frac{Pl}{\frac{\pi}{4}(D)^2 \times E_2} = \frac{4\,Pl}{\pi D^2 E_2}$$

or
$$E_2 = \frac{4\,Pl}{\pi D^2\,.\,\delta l} \qquad \ldots(ii)$$

∴ Percentage error involved (in using the mean diameter to calculate the Young's modulus)

$$= \left(\frac{E_1 - E_2}{E_1}\right) \times 100 = \frac{\left(\frac{4\,Pl}{\pi (D^2 - a^2)\,\delta l}\right) - \left(\frac{4\,Pl}{\pi D^2\,.\,\delta l}\right)}{\frac{4\,Pl}{\pi (D^2 - a^2)\,\delta l}} \times 100$$

$$= \frac{\frac{1}{(D^2 - a^2)} - \frac{1}{D^2}}{\frac{1}{(D^2 - a^2)}} \times 100 = \frac{\frac{D^2 - (D^2 - a^2)}{(D^2 - a^2)(D^2)}}{\frac{1}{(D^2 - a^2)}} \times 100$$

$$= \frac{a^2}{D^2} \times 100 = \left(\frac{10\,a}{D}\right)^2 \quad \textbf{Ans.}$$

**EXAMPLE 3.11.** *Two circular bars A and B of the same material are subjected to the same pull (P) and are deformed by the same amount. What is the ratio of their length, if one of them has a constant diameter of 60 mm and the other uniformly tapers from 80 mm from one end to 40 mm at the other ?*

**SOLUTION.** Given : Modulus of elasticity of bar *A* $(E_A) = E_B$ (because both the bars are of the same material) ; Pull on bar *A* $(P_A) = P_B = P$ ; Deformation in bar *A* $(\delta l_A) = \delta l_B$ ; Diameter of bar *A* $(d_A)$ = 60 mm ; Diameter of bar *B* at section 1 $(d_{B1})$ = 80 mm and diameter of bar *B* at section 2 $(d_{B2})$ = 40 mm.

Let $l_A$ = Length of the bar *A* and

$l_B$ = Length of the bar *B*.

First of all, consider the bar $A$, which has a constant diameter. We know that its deformation.

$$\delta l_A = \frac{P_A\, l_A}{A_A\, E_A} = \frac{P l_A}{\frac{\pi}{4} \times (d)^2 \times E} = \frac{4 P l_A}{\pi\,(60)^2 \times E} = \frac{4 P l_A}{3600\,\pi\, E} \quad ...(i)$$

Now consider the bar $B$, which uniformly tapers from one end to the other. We know that its deformation.

$$\delta l_B = \frac{4 P_B\, l_B}{\pi\, E_B\, d_{B1} \,.\, d_{B2}} = \frac{4 P l_B}{\pi\, E \times 80 \times 40} = \frac{4 P l_B}{3200\,\pi\, E} \quad ...(ii)$$

Since $\delta l_A$ is equal to $\delta l_B$, therefore equating $(i)$ and $(ii)$, we get

$$\frac{4 P l_A}{3600\,\pi\, E} = \frac{4 P l_B}{3200\,\pi\, E} \quad \text{or} \quad \frac{l_A}{3600} = \frac{l_B}{3200}$$

$$\frac{l_A}{l_B} = \frac{3600}{3200} = \frac{9}{8} \quad \text{or} \quad l_A : l_B = 9 : 8 \quad \textbf{Ans.}$$

**EXAMPLE 3.12.** *A round tapered alloy bar 4 m long is subjected to load as shown in Fig. 3.17.*

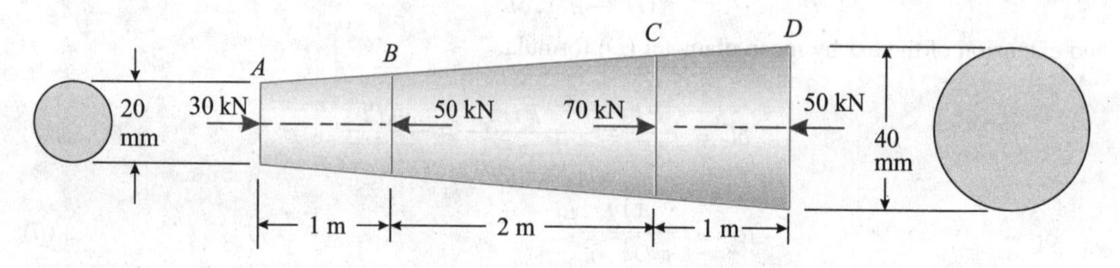


Fig. 3.17

*Find the change in the length of the bar. Take E for the bar material as 120 GPa.*

**SOLUTION.** Given : Length $(L) = 4 \text{ m} = 4 \times 10^3$ mm ; Force $(P_1) = 50 \text{ kN} = 50 \times 10^3$ N ; Force $(P_2) = 70 \text{ kN} = 70 \times 10^3$ N and modulus of elasticity $(E) = 120 \text{ GPa} = 120 \times 10^3 \text{ N/mm}^2$.

From the geometry of the figure, we find that diameter of the bar at $B$.

$$d_B = 20 + (40 - 20) \times \frac{1}{4} = 25 \text{ mm}$$

Similarly, diameter of the bar at $C$.

$$d_C = 25 + (40 - 20) \times \frac{2}{4} = 35 \text{ mm}$$

For the sake of simplification, the forces of 50 kN acting at $B$ may be split up into two forces of 30 kN and 20 kN respectively. Similarly the force of 70 kN acting at $C$ may also be split up into two forces of 20 kN and 50 kN respectively.

Now it will be seen that bar $AB$ subjected to a compressive load of 30 kN and part $BC$ is subjected to a tensile load of 20 kN and part $CD$ is subjected to a compressive load of 50 kN as shown in Fig. 3.18.

We know that shortening of the bar $AB$ due to a compressive force of 30 kN.

$$\delta l_1 = \frac{4 P_A \times l_{AB}}{\pi E d_A \,.\, d_B} = \frac{4 \times (30 \times 10^3) \times (1 \times 10^3)}{\pi \times (120 \times 10^2) \times 20 \times 25} = 0.64 \text{ mm}$$

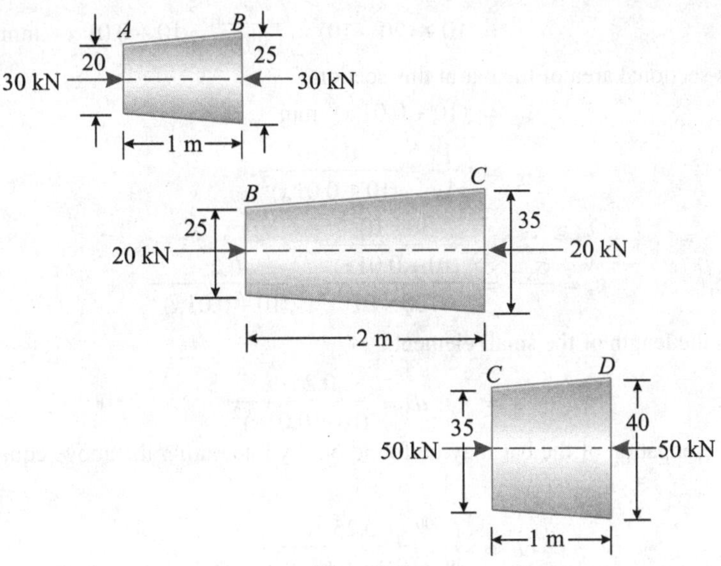


Fig. 3.18

Similarly elongation of the bar $BC$ due to a tensile load of 20 kN

$$\delta l_2 = \frac{4 P_B \times l_{BC}}{\pi E d_B . d_C} = \frac{4 \times (20 \times 10^3) \times (2 \times 10^3)}{\pi \times (120 \times 10^3) \times 25 \times 35} = 0.48 \text{ mm}$$

and shortening of the bar $CD$ due to a compressive load of 50 kN

$$\delta l_3 = \frac{4 P_C \times l_{CD}}{\pi E d_C . d_D} = \frac{4 \times (50 \times 10^3) \times (1 \times 10^3)}{\pi (120 \times 10^3) \times 35 \times 40} = 0.38 \text{ mm}$$

∴ Change in length

$$\delta l = \delta l_1 - \delta l_2 + \delta l_3 = 0.64 - 0.48 + 0.38 = 0.54 \text{ mm (decrease)} \quad \textbf{Ans.}$$

## 3.6. Stresses in the Bars of Uniformly Tapering Rectangular Sections

Sometimes, the uniformly tapering section varies from square section at one end to another square section at the other. Or it may also vary from rectangular section at one end to another rectangular section at the other. In such cases, the stresses should be found out from the fundamentals.

**EXAMPLE 3.13.** *An alloy bar of 1 m length has square section throughout, which tapers from one end of 10 mm × 10 mm to the other end of 20 mm × 20 mm. Find the change in its length due to an axial tensile load of 30 kN. Take E for the alloy as 120 GPa.*

**SOLUTION.** Given : Length of bar ($l$) = 1 m = 1 × $10^3$ mm ; Section at $A$ = 10 mm × 10 mm ; Section at $B$ = 20 mm × 20 mm ; Tensile load ($P$) = 30 kN = 30 × $10^3$ N and modulus of elastictiy ($E$) = 120 GPa = 120 × $10^3$ N/mm$^2$.

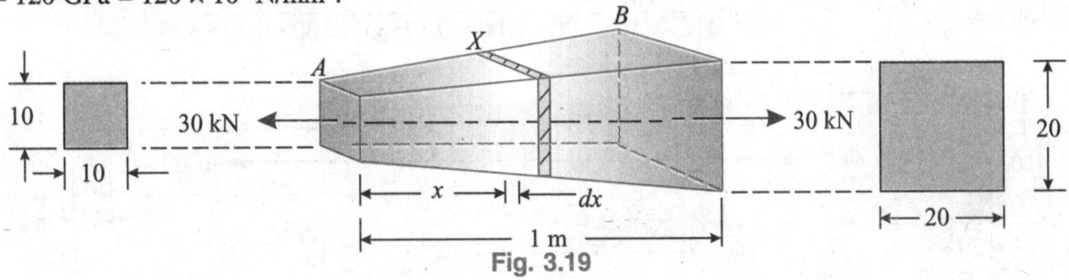


Fig. 3.19

Now consider a small length $dx$ of the bar at a distance $x$ from $A$ as shown in Fig. 3.19. From the geometry of the figure, we find that side of the square at $X$.

$$= 10 + (20 - 10) \times \frac{x}{1 \times 10^3} = 10 + 0.01\,x \quad \text{mm}$$

∴ Cross-sectional area of the bar at this section.

$$A_X = (10 + 0.01\,x)^2 \text{ mm}^2$$

and stress,

$$\sigma_X = \frac{P}{A_X} = \frac{30 \times 10^3}{(10 + 0.01x)^2}$$

∴ Strain,

$$\varepsilon_X = \frac{\sigma_X}{E} = \frac{\dfrac{30 \times 10^3}{(10 + 0.01x)^2}}{120 \times 10^3} = \frac{0.25}{(10 + 0.01x)^2}$$

and increase in the length of the small element.

$$= \varepsilon_X \,.\, dx = \frac{0.25\,dx}{(10 + 0.01x)^2}$$

Now total elongation of the bar may be found out by integrating the above equation between 0 and 1000.

$$\delta l = \int_0^{1000} \frac{0.25\,dx}{(10 + 0.01x)^2}$$

$$= 0.25 \int_0^{1000} (10 + 0.01x)^2 \; dx$$

$$= 0.25 \left[ -\frac{1}{0.01} (10 + 0.01x)^{-1} \right]_0^{1000}$$

$$= -25\,[(20)^{-1} - (10)^{-1}] = -25 \left[ \frac{1}{20} - \frac{1}{10} \right] = 1.25 \text{ mm} \qquad \textbf{Ans.}$$

**EXAMPLE 3.14.** *A steel plate of 20 mm thickness tapers uniformly from 100 mm to 50 mm in a length of 400 mm. What is the elongation of the plate, if an axial force of 80 kN acts on it ? Take E = 200 GPa.*

**SOLUTION.** Given : Plate thickness = 20 mm ; Width at $A$ = 100 mm ; Width at $B$ = 50 mm ; Length ($l$) = 400 mm ; Axial force ($P$) = 80 kN = 80 × $10^3$ N and modulus of elasticity ($E$) = 200 GPa = 200 × $10^3$ N/mm$^2$.

Now consider a small element of length $dx$, of the bar, at a distance $x$ from $A$ as shown in Fig. 3.20. From the geometry of the figure, we find that the width of the plate at a distance $x$ from $A$.

$$= 100 - (100 - 50) \times \frac{x}{400} = 100 - 0.125\,x \text{ mm}$$

∴ Cross-sectional area of the plate at this section.

$$A_X = 20 \times (100 - 0.125\,x)$$

and stress,

$$\sigma_X = \frac{P}{A_X} = \frac{80 \times 10^3}{20 \times (100 - 0.125\,x)} = \frac{4 \times 10^3}{100 - 0.125\,x}$$

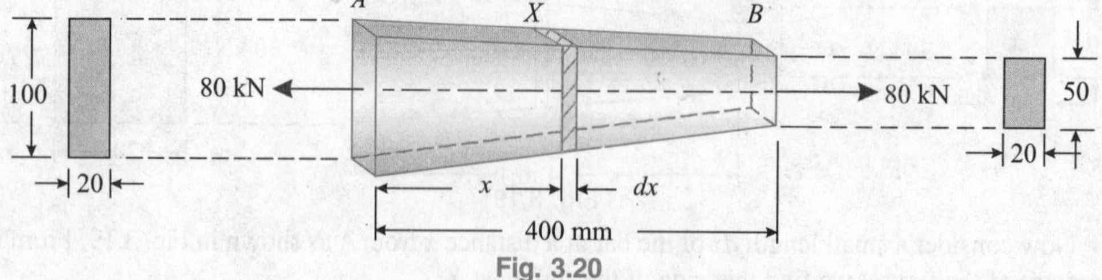

Fig. 3.20

$\therefore$ Strain,
$$\varepsilon_X = \frac{\sigma_X}{E} = \frac{\dfrac{4 \times 10^3}{100 - 0.125x}}{200 \times 10^3} = \frac{1}{50\,(100 - 0.125\,x)}$$

and increase in the length of the small element

$$= \varepsilon_X \,.\, dx = \frac{dx}{50(100 - 0.125\,x)}$$

Now total elongation of the plate may be found out by integrating the above equation between 0 and 400.

$$\therefore \quad \delta l = \int_0^{400} \frac{dx}{50(100 - 0.125\,x)}$$

$$= \frac{1}{50}\int_0^{400} \frac{dx}{(100 - 0.125\,x)}$$

$$= \frac{1}{50(-0.125)}\left[\log_e (100 - 0.125\,x)\right]_0^{400}$$

$$= -\frac{1}{6.25}\left[\log_e (50 - \log_e 100)\right]$$

$$= 0.16\,[\log_e 100 - \log_e 50] \qquad \text{...(Taking minus sign outside)}$$

$$= 0.16 \times \log_e\left(\frac{100}{50}\right) = 0.16 \times \log_e 2 = 0.16 \times 2.3 \log 2$$

$$\text{...}(\because \log_e = 2.3 \log_{10})$$

$$= 0.16 \times 2.3 \times 0.3010 = 0.11 \text{ mm} \qquad \textbf{Ans.}$$

## EXERCISE 3.2

1. A circular bar 2.5 m long tapers uniformly from 25 mm diameter to 12 mm diameter. Determine extension of the rod under a pull of 30 kN. Take E for bar as 200 GPa. [**Ans.** 1.6 mm]
2. A copper rod, circular in cross-section, uniformly tapers from 40 mm to 20 mm in a length of 11 m. Find the magnitude of force, which will deform it by 0.8 mm. Take E = 100 GPa. [**Ans.** 4.56 kN]
3. A circular steel bar 3 m long uniformly tapers from 50 mm diameter from one end to 25 mm at the other. Find the magnitude of force, which will deform it by 0.8 mm. [**Ans.** 52.4 kN]
4. A rectangular bar 2 m long and 12.5 mm thick uniformly tapers from 100 mm at one end to 20 mm at the other. If the bar is subjected to a tensile force of 25 kN, find its deformation. Take *E* as 200 GPa. [**Ans.** 0.4 mm]
5. A steel bar of 100 mm length tapers from 12 mm × 10 mm from one end to 30 mm × 20 mm at the other. If the stress in the bar is not to exceed 100 MPa, find the magnitude of the axial force (*P*). Also find the change in its length. Take *E* as 200 GPa. [**Ans.** 12 kN; 0.2 mm]

## 3.7. Stresses in the Bars of Composite Sections

A bar made up of two or more different materials, joined together is called a composite bar. The bars are joined in such a manner, that the system extends or contracts as one unit, equally, when subjected to tension or compression. Following two points should always be kept in view, while solving example on composite bars :

1. Extension or contraction of the bar being equal, the strain, *i.e.*, deformation per unit length is also equal.
2. The total external load, on the bar, is equal to the sum of the loads carried by the different materials.

Consider a composite bar made up of two different materials as shown in Fig. 3.21.

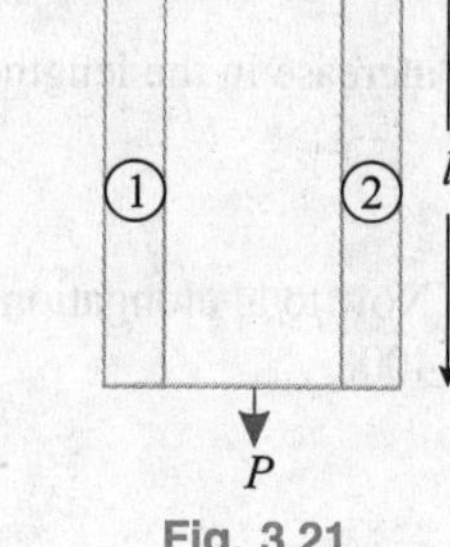


Fig. 3.21

Let

$$P = \text{Total load on the bar,}$$
$$l = \text{Length of the bar,}$$
$$A_1 = \text{Area of bar 1,}$$
$$E_1 = \text{Modulus of elasticity of bar 1.}$$
$$P_1 = \text{Load shared by bar 1, and}$$
$$A_2, E_2, P_2 = \text{Corresponding values for bar 2,}$$

We know that total load on the bar,

$$P = P_1 + P_2 \qquad ...(i)$$

$\therefore$ Stress in bar 1,

$$\sigma_1 = \frac{P_1}{A_1}$$

and strain in bar 1,

$$\varepsilon_1 = \frac{\sigma_1}{E_1} = \frac{P_1}{A_1 E_1}$$

$\therefore$ Elongation,

$$\delta l = \varepsilon_1 . l = \frac{P_1 l}{A_1 E_1} \qquad ...(ii)$$

Similarly, elongation of bar 2,

$$= \frac{P_2 l}{A_2 E_2} \qquad ...(iii)$$

Since both the elongations are equal, therefore equating (*ii*) and (*iii*), we get

$$\frac{P_1 l}{A_1 E_1} = \frac{P_2 l}{A_2 E_2} \quad \text{or} \quad \frac{P_1}{A_1 E_1} = \frac{P_2}{A_2 E_2} \qquad ...(iv)$$

or

$$P_2 = P_1 \times \frac{A_2 E_2}{A_1 E_1}$$

But

$$P = P_1 + P_2 = P_1 + P_1 \times \frac{A_2 E_2}{A_1 E_1}$$

$$= P_1 \left( 1 + \frac{A_2 E_2}{A_1 E_1} \right) = P_1 \left( \frac{A_1 E_1 + A_2 E_2}{A_1 E_1} \right)$$

or

$$P_1 = P \times \frac{A_1 E_1}{A_1 E_1 + A_2 E_2} \qquad ...(v)$$

Similarly,

$$P_2 = P \times \frac{A_2 E_2}{A_1 E_1 + A_2 E_2} \qquad ...(vi)$$

From these equations we can find out the loads shared by the different materials. We have also seen in equation (*iv*) that

$$\frac{P_1}{A_1 E_1} = \frac{P_2}{A_2 E_2}$$

or
$$\frac{\sigma_1}{E_1} = \frac{\sigma_2}{E_2} \qquad \cdots\left(\because \frac{P}{A} = \sigma = \text{Stress}\right)$$

$$\therefore \qquad \sigma_1 = \frac{E_1}{E_2} \times \sigma_2 \qquad \ldots(vii)$$

Similarly,
$$\sigma_2 = \frac{E_2}{E_1} \times \sigma_1 \qquad \ldots(viii)$$

From the above equations, we can find out the stresses in the different materials. We also know that the total load,

$$P = P_1 + P_2 = \sigma_1 A_1 + \sigma_2 A_2$$

From the above equation, we can also find out the stress in the different materials.

**NOTES: 1.** For the sake of simplicity, we have considered the composite bar made up of two different materials only. But this principle may be extended for a bar made up of more than two different materials also.

**2.** If the lengths of the two bars are different, then elongations should be separately calculated and equated.

**3.** The ratio $E_1/E_2$ is known as modulas ratio of the two materials and is denoted by the letter *m*.

**EXAMPLE 3.15.** *A reinforced concrete circular section of 50,000 mm² cross-sectional area carries 6 reinforcing bars whose total area is 500 mm². Find the safe load, the column can carry, if the concrete is not to be stressed more than 3.5 MPa. Take modular ratio for steel and concrete as 18.*

**SOLUTION.** Given : Area of column = 50,000 mm² ; No. of reinforcing bars = 6 ; Total area of steel bars ($A_S$) = 500 mm² ; Max stress in concrete ($\sigma_C$) = 3.5 MPa = 3.5 N/mm² and modular ratio $\left(\frac{E_S}{E_C}\right) = 18$.

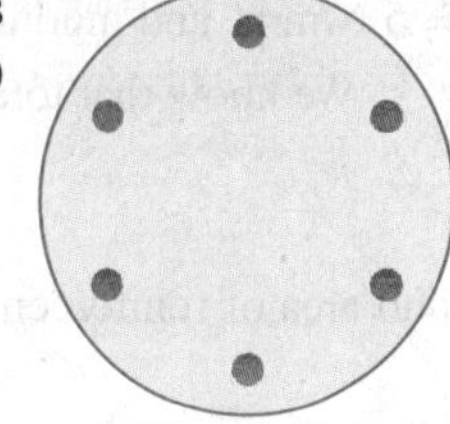

Fig. 3.22

We know that area of concrete,

$$A_C = 50{,}000 - 500 = 49{,}500 \text{ mm}^2$$

and stress in steel,

$$\sigma_S = \frac{E_S}{E_C} \times \sigma_C = 18 \times 3.5 = 63 \text{ N/mm}^2$$

$\therefore$ Safe load, the column can carry,

$$P = (\sigma_S \, . \, A_S) + (\sigma_C \, . \, A_C) = (63 \times 500) + (3.5 \times 49{,}500) \text{ N}$$
$$= 204\,750 \text{ N} = 204.75 \text{ kN} \quad \textbf{Ans.}$$

**EXAMPLE 3.16.** *A reinforced concrete column 500 mm × 500 mm in section is reinforced with 4 steel bars of 25 mm diameter, one in each corner. The column is carrying a load of 1000 kN. Find the stresses in the concrete and steel bars. Take E for steel = 210 GPa and E for concrete = 14 GPa.*

**SOLUTION.** Given : Area of column = 500 × 500 = 2,50,000 mm²; No. of steel bars ($n$) = 4 ; Diameter of steel bars ($d$) = 25 mm ; Load on column ($P$) = 1,000 kN = 1,000 × $10^3$ N ; Modulus of elasticity of steel ($E_S$) = 210 GPa and modulus of elasticity of concrete ($E_C$) = 14 GPa.

Let
$\sigma_S$ = Stress in steel, and
$\sigma_C$ = Stress in concrete.

We know that area of steel bars,

$$A_S = 4 \times \frac{\pi}{4} \times (d)^2 \text{ mm}^2 \qquad \ldots(i)$$

$$= 4 \times \frac{\pi}{4} \times (25)^2 = 1963 \text{ mm}^2$$

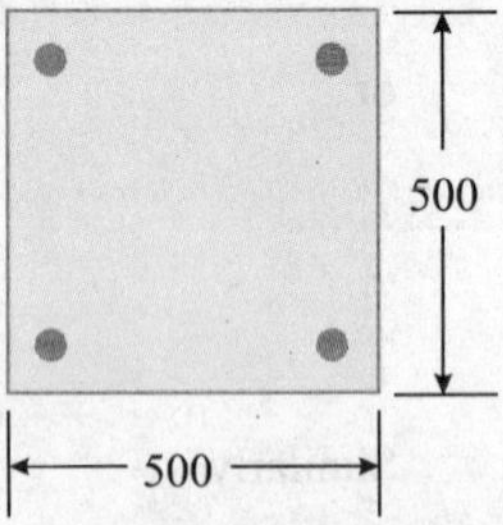

$\therefore$ Area of concrete,
$$A_C = 250,000 - 1963 \text{ mm}^2$$
$$= 248\,037 \text{ mm}^2$$

We also know that stress in steel,

$$\sigma_S = \frac{E_S}{E_C} \times \sigma_C = \frac{210}{14} \times \sigma_C = 15\,\sigma_C \qquad ...(ii)$$

500

500

Fig. 3.23

and total load $(P)$,
$$1,000 \times 10^3 = (\sigma_S \cdot A_S) + (\sigma_C \cdot A_C)$$
$$= (15\,\sigma_C \times 1963) + (\sigma_C \times 248\,037) = 277\,482\,\sigma_C$$

$$\sigma_C = \frac{1,000 \times 10^3}{277\,482} = 3.6 \text{ N/mm}^2 = 3.6 \text{ MPa} \qquad \textbf{Ans.}$$

and
$$\sigma_S = 15\,\sigma_C = 15 \times 3.6 = 54 \text{ MPa} \qquad \textbf{Ans.}$$

**EXAMPLE 3.17.** *A reinforced concrete circular column of 400 mm diameter has 4 steel bars of 20 mm diameter embeded in it. Find the maximum load which the column can carry, if the stresses in steel and concrete are not to exceed 120 MPa and 5 MPa respectively. Take modulus of elasticity of steel as 18 times that of concrete.*

**SOLUTION.** Given : Diameter of column $(D) = 400$ mm ; No. of reinforcing bars = 4 ; Diameter of bars $(d) = 20$ mm ; Maximum stress in steel $(\sigma_{S(max)}) = 120$ MPa $= 120$ N/mm$^2$ ; Maximum stress in concrete $(\sigma_{C(max)}) = 5$ MPa $= 5$ N/mm$^2$ and modulus of elasticity of steel $(E_S) = 18\,E_C$.

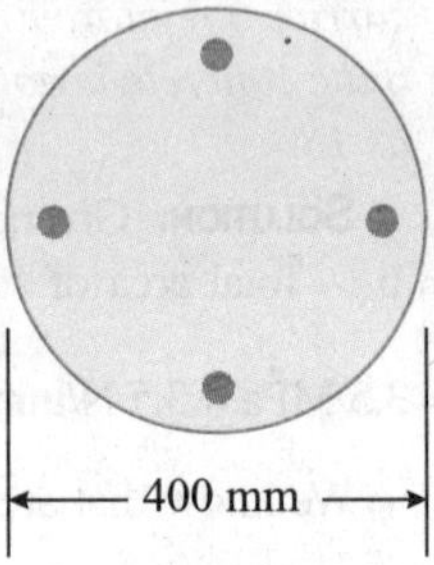

We know that total area of the circular column.

$$= \frac{\pi}{4} \times (D)^2 = \frac{\pi}{4} \times (400)^2 = 125\,660 \text{ mm}^2$$

400 mm

Fig. 3.24

and area of reinforcement (*i.e.*, steel),

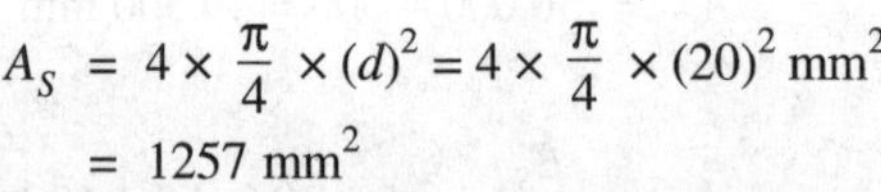

$$A_S = 4 \times \frac{\pi}{4} \times (d)^2 = 4 \times \frac{\pi}{4} \times (20)^2 \text{ mm}^2$$
$$= 1257 \text{ mm}^2$$

$\therefore$ Area of concrete,

$$A_C = 125\,660 - 1257 = 124\,403 \text{ mm}^2$$

First of all let us find out the maximum stresses developed in the steel and concrete. We know that if the stress in steel is 120 N/mm$^2$, then stress in the concrete.

$$\sigma_C = \frac{E_C}{E_S} \times \sigma_S = \frac{1}{18} \times 120 = 6.67 \text{ N/mm}^2 \qquad ...(i)$$

It is more than the stress in the concrete (*i.e.*, 5 N/mm$^2$). Thus these stresses are not accepted. Now if the stress in concrete is 5 N/mm$^2$, then stress in steel,

$$\sigma_S = \frac{E_S}{E_C} \times \sigma_C = 18 \times 5 = 90 \text{ N/mm}^2 \qquad ...(ii)$$

It is less than the stress is steel (*i.e.*, 120 N/mm$^2$). It is thus obvious that stresses in concrete and steel will be taken as 5 N/mm$^2$ and 90 N/mm$^2$ respectively. Therefore maximum load, which the column can carry.

$$P = (\sigma_C \cdot A_C) + (\sigma_S \cdot A_S) = (5 \times 124\,403) + (90 \times 1257) \text{ N}$$
$$= 735\,150 \text{ N} = 735.15 \text{ kN} \qquad \textbf{Ans.}$$

**EXAMPLE 3.18.** *A load of 270 kN is carried by a short concrete column 250 mm × 250 mm in size. The column is reinforced with 8 bars of 16 mm diameter. Find the stresses in concrete and steel, if the modulus of elasticity for the steel is 18 times that of concrete.*

*If the stress in concrete is not to exceed 5 MPa, find the area of steel required, so that the column may carry a load of 500 kN.*

**SOLUTION.** Given : Load on column $(P_1) = 270\ kN = 270 \times 10^3\ N$ ; Area of column $= 250 \times 250 = 62\ 500\ mm^2$, No. of reinforcing bars = 8 ; Diameter of reinforcing bars $(d) = 16$ mm ; Modular ratio $(E_S / E_C) = 18$ ; Maximum stress in concrete $(\sigma_C) = 5$ MPa and load that column may carry $(P_2) = 500\ kN = 500 \times 10^3\ N$.

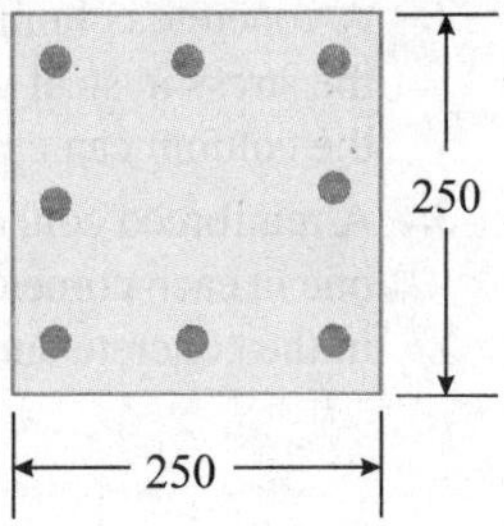

250

250

Fig. 3.25

***Stresses in concrete and steel when the column carries a load of 270 kN***

Let $\sigma_C$ = Stress in concrete, and

$\sigma_S$ = Stress in steel.

We know that area of reinforcement (*i.e.*, steel)

$$A_S = 8 \times \frac{\pi}{4} \times (d)^2 = 8 \times \frac{\pi}{4} \times (16)^2 = 1608\ mm^2$$

∴ Area of concrete $A_C = 62\ 500 - 1608 = 60\ 892\ mm^2$

We also know that stress in steel,

$$\sigma_S = \frac{E_S}{E_C} \times \sigma_C = 18\ \sigma_C \qquad \ldots \left(\because \frac{E_S}{E_C} = 18\right)$$

and total load $(P_1)$,

$$270 \times 10^3 = (\sigma_S \,.\, A_S) + (\sigma_C \,.\, A_C)$$
$$= (18\ \sigma_C \times 1608) + (\sigma_C \times 60\ 892) = 89\ 836\ \sigma_C$$

$$\therefore \quad \sigma_C = \frac{270 \times 10^3}{89\ 836} = 3.0\ N/mm^2 = 3.0\ MPa \qquad \textbf{Ans.}$$

and $\sigma_S = 18 \times 3.0 = 54.0$ MPa **Ans.**

***Area of steel required, so that the column may carry a load of 500 kN***

Let $A_{S1}$ = Area of steel required, if the stress in concrete $(\sigma_C)$ is not to exceed 5 MPa (*i.e.*, 5 N/mm²)

∴ Area of concrete,

$$A_{C1} = 62\ 500 - A_{S1}$$

and total load $(P_2)$

$$500 \times 10^3 = (\sigma_S \cdot A_{S1}) + (\sigma_C \cdot A_{C1})$$
$$= [(18 \times \sigma_C) \times A_{S1}] + [\sigma_C \times (62\ 500 - A_{S1})]$$
$$= [18 \times 5 \times A_{S1}] + [5 \times (62\ 500 - A_{S1})]$$
$$= 90\ A_{S1} + (312.5 \times 10^3) - 5\ A_{S1}$$
$$= (312.5 \times 10^3) + 85\ A_{S1}$$

$$\therefore \quad 85\ A_{S1} = (500 \times 10^3) - (312.5 \times 10^3) = 187.5 \times 10^3$$

$$\text{or} \quad A_{S1} = \frac{187.5 \times 10^3}{85} = 2\ 206\ mm^2 \qquad \textbf{Ans.}$$

## EXERCISE 3.3

1. A reinforced concrete column of 300 mm diameter contain 4 bars of 22 mm diameter. Find the total load, the column can carry, if the stresses in steel and concrete is 50 MPa and 3 MPa respectively. **[Ans.** 283.5 kN]
2. A concrete column of 350 mm diameter is reinforced with four bars of 25 mm diameter. Find the stress in steel when the concrete is subjected to a stress of 4.5 MPa. Also find the safe load the column can carry. Take $E_S/E_C = 18$. **[Ans.** 81 MPa; 583 kN]
3. A reinforced concrete column 300 mm × 300 mm has four reinforcing bars of 20 mm diameter one in each corner. When the column is loaded with 600 kN weight, find the stresses developed in the concrete and steel. Take $E_S/E_C = 15$. **[Ans.** 5.58 MPa ; 83.7 MPa]

## QUESTIONS

1. Define the term bars of varying sections.
2. How will you apply the principle of superposition in a stepped bar ?
3. Obtain a relation for the elongation of a uniformly circular tapering section.
4. Describe the procedure for finding out the stresses in a composite bar.
5. What is a composite section ? Explain the procedure for finding the stresses developed, when a composite section is subjected to an axial load.

## MULTIPLE CHOICE QUESTIONS

1. The total change in length of a bar of different sections is equal to the
   (*a*) sum of changes in the lengths of different sections
   (*b*) average of changes in the lengths of different sections
   (*c*) difference of changes in the lengths of different sections
   (*d*) none of these
2. A circular bar of length ($l$) uniformly tapers from diameter ($d_1$) at one end to diameter ($d_2$) at the other. If the bar is subjected to an axial tensile load ($P$), then its elongation is equal to
   (*a*) $\frac{Pl}{AE}$ (*b*) $\frac{Pl}{A_1 A_2 E}$ (*c*) $\frac{4Pl}{\pi E d_1 d_2}$ (*d*) $\frac{Pl}{4\pi E d_1 d_2}$
3. The maximum stress produced in a bar of tapering sections is at
   (*a*) larger end (*b*) smaller end (*c*) middle (*d*) anywhere
4. In a composite section, the number of different materials is
   (*a*) one only (*b*) two only (*c*) more than two (*d*) all of these
5. A composite section, contains 4 different materials. The stresses in all the different materials will be
   (*a*) zero (*b*) equal
   (*c*) different (*d*) in the ratio of their areas.

## ANSWERS

**1.** (*a*) **2.** (*c*) **3.** (*b*) **4.** (*c*) **5.** (*c*)

Chapter 4

# Stresses and Strains in Statically Indeterminate Structures

## Contents

## 4.1. Introduction

In the previous chapters, we have been discussing the cases, where simple equations of statics were sufficient to solve the examples. But, sometimes, the simple equations are not sufficient to solve such problems. Such problems are called statically indeterminate problems and the structures are called statically indeterminate structures.

For solving statically indeterminate problems, the deformation characteristics of the structure are also taken into account alongwith the statical equilibrium equations. Such equations, which contain the deformation characteristics, are called compatibility equations. The formation of such compatibility equations needs lot of patience and consideration. The solution of such statically indeterminate structures is somewhat different than the solution of simple sections and varying sections as discussed in chapters 2 and 3. So we have to adopt some indirect methods also for solving problems on statically indeterminate structures.

## 4.2. Types of Statically Indeterminate Structures

Though there are many types of statically indeterminate structures in the field of Strength of Materials yet the following are important from the subject point of view :

1. Simple statically indeterminate structures.
2. Indeterminate structures supporting a load.
3. Composite structures of equal lengths.
4. Composite structures of unequal lengths.

Now we shall study the procedures for the stresses and strains in the above mentioned indeterminate structures in the following pages. In order to solve the above mentioned types of statically indeterminate structures, we have to use different types of compatible equations.

## 4.3. Stresses in Simple tatically Indeterminate Structures

The structures in which the stresses can be obtained by forming two or more equations are called simple statically indeterminate structures. The stresses in such structures may be found out with the help of two or three compatible equations.

**EXAMPLE 4.1.** *A square bar of 20 mm side is held between two rigid plates and loaded by an axial force P equal to 450 kN as shown in Fig. 4.1.*

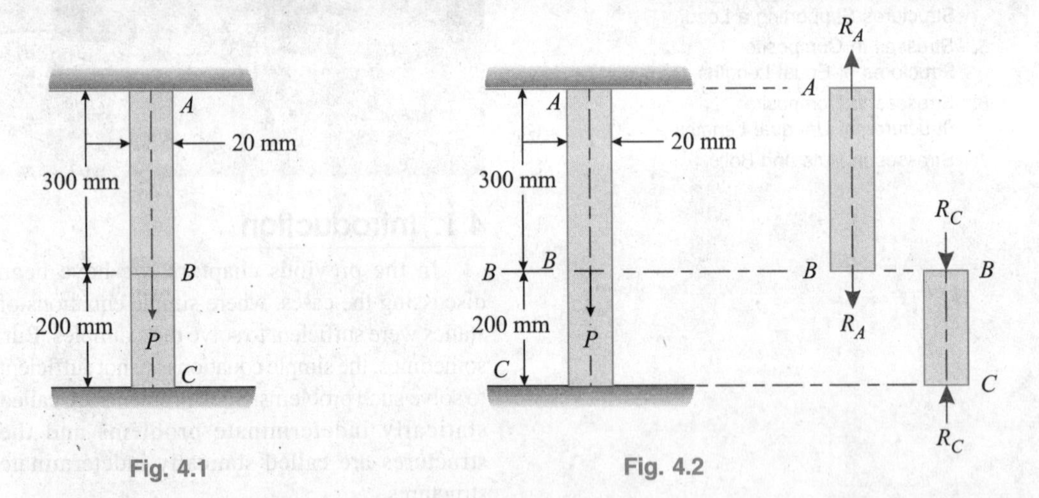


Fig. 4.1

Fig. 4.2

*Find the reactions at the ends A and C and the extension of the portion AB. Take E = 200 GPa.*

**SOLUTION.** Given : Area of bar $(A) = 20 \times 20 = 400\ \text{mm}^2$ ; Axial force $(P) = 450\ \text{kN} = 450 \times 10^3$ N ; Modulus of elasticity $(E) = 200\ \text{GPa} = 200 \times 10^3\ \text{N/mm}^2$ ; Length of $AB$ $(l_{AB}) = 300$ mm and length of $BC$ $(l_{BC}) = 200$ mm.

***Reaction at the ends***

Let $R_A$ = Reaction at $A$, and

$R_C$ = Reaction at $C$.

Since the bar is held between the two rigid plates $A$ and $C$, therefore, the upper portion will be subjected to tension, while the lower portion will be subjected to compression as shown in Fig. 4.2.

Moreover, the increase of portion *AB* will be equal to the decrease of the portion *BC*.

We know that sum of both the reaction is equal to the axial force, *i.e.*,

$$R_A + R_C = 450 \times 10^3 \quad ...(i)$$

Increase in the portion *AB*,

$$\delta l_{AB} = \frac{R_A\, l_{AB}}{AE} = \frac{R_A \times 300}{AE}$$

and decrease in the portion *BC*,

$$\delta l_{BC} = \frac{R_C\, l_{BC}}{AE} = \frac{R_C \times 200}{AE} \quad ...(ii)$$

Since the value $\delta l_{AB}$ is equal to that of $\delta l_{BC}$, therefore equating the equations (*ii*) and (*iii*),

$$\frac{R_A \times 300}{AE} = \frac{R_C \times 200}{AE}$$

$$R_C = \frac{R_A \times 300}{200} = 1.5\, R_A$$

Now substituting the value of $R_C$ in equation (*ii*),

$$R_A + 1.5\, R_A = 450 \quad \text{or} \quad 2.5\, R_A = 450$$

$$\therefore \quad R_A = \frac{450}{2.5} = 180 \text{ kN} \quad \textbf{Ans.}$$

and $$R_C = 1.5\, R_A = 1.5 \times 180 = 270 \text{ kN} \quad \textbf{Ans.}$$

***Extension of the portion AB***

Substituting the value of $R_A$ in equation (*ii*)

$$\delta_{AB} = \frac{R_A \times 300}{AE} = \frac{(180 \times 10^3) \times 300}{400 \times (200 \times 10^3)} = 0.675 \text{ mm} \quad \textbf{Ans.}$$

**EXAMPLE 4.2.** *An aluminium bar 3 m long and 2500 mm² in cross-section is rigidly fixed at A and D as shown in Fig. 4.3.*

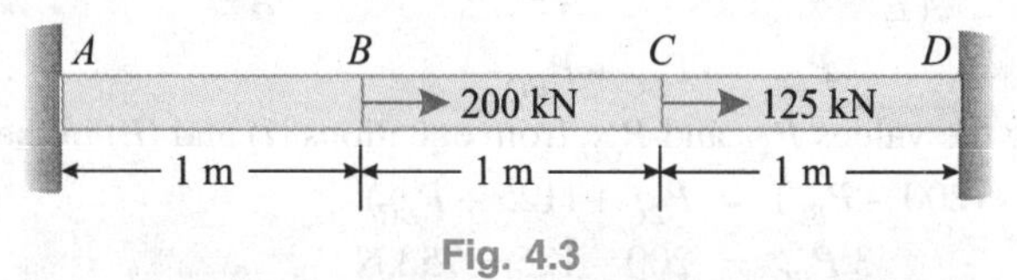


Fig. 4.3

*Determine the loads shared and stresses in each portion and the distances through which the points B and C will move. Take E for aluminium as 80 GPa.*

**SOLUTION.** Given : Total length of bar (*L*) = 3 m ; Area of cross-section A = 2500 mm² ; Modulus of elasticity (*E*) = 80 GPa = $80 \times 10^3$ N/mm² and length of portion *AB* ($l_{AB}$) = $l_{BC} = l_{CD}$ = 1 m = $1 \times 10^3$ mm.

***Loads shared by each portion***

Let $P_{AB}$ = Load shared by the portion *AB*,

$P_{BC}$ = Load shared by the portion *BC* and

$P_{CD}$ = Load shared by the portion *CD*.

Since the bar is rigidly fixed at *A* and *D*, therefore the portion *AB* will be subjected to tension, while the portions *BC* and *CD* will be subjected to compression as shown in Fig. 4.4. Moreover, increase in the portion *AB* will be equal to the sum of the decreases in the portions *BC* and *CD*.

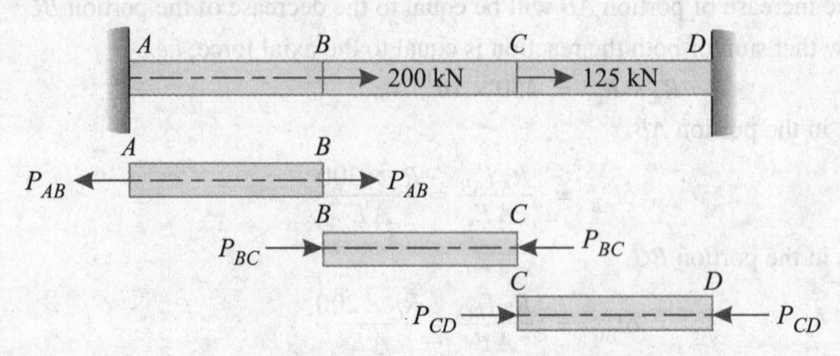


Fig. 4.4

From the geometry of the bar, we find that

$$P_{AB} + P_{BC} = 200 \quad \text{or} \quad P_{AB} = 200 - P_{BC} \qquad ...(i)$$

and

$$P_{CD} - P_{BC} = 125 \quad \text{or} \quad P_{CD} = 125 + P_{BC} \qquad ...(ii)$$

We know that increase in the length of portion *AB*,

$$\delta l_{AB} = \frac{P_{AB}\, l_{AB}}{A\,E} = \frac{P_{AB}\,(1 \times 10^3)}{A\,E} \qquad ...(iii)$$

Similarly, decrease in the length of portion *BC*,

$$\delta l_{BC} = \frac{P_{BC}\, l_{BC}}{A\,E} = \frac{P_{BC}\,(1 \times 10^3)}{A\,E} \qquad ...(iv)$$

and decrease in the length of portion *CD*,

$$\delta l_{CD} = \frac{P_{CD}\, l_{CD}}{A\,E} = \frac{P_{CD}\,(1 \times 10^3)}{A\,E} \qquad ...(v)$$

Since the value of $\delta l_{AB}$ is equal to $\delta l_{BC} + \delta l_{CD}$, therefore

$$\frac{P_{AB} \times (1 \times 10^3)}{A\,E} = \frac{P_{BC} \times (1 \times 10^3)}{A\,E} + \frac{P_{CD} \times (1 \times 10^3)}{A\,E}$$

$$\therefore \quad P_{AB} = P_{BC} + P_{CD}$$

Now substituting the values $P_{AB}$ and $P_{CD}$ from equations (*i*) and (*ii*) in the above equation,

$$(200 - P_{BC}) = P_{BC} + (125 + P_{BC})$$

$$\therefore \quad 3\,P_{BC} = 200 - 125 = 75 \text{ kN}$$

or

$$P_{BC} = \frac{75}{3} = 25 \text{ kN}$$

$$\therefore \quad P_{AB} = 200 - P_{BC} = 200 - 25 = 175 \text{ kN} \qquad \textbf{Ans.}$$

and

$$P_{CD} = 125 + P_{BC} = 125 + 25 = 150 \text{ kN} \qquad \textbf{Ans.}$$

***Stresses in each portion***

We know that stress in *AB*,

$$\sigma_{AB} = \frac{P_{AB}}{A} = \frac{175 \times 10^3}{2500} = 70 \text{ N/mm}^2 = 70 \text{ MPa (tension)} \qquad \textbf{Ans.}$$

Similarly,

$$\sigma_{BC} = \frac{P_{BC}}{A} = \frac{25 \times 10^3}{2500} = 10 \text{ N/mm}^2 = 10 \text{ MPa (compression)} \qquad \textbf{Ans.}$$

and $$\sigma_{CD} = \frac{P_{CD}}{A} = \frac{150 \times 10^3}{2500} = 60 \text{ N/mm}^2 = 60 \text{ MPa (compression)} \quad \textbf{Ans.}$$

***Distance through which the points B and C will move***

Substituting the value of $P_{AB}$ in equation (*iii*), we get

$$\delta l_{AB} = \frac{P_{AB} \times l_{AB}}{AE} = \frac{175 \times 10^3 \times (1 \times 10^3)}{2500 \times (80 \times 10^3)} = 0.875 \text{ mm} \quad \textbf{Ans.}$$

and now substituting the value of $P_{CD}$ in equation (*iv*), we get

$$\delta l_{CD} = \frac{P_{CD} \times l_{CD}}{AE} = \frac{(150 \times 10^3) \times (1 \times 10^3)}{2500 \times (80 \times 10^3)} = 0.75 \text{ mm} \quad \textbf{Ans.}$$

**EXAMPLE 4.3.** *A circular steel bar ABCD, rigidly fixed at A and D is subjected to axial loads of 50 kN and 100 kN at B and C as shown in Fig. 4.5.*

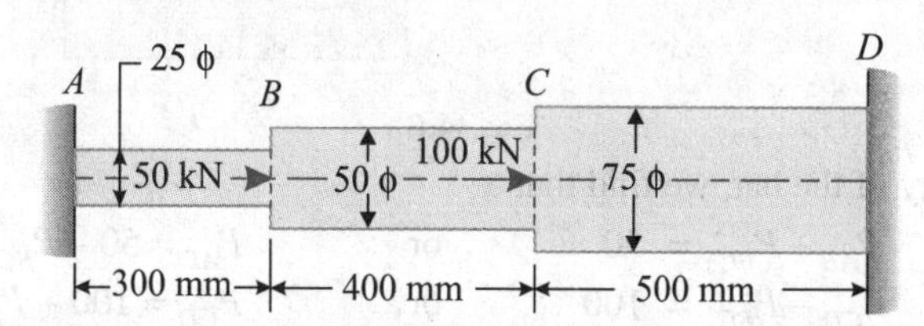


Fig. 4.5

*Find the loads shared by each part of the bar and displacements of the points B and C. Take E for the steel as 200 GPa.*

**SOLUTION.** Given : Axial load at $B$ ($P_1$) = 50 kN = 50 × $10^3$ N ; Axial load at $C$ ($P_2$) = 100 kN = 100 × $10^3$ N ; Diameter of $AB$ ($D_{AB}$) = 25 mm ; length of $AB$ ($l_{AB}$) = 300 mm ; Diameter of $BC$ ($D_{BC}$) = 50 mm ; Length of $BC$ ($l_{BC}$) = 400 mm ; Diameter of $CD$ ($D_{CD}$) = 75 mm ; Length of $CD$ ($l_{CD}$) = 500 mm and modulus of elasticity ($E$) = 200 GPa = 200 × $10^3$ N/mm$^2$.

***Loads shared by each part of the bar***

Let $P_{AB}$ = Load shared by $AB$,

$P_{BC}$ = Load shared by $BC$, and

$P_{CD}$ = Load shared by $CD$.

We know that area of the bar $AB$,

$$A_{AB} = \frac{\pi}{4} \times (D_{AB})^2 = \frac{\pi}{4} \times (25)^2 = 491 \text{ mm}^2$$

Similarly, area of the bar $BC$,

$$A_{BC} = \frac{\pi}{4} \times (D_{BC})^2 = \frac{\pi}{4} \times (50)^2 = 1964 \text{ mm}^2$$

and area of the bar $CD$,

$$A_{CD} = \frac{\pi}{4} \times (D_{CD})^2 = \frac{\pi}{4} (75)^2 = 4418 \text{ mm}^2$$

Since the bar is rigidly fixed at $A$ and $D$, therefore, the portion $AB$ will be subjected to tension, while the portions $BC$ and $CD$ will be subjected to compression as shown in Fig. 4.6. Moreover, increase in the length $AB$ is equal to the sum of decreases in the portions $BC$ and $CD$.

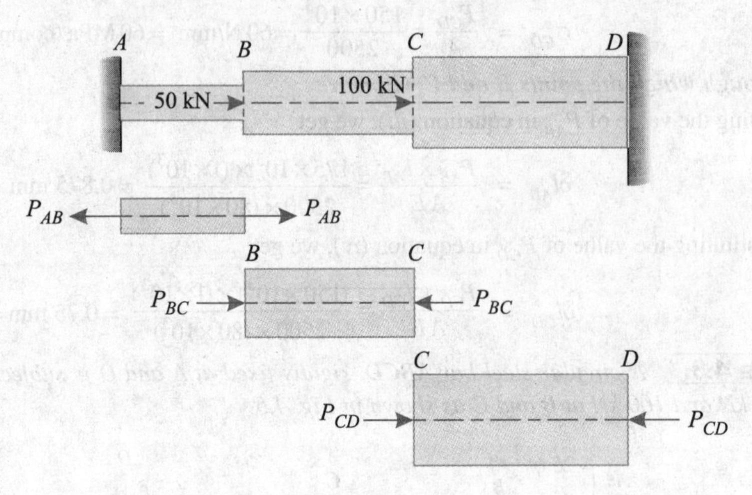


Fig. 4.6

From the geometry of the bar, we find that

$$P_{AB} + P_{BC} = 50 \quad \text{or} \quad P_{AB} = 50 - P_{BC} \quad ...(i)$$

and

$$P_{CD} - P_{BC} = 100 \quad \text{or} \quad P_{CD} = 100 + P_{BC} \quad ...(ii)$$

We know that increase in the length of portion *AB*,

$$\delta l_{AB} = \frac{P_{AB}\, l_{AB}}{A_{AB}\, E} = \frac{P_{AB} \times 300}{491 \times (200 \times 10^3)} = 3.05 \times 10^{-6}\, P_{AB} \text{ mm} \quad ...(iii)$$

Similarly,

$$\delta l_{BC} = \frac{P_{BC}\, l_{BC}}{A_{BC}\, E} = \frac{P_{BC} \times 400}{1964 \times (200 \times 10^3)} = 1.02 \times 10^{-6}\, P_{BC} \text{ mm} \quad ...(iv)$$

and

$$\delta l_{CD} = \frac{P_{CD}\, l_{CD}}{A_{CD}\, E} = \frac{P_{CD} \times 500}{4418 \times (200 \times 10^3)} = 0.57 \times 10^{-6}\, P_{CD} \text{ mm} \quad ...(v)$$

Since the value of $\delta l_{AB}$ is equal to $\delta l_{BC} + \delta l_{CD}$, therefore

$$3.05 \times 10^{-6}\, P_{AB} = 1.02 \times 10^{-6}\, P_{BC} + 0.57 \times 10^{-6}\, P_{CD}$$

$$\therefore \quad 305\, P_{AB} = 102\, P_{BC} + 57\, P_{CD}$$

Now substituting the values of $P_{AB}$ and $P_{CD}$ from equations (*i*) and (*ii*) in the above equation,

$$305\,(50 - P_{BC}) = 102\, P_{BC} + 57\,(100 + P_{BC})$$

$$15\,250 - 305\, P_{BC} = 102\, P_{BC} + 5700 + 57\, P_{BC}$$

$$\therefore \quad 464\, P_{BC} = 9\,550 \quad \text{or} \quad P_{BC} = \frac{9550}{464} = 20.6 \text{ kN} \quad \textbf{Ans.}$$

Similarly,

$$P_{AB} = 50 - P_{BC} = 50 - 20.6 = 29.4 \text{ kN} \quad \textbf{Ans.}$$

and

$$P_{CD} = 100 + P_{BC} = 100 + 20.6 = 120.6 \text{ kN} \quad \textbf{Ans.}$$

***Displacements of the points B and C***

Now substituting the value of $P_{AB}$ in equation (*iii*), we get

$$\delta l_{AB} = 3.05 \times 10^{-6}\, P_{AB} = 3.05 \times 10^{-6} \times (29.4 \times 10^3) = 0.90 \text{ mm} \quad \textbf{Ans.}$$

and now substituting the value of $P_{CD}$ in equation (*v*), we get

$$\delta l_{CD} = 0.57 \times 10^{-6} \times P_{CD} = 0.57 \times 10^{-6} \times (120.6 \times 10^3) = 0.07 \text{ mm} \quad \textbf{Ans.}$$

## EXERCISE 4.1

1. An alloy bar 800 mm long and 200 mm$^2$ in cross-section is held between two rigid plates and is subjected to an axial load of 200 kN as shown in Fig. 4.7.

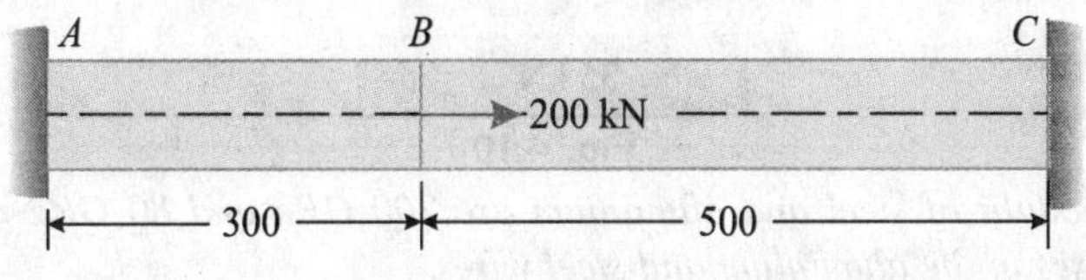


Fig. 4.7

Find the reactions at the two ends *A* and *C* as well as extension of the portion *AB*.

[**Ans.** 125 kN ; 75 kN ; 0.094 mm]

2. A bar *ABC* fixed at both ends *A* and *C* is loaded by an axial load (*P*) at *C*. If the distances *AB* and *BC* are equal to *a* and *b* respectively then find the reactions at the ends *A* and *C*.

3. An axial force of 20 kN is applied to a steel bar *ABC* which is fixed at both ends *A* and *C* as shown in Fig. 4.8.

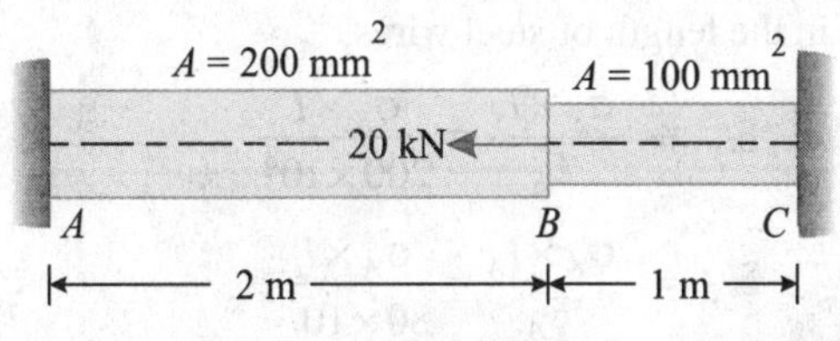


Fig. 4.8

Determine the reactions at both the supports and stresses developed in two parts of the bar. Take $E$ = 200 GPa. [**Ans.** $R_A = R_C$ = 10 kN ; $\sigma_{AB}$ = 50 MPa (*C*); $\sigma_{BC}$ = 100 MPa (*T*)]

4. A prismatic bar *ABCD* has built-in ends *A* and *D*. It is subjected to two point loads $P_1$ and $P_2$ equal to 80 kN and 40 kN at *B* and *C* as shown in Fig. 4.9.

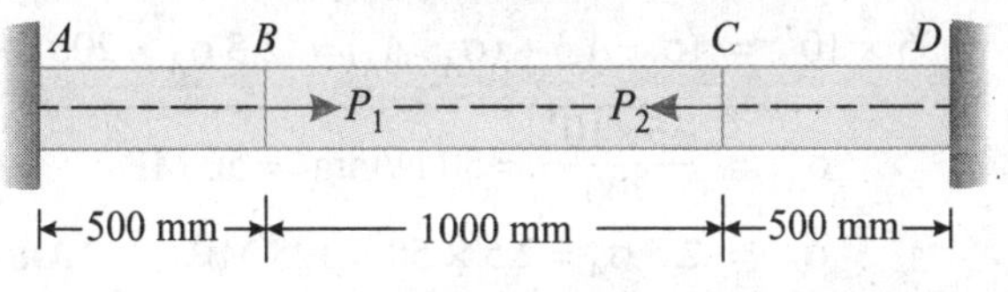


Fig. 4.9

Find the reactions at *A* and *D*. [**Ans.** 70 kN ; 50 kN]

## 4.4. Stresses in Indeterminate Structures Supporting a Load

Sometimes, we come across a set of two or more members supporting a load. In such cases, the deformation of all the members will be the same. If the members are of different cross-sections or have different modulus of elasticity, then the stresses developed in all the members will be different.

**EXAMPLE 4.4.** *A block shown in Fig. 4.10 weighing 35 kN is supported by three wires. The outer two wires are of steel and have an area of 100 mm$^2$ each, whereas the middle wire of aluminium and has an area of 200 mm$^2$.*

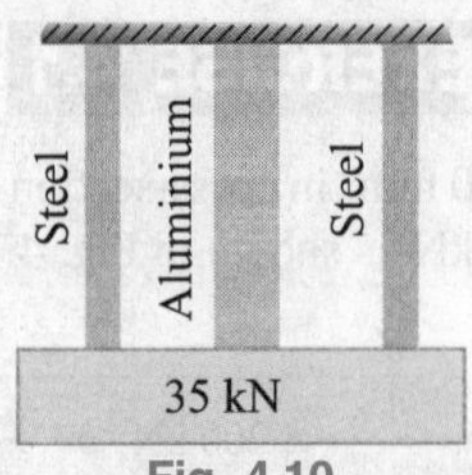

Fig. 4.10

*If the elastic modulii of steel and aluminium are 200 GPa and 80 GPa respectively, then calculate the stresses in the aluminium and steel wires.*

**SOLUTION.** Given: Total load ($P$) = 35 kN = 35 × $10^3$ N ; Total area of steel rods ($A_S$) = 2 × 100 = 200 $mm^2$ ; Area of aluminium rod ($A_A$) = 200 $mm^2$ ; Modulus of elasticity of steel ($E_S$) = 200 GPa = 200 × $10^3$ $N/mm^2$; Modulus of elasticity of aluminium ($E_A$) = 80 GPa = 80 × $10^3$ $N/mm^2$ and load supported by wires ($P$) = 35 kN = 35 × $10^3$ N

Let $\sigma_S$ = Stress in steel wires,

$\sigma_A$ = Stress in aluminium wire and

$l$ = Length of the wires.

We know that increase in the length of steel wires,

$$\delta l_S = \frac{\sigma_S \times l_S}{E_S} = \frac{\sigma_S \times l}{200 \times 10^3}$$

Similarly,

$$\delta l_A = \frac{\sigma_A \times l_A}{E_A} = \frac{\sigma_A \times l}{80 \times 10^3}$$

Since increase in the lengths of steel and aluminium wires is equal, therefore equating equations ($i$) and ($ii$), we get

$$\frac{\sigma_S \times l}{200 \times 10^3} = \frac{\sigma_A \times l}{80 \times 10^3} \quad \text{or} \quad \sigma_S = \frac{200}{80} \times \sigma_A = 2.5\,\sigma_A$$

We also know that load supported by the three wires ($P$),

$$35 \times 10^3 = (\sigma_S \cdot A_S) + (\sigma_A \cdot A_A) = (2.5\,\sigma_A \times 200) + (\sigma_A \times 200) = 700\,\sigma_A$$

$$\therefore \quad \sigma_A = \frac{35 \times 10^3}{700} = 50\ \text{N/mm}^2 = 50\ \text{MPa} \qquad \textbf{Ans.}$$

and $\sigma_S = 2.5\,\sigma_A = 2.5 \times 50 = 125$ MPa **Ans.**

**EXAMPLE 4.5.** *A steel rod of cross-sectional area 800 $mm^2$ and two brass rods each of cross-sectional area 500 $mm^2$ together support a load of 25 kN as shown in Fig. 4.11.*

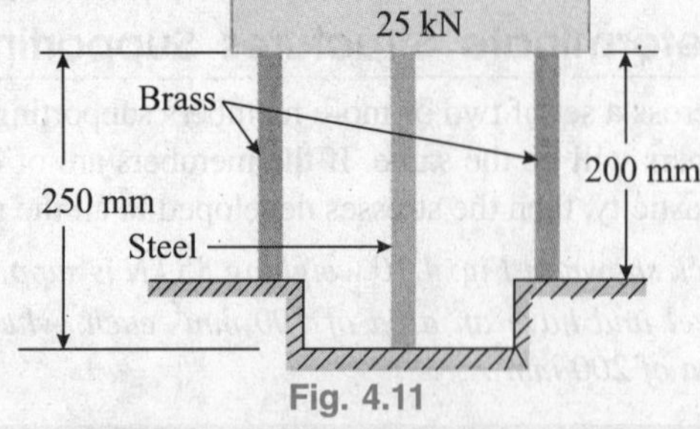

Fig. 4.11

*Calculate the stresses in the rods. Take E for steel as 200 GPa and E for brass as 100 GPa.*

**SOLUTION.** Given : Area of one steel rod, ($A_S$) = 800 $mm^2$ ; Total Area of two brass rods ($A_B$) = 2 × 500 = 1000 $mm^2$ ; Total load ($P$) = 25 kN = 25 × $10^3$ N ; Modulus of elasticity of steel ($E_S$) = 200 GPa : Modulus of elasticity of brass ($E_B$) = 100 GPa ; Length of steel bar ($l_S$) = 250 mm and length of brass rod ($l_B$) = 200 mm.

Let $\sigma_S$ = Stress in steel rod and

$\sigma_B$ = Stress in brass rod.

We know that decrease in the length of the steel rod due to stress,

$$\delta l_S = \frac{\sigma_S \, l_S}{E_S} = \frac{\sigma_S \times 250}{200 \times 10^3} = 1.25 \times 10^{-3} \, \sigma_S$$

and decrease in the length of the brass rods due to stress,

$$\delta l_B = \frac{\sigma_B \, l_B}{E_B} = \frac{\sigma_B \times 200}{100 \times 10^3} = 2 \times 10^{-3} \, \sigma_B$$

Since the value of $\delta l_S$ is equal to that of $\delta l_B$, therefore equating equations (*i*) and (*ii*), we get

$$1.25 \times 10^{-3} \, \sigma_S = 2 \times 10^{-3} \, \sigma_B \quad \text{or} \quad \sigma_S = \frac{2}{1.25} \times \sigma_B = 1.6 \, \sigma_B$$

We also know that total load shared by all the three rods ($P$),

$$25 \times 10^3 = \sigma_S A_S + \sigma_B A_B = (1.6 \, \sigma_B \times 800) + (\sigma_B \times 1000) = 2280 \, \sigma_B$$

$$\therefore \quad \sigma_B = \frac{25 \times 10^3}{2280} = 11.0 \text{ N/mm}^2 = 11.0 \text{ MPa} \qquad \textbf{Ans.}$$

and $\sigma_S = 1.6 \, \sigma_B = 1.6 \times 11.0 = 17.6$ MPa **Ans.**

**EXAMPLE 4.6.** *A load of 80 kN is jointly supported by three rods of 20 mm diameter as shown in Fig. 4.12.*

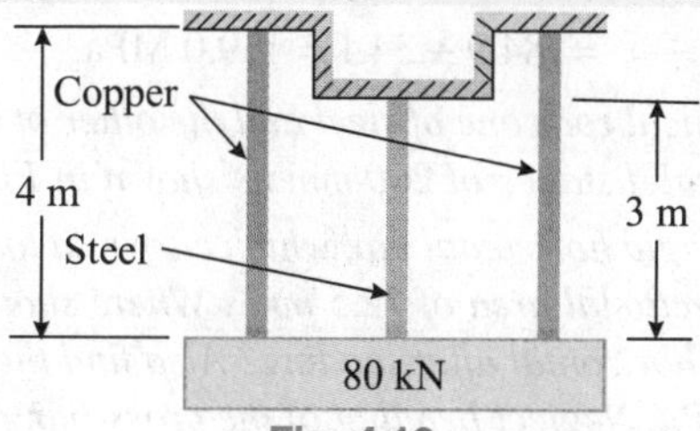


Fig. 4.12

*The rods are adjusted in such a way that they share the load equally. If an additional load of 50 kN is added, find the final stresses in steel and copper. Take E for copper as 100 GPa and for steel as 200 GPa.*

**SOLUTION.** Given : Total load ($P_1$) = 80 kN = 80 × $10^3$ N ; Diameter of each rod ($d$) = 20 mm ; Additional load ($P_2$) = 50 kN = 50 × $10^3$ N ; Modulus of elasticity of copper ($E_C$) = 100 GPa = 100 × $10^3$ N/mm$^2$ and modulus of elasticity of steel ($E_S$) = 200 GPa = 200 × $10^3$ N/mm$^2$.

We know that total area of two copper rods

$$A_C = 2 \times \frac{\pi}{4} \times (d)^2 = 2 \times \frac{\pi}{4} \times (20)^2 = 200 \, \pi \text{ mm}^2$$

and area of one steel rod

$$A_S = \frac{\pi}{4} \times (d)^2 = \frac{\pi}{4} \times (20)^2 = 100 \, \pi \text{ mm}^2$$

First of all consider the 80 kN load only, which is shared equally by all the three rods. We know that initial stress in each rod

$$= \frac{80 \times 10^3}{3 \times 100 \pi} = 84.9 \text{ N/mm}^2 = 84.9 \text{ MPa} \qquad ...(i)$$

Now consider an additional load of 50 kN, which is added to the existing load of 80 kN. We know that this additional load will cause some additional stresses in all the three rods.

Let $\sigma_C$ = Additional stress in copper rods, and

$\sigma_S$ = Additional stress in steel rod

We know that increase in the length of copper rods due to stress,

$$\delta l_C = \frac{\sigma_C \times l_C}{E_C} = \frac{\sigma_C \times (4 \times 10^3)}{100 \times 10^3} = 0.04\ \sigma_C \qquad ...(ii)$$

and increase in the length of steel rod due to stress,

$$\delta l_S = \frac{\sigma_S \times l_S}{E_S} = \frac{\sigma_S \times (3 \times 10^3)}{200 \times 10^3} = 0.015\ \sigma_S \qquad ...(iii)$$

Since the value of $\delta l_C$ is equal to that of $\delta l_S$, therefore equating the equations (*ii*) and (*iii*)

$$0.04\ \sigma_C = 0.015\ \sigma_S \quad \text{or} \quad \sigma_C = 0.375\ \sigma_S$$

We also know that additional load supported by the three rods ($P_2$)

$$50 \times 10^3 = (\sigma_S \cdot A_S) + (\sigma_C \cdot A_C) = (\sigma_S \times 100\ \pi) + (0.375\ \sigma_S \times 200\ \pi)$$

$$= 175\ \pi\ \sigma_S$$

or $$\sigma_S = \frac{50 \times 10^3}{175 \pi} = 90.9 \text{ N/mm}^2 = 90.9 \text{ MPa}$$

and $$\sigma_C = 0.375\ \sigma_S = 0.375 \times 90.9 = 34.1 \text{ MPa}$$

∴ Final stress in the steel

$= 84.9 + 90.9 = 175.8$ MPa **Ans.**

and final stress in copper $= 84.9 + 34.1 = 119.0$ MPa **Ans.**

**EXAMPLE 4.7.** *Two vertical rods one of steel and the other of copper are rigidly fastened at their upper end at a horizontal distance of 200 mm as shown in Fig. 4.13.*

*The lower ends support a rigid horizontal bar, which carries a load of 10 kN. Both the rods are 2.5 m long and have cross-sectional area of 12.5 mm². Where should a load of 10 kN be placed on the bar, so that it remains horizontal after loading? Also find the stresses in each rod. Take $E_S$ = 200 GPa and $E_C$ = 110 GPa. Neglect bending of the cross-bar.*

**SOLUTION.** Given : Distance between the bars = 200 mm ; Total load ($P$) = 10 kN = $10 \times 10^3$ N ; Length of steel rod ($l_S$) = $l_C$ = 2.5 m = $2.5 \times 10^3$ mm ; Area of steel rod ($A_S$) = $A_C$ = 12.5 mm²; Modulus of elasticity of steel ($E_S$) = 200 GPa = $200 \times 10^3$ N/mm² and modulus of elasticity of copper ($E_C$) = 110 GPa = $110 \times 10^3$ N/mm².

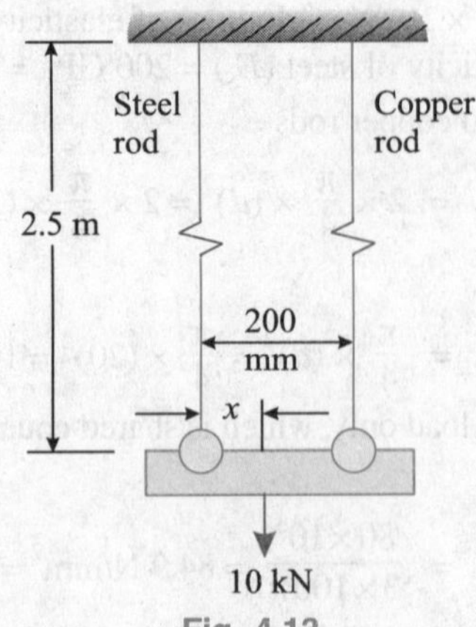


Fig. 4.13

*Position of the load*

Let $x$ = Distance between the load and steel rod in mm

As a matter of fact, the load of 10 kN will be shared by both the rods in such a way that they cause equal extension.

Let $P_S$ = Load shared by the steel rod, and

$P_C$ = Load shared by the copper rod.

$$\therefore \quad P_S + P_C = 10 \text{ kN} \quad ...(i)$$

We know that extension of the steel rod,

$$\delta l_S = \frac{P_S\, l}{A_S\, E_S} = \frac{P_S \times (2.5 \times 10^3)}{12.5 \times (200 \times 10^3)} = \frac{P_S}{1000} \quad ...(ii)$$

and extension of the copper rod,

$$\delta l_C = \frac{P_C\, l}{A_C\, E_C} = \frac{P_C \times (2.5 \times 10^3)}{12.5 \times (110 \times 10^3)} = \frac{P_C}{550} \quad ...(iii)$$

Since both the extensions are equal, therefore equating equations (*ii*) and (*iii*)

$$\frac{P_S}{1000} = \frac{P_C}{550} \quad \text{or} \quad \frac{P_S}{P_C} = \frac{1000}{550} = \frac{20}{11} \quad (iv)$$

Now taking moments of the loads about the steel bar and equating the same,

$$10 \times x = P_C \times 200 \quad \text{or} \quad (P_S + P_C)\, x = 200\, P_C$$

$$P_S . x + P_C . x = 200\, P_C \quad \text{or} \quad P_S . x = 200\, P_C - P_C . x = P_C\,(200 - x)$$

$$\therefore \quad \frac{P_S}{P_C} = \frac{200 - x}{x} \quad ...(v)$$

Now equating two values of $\frac{P_S}{P_C}$ from equations (*iv*) and (*v*),

$$\frac{20}{11} = \frac{200 - x}{x} \quad \text{or} \quad 20x = 2200 - 11x$$

$$\therefore \quad 31x = 2200 \quad \text{or} \quad x = \frac{2200}{31} = 71 \text{ mm} \quad \textbf{Ans.}$$

*Stresses in each rod*

From equation (*iv*), we find that

$$\frac{P_S}{P_C} = \frac{20}{11} \quad \text{or} \quad 11\, P_S - 20\, P_C - 20\,(10 - P_S) = 200 - 20\, P_S$$

$$\therefore \quad 31\, P_S = 200 \quad \text{or} \quad P_S = \frac{200}{31} = 6.45 \text{ kN} = 6.45 \times 10^3 \text{ N}$$

and

$$P_C = 10 - P_S = 10 - 6.45 = 3.5 \text{ kN} = 3.5 \times 10^3 \text{ N}$$

∴ Stress in steel rod,

$$\sigma_S = \frac{P_S}{A_S} = \frac{6.45 \times 10^3}{12.5} = 516 \text{ N/mm}^2 = 516 \text{ MPa} \quad \textbf{Ans.}$$

and stress in copper rod,

$$\sigma_C = \frac{P_C}{A_C} = \frac{3.5 \times 10^3}{12.5} = 280 \text{ N/mm}^2 = 280 \text{ MPa} \quad \textbf{Ans.}$$

**EXAMPLE 4.8.** *A load of 5 kN is suspended by ropes as shown in Fig. 4.14 (a) and (b). In both the cases, the cross-sectional area of the ropes is 200 mm² and the value of E is 1.0 GPa.*

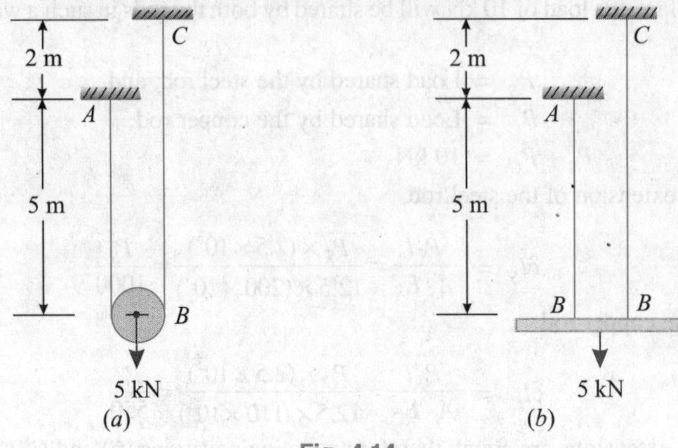


**Fig. 4.14**

*In (a) the rope ABC is continuous over a smooth pulley, from which the load is suspended. In (b) the ropes AB and CB are separate ropes joined to a block, from which the load is suspended in such a way, that both the ropes are stretched by the same amount. Determine, for both the cases, stresses in the ropes and the deflections of the pulley and the block due to the load.*

**SOLUTION.** Given : Total load $(P) = 5\text{ kN} = 5 \times 10^3\text{ N}$ ; Length of $AB\ (l_{AB}) = 5\text{ m} = 5 \times 10^3\text{ mm}$; Length of $BC\ (l_{BC}) = 7\text{ m} = 7 \times 10^3\text{ mm}$ ; Area of each rope $(A) = 200\text{ mm}^2$ and modulus of elasticity $(E) = 1.0\text{ GPa} = 1.0 \times 10^3\text{ N/mm}^2$.

***First case***

We know that the load of 5 kN is suspended from the pulley, therefore load shared by both the ropes is equal. Or in other words, load shared by each rope.

$$P_1 = \frac{5 \times 10^3}{2} = 2.5 \times 10^3\text{ N}$$

$\therefore$ Stress in the ropes,

$$\sigma = \frac{P_1}{A} = \frac{2.5 \times 10^3}{200} = 12.5\text{ N/mm}^2 = 12.5\text{ MPa} \quad \textbf{Ans.}$$

and total elongation of the rope $ABC$,

$$\delta l = \frac{P_1 l_{AB}}{AE} = \frac{P_1 l_{BC}}{AE}$$

$$= \frac{(2.5 \times 10^3) \times (5 \times 10^3)}{200 \times (1.0 \times 10^3)} + \frac{(2.5 \times 10^3) \times (7 \times 10^3)}{200 \times (1.0 \times 10^3)}$$

$$= 62.5 + 87.5 = 150\text{ mm}$$

$\therefore$ Deflection of the pulley

$$= \frac{150}{2} = 75\text{ mm} \quad \textbf{Ans.}$$

***Second case***

Let $\sigma_{AB}$ = Stress in the rope $AB$, and

$\sigma_{BC}$ = Stress in the rope $BC$.

We know that deflection of the rope $AB$,

$$\delta l_{AB} = \frac{\sigma_{AB} \cdot l_{AB}}{E} = \frac{\sigma_{AB} \times (5 \times 10^3)}{1 \times 10^3} = 5\,\sigma_{AB}$$

and deflection of the rope $BC$,

$$\delta l_{BC} = \frac{\sigma_{BC} \cdot l_{BC}}{E} = \frac{\sigma_{BC} \times (5 \times 10^3)}{1 \times 10^3} = 7\,\sigma_{AB}$$

Since both the deflections are equal, therefore equating the value of (*i*) and (*ii*),

$$5\,\sigma_{AB} = 7\,\sigma_{BC} \qquad \text{or} \qquad \sigma_{AB} = \frac{7}{5} \times \sigma_{BC}$$

We also know that the load ($P$) of 5 kN is shared by both the ropes, therefore load ($P$)

$$5 \times 10^3 = \sigma_{AB} \times A + \sigma_{BC} \times A = \left(\frac{7}{5} \times \sigma_{BC} \times 200\right) + (\sigma_{BC} \times 200)$$

$$= 480\,\sigma_{BC}$$

$$\therefore \qquad \sigma_{BC} = \frac{5 \times 10^3}{480} = 10.4 \text{ N/mm}^2 = 10.4 \text{ MPa} \qquad \textbf{Ans.}$$

and

$$\sigma_{AB} = \frac{7}{5}\,\sigma_{BC} = \frac{7}{5} \times 10.4 \;\; 14.56 \text{ MPa} \qquad \textbf{Ans.}$$

Now substituting the value of $\sigma_{AB}$ is equation (*i*),

$$\delta l_{AB} = 5\,\sigma_{AB} = 5 \times 14.56 = 72.8 \text{ mm} \qquad \textbf{Ans.}$$

**Note.** The deflection of the block may also be found out by equating the value of $\sigma_{BC}$ in equation (*ii*),

$$\delta l_{BC} = 7\,\sigma_{BC} = 7 \times 10.4 = 72.8 \text{ mm} \qquad \textbf{Ans.}$$

## EXERCISE 4.2

1. Three long parallel wires equal in length are supporting a rigid bar connected at their bottoms as shown in Fig. 4.15. If the cross-sectional area of each wire is 100 mm², calculate the stresses in each wire. Take $E_B$ = 100 GPa and $E_S$ = 200 GPa. **[Ans.** $\sigma_B$ = 25 MPa ; $\sigma_S$ = 50 MPa]

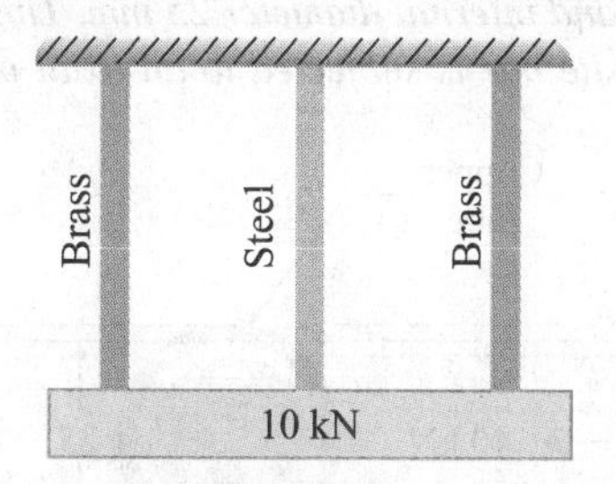


Fig. 4.15

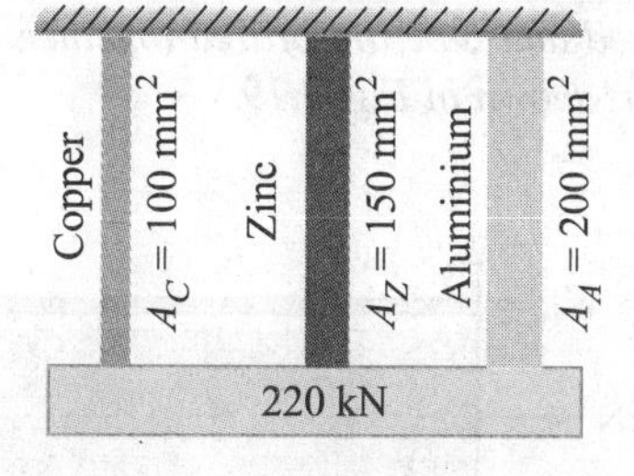


Fig. 4.16

2. Three wires made of copper, zinc and aluminium are of equal lengths and have cross-sectional areas of 100, 150 and 200 square mm respectively. They are rigidly connected at their ends as shown in Fig. 4.16. If this compound member is subjected to a longitudinal pull of 220 kN, estimate the load carried on each wire. Take $E_C$ = 130 GPa, $E_Z$ = 100 GPa and $E_A$ = 80 GPa. **[Ans.** $P_C$ = 65 kN, $P_Z$ = 75 kN, $P_A$ = 80 kN]

3. Two steel rods and one copper rod each of 20 mm diameter together support a load of 50 kN as shown in Fig.4.17. Find the stresses in each rod. Take $E$ for steel and copper as 200 GPa and 100 GPa respectively. **[Ans.** $\sigma_C$ = 39.8 MPa ; $\sigma_S$ = 59.7 MPa]

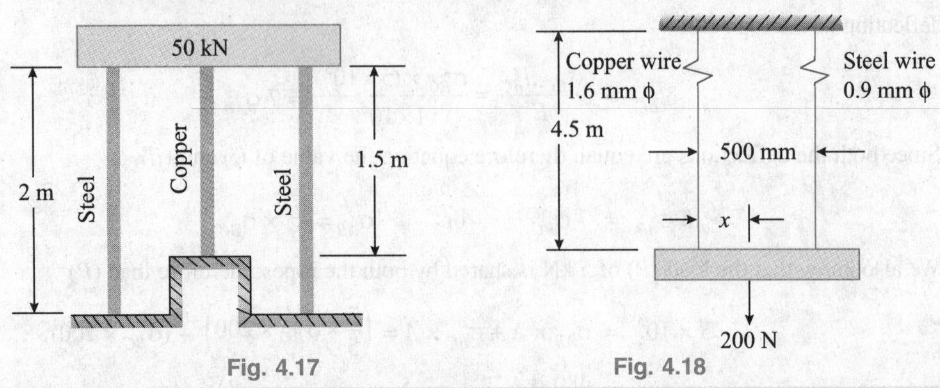


Fig. 4.17 Fig. 4.18

4. Two vertical wires are suspended at a distance of 500 mm apart as shown in Fig. 4.18. Their upper ends are firmly secured and their lower ends support a rigid horizontal bar, which carries a load of 200 N. The left hand wire has a diameter of 1.6 mm and is made of copper, and the right hand wire has a diameter of 0.9 mm and is made of steel. Both wires, initially, are 4.5 metres long. Determine :

(*a*) Position of the line of action of the load, if both the wires extend by the same amount.

(*b*) Slope of the rigid wire, if the load is hung at the centre of the bar. Neglect weight of the bar.

Take $E$ for copper as 100 GPa and $E$ for steel as 200 GPa. **[Ans.** 170 mm ; 0.15°]

## 4.5. Stresses in Composite Structures of Equal Lengths

We have already discussed in Art 3.6 the procedure for stresses in the bars of composite sections. The same principle can be extended to the statically indeterminate structures also. Though there are many types of such structures, yet a rod passing axially through a pipe is an important structure from the subject point of view.

**EXAMPLE 4.9.** *A mild steel rod of 20 mm diameter and 300 mm long is enclosed centrally inside a hollow copper tube of external diameter 30 mm and internal diameter 25 mm. The ends of the rod and tube are brazed together, and the composite bar is subjected to an axial pull of 40 kN as shown in Fig. 4.19.*

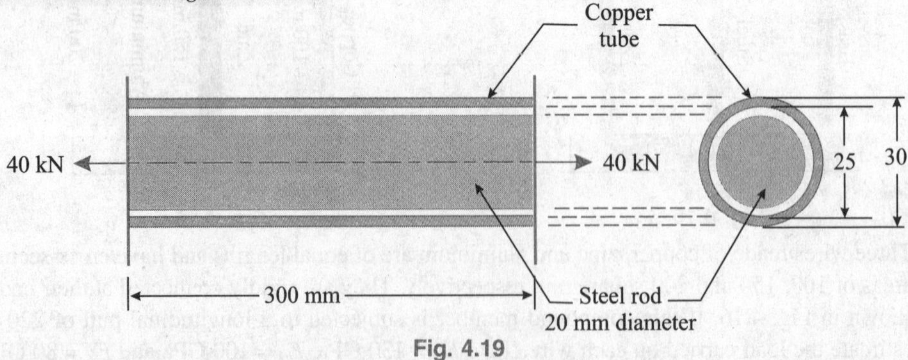


Fig. 4.19

*If E for steel and copper is 200 GPa and 100 GPa respectively, find the stresses developed in the rod and the tube.*

**SOLUTION.** Given : Diameter of steel rod = 20 mm ; External diameter of copper tube = 30 mm; Internal diameter of copper tube = 25 mm ; Total load ($P$) = 40 kN = $40 \times 10^3$ N ; Modulus of elasticity of steel ($E_S$) = 200 GPa and modulus of elasticity of copper ($E_C$) = 100 GPa.

Let $\sigma_S$ = Stress developed in the steel rod and

$\sigma_C$ = Stress developed in the copper tube.

We know that area of steel rod,

$$A_S = \frac{\pi}{4} \times (20)^2 = 314.2 \text{ mm}^2$$

and area of copper tube,

$$A_C = \frac{\pi}{4} [(30)^2 - (25)^2] = 216 \text{ mm}^2$$

We also know that stress in steel,

$$\sigma_S = \frac{E_S}{E_C} \times \sigma_C = \frac{200}{100} \times \sigma_C = 2\,\sigma_C$$

and total load ($P$), $$40 \times 10^3 = (\sigma_S . A_S) + (\sigma_C . A_C)$$
$$= (2\sigma_C \times 314.2) + (\sigma_C \times 216) = 844.4\,\sigma_C$$

$\therefore$ $$\sigma_C = \frac{40 \times 10^3}{844.4} = 47.4 \text{ N/mm}^2 = 47.4 \text{ MPa} \quad \textbf{Ans.}$$

and $$\sigma_S = 2\,\sigma_C = 2 \times 47.4 = 94.8 \text{ MPa} \quad \textbf{Ans.}$$

**EXAMPLE 4.10.** *A composite bar is made up of a brass rod of 25 mm diameter enclosed in a steel tube of 40 mm external diameter and 30 mm internal diameter as shown in Fig. 4.20. The rod and tube, being coaxial and equal in length, are securely fixed at each end. If the stresses in brass and steel are not to exceed 70 MPa and 120 MPa respectively, find the load (P) the composite bar can safely carry.*

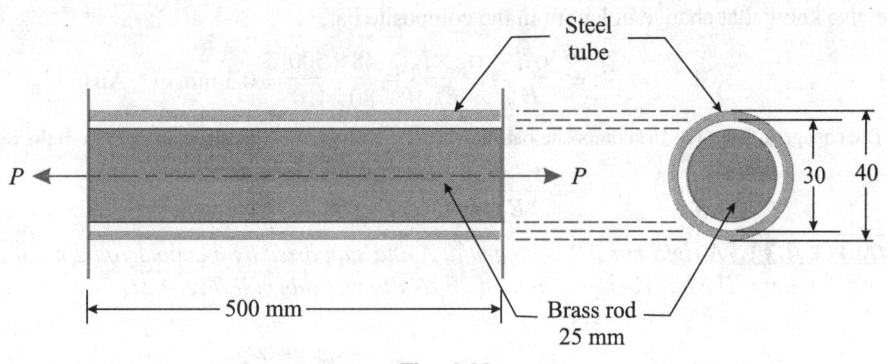


Fig. 4.20

*Also find the change in length, if the composite bar is 500 mm long. Take E for steel tube as 200 GPa and brass rod as 80 GPa respectively.*

**SOLUTION.** Given : Diameter of brass rod = 25 mm ; External diameter of steel tube = 40 mm ; Internal diameter of steel tube = 30 mm ; Maximum stress in brass ($\sigma_{B(max)}$) = 70 MPa = 70 N/mm$^2$ ; Maximum stress in steel ($\sigma_{S(max)}$) = 120 MPa = 120 N/mm$^2$ ; Length of brass rod ($l_B$) = $l_S$ = 500 mm; Modulus of elasticity of steel ($E_S$) = 200 GPa = 200 × 10$^3$ N/mm$^2$ and modulus of elasticity of brass ($E_B$) = 80 GPa = 80 × 10$^3$ N/mm$^2$.

***Load the composite bar can safely carry***

We know that area of brass rod,

$$A_B = \frac{\pi}{4} \times (25)^2 = 491 \text{ mm}^2$$

and area of steel tube, $$A_S = \frac{\pi}{4} \times [(40)^2 - (30)^2] = 550 \text{ mm}^2$$

We also know that as the brass rod and steel tube are securely fixed at each end, therefore strains in both of them will be equal. *i.e.*,

$$\varepsilon_B = \varepsilon_S \quad \text{or} \quad \frac{\sigma_B}{E_B} = \frac{\sigma_S}{E_S}$$

First of all, let us find out the maximum stresses in the brass rod and steel tube. We know that when stress in the brass is 70 N/mm$^2$ (maximum permissible), then stress in the steel tube,

$$\sigma_S = \frac{E_S}{E_B} \times \sigma_B = \frac{200}{80} \times 70 = 175 \text{ N/mm}^2$$

It is more than the permissible stress in the steel (which is given as 120 N/mm$^2$). Therefore we can not accept these values of stresses in brass and steel. Now when the stress in steel tube is 120 N/mm$^2$ (maximum permissible), then stress in the brass rod,

$$\sigma_B = \frac{E_B}{E_S} \times \sigma_S = \frac{80}{200} \times 120 = 48 \text{ N/mm}^2$$

It is less than the permissible stress in brass (which is given as 70 N/mm$^2$). Thus we shall take the stresses in the brass rod ($\sigma_B$) and steel tube ($\sigma_S$) as 48 N/mm$^2$ and 120 N/mm$^2$ respectively. Therefore load which the composite bar can carry,

$$P = (\sigma_B . A_B) + (\sigma_S . A_S) = (48 \times 491) + (120 \times 550) \text{ N}$$

$$= 89\,570 \text{ N} = 89.57 \text{ kN} \quad \textbf{Ans.}$$

***Change in length***

We also know that change in length in the composite bar,

$$\delta l = \frac{\sigma . l}{E} = \frac{\sigma_B \times l_B}{E_B} = \frac{48 \times 500}{80 \times 10^3} = 0.3 \text{ mm} \quad \textbf{Ans.}$$

**NOTE.** The change in length of the composite bar may also be found out by the stress in steel from the relation :

$$\delta l = \frac{\sigma_S \times l_S}{E_S} = \frac{120 \times 500}{200 \times 10^3} = 0.3 \text{ mm.}$$

**EXAMPLE 4.11.** *A rigid bar AB is hinged at A and supported by a copper rod 2 m long and steel rod 1 m long. The bar carries a load of 20 kN at D as shown in Fig. 4.21.*

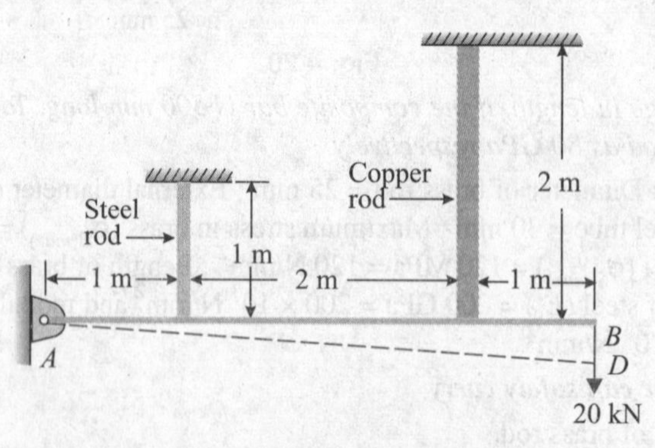


Fig. 4.21

*If the cross-sectional areas of steel and copper rods are 200 mm$^2$ and 400 mm$^2$ respectively, find the stresses developed in each rod. Take the values of E for steel and copper as 200 GPa and 100 GPa respectively.*

**SOLUTION.** Given : Length of copper rod ($l_C$) = 2 cm = 2 × $10^3$ mm ; Length of steel rod ($l_S$) = 1 m = 1 × $10^3$ mm ; Load ($P$) = 20 kN = 20 × $10^3$ N ; Area of steel rod ($A_S$) = 200 mm$^2$ ; Area of copper rod ($A_C$) = 400 mm$^2$ ; Modulus of elasticity of steel ($E_S$) = 200 GPa = 200 × $10^3$ N/mm$^2$ and Modulus of elasticity of copper ($E_C$) = 100 GPa = 100 × $10^3$ N/mm$^2$.

Let $P_S$ = Load shared by the steel rod, and

$P_C$ = Load shared by the copper rod.

Taking moments of the loads about $A$ and equating the same,

$$(P_S \times 1) + (P_C \times 3) = 20 \times 4$$

or
$$P_S + 3P_C = 80 \quad ...(i)$$

We know that deformation of the steel rod due to the load ($P_S$),

$$\delta l_S = \frac{P_S \cdot l_S}{A_S \cdot E_S} = \frac{P_S \times (1 \times 10^3)}{200 \times (200 \times 10^3)} = 0.025 \times 10^{-3}\ P_S \quad ...(ii)$$

and deformation of the copper rod due to the load ($P_C$),

$$\delta l_C = \frac{P_C \cdot l_C}{A_C \cdot E_C} = \frac{P_C \times (2 \times 10^3)}{400 \times (100 \times 10^3)} = 0.05 \times 10^{-3}\ P_C \quad ...(iii)$$

From the geometry of the elongations of the steel rod and copper rod, we find that

$$\frac{\delta l_C}{1} = \frac{\delta l_C}{3} \quad \text{or} \quad \delta l_C = 3\delta l_S$$

Substituting the values of $\delta l_S$ ans $\delta l_C$ from equations (*ii*) and (*iii*) in the above equation,

$$0.05 \times 10^{-3}\ P_C = 3 \times 0.025 \times 10^{-3}\ P_S \quad \text{or} \quad P_C = 1.5\ P_S$$

and now substituting the value of $P_C$ in equation (*i*),

$$P_S + 3 \times (1.5\ P_S) = 80 \quad \text{or} \quad 5.5\ P_S = 80$$

$$\therefore \quad P_S = \frac{80}{5.5} = 14.5 \text{ kN} = 14.5 \times 10^3 \text{ N}$$

and
$$P_C = 1.5\ P_S = 1.5 \times (14.5 \times 10^3) = 21.75 \times 10^3 \text{ N}$$

We know that stress in steel rod,

$$\sigma_S = \frac{P_S}{A_S} = \frac{14.5 \times 10^3}{200} = 72.5 \text{ N/mm}^2 = 72.5 \text{ MPa} \quad \textbf{Ans.}$$

and stress in copper rod,

$$\sigma_C = \frac{P_C}{A_C} = \frac{21.75 \times 10^3}{400} = 54.4 \text{ N/mm}^2 = 54.4 \text{ MPa} \quad \textbf{Ans.}$$

## 4.6. Stresses in Composite Structures of Unequal Lengths

We have already discussed in the last article the procedure for stresses in composite section of equal lengths. But sometimes, the length of one of the member is not equal to the other. In such cases, some of the load (or force) is utilised in extending the member and making its length equal to the other member. Now the remaining load is shared by both the members.

**EXAMPLE 4.12.** *A composite bar ABC, rigidly fixed at A and 1 mm above the lower support, is subjected to an axial load of 50 kN at B as shown in Fig. 4.22.*

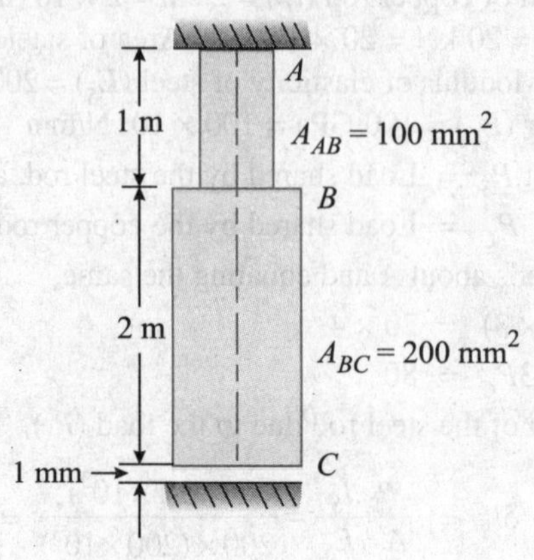


Fig. 4.22

*If the cross-sectional area of the section AB is 100 mm² and that of section BC is 200 mm², find the reactions at both the ends of the bar. Also find the stresses in both the section. Take E = 200 GPa.*

**SOLUTION.** Given : Length of $AB$ ($l_{AB}$) = 1 m = 1 × $10^3$ mm ; Area of $AB$ ($A_{AB}$) = 100 mm²; Length of $BC$ ($l_{BC}$) = 2 m = 2 × $10^3$ mm ; Area of $BC$ ($A_{BC}$) = 200 mm² ; Axial load ($P$) = 50 kN = 50 × $10^3$ N and modulus of elasticity ($E$) = 200 GPa = 200 × $10^3$ N/mm².

***Reactions at both the ends of the bar***

We know that as the bar is rigidly fixed at $A$ and loaded at $B$, therefore, upper portion $AB$ is subjected to tensions. We also know that increase in length of the portion $AB$ due to the load at $B$.

$$\delta l = \frac{P.l_{AB}}{A_{AB}.E} = \frac{(50\times 10^3)\times(1\times 10^3)}{100\times(200\times 10^3)} = 2.5 \text{ mm}$$

From the geometry of the figure, we find that of increase in the length of the portion $AB$ would have been less than 1 mm (*i.e.*, gap between $C$ and lower support), then the lower portion of the bar $BC$ should not have been subjected to any stress. Now it will be interesting to know that as the increase in length $AB$ is 2.5 mm, therefore, first action of the 50 kN load will be to increase the length AB by 1 mm, till the end C touches the lower support. And a part of the load will be required for this increase. Then the remaining load will be shared by both the portions of the bar $AB$ and $BC$ of the bar.

Let $P_1$ = Load required to increase 1 mm length of the bar $AB$,

We know that increase in length,

$$1 = \frac{P_1 \times l_{AB}}{A_{AB}.E} = \frac{P_1\times(1\times 10^3)}{100\times(200\times 10^3)} = 0.05\times 10^{-3}\, P_1$$

$$\therefore \quad P_1 = \frac{1}{0.05\times 10^{-3}} = 20\times 10^3 \text{ N} = 20 \text{ kN}$$

and the remaining loas, which will be shared by the portion $AB$ and $CD$

$= 50 - 20 = 30$ kN

Let $R_A$ = Reaction at $A$ due to 30 kN load, and

$R_C$ = Reaction at $C$ due to 30 kN load.

Thus, $R_A + R_C = 30 \text{ kN} = 30 \times 10^3$ N ...(*i*)

We know that increase in length $AB$ due to reaction $R_A$ (beyond 1 mm),

$$\delta l_1 = \frac{R_A . l_{AB}}{A_{AB} . E} = \frac{R_A \times (1 \times 10^3)}{100 \times (200 \times 10^3)} = 0.05 \times 10^{-3} R_A \qquad ...(ii)$$

and decrease in length $BC$ due to reaction $R_C$,

$$\delta l_2 = \frac{R_C . l_{BC}}{A_{BC} . E} = \frac{R_C \times (2 \times 10^3)}{200 \times (200 \times 10^3)} = 0.05 \times 10^{-3} R_C \qquad ...(iii)$$

Since $\delta l_1$ is equal to $\delta l_2$, therefore equating equations (*i*) and (*ii*),

$$0.05 \times 10^{-3} R_A = 0.05 \times 10^{-3} R_C \quad \text{or} \quad R_A = R_C$$

Now substituting the value of $R_C$ in equation (*i*)

$$R_A + R_A = 30 \quad \text{or} \quad R_A = R_C = \frac{30}{2} = 15 \text{ kN}$$

∴ Total reaction at $A = (20 + 15) = 35$ kN **Ans.**

and total reaction at $C = 15$ kN **Ans.**

***Stresses in both the sections***

We know that stress in the bar $AB$,

$$\sigma_{AB} = \frac{35 \times 10^3}{100} = 350 \text{ N/mm}^2 = 350 \text{ MPa} \qquad \textbf{Ans.}$$

and

$$\sigma_{BC} = \frac{15 \times 10^3}{200} = 75 \text{ N/mm}^2 = 75 \text{ MPa} \qquad \textbf{Ans.}$$

**EXAMPLE 4.13.** *A solid steel bar 500 mm long and 50 mm diameter is placed inside an aluminium tube 75 mm inside diameter and 100 mm outside diameter. The aluminium tube is 0.5 mm longer than the steel bar. An axial load of 600 kN is applied to the bar and cylider through rigid plates as shown in Fig. 4.23.*

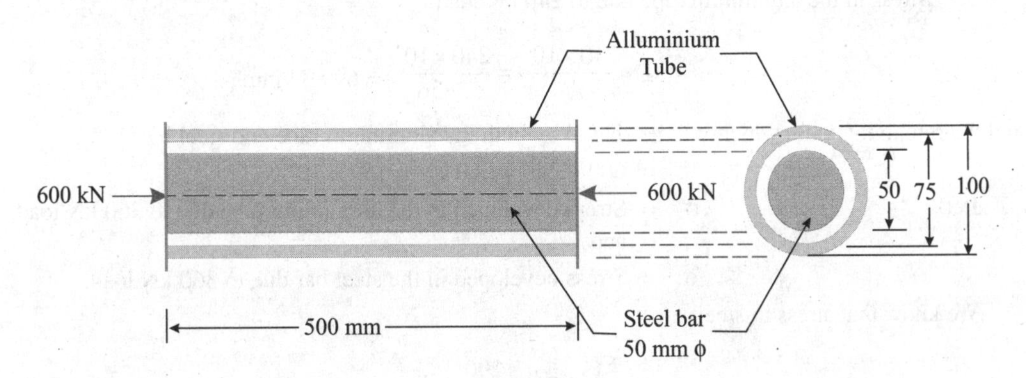


Fig. 4.23

*Find the stresses developed in the steel bar and aluminium tube. Assume E for steel as 200 GPa and E for aluminium is 70 GPa.*

**SOLUTION.** Given : Length of steel bar ($l_S$) = 500 mm ; Diameter of steel bar ($D_S$) = 50 mm; Inside diameter of aluminium tube ($d_A$) = 75 mm ; Outside diameter of aluminium tube ($D_A$) = 100 mm ; Length of aluminium tube ($l_A$) = 500 + 0.5 = 500.5 mm ; Axial load ($P$) = 600 kN = $600 \times 10^3$ N ; Modulus of elasticity of steel ($E_S$) = 200 GPa = $200 \times 10^3$ N/mm² and modulus of elasticity aluminium ($E_A$) = $70 \times 10^3$ N/mm².

We know that area of steel bar,

$$A_S = \frac{\pi}{4} \times (D_S)^2 = \frac{\pi}{4} \times (50)^3 = 1964 \text{ mm}^2$$

and area of aluminium tube,

$$A_S = \frac{\pi}{4} \times [D_A{}^2 - d_A{}^2] = \frac{\pi}{4} \times [(100)^2 - (75)^2] = 3436 \text{ mm}^2$$

We also know that as the aluminium tube is longer than the steel bar by 0.5 mm, therefore the load will first come upon the tube. Therefore decrease in the length of the aluminium tube due to load,

$$\delta l = \frac{P.l_A}{A_A.E_A} = \frac{(600 \times 10^3) \times (500.5)}{3436 \times (70 \times 10^3)} = 1.25 \text{ mm}$$

From the geometry of the figure, we find that if the decrease in the length of the aluminium tube would have been less than 0.5 mm (*i.e.*, difference between the lengths of steel bar and aluminium tube), then the steel bar should not have been subjected to any compressive load. Now it will be interesting to know that as the decrease in the length of aluminium tube is 1.25 mm, therefore, first action of the 600 kN load will be to decrease the length of the aluminium tube by 0.5 mm, till its length becomes equal to that of the steel bar. And a part of the load will be required for this decrease. Then the remaining load will be shared by both the aluminium tube and steel bar.

Let $P_1$ = Load required to decrease 0.5 mm length of the aluminium tube.

We know that decreases in length,

$$0.5 = \frac{P_1.l_A}{A_A.E_A} = \frac{P_1 \times 500.5}{3436 \times (70 \times 10^3)} = 2.08 \times 10^{-6} P_1$$

or

$$P_1 = \frac{0.5}{2.08 \times 10^{-6}} = 240 \times 10^3 \text{ N} = 240 \text{ kN}$$

∴ Stress in the aluminium tube due to 240 kN load

$$= \frac{240 \times 10^3}{A_A} = \frac{240 \times 10^3}{3436} = 69.8 \text{ N/mm}^2$$

and the remaining load, which will be shared by both the aluminium tube and steel bar

$$= 600 - 240 = 360 \text{ kN} = 360 \times 10^3 \text{ N}$$

Let $\sigma_A$ = Stress developed in the aluminium tube due to 360 kN load and

$\sigma_S$ = Stress developed in the steel bar due to 360 kN load.

We know that stress in steel,

$$\sigma_S = \frac{E_S}{E_A} \times \sigma_A = \frac{200}{70} \times \sigma_A = 2.86\, \sigma_A$$

and the load shared by both the aluminium tube and steel bar,

$$360 \times 10^3 = (\sigma_S \,.\, A_S) + (\sigma_A \,.\, A_A)$$

$$= (2.86\, \sigma_A \times 1964) + \sigma_A \times 3436) = 9053\, \sigma_A$$

$$\therefore \quad \sigma_A = \frac{360 \times 10^3}{9053} = 39.8 \text{ N/mm}^2$$

and $\sigma_S = 2.86\,\sigma_A = 2.86 \times 39.8 = 113.8 \text{ N/mm}^2 = 113.8 \text{ MPa}$ **Ans.**

∴ Total stress in aluminium tube

$$= 69.8 + 39.8 = 109.6 \text{ N/mm}^2 = 109.6 \text{ MPa} \quad \textbf{Ans.}$$

## EXERCISE 4.3

1. A composite bar is made up of a brass rod of 25 mm diameter enclosed in a steel tube of 40 mm external diameter, and 35 mm internal diameter. The ends of the rod and tube are securely fixed. Find the stresses developed in the brass rod and steel tube, when the composite bar is subjected to an axial pull of 45 kN. Take E for brass as 80 GPa and E for steel as 200 GPa.

   [**Ans.** 36.6 MPa ; 91.5 MPa]

2. A compound bar consists of a circular rod of steel of diameter 20 mm rigidly fitted into copper tube of internal diameter of 20 mm and external diameter of 30 mm. If the composite bar is 750 mm long and is subjected to a compressive load of 30 kN, find the stresses developed in the steel rod and copper tube. Take $E_C$ = 200 GPa and $E_A$ = 100 GPa. Also find the change in the length of the bar. [**Ans.** 58.8 MPa, 29.4 MPa, 0.22 mm]

3. A uniform rigid block weighing 160 kN is to be supported on three bars as shown in Fig. 4.24.

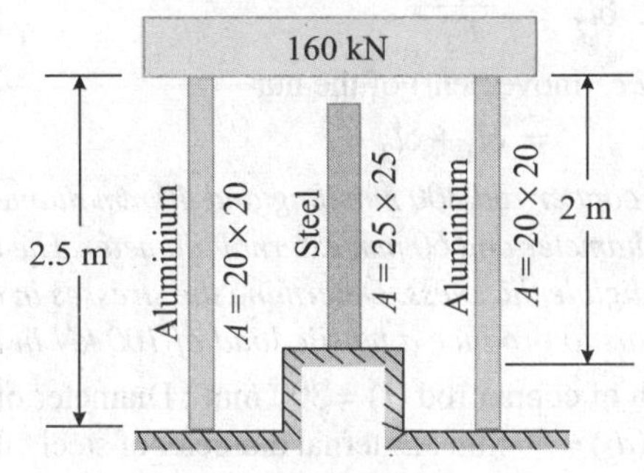


Fig. 4.24

There is 4 mm gap between the block and the top of the steel bar. Find the stresses developed in the bars. Take $E_S$ = 200 GPa and $E_A$ = 80 GPa. [**Ans.** $\sigma_A$ = 148.9 MPa ; $\sigma_S$ = 65.3 MPa]

## 4.7. Stresses in Nuts and Bolts

In our daily life, we use nuts and bolts to tighten the components of a machine or structure. It is generally done by placing washers below the nuts as shown in Fig. 4.25.

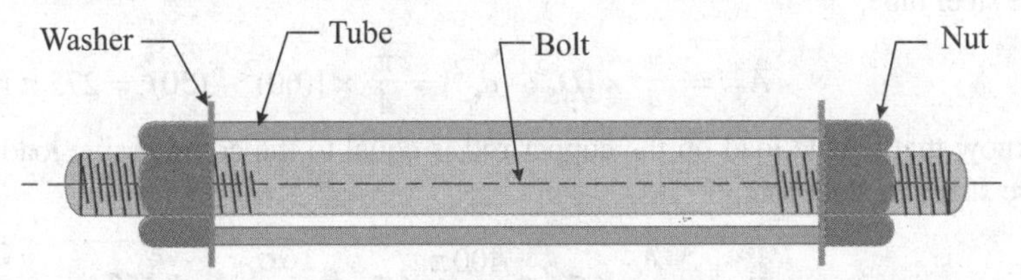


Fig. 4.25

As a matter of fact, a nut can be easily tightened, till the space between the two washers becomes exactly equal to the body placed between them. It will be interesting to know that if we further tighten the nut, it will induce some load in the assembly. As a result of this, bolt will be subjected to some tension, whereas the washers and body between them will be subjected to some compression. And the induced load will be equally shared between the bolt and the body. Now consider an assembly consisting of two nuts and a bolt alongwith a tube as shown in the figure.

Let $P$ = Tensile load induced in the bolt as a result of tightening the nut,

$l$ = Length of the bolt,

$A_1$ = Area of the bolt,

$\sigma_1$ = Stress in the bolt due to induced load,

$E_1$ = Modulus of elasticity for the bolt material.

$A_2, \sigma_2, E_2$ = Corresponding values for the tube

We know that as the tensile load on the bolt is equal to the compressive load on the tube, therefore

$$\sigma_1 . A_1 = \sigma_2 . A_2$$

$$\therefore \quad \sigma_1 = \frac{A_2}{A_1} \times \sigma_2 \qquad \text{Similarly,} \qquad \sigma_2 \frac{A_1}{A_2} \times \sigma_1$$

and the total toad $(P) = \sigma_1 A_1 + \sigma_2 A_2$

We also know that increase in the length of the bolt due to tensile stress in it,

$$\delta l_1 = \frac{\sigma_1 . l}{E_1} \quad \text{...(i)}$$

and decrease in the length of the tube due to compressive stress in it,

$$\delta l_2 = \frac{\sigma_2 . l}{E_2} \quad \text{...(ii)}$$

∴ Axial advancement (*i.e.*, movement) of the nut

$$= \delta l_1 + \delta l_2$$

**EXAMPLE 4.14.** *A solid copper rod 300 mm long and 40 mm diameter passes axially inside a steel tube of 50 mm internal diameter and 60 mm external diameter. The composite bar is tightened by using rigid washers of negligible thickness. Determine the stresses in copper rod and steel tube, when the nut is tightened so as to produce a tensile load of 100 kN in the copper rod.*

**SOLUTION.** Given : Length of copper rod $(l)$ = 300 mm ; Diameter of copper rod $(D_C)$ = 40 mm: Internal diameter of steel tube $(d_S)$ = 50 mm ; External diameter of steel tube $(D_S)$ = 60 mm and tensile load in copper rod $(P)$ = 100 kN = $100 \times 10^3$ N.

Let $\sigma_C$ = Stress in the copper rod and

$\sigma_S$ = Stress in the steel rod.

We know that area of the copper rod,

$$A_C = \frac{\pi}{4} \times (D_C)^2 = \frac{\pi}{4} \times (40)^2 = 400\ \pi\ \text{mm}^2$$

and area of the steel tube,

$$A_S = \frac{\pi}{4} \times [D_S^2 - d_C^2] = \frac{\pi}{4} \times [(60)^2 - (50)^2 = 275\ \pi\ \text{mm}^2$$

We also know that tensile load on the copper rod is equal to the compressive load on the steel tube. Therefore stress in steel rod,

$$\sigma_S = \frac{A_C}{A_S} \times \sigma_C = \frac{400\pi}{275\pi} \times \sigma_C = \frac{16\sigma_C}{11} = 1.455\ \sigma_C$$

and load $(P)$

$$100 \times 10^3 = (\sigma_C . A_C) + (\sigma_S . A_S) = (\sigma_C \times 400\ \pi) + (1.455\ \sigma_C \times 275\ \pi)$$

$$= 800\ \pi\ \sigma_C$$

$$\therefore \quad \sigma_C = \frac{100 \times 10^3}{800\pi} = 39.8\ \text{N/mm}^2 = 39.8\ \text{MPa (tension)} \quad \textbf{Ans.}$$

and

$$\sigma_S = 1.455\ \sigma_C = 1.455 \times 39.8\ \text{N/mm}^2 = 57.9\ \text{N/mm}^2 \quad \textbf{Ans.}$$

$$= 57.9\ \text{MPa (compression)} \quad \textbf{Ans.}$$

**EXAMPLE 4.15.** *A steel bolt of 500 mm length and 18 mm diameter passes coaxially through a steel tube of the same length and 20 mm internal diameter and 30 mm external diameter. The assembly is rigidly fixed at its both ends by washers. If one of the nut is tightened through 45°, find the stresses developed in the steel bolt and steel tube. Take pitch of the threads as 2.4 mm and E for the steel as 200 GPa.*

**SOLUTION.** Given : Length of steel bolt ($l_b$) = 500 mm ; Diameter of steel bolt ($D_b$) = 18 mm ; Length of steel tube ($l_t$) = 500 mm ; Internal diameter of steel tube ($d_t$)= 20 mm ; External diameter of steel tube ($D_t$) = 30 mm ; Angle through which the nut is tightened = 45°; Pitch = 2.4 mm and modulus of elasticity of steel ($E$) = 200 GPa = $200 \times 10^3$ N/mm$^2$.

Let $\sigma_b$ = Stress developed in the steel bolt and

$\sigma_t$ = Stress developed in the steel tube.

We know that area of the bolt,

$$A_b = \frac{\pi}{4} \times (D_b)^2 = \frac{\pi}{4} \times (18)^2 = 81\,\pi \text{ mm}^2$$

and area of tube,

$$A_t = \frac{\pi}{4}[D_t^2 - d_t^2] = \frac{\pi}{4} \times [(30)^2 - (20)^2] = 125\,\pi \text{ mm}^2$$

We also know that tensile load on the steel bolt is equal to the compressive load on the steel tube. Therefore stress in steel tube,

$$\sigma_t = \frac{A_b}{A_t} \times \sigma_b = \frac{81\pi}{125\pi} \times \sigma_b = 0.648\,\sigma_b \qquad ...(i)$$

∴ Decrease in the length of the steel tube,

$$\delta l_t = \frac{\sigma_t . l_t}{E} = \frac{(0.648\sigma_b) \times 500}{200 \times 10^3} = 1.62 \times 10^{-3}\,\sigma_b \qquad ...(ii)$$

and increase in the length of the steel bolt,

$$\delta l_b = \frac{\sigma_b . l_b}{E} = \frac{\sigma_b \times 500}{200 \times 10^3} = 2.5 \times 10^{-3}\,\sigma_b \qquad ...(iii)$$

We know that when the nut is tightened through 45°, then its axial advancement

$$= \frac{45°}{360°} \times \text{Pitch} = \frac{1}{8} \times 2.4 = 0.3 \text{ mm} \qquad ...(iv)$$

Since the axial advancement of the nut is equal to the decrease in the length of the tube plus increase in the length of the bolt, therefore

$$0.3 = (1.62 \times 10^{-3}\,\sigma_b) + (2.5 \times 10^{-3}\,\sigma_b) = 4.12 \times 10^{-3}\,\sigma_b$$

∴ $$\sigma_b = \frac{0.3}{4.12 \times 10^{-3}} = 72.8 \text{ N/mm}^2 = 72.8 \text{ MPa (Tension)} \qquad \textbf{Ans.}$$

and $$\sigma_t = 0.648\,\sigma_b = 0.648 \times 72.8 = 47.2 \text{ MPa (Compression)} \qquad \textbf{Ans.}$$

**EXAMPLE 4.16.** *A steel rod 20 mm diameter passes centrally through a copper tube of 25 mm internal diameter and 35 mm external diameter. Copper tube is 800 mm long and is closed by rigid washers of negligible thickness, which are fastened by nut threaded on the rod as shown in Fig. 4.26.*

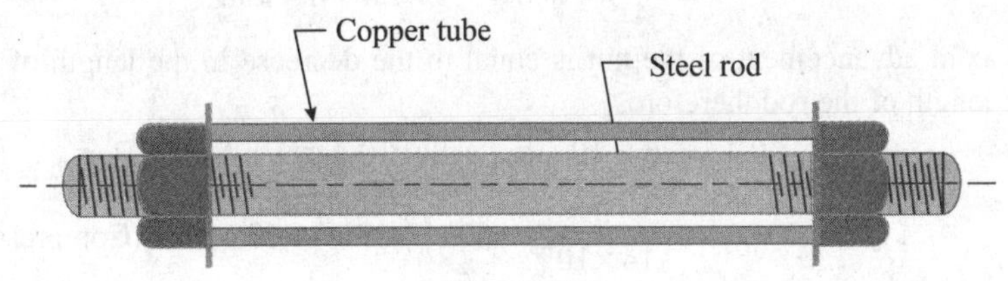


Fig. 4.26

*The nuts are tightened till the load on the assembly is 20 kN. Calculate the initial stresses in the copper tube and steel rod. Also calculate increase in the stresses, when one nut is tightened by one-quarter of a turn relative to the other. Take pitch of the thread as 1.6 mm. Take E for steel and copper as 200 GPa and 100 GPa respectively.*

**SOLUTION.** Given : Diameter of steel rod ($D_S$) = 20 mm ; Internal diameter of copper tube ($d_C$) = 25 mm ; External diameter of copper tube ($D_C$) = 35 mm ; Length of copper tube ($l$) = 800 mm ; Load on assembly ($P$) = 20 kN = $20 \times 10^3$ N ; Pitch = 1.6 mm ; Modulus of elasticity of steel ($E_S$) = 200 GPa = $200 \times 10^3$ N/mm² and modulus of elasticity of copper ($E_C$) = 100 GPa = $100 \times 10^3$ N/mm².

***Initial stress in steel rod and copper tube***

Let $\sigma_S$ = Stress in steel rod, and

$\sigma_C$ = Stress in copper tube.

We know that area of steel rod,

$$A_S = \frac{\pi}{4} \times (D_S)^2 = \frac{\pi}{4} \times (20)^2 = 100\,\pi \text{ mm}^2$$

and area of copper tube,

$$A_C = \frac{\pi}{4} \times \left[D_C^2 - d_C^2\right] = \frac{\pi}{4} \times [(35)^2 - (25)^2] = 150\,\pi \text{ mm}^2$$

We also know that tensile load on the steel rod is equal to the compressive load on the copper tube. Therefore stress in steel rod,

$$\sigma_S = \frac{A_C}{A_S} \times \sigma_C = \frac{150\pi}{100\pi} \times \sigma_C = 1.5\,\sigma_C \quad \text{...(i)}$$

and load ($P$),

$$20 \times 10^3 = \sigma_S \cdot A_S + \sigma_C \cdot A_C = (1.5\,\sigma_C \times 100\,\pi) + (\sigma_C \times 150\,\pi)$$

$$= 300\,\pi\,\sigma_C$$

$$\therefore \quad \sigma_C = \frac{20 \times 10^3}{300\pi} = 21.2 \text{ N/mm}^2 = 21.2 \text{ MPa} \quad \textbf{Ans.}$$

and

$$\sigma_C = 1.5\,\sigma_C = 1.5 \times 21.2 = 31.8 \text{ MPa} \quad \textbf{Ans.}$$

***Increase in stresses when nut is tightened by one-quarter of a turn***

Let $\sigma_{S1}$ = Increase in the stress in the steel rod and

$\sigma_{C1}$ = Increase in the stress in the copper tube.

We know that increase in the length of the steel rod,

$$\delta l_S = \frac{\sigma_{S1} \cdot l}{E_S} = \frac{1.5\,\sigma_{C1} \times 800}{200 \times 10^3} = 6 \times 10^{-3}\,\sigma_{C1} \quad \text{...(ii)}$$

and decrease in the length of the copper tube,

$$\delta l_S = \frac{\sigma_{C1} \cdot l}{E_C} = \frac{\sigma_{C1} \times 800}{100 \times 10^3} = 8 \times 10^{-3}\,\sigma_{C1} \quad \text{...(iii)}$$

We also know that when the nut is tightened by one-quarter of a turn, then its axial advancement

$$= \frac{1}{4} \times \text{Pitch} = \frac{1}{4} \times 1.6 = 0.4 \text{ mm} \quad \text{...(iv)}$$

Since the axial advancement of the nut is equal to the decrease in the length of the tube plus increase in the length of the rod therefore,

$$0.4 = 6 \times 10^{-3}\,\sigma_{C1} + 8 \times 10^{-3}\,\sigma_{C1} = 14 \times 10^{-3}\,\sigma_{C1}$$

$$\therefore \quad \sigma_{C1} = \frac{0.4}{14 \times 10^{-3}} = 28.6 \text{ N/mm}^2 = 28.6 \text{ MPa (Compression)} \quad \textbf{Ans.}$$

and

$$\sigma_{S1} = 1.5\,\sigma_{C1} = 1.5 \times 28.6 = 42.9 \text{ MPa (Tension)} \quad \textbf{Ans.}$$

## EXERCISE 4.4

1. A steel rod of 20 mm diameter and 350 mm long passes centrally through a steel tube of 40 mm external diameter and 30 mm internal diameter. The composite bar is tightened by using rigid washers of negligible thickness, which are fastened by nuts threaded on the rod. Find the stresses developed in the steel tube and rod, when the assembly is subjected to a tensile load of 22 kN. [**Ans.** 20 MPa ; 35 MPa]
2. A steel bolt 25 mm diameter and 400 mm long is surrounded by a copper sleave of 30 mm internal diameter 35 mm external diameter of the same length. The assembly is now rigidly fixed at both ends by washers of negligible length. If pitch of the thread is 1.5 mm and one of the nut is tightened through 60°, calculate the stresses developed in the copper sleeve and steel bolt. Take $E_S = 200$ GPa and $E_C = 100$ GPa. [**Ans.** 82.2 MPa ; 42.7 MPa]

## QUESTIONS

1. What is a statically indeterminate structure ?
2. Give the procedure for solving a statically indeterminate problem.
3. How will you find the load shared by three wires supporting a load at their bottom ?
4. Explain the procedure for finding out stresses developed in a statically indeterminate structure, when one of the support is slightly smaller than the other.
5. Describe the principle for finding out the stresses in the nut and bolt arrangement.

## MULTIPLE CHOICE QUESTIONS

1. Which of the following is a statically indeterminate structure ?
   (*a*) a load supported on one member.
   (*b*) a load supported on two membes.
   (*c*) a load supported on three members.
   (*d*) either '*a*' or '*b*'.
2. A rod is enclosed centrally in a tube and the assembly is tightened by rigid washers. If the assembly is subjected to a compressive load, then
   (*a*) rod is subjected to a compressive load,
   (*b*) tube is subjected to a compressive load,
   (*c*) both are subjected to a compressive load,
   (*d*) rod is subjected to a compressive load, while the tube is subjected to a tensile load.
3. A bolt is made to pass through a tube and both of them are tightly fitted with the help of washers and nuts. If the nut is tightened, then
   (*a*) bolt and tube are subjected to compressive load.
   (*b*) bolt and tube are subjected to tensile load.
   (*c*) bolt is subjected to compressive load, while tube is subjected to tensile load.
   (*d*) bolt is subjected to tensile load while tube is subjected to compressive load.

## ANSWERS

**1.** (*c*) **2.** (*c*) **3.** (*d*)

Chapter **5**

# Thermal Stresses and Strains

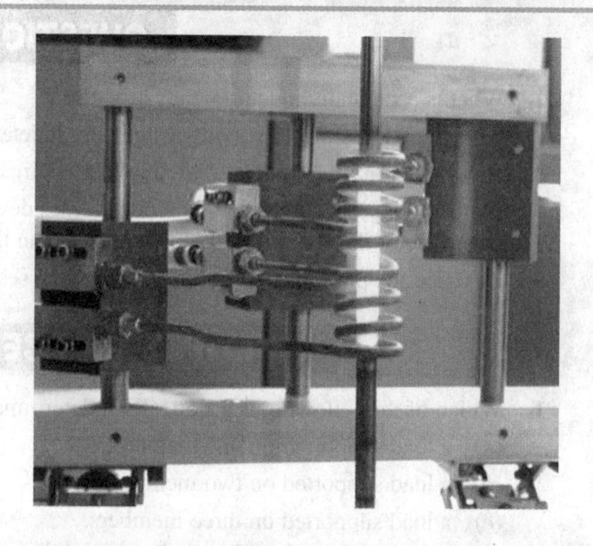

## Contents



## 5.1. Introduction

It has been established since long, that whenever there is some increase or decrease in the temperature of a body, it causes the body to expand or contract. A little consideration will show that if the body is allowed to expand or contract freely, with the rise or fall of the temperature, no stresses are induced in the body. But if the deformation of the body is prevented, some stresses are induced in the body. Such stresses are called thermal stresses or temperature stresses. The corresponding strain are called thermal strains or temperature strains.

## 5.2. Thermal Stresses in Simple Bars

The thermal stresses or strains, in a simple bar, may be found out as discussed below :

1. Calculate the amount of deformation due to change of temperature with the assumption that bar is free to expand or contract.
2. Calculate the load (or force) required to bring the deformed bar to the original length.
3. Calculate the stress and strain in the bar caused by this load.

The thermal stresses or strains may also be found out first by finding out amount of deformation due to change in temperature, and then by finding out the thermal strain due to the deformation. The thermal stress may now be found out from the thermal strain as usual. Now consider a body subjected to an increase in temperature.

Let $l$ = Original length of the body,

$t$ = Increase of temperature and

$\alpha$ = Coefficient of linear expansion.

We know that the increase in length due to increase of temperature.

$$\delta l = l.\alpha.t$$

If the ends of the bar are fixed to rigid supports, so that its expansion is prevented, then compressive strain induced in the bar.

$$\varepsilon = \frac{\delta l}{l} = \frac{l.\alpha.t}{l} = \alpha.t$$

∴ Stress $\sigma = \varepsilon.E = \alpha.t.E.$

**Cor.** If the supports yield by an amount equal to $\Delta$, then the actual expansion that has taken place,

$$\delta l = l\alpha t - \Delta$$

and strain, $$\varepsilon = \frac{\delta l}{l} = \frac{l\alpha t - \Delta}{l} = \left(\alpha t \frac{\Delta}{l}\right)$$

∴ Stress, $$\sigma = \varepsilon.E = \left(\alpha t - \frac{\Delta}{l}\right) E$$

The value of a (*i.e.*, coefficient of linear expansion) of materials in every day use are given below in table 5.1 :

TABLE 5.1

| *S. No.* | *Material* | *Coefficient of linear expansion/°C (α)* |
|---|---|---|
| 1. | Steel | $11.5 \times 10^{-6}$ to $13 \times 10^{-6}$ |
| 2. | Wrought iron, Cast iron | $11 \times 10^{-6}$ to $12 \times 10^{-6}$ |
| 3. | Aluminium | $23 \times 10^{-6}$ to $24 \times 10^{-6}$ |
| 4. | Copper, Brass, Bronze | $17 \times 10^{-6}$ to $18 \times 10^{-6}$ |

**EXAMPLE 5.1.** *A aluminium alloy bar, fixed at its both ends is heated through 20 K. Find the stress developed in the bar. Take modulus of elasticity, and coefficient of linear expansion for the bar material as 80 GPa and 24 × $10^{-6}$/K respectively.*

**SOLUTION.** Given : Increase in temperature ($t$) = 20 K ; Modulus of elasticity ($E$) = 80 GPa = 80 × $10^3$ N/mm$^2$ and Coefficient of linear expansion ($\alpha$) = 24 × $10^{-6}$/K

We know that thermal stress developed in the bar,

$$\sigma = \alpha.t.E = (24 \times 10^{-6}) \times 20 \times (80 \times 10^3) \text{ N/mm}^2$$

$$= 38.4 \text{ N/mm}^2 = 38.4 \text{ MPa} \qquad \textbf{Ans.}$$

**EXAMPLE 5.2.** *A brass rod 2 m long is fixed at both its ends. If the thermal stress is not to exceed 76.5 MPa, calculate the temperature through which the rod should be heated. Take the values of α and E as 17 × 10$^{-6}$/K and 90 GPa respectively.*

**SOLUTION.** Given : * Length ($l$) = 2 m ; Maximum thermal stress ($\sigma_{max}$) = 76.5 MPa = 76.5 N/mm$^2$ ; $\alpha = 17 \times 10^{-6}$/K and $E = 90$ GPa $= 90 \times 10^3$ N/mm$^2$.

Let $t$ = Temperature through which the rod should be heated in K.

We know that maximum stress in the rod ($\sigma_{max}$),

$$76.5 = \alpha.t.E = (17 \times 10^{-6}) \times t \times (90 \times 10^3) = 1.53\,t$$

$$\therefore \quad t = \frac{76.5}{1.53} = 50 \text{ K} \qquad \textbf{Ans.}$$

**EXAMPLE 5.3.** *Two parallel walls 6 m apart are stayed together by a steel rod 25 mm diameter passing through metal plates and nuts at each end. The nuts are tightened home, when the rod is at a temperature of 100°C. Determine the stress in the rod, when the temperature falls down to 60°C, if*

*(a) the ends do not yield, and*

*(b) the ends yield by 1 mm*

*Take E = 200 GPa and α = 12 × 10$^{-6}$/°C*

**SOLUTION.** Given : Length ($l$) = 6 m = 6 × 10$^3$ mm ; ** Diameter ($d$) = 25 mm ; Decrease in temperature ($t$) = 100° – 60° = 40°C ; Amount of yield in ends (Δ) = 1 mm ; Modulus of elasticity ($E$) = 200 GPa = 200 × 10$^3$ N/mm$^2$ and coefficient of linear expansion (α) = 12 × 10$^{-6}$/°C.

***(a) Stress in the rod when the ends do not yield***

We know that stress in the rod when the ends do not yield,

$$\sigma_1 = \alpha.t.E = (12 \times 10^{-6}) \times 40 \times (200 \times 10^3) \text{ N/mm}^2$$

$$= 96 \text{ N/mm}^2 = 96 \text{ MPa} \qquad \textbf{Ans.}$$

***(b) Stress in the rod when the ends yield by 1 mm***

We also know that stress in the rod when the ends yield,

$$\sigma_2 = \left[\alpha t - \frac{\Delta}{l}\right] E = \left[(12 \times 10^{-6})\, 40 - \frac{1}{6 \times 10^3}\right] 200 \times 10^3 \text{ N/mm}^2$$

$$= 62.6 \text{ N/mm}^2 = 62.6 \text{ MPa} \qquad \textbf{Ans.}$$

## 5.3. Thermal Stresses in Bars of Circular Tapering Section

Consider a circular bar of uniformly tapering section fixed at its ends $A$ and $B$ and subjected to an increase of temperature as shown in Fig. 5.1.

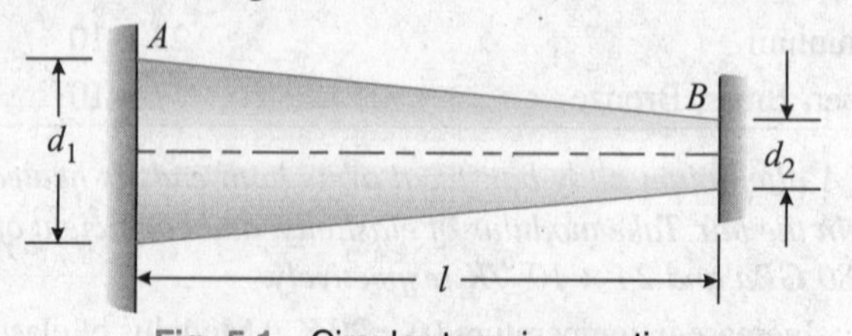


Fig. 5.1. Circular tapering section

* Superfluous data

** Superfluous datar

Let $l$ = Length of the bar.

$d_1$ = Diameter at the bigger end of the bar,

$d_2$ = Diameter at the smaller end of the bar,

$t$ = Increase in temperature and

$a$ = Coefficient of linear expansion.

We know that as a result of the increase in temperature, the bar $AB$ will tend to expand. But since it is fixed at both of its ends, therefore it will cause some compressive stress.

We also know that the increase in length due to increase in temperature,

$$\delta l = l.\alpha.t \qquad ...(i)$$

Now let $P$ = Load (or force) required to bring the deformed bar to the original length.

We know that decrease in the length of the circular bar due to load $P$

$$\delta l = \frac{4Pl}{\pi E d_1 d_2} \qquad ...(ii)$$

Equating equations (*i*) and (*ii*),

$$l.\alpha.t = \frac{4Pl}{\pi E d_1 d_2} \quad \text{or} \quad P = \frac{\pi E d_1 d_2 .\alpha t}{4}$$

$\therefore$ *Max. stress,

$$\sigma_{max} = \frac{P}{\frac{\pi}{4} \times d_2^2} = \frac{\pi E d_1 d_2 .\alpha t}{4 \times \frac{\pi}{4} \times d_2^2} = \frac{\alpha t E d_1}{d_2}$$

**Note.** If we substitute $d_1 = d_2$, the above relation is reduced to

$$\sigma = \alpha.t.E \qquad \text{...(Same as for simple bars)}$$

**Example 5.4.** *A circular bar rigidly fixed at its both ends uniformly tapers from 75 mm at one end to 50 mm at the other end. If its temperature is raised through 26 K, what will be the maximum stress developed in the bar. Take E as 200 GPa and α as 12 × $10^{-6}$/K for the bar material.*

**Solution.** Given : Diameter at end 1 ($d_1$) = 75 mm ; Diameter at end 2 ($d_2$) = 50 mm ; Rise in temperature ($t$) = 26 K ; $E$ = 200 GPa = 200 × $10^3$ N/mm$^2$ and $\alpha$ = 12 × $10^{-6}$/K.

We know that maximum stress developed in the bar,

$$\alpha_{max} = \frac{\alpha t.E.d_1}{d_2} = \frac{(12\times10^{-6})\times 26\times(200\times10^3)\times 75}{50} \text{ N/mm}^2$$

$$= 93.6 \text{ N/mm}^2 = 93.6 \text{ MPa} \qquad \textbf{Ans.}$$

**Example 5.5.** *A rigidly fixed circular bar 1.75 m long uniformly tapers from 125 mm diameter at one end to 100 mm diameter at the other. If the maximum stress in the bar is not to exceed 108 MPa, find the temperature through which it can be heated. Take E and α for the bar material as 100 GPa and 18 × $10^{-6}$ / K respectively.*

**Solution.** Given : ** Length ($l$) = 1.75 m ; Diameter at end 1 ($d_1$) = 125 mm ; Diameter at end 2 ($d_2$) = 100 mm ; Maximum stress ($\sigma_{max}$) = 108 MPa = 108 N/mm$^2$ ; Modulus of elasticity ($E$) = 100 GPa = 100 × $10^3$ N/mm$^2$ and coefficient of linear expansion ($\alpha$) = 18 × $10^{-6}$/K.

Let $t$ = Temperature through which the bar can be heated in K.

We know that maximum stress in the bar ($\sigma_{max}$),

$$108 = \frac{\alpha t.E.d_1}{d_2} = \frac{(18\times10^{-6})\times t\times(100\times10^3)\times 125}{100} = 2.25\, t$$

* The stress will be maximum at $B$, because of lesser areas of cross-section.

* Superfluous data

$$\therefore \qquad t = \frac{108}{2.25} = 48 \text{ K} \qquad \textbf{Ans.}$$

## 5.4. Thermal Stresses in Bars of Varying Section

Consider a bar *ABC* fixed at its ends *A* and *C* and subjected to an increase of temperature as shown in Fig. 5.2.

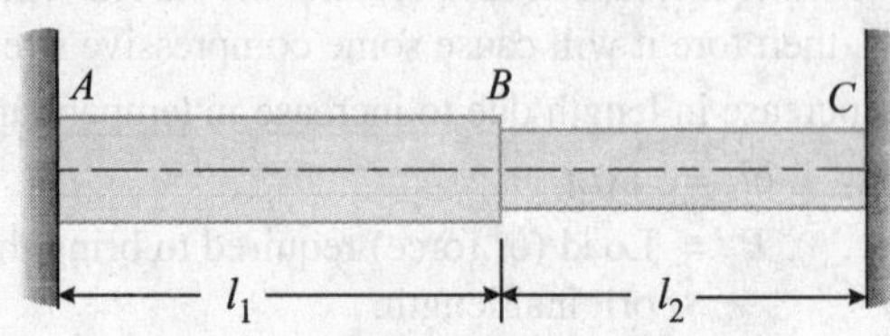


Fig. 5.2. Bar of varying section

Let
$l_1$ = Length of portion *AB*,
$\sigma_1$ = Stress in portion *AB*,
$A_1$ = Cross-sectional area of portion *AB*,
$l_2, \sigma_2, A_2$ = Corresponding values for the portion *BC*,
$\alpha$ = Coefficient of linear expansion and
$t$ = Increase in temperature

We know that as a result of the increase in temperature, the bar *ABC* will tend to expand. But since it is fixed at its ends *A* and *C*, therefore it will cause some compressive stress in the body. Moreover, as the thermal stress is shared equally by both the portions, therefore

$$\sigma_1 A_1 = \sigma_2 A_2$$

Moreover, the total deformation of the bar (assuming it to be free to expand),

$$\delta l = \delta l_1 + \delta l_2 = \frac{\sigma_1 l_1}{E} + \frac{\sigma_2 l_2}{E} = \frac{l}{E}(\sigma_1 l_1 + \sigma_2 l_2)$$

**Note.** Sometimes, the modulus of elasticity is different for different sections. In such cases, the total deformation.

$$\delta l = \left(\frac{\sigma_1 l_1}{E_1} + \frac{\sigma_2 l_2}{E_2}\right)$$

**Example 5.6.** *A steel rod ABC is firmly held between two rigid supports A and C as shown in Fig. 5.3.*

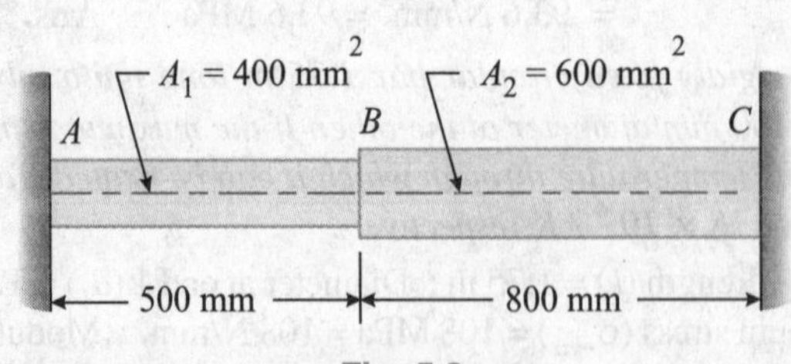


Fig. 5.3

*Find the stresses developed in the two portions of the rod, when it is heated through 15 K. Take $\alpha = 12 \times 10^{-6}$ / K and E = 200 GPa.*

**Solution.** Given : Area of part 1 ($A_1$) = 400 mm$^2$ ; Length of part 1 ($l_1$) = 500 mm ; Area of part 2 ($A_2$) = 600 mm$^2$ ; Length of part 2 ($l_2$) = 800 mm ; Rise in temperature ($t$) = 15K ; Coefficient of linear expansion ($\alpha$) = $12 \times 10^{-6}$/K and modulus of elasticity ($E$) = 200 GPa = $200 \times 10^3$ N/mm$^2$.

Let $\sigma_1$ = Stress developed in the portion $AB$ due to increase in temperatures and

$\sigma_2$ = Stress developed in the portion $BC$ due to increase in temperature

Since the thermal load is shared equally by both the portions, therefore stress developed in the portion $AB$,

$$\sigma_1 = \frac{A_2}{A_1} \times \sigma_2 = \frac{600}{400} \times \sigma_2 \; 1.5\,\sigma_2 \qquad ...(i)$$

We know that free expansion of the part 1 due to increase in temperature,

$$\delta l_1 = l_1 \alpha . t = 500 \times (12 \times 10^{-6}) \times 15 = 0.09 \text{ mm}$$

and
$$\delta l_2 = l_2 . \alpha . t = 800 \times (12 \times 10^{-6}) \times 15 = 0.144 \text{ mm}$$

∴ Total expansion of the rod,

$$\delta l = \delta l_1 + \delta l_2 = 0.09 + 0.144 = 0.234 \text{ mm}$$

Now let us assume a compressive force to be applied at $A$ and $C$, which will cause a contraction of 0.234 mm of the rod (*i.e.*, equal to the total expansion). Therefore,

$$0.234 = \frac{1}{E}(\sigma_1 . l_1 + \sigma_2 . l_2) = \frac{1}{200 \times 10^3}(1.5\,\sigma_2 \times 500 + \sigma_2 \times 800)$$

$$= 7.75 \times 10^{-2} \times \sigma_2$$

$$\therefore \quad \sigma_2 = \frac{0.234}{7.75 \times 10^{-3}} = 30.2 \text{ N/mm}^2 = 30.2 \text{ MPa} \quad \textbf{Ans.}$$

and
$$\sigma_1 = 1.5\,\sigma_2 = 1.5 \times 30.2 = 45.3 \text{ MPa} \quad \textbf{Ans.}$$

**EXAMPLE 5.7.** *A composite bar made up of aluminium and steel, is held between two supports as shown in Fig. 5.4.*

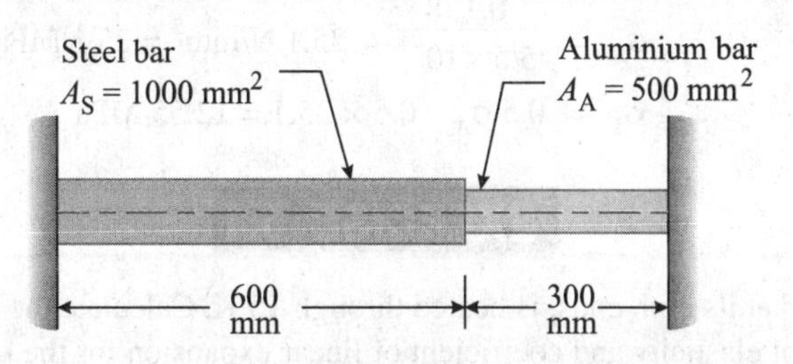


Fig. 5.4

*The bars are stress-free at a temperature of 38°C. What will be the stresses in the two bars, when the temperature is 21°C, if (a) the supports are unyielding, (b) the supports come nearer to each other by 0.1 mm? It can be assumed that the change of temperature is uniform all along the length of the bar.*

*Take E for steel as 200 GPa; E for aluminium as 75 GPa and coefficient of expansion for steel as 11.7 × $10^{-6}$ per °C and coefficient of expansion for aluminium as 23.4 × $10^{-6}$ per °C.*

**SOLUTION.** Given : Length of steel bar ($l_S$) = 600 mm ; Area of steel bar ($A_S$) = 1000 mm$^2$ ; Length of aluminium bar ($l_A$) = 300 mm ; Area of aluminium bar ($A_A$) = 500 mm$^2$ ; Decrease in temperature ($t$) = 38 – 21 = 17°C ; Modulus of elasticity of steel ($E_S$) = 200 GPa = 200 × $10_3$ N/mm$^2$; Modulus of elasticity of aluminium ($E_A$) = 75 GPa = 75 N/mm$^2$ ; Coefficient of expansion for steel ($\alpha_S$) = 11.7 × $10^{-6}$/°C and coefficient of expansion for aluminium ($\alpha_A$) = 23.4 × $10^{-6}$/°C.

Let $\sigma_S$ = Stress in the steel bar, and

$\sigma_A$ = Stress in the aluminium bar.

***(a) Stresses when the supports are unyielding***

$$\sigma_S \cdot A_S = \sigma_A \cdot A_A \quad \text{or} \quad \sigma_S \times 1000 = \sigma_A \times 500$$

$$\therefore \quad \sigma_S = \sigma_A \times 500/1000 = 0.5\ \sigma_A$$

We know that free expansion of steel bar due to increase in temperature,

$$\delta l_S = l_S.\alpha_S.t = 600 \times (11.7 \times 10^{-6}) \times 17 = 0.119 \text{ mm}$$

and

$$\delta l_A = l_A.\alpha_A.t = 300 \times (23.4 \times 10^{-6}) \times 17 = 0.119 \text{ mm}$$

∴ Total contraction of the bar,

$$\delta l = \delta l_S + \delta l_A = 0.119 + 0.119 = 0.238 \text{ mm}$$

Now let us assume a tensile force to be applied at *A* and *C*, which will cause an expansion of 0.238 mm of the rod (*i.e.*, equal to the total contraction). Therefore

$$0.238 = \frac{\sigma_S \cdot l_S}{E_S} + \frac{\sigma_A \cdot l_A}{E_A} = \frac{(0.5\ \sigma_A) \times 600}{200 \times 10^3} + \frac{\sigma_A \times 300}{75 \times 10^3} = 5.5 \times 10^{-3}\ \sigma_A$$

$$\therefore \quad \sigma_A = \frac{0.238}{5.5 \times 10^{-3}} = 43.3 \text{ N/mm}^2 = 43.3 \text{ MPa} \quad \textbf{Ans.}$$

and

$$\sigma_S = 0.5\ \sigma_A = 0.5 \times 43.3 = 21.65 \text{ MPa} \quad \textbf{Ans.}$$

***(b) Stresses when the supports come nearer to each other by 0.1 mm***

In this case, there is an expansion of composite bar equal to 0.238 – 0.1 = 0.138 mm. Now let us assume a tensile force, which will cause an expansion of 0.138 mm. Therefore

$$0.138 = \frac{\sigma_S \cdot l_S}{E_S} + \frac{\sigma_A \cdot l_A}{E_A} = \frac{(0.5\ \sigma_A) \times 600}{200 \times 10^3} + \frac{\sigma_A \times 300}{75 \times 10^3} = 5.5 \times 10^{-3}\ \sigma_A$$

$$\therefore \quad \sigma_A = \frac{0.138}{5.5 \times 10^{-3}} = 25.1 \text{ N/mm}^2 = 25.1 \text{ MPa} \quad \textbf{Ans.}$$

and

$$\sigma_S = 0.5\ \sigma_A = 0.5 \times 25.1 = 12.55 \text{ MPa} \quad \textbf{Ans.}$$

## EXERCISE 5.1

1. A steel bar, fixed at its both ends, is heated through 15 K. Calculate the stress developed in the bar, if modulus of elasticity and coefficient of linear expansion for the bar material is 200 GPa and $12 \times 10^{-6}$/K respectively. [**Ans.** 36 MPa]
2. An alloy bar 2 m long is held between two supports. Find the stresses developed in the bar, when it is heated through 30 K if both the ends (*i*) do not yield; and (*ii*) yield by 1 mm. Take the value of *E* and α for the alloy as 120 GPa and $24 \times 10^{-6}$/K. [**Ans.** 86.4 MPa ; 26.4 MPa]
3. A circular bar rigidly fixed at its both ends is 1.2 m long. It uniformly tapers from 100 mm at one end to 75 mm at the other. What is the maximum stress induced in the bar, when its temperature is raised through 25 K? Take *E* as 200 GPa and α as $12 \times 10^{-6}$/K. [**Ans.** 80 MPa]
4. An alloy circular bar rigidly fixed at its both ends uniformly tapers from 90 mm to 60 mm from one end to another. What will be the maximum stress developed in the bar, when its temperature is raised through 20 K? Take *E* and α for the bar material as 150 GPa and $12 \times 10^{-6}$/K. Also find the maximum stress when the bar is lowered by the same temperature. [**Ans.** 54 MPa (Compn.) ; 54 MPa (Tension)]

**5.** A steel rod *ABC* firmly held at *A* and *C* has a cross-sectional area of 1000 mm$^2$ for 400 mm length and 1500 mm$^2$ for 600 mm length as shown in Fig. 5.5.

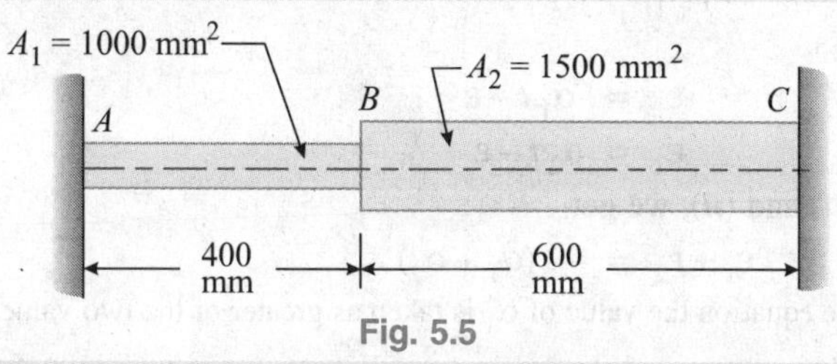


Fig. 5.5

If the rod is heated through 10 K, find the stresses developed in the parts *AB* and *BC*.

[**Ans.** 30 MPa ; 20 MPa]

## 5.5. Thermal Stresses in Composite Bars

Whenever there is some increase or decrease in the temperature of a bar, consisting of two or more different materials, it causes the bar to expand or contract. On account of different coefficients of linear expansions the two materials do not expand or contract by the same amount, but expand or contract by different amounts.

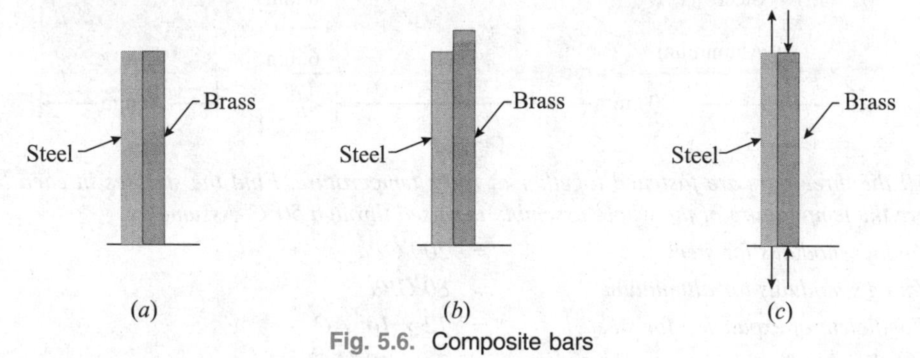


Fig. 5.6. Composite bars

Now consider a composite bar consisting of two members, a bar of steel and another of brass as shown in Fig. 5.6 (*a*).

Let the bar be heated through some temperature. If the component members of the bar (*i.e.*, steel and brass) could have been free to expand, then no internal stresses would have induced. But, since the two members are rigidly fixed, therefore the composite bar, as a whole, will expand by the same amount. We know that the brass expands more than the steel (because the coefficient of linear expansion of the brass is greater than that of the steel). Therefore the free expansion of the brass will be more than that of the steel. But since both the members are not free to expand, therefore the expansion of the composite bar, as a whole, will be less than that of the brass; but more than that of the steel as shown in Fig. 5.6 (*b*). It is thus obvious that the brass will be subjected to compressive force, whereas the steel will be subjected to tensile force as shown in Fig. 5.6 (*c*).

Now let $\sigma_1$ = Stress in brass

$\varepsilon_1$ = Strain in brass,

$\alpha_1$ = Coefficient of linear expansion for brass,

$A_1$ = Cross-sectional area of brass bar,

$\sigma_2, \varepsilon_2, \alpha_2 A_2$ = Corresponding values for steel, and

$\varepsilon$ = Actual strain of the composite bar per unit length.

As the compressive load on the brass is equal to the tensile load on the steel, therefore

$$\sigma_1.A_1 = \sigma_2.A_2$$

Now strain in brass,

$$\varepsilon_1 = \alpha_1.t - \varepsilon \quad ...(i)$$

and strain in steel,

$$\varepsilon_2 = \alpha_2.t - \varepsilon \quad ...(ii)$$

Adding equation (*i*) and (*ii*), we get

$$\varepsilon_1 + \varepsilon_2 = -t\,(\alpha_1 + \alpha_2)$$

NOTES : 1. In the above equation the value of $\alpha_1$ is taken as greater of the two values of $\alpha_1$ and $\alpha_2$.

2. The values of strain ($\varepsilon_1$ and $\varepsilon_2$) may also be found out from the relation $\frac{\text{Stress}}{\text{Modulus of elasticity}}$ or $\frac{\delta l}{l}$.

**EXAMPLE 5.8.** *A flat steel bar 200 mm × 20 mm × 8 mm is placed between two aluminium bars 200 mm × 20 mm × 6 mm so as to form a composite bar as shown in Fig. 5.7.*

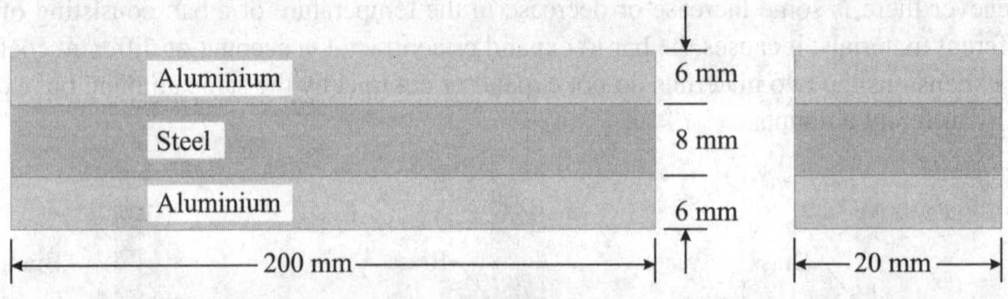


Fig. 5.7

*All the three bars are fastened together at room temperature. Find the stresses in each bar, where the temperature of the whole assembly is raised through 50°C. Assume :*

*Young's modulus for steel* = *200 GPa*
*Young's modulus for aluminium* = *80 GPa*
*Coefficient of expansion for steel* = $12 \times 10^{-6}/°C$
*Coefficient of expansion for aluminium* = $24 \times 10^{-6}/°C$

**SOLUTION.** Given : Size of steel bar = 200 mm × 20 mm × 8 mm ; Size of each aluminium bar = 200 mm × 20 mm × 6 mm ; Rise in temperature ($t$) = 50°C ; Young's modulus for steel ($E_S$) = 200 GPa = $200 \times 10^3$ N/mm$^2$ ; Young's modulus for aluminium ($E_A$)= 80 GPa = $80 \times 10^3$ N/mm$^2$; Coefficient of expansion for steel ($\alpha_S$) = $12 \times 10^{-6}/°C$ and coefficient of expansion for aluminium ($\alpha_A$) = $24 \times 10^{-6}/°C$.

Let $\sigma_S$ = Stress in steel bar and

$\sigma_A$ = Stress in each aluminium bar.

We know that area of steel bar

$$A_S = 20 \times 8 = 160 \text{ mm}^2$$

and total area of two aluminium bars,

$$A_A = 2 \times 20 \times 6 = 240 \text{ mm}^2$$

We also know that when the temperature of the assembly will increase, the free expansion of aluminium bars will be more than that of steel bar (because $\alpha_A$ is more than $\alpha_S$). Thus the aluminium bars will be subjected to compressive stress and the steel bar will be subjected to tensile stress. Since the tensile load on the steel bar is equal to the compressive load on the aluminium bars, therefore stress in steel bar,

$$\sigma_S = \frac{A_A}{A_S} \times \sigma_A = \frac{240}{160} \times \sigma_A = 1.5\,\sigma_A$$

We know that strain in steel bar,

$$\varepsilon_S = \frac{\sigma_S}{E_S} = \frac{\sigma_S}{200 \times 10^3}$$

and

$$\varepsilon_A = \frac{\sigma_A}{E_A} = \frac{\sigma_A}{80 \times 10^3}$$

We also know that total strain,

$$\varepsilon_S + \varepsilon_A = t\,(\alpha_A - \alpha_S)$$

$$\frac{\sigma_S}{200 \times 10^3} + \frac{\sigma_A}{80 \times 10^3} = 50\,[(24 \times 10^{-6}) - (12 \times 10^{-6})]$$

$$\frac{1.5\sigma_A}{200 \times 10^3} + \frac{\sigma_A}{80 \times 10^3} = 50 \times (12 \times 10^{-6})$$

$$20 \times 10^{-6}\,\sigma_A = 600 \times 10^{-6} \quad \text{or} \quad 20\,\sigma_A = 600$$

$$\therefore \quad \sigma_A = \frac{600}{20} = 30 \text{ N/mm}^2 = 30 \text{ MPa} \quad \textbf{Ans.}$$

and

$$\sigma_S = 1.5\,\sigma_A = 1.5 \times 30 \text{ N/mm}^2 = 45 \text{ MPa} \quad \textbf{Ans.}$$

**EXAMPLE 5.9.** *A gun metal rod 20 mm diameter, screwed at the ends, passes through a steel tube 25 mm and 30 mm internal and external diameters respectively. The nuts on the rod are screwed tightly home on the ends of the tube. Find the intensity of stress in each metal, when the common temperature rises by 200°F. Take.*

*Coefficient of expansion for steel = 6 × 10–6/°F*

*Coefficient of expansion for gun metal = $10 \times 10^{-6}$/°F*

*Modulus of elasticity for steel = 200 GPa*

*Modulus of elasticity for gun metal = 100 GPa.*

**SOLUTION.** Given : Diameter of gun metal rod = 20 mm ; Internal diameter of steel tube = 25 mm; External diameter of steel tube = 30 mm ; Rise in temperature ($t$) = 200°F ; Coefficient of expansion for steel ($\alpha_S$) = $6 \times 10^{-6}$/°F ; Coefficient of expansion for gun metals ($\alpha_G$) = $10 \times 10^{-6}$/°F; Modulus of elasticity for steel ($E_S$) = 200 GPa = $200 \times 10^3$ N/mm$^2$ and modulus of elasticity for gun metal ($E_G$) = 100 GPa = $100 \times 10^3$ N/mm$^2$.

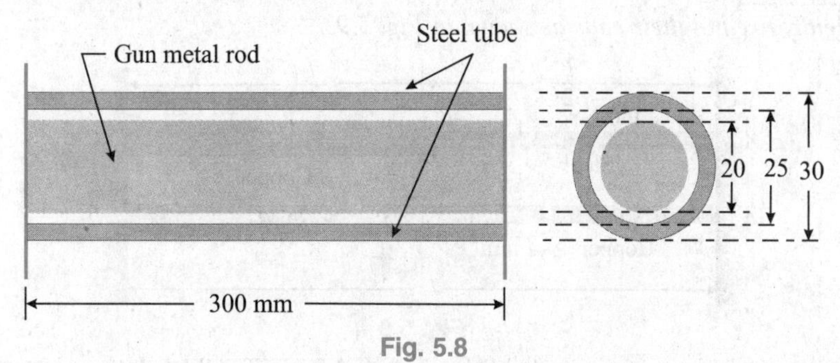


Fig. 5.8

Let $\sigma_G$ = Stress in gun metal rod, and

$\sigma_S$ = Stress in steel tube,

We know that area of gun metal rod,

$$A_G = \frac{\pi}{4} \times (20)^2 = 100\ \pi\ \text{mm}^2$$

and area of steel tube $$A_S = \frac{\pi}{4} [(30)^2 - (25)^2] = 68.75\ \pi\ \text{mm}^2$$

We also know that when the common temperature of the gun metal rod and steel tube will increase, the free expansion of gun metal rod will be more than that of steel tube (because $\alpha_G$ is greater than $\alpha_S$). Thus the gun metal rod will be subjected to compressive stress and the steel tube will be subjected to tensile stress. Since the tensile load on the steel tube is equal to the compressive load on the gun metal rod, therefore stress in steel,

$$\sigma_S = \frac{A_G}{A_S} \times \sigma_S = \frac{100\pi}{68.75\pi} \times \sigma_G = 1.45\ \sigma_G$$

We know that strain in steel tube,

$$\varepsilon_S = \frac{\sigma_S}{E_S} = \frac{\sigma_S}{200 \times 10^3}$$

and $$\varepsilon_G = \frac{\sigma_G}{E_G} = \frac{\sigma_G}{100 \times 10^3}$$

We also know that total strain,

$$\varepsilon_S + \varepsilon_G = t\ (\alpha_G - \alpha_S)$$

$$\frac{\sigma_S}{200 \times 10^3} + \frac{\sigma_G}{100 \times 10^3} = 200\ [(10 \times 10^{-6}) - (6 \times 10^{-6})]$$

$$\frac{1.45\sigma_G}{200 \times 10^3} + \frac{\sigma_G}{100 \times 10^3} = 200 \times (4 \times 10^{-6})$$

$$\frac{3.45\sigma_G}{200 \times 10^3} = 800 \times 10^{-6}$$

$$3.45\ \sigma G = (800 \times 10^{-6}) \times (200 \times 10^3) = 160$$

$$\therefore \quad \sigma_G = \frac{160}{3.45} = 46.4\ \text{N/mm}^2 = 46.4\ \text{MPa} \qquad \textbf{Ans.}$$

and $$\sigma_S = 1.45\ \sigma_G = 1.45 \times 46.4 = 67.3\ \text{MPa} \qquad \textbf{Ans.}$$

**EXAMPLE 5.10.** *A composite bar is made up by connecting a steel member and a copper member, rigidly fixed at their ends as shown in Fig. 5.9.*

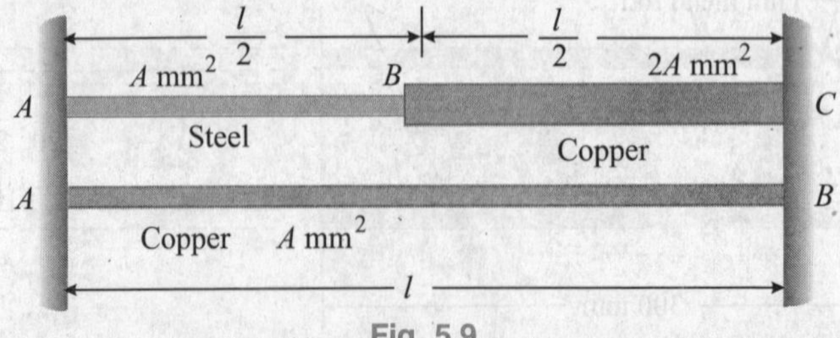


Fig. 5.9

*The cross-sectional area of the steel member is A mm² for half of the length and 2A mm² for the other half of the length ; while that for the copper member is A mm². The coefficients of expansion for steel and copper are α and 1.3 α ; while elastic modulii are E and 0.5 E respectively. Determine the stresses induced in both the members when the composite bar is subjected to a rise of temperature of t degrees.*

**SOLUTION.** Given : Area of steel bar $AB = A$ mm$^2$ ; Area of steel bar $BC = 2A$ mm$^2$ ; Area of copper bar $AB = A$ mm$^2$ ; Coefficient of expansion for steel $(\alpha_S) = \alpha$ ; Coefficient of expansion for copper $(\alpha_C) = 1.3\ \alpha$ ; Young's modulus for steel $(E_S) = E$ and Young's modulus for copper $(E_C) = 0.5$ E.

A little consideration will show that due to rise in temperature, the free expansion of the copper member will be more than that of the steel member (because $\alpha_C$ is more than $\alpha_S$). Hence the copper member will be subjected to compressive stress, whereas the steel member will be subjected to tensile stress.

Let $\sigma_S$ = Stress in the portion $AB$ of the steel bar due to increase in temperature, and

$\sigma_C$ = Stress in the copper bar due to increase in temperature.

Since there is no external load on any member, therefore

$$\sigma = \sigma_S = \sigma_C$$

We know that stress in the portion $BC$ of the steel bar

$$\sigma_S' = \frac{A}{2A} \times \sigma_S = 0.5\ \sigma$$

We also know that elongation of the copper bar due to stress,

$$\sigma l_C = \frac{\sigma_C \times l_C}{E_C} = \frac{\sigma \times l}{0.5 E} = \frac{2\sigma l}{E}$$

and strain in the copper bar,

$$\varepsilon_C = \frac{\delta l_C}{l_C} = \frac{2\sigma l}{E} \times \frac{1}{l} = \frac{2\sigma}{E} \qquad ...(i)$$

Similarly, extension of the steel bar,

$$\delta l_S = \frac{\sigma_S \times l_S}{E_S} = \left(\frac{\sigma_S \times \frac{1}{2}}{E_S}\right) + \left(\frac{\sigma_S' \times \frac{1}{2}}{E_S}\right) = \left(\frac{\sigma \times \frac{l}{2}}{E_S}\right) + \left(\frac{0.5\sigma \times \frac{1}{2}}{E_S}\right)$$

$$= \frac{3\sigma l}{4E}$$

and strain in the steel bar,

$$\varepsilon_S = \frac{\delta l_S}{l_S} = \frac{3\sigma l}{4E} \times \frac{1}{l} = \frac{3\sigma}{4E} \qquad ...(ii)$$

Therefore total strain,

$$\varepsilon_C + \varepsilon_S = t\,(\alpha_C - \alpha_S) \qquad \text{or} \qquad \frac{2\sigma}{E} = \frac{3\sigma}{4E} = t\,(1.3\ \alpha - \alpha)$$

$$\frac{11\sigma}{4E} = 0.3\ \alpha\ t$$

$$\therefore \qquad \sigma = \frac{0.3 \alpha t \times 4E}{11} = 0.109\ \alpha\ t\ E \qquad \textbf{Ans.}$$

and

$$\sigma_S' = 0.5\ \sigma = 0.5 \times 0.109\ \alpha\ t\ E = 0.0545\ \alpha\ t\ E \qquad \textbf{Ans.}$$

## 5.6. Superposition of Thermal Stresses

In the last articles, we have been discussing the thermal stresses in the bars, which were initially free of any type of tensile or compressive stresses. But sometimes, we come across structures, which are subjected to same loading, before their temperature is increased or decreased. Such problems are solved in the following two steps :

1. First of all, find out the stresses caused in its members before there is any change in temperature.
2. Now find out the stresses due to change in temperature and use the principal of superposition on the stresses already obtained.
3. Finally add the two stresses obtained above.

**NOTE :** Such problems are very complicated and need lot of patience in knowing the type of stresses (*i.e.*, tensile or compressive) in both the cases.

**EXAMPLE 5.11.** *A composite bar made up of aluminium bar and steel bar is firmly held between two unyielding supports as shown in Fig. 5.10.*

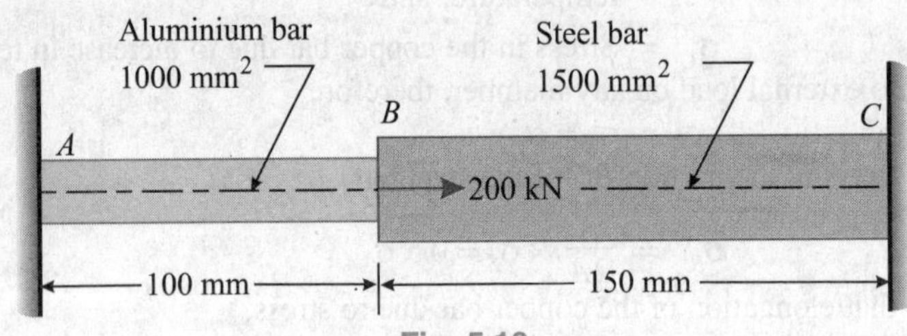


Fig. 5.10

*An axial load of 200 kN is applied at B at 320 K. Find the stresses in each material, when the temperature is 370 K. Take α for aluminium and steel as 70 GPa and 210 GPa respectively. Take α for aluminium and steel as 24 × 10⁻⁶ /K and 12 × 10⁻⁶ /K respectively.*

**SOLUTION.** Given : Length of aluminium bar ($l_A$) = 100 mm ; Area of aluminium bar ($A_A$) = 1000 mm$^2$ ; Length of steel bar ($l_S$) = 150 mm ; Area of steel bar ($A_S$) = 1500 mm$^2$ ; Axial load ($P$) = 200 kN = 200 × $10^3$ N ; Rise in temperature ($t$) = 370 – 320 = 50 K ; Modulus of elasticity of aluminium ($E_A$) = 70 GPa = 70 × $10^3$ N/mm$^2$ ; Modulus of elasticity of steel ($E_S$) = 210 GPa = 210 × $10^3$ N/mm$^2$; Coefficient of expansion of aluminium ($\alpha_A$) = 24 × $10^{-6}$/K and coefficient of expansion of steel ($\alpha_S$) = 12 × $10^{-6}$/K.

First of all, let us find out the stresses developed in the aluminium bar and steel bar due to the application of 200 kN load at *B*.

Let $P_1$ = Load shared by the aluminium bar *AB* in kN.

We know that increase in length of the aluminium bar,

$$\delta l_A = \frac{P_1 . l_A}{A_A . E_A} = \frac{P_1 \times 10^3 \times 100}{1000 \times (70 \times 10^3)} = \frac{P_1}{700} \quad ...(i)$$

Similarly, decrease in length of the steel bar,

$$\delta l_S = \frac{(200 - P_1) \times 10^3 \times l_S}{A_S . E_S} = \frac{(200 - P_1) \times 10^3 \times 150}{1500 \times (210 \times 10^3)} = \frac{(200 - P_1)}{2100} \quad ...(ii)$$

Since the values of $\delta_A$ is equal to that of $\delta_S$, therefore equating equations (*i*) and (*ii*),

$$\frac{P_1}{700} = \frac{200 - P_1}{2100} \quad \text{or} \quad \frac{P_1}{1} = \frac{200 - P_1}{3}$$

$$3P_1 = 200 - P_1 \quad \text{or} \quad P_1 = 200/4 = 50 \text{ kN}$$

∴ Stress in aluminium bar due to axial load

$$\sigma_{A1} = \frac{P_1}{A_A} = \frac{50 \times 10^3}{1000} = 50 \text{ N/mm}^2 \text{ (Tension)} \quad ...(iii)$$

and stress in steel bar due to axial load,

$$\sigma_{S1} = \frac{(200 - P_1) \times 10^3}{A_S} = \frac{(200 - 50) \times 10^3}{1500} \text{ N/mm}^2$$

$$= 100 \text{ N/mm}^2 \text{ (Compression)} \quad ...(iv)$$

Now let us find out the stresses developed in aluminium bar and steel bar due to increase in the temperature. Since the thermal load is shared equally by both the parts, therefore stress in aluminium bar due to increase in temperature,

$$\sigma_{A2} = \frac{A_S}{A_A} \times \sigma_{S2} = \frac{1500}{1000} \times \sigma_{S2} = 1.5\, \sigma_{S2} \quad ...(v)$$

We know that free expansion of the aluminium bar due to increase in temperature,

$$\delta l_{A2} = l_A.\alpha_A.t = 100 \times (24 \times 10^{-6}) \times 50 = 0.12 \text{ mm}$$

and $\delta l_{S2} = l_S.\alpha_S.t = 150 \times (12 \times 10^{-6}) \times 50 = 0.09 \text{ mm}$

∴ Total expansion of the bar,

$$\delta l = \delta l_{A2} + \delta l_{S2} = 0.12 + 0.09 = 0.21 \text{ mm}$$

Now let us assume a tensile force to be applied at *A* and *C*, which will cause a contraction of 0.21 mm of the bar (*i.e.*, equal to the total expansions). Therefore

$$0.21 = \frac{\sigma_{A2}.l_A}{E_A} + \frac{\sigma_{S2}.l_S}{E_S} = \frac{(1.5\, \sigma_{S2}) \times 100}{70 \times 10^3} + \frac{\sigma_{S2} \times 150}{210 \times 10^3}$$

$$= \frac{600\, \sigma_{S2}}{210 \times 10^3}$$

$$\therefore \quad \sigma_{S2} = \frac{0.21 \times (210 \times 10^3)}{600} = 73.5 \text{ N/mm}^2 \text{ (Compression)}$$

and $\sigma_{A2} = 1.5 \times \sigma_{S2} = 1.5 \times 73.5 = 110.3 \text{ N/mm}^2$ (Compression)

∴ Total stress in aluminium,

$$\sigma_A = \sigma_{A1} + \sigma_{A2} = 50 - 110.3 = -60.3 \text{ N/mm}^2$$

$$= 60.3 \text{ MPa (Compression)} \quad \textbf{Ans.}$$

and

$$\sigma_S = \sigma_{S1} + \sigma_{S2} = 100 + 73.5 = 173.5 \text{ N/mm}^2$$

$$= 173.5 \text{ MPa (Compression)} \quad \textbf{Ans.}$$

**EXAMPLE 5.12.** *A steel rod of 20 mm diameter passes centrally through a tight fitting copper tube of external diameter 40 mm. The tube is closed with the help of rigid washers of negligible thickness and nuts threaded on the rod. The nuts are tightened till the compressive load on the tube is 50 kN as shown in Fig. 5.11.*

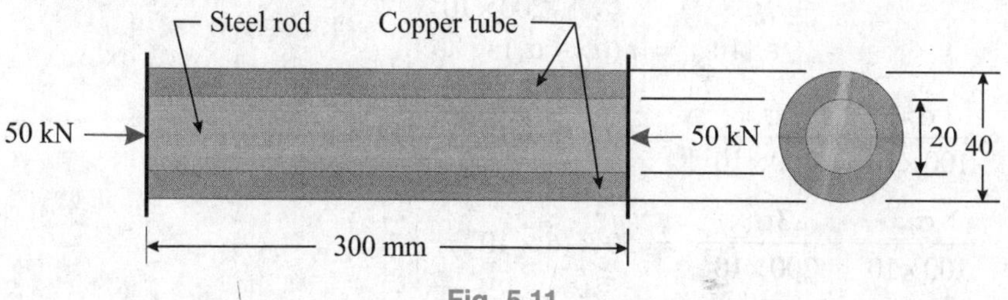


Fig. 5.11

*Determine the stresses in the rod and the tube, when the temperature of the assembly falls, by 50 K. Take E for steel and copper as 200 GPa and 100 GPa respectively. Take coefficient of expansion for steel and copper as 12 × 10⁻⁶ K and 18 × 10⁻⁶ K respectively.*

**SOLUTION.** Given : Diameter of steel rod = 20 mm ; External diameter of copper tube = 40 mm ; Internal diameter of copper tube = 20 mm (because of tight fitting) ; Compressive load ($P$) = 50 kN = $50 \times 10^3$ N ; Fall in temperature ($t$) = 50 K; Modulus of elasticity for steel ($E_S$) = 200 GPa = $200 \times 10^3$ N/mm$^2$ ; Modulus of elasticity for copper ($E_C$) = 100 GPa = $100 \times 10^3$ N/mm$^2$ ; Coefficient of expansion for steel ($\alpha_S$) = $12 \times 10^{-6}$/K and coefficient of expansion for copper ($\alpha_C$) = $18 \times 10^{-6}$/ K.

Let $\sigma_S$ = Stress in steel rod, and

$\sigma_C$ = Stress in copper tube.

We know that area of steel rod,

$$A_S = \frac{\pi}{4}(20)^2 = 100\,\pi \text{ mm}^2$$

and area of copper tube,

$$A_C = \frac{\pi}{4}\left[(40)^2 - (20)^2\right] = 300\,\pi \text{ mm}^2$$

First of all, let us find out the stresses of copper tube and steel rod due to a compressive load of 50 kN. We know that compressive load on the copper tube is equal to tensile load on the steel rod. Therefore stress in the steel rod,

$$\sigma_{S1} = \frac{A_C}{A_S} \times \sigma_{C1} = \frac{300\pi}{100\pi} \times \sigma_{C1} = 3\,\sigma_{C1}$$

and load ($P$) $50 \times 10^3 = (\sigma_{S1} A_S) + (\sigma_{C1} A_C) = (3\sigma_{C1} \times 100\,\pi) + (\sigma_{C1} \times 300\pi)$

$$= 600\,\pi\,\sigma_{C1}$$

$$\therefore \quad \sigma_{C1} = \frac{50 \times 10^3}{600\,\pi} = 26.5 \text{ N/mm}^2 \text{ (Compression)}$$

and $\sigma_{S1} = 3\,\sigma_{C1} = 3 \times 26.5 = 79.5$ N/mm$^2$ (Tension)

Now let us find out the stresses developed in the steel rod and copper tube due to fall in temperature. We know that when temperature of the assembly will fall, the free contraction of the copper tube will be more than that of steel rod (because $\alpha_C$ is greater than $\alpha_S$). Thus the copper tube will be subjected to tension and steel rod will be subjected to compression. Since the tensile load on the copper tube is equal to the compressive load on the steel rod, therefore stress in steel,

$$\sigma_{S2} = 3\sigma_{C2} \qquad \text{... (As obtained earlier)}$$

We know that strain in copper tube,

$$\varepsilon_C = \frac{\sigma_{C2}}{E_C} = \frac{\sigma_{C2}}{100 \times 10^3}$$

and

$$\varepsilon_S = \frac{\sigma_{S2}}{E_S} = \frac{\sigma_{S2}}{200 \times 10^3}$$

$$\therefore \quad \varepsilon_C + \varepsilon_S = t\,(\alpha_C - \alpha_S)$$

$$\frac{\sigma_{C2}}{100 \times 10^3} + \frac{\sigma_{S2}}{200 \times 10^3} = 50\,[(18 \times 10^{-6}) - (12 \times 10^{-6})]$$

$$\frac{\sigma_{C2}}{100 \times 10^3} + \frac{3\sigma_{S2}}{200 \times 10^3} = 50 \times (6 \times 10^{-6})$$

$$\frac{5\sigma_{C2}}{200 \times 10^3} = 300 \times 10^{-6}$$

$$5\,\sigma_{C2} = (300 \times 10^{-6}) \times (200 \times 10^3) = 60$$

or $\sigma_{C2} = 3\,\sigma_{C2} = 12$ N/mm$^2$ (Tension)

and $\sigma_{S2} = 3\,\sigma_{C2} = 3 \times 12 = 36$ N/mm$^2$ (Compression)

∴ Net stress in the copper tube,

$$\sigma_C = \sigma_{C1} + \sigma_{C2} = 26.5 - 12 = 14.5 \text{ N/mm}^2$$

$$= 14.5 \text{ MPa (Compression)} \quad \textbf{Ans.}$$

and $\sigma_S = \sigma_{S1} + \sigma_{S2} = 79.5 - 36 = 43.5$ N/mm$^2$ (Tension) **Ans.**

**EXAMPLE 5.13.** *Two steel rods, each 50 mm diameter are connected end to end by means of a turnbuckle as shown in Fig. 5.12. The other end of each rod is rigidly fixed with a little initial tension in the rods.*

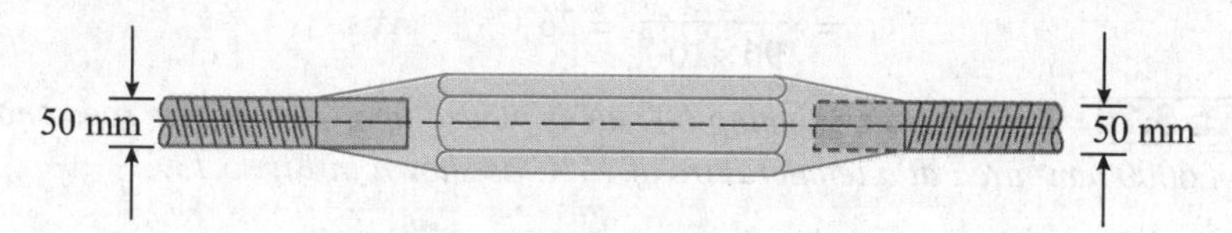


Fig. 5.12

*The length of each rod is 4 m and pitch of the threads on each rod = 5 mm. Neglecting the extension of turnbuckle, calculate the initial tension, when the turnbuckle is tightened by one quarter of a turn. E = 200 GPa. State with reason, whether effect of temperature rise would nullify the increase in tension or add more to it.*

**SOLUTION.** Given : Diameter of each rod $(d) = 50$ mm ; Length of each rod $(l) = 4$ m $= 4 \times 10^3$ mm ; Pitch of the threads = 5 mm and modulus of elasticity $(E) = 200$ GPa $= 200 \times 10^3$ N/mm$^2$.

***Initial tension in the rods, when the turnbuckle is tightened by one-quarter of a turn.***

Let $P$ = Tension in the rods, when the turnbuckle is tightened by one-quarter of a turn in N.

We know that cross-sectional area of the steel rods,

$$A = \frac{\pi}{4} \times (d)^2 = \frac{\pi}{4} \times (50)^2 = 1964 \text{ mm}^2$$

and extension of the first bar,

$$\delta l_1 = \frac{Pl}{AE} = \frac{P \times (4 \times 10^3)}{1964 \times (200 \times 10^3)} = \frac{P}{98.2 \times 10^3} \text{ mm}$$

Similarly, extension of the second bar,

$$\delta l_2 = \frac{Pl}{AE} = \frac{P \times (4 \times 10^3)}{1964 \times (200 \times 10^3)} = \frac{P}{98.2 \times 10^3} \text{ mm}$$

∴ Total extension of both the rods,

$$\delta l = \delta l_1 + \delta l_2 = \frac{P}{98.2 \times 10^3} + \frac{P}{98.2 \times 10^3} = \frac{P}{49.1 \times 10^3} \text{ mm}$$

We also know that the total extension of the two rods, when the turnbuckle is tightened by one-quarter of a turn.

$$\delta l = \left(\frac{1}{4} \times 5\right) + \left(\frac{1}{4} \times 5\right) = 2.5 \text{ mm}$$

Since the total extension of the two rods is equal to the sum of their extensions, therefore,

$$2.5 = \frac{P}{49.1 \times 10^3}$$

or $P = 2.5 \times (49.1 \times 10^3) = 122750 \text{ N} = 122.75 \text{ kN}$ **Ans.**

*Effect of temperature rise*

A little consideration will show that the rise of temperature will increase the length of the bars, whose effect will be to nullify the increase in tension as discussed below:

Let $t$ = Increase of temperature which will nullify the increase in tension in °C.

Since the increase in the length of the two rods due to increase in temperature is equal to the increase in length due to tightening of the turnbuckle, therefore

$$2.5 = l\,\alpha\,t = 2 \times (4 \times 10^3) \times (12 \times 10^{-6}) \times t = 96 \times 10^{-3}\,t$$

( ∵ Standard value of α is $12 \times 10^{-6}$)

$$\therefore \quad t = \frac{2.5}{96 \times 10^{-3}} = 26°C \quad \textbf{Ans.}$$

**EXAMPLE 5.14.** *A rigid slab weighing 600 kN is placed upon two bronze rods and one steel rod each of 6000 mm² area at a temperature of 15°C as shown in Fig. 5.13.*

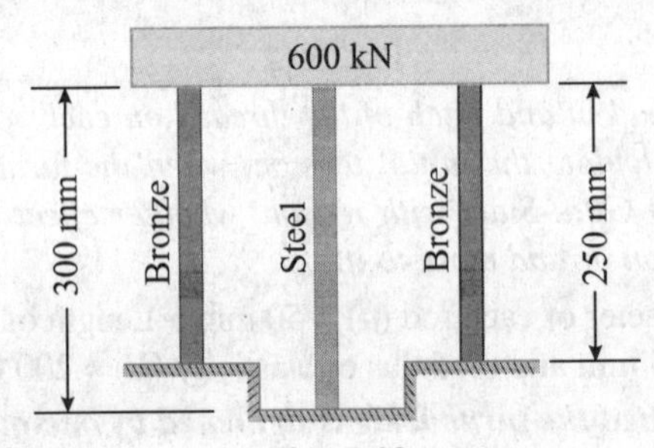


Fig. 5.13

*Find the temperature, at which the stress in steel rod will be zero. Take :*

*Coefficient of expansion for steel* = $12 \times 10^{-6}$/°C
*Coefficient of expansion for bronze* = $18 \times 10^{-6}$/°C
*Young's modulus for steel* = 200 GPa
*Young's modulus for bronze* = 80 GPa.

**SOLUTION.** Given : Weight = 600 kN = $600 \times 10^3$ N ; Area of bronze rod($A_B$) = $A_S$ = 6000 mm²; Coefficient of expansion for steel ($\alpha_S$) = $12 \times 10^{-6}$/°C ; Coefficient of expansion for bronze ($\alpha_B$) = $18 \times 10^{-6}$ /°C ; Modulus of elasticity of steel ($E_S$) = 200 GPa = $200 \times 10^3$ N/mm² and modulus of elasticity of bronze ($E_B$) = 80 GPa = $80 \times 10^3$ N/mm².

Let $t$ = Rise in temperature, when the stress in the steel rod will be zero.

A little consideration will show that due to increase in temperature all the three rods will expand. The expansion of bronze rods will be more than the steel rod (because $\alpha_B$ is greater than $\alpha_S$). If the stress in the steel rod is to be zero, then the entire load should be shared by the two bronze rods. Or in other words, the decrease in the length of two bronze rods should be equal to the difference of the expansion of the bronze rods and steel rod.

We know that free expansion of the steel rod.

$$= l_S.\alpha_S.t = 300 \times 12 \times 10^{-6} \times t = 3.6 \times 10^{-3}\,t$$

Similarly, free expansion of the bronze rods,

$$= l_B.\alpha_B.t = 250 \times 18 \times 10^{-6} \times t = 4.5 \times 10^{-3}\,t$$

∴ Difference in the expansion of the two rods

$$= (4.5 \times 10^{-3})\, t - (3.6 \times 10^{-3})\, t = 0.9 \times 10^{-3}\, t \qquad ...(i)$$

We also know that the contraction of the bronze rods due to load of 600 kN

$$= \frac{Pl}{AE} = \frac{(600 \times 10^3) \times 250}{(2 \times 6000) \times (80 \times 10^3)} = 0.156 \text{ mm} \qquad ...(ii)$$

Now equating equations (*i*) and (*ii*),

$$0.9 \times 10^{-3} \times t = 0.156 \quad \text{or} \quad t = \frac{0.156}{9 \times 10^{-4}} = 173.3°\text{C} \qquad \textbf{Ans.}$$

## EXERCISE 5.2

1. An aluminium rod of 20 mm diameter is completely enclosed in a steel tube of 30 mm external diameter and both the ends of the assembly are rigidly connected. If the composite bar is heated through 50°C, find the stresses developed in the aluminium rod and steel tube. Take:

   Modulus of elasticity for steel = 200 GPa
   Modulus of elasticity for aluminium = 80 GPa
   Coefficient of expansion for steel = $12 \times 10^{-6}$/°C
   Coefficient of expansion for aluminium = $18 \times 10^{-6}$/°C

   [**Ans.** 14.5 MPa (Comp.) ; 18.1 MPa (Tension)]

2. A steel rod of 10 mm diameter passes centrally through a copper tube of external diameter 40 mm and internal diameter 30 mm. The assembly is tightened with the help of washers and nuts. If the whole assembly is heated through 60°C, then find the stresses developed in the steel rod and copper tube. Assume :

   Young's modulus for steel = 200 GPa
   Young's modulus for copper = 100 GPa
   Coefficient of expansion for steel = $11.5 \times 10^{-6}$ /°C
   Coefficient of expansion for copper = $17 \times 10^{-6}$ /°C.

   [**Ans.** 4.4 MPa (Tension) ; 30.8 MPa (Comp.)]

3. A copper bar *ABC* of 500 mm² cross-sectional area is firmly held between two unyielding supports and subjected to an axial load as shown in Fig. 5.14.

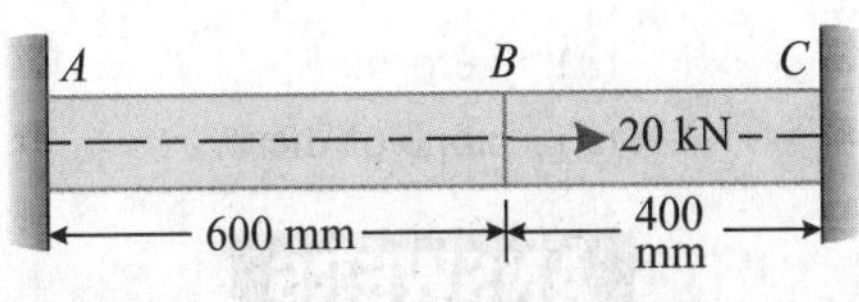


Fig. 5.14

   Calculate the stresses developed in the two portions, when the bar is subjected to an increase of 20 K temperature. Take *E* for the copper as 100 GPa and α for the copper as $18 \times 10^{-6}$/ K.

   [**Ans.** $\sigma_{AB}$ = 20 MPa (Comp.) ; $\sigma_{BC}$ = 60 MPa (Comp.)]

4. A steel rod of 25 mm diameter axially passes through a brass tube of 25 mm internal diameter and 35 mm external diameter when the nut on the rod is tightened, initial stress of 10 MPa is developed in the rod. The temperature of the tube is then raised by 60°C. Calculate the final stresses in the rod and tube. Take $E_S$ = 200 GPa, $E_B$ = 80 GPa, $\alpha_S = 11.7 \times 10^{-6}$ /°C and $\alpha_B = 19 \times 10^{-6}$ /°C.

   [**Ans.** 34.2 MPa ; 35.6 MPa]

## QUESTIONS

1. Define thermal stress and thermal strain.
2. Explain the procedure for finding out the stresses developed in a body due to change of temperature.
3. Obtain the relation for the thermal stress in a circular bar of uniformly tapering section.
4. What is the effect of thermal stresses of a body, when its ends (*i*) do not yield and (*ii*) yield by a small amount ?
5. Describe the methods for finding out the stresses in a bar of varying section, when it is made up of (*a*) one material throughout, (*b*) two different materials.
6. Explain clearly the effect of change of temperature in a composite bar.

## OBJECTIVE TYPE QUESTIONS

1. Thermal stress is caused, when the temperature of a body

   (*a*) is increased (*b*) is decreased

   (*c*) remains constant (*d*) either '*a*' or '*b*'

2. When the temperature of a body is increased, the stress induced will be

   (*a*) tension (*b*) compression

   (*c*) both '*a*' and '*b*' (*d*) neither '*a*' nor '*b*'

3. If the ends of a body yield, the magnitude of thermal stress will

   (*a*) increase (*b*) decrease

   (*c*) remain the same (*d*) none of these

4. The maximum thermal stress in a circular tapering section is

   (*a*) directly proportional to the bigger diameter

   (*b*) directly proportional to the smaller diameter

   (*c*) inversely proportional to the bigger diameter

   (*d*) both '*b*' and '*c*'

5. If a composite bar is cooled, then the nature of stress in the part with high coefficient of thermal expansion will be

   (*a*) tensile (*b*) zero

   (*c*) compressive (*d*) none of these.

## ANSWERS

**1.** (*d*) **2.** (*b*) **3.** (*c*) **4.** (*a*) **5.** (*a*)

Chapter 6

# Elastic Constants

## Contents



## 6.1. Introduction

In the previous chapter, we have discussed the axial deformation of a body, when it is subjected to a direct tensile or compressive stress. But we have not discussed the lateral or side effects of the pulls or pushes. It has been experimentally found, that the axial strain of a body is always followed by an opposite kind of strain in all directions at right angle to it. Thus, in general, there is always a set of the following two types of strains in a body, when it is subjected to a direct stress.

1. Primary or linear strain, and
2. Secondary or lateral strain.

## 6.2. Primary or Linear Strain

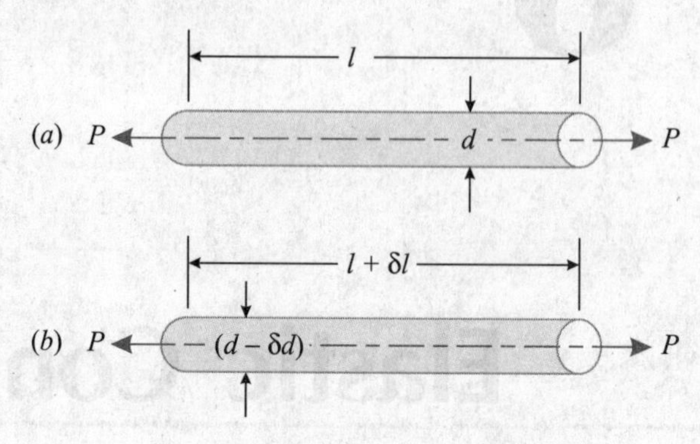


Fig. 6.1. Linear and Lateral strain

We have already discussed in Art 2.2 that whenever some external force acts on a body, it undergoes some deformation. Now consider a circular bar subjected to a tensile force as shown in Fig. 6.1 (*a*).

Let $l$ = Length of the bar,

$d$ = Diameter of the bar,

$P$ = Tensile force acting on the bar, and

$dl$ = Increase in the length of the bar, as a result of the tensile force.

The deformation of the bar per unit length in the direction of the force, *i.e.*, $\frac{\delta l}{l}$ is known as primary or linear strain.

## 6.3. Secondary or Lateral Strain

We have already discussed in the last article the linear deformation of a circular bar of length $l$ and diameter $d$ subjected to a tensile force $P$. If we actually study the deformation of the bar, we will find that bar has extended through a length $\delta l$, which will be followed by the decrease of diameter from $d$ to $(d - \delta d)$ as shown in Fig. 6.1 (*b*). Similarly, if the bar is subjected to a compressive force, the length of the bar will decrease by $\delta l$ which will be followed by the increase of diamcter from $d$ to $(d + \delta d)$.

It is thus obvious that every direct stress is always accompanied by a strain in its own direction and an opposite kind of strain in every direction at right angles to it. Such a strain is known as secondary or lateral strain.

## 6.4. *Poisson's Ratio

It has been experimentally found, that if a body is stressed within its elastic limit, the lateral strain bears a constant ratio to the linear strain. Mathematically :

$$\frac{\text{Lateral strain}}{\text{Linear strain}} = \text{(constant)}$$

This constant is known as Poisson's ratio and is denoted by $\frac{1}{m}$ or $\mu$. Mathematically,

$$\text{Lateral strain} = \frac{1}{m} \times \varepsilon = \mu\,\varepsilon$$

* Named after French mathematician Poisson, who first predicted its existence and value by using the molcular theory of structure of the material. He found this value for many isotropic materials (*i.e.*, the materials which have the same properties in all directions).

The corresponding change in the lateral length may be found out, as usual, *i.e.*, by multiplying the lateral length (*i.e.*, width or thickness).

NOTE. The value of Poisson's ratio is the same in tension and compression.

**Table 6.1.**

**The value of Poisson's ratio of materials, in every day use, are given below :**

| *S. No.* | *Material* | *Poisson's ratio* $\left(\frac{1}{m} \text{ or } \mu\right)$ |
|---|---|---|
| **1.** | Steel | 0.25 to 0.33 |
| **2.** | Cast iron | 0.23 to 0.27 |
| **3.** | Copper | 0.31 to 0.34 |
| **4.** | Brass | 0.32 to 0.42 |
| **5.** | Aluminium | 0.32 to 0.36 |
| **6.** | Concrete | 0.08 to 0.18 |
| **7.** | Rubber | 0.45 to 0.50 |

**EXAMPLE 6.1.** *A steel bar 2 m long, 40 mm wide and 20 mm thick is subjected to an axial pull of 160 kN in the direction of its length. Find the changes in length, width and thickness of the bar. Take E = 200 GPa and Poisson's ratio = 0.3.*

**SOLUTION.** Given : Length $(l) = 2 \text{ m} = 2 \times 10^3$ mm ; Width $(b) = 40$ mm ; Thickness $(t) = 20$ mm; Axial pull $(P) = 160 \text{ kN} = 160 \times 10^3$ N ; Modulus of elasticity $(E) = 200 \text{ GPa} = 200 \times 10^3 \text{ N/mm}^2$ and poisson's ratio $(\frac{1}{m}) = 0.3$.

***Change in length***

We know that change in length,

$$\delta l = \frac{Pl}{AE} = \frac{(160 \times 10^3) \times (2 \times 10^3)}{(40 \times 20) \times (200 \times 10^3)} = 2 \text{ mm} \quad \textbf{Ans.}$$

***Change in width***

We know that linear strain,

$$\varepsilon = \frac{\delta l}{l} = \frac{2}{2 \times 10^3} = 0.001$$

and lateral strain

$$= \frac{1}{m} \times \varepsilon = 0.3 \times 0.01 = 0.0003$$

$\therefore$ Change in width,

$$\delta b = b \times \text{Lateral strain} = 40 \times 0.0003 = 0.012 \text{ mm} \quad \textbf{Ans.}$$

***Change in thickness***

We also know that change in thickness,

$$\delta t = t \times \text{Lateral strain} = 20 \times 0.0003 = 0.006 \text{ mm} \quad \textbf{Ans.}$$

**EXAMPLE 6.2.** *A metal bar 50 mm × 50 mm in section is subjected to an axial compressive load of 500 kN. If the contraction of a 200 mm gauge length was found to be 0.5 mm and the increase in thickness 0.04 mm, find the values of Young's modulus and Poisson's ratio for the bar material.*

**SOLUTION.** Given : Width $(b) = 50$ mm ; Thickness $(t) = 50$ mm ; Axial compressive load $(P) = 500 \text{ kN} = 500 \times 10^3$ N ; Length $(l) = 200$ mm ; Change in length $(\delta l) = 0.5$ mm and change in thickness $(\delta t) = 0.04$ mm.

***Value of Young's modulus for the bar material***

Let $E$ = Value of Young's modulus for the bar material.

We know that contraction of the bar ($\delta l$),

$$0.5 = \frac{P.l}{A.E} = \frac{(500 \times 10^3) \times 200}{(50 \times 50) \times E} = \frac{40 \times 10^3}{E}$$

$$\therefore \quad E = \frac{40 \times 10^3}{0.5} = 80 \times 10^3 \text{ N/mm}^2 = 80 \text{ GPa} \qquad \textbf{Ans.}$$

Value of Poisson's ratio for the bar material

Let $\frac{1}{m}$ = Value of Poisson's ratio for the bar material.

We know that linear strain,

$$\varepsilon = \frac{\delta l}{l} = \frac{0.5}{200} = 0.0025$$

and lateral strain $= \frac{1}{m} \times \text{Linear strain} = \frac{1}{m} \times 0.0025$

We also know that increase in thickness ($\delta t$),

$$0.04 = t \times \text{Lateral strain} = 50 \times \frac{1}{m} \times 0.0025 = \frac{0.125}{m}$$

$$\therefore \quad \frac{1}{m} = \frac{0.04}{0.125} = 0.32 \qquad \textbf{Ans.}$$

## 6.5. Volumetric Strain

We have already discussed that whenever a body is subjected to a single force (or a system of forces), it undergoes some changes in its dimensions. A little consideration will show, that the change in dimensions of a body will cause some changes in its volume. The ratio of change in volume, to the original volume, is known as volumetric strain. Mathematically volumetric strain,

$$\varepsilon_V = \frac{\delta V}{V}$$

where $\delta V$ = Change in volume, and

$V$ = Original volume.

Though there are numerous ways, in which a force (or a system of forces) may act, yet the following are important from the subject point of view :

1. A rectangular body subjected to an axial force.
2. A rectangular body subjected to three mutually perpendicular forces.

Now we shall discuss the volumetric strains on all the types of bodies one by one in the following pages :

## 6.6. Volumetric Strain of a Rectangular Body Subjected to an Axial Force

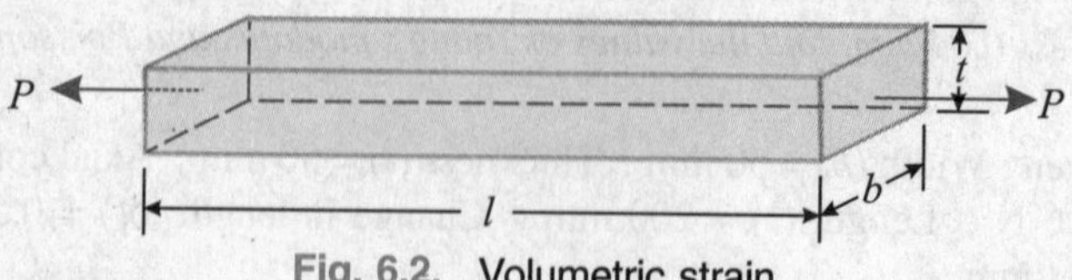


Fig. 6.2. Volumetric strain

Consider a bar, rectangular in section, subjected to an axial tensile force as shown in Fig. 6.2.

Let $l$ = Length of the bar,
$b$ = Breadth of the bar,
$t$ = Thickness of the bar,
$P$ = Tensile force acting on the bar,
$E$ = Modulus of elasticity and
$\frac{1}{m}$ = Poisson's ratio.

We know that change in length,

$$\delta l = \frac{Pl}{AE} = \frac{Pl}{bt\,E} \quad ...(i)$$

and linear stress,

$$\sigma = \frac{\text{Force}}{\text{Area}} = \frac{P}{b\,t}$$

∴ Linear strain $= \frac{\text{Stress}}{E} = \frac{P}{bt\,E}$

and lateral strain $= \frac{1}{m} \times$ Linear strain $= \frac{1}{m} \times \frac{P}{bt\,E}$

∴ Change in thickness,

$$\delta t = t \times \frac{1}{m} \times \frac{P}{bt\,E} = \frac{P}{mb\,E} \quad ...(ii)$$

and change in breadth,

$$\delta b = b \times \frac{1}{m} \times \frac{P}{bt\,E} = \frac{P}{mt\,E} \quad ...(iii)$$

As a result of this tensile force, let the final length

$= l + \delta l$

Final breadth $= b - \delta b$ ...(Minus sign due to compression)

and final thickness $= t - \delta t$ ...(Minus sign due to compression)

We know that original volume of the body,

$$V = l.b.t.$$

and final volume $= (l + \delta l)(b - \delta b)(t - \delta t)$

$$= lbt\left(1 + \frac{\delta l}{l}\right)\left(1 - \frac{\delta b}{b}\right)\left(1 - \frac{\delta t}{t}\right)$$

$$= lbt\left[1 + \frac{\delta l}{l} - \frac{\delta b}{b} - \frac{\delta t}{t} - \left(\frac{\delta l}{l} \times \frac{\delta b}{b}\right) - \left(\frac{\delta l}{l} \times \frac{\delta t}{t}\right) + \left(\frac{\delta l}{b} \times \frac{\delta t}{t}\right) + \left(\frac{\delta l}{l} \times \frac{\delta b}{b} \times \frac{\delta t}{t}\right)\right]$$

$$= lbt\left[1 + \frac{\delta l}{l} - \frac{\delta b}{b} - \frac{\delta t}{t}\right] \quad \text{...(Ignoring other negligible values)}$$

∴ Change in volume,

$$\delta V = \text{Final volume} - \text{Original volume}$$

$$= lbt\left(1 + \frac{\delta l}{l} - \frac{\delta b}{b} - \frac{\delta t}{t}\right) - lbt = lbt\left(\frac{\delta l}{l} - \frac{\delta b}{b} - \frac{\delta t}{t}\right)$$

$$= lbt\left[\frac{\frac{Pl}{btE}}{l} - \frac{\frac{P}{mtE}}{b} - \frac{\frac{P}{mbE}}{t}\right] = lbt\left(\frac{P}{btE} - \frac{P}{mbtE} - \frac{P}{mbtE}\right)$$

$$= V \times \frac{P}{btE}\left(1 - \frac{2}{m}\right)$$

and volumetric strain,

$$\frac{\delta V}{V} = \frac{V \times \frac{P}{btE}\left(1 - \frac{2}{m}\right)}{V} = \frac{P}{btE}\left(1 - \frac{2}{m}\right)$$

$$= \varepsilon\left(1 - \frac{2}{m}\right) \qquad \ldots\left(\because \frac{P}{btE} = \varepsilon = \text{Strain}\right)$$

**NOTE.** The above formula holds good for compressive force also.

**EXAMPLE 6.3.** *A steel bar 2 m long, 20 mm wide and 15 mm thick is subjected to a tensile load of 30 kN. Find the increase in volume, if Poisson's ratio is 0.25 and Young's modulus is 200 GPa.*

**SOLUTION.** Given : Length ($l$) = 2 m = $2 \times 10^3$ mm ; Width ($b$) = 20 mm ; Thickness ($t$) = 15 mm; Tensile load ($P$) = 30 kN = $30 \times 10^3$ N ; Poisson's ratio $\left(\frac{1}{m}\right)$ = 0.25 or $m$ = 4 and Young's modulus of elasticity ($E$) = 200 GPa = $200 \times 10^3$ N/mm$^2$.

Let $\delta V$ = Increase in volume of the bar.

We know that original volume of the bar,

$$V = l.b.t = (2 \times 10^3) \times 20 \times 15 = 600 \times 10^3 \text{ mm}^3$$

and

$$\frac{\delta V}{V} = \frac{P}{btE}\left(1 - \frac{2}{m}\right) = \frac{30 \times 10^3}{20 \times 15 \times (200 \times 10^3)}\left(1 - \frac{2}{4}\right) = 0.000\ 25$$

$\therefore$

$$\delta V = 0.000\ 25 \times V = 0.000\ 25 \times (600 \times 10^3) = 150 \text{ mm}^3 \qquad \textbf{Ans.}$$

**EXAMPLE 6.4.** *A copper bar 250 mm long and 50 mm × 50 mm in cross-section is subjected to an axial pull in the direction of its length. If the increase in volume of the bar is 37.5 mm$^3$, find the magnitude of the pull. Take m = 4 and E = 100 GPa.*

**SOLUTION.** Given: Length ($l$) = 250 mm ; Width ($b$) = 50 mm ; Thickness ($t$) = 50 mm ; Increase in volume ($\delta V$) = 37.5 mm$^3$ ; ($m$) = 4 and modulus of elasticity ($E$) = 100 GPa = $100 \times 10^3$ N/mm$^2$.

Let $P$ = Magnitude of the pull in kN.

We know that original volume of the copper bar,

$$V = l.b.t = (250 \times 50 \times 50) = 625 \times 10^3 \text{ mm}^3$$

and

$$\frac{\delta V}{V} = \frac{P}{btE}\left(1 - \frac{2}{m}\right) = \frac{P}{50 \times 10 \times (100 \times 10^3)}\left(1 = \frac{2}{4}\right)$$

or

$$\frac{37.5}{625 \times 10^3} = \frac{P}{500 \times 10^6}$$

$\therefore$

$$P = \frac{37.5 \times (500 \times 10^6)}{625 \times 10^3} = 30 \times 10^3 \text{ N} = 30 \text{ kN} \qquad \textbf{Ans.}$$

**EXAMPLE 6.5.** *A steel bar 50 mm × 50 mm in cross-section is 1.2 m long. It is subjected to an axial pull of 200 kN. What are the changes in length, width and volume of the bar, if the value of Poisson's ratio is 0.3? Take E as 200 GPa.*

**Solution.** Given : Width ($b$) = 50 mm ; Thickness ($t$) = 50 mm ; Length ($l$) = 1.2 m = $1.2 \times 10^3$ mm ; Axial pull ($P$) = 200 kN = $200 \times 10^3$ N ; Poisson's ratio $\left(\frac{1}{m}\right)$ = 0.3 and modulus of elasticity ($E$) = 200 GPa = $200 \times 10^3$ N/mm$^2$.

***Change in length***

We know that change in length,

$$\delta l = \frac{Pl}{AE} = \frac{200 \times 10^3 \times (1.2 \times 10^3)}{(50 \times 50) \times (200 \times 10^3)} = 0.48 \text{ mm} \qquad \textbf{Ans.}$$

***Change in width***

We know that linear strain,

$$\varepsilon = \frac{\delta l}{l} = \frac{0.48}{1.2 \times 10^3} = 0.0004$$

and lateral strain $= \frac{1}{m} \times \varepsilon = 0.3 \times 0.0004 = 0.000\,12$

$\therefore$ Change in width, $\delta b = b \times \text{Lateral strain} = 50 \times 0.000\,12 = 0.006$ mm **Ans.**

***Change in volume***

We also know that volume of the bar,

$$V = l.b.t = (1.2 \times 10^3) \times 50 \times 50 = 3 \times 10^6 \text{ mm}^3$$

and

$$\frac{\delta V}{V} = \frac{P}{btE}\left(1 - \frac{2}{m}\right) = \frac{200 \times 10^3}{50 \times 50 \times (200 \times 10^3)}[1 - (2 \times 0.3)]$$

$$= 0.000\,16$$

$\therefore$

$$\delta V = 0.000\,16\,V = 0.00016 \times (3 \times 10^6) = 480 \text{ mm}^3 \qquad \textbf{Ans.}$$

## 6.7. Volumetric Strain of a Rectangular Body Subjected to Three Mutually Perpendicular Forces

Consider a rectangular body subjected to direct tensile stresses along three mutually perpendicular axes as shown in Fig. 6.3.

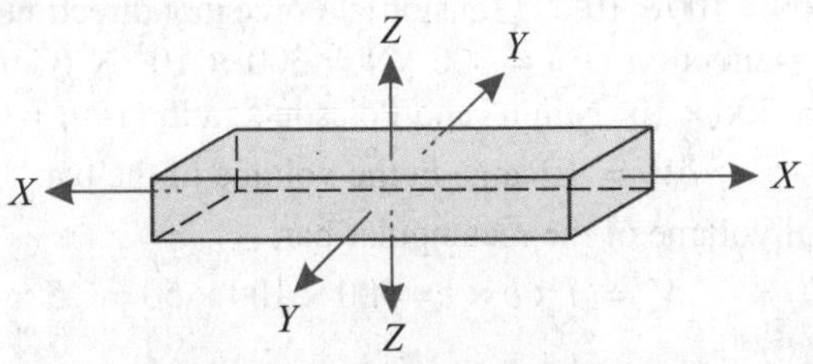


Fig. 6.3. Volumetric strain

Let $\sigma_x$ = Stress in $x$-$x$ direction,

$\sigma_y$ = Stress in $y$-$y$ direction,

$\sigma_z$ = Stress in $z$-$z$ direction and

$E$ = Young's modulus of elasticity.

$\therefore$ Strain in $x$-$x$ direction due to stress $\sigma_x$,

$$\varepsilon_x = \frac{\sigma_x}{E}$$

Similarly,

$$\varepsilon_y = \frac{\sigma_y}{E} \quad \text{and} \quad \varepsilon_z = \frac{\sigma_z}{E}$$

The resulting strains in the three directions, may be found out by the principle of superposition, *i.e.*, by adding algebraically the strains in each direction due to each individual stress.

For the three tensile stresses shown in Fig. 6.3. (taking tensile strains as +ve and compressive strains as –ve) the resultant strain in *x-x* direction,

$$\varepsilon_x = \frac{\sigma_x}{E} - \frac{\sigma_y}{mE} - \frac{\sigma_z}{mE} = \frac{1}{E}\left[\sigma_x - \frac{\sigma_y}{m} - \frac{\sigma_z}{m}\right]$$

Similarly,

$$\varepsilon_y = \frac{\sigma_y}{E} - \frac{\sigma_x}{mE} - \frac{\sigma_z}{mE} = \frac{1}{E}\left[\sigma_y - \frac{\sigma_x}{m} - \frac{\sigma_z}{m}\right]$$

and

$$\varepsilon_z = \frac{\sigma_z}{E} - \frac{\sigma_x}{mE} - \frac{\sigma_y}{mE} = \frac{1}{E}\left[\sigma_z - \frac{\sigma_x}{m} - \frac{\sigma_y}{m}\right]$$

The volumetric strain may then be found by the relation;

$$\frac{\delta V}{V} = \varepsilon_x + \varepsilon_y + \varepsilon_z$$

**Note.** In the above relation, the values of $\varepsilon_x$, $\varepsilon_y$ and $\varepsilon_z$ should be taken tensile as positive and compressive as negative.

**Example 6.6.** *A rectangular bar 500 mm long and 100 mm × 50 mm in cross-section is subjected to forces as shown in Fig. 6.4.*

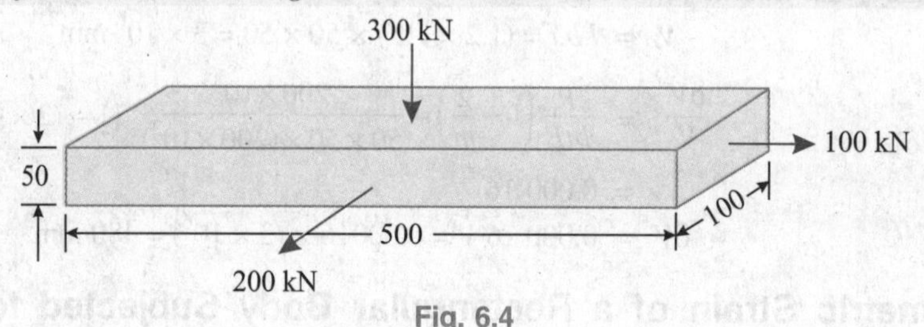


**Fig. 6.4**

*What is the change in the volume of the bar? Take modulus of elasticity for the bar material as 200 GPa and Poisson's ratio as 0.25.*

**Solution.** Given : Length ($l$) = 500 mm ; Width ($b$) = 100 mm ; Thickness ($t$) = 50 mm ; Force in *x*-direction ($P_x$) = 100 kN = $100 \times 10^3$ N (Tension) ; Force in *y*-direction ($P_y$) = 200 kN = $200 \times 10^3$ N (Tension) ; Force in *z*-direction ($P_z$) = 300 kN = $300 \times 10^3$ N (Compression) ; Modulus of elasticity ($E$) = 200 GPa = $200 \times 10^3$ N/mm$^2$ and Poisson's ratio ($1/m$) = 0.25 or $m = 4$.

Let $\delta V$ = Change in the volume of the bar.

We know that original volume of the rectangular bar,

$$V = l \times b \times t = 500 \times 100 \times 50 = 2.5 \times 10^6 \text{ mm}^3$$

and stress in *x-x* direction,

$$\sigma_x = \frac{P_x}{A_x} = \frac{100 \times 10^3}{100 \times 50} = 20 \text{ N/mm}^2 \text{ (Tension)}$$

Similarly,

$$\sigma_y = \frac{P_y}{A_y} = \frac{200 \times 10^3}{500 \times 50} = 8 \text{ N/mm}^3 \text{ (Tension)}$$

and

$$\sigma_z = \frac{P_z}{A_z} = \frac{300 \times 10^3}{500 \times 100} = 6 \text{ N/mm}^2 \text{ (Compression)}$$

We also know that resultant strain in *x-x* direction considering tension as positive and compression as negative

$$\varepsilon_x = +\frac{\sigma_x}{E} - \frac{\sigma_y}{mE} + \frac{\sigma_z}{mE} = +\frac{20}{E} - \frac{8}{4E} + \frac{6}{4E} = \frac{19.5}{E}$$

Similarly
$$\varepsilon_y = +\frac{\sigma_y}{E} - \frac{\sigma_x}{mE} + \frac{\sigma_z}{mE} = +\frac{8}{E} - \frac{20}{4E} + \frac{6}{4E} = \frac{4.5}{E}$$

and
$$\varepsilon_z = -\frac{\sigma_z}{E} - \frac{\sigma_x}{mE} - \frac{\sigma_y}{mE} = -\frac{6}{E} - \frac{20}{4E} - \frac{8}{4E} = -\frac{13}{E}$$

We also know that volumetric strain,

$$\frac{\delta V}{V} = \varepsilon_x + \varepsilon_y + \varepsilon_z$$

$$\frac{\delta V}{2.5 \times 10^6} = \frac{19.5}{E} + \frac{4.5}{E} - \frac{13}{E} = \frac{11}{E} = \frac{11}{200 \times 10^3} = 0.055 \times 10^{-3}$$

$$\therefore \quad \delta V = (0.055 \times 10^{-3}) \times (2.5 \times 10^6) = 137.5 \text{ mm}^3 \qquad \textbf{Ans.}$$

**EXAMPLE 6.7.** *A steel cube block of 50 mm side is subjected to a force of 6 kN (Tension), 8 kN (Compression) and 4 kN (Tension) along x, y and z direction respectively. Determine the change in volume of the block. Take E as 200 GPa and m as 10/3.*

**SOLUTION.** Given : Side of the cube = 50 mm ; Force in x-direction ($P_x$) = 6 kN = $6 \times 10^3$ N (Tension) ; Force in y-direction ($P_y$) = 8 kN = $8 \times 10^3$ N (Compression) ; Force in z-direction ($P_z$) = 4 kN = $4 \times 10^3$ N (Tension) and modulus of elasticity ($E$) = 200 GPa = $200 \times 10^3$ N/mm² and $m = \frac{10}{3}$ or $\frac{1}{m} = \frac{3}{10}$

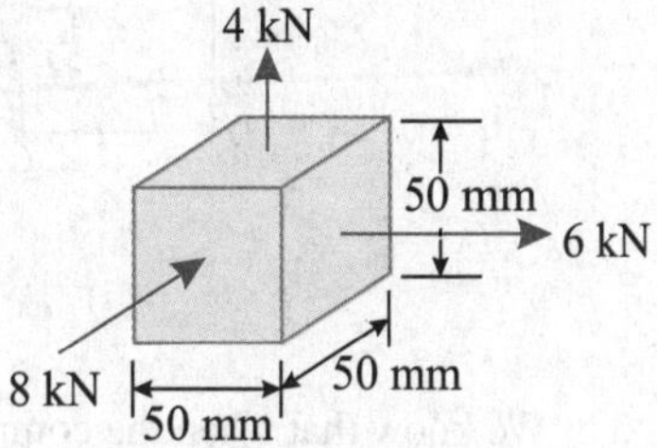

Fig. 6.5

Let $\delta V$ = Change in volume of the block.

We know that original volume of the steel cube,

$$V = 50 \times 50 \times 50 = 125 \times 10^3 \text{ mm}^3$$

and stress in x-x direction,

$$\sigma_x = \frac{P_x}{A} = \frac{6 \times 10^3}{2500} = 2.4 \text{ N/mm}^2 \text{ (Tension)}$$

Similarly
$$\sigma_y = \frac{P_y}{A} = \frac{8 \times 10^3}{2500} = 3.2 \text{ N/mm}^2 \text{ (Compression)}$$

and
$$\sigma_z = \frac{P_z}{A} = \frac{4 \times 10^3}{2500} = 1.6 \text{ N/mm}^2 \text{ (Tension)}$$

We also know that resultant strain in x-x direction considering tension as positive and compression as negative,

$$\varepsilon_x = \frac{\sigma_x}{E} + \frac{\sigma_y}{mE} - \frac{\sigma_z}{mE} = \frac{2.4}{E} + \frac{3.2 \times 3}{10E} - \frac{1.6 \times 3}{10E} = \frac{2.88}{E}$$

Similarly,
$$\varepsilon_y = -\frac{\sigma_y}{E} - \frac{\sigma_x}{mE} - \frac{\sigma_z}{mE} = -\frac{3.2}{E} - \frac{2.4 \times 3}{10E} - \frac{1.6 \times 3}{10E} = -\frac{4.4}{E}$$

and
$$\varepsilon_z = \frac{\sigma_z}{E} - \frac{\sigma_x}{mE} + \frac{\sigma_y}{mE} = \frac{1.6}{E} - \frac{2.4 \times 3}{10E} + \frac{3.2 \times 3}{10E} = \frac{1.84}{E}$$

We also know that volumetric strain,

$$\frac{\delta V}{V} = \varepsilon_x + \varepsilon_y + \varepsilon_z$$

$$\frac{\delta V}{125 \times 10^3} = \frac{2.88}{E} - \frac{4.4}{E} + \frac{1.84}{E} = \frac{0.32}{E} = \frac{0.32}{200 \times 10^3}$$

$$\therefore \quad \delta V = 125 \times 10^3 \times \frac{0.32}{200 \times 10^3} = 0.2 \text{ mm}^3 \quad \textbf{Ans.}$$

**EXAMPLE 6.8.** *A cubical block is subjected to a compressive load (P) in one of the directions. If the lateral strains, in other two directions are to be completely prevented, by the application of another compressive load ($P_1$), then find the value of $P_1$ in terms of P.*

**SOLUTION.** Given : A cubical block *ABCDEFGH* and load on two opposite faces *ADHE* and *BCGF* = *P*.

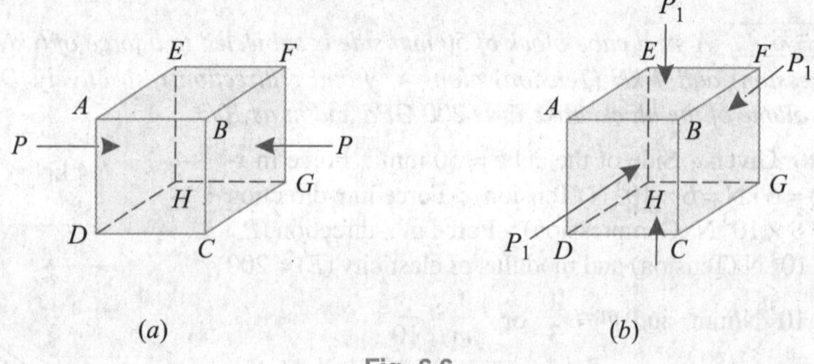


**Fig. 6.6**

We know that when the compressive load (*P*) is applied on the faces *ADHE* and *BCGF*, the other two faces will be subjected to lateral tensile stress as shown in Fig. 6.6 (*a*). Now in order to prevent the lateral strains in the other two directions, we have to apply a compressive load ($P_1$) as shown in Fig. 6.6 (*b*).

We also know that lateral strain ($\varepsilon_y$)

$$0 = \frac{1}{E} \times \left[\sigma_y - \frac{\sigma_x}{m} - \frac{\sigma_z}{m}\right] = \frac{1}{E} \times \left[P_1 - \frac{P}{m} - \frac{P_1}{m}\right]$$

...($\because$ Stresses are directly proportional to loads)

$$\therefore \quad P_1 - \frac{P}{m} - \frac{P_1}{m} = 0 \quad \text{or} \quad \frac{P}{m} = P_1 - \frac{P_1}{m} = P_1\left(1 - \frac{1}{m}\right) = P_1\left(\frac{m-1}{m}\right)$$

$$\text{or} \quad P_1 = \frac{P}{m} \times \frac{m}{m-1} = \frac{P}{m-1} \quad \textbf{Ans.}$$

## EXERCISE 6.1

1. A steel rod 1.5 m long and 20 mm diameter is subjected to an axial pull of 100 kN. Find the change in length and diameter of the rod, if $E$ = 200 GPa and 1/m = 0.32.

   [**Ans.** 2.4 mm ; 0.01 mm]

2. Determine the changes in length, breadth and thickness of a steel bar 4 m long, 30 mm wide and 20 mm thick, when subjected to an axial pull of 120 kN in the direction of its length. Take $E$ = 200 GPa and Poisson's ratio 0.3. [**Ans.** 4 mm ; 0.009 mm ; 0.006 mm]

3. A steel bar 1.2 m long, 50 mm wide and 40 mm thick is subjected to an axial pull of 150 kN in the direction of its length. Determine the change in volume of the bar. Take $E$ = 200 GPa and m = 4. **[Ans.** 450 mm$^3$]
4. A steel block 200 mm × 20 mm × 20 mm is subjected to a tensile load of 40 kN in the direction of its length. Determine the change in volume, if $E$ is 205 GPa and 1/m = 0.3. **[Ans.** 15.6 mm$^3$]
5. A rectangular bar is subjected to an axial stress $\sigma_1$, $\sigma_2$ and $\sigma_3$ on its sides. Show that the volumetric strain,

$$\frac{\delta V}{V} = (\sigma_1 + \sigma_2 + \sigma_3) \times \frac{1}{E}\left(1 - \frac{2}{m}\right)$$

## 6.8. Bulk Modulus

When a body is subjected to three mutually perpendicular stresses, of equal intensity, the ratio of direct stress to the corresponding volumetric strain is known as bulk modulus. It is, denoted by $K$. Mathematically bulk modulus,

$$K = \frac{\text{Direct stress}}{\text{Volumetric strain}} = \frac{\sigma}{\frac{\delta V}{V}}$$

## 6.9. Relation Between Bulk Modulus and Young's Modulus

Consider a cube $ABCD\ A_1B_1C_1D_1$ as shown in Fig. 6.7. Let the cube be subjected to three mutually perpendicular tensile stresses of equal intensity.

Let $\sigma$ = Stress on the faces.

$l$ = Length of the cube, and

$E$ = Young's modulus for the material of the block.

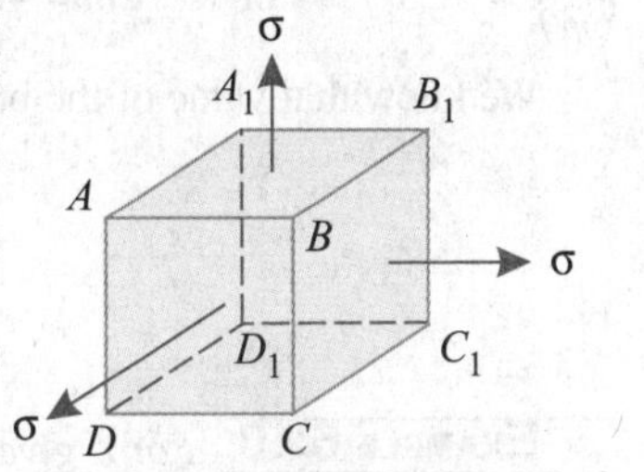

Fig. 6.7. Cube $ABCD\ A_1B_1C_1D_1$

Now consider the deformation of one side of cube (say $AB$) under the action of the three mutually perpendicular stresses. We know that this side will suffer the following strains due to the pair of stresses:

1. Tensile strain equal to $\frac{\sigma}{E}$ due to stresses on the faces $BB_1\ CC_1$ and $AA_1\ DD_1$.
2. Compressive lateral strain equal to due to stresses on faces $AA_1\ BB_1$ and $DD_1\ CC_1$.
3. Compressive lateral strain equal to $\frac{1}{m} \times \frac{\sigma}{E}$ due to stresses on faces $ABCD$ and $A_1\ B_1\ C_1\ D_1$.

Therefore net tensile strain, which the side $AB$ will suffer, due to these stresses,

$$\frac{\delta l}{l} = \frac{\sigma}{E} - \left(\frac{1}{m} \times \frac{\sigma}{E}\right) - \left(\frac{1}{m} \times \frac{\sigma}{E}\right) = \frac{\sigma}{E}\left(1 - \frac{2}{m}\right) \quad ...(i)$$

We know that the original volume of the cube,

$$V = l^3$$

Differentiating the above equation with respect to $l$,

$$\frac{\delta V}{\delta l} = 3\,l^2$$

or $$\delta V = 3\,l^2\,.\,\delta l = 3\,l^3 \times \frac{\delta l}{l}$$

Substituting the value of $\frac{\delta l}{l}$ from equation (*i*)

$$\delta V = 3\,l^3 \times \frac{\sigma}{E}\left(1 - \frac{2}{m}\right)$$

or $$\frac{\delta V}{V} = \frac{3l^3}{l^3} \times \frac{\sigma}{E}\left(1 - \frac{2}{m}\right) = \frac{3\sigma}{E}\left(1 - \frac{2}{m}\right)$$

$\therefore$ $$\frac{\sigma}{\frac{\delta V}{V}} = \frac{E}{3} \times \frac{1}{\left(1 - \frac{2}{m}\right)} = \frac{E}{3} \times \frac{1}{\left(\frac{m-2}{m}\right)}$$

or $$K = \frac{mE}{3\,(m-2)}$$

**EXAMPLE 6.9.** *If the values of modulus of elasticity and Poisson's ratio for an alloy body is 150 GPa and 0.25 respectively, determine the value of bulk modulus for the alloy.*

**SOLUTION.** Given: Modulus of elasticity $(E)$ = 150 GPa = $150 \times 10^3$ N/mm² and Poisson's ratio $\left(\frac{1}{m}\right) = 0.25$ or $m = 4$.

We know that value of the bulk modulus for the alloy,

$$K = \frac{mE}{3\,(m-2)} = \frac{4 \times (150 \times 10^3)}{3\,(4-2)} = 100 \times 10^3 \text{ N/mm}^2$$

$$= 100 \text{ GPa} \quad \textbf{Ans.}$$

**EXAMPLE 6.10.** *For a given material, Young's modulus is 120 GPa and modulus of rigidity is 40 GPa. Find the bulk modulus and lateral contraction of a round bar of 50 mm diameter and 2.5 m long, when stretched 2.5 mm. Take poisson's ratio as 0.25.*

**SOLUTION.** Given : Young's modulus $(E)$ = 120 GPa = $120 \times 10^3$ N/mm² ; Modulus of rigidity $(C)$ = 40 GPa = $40 \times 10^3$ N/mm² ; Diameter $(d)$ = 50 mm ; Length $(l)$ = 2.5 m = $2.5 \times 10^3$ mm ; Linear stretching or change in length $(\delta l)$ = 2.5 mm and poisson's ratio = 0.25 or m = 4.

***Bulk modulus of the bar***

We know that bulk modulus of the bar,

$$K = \frac{mE}{3\,(m-2)} = \frac{4 \times (120 \times 10^3)}{3\,(4-2)} = 80 \times 10^3 \text{ N/mm}^2$$

$$= 80 \text{ GPa} \quad \textbf{Ans.}$$

***Lateral contraction of the bar***

Let $\delta d$ = Lateral contraction of the bar (or change in diameter)

We know that linear strain,

$$\varepsilon = \frac{\delta l}{l} = \frac{2.5}{2.5 \times 10^3} = \frac{1}{1000} = 0.001$$

and lateral strain, $\frac{\delta d}{d} = \frac{1}{m} \times \varepsilon = 0.25 \times 0.001 = 0.25 \times 10^{-3}$

$\therefore$ $\delta d = d \times (0.25 \times 10^{-3}) = 50 \times (0.25 \times 10^{-3}) = 0.0125 \text{ mm}$ **Ans.**

## 6.10. Shear Stress

When a section is subjected to two equal and opposite forces, acting tangentially across the resisting section, as a result of which the body tends to shear off across the section as shown in Fig. 6.8, the stress induced is called shear stress. The corresponding strain is called shear strain.

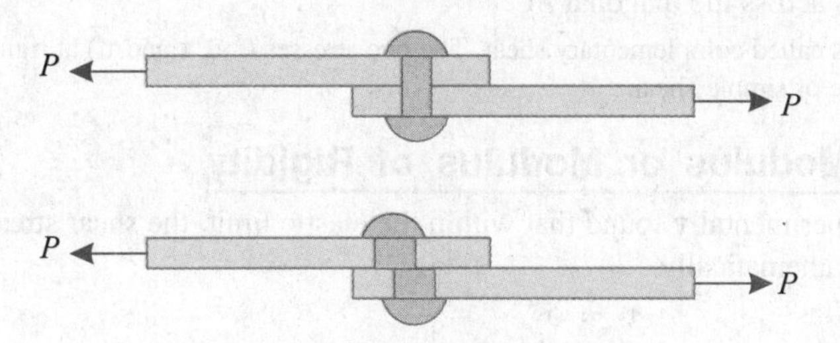


Fig. 6.8. Shear stress on a rivet.

Consider a cube of length l fixed at the bottom face *AB*. Let a force *P* be applied at the face *DC*, tangentially to the face *AB*. As a result of the force, let the cube be distorted from *ABCD* to $AB\,C_1\,D_1$ through an angle $\phi$ as shown in Fig. 6.9. We know that

$$\text{Shear strain} = \frac{\text{Deformation}}{\text{Original length}}$$

$$= \frac{CC_1}{l} = \phi$$

and shere stress, $\tau = \frac{P}{AB}$

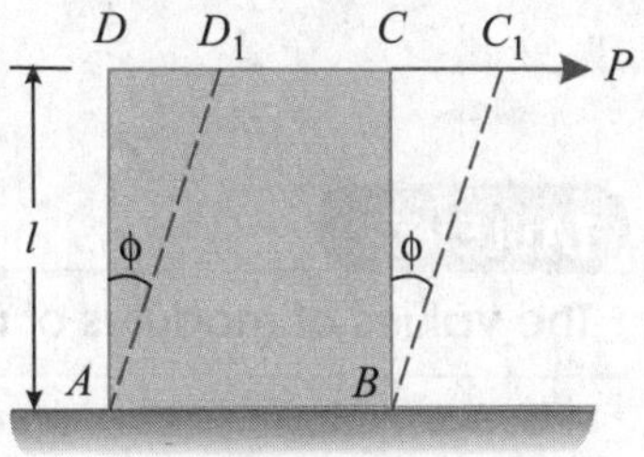

Fig. 6.9. Shear strain.

## 6.11. Principle of Shear Stress

It states, "*A shear stress across a plane, is always accompanied by a balancing shear stress across the plane and normal to it*."

### Proof

Consider a rectangular block *ABCD*, subjected to a shear stress of intensity $\tau$ on the faces *AD* and *CB* as shown in Fig. 6.10. Now consider a unit thickness of the block. Therefore force acting on the faces *AD* and *CB*,

$$P = \tau \times AD = \tau \times CB$$

A little consideration will show that these forces will form a couple, whose moment is equal to $\tau \times AD \times AB$ *i.e.*, force × distance. If the block is in equilibrium, there must be a restoring couple, whose moment must be equal to this couple. Let the shear stress of intensity $\tau$ be set up on the faces *AB* and *CD* as shown in Fig. 6.10. Therefore forces acting on the faces *AB* and *CD*,.

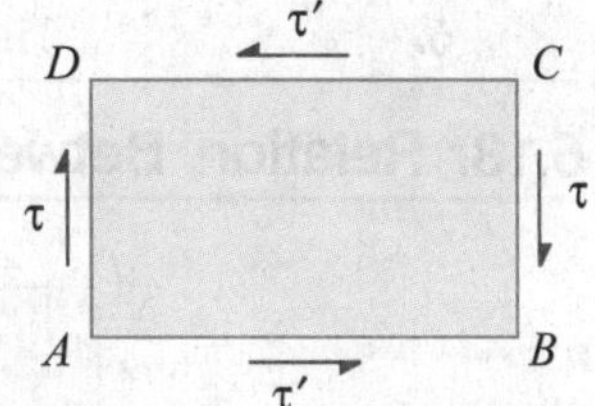


Fig. 6.10. Princciple of shear stress

$$P = \tau' \times AB = \tau' \times CD$$

We see that these forces will also form a couple, whose moment is equal to $\tau' \times AD \times AB$ *i.e.*, force × distance. Equating these two moments, we get

$$\tau \times AD \times AB = \tau' \times AD \times AB$$

or

$$\tau = \tau'$$

As a result of the two couples formed by the shear forces, the diagonal *BD* of the block will be subjected to tension whereas the diagonal *AC* will be subjected to compression. A little consideration will show that if the block material is poor in tension, it will fail due to excessive tensile stress across the diagonal *BD*. Similarly if the block material is poor in compression, it will fail due to excessive compressive forces across the diagonal *AC*.

**Note.** The stress $\tau'$ is called complementary shear. The two stresses (*i.e.*, $\tau$ and $\tau'$) at right angles to each other constitute a state of simple shear.

## 6.12. Shear Modulus or Modulus of Rigidity

It has been experimentally found that within the elastic limit, the shear stress is proportional to the shear strain. Mathematically

$$\tau \propto \phi$$

or

$$\tau = C \times \phi$$

or

$$\frac{\tau}{\phi} = C \text{ (or } G \text{ or } N)$$

where $\tau$ = Shear stress,

$\phi$ = Shear strain, and

$C$ = A constant, known as shear modulus or modulus of rigidity. It is also denoted by *G* or *N*.

**TABLE 6.2.**

**The values of modulus of rigidity of materials in every day use are given below :**

| *S. No.* | *Material* | *Modulus of rigidity (C) in GPa i.e., GN/m² or kN/mm²* |
|---|---|---|
| 1. | Steel | 80 to 100 |
| 2. | Wrought iron | 80 to 90 |
| 3. | Cast iron | 40 to 50 |
| 4. | Copper | 30 to 50 |
| 5. | Brass | 30 to 60 |
| 6. | Timber | 10 |

## 6.13. Relation Between Modulus of Elasticity and Modulus of Rigidity

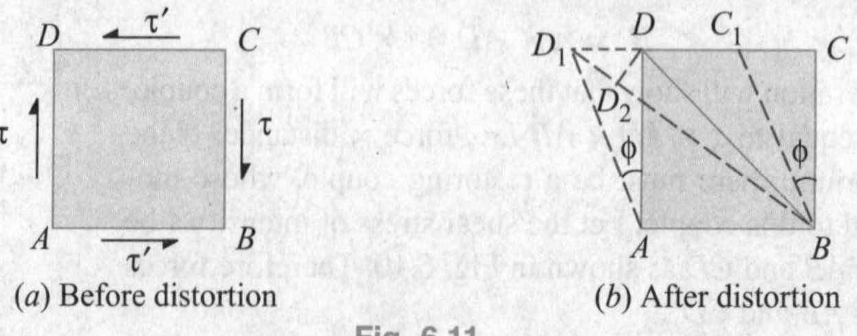


(*a*) Before distortion (*b*) After distortion

Fig. 6.11

Consider a cube of length *l* subjected to a shear stress of $\tau$ as shown in Fig. 6.11 (*a*). A little consideration will show that due to these stresses the cube is subjected to some distortion, such that the diagonal $BD$ will be elongated and the diagonal $AC$ will be shortened. Let this shear stress t cause shear strain $\phi$ as shown in Fig. 6.11 (*b*). We see that the diagonal $BD$ is now distorted to $BD_1$.

$$\therefore \quad \text{Strain of } BD = \frac{BD_1 - BD}{BD} \qquad \ldots\left(\because \text{ Strain} = \frac{\delta l}{l}\right)$$

$$= \frac{D_1 D_2}{BD} = \frac{DD_1 \cos 45°}{AD\sqrt{2}} = \frac{DD_1}{2\,AD} = \frac{\phi}{2}$$

Thus we see that the linear strain of the diagonal $BD$ is half of the shear strain and is tensile in nature. Similarly it can be proved that the linear strain of the diagonal $AC$ is also equal to half of the shear strain, but is compressive in nature. Now this linear strain of the diagonal $BD$.

$$= \frac{\phi}{2} = \frac{\tau}{2C} \qquad \ldots(i)$$

where $\tau$ = Shear stress and

$C$ = Modulus of rigidity.

Let us now consider this shear stress t acting on the sides $AB$, $CD$, $CB$ and $AD$. We know that the effect of this stress is to cause tensile stress on the diagonal $BD$ and compressive stress on the diagonal $AC$. Therefore tensile strain on the diagonal $BD$ due to tensile stress on the diagonal $BD$

$$= \frac{\tau}{E} \qquad \ldots(ii)$$

and the tensile strain on the diagonal $BD$ due to compressive stress on the diagonal $AC$

$$= \frac{1}{m} \times \frac{\tau}{E} \qquad \ldots(iii)$$

The combined effect of the above two stresses on the diagonal $BD$

$$= \frac{\tau}{E} + \frac{1}{m} \times \frac{\tau}{E} = \frac{\tau}{E}\left(1 + \frac{1}{m}\right) = \frac{\tau}{E}\left(\frac{m+1}{m}\right) \qquad \ldots(iv)$$

Equating equations (*i*) and (*iv*),

$$\frac{\tau}{2C} = \frac{\tau}{E}\left(\frac{m+1}{m}\right) \quad \text{or} \quad C = \frac{mE}{2(m+1)}$$

**EXAMPLE 6.11.** *An alloy specimen has a modulus of elasticity of 120 GPa and modulus of rigidity of 45 GPa. Determine the Poisson's ratio of the material.*

**SOLUTION.** Given : Modulus of elasticity ($E$) = 120 GPa and modulus of rigidity ($C$) = 45 GPa.

Let $\frac{1}{m}$ = Poisson's ratio of the material.

We know that modulus of rigidity ($C$),

$$45 = \frac{mE}{2(m+1)} = \frac{m \times 120}{2\,(m+1)} = \frac{120m}{2m+2}$$

$$90\,m + 90 = 120\,m \quad \text{or} \quad 30\,m = 90$$

$$\therefore \quad m = \frac{90}{30} = 3 \quad \text{or} \quad \frac{1}{m} = \frac{1}{3} \quad \textbf{Ans.}$$

**EXAMPLE 6.12.** *In an experiment, a bar of 30 mm diameter is subjected to a pull of 60 kN. The measured extension on gauge length of 200 mm is 0.09 mm and the change in diameter is 0.0039 mm. Calculate the Poisson's ratio and the values of the three moduli.*

**SOLUTION.** Given : Diameter ($d$) = 30 mm ; Pull ($P$) = 60 kN = $60 \times 10^3$ N ; Length ($l$) = 200 mm; Extension ($\delta l$) = 0.09 mm and change in diameter ($\delta d$) = 0.0039 mm.

***Poisson's ratio***

We know that linear strain,

$$\varepsilon = \frac{\delta l}{l} = \frac{0.09}{200} = 0.000\,45$$

and lateral strain $= \frac{\delta d}{d} = \frac{0.0039}{30} = 0.000\,13$

We also know that Poisson's ratio,

$$\frac{1}{m} = \frac{\text{Lateral strain}}{\text{Linear strain}} = \frac{0.00013}{0.00045} = 0.289 \quad \textbf{Ans.}$$

***Values of three moduli***

Let $E$ = Value of Young's modulus.

We know that area of the bar,

$$A = \frac{\pi}{4} \times (d)^2 = \frac{\pi}{4}(30)^2 = 706.9 \text{ mm}^2$$

and extension of the bar ($\delta l$),

$$0.09 = \frac{P.l}{A.E} = \frac{(60 \times 10^3) \times 200}{706.9E} = \frac{17 \times 10^3}{E}$$

$$\therefore \quad E = 17 \times 10^3/0.09 = 188.9 \times 10^3 \text{ N/mm}^2 = 188.9 \text{ GPa} \quad \textbf{Ans.}$$

We know from the value of Poisson's ratio that

$$m = \frac{1}{0.289} = 3.46$$

and value of modulus of rigidity,

$$C = \frac{m.E}{2(m+1)} = \frac{3.46 \times (188.9 \times 10^3)}{2(3.46+1)} \text{ N/mm}^2$$

$$= 73.3 \times 10^3 \text{ N/mm}^2 = 149.2 \text{ GPa} \quad \textbf{Ans.}$$

We also know that the value of bulk modulus,

$$K = \frac{m.E}{3(m-2)} = \frac{3.46 \times (188.9 \times 10^3)}{2(3.46-2)} \text{ N/mm}^2$$

$$= 149.2 \times 10^3 \text{ N/mm}^2 = 149.2 \text{ GPa} \quad \textbf{Ans.}$$

## EXERCISE 6.2

1. A steel plate has modulus of elasticity as 200 GPa and Poisson's ratio as 0.3. What is the value of bulk modulus for the steel plate? **[Ans.** 166.7 GPa]
2. In an experiment an alloy bar 1 m long and 20 mm × 20 mm in section was tested to increase through 0.1 mm, when subjected to an axial tensile load of 6.4 kN. If the value of bulk modulus for the bar is 133 GPa, find the value of Poisson's ratio. **[Ans.** 0.3]
3. What is the value of modulus of rigidity of a steel alloy, if its modulus of elasticity is 180 GPa and Poisson's ratio is 0.25? **[Ans.** 72 GPa]
4. An alloy bar has bulk modulus as 150 GPa and Poisson's ratio as 0.3. Find its modulus of rigidity. **[Ans.** 69.2 GPa]
5. A round bar 40 mm diameter is subjected to an axial pull of 80 kN and reduction in diameter was found to be 0.007 75 mm. Find Poisson's ratio and Young's modulus for the material of the bar. Take value of shear modulus as 40 GPa. **[Ans.** 0.322 ; 105.7 GPa]

## QUESTIONS

1. Explain the difference between 'primary strain' and 'secondary strain'.
2. Define Poisson's ratio.
3. Derive a relation for the volumetric strain of a body.
4. Prove

$$E = 3K\left(1 - \frac{2}{m}\right)$$

where $E$ = Young's modulus,
K = Bulk modulus, and
$\frac{1}{m}$ = Poisson's ratio

5. Define shear stress and state the principle of shear stress.
6. Explain clearly the term modulus of rigidity.
7. Derive a relation between modulus of elasticity and modulus of rigidity.

## OBJECTIVE TYPE QUESTIONS

1. The ratio of lateral strain to the linear strain is called
   (*a*) modulus of elasticity (*b*) modulus of rigidity
   (*c*) bulk modulus (*d*) Poisson's ratio
2. The value of Poisson's ratio for steel varies from
   (*a*) 0.20 to 0.25 (*b*) 0.25 to 0.35 (*c*) 0.35 to 0.40 (*d*) 0.40 to 0.55
3. When a rectangular bar is subjected to a tensile stress, then the volumetric strain is equal to
   (*a*) $\varepsilon\left[1 - \frac{2}{m}\right]$ (*b*) $\varepsilon\left[1 + \frac{2}{m}\right]$ (*c*) $\varepsilon\left[2 - \frac{1}{m}\right]s$ (*d*) $\varepsilon\left[2 - \frac{1}{m}\right]$
   where ε = Linear strain for the bar, and
   1/m = Poisson's ratio for the bar material.
4. The bulk modulus of a body is equal to
   (*a*) $\frac{mE}{3(m-2)}$ (*b*) $\frac{mE}{3(m+2)}$ (*c*) $\frac{mE}{2(m-2)}$ (*d*) $\frac{mE}{2(m+2)}$

## ANSWERS

1. (*a*) 2. (*b*) 3. (*a*) 4. (*a*)

Chapter 7

# Principal Stresses and Strains

## Contents



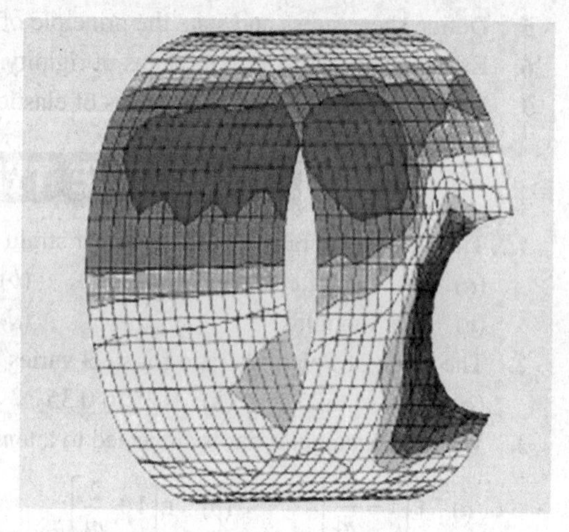

## 7.1. Introduction

In the previous chapters, we have studied in detail, the direct tensile and compressive stress as well as simple shear. In these chapters, we have always referred the stress in a plane, which is at right angles to the line of action of the force (in case of direct tensile or compressive stress). Moreover, we have considered at a time one type of stress, acting in one direction only. But the majority of engineering, component and structures are subjected to such loading conditions (or sometimes are of such shapes) that there exists a complex state of stresses; involving direct tensile and compressive stress as well as shear stress in various directions. Now in this chapter we shall study the nature and intensity of stresses on planes, other than that, which is at right angles to the line of action of the force.

## 7.2. Principal Planes

It has been observed that at any point in a strained material, there are three planes, mutually perpendicular to each other, which carry direct stresses only, and no shear stress. A little consideration will show that out of these three direct stresses one will be maximum, the other minimum, and the third an intermediate between the two. These particular planes, which have no shear stress, are known as *principal planes*.

## 7.3. Principal Stress

The magnitude of direct stress, across a principal plane, is known as principal stress. The determination of principal planes, and then principal stress is an important factor in the design of various structures and machine components.

## 7.4. Methods for the Stresses on an Oblique Section of a Body

The following two methods for the determination of stresses on an oblique section of a strained body are important from the subject point of view :

**1.** Analytical method and **2.** Graphical method.

## 7.5. Analytical Method for the Stresses on an Oblique Section of a Body

Here we shall first discuss the analytical method for the determination of stresses on an oblique section in the following cases, which are important from the subject point of view :

1. A body subjected to a direct stress in one plane.
2. A body subjected to direct stresses in two mutually perpendicular directions.

## 7.6. Sign Conventions for Analytical Method

Though there are different sign conventions, used in different books, yet we shall adopt the following sign conventions, which are widely used and internationally recognised :

1. All the tensile stresses and strains are taken as positive, whereas all the compressive stresses and strains are taken as negative.
2. The well established principles of mechanics is used for the shear stress. The shear stress which tends to rotate the element in the clockwise direction is taken as positive, whereas that which tends to rotate in an anticlockwise direction as negative.

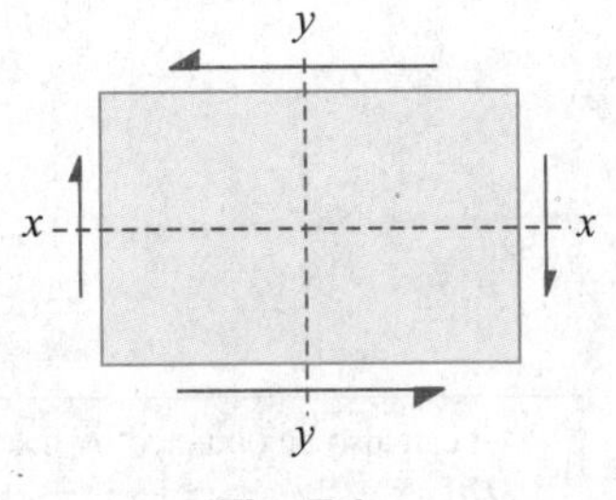

Fig. 7.1

In the element shown in Fig. 7.1, the shear stress on the vertical faces (or *x-x* axis) is taken as positive, whereas the shear stress on the horizontal faces (or *y-y* axis) is taken as negative.

## 7.7. Stresses on an Oblique Section of a Body Subjected to a Direct Stress in One Plane

Consider a rectangular body of uniform cross-sectional area and unit thickness subjected to a direct tensile stress along *x-x* axis as shown in Fig. 7.2 (*a*). Now let us consider an oblique section *AB* inclined with the $x$-$x$ axis (*i.e.*, with the line of action of the tensile stress on which we are required to find out the stresses as shown in the figure).

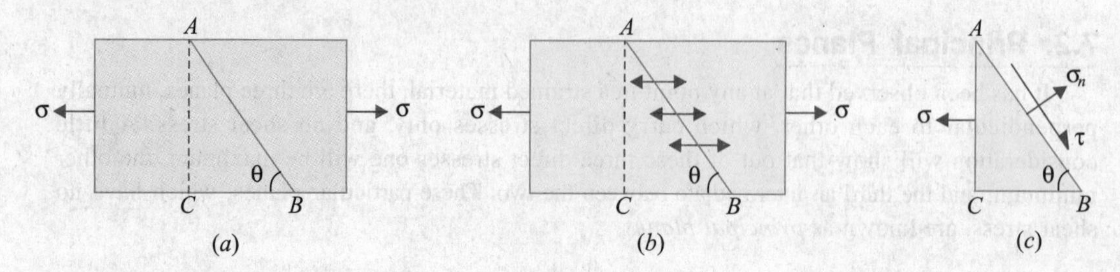


Fig. 7.2

Let $\sigma$ = Tensile stress across the face $AC$ and

$\theta$ = Angle, which the oblique section $AB$ makes with $BC$ *i.e.* with the $x$-$x$ axis in the clockwise direction.

First of all, consider the equilibrium of an element or wedge $ABC$ whose free body diagram is shown in fig 7.2 (*b*) and (*c*). We know that the horizontal force acting on the face $AC$,

$$P = \sigma \, . \, AC \; (\leftarrow)$$

Resolving the force perpendicular or normal to the section $AB$

$$P_n = P \sin\theta = \sigma \, . \, AC \sin\theta \quad \text{....}(i)$$

and now resolving the force tangential to the section $AB$,

$$P_t = P \cos\theta = \sigma \, . \, AC \cos\theta \quad \text{....}(ii)$$

We know that normal stress across the section $AB$*,

$$\sigma_n = \frac{P_n}{AB} = \frac{\sigma AC \sin\theta}{AB} = \frac{\sigma . AC \sin\theta}{\dfrac{AC}{\sin\theta}} = \sigma \sin^2\theta$$

$$= \frac{\sigma}{2}(1 - \cos 2\theta) = \frac{\sigma}{2} - \frac{\sigma}{2}\cos 2\theta \quad \text{...}(iii)$$

and shear stress (*i.e.*, tangential stress) across the section $AB$,

$$\tau = \frac{P_t}{AB} = \frac{\sigma . AC \cos\theta}{AB} = \frac{\sigma . AC \cos\theta}{\dfrac{AC}{\sin\theta}} = \sigma \sin\theta \cos\theta$$

$$= \frac{\sigma}{2}\sin 2\theta \quad \text{...}(iv)$$

---

* It can also be obtained by resolving the stress along the normal and across the section $AB$ as shown in Fig. 7.2. (*b*).

We know that the stress across the section $AB$

$$= \sigma \cos\theta$$

Now resolving the stress normal to the section $AB$,

$$\sigma_n = \sigma \cos\theta \, . \, \cos\theta = \sigma \cos^2\theta$$

and now resolving the stress along the section $AB$

$$\tau = \sigma \sin\theta \, . \, \cos\theta$$

It will be interesting to know from equation (*iii*) above that the normal stress across the section *AB* will be maximum, when $\sin^2 \theta = 1$ or $\sin \theta = 1$ or $\theta = 90°$. Or in other words, the face *AC* will carry the maximum direct stress. Similarly, the shear stress across the section *AB* will be maximum when $\sin 2\theta = 1$ or $2\theta = 90°$ or $270°$. Or in other words, the shear stress will be maximum on the planes inclined at 45° and 135° with the line of action of the tensile stress. Therefore maximum shear stress when $\theta$ is equal to 45°,

$$\tau_{max} = \frac{\sigma}{2} \sin 90° = \frac{\sigma}{2} \times 1 = \frac{\sigma}{2}$$

and maximum shear stress, when $\theta$ is equal to 135°,

$$\tau_{max} = -\frac{\sigma}{2} \sin 270° = -\frac{\sigma}{2}(-1) = \frac{\sigma}{2}$$

It is thus obvious that the magnitudes of maximum shear stress is half of the tensile stress. Now the resultant stress may be found out from the relation :

$$\sigma_R = \sqrt{\sigma_n^2 + \tau^2}$$

**Note :** The planes of maximum and minimum normal stresses (*i.e.* principal planes) may also be found out by equating the shear stress to zero. This happens as the normal stress is either maximum or minimum on a plane having zero shear stress. Now equating the shear stress to zero,

$$\sigma \sin \theta \cos \theta = 0$$

It will be interesting to know that in the above equation either $\sin \theta$ is equal to zero or $\cos \theta$ is equal to zero. We know that if sin is zero, then $\theta$ is equal to 0°. Or in other words, the plane coincides with the line of action of the tensile stress. Similarly, if $\cos \theta$ is zero, then $\theta$ is equal to 90°. Or in other words, the plane is at right angles to the line of action of the tensile stress. Thus we see that there are two principal planes, at right angles to each other, one of them coincides with the line of action of the stress and the other at right angles to it.

**Example 7.1.** *A wooden bar is subjected to a tensile stress of 5 MPa. What will be the values of normal and shear stresses across a section, which makes an angle of 25° with the direction of the tensile stress.*

**Solution.** Given : Tensile stress ($\sigma$) = 5 MPa and angle made by section with the direction of the tensile stress ($\theta$) = 25°.

***Normal stress across the section***

We know that normal stress across the section

$$\sigma_n = \frac{\sigma}{2} - \frac{\sigma}{2} \cos 2\theta = \frac{5}{2} - \frac{5}{2} \cos (2 \times 25°) \text{ MPa}$$

$$= 2.5 - 2.5 \cos 50° = 2.5 - (2.5 \times 0.6428) \text{ MPa}$$

$$= 2.5 - 1.607 = 0.89 \text{ MPa}$$ **Ans.**

***Shear stress across the section***

We also know that shear stress across the section,

$$\tau = \frac{\sigma}{2} \sin 2\theta = \frac{\sigma}{2} \sin (2 \times 25°) = 2.5 \sin 50° \text{ MPa}$$

$$= 2.5 \times 0.766 = 1.915 \text{ MPa}$$ **Ans.**

**EXAMPLE 7.2.** *Two wooden pieces 100 mm × 100 mm in cross-section are joined together along a line AB as shown in Fig. 7.3.*

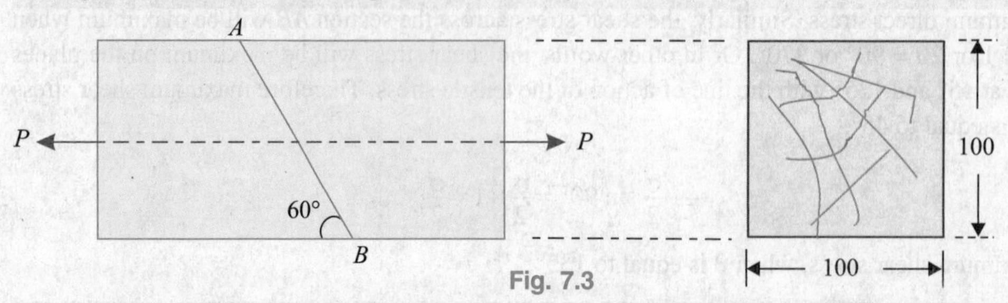


Fig. 7.3

*Find the maximum force (P), which can be applied if the shear stress along the joint AB is 1.3 MPa.*

**SOLUTION.** Given : Section = 100 mm × 100 mm ; Angle made by section with the direction of tensile stress ($\theta$) = 60° and permissible shear stress ($\tau$) = 1.3 MPa = 1.3 N/mm$^2$.

Let $\sigma$ = Safe tensile stress in the member

We know that cross- sectional area of the wooden member,

$$A = 100 \times 100 = 10\,000 \text{ mm}^2$$

and shear stress ($\tau$),

$$1.3 = \frac{\sigma}{2} \sin 2\theta = \frac{\sigma}{2} \sin (2 \times 60°) = \frac{\sigma}{2} \sin 120° = \frac{\sigma}{2} \times 0.866$$

$$= 0.433\,\sigma$$

or

$$\sigma = \frac{1.3}{0.433} = 3.0 \text{ N/mm}^2$$

$\therefore$ Maximum axial force, which can be applied,

$$P = \sigma.A = 3.0 \times 10\,000 = 30\,000 \text{ N} = 30 \text{ kN}$$ **Ans.**

**EXAMPLE 7.3.** *A tension member is formed by connecting two wooden members 200 mm × 100 mm as shown in the figure given below:*

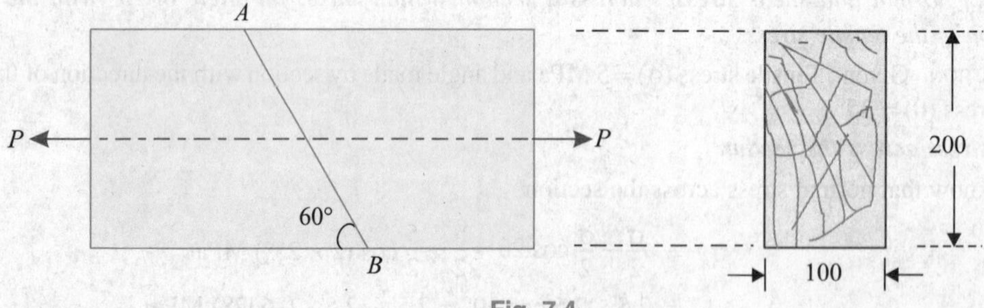


Fig. 7.4

*Determine the safe value of the force (P), if permissible normal and shear stresses in the joint are 0.5 MPa and 1.25 MPa respectively.*

**SOLUTION.** Given : Section = 200 mm × 100 mm ; Angle made by section *AB* with the direction of the tensile stress ($\sigma$) = 60° ; Permissible normal stress ($\sigma_n$) = 0.5 MPa = 0.5 N/mm$^2$ and permissible shear stress ($\tau$) = 1.25 MPa = 1.25 N/mm$^2$.

Let $\sigma$ = Safe stress in the joint in N/mm$^2$.

We know that cross-sectional area of the member

$$A = 200 \times 100 = 20\,000 \text{ mm}^2$$

We also know that normal stress ($\sigma_n$),

$$0.5 = \frac{\sigma}{2} - \frac{\sigma}{2}\cos 2\theta = \frac{\sigma}{2} - \frac{\sigma}{2}\cos(2 \times 60°)$$

$$= \frac{\sigma}{2} - \frac{\sigma}{2}\cos 120° = \frac{\sigma}{2} - \frac{\sigma}{2}(-0.5) = 0.75\,\sigma$$

$$\therefore \quad \sigma = \frac{0.5}{0.75} = 0.67 \text{ N/mm}^2 \quad ...(i)$$

and shear stress ($\tau$)

$$1.25 = \frac{\sigma}{2}\sin 2\theta = \frac{\sigma}{2}\sin(2 \times 60°) = \frac{\sigma}{2}\sin 120° = \frac{\sigma}{2} \times 0.866 = 0.433\sigma$$

$$\sigma = \frac{1.25}{0.433} = 2.89 \text{ N/mm}^2 \quad ...(ii)$$

From the above two values, we find that the safe stress is least of the two values, *i.e.* 0.67 N/mm$^2$. Therefore safe value of the force

$$P = \sigma \,.\, A = 0.67 \times 20\,000 = 13\,400 \text{ N} = 13.4 \text{ kN} \quad \textbf{Ans.}$$

## 7.8. Stresses on an Oblique Section of a Body Subjected to Direct Stresses in Two Mutually Perpendicular Directions

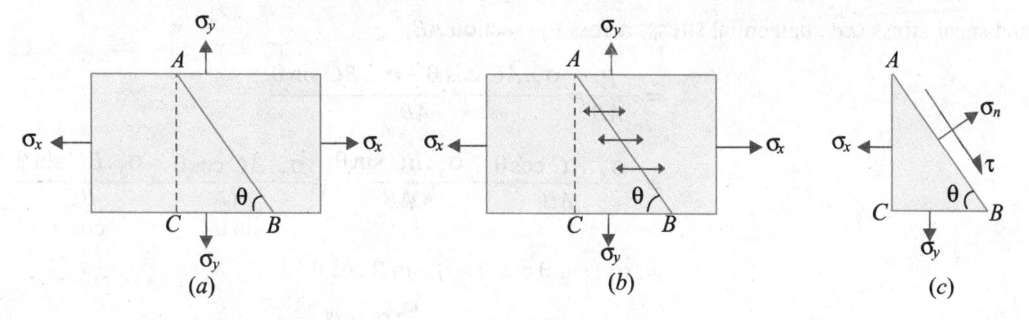


Fig. 7.5

Consider a rectangular body of uniform cross-sectional area and unit thickness subjected to direct tensile stresses in two mutually perpendicular directions along *x-x* and *y-y* axes as shown in Fig. 7.5. Now let us consider an oblique section *AB* inclined with *x-x* axis (*i.e.* with the line of action of the stress along *x-x* axis, termed as a major tensile stress on which we are required to find out the stresses as shown in the figure).

Let $\sigma_x$ = Tensile stress along *x-x* axis (also termed as major tensile stress),

$\sigma_y$ = Tensile stress along *y-y* axis (also termed as a minor tensile stress), and

$\theta$ = Angle which the oblique section *AB* makes with *x-x* axis in the clockwise direction.

First of all, consider the equilibrium of the wedge *ABC*. We know that horizontal force acting on the face *AC* (or *x-x* axis).

$$P_x = \sigma_x \,.\, AC \; (\leftarrow)$$

and vertical force acting on the face *BC* (or *y-y* axis),

$$P_y = \sigma_y \,.\, BC \; (\downarrow)$$

Resolving the forces perpendicular or normal to the section $AB$,

$$P_n = P_x \sin\theta + P_y \cos\theta = \sigma_x . AC \sin\theta + \sigma_y . BC \cos\theta \quad ...(i)$$

and now resolving the forces tangential to the section $AB$,

$$P_t = P_x \cos\theta - P_y \sin\theta = \sigma_x . AC \cos\theta - \sigma_y . BC \sin\theta \quad ....(ii)$$

We know that normal stress across the section $AB$,

$$\sigma_n = \frac{P_n}{AB} = \frac{\sigma_x . AC \sin\theta + \sigma_y\, BC \cos\theta}{AB}$$

$$= \frac{\sigma_x . AC \sin\theta}{AB} + \frac{\sigma_y . BC \cos\theta}{AB} = \frac{\sigma_x . AC \sin\theta}{\frac{AC}{\sin\theta}} + \frac{\sigma_y . BC \cos\theta}{\frac{BC}{\cos\theta}}$$

$$= \sigma_x \sin^2\theta + \sigma_y . \cos^2\theta = \frac{\sigma_x}{2}(1 - \cos 2\theta) + \frac{\sigma_y}{2}(1 + \cos 2\theta)$$

$$= \frac{\sigma_x}{2} - \frac{\sigma_x}{2}\cos 2\theta + \frac{\sigma_y}{2} + \frac{\sigma_y}{2}\cos 2\theta$$

$$= \frac{\sigma_x + \sigma_y}{2} - \frac{\sigma_x - \sigma_y}{2}\cos 2\theta \quad ...(iii)$$

and shear stress (*i.e.*, tangential stress) across the section $AB$,

$$\tau = \frac{P_t}{AB} = \frac{\sigma_x . AC \cos\theta - \sigma_y . BC \sin\theta}{AB}$$

$$= \frac{\sigma_x . AC \cos\theta}{AB} - \frac{\sigma_y . BC \sin\theta}{AB} = \frac{\sigma_x . AC \cos\theta}{\frac{AC}{\sin\theta}} - \frac{\sigma_y . BC \sin\theta}{\frac{BC}{\cos\theta}}$$

$$= \sigma_x . \sin\theta\cos\theta - \sigma_y \sin\theta\cos\theta$$

$$= (\sigma_x - \sigma_y)\sin\theta\cos\theta = \frac{\sigma_x - \sigma_y}{2}\sin 2\theta \quad ...(iv)$$

It will be interesting to know from equation (*iii*) the shear stress across the section $AB$ will be maximum when $\sin 2\theta = 1$ or $2\theta = 90°$ or $\theta = 45°$. Therefore maximum shear stress,

$$\tau_{max} = \frac{\sigma_x - \sigma_y}{2}$$

Now the resultant stress may be found out from the relation :

$$\sigma_R = \sqrt{\sigma_n^2 + \tau^2}$$

**EXAMPLE 7.4.** *A point in a strained material is subjected to two mutually perpendicular tensile stresses of 200 MPa and 100 MPa. Determine the intensities of normal, shear and resultant stresses on a plane inclined at 30° with the axis of minor tensile stress.*

**SOLUTION.** Given : Tensile stress along *x-x* axis ($\sigma_x$) =150 MPa ; Tensile stress along *y-y* axis ($\sigma_y$) = 100 MPa and angle made by plane with the axis of tensile stress $\theta = 30°$

***Normal stress on the inclined plane***

We know that normal stress on the inclined plane,

$$\sigma_n = \frac{\sigma_x + \sigma_y}{2} - \frac{\sigma_x - \sigma_y}{2}\cos 2\theta$$

$$= \frac{200+100}{2} - \frac{20-100}{2} \cos (2 \times 30°) \text{ MPa}$$

$$= 150 - (50 \times 0.5) = 125 \text{ MPa} \qquad \textbf{Ans.}$$

***Shear stress on the inclined plane***

We know that shear stress on the inclined plane,

$$\tau = \frac{\sigma_x - \sigma_y}{2} \sin 2\theta = \frac{200-100}{2} \times \sin (2 \times 30°) \text{ MPa}$$

$$= 50 \sin 60° = 50 \times 0.866 = 43.3 \text{ MPa} \qquad \textbf{Ans.}$$

***Resultant stress on the inclined plane***

We also know that resultant stress on the inclined plane,

$$\sigma_R = \sqrt{\sigma_n^2 + \tau^2} = \sqrt{(125)^2 + (43.3)^2} = 132.3 \text{ MPa} \qquad \textbf{Ans.}$$

**EXAMPLE 7.5.** *The stresses at point of a machine component are 150 MPa and 50 MPa both tensile. Find the intensities of normal, shear and resultant stresses on a plane inclined at an angle of 55° with the axis of major tensile stress. Also find the magnitude of the maximum shear stress in the component.*

**SOLUTION.** Given : Tensile stress along *x-x* axis ($\sigma_x$) = 150 MPa ; Tensile stress along *y-y* axis ($\sigma_y$) = 50 MPa and angle made by the plane with the major tensile stress ($\theta$) = 55°.

***Normal stress on the inclined plane***

We know that the normal stress on the inclined plane,

$$\sigma_n = \frac{\sigma_x + \sigma_y}{2} - \frac{\sigma_x - \sigma_y}{2} \cos 2\theta$$

$$= \frac{150+50}{2} - \frac{150-50}{2} \cos (2 \times 55°) \text{ MPa}$$

$$= 100 - 50 \cos 110° = 100 - 50 (-0.342) \text{ MPa}$$

$$= 10 + 17.1 = 117.1 \text{ MPa} \qquad \textbf{Ans.}$$

***Shear stress on the inclined plane***

We know that the shear stress on the inclined plane,

$$\tau = \frac{\sigma_x - \sigma_y}{2} \sin 2\theta = \frac{150-50}{2} \times \sin (2 \times 55°) \text{ MPa}$$

$$= 50 \sin 110° = 50 \times 0.9397 = 47 \text{ MPa} \qquad \textbf{Ans.}$$

***Resultant stress on the inclined plane***

We know that resultant stress on the inclined plane,

$$\sigma_R = \sqrt{\sigma_n^2 + \tau^2} = \sqrt{(117.1)^2 + (47.0)^2} = 126.2 \text{ MPa} \qquad \textbf{Ans.}$$

***Magnitude of the maximum shear stress in the component***

We also know that the magnitude of the maximum shear stress in the component,

$$\tau_{max} = \pm \frac{\sigma_x - \sigma_y}{2} = \pm \frac{150-50}{2} = \pm 50 \text{ MPa} \qquad \textbf{Ans.}$$

**EXAMPLE 7.6.** *The stresses at a point in a component are 100 MPa (tensile) and 50 MPa (compressive). Determine the magnitude of the normal and shear stresses on a plane inclined at an angle of 25° with tensile stress. Also determine the direction of the resultant stress and the magnitude of the maximum intensity of shear stress.*

**Solution.** Given : Tensile stress along *x-x* axis ($\sigma_x$) 100 MPa ; Compressive stress along *y-y* axis ($\sigma_y$) = –50 MPa ( Minus sign due to compression ) and angle made by the plane with tensile stress ($\theta$) = 25°.

***Normal stress on the inclined plane***

We know that the normal stress on the inclined plane,

$$\sigma_n = \frac{\sigma_x + \sigma_y}{2} - \frac{\sigma_x - \sigma_y}{2} \cos 2\theta$$

$$= \frac{100 + (-50)}{2} - \frac{100 - (-50)}{2} \cos (2 \times 25°) \text{ MPa}$$

$$= 25 - 75 \cos 50° = 25 - (75 \times 0.6428) = -23.21 \text{ MPa} \quad \textbf{Ans.}$$

***Shear stress on the inclined plane***

We know that the shear stress on the inclined plane,

$$\tau = \frac{\sigma_x - \sigma_y}{2} \sin 2\theta = \frac{100 - (-50)}{2} \sin (2 \times 25°) \text{ MPa}$$

$$= 75 \sin 50° = 75 \times 0.766 = 57.45 \text{ MPa} \quad \textbf{Ans.}$$

***Direction of the resultant stress***

Let $\theta$ = Angle, which the resultant stress makes with *x-x* axis.

We know that $\tan \theta = \frac{\tau}{\sigma_n} = \frac{57.45}{-23.21} = -2.4752$ or $\theta = -68°$ **Ans.**

***Magnitude of the maximum shear stress***

We also know that magnitude of the maximum shear stress,

$$\tau_{max} = \pm \frac{\sigma_x - \sigma_y}{2} = \pm \frac{100 - (-50)}{2} = \pm 75 \text{ MPa} \quad \textbf{Ans.}$$

## 7.9. Stresses on an Oblique Section of a Body Subjected to a Simple Shear stress

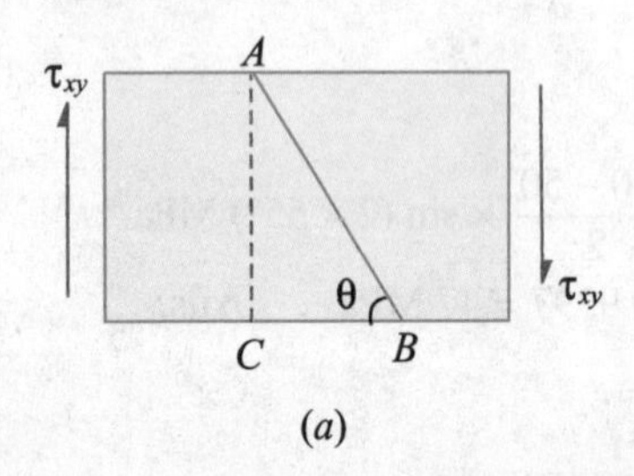

(*a*)

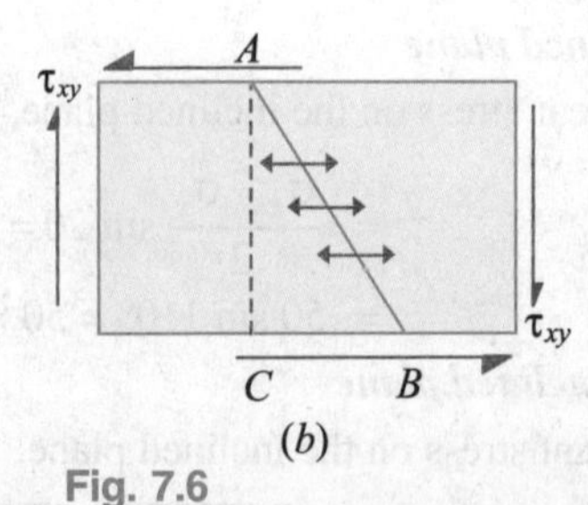

(*b*)

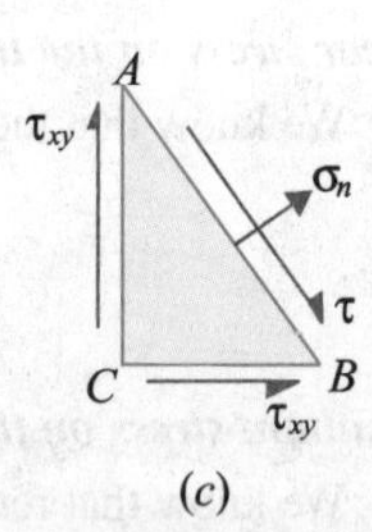

(*c*)

**Fig. 7.6**

Consider a rectangular body of uniform cross-sectional area and unit thickness subjected to a positive (*i.e.*, clockwise) shear stress along *x-x* axis as shown in Fig.7.6 (*a*). Now let us consider an oblique section *AB* inclined with *x-x* axis on which we are required to find out the stresses as shown in the figure 7.6 (*b*).

Let $\tau_{xy}$ = Positive (*i.e.*, clockwise) shear stress along *x-x* axis, and

$\theta$ = Angle , which the oblique section *AB* makes with *x-x* axis in the anticlockwise direction.

First of all, consider the equilibrium of the wedge *ABC*. We know that as per the principle of simple shear, the face *BC*, of the wedge will be subjected to an anticlockwise shear stress equal to $\tau_{xy}$ as shown in the Fig. 7.6 (*b*). We know that vertical force acting on the face *AC*,

$$P_1 = \tau_{xy} \cdot AC \; (\uparrow)$$

and horizontal force acting on the face *BC*,

$$P_2 = \tau_{xy} \,.\, BC\ (\rightarrow)$$

Resolving the forces perpendicular or normal to the *AB*,

$$P_n = P_1 \cos\theta + P_2 \sin\theta = \tau_{xy} \,.\, AC \cos\theta + \tau_{xy} \,.\, BC \sin\theta$$

and now resolving the forces tangential to the section *AB*,

$$P_t = P_2 \sin\theta - P_1 \cos\theta = \tau_{xy} \,.\, BC \sin\theta - \tau_{xy} \,.\, AC \cos\theta$$

We know that normal stress across the section *AB*,

$$\sigma_n = \frac{P_n}{AB} = \frac{\tau_{xy} \,.\, AC \cos\theta + \tau_{xy} \,.\, BC \sin\theta}{AB}$$

$$= \frac{\tau_{xy} \,.\, AC \cos\theta}{AB} + \frac{\tau_{xy} \,.\, BC \sin\theta}{AB}$$

$$= \frac{\tau_{xy} \,.\, AC \cos\theta}{\dfrac{AC}{\sin\theta}} + \frac{\tau_{xy} \,.\, BC \sin\theta}{\dfrac{BC}{\cos\theta}}$$

$$= \tau_{xy} \,.\, \sin\theta \cos\theta + \tau_{xy} \,.\, \sin\theta \cos\theta$$

$$= 2\,\tau_{xy} \,.\, \sin\theta \cos\theta = \tau_{xy} \,.\, \sin 2\theta$$

and shear stress (*i.e.* tangential stress) across the section *AB*

$$\tau = \frac{P_t}{AB} = \frac{\tau_{xy} \,.\, BC \sin\theta - \tau_{xy} \,.\, AC \cos\theta}{AB}$$

$$= \frac{\tau_{xy} \,.\, BC \sin\theta}{AB} - \frac{\tau_{xy} \,.\, AC \cos\theta}{AB} = \frac{\tau_{xy} \,.\, BC \sin\theta}{\dfrac{BC}{\sin\theta}} - \frac{\tau_{xy} \,.\, AC \cos\theta}{\dfrac{AC}{\cos\theta}}$$

$$= \tau_{xy} \sin^2\theta - \tau_{xy} \cos^2\theta$$

$$= \frac{\tau_{xy}}{2}(1 - \cos 2\theta) - \frac{\tau_{xy}}{2}(1 + \cos 2\theta)$$

$$= \frac{\tau_{xy}}{2} - \frac{\tau_{xy}}{2}\cos 2\theta - \frac{\tau_{xy}}{2} - \frac{\tau_{xy}}{2}\cos 2\theta$$

$$= -\,\tau_{xy} \cos 2\theta$$

...(Minus sign means that normal stress is opposite to that across *AC*)

Now the planes of maximum and minimum normal stresses (*i.e.*, principal planes ) may be found out by equating the shear stress to zero *i.e.*

$$-\,\tau_{xy} \cos 2\theta = 0$$

The above equation is possible only if $2\theta = 90°$ or $270°$ (because $\cos 90° $ or $\cos 270° = 0$) or in other words, $\theta = 45°$ or $135°$.

## 7.10. Stresses on an Oblique Section of a Body Subjected to a Direct Stress in One Plane and Accompanied by a Simple Shear Stress

Consider a rectangular body of uniform cross-sectional area and unit thickness subjected to a tensile stress along *x-x* axis accompanied by a positive (*i.e.* clockwise ) shear stress along *x-x* axis as shown in Fig. 7.7 (*a*). Now let us consider an oblique section *AB* inclined with *x-x* axis on which we are required to find out the stresses as shown in the figure.

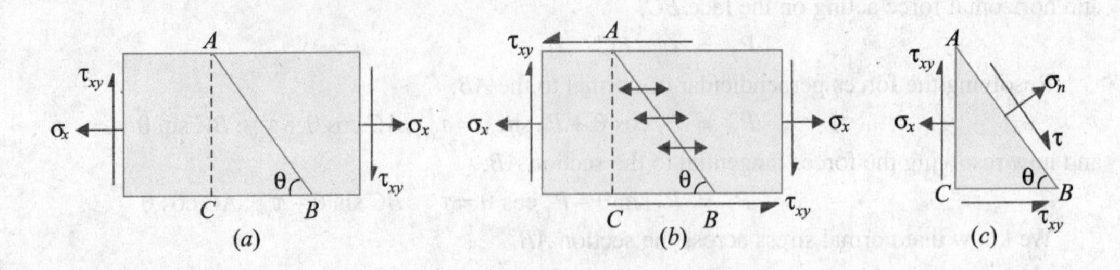


Fig. 7.7

Let $\sigma_x$ = Tensile stress along *x-x* axis,

$\tau_{xy}$ = Positive (*i.e.* clockwise) shear stress along *x-x* axis, and

$\theta$ = Angle which the oblique section *AB* makes with *x-x* axis in clockwise direction.

First of all, consider the equilibrium of the wedge *ABC*. We know that as per the principle of simple shear, the face *BC* of the wedge will be subjected to an anticlockwise shear stress equal to $\tau_{xy}$ as shown in Fig. 7.7 (*b*). We know that horizontal force acting on the face *AC*,

$$P_x = \sigma_x \cdot AC \ (\leftarrow) \qquad \text{...(i)}$$

Similarly, vertical force acting on the face *AC*,

$$P_y = \tau_{xy} \cdot AC \ (\uparrow) \qquad \text{... (ii)}$$

and horizontal force acting on the face *BC*,

$$P = \tau_{xy} \cdot BC \ (\rightarrow) \qquad \text{...(iii)}$$

Resolving the forces perpendicular to the section *AB*,

$$P_n = P_x \sin\theta - P_y \cos\theta - P \sin\theta$$

$$= \sigma_x \cdot AC \sin\theta - \tau_{xy} \cdot AC \cos\theta - \tau_{xy} \cdot BC \sin\theta$$

and now resolving the forces tangential to the section *AB*,

$$P_t = P_x \cos\theta + P_y \sin\theta - P \cos\theta$$

$$= \sigma_x \cdot AC \cos\theta + \tau_{xy} \cdot AC \sin\theta - \tau_{xy} \cdot BC \cos\theta$$

We know that normal stress across the section *AB*,

$$\sigma_n = \frac{P_n}{AB} = \frac{\sigma_x . AC \sin\theta - \tau_{xy} . AC \cos\theta - \tau_{xy} . BC \sin\theta}{AB}$$

$$= \frac{\sigma_x . AC \sin\theta}{AB} - \frac{\tau_{xy} . AC \cos\theta}{AB} - \frac{\tau_{xy} . BC \sin\theta}{AB}$$

$$= \frac{\sigma_x . AC \sin\theta}{\dfrac{AC}{\sin\theta}} - \frac{\tau_{xy} . AC \cos\theta}{\dfrac{AC}{\sin\theta}} - \frac{\tau_{xy} . BC \sin\theta}{\dfrac{BC}{\cos\theta}}$$

$$= \sigma_x . \sin^2\theta - \tau_{xy} \sin\theta \cos\theta - \tau_{xy} \sin\theta \cos\theta$$

$$= \frac{\sigma_x}{2} (1 - \cos 2\theta) - 2\,\tau_{xy} \sin\theta \cos\theta$$

$$= \frac{\sigma_x}{2} - \frac{\sigma_x}{2} \cos 2\theta - \tau_{xy} \sin 2\theta \qquad \text{...(iv)}$$

and shear stress (*i.e.*, tangential stress) across the section *AB*,

$$\tau = \frac{P_t}{AB} = \frac{\sigma_x . AC \cos\theta + \tau_{xy} . AC \sin\theta - \tau_{xy} . BC \cos\theta}{AB}$$

$$= \frac{\sigma_x . AC \cos\theta}{AB} + \frac{\tau_{xy}\ AC \sin\theta}{AB} - \frac{\tau_{xy} . BC \cos\theta}{AB}$$

$$= \frac{\sigma_x . AC \cos\theta}{\frac{AC}{\sin\theta}} + \frac{\tau_{xy}\ AC \sin\theta}{\frac{AC}{\sin\theta}} - \frac{\tau_{xy} . BC \cos\theta}{\frac{BC}{\cos\theta}}$$

$$= \sigma_x \sin\theta \cos\theta + \tau_{xy} \sin^2\theta - \tau_{xy} \cos^2\theta$$

$$= \frac{\sigma_x}{2} \sin 2\theta + \frac{\tau_{xy}}{2}(1 - \cos 2\theta) - \frac{\tau_{xy}}{2}(1 + \cos 2\theta)$$

$$= \frac{\sigma_x}{2} \sin 2\theta + \frac{\tau_{xy}}{2} - \frac{\tau_{xy}}{2} \cos 2\theta - \frac{\tau_{xy}}{2} - \frac{\tau_{xy}}{2} \cos 2\theta$$

$$= \frac{\sigma_x}{2} \sin 2\theta - \tau_{xy} \cos 2\theta \quad ...(v)$$

Now the planes of maximum and minimum normal stresses (*i.e.*, principal planes) may be found out by equating the shear stress to zero *i.e.*, from the above equation, we find that the shear stress on any plane is a function of $\sigma_x$, $\tau_{xy}$ and $\theta$. A little consideration will show that the values of $\sigma_x$ and $\tau_{xy}$ are constant and thus the shear stress varies with the angle $\theta$. Now let $\theta_p$ be the value of the angle for which the shear stress is zero.

$$\therefore \quad \frac{\sigma_x}{2} \sin 2\theta_p - \tau_{xy} \cos 2\theta_p = 0 \quad \text{or} \quad \frac{\sigma_x}{2} \sin 2\theta_p = \tau_{xy} \cos 2\theta_p$$

$$\therefore \quad \tan 2\theta_p = \frac{2\tau_{xy}}{\sigma_x}$$

From the above equation we find that the following two cases satisfy this condition as shown in Fig 7.8 (*a*) and (*b*)

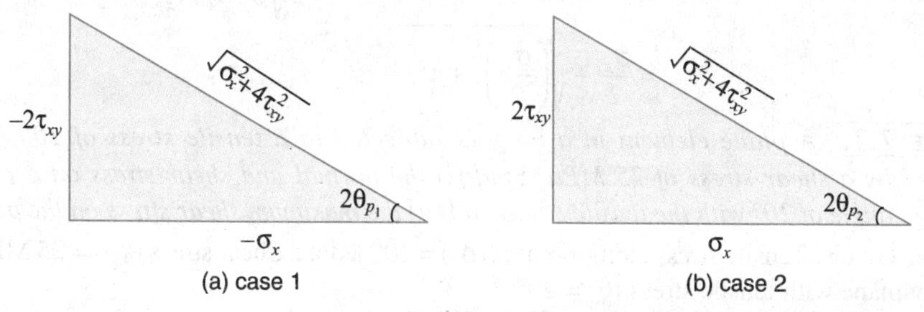


**Fig. 7.8**

Thus we find that these are two principal planes at right angles to each other, their inclination with *x*-*x* axis being $\theta_{p_1}$ and $\theta_{p_2}$.

Now for case 1,

$$\sin 2\theta_{p_1} = \frac{-2\tau_{xy}}{\sqrt{\sigma_x^2 + 4\tau_{xy}^2}} \quad \text{and} \quad \cos 2\theta_{p_1} = \frac{-\sigma_x}{\sqrt{\sigma_x^2 + 4\tau_{xy}^2}}$$

Similarly for case 2,

$$\sin 2\theta_{p_2} = \frac{2\tau_{xy}}{\sqrt{\sigma_x^2 + 4\tau_{xy}^2}} \quad \text{and} \quad \cos 2\theta_{p_2} = \frac{\sigma_x}{\sqrt{\sigma_x^2 + 4\tau_{xy}^2}}$$

Now the values of principal stresses may be found out by substituting the above values of $2\theta_{p_1}$ and $2\theta_{p_2}$ in equation (*iv*).

Maximum principal stress,
$$\sigma_{p_1} = \frac{\sigma_x}{2} - \frac{\sigma_x}{2}\cos 2\theta - \tau_{xy}\sin 2\theta$$
$$= \frac{\sigma_x}{2} - \frac{\sigma_x}{2} \times \frac{-\sigma_x}{\sqrt{\sigma_x^2 + 4\tau_{xy}^2}} - \tau_{xy} \times \frac{-2\tau_{xy}}{\sqrt{\sigma_x^2 + 4\tau_{xy}^2}}$$
$$= \frac{\sigma_x}{2} + \frac{\sigma_x^2}{2\sqrt{\sigma_x^2 + 4\tau_{xy}^2}} + \frac{2\tau_{xy}^2}{\sqrt{\sigma_x^2 + 4\tau_{xy}^2}}$$
$$= \frac{\sigma_x}{2} + \frac{\sigma_x^2 + 4\tau_{xy}^2}{2\sqrt{\sigma_x^2 + 4\tau_{xy}^2}} = \frac{\sigma_x}{2} + \frac{\sqrt{\sigma_x^2 + 4\tau_{xy}^2}}{2}$$
$$= \frac{\sigma_x}{2} + \sqrt{\left(\frac{\sigma_x^2}{2}\right) + \tau_{xy}^2}$$

Minimum principal stress,
$$\sigma_{p_2} = \frac{\sigma_x}{2} - \frac{\sigma_x}{2}\cos 2\theta - \tau_{xy}\sin 2\theta$$
$$= \frac{\sigma_x}{2} - \frac{\sigma_x}{2} \times \frac{\sigma_x}{\sqrt{\sigma_x^2 + 4\tau_{xy}^2}} - \tau_{xy} \times \frac{2\tau_{xy}}{\sqrt{\sigma_x^2 + 4\tau_{xy}^2}}$$
$$= \frac{\sigma_x}{2} - \frac{\sigma_x^2}{2\sqrt{\sigma_x^2 + 4\tau_{xy}^2}} - \frac{2\tau_{xy}^2}{\sqrt{\sigma_x^2 + 4\tau_{xy}^2}}$$
$$= \frac{\sigma_x}{2} - \frac{\sqrt{\sigma_x^2 + 4\tau_{xy}^2}}{2} = \frac{\sigma_x}{2} - \frac{\sigma_x^2 + 4\tau_{xy}^2}{2\sqrt{\sigma_x^2 + 4\tau_{xy}^2}}$$
$$= \frac{\sigma_x}{2} - \sqrt{\left(\frac{\sigma_x}{2}\right)^2 + \tau_{xy}^2}$$

**EXAMPLE 7.7.** *A plane element in a body is subjected to a tensile stress of 100 MPa accompanied by a shear stress of 25 MPa. Find (i) the normal and shear stress on a plane inclined at an angle of 20° with the tensile stress and (ii) the maximum shear stress on the plane.*

**SOLUTION.** Given : Tensile stress along *x-x* axis ($\sigma_x$) = 100 MPa ; Shear stress ($\tau_{xy}$) = 25 MPa and angle made by plane with tensile stress ($\theta$) = 20°.

***(i) Normal and shear stresses***

We know that the normal stress on the plane,
$$\sigma_n = \frac{\sigma_x}{2} - \frac{\sigma_x}{2}\cos 2\theta - \tau_{xy}\sin 2\theta$$
$$= \frac{100}{2} - \frac{100}{2}\cos(2 \times 20°) - 25\sin(2 \times 20°) \text{ MPa}$$
$$= 50 - 50\cos 40° - 25\sin 40° \text{ MPa}$$
$$= 50 - (50 \times 0.766) - (25 \times 0.6428) \text{ MPa}$$
$$= 50 - 38.3 - 16.07 = -4.37 \text{ MPa} \qquad \textbf{Ans.}$$

and shear stress on the plane,
$$\tau = \frac{\sigma_x}{2}\sin 2\theta - \tau_{xy}\cos 2\theta$$
$$= \frac{100}{2}\sin(2 \times 20°) - 25\cos(2 \times 20°) \text{ MPa}$$
$$= 50\sin 40° - 25\cos 40° \text{ MPa}$$
$$= (50 \times 0.6428) - (25 \times 0.766) \text{ MPa}$$
$$= 32.14 - 19.15 = 12.99 \text{ MPa} \qquad \textbf{Ans.}$$

*(ii) Maximum shear stress on the plane*

We also know that maximum shear stress on the plane,

$$\tau_{max} = \sqrt{\left(\frac{\sigma_x}{2}\right)^2 + \tau_{xy}^2} = \sqrt{\left(\frac{100}{2}\right)^2 + (25)^2} = 55.9 \text{ MPa} \qquad \textbf{Ans.}$$

**EXAMPLE 7.8.** *An element in a strained body is subjected to a tensile stress of 150 MPa and a shear stress of 50 MPa tending to rotate the element in an anticlockwise direction. Find (i) the magnitude of the normal and shear stresses on a section inclined at 40° with the tensile stress; and (ii) the magnitude and direction of maximum shear stress that can exist on the element.*

**SOLUTION.** Given : Tensile stress along horizontal *x-x* axis ($\sigma_x$) = 150 MPa ; Shear stress ($\tau_{xy}$) – 50 MPa (Minus sign due to anticlockwise) and angle made by section with the tensile stress ($\theta$) = 40°.

*(i) Magnitude of the normal and shear stress on the section*

We know that magnitude of the normal stress on the section,

$$\sigma_n = \frac{\sigma_x}{2} - \frac{\sigma_x}{2}\cos 2\theta - \tau_{xy}\sin 2\theta$$

$$= \frac{150}{2} - \frac{150}{2}\cos(2\times 40°) - (-50)\sin(2\times 40°) \text{ MPa}$$

$$= 75 - (75\times 0.1736) + (50\times 0.9848) \text{ MPa}$$

$$= 75 - 13.02 + 49.24 = 111.22 \text{ MPa} \qquad \textbf{Ans.}$$

and shear stress on the section

$$\tau = \frac{\sigma_x}{2}\sin 2\theta - \tau_{xy}\cos 2\theta$$

$$= \frac{150}{2}\sin(2\times 40°) - (-50)\cos(2\times 40°) \text{ MPa}$$

$$= (75\times 0.9848) + (50\times 0.1736) \text{ MPa}$$

$$= 73.86 + 8.68 = 82.54 \text{ MPa} \qquad \textbf{Ans.}$$

*(ii) Magnitude and direction of the maximum shear stress that can exist on the element*

We know that magnitude of the maximum shear stress.

$$\tau_{max} = \pm\sqrt{\left(\frac{\sigma_x}{2}\right)^2 + \tau_{xy}^2} = \pm\sqrt{\left(\frac{150}{2}\right)^2 + (-50)^2} = \pm 90.14 \text{ MPa} \quad \textbf{Ans.}$$

Let $\theta_x$ = Angle which plane of maximum shear stress makes with *x-x* axis.

We know that, $\tan 2\theta_s = \frac{\sigma_x}{2\tau_{xy}} = \frac{150}{2\times 50} = 1.5$ or $2\theta_s = 56.3°$

$\therefore$ $\theta_s = 28.15°$ or $118.15°$ **Ans.**

**EXAMPLE 7.9.** *An element in a strained body is subjected to a compressive stress of 200 MPa and a clockwise shear stress of 50 MPa on the same plane. Calculate the values of normal and shear stresses on a plane inclined at 35° with the compressive stress. Also calculate the value of maximum shear stress in the element.*

**SOLUTION.** Given : Compressive stress along horizontal *x-x* axis ($\sigma_x$) = – 200 MPa (Minus sign due to compressive stress) ; Shear stress ($\tau_{xy}$) = 50 MPa and angle made by the plane with the compressive stress ($\theta$) = 35°

*Values of normal and shear stresses*

We know that normal stress on the plane,

$$\sigma_n = \frac{\sigma_x}{2} - \frac{\sigma_x}{2}\cos 2\theta - \tau_{xy}\sin 2\theta$$

$$= \frac{-200}{2} - \frac{-200}{2}\cos(2\times 35°) - 50\sin(2\times 35°)\text{ MPa}$$

$$= -100 + (10\times 0.342) - (50\times 0.94)\text{ MPa}$$

$$= -100 + 34.2 - 46.9 = -112.9\text{ MPa}$$ **Ans.**

and shear stress on the plane,

$$\tau = \frac{\sigma_x}{2}\sin 2\theta - \tau_{xy}\cos 2\theta$$

$$= \frac{-200}{2}\sin(2\times 35°) - 50\cos(2\times 35°)\text{ MPa}$$

$$= (-100\times 0.9397) - (50\times 0.342)\text{ MPa}$$

$$= -93.97 - 17.1 = -111.07\text{ MPa}$$ **Ans.**

*Values of maximum shear stress in the element*

We also know that value of maximum shear stress in the element,

$$\tau_{max} = \sqrt{\left(\frac{\sigma_x}{2}\right)^2 + \tau_{xy}^2} = \sqrt{\left(\frac{-200}{2}\right)^2 + (50)^2} = 111.8\text{ MPa}$$ **Ans.**

## 7.11. Stresses on an Oblique Section of a Body Subjected to Direct Stresses in Two Mutually Perpendicular Directions Accompanied by a Simple Shear Stress

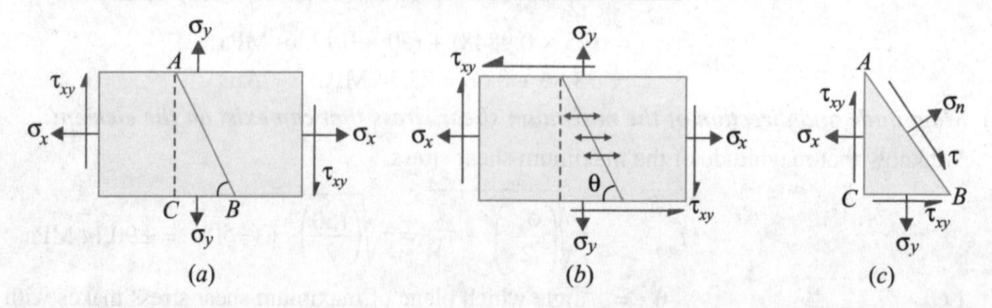


Fig. 7.9

Consider a rectangular body of uniform cross-sectional area and unit thickness subjected to tensile stresses along *x-x* and *y-y* axes and accompanied by a positive ( *i.e.*, clockwise) shear stress along *x-x* axis as shown in Fig.7.9 (*b*). Now let us consider an oblique section *AB* inclined with *x-x* axis on which we are required to find out the stresses as shown in the figure.

Let $\sigma_x$ = Tensile stress along *x-x* axis,

$\sigma_y$ = Tensile stress along *y-y* axis,

$\tau_{xy}$ = Positive (*i.e.* clockwise) shear stress along *x-x* axis, and

$\theta$ = Angle, which the oblique section *AB* makes with *x-x* axis in an anticlockwise direction.

First of all, consider the equilibrium of the wedge *ABC*. We know that as per the principle of simple shear, the face *BC* of the wedge will be subjected to an anticlockwise shear stress equal to $\tau_{xy}$

as shown in Fig. 7.9 (*b*). We know that horizontal force acting on the face *AC*,

$$P_1 = \sigma_x . AC \ (\leftarrow) \qquad \text{... (i)}$$

and vertical force acting on the face *AC*,

$$P_2 = \tau_{xy} . AC \ (\uparrow) \qquad \text{...(ii)}$$

Similarly, vertical force acting on the face *BC*,

$$P_3 = \sigma_y . BC \ (\downarrow) \qquad \text{...(iii)}$$

and horizontal force on the face *BC*,

$$P_4 = \tau_{xy} . BC \ (\rightarrow) \qquad \text{...(iv)}$$

Now resolving the forces perpendicular to the section *AB*,

$$P_n = P_1 \sin\theta - P_2 \cos\theta + P_3 \cos\theta - P_4 \sin\theta$$

$$= \sigma_x . AC \sin\theta - \tau_{xy} . AC \cos\theta + \sigma_y . BC \cos\theta - \tau_{xy} . BC \sin\theta$$

and now resolving the forces tangential to *AB*,

$$P_t = P_1 \cos\theta + P_2 \sin\theta - P_3 \sin\theta - P_4 \cos\theta$$

$$= \sigma_x . AC \cos\theta + \tau_{xy} . AC \sin\theta - \sigma_y . BC \sin\theta - \tau_{xy} . BC \cos\theta$$

**Normal Stress (across the section *AB*)**

$$\sigma_n = \frac{P_n}{AB} = \frac{\sigma_x . AC \sin\theta - \tau_{xy} . AC \cos\theta + \sigma_y . BC \cos\theta - \tau_{xy} . BC \sin\theta}{AB}$$

$$= \frac{\sigma_x . AC \sin\theta}{AB} - \frac{\tau_{xy} . AC \cos\theta}{AB} + \frac{\sigma_y . BC \cos\theta}{AB} - \frac{\tau_{xy} . BC \sin\theta}{AB}$$

$$= \frac{\sigma_x . AC \sin\theta}{\dfrac{AC}{\sin\theta}} - \frac{\tau_{xy} . AC \cos\theta}{\dfrac{AC}{\sin\theta}} + \frac{\sigma_y . BC \cos\theta}{\dfrac{BC}{\cos\theta}} - \frac{\tau_{xy} . BC \sin\theta}{\dfrac{BC}{\cos\theta}}$$

$$= \sigma_x . \sin^2\theta - \tau_{xy} \sin\theta \cos\theta + \sigma_y . \cos^2\theta - \tau_{xy} . \sin\theta \cos\theta$$

$$= \frac{\sigma_x}{2} (1 - \cos 2\theta) + \frac{\sigma_y}{2} (1 + \cos 2\theta) - 2\tau_{xy} . \sin\theta \cos\theta$$

$$= \frac{\sigma_x}{2} - \frac{\sigma_x}{2} \cos 2\theta + \frac{\sigma_y}{2} + \frac{\sigma_y}{2} \cos 2\theta - \tau_{xy} \sin 2\theta$$

or

$$\sigma_n = \frac{\sigma_x + \sigma_y}{2} - \frac{\sigma_x - \sigma_y}{2} \cos 2\theta - \tau_{xy} \sin 2\theta \qquad \text{...(v)}$$

**Shear Stress or Tangential Stress (across the section *AB*)**

$$\tau = \frac{P_t}{AB} = \frac{\sigma_x . AC \cos\theta + \tau_{xy} . AC \sin\theta - \sigma_y . BC \sin\theta - \tau_{xy} BC \cos\theta}{AB}$$

$$= \frac{\sigma_x . AC \cos\theta}{AB} + \frac{\tau_{xy} . AC \sin\theta}{AB} - \frac{\sigma_y . BC \sin\theta}{AB} - \frac{\tau_{xy} BC \cos\theta}{AB}$$

$$= \frac{\sigma_x . AC \cos\theta}{\dfrac{AC}{\sin\theta}} + \frac{\tau_{xy} . AC \sin\theta}{\dfrac{AC}{\sin\theta}} - \frac{\sigma_y . BC \sin\theta}{\dfrac{BC}{\cos\theta}} - \frac{\tau_{xy} . BC \cos\theta}{\dfrac{BC}{\cos\theta}}$$

$$= \sigma_x \sin\theta \cos\theta + \tau_{xy} \sin^2\theta - \sigma_y \sin\theta \cos\theta - \tau_{xy} \cos^2\theta$$

$$= (\sigma_x - \sigma_y) \sin\theta \cos\theta + \frac{\tau_{xy}}{2} (1 - \cos 2\theta) - \frac{\tau_{xy}}{2} (1 + \cos 2\theta)$$

or

$$\tau = \frac{\sigma_x - \sigma_y}{2} \sin 2\theta - \tau_{xy} \cos 2\theta \qquad \text{...(vi)}$$

Now the planes of maximum and minimum normal stresses (*i.e.* principal planes) may be found out by equating the shear stress to zero. From the above equations, we find that the shear stress to any plane is a function of $\sigma_y$, $\sigma_x$, $\tau_{xy}$ and $\theta$. A little consideration will show that the values of $\sigma_y$, $\sigma_x$ and $\tau_{xy}$ are constant and thus the shear stress varies in the angle $\theta$. Now let $\theta_p$ be the value of the angle for which the shear stress is zero.

$$\therefore \quad \frac{\sigma_x - \sigma_y}{2} \sin 2\theta_p - \tau_{xy} \cos 2\theta_p = 0$$

$$\text{or} \quad \frac{\sigma_x - \sigma_y}{2} \sin 2\theta_p = \tau_{xy} \cos 2\theta_p \quad \text{or} \quad \tan 2\theta_p = \frac{2\tau_{xy}}{\sigma_x - \sigma_y}$$

From the above equation, we find that the following two cases satisfy this condition as shown in Fig 7.10 (*a*) and (*b*).

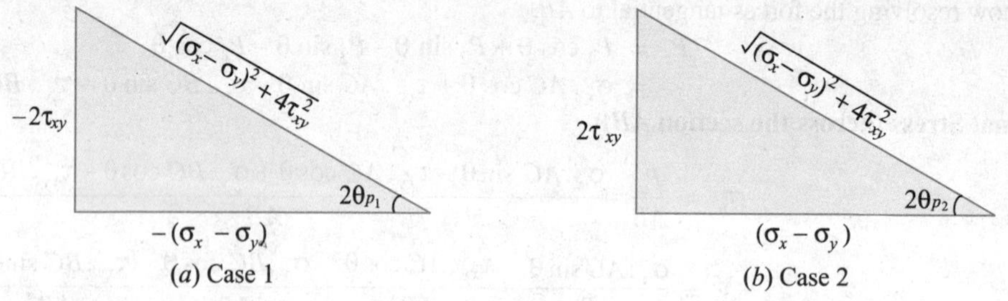


(*a*) Case 1 (*b*) Case 2

**Fig. 7.10**

Thus we find that there are two principal planes, at right angles to each other, their inclinations with *x-x* axis being $\theta_{p_1}$ and $\theta_{p_2}$.

Now for case 1,

$$\sin 2\theta_{p_1} = \frac{-2\tau_{xy}}{\sqrt{(\sigma_x - \sigma_y)^2 + 4\tau_{xy}^2}} \quad \text{and} \quad \cos 2\theta_{p_1} = \frac{-(\sigma_x - \sigma_y)}{\sqrt{(\sigma_x - \sigma_y)^2 + 4\tau_{xy}^2}}$$

Similarly for case 2,

$$\sin 2\theta_{p_2} = \frac{2\tau_{xy}}{\sqrt{(\sigma_x - \sigma_y)^2 + 4\tau_{xy}^2}} \quad \text{and} \quad \cos 2\theta_{p_2} = \frac{(\sigma_x - \sigma_y)}{\sqrt{(\sigma_x - \sigma_y)^2 + 4\tau_{xy}^2}}$$

Now the values of principal stresses may be found out by substituting the above values of $2\theta_{p_1}$ and $2\theta_{p_2}$ in equation (*v*).

**Maximum Principal Stress,**

$$\sigma_{p_1} = \frac{\sigma_x + \sigma_y}{2} - \frac{\sigma_x - \sigma_y}{2} \cos 2\theta - \tau_{xy} \sin 2\theta$$

$$= \frac{\sigma_x + \sigma_y}{2} - \left( \frac{\sigma_x - \sigma_y}{2} \times \frac{-(\sigma_x - \sigma_y)}{\sqrt{(\sigma_x - \sigma_y^2) + 4\tau_{xy}^2}} \right) - \left( \tau_{xy} \times \frac{-2\tau_{xy}}{\sqrt{(\sigma_x - \sigma_y)^2 + 4\tau_{xy}^2}} \right)$$

$$= \frac{\sigma_x + \sigma_y}{2} + \frac{(\sigma_x - \sigma_y)^2 + 4\tau_{xy}^2}{2\sqrt{\sigma_x - \sigma_y)^2 + 4\tau_{xy}^2}} = \frac{\sigma_x + \sigma_y}{2} + \frac{\sqrt{(\sigma_x - \sigma_y)^2 + 4\tau_{xy}^2}}{2}$$

$$\text{or} \quad \sigma_{p_1} = \frac{\sigma_x + \sigma_y}{2} + \sqrt{\left( \frac{\sigma_x - \sigma_y}{2} \right)^2 + \tau_{xy}^2}$$

**Minimum Principal Stress**

$$\sigma_{p2} = \frac{\sigma_x+\sigma_y}{2} - \frac{(\sigma_x-\sigma_y)}{2}\cos 2\theta - \tau_{xy}\sin 2\theta$$

$$= \frac{\sigma_x+\sigma_y}{2} - \left(\frac{\sigma_x-\sigma_y}{2} \times \frac{(\sigma_x-\sigma_y)}{\sqrt{(\sigma_x-\sigma_y)^2+4\tau_{xy}^2}}\right) - \left(\tau_{xy} \times \frac{2\tau_{xy}}{\sqrt{(\sigma_x-\sigma_y)^2+4\tau_{xy}^2}}\right)$$

$$= \frac{\sigma_x+\sigma_y}{2} - \frac{(\sigma_x-\sigma_y)^2+4\tau_{xy}^2}{2\sqrt{(\sigma_x-\sigma_y)^2+4\tau_{xy}^2}} = \frac{\sigma_x-\sigma_y}{2} - \frac{\sqrt{(\sigma_x-\sigma_y)^2+4\tau_{xy}^2}}{2}$$

or

$$\sigma_{p_2} = \frac{\sigma_x+\sigma_y}{2} - \sqrt{\left(\frac{\sigma_x-\sigma_y}{2}\right)^2+\tau_{xy}^2}$$

**EXAMPLE 7.10.** *A point is subjected to a tensile stress of 250 MPa in the horizontal direction and another tensile stress of 100 MPa in the vertical direction. The point is also subjected to a simple shear stress of 25 MPa, such that when it is associated with the major tensile stress, it tends to rotate the element in the clockwise direction. What is the magnitude of the normal and shear stresses on a section inclined at an angle of 20° with the major tensile stress?*

**SOLUTION.** Given : Tensile stress in horizontal *x-x* direction ($\sigma_x$) = 250 MPa ; Tensile stress in vertical *y-y* direction ($\sigma_y$) = 100 MPa ; Shear stress ($\tau_{xy}$) = 25 MPa and angle made by section with the major tensile stress ($\theta$) = 20°.

***Magnitude of normal stress***

We know that magnitude of normal stress,

$$\sigma_n = \frac{\sigma_x+\sigma_y}{2} - \frac{\sigma_x-\sigma_y}{2}\cos 2\theta - \tau_{xy}\sin 2\theta$$

$$= \frac{250+100}{2} - \frac{250-100}{2}\cos(2\times 20°) - 25\sin(2\times 20°)$$

$$= 175 - 75\cos 40° - 25\sin 40° \text{ MPa}$$

$$= 175 - (75\times 0.766) - (25\times 0.6428) \text{ MPa}$$

$$= 175 - 57.45 - 16.07 = 101.48 \text{ MPa}$$ **Ans.**

***Magnitude of shear stress***

We also know that magnitude of shear stress,

$$\tau = \frac{\sigma_x-\sigma_y}{2}\sin 2\theta - \tau_{xy}\cos 2\theta$$

$$= \frac{250-100}{2}\sin(2\times 20°) - 25\cos(2\times 20°)$$

$$= 75\sin 40° - 25\cos 40° \text{ MPa}$$

$$= (75\times 0.6428) - (25\times 0.766) \text{ MPa}$$

$$= 48.21 - 19.15 = 29.06 \text{ MPa}$$ **Ans.**

**EXAMPLE 7.11.** *A plane element in a boiler is subjected to tensile stresses of 400 MPa on one plane and 150 MPa on the other at right angles to the former. Each of the above stresses is accompanied by a shear stress of 100 MPa such that when associated with the minor tensile stress tends to rotate the element in anticlockwise direction. Find*

*(a) Principal stresses and their directions.*

*(b) Maximum shearing stresses and the directions of the plane on which they act.*

**SOLUTION.** Given : Tensile stress along *x-x* axis ($\sigma_x$) = 400 MPa ; Tensile stress along *y-y* axis ($\sigma_y$) = 150 MPa and shear stress ($\tau_{xy}$) = – 100 MPa (Minus sign due to anticlockwise on *x-x* direction).

(*a*) ***Principal stresses and their directions***

We know that maximum principal stress,

$$\sigma_{max} = \frac{\sigma_x + \sigma_y}{2} + \sqrt{\left(\frac{\sigma_x - \sigma_y}{2}\right)^2 + \tau_{xy}^2}$$

$$= \frac{400 + 150}{2} + \sqrt{\left(\frac{400 - 150}{2}\right)^2 + (-100)^2} \text{ MPa}$$

$$= 275 + 160.1 = 435.1 \text{ MPa} \qquad \textbf{Ans.}$$

and minimum principal stress,

$$\sigma_{min} = \frac{\sigma_x + \sigma_y}{2} - \sqrt{\left(\frac{\sigma_x - \sigma_y}{2}\right)^2 + \tau_{xy}^2}$$

$$= \frac{400 + 150}{2} - \sqrt{\left(\frac{400 - 150}{2}\right)^2 + (-100)^2} \text{ MPa}$$

$$= 275 - 160.1 = 114.9 \text{ MPa} \qquad \textbf{Ans.}$$

Let $\theta_p$ = Angle which plane of principal stress makes with *x-x* axis.

We know that,
$$\tan 2\theta_p = \frac{2\tau_{xy}}{\sigma_x - \sigma_y} = \frac{2 \times 100}{400 - 150} = 0.8 \quad \text{or} \quad 2\theta_p = 38.66°$$

$$\therefore \quad \theta_p = 19.33° \quad \text{or} \quad 109.33° \qquad \textbf{Ans.}$$

(*b*) ***Maximum shearing stresses and their directions***

We also know that maximum shearing stress

$$\tau_{max} = \sqrt{\left(\frac{\sigma_x - \sigma_y}{2}\right)^2 + \tau_{xy}^2} = \sqrt{\left(\frac{400 - 150}{2}\right)^2 + (-100)^2}$$

$$= 160.1 \text{ MPa} \qquad \textbf{Ans.}$$

Let $\theta_s$ = Angle which plane of maximum shearing stress makes with *x-x* axis.

We know that,
$$\tan 2\theta_s = \frac{\sigma_x - \sigma_y}{2\tau_{xy}} = \frac{400 - 150}{2 \times 100} = 1.25 \quad \text{or} \quad 2\theta_s = 51.34°$$

$$\theta_s = 25.67° \quad \text{or} \quad 115.67° \qquad \textbf{Ans.}$$

**EXAMPLE 7.12.** *A point in a strained material is subjected to the stresses as shown in Fig. 7.11.*

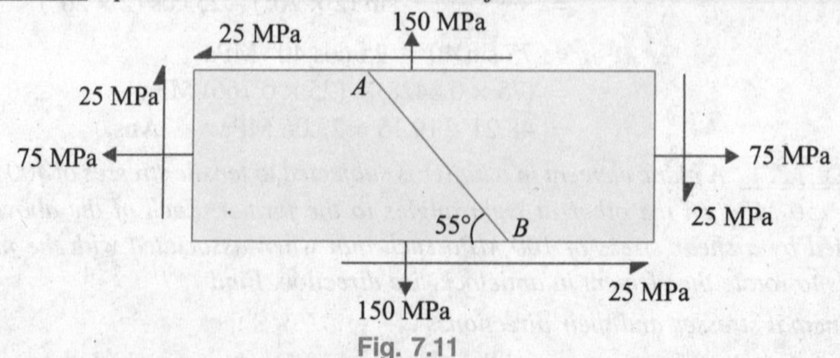


Fig. 7.11

*Find graphically, or otherwise, the normal and shear stresses on the section AB.*

**SOLUTION.** Given : Tensile stress along horizontal *x-x* axis ($\sigma_x$) = 75 MPa ; Tensile stress along vertical *y-y* axis ($\sigma_y$) = 150 MPa ; Shear stress ($\tau_{xy}$) = 25 MPa and angle made by section with the horizontal direction ($\theta$) = 55°.

***Normal stress on the section AB***

We know that normal stress on the section *AB*,

$$\sigma_n = \frac{\sigma_x - \sigma_y}{2} - \frac{\sigma_x - \sigma_y}{2} \cos 2\theta - \tau_{xy} \sin 2\theta$$

$$= \frac{75+150}{2} - \frac{75-150}{2} \cos (2 \times 55°) - 25 \sin (2 \times 55°) \text{ MPa}$$

$$= 112.5 + 37.5 \cos 110° - 25 \sin 110° \text{ MPa}$$

$$= 11.25 + 37.5 \times (-0.342) - (25 \times 0.9397) \text{ MPa}$$

$$= 112.5 - 12.83 - 23.49 = 76.18 \text{ MPa} \quad \textbf{Ans.}$$

***Shear stress on the section AB***

We also know that shear stress on the section *AB*.

$$\tau = \frac{\sigma_x - \sigma_y}{2} \sin 2\theta - \tau_{xy} \cos 2\theta$$

$$= \frac{75-150}{2} \sin (2 \times 55°) - 25 \cos (2 \times 55°) \text{ MPa}$$

$$= -37.5 \sin 110° - 25 \cos 110° \text{ MPa}$$

$$= -37.5 \times 0.9397 - 25 \times (-0.342) \text{ MPa}$$

$$= -35.24 + 8.55 = -26.69 \text{ MPa} \quad \textbf{Ans.}$$

**EXAMPLE 7.13.** *A plane element of a body is subjected to a compressive stress of 300 MPa in x-x direction and a tensile stress of 200 MPa in the y-y direction. Each of the above stresses is subjected to a shear stress of 100 MPa such that when it is associated with the compressive stress, it tends to rotate the element in an anticlockwise direction. Find graphically, or otherwise, the normal and shear stresses on a plane inclined at an angle of 30° with the x-x axis.*

**SOLUTION.** Given : Compressive stress in *x-x* direction ($\sigma_x$) = – 300 MPa (Minus sign due to compressive stress) ; Tensile stress in *y-y* direction ($\sigma_y$) = 200 MPa ; Shear stress ($\tau_{xy}$) = – 100 MPa (Minus sign due to anticlockwise direction along the compressive stress *i.e.*, $\sigma_x$) and angle made by section with the *x-x* axis ($\theta$) = 30°.

***Normal stress on the plane***

We know that normal stress on the plane,

$$\sigma_n = \frac{\sigma_x + \sigma_y}{2} - \frac{\sigma_x - \sigma_y}{2} \cos 2\theta - \tau_{xy} \sin 2\theta$$

$$= \frac{-300+200}{2} - \frac{-300-200}{2} \cos (2 \times 30°) - [-100 \sin (2 \times 30°]$$

$$= -50 - (-250 \cos 60°) + 100 \sin 60° \text{ MPa}$$

$$= -50 + (250 \times 0.5) + (10 \times 0.866) \text{ MPa}$$

$$= -50 + 125 + 86.6 = 161.6 \text{ MPa} \quad \textbf{Ans.}$$

***Shear stress on the plane***

We also know that shear stress on the plane.

$$\tau = \frac{\sigma_x - \sigma_y}{2} \sin 2\theta - \tau_{xy} \cos 2\theta$$

$$= \frac{-300-200}{2} \sin(2 \times 30°) - [-100 \cos(2 \times 30°)] \text{ MPa}$$
$$= -250 \sin 60° + 100 \cos 60° \text{ MPa}$$
$$= -250 \times 0.866 + 100 \times 0.5 \text{ MPa}$$
$$= -216.5 + 50 = -166.5 \text{ MPa} \quad \textbf{Ans.}$$

**EXAMPLE 7.14.** *A machine component is subjected to the stresses as shown in the figure given below :*

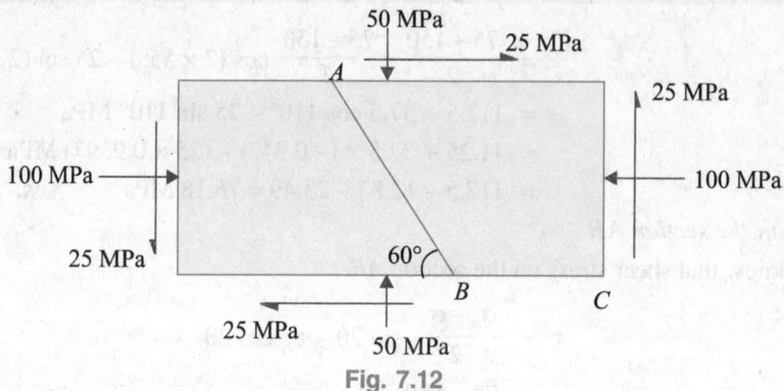


Fig. 7.12

*Find the normal and shearing stresses on the section AB inclined at an angle of 60° with x-x axis. Also find the resultant stress on the section.*

**SOLUTION.** Given : Compressive stress along horizontal *x-x* axis ($\sigma_x$) = – 100 MPa (Minus sign due to compressive stress) ; Compressive stress along vertical *y-y* axis ($\sigma_y$) = – 50 MPa (Minus sign due to compressive stress) ; Shear stress ($\tau_{xy}$) = – 25 MPa (Minus sign due to anticlockwise on *x-x* axis) and angle made by section *AB* with *x-x* axis ($\theta$) = 60°.

***Normal stress on the section AB***

We know that normal stress on the section *AB*,

$$\sigma_n = \frac{\sigma_x + \sigma_y}{2} - \frac{\sigma_x - \sigma_y}{2} \cos 2\theta - \tau_{xy} \sin 2\theta$$
$$= \frac{-100+(-50)}{2} - \frac{-100-(-50)}{2} \cos(2\times 60°) - [-25 \sin(2 \times 60°)]$$
$$= -75 + 25 \cos 120° + 25 \sin 120° \text{ MPa}$$
$$= -75 + [25 \times (-0.5)] + (25 \times 0.866) \text{ MPa}$$
$$= -75 - 12.5 + 21.65 = -65.85 \text{ MPa} \quad \textbf{Ans.}$$

***Shearing stress on the section AB***

We know that shearing stress on the section *AB*,

$$\tau = \frac{\sigma_x - \sigma_y}{2} \sin 2\theta - \tau_{xy} \cos 2\theta$$
$$= \frac{-100-(-50)}{2} \sin(2 \times 60°) - [-25 \cos(2 \times 60°)]$$
$$= -25 \sin 120° + 25 \cos 120° = -25 \times 0.866 + [25 \times (-0.5)] \text{ MPa}$$
$$= -21.65 - 12.5 = -34.15 \text{ MPa} \quad \textbf{Ans.}$$

***Resultant stress on the section AB***

We also know that resultant stress on the section *AB*,

$$\sigma_R = \sqrt{\sigma_n^2 + \tau^2} = \sqrt{(-65.85)^2 + (-34.15)^2} = 74.2 \text{ MPa} \quad \textbf{Ans.}$$

## EXERCISE 7.1

1. A bar is subjected to a tensile stress of 100 MPa, Determine the normal and tangential stresses on a plane making an angle of 30° with the direction of the tensile stress.
   (**Ans.** 75 MPa ; 43.3 MPa)
2. A point in a strained material is subjected to a tensile stress of 50 MPa. Find the normal and shear stress at an angle of 50° with the direction of the stress. (**Ans.** 29.34 MPa ; 24.62 MPa)
3. At a point in a strained material, the principal stresses are 100 MPa and 50 MPa both tensile. Find the normal and shear stresses at a section inclined at 30° with the axis of the major principal stress. (**Ans.** 87.5 MPa ; 21.65 MPa)
4. A point in a strained material is subjected to a tensile stress of 120 MPa and a clockwise shear stress of 40 MPa. What are the values of normal and shear stresses on a plane inclined at 45° with the normal to the tensile stress. (**Ans.** 20 MPa ; 60 MPa)
5. The principal stresses or a point in the section of a member are 50 MPa or 20 MPa both tensile. If there is a clockwise shear stress of 30 MPa, find the normal and shear stresses on a section inclined at an angle of 15° with the normal to the major tensile stress.
   (**Ans.** 32.99 MPa ; 33.48 MPa)

## 7.12. Graphical Method for the Stresses on an Oblique Section of a Body

In the previous articles, we have been discussing the analytical method for the determination of normal, shear and resultant stresses across a section. But we shall now discuss a graphical method for this purpose. This is done by drawing a Mohr's Circle of Stresses. The construction of Mohr's Circle of Stresses as well as determination of normal, shear and resultant stresses is very easier than the analytical method. Moreover, there is a little chance of committing any error in this method. In the following pages, we shall draw the Mohr's Circle of Stresses for the following cases :

1. A body subjected to a direct stress in one plane.
2. A body subjected to direct stresses in two mutually perpendicular directions.
3. A body subjected to a simple shear stress.
4. A body subjected to a direct stress in one plane accompanied by a simple shear stress.
5. A body subjected to direct stresses in two mutually perpendicular directions accompanied by a simple shear stress.

## 7.13. Sign Conventions for Graphical Method

Though there are different sign conventions used in different books for graphical method also, yet we shall adopt the following sign conventions, which are widely used and internationally recognised :

1. The angle is taken with reference to the *X-X* axis. All the angles traced in the anticlockwise direction to the *X-X* axis are taken as negative, whereas those in the clockwise direction as positive as shown in Fig. 7.13 (*a*). The value of angle $\theta$, until and unless mentioned is taken as positive and drawn clockwise.
2. The measurements above *X-X* axis and to the right of *Y-Y* axis are taken as positive, whereas those below *X-X* axis and to the left of *Y-Y* axis as negative as shown in Fig 7.13 (*b*) and (*c*).
3. Sometimes there is a slight variation in the results obtained by analytical method and graphical method. The values obtained by graphical method are taken to be correct if they agree upto the first decimal point with values obtained by analytical method, *e.g.*, 8.66 (Analytical) = 8.7 (Graphical), similarly 4.32 (Analytical) = 4.3 (Graphical)

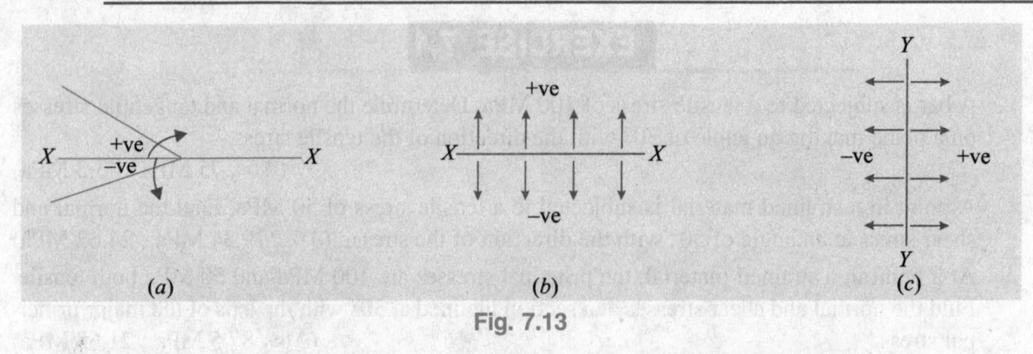


Fig. 7.13

## 7.14. Mohr's Circle for Stresses on an Oblique Section of a Body Subjected to a Direct Stress in One Plane

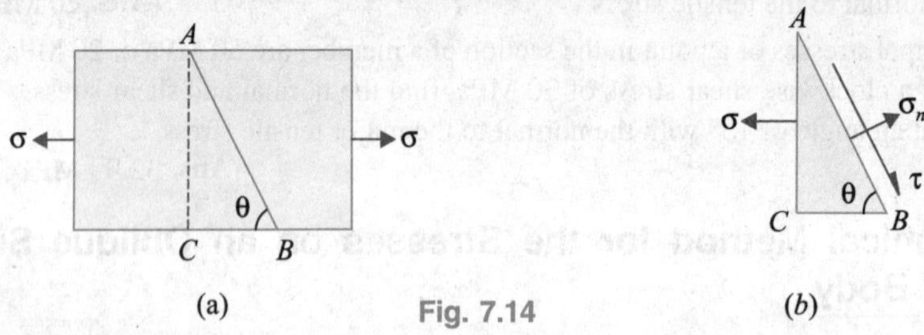


Fig. 7.14

Consider a rectangular body of uniform cross-sectional area and unit thickness subjected to a direct tensile stress along *X–X* axis as shown in Fig 7.14 (*a*) and (*b*). Now let us consider an oblique section *AB* inclined with *X–X* axis, on which we are required to find out the stresses as shown in the figure.

Let $\sigma$ = Tensile stress, in *x-x* direction and

$\theta$ = Angle which the oblique section *AB* makes with the *x-x* axis in clockwise direction.

First of all, consider the equilibrium of the wedge *ABC*. Now draw the Mohr's* Circle of Stresses as shown in Fig.7.15 and as discussed below :

1. First of all, take some suitable point *O* and through it draw a horizontal line *XOX*.
2. Cut off *OJ* equal to the tensile stress ($\sigma$) to some suitable scale and towards right (because $\sigma$ is tensile). Bisect *OJ* at *C*. Now the point *O* represents the stress system on plane *BC* and the point *J* represents the stress system on plane *AC*.
3. Now with *C* as centre and radius equal to *CO* and or *CJ* draw a circle. It is known as Mohr's Circle for Stresses.

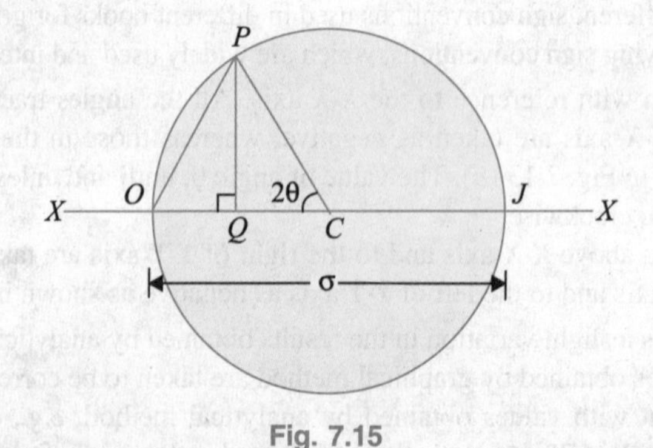


Fig. 7.15

---

* The diagram was first presented by German Scientist Otto Mohr in 1982.

4. Now through $C$ draw a line $CP$ making an angle of $2\theta$ with $CO$ in the clockwise direction meeting the circle at $P$. The point $P$ represents the section $AB$.
5. Through $P$, draw $PQ$ perpendicular to $OX$. Join $OP$.
6. Now $OQ$, $QP$ and $OP$ will give the normal stress, shear stress and resultant stress respectively to the scale. And the angle $POJ$ is called the angle of obliquity ($\theta$).

**Proof**

From the geometry of the Mohr's Circle of Stresses, we find that,

$$OC = CJ = CP = \sigma/2 \qquad \text{... (Radius of the circle)}$$

$\therefore$ Normal Stress.

$$\sigma_n = OQ = OC - QC = \left(\frac{\sigma}{2}\right) - \left(\frac{\sigma}{2}\right)\cos 2\theta \qquad \text{...(Same as in Art. 7.7)}$$

and shear stress

$$\tau = QP = CP \sin 2\theta = \frac{\sigma}{2}\sin 2\theta \qquad \text{...(Same as in Art. 7.7)}$$

We also find that maximum shear stress will be equal to the radius of the Mohr's Circle of Stresses *i.e.*, $\frac{\sigma}{2}$. It will happen when $2\theta$ is equal to 90° or 270° *i.e.*, $\theta$ is equal to 45° or 135°.

However when $\theta = 45°$ then the shear stress is equal to $\frac{\sigma}{2}$.

And when $\theta = 135°$ then the shear stress is equal to $-\frac{\sigma}{2}$.

## 7.15. Mohr's Circle for Stresses on an Oblique Section of a Body Subjected to Direct Stresses in Two Mutually Perpendicular Directions

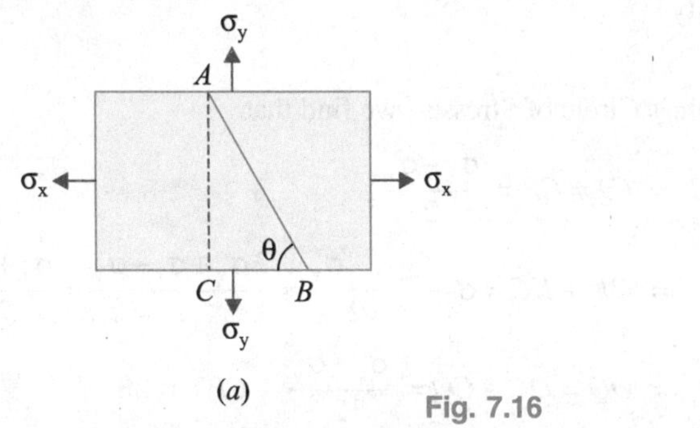


Fig. 7.16

Consider a rectangular body of uniform cross-sectional area and unit thickness subjected to direct tensile stresses in two mutually perpendicular directions along *x-x* and *y-y* axis as shown in Fig 7.16 (*a*) and (*b*). Now let us consider an oblique section $AB$ inclined with *x-x* axis on which we are required to find out the stresses as shown in the figure.

Let $\sigma_x$ = Tensile stress in *x-x* direction (also termed as major tensile stress),

$\sigma_y$ = Tensile stress in *y-y* direction (also termed as minor tensile stress). and

$\theta$ = Angle which the oblique section $AB$ makes with *x-x* axis in clockwise direction.

First of all consider the equilibrium of the wedge $ABC$. Now draw the Mohr's Circle of Stresses as shown in Fig. 7.17 and as discussed below :

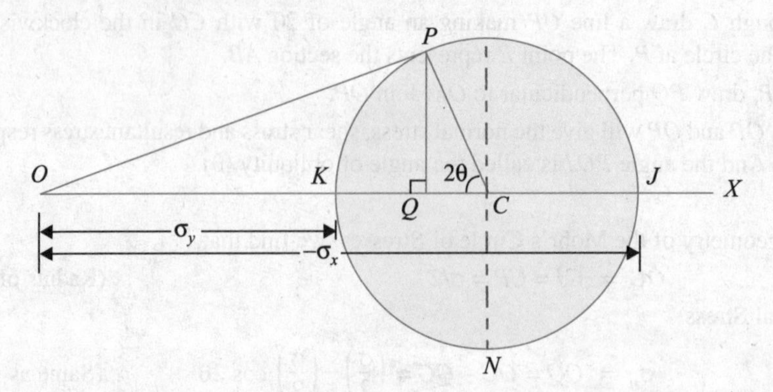


Fig. 7.17

1. First of all, take some suitable point $O$ and draw a horizontal line $OX$.
2. Cut off $OJ$ and $OK$ equal to the tensile stresses $\sigma_x$ and $\sigma_y$ to some suitable scale towards right (because both the stresses are tensile). The point $J$ represents the stress system on plane $AC$ and the point $K$ represents the stress system on plane $BC$. Bisect $JK$ at $C$.
3. Now with $C$ as centre and radius equal to $CJ$ or $CJ$ draw a circle. It is known as Mohr's Circle of Stresses.
4. Now through $C$, draw a line $CP$ making an angle of $2\theta$ with $CK$ in clockwise direction meeting the circle at $P$. The point $P$ represents the stress systems on the section $AB$.
5. Through $P$, draw $PQ$ perpendicular to the line $OX$. Join $OP$.
6. Now $OQ$, $QP$ and $OP$ will give the normal stress, shear stress and resultant stress respectively to the scale. Similarly $CM$ or $CN$ will give the maximum shear stress to the scale. The angle $POC$ is called the angle of obliquity.

**Proof**

From the geometry of the Mohr's Circle of Stresses, we find that

$$KC = CJ = CP = \frac{\sigma_x - \sigma_y}{2}$$

or

$$OC = OK + KC = \sigma_y + \frac{\sigma_x - \sigma_y}{2} = \frac{2\sigma_y + \sigma_x - \sigma_y}{2} = \frac{\sigma_x + \sigma_y}{2}$$

$\therefore$ Normal stress,

$$\sigma_n = OQ = OC - CQ = \frac{\sigma_x - \sigma_y}{2} - CP\cos 2\theta$$

$$= \frac{\sigma_x + \sigma_y}{2} - \frac{\sigma_x - \sigma_y}{2}\cos 2\theta \qquad \text{...(Same as Art. 7.8)}$$

and shear stress,

$$\tau = QP = CP\sin 2\theta$$

$$= \frac{\sigma_x + \sigma_y}{2}\sin 2\theta \qquad \text{...(Same as Art. 7.8)}$$

We also find that the maximum shear stress will be equal to the radius of the Mohr's Circle of Stresses. *i.e.*, $\frac{\sigma_x - \sigma_y}{2}$. It will happen when $2\theta$ is equal to 90° or 270° *i.e.*, when $\theta$ is equal to 45° or 135°.

However when $\theta = 45°$ then the shear stress is equal to $\dfrac{\sigma_x - \sigma_y}{2}$

And when $\theta = 135°$ then the shear stress will be equal to $\dfrac{-(\sigma_x - \sigma_y)}{2}$ or $\dfrac{\sigma_y - \sigma_x}{2}$.

**EXAMPLE 7.15.** *The stresses at a point of a machine component are 150 MPa and 50 MPa both tensile. Find the intensities of normal, shear and resultant stresses on a plane inclined at an angle of 55° with the axis of major tensile stress.*

*Also find the magnitude of the maximum shear stresses in the component.*

***SOLUTION.** Given : Tensile stress along horizontal *x-x* axis ($\sigma_x$) = 150 MPa ; Tensile stress along vertical *y-y* axis ($\sigma_y$) = 50 MPa and angle made by the plane with the axis of major tensile stress ($\theta$) = 55°.

The given stresses on the planes *AC* and *BC* in the machine component are shown in Fig. 7.18 (*a*). Now draw the Mohr's Circle of Stresses as shown in Fig. 7.18 (*b*) and as discussed below :

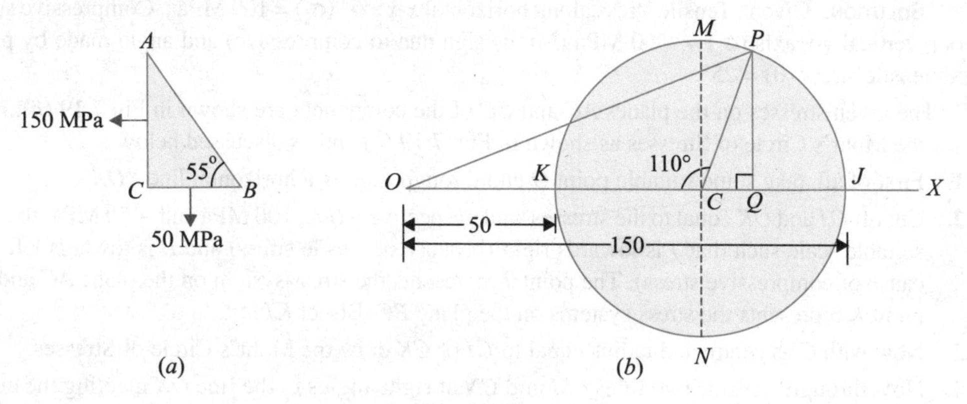


Fig. 7.18

1. First of all, take some suitable point *O* and draw a horizontal line *OX*.
2. Cut off *OJ* and *OK* equal to the tensile stresses $\sigma_x$ and $\sigma_y$ respectively (*i.e.* 150 MPa and 50 MPa) to some suitable scale towards right. The point *J* represents the stress system on the plane *AC* and the point *K* represents the stress system on the plane *BC*. Bisect *KJ* at *C*.
3. Now with *C* as centre and radius equal to *CJ* or *CK* draw the Mohr's Circle of Stresses.
4. Now through *C* draw two lines *CM* and *CN* at right angles to the line *OX* meeting the circle at *M* and *N*. Also through *C* draw a line *CP* making an angle of 2 × 55° = 110° with *CK* in clockwise direction meeting the circle at *P*. The point *P* represents the stress system on the plane *AB*.
5. Through *P*, draw *PQ* perpendicular to the line *OX*. Join *OP*.

   By measurement, we find that the normal stress ($\sigma_n$) = *OQ* = 117.1 MPa ; Shear stress ($\tau$) = *QP* = 47.0 MPa ; Resultant stress ($\sigma_R$) = *OP* = 126.2 MPa and maximum shear stress ($\tau_{max}$) = *CM* = ± 50 MPa **Ans.**

**EXAMPLE 7.16.** *The stresses at a point in a component are 100 MPa (tensile) and 50 MPa (compressive). Determine the magnitude of the normal and shear stresses on a plane inclined at an angle of 25° with tensile stress. Also determine the direction of the resultant stress and the magnitude of the maximum intensity of shear stress.*

* We have already solved this question analytically, as example 7.5.

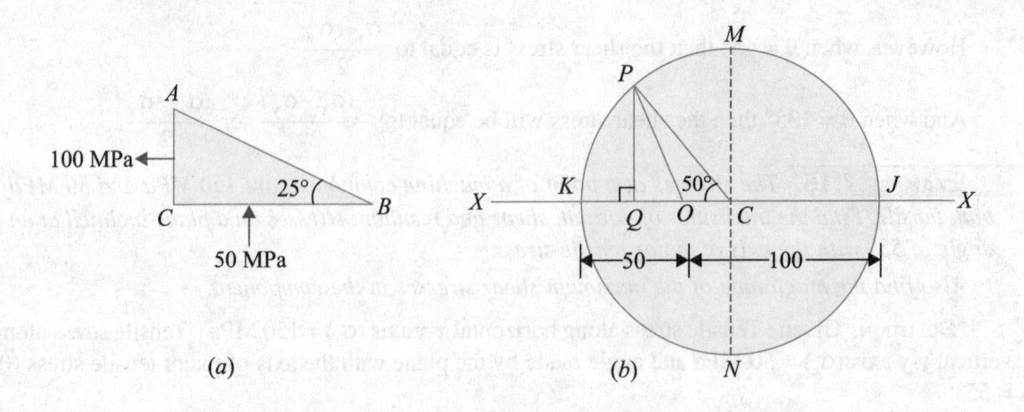


Fig. 7.19

***SOLUTION.** Given : Tensile stress along horizontal *x-x* axis ($\sigma_x$) = 100 MPa ; Compressive stress along vertical *y-y* axis ($\sigma_y$) = – 50 MPa (Minus sign due to compressive) and angle made by plane with tensile stress ($\theta$) = 25°.

The given stresses on the planes *AC* and *BC* of the component are shown in Fig 7.19 (*a*). Now draw the Mohr's Circle of Stresses as shown in Fig. 7.19 (*b*) and as discussed below :

1. First of all, take some suitable point *O* and through it draw a horizontal line *XOX*.
2. Cut off *OJ* and *OK* equal to the stresses and respectively (*i.e.*, 100 MPa and – 50 MPa) to some suitable scale such that *J* is towards right (because of tensile stress) and *B* is towards left (because of compressive stress). The point *J* represents the stress system on the plane *AC* and the point *K* represents the stress systems on the plane *BC*. Bisect *KJ* at *C*.
3. Now with *C* as centre and radius equal to *CJ* or *CK* draw the Mohr's Circle of Stresses.
4. Now through *C*, draw two lines *CM* and *CN* at right angles to the line *OX* meeting the circle at *M* and *N*. Also through *C*, draw a line *CP* making an angle of 2 × 25° = 50° with *CK* in clockwise direction meeting the circle at *P*. The point *P* represents the stress system on the plane *AB*.
5. Through *P*, draw *PQ* perpendicular to the line *OX*. Join *OP*.

By measurement, we find that the normal stress ($\sigma_n$) = – 23.2 MPa ; Shear stress ($\tau$) = *PQ* = 57.45 MPa; Direction of the resultant stress $\angle POQ$ = 68.1° and maximum shear stress ($\tau_{max}$) = *CM* = *CN* = ± 75 MPa **Ans.**

## 7.16. Mohr's Circle for Stresses on an Oblique Section of a Body Subjected to a Direct Stresses in One Plane Accompanied by a Simple Shear Stress

Consider a rectangular body of uniform cross-sectional area and unit thickness subjected to a direct tensile stress along *X-X* axis accompanied by a positive (*i.e.* clockwise ) shear stress along *X-X* axis as shown in Fig 7.20 (*a*) and (*b*). Now let us consider an oblique section *AB* inclined with *x-x* axis on which we are required to find out the stresses as shown in the figure 7.20.

* We have already solved this question analytically, as example 7.6.

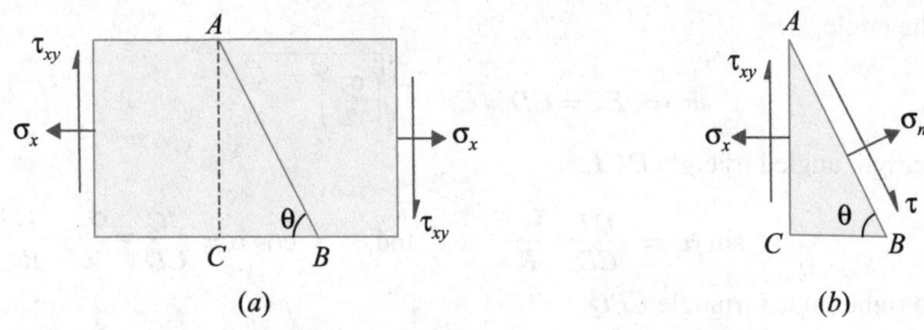


Fig. 7.20

Let $\sigma_x$ = Tensile stress in $x$-$x$ direction,

$\tau_{xy}$ = Positive (*i.e.*, clockwise) shear stress along $x$-$x$ axis, and

$\theta$ = Angle which oblique section $AB$ makes with $x$-$x$ axis in clock wise direction.

First of all consider the equilibrium of the wedge $ABC$. **We know that as per the principle of simple shear the face $BC$ of the wedge will also be subjected to an anticlockwise shear stress.** Now draw the Mohr's Circle of Stresses as shown in Fig.7.21 and as discussed below :

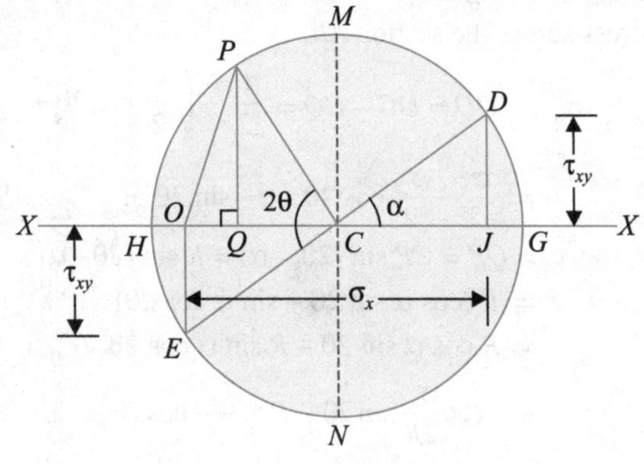


Fig. 7.21

1. First of all, take some suitable point $O$ and through it draw a horizontal line $XOX$.
2. Cut off $OJ$ equal to the tensile stress $\sigma_x$ to some suitable scale and towards right (because $\sigma_x$ is tensile).
3. Now erect a perpendicular at $J$ above the line $X$-$X$ (because $\tau_{xy}$ is positive along $x$-$x$ axis) and cut off $JD$ equal to the shear stress $\tau_{xy}$ to the scale. The point $D$ represents the stress system on plane $AC$. Similarly, erect a perpendicular below the line $x$-$x$ (**because $\tau_{xy}$ is negative along $y$-$y$ axis**) and cut off $OE$ equal to the shear stress $\tau_{xy}$ to the scale. The point $E$ represents the stress system on plane $BC$. Join $DE$ and bisect it at $C$.
4. Now with $C$ as centre and radius equal to $CD$ or $CE$ draw a circle. It is known as Mohr's Circle of Stresses.
5. Now through $C$, draw a line $CP$ making an angle $2\theta$ with $CE$ in clockwise direction meeting the circle at $P$. The point $P$ represents the stress system on the section $AB$.
6. Through $P$, draw $PQ$ perpendicular to the line $OX$. Join $OP$.
7. Now $OQ$, $QP$ and $OP$ will give the normal, shear and resultant stresses to the scale. And the angle $POC$ is called the angle of obliquity.

**Proof**

From the geometry of the Mohr's Circle of Stresses, we find that

$$OC = \frac{\sigma_x}{2}$$

and radius of the circle,

$$R = EC = CD = CP = \sqrt{\left(\frac{\sigma_x}{2}\right)^2 + \tau_{xy}^2}$$

Now in the right angled triangle *DCJ*,

$$\sin\alpha = \frac{DJ}{CD} = \frac{\tau_{xy}}{R} \quad \text{and} \quad \cos\alpha = \frac{JC}{CD} = \frac{\sigma_x}{2} \times \frac{1}{R} = \frac{\sigma_x}{2R}$$

and similarly in right angled triangle *CPQ*,

$$\angle PCQ = (2\theta - \alpha)$$

$$\therefore \quad CQ = CP\cos(2\theta - \alpha) = R[\cos(2\theta - \alpha)]$$

$$= R[\cos\alpha\cos 2\theta + \sin\alpha\sin 2\theta]$$

$$= R\cos\alpha\cos 2\theta + R\sin\alpha\sin 2\theta$$

$$= R \times \frac{\sigma_x}{2R}\cos 2\theta + R \times \frac{\tau_{xy}}{R}\sin 2\theta$$

$$= \frac{\sigma_x}{2}\cos 2\theta + \tau_{xy}\sin 2\theta$$

We know that normal stress across the section *AB*,

$$\sigma_n = OQ = OC - CQ = \frac{\sigma_x}{2} - \left(\frac{\sigma_x}{2}\cos 2\theta + \tau_{xy}\sin 2\theta\right)$$

$$= \frac{\sigma_x}{2} - \frac{\sigma_x}{2}\cos 2\theta - \tau_{xy}\sin 2\theta \qquad \text{...(Same as in Art. 7.10)}$$

and shear stress,

$$\tau = QP = CP\sin(2\theta - \alpha) = R\sin(2\theta - \alpha)$$

$$= R(\cos\alpha\sin 2\theta - \sin\alpha\cos 2\theta)$$

$$= R\cos\alpha\sin 2\theta - R\sin\alpha\cos 2\theta$$

$$= R \times \frac{\sigma_x}{2R}\sin 2\theta - R \times \frac{\tau_{xy}}{2}\cos 2\theta$$

$$= \frac{\sigma_x}{2}\sin 2\theta - \tau_{xy}\cos 2\theta \qquad \text{...(Same as in Art. 7.10)}$$

We also know that maximum stress,

$$\sigma_{max} = OG = OC + CG = \frac{\sigma_x}{2} + \sqrt{\left(\frac{\sigma_x}{2}\right)^2 + \tau_{xy}^2}$$

and minimum stress

$$\sigma_{min} = OH = OC - CH = \frac{\sigma_x}{2} - \sqrt{\left(\frac{\sigma_x}{2}\right)^2 + \tau_{xy}^2}$$

We also find that the maximum shear stress will be equal to the radius of the Mohr's circle of stresses *i.e.*, $\sqrt{\left(\frac{\sigma_x}{2}\right)^2 + \tau_{xy}^2}$. It will happen when $(2\theta - \alpha)$ is equal to 90° or 270°.

However when $(2\theta - \alpha)$ is equal to 90° then the shear stress is equal to $+\sqrt{\left(\frac{\sigma_x}{2}\right)^2 + \tau_{xy}^2}$.

And when $(2\theta - \alpha) = 270°$ then the shear stress is equal to $-\sqrt{\left(\frac{\sigma_x}{2}\right)^2 + \tau_{xy}^2}$.

**EXAMPLE 7.17.** *A plane element in a body is subjected to a tensile stress of 100 MPa accompanied by a clockwise shear stress of 25 MPa. Find (i) the normal and shear stress on a plane inclined at an angle of 20° with the tensile stress ; and (ii) the maximum shear stress on the plane.*

***SOLUTION.** Given : Tensile stress along horizontal *x-x* axis ($\sigma_x$) = 100 MPa ; Shear stress ($\tau_{xy}$) = 25 MPa and angle made by plane with tensile stress ($\theta$) = 20°.

The given stresses on the element and a complimentary shear stress on the *BC* plane are shown in Fig. 7.22 (*a*). Now draw the Mohr's Circle of Stresses as shown in Fig 7.22 (*b*) and as discussed below :

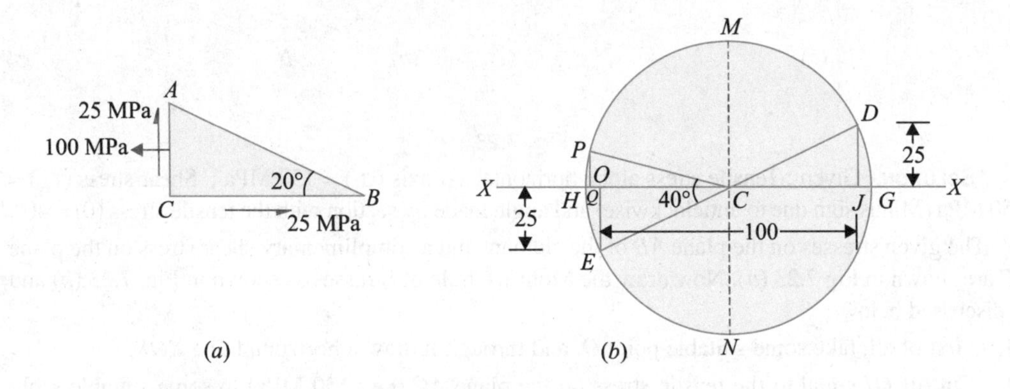


Fig. 7.22

1. First of all, take some suitable point *O*, and through it draw a horizontal line *XOX*.
2. Cut off *OJ* equal to the tensile stress on the plane *AC* (*i.e.,* 100 MPa) to some suitable scale towards right.
3. Now erect a perpendicular at *J* above the line *X-X* and cut off *JD* equal to the positive shear stress on the plane *BC* (*i.e.,* 25 MPa) to the scale. The point *D* represents the stress system on the plane *AC*. Similarly erect a perpendicular at *O* below the line *X-X* and cut off *OE* equal to the negative shear stress on the plane *BC* (*i.e.,* 25 MPa) to the scale. The point *E* represents the stress system on the plane *BC*. Join *DE* and bisect it at *C*.
4. Now with *C* as centre and radius equal to *CD* or *CE* draw the Mohr's Circle of Stresses.
5. Now through *C*, draw two lines *CM* and *CN* at right angle to the line *OX* meeting the circle at *M* and *N*. Also through *C*, draw a line *CP* making an angle of 2 × 20° = 40° with *CE* in clockwise direction meeting the circle at *P*. The point *P* represents the stress system on the section *AB*.
6. Through *P*, draw *PQ* perpendicular to the line *OX*.

   By measurement, we find that the normal stress ($\sigma_n$) = *OQ* = 4.4 MPa (compression) ; Shear stress ($\tau$) = *QP* = 13.0 MPa and maximum shear stress ($\tau_{max}$) = *CM* = 55.9 MPa **Ans.**

**EXAMPLE 7.18.** *An element in a strained body is subjected to a tensile stress of 150 MPa and a shear stress of 50 MPa tending to rotate the element in an anticlockwise direction. Find (i) the magnitude of the normal and shear stresses on a section inclined at 40° with the tensile stress ; and (ii) the magnitude and direction of maximum shear stress that can exist on the element.*

* We have already solved this question analytically, as example 7.7.

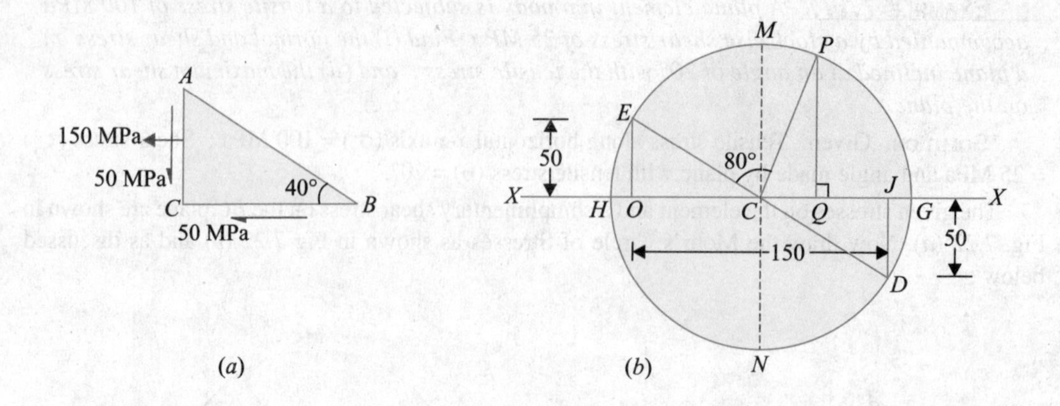


Fig. 7.23

*SOLUTION. Given : Tensile stress along horizontal $x$-$x$ axis ($\sigma_x$) = 150 MPa ; Shear stress ($\tau_{xy}$) = – 50 MPa (Minus sign due to anticlockwise) and angle made by section with the tensile stress ($\theta$) = 40°.

The given stresses on the plane $AB$ of the element and a complimentary shear stress on the plane $BC$ are shown in Fig 7.23 (*a*). Now draw the Mohr's Circle of Stresses as shown in Fig. 7.23 (*b*) and as discussed below :

1. First of all, take some suitable point $O$, and through it draw a horizontal line $XOX$.
2. Cut off $OJ$ equal to the tensile stress on the plane $AC$ (*i.e.*, 150 MPa) to some suitable scale towards right.
3. Now erect a perpendicular at $J$ below the line $X$-$X$ and cut off $JD$ equal to the negative shear stress on the plane $AC$ (*i.e.*, 50 MPa) to the scale. The point $D$ represents the stress system on the plane $AC$. Similarly, erect a perpendicular at $O$ above the line $X$-$X$ and cut off $OE$ equal to the positive shear stress on the plane $BC$ (*i.e.*, 50 MPa) to the scale. The point $E$ represents the stress system on the plane $BC$. Join $DE$ and bisect it at $C$.
4. Now with $C$ as centre and radius equal to $CD$ or $CE$ draw the Mohr's Circle of Stresses meeting the line $X$-$X$ at $G$ and $H$.
5. Through $C$, draw two lines $CM$ and $CN$ at right angles to the line $X$-$X$ meeting the circle at $M$ and $N$. Also through $C$, draw a line $CP$ making an angle of 2 × 40° = 80° with $CE$ in clockwise direction meeting the circle at $P$. The point $P$ represents the stress system on the section $AB$.
6. Through $P$, draw $PQ$ perpendicular to the line $OX$.

   By measurement, we find that the Normal stress ($\sigma_n$) = $OQ$ = 112.2 MPa ; Shear stress ($\tau$) = $QP$ = 82.5 MPa and maximum shear stress, that can exist on element ($\tau_{max}$) = ± $CM$ = $CN$ = 90.14 MPa **Ans.**

**EXAMPLE 7.19.** *An element in a strained body is subjected to a compressive stress of 200 MPa and a clockwise shear stress of 50 MPa on the same plane. Calculate the values of normal and shear stresses on a plane inclined at 35° with the compressive stress. Also calculate the value of maximum shear stress in the element.*

**SOLUTION. Given : Compressive stress along horizontal $x$-$x$ axis ($\sigma_x$) = – 200 MPa (Minus sign due to compressive stress) ; Shear stress ($\tau_{xy}$) = 50 MPa ; and angle made by plane with the compressive stress ($\theta$) = 35°.

* We have already solved this question analytically, as example 7.8.
** We have already solved this question analytically, as example 7.9.

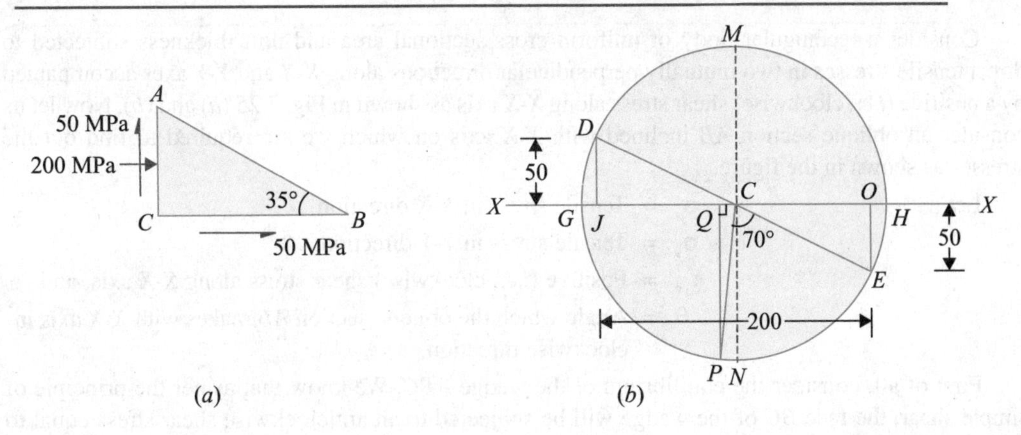


Fig. 7.24

The given stresses on the plane $AC$ of the element and a complimentary shear stress on the plane $BC$ are shown in Fig. 7.24 (*a*). Now draw the Mohr's Circle of Stresses as shown in Fig.7.24 (*b*) and as discussed below :

1. First of all, take some suitable point $O$, and through it draw a horizontal line $XOX$.
2. Cut off $OJ$ equal to the compressive stress on the plane $AC$ (*i.e.*, 200 MPa) to some suitable scale towards left .
3. Now erect a perpendicular at $J$ above the line $X$-$X$ and cut off $JD$ equal to the positive shear stress on the plane $AC$ (*i.e.*, 50 MPa) to the scale. The point $D$ represents the stress system on the plane $AC$. Similarly, erect a perpendicular at $O$ below the line $X$-$X$ and cut off $OE$ equal to the negative shear stress on the plane $BC$ (*i.e.*, 50 MPa) to the scale. The point $E$ represents the stress system on the plane $BC$. Join $DE$ and bisect it at $C$.
4. Now with $C$ as centre and radius equal to $CD$ or $CE$ draw the Mohr's Circle of Stresses. Meeting the line $X$-$X$ at $G$ and $H$.
5. Through $C$, draw two lines $CM$ and $CN$ at right angles to the line $X$-$X$ meeting the circle at $M$ and $N$. Also through $C$ draw a line $CP$ making an angle of $2 \times 35° = 70°$ with $CE$ in clockwise direction meeting the circle at $P$. The point $P$ represents the stress system on the plane $AB$.
6. Through $P$, draw $PQ$ perpendicular to the line $OX$.
   By measurement, we find that the Normal stress $(\sigma_n) = OQ = -112.8$ MPa ; Shear stress $(\tau) = QP = -111.1$ MPa and maximum shear stress in the element $(t_{max}) = \pm CM = CN = 112.1$ MPa **Ans.**

## 7.17. Mohr's Circle for Stresses on an Oblique Section of a Body Subjected to Direct Stresses in Two Mutually Perpendicular Directions Accompanied by a Simple Shear Stress

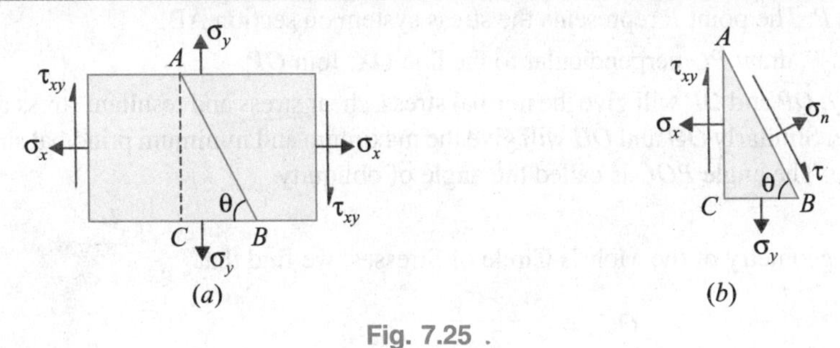


Fig. 7.25 .

Consider a rectangular body of uniform cross-sectional area and unit thickness subjected to direct tensile stresses in two mutually perpendicular directions along *X-X* and *Y-Y* axes accompanied by a positive (*i.e.*, clockwise) shear stress along *X-X* axis as shown in Fig. 7.25 (*a*) and (*b*). Now let us consider an oblique section *AB* inclined with *X-X* axis on which we are required to find out the stresses as shown in the figure.

Let $\sigma_x$ = Tensile stress in *X-X* direction,

$\sigma_y$ = Tensile stress in *Y-Y* direction,

$\tau_{xy}$ = Positive (*i.e.*, clockwise) shear stress along *X-X* axis, and

$\theta$ = Angle which the oblique section *AB* makes with *X-X* axis in clockwise direction.

First of all, consider the equilibrium of the wedge *ABC*. We know that as per the principle of simple shear, the face *BC* of the wedge will be subjected to an anticlockwise shear stress equal to $\tau_{xy}$ as shown in Fig. 7.25 (*b*). Now draw the Mohr's Circle of Stresses as shown in Fig. 7.26 and as discussed below :

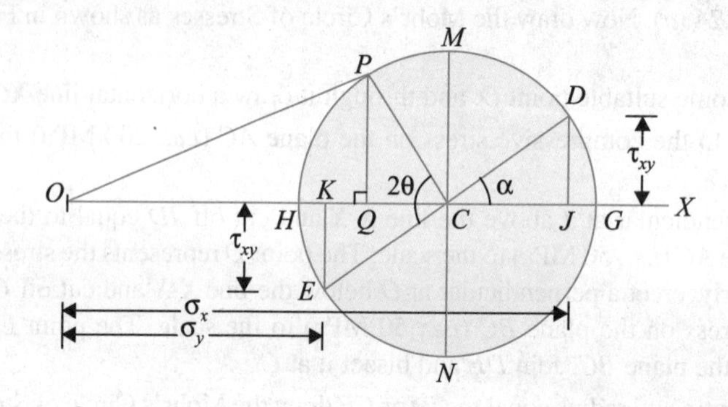


Fig. 7.26

1. First of all, take some suitable point *O* and through it draw a horizontal line *OX*.
2. Cut off *OJ* and *OK* equal to the tensile stresses $\sigma_x$ and $\sigma_y$ respectively to some suitable scale and towards right (because both the stresses are tensile).
3. Now erect a perpendicular at *J* above the line *X-X* (because $\tau_{xy}$ is positive along *X-X* axis) and cut off *JD* equal to the shear stress $\tau_{xy}$ to the scale. The point *D* represents the stress system on plane *AC*. Similarly, erect perpendicular below the line *X-X* (because $\tau_{xy}$ is negative along *Y-Y* axis) and cut off *KE* equal to the shear stress $\tau_{xy}$ to the scale. The point *E* represents the plane *BC*. Join *DE* and bisect it at *C*.
4. Now with *C* as centre and radius equal to *CD* or *CE* draw a circle. It is known as Mohr's Circle of Stresses.
5. Now through *C*, draw a line *CP* making an angle 2θ with *CE* in clockwise direction meeting the circle at *P*. The point *P* represents the stress system on section AB.
6. Through *P*, draw *PQ* perpendicular to the line *OX*. Join *OP*.
7. Now *OQ*, *QP* and *OP* will give the normal stress, shear stress and resultant stress respectively to the scale. Similarly *OG* and *OH* will give the maximum and minimum principal shear stresses to the scale. The angle *POC* is called the angle of obliquity.

**Proof**

From the geometry of the Mohr's Circle of Stresses, we find that

$$OC = \frac{\sigma_x + \sigma_y}{2}$$

and radius of the circle

$$R = EC = CD = CP = \sqrt{\left(\frac{\sigma_x - \sigma_y}{2}\right)^2 + \tau_{xy}^2}$$

Now in the right angled triangle $DCJ$

$$\sin\alpha = \frac{JD}{DC} = \frac{\tau_{xy}}{R} \quad \text{and} \quad \cos\alpha = \frac{JD}{DC} = \frac{\sigma_x - \sigma_y}{2} \times \frac{1}{R} = \frac{\sigma_x - \sigma_y}{2R}$$

Similarly in right angled triangle $CPQ$

$$\therefore \quad \angle PCQ = (2\theta - \alpha)$$

$$CQ = CP \cos 2\theta - \alpha$$

$$= R\,[\cos(2\theta - \alpha)]$$

$$= R\,[\cos\alpha\cos 2\theta + \sin\alpha\sin 2\theta]$$

$$= R\cos\alpha\cos 2\theta + R\sin\alpha\sin 2\theta$$

$$= R \times \frac{\sigma_x - \sigma_y}{2R}\cos 2\theta + R \times \frac{\tau_{xy}}{R}\sin 2\theta$$

$$= \frac{\sigma_x - \sigma_y}{2}\cos 2\theta + \tau_{xy}\sin 2\theta$$

**Normal Stress (across the section *AB*)**

$$\sigma_n = OQ = OC - CQ$$

or

$$\sigma_n = \frac{\sigma_x + \sigma_y}{2} - \frac{\sigma_x - \sigma_y}{2}\cos 2\theta - \tau_{xy}\sin 2\theta \quad \text{...(Same as in Art. 7.11)}$$

**Shear Stress or Tangential Stress (across the section *AB*)**

$$\tau = QP = CP\sin[(2\theta - \alpha)] = R\sin(2\theta - \alpha)$$

$$= R(\cos\alpha\sin 2\theta - \sin\alpha\cos 2\theta)$$

$$= R\cos\alpha\sin 2\theta - R\sin\alpha\cos 2\theta$$

$$= R \times \frac{\sigma_x - \sigma_y}{2R}\sin 2\theta - R \times \frac{\tau_{xy}}{R}\cos 2\theta$$

or

$$\tau = \frac{\sigma_x - \sigma_y}{2}\sin 2\theta - \tau_{xy}\cos 2\theta \quad \text{...(Same as in Art. 7.11)}$$

**Maximum Principal Stress**

$$\sigma_{max} = OG = OC + CG = \frac{\sigma_x + \sigma_y}{2} + \sqrt{\left(\frac{\sigma_x - \sigma_y}{2}\right)^2 + \tau_{xy}^2}$$

**Minimum Principal Stress**

$$\sigma_{min} = OH = OC - CH = \frac{\sigma_x + \sigma_y}{2} - \sqrt{\left(\frac{\sigma_x - \sigma_y}{2}\right)^2 + \tau_{xy}^2}$$

We also find the maximum shear stress will be equal to the radius of the Mohr's circle of Stresses. *i.e.*, $\sqrt{\left(\frac{\sigma_x - \sigma_y}{2}\right)^2 + \tau_{xy}^2}$. It will happen when $(2\theta - \alpha)$ is equal to 90° or 270°.

However when $(2\theta - \alpha) = 90°$ then the shear stress is equal to $+\sqrt{\left(\frac{\sigma_x - \sigma_y}{2}\right)^2 + \tau_{xy}^2}$.

And when $(2\theta - \alpha) = 270°$ then the shear stress is equal to $-\sqrt{\left(\frac{\sigma_x - \sigma_y}{2}\right)^2 + \tau_{xy}^2}$.

**EXAMPLE 7.20.** *A point is subjected to a tensile stress of 250 MPa in the horizontal direction and another tensile stress of 100 MPa in the vertical direction. The point is also subjected to a simple shear stress of 25 MPa, such that when it is associated with the major tensile stress, it tends to rotate the element in the clockwise direction. What is the magnitude of the normal and shear stresses inclined on a section at an angle of 20° with the major tensile stress ?*

***SOLUTION.** Given : Tensile stress in horizontal direction ($\sigma_x$) = 250 MPa ; Tensile stress in vertical direction ($\sigma_y$) = 100 MPa ; Shear stress ($\tau$) = 25 MPa and angle made by section with major tensile stress ($\theta$) = 20°.

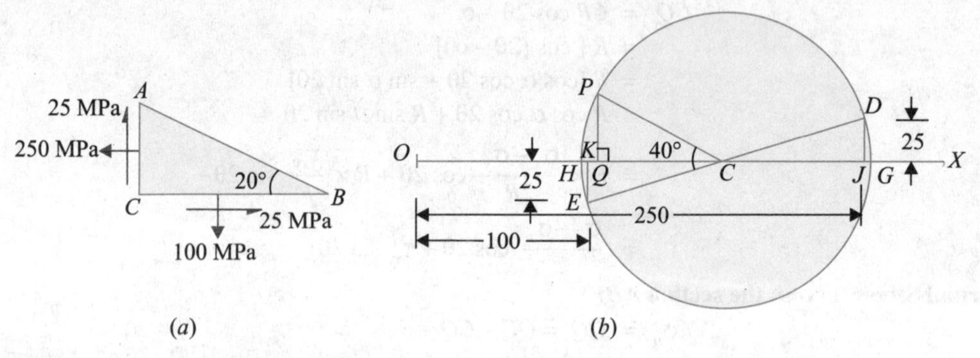


Fig. 7.27

The given stresses on the face *AC* of the point alongwith a tensile stress on the plane *BC* and a complimentary shear stress on the plane *BC* are shown in Fig 7.27 (*a*). Now draw the Mohr's Circle of Stresses as shown in Fig. 7.27 (*b*) and as discussed below :

1. First of all, take some suitable point *O*, and through it draw a horizontal line *OX*.
2. Cut off *OJ* and *OK* equal to the tensile stresses $\sigma_x$ and $\sigma_y$ respectively (*i.e.*, 250 MPa and 100 MPa) to some suitable scale towards right.
3. Now erect a perpendicular at *J* above the line *OX* and cut off *JD* equal to the positive shear stress on the plane *AC* (*i.e.*, 25 MPa) to the scale. The point *D* represents the stress system on the plane *AC*. Similarly, erect a perpendicular at *K* below the *OX* and cut off *KE* equal to the negative shear stress on the plane *BC* (*i.e.*, 25 MPa) to the scale. The point *E* represents the stress system on the plane *BC*. Join *DE* and bisect it at *C*.
4. Now with *C* as centre and radius equal to *CD* or *CE* draw the Mohr's Circle of Stresses.
5. Now through *C* draw a line *CP* making an angle of 2 × 20° = 440° with *CE* in clockwise direction meeting the circle at *P*. The point *P* represents the stress system on the section to *AB*.
6. Through *P*, draw *PQ* perpendicular to the line *OX*.

   By measurement, we find that the normal stress, ($\sigma_x$) = *OQ* = 101.5 MPa and shear stress $\tau$ = *QP* = 29.0 MPa **Ans.**

**EXAMPLE 7.21.** *A plane element in a boiler is subjected to tensile stresses of 400 MPa on one plane and 150 MPa on the other at right angle to the former. Each of the above stresses is accompanied by a shear stress of 100 MPa such that when associated with the major tensile stress tends to rotate the element in an anticlockwise direction. Find (a) Principal stresses and their directions. (b) Maximum shearing stresses and directions of the plane on which they act.*

* We have already solved this question analytically, as example 7.10.

***SOLUTION.** Given : Tensile stress along horizontal *x-x* axis ($\sigma_x$) = 400 MPa ; Tensile stress along vertical *y-y* axis ($\sigma_y$) = 150 MPa and Shear stress ($\tau_{xy}$) = – 100 MPa (Minus sign due to anticlockwise on *x-x* axis).

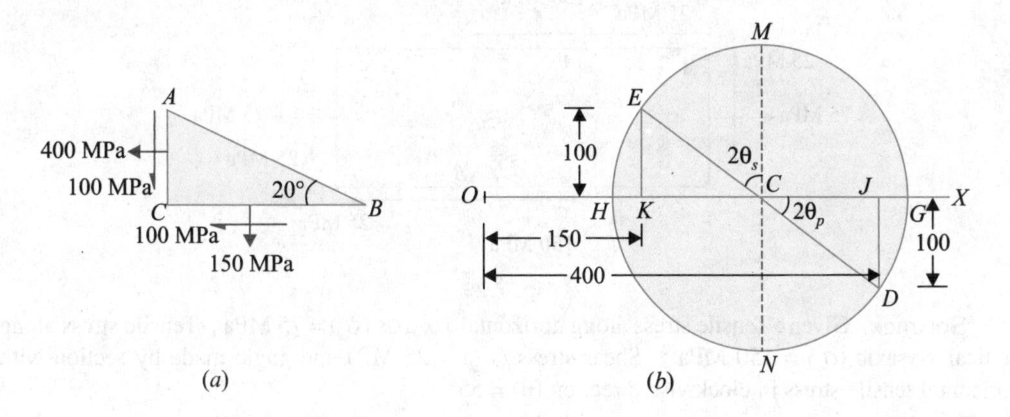


Fig. 7.28

The given stresses on the plane *AC* and *BC* of the element along with a complimentary shear stress on the plane *BC* are shown in Fig. 7.28 (*a*). Now Draw the Mohr's Circle of Stresses as shown in Fig 7.28 (*b*) and as discussed below :

1. First of all, take some suitable point *O*, and draw a horizontal line *OX*.
2. Cut off *OJ* and *OK* equal to the tensile stresses $\sigma_x$ and $\sigma_y$ respetitively (*i.e*,. 400 MPa and 150 MPa) to some suitable scale towards right.
3. Now erect a perpendicular at *J* below the line *OX* and cut off *JD* equal to the negative shear stress on the plane *AC* (*i.e.*, 100 MPa) to the scale. The point *D* represents the stress systems on the plane *AC*. Similarly, erect a perpendicular at *K* above the line *OX* and cut off *KE* equal to the positive shear stress on the plane *BC* (*i.e.*, 100 MPa) to the scale. The point *E* represents the stress system on the plane *BC*. Join *DE* and bisect it at *C*.
4. Now with *C* as centre and radius equal to *CD* or *CE* draw the Mohr's Circle of Stresses meeting the line *OX* at *G* and *H*.
5. Through *C* draw two lines *CM* and *CN* at right angles to the line *OX* meeting the circle at *M* and *N*.

By measurement, we find that maximum principal stress ($\sigma_{max}$) = *OG* = 435.0 MPa ; Minimum principal stress ($\sigma_{min}$) = *OH* = 115.0 MPa ; By measurement ∠*JCD* therefore angle which the plane of principal stress makes with *x-x* axis ($\theta_p$) = $\frac{\angle JCD}{2} = \frac{38.66°}{2}$ = 19.33° ; Maximum shearing stress ($\tau_{max}$) = *CM* = 160.0 MPa ; By measurement ∠ *MCE* = $2\theta_s$ = 51.34°, therefore angle which the plane of maximum shearing stress makes with *x-x* axis ($\theta_s$) = $\frac{51.34°}{2}$ = 25.7° **Ans.**

* We have already solved this question analytically, as example 7.11.

**EXAMPLE 7.22.** *A point in a strained material is subjected to the stresses as shown in Fig. 7.29. Find graphically, or otherwise, the normal and shear stresses on the section AB.*

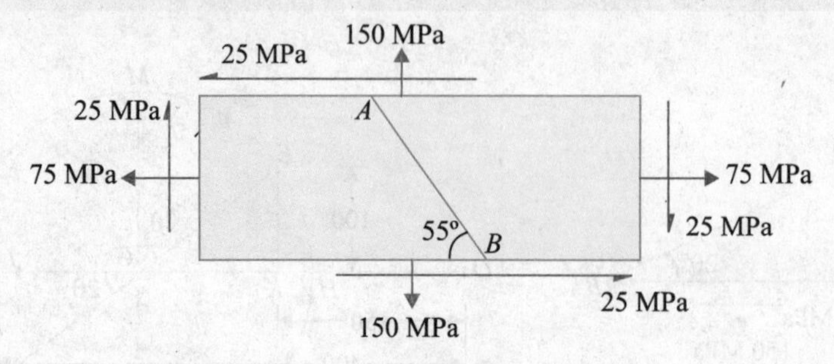


Fig. 7.29

***SOLUTION.** Given : Tensile stress along horizontal *x-x* axis ($\sigma_x$) = 75 MPa ; Tensile stress along vertical *y-y* axis ($\sigma_y$) = 150 MPa ; Shear stress ($\tau_{xy}$) = 25 MPa and angle made by section with horizontal tensile stress in clockwise direction ($\theta$) = 55°.

The given stresses on the planes *AC* and *BC* are shown in Fig.7.30 (*a*). Now draw the Mohr's Circle of Stresses as shown in Fig. 7.30 (*b*) and as discussed below :

1. First of all, take some suitable point *O*, and draw a horizontal line *OX*.
2. Cut off *OJ* and *OK* equal to the tensile stresses $\sigma_x$ and $\sigma_y$ respectively (*i.e.*,75 MPa and 150 MPa) to some suitable scale towards right.

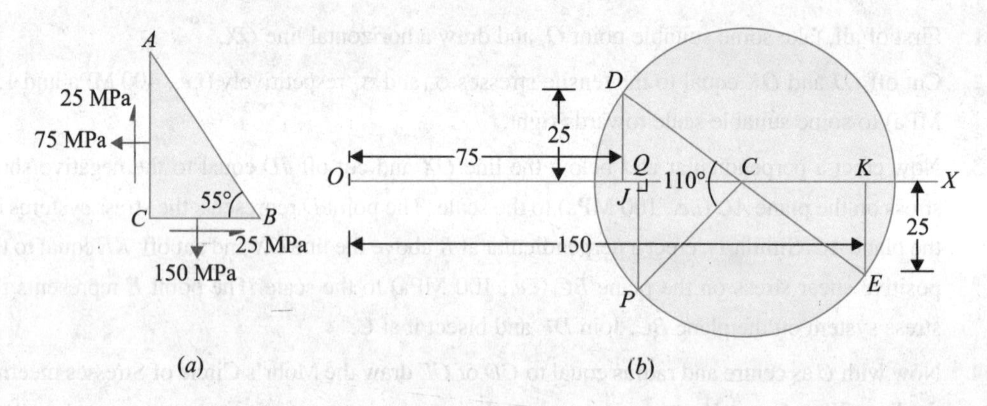


Fig. 7.30

3. Now erect a perpendicular at J above the line *OX* and cut off *JD* equal to the positive shear stress on the plane *AC* (*i.e.*, 25 MPa) to the scale. The point *D* represents the stress system on the plane *AC*. Similarly, erect a perpendicular at *K* below the line *OX* and cut off *KE* equal to the negative shear stress on the plane *BC* (*i.e.*, 25 MPa) to the scale. The point *E* represents the stress system on the plane *BC*. Join *DE* and bisect it at *C*.
4. Now with *C* as centre and radius equal to *CD* or *CE* draw the Mohr's Circle of Stresses.
5. Now through *C* draw a line *CP* making an angle of 2 × 55° = 110° with *CD* in an anticlockwise direction meeting the circle at *P*. The point *P* represents the stress system on the section *AB*.

   By measurement, we find that the normal stress ($\sigma_n$) = *OQ* = 76.1 MPa and shear stress ($\tau$) = *PQ* = – 26.7 MPa. **Ans.**

* We have already solved this question analytically, as example 7.12.

**EXAMPLE 7.23.** *A plane element of a body is subjected to a compressive stress of 300 MPa in x-x direction and a tensile stress of 200 MPa in the y-y direction. Each of the above stresses is subjected to a shear stress of 100 MPa such that when it is associated with the compressive stress, it tends to rotate the element in an anticlockwise direction.*

*Find graphically, or otherwise, the normal and shear stresses on a plane inclined at an angle of 30° with the x-x axis.*

***SOLUTION.** Given : Compressive stress in *x-x* direction ($\sigma_x$) = – 300 MPa (Minus sign due to compressive). Tensile stress in *y-y* direction ($\sigma_y$) = 200 MPa ; Shear stress ($\tau_{xy}$) = 100 MPa (Minus sign due to anticlockwise direction along the compressive stress *i.e.*, $\sigma_x$) and angle of plane with *x-x* axis ($\theta$) = 30°.

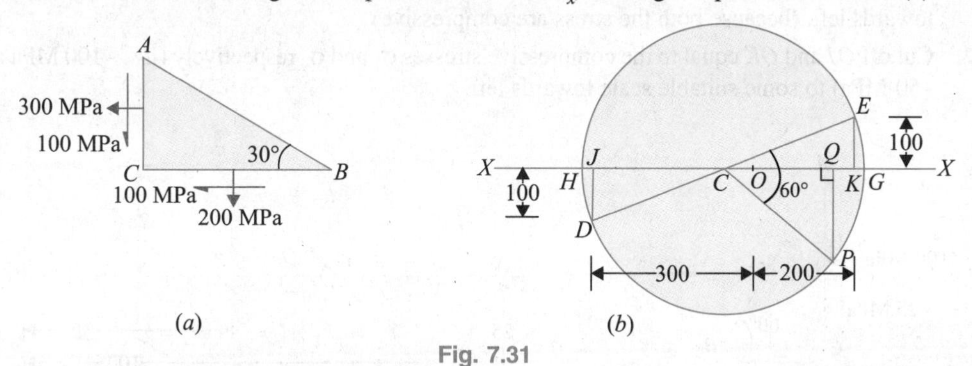


Fig. 7.31

The given stresses on the plane *AC* of the element alongwith a tensile stress on the plane *BC* and a complimentary shear stress on the plane *BC* are shown in Fig. 7.31 (*a*). Now draw the Mohr's Circle of Stresses as shown in Fig. 7.31 (*b*) and as discussed below :

1. First of all, take some suitable point *O*, and through it draw horizontal line *XOX*.
2. Cut off *OJ* and *OK* equal to the stresses $\sigma_x$ and $\sigma_y$ respectively (*i.e.*, – 300 MPa and 200 MPa) to some suitable scale such that *J* is towards left (because of compressive) and *K* is towards right (because of tensile).
3. Now erect a perpendicular at *J* below the line *XOX* and cut off *JD* equal to the negative shear stress on the plane *AC* (*i.e.*, 100 MPa) to the scale. The point *D* represents the stress system on the plane *AC*. Similarly, erect a perpendicular at *K* above the line *XOX* and cut off *KE* equal to the positive shear stress on the plane *BC* (*i.e.*, 100 MPa) to the scale. The point *E* represents the stress system on the plane *BC*. Join *DE* and bisect it at *C*.
4. Now with *C* as centre and radius equal to *CD* or *CE* draw the Mohr's Circle of Stresses.
5. Now through *C* draw a line *CP* making an angle of 2 × 30° = 60° with *CE* in clockwise direction meeting the circle at *P*. The point *P* represents the stress system on plane *AB*.
6. Through, *P*, draw *PQ* perpendicular to the line *OX*.

   By measurement, we find that the normal stress ($\sigma_n$) = *OQ* = 161.6 MPa ; and shear stress ($\tau$) = *QP* = – 166.5 MPa **Ans.**

**EXAMPLE 7.24.** *A machine component is subjected to the stresses as shown in Fig. 7.32.*

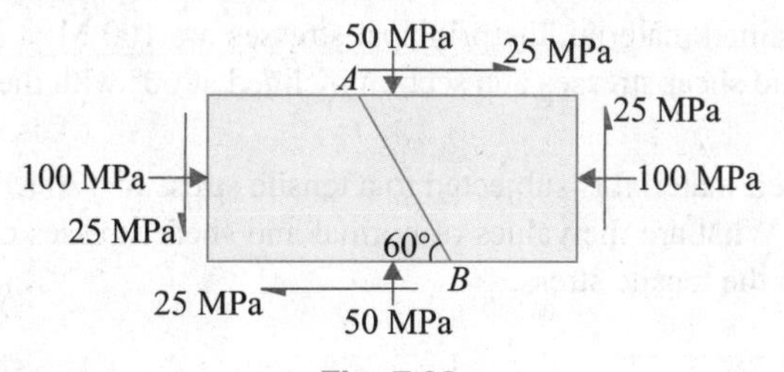


Fig. 7.32

* We have already solved this question analitically, as example 7.13.

*Find the normal and shearing stresses on the section AB inclined at an angle of 60° with x-x axis. Also find the resultant stress on the section.*

*SOLUTION. Given : Compressive stress along horizontal x-x axis ($\sigma_x$) = – 100 MPa (Minus sign due to compressive) ; Compressive stress along vertical y-y axis ($\sigma_y$) = – 50 MPa (Minus sign due to compressive) ; Shear stress ($\tau_{xy}$) = – 25 MPa (Minus sign due to anticlockwise on x-x axis and angle between section and horizontal x-x axis ($\theta$) = 60°.

The given stresses on the planes *AC* and *BC* are shown in Fig. 7.33 (*a*). Now draw the Mohr's Circle of Stresses as shown in Fig. 7.33 (*b*) and as discussed below :

1. First of all, take some suitable point *O* and through it draw a horizontal line, such that *X* is towards left. (because both the stress are compressive)
2. Cut off *OJ* and *OK* equal to the compressive stresses $\sigma_x$ and $\sigma_y$ respectively (*i.e.*, –100 MPa and –50 MPa) to some suitable scale towards left.

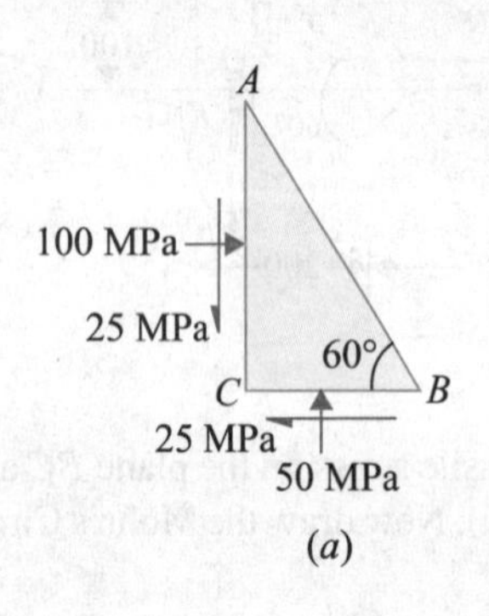

(*a*)

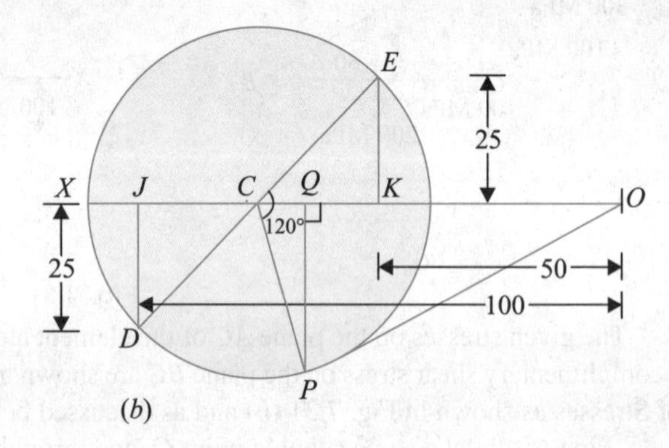

(*b*)

Fig. 7.33

3. Now erect a perpendicular at *J* below the line *XO* and cut off *JD* equal to the negative shear stress on the plane *AC* (*i.e.*, 25 MPa) to the scale. The point *D* represents the stress system on the plane *AC*. Similarly, erect a perpendicular at *K* above the line *XO* and cut off *KE* equal to the positive shear stress on the plane *BC* (*i.e.*, 25 MPa) to the scale. The point *E* represents the stress system on the plane *BC*. Join *DE* and bisect it at *C*.
4. Now with *C* as centre and radius equal to *CD* or *CE* draw the Mohr's Circle of Stresses.
5. Now through *C*, draw a line *CP* making an angle of 2 × 60° = 120° with *CE* in clockwise direction meeting the circle at *P*. The point *P* represents the stress system on the section *AB*.
6. Through *P*, draw *PQ* perpendicular to the line *XO*. Join *OP*.

   By measurement, we find that the normal stress ($\sigma_n$) = *OQ* = – 65.8 MPa ; Shear stress ($\tau$) = QP = – 34.1 MPa and resultant stress ($\sigma_R$) = *OP* = 74 MPa **Ans.**

## EXERCISE 7.2

1. At a point in a strained material, the principal stresses are 100 MPa and 50 MPa both tensile. Find the normal and shear stresses at a section inclined at 60° with the axis of the major principal stress. (**Ans.** 87.5 MPa ; 21.65 MPa)
2. A point in a strained material is subjected to a tensile stress of 120 MPa and a clockwise shear stress of 40 MPa. What are the values of normal and shear stresses on a plane inclined at 25° with the normal to the tensile stress. (**Ans.** 20 MPa ; 60 MPa)

* We have already solved this question analytically, as example 7.14.

3. The principal stresses at a point in the section of a member are 50 MPa and 20 MPa both tensile. If there is a clockwise shear stress of 30 MPa, find graphically the normal and shear stresses on a section inclined at an angle of 15° with the normal to the major tensile stress.
(**Ans.** 32.99 MPa ; 33.48 MPa)
4. A point is subjected to tensile stresses of 200 MPa and 150 MPa on two mutually perpendicular planes and an anticlockwise shear stress of 30 MPa. Determine by any method the values of normal and shear stresses on a plane inclined at 60° with the minor tensile stress.
(**Ans.** 188.48 MPa ; 36.65 MPa)
5. At a point in a stressed element, the normal stresses in two mutually perpendicular directions are 45 MPa and 25 MPa both tensile. The complimentary shear stress is these directions is 15 MPa. By using Mohr's circle method, or otherwise, determine the maximum and minimum principal stresses. (**Ans.** 188.48 MPa ; 36.65 MPa)

## QUESTIONS

1. Define principal planes and principal stresses and explain their uses.
2. Derive an expression for the stresses on an oblique section of a rectangular body, when it is subjected to (*a*) a direct stress in one plane only and (*b*) direct stresses in two mutually perpendicular directions.
3. Obtain an expression for the major and minor principal stresses on a plane, when the body is subjected to direct stresses in two mutually perpendicular directions accompanied by a shear stress.
4. How will you find out graphically the resultant stress on an oblique section when the body is subjected to direct stresses in two mutually perpendicular directions?

## OBJECTIVE TYPE QUESTIONS

1. When a body is subjected to a direct tensile stress (σ) in one plane, then normal stress on an oblique section of body inclined at an angle to the normal of the section is equal to
(*a*) $\sigma \sin \theta$ (*b*) $\sigma \cos \theta$ (*c*) $\sigma \sin^2 \theta$ (*d*) $\sigma \cos^2 \theta$
2. When a body is subjected to a direct tensile stress (σ) in one plane, then the tangential stress on an oblique section of the body inclined at an angle (θ) to normal of the section is equal to
(*a*) $p \sin 2\theta$ (*b*) $p \cos 2\theta$ (*c*) $\frac{P}{2} \sin 2\theta$ (*d*) $\frac{P}{2} \cos 2\theta$
3. When a body is subjected to a direct tensile stress (σ) in one plane and accompanied by a single shear stress (τ), the maximum normal stress is
(*a*) $\frac{\sigma}{2} + \frac{1}{2}\sqrt{\sigma^2 + 4\tau^2}$ (*b*) $\frac{\sigma}{2} - \frac{1}{2}\sqrt{\sigma^2 + 4\tau^2}$
(*c*) $\frac{\sigma}{2} + \sqrt{\sigma^2 - 4\tau^2}$ (*d*) $\frac{\sigma}{2} - \frac{1}{2}\sqrt{\sigma^2 - 4\tau^2}$
4. When a body is subjected to the mutually perpendicular stress ($\sigma_x$ and $\sigma_y$) then the centre of the Mohr's circle from y-axis is taken as
(*a*) $\frac{\sigma_x + \sigma_y}{2}$ (*b*) $\frac{\sigma_x - \sigma_y}{2}$ (*c*) $\frac{\sigma_x - \sigma_y}{2} + \tau_{xy}$ (*d*) $\frac{\sigma_x - \sigma_y}{2} - \tau_{xy}$

## ANSWERS

1. (*d*) 2. (*c*) 3. (*a*) 4. (*b*)

Chapter **8**

# Strain Energy and Impact Loading

### Contents

## 8.1. Introduction

We have studied in Chapter 2 that whenever some load is attached to a hanging wire, it extends and the load moves downwards by an amount equal to the extension of the wire. A little consideration will show that when the load moves downwards, it loses its *potential energy. This energy is absorbed ( or stored ) in the stretched wire, which may be released by removing the load. On removing the load, the wire will spring back to its original position. This energy, which is absorbed in a body, when strained within its elastic

* It is the energy possessed by a body by virtue of its position.

limit, is known as strain energy. It has been experimentally found that this strain energy is always capable of doing some work. The amount of strain energy, in a body is found out by the principal of work. Mathematically

Strain energy = Work done

## 8.2. Resilience

It is a common term used for the total strain energy stored in a body. Sometimes the resilience is also defined as the capacity of a strained body for doing work (when it springs back) on the removal of the straining force.

## 8.3. Proof Resilience

It is also a common term, used for the maximum strain energy, which can be stored in a body. (This happens when the body is stressed up to the elastic limit). The corresponding stress is known as proof stress.

## 8.4. Modulus of Resilience

The proof resilience per unit volume of a material, is known as modulus of resilience and is an important property of the material.

## 8.5. Types of Loading

In the previous chapter, we have solved the problems on the assumption that the load applied was gradual. But in actual practice, it is not always possible that the load may act gradually. As a matter of fact, a load may act in either of the following three ways:

**1.** gradually. **2.** suddenly. **3.** with impact.

Now in the succeeding pages, we shall discuss the work done, or in other words strain energy stored in a body, when loaded in any one of the above mentioned loadings.

## 8.6. Strain Energy Stored in a Body, when the Load is Gradually Applied

It is the most common and practical way of loading a body, in which the loading starts from zero and increases gradually till the body is fully loaded. *e.g.*, when we lower a body with the help of a crane, the body first touches the platform on which it is to be placed. On further releasing the chain, the platform goes on loading till it is fully loaded by the body. This is the case of a gradually applied load. Now consider a metallic bar subjected to a gradual load.

Let $P$ = Load gradually applied,

$A$ = Cross-sectional area of the bar,

$l$ = Length of the bar,

$E$ = Modulus of elasticity of the bar material and

$\delta$ = Deformation of the bar due to load.

Since the load applied is gradual, and varies from zero to $P$, therefore the average load is equal to $\frac{P}{2}$

$$\therefore \quad \text{Work done} = \text{Force} \times \text{Distance}$$

$$= \text{Average load} \times \text{Deformation}$$

$$= \frac{P}{2} \times \delta l = \frac{P}{2}(\varepsilon . l) \qquad ...(\because \delta l = \varepsilon . l)$$

$$= \frac{1}{2}\sigma . \varepsilon A . l \qquad ...(\because P = \sigma A)$$

$$= \frac{1}{2} \times (\text{Stress} \times \text{Strain} \times \text{Volume})$$

$$= \frac{1}{2} \times \sigma . \frac{\sigma}{E} . Al \qquad ...(\because \varepsilon = \frac{\sigma}{E})$$

$$= \frac{1}{2} \times \frac{\sigma^2}{E} \times Al = \frac{\sigma^2}{2E} \times V \qquad ...(\because Al = \text{Volume} = V)$$

Since the energy stored is also equal to the work done, therefore strain energy stored,

$$U = \frac{\sigma^2}{2E} \times V$$

We also know that modulus of resilience

$$= \text{Strain energy per unit volume}$$

$$= \frac{\sigma^2}{2E}$$

**EXAMPLE 8.1.** *Calculate the strain energy strored in a bar 2 m long, 50 mm wide and 40 mm thick when it is subjected to a tensile load of 60kN. Take E as 200 GPa.*

**SOLUTION.** Given : Length of bar ($l$) = 2 m = $2 \times 10^3$ mm ; Width of bar ($b$) = 50 mm ; Thickness of bar ($t$) = 40 mm ; Tensile load on bar ($P$) = 60 kN = $60 \times 10^3$ N and modulus of elasticity ($E$) = 200 GPa = $200 \times 10^3$ N/mm$^2$

We know that stress in the bar,

$$\sigma = \frac{P}{A} = \frac{60 \times 10^3}{50 \times 40} = 30 \text{ N/mm}^2$$

∴ Strain energy stored in the bar,

$$U = \frac{\sigma^2}{2E} \times V = \frac{(30)^2}{2 \times (200 \times 10^3)} \times 4 \times 10^6 \text{ N-mm}$$

$$= 9 \times 10^3 \text{ N-mm} = 9 \text{ kN-mm} \qquad \textbf{Ans.}$$

## 8.7. Strain Energy Stored in a Body when the Load is Suddenly Applied

Sometimes in factories and workshops, the load is suddenly applied on a body. *e.g.*, when we lower a body with the help of a crane, the body is, first of all, just above the platform on which it is to be placed. If the chain breaks at once at this moment the whole load of the body begins to act on the platform. This is the case of a suddenly applied load. Now consider a bar subjected to a sudden load.

$P$ = Load applied suddenly,
$A$ = Cross-sectional area of the bar,
$l$ = Length of the bar,
$E$ = Modulus of elasticity of the material,
$\delta$ = Deformation of the bar, and
$\sigma$ = Stress induced by the application of the sudden load

Since the load is applied suddenly, therefore the load ($P$) is constant throughout the process of deformation of the bar.

$$\therefore \text{ Work done} = \text{Force} \times \text{Distance} = \text{Load} \times \text{Deformation} \qquad ...(i)$$

$$= P \times \delta l$$

We know that strain energy stored,

$$U = \frac{\sigma^2}{2E} \times Al \qquad ...(ii)$$

Since the energy stored is equal to the work done, therefore

$$\frac{\sigma^2}{2E} \times Al = P \times \delta l = P \times \frac{\sigma}{E} l \qquad ...\left(\delta l = \frac{\sigma}{E}.l\right)$$

or

$$\sigma = 2 \times \frac{P}{A}$$

It means that the stress induced in this case is twice the stress induced when the same load is applied gradually. Once the stress ($\sigma$), is obtained, the corresponding instantaneous deformation ($\delta l$) and the strain energy may be found out as usual.

**EXAMPLE 8.2.** *An axial pull of 20 kN is suddenly applied on a steel rod 2.5 m long and 1000 mm² in cross-section. Calculate the strain energy, which can be absorbed in the rod. Take E = 200 GPa.*

**SOLUTION.** Given : Axial pull on the rod ($P$) = 20 kN = $20 \times 10^3$ N ; Length of rod ($l$) = 2.5 m = $2.5 \times 10^3$ mm ; Cross-sectional area of rod ($A$) = 1000 mm² and modulus of elasticity ($E$) = 200 GPa = $200 \times 10^3$ N/mm².

We know that stress in the rod, when the load is suddenly applied,

$$\sigma = 2 \times \frac{P}{A} = 2 \times \frac{20 \times 10^2}{1000} = 440 \text{ N/mm}^2$$

and volume of the rod,

$$V = l \,.\, A = (2.5 \times 10^3) \times 1000 = 2.5 \times 10^6 \text{ mm}^3$$

$\therefore$ Strain energy which can be absorbed in the rod,

$$U = \frac{\sigma^2}{2E} \times V = \frac{(40)^2}{2 \times (200 \times 10^3)} \times (2.5 \times 10^6) \text{ N-mm}$$

$$= 10 \times 10^3 \text{ N-mm} = 10 \text{ kN-mm} \qquad \textbf{Ans.}$$

**EXAMPLE 8.3** *A steel rod of 28 mm diameter is 2.5 m long. Find the maximum instantaneous stress and work done at maximum elongation, when an axial load of 50 kN is suddenly applied to it. Also calculate the maximum dynamic force in the rod. Take E = 200 GPa.*

**SOLUTION.** Given : Diameter of rod ($d$) = 28 mm ; Length of rod ($l$) = 2.5 m = $2.5 \times 10^3$ mm ; Axial load on rod ($P$) = 50 kN = $50 \times 10^3$ N and modulus of elasticity ($E$) = 200 GPa = $200 \times 10^3$ N/mm².

***Maximum Instantaneous stress***

We know that cross-sectional area of rod,

$$A = \frac{\pi}{4} \times (d)^2 = \frac{\pi}{4} \times (28)^2 = 615.8 \text{ mm}^2$$

and maximum instantaneous stress, when the load is suddenly applied,

$$\sigma_{max} = 2 \times \frac{P}{A} = 2 \times \frac{50 \times 10^3}{615.8} = 162.4 \text{ N/mm}^2 = 162.4 \text{ MPa} \qquad \textbf{Ans.}$$

***Work done at maximum elongation***

We know that maximum elongation,

$$\delta l = \frac{\sigma_{max} \times 1}{E} = \frac{162.4 \times (2.5 \times 10^3)}{200 \times 10^3} = 2.03 \text{ mm}$$

and work done $= P \times \delta l = (50 \times 10^3) \times 2.03 = 101.5 \times 10^3$ N-mm

$= 101.5$ kN-mm **Ans.**

***Maximum dynamic force***

We also know that maximum dynamic force,

$= A \times \sigma_{max} = 615.8 \times 162.4 = 100 \times 10^3$ N $= 100$ kN **Ans.**

## 8.8. Strain Energy Stored in a Body, when the Load is Applied with Impact

Sometimes in factories and workshops, the impact load is applied on a body *e.g.*, when we lower a body with the help of a crane, and the chain breaks while the load is being lowered the load falls through a distance, before it touches the platform. This is the case of a load applied with impact.

Now consider a bar subject to a load applied with impact as shown in Fig 8.1.

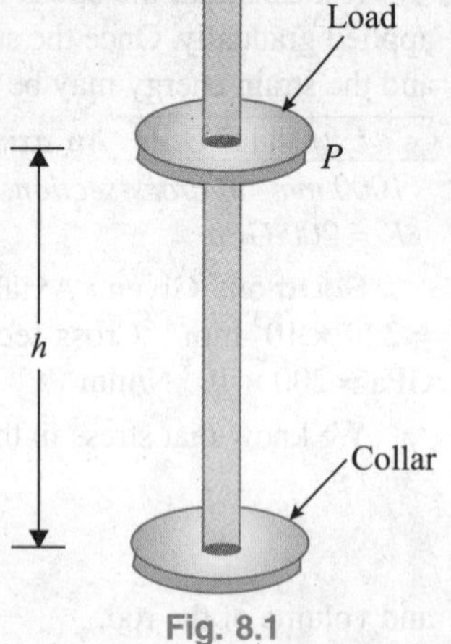

Fig. 8.1

Let $P$ = Load applied with impact,

$A$ = Cross-sectional area of the bar,

$E$ = Modulus of elasticity of the bar material,

$l$ = Length of the bar,

$\delta l$ = Deformation of the bar, as a result of this load,

$\sigma$ = Stress induced by the application of this load with impact, and

$h$ = Height through which the load will fall, before impacting on the collar of the bar.

$\therefore$ Work done = Load × Distance moved

$= P(h + \delta l)$

and energy stored, $U = \dfrac{\sigma^2}{2E} \times Al$

Since energy stored is equal to the work done, therefore

$$\frac{\sigma^2}{2E} \times Al = P(h + \delta l) = P\left(h + \frac{\sigma}{E}.l\right) \quad ...\left(\delta l = \frac{\sigma}{E}.l\right)$$

$$\frac{\sigma^2}{2E} \times Al = Ph + \frac{P\sigma l}{E}$$

$$\therefore \quad \sigma^2\left(\frac{Al}{2E}\right) - \sigma\left(\frac{Pl}{E}\right) - Ph = 0$$

Multiplying both sides by $\left(\dfrac{E}{Al}\right)$,

$$\frac{\sigma^2}{2} - \sigma\left(\frac{P}{A}\right) - \frac{PEh}{Al} = 0$$

This is a quadratic equation. We know that

$$\sigma = \frac{P}{A} \pm \sqrt{\left(\frac{P}{A}\right)^2 + \left(4 \times \frac{1}{2}\right)\left(\frac{PEh}{Al}\right)}$$

$$= \frac{P}{A}\left[1 \pm \sqrt{1+\frac{2AEh}{Pl}}\right]$$

Once the stress ($\sigma$)is obtained, the corresponding instantaneous deformation ($\delta l$) or the strain energy stored may be found out as usual.

**Cor.** When $\delta$ is very small as compared to $h$, then

$$\text{Work done} = Ph$$

$$\therefore \quad \frac{\sigma^2}{2E} Al = Ph$$

$$\text{or} \quad \sigma^2 = \frac{2EPh}{Al}$$

$$\therefore \quad \sigma = \sqrt{\frac{2EPh}{Al}}$$

**EXAMPLE 8.4.** *A 2 m long alloy bar of 1500 mm² cross-sectional area hangs vertically and has a collar securely fixed at its lower end. Find the stress induced in the bar, when a weight of 2 kN falls from a height of 100 mm on the collar. Take E = 120 GPa. Also find the strain energy stored in the bar.*

**SOLUTION.** Given : Length of bar ($l$) = 2 m = $2 \times 10^3$ mm ; Cross-sectional area of bar ($A$) = 1500 mm² ; Weight falling on collar of bar ($P$) = 2 kN = $2 \times 10^3$ N ; Height from which weight falls ($h$) = 100 mm and modulus of elasticity ($E$) = 120 GPa = $120 \times 10^3$ N/mm².

***Stress induced in the bar***

We know that in this case, extension of the bar will be small and negligible as compared to the height ($h$) from where the weight falls on the collar (due to small value of weight *i.e.*, 2 kN and a large value of $h$ *i.e.*, 100 mm). Therefore stress induced in the bar

$$\sigma = \sqrt{\frac{2EPh}{A.l}} = \sqrt{\frac{2\times(120\times10^3)\times(2\times10^3)\times100}{1500\times(2\times10^3)}} \text{ N/mm}^2$$

$$= 126.5 \text{ N/mm}^2 = 126.5 \text{ MPa} \quad \textbf{Ans.}$$

***Strain energy stored in the bar***

We also know that volume of the bar,

$$V = l.A = (2 \times 10^3) \times 1500 = 3 \times 10^6 \text{ mm}^3$$

and strain energy stored in the bar,

$$U = \frac{\sigma^2}{2E}\times V = \frac{(126.5)^2}{2\times(120\times10^2)} \times (3 \times 10^6) \text{ N-mm}$$

$$= 200 \times 10^3 \text{ N-mm} = 200 \text{ N-m} \quad \textbf{Ans.}$$

**EXAMPLE 8.5.** *A steel bar 3 m long and 2500 mm² in area hangs vertically, which is securely fixed on a collar at its lower end. If a weight of 15 kN falls on the collar from a height of 10 mm, determine the stress developed in the bar. What will be the strain energy stored in the bar? Take E as 200 GPa.*

**SOLUTION.** Given : Length of bar ($l$) = 3 m = $3 \times 10^3$ mm ; Area of bar ($A$) = 2500 mm² ; Weight falling on collar of bar ($P$) = 15 kN = $15 \times 10^3$ N ; Height from which weight falls ($h$) = 10 mm and modulus of elasticity ($E$) = 200 GPa = $20 \times 10^3$ N/mm².

***Stress developed in the bar***

We know that in this case, extension of the bar will be considerable as compared to the height ($h$) from where the weight falls on the collar (due to a large value of weight *i.e.*, 15 kN and a small value of $h = 10$ mm). Therefore stress developed in the bar,

$$\sigma = \frac{P}{A}\left[1+\sqrt{1+\frac{2AEh}{Pl}}\right]$$

$$= \frac{15\times10^3}{2500}\left[1+\sqrt{1+\frac{2\times2500\times(200\times10^3)\times10}{(15\times10^3)\times(3\times10^3)}}\right] \text{N/mm}^2$$

$$= 6\,(1 + 14.9) = 95.4 \text{ N/mm}^2 = 95.4 \text{ MPa} \qquad \textbf{Ans.}$$

***Strain energy stored in the bar***

We know that volume of the bar,

$$V = l\,.\,A = (3\times10^3)\times2500 = 7.5\times10^6 \text{ mm}^3$$

and strain energy stored in the bar,

$$U = \frac{\sigma^2}{2E}\times V = \frac{(95,4)^2}{2\times(200\times10^3)}\times7.5\times10^6 \text{ N-mm}$$

$$= 170.6\times10^3 \text{ N-mm} = 170.6 \text{ N-m} \qquad \textbf{Ans.}$$

**EXAMPLE 8.6.** *A copper bar of 12 mm diameter gets stretched by 1 mm under a steady load of 4 kN. What stress would be produced in the bar by a weight 500 N, the weight falls through 80 mm before striking the collar rigidly fixed to the lower end of the bar ? Take Young's modulus for the bar material as 100 GPa.*

**SOLUTION.** Given : Diameter of bar $(d) = 12$ mm ; Change in length of bar $(\delta l) = 1$ mm ; Load on bar $(P_1) = 4$ kN $= 4\times10^3$ N ; Weight falling on collar $(P_2) = 500$ N ; Height from which weight falls $(h) = 80$ mm and modulus of elasticity $(E) = 100$ GPa $= 100\times10^3$ N/mm$^2$

Let $\qquad l =$ Length of the copper bar.

We know that cross-sectional area of the bar,

$$A = \frac{\pi}{4}\times(d)^2 = \frac{\pi}{4}\times(12)^2 = 113.1 \text{ mm}^2$$

and stretching of the bar $(\delta l)$,

$$l = \frac{P.l}{A.E} = \frac{(4\times10^3)}{113.1\times(100\times10^3)} = \frac{l}{2.83\times10^3}$$

$$\therefore \qquad l = 1\times(2.83\times10^3) = 2.83\times10^3 \text{ mm}$$

We also know that stress produced in the bar by the falling weight.

$$\sigma = \frac{P_2}{A}\left(1+\sqrt{1+\frac{2AEh}{P_2 l}}\right)$$

$$= \frac{1500}{113.1}\left(1+\sqrt{1+\frac{2\times113.1\times(100\times10^3)\times80}{500\times(2.83\times10^3)}}\right) \text{N/mm}^2$$

$$= 4.2\,(1 + 35.77) = 162.52 \text{ N/mm}^2 = 162.52 \text{ MPa} \qquad \textbf{Ans.}$$

**EXAMPLE 8.7.** *An unknown weight falls through 10 mm on a collar rigidly attached to the lower end of a vertical bar 4 m long and 600 mm$^2$ in section. If the maximum instantaneous extension is known to be 2 mm, what is the corresponding stress and the value of unknown weight. Take E = 200 GPa.*

**SOLUTION.** Given : Height from which weight falls $(h) = 10$ mm ; Length $(l) = 4$ m $= 4\times10^3$ mm; Cross-sectional area of bar $(A) = 600$ mm$^2$ ; Instananeous extension $(\delta l) = 2$ mm and modulus of elasticity $(E) = 200$ GPa $= 200\times10^3$ N/mm$^2$.

***Stress in the bar***

Let $\sigma$ = Stress in the bar in N/mm$^2$.

We know that instantaneous extension of the bar ($\delta l$),

$$2 = \frac{\sigma . l}{E} = \frac{\sigma \times (4 \times 10^3)}{200 \times 10^3} = \frac{\sigma}{50}$$

$$\sigma = 2 \times 50 = 100 \text{ N/mm}^2 = 100 \text{ MPa} \qquad \textbf{Ans.}$$

***Value of unknown weight***

Let $P$ = Value of the unknown weight in N.

We also know that the stress ($\sigma$),

$$100 = \frac{P}{A}\left(1 + \sqrt{1 + \frac{2AEh}{Pl}}\right)$$

$$= \frac{P}{600}\left(1 + \sqrt{1 + \frac{2 \times 600 \times (200 \times 10^3) \times 10}{P \times (4 \times 10^3)}}\right)$$

$$\frac{100 \times 600}{P} = 1 + \sqrt{1 + \frac{600 \times 10^3}{P}}$$

$$\frac{60 \times 10^3}{P} - 1 = \sqrt{1 + \frac{600 \times 10^3}{P}}$$

Squaring both sides of the equation,

$$\frac{3600 \times 10^6}{P^2} + 1 - \frac{120 \times 10^3}{P} = 1 + \frac{600 \times 10^3}{P}$$

$$\frac{3600 \times 10^3}{P} = 600 + 120 = 720$$

$$P = \frac{(3600 \times 10^3)}{720} = 5 \times 10^3 \text{ N} = 5 \text{ kN} \qquad \textbf{Ans.}$$

## EXERCISE 8.1

1. Calculate the strain energy that can be stored in a steel bar 2.4 m long and 1000 mm$^2$ cross-sectional area, when subjected to a tensile stress of 50 MPa. Take $E$ = 200 GPa. [**Ans.** 15 kN-mm]
2. A mild steel rod 1 m long and 20 mm diameter is subjected to an axial pull of 62.5 kN. What is the elongation of the rod, when the load is applied (*i*) gradually. and (*ii*) suddenly. Take $E$ as 200 GPa [**Ans.** 1mm ; 2mm]
3. Find the maximum stress and strain energy stored in a 2 m long and 25 mm diameter bar, when an axial pull of 15 kN is suddenly applied on it. Take $E$ as 100 GPa. [**Ans.** 61.1 MPa ; 18.3 kN-mm]
4. A steel bar 3 m long is 500 mm$^2$ in cross-sectional area. What is the instantaneous stress produced in the bar, due to an axial pull, when its extension was observed to be 1.5 mm? Also find magnitude of the axial pull. Take modulus of elasticity as 200 GPa. [**Ans.** 100 MPa ; 25 kN]
5. An alloy bar 1.5 m long and of 1206 mm$^2$ cross-sectional area has a collar securely fixed at its lower end. Find the stress induced in the bar, when a load of 500 N falls from a height of 100 mm on the collar. Take $E$ = 150 GPa. [**Ans.** 91.3 MPa]

   **Hint :** Extension of the bar will be negligible as compared to the height (100 mm).

6. A load of 10 kN falls freely through a height of 12.5 mm on to a collar attached to the end of a vertical rod 50 mm diameter and 3 m long, the upper end being fixed to the ceiling. What is the maximum stress induced in the bar? Take $E$ for the rod material as 120 GPa. [**Ans.** 76.6 MPa]

## 8.9. Strain Energy Stored in a Body of Varying Section

Sometimes, we come across bodies of varying section. The strain energy in such a body is obtained by adding the strain energies stored in different parts of the body. Mathematically total strain energy stored in a body.

$$U = U_1 + U_2 + U_3 + \ldots\ldots$$

Where $U_1$ = Strain energy stored in part 1,

$U_2$ = Strain energy stored in part 2,

$U_3$ = Strain energy stored in part 3.

**Note.** The above relation is also used for finding strain energy stored in a composite body.

**Example 8.8.** *A non-uniform tension bar 5 m long is made up of two parts as shown in Fig 8.2.*

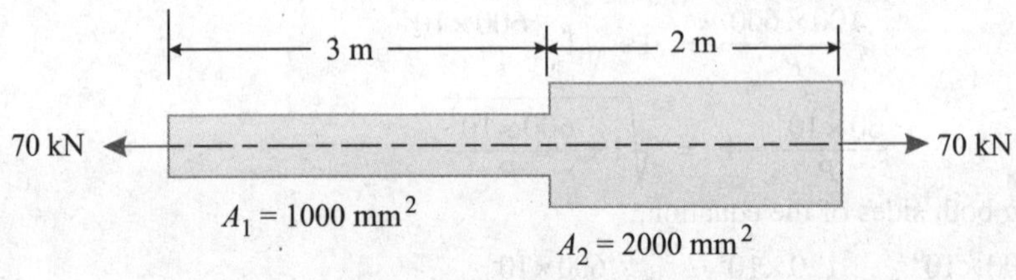


Fig. 8.2

*Find the total strain energy stored in the bar, when it is subjected to a gradual load of 70 kN. Also find the total strain energy stored in the bar, when the bar is made of uniform cross-section of the same volume under the same load. Take E = 200 GPa.*

**Solution.** Given : Total length of bar ($L$) = 5 m = $5 \times 10^3$ mm ; Length of part 1 ($L_1$) = 3 m = $3 \times 10^3$ mm ; Length of part 2 ($L_2$) = 2 m = $2 \times 10^3$ mm ; Area of part 1 ($A_1$) = 1000 mm$^2$ ; Area of part 2 ($A_2$) = 2000 mm$^2$ ; Pull ($P$) = 70 kN = $70 \times 10^3$ N and modulus of elasticity ($E$) = 200 GPa = $200 \times 10^3$ N/mm$^2$

***Total strain energy stored in the non-uniform bar***

We know that stress in the first part,

$$\sigma_1 = \frac{P}{A_1} = \frac{70 \times 10^3}{1000} = 70 \text{ N/mm}^2$$

and volume of the first part,

$$V_1 = (3 \times 10^3) \times 1000 = 3 \times 10^6 \text{ mm}^3$$

∴ Strain energy stored in the first part,

$$U_1 = \frac{\sigma_1^2}{2E} \times V_1 = \frac{(70)^2}{2 \times (200 \times 10^3)} \times (3 \times 10^6) = 36.75 \times 10^3 \text{ N-mm} \quad \ldots(i)$$

Similarly, stress in the second part,

$$\sigma_2 = \frac{P}{A_2} = \frac{70 \times 10^3}{2000} = 35 \text{ N/mm}^2$$

and volume of the second part,

$$V_2 = (2 \times 10^3) \times 2000 = 4 \times 10^6 \text{ mm}^3$$

∴ Strain energy stored in the second part,

$$U_2 = \frac{\sigma_2^2}{2E} \times V_2 = \frac{(35)^2}{2 \times (200 \times 10^3)} \times (4 \times 10^6) = 12.25 \times 10^3 \text{ N-mm} \quad ...(ii)$$

and total strain energy stored in the non-uniform bar,

$$U = U_1 + U_2 = (36.75 \times 10^3) + (12.25 \times 10^3) = 49 \times 10^3 \text{ N=mm} = 49 \text{ N-m} \quad \textbf{Ans.}$$

***Total strain energy in the uniform bar***

We know that total volume of the bar,

$$V = V_1 + V_2 = (3 \times 10^6) + (4 \times 10^6) = 7 \times 10^6 \text{ mm}^3$$

and cross-sectional area of the circular bar,

$$A = \frac{\text{Volume of the bar}}{\text{Length of the bar}} = \frac{7 \times 10^6}{5 \times 10^3} = 1400 \text{ mm}^2$$

∴ Stress in the bar

$$\sigma = \frac{70 \times 10^3}{1400} = 50 \text{ N/mm}^2$$

and strain energy storad in the uniform bar,

$$U = \frac{\sigma^2}{2E} \times V = \frac{(50)^2}{2 \times (200 \times 10^3)} \times (7 \times 10^6) = 43.75 \times 10^3 \text{ N-mm}$$

$$= 43.75 \text{ N-m} \quad \textbf{Ans.}$$

**EXAMPLE 8.9.** *Two similar round bars A and B are each 300 mm long as shown in Fig 8.3. The bar A receives an axial blow, which produces a maximum stress of 100 MPa. Find the maximum stress produced by the same blow on the bar B. If the bar B is also stressed to 100 MPa, determine the ratio of energies stored by the bars B and A.*

**SOLUTION.** Given : Total Length of the bars = 300 mm ; Maximum stress in bar *A* in 20 mm diameter portion ($\sigma_A$) 100 MPa = 100 N/mm$^2$.

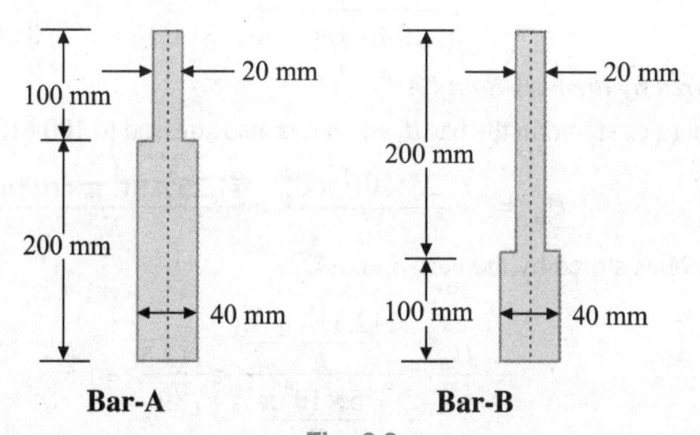


Fig. 8.3

***Maximum stress in the bar B***

Let $\sigma_B$ = Maximum stress produced in the bar *B* (in 20 mm diameter portion)

$E$ = Young's modulus for both the bars.

We know that the area of 20 mm diameter portion,

$$A_1 = \frac{\pi}{4} \times (20)^2 = 100\,\pi\,\text{mm}^2$$

and area of 40 mm diameter portion,

$$A_2 = \frac{\pi}{4} \times (40)^2 = 400\,\pi\,\text{mm}^2$$

From the geometry of the figure, we find that stress in the 40 mm diameter of the bar *A*,

$$= \frac{100}{4} = 25\ \text{N/mm}^2$$

and stress in the 40 mm diameter of the bar *B*

$$= \frac{\sigma_B}{4}\ \text{N/mm}^2$$

We know that energy stored in the bar *A*,

$$U_A = \frac{\sigma^2}{2E} \times V = \frac{\sigma^2}{2E}\,(l\,.\,A)$$

$$= \frac{(100)^2}{2E}\left[100 \times (100\,\pi)\right] + \frac{(25)^2}{2E}\left[200 \times (400\,\pi)\right]$$

$$= \frac{75 \times 10^6\,\pi}{E} \qquad ...(i)$$

and energy stored in the bar *B*,

$$U_B = \frac{\sigma_B^2}{2E}\left[200 \times 100\,\pi\right] + \frac{(\sigma_B/4)^2}{2E}\left[100 \times (400\,\pi)\right]$$

$$= \frac{11.25 \times 10^3\,\pi\,\sigma_B^2}{E} \qquad ...(ii)$$

Since the blow on both the bars *A* and *B* is the same, therefore energies stored in both the bars is equal. Now equating equation (*i*) and (*ii*),

$$\frac{75 \times 10^6\,\pi}{E} = \frac{11.25 \times 10^3\,\pi\,\sigma_B^2}{E}$$

or $\quad 75 \times 10^3 = 11.25\,\sigma_B^2$

$$\therefore \quad \sigma_B = \sqrt{(75 \times 10^3)/11.25} = 81.6\ \text{N/mm}^2 = 81.6\ \text{MPa} \qquad \textbf{Ans.}$$

***Ratio of energies stored by the bars B and A***

We know that energies stored in the bar *B*, when it is also stressed to 100 MPa (*i.e.*, 100 N/mm²).

$$U_B = \frac{11.25 \times 10^3\,\pi\,\sigma_B^2}{E} = \frac{11.25 \times 10^3\,\pi \times (100)^2}{E} = \frac{112.5 \times 10^6\,\pi}{E}$$

∴ Ratio of energies stored by the bars *B* and *A*,

$$= \frac{U_B}{U_A} = \frac{\dfrac{112.5 \times 10^6\,\pi}{E}}{\dfrac{75 \times 10^6\,\pi}{E}} = \frac{112.5}{75} = 1.5 \qquad \textbf{Ans.}$$

**EXAMPLE 8.10.** *A vertical tie fixed rigidly at the top, consists of a steel rod 2.5 m long and 20 mm diameter encased throughout in a brass tube 20 mm internal diameter and 30 mm external diameter. The rod and casing are fixed together at both ends. The compound rod is suddenly loaded in tension by a weight of 10 kN falling through 3 mm before being arrested by the tie.*

*Calculate the maximum stress in steel and brass. Take* $E_S = 200$ *GPa and* $E_B = 100$ *GPa.*

**SOLUTION.** Given : Length of rod $(l) = 2.5$ m $= 2.5 \times 10^3$ mm ; Load $(P) = 10$ kN $= 10 \times 10^3$ N; Height through which load falls $(h) = 3$ mm ; Modulus of elasticity of steel $(E_S) = 200$ GPa $= 200 \times 10^3$ N/mm² and modulus of elasticity of brass $(E_B) = 100$ GPa $= 100$ N/mm².

Let $\sigma_S$ = Maximum stress in steel and

$\sigma_B$ = Maximum stress in brass

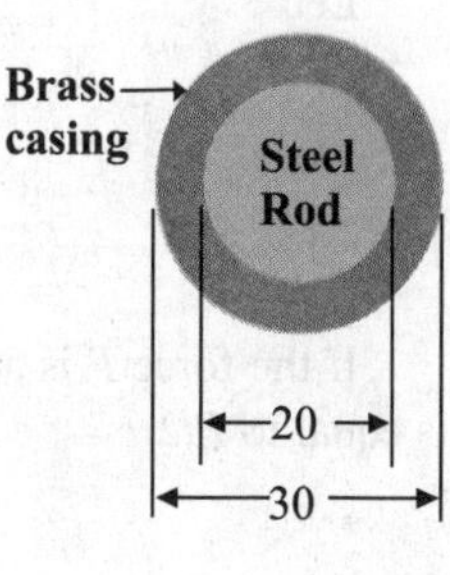


Fig. 8.4

We know that area of steel rod,

$$A_S = \frac{\pi}{4}(20)^2 = 100\,\pi \text{ mm}^2$$

and area of brass tube,

$$A_B = \frac{\pi}{4}[(30)^2 - (20)^2] = 125\,\pi \text{ mm}^2$$

We also know that stress in steel,

$$\sigma_S = \frac{E_S}{E_B} \times \sigma_B = \frac{200}{100} \times \sigma_B = 2\sigma_B$$

∴ Strain energy stored in the steel rod,

$$U_S = \frac{\sigma_S^2}{2E_S} \times V_S = \frac{(2\sigma_B)^2}{2 \times (200 \times 10^2)}\left[(100\,\pi) \times (2.5 \times 10^3)\right]$$

$$= 7.854\,\sigma_B^2$$

and strain energy stored in the brass tube,

$$U_B = \frac{\sigma_B^2}{2E_B} \times V_B = \frac{\sigma_B^2}{2 \times (100 \times 10^3)}\left[(125\,\pi) \times (2.5 \times 10^3)\right]$$

$$= 4.909\,\sigma_B^2$$

We know that work done by the falling weight (or loss of potential energy of the falling weight)

$$= P(h + \delta l) = 10 \times 10^3 \left(3 + \frac{\sigma_B \times (2.5 \times 10^3)}{100 \times 10^3}\right)$$

$$= (30 \times 10^3) + (250 \times \sigma_B)$$

We also know as per the principle of work that work done by the falling weight = Energy stored by steel + Energy stored by brass,

$$(30 \times 10^3) + (250 \times \sigma_B) = 7.854\,\sigma_B^2 + 4.909\,\sigma_B^2 = 12.763\,\sigma_B^2$$

$$12.763\,\sigma_B^2 - 250\,\sigma_B - 30 \times 10^3 = 0$$

or $$\sigma_B^2 - 19.6\,\sigma_B - 2.35 \times 10^3 = 0$$

This is a quadratic equation. Therefore

$$\sigma_B = \frac{19.6 \pm \sqrt{(-19.6)^2 - 4 \times (-2.35 \times 10^3)}}{2} = \frac{19.6 + 99}{2} \text{ N/mm}^2$$

$$= 59.3 \text{ N/mm}^2 = 59.3 \text{ MPa}$$ **Ans.**

$$\sigma_S = 2\sigma_B = 2 \times 59.3 = 118.6 \text{ MPa}$$ **Ans.**

## 8.10. Strain Energy Stored in a Body due to Shear Stress

Consider a cube *ABCD* of length $l$ fixed at the bottom face *AB* as shown in Fig 8.5.

Let $P$ = Force applied tangentially on the face *DC*,

$\tau$ = Shear stress

$\phi$ = Shear strain, and

N = Modulus of rigidity or shear modulus.

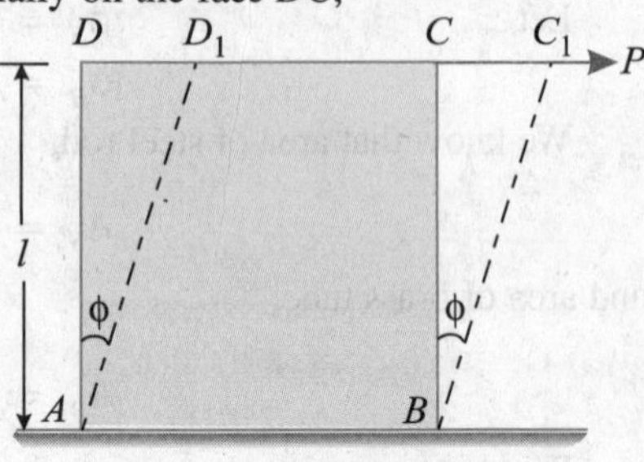


Fig. 8.5. Strain energy due to shear stress

If the force $P$ is applied gradually then the average force is equal to $P/2$.

$\therefore$ Work done = Average force × Distance

$$= \frac{P}{2} \times DD_1$$

$$= \frac{1}{2} \times P \times AD \times \phi \qquad ...(\because DD_1 = AD \times \phi)$$

$$= \frac{1}{2} \times \tau \times DC \times l \times AD \times \phi \qquad ...(\because P = \tau \times DC \times l)$$

$$= \frac{1}{2} \times \tau \times \phi \times DC \times AD \times l$$

$$= \frac{1}{2} \text{ (stress × strain × volume)}$$

$$= \frac{1}{2} \times \tau \times \frac{\tau}{N} \times V \qquad ...\left(\because \phi = \frac{\tau}{N}\right)$$

$$= \frac{\tau^2}{2N} \times V \qquad \text{...(where } V \text{ is the volume)}$$

Since energy stored is also equal to the work done, therefore energy stored,

$$U = \frac{\tau^2}{2N} \times V$$

We also know that modulus of resilience

= Strain energy per unit volume

$$= \frac{\tau^2}{2N}$$

**Example 8.11.** *A rectangular body 500 mm long, 100 mm wide and 50 mm thick is subjected to a shear stress of 80 MPa. Determine the strain energy stored in the body. Take N = 85 GPa.*

**Solution.** Given : Length of rectangular body ($l$) = 500 mm ; Width of rectangular body ($b$) = 100 mm ; Thickness of rectangular body ($t$) = 50 mm ; Shear stress ($\tau$) = 80 MPa = 80 N/mm$^2$ and modulus of rigidity ($N$) = 85 N/mm$^2$.

We know that volume of the bar,

$$V = l.b.t = 500 \times 100 \times 50 = 2.5 \times 10^6 \text{ mm}^3$$

and strain energy stored in the body,

$$U = \frac{\tau^2}{2N} \times V = \frac{(80)^2}{2 \times (85 \times 10^3)} \times 2.5 \times 10^6 \text{ N-mm}$$

$$= 94.1 \times 10^3 \text{ N-mm} = 94.1 \text{ N-m} \qquad \textbf{Ans.}$$

## EXERCISE 8.2

1. Find the ratio of strain energies stored in bars *A* and *B* of the same material and subjected to the same axial tensile loads. The bar *A* is of 50 mm diameter throughout its length, while the bar *B* though of the same length as of *A* but has diameter of 25 mm for the middle one-third of its length and the remainder is of 50 mm diameter. **[Ans.** 1 : 2]
2. A rectangular body 400 mm × 50 mm × 40 mm is subjected to a shear stress of 60 MPa. Calculate the strain energy stored in the body. Take $N$ = 80 GPa. **[Ans.** 18 N-m]

## QUESTIONS

1. Define strain energy and explain how it is stored in a body ?
2. Write short notes on :
   (*a*) Resilience,
   (*b*) Proof resilience,
   (*c*) Modulus of resilience.
3. From first principle, derive an equation for the energy stored in a strained body.
4. Show that in a bar, subjected to an axial load, the instantaneous stress due to sudden application of a load is twice the stress caused by the gradual application of load.
5. Obtain a relation for the stress induced in a body, if a load *P* is applied with an impact.
6. Derive an equation for the energy stored due to shear resilience.

## OBJECTIVE TYPE QUESTIONS

1. Strain energy is the
   (*a*) maximum energy which can be stored in a body
   (*b*) energy stored in a body when stressed in the elastic limit
   (*c*) energy stored in a body when stressed up to the breaking point
   (*d*) none of the above
2. The total strain energy stored in a body is known as
   (*a*) impact energy (*b*) resilience
   (*c*) proof resilience (*d*) modulus of resilience
3. The strain energy stored in a body, when the load is gradually applied, is equal to
   (*a*) $\frac{\sigma^2}{2E} \times V$ (*b*) $\frac{\sigma^2}{E} \times V$ (*c*) $\frac{\sigma^2}{2V} \times E$ (*d*) $\frac{\sigma^2}{V} \times E$
   where $\sigma$ = Stress in the body
   $E$ = Modulus of elasticity for the meterial and
   $V$ = Volume of the body.
4. The stress in a body if suddenly loaded is ...... the stress induced, when the same load is applied gradually.
   (*a*) One-half (*b*) euqal to (*c*) twice (*d*) four times.

## ANSWERS

1. (*b*) 2. (*b*) 3. (*a*) 4. (*c*)

Chapter **9**

# Centre of Gravity

### Contents

## 9.1. Introduction

It has been established, since long, that every particle of a body is attracted by the earth towards its centre. The force of attraction, which is proportional to the mass of the particle, acts vertically downwards and is known as weight of the body. As the *distance between the different particles of a body and the centre of the earth is the same, therefore these forces may be taken to act along parallel lines.

We have already discussed in Art. 4.6 that a point may be found out in a body, through which

* Strictly speaking, this distance is not the same. But it is taken to the same, because of the very small size of the body as compared to the earth.

the resultant of all such parallel forces act. This point, through which the whole weight of the body acts, irrespect of its position, is known as centre of gravity (briefly written as C.G.). It may be noted that every body has one and only one centre of gravity.

## 9.2. Centroid

The plane figures (like triangle, quadrilateral, circle etc.) have only areas, but no mass. The centre of area of such figures is known as *centroid*. The method of finding out the centroid of a figure is the same as that of finding out the centre of gravity of a body. In many books, the authors also write centre of gravity for centroid and *vice-versa*.

## 9.3. Methods for Centre of Gravity

The centre of gravity (or centroid) may be found out by any one of the following two methods:

1. By geometrical considerations
2. By moments
3. By graphical method

As a matter of fact, the graphical method is a tedious and cumbersome method for finding out the centre of gravity of simple figures. That is why, it has academic value only. But in this book, we shall discuss the procedure for finding out the centre of gravity of simple figures by geometrical considerations and by moments one by ones.

## 9.4. Centre of Gravity by Geometrical Considerations

The centre of gravity of simple figures may be found out from the geometry of the figure as given below.

**1.** The centre of gravity of uniform rod is at its middle point.

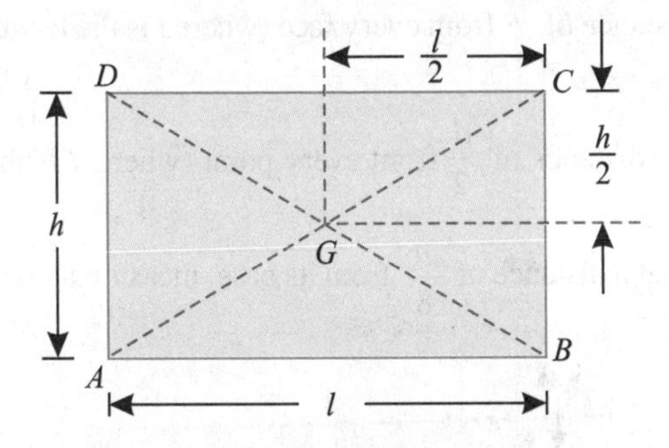


Fig. 9.1. Rectangle

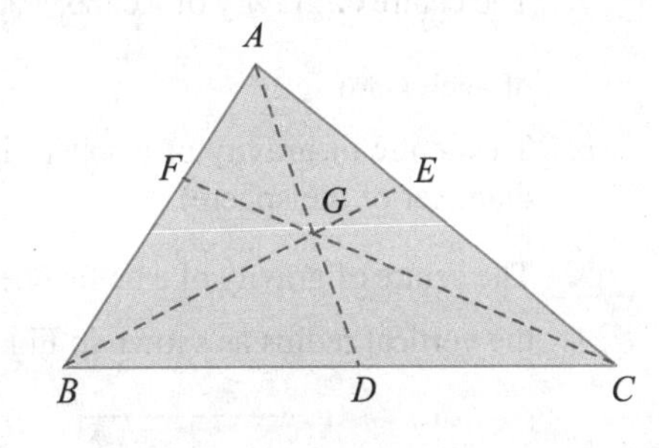


Fig. 9.2. Triangle

**2.** The centre of gravity of a rectangle (or a parallelogram) is at the point, where its diagonals meet each other. It is also a middle point of the length as well as the breadth of the rectangle as shown in Fig. 9.1.

**3.** The centre of gravity of a triangle is at the point, where the three medians (a median is a line connecting the vertex and middle point of the opposite side) of the triangle meet as shown in Fig. 9.2.

**4.** The centre of gravity of a trapezium with parallel sides $a$ and $b$ is at a distance of $\frac{h}{3} \times \left(\frac{b + 2a}{b + a}\right)$ measured form the side $b$ as shown in Fig. 9.3.

5. The centre of gravity of a semicircle is at a distance of $\frac{4r}{3\pi}$ from its base measured along the vertical radius as shown in Fig. 9.4.

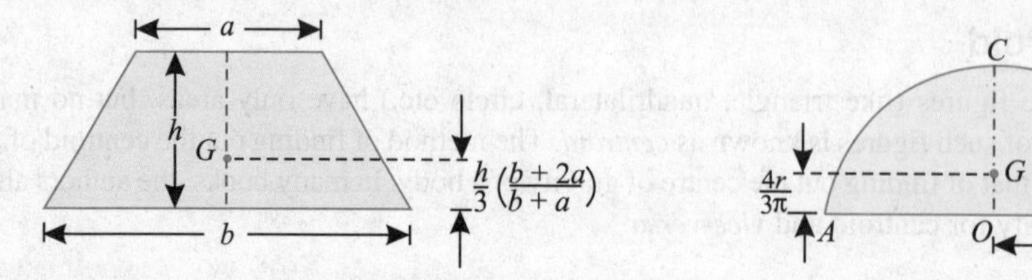


Fig. 9.3. Trapezium

Fig. 9.4. Semicircle

6. The centre of gravity of a circular sector making semi-vertical angle α is at a distance of $\frac{2r}{3}\frac{\sin\alpha}{\alpha}$ from the centre of the sector measured along the central axis as shown in Fig. 9.5.

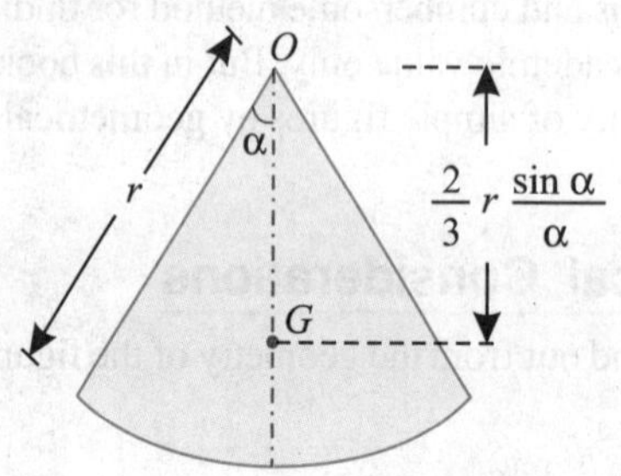


Fig. 9.5. Circular sector

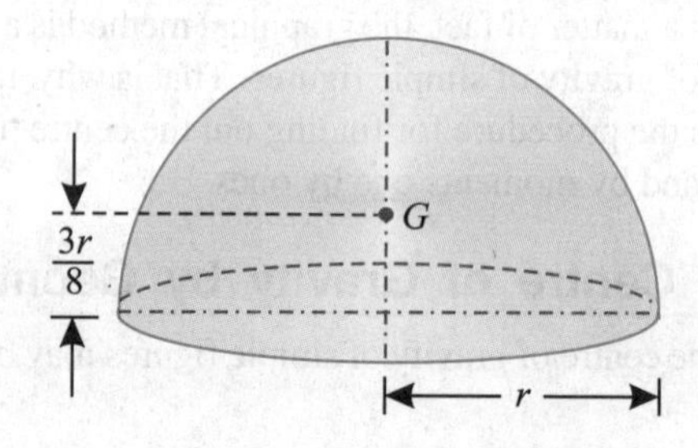


Fig. 9.6. Hemisphere

7. The centre of gravity of a cube is at a distance of $\frac{l}{2}$ from every face (where $l$ is the length of each side).

8. The centre of gravity of a sphere is at a distance of $\frac{d}{2}$ from every point (where $d$ is the diameter of the sphere).

9. The centre of gravity of a hemisphere is at a distance of $\frac{3r}{8}$ from its base, measured along the vertical radius as shown in Fig. 9.6.

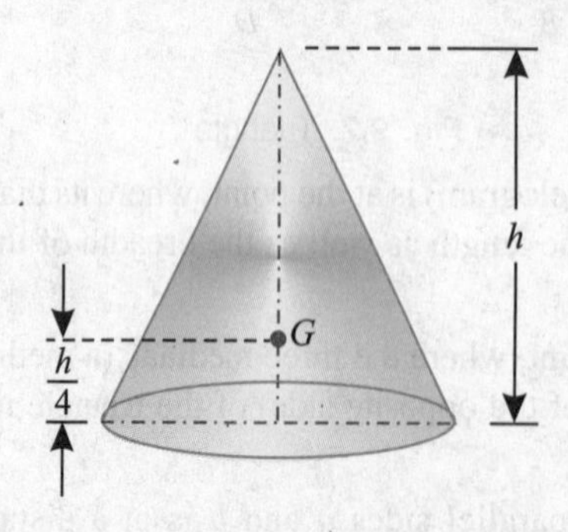


Fig. 9.7. Right circular solid cone

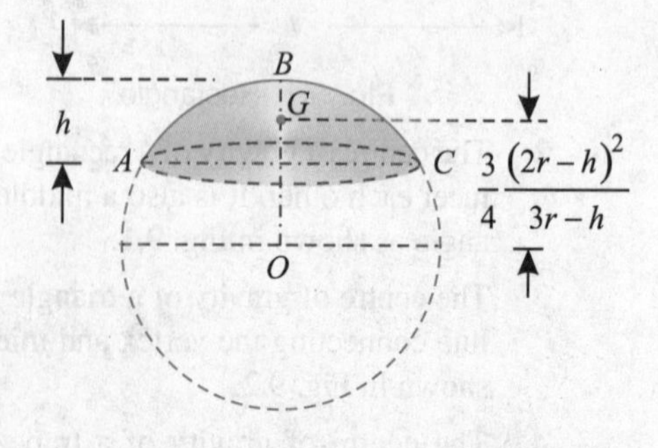


Fig. 9.8. Segment of a sphere

**10.** The centre of gravity of right circular solid cone is at a distance of $\frac{h}{4}$ from its base, measured along the vertical axis as shown in Fig. 9.7.

**11.** The centre of gravity of a segment of sphere of a height $h$ is at a distance of $\frac{3}{4}\frac{(2r-h)^2}{(3r-h)}$ from the centre of the sphere measured along the height. as shown in Fig. 9.8.

## 9.5. Centre of Gravity by Moments

The centre of gravity of a body may also be found out by moments as discussed below:

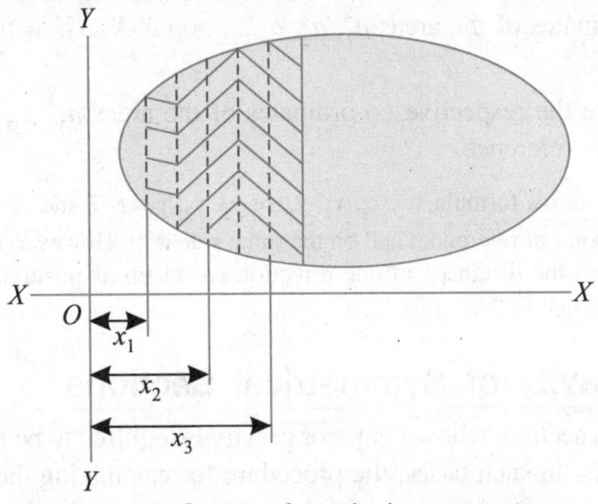


Fig. 9.9. Centre of gravity by moments

Consider a body of mass $M$ whose centre of gravity is required to be found out. Divide the body into small masses, whose centres of gravity are known as shown in Fig. 9.9. Let $m_1$, $m_2$, $m_3$....; etc. be the masses of the particles and $(x_1, y_1)$, $(x_2, y_2)$, $(x_3, y_3)$, ...... be the co-ordinates of the centres of gravity from a fixed point $O$ as shown in Fig. 9.9.

Let $\bar{x}$ and $\bar{y}$ be the co-ordinates of the centre of gravity of the body. From the principle of moments, we know that

$$M\,\bar{x} = m_1 x_1 + m_2 x_2 + m_3 x_3 \,.....$$

or
$$\bar{x} = \frac{\Sigma mx}{M}$$

Similarly
$$\bar{y} = \frac{\Sigma my}{M},$$

where
$$M = m_1 + m_2 + m_3 + .....$$

## 9.6. Axis of Reference

The centre of gravity of a body is always calculated with reference to some assumed axis known as axis of reference (or sometimes with reference to some point of reference). The axis of reference, of plane figures, is generally taken as the lowest line of the figure for calculating $\bar{y}$ and the left line of the figure for calculating $\bar{x}$.

## 9.7. Centre of Gravity of Plane Figures

The plane geometrical figures (such as *T*-section, *I*-section, *L*-section etc.) have only areas but no mass. The centre of gravity of such figures is found out in the same way as that of solid bodies. The centre of area of such figures is known as centroid, and coincides with the centre of gravity of the figure. It is a common practice to use centre of gravity for centroid and vice versa.

Let $\bar{x}$ and $\bar{y}$ be the co-ordinates of the centre of gravity with respect to some axis of reference, then

$$\bar{x} = \frac{a_1 x_1 + a_2 x_2 + a_3 x_3 + \ldots\ldots}{a_1 + a_2 + a_3}$$

and
$$\bar{y} = \frac{a_1 y_1 + a_2 y_2 + a_3 y_3 + \ldots\ldots}{a_1 + a_2 + a_3 + \ldots}$$

where $a_1$, $a_2$, $a_3$........ etc., are the areas into which the whole figure is divided $x_1$, $x_2$, $x_3$ ..... etc., are the respective co-ordinates of the areas $a_1$, $a_2$, $a_3$....... on *X-X* axis with respect to same axis of reference.

$y_1$, $y_2$, $y_3$....... etc., are the respective co-ordinates of the areas $a_1$, $a_2$, $a_3$....... on *Y-Y* axis with respect to same axis of the reference.

**NOTE.** While using the above formula, $x_1$, $x_2$, $x_3$ ..... or $y_1$, $y_2$, $y_3$ or $\bar{x}$ and $\bar{y}$ must be measured from the same axis of reference (or point of reference) and on the same side of it. However, if the figure is on both sides of the axis of reference, then the distances in one direction are taken as positive and those in the opposite directions must be taken as negative.

## 9.8. Centre of Gravity of Symmetrical Sections

Sometimes, the given section, whose centre of gravity is required to be found out, is symmetrical about *X-X* axis or *Y-Y* axis. In such cases, the procedure for calculating the centre of gravity of the body is very much simplified; as we have only to calculate either $\bar{x}$ or $\bar{y}$. This is due to the reason that the centre of gravity of the body will lie on the axis of symmetry.

**EXAMPLE 9.1.** *Find the centre of gravity of a 100 mm × 150 mm × 30 mm T-section.*

**SOLUTION.** As the section is symmetrical about *Y-Y* axis, bisecting the web, therefore its centre of gravity will lie on this axis. Split up the section into two rectangles *ABCH* and *DEFG* as shown in Fig. 9.10.

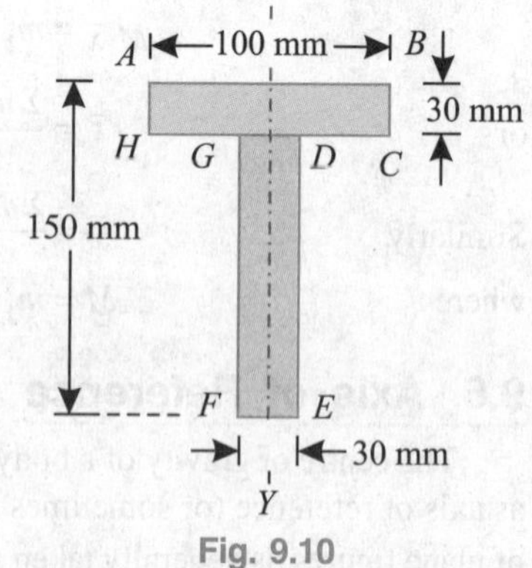

Fig. 9.10

Let bottom of the web *FE* be the axis of reference.

(*i*) *Rectangle ABCH*

$$a_1 = 100 \times 30 = 3000 \text{ mm}^2$$

and
$$y_1 = \left(150 - \frac{30}{2}\right) = 135 \text{ mm}$$

(*ii*) *Rectangle DEFG*

$$a_2 = 120 \times 30 = 3600 \text{ mm}^2$$

and
$$y_2 = \frac{120}{2} = 60 \text{ mm}$$

We know that distance between centre of gravity of the section and bottom of the flange *FE*,

$$\bar{y} = \frac{a_1 y_1 + a_2 y_2}{a_1 + a_2} = \frac{(3000 \times 135) + (3600 \times 60)}{3000 + 3600} \text{ mm}$$

$$= 94.1 \text{ mm} \quad \textbf{Ans.}$$

**EXAMPLE 9.2.** *Find the centre of gravity of a channel section 100 mm × 50 mm × 15 mm.*

**SOLUTION.** As the section is symmetrical about *X-X* axis, therefore its centre of gravity will lie on this axis. Now split up the whole section into three rectangles *ABFJ*, *EGKJ* and *CDHK* as shown in Fig. 9.11.

Let the face *AC* be the axis of reference.

(*i*) *Rectangle ABFJ*

$$a_1 = 50 \times 15 = 750 \text{ mm}^2$$

and $$x_1 = \frac{50}{2} = 25 \text{ mm}$$

(*ii*) *Rectangle EGKJ*

$$a_2 = (100 - 30) \times 15 = 1050 \text{ mm}^2$$

and $$x_2 = \frac{15}{2} = 7.5 \text{ mm}$$

(*iii*) *Rectangle CDHK*

$$a_3 = 50 \times 15 = 750 \text{ mm}^2$$

and $$x_3 = \frac{50}{2} = 25 \text{ mm}$$

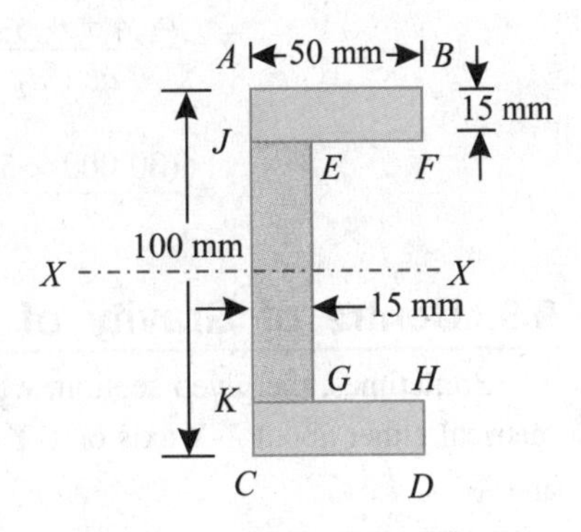

Fig. 9.11

We know that distance between the centre of gravity of the section and left face of the section *AC*,

$$\bar{x} = \frac{a_1 x_1 + a_2 x_2 + a_3 x_3}{a_1 + a_2 + a_3}$$

$$= \frac{(750 \times 25) + (1050 \times 7.5) + (750 \times 25)}{750 + 1050 + 750} = 17.8 \text{ mm} \quad \textbf{Ans.}$$

**EXAMPLE 9.3.** *An I-section has the following dimensions in mm units :*

*Bottom flange = 300 × 100*

*Top flange = 150 × 50*

*Web = 300 × 50*

*Determine mathematically the position of centre of gravity of the section.*

**SOLUTION.** As the section is symmetrical about *Y-Y* axis, bisecting the web, therefore its centre of gravity will lie on this axis. Now split up the section into three rectangles as shown in Fig. 9.12.

Let bottom of the bottom flange be the axis of reference.

(*i*) *Bottom flange*

$$a_1 = 300 \times 100 = 30\,000 \text{ mm}^2$$

and $$y_1 = \frac{100}{2} = 50 \text{ mm}$$

(*ii*) *Web*

$$a_2 = 300 \times 50 = 15\,000 \text{ mm}^2$$

and $$y_2 = 100 + \frac{300}{2} = 250 \text{ mm}$$

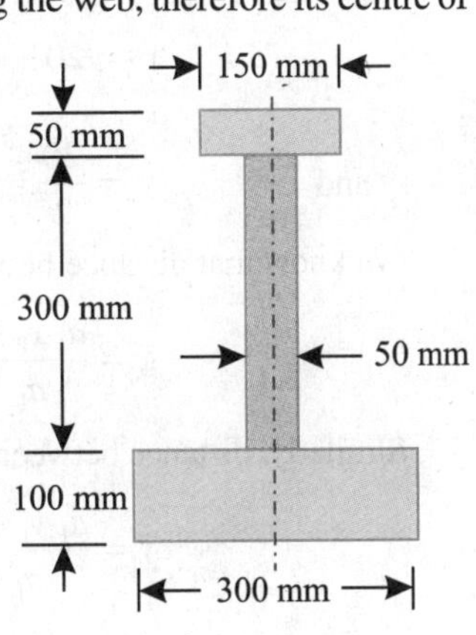

Fig. 9.12

(*iii*) *Top flange*

$$a_3 = 150 \times 50 = 7500 \text{ mm}^2$$

and $$y_3 = 100 + 300 + \frac{50}{2} = 425 \text{ mm}$$

We know that distance between centre of gravity of the section and bottom of the flange,

$$\bar{y} = \frac{a_1 y_1 + a_2 y_2 + a_3 y_3}{a_1 + a_2 + a_3}$$

$$= \frac{(30\,000 \times 50) + (15\,000 \times 250) + (7500 \times 425)}{30\,000 + 15\,000 + 7500} = 160.7 \text{ mm} \quad \textbf{Ans.}$$

## 9.9. Centre of Gravity of Unsymmetrical Sections

Sometimes, the given section, whose centre of gravity is required to be found out, is not symmetrical either about *X-X* axis or *Y-Y* axis. In such cases, we have to find out both the values of $\bar{x}$ and $\bar{y}$

**EXAMPLE 9.4.** *Find the centroid of an unequal angle section 100 mm × 80 mm × 20 mm.*

**SOLUTION.** As the section is not symmetrical about any axis, therefore we have to find out the values of $\bar{x}$ and $\bar{y}$ for the angle section. Split up the section into two rectangles as shown in Fig. 9.13.

Let left face of the vertical section and bottom face of the horizontal section be axes of reference.

(*i*) *Rectangle* 1

$$a_1 = 100 \times 20 = 2000 \text{ mm}^2$$

$$x_1 = \frac{20}{2} = 10 \text{ mm}$$

and $$y_1 = \frac{100}{2} = 50 \text{ mm}$$

(*ii*) *Rectangle* 2

$$a_2 = (80 - 20) \times 20 = 1200 \text{ mm}^2$$

$$x_2 = 20 + \frac{60}{2} = 50 \text{ mm}.$$

and $$y_2 = \frac{20}{2} = 10 \text{ mm}$$

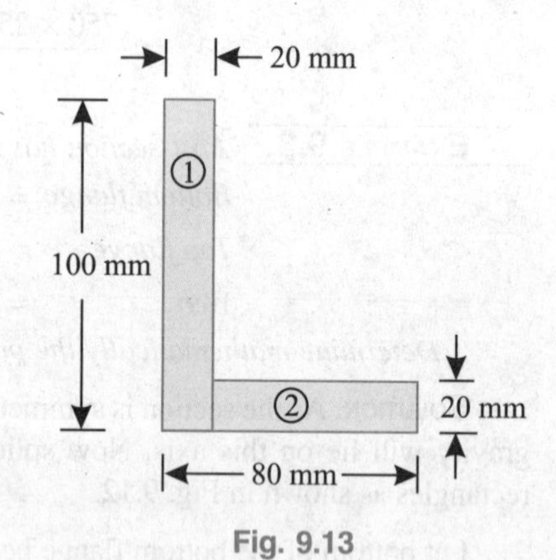


Fig. 9.13

We know that distance between centre of gravity of the section and left face,

$$\bar{x} = \frac{a_1 x_1 + a_2 x_2}{a_1 + a_2} = \frac{(2000 \times 10) + (1200 \times 50)}{2000 + 1200} = 25 \text{ mm} \quad \textbf{Ans.}$$

Similarly, distance between centre of gravity of the section and bottom face,

$$\bar{y} = \frac{a_1 y_1 + a_2 y_2}{a_1 + a_2} = \frac{(2000 \times 50) + (1200 \times 10)}{2000 + 1200} = 35 \text{ mm} \quad \textbf{Ans.}$$

**EXAMPLE 9.5.** *A uniform lamina shown in Fig. 9.14 consists of a rectangle, a circle and a triangle.*

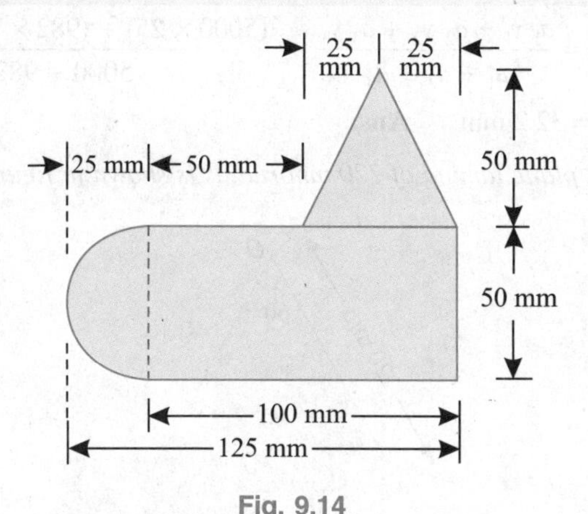


Fig. 9.14

*Determine the centre of gravity of the lamina. All dimensions are in mm.*

**SOLUTION.** As the section is not symmetrical about any axis, therefore we have to find out the values of both $\bar{x}$ and $\bar{y}$ for the lamina.

Let left edge of circular portion and bottom face rectangular portion be the axes of reference.

(*i*) *Rectangular portion*

$$a_1 = 100 \times 50 = 5000 \text{ mm}^2$$

$$x_1 = 25 + \frac{100}{2} = 75 \text{ mm}$$

and $$y_1 = \frac{50}{2} = 25 \text{ mm}$$

(*ii*) *Semicircular portion*

$$a_2 = \frac{\pi}{2} \times r^2 = \frac{\pi}{2}(25)^2 = 982 \text{ mm}^2$$

$$x_2 = 25 - \frac{4r}{3\pi} = 25 - \frac{4 \times 25}{3\pi} = 14.4 \text{ mm}$$

and $$y_2 = \frac{50}{2} = 25 \text{ mm}$$

(*iii*) *Triangular portion*

$$a_3 = \frac{50 \times 50}{2} = 1250 \text{ mm}^2$$

$$x_3 = 25 + 50 + 25 = 100 \text{ mm}$$

and $$y_3 = 50 + \frac{50}{3} = 66.7 \text{ mm}$$

We know that distance between centre of gravity of the section and left edge of the circular portion,

$$\bar{x} = \frac{a_1 x_1 + a_2 x_2 + a_3 x_3}{a_1 + a_2 + a_3} = \frac{(5000 \times 75) + (982 \times 14.4) + (1250 \times 100)}{5000 + 982 + 1250}$$

$$= 71.1 \text{ mm} \quad \textbf{Ans.}$$

Similarly, distance between centre of gravity of the section and bottom face of the rectangular portion,

$$\bar{y} = \frac{a_1 y_1 + a_2 y_2 + a_3 y_3}{a_1 + a_2 + a_3} = \frac{(5000 \times 25) + (982 \times 25) + (1250 \times 66.7)}{5000 + 982 + 1250} \text{ mm}$$

$= 32.2$ mm **Ans.**

**EXAMPLE 9.6.** *A plane lamina of 220 mm radius is shown in figure given below*

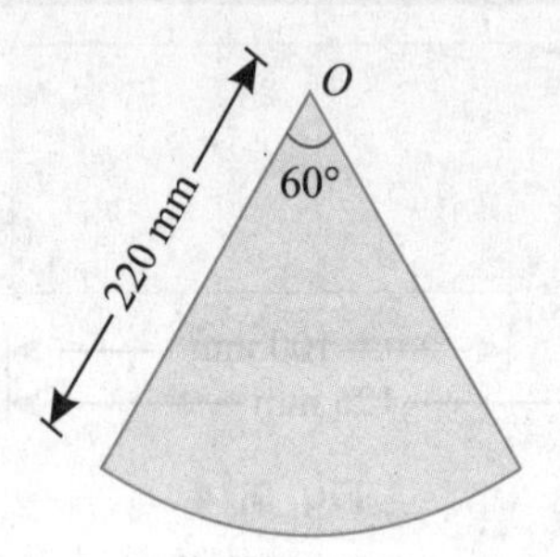


Fig. 9.15

*Find the centre of gravity of lamina from the point O.*

**SOLUTION.** As the lamina is symmetrical about *y-y* axis, bisecting the lamina, therefore its centre of gravity lies on this axis. Let *O* be the reference point. From the geometry of the lamina. We find that semi-vertical angle of the lamina

$$\alpha = 30° = \frac{\pi}{6} \text{ rad}$$

We know that distance between the reference point *O* and centre of gravity of the lamina,

$$\bar{y} = \frac{2r}{3} \frac{\sin \alpha}{\alpha} = \frac{2 \times 220}{3} \times \frac{\sin 30°}{\left(\frac{\pi}{6}\right)} = \frac{440}{3} \times \frac{0.5}{\left(\frac{\pi}{6}\right)} = 140 \text{ mm} \quad \textbf{Ans.}$$

## EXERCISE 9.1

1. Find the centre of gravity of a *T*-section with flange 150 mm × 10 mm and web also 150 mm × 10 mm. [**Ans.** 115 mm for bottom of the web]
2. Find the centre of gravity of an inverted *T*-section with flange 60 mm × 10 mm and web 50 mm × 10 mm [**Ans.** 18.6 mm from bottom of the flange]
3. A channel section 300 mm × 10 mm is 20 mm thick. Find the centre of gravity of the section from the back of the web. [**Ans.** 27.4 mm]
4. Find the centre of gravity of an *T*-section with top flange 100 mm × 20 mm, web 200 mm × 30 mm and bottom flange 300 mm × 40 mm. [**Ans.** 79 mm from bottom of lower flange]
5. Find the position of the centre of gravity of an unequal angle section 10 cm × 16 cm × 2cm. [**Ans.** 5.67 cm and 2.67 cm]

6. A figure consists of a rectangle having one of its sides twice the other, with an equilateral triangle described on the larger side. Show that centre of gravity of the section lies on the line joining the rectangle and triangle.
7. A plane lamina of radius 100 mm as shown in fig 9.16 given below:

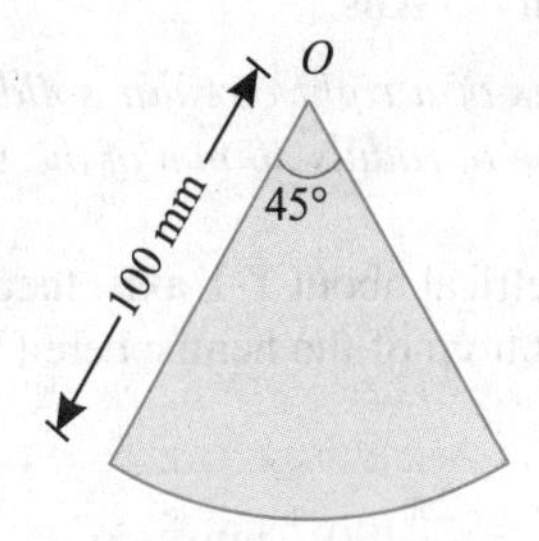


Fig. 9.16

Find the centre of gravity of lamina from the point $O$. **[Ans.** 65 mm]

## 9.10. Centre of Gravity of Solid Bodies

The centre of gravity of solid bodies (such as hemispheres, cylinders, right circular solid cones etc.) is found out in the same way as that of plane figures. The only difference, between the plane figures and solid bodies, is that in the case of solid bodies, we calculate volumes instead of areas. The volumes of few solid bodies are given below :

1. Volume of cylinder $= \pi \times r^2 \times h$
2. Volume of hemisphere $= \dfrac{2\pi}{3} \times r^3$
3. Volume of right circular solid cone $= \dfrac{\pi}{3} \times r^2 \times h$

where $r$ = Radius of the body, and
$h$ = Height of the body.

**Note.** Sometimes the densities of the two solids are different. In such a case, we calculate the weights instead of volumes and the centre of gravity of the body is found out as usual.

**Example 9.7.** *A solid body formed by joining the base of a right circular cone of height H to the equal base of a right circular cylinder of height h. Calculate the distance of the centre of mass of the solid from its plane face, when H = 120 mm and h = 30 mm.*

**Solution.** As the body is symmetrical about the vertical axis, therefore its centre of gravity will lie on this axis as shown in Fig. 9.17. Let $r$ be the radius of the cylinder base in cm. Now let base of the cylinder be the axis of reference.

(*i*) *Cylinder*

$$v_1 = \pi \times r^2 \times 30 = 30\,\pi\, r^2 \text{ mm}^3$$

and $$y_1 = \frac{30}{2} = 15 \text{ mm}$$

(*ii*) *Right circular cone*

$$v_2 = \frac{\pi}{3} \times r^2 \times h = \frac{\pi}{3} \times r^2 \times 120 \text{ mm}^3$$
$$= 40\,\pi r^2 \text{ mm}^3$$

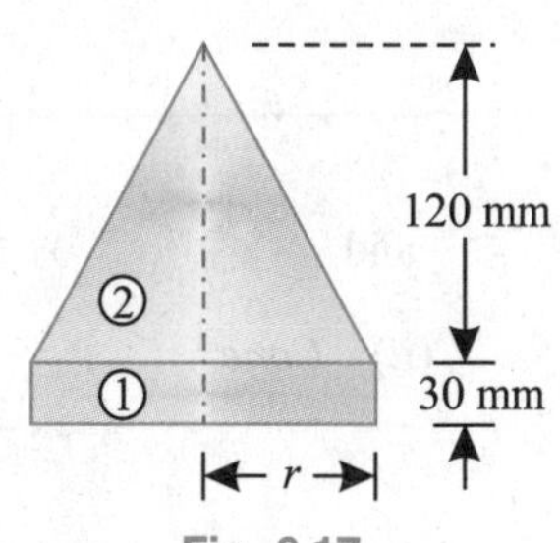


Fig. 9.17

and $$y_2 = 30 + \frac{120}{4} = 60 \text{ mm}$$

We know that distance between centre of gravity of the section and base of the cylinder,

$$\bar{y} = \frac{v_1 y_1 + v_2 y_2}{v_1 + v_2} = \frac{(30\pi r^2 \times 15) + (40\pi r^2 \times 60)}{30\pi r^2 + 40\pi r^2} = \frac{2850}{70} \text{ mm}$$

$$= 40.7 \text{ mm} \quad \textbf{Ans.}$$

**EXAMPLE 9.8.** *A body consists of a right circular solid cone of height 40 mm and radius 30 mm placed on a solid hemisphere of radius 30 mm of the same material. Find the position of centre of gravity of the body.*

**SOLUTION.** As the body is symmetrical about *Y-Y* axis, therefore its centre of gravity will lie on this axis as shown in Fig. 9.18. Let bottom of the hemisphere (*D*) be the point of reference.

(*i*) *Hemisphere*

$$v_1 = \frac{2\pi}{3} \times r^3 = \frac{2\pi}{3}(30)^3 \text{ mm}^3$$

$$= 18\,000\,\pi \text{ mm}^3$$

and
$$y_1 = r - \frac{3r}{8} = \frac{5r}{8} = \frac{5 \times 30}{8} = 18.75 \text{ mm}$$

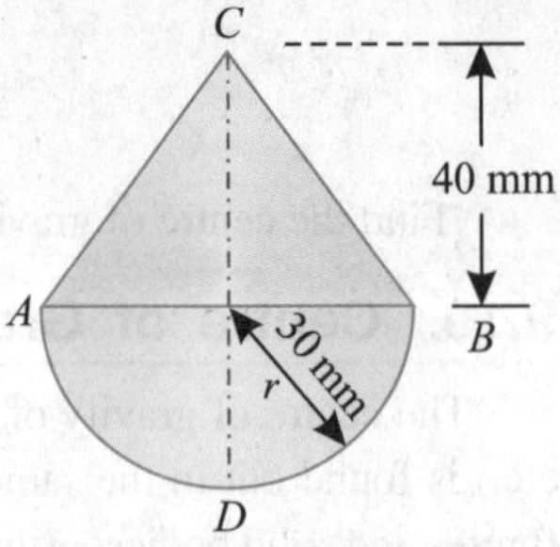

(*ii*) *Right circular cone*

$$v_2 = \frac{\pi}{3} \times r^2 \times h = \frac{\pi}{3} \times (30)^2 \times 40 \text{ mm}^3$$

$$= 12\,000\,\pi \text{ mm}^3$$

and
$$y_2 = 30 + \frac{40}{4} = 40 \text{ mm}$$

Fig. 9.18

We know that distance between centre of gravity of the body and bottom of hemisphere *D*,

$$\bar{y} = \frac{v_1 y_1 + v_2 y_2}{v_1 + v_2} = \frac{(18\,000\pi \times 18.75) + (12\,000\pi \times 40)}{18\,000\pi + 12\,000\pi} \text{ mm}$$

$$= 27.3 \text{ mm} \quad \textbf{Ans.}$$

**EXAMPLE 9.9.** *A body consisting of a cone and hemisphere of radius r fixed on the same base rests on a table, the hemisphere being in contact with the table. Find the greatest height of the cone, so that the combined body may stand upright.*

**SOLUTION.** As the body is symmetrical about *Y-Y* axis, therefore its centre of gravity will lie on this axis as shown in Fig. 9.19. Now consider two parts of the body *viz*., hemisphere and cone. Let bottom of the hemisphere (*D*) be the axis of reference.

(*i*) *Hemisphere*

$$v_1 = \frac{2\pi}{3} \times r^3$$

and
$$y_1 = \frac{5r}{8}$$

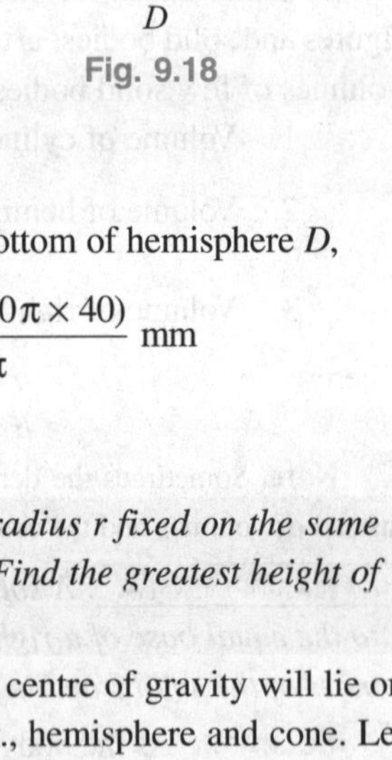

(*ii*) *Cone*

$$v_2 = \frac{\pi}{3} \times r^2 \times h$$

and
$$y_2 = r + \frac{h}{4}$$

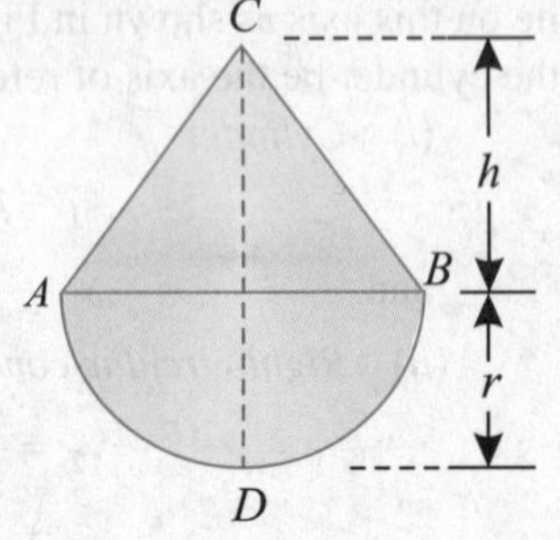

Fig. 9.19

We know that distance between centre of gravity of the body and bottom of hemisphere $D$,

$$\bar{y} = \frac{v_1 y_1 + v_2 y_2}{v_1 + v_2} = \frac{\left(\frac{2\pi}{3} \times r^3 \times \frac{5r}{8}\right) + \left(\frac{\pi}{3} \times r^2 \times h\right)\left(r + \frac{h}{4}\right)}{\left(\frac{2\pi}{3} \times r^3\right) + \left(\frac{\pi}{3} \times r^2 \times h\right)}$$

Now for stable equilibrium, we know that the centre of gravity of the body should preferably be below the common face $AB$ or maximum may coincide with it. Therefore substituting $\bar{y}$ equal to $r$ in the above equation,

$$r = \frac{\left(\frac{2\pi}{3} \times r^3 \times \frac{5r}{8}\right) + \left(\frac{\pi}{3} \times r^2 \times h\right)\left(r + \frac{h}{4}\right)}{\left(\frac{2\pi}{3} \times r^3\right) + \left(\frac{\pi}{3} \times r^2 \times h\right)}$$

or
$$\left(\frac{2\pi}{3} \times r^4\right) + \left(\frac{\pi}{3} \times r^3 h\right) = \left(\frac{5\pi}{12} \times r^4\right) + \left(\frac{\pi}{3} \times r^3 \times h\right) + \left(\frac{\pi}{12} \times r^2 \times h^2\right)$$

Dividing both sides by $\pi r^2$,

$$\frac{2r^2}{3} + \frac{rh}{3} = \frac{5r^2}{12} + \frac{rh}{3} + \frac{h^2}{12} \quad \text{or} \quad \frac{3r^2}{12} = \frac{h^2}{12}$$

$$3r^2 = h^2 \quad \text{or} \quad h = 1.732\ r \quad \textbf{Ans.}$$

**EXAMPLE 9.10.** *A right circular cylinder of 12 cm diameter is joined with a hemisphere of the same diameter face to face. Find the greatest height of the cylinder, so that centre of gravity of the composite section coincides with the plane of joining the two sections. The density of the material of hemisphere is twice that the material of cylinder.*

**SOLUTION.** As the body is symmetrical about the vertical axis, therefore its centre of gravity will lie on this axis. Now let the vertical axis cut the plane joining the two sections at O as shown in Fig. 9.20. Therefore centre of gravity of the section is at a distance of 60 mm from $P$ *i.e.*, bottom of the hemisphere.

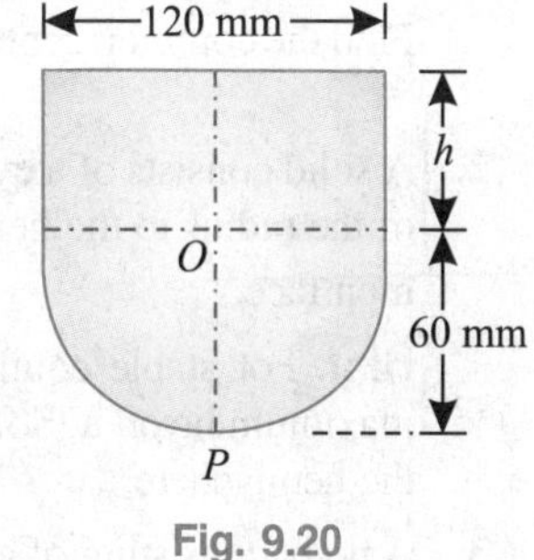

Fig. 9.20

Let $\quad h$ = Height of the cylinder in mm.

(*i*) *Right circular cylinder*

Weight $\quad (w_1) = \rho_1 \times \frac{\pi}{4} \times d^2 \times h$

$$= \rho_1 \times \frac{\pi}{4} \times (120)^2 \times h = 3\,600\pi\rho_1 h$$

and $\quad y_1 = 60 + \frac{h}{2} = 60 + 0.5h$ mm

(*ii*) *Hemisphere*

Weight $\quad (w_2) = \rho_2 \times \frac{2\pi}{3} \times r^3 = 2\rho_1 \times \frac{2\pi}{3} \times (60)^3 \quad ...(\because \rho_2 = 2\rho_1)$

$$= 288\,000\,\pi\,\rho_1$$

and $\quad y_2 = \frac{5r}{8} = \frac{5 \times 60}{8} = \frac{300}{8} = 37.5$ mm

We know that distance between centre of gravity of the combined body from $P(\bar{y})$,

$$60 = \frac{w_1 y_1 + w_2 y_2}{w_1 + w_2} = \frac{3\,600\pi\rho_1 h(60 + 0.5h) + (288\,000\pi\rho_1 \times 37.5)}{3\,600\pi\rho_1 h + 288\,000\pi\rho_1}$$

$$= \frac{216\,000h + 1800\,h^2 + 10\,800\,000}{3\,600h + 288\,000}$$

$$216\,000\,h + 17\,280\,000 = 216\,000\,h + 1\,800\,h^2 + 10\,800\,000$$

$$1\,800\,h^2 = 17\,280\,000 - 10\,800\,000 = 6\,480\,000$$

$$h = \sqrt{\frac{6\,480\,000}{1\,800}} = \sqrt{3\,600} = 60 \text{ mm} \quad \textbf{Ans.}$$

**EXAMPLE 9.11.** *Find the centre of gravity of a segment of height 30 mm of a sphere of radius 60 mm.*

**SOLUTION.** Let $O$ be the centre of the given sphere and $ABC$ is the segment of this sphere as shown in Fig. 9.21

As the section is symmetrical about $X$-$X$ axis, therefore its centre of gravity lies on this axis.

Let $O$ be the reference point.

We know that centre of gravity of the segment of sphere

$$\bar{x} = \frac{3(2r - h)^2}{4(3r - h)} = \frac{3(2 \times 60 - 30)^2}{4(3 \times 60 - 30)}$$

$$= \frac{3 \times (90)^2}{4 \times 150} = 40.5 \text{ mm.} \quad \textbf{Ans.}$$

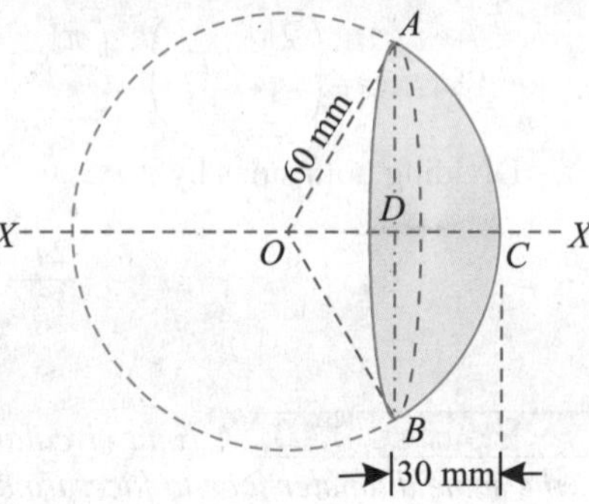

Fig. 9.21

## EXERCISE 9.2

1. A hemisphere of 60 mm diameter is placed on the top of the cylinder having 60 mm diameter. Find the common centre of gravity of the body from the base of cylinder, if its height is 100 mm. [**Ans.** 60.2 mm]
2. A solid consists of a cylinder and a hemisphere of equal radius fixed base to base. Find the ratio of the radius to the height of the cylinder, so that the solid has its centre of gravity at the common face. [**Ans.** $\sqrt{2} : 1$]

   **Hint.** For stable equilibrium, the centre of the body should be below the common face or maximum lie on it. So take the centre of gravity of the body at a distance ($a$) from the bottom of the hemisphere.
3. A body consisting of a cone and hemisphere of radius ($r$) on the same base rests on a table, the hemisphere being in contact with the table. Find the greatest height of the cone, so that the combined solid may be in stable equilibrium. [**Ans.** 1.732 $r$]
4. Find the centre of gravity of a segment of height 77 mm of a sphere of radius 150 mm. [**Ans.** 100 mm]

## 9.11. Centre of Gravity of Sections with Cut out Holes

The centre of gravity of such a section is found out by considering the main section, first as a complete one, and then deducting the area of the cut out hole *i.e.*, by taking the area of the cut out hole as negative. Now substituting $a_2$ (*i.e.*, the area of the cut out hole) as negative, in the general equation for the centre of gravity, we get

$$\bar{x} = \frac{a_1 x_1 - a_2 x_2}{a_1 - a_2} \quad \text{and} \quad \bar{y} = \frac{a_1 y_1 - a_2 y_2}{a_1 - a_2}$$

**Note.** In case of circle the section will be symmeterical along the line joining the centres of the bigger and the cut out circle.

**Example 9.12.** *A square hole is punched out of circular lamina, the digonal of the square being the radius of the circle as shown in Fig.9.22. Find the centre of gravity of the remainder, if r is the radius of the circle.*

**Solution.** As the section is symmetrical about *X-X* axis, therefore its centre of gravity will lie on this axis. Let *A* be the point of reference.

(*i*) *Main circle*

$$a_1 = \pi r^2$$

and $$x_1 = r$$

(*ii*) *Cut out square*

$$a_2 = \frac{r \times r}{2} = 0.5\, r^2$$

and $$x_2 = r + \frac{r}{2} = 1.5\, r$$

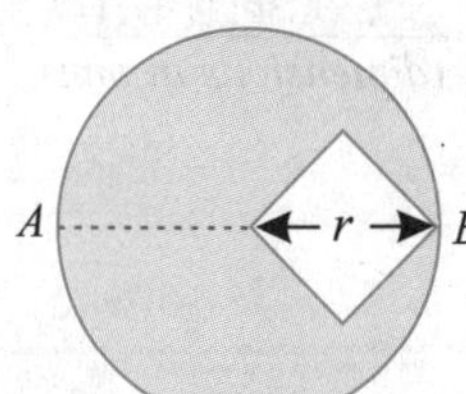

Fig. 9.22

We know that distance between centre of gravity of the section and *A*,

$$\bar{x} = \frac{a_1 x_1 - a_2 x_2}{a_1 - a_2} = \frac{(\pi r^2 \times r) - (0.5 r^2 \times 1.5 r)}{\pi r^2 - 0.5 r^2}$$

$$= \frac{r^3 (\pi - 0.75)}{r^2 (\pi - 0.5)} = \frac{r(\pi - 0.75)}{\pi - 0.5} \quad \textbf{Ans.}$$

**Example 9.13.** *A semicircle of 90 mm radius is cut out from a trapezium as shown in Fig. 9.23*

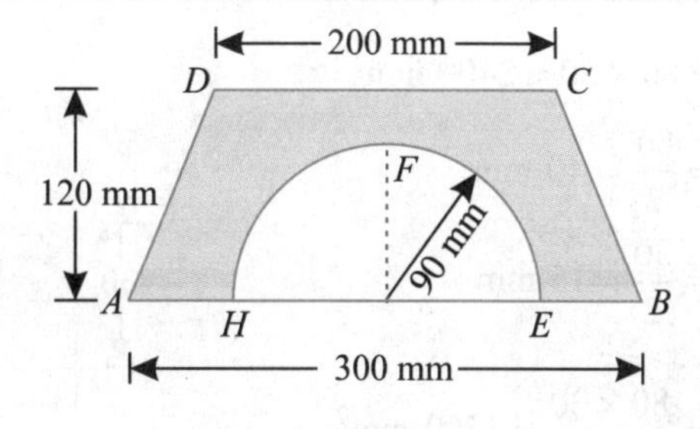

Fig. 9.23

*Find the position of the centre of gravity of the figure.*

**Solution.** As the section is symmetrical about *Y-Y* axis, therefore its centre of gravity will lie on this axis. Now consider two portions of the figure *viz.*, trapezium *ABCD* and semicircle *EFH*.

Let base of the trapezium *AB* be the axis of reference.

(*i*) *Trapezium ABCD*

$$a_1 = 120 \times \frac{200 + 300}{2} = 30\,000 \text{ mm}^2$$

and $$y_1 = \frac{120}{3} \times \left( \frac{300 + 2 \times 200}{300 + 200} \right) = 56 \text{ mm}$$

(*ii*) *Semicircle*

$$a_2 = \frac{1}{2} \times \pi r^2 = \frac{1}{2} \times \pi \times (90)^2 = 4050\pi \text{ mm}^2$$

and $$y_2 = \frac{4r}{3\pi} = \frac{4 \times 90}{3\pi} = \frac{120}{\pi} \text{ mm}$$

We know that distance between centre of gravity of the section and *AB*,

$$\bar{y} = \frac{a_1 y_1 - a_2 y_2}{a_1 - a_2} = \frac{(30\,000 \times 56) - \left(4050\pi \times \frac{120}{\pi}\right)}{30\,000 - 4050\pi} \text{ mm}$$

$$= 69.1 \text{ mm} \quad \textbf{Ans.}$$

**EXAMPLE 9.14.** *A semicircular area is removed from a trapezium as shown in Fig.9.24 (dimensions in mm)*

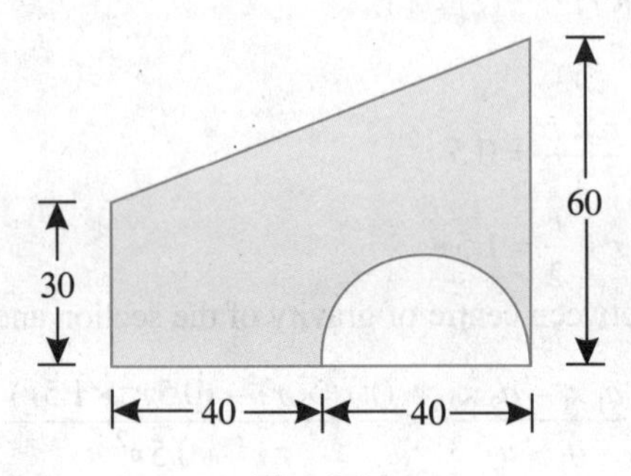


Fig. 9.24

*Determine the centroid of the remaining area (shown hatched).*

**SOLUTION.** As the section in not symmetrical about any axis, therefore we have to find out the values of $\bar{x}$ and $\bar{y}$ for the area. Split up the area into three parts as shown in Fig. 9.25. Let left face and base of the trapezium be the axes of reference.

(*i*) *Rectangle*

$$a_1 = 80 \times 30 = 2400 \text{ mm}^2$$

$$x_1 = \frac{80}{2} = 40 \text{ mm}$$

and $$y_1 = \frac{30}{2} = 15 \text{ mm}$$

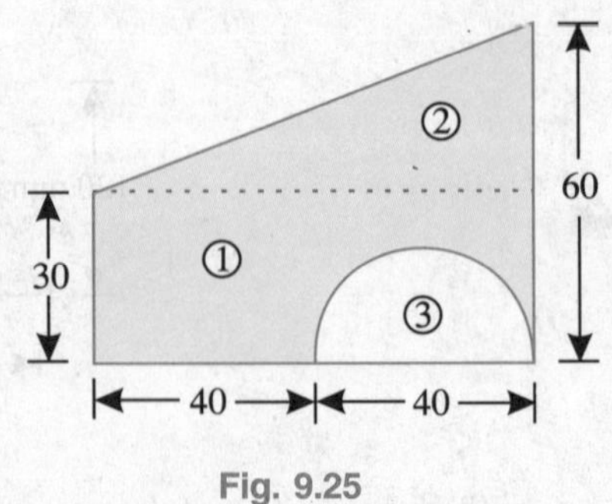


Fig. 9.25

(*ii*) *Triangle*

$$a_2 = \frac{80 \times 30}{2} = 1200 \text{ mm}^2$$

$$x_2 = \frac{80 \times 2}{3} = 53.3 \text{ mm}$$

and $$y_2 = 30 + \frac{30}{3} = 40 \text{ mm}$$

(*iii*) *Semicircle*

$$a_3 = \frac{\pi}{2} \times r^2 = \frac{\pi}{2}(20)^2 = 628.3 \text{ mm}^2$$

$$x_3 = 40 + \frac{40}{2} = 60 \text{ mm}$$

and $$y_3 = \frac{4r}{3\pi} = \frac{4 \times 20}{3\pi} = 8.5 \text{ mm}$$

We know that distance between centre of gravity of the area and left face of trapezium,

$$\bar{x} = \frac{a_1 x_1 + a_2 x_2 - a_3 x_3}{a_1 + a_2 - a_3} = \frac{(2400 \times 40) + (1200 \times 53.3) - (628.3 \times 60)}{2400 + 1200 - 628.3}$$

$$= 41.1 \text{ mm} \quad \textbf{Ans.}$$

Similarly, distance between centre of gravity of the area and base of the trapezium,

$$\bar{y} = \frac{a_1 y_1 + a_2 y_2 - a_3 y_3}{a_1 + a_2 - a_3} = \frac{(2400 \times 15) + (1200 \times 40) - (628.3 \times 8.5)}{2400 + 1200 - 628.3}$$

$$= 26.5 \text{ mm} \quad \textbf{Ans.}$$

**EXAMPLE 9.15.** *A circular sector of angle 60° is cut from the circle of radius r as shown in Fig. 9.26 :*

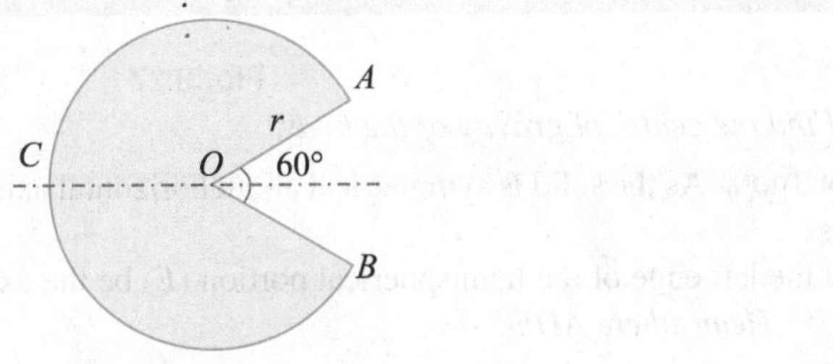


Fig. 9.26

*Determine the centre of gravity of the remainder.*

**SOLUTION.** As the section is symmetrical about $X$-$X$ axis, therefore its centre of gravity will lie on this axis.

Let $C$ be the reference point.

(*i*) *Main circle*

$$a_1 = \pi r^2$$

and $$x_1 = r$$

(*ii*) *Cut out sector*

$$a_2 = \frac{\pi r^2 \theta}{360°} = \frac{\pi r^2 \times 60°}{360°} = \frac{\pi r^2}{6}$$

and $$x_2 = r + \frac{2r}{\pi}$$

We know that distance between the centre of gravity of the section and $C$

$$\bar{x} = \frac{a_1 x_1 - a_2 x_2}{a_1 - a_2} = \frac{(\pi r^2 \times r) - \left[\frac{\pi r^2}{6} \times \left(r + \frac{2r}{\pi}\right)\right]}{\pi r^2 - \frac{\pi r^2}{6}}$$

$$= \frac{\pi r^2 \left[r - \frac{1}{6}\left(r + \frac{2r}{\pi}\right)\right]}{\pi r^2 \left(1 - \frac{1}{6}\right)} = \frac{r - \left[\frac{1}{6} \times \left(r + \frac{2r}{\pi}\right)\right]}{1 - \frac{1}{6}}$$

$$= \frac{6}{5}\left[r - \left(\frac{r}{6} + \frac{2r}{6\pi}\right)\right] = \frac{6}{5}\left[r - \frac{r}{6} - \frac{r}{3\pi}\right]$$

$$= \frac{6}{5}\left(\frac{5}{6}r - \frac{r}{3\pi}\right) = r - \frac{2r}{5\pi} \quad \textbf{Ans.}$$

**EXAMPLE 9.16.** *A solid consists of a right circular cylinder and a hemisphere with a cone cut out from the cylinder as shown in Fig. 9.27.*

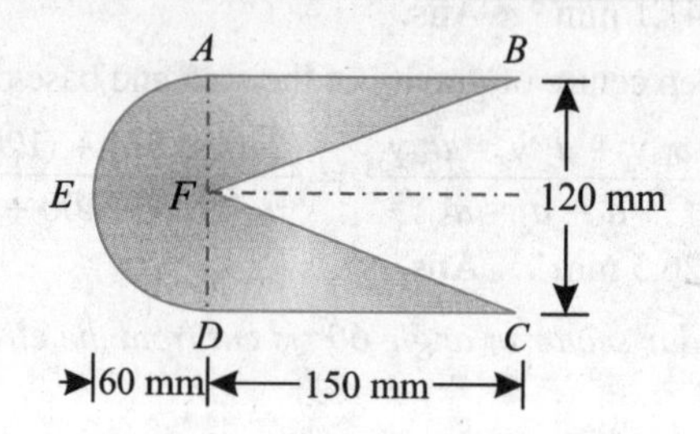


Fig. 9.27

*Find the centre of gravity of the body.*

**SOLUTION.** As the solid is symmetrical about horizontal axis, therefore its centre of gravity lie on this axis.

Let the left edge of the hemispherical portion (*E*) be the axis of reference.

(*i*) *Hemisphere ADE*

$$v_1 = \frac{2\pi}{3} \times r^3 = \frac{2\pi}{3} \times (60)^3 = 144\ 000\ \pi\ \text{mm}^3$$

and $$x_1 = \frac{5r}{8} = \frac{5 \times 60}{8} = 37.5\ \text{mm}$$

(*ii*) *Right circular cylinder ABCD*

$$v_2 = \pi \times r^2 \times h = \pi \times (60)^2 \times 150 = 540\ 000\ \pi\ \text{mm}^3$$

and $$x_2 = 60 + \frac{150}{2} = 135\ \text{mm}$$

(*iii*) *Cone BCF*

$$v_3 = \frac{\pi}{3} \times r^2 \times h = \frac{\pi}{3} \times (60)^2 \times 150 = 180\ 000\ \pi\ \text{mm}^3$$

and $$x_3 = 60 + 150 \times \frac{3}{4} = 172.5\ \text{mm}$$

We know that distance between centre of gravity of the solid and left edge *E* of hemisphere,

$$\bar{x} = \frac{v_1 x_1 + v_2 x_2 - v_3 x_3}{v_1 + v_2 - v_3}$$

$$= \frac{(144\ 000\ \pi \times 37.5) + (540\ 000\ \pi \times 135) - (180\ 000\ \pi \times 172.5)}{144\ 000\ \pi + 540\ 000\ \pi - 180\ 000\ \pi}$$

$$= 93.75\ \text{mm} \quad \textbf{Ans.}$$

**EXAMPLE 9.17.** *A frustum of a solid right circular cone has an axial hole of 50 cm diameter as shown in Fig. 9.28.*

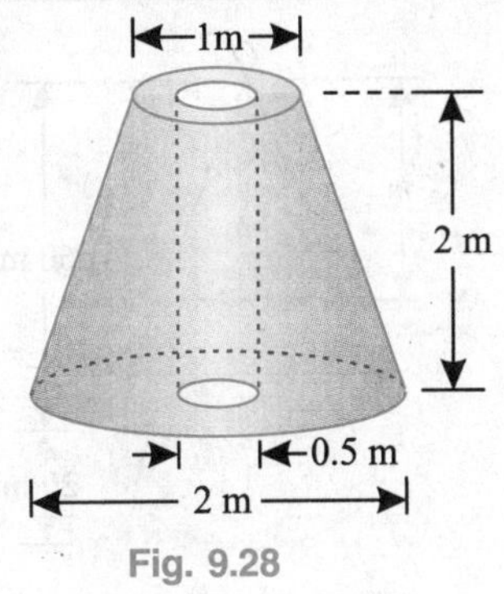


Fig. 9.28

*Determine the centre of gravity of the body.*

**SOLUTION.** As the body is symmetrical about vertical axis, therefore its centre of geravity lie on this axis. For the sake of simplicity, let us assume a right circular cone *OCD*, from which a right circulr cone *OAB* is cut off as shown in Fig. 9.29.

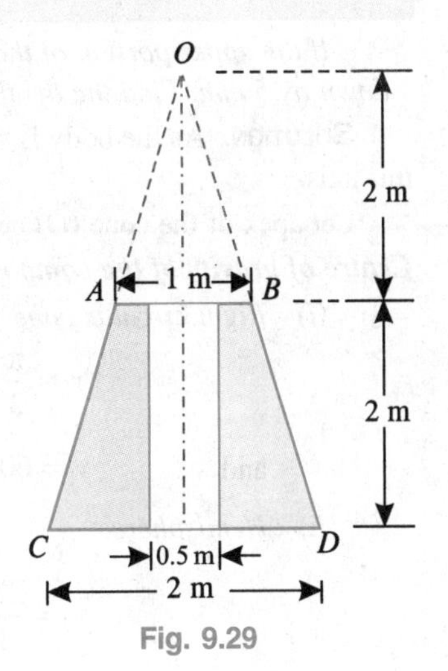


Fig. 9.29

Let base of cone *CD* be the axis of reference.

*(i)* ***Right circular cone OCD***

$$v_1 = \frac{\pi}{3} \times R^2 \times H$$

$$= \frac{\pi}{3} \times (1)^2 \times 4 = \frac{4\pi}{3} \text{ m}^3$$

and $$y_1 = \frac{4}{4} = 1 \text{m}$$

*(ii)* ***Right circular cone OAB***

$$v_2 = \frac{\pi}{3} \times r^2 \times h$$

$$= \frac{\pi}{3} \times \left(\frac{2}{4}\right)^2 \times 2 = \frac{\pi}{6} \text{ m}^3$$

and $$y_2 = 2 + \frac{2}{4} = \frac{5}{2} \text{m}$$

*(iii)* ***Circular hole***

$$v_3 = \frac{\pi}{4} \times d^2 \times h = \frac{\pi}{4} \times (0.5)^2 \times 2 = \frac{\pi}{8} \text{ m}^3$$

and $$y_2 = \frac{2}{2} = 1 \text{ m}$$

We know that distance between centre of gravity of the body and the base of the cone,

$$\bar{y} = \frac{v_1 y_1 - v_2 y_2 - v_3 y_3}{v_1 - v_2 - v_3}$$

$$= \frac{\left(\frac{4\pi}{3} \times 1\right) - \left(\frac{\pi}{6} \times \frac{5}{2}\right) - \left(\frac{\pi}{8} \times 1\right)}{\frac{4\pi}{3} - \frac{\pi}{6} - \frac{\pi}{8}} = \frac{\frac{4}{3} - \frac{5}{12} - \frac{1}{8}}{\frac{4}{3} - \frac{1}{6} - \frac{1}{8}} = \frac{19}{25} = 0.76 \text{ m} \quad \textbf{Ans.}$$

**EXAMPLE 9.18.** *A solid hemisphere of 20 mm radius supports a solid cone of the same base and 60 mm height as shown in Fig. 9.30. Locate the centre of gravity of the composite section.*

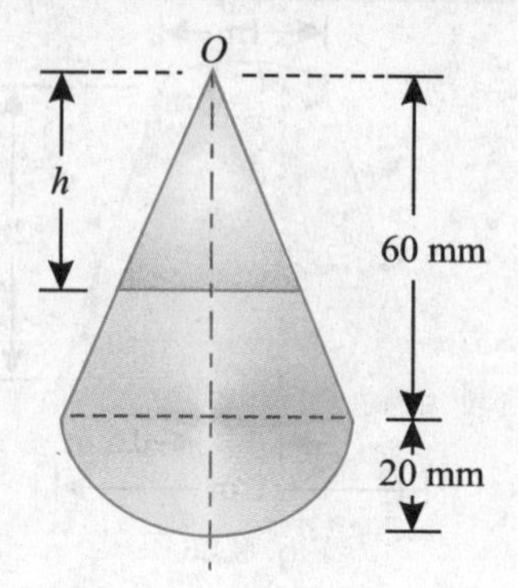


Fig. 9.30

*If the upper portion of the cone is removed by a certain section, the centre of gravity lowers down by 5 mm. Find the depth of the section plane (h) below the apex.*

**SOLUTION.** As the body is symmetrical about *Y-Y* axis, therefore its centre of gravity will lie on this axis.

Let apex of the cone (*O*) be the axis of reference.

***Centre of gravity of the composite section***

(*i*) *Right circular cone*

$$v_1 = \frac{\pi}{3} \times r^2 \times h = \frac{\pi}{3} \times (20)^2\, 60 = 25\,133 \text{ mm}^3$$

and $$y_1 = 60 \times \frac{3}{4} = 45 \text{ mm}$$

(*ii*) *Hemisphere*

$$v_2 = \frac{2\pi}{3} \times r^2 = \frac{2\pi}{3} \times (20)^3 = 16\,755 \text{ mm}^3$$

and $$y_2 = 60 + \frac{3 \times 20}{8} = 67.5 \text{ mm}$$

We know that distance between centre of gravity of the body and apex of the cone,

$$\bar{y} = \frac{v_1 y_1 + v_2 y_2}{v_1 + v_2} = \frac{(25\,133 \times 45) + (16\,755 \times 67.5)}{25\,133 + 16\,755} \text{ mm}$$

$$= \frac{2\,261\,950}{41\,888} = 54 \text{ mm} \quad \textbf{Ans.}$$

***Depth of the section plane below the apex***

We know that the radius of the cut out cone,

$$r = \frac{h}{3} \qquad \ldots \left(\because \frac{r}{20} = \frac{h}{60}\right)$$

∴ Volume of the cut out cone,

$$v_3 = \frac{\pi}{3} \times r^2 \times h = \frac{\pi}{3}\left(\frac{h}{3}\right)^2 \times h = 0.1164\, h^2 \text{ mm}^3$$

and distance between centre of gravity of the cut out cone and its apex,

$$y_3 = \frac{3h}{4} = 0.75\ h$$

We also know that distance between the centre of gravity of the body and apex of the cone (*i.e.* 54 + 5 = 59 mm),

$$\bar{y} = \frac{v_1\, y_1 + v_2\, y_2 - v_3\, y_3}{v_1 + v_2 - v_2}$$

$$\therefore \quad 59 = \frac{(25\,133 \times 45) + (16\,755 \times 67.5) - 0.1164\ h^3 \times 0.75\ h}{25\,133 + 16\,755 - 0.1164\ h^3}$$

$$= \frac{2\,261\,950 - 0.0873 h^4}{41\,888 - 0.1164\ h^3}$$

$$2\,471\,400 - 6.868\ h^3 = 2\,261\,950 - 0.0873\ h^4$$

$$0.0873\ h^4 - 6.868\ h^3 = -\ 209\,450$$

Dividing both sides by 0.0873,

$$h^4 - 78.67\ h^3 = -2\,399\,200 \qquad ...(i)$$

We shall solve this equation by trial and error. First of all, let us substitute $h$ = 10 mm in the left hand side of equation (*i*). We find

$$(10)^4 - 78.67\ (10)^3 = -\ 68\,670$$

We find that answer obtained does not tally with the value of right hand side of equation (*i*), and is much less than that. Now let us substitute $h$ = 20 mm in the left hand side of equation (*i*),

$$(20)^4 - 78.67\ (20)^3 = -\ 469\,360$$

We again find that the answer obtained does not tally with the right hand side of equation (*i*), But it is closer to the value of right hand side than the first case (*i.e.* when we substituted $h$ = 10 mm.) Or in other words, the value obtained is still less than the right hand side of equation (*i*). But the difference has reduced. Now let us substitute $h$ = 30 mm in the left hand side of equation (*i*).

$$(30)^4 - 78.67\ (30)^3 = 1\,314\,100$$

We again find the answer obtained does not tally with the right hand side of equation (*i*), But it is more close to the right hand side than the previous case *i.e.* when we substituted $h$ = 20 mm. Now let us substitute $h$ = 40 mm in the left hand side of the equation (*i*).

$$(40)^4 - 78.67\ (40)^3 = 2474900$$

Now we find that the answer obtained does not tally with the right hand side of equation (*i*). But its value is more than the right hand side of equation (*i*), In the previous cases, the value of the answer obtained was less. Thus we find that the value of ($h$) is less than 40 mm.

A little consideration will show, that as the value of the answer is slightly more than the right hand side of equation (*i*). (as compared to the previous answers), the value of ($h$) is slightly less than 40 mm. Now let us substitude $h$ = 39 mm in the left hand side of the equation (*i*).

$$(39)^4 - 78.67\ (39)^3 = -\ 2\,153\,200$$

Now we find that the answer obtained is less than the right hand side of equation (*i*). Thus the value of ($h$) is more than 39 mm. Or in other words it is within 39 and 40 mm. This is due to the reason that when we substitude $h$ = 39 mm, the answer is less and when we substitute $h$ = 40 mm, answer is more than the right hand side of equation (*i*), Now let us substitute $h$ = 39.5 mm in the left hand side of the equation (*i*).

$$(39.5)^4 - 78.67\ (39.5)^3 = -\ 2\,414\,000$$

Now we find that the answer obtained is more than the right hand side of equation (*i*). Thus the value of (*h*) is less than 39.5 mm. Now let us substitute the $h$ = 39.4 mm in the left hand side of equation, (*i*).

$$(39.4)^4 - 78.67\,(39.4)^3 = -\,2\,401\,900$$

We find that is answer is very close to the right hand side of the equation and there is no need of further calculations. Thus the value of $h$ = 39.4 mm **Ans.**

## EXERCISE 9.3

1. A circular hole of 50 mm diameter is cut out from a circular disc of 100 mm diameter as shown in Fig. 9.31. Find the centre of gravity of the section from *A*. [**Ans.** 41.7 mm]

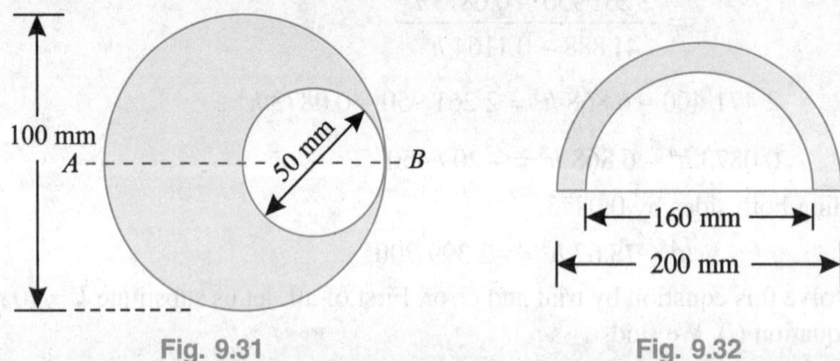


Fig. 9.31 Fig. 9.32

2. Find the centre of gravity of a semicircular section having outer and inner diameters of 200 mm and 160 mm respectively as shown in Fig. 9.32. [**Ans.** 57.5 mm from the base]
3. A circular sector of angle 45° is cut from the circle of radius 220 mm Determine the centre of gravity of the remainder from the centre of the sector. [**Ans.** 200 mm]
4. A hemisphere of diameter 80 mm is cut out from a right circular cylinder of diameter 80 mm and height 160 mm as shown in Fig. 9.33. Find the centre of gravity of the body from the base *AB*. [**Ans.** 77.2 mm]

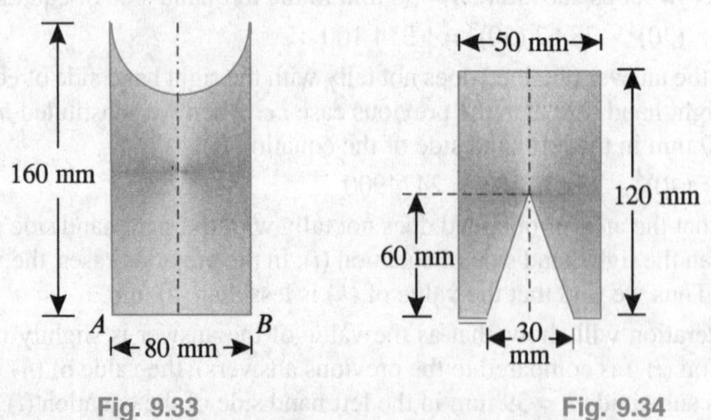


Fig. 9.33 Fig. 9.34

5. A right circular cone of 30 mm diameter and 60 mm height is cut from a cylinder of 50 mm diameter at 120 mm height as shown in Fig. 9.34. Find the position of the centre of gravity of the body from its base. [**Ans.** 60.7 mm]

## QUESTIONS

1. Define the terms 'centre of gravity'.

2. Distinguish between centre of gravity and centroid.
3. How many centres of gravity a body has?
4. Describe the various methods of finding out the centre of gravity of a body.
5. How would you find out the centre of gravity of a section, with a cut out hole?

## OBJECTIVE TYPE QUESTIONS

1. The centre of gravity of an equilateral triangle with each side (*a*) is ...... from any of the three sides.

   (*a*) $\frac{a\sqrt{3}}{2}$ (*b*) $\frac{a\sqrt{2}}{3}$ (*c*) $\frac{a}{2\sqrt{3}}$ (*d*) $\frac{a}{3\sqrt{2}}$

2. The centre of gravity of hemisphere lies at a distance of ......form its base measured along the vertical radius.

   (*a*) $\frac{3r}{8}$ (*b*) $\frac{3}{8r}$ (*c*) $\frac{8r}{3}$ (*d*) $\frac{8}{3r}$

3. The centre of gravity of a right circular cone of diameter (*d*) and height (*h*) lies at a distance of ...... from the base measured along the vertical radius.

   (*a*) $\frac{h}{2}$ (*b*) $\frac{h}{3}$ (*c*) $\frac{h}{4}$ (*d*) $\frac{h}{6}$

4. A circular hole of radius (*r*) is cut out from a circular disc of radius (2*r*) in such a way that the diagonal of the hole is the radius of the disc. The centre of gravity of the section lies at

   (*a*) Centre of a disc (*b* Centre of the hole
   (*c*) Somewhere in the disc (*d*) Somewhere in the hole

## ANSWERS

1. (*c*) 2. (*a*) 3. (*c*) 4. (*c*)

Chapter **10**

# Moment of Inertia

## Contents

## 10.1. Introduction

We have already discussed in Art. 3.2 that the moment of a force ($P$) about a point, is the product of the force and perpendicular distance ($x$) between the point and the line of action of the force (*i.e.* $P.x$). This moment is also called first moment of force. If this moment is again multiplied by the perpendicular distance ($x$) between the point and the line of action of the force *i.e.* $P.x\,(x) = Px^2$, then this quantity is called moment of the moment of a force or second moment of force or moment of inertia (briefly written as M.I.).

Sometimes, instead of force, area or mass of a figure or body is taken into consideration. Then the second moment is known as second moment of area or second moment of mass. But all such second moments are broadly termed as moment of inertia. In this chapter, we shall discuss the moment of inertia of plane areas only.

## 10.2. Moment of Inertia of a Plane Area

Consider a plane area, whose moment of inertia is required to be found out. Split up the whole area into a number of small elements.

Let $a_1, a_2, a_3, \ldots =$ Areas of small elements, and

$r_1, r_2, r_3, \ldots =$ Corresponding distances of the elements from the line about which the moment of inertia is required to be found out.

Now the moment of inertia of the area,

$$I = a_1 r_1^2 + a_2 r_2^2 + a_3 r_3^2 + \ldots$$

$$= \Sigma a r^2$$

## 10.3. Units of Moment of Inertia

As a matter of fact the units of moment of inertia of a plane area depend upon the units of the area and the length. *e.g.*,

1. If area is in $m^2$ and the length is also in m, the moment of inertia is expressed in $m^4$.
2. If area in $mm^2$ and the length is also in mm, then moment of inertia is expressed in $mm^4$.

## 10.4. Methods for Moment of Inertia

The moment of inertia of a plane area (or a body) may be found out by any one of the following two methods :

1. By Routh's rule    2. By Integration.

**NOTE :** The Routh's Rule is used for finding the moment of inertia of a plane area or a body of uniform thickness.

## 10.5. Moment of Inertia by Routh's Rule

The Routh's Rule states, if a body is symmetrical about three mutually perpendicular axes*, then the moment of inertia, about any one axis passing through its centre of gravity is given by:

$$I = \frac{A \text{ (or } M) \times S}{3} \qquad \text{... (For a Square or Rectangular Lamina)}$$

$$I = \frac{A \text{ (or } M) \times S}{4} \qquad \text{... (For a Circular or Elliptical Lamina)}$$

$$I = \frac{A \text{ (or } M) \times S}{5} \qquad \text{... (For a Spherical Body)}$$

where $A =$ Area of the plane area

$M =$ Mass of the body, and

$S =$ Sum of the squares of the two semi-axis, other than the axis, about which the moment of inertia is required to be found out.

**NOTE :** This method has only academic importance and is rarely used in the field of science and technology these days. The reason for the same is that it is equally convenient to use the method of integration for the moment of inertia of a body.

---

* *i.e.*, *X-X* axis, *Y-Y* axis and *Z-Z* axis.

## 10.6. Moment of Inertia by Integration

The moment of inertia of an area may also be found out by the method of integration as discussed below:

Consider a plane figure, whose moment of inertia is required to be found out about $X$-$X$ axis and $Y$-$Y$ axis as shown in Fig 10.1. Let us divide the whole area into a no. of strips. Consider one of these strips.

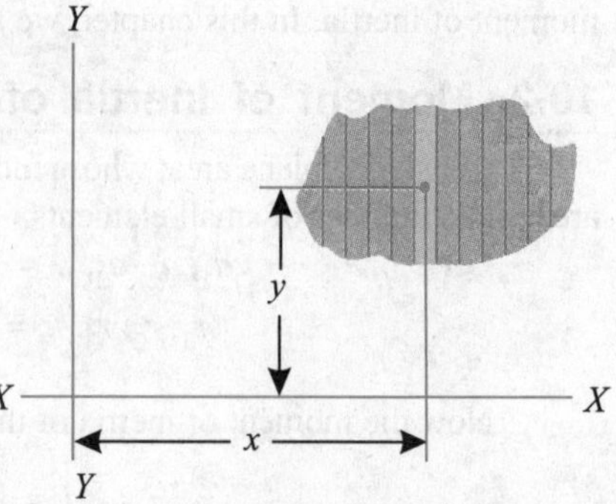

Let $dA$ = Area of the strip

$x$ = Distance of the centre of gravity of the strip on $X$-$X$ axis and

$y$ = Distance of the centre of gravity of the strip on $Y$-$Y$ axis.

Fig. 10.1. Moment of inertia by integration.

We know that the moment of inertia of the strip about $Y$-$Y$ axis

$$= dA \cdot x^2$$

Now the moment of inertia of the whole area may be found out by integrating above equation. *i.e.*,

$$I_{YY} = \Sigma\, dA \cdot x^2$$

Similarly $\quad I_{XX} = \Sigma\, dA \cdot y^2$

In the following pages, we shall discuss the applications of this method for finding out the moment of inertia of various cross-sections.

## 10.7. Moment of Inertia of a Rectangular Section

Consider a rectangular section $ABCD$ as shown in Fig. 10.2 whose moment of inertia is required to be found out.

Let $b$ = Width of the section and

$d$ = Depth of the section.

Now consider a strip $PQ$ of thickness $dy$ parallel to $X$-$X$ axis and at a distance $y$ from it as shown in the figure

∴ Area of the strip

$$= b.dy$$

We know that moment of inertia of the strip about $X$-$X$ axis,

$$= \text{Area} \times y^2 = (b.\, dy)\, y^2 = b.\, y^2.\, dy$$

Now *moment of inertia of the whole section may be found out by integrating the above equation for the whole length of the lamina *i.e.* from $-\dfrac{d}{2}$ to $+\dfrac{d}{2}$,

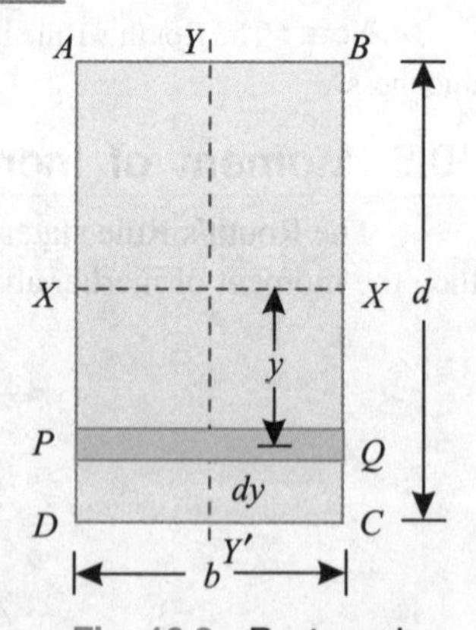

Fig. 10.2. Rectangular section.

* This may also be obtained by Routh's rule as discussed below :

$$I_{XX} = \frac{AS}{3} \qquad \text{...(for rectangular section)}$$

where area, $\quad A = b \times d$ and sum of the square of semi axes $Y$-$Y$ and $Z$-$Z$,

$$S = \left(\frac{d}{2}\right)^2 + 0 = \frac{d^2}{4}$$

$$\therefore \quad I_{XX} = \frac{AS}{3} = \frac{(b \times d) \times \dfrac{d^2}{4}}{3} = \frac{bd^3}{12}$$

$$I_{xx} = \int_{-\frac{d}{2}}^{+\frac{d}{2}} b.y^2.dy = b \int_{-\frac{d}{2}}^{+\frac{d}{2}} y^2 . dy$$

$$= b\left[\frac{y^3}{3}\right]_{-\frac{d}{2}}^{+\frac{d}{2}} = b\left[\frac{(d/2)^3}{3} - \frac{(-d/2)^3}{3}\right] = \frac{bd^3}{12}$$

Similarly, $$I_{YY} = \frac{db^3}{12}$$

**NOTE.** Cube is to be taken of the side, which is at right angles to the line of reference.

**EXAMPLE 10.1.** *Find the moment of inertia of a rectangular section 30 mm wide and 40 mm deep about X-X axis and Y-Y axis.*

**SOLUTION.** Given: Width of the section ($b$) = 30 mm and depth of the section ($d$) = 40 mm.

We know that moment of inertia of the section about an axis passing through its centre of gravity and parallel to $X$-$X$ axis,

$$I_{XX} = \frac{bd^3}{12} = \frac{30 \times (40)^3}{12} = 160 \times 10^3 \text{ mm}^4 \quad \textbf{Ans.}$$

Similarly $$I_{YY} = \frac{db^3}{12} = \frac{40 \times (30)^3}{12} = 90 \times 10^3 \text{ mm}^4 \quad \textbf{Ans.}$$

## 10.8. Moment of Inertia of a Hollow Rectangular Section

Consider a hollow rectangular section, in which $ABCD$ is the main section and $EFGH$ is the cut out section as shown in Fig 10.3

Let $b$ = Breadth of the outer rectangle,

$d$ = Depth of the outer rectangle and

$b_1, d_1$ = Corresponding values for the cut out rectangle.

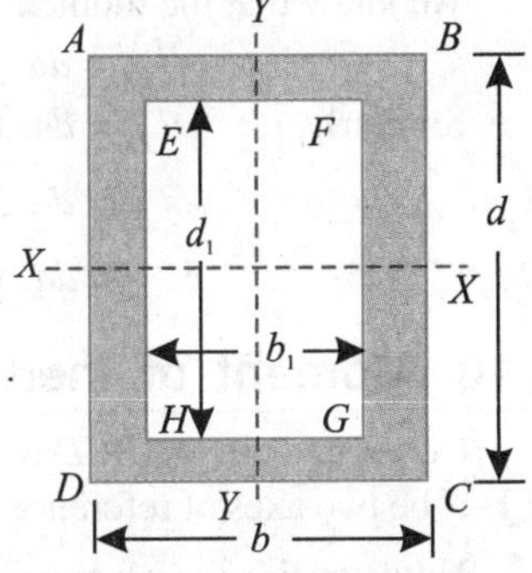

Fig. 10.3. Hollow rectangular section.

We know that the moment of inertia, of the outer rectangle $ABCD$ about $X$-$X$ axis

$$= \frac{bd^3}{12} \quad ...(i)$$

and moment of inertia of the cut out rectangle $EFGH$ about $X$-$X$ axis

$$= \frac{b_1 d_1^3}{12} \quad ...(ii)$$

∴ M.I. of the hollow rectangular section about $X$-$X$ axis,

$I_{XX}$ = M.I. of rectangle $ABCD$ – M.I. of rectangle $EFGH$

$$= \frac{bd^3}{12} - \frac{b_1 d_1^3}{12}$$

Similarly, $$I_{yy} = \frac{db^3}{12} - \frac{d_1 b_1^3}{12}$$

**NOTE :** This relation holds good only if the centre of gravity of the main section as well as that of the cut out section coincide with each other.

**EXAMPLE 10.2.** *Find the moment of inertia of a hollow rectangular section about its centre of gravity if the external dimensions are breadth 60 mm, depth 80 mm and internal dimensions are breadth 30 mm and depth 40 mm respectively.*

**Solution.** Given: External breadth ($b$) = 60 mm; External depth ($d$) = 80 mm ; Internal breadth ($b_1$) = 30 mm and internal depth ($d_1$) = 40 mm.

We know that moment of inertia of hollow rectangular section about an axis passing through its centre of gravity and parallel to $X$-$X$ axis,

$$I_{XX} = \frac{bd^3}{12} - \frac{b_1 d_1^3}{12} = \frac{60\,(80)^3}{12} - \frac{30\,(40)^3}{12} = 2400 \times 10^3 \text{ mm}^4 \quad \textbf{Ans.}$$

Similarly,
$$I_{YY} = \frac{db^3}{12} - \frac{d_1 b_1^3}{12} = \frac{80\,(60)^3}{12} - \frac{40\,(30)^3}{12} = 1350 \times 10^3 \text{ mm}^4 \quad \textbf{Ans.}$$

## 10.9. Theorem of Perpendicular Axis

It states, *If $I_{XX}$ and $I_{YY}$ be the moments of inertia of a plane section about two perpendicular axis meeting at O, the moment of inertia $I_{ZZ}$ about the axis Z-Z, perpendicular to the plane and passing through the intersection of X-X and Y-Y is given by:*

$$I_{ZZ} = I_{XX} + I_{YY}$$

**Proof :**

Consider a small lamina ($P$) of area $da$ having co-ordinates as $x$ and $y$ along $OX$ and $OY$ two mutually perpendicular axes on a plane section as shown in Fig. 10.4.

Now consider a plane $OZ$ perpendicular to $OX$ and $OY$. Let ($r$) be the distance of the lamina ($P$) from $Z$-$Z$ axis such that $OP = r$.

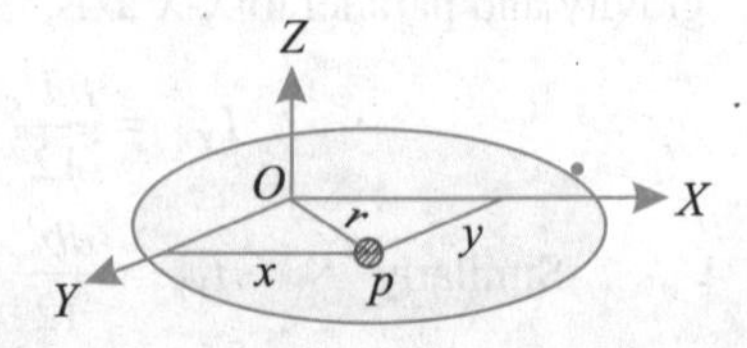


Fig. 10.4. Theorem of perpendicular axis.

From the geometry of the figure, we find that

$$r^2 = x^2 + y^2$$

We know that the moment of inertia of the lamina $P$ about $X$-$X$ axis,

$$I_{XX} = da.\ y^2 \qquad ...[\because I = \text{Area} \times (\text{Distance})^2]$$

Similarly,
$$I_{YY} = da.\ x^2$$

and
$$I_{ZZ} = da.\ r^2 = da\ (x^2 + y^2) \qquad ...(\because r^2 = x^2 + y^2)$$
$$= da.\ x^2 + da.\ y^2 = I_{YY} + I_{XX}$$

## 10.10. Moment of Inertia of a Circular Section

Consider a circle $ABCD$ of radius ($r$) with centre $O$ and $X$-$X'$ and $Y$-$Y'$ be two axes of reference through $O$ as shown in Fig. 10.5.

Now consider an elementary ring of radius $x$ and thickness $dx$. Therefore area of the ring,

$$da = 2\,\pi\,x.\ dx$$

and moment of inertia of ring, about $X$-$X$ axis or $Y$-$Y$ axis

$$= \text{Area} \times (\text{Distance})^2$$
$$= 2\,\pi\,x.\ dx \times x^2$$
$$= 2\,\pi\,x^3.\ dx$$

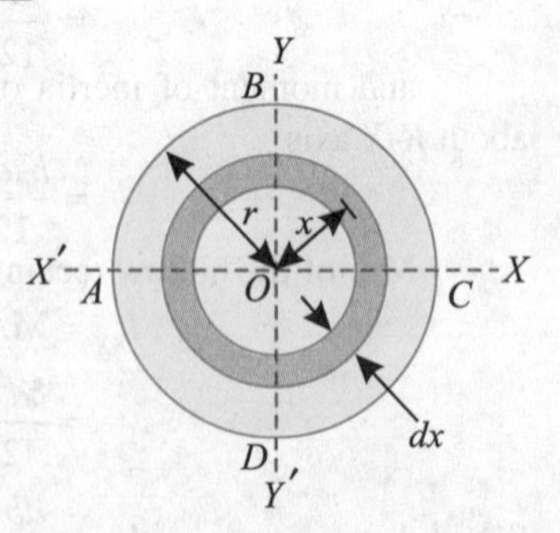


Fig. 10.5. Circular section.

Now moment of inertia of the whole section, about the central axis, can be found out by integrating the above equation for the whole radius of the circle *i.e.*, from 0 to $r$.

$$\therefore \quad I_{ZZ} = \int_0^r 2\pi x^3 .\ dx = 2\pi \int_0^r x^3 .\ dx$$

$$I_{ZZ} = 2\pi \left[ \frac{x^4}{4} \right]_0^r = \frac{\pi}{2} (r)^4 = \frac{\pi}{32} (d)^4 \qquad \text{...}\left( \text{substituting } r = \frac{d}{2} \right)$$

We know from the Theorem of Perpendicular Axis that

$$I_{XX} + I_{YY} = I_{ZZ}$$

$$\therefore \qquad {}^{*}I_{XX} = I_{YY} = \frac{I_{ZZ}}{2} = \frac{1}{2} \times \frac{\pi}{32} (d)^4 = \frac{\pi}{64} (d)^4$$

**EXAMPLE 10.3.** *Find the moment of inertia of a circular section of 50 mm diameter about an axis passing through its centre.*

**SOLUTION.** Given: Diameter $(d) = 50$ mm

We know that moment of inertia of the circular section about an axis passing through its centre,

$$I_{XX} = \frac{\pi}{64} (d)^4 = \frac{\pi}{64} \times (50)^4 = 307 \times 10^3 \text{ mm}^4 \quad \textbf{Ans.}$$

## 10.11. Moment of Inertia of a Hollow Circular Section

Consider a hollow circular section as shown in Fig.10.6, whose moment of inertia is required to be found out.

Let $D$ = Diameter of the main circle, and

$d$ = Diameter of the cut out circle.

We know that the moment of inertia of the main circle about $X$-$X$ axis

$$= \frac{\pi}{64} (D)^4$$

and moment of inertia of the cut-out circle about $X$-$X$ axis

$$= \frac{\pi}{64} (d)^4$$

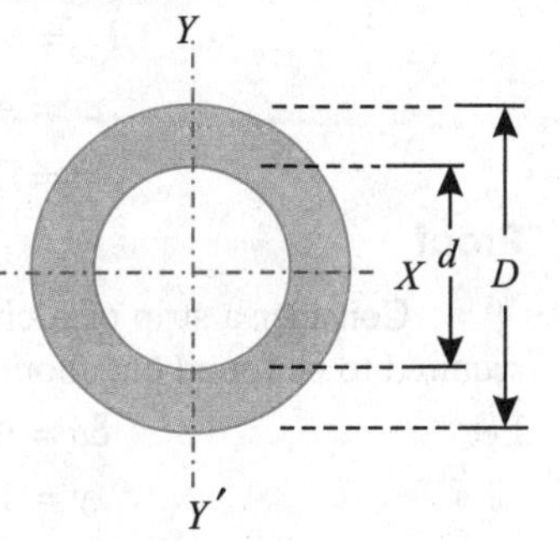


Fig. 10.6. Hollow circular section.

$\therefore$ Moment of inertia of the hollow circular section about $X$-$X$ axis,

$I_{XX}$ = Moment of inertia of main circle – Moment of inertia of cut out circle,

$$= \frac{\pi}{64} (D)^4 - \frac{\pi}{64} (d)^4 = \frac{\pi}{64} (D^4 - d^4)$$

Similarly, $$I_{YY} = \frac{\pi}{64} (D^4 - d^4)$$

**NOTE :** This relation holds good only if the centre of the main circular section as well as that of the cut out circular section coincide with each other.

* This may also be obtained by Routh's rule as discussed below

$$I_{XX} = \frac{AS}{4} \qquad \text{(for circular section)}$$

where area, $A = \frac{\pi}{4} \times d^2$ and sum of the square of semi axis $Y$-$Y$ and $Z$-$Z$,

$$S = \left( \frac{d}{2} \right)^2 + 0 = \frac{d^2}{4}$$

$$\therefore \qquad I_{XX} = \frac{AS}{4} = \frac{\left[ \frac{\pi}{4} \times d^2 \right] \times \frac{d^2}{4}}{4} = \frac{\pi}{64} (d)^4$$

**EXAMPLE 10.4.** *A hollow circular section has an external diameter of 80 mm and internal diameter of 60 mm. Find its moment of inertia about the horizontal axis passing through its centre.*

**SOLUTION.** Given : External diameter ($D$) = 80 mm and internal diameter ($d$) = 60 mm.

We know that moment of inertia of the hollow circular section about the horizontal axis passing through its centre,

$$I_{XX} = \frac{\pi}{64}(D^4 - d^4) = \frac{\pi}{64}[(80)^4 - (60)^4] = 1374 \times 10^3 \text{ mm}^4 \quad \textbf{Ans.}$$

## 10.12. Theorem of Parallel Axis

It states, *If the moment of inertia of a plane area about an axis through its centre of gravity is denoted by $I_G$, then moment of inertia of the area about any other axis AB, parallel to the first, and at a distance h from the centre of gravity is given by:*

$$I_{AB} = I_G + ah^2$$

where $I_{AB}$ = Moment of inertia of the area about an axis *AB*,

$I_G$ = Moment of Inertia of the area about its centre of gravity

$a$ = Area of the section, and

$h$ = Distance between centre of gravity of the section and axis *AB*.

**Proof**

Consider a strip of a circle, whose moment of inertia is required to be found out about a line *AB* as shown in Fig. 10.7.

Let $\delta a$ = Area of the strip

$y$ = Distance of the strip from the centre of gravity the section and

$h$ = Distance between centre of gravity of the section and the axis *AB*.

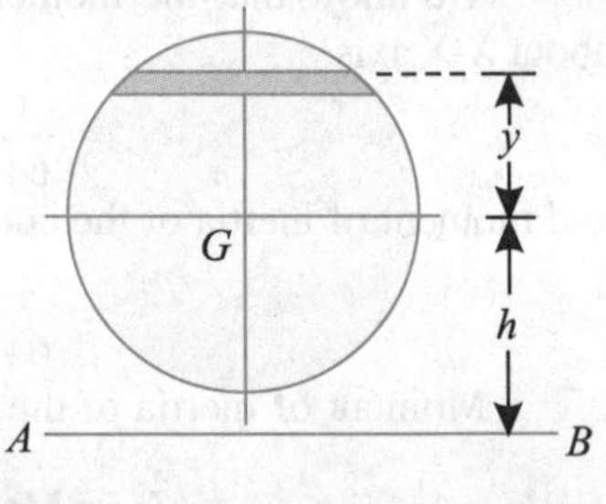

Fig. 10.7. Theorem of parallel axis.

We know that moment of inertia of the whole section about an axis passing through the centre of gravity of the section

$$= \delta a.\ y^2$$

and moment of inertia of the whole section about an axis passing through its centre of gravity,

$$I_G = \Sigma\ \delta a.\ y^2$$

$\therefore$ Moment of inertia of the section about the axis *AB*,

$$I_{AB} = \Sigma\ \delta a\ (h + y)^2 = \Sigma\ \delta a\ (h^2 + y^2 + 2\,h\,y)$$

$$= (\Sigma\ h^2.\ \delta a) + (\Sigma\ y^2.\ \delta a) + (\Sigma\ 2\,h\,y\ .\ \delta a)$$

$$= a\,h^2 + I_G + 0$$

It may be noted that $\Sigma\ h^2\ .\ \delta a = a\,h^2$ and $\Sigma\ y^2\ .\ \delta a = I_G$ [as per equation (*i*) above] and $\Sigma\ \delta a.y$ is the algebraic sum of moments of all the areas, about an axis through centre of gravity of the section and is equal to $a.\bar{y}$, where $\bar{y}$ is the distance between the section and the axis passing through the centre of gravity, which obviously is zero.

## 10.13. Moment of Inertia of a triangular Section

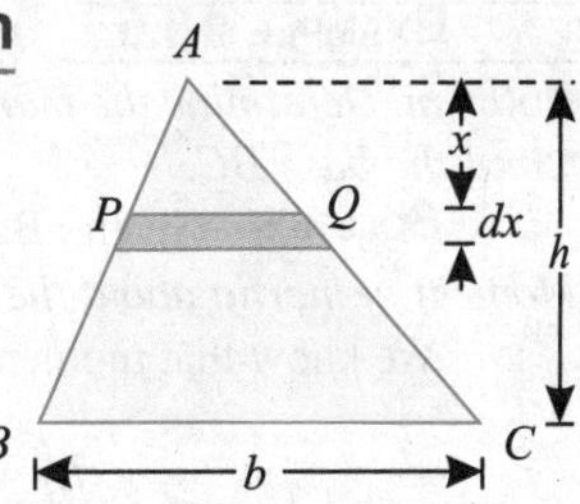

Fig. 10.8. Triangular section.

Consider a triangular section *ABC* whose moment of inertia is required to be found out.

Let $b$ = Base of the triangular section and

$h$ = Height of the triangular section.

Now consider a small strip *PQ* of thickness *dx* at a distance of *x* from the vertex *A* as shown in Fig. 10.8. From the geometry of the figure, we find that the two triangles *APQ* and *ABC* are similar. Therefore

$$\frac{PQ}{BC} = \frac{x}{h} \quad \text{or} \quad PQ = \frac{BC.x}{h} = \frac{bx}{h} \qquad (\because BC = \text{base} = b)$$

We know that area of the strip *PQ*

$$= \frac{bx}{h} \cdot dx$$

and moment of inertia of the strip about the base *BC*

$$= \text{Area} \times (\text{Distance})^2 = \frac{bx}{h}\, dx\, (h - x)^2 = \frac{bx}{h}\, (h - x)^2\, dx$$

Now moment of inertia of the whole triangular section may be found out by integrating the above equation for the whole height of the triangle *i.e.*, from 0 to *h*.

$$I_{BC} = \int_0^h \frac{b\,x}{h}\, (h - x)^2\, dx$$

$$= \frac{b}{h} \int_0^h x\, (h^2 + x^2 - 2hx)\, dx$$

$$= \frac{b}{h} \int_0^h (xh^2 + x^3 - 2hx^2)\, dx$$

$$= \frac{b}{h} \left[ \frac{x^2 h^2}{2} + \frac{x^4}{4} - \frac{2hx^3}{3} \right]_0^h = \frac{bh^3}{12}$$

We know that distance between centre of gravity of the triangular section and base *BC*,

$$d = \frac{h}{3}$$

∴ Moment of inertia of the triangular section about an axis through its centre of gravity and parallel to *X-X* axis,

$$I_G = I_{BC} - ad^2 \qquad ...(\because I_{XX} = I_G + a\,h^2)$$

$$= \frac{bh^3}{12} - \left(\frac{bh}{2}\right)\left(\frac{h}{3}\right)^2 = \frac{bh^3}{36}$$

**Notes : 1.** The moment of inertia of section about an axis through its vertex and parallel to the base

$$= I_G + a\,d^2 = \frac{bh^3}{36} + \left(\frac{bh}{2}\right)\left(\frac{2h}{3}\right)^2 = \frac{9bh^3}{36} = \frac{bh^3}{4}$$

**2.** This relation holds good for any type of triangle.

**EXAMPLE 10.5.** *An isosceles triangular section ABC has base width 80 mm and height 60 mm. Determine the moment of inertia of the section about the centre of gravity of the section and the base BC.*

**SOLUTION.** Given : Base width ($b$) = 80 mm and height ($h$) = 60 mm.

*Moment of inertia about the centre of gravity of the section*

We know that moment of inertia of triangular section about its centre of gravity,

$$I_G = \frac{b\,h^3}{36} = \frac{80 \times (60)^3}{36} = 480 \times 10^3 \text{ mm}^4$$

*Moment of inertia about the base BC*

We also know that moment of inertia of triangular section about the base $BC$,

$$I_{BC} = \frac{b\,h^3}{12} = \frac{80 \times (60)^3}{12} = 1440 \times 10^3 \text{ mm}^4$$

**EXAMPLE 10.6.** *A hollow triangular section shown in Fig. 10.9 is symmetrical about its vertical axis.*

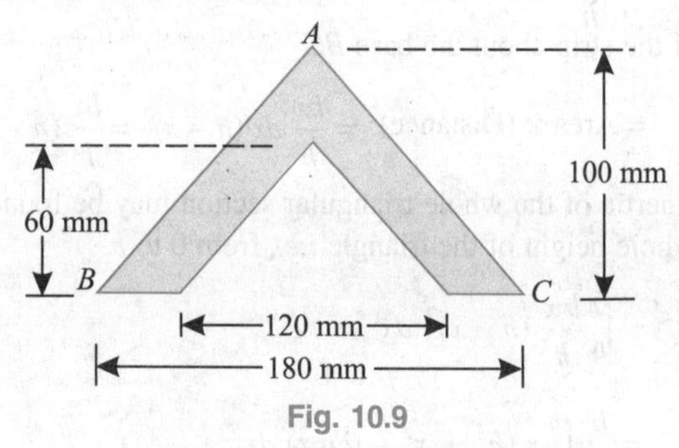


Fig. 10.9

*Find the moment of inertia of the section about the base BC.*

**SOLUTION.** Given : Base width of main triangle ($B$) = 180 mm; Base width of cut out triangle ($b$) = 120 mm; Height of main triangle ($H$) = 100 mm and height of cut out triangle ($h$) = 60 mm.

We know that moment of inertia of the triangular, section about the base $BC$,

$$I_{BC} = \frac{BH^3}{12} - \frac{bh^3}{12} = \frac{180 \times (100)^3}{12} - \frac{120 \times (60)^3}{12} \text{ mm}^4$$

$$= (15 \times 10^6) - (2.16 \times 10^6) = 12.84 \times 10^6 \text{ mm}^4$$ **Ans.**

## 10.14. Moment of Inertia of a Semicircular Section

Consider a semicircular section $ABC$ whose moment of inertia is required to be found out as shown in Fig. 10.10.

Let $r$ = Radius of the semicircle.

We know that moment of inertia of the semicircular section about the base $AC$ is equal to half the moment of inertia of the circular section about $AC$. Therefore moment of inertia of the semicircular section $ABC$ about the base $AC$,

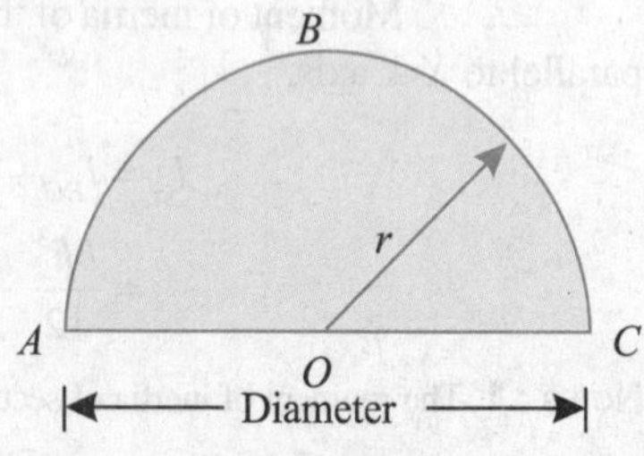


Fig. 10.10. Semicircular section ABC.

$$I_{AC} = \frac{1}{2} \times \frac{\pi}{64} \times (d)^4 = 0.393\, r^4$$

We also know that area of semicircular section,

$$a = \frac{1}{2} \times \pi r^2 \; \frac{\pi r^2}{2}$$

and distance between centre of gravity of the section and the base $AC$,

$$h = \frac{4r}{3\pi}$$

∴ Moment of inertia of the section through its centre of gravity and parallel to $x$-$x$ axis,

$$I_G = I_{AC} - ah^2 = \left[\frac{\pi}{8} \times (r)^4\right] - \left[\frac{\pi r^2}{2}\left(\frac{4r}{3\pi}\right)^2\right]$$

$$= \left[\frac{\pi}{8} \times (r)^4\right] - \left[\frac{8}{9\pi} \times (r)^4\right] = 0.11\, r^4$$

**Note.** The moment of inertia about $y$-$y$ axis will be the same as that about the base $AC$ *i.e.*, $0.393\, r^4$.

**Example 10.7.** *Determine the moment of inertia of a semicircular section of 100 mm diameter about its centre of gravity and parallel to X-X and Y-Y axes.*

**Solution.** Given: Diameter of the section ($d$) = 100 mm or radius ($r$) = 50 mm

*Moment of inertia of the section about its centre of gravity and parallel to X-X axis*

We know that moment of inertia of the semicircular section about its centre of gravity and parallel to $X$-$X$ axis,

$$I_{XX} = 0.11\, r^4 = 0.11 \times (50)^4 = 687.5 \times 10^3 \text{ mm}^4 \quad \textbf{Ans.}$$

*Moment of inertia of the section about its centre of gravity and parallel to Y-Y axis.*

We also know that moment of inertia of the semicircular section about its centre of gravity and parallel to $Y$-$Y$ axis.

$$I_{YY} = 0.393\, r^4 = 0.393 \times (50)^4 = 2456 \times 10^3 \text{ mm}^4 \quad \textbf{Ans.}$$

**Example 10.8.** *A hollow semicircular section has its outer and inner diameter of 200 mm and 120 mm respectively as shown in Fig. 10.11.*

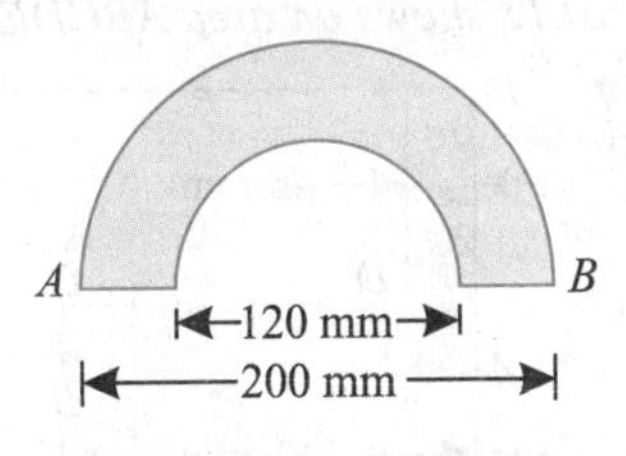


Fig. 10.11

*What is its moment of inertia about the base AB ?*

**Solution.** Given: Outer diameter ($D$) = 200 mm or Outer Radius ($R$) = 100 mm and inner diameter ($d$) = 120 mm or inner radius ($r$) = 60 mm.

We know that moment of inertia of the hollow semicircular section about the base $AB$,

$$I_{AB} = 0.393\,(R^4 - r^4) = 0.393\,[(100)^4 - (60)^4] = 34.21 \times 10^6 \text{ mm}^4 \quad \textbf{Ans.}$$

## EXERCISE 10.1

1. Find the moment of inertia of a rectangular section 60 mm wide and 40 mm deep about its centre of gravity. [**Ans.** $I_{XX} = 320 \times 10^3$ mm$^4$ ; $I_{YY} = 720 \times 10^3$ mm$^4$]
2. Find the moment of inertia of a hollow rectangular section about its centre of gravity, if the external dimensions are 40 mm deep and 30 mm wide and internal dimensions are 25 mm deep and 15 mm wide. [**Ans.** $I_{XX} = 140\,470$ mm$^4$ : $I_{YY} = 82\,970$ mm$^4$]

3. Find the moment of inertia of a circular section of 20 mm diameter through its centre of gravity. [**Ans.** 7854 $mm^4$]
4. Calculate the moment of inertia of a hollow circular section of external and internal diameters 100 mm and 80 mm respectively about an axis passing through its centroid. [**Ans.** $2.898 \times 10^6$ $mm^4$]
5. Find the moment of inertia of a triangular section having 50 mm base and 60 mm height about an axis through its centre of gravity and base. [**Ans.** $300 \times 10^3$ $mm^4$: $900 \times 10^3$ $mm^4$]
6. Find the moment of inertia of a semicircular section of 30 mm radius about its centre of gravity and parallel to *X-X* and *Y-Y* axes. [**Ans.** 89 100 $mm^4$ : 381 330 $mm^4$]

## 10.15. Moment of Inertia of a Composite Section

The moment of inertia of a composite section may be found out by the following steps :

1. First of all, split up the given section into plane areas (*i.e.*, rectangular, triangular, circular etc., and find the centre of gravity of the section).
2. Find the moments of inertia of these areas about their respective centres of gravity.
3. Now transfer these moment of inertia about the required axis (*AB*) by the Theorem of Parallel Axis, *i.e.*,

$$I_{AB} = I_G + ah^2$$

where $I_G$ = Moment of inertia of a section about its centre of gravity and parallel to the axis.
$a$ = Area of the section,
$h$ = Distance between the required axis and centre of gravity of the section.

4. The moments of inertia of the given section may now be obtained by the algebraic sum of the moment of inertia about the required axis.

**EXAMPLE 10.9.** *Figure 10.12 shows an area ABCDEF.*

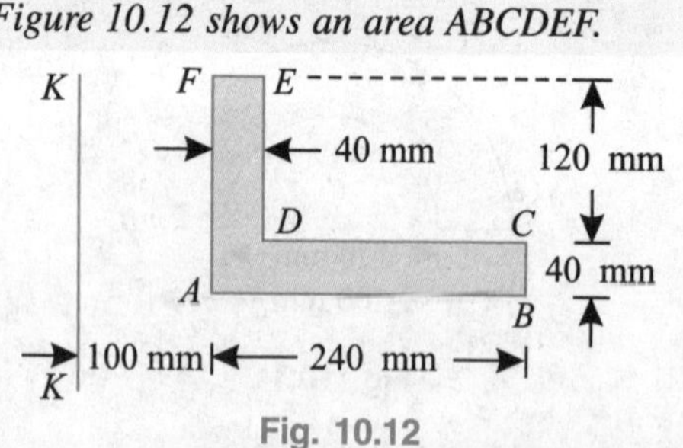


Fig. 10.12

*Compute the moment of inertia of the above area about axis K-K.*

**SOLUTION.** As the moment of inertia is required to be found out about the axis *K-K*, therefore there is no need of finding out the centre of gravity of the area.

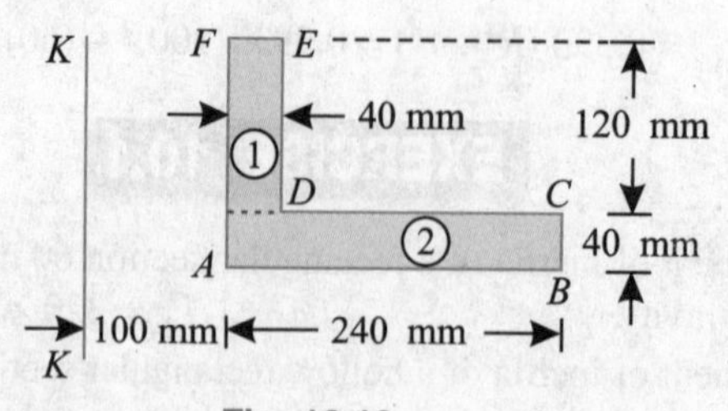


Fig. 10.13

Let us split up the area into two rectangles 1 and 2 as shown in Fig. 10.13.

We know that moment of inertia of section (1) about its centre of gravity and parallel to axis *K-K*,

$$I_{G1} = \frac{120 \times (40)^3}{12} = 640 \times 10^3 \text{ mm}^4$$

and distance between centre of gravity of section (1) and axis *K-K*,

$$h_1 = 100 + \frac{40}{2} = 120 \text{ mm}$$

∴ Moment of inertia of section (1) about axis *K-K*

$$= I_{G1} + a_1 h_1^2 = (640 \times 10^3) + [(120 \times 40) \times (120)^2] = 69.76 \times 10^6 \text{ mm}^4$$

Similarly, moment of inertia of section (2) about its centre of gravity and parallel to axis *K-K*,

$$I_{G2} = \frac{40 \times (240)^3}{12} = 46.08 \times 10^6 \text{ mm}^4$$

and distance between centre of gravity of section (2) and axis *K-K*,

$$h_2 = 100 + \frac{240}{2} = 220 \text{ mm}$$

∴ Moment of inertia of section (2) about the axis *K-K*,

$$= I_{G2} + a_2 h_2^2 = (46.08 \times 10^6) + [(240 \times 40) \times (220)^2] = 510.72 \times 10^6 \text{ mm}^4$$

Now moment of inertia of the whole area about axis *K-K*,

$$I_{KK} = (69.76 \times 10^6) + (510.72 \times 10^6) = 580.48 \times 10^6 \text{ mm}^4 \quad \textbf{Ans.}$$

**EXAMPLE 10.10.** *Find the moment of inertia of a T-section with flange as 150 mm × 50 mm and web as 150 mm × 50 mm about X-X and Y-Y axes through the centre of gravity of the section.*

**SOLUTION.** The given *T*-section is shown in Fig. 10.14.

First of all, let us find out centre of gravity of the section. As the section is symmetrical about *Y-Y* axis, therefore its centre of gravity will lie on this axis. Split up the whole section into two rectangles *viz*., 1 and 2 as shown in figure. Let bottom of the web be the axis of reference.

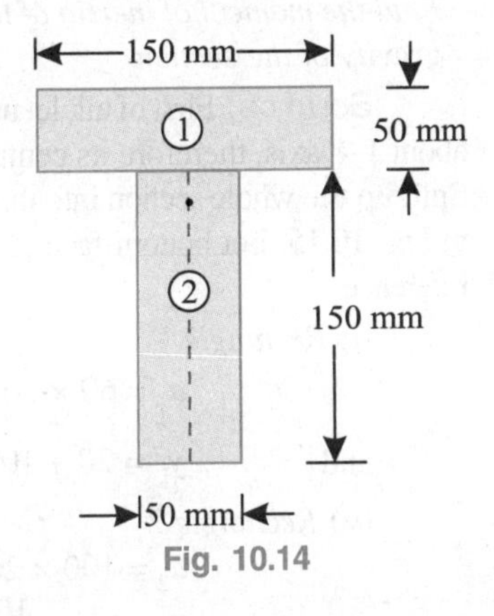

Fig. 10.14

(*i*) *Rectangle* (1)

$$a_1 = 150 \times 50 = 7500 \text{ mm}^2$$

and $$y_1 = 150 + \frac{50}{2} = 175 \text{ mm}$$

(*ii*) *Rectangle* (2)

$$a_2 = 150 \times 50 = 7500 \text{ mm}^2$$

and $$y_2 = \frac{150}{2} = 75 \text{ mm}$$

We know that distance between centre of gravity of the section and bottom of the web,

$$\bar{y} = \frac{a_1 y_1 + a_2 y_2}{a_1 + a_2} = \frac{(7500 \times 175) + (7500 \times 75)}{7500 + 7500} = 125 \text{ mm}$$

*Moment of inertia about X-X axis*

We also know that M.I. of rectangle (1) about an axis through its centre of gravity and parallel to *X-X* axis.

$$I_{G1} = \frac{150\,(50)^3}{12} = 1.5625 \times 10^6 \text{ mm}^4$$

and distance between centre of gravity of rectangle (1) and *X-X* axis,

$$h_1 = 175 - 125 = 50 \text{ mm}$$

∴ Moment of inertia of rectangle (1) about $X$-$X$ axis

$$I_{G1} + a_1 h_1^2 = (1.5625 \times 10^6) + [7500 \times (50)^2] = 20.3125 \times 10^6 \text{ mm}^4$$

Similarly, moment of inertia of rectangle (2) about an axis through its centre of gravity and parallel to $X$-$X$ axis,

$$I_{G2} = \frac{50\,(150)^3}{12} = 14.0625 \times 10^6 \text{ mm}^4$$

and distance between centre of gravity of rectangle (2) and $X$-$X$ axis,

$$h_2 = 125 - 75 = 50 \text{ mm}$$

∴ Moment of inertia of rectangle (2) about $X$-$X$ axis

$$= I_{G2} + a_2 h_2^2 = (14.0625 \times 10^6) + [7500 \times (50)^2] = 32.8125 \times 10^6 \text{ mm}^4$$

Now moment of inertia of the whole section about $X$-$X$ axis,

$$I_{XX} = (20.3125 \times 10^6) + (32.8125 \times 10^6) = 53.125 \times 10^6 \text{ mm}^4 \quad \textbf{Ans.}$$

*Moment of inertia about Y-Y axis*

We know that M.I. of rectangle (1) about $Y$-$Y$ axis

$$= \frac{50\,(150)^3}{12} = 14.0625 \times 10^6 \text{ mm}^4$$

and moment of inertia of rectangle (2) about $Y$-$Y$ axis,

$$= \frac{150\,(50)^3}{12} = 1.5625 \times 10^6 \text{ mm}^4$$

Now moment of inertia of the whole section about $Y$-$Y$ axis,

$$I_{YY} = (14.0625 \times 10^6) + (1.5625 \times 10^6) = 15.625 \times 10^6 \text{ mm}^4 \quad \textbf{Ans.}$$

**EXAMPLE 10.11.** *An I-section is made up of three rectangles as shown in Fig. 10.15. Find the moment of inertia of the section about the horizontal axis passing through the centre of gravity of the section.*

**SOLUTION.** First of all, let us find out centre of gravity of the section. As the section is symmetrical about $Y$-$Y$ axis, therefore its centre of gravity will lie on this axis.

Split up the whole section into three rectangles 1, 2 and 3 as shown in Fig. 10.15. Let bottom face of the bottom flange be the axis of reference.

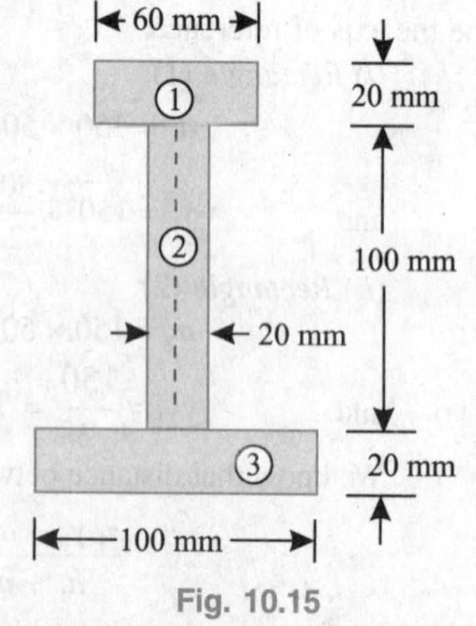

Fig. 10.15

(i) *Rectangle* 1

$$a_1 = 60 \times 20 = 1200 \text{ mm}$$

and $$y_1 = 20 + 100 + \frac{20}{2} = 130 \text{ mm}$$

(ii) *Rectangle* 2

$$a_2 = 100 \times 20 = 2000 \text{ mm}^2$$

and $$y_2 = 20 + \frac{100}{2} = 70 \text{ mm}$$

(iii) *Rectangle* 3

$$a_3 = 100 \times 20 = 2000 \text{ mm}^2$$

and $$y_3 = \frac{20}{2} = 10 \text{ mm}$$

We know that the distance between centre of gravity of the section and bottom face,

$$\bar{y} = \frac{a_1 y_1 + a_2 y_2 + a_3 y_3}{a_1 + a_2 + a_3} = \frac{(1200 \times 130) + (2000 \times 70) + (2000 \times 10)}{1200 + 2000 + 2000} \text{ mm}$$

$$= 60.8 \text{ mm}$$

We know that moment of inertia of rectangle (1) about an axis through its centre of gravity and parallel to *X-X* axis,

$$I_{G1} = \frac{60 \times (20)^3}{12} = 40 \times 10^3 \text{ mm}^4$$

and distance between centre of gravity of rectangle (1) and *X-X* axis,

$$h_1 = 130 - 60.8 = 69.2 \text{ mm}$$

∴ Moment of inertia of rectangle (1) about *X-X* axis,

$$= I_{G1} + a_1 h_1^2 = (40 \times 10^3) + [1200 \times (69.2)^2] = 5786 \times 10^3 \text{ mm}^4$$

Similarly, moment of inertia of rectangle (2) about an axis through its centre of gravity and parallel to *X-X* axis,

$$I_{G2} = \frac{20 \times (100)^3}{12} = 1666.7 \times 10^3 \text{ mm}^4$$

and distance between centre of gravity of rectangle (2) and *X-X* axis,

$$h_2 = 70 - 60.8 = 9.2 \text{ mm}$$

∴ Moment of inertia of rectangle (2) about *X-X* axis,

$$= I_{G2} + a_2 h_2^2 = (1666.7 \times 10^3) + [2000 \times (9.2)^2] = 1836 \times 10^3 \text{ mm}^4$$

Now moment of inertia of rectangle (3) about an axis through its centre of gravity and parallel to *X-X* axis,

$$I_{G3} = \frac{100 \times (20)^3}{12} = 66.7 \times 10^3 \text{ mm}^4$$

and distance between centre of gravity of rectangle (3) and *X-X* axis,

$$h_3 = 60.8 - 10 = 50.8 \text{ mm}$$

∴ Moment of inertia of rectangle (3) about *X-X* axis,

$$= I_{G3} + a_3 h_3^2 = (66.7 \times 10^3) + [2000 \times (50.8)^2] = 5228 \times 10^3 \text{ mm}^4$$

Now moment of inertia of the whole section about *X-X* axis,

$$I_{XX} = (5786 \times 10^3) + (1836 \times 10^3) + (5228 \times 10^3) = 12\,850 \times 10^3 \text{ mm}^4$$ **Ans.**

**EXAMPLE 10.12.** *Find the moment of inertia about the centroidal X-X and Y-Y axes of the angle section shown in Fig. 10.16.*

**SOLUTION.** First of all, let us find the centre of gravity of the section. As the section is not symmetrical about any section, therefore we have to find out the values of $\bar{x}$ and $\bar{y}$ for the angle section. Split up the section into two rectangles (1) and (2) as shown in Fig. 10.16.

*Moment of inertia about centroidal X-X axis*

Let bottom face of the angle section be the axis of reference.

*Rectangle* (1)

$$a_1 = 100 \times 20 = 2000 \text{ mm}^2$$

and $$y_1 = \frac{100}{2} = 50 \text{ mm}$$

*Rectangle* (2)

$$a_2 = (80 - 20) \times 20 = 1200 \text{ mm}^2$$

and $$y_2 = \frac{20}{2} = 10 \text{ mm}$$

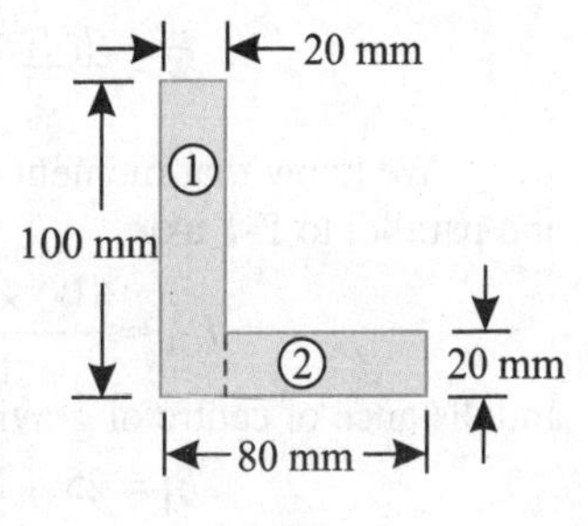


Fig. 10.16

We know that distance between the centre of gravity of the section and bottom face,

$$\bar{y} = \frac{a_1 y_1 + a_2 y_2}{a_1 + a_2} = \frac{(2000 \times 50) + (1200 \times 10)}{2000 + 1200} = 35 \text{ mm}$$

We know that moment of inertia of rectangle (1) about an axis through its centre of gravity and parallel to *X-X* axis,

$$I_{G1} = \frac{20 \times (100)^3}{12} = 1.667 \times 10^6 \text{ mm}^4$$

and distance of centre of gravity of rectangle (1) from *X-X* axis,

$$h_1 = 50 - 35 = 15 \text{ mm}$$

∴ Moment of inertia of rectangle (1) about *X-X* axis

$$= I_{G1} + a h_1^2 = (1.667 \times 10^6) + [2000 \times (15)^2] = 2.117 \times 10^6 \text{ mm}^4$$

Similarly, moment of inertia of rectangle (2) about an axis through its centre of gravity and parallel to *X-X* axis,

$$I_{G2} = \frac{60 \times (20)^3}{12} \; 0.04 \times 10^6 \text{ mm}^4$$

and distance of centre of gravity of rectangle (2) from *X-X* axis,

$$h_2 = 35 - 10 = 25 \text{ mm}$$

∴ Moment of inertia of rectangle (2) about *X-X* axis

$$= I_{G2} + a h_2^2 = (0.04 \times 10^6) + [1200 \times (25)^2] = 0.79 \times 10^6 \text{ mm}^4$$

Now moment of inertia of the whole section about *X-X* axis,

$$I_{XX} = (2.117 \times 10^6) + (0.79 \times 10^6) = 2.907 \times 10^6 \text{ mm}^4 \qquad \textbf{Ans.}$$

*Moment of inertia about centroidal Y-Y axis*

Let left face of the angle section be the axis of reference.

*Rectangle* (1)

$$a_1 = 2000 \text{ mm}^2 \qquad \text{...(As before)}$$

and $x_1 = \frac{20}{2} = 10$ mm

*Rectangle* (2)

$$a_2 = 1200 \text{ mm}^2 \qquad \text{...(As before)}$$

and $x_2 = 20 + \frac{60}{2} = 50$ mm

We know that distance between the centre of gravity of the section and left face,

$$\bar{x} = \frac{a_1 x_1 + a_2 x_2}{a_1 + a_2} = \frac{(2000 \times 10) + (1200 \times 50)}{2000 + 1200} = 25 \text{ mm}$$

We know that moment of inertia of rectangle (1) about an axis through its centre of gravity and parallel to *Y-Y* axis,

$$I_{G1} = \frac{100 \times (20)^3}{12} = 0.067 \times 10^6 \text{ mm}^4$$

and distance of centre of gravity of rectangle (1) from *Y-Y* axis,

$$h_1 = 25 - 10 = 15 \text{ mm}$$

∴ Moment of inertia of rectangle (1) about *Y-Y* axis

$$= I_{G1} + a_1 h_1^2 = (0.067 \times 10^6) + [2000 \times (15)^2] = 0.517 \times 10^6 \text{ mm}^4$$

Similarly, moment of inertia of rectangle (2) about an axis through its centre of gravity and parallel to *Y-Y* axis,

$$I_{G2} = \frac{20 \times (60)^3}{12} = 0.36 \times 10^6 \text{ mm}^4$$

and distance of centre of gravity of rectangle (2) from *Y-Y* axis,

$$h_2 = 50 - 25 = 25 \text{ mm},$$

∴ Moment of inertia of rectangle (2) about *Y-Y* axis

$$= I_{G2} + a_2 h_2^2 = 0.36 \times 10^6 + [1200 \times (25)^2] = 1.11 \times 10^6 \text{ mm}^4$$

Now moment of inertia of the whole section about *Y-Y* axis,

$$I_{YY} = (0.517 \times 10^6) + (1.11 \times 10^6) = 1.627 \times 10^6 \text{ mm}^4 \qquad \textbf{Ans.}$$

**EXAMPLE 10.13.** *Figure 10.17 shows the cross-section of a cast iron beam.*

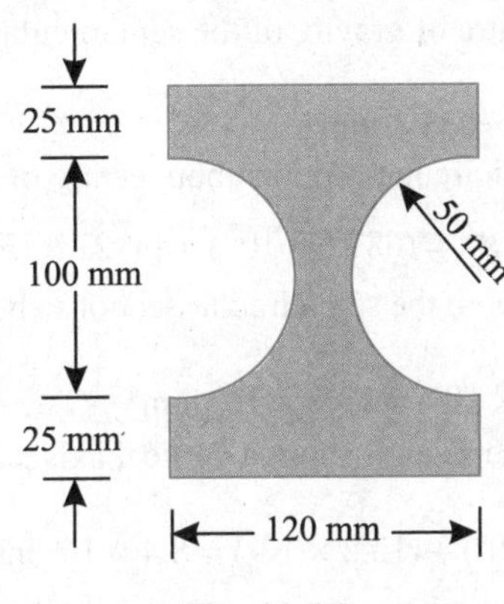


Fig. 10.17

*Determine the moments of inertia of the section about horizontal and vertical axes passing through the centroid of the section.*

**SOLUTION.** As the section is symmetrical about its horizontal and vertical axes, therefore centre of gravity of the section will lie at the centre of the rectangle. A little consideration will show that when the two semicircles are placed together, it will form a circular hole with 50 mm radius or 100 mm diameter.

*Moment of inertia of the section about horizontal axis passing through the centroid of the section.*

We know that moment of inertia of the rectangular section about its horizontal axis passing through its centre of gravity,

$$= \frac{bd^3}{12} = \frac{120 \times (150)^3}{12} = 33.75 \times 10^6 \text{ mm}^4$$

and moment of inertia of the circular section about a horizontal axis passing through its centre of gravity,

$$= \frac{\pi}{4}(r)^4 = \frac{\pi}{4}(50)^4 = 4.91 \times 10^6 \text{ mm}^4$$

∴ Moment of inertia of the whole section about horizontal axis passing through the centroid of the section,

$$I_{XX} = (33.75 \times 10^6) - (4.91 \times 10^6) = 28.84 \times 10^6 \text{ mm}^4 \qquad \textbf{Ans.}$$

*Moment of inertia of the section about vertical axis passing through the centroid of the section*

We know that moment of inertia of the rectangular section about the vertical axis passing through its centre of gravity,

$$I_{G1} = \frac{db^3}{12} = \frac{150 \times (120)^3}{12} = 21.6 \times 10^6 \text{ mm}^4 \quad \ldots(i)$$

and area of one semicircular section with 50 mm radius,

$$a = \frac{\pi r^2}{2} = \frac{\pi (50)^2}{2} = 3927 \text{ mm}^2$$

We also know that moment of inertia of a semicircular section about a vertical axis passing through its centre of gravity,

$$I_{G2} = 0.11\, r^4 = 0.11 \times (50)^4 = 687.5 \times 10^3 \text{ mm}^4$$

and distance between centre of gravity of the semicircular section and its base

$$= \frac{4r}{3\pi} = \frac{4 \times 50}{3\pi} = 21.2 \text{ mm}$$

∴ Distance between centre of gravity of the semicircular section and centre of gravity of the whole section,

$$h_2 = 60 - 21.2 = 38.8 \text{ mm}$$

and moment of inertia of one semicircular section about centre of gravity of the whole section,

$$= I_{G2} + a_2 h_2^2 = (687.5 \times 10^3) + [3927 \times (38.8)^2] = 6.6 \times 10^6 \text{ mm}^4$$

∴ Moment of inertia of both the semicircular sections about centre of gravity of the whole section,

$$= 2 \times (6.6 \times 10^6) = 13.2 \times 10^6 \text{ mm}^4 \quad \ldots(ii)$$

and moment of inertia of the whole section about a vertical axis passing through the centroid of the section,

$$= (21.6 \times 10^6) - (13.2 \times 10^6) = 8.4 \times 10^6 \text{ mm}^4 \quad \textbf{Ans.}$$

**EXAMPLE 10.14.** *Find the moment of inertia of a hollow section shown in Fig. 10.18. about an axis passing through its centre of gravity or parallel X-X axis.*

**SOLUTION.** As the section is symmentrical about $Y$-$Y$ axis, therefore centre of a gravity of the section will lie on this axis. Let $\bar{y}$ be the distance between centre of gravity of the section from the bottom face.

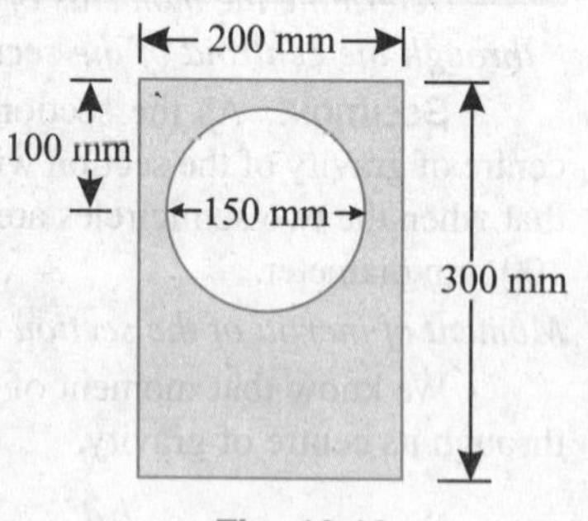

Fig. 10.18

*(i) Rectangle*

$$a_1 = 300 \times 200 = 60\,000 \text{ mm}^2$$

and $$y_1 = \frac{300}{2} = 150 \text{ mm}$$

*(ii) Circular hole*

$$a_2 = \frac{\pi}{4} \times (150)^2 = 17\,670 \text{ mm}^2$$

and $$y_2 = 300 - 100 = 200 \text{ mm}$$

We know that distance between the centre of gravity of the section and its bottom face,

$$\bar{y} = \frac{a_1 y_1 - a_2 y_2}{a_1 - a_2} = \frac{(60000 \times 150) - (17670 \times 200)}{60000 - 17670} = 129.1 \text{ mm}$$

∴ Moment of inertia of rectangular section about an axis through its centre of gravity and parallel to $X$-$X$ axis,

$$I_{G1} = \frac{200 \times (300)^3}{12} = 450 \times 10^6 \text{ mm}^4$$

and distance of centre of gravity of rectangular section and *X-X* axis,

$$h_1 = 150 - 129.1 = 20.9 \text{ mm}$$

∴ Moment of inertia of rectangle about *X-X* axis

$$= I_{G1} + ah^2 = (450 \times 10^6) + [(300 \times 200) \times (20.9)]^2 = 476.21 \times 10^6 \text{ mm}^4$$

Similarly, moment of inertia of circular section about an axis through its centre of gravity and parallel to *X-X* axis,

$$I_{G2} = \frac{\pi}{64} \times (150)^4 = 24.85 \times 10^6 \text{ mm}^4$$

and distance between centre of gravity of the circular section and *X-X* axis,

$$h_2 = 200 - 129.1 = 70.9 \text{ mm}$$

∴ Moment of inertia of the circular section about *X-X* axis,

$$= I_{G2} + ah^2 = (24.85 \times 10^6) + [(17\,670) \times (70.9)^2] = 113.67 \times 10^6 \text{ mm}^4$$

Now moment of inertia of the whole section about *X-X* axis

$$= (476.21 \times 10^6) - (113.67 \times 10^6) = 362.54 \times 10^6 \text{ mm}^4$$ **Ans.**

**EXAMPLE 10.15.** *A rectangular hole is made in a triangular section as shown in Fig. 10.19.*

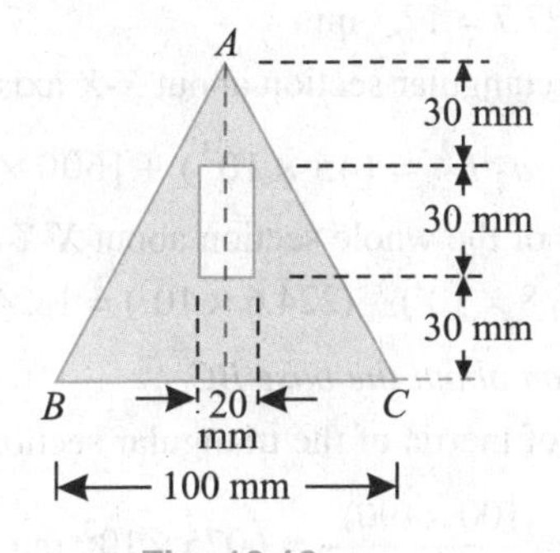


Fig. 10.19

*Determine the moment of inertia of the section about X-X axis passing through its centre of gravity and the base BC.*

**SOLUTION.** As the section is symmetrical about *Y-Y* axis, therefore centre of gravity of the section will lie on this axis. Let $\bar{y}$ be the distance between the centre of gravity of the section and the base *BC*.

*(i) Triangular section*

$$a_1 = \frac{100 \times 90}{2} = 4500 \text{ mm}^2$$

and $$y_1 = \frac{90}{3} = 30 \text{ mm}$$

*(ii) Rectangular hole*

$$a_2 = 30 \times 20 = 600 \text{ mm}^2$$

and $$y_2 = 30 + \frac{30}{2} = 45 \text{ mm}$$

We know that distance between the centre of gravity of the section and base *BC* of the triangle,

$$\bar{y} = \frac{a_1 y_1 - a_2 y_2}{a_1 - a_2} = \frac{(4500 \times 30) - (600 \times 45)}{4500 - 600} = 27.7 \text{ mm}$$

*Moment of inertia of the section about X-X axis.*

We also know that moment of inertia of the triangular section through its centre of gravity and parallel to $X$-$X$ axis,

$$I_{G1} = \frac{bd^3}{36} = \frac{100 \times (90)^3}{36} = 2025 \times 10^3 \text{ mm}^4$$

and distance between the centre of gravity of the section and $X$-$X$ axis,

$$h_1 = 30 - 27.7 = 2.3 \text{ mm}$$

∴ Moment of inertia of the triangular section about $X$-$X$ axis

$$= I_{G1} + a_2 h_1^2 = 2025 \times 10^3 + [4500 \times (2.3)^2] = 2048.8 \times 10^3 \text{ mm}^4$$

Similarly moment of inertia of the rectangular hole through its centre of gravity and parallel to the $X$-$X$ axis

$$I_{G2} = \frac{bd^3}{12} = \frac{20 \times (30)^3}{12} = 45 \times 10^3 \text{ mm}^4$$

and distance between the centre of gravity of the section and $X$-$X$ axis

$$h_2 = 45 - 27.7 = 17.3 \text{ mm}$$

∴ Moment of inertia of rectangular section about $X$-$X$ axis

$$= I_{G2} + a_2 h_2^2 = (45 \times 10^3) + [600 \times (17.3)^2] = 224.6 \times 10^3 \text{ mm}^4$$

Now moment of inertia of the whole section about $X$-$X$ axis.

$$I_{xx} = (2048.8 \times 10^3) - (224.6 \times 10^3) = 1824.2 \times 10^3 \text{ mm}^4 \quad \textbf{Ans.}$$

*Moment of inertia of the section about the base BC*

We know that moment of inertia of the triangular section about the base $BC$

$$I_{G1} = \frac{bd^3}{12} = \frac{100 \times (90)^3}{12} = 6075 \times 10^3 \text{ mm}^4$$

Similarly moment of inertia of the rectangular hole through its centre of gravity and parallel to $X$-$X$ axis,

$$I_{G2} = \frac{bd^3}{12} = \frac{20 \times (30)^3}{12} = 45 \times 10^3 \text{ mm}^4$$

and distance between the centre of gravity of the section about the base $BC$,

$$h_2 = 30 + \frac{30}{2} = 45 \text{ mm}$$

∴ Moment of inertia of rectangular section about the base $BC$,

$$= I_{G2} + a_2 h_2^2 = (45 \times 10^3) + [600 \times (45)^2] = 1260 \times 10^3 \text{ mm}^4$$

Now moment of inertia of the whole section about the base $BC$,

$$I_{BC} = (6075 \times 10^3) - (1260 \times 10^3) = 4815 \times 10^3 \text{ mm}^4 \quad \textbf{Ans.}$$

## 10.16. Moment of Inertia of a Built-up Section

A built-up section consists of a number of sections such as rectangular sections, channel sections, I-sections etc., A built-up section is generally made by symmetrically placing and then fixing these section by welding or riveting. It will be interesting to know that a built-up section behaves as one unit. The moment of inertia of such a section is found out by the following steps.

1. Find out the moment of inertia of the various sections about their respective centres of gravity as usual.
2. Now transfer these moments of inertia about the required axis (say *X-X* axis or *Y-Y* axis) by the Theorem of Parallel Axis.

**NOTE.** In most of the standard sections, their moments of inertia of about their respective centres of gravity is generally given. However, if it is not given then we have to calculate it before transferring it to the required axis.

**EXAMPLE 10.16.** *A compound beam is made by welding two steel plates 160 mm × 12 mm one on each flange of an ISLB 300 section as shown in Fig 10.20.*

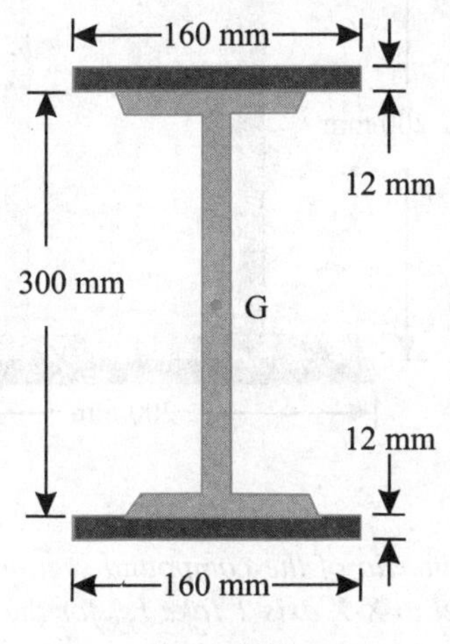


Fig. 10.20

*Find the moment of inertia the beam section about an axis passing through its centre of gravity and parallel to X-X axis. Take moment of inertia of the ISLB 300 section about X-X axis as 73.329 × 10⁶ mm⁴.*

**SOLUTION.** Given: Size of two steel plates = 160 mm × 12 mm and moment of inertia of ISLB 300 section about *X-X* axis = 73.329

From the geometry of the compound section, we find that it is symmetrical about both the *X-X* and *Y-Y* axes. Therefore centre of gravity of the section will lie at *G i.e.* centre of gravity of the beam section.

We know that moment of inertia of one steel plate section about an axis passing through its centre of gravity and parallel to *X-X* axis.

$$I_G = \frac{160 \times (12)^3}{12} = 0.023 \times 10^6 \text{ mm}^4$$

and distance between the centre of gravity of the plate section and *X-X* axis,

$$h = 150 + \frac{12}{2} = 156 \text{ mm}$$

∴ Moment of inertia of one plate section about *X-X* axis,

$$= I_G + a\,h^2 = (0.023 \times 10^6) + [(160 \times 12) \times (156)^2] = 46.748 \times 10^6 \text{ mm}^4$$

and moment of inertia of the compound beam section about $X$-$X$ axis,

$I_{XX}$ = Moment of inertia of ISLB section

+ Moment of inertia of two plate sections.

$= (73.329 \times 10^6) + 2\,(46.748 \times 10^6) = 166.825 \times 10^6 \text{ mm}^4$ **Ans.**

**EXAMPLE 10.17.** *A compound section is built-up by welding two plates 200 mm × 15 mm on two steel beams ISJB 200 placed symmetrically side by side as shown in Fig. 10.21.*

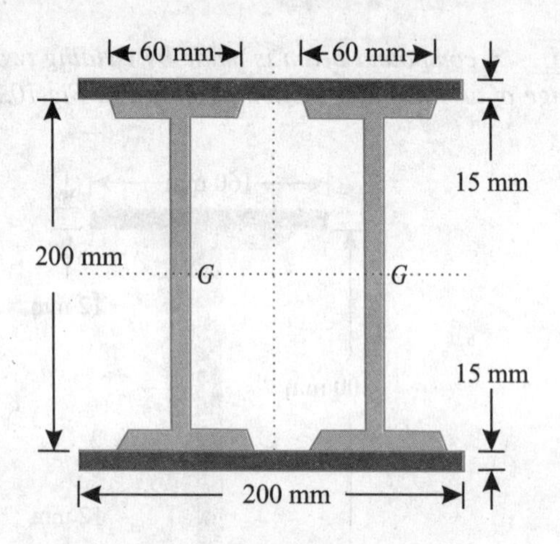


Fig. 10.21

*What is the moment of inertia of the compound section about an axis passing through its centre of gravity and parallel to X-X axis ? Take $I_{XX}$ for the ISJB section as $7.807 \times 10^6$ mm⁴.*

**SOLUTION.** Given: Size of two plates = 200 mm × 15 mm and moment of inertia of ISJB 200 section about $X$-$X$ axis $= 7.807 \times 10^6 \text{ mm}^4$.

From the geometry of the compound section, we find that it is symmetrical about both the $X$-$X$ and $Y$-$Y$ axis. Therefore centre of gravity of the section will lie at $G$ *i.e.*, centre of gravity of the beam sections.

We know that moment of inertia of one plate section about an axis passing through its centre of gravity and parallel to $X$-$X$ axis,

$$I_G = \frac{200 \times (15)^3}{12} = 0.056 \times 10^6 \text{ mm}^4$$

and distance between the centre of gravity of the plate section and $X$-$X$ axis,

$$h = 100 + \frac{15}{2} = 107.5 \text{ mm}$$

∴ Moment of inertia of the plate section about $x$-$x$ axis

$= I_G + a\,h^2 = (0.056 \times 10^6) + (200 \times 15) \times (107.5)^2 = 34.725 \times 10^6 \text{ mm}^4$

and moment of inertia of the compound section about $x$-$x$ axis,

$I_{XX}$ = Moment of inertia of two ISJB sections

+ Moment of inertia of two plate sections

$= [2 \times (7.807 \times 10^6) + 2 \times (34.725 \times 10^6)] = 85.064 \times 10^6 \text{ mm}^4$ **Ans.**

**EXAMPLE 10.18.** *A built up section is made by needing too stable and two channel sections as shown in Fig. 10.22.*

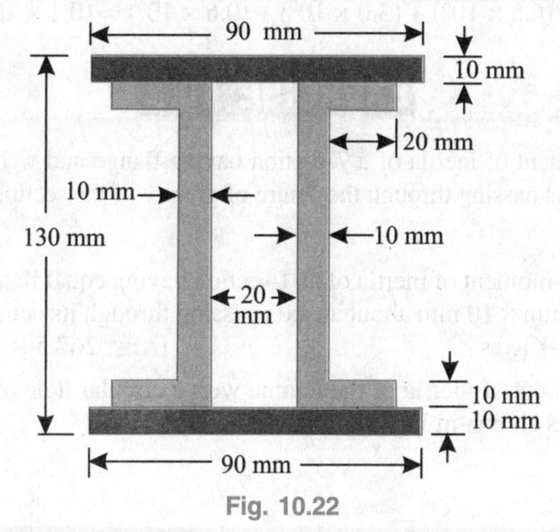


Fig. 10.22

*Determine moment of inertia of a built up section about X-X axis passing through centre of gravity of the section.*

**SOLUTION.** As the section is symmetrical about *X-X* axis and *Y-Y* axis therefore centre of gravity of the section will coincide with the geometrical centre of section.

We know that the moment of inertia of one top or bottom plate about an axis through its centre os gravity and parallel to *X-X* axis,

$$I_{G1} = \frac{90 \times (10)^3}{12} = 7500 \text{ mm}^4$$

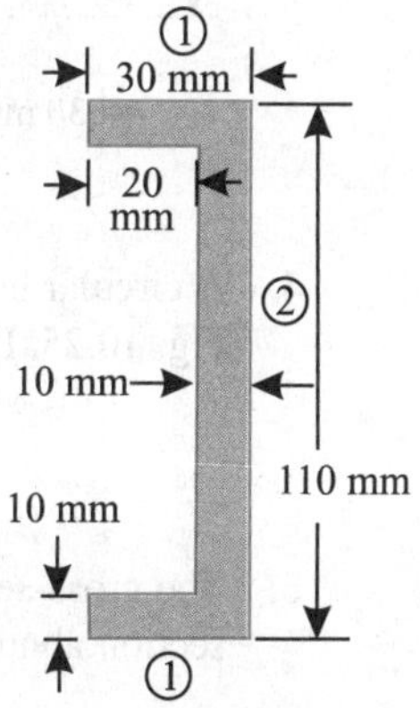

and distance between centre of gravity of the plates from *X-X* axis,

$$h_1 = 65 - 5 = 60 \text{ mm}$$

∴ Moment of inertia of top and bottom plates about *X-X* axis,

$$= I_{G1} + a\,h^2 = 2\,[7500 + (90 \times 10) \times (60)^2] \text{ mm}^4$$

(because of two plates)

$$= 6.5 \times 10^6 \text{ mm}^4$$

Fig. 10.23

Now moment of inertia of part (1) of one channel section about an axis through its centre of gravity and parallel to *X-X* axis,

$$I_{G2} = \frac{30 \times (10)^3}{12} = 2500 \text{ mm}^4$$

and distance of centre of gravity of this part from *X-X* axis,

$$h_2 = 55 - 5 = 50 \text{ mm}$$

∴ Moment of inertia of part (1) about *X-X* axis,

$$= I_{G2} + a\,h^2 = 4\,[2500 + (30 \times 10) \times (50)^2 \text{ mm}^4$$ ...(because of four plates)

$$= 3.0 \times 10^6 \text{ mm}^4$$

Similarly moment of inertia of part (2) of the channel about an axis through its centre of g avity and parallel to *X-X* axis,

$$I_{G3} = 2\left[\frac{10 \times (90)^3}{12}\right] = 0.6 \times 10^6 \text{ mm}^4$$ ...(because of two plates)

Now moment of inertia of the whole built-up section about an axis through its centre of gravity and parallel to X-X axis,

$$I_{XX} = (6.5 \times 10^6) + (3.0 \times 10^6) + (0.6 \times 10^6) = 10.1 \times 10^6 \text{ mm}^4 \qquad \textbf{Ans.}$$

## EXERCISE 10.2

1. Find the moment of inertia of a *T*-section having flange and web both 120 mm × 30 mm about *X-X* axis passing through the centre of gravity of the section.
   [**Ans.** 14 715 × $10^3$ $mm^4$]

2. Calculate the moment of inertia of an I-section having equal flanges 30 mm × 10 mm and web also 30 mm × 10 mm about an axis passing through its centre of gravity and parallel to *X-X* and *Y-Y* axes. [**Ans.** 267.5 × $10^3$ $mm^4$; 47 × $10^3$ $mm^4$]

3. Find the moment of inertia of the lamina with a circular hole of 30 mm diameter about the axis *AB* as shown in Fig. 10.24. [**Ans.** 638.3 × $10^3$ $mm^4$]

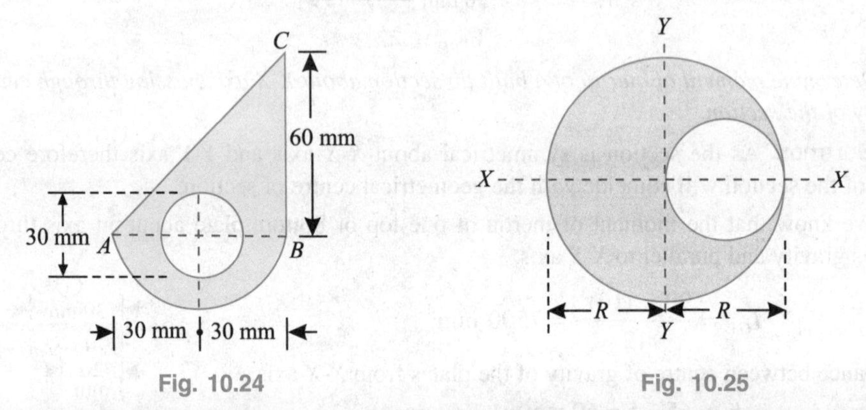


Fig. 10.24

Fig. 10.25

4. A circular hole of diameter *R* is punched out from a circular plate of radius *R* shown in Fig. 10.25. Find the moment of inertia about both the centroidal axes.

$$\left[\textbf{Ans. } I_{XX} = \frac{15\pi R^4}{64};\ I_{YY} = \frac{29\pi R^4}{192}\right]$$

5. The cross-section of a beam is shown in Fig. 10.26. Find the moment of inertia of the section about the horizontal centroidal axis. [**Ans.** 1.354 × $10^6$ $mm^4$]

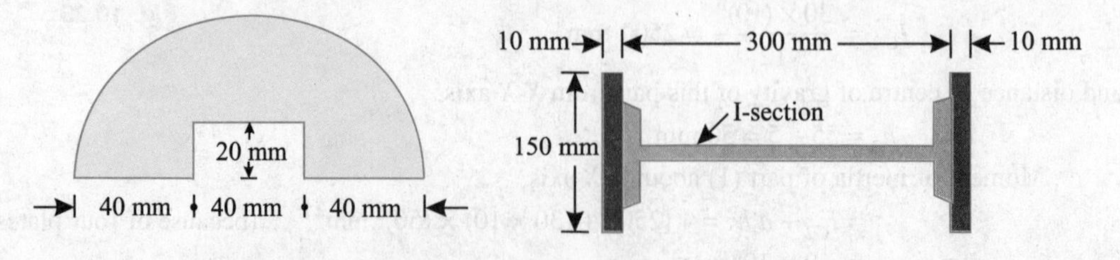


Fig. 10.26

Fig. 10.27

6. A built-up section consists of an I-section and two plates as shown in Fig 10.27. Find values of $I_{XX}$ and $I_{YY}$ of the section. Take values of $I_{XX}$ as 3.762 × $10^6$ $mm^4$ and $I_{YY}$ as 73.329 × $10^6$ $mm^6$ respectively for the I-section.

   [**Ans.** $I_{XX}$ = 17.095 × $10^6$ $mm^4$ ; $I_{YY}$ = 169.46 × $10^6$ $mm^4$]

## QUESTIONS

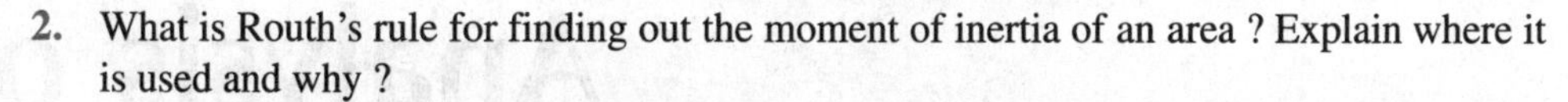

1. How would you find out the moment of inertia of a plane area ?
2. What is Routh's rule for finding out the moment of inertia of an area ? Explain where it is used and why ?
3. Derive an equation for moment of inertia of the following sections about centroidal axis:
   (*a*) a rectangular section,
   (*b*) a hollow rectangular section,
   (*c*) a circular section, and
   (*d*) a hollow circular section.
4. State and prove the theorem of perpendicular axis applied to moment of inertia.
5. Prove the parallel axis theorem in the determination of moment of inertia of areas with the help of a neat sketch.
6. Describe the method of finding out the moment of inertia of a composite section.

## OBJECTIVE TYPE QUESTIONS

1. If the area of a section is in mm$^2$ and the distance of the centre of area from a lines is in mm, then units of the moment of inertia of the section about the line is expressed in
   (*a*) mm$^2$ (*b*) mm$^3$ (*c*) mm$^4$ (*d*) mm$^5$
2. Theorem of perpendicular axis is used in obtaining the moment of inertia of a
   (*a*) triangular lamina (*b*) square lamina
   (*c*) circular lamina (*d*) semicircular lamina
3. The moment of inertia of a circular section of diameter (*d*) is given by the relation
   (*a*) $\frac{\pi}{16}(d)^4$ (*b*) $\frac{\pi}{32}(d)^4$ (*c*) $\frac{\pi}{64}(d)^4$ (*d*) $\frac{\pi}{96}(d)^4$
4. The moment of inertia of a triangular section of base (*b*) and height (*h*) about an axis through its *c.g.* and parallel to the base is given by the relation.
   (*a*) $\frac{bh^3}{12}$ (*b*) $\frac{bh^3}{24}$ (*c*) $\frac{bh^3}{36}$ (*d*) $\frac{bh^3}{48}$
5. The moment of inertia of a triangular section of base (*b*) and height (*h*) about an axis passing through its vertex and parallel to the base is ... as that passing through its C.G. and parallel to the base.
   (*a*) twelve times (*b*) nine times
   (*c*) six times (*d*) four times

## ANSWERS

**1.** (*c*) **2.** (*b*) **3.** (*c*) **4.** (*c*) **5.** (*b*)

Chapter

# 11

# Analysis of Perfect Frames (Analytical Method)

***Contents***



## 11.1. Introduction

A frame may be defined as a structure, made up of several bars, riveted or welded together. these are made up of angle irons or channel sections, and are called members of the frame or framed structure. though these members are welded or riveted together, at their joints, yet for calculation purposes, the joints are assumed to be hinged or pin-jointed the determination of force in a frame is an important problem in engineering- science, which can be solved by the application of the principles of either statics or graphics. in this chapter, we shall be using the principles of statics for determining the forces in frames.

## 11.2. Types of Frames

Though there are many types of frames, yet from the analysis point of view, the frames may be classified into the following two groups:

1. Perfect frame. 2. Imperfect frame.

## 11.3. Perfect Frame

A perfect frame is that, which is made up of members just sufficient to keep it in equilibrium, when loaded, without any change in its shape.

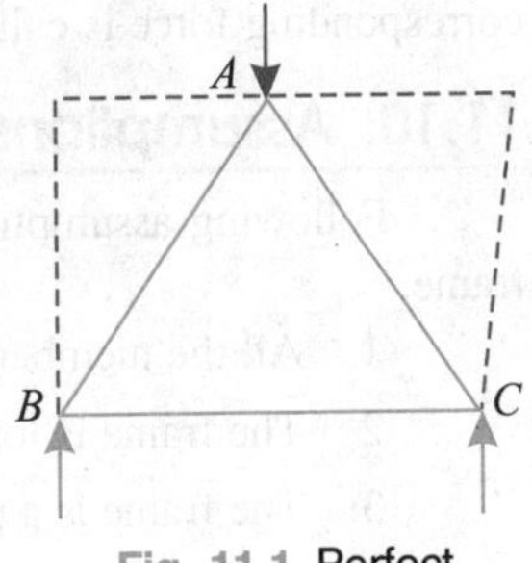


Fig. 11.1. Perfect Frame.

The simplest perfect frame is a triangle, which contains three members and three joints as shown in Fig. 11.1. It will be intersting to know that if such a structure is loaded, its shape will not be distorted. Thus, for three jointed frame, there should be three members to prevent any distortion. It will be further noticed that if we want to increase a joint, to a triangular frame, we require two members as shown by dotted lines in Fig. 11.1. Thus we see that for every additional joint, to a triangular frame, two members are required.

The no. of members, in a perfect frame, may also be expressed by the relation :

$$n = (2j - 3)$$

$n$ = No. of members, and

$j$ = No. of joints.

## 11.4. Imperfect Frame

An imperfect frame is that which does not satisfy the equation :

$$n = (2j - 3)$$

Or in other words, it is a frame in which the no. of members are *more* or *less* than $(2j - 3)$. The imperfect frames may be further classified into the following two types :

1. Deficient frame. 2. Redundant frame.

## 11.5. Deficient Frame

A deficient frame is an imperfect frame, in which the no. of members are less than $(2j - 3)$.

## 11.6. Redundant Frame

A redundant frame is an imperfect frame, in which the no. of members are more than $(2j - 3)$. In this chapter, we shall discuss only perfect frames.

## 11.7. Stress

When a body is acted upon by a force, the internal force which is transmitted through the body is known as stress. Following two types of stress are important from the subject point of view :

1. Tensile stress. 2. Compressive stress.

## 11.8. Tensile Stress

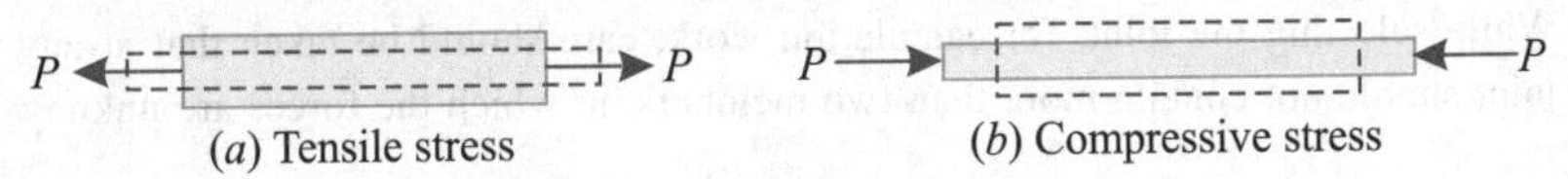


(*a*) Tensile stress (*b*) Compressive stress

Fig. 11.2

Sometimes, a body is pulled outwards by two equal and opposite forces and the body tends to extend, as shown in Fig 11.2. (*a*). The stress induced is called tensile stress and corresponding force is called tensile force.

## 11.9. Compressive Stress

Sometimes, a body is pushed inwards by two equal and opposite forces and the body tends to shorten its length as shown in Fig. 11.2 (*b*). The stress induced is called compressive stress and the corresponding force is called compressive force.

## 11.10. Assumptions for Forces in the Members of a Perfect Frame

Following assumptions are made, while finding out the forces in the members of a perfect frame:

1. All the members are pin-jointed.
2. The frame is loaded only at the joints.
3. The frame is a perfect one.
4. The weight of the members, unless stated otherwise, is regarded as negligible in comparison with the other external forces or loads acting on the truss.

The forces in the members of a perfect frame may be found out either by analytical method or graphical method. But in this chapter, we shall discuss the analytical method only.

## 11.11. Analytical Methods for the Forces

The following two analytical methods for finding out the forces, in the members of a perfect frame, are important from the subject point of view :

1. Method of joints. 2. Method of sections.

## 11.12. Method of Joints

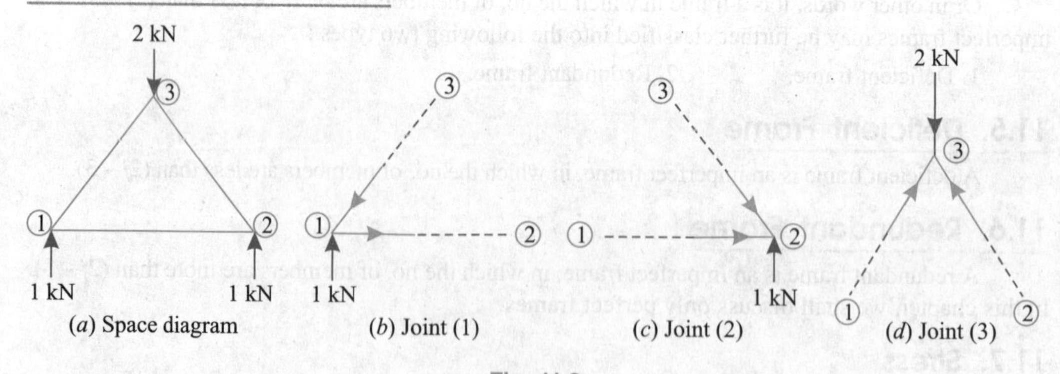


Fig. 11.3

In this method, each and every joint is treated as a free body in equilibrium as shown in Fig. 11.3 (*a*), (*b*), (*c*) and (*d*). The unknown forces are then determined by equilibrium equations viz., $\Sigma V = 0$ and $\Sigma H = 0$. *i.e.,* Sum of all the vertical forces and horizontal forces is equated to zero.

**Notes: 1.** The members of the frame may be named either by Bow's methods or by the joints at their ends.

**2.** While selecting the joint, for calculation work, care should be taken that at any instant, the joint should not contain more than two members, in which the forces are unknown.

## 11.13. Method of Sections (or Method of Moments)

This method is particularly convenient, when the forces in a few members of a frame are required to be found out. In this method, a section line is passed through the member or members, in which the forces are required to be found out as shown in Fig. 11.4 (*a*). A part of the structure, on any one side of the section line, is then treated as a free body in equilibrium under the action of external forces as shown in Fig. 11.4 (*b*) and (*c*).

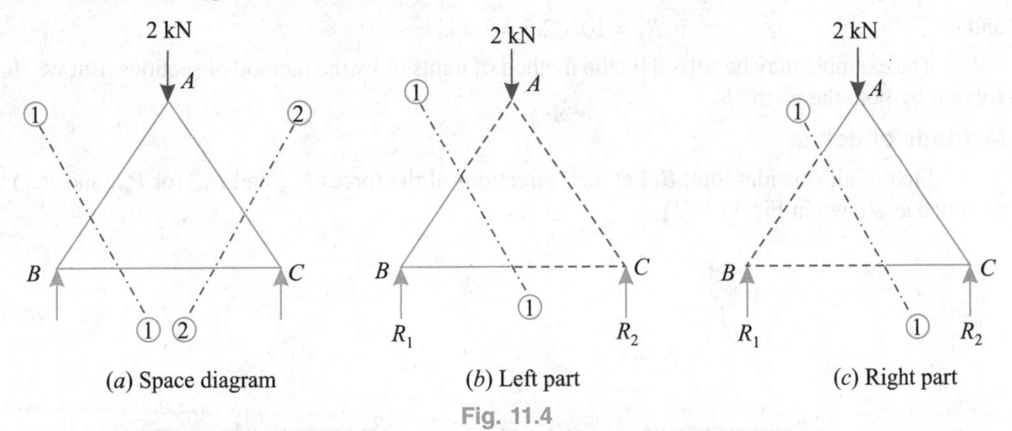


(*a*) Space diagram (*b*) Left part (*c*) Right part

Fig. 11.4

The unknown forces are then found out by the application of equilibrium or the principles of statics *i.e.*, $\Sigma M = 0$.

**Notes:** 1. To start with, we have shown section line 1-1 cutting the members *AB* and *BC*. Now in order to find out the forces in the member *AC*, section line 2-2 may be drawn.

2. While drawing a section line, care should always be taken not to cut more than three members, in which the forces are unknown.

## 11.14. Force Table

Finally, the results are tabulated showing the members, magnitudes of forces and their nature. Sometimes, tensile force is represented with a + ve sign and compressive force with a – ve sign.

**Note:** The force table is generally prepared, when force in all the members of a truss are required to be found out.

**Example 11.1.** *The truss ABC shown in Fig. 11.5 has a span of 5 metres. It is carrying a load of 10 kN at its apex.*

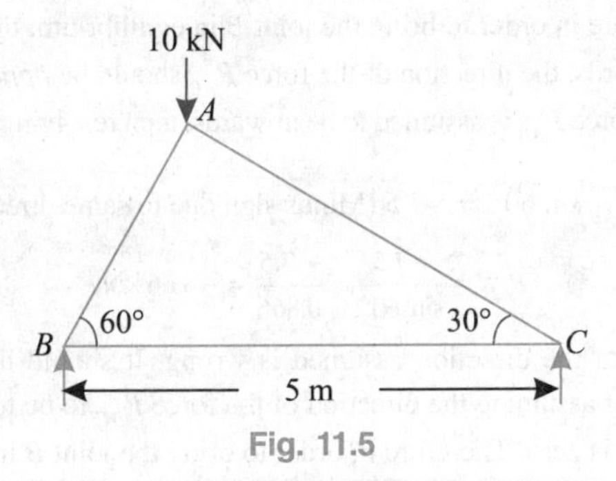


Fig. 11.5

*Find the forces in the members AB, AC and BC.*

**Solution.** From the geometry of the truss, we find that the load of 10 kN is acting at a distance 1.25 m from the left hand support *i.e.*, *B* and 3.75 m from *C*. Taking moments about *B* and equating the same,

$$R_C \times 5 = 10 \times 1.25 = 12.5$$

$\therefore$ $$R_C = \frac{12.5}{5} = 2.5 \text{ kN}$$

and $$R_B = 10 - 2.5 = 7.5 \text{ kN}$$

The example may be solved by the method of joints or by the method of sections. But we shall solve it by both the methods.

**Methods of Joints**

First of all consider joint *B*. Let the *directions of the forces $P_{AB}$ and $P_{BC}$ (or $P_{BA}$ and $P_{CB}$) be assumed as shown in Fig 11.6 (*a*).

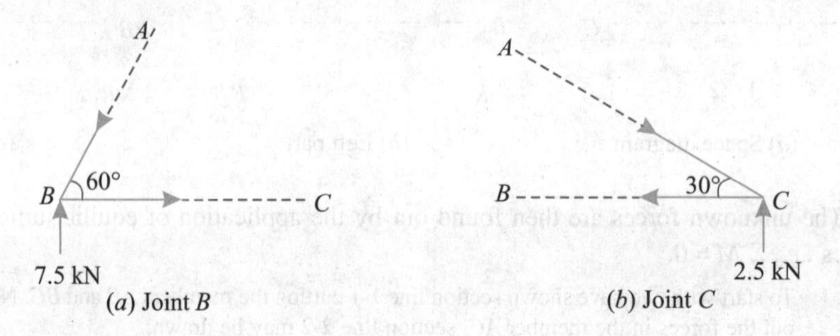


(*a*) Joint *B* (*b*) Joint *C*

**Fig. 11.6**

Resolving the forces vertically and equating the same,

$$P_{AB} \sin 60° = 7.5$$

or $$P_{AB} = \frac{7.5}{\sin 60°} = \frac{7.5}{0.866} = 8.66 \text{ kN (Compression)}$$

and now resolving the forces horizontally and equating the same,

$$P_{BC} = P_{AB} \cos 60° = 8.66 \times 0.5 = 4.33 \text{ kN (Tension)}$$

Now consider the joint *C*. Let the *directions of the forces $P_{AC}$ and $P_{BC}$ (or $P_{CA}$ and $P_{CB}$) be

---

* The idea, of assuming the direction of the force $P_{AB}$ to be downwards, is that the vertical component of the force $P_{BC}$ is zero. Therefore in order to bring the joint *B* in equilibrium, the direction of the force $P_{AB}$ must be downwards, or in other words, the direction of the force $P_{AB}$ should be *opposite* to that of the reaction $R_B$. If, however the direction of the force $P_{AB}$ is assumed to be upwards, then resolving the forces vertically and equating the same,

$$P_{AB} \sin 60° = -7.5 \text{ (Minus sign due to same direction of } R_B \text{ and } P_{AB}.)$$

$\therefore$ $$P_{AB} = \frac{-7.5}{\sin 60°} = \frac{-7.5}{0.866} = -8.66 \text{ kN}$$

Minus sign means that the direction assumed is wrong. It should have been downwards instead of upwards. Similarly, the idea of assuming the direction of the force $P_{BC}$ to be towards right is that the horizontal component of the reaction $R_B$ is zero. Therefore in order to bring the joint *B* in equilibrium, the direction of the force $P_{AB}$ must be towards right (because the direction of the horizontal component of the force $P_{AB}$ is towards left).

assumed as shown in Fig. 11.6 (*b*). Resolving the forces vertically and equating the same,

$$P_{AC} \sin 30° = 2.5$$

$$\therefore \quad P_{AC} = \frac{2.5}{\sin 30°} = \frac{2.5}{0.5} = 5.0 \text{ kN (Compression)}$$

and now resolving the forces horizontally and equating the same,

$$P_{BC} = P_{AC} \cos 30° = 5.0 \times 0.866 = 4.33 \text{ kN (Tension).}$$

...(As already obtained)

## Method of Sections

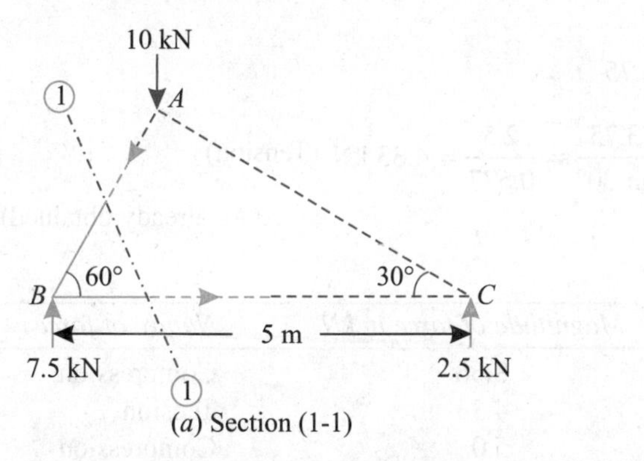

(*a*) Section (1-1)

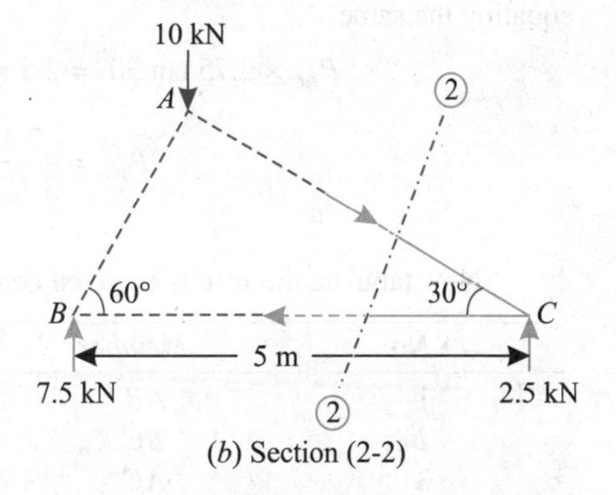

(*b*) Section (2-2)

**Fig. 11.7**

First of all, pass section (1-1) cutting the truss into two parts (one part shown by firm lines and the other by dotted lines) through the members *AB* and *BC* of the truss as shown in Fig 11.7 (*a*). Now consider equilibrium of the left part of the truss (because it is smaller than the right part). Let the directions of the forces $P_{AB}$ and $P_{AC}$ be assumed as shown in Fig 11.7 (*a*).

Taking** moments of the forces acting in the left part of the truss only about the joint *C* and equating the same,

$$P_{AB} \times 5 \sin 60° = 7.5 \times 5$$

$$\therefore \quad P_{AB} = \frac{7.5 \times 5}{5 \sin 60°} = \frac{7.5}{0.866} = 8.66 \text{ kN (Compression)}$$

and now taking moments of the forces acting in the left part of the truss only about the joint *A* and equating the same,

$$P_{BC} \times 1.25 \tan 60° = 7.5 \times 1.25$$

$$\therefore \quad P_{BC} = \frac{7.5 \times 1.25}{1.25 \tan 60°} = \frac{7.5}{1.732} = 4.33 \text{ kN (Tension)}$$

* For details, please refer to the foot note on last page.

** The moment of the force $P_{AB}$ about the joint *C* may be obtained in any one of the following two ways :

1. The vertical distance between the member *AB* and the joint *C* (*i.e.*, *AC* in this case) is equal to 5 sin 60° m. Therefore moment about *C* is equal to $P_{AB} \times 5 \sin 60°$ kN-m.
2. Resolve the force $P_{AB}$ vertically and horizontally at *B*. The moment of horizontal component about *C* will be zero. The moment of vertical component (which is equal to $P_{AB} \times \sin 60°$) is equal to $P_{AB} \times \sin 60° \times 5 = P_{AB} \times 5 \sin 60°$ kN-m.

Now pass section (2-2) cutting the truss into two parts through the members $AC$ and $BC$. Now consider the equilibrium of the right part of the truss (because it is smaller than the left part). Let the †direction of the forces $P_{AC}$ and $P_{BC}$ be assumed as shown in Fig 11.7 (*b*).

Taking moments of the force acting in the right part of the truss only about the joint $B$ and equating the same,

$$P_{AC} \times 5 \sin 30° = 2.5 \times 5$$

$$\therefore \qquad P_{AC} = \frac{2.5}{\sin 30°} = \frac{2.5}{0.5} = 5 \text{ kN (Compression)}$$

and now taking moments of the forces acting in the right part of the truss only about the joint $A$ and equating the same,

$$P_{BC} \times 3.75 \tan 30° = 2.5 \times 3.75$$

$$\therefore \qquad P_{BC} = \frac{2.5 \times 3.75}{3.75 \tan 30°} = \frac{2.5}{0.577} = 4.33 \text{ kN (Tension)}$$

...(As already obtained)

Now tabulate the results as given below :

| *S.No.* | *Member* | *Magnitude of force in kN* | *Nature of force* |
|---|---|---|---|
| 1 | *AB* | 8.66 | Compression |
| 2 | *BC* | 4.33 | Tension |
| 3 | *AC* | 5.0 | Compression |

**EXAMPLE 11.2.** *Fig 11.8 shows a Warren girder consisting of seven members each of 3 m length freely supported at its end points.*

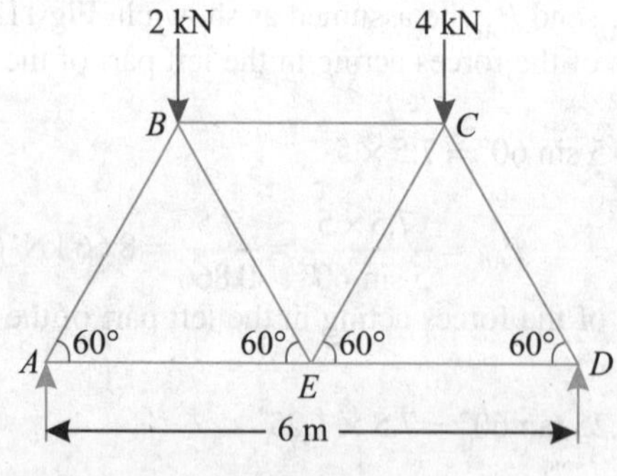


Fig. 11.8

*The girder is loaded at B and C as shown. Find the forces in all the members of the girder, indicating whether the force is compressive or tensile.*

**SOLUTION.** Taking moments about $A$ and equating the same,

$$R_D \times 6 = (2 \times 1.5) + (4 \times 4.5) = 21$$

$$\therefore \qquad R_D = \frac{21}{6} = 3.5 \text{ kN}$$

and $$R_A = (2 + 4) - 3.5 = 2.5 \text{ kN}$$

† For details, please refer to the foot note on last page.

The example may be solved by the method of joints or method of sections. But we shall solve it by both the methods.

**Method of Joints**

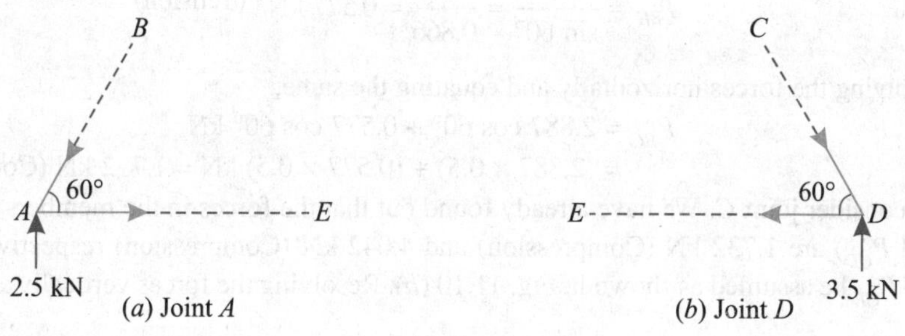


(*a*) Joint *A* (*b*) Joint *D*

Fig. 11.9

First of all, consider the joint *A*. Let the directions of $P_{AB}$ and $P_{AE}$ be assumed as shown in Fig. 13.9 (*a*) Resolving the forces vertically and equating the same,

$$P_{AB} \sin 60° = 2.5$$

$$\therefore \quad P_{AB} = \frac{2.5}{\sin 60°} = \frac{2.5}{0.866} = 2.887 \text{ kN (Compression)}$$

and now resolving the forces horizontally and equating the same,

$$P_{AE} = P_{AB} \cos 60° = 2.887 \times 0.5 = 1.444 \text{ kN (Tension)}$$

Now consider the joint *D*. Let the directions of the forces $P_{CD}$ and $P_{ED}$ be assumed as shown in Fig. 11.9 (*b*).

Resolving the forces vertically and equating the same,

$$P_{CD} \times \sin 60° = 3.5$$

$$\therefore \quad P_{CD} = \frac{3.5}{\sin 60°} = \frac{3.5}{0.866} = 4.042 \text{ kN (Compression)}$$

and now resolving the forces horizontally and equating the same,

$$P_{DE} = P_{CD} \cos 60° = 4.042 \times 0.5 = 2.021 \text{ kN (Tension)}$$

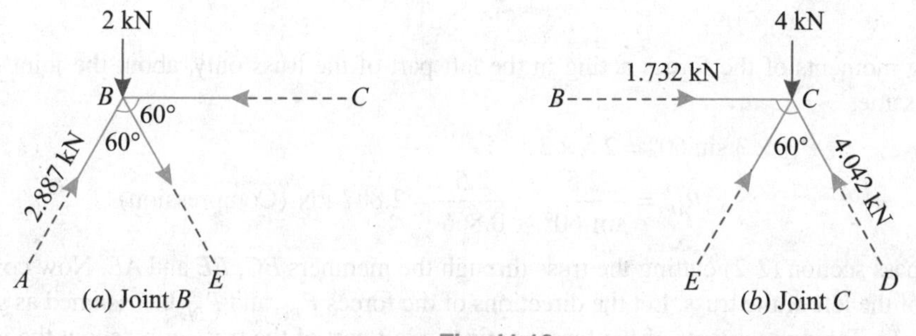


(*a*) Joint *B* (*b*) Joint *C*

Fig. 11.10

Now consider the joint *B*. We have already found that force in member *AB i.e.*, $P_{AB}$ is 2.887 kN (Compression). Let the direction of the forces $P_{BC}$ and $P_{BE}$ be assumed as shown in Fig.13.10 (*a*).

Resolve the forces vertically and equating the same,

$$P_{BE} \sin 60° = P_{AB} \sin 60° - 2.0 = 2.887 \times 0.866 - 2.0 = 0.5 \text{ kN}$$

$$\therefore \quad P_{BE} = \frac{0.5}{\sin 60°} = \frac{0.5}{0.866} = 0.577 \text{ kN (Tension)}$$

and now resolving the forces horizontally and equating the same,

$$P_{BC} = 2.887 \cos 60° + 0.577 \cos 60° \text{ kN}$$

$$= (2.887 \times 0.5) + (0.577 \times 0.5) \text{ kN} = 1.732 \text{ kN (Compression)}$$

Now consider joint *C*. We have already found out that the forces in the members *BC* and *CD* (*i.e.* $P_{BC}$ and $P_{CD}$) are 1.732 kN (Compression) and 4.042 kN (Compression) respectively. Let the directions of $P_{CE}$ be assumed as shown in Fig. 11.10 (*b*). Resolving the forces vertically and equating the same,

$$P_{CE} \sin 60° = 4 - P_{CD} \sin 60° = 4 - (4.042 \times 0.866) = 0.5$$

$$\therefore \quad P_{CE} = \frac{0.5}{\sin 60°} = \frac{0.5}{0.866} = 0.577 \text{ kN (Compression)}$$

**Method of sections**

First of all, pass section (1-1) cutting the truss through the members *AB* and *AE*. Now consider equilibrium of the left part of the truss. Let the directions of the forces $P_{AB}$ and $P_{AE}$ be assumed as shown in Fig. 11.11 (*a*).

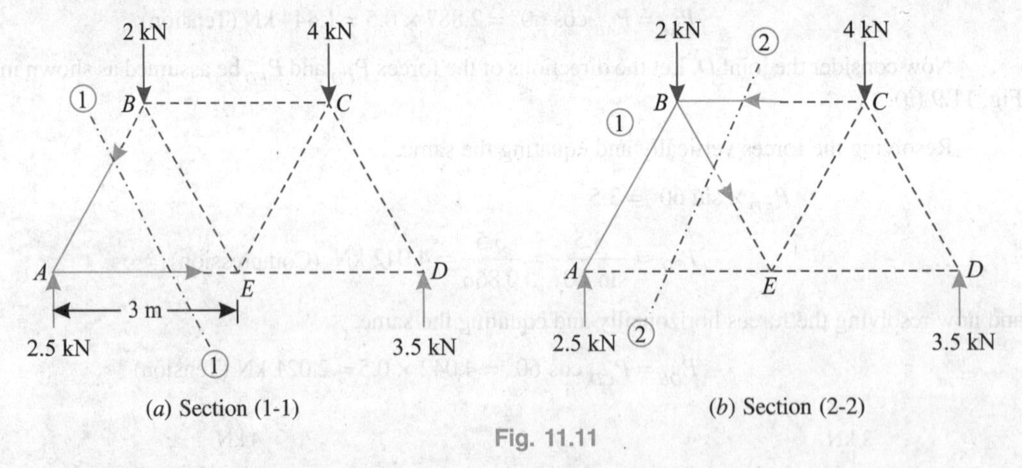


(*a*) Section (1-1) (*b*) Section (2-2)

**Fig. 11.11**

Taking moments of the forces acting in the left part of the truss only, about the joint *E* and equating the same,

$$P_{AB} \times 3 \sin 60° = 2.5 \times 3$$

$$P_{AB} = \frac{2.5}{\sin 60°} = \frac{2.5}{0.866} = 2.887 \text{ kN (Compression)}$$

Now pass section (2-2) cutting the truss through the members *BC*, *BE* and *AE*. Now consider equilibrium of the left of the truss. Let the directions of the forces $P_{BC}$ and $P_{BE}$ be assumed as shown in Fig. 11.11 (*b*). Taking moments of the forces acting in left part of the truss only, about the joint *E* and equating the same,

$$P_{BC} \times 3 \sin 60° = (2.5 \times 3) - (2 \times 1.5) = 4.5$$

$$\therefore \quad P_{BC} = \frac{4.5}{3 \sin 60°} = \frac{4.5}{3 \times 0.866} = 1.732 \text{ kN (Compression)}$$

and now taking moments of the forces acting in the left part of the truss only about the joint *A* and equating the same,

$$P_{BE} \times 3 \sin 60° = (P_{BC} \times 3 \sin 60°) - (2 \times 1.5) = (1.732 \times 3 \times 0.866) - 3.0 = 1.5$$

$$P_{BE} = \frac{1.5}{3 \sin 60°} = \frac{1.5}{3 \times 0.866} = 0.577 \text{ kN (Tension)}$$

Now pass section (3-3) cutting the truss through the members *BC*, *CE* and *ED*. Now consider the equilibrium of the right part of the truss. Let the directions of the forces $P_{CE}$ and $P_{DE}$ be assumed as shown in Fig. 11.12 (*a*) Taking moments of the forces in the right part of the truss only, about the joint *D* and equating the same,

$$P_{CE} \times 3 \sin 60° = (4 \times 1.5) - (P_{BC} \times 3 \sin 60°)$$

$$= 6.0 - (1.732 \times 3 \times 0.866) = 1.5$$

$$\therefore \quad P_{CE} = \frac{1.5}{3 \sin 60°} = \frac{1.5}{3 \times 0.866} = 0.577 \text{ kN (Compression)}$$

and now taking moments of the forces in the right part of the truss only about the joint *C* and equating the same,

$$P_{DE} \times 3 \sin 60° = 3.5 \times 1.5 = 5.25$$

$$\therefore \quad P_{DE} = \frac{5.25}{3 \sin 60°} = \frac{5.25}{3 \times 0.866} = 2.021 \text{ kN (Tension)}$$

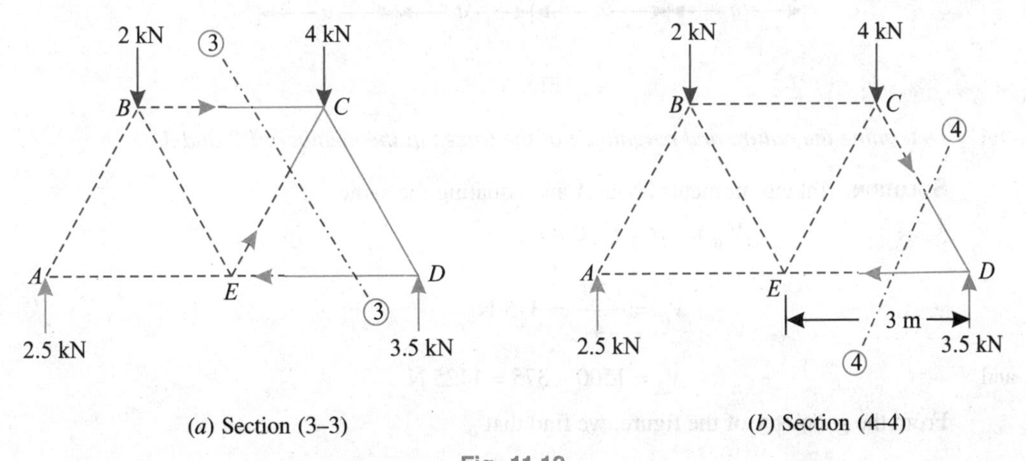


(*a*) Section (3–3)

(*b*) Section (4–4)

**Fig. 11.12**

Now pass section (4-4) cutting the truss through the members *CD* and *DE*. Let the directions of the forces $P_{CD}$ be assumed as shown in Fig 11.12 (*b*). Taking moments of the forces acting in the right part of the truss only about the joint *E* and equating the same,

$$P_{CD} \times 3 \sin 60° = 3.5 \times 3$$

$$P_{CD} = \frac{3.5}{\sin 60°} = \frac{3.5}{0.866} = 4.042 \text{ kN (Compression)}$$

Now tabulate the results as given below :

| S.No. | Member | Magnitude of force in kN | Nature of force |
|---|---|---|---|
| 1 | AB | 2.887 | Compression |
| 2 | AE | 1.444 | Tension |
| 3 | CD | 4.042 | Compression |
| 4 | DE | 2.021 | Tension |
| 5 | BE | 0.577 | Tension |
| 6 | BC | 1.732 | Compression |
| 7 | CE | 0.577 | Compression |

**EXAMPLE 11.3.** *A plane is loaded and supported as shown in Fig 11.13.*

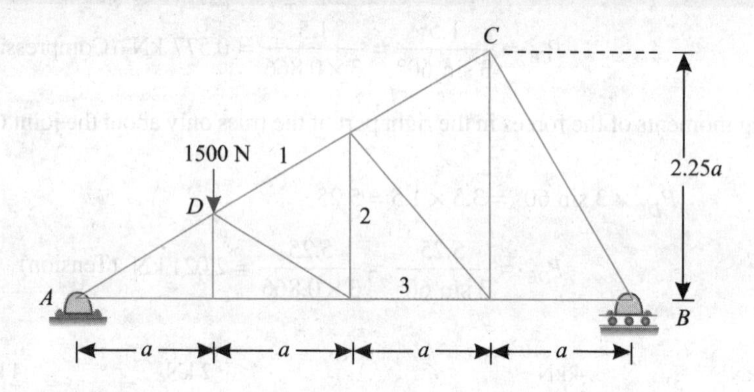


Fig. 11.13

*Determine the nature and magnitude of the forces in the members 1,2 and 3.*

**SOLUTION.** Taking moments about $A$ and equating the same,

$$V_B \times 4a = 1500 \times a$$

$$\therefore \quad V_B = \frac{1500}{4} = 375 \text{ N}$$

and $$V_A = 1500 - 375 = 1125 \text{ N}$$

From the geometry of the figure, we find that

$$\tan\theta = \frac{2.25\,a}{3\,a} = 0.75$$

and $$\sin\theta = \frac{3}{5} = 0.6 \text{ and } \cos\theta = \frac{4}{5} = 0.8$$

The example may be solved by any method. But we shall solve it by the method of sections, as one section line can cut the members 1, 2 and 3 in which the forces are required to be found out. Now let us pass section (1-1) cutting the truss into two parts as shown in Fig 11.14.

Now consider the equilibrium of the right part of the truss. Let the directions of $P_1$, $P_2$ and $P_3$ be assumed as shown in Fig. 11.14.

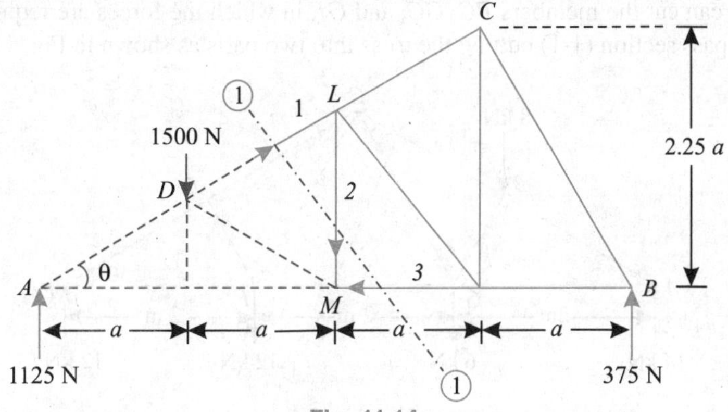


Fig. 11.14

Taking moments about joint $M$ and equating the same,

$$P_1 \times 2a \sin\theta = 375 \times 2a$$

$$\therefore \quad P_1 = \frac{375}{\sin\theta} = \frac{375}{0.6} = 625\,\text{N (Compression)}$$

Similarly, taking moments about joint $A$ and equating the same,

$$P_2 \times 2a = 375 \times 4a = 1500a$$

$$\therefore \quad P_2 = \frac{1500a}{2a} = 750\,\text{N (Tension)}$$

and now taking moments about the joint $L$, and equating the same,

$$P_3 \times \frac{3a}{2} = 375 \times 2a = 750a$$

$$\therefore \quad P_3 = \frac{750}{1.5} = 500\,\text{N (Tension)}$$

**EXAMPLE 11.4.** *An inclined truss shown in Fig 11.15 is loaded as shown.*

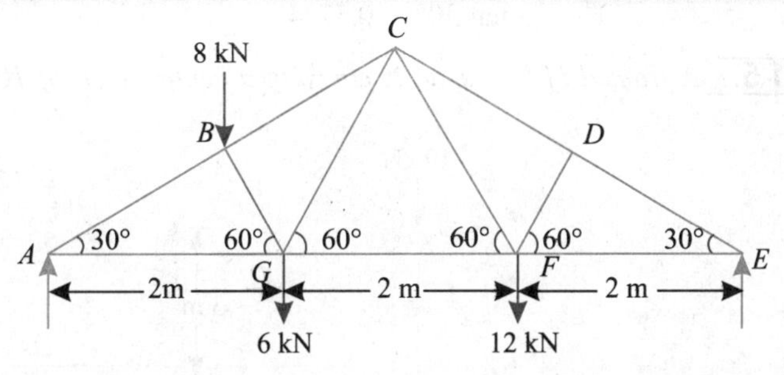


Fig. 11.15

*Determine the nature and magnitude of the forces in the members BC, GC and GF of the truss.*

**SOLUTION.** From the geometry of the figure, we find that the load 8 kN at $B$ is acting at a distance of 1.5 m from the joint $A$. Taking moments about $A$ and equating the same,

$$R_E \times 6 = (8 \times 1.5) + (6 \times 2) + (12 \times 4) = 72$$

$$\therefore \quad R_E = \frac{72}{6} = 12\,\text{kN}$$

$$R_A = (8 + 6 + 12) - 12 = 14 \text{ kN}$$

The example may be solved by any method. But we shall solve it by the method of sections, as one section line can cut the members *BC*, *GC*, and *GF* in which the forces are required to be found out. Now let us pass section (1-1) cutting the truss into two parts as shown in Fig. 11.16

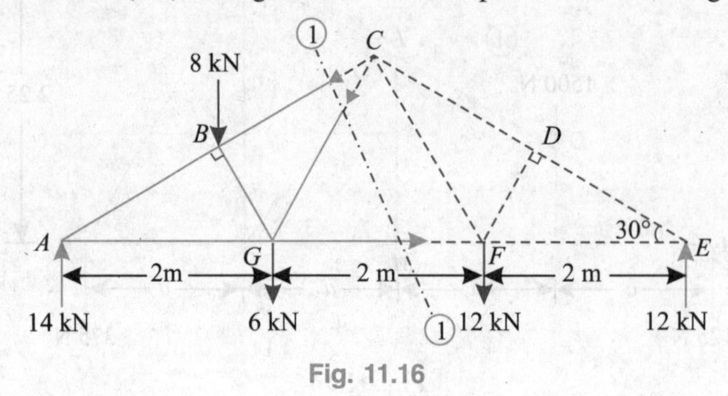


Fig. 11.16

Now consider equilibrium of the left part of the truss. Let the directions of the force $P_{BC}$, $P_{GC}$ and $P_{GF}$ be assumed as shown in Fig 11.16. Taking moments about the joint *G* and equating the same,

$$P_{BC} \times 2 \sin 30° = (14 \times 2) - (8 \times 0.5) = 24$$

$$\therefore \quad P_{BC} = \frac{24}{2 \sin 30°} = \frac{24}{2 \times 0.5} = 24 \text{ kN (Compression)}$$

Similarly, taking moments about the joint *B* and equating the same,

$$P_{GC} \times 1 \cos 30° = (14 \times 1.5) + (6 \times 0.5) = 24 \text{ kN}$$

$$P_{GC} = \frac{24}{\cos 30°} = \frac{24}{0.866} = 27.7 \text{ kN (Compression)}$$

and now taking moments about the joint *C* and equating the same,

$$P_{GF} \times 3 \tan 30° = (14 \times 3) - (6 \times 1) = 36$$

$$\therefore \quad P_{GF} = \frac{36}{3 \tan 30°} = \frac{12}{0.5774} = 20.8 \text{ kN (Tension)}$$

**EXAMPLE 11.5.** *A framed of 6 m span is carrying a central load of 10 kN as shown in Fig. 11.17.*

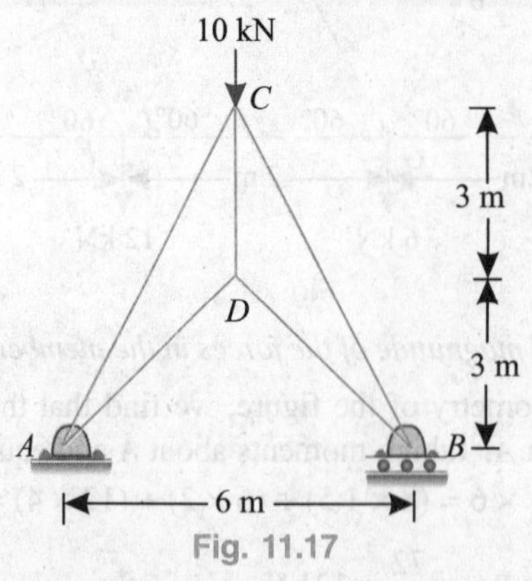


Fig. 11.17

*Find by any method, the magnitude and nature of forces in all members of the structure and tabulate the results.*

**SOLUTION.** Since the structure is symmetrical in geometry and loading, therefore reaction at $A$,

$$R_A = R_B = 5 \text{ kN}$$

From the geometry of the structure, shown in Fig. 11.18 (*a*). we find that

$$\tan \theta = \frac{3}{3} = 1.0 \quad \text{or} \quad \theta = 45°$$

$$\tan \alpha = \frac{6}{3} = 2.0 \quad \text{or} \quad \alpha = 63.4°$$

The example may be solved either by the method of joints or method of sections. But we shall solve it by the method of joints only.

First of all, consider the joint $A$. Let the directions of the forces $P_{AC}$ and $P_{AD}$ be assumed as shown in Fig 11.18 (*a*). Resolving the forces horizontally and equating the same,

$$P_{AC} \cos 63.4° = P_{AD} \cos 45°$$

$$\therefore \quad P_{AC} = \frac{P_{AD} \cos 45°}{\cos 63.4°} = \frac{P_{AD} \times 0.707}{0.4477} = 1.58\, P_{AD}$$

and now resolving the forces vertically and equating the same,

$$P_{AC} \sin 63.4° = 5 + P_{AD} \sin 45°$$

$$1.58\, P_{AD} \times 0.8941 = 5 + P_{AD} \times 0.707 \quad \text{...}(\because P_{AC} = 1.58\, P_{AD})$$

$$\therefore \quad 0.7056\, P_{AD} = 5$$

$$P_{AD} = \frac{5}{0.7056} = 7.08 \text{ kN (Tension)}$$

$$P_{AC} = 1.58 \times P_{AD} = 1.58 \times 7.08 = 11.19 \text{ kN (Compression)}$$

Now consider the joint $D$. Let the directions of the forces $P_{CD}$ and $P_{BD}$ be assumed as shown in Fig. 11.18 (*b*). Resolving the forces vertically and equating the same,

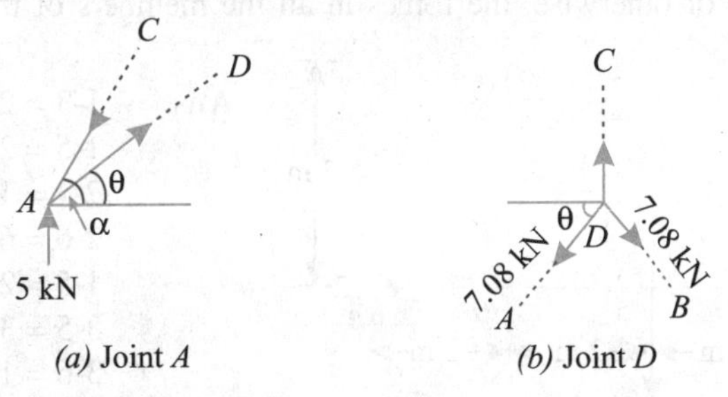


(*a*) Joint *A* (*b*) Joint *D*

**Fig. 11.18**

$$P_{CD} = P_{AD} \sin 45° + P_{BD} \sin 45° = 2\, P_{AD} \sin 45° \quad \text{...}(\because P_{BD} = P_{AD})$$

$$= 2 \times 7.08 \times 0.707 = 10.0 \text{ kN (Tension)}$$

Now tabulate these results as given below :

| *S.No.* | *Member* | *Magnitude of force in kN* | *Nature of force* |
|---|---|---|---|
| 1 | *AD, DB* | 7.08 | Tension |
| 2 | *AC, CB* | 11.19 | Compression |
| 3 | *CD* | 10.0 | Tension |

## EXERCISE 11.1

**1.** A truss of span 10 meters is loaded as shown in Fig. 11.19. Find the forces in all the members of the truss.

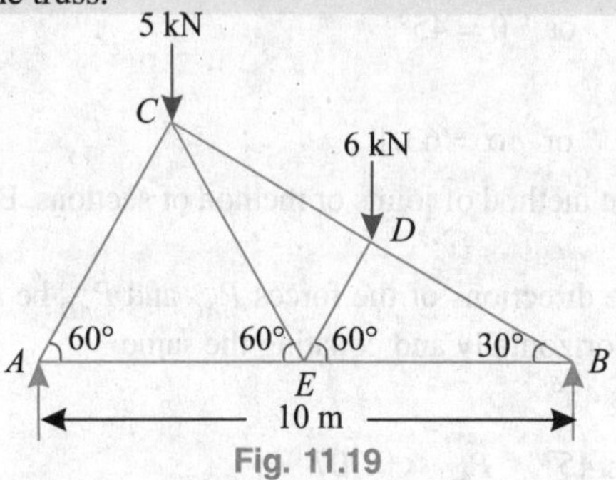

Fig. 11.19

**Ans.** $AC$ = 6.92 kN (Compression)
$AE$ = 3.46 kN (Tension)
$BD$ = 10.0 kN (Compression)
$BE$ = 8.66 kN (Tension)
$CD$ = 7.0 kN (Compression)
$ED$ = 5.2 kN (Compression)
$CE$ = 5.2 kN (Tension)

**2.** A king post truss of 8 m span is loaded as shown in Fig 11.20. Find the forces in each member of the truss and tabulate the results.

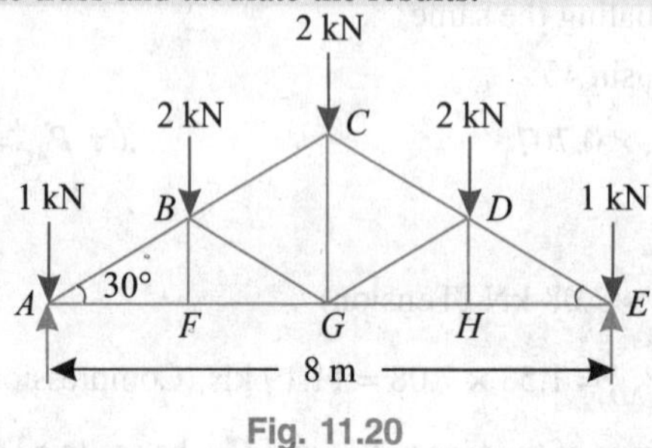

Fig. 11.20

**Ans.** $AB, DE$ = 6.0 kN (Compression)
$AF, EH$ = 5.2 kN (Tension)
$FG, GH$ = 5.2 kN (Tension)
$BF, DH$ = 0
$BG, DG$ = 2.0 kN (Compression)
$BC, CD$ = 4.0 kN (Compression)
$CG$ = 2.0 kN (Tension)

**3.** A plane truss of 6 m span is subjected to a point load of 30 kN as shown in the figure 11.21. Find graphically, or otherwise, the forces in all the members of the truss and tabulate the results.

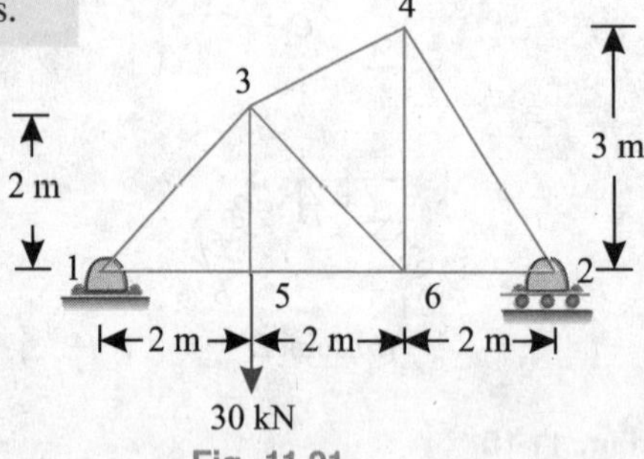

Fig. 11.21

**Ans.** 1-3 = 28.3 kN (Compression)
1-5 = 20.0 kN (Tension)
2-4 = 12.0 kN (Compression)
2-6 = 6.7 kN (Tension)
1-5 = 20.0 kN (Tension)
3-5 = 30.0 kN (Tension)
3-6 = 18.8 kN (Compression)
4-6 = 13.3 kN (Tension)
3-4 = 7.5 kN (Compression)

**4.** A 9 m span truss is loaded as shown in Fig 11.22. Find the forces in the members $BC$, $CH$ and $HG$ of the truss.

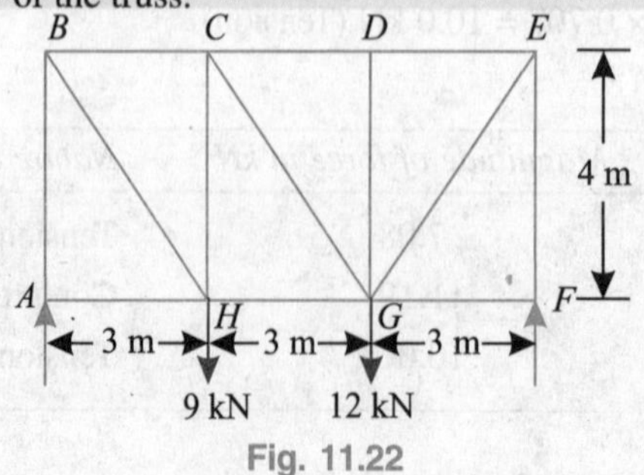

Fig. 11.22

**Ans.** $BC$ = 7.5 kN (Compression)
$CH$ = 1.0 kN (Compression)
$GH$ = 7.5 kN (Tension)

5. The roof truss shown in Fig. 11.23 is supported at $A$ and $B$ and carries vertical loads at each of the upper chord points.

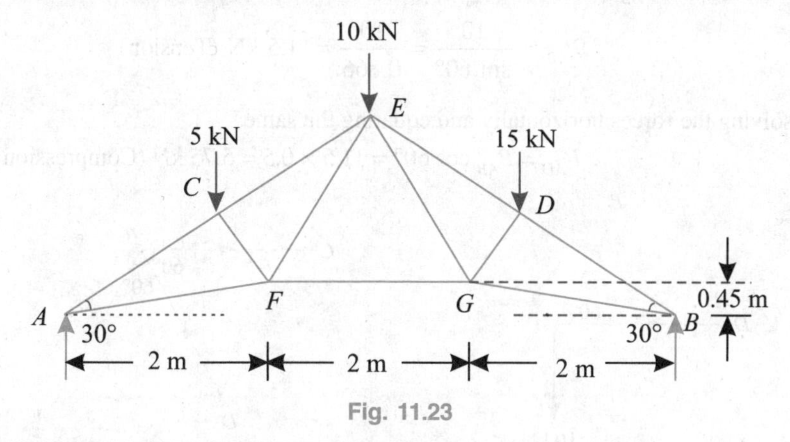


Fig. 11.23

Using the method of sections, determine the forces in the members $CE$ and $FG$ of truss, stating whether they are in tension or compression.

[**Ans.** 38.5 kN (Compression); 24.2 kN (Tension)]

## 11.15. Cantilever Trusses

A truss, which is connected to a wall or a column at one end, and free at the other is known as a cantilever truss. In the previous examples, the determination of support reactions was absolutely essential to start the work. But in the case of cantilever trusses, determination of support reaction is not essential, as we can start the calculation work from the free end of the cantilever.

**EXAMPLE 11.6.** *A cantilever truss of 3 m span is loaded as shown in Fig 11.24.*

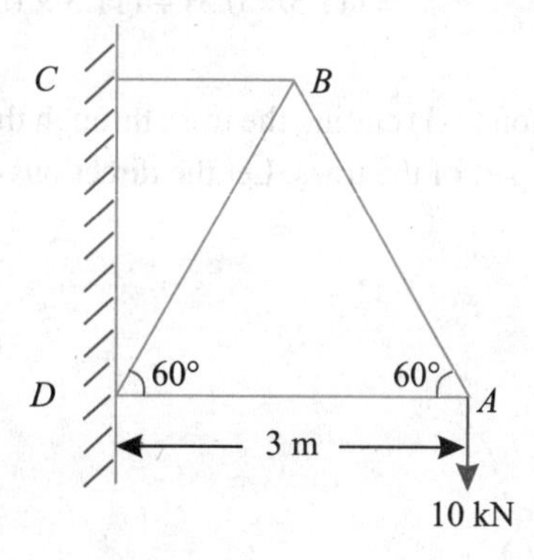


Fig. 11.24

*Find the forces in the various members of the framed truss, and tabulate the results.*

**SOLUTION.** The example may be solved either by the method of joints or method of sections. But we shall solve it by both the methods one by one.

### Method of joints

First of all, consider the joint $A$, Let the directions of the forces $P_{AB}$ and $P_{AD}$ be assumed as shown Fig 11.25 ($a$).

Resolving the forces vertically and equating the same,

$$P_{AB} \sin 60° = 10$$

$$\therefore \quad P_{AB} = \frac{10}{\sin 60°} = \frac{10}{0.866} = 11.5 \text{ kN (Tension)}$$

and now resolving the forces horizontally and equating the same,

$$P_{AD} = P_{AB} \cos 60° = 11.5 \times 0.5 = 5.75 \text{ kN (Compression)}$$

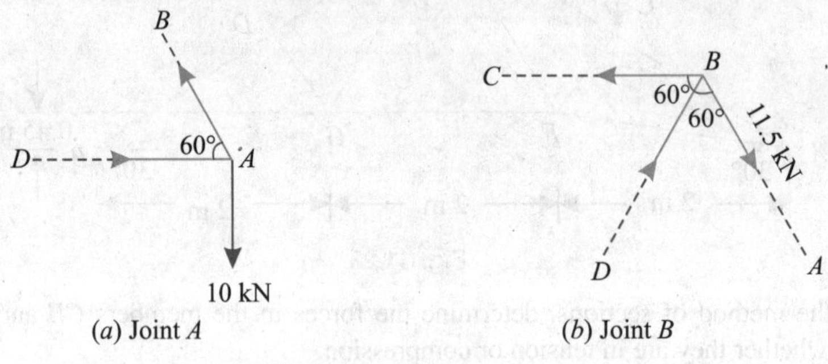


(*a*) Joint *A* (*b*) Joint *B*

Fig. 11.25

Now consider the joint *B*. Let the directions of $P_{BD}$ and $P_{BC}$ be assumed as shown in Fig 11.25 (*b*). We have already found out that the force in member *AB* is 11.5 kN (Tension) as shown in the figure 11.25 (*b*). Resolving the forces vertically and equating the same,

$$P_{BD} \sin 60° = P_{AB} \sin 60° = 11.5 \sin 60°$$

$$\therefore \quad P_{BD} = P_{AB} = 11.5 \text{ kN (Compression)}$$

and now resolving the forces horizontally and equating the same,

$$P_{BC} = P_{AB} \cos 60° + P_{BD} \cos 60°$$

$$= (11.5 \times 0.5) + (11.5 \times 0.5) = 11.5 \text{ kN (Tension)}$$

### Method of sections

First of all, pass section (1-1) cutting the truss through the members *AB* and *AD*. Now consider the equilibrium of the right part of the truss. Let the directions of the forces $P_{AB}$ and $P_{AD}$ be assumed as shown in Fig 11.26 (*a*).

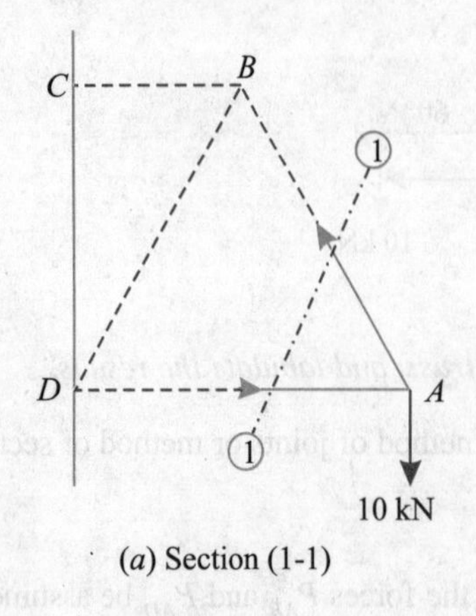


(*a*) Section (1-1)

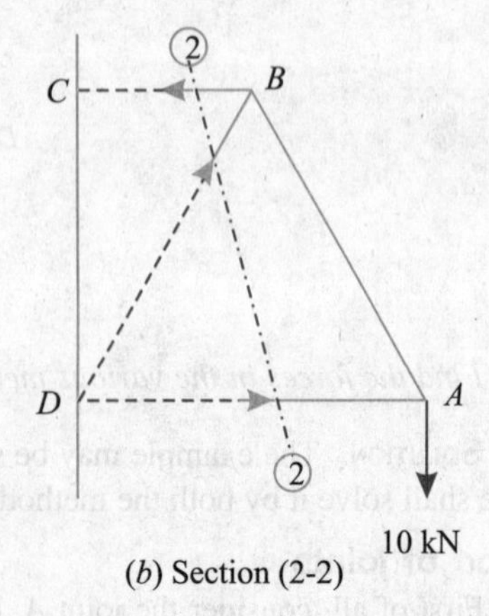


(*b*) Section (2-2)

Fig. 11.26

Taking moments of the forces acting on right part of the truss only, about the joint $D$ and equating the same,

$$P_{AB} \times 3 \sin 60° = 10 \times 3$$

$$\therefore \quad P_{AB} = \frac{10}{\sin 60°} = \frac{10}{0.866} = 11.5 \text{ kN (Tension)}$$

and now taking moments of the forces in the right part of the truss only about the joint $B$ and equating the same,

$$P_{AD} \times 3 \sin 60° = 10 \times 1.5 = 15$$

$$\therefore \quad P_{AD} = \frac{15}{3 \sin 60°} = \frac{15}{3 \times 0.866} = 5.75 \text{ kN (Compression)}$$

Now pass section (2-2) cutting the truss through the members $BC$, $BD$ and $AD$. Now consider the equilibrium of the right part of the truss. Let the directions of the forces $P_{BC}$ and $P_{BD}$ be assumed as shown in Fig. 11.26 (*b*)

Taking moments of the forces acting on the right part of the truss only, about the joint $D$ and equating the same,

$$P_{BC} \times 3 \sin 60° = 10 \times 3$$

$$\therefore \quad P_{BC} = \frac{10}{\sin 60°} = \frac{10}{0.866} = 11.5 \text{ kN (Tension)}$$

and now taking moments of the forces in the right part of the truss only, about the joint $C$ and equating the same,

$$P_{BD} \times 1.5 \sin 60° = (10 \times 3) - P_{AD} \times 3 \sin 60° = 30 - (5.75 \times 3 \times 0.866) = 15$$

$$P_{BD} = \frac{15}{1.5 \sin 60°} = \frac{15}{1.5 \times 0.866} = 11.5 \text{ kN (Compression)}$$

Now tabulate the results as given below :

| *S.No.* | *Members* | *Magnitude of force in kN* | *Nature of force* |
|---|---|---|---|
| 1 | *AB* | 11.5 | Tension |
| 2 | *AD* | 5.75 | Compression |
| 3 | *BD* | 11.5 | Compression |
| 4 | *BC* | 11.5 | Tension |

**EXAMPLE 11.7.** *A cantilever truss is loaded as shown in Fig 11.27.*

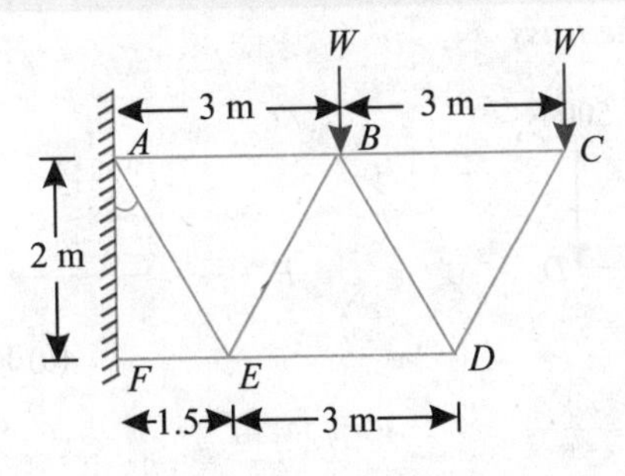


Fig. 11.27

*Find the value W, which would produce the force of magnitude 15 kN in the member AB.*

**SOLUTION.** The example may be solved either by the method of joints or method of sections. But we shall solve it by the method of section only as we have to find out the force in member $AB$ only.

First of all, let us find out the force in the member $AB$ of the truss in terms of $W$. Now pass section (1-1) cutting the truss through the members $AB$, $BE$ and $ED$ as shown in Fig. 11.28.

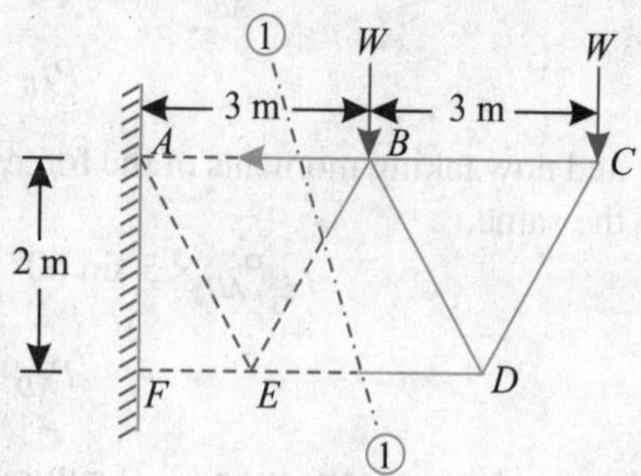

Fig. 11.28

Now consider the equilibrium of the right part of the truss. Let the direction $P_{AB}$ be assumed as shown in Fig 11.28. Taking moments of the forces in the right part of the truss only, about the joint $E$ and equating the same,

$$P_{AB} \times 2 = (W \times 1.5) + (W \times 4.5) = 6\,W$$

$$P_{AB} = \frac{6W}{2} = 3W$$

Thus the value of $W$, which would produce the force of 15 kN in the member $AB$

$$= \frac{W}{3W} \times 15 = 5 \text{ kN} \quad \textbf{Ans.}$$

**EXAMPLE 11.8.** *Figure 11.29 shows a cantilever truss having a span of 4.5 meters. It is hinged at two joints to a wall and is loaded as shown.*

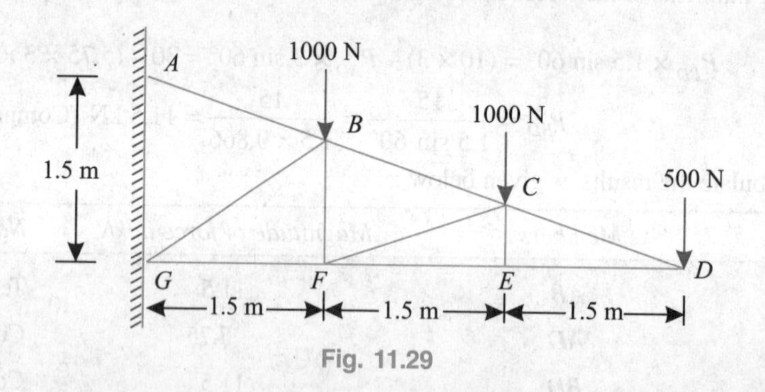

Fig. 11.29

*Find the forces in all the member of the truss.*

**SOLUTION.** The example may be solved either by the method of joints or method of sections. But we shall solve it by the method of joints as we have to find out forces in all members of the truss.

*Force in all the members of the truss*

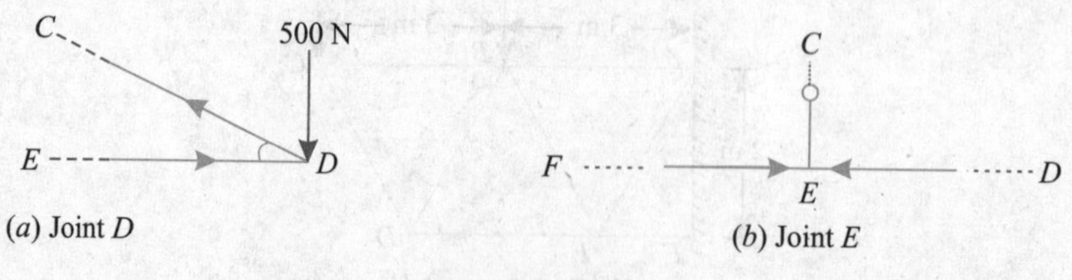

Fig. 11.30

First of all, consider the joint $D$. Let the directions of $P_{CD}$ and $P_{DE}$ be assumed as shown in Fig. 11.30 (*a*).

From the geometry of the figure, we find that

$$\tan \angle CDE = \frac{1.5}{4.5} = 0.3333 \quad \text{or} \quad \angle CDE = 18.4°$$

Resolving the forces vertically at $D$

$$P_{CD} \sin \angle CDE = 500 \quad \text{or} \quad P_{CD} \sin 18.4° = 500$$

$$\therefore \quad P_{CD} = \frac{500}{\sin 18.4°} = \frac{500}{0.3156} = 1584 \text{ N (Tension)}$$

and now resolving the forces horizontally at $D$

$$P_{DE} = P_{CD} \cos \angle CDE = 1584 \cos 18.4°$$

$$\therefore \quad P_{DE} = 1584 \times 0.9488 = 1503 \text{ N (Compression)}$$

Now consider the joint $E$. A little consideration will show that the value of the force $P_{FE}$ will be equal to the force $P_{ED}$ *i.e.*, 1503 N (Compression). Since the vertical components of the forces $P_{FE}$ and $P_{ED}$ are zero, therefore the value of the force $P_{CE}$ will also be zero.

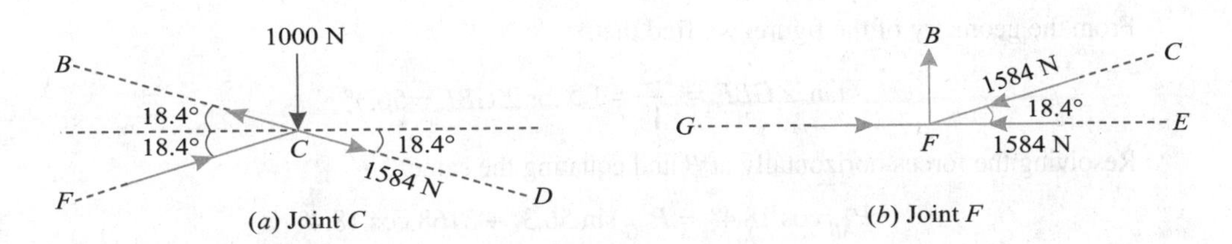


Fig. 11.31

Now consider the joint $C$. Let the directions of $P_{BC}$ and $P_{FC}$ be assumed as shown in Fig. 11.31 (*a*). From the geometry of the figure, we find that the members $CD$, $BC$ and $FC$ make angle of 18.4° with the horizontal. Resolving the forces horizontally and equating the same,

$$P_{BC} \cos 18.4° = 1584 \cos 18.4° + P_{FC} \cos 18.4°$$

$$\text{or} \quad P_{BC} = 1584 + P_{FC} \quad \ldots(i)$$

and now resolving the forces vertically and equating the same,

$$1000 + 1584 \sin 18.4° = P_{FC} \sin 18.4° + P_{BC} \sin 18.4°$$

$$1000 + (1584 \times 0.3156) = (P_{FC} \times 0.3156) + (P_{BC} \times 0.3156)$$

$$1000 + (1581 \times 0.3156) = 0.3156\, P_{FC} + (1584 + P_{FC}) \times 0.3156$$

$$\ldots(\because P_{BC} = 1584 + P_{FC})$$

$$1000 + (1581 \times 0.3156) = 0.3156\, P_{FC} + (1584 \times 0.3156) + 0.3156\, P_{FC}$$

$$\therefore \quad P_{FC} = \frac{1000}{0.6312} = 1584 \text{ N (Compression)}$$

Substituting the value of $P_{FC}$ in equation (*i*)

$$P_{BC} = 1584 + 1584 = 3168 \text{ N (Tension)}$$

Now consider the joint $F$. Let the directions of the forces $P_{GF}$ and $P_{FB}$ be assumed as shown in Fig 11.31 (*b*). Resolving the forces horizontally,

$$P_{GF} = 1584 + 1584 \cos 18.4° = 1584 + (1584 \times 0.9488) \text{ N}$$

$$= 1584 + 1503 = 3087 \text{ N (Compression)}$$

and now resolving the forces vertically and equating the same,

$$P_{BF} = 1584 \sin 18.4° = 1584 \times 0.3156 = 500 \text{ N (Tension)}$$

Now consider the joint $B$. Let the direction of $P_{BG}$ and $P_{AB}$ be assumed as shown in Fig 11.32.

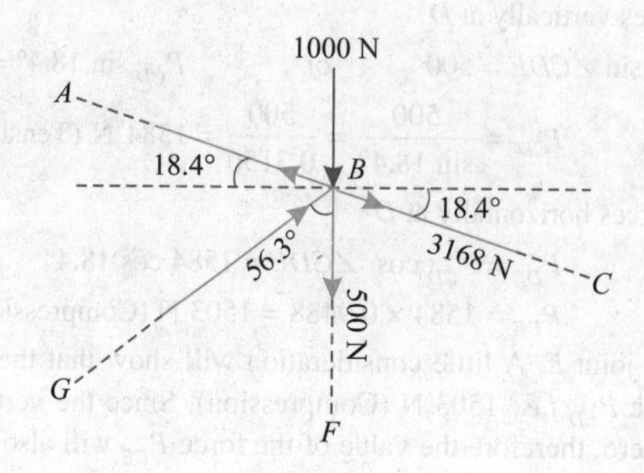


Fig. 11.32

From the geometry of the figure, we find that

$$\tan \angle GBF = \frac{1.5}{1} = 1.5 \text{ or } \angle GBF = 56.3°$$

Resolving the forces horizontally at $B$ and equating the same,

$$P_{AB} \cos 18.4° = P_{BG} \sin 56.3° + 3168 \cos 18.4°$$

$$P_{AB} \times 0.9488 = P_{BG} \times 0.832 + 3168 \times 0.9488$$

$$\therefore \quad 0.9488\, P_{AB} = 0.832\, P_{BG} + 3000 \quad \text{....(ii)}$$

Dividing the above equation by 3,

$$0.3156\, P_{AB} = 0.2773\, P_{BG} + 1000 \quad \text{....(iii)}$$

and now resolving the forces vertically at $B$ and equating the same,

$$P_{AB} \sin 18.4° + P_{BG} \cos 56.3° = 1000 + 500 + 3168 \sin 18.4°$$

$$= 1500 + (3168 \times 0.3156)$$

$$P_{AB} \times 0.3156 + P_{BG} \times 0.5548 = 1500 + 1000$$

$$0.3156\, P_{AB} + 0.5548\, P_{BG} = 2500 \quad \text{...(iv)}$$

Substracting equation (*iii*) from equation (*iv*),

$$0.8321\, P_{BG} = 1500$$

or

$$P_{BG} = \frac{1500}{0.8321} = 1801 \text{ N (Compression)}$$

Substituting the value of $P_{BG}$ in equation (*iii*)

$$0.3156\, P_{AB} = (0.2773 \times 1801) + 1000$$

$$0.3156\, P_{AB} = 500 + 1000 = 500$$

$$P_{AB} = \frac{1500}{0.3156} = 4753 \text{ N (Tension)}$$

Now tabulate the results as given below :

| S.No. | Member | Magnitude of force in kN | Nature of force |
|---|---|---|---|
| 1 | AB | 4753 | Tension |
| 2 | BC | 3168 | Tension |
| 3 | CD | 1584 | Tension |
| 4 | DE | 1503 | Compression |
| 5 | CE | 0 | — |
| 6 | FE | 1503 | Compression |
| 7 | FC | 1584 | Compression |
| 8 | BF | 500 | Tension |
| 9 | GF | 3087 | Compression |
| 10 | BG | 1801 | Compression |

**EXAMPLE 11.9.** *A truss shown in Fig 11.33 is carrying a point load of 5 kN at E.*

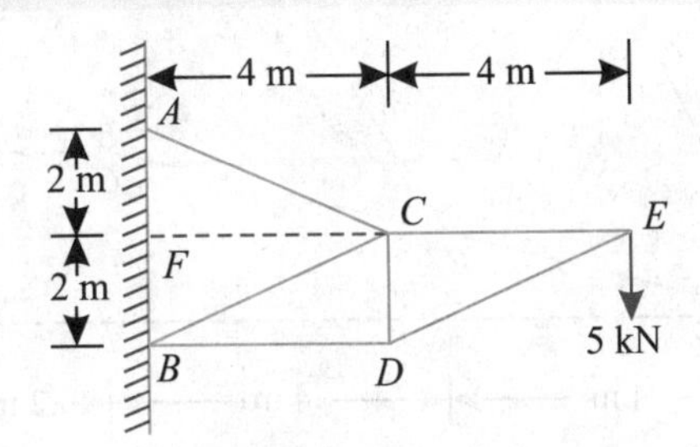


Fig. 11.33

*Find graphically, or otherwise, the force in the members CE, CD and BD of the truss.*

**SOLUTION.** The example may be solved either by the method of joints or method of sections. But we shall solve it by the method of sections, as one section line can cut the members *CE*, *CD* and *BD* in which the forces are required to be found out. Now let us pass section (1-1) cutting truss into two parts as shown in Fig. 11.34.

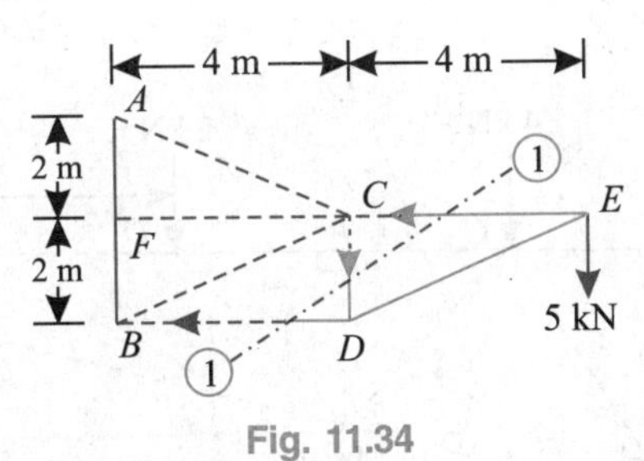


Fig. 11.34

Now consider equilibrium of the right parts of the truss. Let the directions of the force $P_{CE}$ $P_{CD}$ and $P_{BD}$ be assumed as shown in Fig. 11.34. Taking moments about the joint *D* and equating the same,

$$P_{CE} \times 2 = 5 \times 4 = 20$$

$$\therefore \qquad P_{CE} = \frac{20}{2} = 10 \text{ kN (Tension)}$$

Similarly, taking moments about the joint $B$ and equating the same,

$$P_{CD} \times 4 = (5 \times 8) - (P_{CE} \times 2) = 40 - (10 \times 2) = 20$$

$$\therefore \quad P_{CD} = \frac{20}{4} = 5\,\text{kN (Compession)}$$

and now taking moments about the joint $C$ and equating the same,

$$P_{BD} \times 2 = 5 \times 4 = 20$$

$$\therefore \quad P_{BD} = \frac{20}{2} = 10\ \text{kN (Tension)}$$

**EXAMPLE 11.10.** *A pin-joined cantilever frame is hinged to a vertical wall at A and E and is loaded as shown in Fig 11.35.*

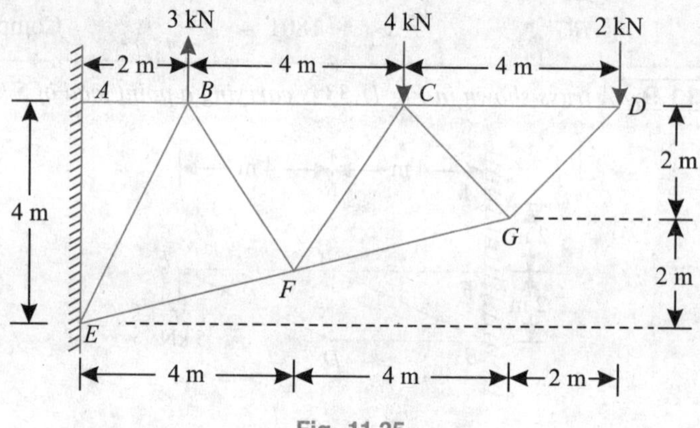


Fig. 11.35

*Determine the forces in the members CD, CG and FG.*

**SOLUTION.** First of all, extend the lines through the joints $B$, $C$ and $D$ as $E$, $F$ and $G$ meeting at $O$. Through $G$, draw $GP$ perpendicular to $CD$. Similarly, through $C$, draw $CQ$ perpendicular to $FG$.

Now extend the line of action of the member $CG$, and through $O$, draw a perpendicular to this line meeting at $R$ as shown in Fig. 11.36.

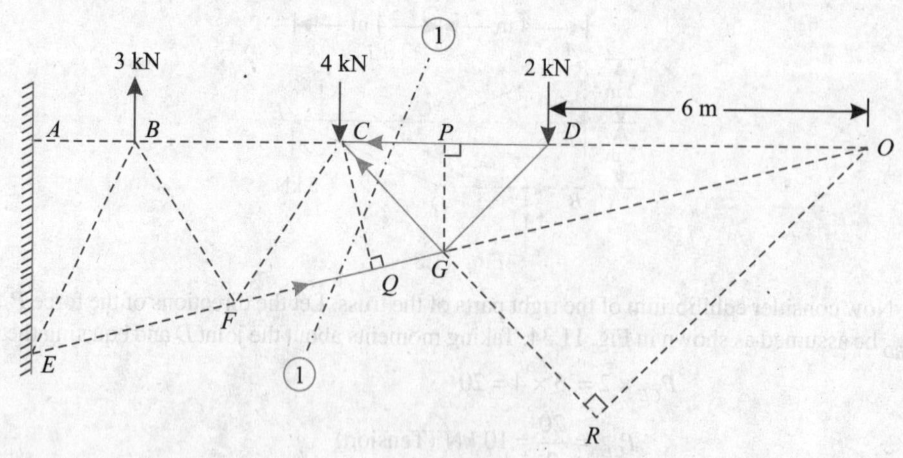


Fig. 11.36

We know that in similar triangles $OPG$ and $OAE$,

$$\frac{AO}{AE} = \frac{AP}{PG} \qquad \text{or} \qquad \frac{AO}{4} = \frac{8}{2} = 4$$

$$\therefore \qquad AO = 4 \times 4 = 16 \text{ m}$$

and $$DO = 16 - 10 = 6 \text{ m}$$

Now in triangle $CGP$, we find that

$$\tan \angle GCP = \frac{2}{2} = 1 \qquad \text{or} \qquad \angle GCP = 45°$$

$$\therefore \qquad \angle COR = 90° - 45° = 45°$$

and $$OR = OC \cos 45° = 10 \times 0.707 \text{ m} = 7.07 \text{ m}$$

From the geometry of the triangle $OPG$, we find that

$$\tan \angle GOP = \frac{2}{8} = 0.25 \qquad \text{or} \qquad \angle GOP = 14°$$

Similarly, in triangle $OCQ$, we find that

$$CQ = CO \sin 14° = 10 \times 0.2425 = 2.425 \text{ m}$$

Now pass section (1-1) cutting the frame through the members $CD$, $CG$ and $FG$. Let the directions of the forces $P_{CD}$, $P_{CG}$ and $P_{FG}$ be assumed as shown in Fig. 11.36. Taking moments of the forces acting on right part of the frame only, about the joint $G$ and equating the same,

$$P_{CD} \times 2 = 2 \times 2 \qquad \text{or} \qquad P_{CD} = 2 \text{ kN (Tension)} \qquad \textbf{Ans.}$$

Similarly, taking moments of the forces acting in the right part of the truss only about the imaginary joint $O$ and equating the same,

$$P_{CG} \times 7.07 = 2 \times 6$$

or $$P_{CG} = \frac{12}{7.07} = 1.7 \text{ kN (Tension)} \qquad \textbf{Ans.}$$

and now taking moments of the forces acting in the right part of the truss only about the joint $C$ and equating the same,

$$P_{FG} \times 2.425 = 2 \times 4 = 8$$

$$\therefore \qquad P_{FG} = \frac{8}{2.425} = 3.3 \text{kN (Compression)}$$

## EXERCISE 11.2

**1.** Determine the forces in the various members of a pin-joined frame as shown in Fig. 11.37. Tabulate the result stating whether they are in tension or compression.

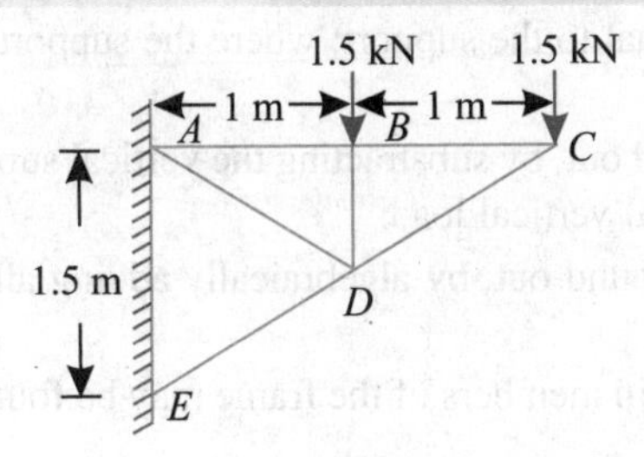


Fig. 11.37

**Ans.** $CD$ = 2.5 kN (Compression)
$BC$ = 2.0 kN (Tension)
$AB$ = 2.0 kN (Tension)
$BD$ = 1.5 kN (Compression)
$AD$ = 1.25 kN (Tension)
$ED$ = 3.75 kN (Compression)

2. A cantilever truss of 4 m span is carrying two point loads of 1.5 kN each as shown in Fig. 11.38 Find the stresses in the members *BC* and *BD* of the truss.

**Ans.** 2.52 kN (Tension) ; zero

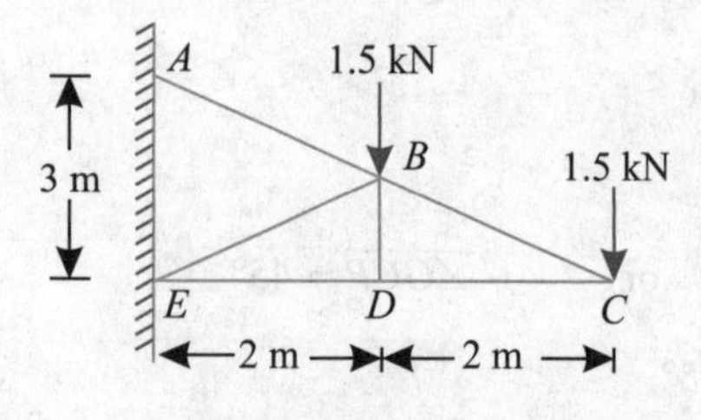


Fig. 11.38

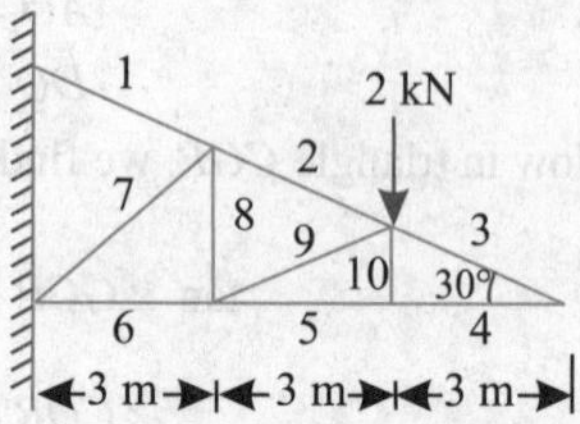

1
2 kN
2
7
8
9
3
10
30°
6
5
4
3 m
3 m
3 m

Fig. 11.39

3. A cantilever truss carries two vertical load as shown in the Fig. 11.39. Find the magnitude and nature of strees in the members 2, 9, 5 and 10 of the truss.

**Ans.** $P_2 = 6.0$ kN (Tension)
$P_9 = 2.9$ kN (Compression)
$P_5 = 3.46$ kN (Compression)
$P_{10} = 0$

4. A cantilever truss is subjected to two point loads of 3 kN each at *B* and *C* as shown in Fig 11.40. Find by any method the forces in the members *AB*. *BE* and *ED* of the truss.

**Ans.** *AB* = 8.6 kN (Tension)
*BE* = 2.0 kN (Tension)
*ED* = 2.0 kN (Compression)

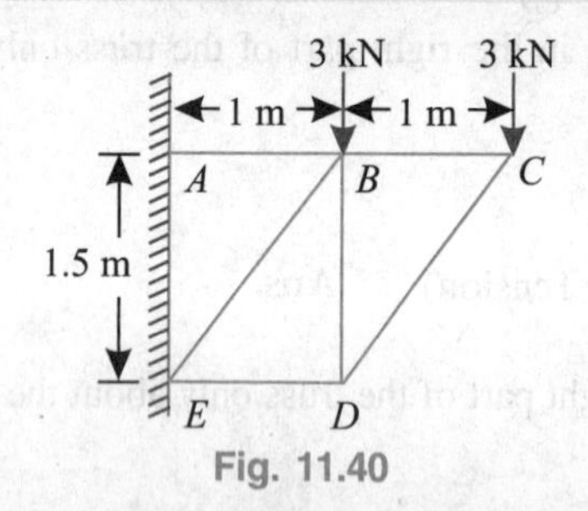


Fig. 11.40

## 11.16. Structures With One End Hinged (or Pin-jointed) and the Other Freely Supported on Rollers and Carrying Horizontal Loads

Sometimes, a structure is hinged or pin-jointed at one end, and freely supported on rollers at the other end. If such a truss carries vertical loads only, it does not present any special features. Such a structure may be solved just as a simply supported structure.

But, if such a structure carries horizontal loads (with or without vertical loads) the support reaction at the roller supported end will be normal to the support; where the support reaction at the hinged end will consist of :

1. Vertical reaction, which may be found out, by substracting the vertical support reaction at the roller supported end from the total vertical load.
2. Horizontal reaction, which may be found out, by algebraically adding all the horizontal loads.

After finding out the reactions, the forces in members of the frame may be found out as usual.

**EXAMPLE. 11.11.** *Figure 11.41 shows a framed of 4 m span and 1.5 m height subjected to two point loads at B and D.*

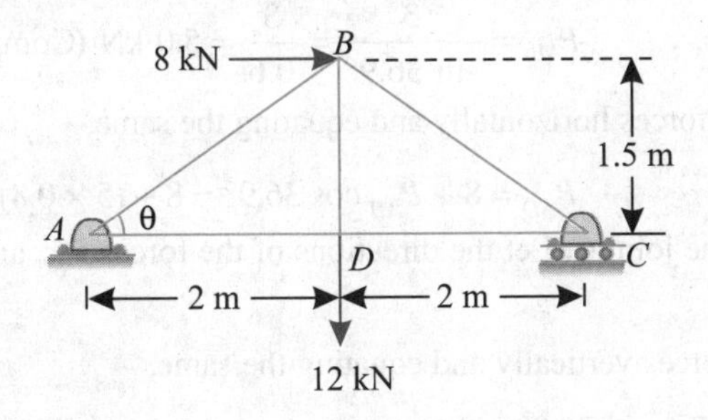


Fig. 11.41

*Find graphically, or otherwise, the forces in all the members of the structure.*

**SOLUTION.** Since the structure is supported on rollers at the right hand support (*C*), therefore the reaction at this support will be vertical (because of horizontal support). The reaction at the left hand support (*A*) will be the resultant of vertical and horizontal forces and inclined with the vertical.

Taking moments about *A* and equating the same,

$$V_C \times 4 = (8 \times 1.5) + (12 \times 2) = 36$$

$$V_C = \frac{36}{4} = 9 \text{ kN} (\uparrow)$$

$$V_A = 12 - 9 = 3 \text{ kN} (\uparrow) \quad \text{and} \quad H_A = 8 \text{ kN} (\leftarrow)$$

From the geometry of the figure, we find that

$$\tan \theta = \frac{1.5}{2} = 0.75 \quad \text{or} \quad \theta = 36.9°$$

Similarly $\quad \sin \theta = \sin 36.9° = 0.6 \quad \text{and} \quad \cos \theta = \cos 36.9° = 0.8$

The example may be solved either by the method of joints or method of sections. But we shall solve it by the method of joints as we have to find forces in all the members of the structure.

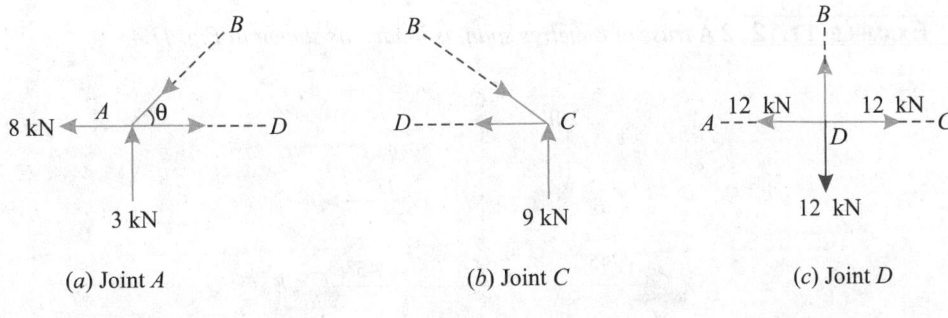


Fig. 11.42

First of all, consider joint *A*. Let directions of the forces $P_{AB}$ and $P_{AD}$ be assumed as shown in Fig. 11.42 (*a*). We have already found that a horizontal force of 8 kN is acting at *A* as shown in Fig. 11.42 (*a*).

Resolving the forces vertically and equating the same,

$$P_{AB} \sin 36.9° = 3$$

$$\therefore \qquad P_{AB} = \frac{3}{\sin 36.9°} = \frac{3}{0.6} = 5.0 \text{ kN (Compression)}$$

and now resolving the forces horizontally and equating the same,

$$P_{AD} = 8 + P_{AB} \cos 36.9° = 8 + (5 \times 0.8) = 12.0 \text{ kN (Tension)}$$

Now consider the joint *C*. Let the directions of the forces $P_{BC}$ and $P_{CD}$ be assumed as shown in Fig. 11.42 (*b*).

Resolving the forces vertically and equating the same,

$$P_{BC} \sin 36.9° = 9$$

$$P_{BC} = \frac{9}{\sin 36.9°} = \frac{9}{0.6} = 15 \text{ kN (Compression)}$$

and now resolving the forces horizontally and equating the same,

$$P_{CD} = P_{BC} \cos 36.9° = 15 \times 0.8 = 12.0 \text{ kN (Tension)}$$

Now consider the joint *D*. A little consideration will show that the value of the force $P_{BD}$ will be equal to the load 12 kN (Tension) as shown in Fig 11.42. (*c*). This will happen as the vertical components of the forces $P_{AD}$ and $P_{CD}$ will be zero.

Now tabulate the results as given below :

| *S.No.* | *Member* | *Magnitude of force in kN* | *Nature of force* |
|---|---|---|---|
| 1 | *AB* | 5.0 | Compression |
| 2 | *AD* | 12.0 | Tension |
| 3 | *BC* | 15.0 | Compression |
| 4 | *CD* | 12.0 | Tension |
| 5 | *BD* | 12.0 | Tension |

**EXAMPLE 11.12.** *2 A truss of 8 metres span, is loaded as shown in Fig. 11.43.*

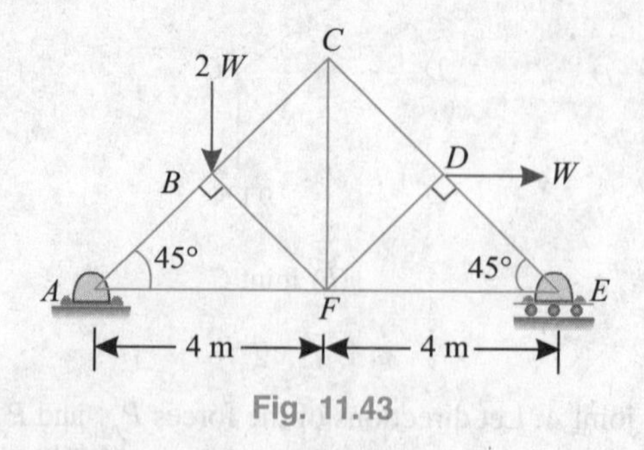


**Fig. 11.43**

*Find the forces in the members CD, FD and FE of the truss.*

**SOLUTION.** Since the truss is supported on rollers at the right hand support ($E$), therefore the reaction at this support will be vertical (because of horizontal support). The reaction at the left hand support ($A$) will be the resultant of vertical and horizontal forces and inclined with vertical.

Taking moments about $A$ and equating same,

$$V_E \times 8 = (2\,W \times 2) + (W \times 2) = 6\,W$$

$$\therefore \qquad V_E = \frac{6W}{8} = 0.75W\ (\uparrow)$$

and $\qquad {}^*V_A = 2\,W - 0.75\,W = 1.25\,W\,(\uparrow) \qquad$ and $\qquad H_A = W\,(\leftarrow)$

The example may be solved either by the method of joints or method of sections. But we shall solve it by the method of sections, as one section line can cut the members $CD$, $FD$ and $FE$ in which the forces are required to be found out. Now let us pass section (1-1) cutting the truss into two parts as shown in Fig. 11.44.

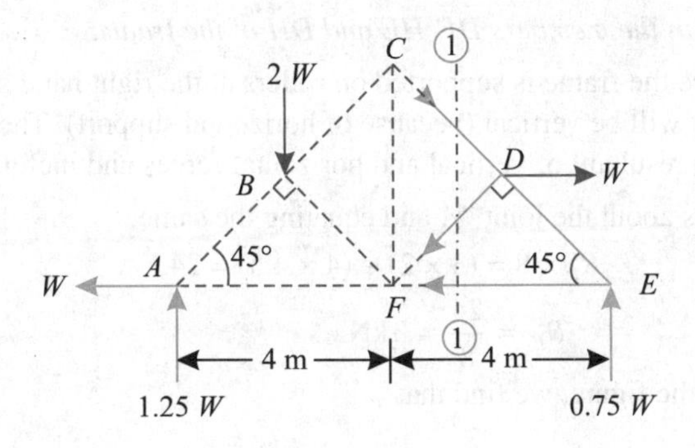


Fig. 11.44

Now consider equilibrium of the right part of the truss. Let the directions of the forces $P_{CD}$, $P_{FD}$ and $P_{FE}$ be assumed as shown in Fig. 11.44. Taking moments about the joint $F$ and equating the same,

$$P_{CD} \times 4 \sin 45° = (0.75\,W \times 4) - (W \times 2) = W$$

$$\therefore \qquad P_{CD} = \frac{W}{4 \sin 45°} = \frac{W}{4 \times 0.707} = 0.354\,W \text{ (Compression)}$$

Similarly, taking moments about the joint $E$ and equating the same,

$$P_{FD} \times 4 \cos 45° = W \times 2 = 2\,W$$

$$\therefore \qquad P_{FD} = \frac{2W}{4 \cos 45°} = \frac{2W}{4 \times 0.707} = 0.707\,W \text{ (Tension)}$$

and now taking moments about the joint $D$ and equating the same,

$$P_{FE} \times 2 = 0.75\,W \times 2 = 1.5\,W$$

$$\therefore \qquad P_{FE} = \frac{1.5W}{2} = 0.75\,W \text{ (Tension)}$$

---

* There is no need of finding out the vertical and horizontal reaction at A, as we are not considering this part of the truss.

**EXAMPLE 11.13.** *Figure 11.45 shows a pin-jointed frame carrying a vertical load at B and a horizontal load at D*

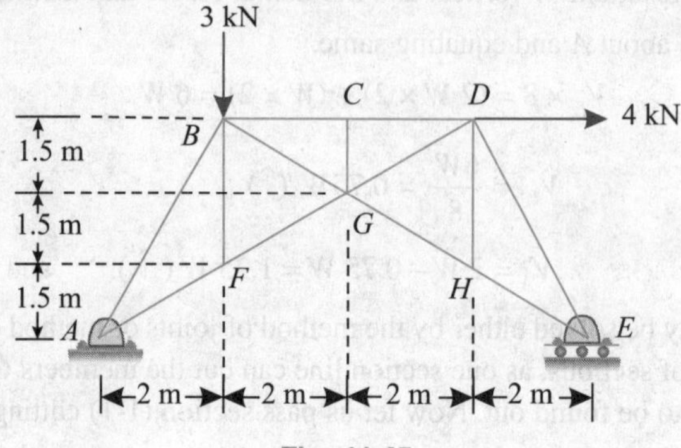


Fig. 11.45

*Find the forces in the members DF, HE and DH of the frame.*

**SOLUTION.** Since the frame is supported on rollers at the right hand support ($E$), therefore the reaction at this support will be vertical (because of horizontal support). The reaction at the left hand support ($A$) will be the resultant of vertical and horizontal forces and inclined with the vertical.

Taking moments about the joint* $A$ and equating the same,

$$R_E \times 8 = (3 \times 2) + (4 \times 4.5) = 24$$

$$\therefore \qquad R_E = \frac{24}{8} = 3 \text{ kN}$$

From the geometry of the figure, we find that

$$\tan \theta = \frac{3}{4} = 0.75 \qquad \text{or} \qquad \theta = 36.9°$$

$$\tan \alpha = \frac{4.5}{2} = 2.25 \qquad \text{or} \qquad \alpha = 66°$$

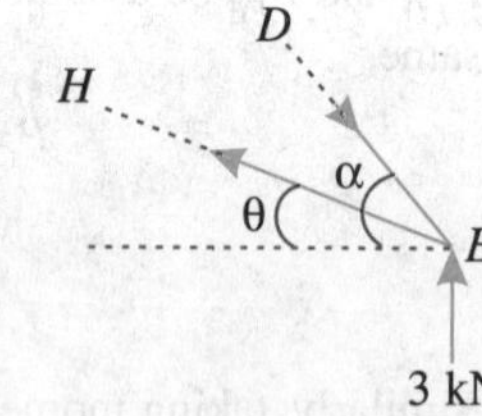

The example may be solved either by the method of joints or method of sections. But we shall solve it by the method of joints, as we can resolve the force in the members at joint $E$ in which the force are required to be found out. Now consider the point $E$. Let the directions of the forces $P_{DE}$ and $P_{HE}$ be assumed as shown in Fig. 11.46.

Fig. 11.46

Resolving the forces horizontally and equating the same,

$$P_{DE} \cos 66° = P_{HE} \cos 36.9° = P_{HE} \times 0.8$$

$$\therefore \qquad P_{DE} = \frac{P_{HE} \times 0.8}{\cos 66°} = \frac{P_{HE} \times 0.8}{0.4062} = 1.97\, P_{HE}$$

and now resolving the forces vertically and equating the same,

$$P_{DE} \sin 66° = P_{HE} \sin 36.9° + 3$$

$$1.97\, P_{HE} \times 0.9137 = (P_{HE} \times 0.6) + 3$$

$$1.2\, P_{HE} = 3$$

$$\text{or} \qquad P_{HE} = \frac{3}{1.2} = 2.5 \text{ kN (Tension)}$$

$$\text{and} \qquad P_{DE} = 1.97\, P_{HE} = 1.97 \times 2.5 = 4.93 \text{ (Compression)}$$

* There are no need of finding out the vertical and horizontal reaction at A, as we are not considering this part of the truss.

Now consider the joint $H$. We have already found out that $P_{HE}$ = 2.5 kN (Tension). It will be interesting to know that the force $P_{DH}$ will be zero, as there is no other member at joint $H$ to balance the component of this forces (if any) at right angle to the member $GHE$.

## 11.17. Structures With One End Hinged (or Pin-Jointed) and the other Freely Supported on Rollers and Carrying Inclined Loads

We have already discussed in the last article that if a structure is hinged at one end, freely supported on rollers at the other, and carries horizontal loads (with or without vertical loads), the support reaction at the roller- supported end will be normal to the support. The same principle is used for structures carrying inclined loads also. In such a case, the support reaction at the hinged end will be the resultant of :

1. Vertical reaction, which may be found out by subtracting the vertical component of the support reaction at the roller supported end from the total vertical loads.
2. Horizontal reaction, which may be found out algebraically by adding all the horizontal loads.

**EXAMPLE 11.14.** *Figure 11.47 represents a north-light roof truss with wind loads acting on it.*

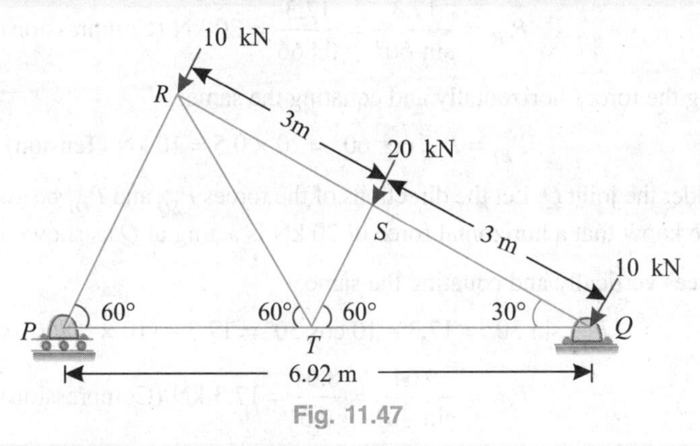


Fig. 11.47

*Find graphically, or otherwise, the forces in all the members of the truss Give your results in a tabulated form.*

**SOLUTION.** Since the truss is supported on rollers at $P$, therefore the reaction at this end will be vertical (because of horizontal support). Moreover, it is hinged at $Q$, therefore the reaction at this end will be the resultant of horizontal and vertical forces and inclined with the vertical.

Taking moments about $Q$ and equating the same,

$$V_P \times 6.92 = (20 \times 3) + (10 \times 6) = 120$$

$$\therefore \qquad V_P = \frac{120}{6.92} = 17.3\,\text{kN}$$

We know that total wind loads on the truss

$$= 10 + 20 + 10 = 40 \text{ kN}$$

∴ Horizontal component of wind load,

$$H_Q = 40 \cos 60° = 40 \times 0.5 = 20 \text{ kN } (\rightarrow)$$

and vertical component of the wind load

$$= 40 \sin 60° = 40 \times 0.866 = 34.6 \text{ kN } (\downarrow)$$

∴ Vertical reaction at $Q$,

$$V_Q = 34.6 - 17.3 = 17.3 \text{ kN } (\uparrow)$$

The example may be solved either by the method of joints or method of sections. But we shall solve it by the method of joints, as we have to find out the forces in all the members of the truss.

First of all, consider the joint *P*. Let the directions of the forces $P_{PR}$ and $P_{PT}$ be assumed as shown in Fig 11.48(*a*). We know that a horizontal force of 20 kN is acting at *Q* as shown in Fig. 11.48 (*b*).

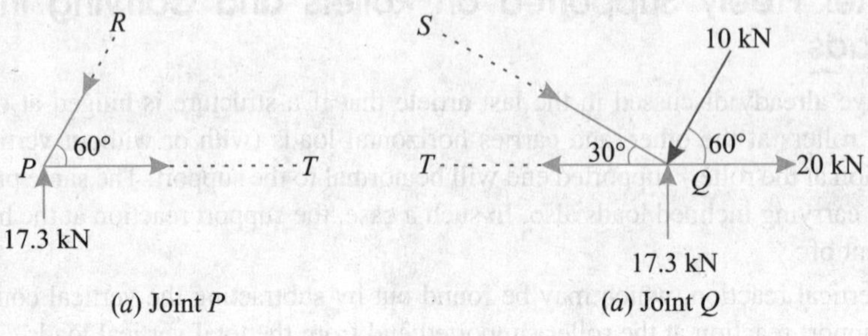


(*a*) Joint *P* (*a*) Joint *Q*

**Fig. 11.48**

Resolving the forces vertically and equating the same,

$$P_{PR} \sin 60° = 17.3$$

$$\therefore \quad P_{PR} = \frac{17.3}{\sin 60°} = \frac{17.3}{0.866} = 20 \text{ kN (Compression)}$$

and now resolving the forces horizontally and equating the same,

$$P_{PT} = P_{PR} \cos 60° = 20 \times 0.5 = 10 \text{ kN (Tension)}$$

Now consider the joint *Q*. Let the directions of the forces $P_{SQ}$ and $P_{QT}$ be assumed as shown in Fig. 11.48 (*b*). We know that a horizontal force of 20 kN is acting at *Q* as shown in Fig 11.48 (*b*).

Resolving the forces vertically and equating the same,

$$P_{SQ} \sin 30° = 17.3 - 10 \cos 30° = 17.3 - (10 \times 0.866) = 8.64$$

$$\therefore \quad P_{SQ} = \frac{8.64}{\sin 30°} = \frac{8.64}{0.5} = 17.3 \text{ kN (Compression)}$$

and now resolving the forces horizontally and equating the same,

$$P_{QT} = P_{SQ} \cos 30° + 20 - 10 \sin 30°$$
$$= (17.3 \times 0.866) + 20 - (10 \times 0.5) = 30 \text{ kN (Tension)}$$

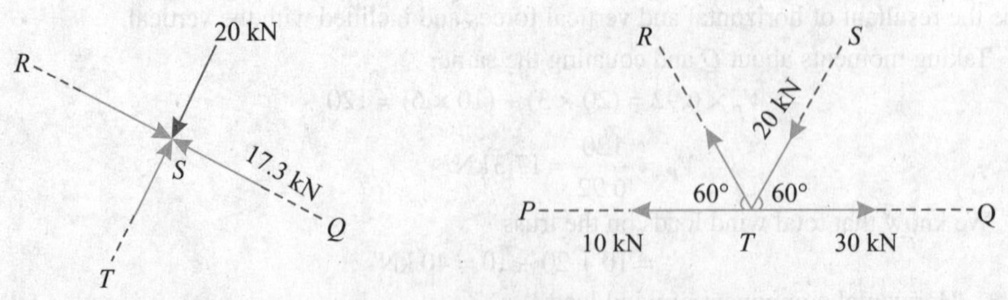


**Fig. 11.49**

Now consider the joint *S*. We have already found out that $P_{SQ}$ = 17.3 kN (Compression). A little consideration will show that the value of the force $P_{TS}$ will be equal to the force 20 kN (Compression). Similarly, the value of the force $P_{RS}$ will be equal to $P_{SQ}$ *i.e.*, 17.3 kN (Compression) as shown in Fig. 11.49 (*a*).

Now consider the joint $T$. Let the directions of the force $P_{RT}$ be assumed as shown in Fig. 11.49 (*b*). We have already found out that $P_{ST} = 20$ kN (Compression).

Resolving the forces vertically and equating the same,

$$P_{RT} \sin 60° = P_{ST} \sin 60° = 20 \sin 60°$$

or
$$P_{RT} = 20 \text{ kN (Tension)}$$

Now tabulate the results as given below:

| *S.No.* | *Member* | *Magnitude of force in kN* | *Nature of force* |
|---|---|---|---|
| 1 | *PR* | 20.0 | Compression |
| 2 | *PT* | 10.0 | Tension |
| 3 | *SQ* | 17.3 | Compression |
| 4 | *QT* | 30.0 | Tension |
| 5 | *ST* | 20.0 | Compression |
| 6 | *RS* | 17.3 | Compression |
| 7 | *RT* | 20.0 | Tension |

**EXAMPLE 11.15.** *A truss of 12 m span is loaded as shown in Fig 11.50.*

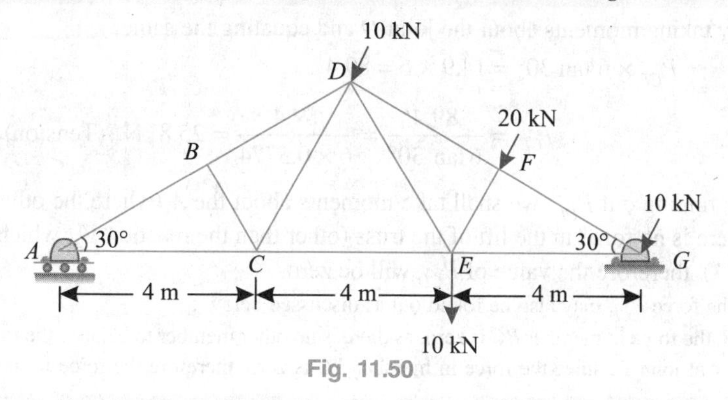


Fig. 11.50

*Determine the force in the members BD, CE and CD of the truss.*

**SOLUTION.** Since the truss is supported on rollers on the left end (*A*), therefore the reaction at this end will be vertical (because of horizontal support). Moreover, it is hinged at the right hand support (*G*), therefore the reaction at this end will be the resultant of horizontal and vertical forces and will be inclined with the vertical.

Taking * moments about $G$ and equating the same,

$$V_A \times 12 = (10 \times 4)\ (20 \times 4 \cos 30°) + (10 \times 8 \cos 30°)$$

$$= 40 + (80 \times 0.866) + (80 \times 0.866) = 178.6$$

$$\therefore \qquad V_A = \frac{178.6}{12} = 14.9 \text{ kN}$$

The example may be solved either by the method of joints or method of sections. But we shall solve it by the method of sections, as one section line can cut the members *BD*, *CE* and *CD* in which forces are required to be found out.

* There is no need of finding out the vertical and horizontal reaction at *G*, as we are not considering this part of the truss.

Now let us pass section (1-1) cutting the truss into two parts as shown in Fig 13.51.

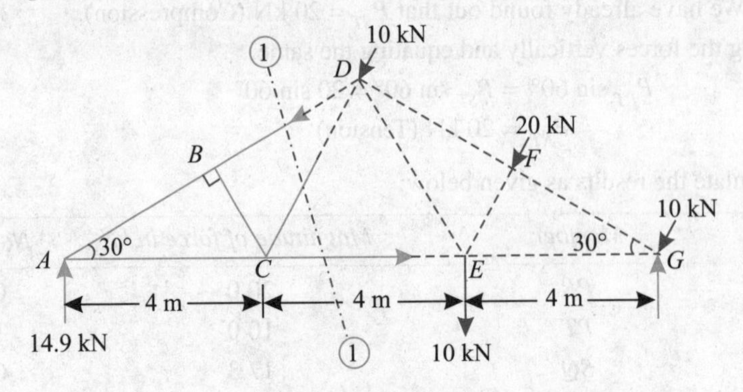


Fig. 11.51

Now consider equilibrium of the left part of the truss. Let the directions of the forces $P_{BD}$, $P_{CE}$ and $P_{CD}$ be assumed as shown in Fig 11.51. Taking moments about the joint $C$ and equating the same,

$$P_{BD} \times 2 = 14.9 \times 4 = 59.6$$

$$\therefore \qquad P_{BD} = \frac{59.6}{2} = 29.8 \text{ kN (Compression)}$$

Similarly taking moments about the joint $D$ and equating the same,

$$P_{CE} \times 6 \tan 30° = 14.9 \times 6 = 89.4$$

$$\therefore \qquad P_{CE} = \frac{89.4}{6 \tan 30°} = \frac{89.4}{6 \times 0.5774} = 25.8 \text{kN} \quad \text{(Tension)}$$

Now for finding out $P_{CD}$, we shall take moments about the $A$ (where the other two members meet). Since there is no force in the lift of the truss (other than the reaction $V_A$, which will have zero moment about $A$), therefore the value of $P_{CD}$ will be zero.

**Note:** The force $P_{CD}$ may also be found out as discussed below :

At joint $B$, the force in member $BC$ is zero, as there is no other member to balance the force (if any) in the member $BC$. Now at joint $C$, since the force in member $BC$ is zero, therefore the force in member $CD$ is also equal to zero.

**Example 11.16.** *A truss hinged at A and supported on rollers at D, is loaded as shown in Fig. 11.52.*

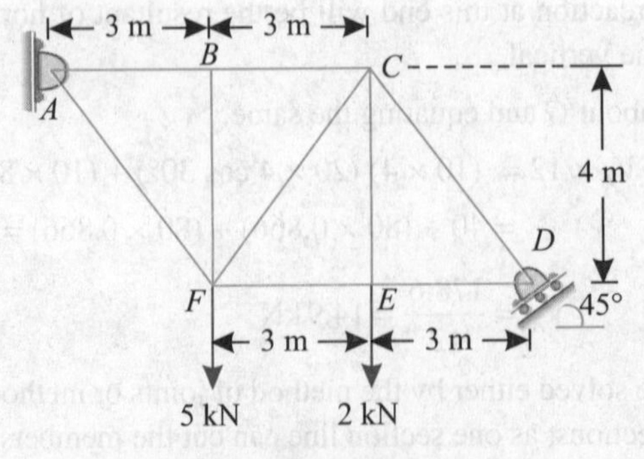


Fig. 11.52

*Find the forces in the members BC, FC, FE of the truss.*

**SOLUTION.** Since the truss is supported on rollers at the right end $D$, therefore the reaction at this support will be normal to the support *i.e.*, inclined at 45° with the horizontal. The reaction at $A$ will be the resultant of horizontal and vertical forces. It will be interesting to know that as the reaction at $D$ is inclined at 45° with the horizontal, therefore horizontal component ($R_{DH}$) and vertical component ($R_{DV}$) of this reaction will be equal. Mathematically $R_{DH} = R_{DV}$.

Taking moments about $A$ and equating the same,

$$(R_{DV} \times 9) - (R_{DH} \times 4) = (5 \times 3) + (2 \times 6)$$

$$5\,R_{DH} = 27 \qquad [\because R_{DH} = R_{DV}]$$

$$\therefore \quad R_{DH} = \frac{27}{5} = 5.4 \text{ kN } (\leftarrow)$$

and

$$R_{DV} = 5.4 \text{ kN } (\uparrow)$$

The example may be solved either by the method of joints or method of sections. But we shall solve it by the method of sections, as one section line can cut the members $BC$, $FE$ and $FC$ and in which forces are required to be found out.

Now let us pass section (1-1) cutting the truss into two parts as shown in Fig. 11.53.

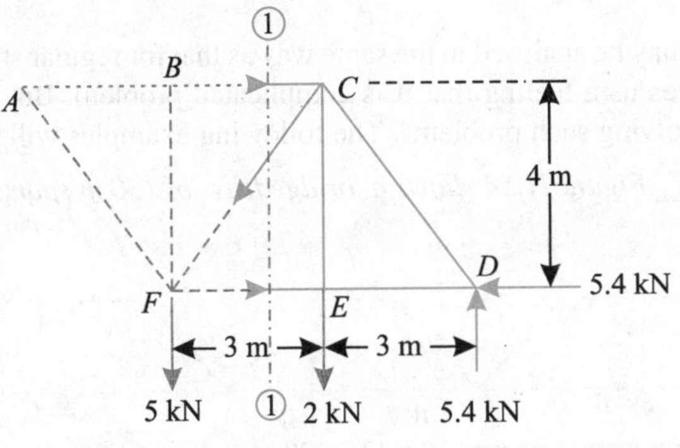


Fig. 11.53

Now consider equilibrium of right part of the truss. Let the directions of the forces $P_{BC}$ and $P_{FE}$ be assumed as shown in Fig 11.53. Taking moments about the joint $F$ and equating the same,

$$P_{BC} \times 4 = (5.4 \times 6) - (2 \times 3) = 26.4$$

$$\therefore \quad P_{BC} = \frac{26.4}{4} = 6.6 \text{ kN (Compression)}$$

Similarly, taking moments about the joint $C$ and equating the same,

$$P_{FE} \times 4 = (5.4 \times 4) - (5.4 \times 3) = 5.4$$

$$\therefore \quad P_{FE} = \frac{5.4}{4} = 1.35 \text{kN (Compression)}$$

and now taking moments about the joint $B$ and equating the same,

$$P_{FC} \times 2.4 = (P_{FE} \times 4) - (2 \times 3) + (5.4 \times 6) - (5.4 \times 4)$$

$$= (1.35 \times 4) - 6 + 32.4 - 21.6 = 10.2$$

$$\therefore \quad P_{FC} = \frac{10.2}{2.4} = 4.25 \text{ kN (Tension)}$$

## 11.18. Miscellaneous Structures

In the previous articles we have been analysing the regular frames subjected to vertical, horizontal and inclined loads. We have been solving such examples by the methods of joints and sections. But sometimes we come across irregular structures.

Such structures may be analysed in the same way as that for regular structures. The casual look at such a structure, gives us a feeling that it is complicated problem. But a patient and thoughtful procedure helps us in solving such problems. The following examples will illustrate this point.

**EXAMPLE 11.17.** *Figure 11.54 shows a bridge truss of 130 m span subjected to two points loads.*

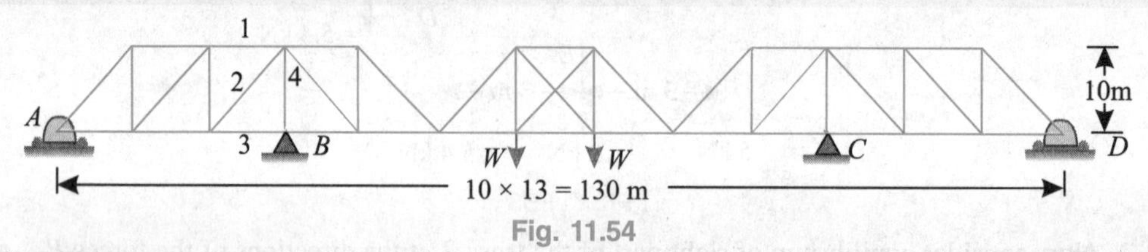


Fig. 11.54

*Determine the forces in the members 1, 2 and 3 of the bridge truss by any suitable method.*

**SOLUTION.** The whole structure may be considered to consist of two cantilever trusses supporting an intermediate truss. As a matter of fact, the two point loads acting at the intermediate truss are transferred to the ends of the cantilever trusses.

Since the two cantilever trusses are symmetrical and the point loads on the intermediate truss are also symmetrical, therefore each cantilever truss is subjected to a point load as shown in Fig. 11.55 (*a*).

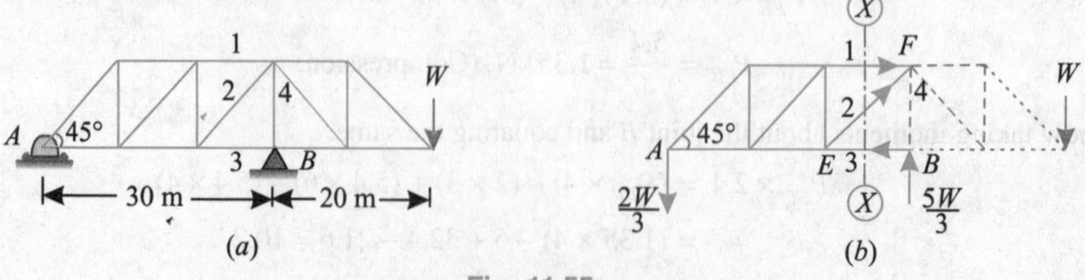


Fig. 11.55

Let $\quad V_B$ = Vertical reaction at the support $B$.

Taking moments about the support $A$ and equating the same,

$$V_B \times 30 = W \times 50 = 50\ W$$

$$V_B = \frac{50\,W}{30} = \frac{5\,W}{3}\ (\uparrow)$$

and
$$V_A = \frac{5\,W}{3} - W = \frac{2\,W}{3}\ (\downarrow)$$

First of all, pass section (*X-X*) cutting the truss into two parts and consider the equilibrium of the left part of the truss as shown in Fig. 11.55 (*b*). Now let the directions of the forces $P_1$, $P_2$ and $P_3$ be assumed as shown in Fig 11.55 (*b*). First of all, let us consider the joint *B*. A little consideration will show that the magnitude of the force $P_4$ will be equal to the reaction $V_B$ *i.e.*, 5*W*/3 (Compression). This will happen as the vertical components of the horizontal members at *B* will be zero.

Now resolving the forces vertically and equating the same,

$$P_2 \times \cos 45° = \frac{2W}{3}$$

or
$$P_2 = \frac{2W}{3} \times \frac{1}{\cos 45°} = \frac{2W}{3 \times 0.707} = 0.943\ W\ \text{(Tension)}$$

Taking moments of the forces acting on the left part of the truss only about the joint *E* and equating the same,

$$P_1 \times 10 = \frac{2W}{3} \times 20 = \frac{40W}{3}$$

$$\therefore \quad P_1 = \frac{40W}{3} \times \frac{1}{10} = \frac{4W}{3}\ \text{(Tension)} \qquad \textbf{Ans.}$$

and now taking moments of the forces acting on the left part of the truss only about the joint *F* and equating the same,

$$P_3 \times 10 = \frac{2W}{3} \times 30 = 20W$$

$$\therefore \quad P_3 = \frac{20W}{10} = 2W\ \text{(Compression)} \qquad \textbf{Ans.}$$

**EXAMPLE 11.18.** *A pin-jointed frame shown in Fig 11.56 is hinged at A and loaded at D. A horizontal chain is attached to C and pulled so that AD is horizontal.*

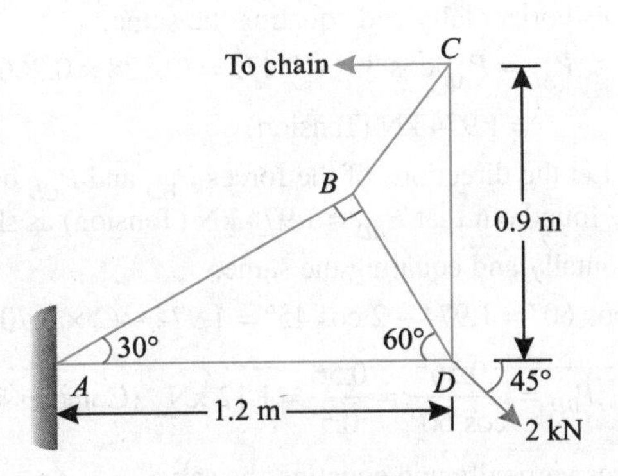


Fig. 11.56

*Determine the pull in the chain and also the force in each member. Tabulate the results.*

**SOLUTION.** The example may be solved either by the method of joints or method of sections. But we shall solve it by the method of joints, as we have to find the force in each member.

*Pull in the chain*

Let $P$ = Pull in the chain.

Taking moments about the joint $A$ and equating the same,

$$P \times 0.9 = 2 \cos 45° \times 1.2 = 2 \times 0.707 \times 1.2 = 1.7$$

$$\therefore \quad P = \frac{1.7}{0.9} = 1.889 \text{ kN} \quad \textbf{Ans.}$$

*Force in each member*

We know that horizontal reaction at $A$,

$$H_A = 1.889 - (2 \cos 45°) = 1.889 - (2 \times 0.707) = 0.475 \text{ kN} (\rightarrow)$$

and vertical reaction at $A$,

$$V_A = 2 \sin 45° = 2 \times 0.707 = 1.414 \text{ kN} (\uparrow)$$

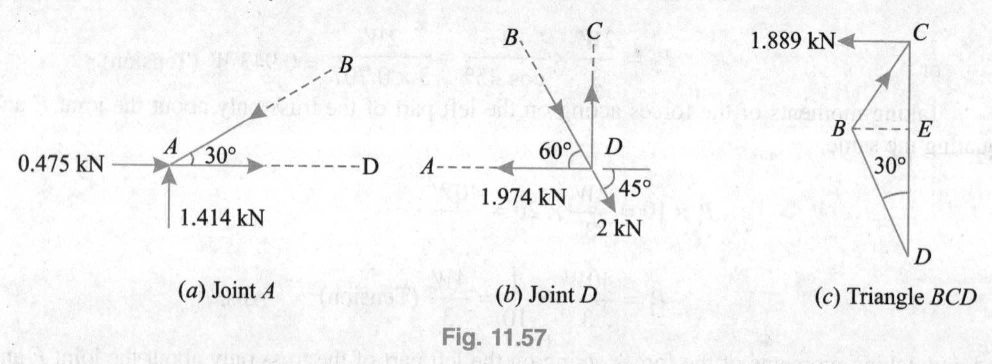


(*a*) Joint $A$ (*b*) Joint $D$ (*c*) Triangle $BCD$

**Fig. 11.57**

First of all, consider the joint $A$. Let the directions of the forces $P_{AB}$ and $P_{AD}$ be assumed as shown in Fig 11.57 (*a*). We have already found out that zthe horizontal and vertical reactions at $A$ are 0.475 kN and 1.414 kN repectively as shown in the figure.

Resolving the forces vertically and equating the same,

$$P_{AB} \sin 30° = 1.414$$

$$P_{AB} = \frac{1.414}{\sin 30°} = \frac{1.414}{0.5} = 2.828 \text{ kN (Compression)}$$

and now resolving the forces horizontally and equating the same,

$$P_{AD} = P_{AB} \cos 30° - 0.475 = (2.828 \times 0.866) - 0.475$$

$$= 1.974 \text{ kN (Tension)}$$

Now consider the joint D. Let the directions of the forces $P_{BD}$ and $P_{CD}$ be assumed as shown in Fig 13.57 (*b*). We have already found out that $P_{AD}$ = 1.974 kN (Tension) as shown in the figure.

Resolving the forces horizontally and equating the same,

$$P_{BD} \cos 60° = 1.974 - 2 \cos 45° = 1.974 - (2 \times 0.707) = 0.56 \text{ kN}$$

$$\therefore \quad P_{BD} = \frac{0.56}{\cos 60°} = \frac{0.56}{0.5} = 1.12 \text{ kN (Compression)}$$

and now resolving the forces vertically and equating the same,

$$P_{CD} = P_{BD} \sin 60° + 2 \sin 45°$$

$$= (1.12 \times 0.866) + (2 \times 0.707) = 2.384 \text{ kN (Tension)}$$

Now consider the triangle $BCD$. From $B$, draw $BE$ perpendicular to $CD$. Let the direction of $P_{BC}$ be assumed as shown in Fig 11.57 (*c*).

From the geometry of this triangle, we find that

$$BD = AD \sin 30° = 1.2 \times 0.5 = 0.6 \text{ m}$$

and $$BE = BD \sin 30° = 0.6 \times 0.5 = 0.3 \text{ m}$$

$\therefore$ $$DE = BD \cos 30° = 0.6 \times 0.866 = 0.52 \text{ m}$$

and $$CE = DC - DE = 0.9 - 0.52 = 0.38 \text{ m}$$

$\therefore$ $$\tan \angle BCE = \frac{BE}{CE} = \frac{0.3}{0.38} = 0.7895$$

or $$\angle BCE = 38.3°$$

Resolving the forces horizontally at $C$ and equating the same,

$$P_{BC} \sin 38.3° = 1.889$$

$\therefore$ $$P_{BC} = \frac{1.889}{\sin 38.3°} = \frac{1.889}{0.6196} = 3.049 \text{ kN (Compression)}$$

Now tabulate the results as given below :

| *S.No.* | *Member* | *Magnitude of force in kN* | *Nature of force* |
|---|---|---|---|
| 1 | *AB* | 2.828 | Compression |
| 2 | *AD* | 1.974 | Tension |
| 3 | *BD* | 1.12 | Compression |
| 4 | *CD* | 2.384 | Tension |
| 5 | *BC* | 3.049 | Compression |

**EXAMPLE 11.19.** *The truss shown in the Fig. 11.58 is made up of three equilateral triangles loaded at each of the lower panel pains.*

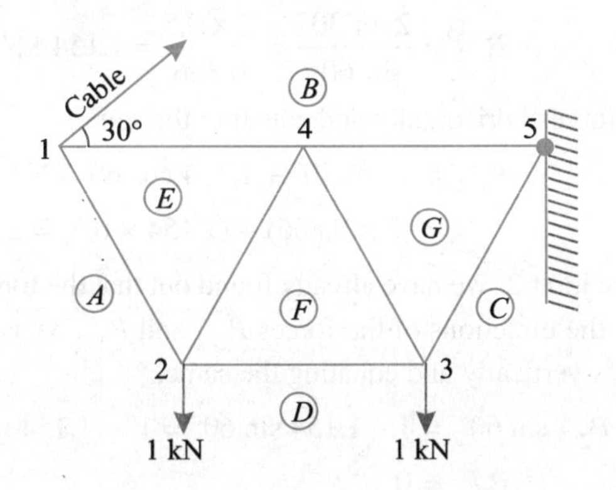


Fig. 11.58

*It is supported at the wall on the right hand side and by a cable on the left as shown. Determine (a) the tension in the cable (b) the reaction at the wall and (c) the nature and magnitude of the force in each bar.*

**SOLUTION.** The example may be solved either by the method of joints or method of sections. But we shall solve it by the method of joints, as we have to find out the forces in all the members of the truss.

*(a) Tension in the cable*

Let $T$ = Tension in the cable and

$a$ = Length of each side of the equilateral triangle.

Taking moments about the joint 5 and equating the same,

$$(T \cos 60°) \times 2a = (1 \times 1.5\,a) + (1 \times 0.5\,a)$$

$$(T \times 0.5)\,2a = 2a$$

$$\therefore \quad T = 2 \text{ kN} \quad \textbf{Ans.}$$

*(b) Nature and magnitude of the force in each bar*

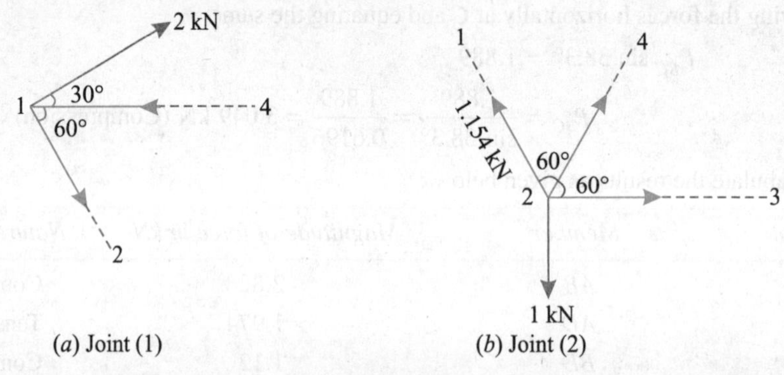


(*a*) Joint (1) (*b*) Joint (2)

**Fig. 11.59**

First of all consider the joint 1. We have already found out that tension in the cable is 2 kN as shown in the figure. Let the directions of $P_{1-2}$ and $P_{1-4}$ be assumed as shown in Fig. 11.59 (*a*). Resolving the forces vertically and equating the same,

$$P_{1-2} \sin 60° = 2 \sin 30°$$

$$\therefore \quad P_{1-2} = \frac{2 \sin 30°}{\sin 60°} = \frac{2 \times 0.5}{0.866} = 1.154 \text{ kN (Tension)}$$

and now resolving the forces horizontally and equating the same,

$$P_{1-4} = 2 \cos 30° + 1.154 \cos 60° \text{ kN}$$

$$= (2 \times 0.866) + (1.154 \times 0.5) = 2.309 \text{ kN (Compression)}$$

Now consider the joint 2. We have already found out that the force in member 1-2 (*i.e.* $P_{1-2}$) is 1.54 kN (Tension). Let the directions of the forces $P_{2-4}$ and $P_{2-3}$ be assumed as shown in Fig 11.59 (*b*). Resolving the forces vertically and equating the same,

$$P_{2-4} \sin 60° = 1 - 1.154 \sin 60° = 1 - (1.154 \times 0.866) = 0$$

$$\therefore \quad P_{2-4} = 0$$

and now resolving the forces horizontally and equating the same,

$$P_{2-3} = 1.154 \cos 60° = 1.154 \times 0.5 = 0.577 \text{ kN (Tension)}$$

Now consider the joint 4. A little consideration will show that the force $P_{3-4}$ will be zero. This will happen as the force $P_{2-4}$ is zero and the vertical components of the forces $P_{1-4}$ and $P_{4-5}$ are also zero. Moreover, the force $P_{4-5}$ will be equal to the force $P_{1-4}$ *i.e.*, 2.309 kN (Compression). This will happen as the forces $P_{2-4}$ and $P_{2-5}$ (being zero) will have their vertical components as zero.

Now consider the joint 3. Let the direction of the force $P_{3-5}$ be assumed as shown in Fig. 11.60 (*b*). We have already found out that the force $P_{2-3}$ is 0.577 kN (Tension) and force $P_{3-4}$ is zero.

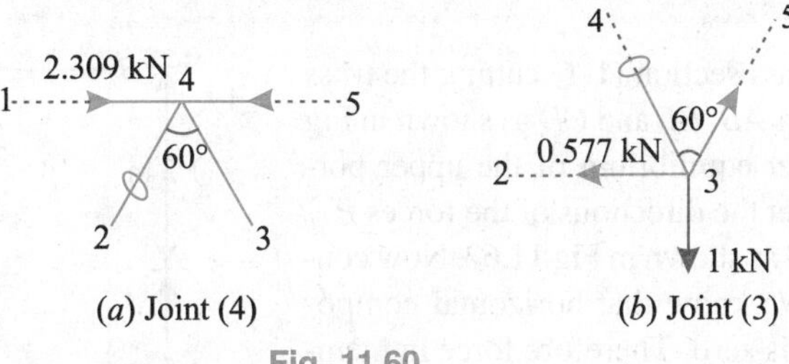


Fig. 11.60

Resolving the forces vertically and equating the same,

$$P_{3-5} \cos 30° = 1$$

$$\therefore \quad P_{3-5} = \frac{1}{\cos 30°} = \frac{1}{0.866} = 1.154 \text{ kN (Tension)}$$

Now tabulate the results as given below :

| *S.No.* | *Member* | *Magnitude of force in kN* | *Nature of force* |
|---|---|---|---|
| 1 | 1-2 (*AE*) | 1.154 | Tension |
| 2 | 1-4 (*BE*) | 2.309 | Compression |
| 3 | 2-4 (*EF*) | 0 | — |
| 4 | 2-3 (*FD*) | 0.577 | Tension |
| 5 | 3-4 (*FG*) | 0 | — |
| 6 | 4-5 (*BG*) | 2.309 | Compression |
| 7 | 3-5 (*GD*) | 1.154 | Tension |

**(*C*) *Reaction at the wall***

We know that the reaction at the wall will be the resultant of the forces $P_{4-5}$ (*i.e.*, 2.309 kN Compression) and $P_{3-5}$ (*i.e.*, 1.154 kN Tension). This can be easily found out by the parallelogram law of forces *i.e.*,

$$R = \sqrt{(1.154)^2 + (2.309)^2 + 2 \times 1.154 \times 2.309 \cos 120°}$$

$$= \sqrt{1.332 + 5.331 + 5.329(-0.5)} = 2 \text{ kN} \qquad \textbf{Ans.}$$

**EXAMPLE 11.20.** *A frame ABCD is hinged at A and supported on rollers at D as shown in Fig. 11.61.*

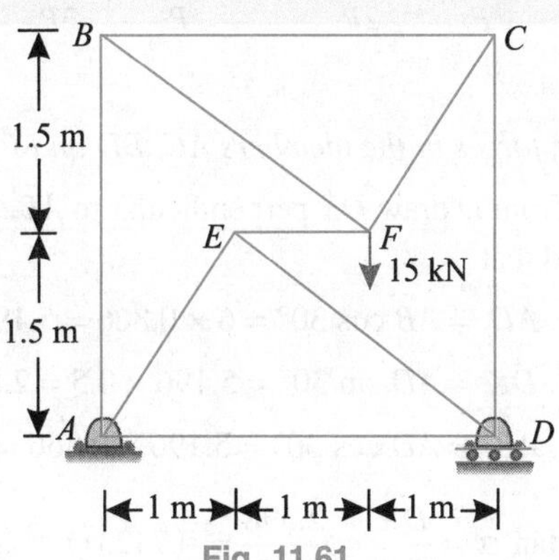


Fig. 11.61

*Determine the forces in the member AB, CD and EF,.*

**Solution.** The example may be solved either by the method of joints or method of sections. But we shall solve it by the method of sections, as we have to determine forces in three members of the frame only.

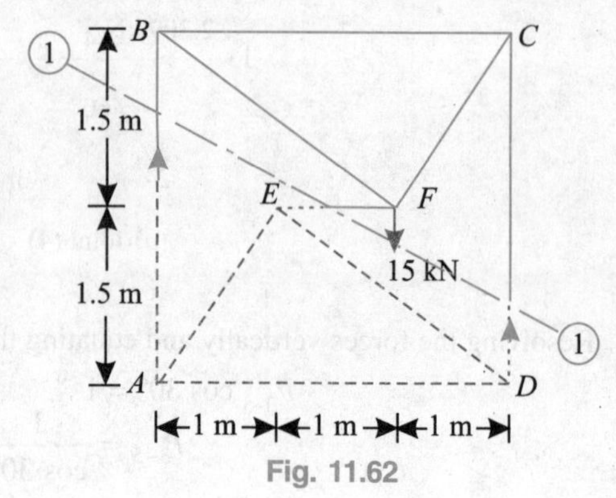

Fig. 11.62

First of all pass section (1-1) cutting the truss through the members *AB*, *EF* and *CD* as shown in Fig 11.62. Now consider equilibrium of the upper portion of the frame. Let the directions of the forces $P_{AB}$ and $P_{CD}$ be assumed as shown in Fig 11.62. Now consider the joint *F*. We know that horizontal component of 15 kN load is zero. Therefore force in member *EF* is also zero. **Ans.**

Now taking moments of the forces acting on the upper portion of the frame about the joint *A* and equating the same,

$$P_{CD} \times 3 = 15 \times 2 = 30$$

or
$$P_{CD} = \frac{30}{3} = 10 \text{ kN} \quad \textbf{Ans.}$$

and now taking moments of the forces about the joint *D* and equating the same,

$$P_{AB} \times 3 = 15 \times 1 = 15$$

or
$$P_{AB} = \frac{15}{3} = 5 \text{ kN} \quad \textbf{Ans.}$$

**Example 11.21.** *A framed structure of 6 m span is carrying point loads as shown in Fig 11.63.*

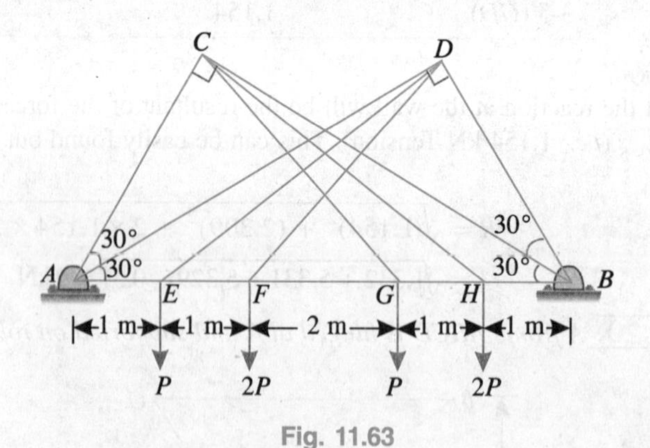

Fig. 11.63

*Find by any method the forces in the members AC, BD and FG of the structure.*

**Solution.** First of all, from *D* draw *DK* perpendicular to *AB* as shown in Fig 11.63. From the geometry of the figure, we find that

$$AD = AB \cos 30° = 6 \times 0.866 = 5.196 \text{ m}$$

and
$$DK = AD \sin 30° = 5.196 \times 0.5 = 2.598 \text{ m}$$

Similarly
$$AK = AD \cos 30° = 5.196 \times 0.866 = 4.5 \text{ m}$$

$\therefore$
$$\tan \alpha = \frac{DK}{EK} = \frac{2.598}{3.5} = 0.7423 \quad \text{or} \quad \alpha = 36.6°$$

and $\tan \beta = \dfrac{DK}{FK} = \dfrac{2.598}{2.5} = 1.0392$ or $\beta = 46.1°$

Taking moments about B and equating the same,

$$R_A \times 6 = (P \times 5) + (2\,P \times 4) + (P \times 2) + (2\,P \times 1) = 17\,P$$

$$\therefore \quad R_A = \frac{17\,P}{6} = 2.83\,P.$$

Let the directions of the various forces be assumed as shown in Fig 11.64. Now resolving the forces vertically at $E$ and equating the same,

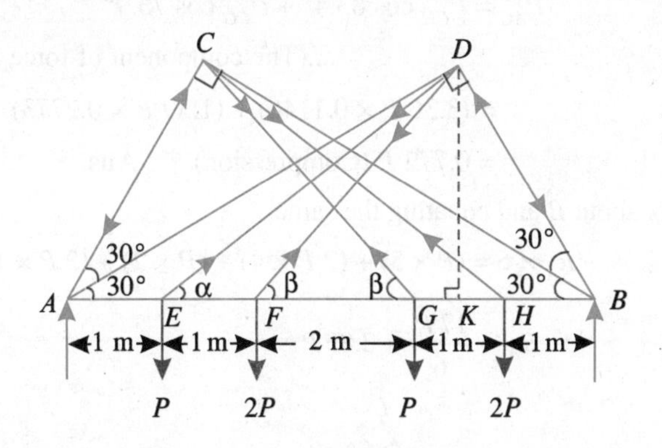


Fig. 11.64

$$P_{ED} \sin 36.6° = P$$

$$\therefore \quad P_{ED} = \frac{P}{\sin 36.6°} = \frac{P}{0.5960} = 1.68\ P \text{ (Tension)}$$

and now resolving the forces vertically at $F$ and equating the same,

$$P_{FD} \sin 46.1° = 2\,P$$

$$\therefore \quad P_{FD} = \frac{2\,P}{\sin 46.1°} = \frac{2\,P}{0.7206} = 2.78\ P \text{ (Tension)}$$

Similarly, resolving the forces vertically at $G$ and equating the same,

$$P_{CG} \sin 46.1° = P$$

$$\therefore \quad P_{CG} = \frac{P}{\sin 46.1°} = \frac{P}{0.7206} = 1.39\ P \text{ (Tension)}$$

and now resolving the forces vertically at $H$ and equating the same,

$$P_{CH} \sin 36.6° = 2\,P$$

$$\therefore \quad P_{CH} = \frac{2\,P}{\sin 36.6°} = \frac{2\,P}{0.5960} = 3.36\ P \text{ (Tension)}$$

From the geometry of the figure, we also find that

$$\angle EDB = \angle ACH = 180° - (36.6° + 60°) = 83.4°$$

and $\angle FDB = \angle ACG = 180° - (46.1° + 60°) = 73.9°$

Now at $D$, resolving the forces along $BD$ and equating the same,

$$P_{BD} = P_{ED} \cos 83.4° + P_{FD} \cos 73.9°$$

....(The component of force $P_{AD}$ about $BD$ is zero)

$$= (1.68\ P \times 0.1146) + (2.78\ P \times 0.2773)$$

$$= 0.963\ P \text{ (Compression)} \qquad \textbf{Ans.}$$

and at $C$ resolving the forces along $AC$ and equating the same,

$$P_{AC} = P_{CH} \cos 83.4° + P_{CG} \cos 73.9°$$

....(The component of force $P_{BC}$ about $AC$ is zero)

$$= (3.36\ P \times 0.1146) + (1.39\ P \times 0.2773)$$

$$= 0.772\ P \text{ (Compression)} \qquad \textbf{Ans.}$$

Taking moments about $B$ and equating the same,

$$R_A \times 6 = (P \times 5) + (2\ P \times 4) + (P \times 2) + (2\ P \times 1) = 17\ P$$

$$R_A = \frac{17\ P}{6} = 2.83\ P$$

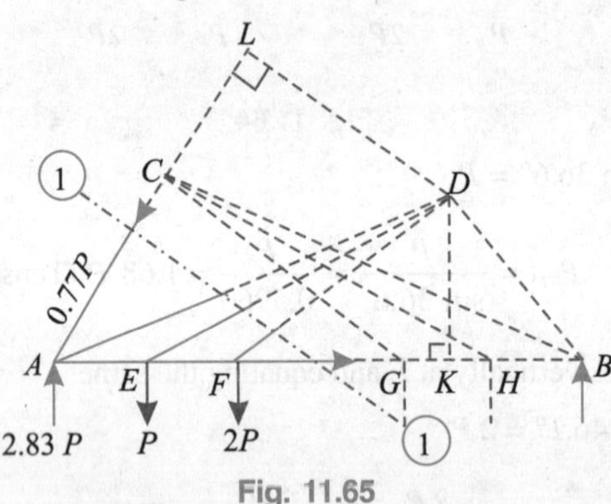


Fig. 11.65

Now pass section (1-1) cutting the truss into two parts as shown in Fig 11.65. Let us extend the line $AC$ and through $D$ draw $DL$ perpendicular to $AC$. From the geometry of the figure, we find that

$$DL = AD \sin 30° = 5.196 \times 0.5 = 2.598 \text{ m}$$

Taking moments of the forces in the left part of the truss about $D$ and equating the same,

$$2.83\ P \times 4.5 = (0.772\ P \times 2.598) + (P \times 3.5)$$
$$+ (2\ P \times 2.5) + (P_{FG} \times 2.598)$$

$$12.74\ P = 10.5\ P + (P_{FG} \times 2.598)$$

$$\therefore \quad 2.598\ P_{IG} = 12.74\ P - 10.5\ P = 2.24\ P$$

or

$$P_{FG} = \frac{2.24P}{2.598} = 0.862\ P \text{ (Tension)} \qquad \textbf{Ans.}$$

## EXERCISE 11.3

1. A truss shown in Fig. 11.66 is subjected to two points loads at $B$ and $F$. Find the forces in all the members of the truss and tabulate the results.

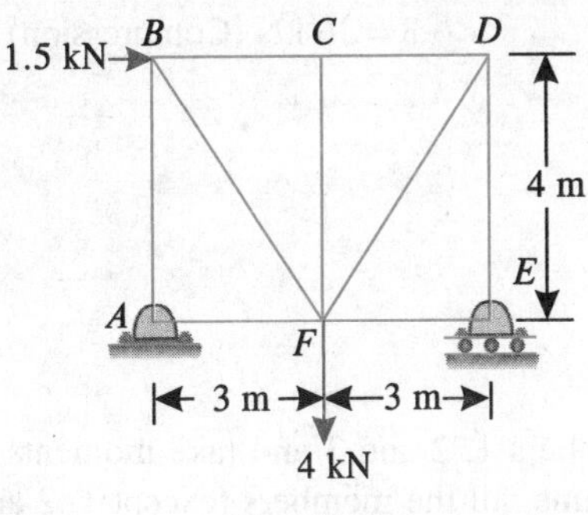

Fig. 11.66

**Ans.** $AB$ = 1.0 kN (Compression)
$AF$ = 1.5 kN (Tension)
$AE$ = 3.0 kN (Compression)
$EF$ = 0
$BF$ = 1.25 kN (Tension)
$BC$ = 2.25 kN (Compression)
$DF$ = 3.75 kN (Tension)
$CD$ = 2.25 kN (Compression)
$CF$ = 0

2. A cantilever braced truss supported on rollers at E and hinged at A is loaded as shown in Fig 11.67. Determine graphically or otherwise, the forces in the members of the truss, also determine the reactions at $A$ and $E$.

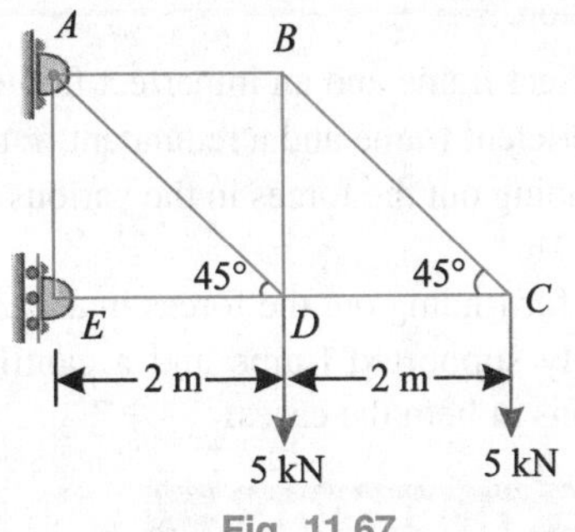

Fig. 11.67

**Ans.** $BC$ = 7.1 kN (Compression)
$CD$ = 5.0 kN (Tension)
$AB$ = 5.0 kN (Compression)
$BD$ = 5.0 kN (Tension)
$AD$ = 14.1 kN (Tension)
$ED$ = 15.0 kN (Compression)
$R_E$ = 15 kN
$R_E$ = 18 kN

**NOTE:** Since the truss is freely supported on rollers at $E$, therefore the reaction at this support will be horizontal (because of vertical support).

3. A truss of 5 m span and 2.5 m height is subjected to wind load as shown in Fig. 11.68. Find by any method the magnitude of forces in all the members of the truss. Also state their nature.

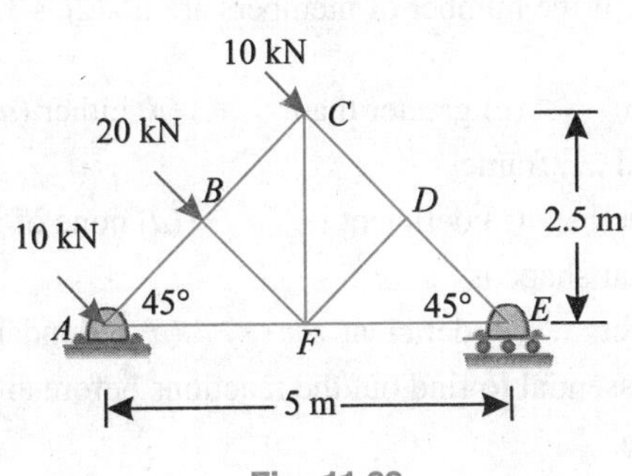

Fig. 11.68

**Ans.** $AB$ = 10.0 kN (Compression)
$AF$ = 28.28 kN (Tension)
$DE$ = 20.0 kN (Compression)
$EF$ = 14.14 kN (Tension)
$BF$ = 20.0 kN (Compression)
$BC$ = 10.0 kN (Compression)
$CF$ = 14.11 kN (Tension)
$CD$ = 20.0 kN (Compression)
$DF$ = 0

4. A truss 15 m long is subjected to a point load of 10 kN as shown in Fig. 11.69. Find the forces in the members 1, 2 and 3 of the truss.

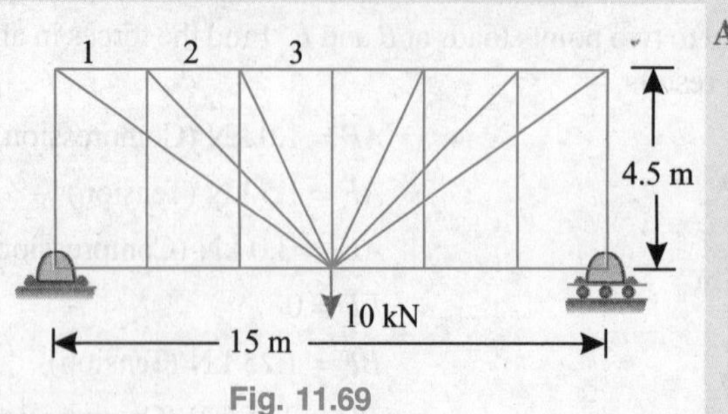

Fig. 11.69

**Ans.** 1 = 40 kN (Compression)
2 = 10 kN (Compression)
3 = 10 kN (Compression)

**Hint.** Pass vertical sections cutting the members 1, 2 and 3 and take moments about the joint containing 100 kN load. Each time, all the members (except 1, 2 and 3) pass through the joint about which moments are taken.

## QUESTIONS

1. What is a 'frame' ? Discuss its classification.
2. State clearly the difference between a perfect frame and an imperfect frame.
3. How would you distinguish between a deficient frame and a redundant frame ?
4. What are the assumptions made, while finding out the forces in the various members of a framed structure ?
5. Name the methods, which are employed, for finding out the forces in a frame.
6. What is the difference between a simply supported frame and a cantilever frame ? Discuss the method of finding out reactions in both the cases.

## OBJECTIVE TYPE QUESTIONS

1. A framed structure is perfect, if the number of members are .....$(2j - 3)$, where $j$ is the number of joints.
   (*a*) less than (*b*) equal to (*c*) greater than (*d*) either (*a*) or (*c*)
2. A framed structure is imperfect, if the number of members are .....$(2j - 3)$, where $j$ is the number of joints.
   (*a*) less than (*b*) equal to (*c*) greater than (*d*) either (*a*) or (*c*)
3. A redundant frame is also called ......frame
   (*a*) perfect (*b*) imperfect (*c*) deficient (*d*) none of these
4. A framed structure of a triangular shape is
   (*a*) perfect (*b*) imperfect (*c*) deficient (*d*) redundant
5. In a cantilever truss, it is very essential to find out the reactions before analyzing it.
   (*a*) agree (*b*) disagree

## ANSWERS

1. (*b*) 2. (*d*) 3. (*b*) 4. (*a*) 5. (*b*)

Chapter

# 12

# Analysis of Perfect Frames (Graphical Method)

## Contents

## 12.1. Introduction

In the previous chapter, we have discussed the analytical methods for determining the forces in perfect frames. We have seen that the method of joints involves a long process, whereas the method of sections is a tedious one. Moreover, there is a possibility of committing some mathematical mistake, while finding out the forces in the various members of truss. The graphical method, for determining the forces in the members of a perfect frame, is a simple and comparatively fool-proof method. The graphical solution of a frame is done in the following steps:

1. Construction of space diagram,
2. Construction of vector diagram and
3. Preparation of the table.

## 12.2. Construction of Space Diagram

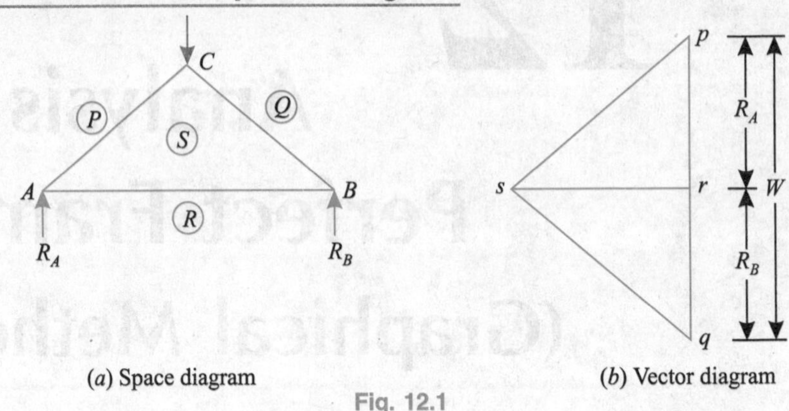


(*a*) Space diagram (*b*) Vector diagram

Fig. 12.1

It means the construction of a diagram of the given frame to a suitable linear scale, alongwith the loads it carries. The magnitude of support reactions is also found out and shown in the space diagram. Now name the various members and forces according to Bow's notations as shown in Fig. 12.1 (*a*).

In the space diagram of the truss *ABC* shown in Fig. 12.1 (*a*), the members *AB*, *BC* and *CA* are represented by *SR* (or *RS*), *SQ* (or *QS*) and *PS* (or *SP*) respectively. Similarly, load at *C* and reactions at *A* and *B* are represented by *PQ*, *RP* and *QR* respectively.

**NOTE:** The reactions are generally found out by analytical method as discussed in the last chapter.

## 12.3. Construction of Vector Diagram

After drawing the space diagram and naming the various members of the frame according to Bow's notations, as discussed in the last article, the next step is the construction of vector diagram. It is done in the following steps :

1. Select a suitable point *p* and draw *pq* parallel to *PQ* (*i.e.*, vertically downwards) and equal to the load *W* at *C* to some suitable scale.
2. Now cut off *qr* parallel to *QR* (*i.e.*, vertically upwards) equal to the reaction $R_B$ to the scale.
3. Similarly, cut off *rp* parallel to *RP* (*i.e.*, vertically upwards) equal to the reaction $R_A$ to the scale. Thus we see that in the space diagram, we started from *P* and returned to *P* after going for *P-Q-R-P* (*i.e.*, considering the loads and reactions only).
4. Now through *p* draw a line *ps* parallel to *PS* and throgh *r* draw *rs* parallel to *RS*, meeting the first line at *s* as shown in Fig. 12.1 (*b*). Thus *psrp* is the vector diagram for the joint (*A*).
5. Similarly, draw the vector diagram *qrsq* for the joint (*B*) and *pqsp* is the vector diagram for the joint (*C*) as shown in Fig. 12.1 (*b*).

**NOTES:** 1. While drawing the vector diagram, for a joint, care should be taken that the joint under consideration does not contain more than two members whose forces are unknown. if the joint, under consideration contains more than two such members whose forces are unknown, then some other joint which does not contain more than two unknown force members, should be considered for drawing the vector diagram.

2. If at any stage (which normally does not arise in a perfect frame) the work of drawing the vector diagram is held up at some joint, it will be then necessary to determine the force at such a joint by some other method *i.e.*, method of sections or method of joints.

## 12.4. Force Table

After drawing the vector diagram, the next step is to measure the various sides of the vector diagram and tabulate the forces in the members of the frame. For the preparation of the table, we require :

1. Magnitude of forces, and 2. Nature of forces.

## 12.5. Magnitude of Force

Measure all the sides of the vector diagram, whose lengths will give the forces in the corresponding members of the frame to the scale *e.g.*, the length *ps* of the vector diagram will give the force in the member *PS* of the frame to the scale. Similarly, the length *sr* will give the force in the member *SR* to the scale and so on as shown in Fig. 12.2. (*b*).

If any two points in the vector diagram coincide in the each other, then force in the member represented by the two letters will be zero.

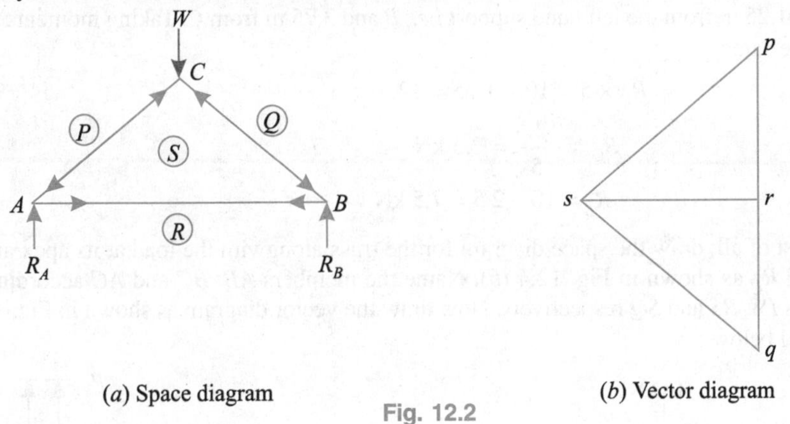


(*a*) Space diagram (*b*) Vector diagram

Fig. 12.2

## 12.6. Nature of Force

The nature of forces in the various members of a frame is determined by the following steps:

1. In the space diagram, go round a joint in a clockwise direction and note the order of the two letters by which the members are named *e.g.*, in Fig. 12.2 (*a*) the members at joint (*A*) are *RP*, *PS* and *SR*. Similarly, the members at joint (*B*) are *QR*, *RS* and *SQ*. And the members at joint (*C*) are *PQ*, *QS* and *SP*.
2. Now consider a joint of the space diagram and note the order of the letters of all the members (as stated above). Move on the vector diagram in the order of the letters noted on the space diagram.
3. Make the arrows on the members of the space diagram, near the joint, under consideration, which should show the direction of movement on the vector diagram. Put another arrow in the opposite direction on the other end of the member, so as to indicate the equilibrium of the method under the action of the internal stress.
4. Similarly, go round all the joints and put arrows.
5. Since these arrows indicates the direction of the internal forces only, thus the direction of the actual force in the member will be in opposite direction of the arrows, *e.g.*, a member with arrows pointing outwards *i.e.*, towards the joints [as member *PS* and *SQ* of Fig. 12.2 (*a*)] will be in compression; whereas a member with arrow pointing inwards *i.e.*, away from the joints [as member *SR* in Fig. 12.2 (*b*)] will be in tension.

**EXAMPLE 12.1.** *The truss ABC shown in Fig. 12.3 has a span of 5 metres. It is carrying a load of 10 kN at its apex.*

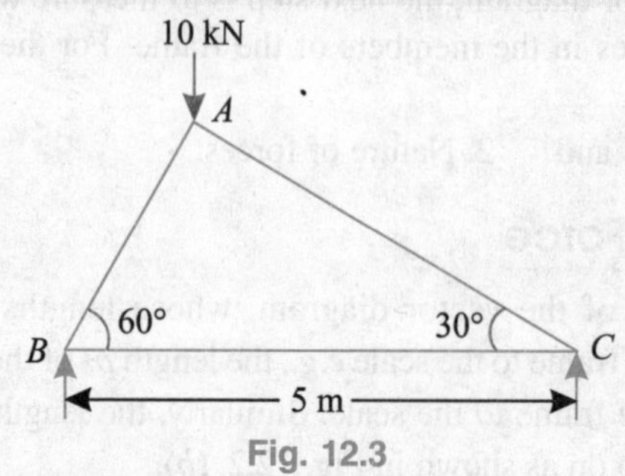


**Fig. 12.3**

*Find the forces in the members AB, AC and BC*

**SOLUTION*.** From the geometry of the truss, we find that the load of 10 kN is acting at a distance of 1.25 m from the left hand support *i.e.*, *B* and 3.75 m from *C*. Taking moments about *B* and equating the same,

$$R_C \times 5 = 10 \times 1.25 = 12.5$$

$$\therefore \qquad R_C = \frac{12.5}{5} = 2.5 \text{ kN}$$

and

$$R_B = 10 - 2.5 = 7.5 \text{ kN}$$

†First of all, draw the space diagram for the truss alongwith the load at its apex and the reaction $R_B$ and $R_C$ as shown in Fig. 12.4 (*b*). Name the members *AB*, *BC* and *AC* according to Bow's notations as *PS*, *RS* and *SQ* respectively. Now draw the vector diagram as shown in Fig. 12.4 (*b*) and as discussed below :

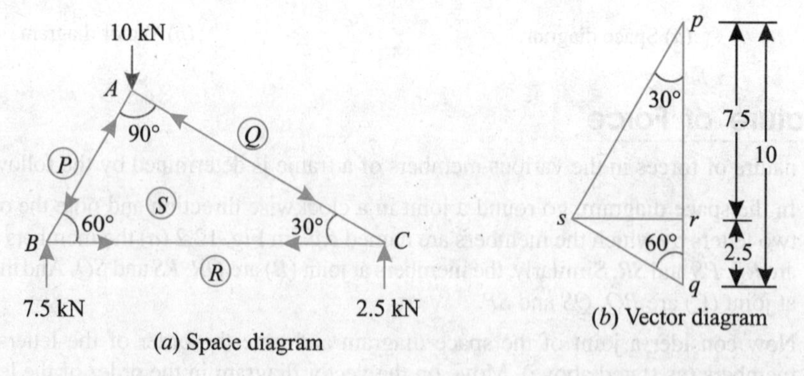


(*a*) Space diagram (*b*) Vector diagram

**Fig. 12.4**

1. Select some suitable point *p* and draw a vertical line *pq* equal to 10 kN to some suitable scale to represent the load *PQ* at joint *A*.
2. Now cut off *qr* equal to 2.5 kN to the scale to represent the reaction $R_C$ at *C*. This *rp* will represent the reaction $R_B$ to the scale.
3. Now draw the vector diagram for the joint *B*. For doing so, through *p* draw *ps* parallel to *PS* and through *r* draw *rs* parallel to *RS* meeting the first line at *s*. Now *psrp* is the vector diagram for the joint *B*, whose directions follow *p-s*; *s-r* and *r-p*.

* We have already solved this example analylically in the last chapter.

† As a matter of fact, this is the advantage of graphical method, that the previous work is checked. If at any stage some error is noticed, the complete vector diagram should be drawn once again.

4. Similarly, draw vector diagram for the joint $C$, whose directions follow $q$-$r$; $r$-$s$ and $s$-$q$ shown Fig.12.4 (*a*) and (*b*). Now check the vector diagram for the joint $A$, whose directions follow $p$-$q$ ; $q$-$s$ and $s$-$p$.

Now measuring† the various sides of the vector diagram and keeping due note of the directions of the arrow heads, the results are tabulated here :

| *S.No.* | *Member* | *Magnitude of force in kN* | *Nature of force* |
|---|---|---|---|
| 1 | *AB (PS)* | 8.7 | Compression |
| 2 | *BC (RS)* | 4.3 | Tension |
| 3 | *AC (SQ)* | 5.0 | Compression |

**EXAMPLE 12.2.** *A truss of span 10 metres is loaded as shown in Fig. 12.5.*

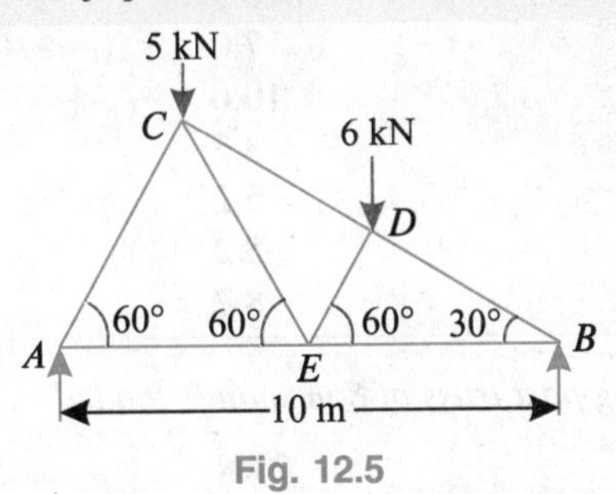


Fig. 12.5

*Find the reactions and forces in the members of the truss.*

**SOLUTION.** From the geometry of the figure, we find the load 5 kN is acting at a distance of 2.5 metres and the load of 6 kN at a distance 6.25 metres from the left hand support.

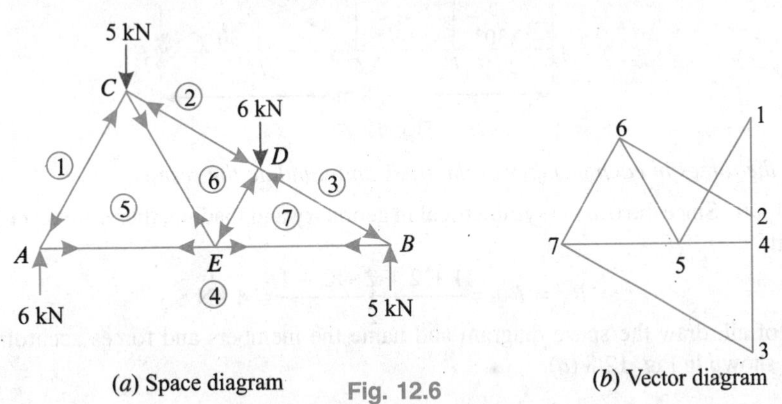


(*a*) Space diagram Fig. 12.6 (*b*) Vector diagram

Taking moments about the left hand support and equating the same.

$$R_B \times 10 = (5 \times 2.5) + (6 \times 6.25) = 50$$

$$\therefore \qquad R_B = \frac{50}{10} = 5 \text{ kN}$$

and

$$R_A = (5 + 6) - 5 = 6 \text{ kN}$$

First of all, draw space diagram for the truss alongwith loads and reactions as shown in Fig. 12.6 (*a*). Name the various members of the truss and forces according to Bow's notations.

† Sometimes, there is a slight variation in the results obtained by the analytical method and graphical method. The values obtained by graphical method are taken to be correct, if they agree upto the first decimal point with the values obtained by analytical method, *e.g.*, 8.66 (Analytical) = 8.7 (graphical). Similarly, 4.32 (Analytical) = 4.3 (graphical).

Now draw vector diagram as shown in Fig. 12.6 (*b*) and as discussed below :

1. Select some suitable point 1 and draw a vertical line 1-2 equal to 5 kN to some suitable scale to represent the load 5 kN at *C*. Similarly, draw 2-3 equal to 6 kN to the scale to represent the load 6 kN at *D*.
2. Now cut off 3-4 equal to 5 kN to the scale to represent the reaction $R_B$. Thus 4-1 will represent the reaction $R_A$ to the scale.
3. Now draw vector diagram for the joint *A*. For doing so through 1, draw 1-5 parallel to *AC* and through 4, draw 4-5 parallel to *AE* meeting the first line at 5. Now 1-5-4-1 is the vector diagram for joint *A*, whose directions follow 1-5, 5-4 and 4-1. Similarly, draw vector diagrams for the joints *B*, *C*, *D* and *E* as shown in Fig. 12.6 (*b*).

Now measuring the various sides of the vector diagram, the results are tabulated here :

| *S. No.* | *Member* | *Magnitude of force in kN* | *Nature of force* |
|---|---|---|---|
| 1 | *AC* (1-5) | 6.9 | Compression |
| 2 | *CD* (2-6) | 7.0 | Compression |
| 3 | *BD* (3-7) | 10.0 | Compression |
| 4 | *AE* (4-5) | 3.5 | Tension |
| 5 | *CE* (5-6) | 5.2 | Tension |
| 6 | *DE* (6-7) | 5.2 | Compression |
| 7 | *BE* (4-7) | 8.7 | Tension |

**EXAMPLE 12.3.** *A king post truss of 8 m span is loaded as shown in Fig. 12.7.*

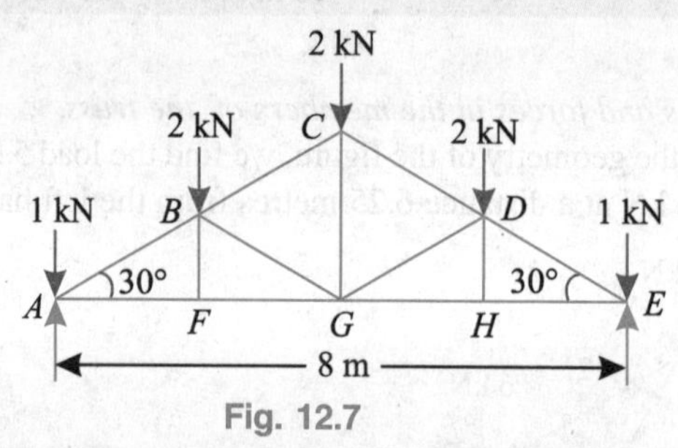


Fig. 12.7

*Find the forces in each member of the truss and tabulate the results.*

**SOLUTION.** Since the truss is symmetrical in geometry and loading, therefore reaction at the left hand support,

$$R_A = R_E = \frac{1 + 2 + 2 + 2 + 1}{2} = 4 \text{ kN}$$

First of all, draw the space diagram and name the members and forces according to Bow's notations as shown in Fig. 12.8 (*a*).

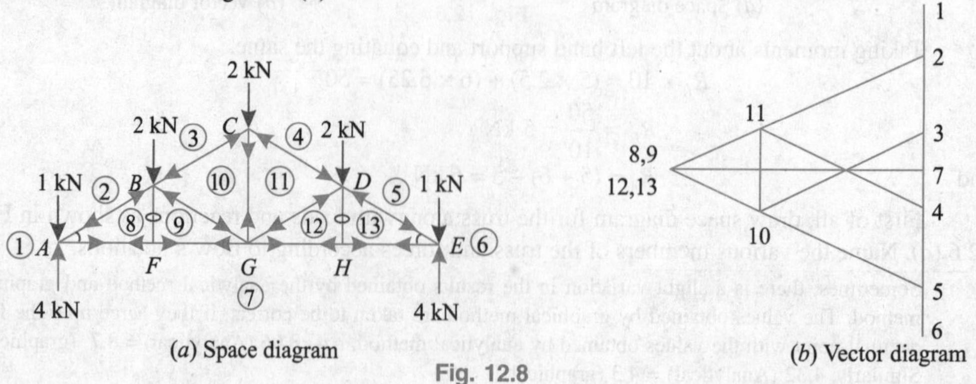


Fig. 12.8

Now draw the vector diagram as shown in Fig. 12.8 (*b*). Measuring various sides of the vector diagram, the result are tabulated here :

| *S. No.* | *Member* | *Magnitude of force in kN* | *Nature of force* |
|---|---|---|---|
| 1 | *AB, DE* | 6.0 | Compression |
| 2 | *AF, EH* | 5.2 | Tension |
| 3 | *FG, GH* | 5.2 | Tension |
| 4 | *BF, DH* | 0 | — |
| 5 | *BG, DG* | 2.0 | Compression |
| 6 | *BC, CD* | 4.0 | Compression |
| 7 | *CG* | 2.0 | Tension |

**EXAMPLE 12.4.** *A horizontal link AB is divided into three equal parts AC, CD and DB and above each, an equilateral triangle is drawn. The apices E, F and G of the triangles on AC, CD and DB respectively are also jointed.*

*The figure is then represented by centre lines, a framework simply at its ends A and B. Vertical loads each equal to W are carried at E and C as shown in Fig. 12.9.*

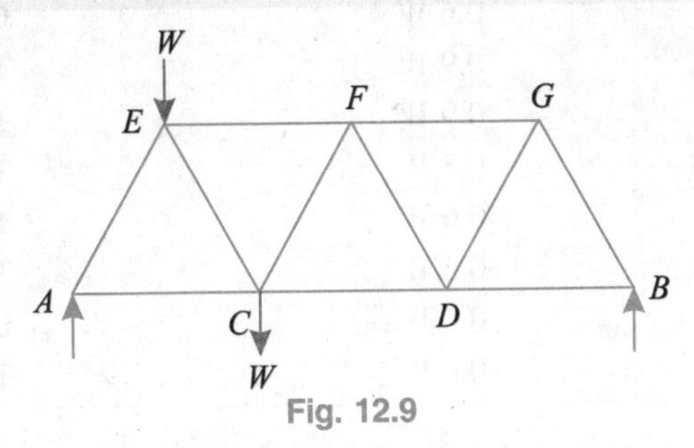


Fig. 12.9

*Find the nature and magnitude of forces in each of the member and write them upon the members of your diagram or in a table.*

**SOLUTION.** Taking moments about *A* and equating the same,

$$R_B \times 3 = W \times \frac{1}{2} + W \times 1 = \frac{3}{2} W$$

$$\therefore \qquad R_B = \frac{3}{2} \times W \times \frac{1}{3} = \frac{W}{2}$$

and

$$R_A = (W + W) - \left(\frac{W}{2}\right) = \frac{3W}{2}$$

First of all, draw the space diagram for the truss and name the various members according to Bow's notations as shown in Fig. 12.10 (*a*).

Now draw the vector diagram as shown in Fig. 12.10 (*b*). Measuring the various sides of the vector diagram the results are tabulated here :

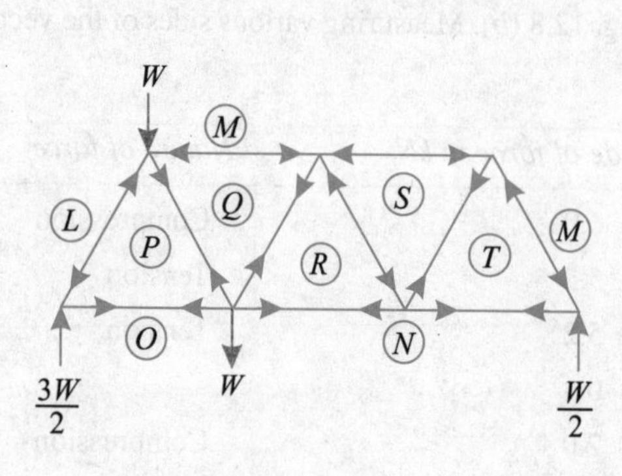


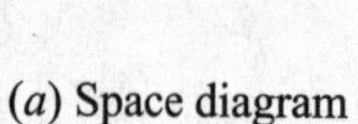

(a) Space diagram

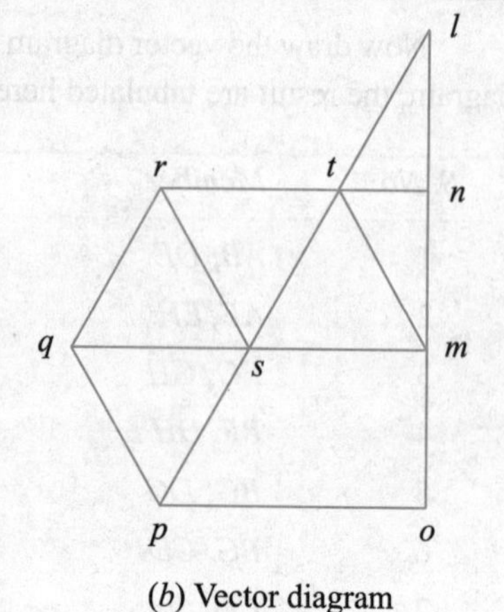


(b) Vector diagram

Fig. 12.10

| S.No | Member | Force | Nature |
|---|---|---|---|
| 1 | AE | 1.7 W | Compression |
| 2 | EF | 1.2 W | Compression |
| 3 | FG | 0.6 W | Compression |
| 4 | GB | 0.6 W | Compression |
| 5 | AC | 0.9 W | Tension |
| 6 | CD | 0.9 W | Tension |
| 7 | DB | 0.3 W | Tension |
| 8 | EC | 0.6 W | Tension |
| 9 | FC | 0.6 W | Tension |
| 10 | FD | 0.6 W | Compression |
| 11 | GD | 0.6 W | Tension |

**EXAMPLE 12.5.** *A truss of 32 metres span is loaded as shown in Fig. 12.11.*

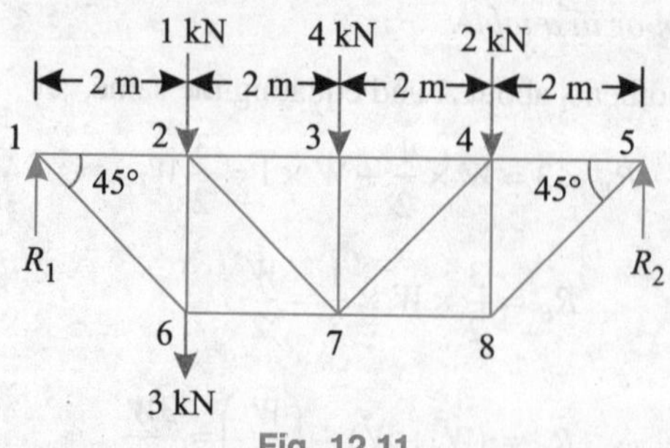


Fig. 12.11

*Find graphically, or otherwise, the magnitude and nature of forces in all the members of the truss.*

**SOLUTION.** Taking moments about the left end support and equating the same,

$$\therefore \quad R_5 \times 8 = (1 \times 2) + (4 \times 4) + (2 \times 6) + (3 \times 2) = 36$$

and

$$R_5 = \frac{36}{8} = 4.5 \text{ kN}$$

$$R_1 = (1 + 4 + 2 + 3) - 4.5 = 5.5 \text{ kN}$$

First of all, draw the space diagram and name all the members and forces according to Bow's notations as shown in Fig. 12.12 (*a*).

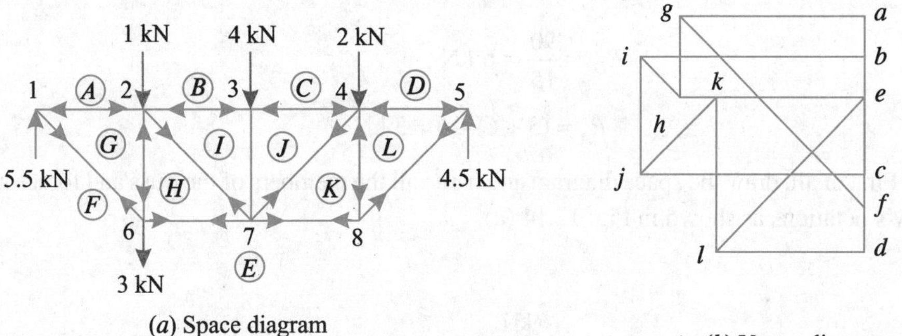


(*a*) Space diagram

Fig. 12.12

(*b*) Vector diagram

Now draw the vector diagram as shown in Fig. 12.12 (*b*). Measuring the various sides of the vector diagram, the results are tabulated here :

| *S. No.* | *Member* | *Magnitude of force in kN* | *Nature of force* |
|---|---|---|---|
| 1 | 1-2 (*AG*) | 5.5 | Compression |
| 2 | 2-3 (*BI*) | 7.0 | Compression |
| 3 | 3-4 (*CJ*) | 7.0 | Compression |
| 4 | 4-5 (*DL*) | 4.5 | Compression |
| 5 | 1-6 (*FG*) | 7.8 | Tension |
| 6 | 2-6 (*GH*) | 2.5 | Compression |
| 7 | 6-7 (*EH*) | 5.5 | Tension |
| 8 | 2-7 (*HI*) | 2.1 | Tension |
| 9 | 3-7 (*IJ*) | 4.0 | Compression |
| 10 | 4-7 (*JK*) | 3.5 | Tension |
| 11 | 7-8 (*EK*) | 4.5 | Tension |
| 12 | 4-8 (*KL*) | 4.5 | Compression |
| 13 | 5-8 (*EL*) | 6.4 | Tension |

**EXAMPLE 12.6.** *Find graphically or otherwise, the magnitude and nature of the forces in the truss shown in Fig. 12.13.*

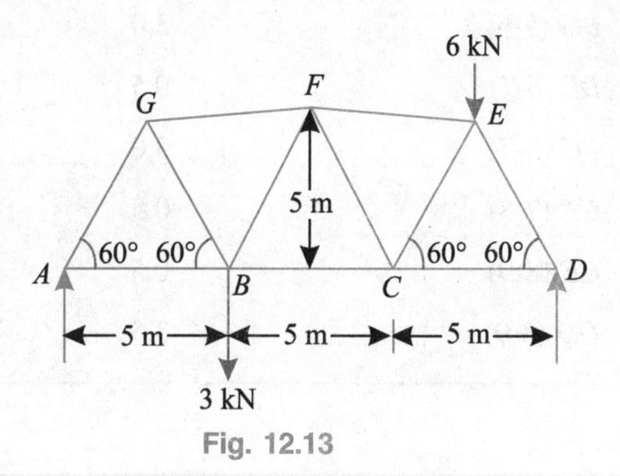


Fig. 12.13

*Also Indicate the results in a tabular form.*

**SOLUTION.** Taking moments about $A$ and equating the same,

$$R_D \times 15 = (3 \times 5) + (6 \times 12.5) = 90$$

$$\therefore \qquad R_D = \frac{90}{15} = 6 \text{ kN}$$

and

$$R_A = (3 + 6) - 6 = 3 \text{ kN}$$

First of all, draw the space diagram and name all the members of the truss and forces according to Bow's notations as shown in Fig. 12.14 (*a*).

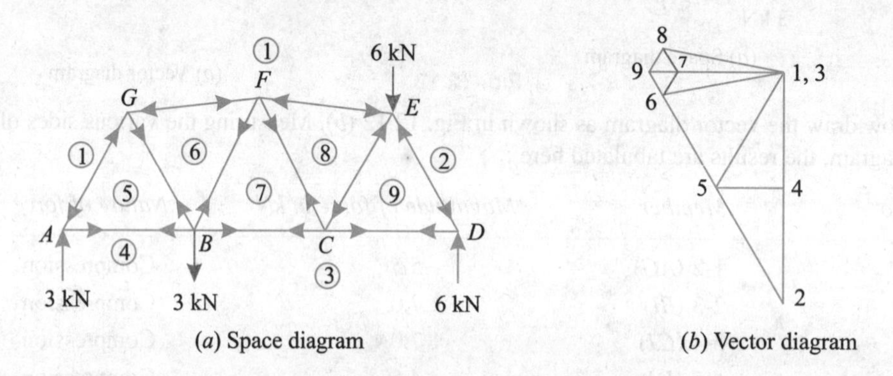


(*a*) Space diagram (*b*) Vector diagram

**Fig. 12.14**

Now draw vector diagram as shown in Fig. 12.14 (*b*). Measuring various sides of the vector diagram, the results are tabulated here :

| *S. No.* | *Member* | *Magnitude of force in kN* | *Nature of force* |
|---|---|---|---|
| 1 | *AG* (1-5) | 3.5 | Compression |
| 2 | *FG* (1-6) | 3.2 | Compression |
| 3 | *FE* (1-8) | 3.2 | Compression |
| 4 | *ED* (2-9) | 7.0 | Compression |
| 5 | *AB* (4-5) | 1.7 | Tension |
| 6 | *BG* (5-6) | 3.0 | Tension |
| 7 | *BF* (6-7) | 0.5 | Tension |
| 8 | *BC* (3-7) | 3.0 | Tension |
| 9 | *CF* (7-8) | 0.5 | Tension |
| 10 | *CE* (8-9) | 0.5 | Compression |
| 11 | *CD* (3-9) | 3.5 | Tension |

**EXAMPLE 12.7.** *A framed structure of 6 m span is carrying a central point load of 10 kN as shown in Fig 12.15.*

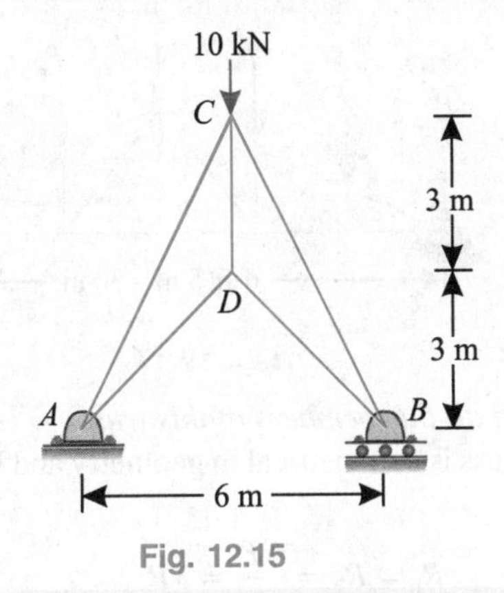


Fig. 12.15

*Find by any method the magnitude and nature of forces in all members of the sturcture.*

*SOLUTION. Since the structure is symmetrical in geometry and loading, therefore the reaction at *A*,

$$R_A = R_B = \frac{10}{2} = 5 \text{ kN}$$

First of all, draw the space diagram and name the members and forces according to Bow's notations as shown in Fig. 12.16 (*a*).

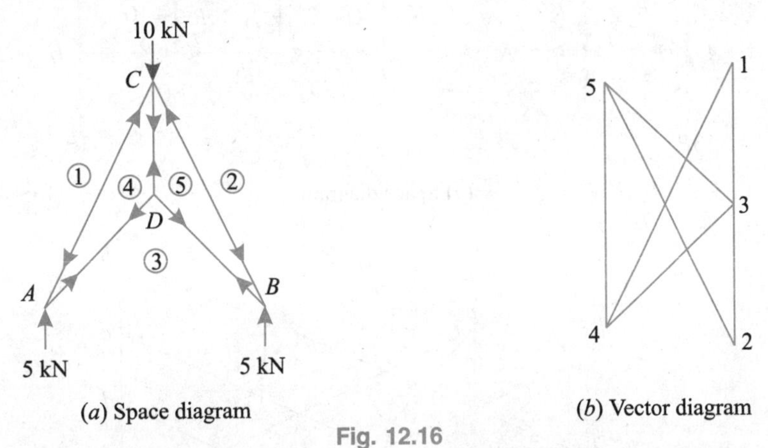


(*a*) Space diagram (*b*) Vector diagram

Fig. 12.16

Now draw the vector diagarm as shown in Fig. 12.16 (*b*). Measuring the various sides of the vector diagram, the results are tabulated here :

| *S. No.* | *Member* | *Magnitude of force in kN* | *Nature of force* |
|---|---|---|---|
| 1 | *AC, CB* | 11.2 | Compression |
| 2 | *AD, DB* | 7.1 | Tension |
| 3 | *CD* | 10.0 | Tension |

* We have already solved this example analytically in the last chapter.

**Example 12.8.** *Construct a vector diagram for the truss shown in Fig. 12.17.*

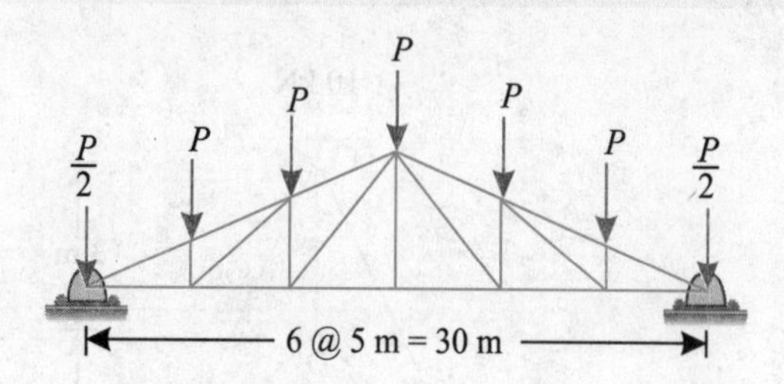


Fig. 12.17

*Determine the forces in all the members of this truss.*

**Solution.** Since the truss is symmetrical in geometry and loading, therefore the reaction at the left hand support,

$$R_1 = R_2 = \frac{6P}{2} = 3P$$

First of all, draw the space diagram and name the members and forces according to Bow's notations as shown in Fig. 12.18 (*a*).

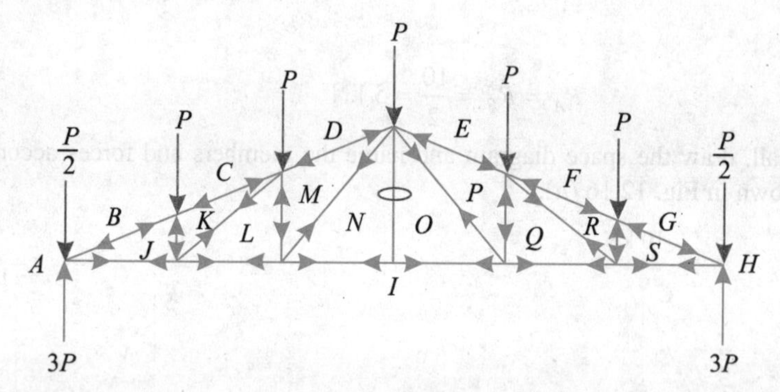


(*a*) Space diagram

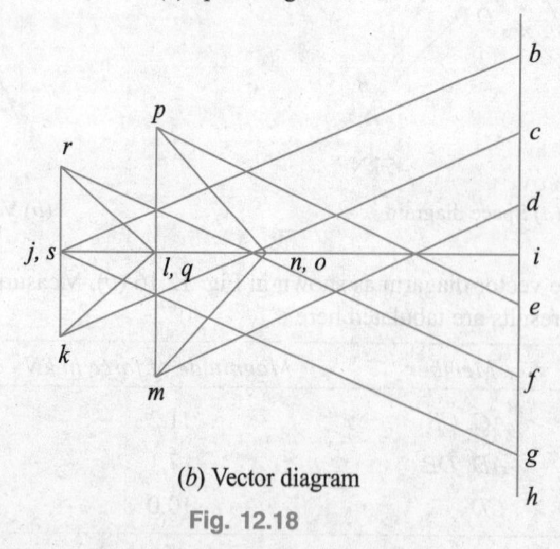


(*b*) Vector diagram

Fig. 12.18

Now draw the vector (*i.e.*, stress) diagram as shown in Fig. 12.18 (*b*). Measuring the various sides of the vector diagram, the results are tabulated here :

| S.No. | Member | Magnitude of force in terms of P | Nature of force (stress) |
|---|---|---|---|
| 1 | *BJ, GS* | 6.73 | Compression |
| 2 | *JI, IS* | 6.25 | Tension |
| 3 | *JK, RS* | 1.00 | Compression |
| 4 | *CK, RF* | 6.73 | Compression |
| 5 | *KL, QR* | 1.60 | Tension |
| 6 | *LI, IQ* | 1.00 | Tension |
| 7 | *LM, PQ* | 1.50 | Compression |
| 8 | *DM, EP* | 5.40 | Compression |
| 9 | *MN, OP* | 1.95 | Tension |
| 10 | *NI, IO* | 4.75 | Tension |
| 11 | *NO* | 0 | — |

## EXERCISE 12.1

1. Figure 12.19, shows a warren girder consisting of seven members each of 3 m length freely supported at its end points. The girder is loaded at B and C as shown. Find the forces in all the members of the girder, indicating whether the force is compressive or tensile.

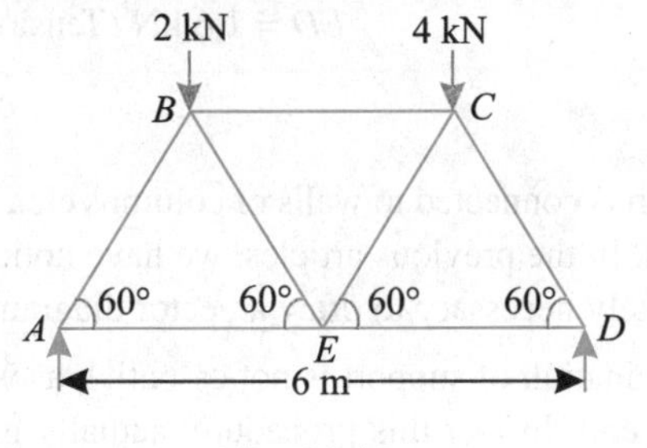


Fig. 12.19

**Ans.** $AB$ = 2.9 kN (Compression)
$AE$ = 1.4 kN (Tension)
$CD$ = 4.0 kN (Compression)
$DE$ = 2.0 kN (Tension)
$BE$ = 0.6 kN (Tension)
$BC$ = 1.7 kN (Compression)
$CE$ = 0.6 kN (Compression)

2. Figure 12.20 shows a framed structure of 5 m span. The structure carries vertical loads as shown in the figure. Find the forces in the members of the structure and tabulate the results.

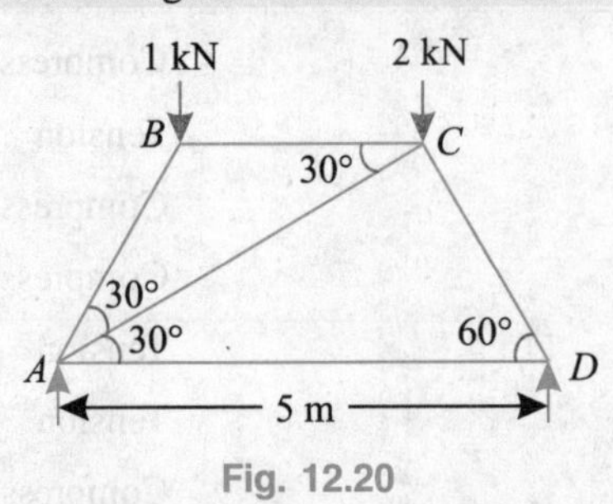


Fig. 12.20

**Ans.** $AB$ = 1.2 kN (Compression)
$BC$ = 0.6 kN (Compression)
$CD$ = 2.0 kN (Compression)
$AC$ = 0.5 kN (Compression)
$AD$ = 1.0 kN (Tension)

3. A pin-jointed frame is supported at $F$ and $E$ and loaded as shown in Fig. 12.21. Find the forces in all the members of the frame and state in each case, whether the member is in tension or compression.

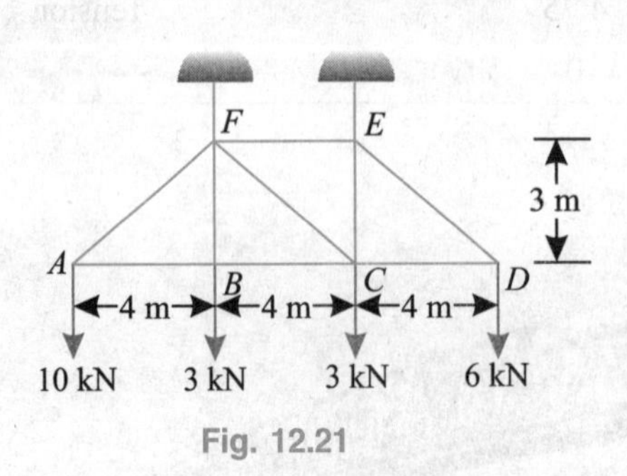


Fig. 12.21

**Ans.** $AF$ = 16.7 kN (Tension)
$FE$ = 8.0 kN (Tension)
$ED$ = 10.0 kN (Tenison)
$AB$ = 13.3 kN (Compression)
$BF$ = 3.0 kN (Tension)
$BC$ = 13.3 kN (Compression)
$FC$ = 6.7 kN (Tension)
$EC$ = 1.0 kN (Compression)
$CD$ = 8.0 kN (Compression)

4. A pin-jointed truss is subjected to three points loads at $A$, $B$ and $C$ as shown in Fig. 12.22. Find by any method, the forces in all the members of the truss.

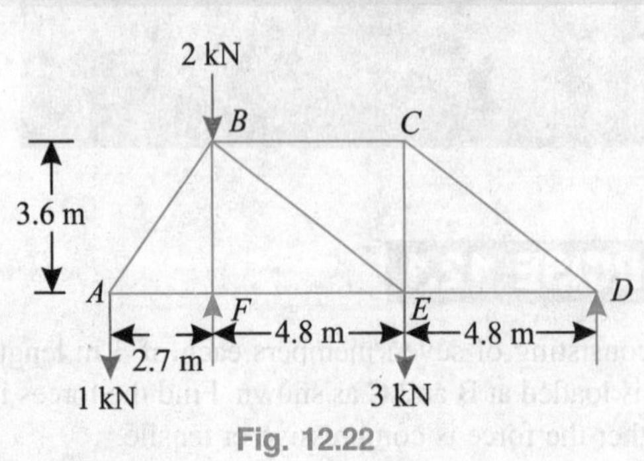


Fig. 12.22

**Ans.** $AB$ = 1.25 kN (Tension)
$BC$ = 1.6 kN (Compression)
$CD$ = 2.0 kN (Compression)
$AF$ = 0.75 kN (Compression)
$BF$ = 4.8 kN (Compression)
$FE$ = 0.75 kN (Compression)
$BE$ = 3.0 kN (Tension)
$CE$ = 1.2 kN (Tension)
$ED$ = 1.6 kN (Tension)

## 12.7. Cantilever Trusses

We have already discussed that a truss which is connected to walls or columns etc., at one end, and free at the other is known as a cantilever truss. In the previous articles, we have noticed that the determination of the support reactions was absolutely necessary to draw a vector diagram.

But in the case of cantilever trusses, determination of support is not essential, as we can start the construction of vector diagram from the free end. In fact this procedure, actually gives us the reactions at the connected ends of the truss.

**EXAMPLE 12.9.** *Figure 12.23 shows a cantilever truss with two vertical loads of 1 kN each.*

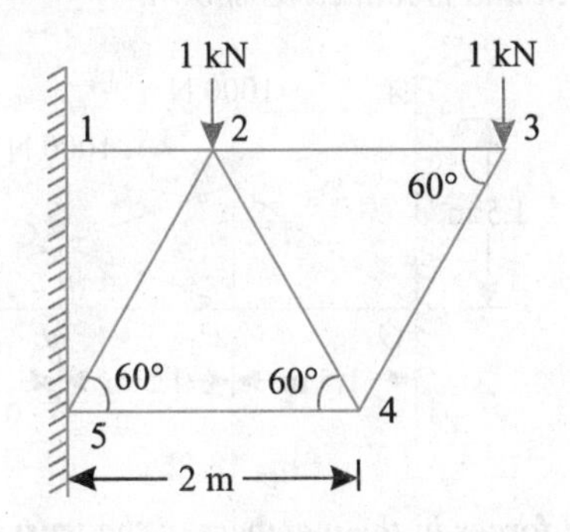


Fig. 12.23

*Find the reactions and forces in all the members of the truss.*

**SOLUTION.** First of all, draw the space diagram and name all the members and forces according to Bow's notations as shown in Fig. 12.24 (*a*).

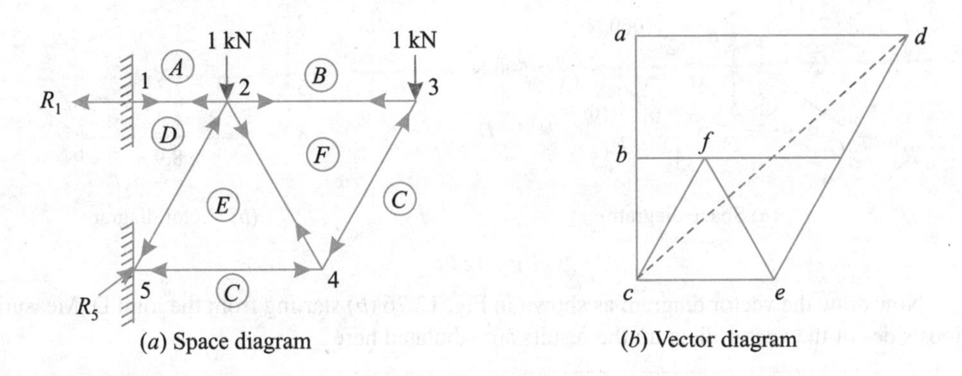


(*a*) Space diagram (*b*) Vector diagram

Fig. 12.24

Now draw the vector diagram, starting from the free joint (3) as shown in Fig. 13.24 (*b*). Measuring the various sides of the vector diagram, the results are tabulated here :

| *S.No.* | *Member* | *Magnitude of force in kN* | *Nature of force* |
|---|---|---|---|
| 1 | 1-2 (*AD*) | 2.3 | Tension |
| 2 | 2-3 (*BF*) | 0.6 | Tension |
| 3 | 3-4 (*CF*) | 1.15 | Compression |
| 4 | 2-4 (*EF*) | 1.15 | Tension |
| 5 | 4-5 (*CE*) | 1.15 | Compression |
| 6 | 2-5 (*DE*) | 2.3 | Compression |

***Reactions***

Upper $R_1$ (*ad*) = 2.3 kN;

Lower $R_5$ (*cd*) = 3.05 kN **Ans.**

**EXAMPLE 12.10.** *Figure 13.25 shows a cantilever truss having a span of 4.5 metres. It is hinged at two joints to a wall and is loaded as shown.*

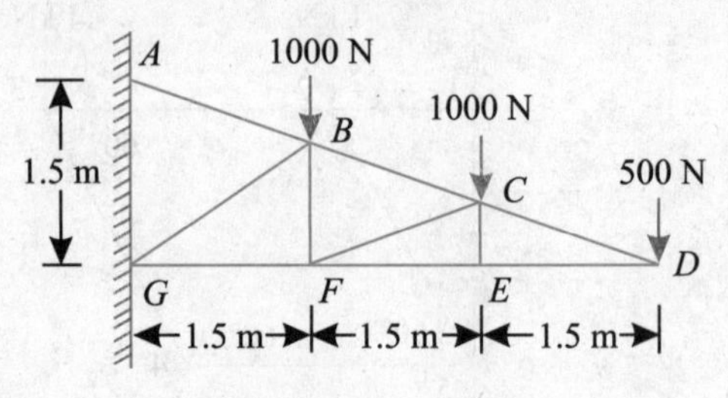


Fig. 12.25

*Find the reactions and forces in the members of the truss.*

**SOLUTION.** First of all, draw the space diagram and name all the members and forces according to Bow's notations as shown in Fig. 12.26 (*a*).

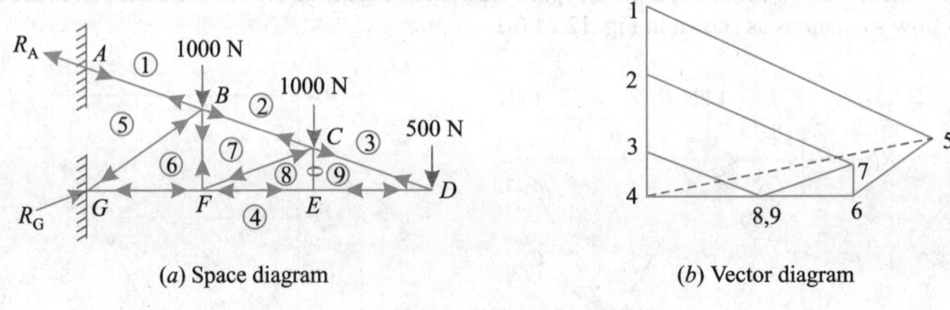


(*a*) Space diagram (*b*) Vector diagram

Fig. 12.26

Now draw the vector diagram as shown in Fig. 12.26 (*b*) starting from the joint D. Measuring various sides of the vector diagram the results are tabulated here :

| *S.No.* | *Member* | *Magnitude of force in kN* | *Nature of force* |
|---|---|---|---|
| 1 | *AB* (1-5) | 4750 | Tension |
| 2 | *BC* (2-7) | 3160 | Tension |
| 3 | *CD* (3-9) | 1580 | Tension |
| 4 | *DE* (4-9) | 1500 | Compression |
| 5 | *CE* (8-9) | 0 | — |
| 6 | *EF* (4-8) | 1500 | Compression |
| 7 | *CF* (7-8) | 1580 | Compression |
| 8 | *BF* (6-7) | 500 | Tension |
| 9 | *FG* (4-6) | 3080 | Compression |
| 10 | *BG* (5-6) | 1800 | Compression |

***Reaction***

Upper $R_A$ (1-5) = 4750 kN

Lower $R_G$ (4-5) = 4600 kN

**EXAMPLE 12.11.** *A truss shown in Fig. 12.27 is carrying point load of 5 kN at E.*

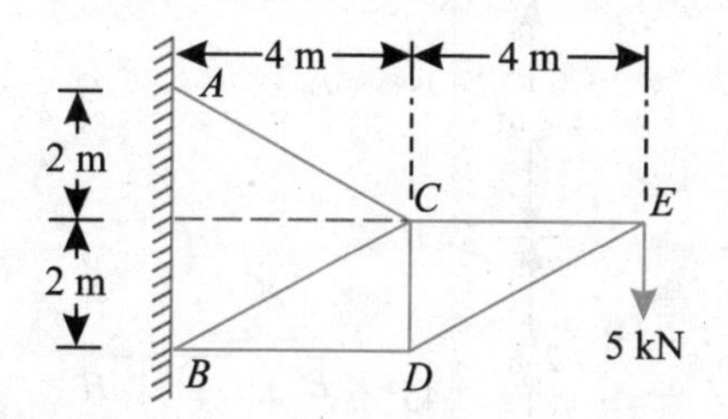


Fig. 12.27

*Find graphically, or otherwise, the forces in all the members of the truss and indicate results in a tabular form.*

**SOLUTION.** First of all, draw the space diagram and name all the various members according to Bow's notations as shown in Fig. 12.28 (*a*).

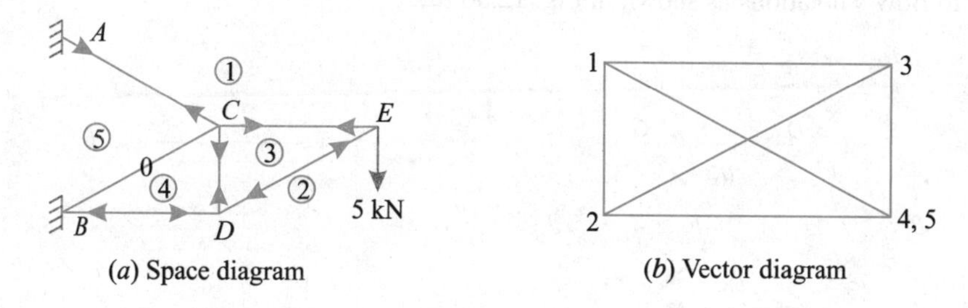


Fig. 12.28

Now draw the vector diagram as shown in Fig. 12.28 (*b*), starting from the joint *E*. Measuring the various sides of the vector diagram, the results are tabulated here :

| *No.* | *Member* | *Magnitude of force in kN* | *Nature of force* |
|---|---|---|---|
| 1 | *CE* (1-3) | 10 | Tension |
| 2 | *DE* (2-3) | 11.2 | Compression |
| 3 | *CD* (4-3) | 5.0 | Tension |
| 4 | *BD* (2-4) | 10 | Compression |
| 5 | *BC* (4-5) | 0 | — |
| 6 | *AC* (1-5) | 11.2 | Tension |

**EXAMPLE 12.12.** *A cantilever truss shown in Fig. 12.29 is carrying a point load of 15 kN.*

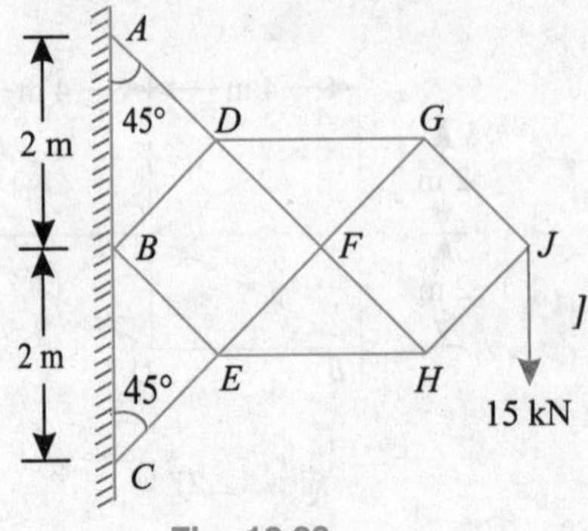


Fig. 12.29

*Find the forces in all the members of the truss. All the inclined members are at 45° with the horizontal.*

*SOLUTION. First of all, draw the space diagram and name all the members and forces according to Bow's notations as shown in Fig. 12.30 (*a*).

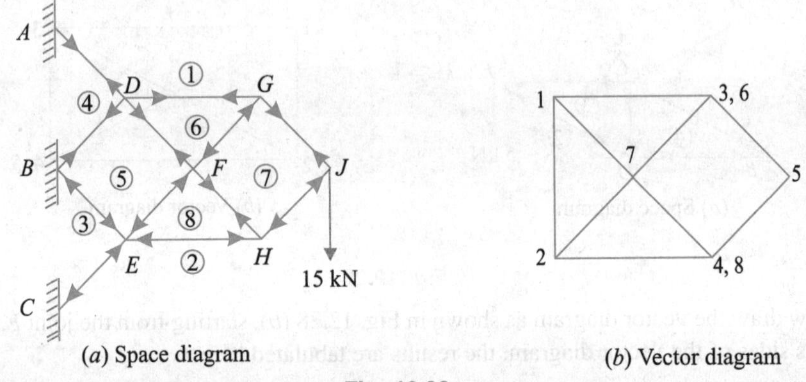


(*a*) Space diagram (*b*) Vector diagram

Fig. 12.30

Now draw the vector diagram as shown in Fig. 12.30 (*b*), starting from the joint *J*, Measuring the various sides of the vector diagram, the results are tabulated here :

| S.No. | Member | Magnitude of force in kN | Nature of force |
|---|---|---|---|
| 1 | *GJ* (1-7) | 10.6 | Tension |
| 2 | *HJ* (2-7) | 10.6 | Compression |
| 3 | *DG* (1-6) | 15.0 | Tension |
| 4 | *FG* (6-7) | 10.6 | Compression |
| 5 | *EH* (2-8) | 15.0 | Compression |
| 6 | *FH* (8-7) | 10.6 | Tension |
| 7 | *EF* (5-8) | 10.6 | Compression |
| 8 | *DF* (5-6) | 10.6 | Tension |
| 9 | *DA* (1-4) | 21.2 | Tension |
| 10 | *BD* (4-5) | 10.6 | Tension |
| 11 | *CE* (2-3) | 21.2 | Compression |
| 12 | *BE* (3-5) | 10.6 | Compression |

* We have already solved this example analytically in the last chapter.

**EXAMPLE 12.13.** *A frame is supporting two loads of 5 kN each at D and E as shown in Fig. 14.31.*

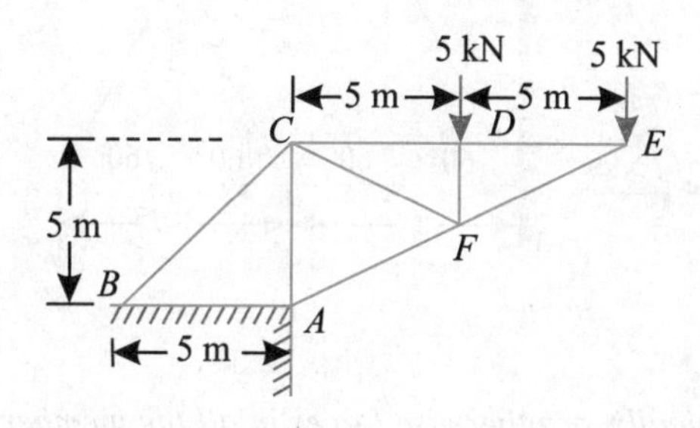


Fig. 12.31

*Find the forces in the members of the frame and the reactions at A and B.*

**SOLUTION.** First of all, draw the space diagram for the frame and name the members according to Bow's notations as shown in Fig. 12.31 (*a*).

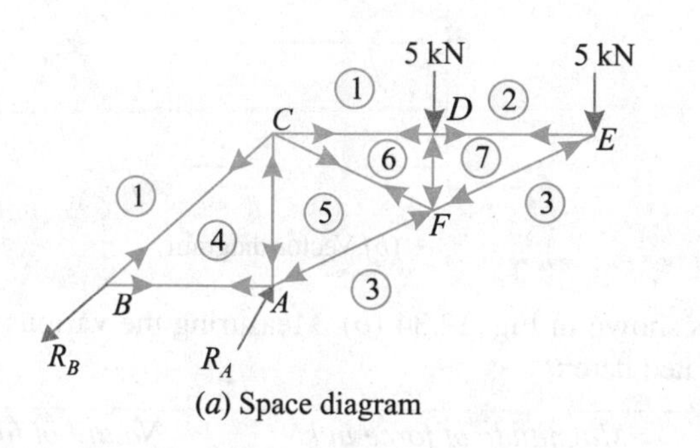


(*a*) Space diagram

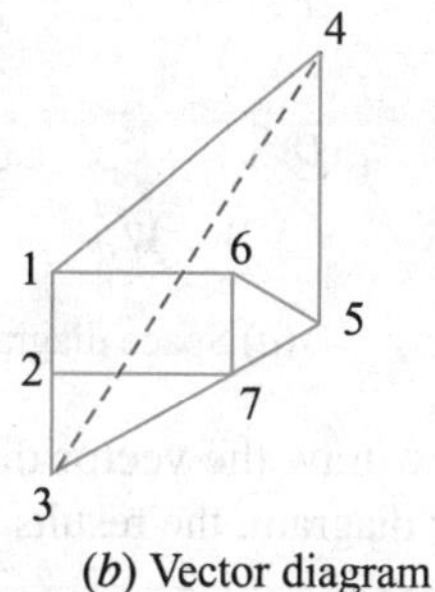


(*b*) Vector diagram

Fig. 12.32

Now draw the vector diagram for the frame as shown in Fig. 12.32 (*b*), starting from the joint *E*. Measuring the various sides of the vector diagram, the results are tabulated here :

| *S.No.* | *Member* | *Magnitude of force in kN* | *Nature of force* |
|---|---|---|---|
| 1 | *EF* | 11.2 | Compression |
| 2 | *ED* | 10.0 | Tension |
| 3 | *DF* | 5.0 | Compression |
| 4 | *CD* | 10.6 | Tension |
| 5 | *CF* | 5.6 | Tension |
| 6 | *FA* | 16.75 | Compression |
| 7 | *AC* | 17.5 | Compression |
| 8 | *CB* | 21.2 | Tension |

Reactions at $A = R_A$ (3-4) = 29.2 kN **Ans.**

and reaction at $B = R_B$ (1-4) = 21.2 kN **Ans.**

**EXAMPLE 12.14.** *A cantilever truss of span 2l is carrying loads as shown in Fig. 14.33.*

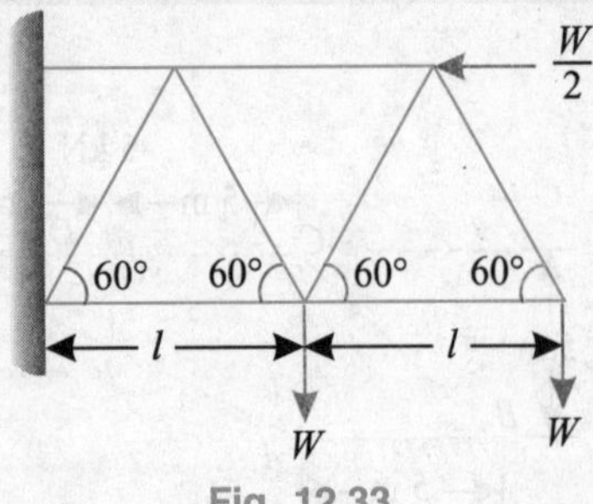


Fig. 12.33

*Determine graphically, or otherwise forces in all the members of the truss.*

**SOLUTION.** First of all, draw the space diagram, and name all the members according to Bow's notations as shown in Fig. 12.34 (*a*).

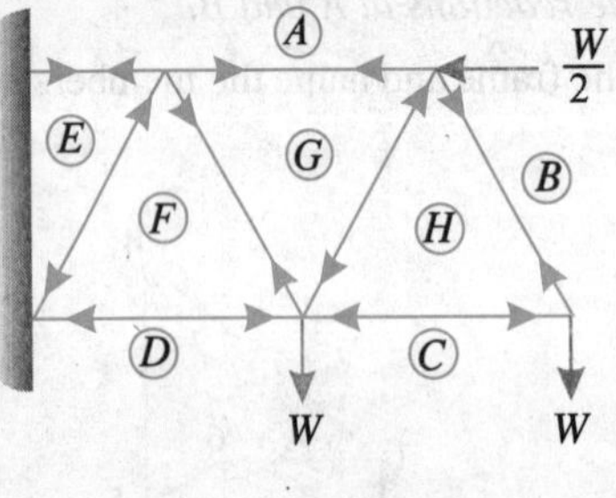


(*a*) Space diagram

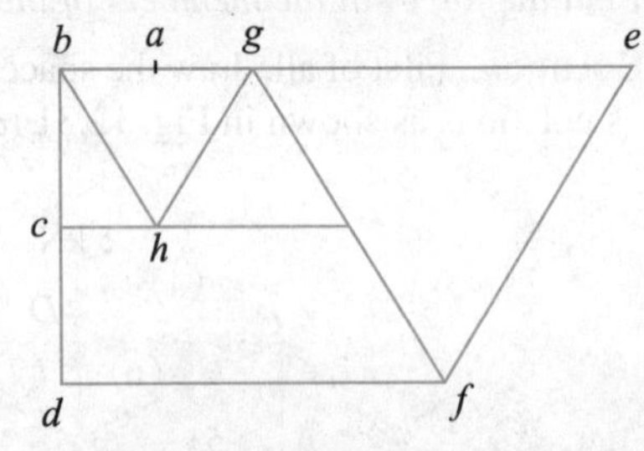


(*b*) Vector diagram

Fig. 12.34

Now draw the vector diagram as shown in Fig. 12.34 (*b*). Measuring the various sides of the vector diagram, the results are tabulated here :

| *S.No.* | *Member* | *Magnitude of force in kN* | *Nature of force* |
|---|---|---|---|
| 1 | *HB* | 1.2 | Tension |
| 2 | *CH* | 0.6 | Compression |
| 3 | *GH* | 1.2 | Compression |
| 4 | *AG* | 0.6 | Tension |
| 5 | *GF* | 2.3 | Tension |
| 6 | *DF* | 2.3 | Compression |
| 7 | *EF* | 2.3 | Compression |
| 8 | *AE* | 2.9 | Tension |

## EXERCISE 12.2

1. Determine the forces in the various members of a pin-jointed frame shown in Fig. 12.35. Tabulate the results stating whether they are in tension or compression.

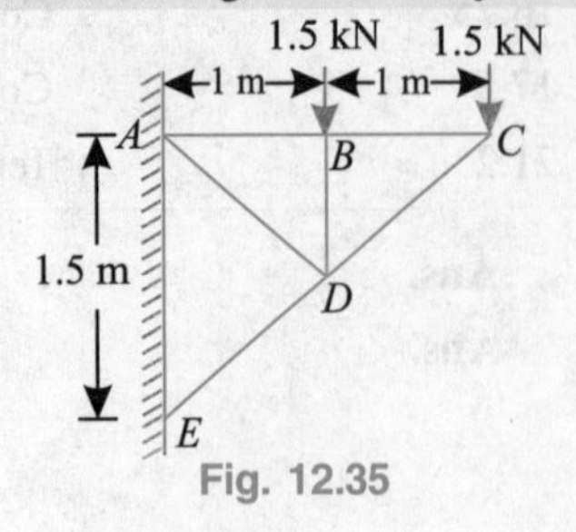


Fig. 12.35

**Ans.**
$AB$ = 2.0 kN (Tension)
$BC$ = 2.0 kN (Tension)
$CD$ = 2.5 kN (Compression)
$DE$ = 3.75 kN (Compression)
$BD$ = 1.5 kN (Compression)
$AD$ = 1.72 kN (Tension)

**2.** Find the forces in all the members of a cantilever truss shown in Fig. 12.36.

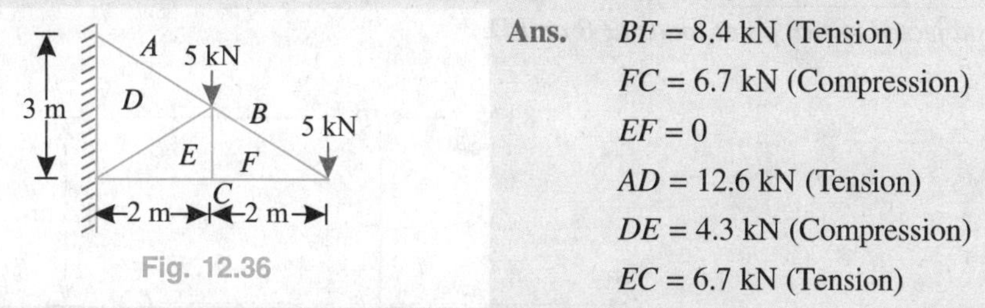


Fig. 12.36

**Ans.** $BF$ = 8.4 kN (Tension)
$FC$ = 6.7 kN (Compression)
$EF$ = 0
$AD$ = 12.6 kN (Tension)
$DE$ = 4.3 kN (Compression)
$EC$ = 6.7 kN (Tension)

**3.** Find graphically or otherwise the forces in the members 2, 5, 9 and 10 of the truss shown in Fig 12.37. Also state whether they are in tension or compression.

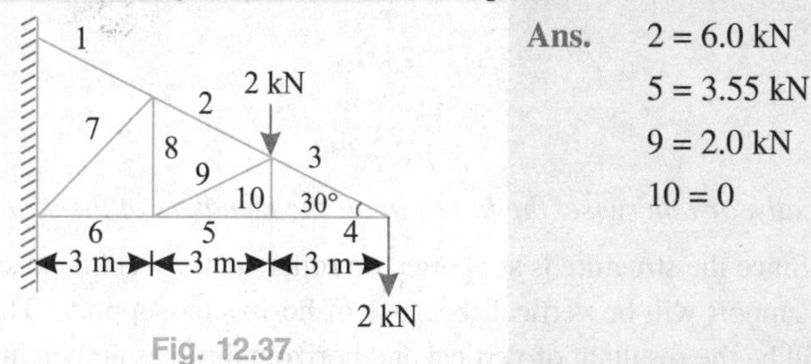


Fig. 12.37

**Ans.** 2 = 6.0 kN
5 = 3.55 kN
9 = 2.0 kN
10 = 0

**4.** Find the forces in the members of the frame given in Fig. 12.38.

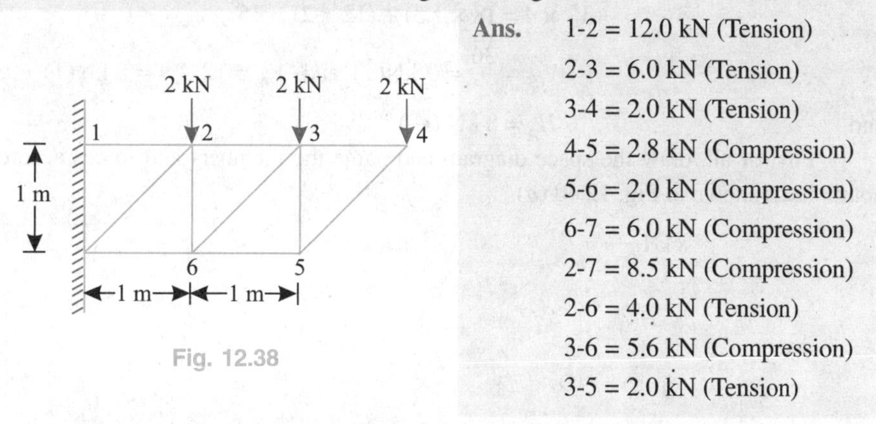


Fig. 12.38

**Ans.** 1-2 = 12.0 kN (Tension)
2-3 = 6.0 kN (Tension)
3-4 = 2.0 kN (Tension)
4-5 = 2.8 kN (Compression)
5-6 = 2.0 kN (Compression)
6-7 = 6.0 kN (Compression)
2-7 = 8.5 kN (Compression)
2-6 = 4.0 kN (Tension)
3-6 = 5.6 kN (Compression)
3-5 = 2.0 kN (Tension)

## 12.8. Structures with one end hinged (or Pin-jointed) and the other freely supported on rollers and carrying horizontal loads

We have already discussed in Art 14.16 that sometimes a structure is hinged or pin-jointed at one end and freely supported on rollers at the others end. If such a structure carries vertical loads only, the problem does not present any special features. Such a problem may be solved just as a simply supported structure.

But, if such a structure carries horizontal loads (with or without vertical loads) the support reaction at the roller supported end will be normal to the support; whereas the support reaction at the hinged end will consist of :

1. Vertical reaction, which may be found out by subtracting the vertical support reaction at the roller supported end from the total vertical load.
2. Horizontal reaction, which may be found out by algebraically adding all the horizontal loads. After finding out the reactions, the space and vector diagram may be drawn as usual.

**EXAMPLE 12.15.** *Figure 14.39 shows a framed structure of 4 m span and 1.5 m height subjected to two point loads at B and D.*

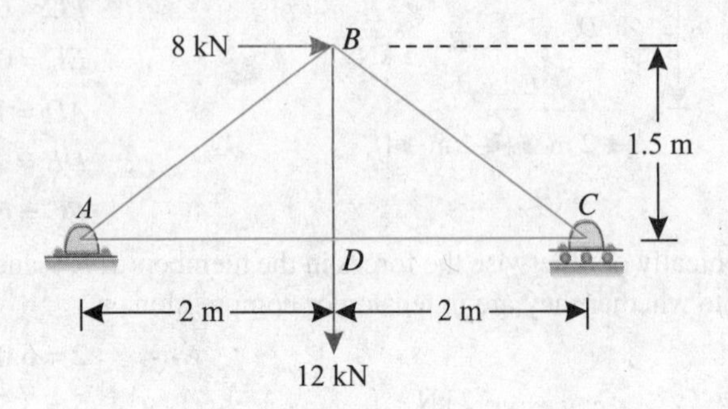


Fig. 12.39

*Find graphically, or otherwise, the forces in all the members of the structure.*

**SOLUTION.** *Since the structure is supported on rollers at the right hand support ($C$), therefore the reaction at this support will be vertical (because of horizontal support). The reaction at the left hand support ($A$) will be the resultant of vertical and horizontal forces and inclined with the vertical.

Taking moments about A and equating the same,

$$V_C \times 4 = (8 \times 1.5) + (12 \times 2) = 36$$

$$\therefore \quad V_C = \frac{36}{4} = 9 \text{ kN}(\uparrow) \text{ and } V_A = 12 - 9 = 3 \text{ kN}(\uparrow)$$

and

$$H_A = 8 \text{ kN } (\leftarrow)$$

First of all, draw the space diagram and name the members and forces according to Bow's notations as shown in Fig. 12.40 (*a*).

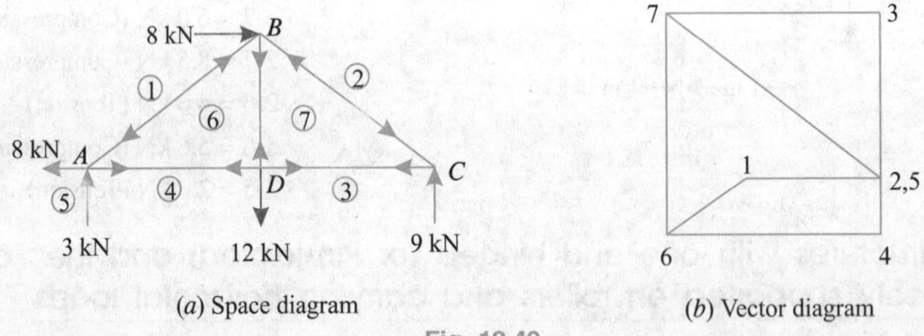


(*a*) Space diagram (*b*) Vector diagram

Fig. 12.40

Now draw the vector diagram as shown in Fig. 12.40 (*b*). Measuring the various sides of the vector diagram the results are tabulated here :

| *S.No.* | *Member* | *Magnitude of force in kN* | *Nature of force* |
|---|---|---|---|
| 1 | *AB* (1-6) | 5.0 | Compression |
| 2 | *BC* (2-7) | 15.0 | Compression |
| 3 | *AD* (4-6) | 12.0 | Tension |
| 4 | *BD* (6-7) | 12.0 | Tension |
| 5 | *DC* (3-7) | 12.0 | Tension |

* We have already solved this example analytically in the last chapter.

**EXAMPLE 12.16.** *A truss of 8 m span and 4 m height is loaded as shown in Fig. 12.41.*

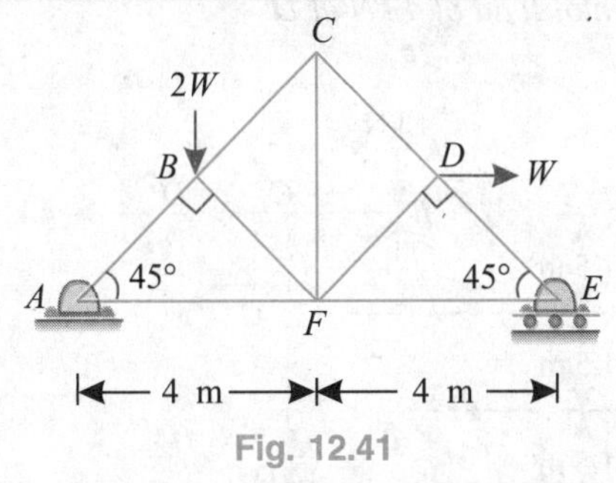


Fig. 12.41

*Find the forces in all the members of the truss and mention their nature in each case.*

***SOLUTION.** Since the truss is supported on rollers at the right hand support ($E$), therefore the reaction at this support will be vertical (because of horizontal support). The reaction at $A$ will be the resultant of vertical and horizontal forces.

Taking moments about $A$,

$$V_E \times 8 = (2\,W \times 2) + (W \times 2) = 6\,W$$

$$\therefore \qquad V_E = \frac{6W}{8} = 0.75\,W(\uparrow) \text{ and } V_A = 2W - 0.75W = 1.25\,W(\uparrow)$$

and
$$H_A = W\,(\leftarrow)$$

First of all, draw the space diagram and name all the members and forces according to Bow's notations as shown in Fig. 12.42 ($a$).

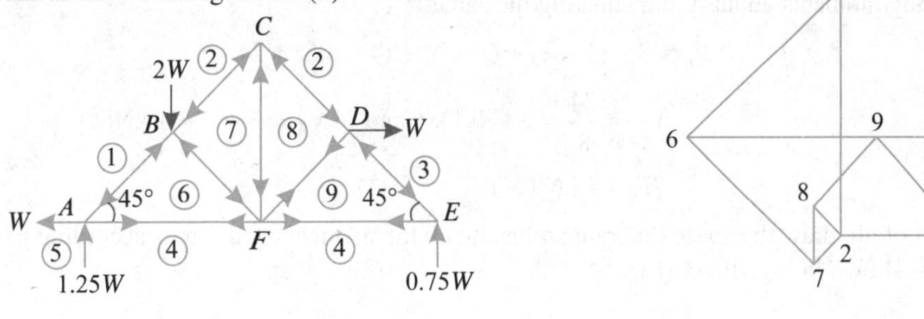


($a$) Space diagram ($b$) Vector diagram

Fig. 12.42

Now draw the vector diagram as shown in Fig. 12.42 ($b$). Measuring the various sides of the vector diagram, the results are tabulated here :

| *S.No.* | *Member* | *Magnitude of force in kN* | *Nature of force* |
|---|---|---|---|
| 1 | *AB* (1-6) | 1.77 | Compression |
| 2 | *BC* (2-7) | 0.35 | Compression |
| 3 | *CD* (2-8) | 0.35 | Compression |
| 4 | *DE* (3-9) | 1.06 | Compression |
| 5 | *AF* (4-6) | 2.25 | Tension |
| 6 | *BF* (6-7) | 1.41 | Compression |
| 7 | *CF* (7-8) | 0.5 | Tension |
| 8 | *FD* (8-9) | 0.71 | Tension |
| 9 | *FE* (4-9) | 0.75 | Tension |

* We have already solved this example analytically in the last chapter.

**Example 12.17.** *Figure 12.43 shows a pin-jointed frame carrying vertical loads of 1 kN each at B and G and horizontal load of 4 kN at D.*

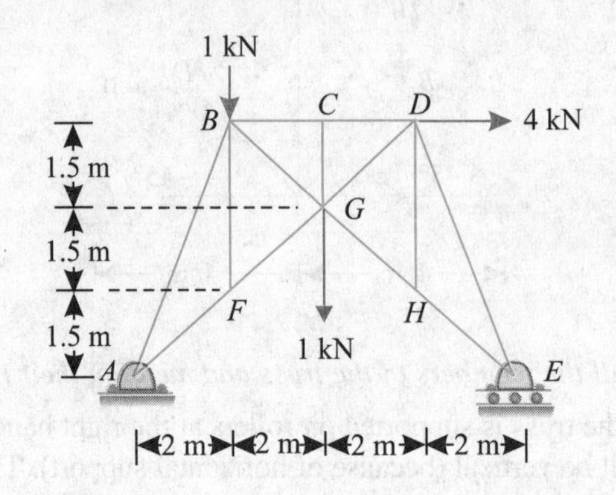


Fig. 12.43

*Find graphically, or otherwise, force in the various members of the truss. Also prepare a table stating the nature of forces.*

**Solution.** Since the frame is supported on rollers at the right hand support ($E$), therefore the reaction at this support will be vertical (because of horizontal support). The reaction at the left hand support ($A$) will be the resultant of vertical and horizontal forces.

Taking moments about $A$ and equating the same,

$$V_E \times 8 = (1 \times 2) + (1 \times 4) + (4 \times 4.5) = 24$$

$$\therefore \qquad V_E = \frac{24}{8} = 3 \text{ kN}(\uparrow) \quad \text{and} \quad V_A = 3 - 2 = 1 \text{ kN}(\downarrow)$$

and

$$H_A = 4 \text{ kN } (\leftarrow)$$

First of all, draw the space diagram and name all the members and forces according to Bow's notations as shown in Fig. 12.44 ($a$).

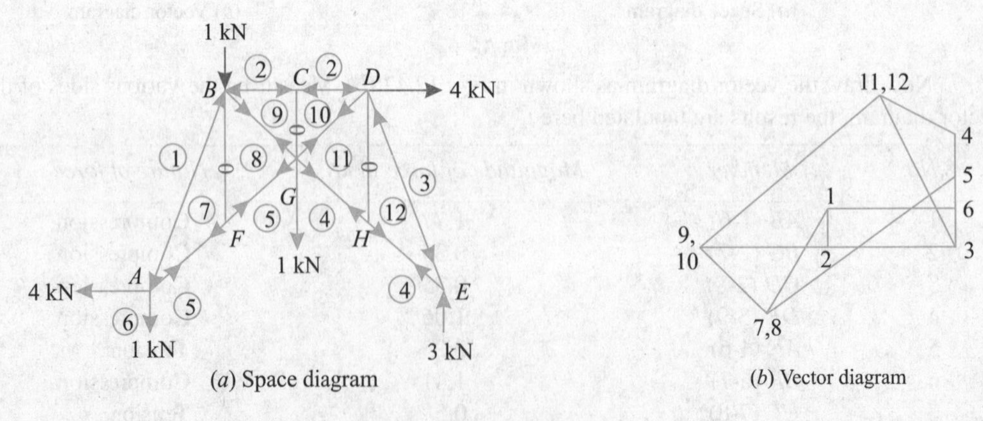


($a$) Space diagram ($b$) Vector diagram

Fig. 12.44

Now draw the vector diagram as shown in Fig. 12.44 ($b$). Measuring the various sides of the vector diagram, the results are tabulated here :

| S.No. | Member | Magnitude of force in kN | Nature of force |
|---|---|---|---|
| 1 | AB (1-7) | 3.3 | Compression |
| 2 | BC (2-9) | 4.0 | Copression |
| 3 | CD (2-10) | 4.0 | Compression |
| 4 | DE (3-12) | 4.9 | Compression |
| 5 | EH (4-12) | 2.5 | Tension |
| 6 | HG (4-11) | 2.5 | Tension |
| 7 | GF (5-8) | 6.7 | Tension |
| 8 | FA (5-7) | 6.7 | Tension |
| 9 | BF (7-8) | 0 | — |
| 10 | BG (8-9) | 3.3 | Tension |
| 11 | CG (9-10) | 0 | — |
| 12 | GD (10-11) | 7.5 | Tension |
| 13 | DH (11-12) | 0 | — |

## 12.9. Structures with one end Hinged (or Pin-jointed) and the Other Freely Supported on Rollers and Carrying Inclined Loads

We have already discussed in Art 14.8 that if a structure is hinged at one end, freely supported on rollers at the other and carries inclined loads (with or without vertical loads), the support reaction at the roller supported end will be normal to the support. The support reaction at the hinged end will be the resultant of :

1. Vertical reaction, which may be found out by subtracting the vertical component of the support reaction at the roller supported end from the total vertical load.
2. Horizontal reaction, which may be found out by algebraically adding all the horizontal loads.

**EXAMPLE 12.18.** *Figure 12.45 shows a north-light roof truss with wind loads acting on it.*

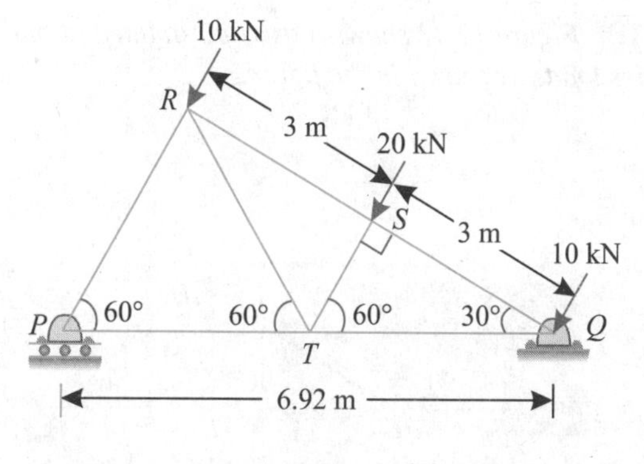


Fig. 12.45

*Find graphically, or otherwise, the forces in all the members of the truss. Give your result in a tabulated form.*

*SOLUTION. Since the truss is supported on rollers at $P$, threfore the reaction at this end will be vertical (because of horizontal support). Moreover, it is hinged at $Q$, therefore the reaction at this end will be resultant of horizontal and vertical forces and inclined with the vertical.

Taking moments about $Q$ and equating the same,

$$V_P \times 6.92 = (20 \times 3) + (10 \times 6) = 120$$

$$\therefore \quad V_P = \frac{120}{6.92} = 17.3 \text{ kN} (\uparrow) \quad \text{and} \quad V_Q = [(10 + 20 + 10) \sin 60°] - 17.3 = 17.3 \text{ kN} (\uparrow)$$

and
$$H_Q = (10 + 20 + 10) \cos 60° = 40 \times 0.5 = 20 \text{ kN} (\rightarrow)$$

First of all, draw the space diagram and name the members and forces according to Bow's notations as shown in Fig. 12.46 (*a*).

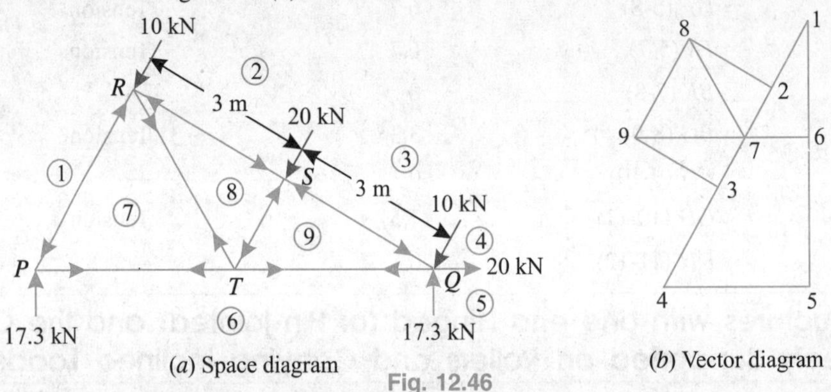


(*a*) Space diagram (*b*) Vector diagram

Fig. 12.46

Now draw the vector diagram as shown in Fig. 12.46 (*b*). Measuring the various sides of the vector diagram, the results are tabulated here :

| S.No. | Member | Magnitude of force in kN | Nature of force |
|---|---|---|---|
| 1 | *PR* (1-7) | 20.0 | Compression |
| 2 | *RS* (2-8) | 17.3 | Compression |
| 3 | *SQ* (3-9) | 17.3 | Compression |
| 4 | *QT* (6-9) | 30.0 | Tension |
| 5 | *PT* (6-7) | 10.0 | Tension |
| 6 | *RT* (7-8) | 20.0 | Tension |
| 7 | *ST* (8-9) | 20.0 | Compression |

**EXAMPLE 12.19.** *Figure 12.47 shows a truss pin-jointed at one end, and freely supported at the other. It carries loads as shown in the figure.*

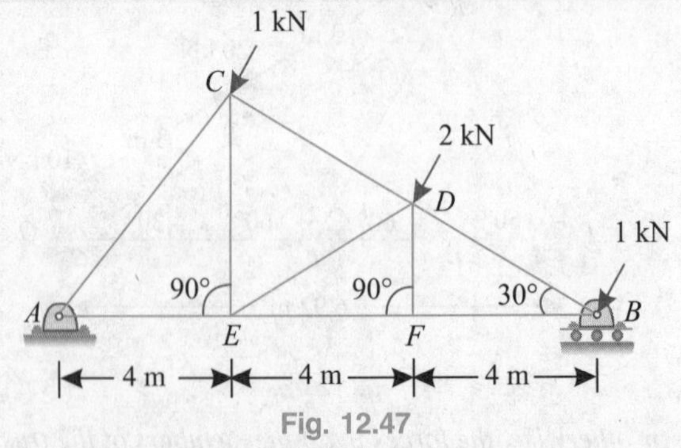


Fig. 12.47

*Determine forces in all the members of the truss and state their nature.*

* We have already solved this example analytically in the last chapter.

**SOLUTION.** Since the truss is supported on rollers at the right end, therefore the reaction at this end will be vertical. Moreover, as the truss is hinged at the left end, therefore the reaction at this end will be inclined with the vertical.

1. First of all draw the space diagram for the roof truss and name the various forces and reactions according to Bow's notations.
2. Compound all the forces together and assume them to act through the centre of gravity of the forces, *i.e.*, along the line of action of 2 kN force.
3. Produce the line of action of the resultant force (compound together as per item 2) to meet the line of action of the roller support (which will be vertical due to support on rollers) at *O*.
4. Join *OA*. From *O* cut off *OM* equal to the total compound load (*i.e.*, 1 + 2 + 1 = 4 kN) according to some scale, along the line of action of the resultant load.
5. Complete the parallelogram *OLMN* with *OM* as diagonal.
6. Measure *OL* and *ON*. The length *ON* gives the magnitude and direction of the reaction $R_A$. The length *OL* gives the magnitude of the reaction $R_B$.
7. By measurement, we find that
   $R_1 = 2.52$ kN, $R_2 = 1.92$ kN and $\theta = 51°$ **Ans.**

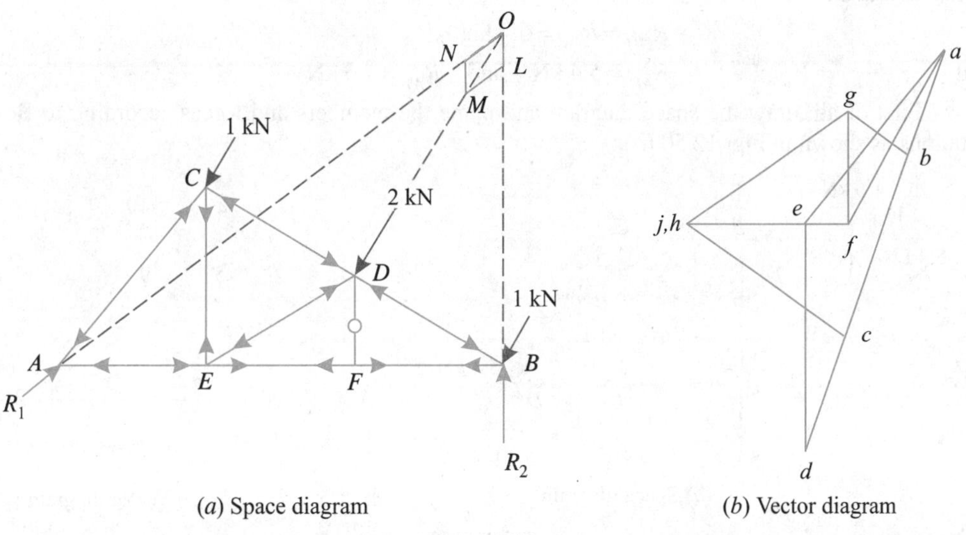


(*a*) Space diagram (*b*) Vector diagram

Fig. 12.48

Now draw the vector diagram as shown in Fig. 12.48 (*b*). Measuring the various sides of the vector diagram, the results are tabulated here :

| *S.No.* | *Member* | *Magnitude of force in kN* | *Nature of force* |
|---|---|---|---|
| 1 | *EJ* | 1.3 | Tension |
| 2 | *JC* | 2.1 | Compression |
| 3 | *HJ* | 0 | — |
| 4 | *HE* | 1.3 | Tension |
| 5 | *HG* | 2.3 | Compression |
| 6 | *GB* | 0.9 | Compression |
| 7 | *FG* | 1.2 | Tension |
| 8 | *FE* | 0.7 | Compression |
| 9 | *AF* | 2.0 | Compression |

**EXAMPLE 12.20.** *A truss hinged at A and supported on rollers at D is loaded as shown in Fig. 12.49.*

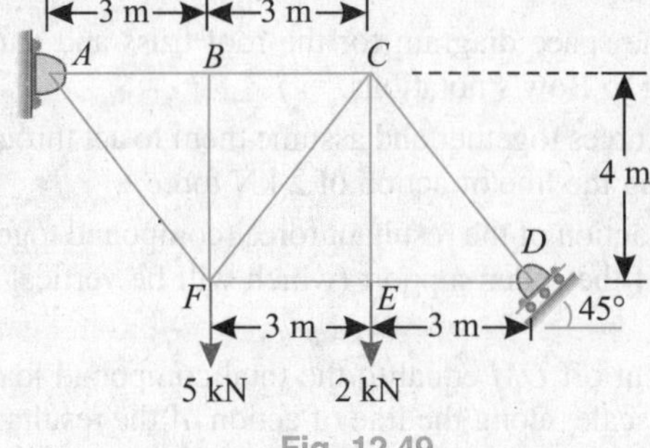


Fig. 12.49

*Find by any method the forces in all the members of the truss and mention the nature of forces.*

***SOLUTION.** Since the truss is supported on rollers at the right end $D$, therefore reaction at this support will be inclined at 45°, with the vertical (because the support is inclined at 45° with the horizontal). Now find out the reactions as done in example 12.17. We know that horizontal component of reaction at $D$.

$$R_{DH} = R_{DV} = 5.4 \text{ kN}$$

and $$R_{AH} = 5.4 \text{ kN} \quad \text{and} \quad R_{AV} = 1.6 \text{ kN}$$

First of all, draw the space diagram and name the members and forces according to Bow's notations as shown in Fig. 12.50 (*a*).

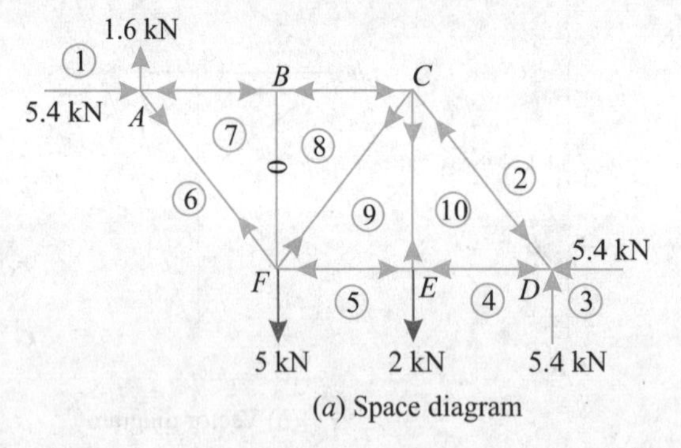


(*a*) Space diagram

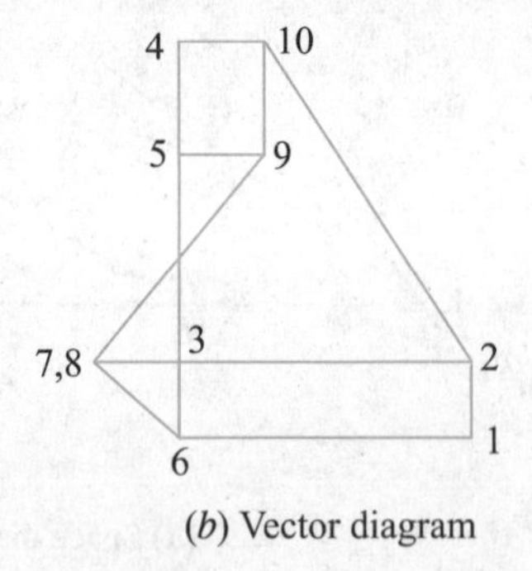


(*b*) Vector diagram

Fig. 12.50

Now draw the vector diagram as shown in Fig. 12.50 (*b*). Measuring the various sides of the vector diagram, the results are tabulated here :

| S.No. | Member | Magnitude of force in kN | Nature of force |
|---|---|---|---|
| 1 | *AB* (2-7) | 6.6 | Compression |
| 2 | *BC* (2-8) | 6.6 | Compression |
| 3 | *CD* (2-10) | 6.75 | Compression |
| 4 | *DE* (4-10) | 1.35 | Compression |
| 5 | *EF* (5-9) | 1.35 | Compression |
| 6 | *FA* (6-7) | 2.0 | Tension |
| 7 | *BF* (7-8) | 0 | — |
| 8 | *CF* (8-9) | 4.25 | Tension |
| 9 | *CE* (9-10) | 2.0 | Tension |

* We have already solved this example analytically in the last chapter.

## 12.10. Frames with both ends fixed

Sometimes, a frame or a truss is fixed or built-in at its both ends. In such a case, the reactions at both the supports can not be determined, unless some assumption is made. The assumptions usually made are :

1. The reactions are parallel to the direction of the loads and
2. In case of inclined loads, the horizontal thrust is equally shared by the two reactions.

Generally, the first assumption is made and the reactions are determined as usual by taking moments about one of the supports.

**EXAMPLE 12.21.** *Figure 12.51 shows as roof truss with both ends fixed. The truss is subjected to wind loads normal to the main rafter.*

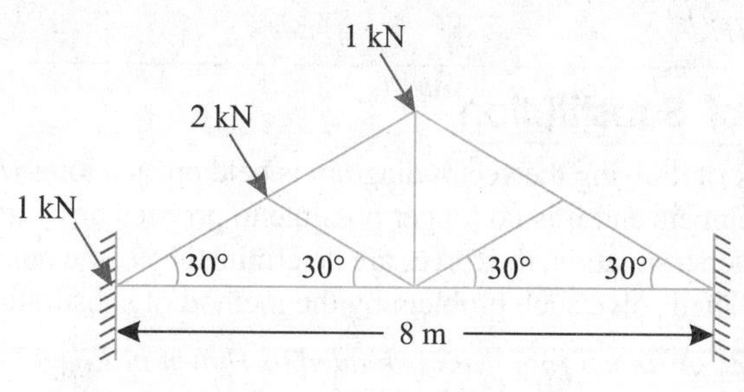


Fig. 12.51

*Find the force in various members of the truss.*

**SOLUTION.** The reactions may be obtained by any one assumption as mentioned. With the help of first assumption the reactions have been found out as shown in Fig. 12.52 (*a*).

Equating the anticlockwise moments and the clockwise moments about *A*,

$$R_1 \times 8 \sin 60° = \frac{2 \times 2}{\cos 30°} + \frac{1 \times 4}{\cos 30°} = \frac{8}{0.866} = 9.24 \text{ kN}$$

$$\therefore \qquad R_1 = \frac{9.24}{8 \sin 60°} = \frac{9.24}{8 \times 0.866} = 1.33 \text{ kN}$$

and

$$R_2 = (1 + 2 + 1) - 1.33 = 2.67 \text{ kN}$$

First of all, draw the space diagram and name the members according to Bow's notations as shown in Fig. 12.52 (*a*).

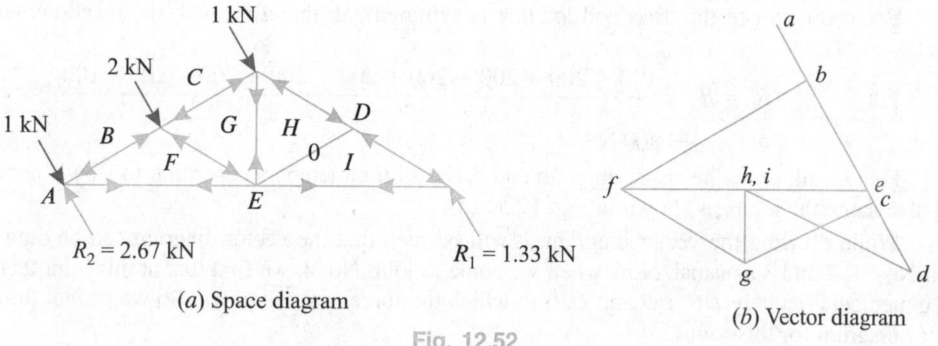


Fig. 12.52

Now draw the vector diagram as shown in Fig. 12.52 (*b*). Measuring the various sides of the vector diagram, the results are tabulated here :

| S.No. | Member | Magnitude of force in kN | Nature of force |
|---|---|---|---|
| 1 | BF | 2.9 | Compression |
| 2 | FE | 3.3 | Tension |
| 3 | CG | 1.9 | Compression |
| 4 | FG | 2.3 | Compression |
| 5 | GH | 1.15 | Tension |
| 6 | HD | 2.3 | Compression |
| 7 | HI | 0 | — |
| 8 | ID | 2.3 | Compression |
| 9 | IE | 1.33 | Tension |

## 12.11. Method of Substitution

Sometimes work of drawing the vector diagram is held up, at a joint which contains more than two unknown force members and it is no longer possible to proceed any further for the construction of vector diagram. In such a situation, the forces are determined by some other method. Here we shall discuss such cases and shall solve such problem by the method of substitution.

**EXAMPLE 12.22.** *A french roof truss is loaded as shown in Fig. 12.53.*

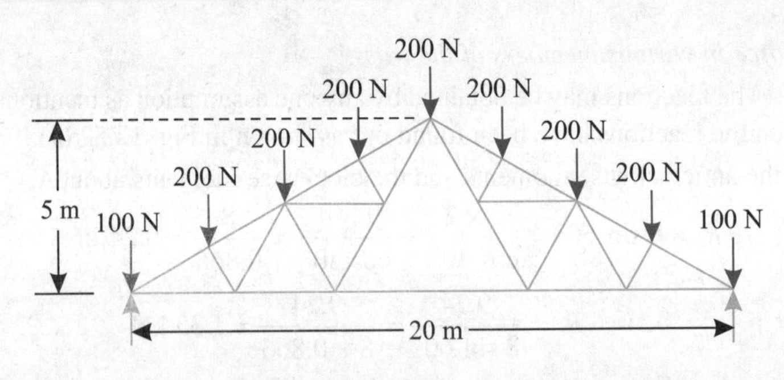


Fig. 12.53

*Find the forces in all the members of the truss, indicating whether the member is in tension or compression.*

**SOLUTION.** Since the truss and loading is symmetrical, therefore both the reactions will be equal.

$$\therefore \quad R_1 = R_2 = \frac{100 + 200 + 200 + 200 + 200 + 200 + 200 + 200 + 100}{2} \text{ N}$$

$$= 800 \text{ N}$$

First of all, draw the space diagram and name all the members according to Bow's notations and also name the joints as shown in Fig. 12.54 (*a*).

While drawing the vector diagram, it will be seen that the vector diagram can be drawn for joint Nos. 1, 2 and 3 as usual. Now when we come to joint No. 4, we find that at this joint there are three members (namely *DP*, *PO* and *ON*) in which the forces are unknown. So we cannot draw the vector diagram for this joint.

Now, as an alternative attempt, we look to joint No. 5. We again find that there are also three members (namely *NO*, *OR* and *RK*) in which the forces are unknown. So we can not draw the vector diagram for this joint also. Thus we find that the work of drawing vector diagram is held up beyond joint No. 3. In such cases, we can proceed by the substitution of an imaginary member.

Now, consider (for the time being only) the members *OP* and *PQ* as removed and substitute an imaginary member joining the joints 5 and 6 (as shown by the dotted line) as shown in Fig. 14.54. (*a*). Now we find that this substitution reduces the unknown force members at joint 4, from three to two (*i.e.*, members *DI* and *IN*; assuming the letter *I* in place of *P* and *O*) and thus we can draw the vector diagram for this joint (*i.e.*, No. 4).

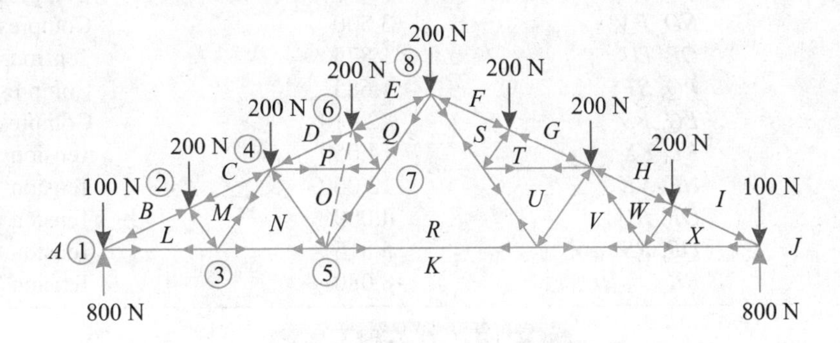


Fig. 12.54 (*a*)

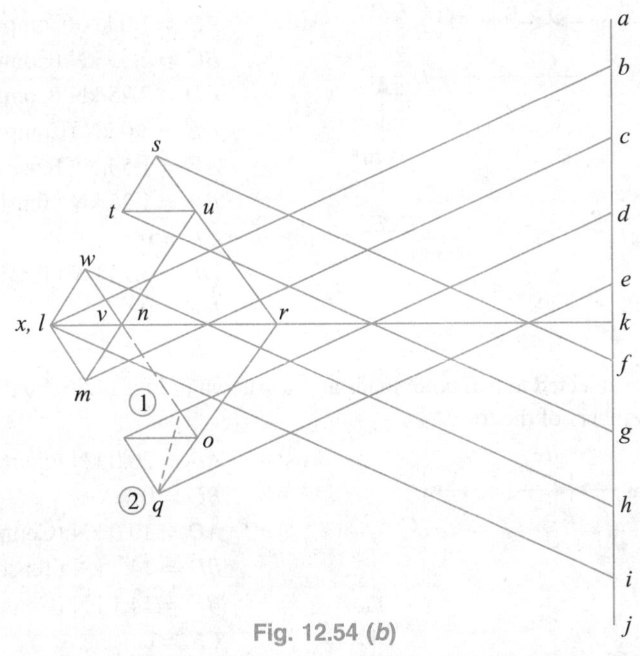


Fig. 12.54 (*b*)

Now after drawing the vector diagram for joint 4, proceed to joint 6 at which there are only two members (*i.e.*, *EQ* and *QI*) in which the forces are unknown. The vector diagram, at this joint will give the forces in *EQ* by the side *eq* of the vector diagram.

After drawing vector diagram at joint 6 and determining the forces in *EQ* (*i.e.*, *eq*) replace the imaginary member by the original members *PQ* and *PO* and again draw vector diagram for the joint No. 6 as shown in Fig. 14.54 (*b*). This will give the force in the member *PO*.

Now proceed to joint No. 5 as usual and complete the whole vector diagram as shown in Fig. 14.54 (*b*). Meausring the various sides of the vector diagram, the results are tabulated here :

| S. No. | Member | Magnitude of force in kN | Nature of force |
|---|---|---|---|
| 1 | BL, IX | 15,720 | Compression |
| 2 | LM, WX | 1,750 | Compression |
| 3 | CM, HW | 14,750 | Compression |
| 4 | MN, VW | 2,000 | Tension |
| 5 | DP, GT | 13,780 | Compression |
| 6 | NO, UV | 3,500 | Compression |
| 7 | OP, TU | 1,875 | Tension |
| 8 | PQ, ST | 1,685 | Compression |
| 9 | EQ, FS | 12,810 | Compression |
| 10 | KL, KX | 14,050 | Tension |
| 11 | NK, VK | 12,060 | Tension |
| 12 | OR, RU | 4,000 | Tension |
| 13 | QR, RS | 5,815 | Tension |
| 14 | RK | 8,080 | Tension |

## EXERCISE 12.3

**1.** A truss shown in Fig. 14.55 is subjected to two point loads at *B* and *F*. Find the forces in all the members of the truss and tabulate the results.

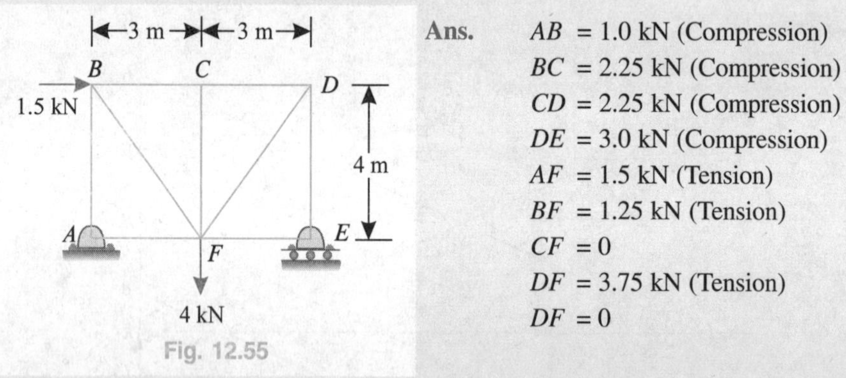


Fig. 12.55

**Ans.**
$AB$ = 1.0 kN (Compression)
$BC$ = 2.25 kN (Compression)
$CD$ = 2.25 kN (Compression)
$DE$ = 3.0 kN (Compression)
$AF$ = 1.5 kN (Tension)
$BF$ = 1.25 kN (Tension)
$CF$ = 0
$DF$ = 3.75 kN (Tension)
$DF$ = 0

**2.** A truss is subjected to two point loads at *A* as shown in Fig. 14.56. Find by any method, the forces in all the members of the truss.

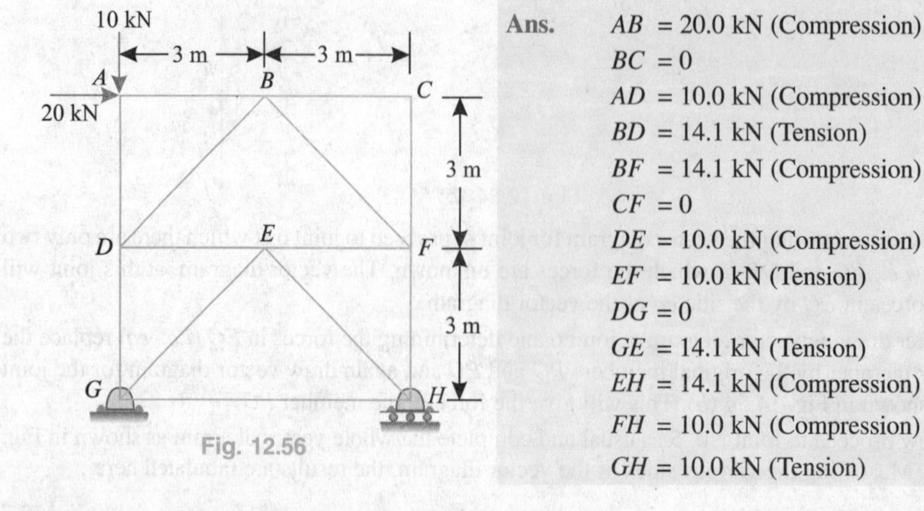


Fig. 12.56

**Ans.**
$AB$ = 20.0 kN (Compression)
$BC$ = 0
$AD$ = 10.0 kN (Compression)
$BD$ = 14.1 kN (Tension)
$BF$ = 14.1 kN (Compression)
$CF$ = 0
$DE$ = 10.0 kN (Compression)
$EF$ = 10.0 kN (Tension)
$DG$ = 0
$GE$ = 14.1 kN (Tension)
$EH$ = 14.1 kN (Compression)
$FH$ = 10.0 kN (Compression)
$GH$ = 10.0 kN (Tension)

3. Fig. 14.57 shows a truss pin-joint at one end, and freely supported at the other. It carries loads as shown in the figure. Determine forces in all the members of the truss and state their nature.

**Ans.** $AB = 2.0$ kN (Compression)
$BC = 0.9$ kN (Compression)
$CD = 2.1$ kN (Compression)
$AF = 0.7$ kN (Compression)
$BF = 1.2$ kN (Tension)
$CF = 2.3$ kN (Compression)
$FE = 1.3$ kN (Tension)
$CE = 0$
$ED = 1.3$ kN (Tension)

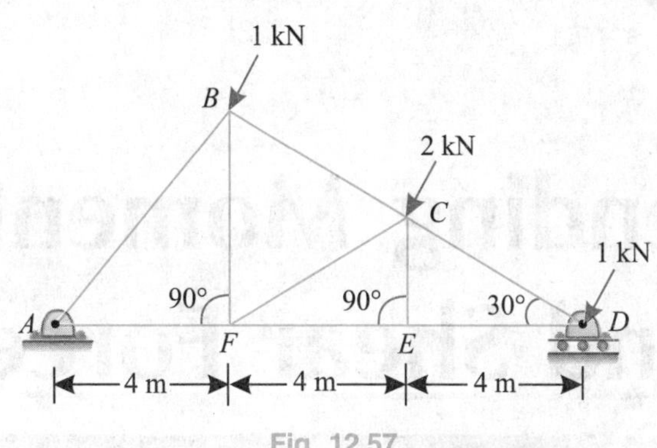


Fig. 12.57

## QUESTIONS

1. Discuss the procedure for drawing the vector diagram of a frame.
2. How will you find out (*i*) magnitude of a force, and (*ii*) nature of a force from the vector diagram?
3. What is a cantilever truss? How will you find out its reactions?
4. Explain why it is not essential to obtain the reactions of a cantilever truss before drawing the vector diagram ?
5. Describe the procedure for drawing the vector diagram of a truss subjected to horizontal loads.

## OBJECTIVE QUESTIONS

1. The space diagram of a framed structure must have all the
   (*a*) loads (*b*) reactions (*c*) both (*a*) and (*b*)
2. The Bow's notations is used only in case of
   (*a*) simply supported structure
   (*b*) cantilever structure
   (*c*) structures with one end hinged and the other supported on rollers.
   (*d*) all of the above.
3. If in a vector diagram, any two points coincide, then the force in the member represented by the two letters is zero.
   (*a*) True (*b*) False
4. In a graphical method, for analysing the perfect frames, it is possible to check the previous work in any subsequent step.
   (*a*) Yes (*b*) No

## ANSWERS

1. (*c*) 2. (*d*) 3. (*a*) 4. (*a*)

Chapter

# 13

# Bending Moment and Shear Force

## Contents

## 13.1. Introduction

We see that whenever a horizontal beam is loaded with vertical loads, sometimes, it bends (*i.e.*, deflects) due to the action of the loads. The amount with which a beam bends, depends upon the amount and type of the loads, length of the beam, elasticity of the beam and type of the beam. The scientific way of studying the deflection or any other effect is to draw and analyse the shear force or bending moment diagrams of a beam. In general, the beams are classified as under:

1. Cantilever beam,
2. Simply supported beam,
3. Overhanging beam,
4. Rigidly fixed or built-in-beam and
5. Continuous beam.

**Note.** In this chapter, we shall study the first three types of beams only.

## 13.2. Types of Loading

A beam may be subjected to either or in combination of the following types of loads:

1. Concentrated or point load,
2. Uniformly distributed load and
3. Uniformly varying load.

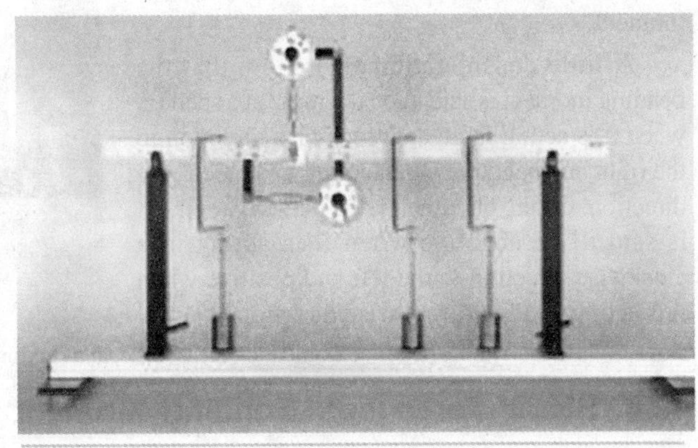

Shearing force

## 13.3. Shear Force

The shear force (briefly written as S.F.) at the cross-section of a beam may be defined as the unbalanced vertical force to the right or left of the section.

## 13.4. Bending Moment

The bending moment (briefly written as B.M.) at the cross-section of a beam may be defined as the algebraic sum of the moments of the forces, to the right or left of the section.

**Note.** While calculating the shear force or bending moment at a section, the end reactions must also be considered alongwith other external loads.

## 13.5. Sign Conventions

We find different sign conventions in different books, regarding shear force and bending moment at a section. But in this book the following sign conventions will be used, which are widely followed and internationally recognised.

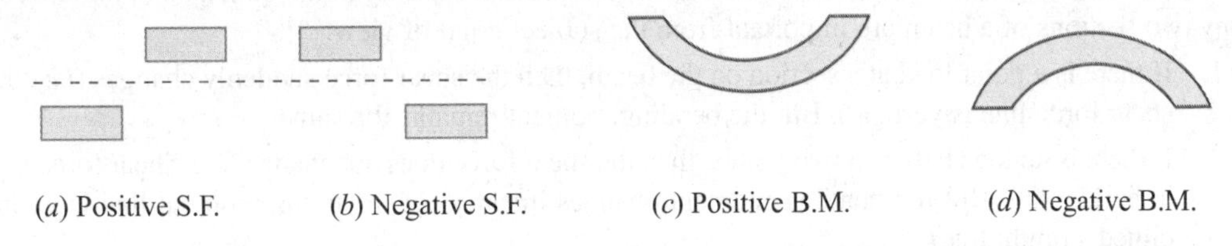

(*a*) Positive S.F. (*b*) Negative S.F. (*c*) Positive B.M. (*d*) Negative B.M.

Fig. 13.1

1. ***Shear Force.*** We know that as the shear force is the unbalanced vertical force, therefore it tends to slide one portion of the beam, upwards or downwards with respect to the other. The shear force is said to be positive, at a section, when the left hand portion tends to slide downwards or the right hand portion tends to slide upwards shown in Fig. 13.1 (*a*). Or in other words, all the downward forces to the left of the section cause positive shear and those acting upwards cause negative shear as shown in Fig. 13.1 (*a*).

Similarly, the shear force, is said to be negative at a section when the left hand portion tends to slide upwards or the right hand portion tends to slide downwards as shown in Fig. 13.1 (*b*). Or in other words, all the upward forces to the left of the section cause negative shear and those acting downwards cause positive shear as shown in Fig. 13.1 (*b*).

2. ***Bending Moment.*** At sections, where the bending moment, is such that it tends to bend the beam at that point to a curvature having concavity at the top, as shown in Fig. 13.1 (*c*) is taken as

positive. On the other hand, where the bending moment is such that it tends to bend the beam at that point to a curvature having convexity at the top, as shown in Fig. 13.1 (*d*) is taken as negative. The positive bending moment is often called sagging moment and negative as hogging moment.

A little consideration will show that the bending moment is said to be positive, at a section, when it is acting in an anticlockwise direction to the right and negative when acting in a clockwise direction. On the other hand, the bending moment is said to be negative when it is acting in a clockwise direction to the left and positive when it is acting in an anticlockwise direction.

Bending test of resin concrete

NOTE. While calculating bending moment or shear force, at a section the beam will be assumed to be weightless.

## 13.6. Shear Force and Bending Moment Diagrams

The shear force and bending moment can be calculated numerically at any particular section. But sometimes, we are interested to know the manner, in which these values vary, along the length of the beam. This can be done by plotting the shear force or the bending moment as ordinate and the position of the cross as abscissa. These diagrams are very useful, as they give a clear picture of the distribution of shear force and bending moment all along the beam.

NOTE. While drawing the shear force or bending moment diagrams, all the positive values are plotted above the base line and negative values below it.

## 13.7. Relation between Loading, Shear Force and Bending Moment

The following relations between loading, shear force and bending moment at a point or between any two sections of a beam are important from the subject point of view:

1. If there is a point load at a section on the beam, then the shear force suddenly changes (*i.e.*, the shear force line is vertical). But the bending moment remains the same.
2. If there is no load between two points, then the shear force does not change (*i.e.*, shear force line is horizontal). But the bending moment changes linearly (*i.e.*, bending moment line is an inclined straight line).
3. If there is a uniformly distributed load between two points, then the shear force changes linearly (*i.e.*, shear force line is an inclined straight line). But the bending moment changes according to the parabolic law. (*i.e.*, bending moment line will be a parabola).
4. If there is a uniformly varying load between two points then the shear force changes according to the parabolic law (*i.e.*, shear force line will be a parabola). But the bending moment changes according to the cubic law.

## 13.8. Cantilever with a Point Load at its Free End

Consider a *cantilever *AB* of length *l* and carrying a point load *W* at its free end *B* as shown in Fig. 13.2 (*a*). We know that shear force at any section *X*, at a distance *x* from the free end, is equal to the total unbalanced vertical force. *i.e.*,

$$F_x = -W \qquad \text{...(Minus sign due to right downward)}$$

* It is a beam fixed at one end and free at the other.

and bending moment at this section,

$$M_x = -W.x \quad \text{...(Minus sign due to hogging)}$$

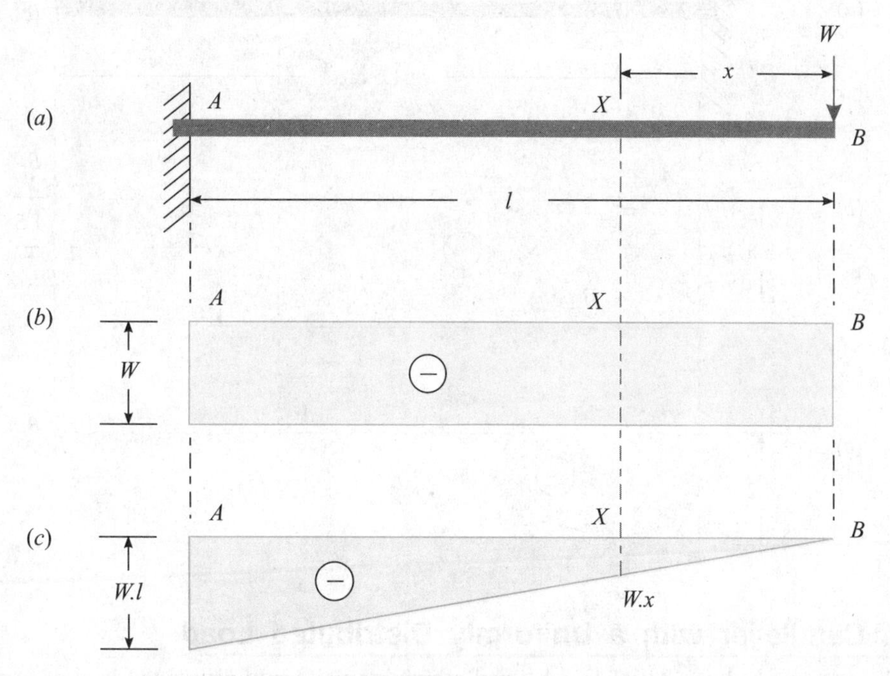


Fig. 13.2. Cantilever with a point load

Thus from the equation of shear force, we see that the shear force is constant and is equal to – $W$ at all sections between $B$ and $A$. And from the bending moment equation, we see that the bending moment is zero at $B$ (where $x = 0$) and increases by a straight line law to – $Wl$; . at (where $x = l$). Now draw the shear force and bending moment diagrams as shown in Fig. 13.2 (*b*) and 13.2 (*c*) respectively.

**EXAMPLE 13.1.** *Draw shear force and bending moment diagrams for a cantilever beam of span 1.5 m carrying point loads as shown in Fig. 13.3 (a).*

**SOLUTION.** Given : Span ($l$) = 1.5 m ; Point load at $B$ ($W_1$) = 1.5 kN and point load at $C$ ($W_2$) = 2 kN.

***Shear force diagram***

The shear force diagram is shown in Fig. 13.3 (*b*) and the values are tabulated here:

$$F_B = -W_1 = -1.5 \text{ kN}$$

$$F_C = -(1.5 + W_2) = -(1.5 + 2) = -3.5 \text{ kN}$$

$$F_A = -3.5 \text{ kN}$$

***Bending moment diagram***

The bending moment diagram is shown in Fig. 13.3 (*c*) and the values are tabulated here:

$$M_B = 0$$

$$M_C = -[1.5 \times 0.5] = -0.75 \text{ kN-m}$$

$$M_A = -[(1.5 \times 1.5) + (2 \times 1)] = -4.25 \text{ kN-m}$$

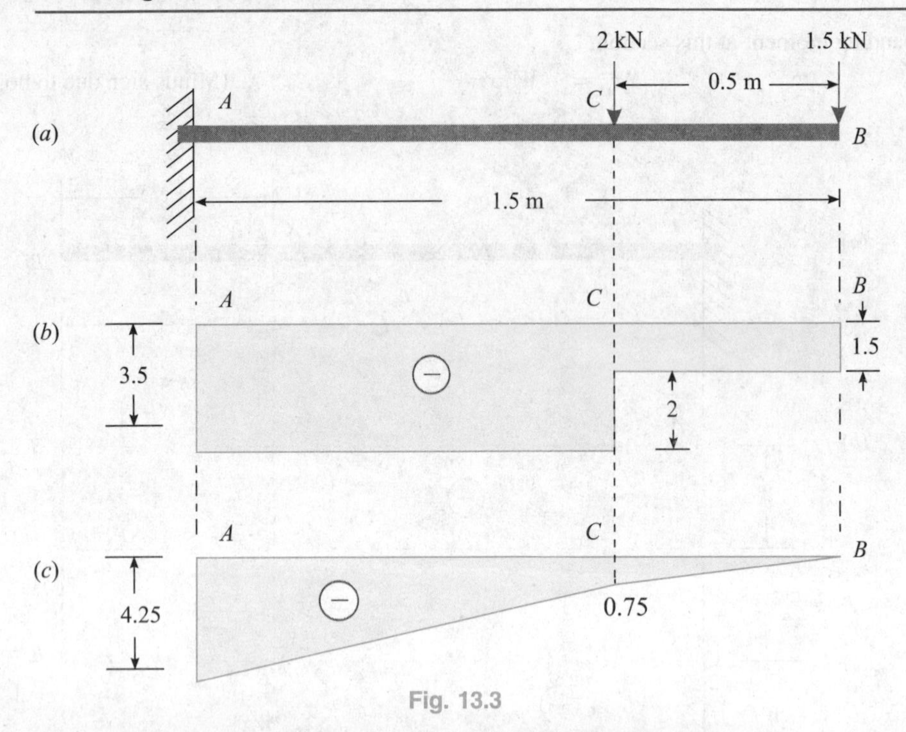


Fig. 13.3

## 13.9. Cantilever with a Uniformly Distributed Load

Consider a cantilever $AB$ of length $l$ and carrying a uniformly distributed load of $w$ per unit length, over the entire length of the cantilever as shown in Fig. 13.4 ($a$).

We know that shear force at any section $X$, at a distance $x$ from $B$,

$$F_x = -w \cdot x \qquad \text{... (Minus sign due to right downwards)}$$

Thus we see that shear force is zero at $B$ (where $x = 0$) and increases by a straight line law to $-wl$ at $A$ as shown in Fig. 13.4 ($b$).

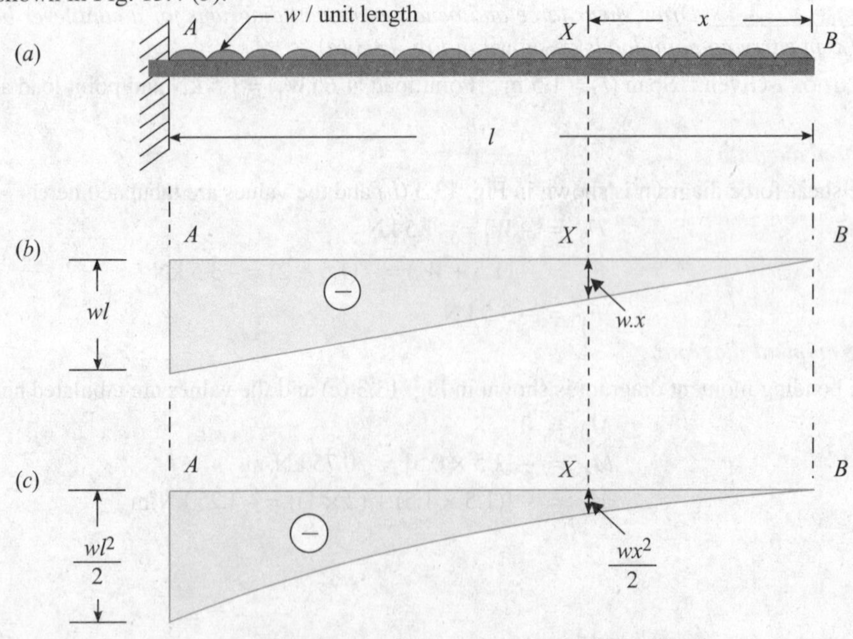


Fig. 13.4. Cantilever with a uniformly distributed load

We also know that bending moment at $X$,

$$M_x = -wx \cdot \frac{x}{2} = -\frac{wx^2}{2} \qquad \text{...(Minus sign due to hogging)}$$

Thus we also see that the bending moment is zero at $B$ (where $x = 0$) and increases in the form of a parabolic curve to $-\frac{wl^2}{2}$ at $B$ (where $x = 1$) as shown in Fig. 13.4 (*c*).

**EXAMPLE 13.2.** *A cantilever beam AB, 2 m long carries a uniformly distributed load of 1.5 kN/m over a length of 1.6 m from the free end. Draw shear force and bending moment diagrams for the beam.*

**SOLUTION.** Given : span ($l$) = 2 m ; Uniformly distributed load ($w$) = 1.5 kN/m and length of the cantilever $CB$ carrying load ($a$) = 1.6 m.

***Shear force diagram***

The shear force diagram is shown in Fig. 13.5 (*b*) and the values are tabulated here:

$$F_B = 0$$

$$F_C = -w \cdot a = -1.5 \times 1.6 = -2.4 \text{ kN}$$

$$F_A = -2.4 \text{ kN}$$

***Bending moment diagram***

The bending moment diagram is shown in Fig. 13.5 (*c*) and the values are tabulated here:

$$M_B = 0$$

$$M_C = -\frac{wa^2}{2} = \frac{1.5 \times (1.6)^2}{2} = -1.92 \text{ kN-m}$$

$$M_A = -\left[(1.5 \times 1.6)\left(0.4 + \frac{1.6}{2}\right)\right] = -2.88 \text{ kN-m}$$

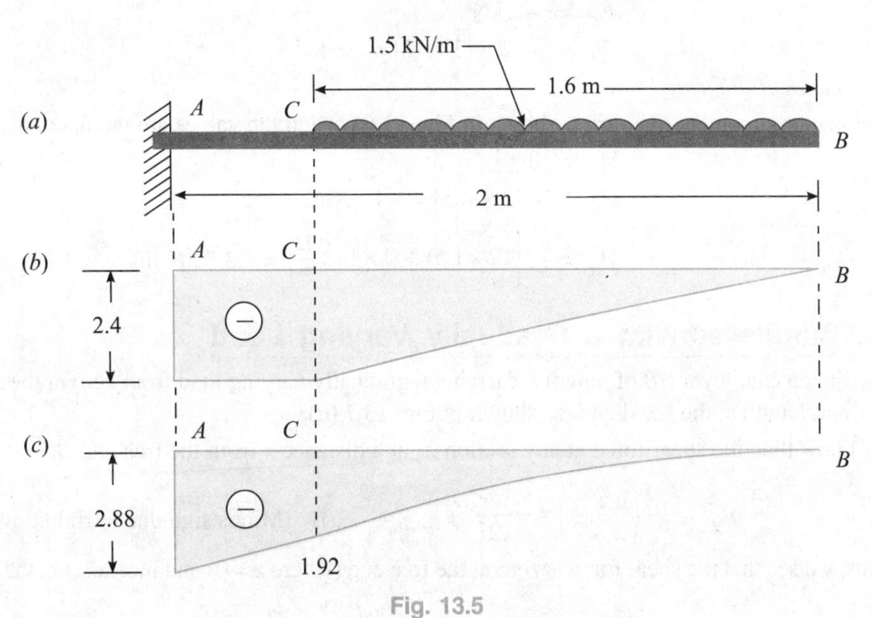


Fig. 13.5

**NOTE.** The bending moment at A is the moment of the load between $C$ and $B$ (equal to $1.5 \times 1.6 = 2.4$ kN) about $A$. The distance between the centre of the load and $A$ is $0.4 + \frac{1.6}{2} = 1.2$ m.

**Example 13.3.** *A cantilever beam of 1.5 m span is loaded as shown in Fig. 13.6 (a). Draw the shear force and bending moment diagrams.*

**Solution.** Given : Span ($l$) = 1.5 m ; Point load at $B$ ($W$) = 2 kN ; Uniformly distributed load ($w$) = 1 kN/m and length of the cantilever $AC$ carrying the load ($a$) = 1 m.

***Shear force diagram***

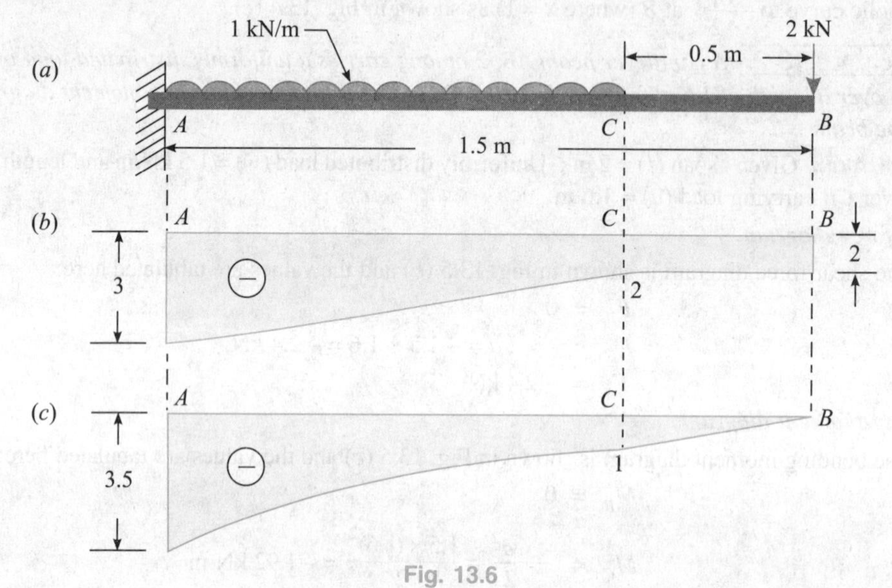


Fig. 13.6

The shear force diagram is shown in Fig. 13.6 ($b$) and the values are tabulated here:

$$F_B = -W = -2 \text{ kN}$$

$$F_C = -2 \text{ kN}$$

$$F_A = -[2 + (1 \times 1)] = -3 \text{ kN}$$

***Bending moment diagram***

The bending moment diagram is shown in Fig. 13.6 ($c$) and the values are tabulated here:

$$M_B = 0$$

$$M_C = -[2 \times 0.5] = -1 \text{ kN-m}$$

$$M_A = -\left[(2 \times 1.5) + (1 \times 1) \times \frac{1}{2}\right] = -3.5 \text{ kN-m}$$

## 13.10. Cantilever with a Gradually Varying Load

Consider a cantilever $AB$ of length $l$, carrying a gradually varying load from zero at the free end to w per unit length at the fixed end, as shown in Fig. 13.7 ($a$).

We know that, the shear force at any section $X$, at a distance $x$ from the free end $B$,

$$F_X = -\left(\frac{wx}{l} \cdot \frac{x}{2}\right) = -\frac{wx^2}{2l} \qquad ...(i) \text{ (Minus sign due to right downward)}$$

Thus, we see that the shear force is zero at the free end (where $x = 0$) and increases in the form of a parabolic curve [as given by equation ($i$) above] to $-\dfrac{wl^2}{2l} = -\dfrac{wl}{2}$ = at $A$ (where $x = l$) as shown in Fig. 13.7 ($b$).

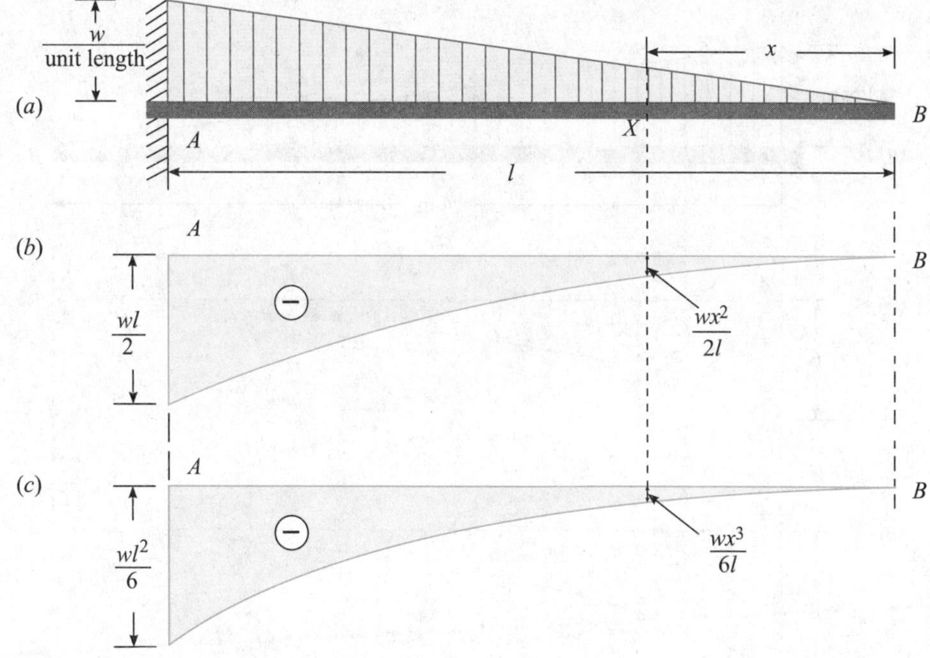


Fig. 13.7

We also know that the bending moment at $X$,

$$M_X = -\frac{wx^2}{2l} \times \frac{x}{3} = -\frac{wx^2}{6l} \quad \text{...(ii) (Minus sign due to hogging)}$$

Thus, we see that the bending moment is zero at the free end (where $x = 0$) and increases in the form of a *cubic parabolic curve* [as given by equation (*ii*) above] to $-\frac{wl^3}{6l} = -\frac{wl^2}{6}$ at $A$ (where $x = l$) as shown in Fig. 13.7 (*c*).

**EXAMPLE 13.4.** *A cantilever beam 4 m long carries a gradually varying load, zero at the free end to 3 kN/m at the fixed end. Draw B.M. and S.F. diagrams for the beam.*

**SOLUTION.** Given : Span ($l$) = 4 m and gradually varying load at $A$ ($w$) = 3 kN/m

The cantilever beam is shown in Fig. 13.8 (*a*).

***Shear force diagram***

The shear force diagram is shown in Fig. 13.8 (*b*) and the values are tabulated here:

$$F_B = 0$$

$$F_A = -\frac{3 \times 4}{2} = -6 \text{ kN}$$

***Bending moment diagram***

The bending moment diagram is shown in Fig. 13.8 (*c*) and the values are tabulated here:

$$M_B = 0$$

$$M_A = -\frac{3 \times (4)^2}{6} = -8 \text{ kN-m}$$

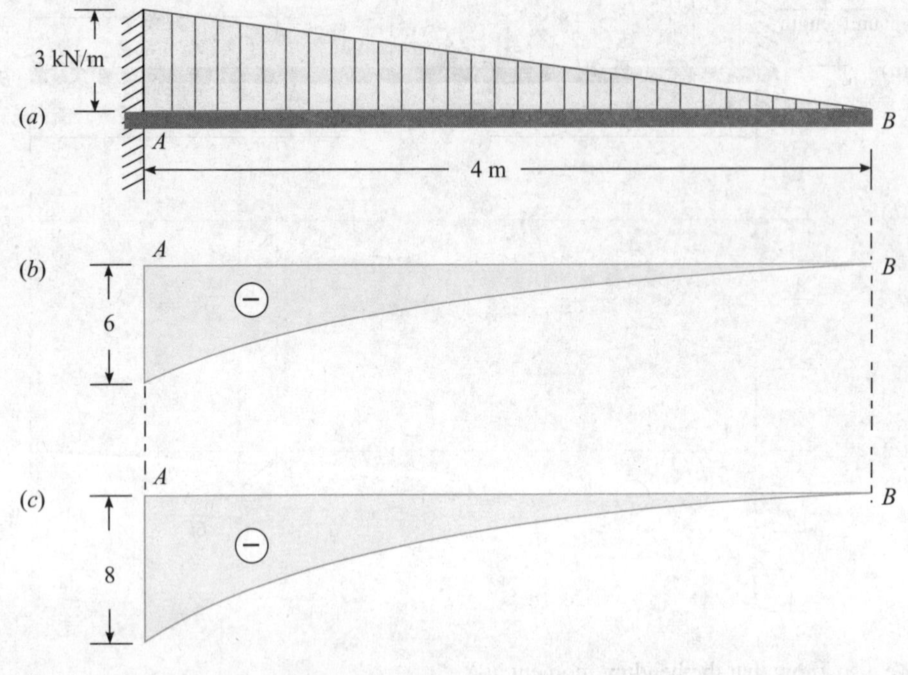


Fig. 13.8

**EXAMPLE 13.5.** *A cantilever beam of 2 m span is subjected to a gradually varying load from 2 kN/m to 5 kN/m as shown in Fig. 13.9.*

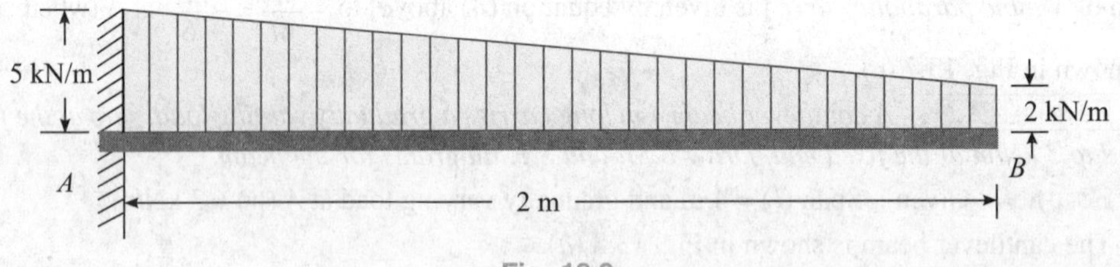


Fig. 13.9

*Draw the shear force and bending moment diagrams for the beam.*

**SOLUTION.** Given : Span ($l$) = 2 m ; Gradually varying load at $A$ ($w_A$) = 5 kN/m and gradually varying load at $B$ ($w_B$) = 2 kN/m.

The load may be assumed to be split up into (*i*) a uniformly distributed load ($w_1$) of 2 kN/m over the entire span and (*ii*) a gradually varying load ($w_1$) from zero at $B$ to 3 kN/m at $A$ as shown in Fig. 13.10 (*a*)

***Shear force diagram***

The shear force diagram is shown in Fig. 13.10 (*b*) and the values are tabulated here:

$$F_B = 0$$

$$F_A = -\left[(2 \times 2) + \left(\frac{3 \times 2}{2}\right)\right] = -7 \text{ kN}$$

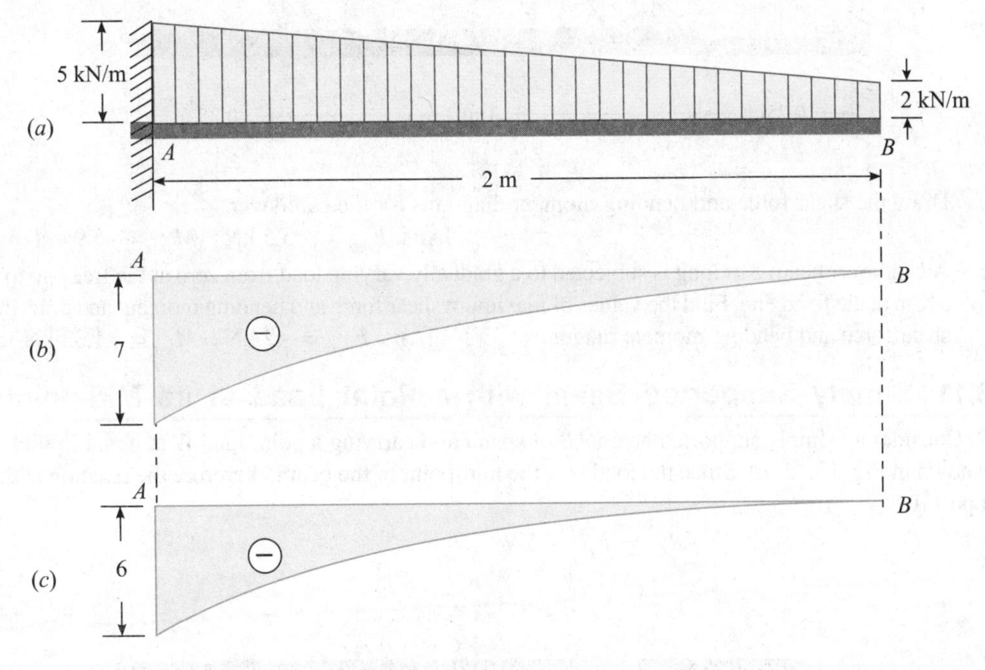


Fig. 13.10

***Bending moment diagram***

The bending moment diagram is shown in Fig. 13.10 (*c*) and the values are tabulated here:

$$M_B = 0$$

$$M_A = -\left[\left(\frac{2\times(2)^2}{2}\right)+\left(\frac{3(2)^2}{6}\right)\right] = -6 \text{ kN-m}$$

## EXERCISE 13.1

1. A cantilever beam 2 m long carries a point load of 1.8 kN at its free end. Draw shear force and bending moment diagrams for the cantilever. **[Ans.** $F_{max}$ = – 1.8 kN ; $M_{max}$ = – 3.6 kN-m]
2. A cantilever beam 1.5 m long carries point loads of 1 kN, 2 kN and 3 kN at 0.5 m, 1.0 m and 1.5 m from the fixed end respectively. Draw the shear force and bending moment diagrams for the beam.
   **[Ans.** $F_{max}$ = – 6 kN ; $M_{max}$ = – 7 kN-m]
3. A cantilever beam of 1.4 m length carries a uniformly distributed load of 1.5 kN/m over its entire length. Draw S.F. and B.M. diagrams for the cantilever.
   **[Ans.** $F_{max}$ = – 2.1 kN ; $M_{max}$ = – 1.47 kN-m]
4. A cantilever *AB* 1.8 m long carries a point load of 2.5 kN at its free end and a uniformly distributed load of 1 kN/m from *A* to *B*. Draw the shear force the bending moment diagrams for the beam.
   **[Ans.** $F_{max}$ = – 4.3 kN ; $M_{max}$ = – 6.12 kN-m]
5. A cantilever 1.5 m long is loaded with a uniformly distributed load of 2 kN/m and a point load of 3 kN as shown in Fig. 13.11

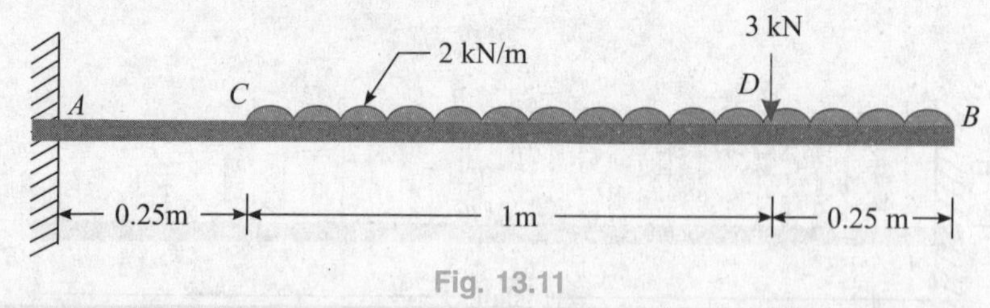


Fig. 13.11

Draw the shear force and bending moment diagrams for the cantilever.

[**Ans.** $F_{max}$ = – 5.5 kN ; $M_{max}$ = – 5.94 kN-m]

6. A cantilever beam 2 m long is subjected to a gradually varying load from zero at the free end to 2 kN/m at the fixed end. Find the values of maximum shear force and bending moment and draw the shear force and bending moment diagrams. [**Ans.** $F_{max}$ = – 2 kN ; $M_{max}$ = – 1.33 kN-m]

## 13.11. Simply Supported Beam with a Point Load at its Mid-point

Consider a *simply supported beam *AB* of span *l* and carrying a point load *W* at its mid-point *C* as shown in Fig. 13.12 (*a*). Since the load is at the mid-point of the beam, therefore the reaction at the support *A*,

$$R_A = R_B = 0.5\ W$$

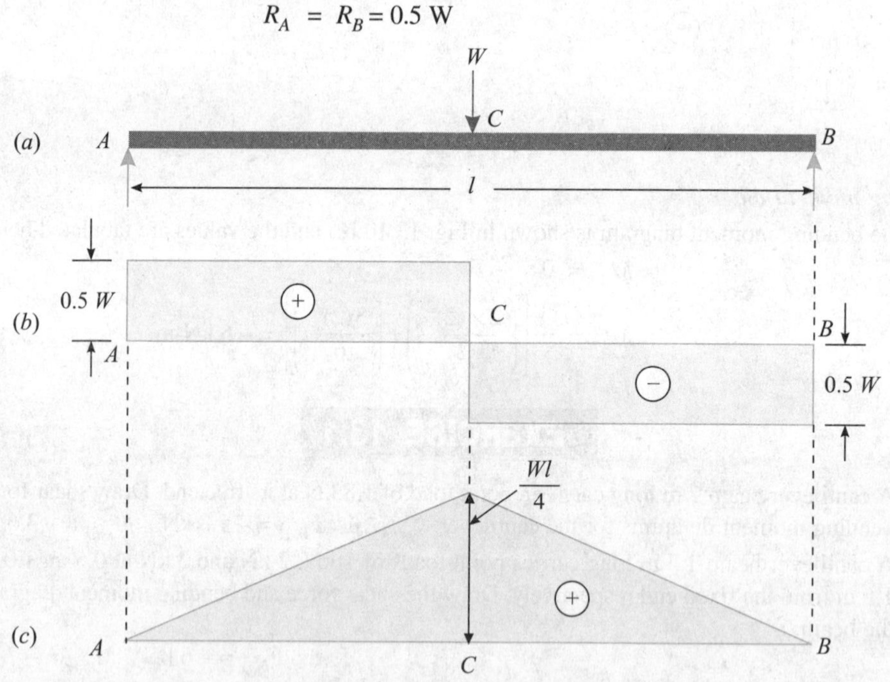


Fig. 13.12. Simply supported beam with a point load

Thus we see that the shear force at any section between *A* and *C* (*i.e.*, up to the point just before the load *W*) is constant and is equal to the unbalanced vertical force, *i.e.*, + 0.5 *W*. Shear force at any section between *C* and *B* (*i.e.*, just after the load *W*) is also constant and is equal to the unbalanced vertical force, *i.e.*, – 0.5 *W* as shown in Fig. 13.12 (*b*).

We also see that the bending moment at *A* and *B* is zero. It increases by a straight line law and is maximum at centre of beam, where shear force changes sign as shown in Fig. 13.12 (*c*).

* It is beam supported or resting freely on the walls or columns on both ends.

Therefore bending moment at $C$,

$$M_C = \frac{W}{2} \times \frac{1}{2} = \frac{Wl}{4} \quad \text{...(Plus sign due to sagging)}$$

**NOTE.** If the point load does not act at the mid-point of the beam, then the two reactions are obtained and the diagrams are drawn as usual.

**EXAMPLE 13.6.** *A simply supported beam AB of span 2.5 m is carrying two point loads as shown in Fig. 13.13.*

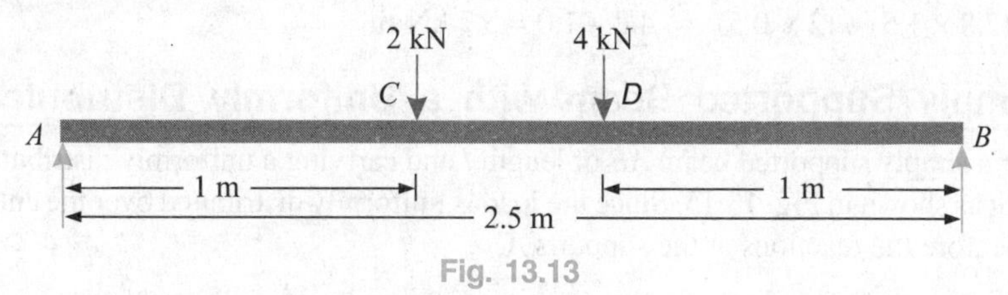


Fig. 13.13

*Draw the shear force and bending moment diagrams for the beam.*

**SOLUTION.** Given : Span $(l) = 2.5$ m ; Point load at $C\ (W_1) = 2$ kN and point load at $B\ (W_2) = 4$ kN.

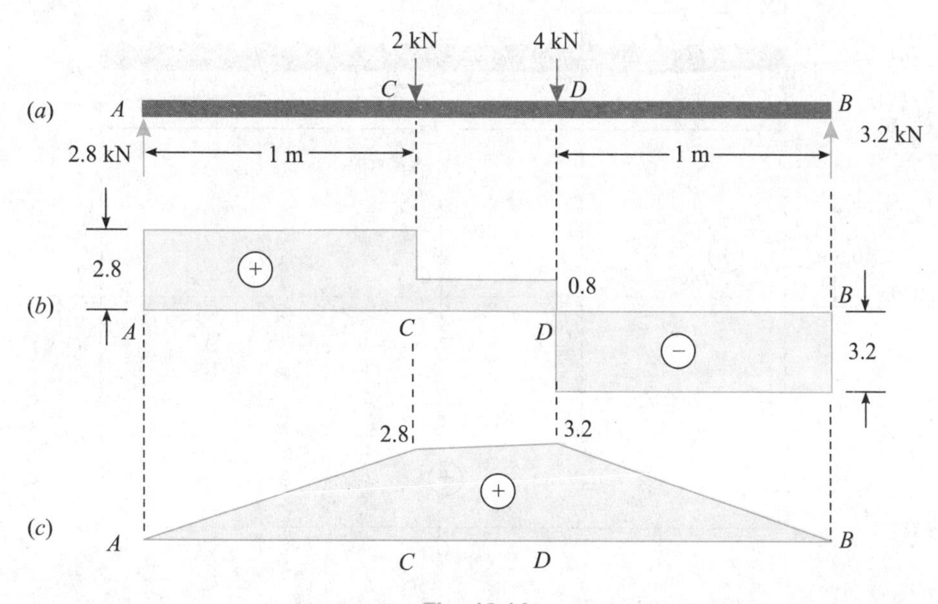


Fig. 13.14

First of all let us find out the reactions $R_A$ and $R_B$. Taking moments about $A$ and equating the same,

$$R_B \times 2.5 = (2 \times 1) + (4 \times 1.5) = 8$$

$$R_B = 8/2.5 = 3.2 \text{ kN}$$

and

$$R_A = (2 + 4) - 3.2 = 2.8 \text{ kN}$$

***Shear force diagram***

The shear force diagram is shown in Fig. 13.14 (*b*) and the values are tabulated here:

$$F_A = +R_A = 2.8 \text{ kN}$$

$$F_C = +2.8 - 2 = 0.8 \text{ kN}$$

$$F_D = 0.8 - 4 = -3.2 \text{ kN}$$

$$F_B = -3.2 \text{ kN}$$

***Bending moment diagram***

The bending moment diagram is shown in Fig. 13.14 (*c*) and the values are tabulated here:

$$M_A = 0$$
$$M_C = 2.8 \times 1 = 2.8 \text{ kN-m}$$
$$M_D = 3.2 \times 1 = 3.2 \text{ kN-m}$$
$$M_B = 0$$

NOTE. The value of $M_D$ may also be found and from the reaction $R_A$. *i.e.*,

$$M_D = (2.8 \times 1.5) - (2 \times 0.5) = 4.2 - 1.0 = 3.2 \text{ kN-m}$$

## 13.12. Simply Supported Beam with a Uniformly Distributed Load

Consider a simply supported beam *AB* of length *l* and carrying a uniformly distributed load of *w* per unit length as shown in Fig. 13.15. Since the load is uniformly distributed over the entire length of the beam, therefore the reactions at the supports *A*,

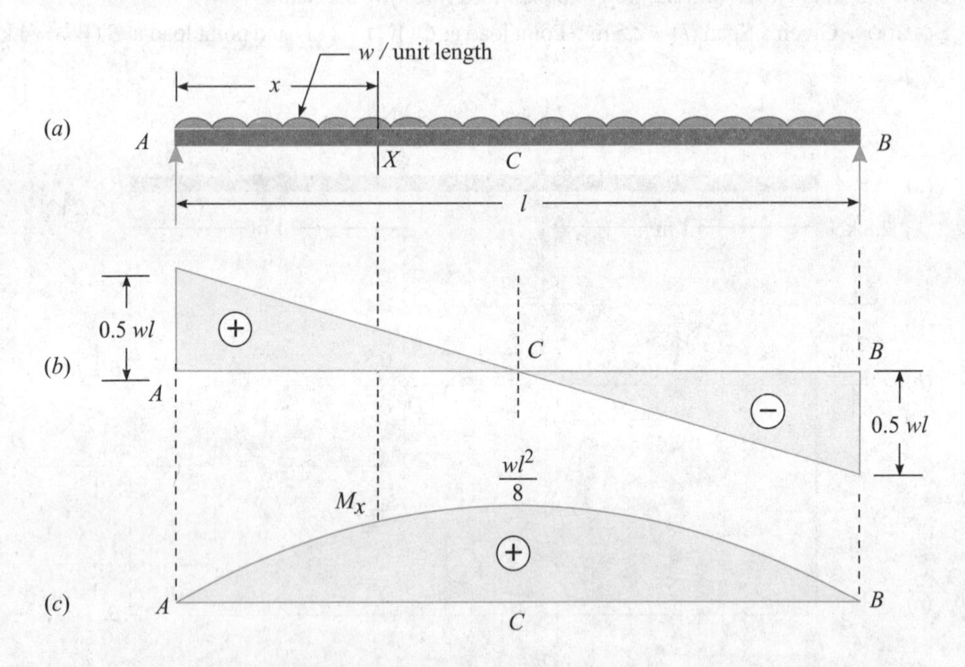


Fig. 13.15. Simply supported beam with a uniformly distributed load

$$R_A = R_B = \frac{wl}{2} = 0.5\,wl$$

We know that shear force at any section *X* at a distance *x* from *A*,

$$F_x = R_A - wx = 0.5\,wl - wx$$

We see that the shear force at *A* is equal to $R_A = 0.5\,wl$, where $x = 0$ and decreases uniformly by a straight line law, to zero at the mid-point of the beam ; beyond which it continues to decrease uniformly to $-\,0.5\,wl$ at *B i.e.*, $R_B$ as shown in Fig. 13.15 (*b*). We also know that bending moment at any section at a distance *x* from *A*,

$$M_x = R_A \, . \, x - \frac{wx^2}{2} = \frac{wl}{2}\,x - \frac{wx^2}{2}$$

We also see that the bending moment is zero at $A$ and $B$ (where $x = 0$ and $x = l$) and increases in the form of a parabolic curve at $C$, *i.e.*, mid-point of the beam where shear force changes sign as shown in Fig. 13.15 (*c*). Thus bending moment at $C$,

$$M_C = \frac{wl}{2}\left(\frac{l}{2}\right) - \frac{w}{2}\left(\frac{l}{2}\right)^2 = \frac{wl^2}{4} - \frac{wl^2}{8} = \frac{wl^2}{8}$$

**EXAMPLE 13.7.** *A simply supported beam 6 m long is carrying a uniformly distributed load of 5 kN/m over a length of 3 m from the right end. Draw the S.F. and B.M. diagrams for the beam and also calculate the maximum B.M. on the section.*

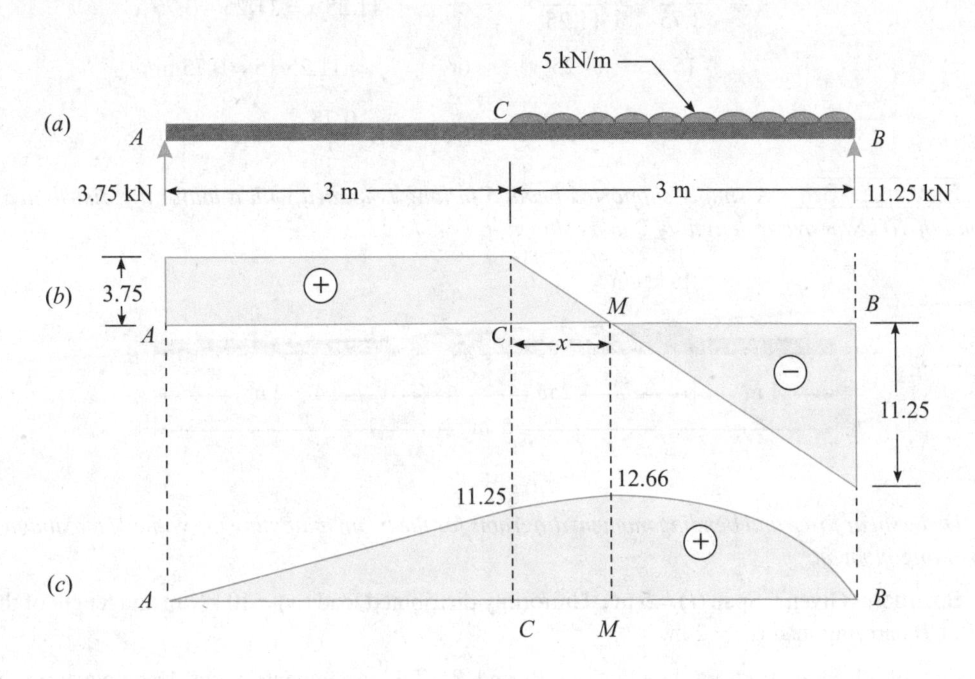


Fig. 13.16

**SOLUTION.** Given : Span ($l$) = 6 m ; Uniformly distributed load ($w$) = 5 kN/m and length of the beam $CB$ carrying load ($a$) = 3 m.

First of all, let us find out the reactions $R_A$ and $R_B$. Taking moments about $A$ and equating the same,

$$R_B \times 6 = (5 \times 3) \times 4.5 = 67.5$$

$$\therefore \qquad R_B = \frac{67.5}{6} = 11.25 \text{ kN}$$

and $$R_A = (5 \times 3) - 11.25 = 3.75 \text{ kN}$$

***Shear force diagram***

The shear force diagram is shown in Fig. 13.16 (*b*) and the values are tabulated here:

$$F_A = +R_A = +3.75 \text{ kN}$$

$$F_C = +3.75 \text{ kN}$$

$$F_B = +3.75 - (5 \times 3) = -11.25 \text{ kN}$$

***Bending moment diagram***

The bending moment is shown in Fig. 13.16 (*c*) and the values are tabulated here:

$$M_A = 0$$

$$M_C = 3.75 \times 3 = 11.25 \text{ kN}$$

$$M_B = 0$$

We know that the maximum bending moment will occur at *M*, where the shear force changes sign. Let *x* be the distance between *C* and *M*. From the geometry of the figure between *C* and *B*, we find that

$$\frac{x}{3.75} = \frac{3-x}{11.25} \quad \text{or} \quad 11.25\,x = 11.25 - 3.75\,x$$

$$15\,x = 11.25 \quad \text{or} \quad x = 11.25/15 = 0.75 \text{ m}$$

$$\therefore \quad M_M = 3.75 \times (3 + 0.75) - 5 \times \frac{0.75}{2} = 12.66 \text{ kN-m}$$

**EXAMPLE 13.8.** *A simply supported beam 5 m long is loaded with a uniformly distributed load of 10 kN/m over a length of 2 m as shown in Fig. 13.17.*

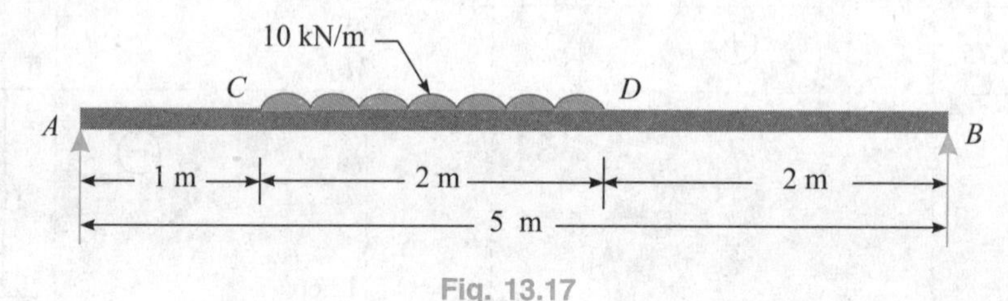


Fig. 13.17

*Draw shear force and bending moment diagrams for the beam indicating the value of maximum bending moment.*

**SOLUTION.** Given : Span ($l$) = 5 m ; Uniformly distributed load ($w$) = 10 kN/m and length of the beam *CD* carrying load ($a$) = 2 m.

First of all, let us find out the reactions $R_A$ and $R_B$. Taking moments about *A* and equating the same,

$$R_B \times 5 = (10 \times 2) \times 2 = 40$$

$$\therefore \quad R_B = 40/5 = 8 \text{ kN}$$

$$\text{and} \quad R_A = (10 \times 2) - 8 = 12 \text{ kN}$$

***Shear force diagram***

The shear force diagram is shown in Fig. 13.18 (*b*) and the values are tabulated here:

$$F_A = +R_A = +12 \text{ kN}$$

$$F_C = +12 \text{ kN}$$

$$F_D = +12 - (10 \times 2) = -8 \text{ kN}$$

$$F_B = -8 \text{ kN}$$

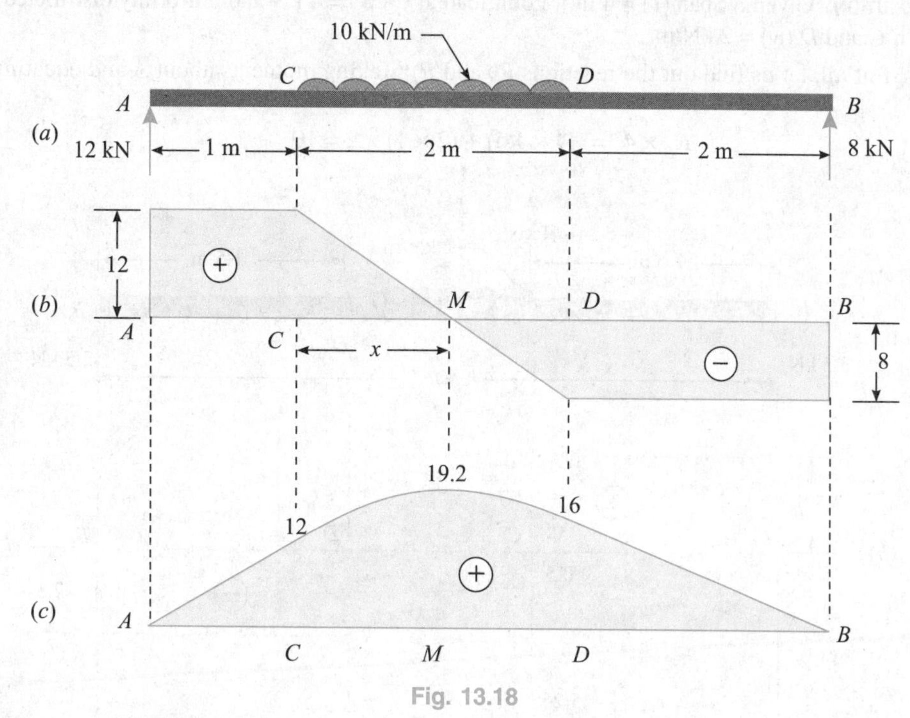


Fig. 13.18

***Bending moment diagram***

The bending moment diagram is shown in Fig. 13.18 (*c*) and the values are tabulated here:

$$M_A = 0$$

$$M_C = 12 \times 1 = 12 \text{ kN-m}$$

$$M_D = 8 \times 2 = 16 \text{ kN-m}$$

We know that maximum bending moment will occur at $M$, where the shear force changes sign. Let $x$ be the distance between $C$ and $M$. From the geometry of the figure between $C$ and $D$, we find that

$$\frac{x}{12} = \frac{2-x}{8} \quad \text{or} \quad 8x = 24 - 12x$$

$$20x = 24 \quad \text{or} \quad x = 24/20 = 1.2 \text{ m}$$

$$M_M = 12(1 + 1.2) - 10 \times 1.2 \times \frac{1.2}{2} = 19.2 \text{ kN-m}$$

**EXAMPLE 13.9.** *A simply supported beam of 4 m span is carrying loads as shown in Fig. 13.19.*

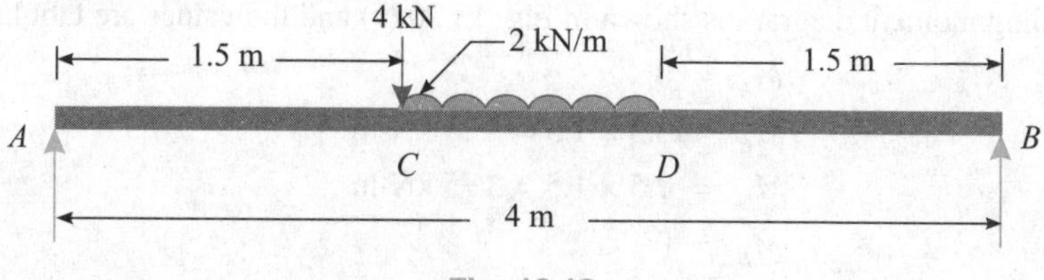


Fig. 13.19

*Draw shear force and bending moment diagrams for the beam.*

**Solution.** Given : Span $(l) = 4$ m ; Point load at $C$ $(W) = 4$ kN and uniformly distributed load between $C$ and $D$ $(w) = 2$ kN/m.

First of all, let us find out the reactions $R_A$ and $R_B$. Taking moments about $A$ and equating the same,

$$R_B \times 4 = (4 \times 1.5) + (2 \times 1) \times 2 = 10$$

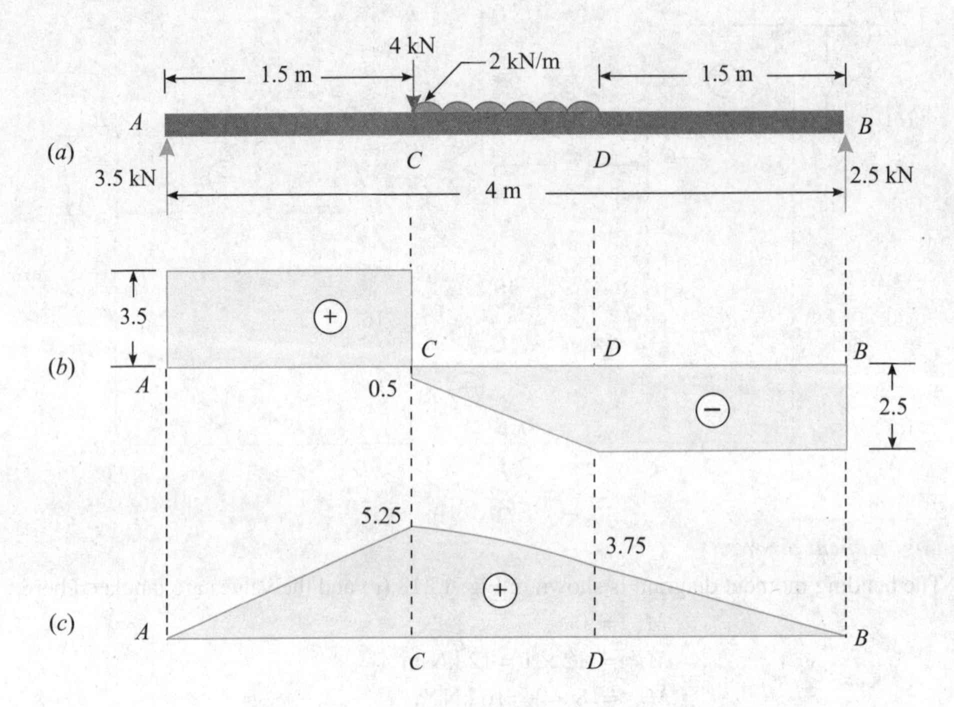


Fig. 13.20

$$R_B = 10/4 = 2.5 \text{ kN}$$

and
$$R_A = 4 + (2 \times 1) - 2.5 = 3.5 \text{ kN}$$

***Shear force diagram***

The shear force diagram is shown in Fig. 13.20 (*b*) and the values are tabulated here:

$$F_A = +R_A = +3.5 \text{ kN}$$
$$F_C = +3.5 - 4 = -0.5 \text{ kN}$$
$$F_D = -0.5 - (2 \times 1) = -2.5 \text{ kN}$$
$$F_B = -2.5 \text{ kN}$$

***Bending moment diagram***

The bending moment diagram is shown in Fig. 13.20 (*c*) and the values are tabulated here:

$$M_A = 0$$
$$M_C = 3.5 \times 1.5 = 5.25 \text{ kN-m}$$
$$M_D = 2.5 \times 1.5 = 3.75 \text{ kN-m}$$
$$M_B = 0$$

We know that the maximum bending moment will occur at $C$, where the shear force changes sign, *i.e.*, at $C$ as shown in the figure.

**EXAMPLE 13.10.** *A simply supported beam AB, 6 m long is loaded as shown in Fig. 13.21.*

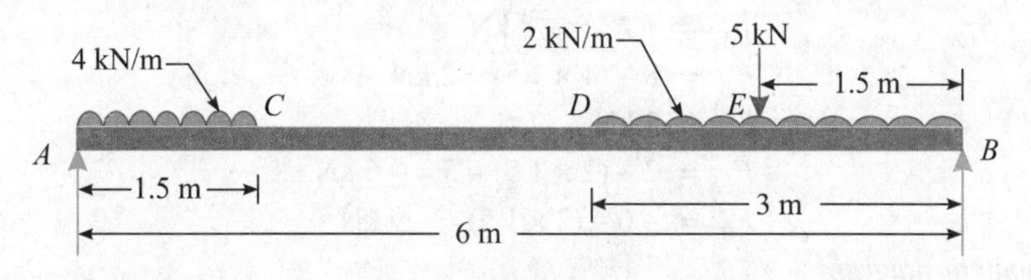


Fig. 13.21

*Construct the shear force and bending moment diagrams for the beam and find the position and value of maximum bending moment.*

**SOLUTION.** Given : Span ($l$) = 6 m ; Point load at $E$ ($W$) = 5 kN ; Uniformly distributed load between $A$ and $C$ ($w_1$) = 4 kN/m and uniformly distributed load between $D$ and $B$ = 2 kN/m.

First of all, let us find out the reactions $R_A$ and $R_B$. Taking moments about $A$ and equating the same,

$$R_B \times 6 = (4 \times 1.5 \times 0.75) + (2 \times 3 \times 4.5) + (5 \times 4.5) = 54$$

$$R_B = 54/6 = 9 \text{ kN}$$

and

$$R_A = (4 \times 1.5) + (2 \times 3) + 5 - 9 = 8 \text{ kN}$$

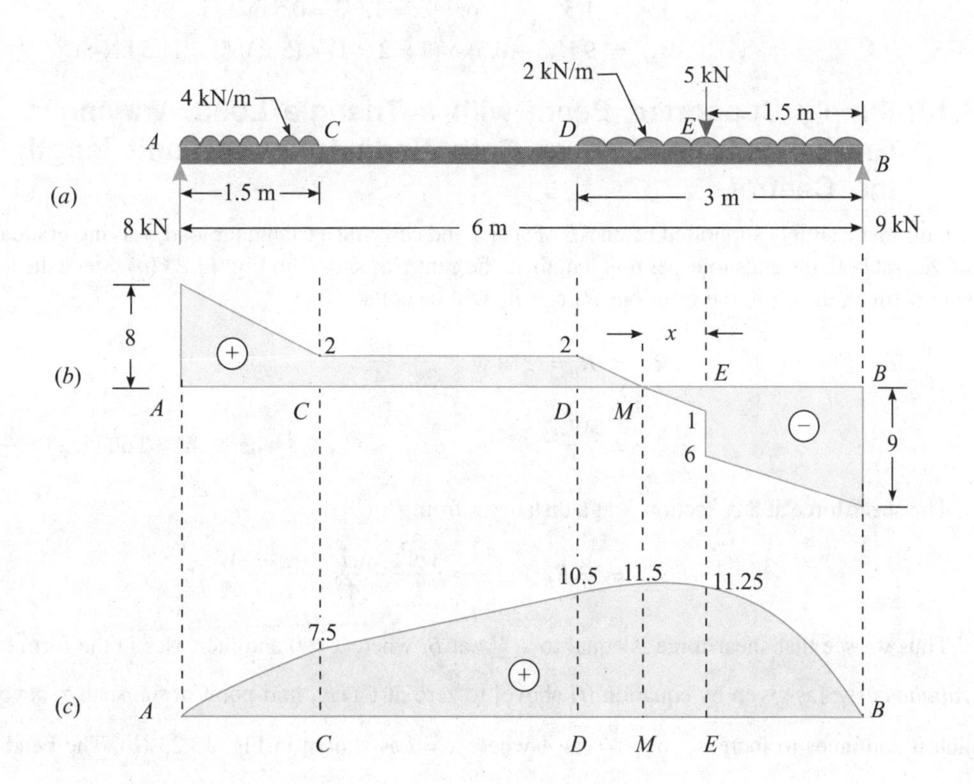


Fig. 13.22

***Shear force diagram***

The shear force diagram is shown in Fig. 13.22 (*b*) and the values are tabulated here:

$$F_A = +R_A = +8 \text{ kN}$$
$$F_C = 8 - (4 \times 1.5) = 2 \text{ kN}$$
$$F_D = 2 \text{ kN}$$
$$F_E = 2 - (2 \times 1.5) - 5 = -6 \text{ kN}$$
$$F_B = -6 - (2 \times 1.5) = -9 \text{ kN}$$

***Bending moment diagram***

The bending moment diagram is shown in Fig. 13.22 (c) and the values are tabulated here:

$$M_A = 0$$
$$M_C = (8 \times 1.5) - (4 \times 1.5 \times 0.75) = 7.5 \text{ kN-m}$$
$$M_D = (8 \times 3) - (4 \times 1.5 \times 2.25) = 10.5 \text{ kN-m}$$
$$M_E = (9 \times 1.5) - (2 \times 1.5 \times 0.75) = 11.25 \text{ kN-m}$$
$$M_B = 0$$

We know that maximum bending moment will occur at $M$, where the shear force changes sign. Let $x$ be the distance between $E$ and $M$. From the geometry of the figure between $D$ and $E$, we find that

$$\frac{x}{1} = \frac{1.5 - x}{2} \quad \text{or} \quad 2x = 1.5 - x$$

$$3x = 1.5 \quad \text{or} \quad x = 1.5/3 = 0.5 \text{ m}$$

$$\therefore \quad M_M = 9(1.5 + 0.5) - (2 \times 2 \times 1) - (5 \times 0.5) = 11.5 \text{ kN-m}$$

## 13.13. Simply Supported Beam with a Triangle Load, Varying Gradually from Zero at Both Ends to *w* per unit length at the Centre

Consider a simply supported beam *AB* of span $l$ and carrying a triangular load, varying gradually from zero at both the ends to $w$ per unit length, at the centre as shown in Fig. 13.23 (*a*). Since the load is symmetrical, therefore the reactions $R_A$ and $R_B$ will be equal.

$$\text{or} \quad R_A = R_B = \frac{1}{2} \times w \times \frac{1}{2} = \frac{wl}{4}$$

$$= \frac{W}{2} \quad \quad \text{...}\left(\text{where } W = \text{Total load} = \frac{wl}{2}\right)$$

The shear force at any section $X$ at a distance $x$ from $B$,

$$F_X = -R_B + \frac{wx^2}{l} = \frac{wx^2}{l} - \frac{wl}{4} = \frac{wx^2}{l} - \frac{W}{2} \quad \text{...}(i)$$

Thus we see that shear force is equal to $-\frac{W}{2}$ at $B$, where $x = 0$ and increases in the form of a *parabolic curve* [as given by equation (*i*) above] to zero at $C$, *i.e.*, mid-point of the span ; beyond which it continues to increase to $+\frac{W}{2}$ at $A$ where $x = l$ as shown in Fig. 13.23 (*b*). The bending moment at any section $X$ at a distance $x$ from $B$,

$$M_X = R_B \cdot x - \frac{wx}{\frac{l}{2}} \times \frac{x}{2} \times \frac{x}{3} = \frac{wlx}{4} - \frac{wx^3}{3l} \quad \text{...}(ii)$$

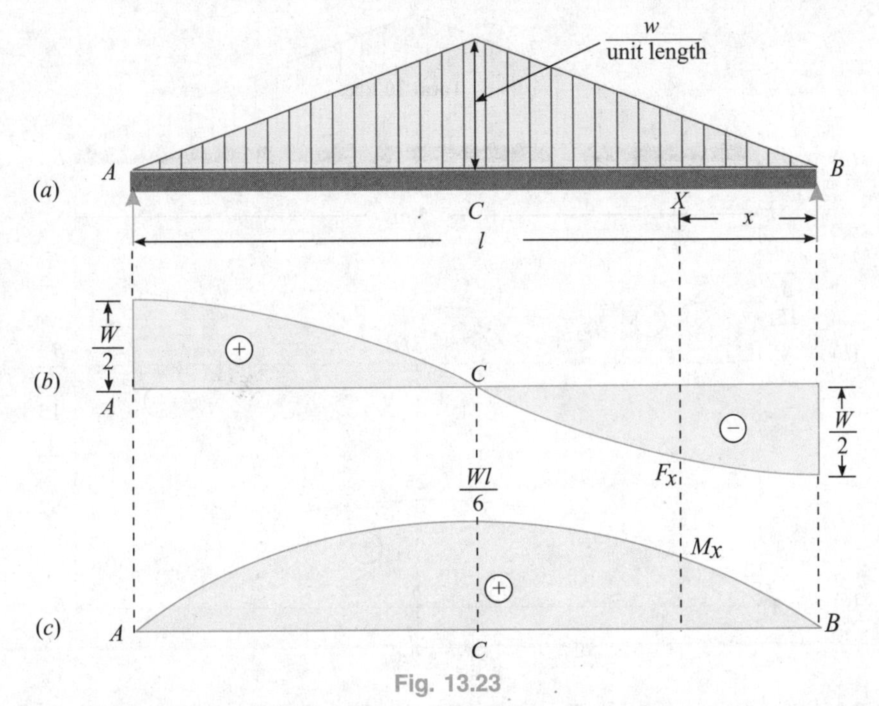


Fig. 13.23

Thus we see that the bending moment at $A$ and $B$ is zero and increases in the form of a cubic curve [as given by the equation (*ii*) above] at $C$, *i.e.*, mid-point of the beam, where bending moment will be maximum because shear force changes sign.

$$\therefore \qquad M_M = \frac{wl}{4}\left(\frac{l}{2}\right) - \frac{w}{3l}\left(\frac{l}{2}\right)^3 = \frac{wl^2}{12}$$

$$= \frac{Wl}{6} \qquad \ldots \left(\text{where } W = \text{Total load} = \frac{wl}{2}\right)$$

**EXAMPLE 13.11.** *A simply supported beam of 5 m span carries a triangular load of 30 kN. Draw S.F. and B.M. diagrams for the beam.*

**SOLUTION.** Given : Span ($l$) = 5 m and total triangular load ($W$) = 30 kN

By symmetry, $\qquad R_A = R_B = \dfrac{30}{2} = 15 \text{ kN}$

***Shear force diagram***

The shear force diagram is shown in Fig. 13.24 (*b*) and the values are tabulated here:

$$F_A = +R_A = +15 \text{ kN}$$

$$F_B = -R_B = -15 \text{ kN}$$

***Bending moment diagram***

The bending moment diagram is shown in Fig. 13.24 (*c*). It is zero at $A$ and $B$ and the maximum bending moment will occur at the centre *i.e.*, at $M$, where the shear force changes sign. We know that maximum bending moment,

$$M_M = \frac{Wl}{6} = \frac{30 \times 5}{6} = 25 \text{ kN-m}$$

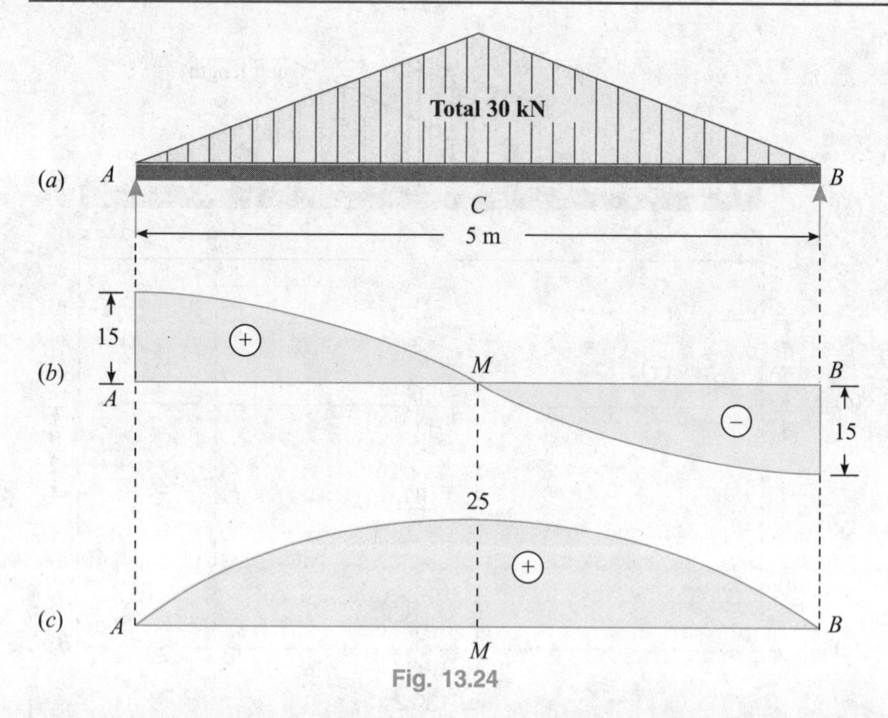


Fig. 13.24

## 13.14. Simply Supported Beam with a Gradually Varying Load from Zero at One End to *w* per unit length at the Other End

Consider a simply supported beam *AB* of length *l* and carrying a gradually varying load zero at one end and *w* per unit length at the other as shown in Fig. 13.25 (*a*). Since the load is varying gradually from zero at one end to *w* per unit length at the other, therefore both the reactions at *A* and *B* will have to be first calculated.

Taking moments about *A*,

$$R_B \times 1 = \left(\frac{0+w}{2}\right) \times l \times \frac{l}{3} = \frac{wl^2}{6}$$

$$\therefore \quad R_B = \frac{wl^2}{6} \times \frac{1}{l} = \frac{wl}{6}$$

$$= \frac{W}{3} \qquad \ldots\left(\text{where } W = \text{Total load} = \frac{wl}{2}\right)$$

and

$$R_A = \frac{wl}{2} - \frac{wl}{6} = \frac{wl}{3}$$

$$= \frac{2W}{3} \qquad \ldots\left(\text{where } W = \frac{wl}{2}\right)$$

We know that the shear force at any section *X* at a distance *x* from *B*,

$$F_X = -R_B + \frac{wx^2}{2l} = \frac{wx^2}{2l} - \frac{W}{3} \qquad \ldots(i)$$

Thus we see that the shear force is equal to $-\frac{W}{3}$ at *B* (where $x = 0$) and increases in the form of a *parabolic curve* [as is given by the equation (*i*) above] to zero at *M* ; beyond which it continues to increase to $+\frac{2W}{3}$ at *A* (where $x = l$) as shown in Fig. 13.25 (*b*). The bending moment at any section *X* at a distance *x* from *B*,

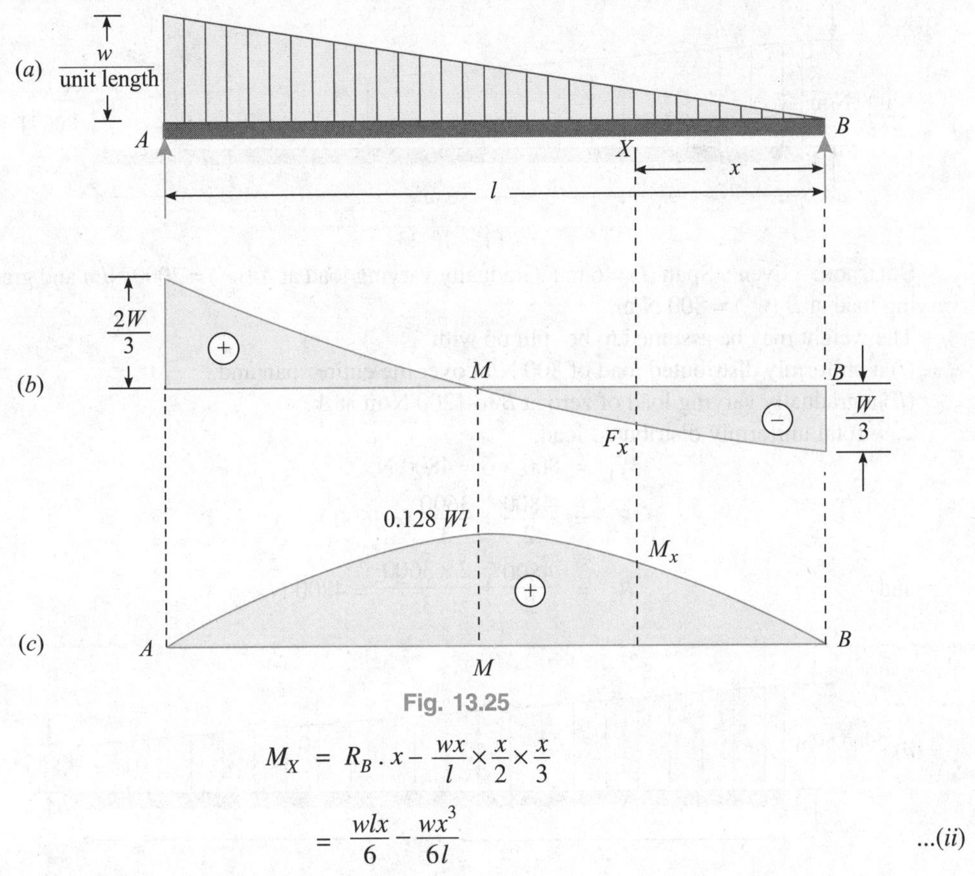


Fig. 13.25

$$M_X = R_B \, . \, x - \frac{wx}{l} \times \frac{x}{2} \times \frac{x}{3}$$

$$= \frac{wlx}{6} - \frac{wx^3}{6l} \qquad \ldots(ii)$$

Thus bending moment at $A$ and $B$ is zero and it increases in the form of a cubic curve [as given by the equation (*ii*) above] at $M$, where the shear force changes sign. To find out the position $M$, let us equate the equation (*i*) to zero, *i.e.*,

$$\frac{wx^2}{2l} - \frac{wl}{6} = 0 \quad \text{or} \quad \frac{wx^2}{2l} - \frac{wl}{6}$$

$$\therefore \quad x^2 = \frac{l^2}{3} \quad \text{or} \quad x = \frac{l}{\sqrt{3}} = 0.577\, l$$

$$\therefore \quad M_M = \frac{wl}{6}\left(\frac{l}{\sqrt{3}}\right) - \frac{w}{6l}\left(\frac{l}{\sqrt{3}}\right)^3 = \frac{wl^2}{9\sqrt{3}}$$

$$= \frac{2\,Wl}{9\sqrt{3}} = 0.128\, Wl \qquad \ldots\left(\text{where } W = \frac{wl}{2}\right)$$

**Note.** In such cases the different values of shear force and bending moment should be calculated at intervals of 0.5 m or 1 m [as per equations (*i*) and (*ii*) above] and then the diagrams should be drawn.

**Example 13.12.** *The intensity of loading on a simply supported beam of 6 m span increases gradually from 800 N/m run at one end to 2000 N/m run at the other as shown in Fig. 13.26.*

*Find the position and amount of maximum bending moment. Also draw the shear force and bending moment diagrams.*

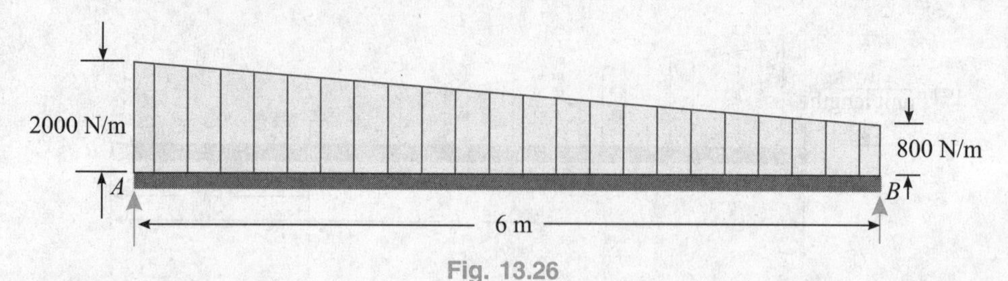


Fig. 13.26

**Solution.** Given : Span ($l$) = 6 m ; Gradually varying load at $A$ ($w_A$) = 2000 N/m and gradually varying load at $B$ ($w_B$) = 800 N/m.

The weight may be assumed to be split up with

(*i*) a uniformly distributed load of 800 N/m over the entire span and

(*ii*) a gradually varying load of zero at $B$ to 1200 N/m at $A$.

∴ Total uniformly distributed load,

$$W_1 = 800 \times 6 = 4800 \text{ N}$$

∴
$$R_B = \frac{4800}{2} + \frac{3600}{3} = 3600 \text{ N}$$

and
$$R_A = \frac{4800}{2} + \frac{2 \times 3600}{3} = 4800 \text{ N.}$$

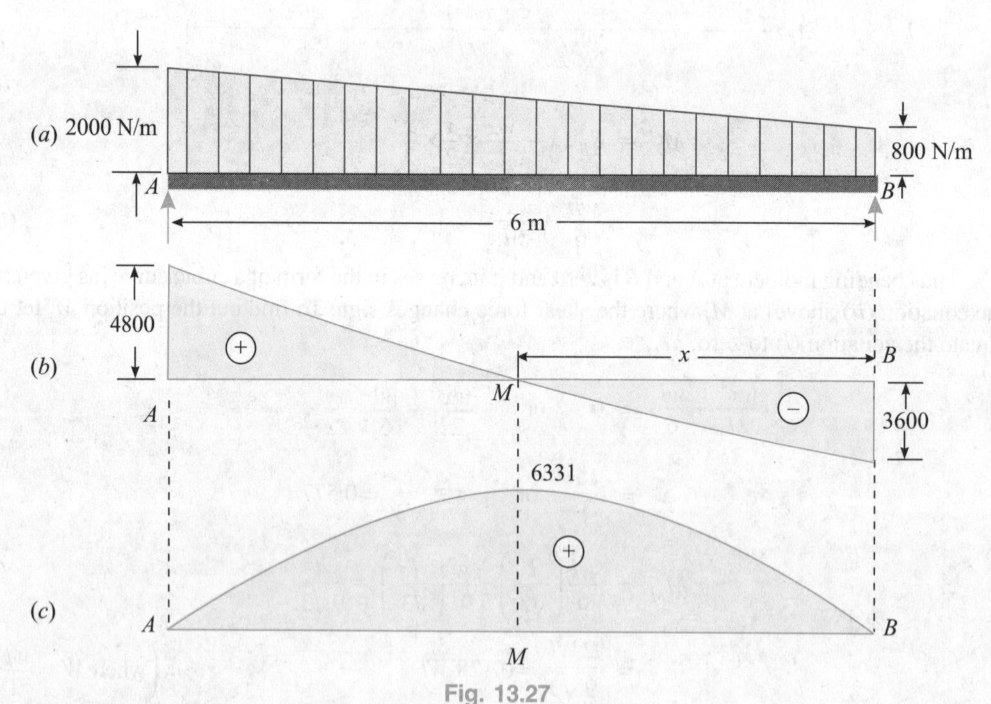


Fig. 13.27

***Shear force diagram***

The shear force diagram is shown in Fig. 13.27 (*b*), and the values are tabulated here:

$$F_A = +R_A = 4800 \text{ N}$$

$$F_B = -R_B = -3600 \text{ N}$$

***Bending moment diagram***

The bending moment diagram is shown in Fig. 13.27 (*c*). It is zero at $A$ and $B$ and the maximum bending moment will occur at $M$, where the shear force changes sign.

***Maximum bending moment***

We know that maximum bending moment will occur at a point (*M*), where shear force changes sign. Let *x* be the distance between *B* and *M*. We also know that shear force at a distance *x* from *M*,

$$= -3600 + 800\,x + \frac{1}{2} \times 1200\,x \times \frac{x}{6}$$

$$= -3600 + 800\,x + 100\,x^2$$

$$= 100\,x^2 + 800\,x - 3600$$

Now to find the position of *M* (*i.e.*, the point where shear force changes sign), let us equate the above equation to zero. *i.e.*,

$$100\,x^2 + 800\,x - 3600 = 0$$

or $\quad x^2 + 8\,x - 36 = 0$

This is a quadratic equation. Therefore

$$\therefore \quad x = \frac{-8 \pm \sqrt{(8)^2 + (4 \times 36)}}{2} = 3.21 \text{ m}$$

and bending moment at *M*,

$$M_M = 3600\,x - \left(800 \times \frac{x^2}{2}\right) - \left(\frac{1}{2} \times 1200\,x \times \frac{x}{6} \times \frac{x}{3}\right)$$

$$= 3600\,x - 400\,x^2 - \frac{100}{3}\,x^3$$

$$= (3600 \times 3.21) - 400 \times (3.21)^2 - \frac{100}{3} \times (3.21)^3 \text{ N-m}$$

$$= 11556 - 4122 - 1102 = 6332 \text{ N-m}$$

## EXERCISE 13.2

1. A simply supported beam of 3 m span carries two loads of 5 kN each at 1 m and 2 m from the left hand support. Draw the shear force and bending moment diagrams for the beam. **[Ans.** $M_{max}$ = 5 kN-m]
2. A simply supported beam of span 4.5 m carries a uniformly distributed load of 3.6 kN/m over a length of 2 m from the left end *A*. Draw the shear force and bending moment diagrams for the beam. **[Ans.** $M_{max}$ = 4.36 kN-m at 1.56 m from *A*]
3. A simply supported beam *ABCD* is of 5 m span, such that *AB* = 2 m, *BC* = 1 m and *CD* = 2 m. It is loaded with 5 kN/m over *AB* and 2 kN/m over *CD*. Draw shear force and bending moment diagrams for the beam. **[Ans.** $M_{max}$ = 7.74 kN-m at 1.76 m from *A*]
4. Draw shear force and bending moment diagrams for a simply supported beam, loaded as shown in Fig. 13.28.

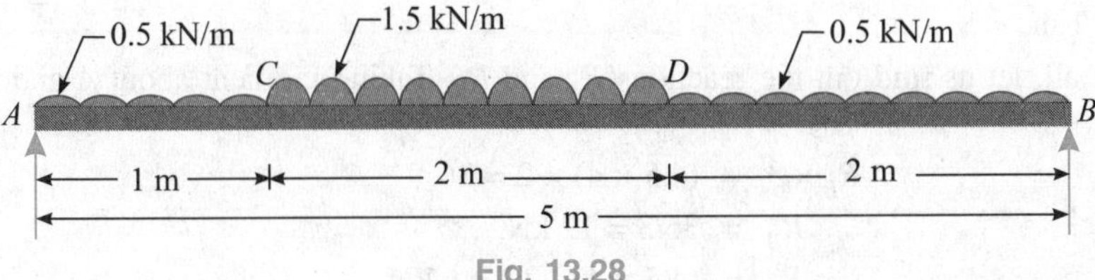


Fig. 13.28

Find the position and value of the maximum bending moment that will occur in the beam. **[Ans.** 3.47 kN-m at 1.3 m from *C*]

5. A simply supported beam $AB$, 6 m long is loaded as shown in Fig. 13.29.

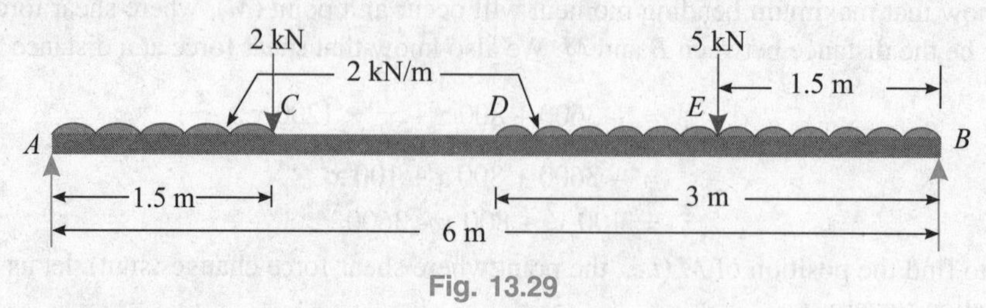


Fig. 13.29

Draw the shear force and bending moment diagrams for the beam.

[*Ans.* $M_{max}$ = 11.75 kN-m at 0.56 m from $E$]

6. A simply supported beam 3 m long carries a triangular load of 12 kN. Draw the S.F and B.M. diagrams for the beam. [**Ans.** $M_{max}$ = 6 kN-m]

## 13.15. Overhanging Beam

It is a simply supported beam which overhangs (*i.e.*, extends in the form of a cantilever) from its support.

For the purposes of shear force and bending moment diagrams, the overhanging beam is analysed as a combination of a simply supported beam and a cantilever. An overhanging beam may overhang on one side only or on both sides of the supports.

## 13.16. Point of Contraflexure

We have already discussed in the previous article that an overhanging beam is analysed as a combination of simply supported beam and a cantilever. In the previous examples, we have seen that the bending moment in a cantilever is negative, whereas that in a simply supported beam is positive. It is thus obvious that in an overhanging beam, there will be a point, where the bending moment will change sign from negative to positive or *vice versa*. Such a point, where the bending moment changes sign, is known as a point of contraflexure.

**EXAMPLE 13.13.** *An overhanging beam ABC is loaded as shown in Fig. 13.30.*

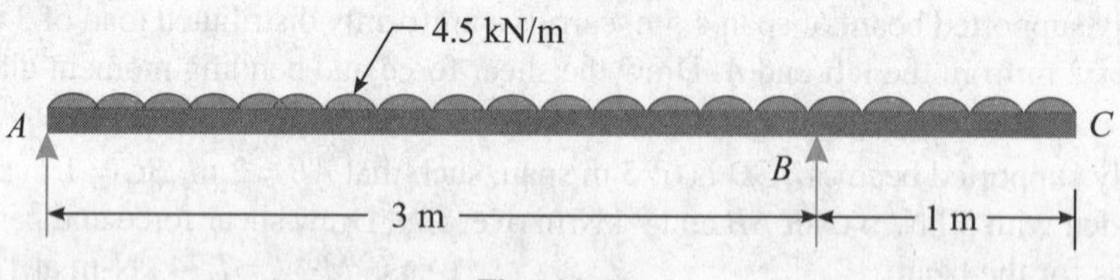


Fig. 13.30

*Draw the shear force and bending moment diagrams and find the point of contraflexure, if any.*

**SOLUTION.** Given : Span ($l$) = 4 m ; Uniformly distributed load ($w$) = 4.5 kN/m and overhanging length ($c$) = 1 m.

First of all, let us find out the reactions $R_A$ and $R_B$. Taking moment about $A$ and equating the same,

$$R_B \times 3 = (4.5 \times 4) \times 2 = 36$$

$$\therefore \quad R_B = 36/3 = 12 \text{ kN}$$

and $$R_A = (4.5 \times 4) - 12 = 6 \text{ kN}$$

***Shear force diagram***

The shear force diagram is shown in Fig. 13.31 ($b$) and the values are tabulated here:

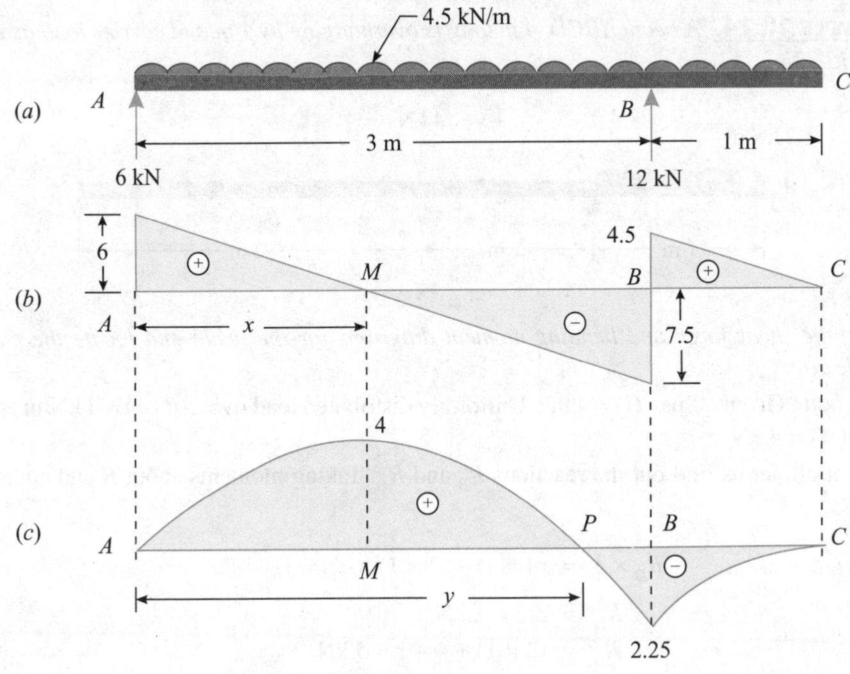


Fig. 13.31

$$F_A = +R_A = +6 \text{ kN}$$
$$F_B = +6 - (4.5 \times 3) + 12 = 4.5 \text{ kN}$$
$$F_C = +4.5 - (4.5 \times 1) = 0$$

***Bending moment diagram***

The bending moment diagram is shown in Fig. 13.31 (*c*) and the values are tabulated here:

$$M_A = 0$$

$$M_B = -\left(4.5 \times 1 \times \frac{1}{2}\right) = -2.25 \text{ kN-m}$$

$$M_C = 0$$

We know that maximum bending moment will occur at *M*, where the shear force changes sign. Let *x* be the distance between *A* and *M*. From the geometry of the figure between *A* and *B*, we find that

$$\frac{x}{6} = \frac{3-x}{7.5} \quad \text{or} \quad 7.5\,x = 18 - 6\,x$$

$$13.5\,x = 18 \quad \text{or} \quad x = 18/13.5 = 1.33 \text{ m}$$

$$\therefore \quad M_M = (6 \times 1.33) - 4.5 \times 1.33 \times \frac{1.33}{2} = 4 \text{ kN-m}$$

***Point of contraflexure***

Let *P* be the point of contraflexure at a distance *y* from the support *A*. We know that bending moment at *P*.

$$M_P = 6 \times y - 4.5 \times y \times \frac{y}{2} = 0$$

$$2.25\,y^2 - 6\,y = 0 \quad \text{or} \quad 2.25\,y = 6$$

$$\therefore \quad y = 6/2.25 = 2.67 \text{ m} \qquad \textbf{Ans.}$$

**EXAMPLE 13.14.** *A beam ABCD, 4 m long is overhanging by 1 m and carries load as shown in Fig. 13.32.*

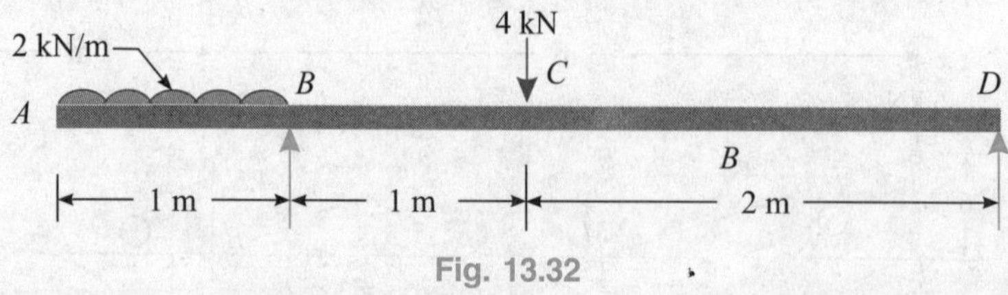


Fig. 13.32

*Draw the shear force and bending moment diagrams for the beam and locate the point of contraflexure.*

**SOLUTION.** Given : Span ($l$) = 4 m ; Uniformly distributed load over $AB$ ($w$) = 2 kN/m and point load at $C$ ($W$) = 4 kN.

First of all, let us find out the reactions $R_B$ and $R_D$. Taking moments about $B$ and equating the same,

$$R_D \times 3 = (4 \times 1) - (2 \times 1) \times \frac{1}{2} = 3$$

$$\therefore \quad R_D = 3/3 = 1 \text{ kN}$$

and

$$R_B = (2 \times 1) + 4 - 1 = 5 \text{ kN}$$

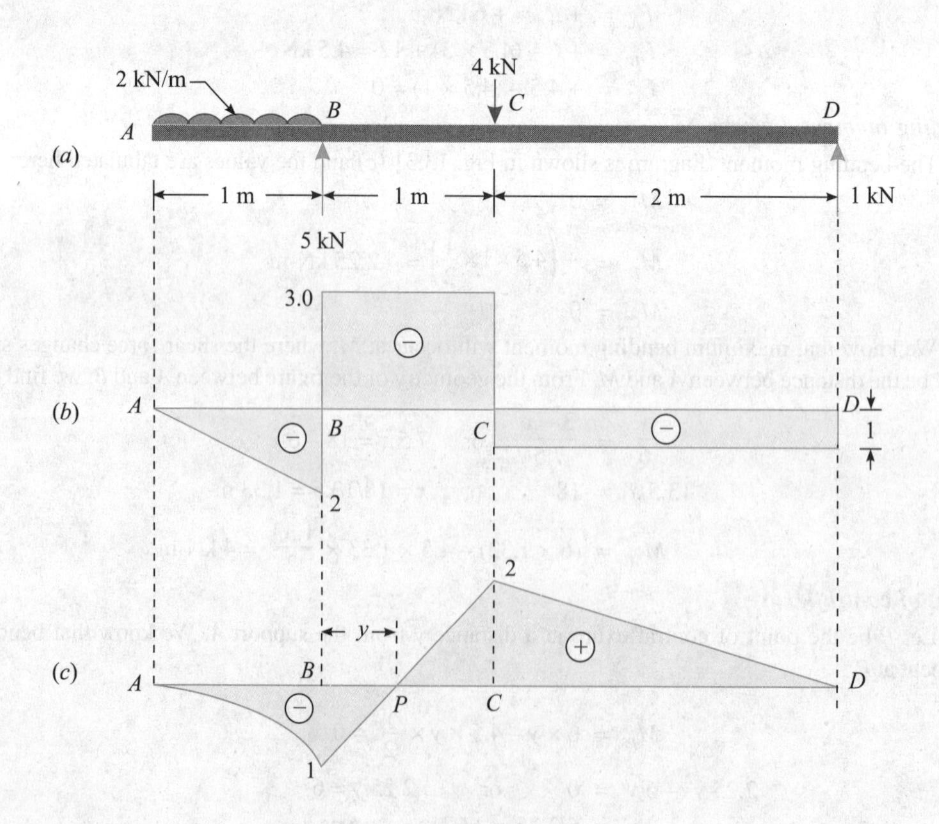


Fig. 13.33

***Shear force diagram***

The shear force diagram is shown in Fig. 13.33 (*b*) and the values are tabulated here:

$$F_A = 0$$

$$F_B = 0 - (2 \times 1) + 5 = +3 \text{ kN}$$

$$F_C = +3 - 4 = -1 \text{ kN}$$

$$F_D = 1 \text{ kN}$$

***Bending moment diagram***

The bending moment diagram is shown in Fig. 13.33 (*c*) and the values are tabulated here:

$$M_A = 0$$

$$M_B = -(2 \times 1)\, 0.5 = -1 \text{ kN-m}$$

$$M_C = 1 \times 2 = +2 \text{ kN}$$

$$M_D = 0$$

We know that maximum bending moment occurs either at *B* or *C*, where the shear force changes sign. From the geometry of the bending moment diagram, we find that maximum negative bending moment occurs at *B* and maximum positive bending moment occurs at *C*.

***Point of contraflexure***

Let *P* be the point of contraflexure at a distance *y* from the support *B*. From the geometry of the figure between *B* and *C*, we find that

$$\frac{y}{1.0} = \frac{1-y}{2.0}$$

$$2y = 1 - y \quad \text{or} \quad 3y = 1$$

or $$y = 1/3 = 0.33 \text{ m} \quad \textbf{Ans.}$$

**EXAMPLE 13.15.** *Draw shear force and bending moment diagrams for the beam shown in Fig. 13.34. Indicate the numerical values at all important sections.*

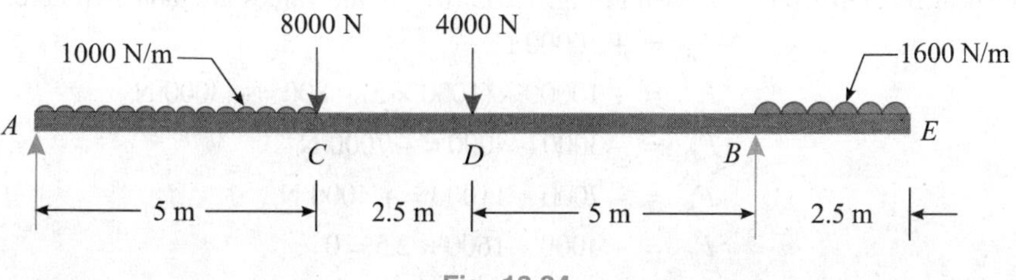


Fig. 13.34

**SOLUTION.** Given : Span ($l$) = 15 m ; Uniformly distributed load between *A* and *B* ($w_1$) = 1000 N/m ; Point load at *C* ($W_1$) = 8000 N ; Point load at *D* ($W_2$) = 4000 N and uniformly distributed load between *B* and *E* ($w_2$) = 1600 N/m.

First of all, let us find out the reactions $R_A$ and $R_B$.

Taking moments about *A* and equating the same,

$$R_B \times 12.5 = (1600 \times 2.5) \times 13.75 + (4000 \times 7.5) + (8000 \times 5) + (1000 \times 5) \times 2.5$$

$$= 137500$$

$$R_B = \frac{137500}{12.6} = 110000 \text{ N}$$

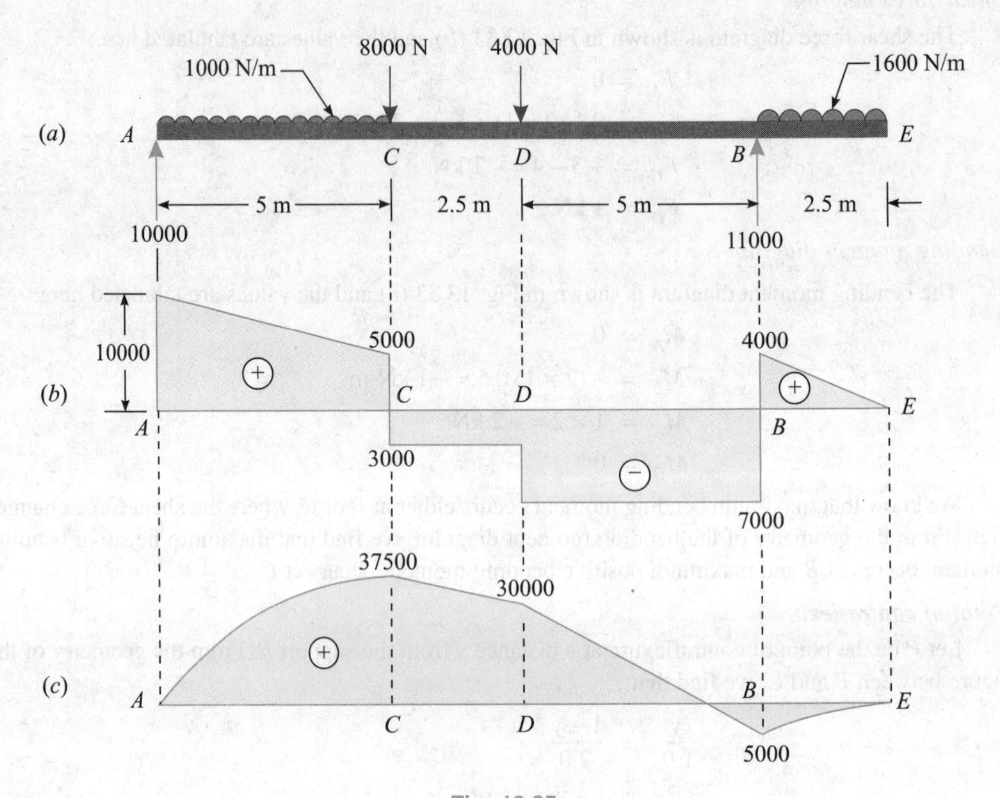


Fig. 13.35

and $$R_A = (1000 \times 5 + 8000 + 4000 + 1600 \times 2.5) - 11000 \text{ N}$$
$$= 10000 \text{ N}$$

***Shear force***

The shear force diagram is shown in Fig. 13.35 (*b*) and the values are tabulated here:

$$F_A = +10000 \text{ N}$$
$$F_C = +10000 - (1000 \times 5) - 800 = -3000 \text{ N}$$
$$F_D = -3000 - 4000 = -7000 \text{ N}$$
$$F_B = -7000 + 11000 = +4000 \text{ N}$$
$$F_E = +4000 - 1600 \times 2.5 = 0$$

***Bending moment***

The bending moment diagram is shown in Fig. 13.35 (*c*), and the values are tabulated here:

$$M_A = 0$$
$$M_C = (10000 \times 5) - (1000) \times (5 \times 2.5) = 37500 \text{ N-m}$$
$$M_D = (10000 \times 7.5) - (1000 \times 5 \times 5) - (8000 \times 2.5) \text{ N-m}$$
$$= 30000 \text{ N-m}$$
$$M_B = -1600 \times 2.5 \times \frac{2.5}{2} = -5000 \text{ N-m}$$

*Maximum bending moment*

The maximum bending moment, positive or negative will occur at $C$ or at $B$ because the shear force changes sign at both these points. But from the bending moment diagram, we see that the maximum positive bending moment occurs at $C$ and the maximum negative bending moment occurs at $B$.

**EXAMPLE 13.16.** *Draw the complete shear force diagram for the overhanging beam shown in Fig. 13.36.*

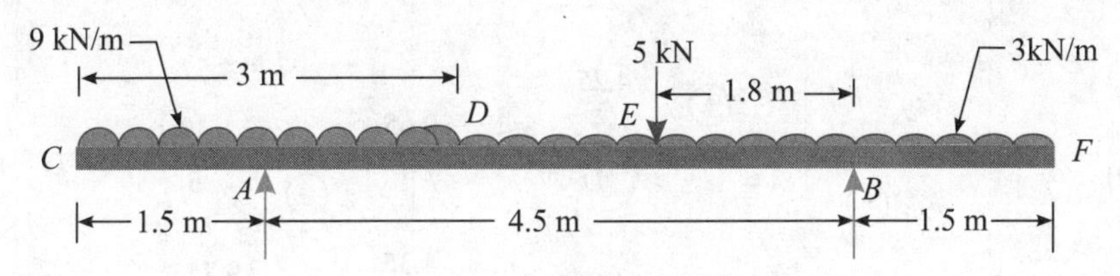


Fig. 13.36

*Hence, determine the position in the central bay, at which the positive bending moment occurs. Find also magnitude of the maximum positive and negative bending moment.*

**SOLUTION.** Given : Span ($l$) = 7.5 m ; Uniformly distributed load between $C$ and $D$ = 9 kN/m ; Point load at $E$ ($W$) = 5 kN ; Uniformly distributed load between $D$ and $F$ ($w_2$) = 3 kN/m and overhanging on both sides = 1.5 m.

Taking moments about $A$,

$$R_B \times 4.5 = (3 \times 4.5) \times 3,75 + (5 \times 2.7) = 64.125$$

($\because$ U.D.L. of 9 kN/m will have zero moment about $A$)

$$\therefore \quad R_B = \frac{64.125}{4.5} = 14.25 \text{ kN}$$

and

$$R_A = (9 \times 3 + 5 + 3 \times 4.5) - 14.25 = 31.25 \text{ kN}$$

*Shear force diagram*

The shear force diagram is shown in Fig. 13.37 ($b$) and the values are tabulated here:

$$F_C = 0$$

$$F_A = 0 - 9 \times 1.5 + 31.25 = +\ 17.75 \text{ kN}$$

$$F_D = +\ 17.75 - 9 \times 1.5 = +\ 4.25 \text{ kN}$$

$$F_E = +\ 4.25 - 3 \times 1.2 - 5.0 = -\ 4.35 \text{ kN}$$

$$F_B = -\ 4.35 - 3 \times 1.8 + 14.25 = +\ 4.5 \text{ kN}$$

$$F_F = +\ 4.5 - 3 \times 1.5 = 0$$

*Bending moment diagram*

The bending moment diagram is shown in Fig. 13.37 ($c$) and the values are tabulated here:

$$M_C = 0$$

$$M_A = -\frac{9 \times (1.5)^2}{2} = -\ 10.125 \text{ kN-m}$$

$$M_D = -\frac{9 \times (3)^2}{2} + 31.25 \times 1.5 = 6.375 \text{ kN-m}$$

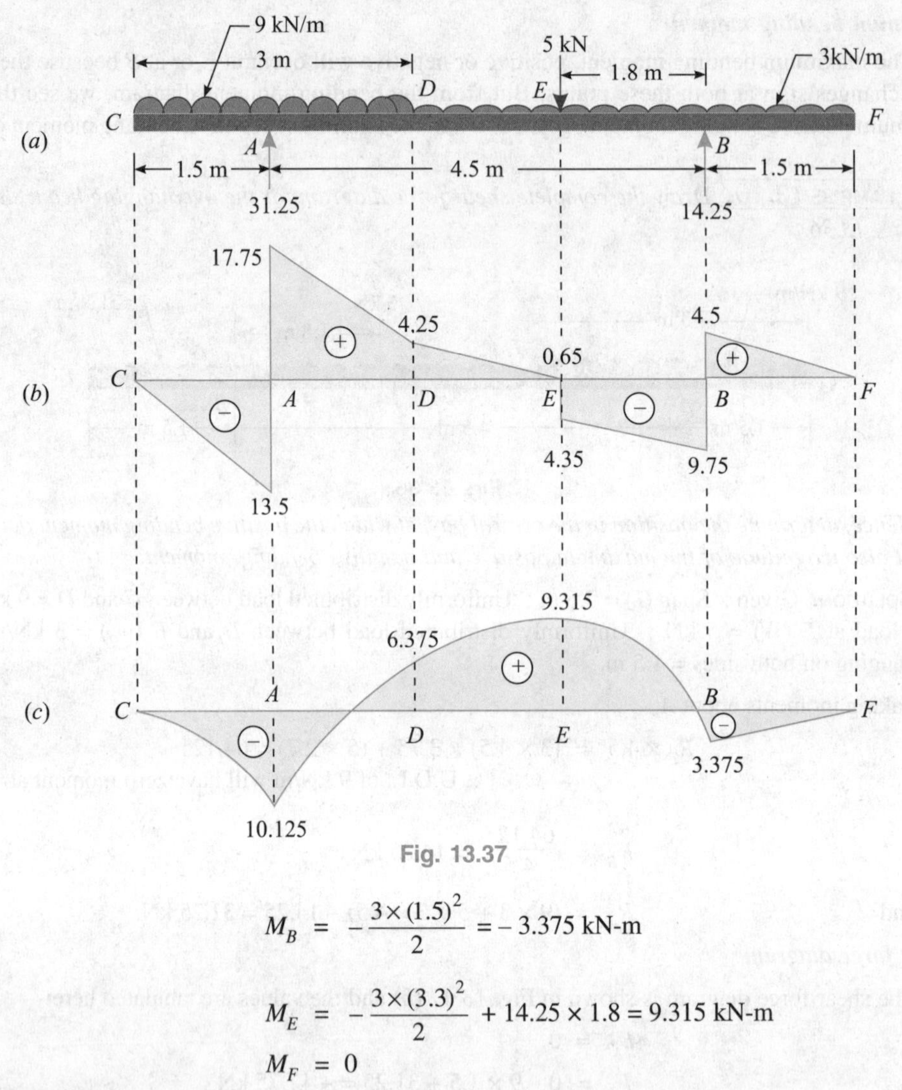


Fig. 13.37

$$M_B = -\frac{3\times(1.5)^2}{2} = -3.375 \text{ kN-m}$$

$$M_E = -\frac{3\times(3.3)^2}{2} + 14.25 \times 1.8 = 9.315 \text{ kN-m}$$

$$M_F = 0$$

***Maximum bending moment***

The maximum bending moment, positive or negative will occur at $A$, $E$ or $B$, because the shear force changes sign at all these three points. But from the bending moment diagram, we see that the maximum negative bending moment occurs at $A$ and the maximum positive bending moment occurs at $E$.

**Example 13.17.** *A simply supported beam with over-hanging ends carries transverse loads as shown in Fig. 13.38.*

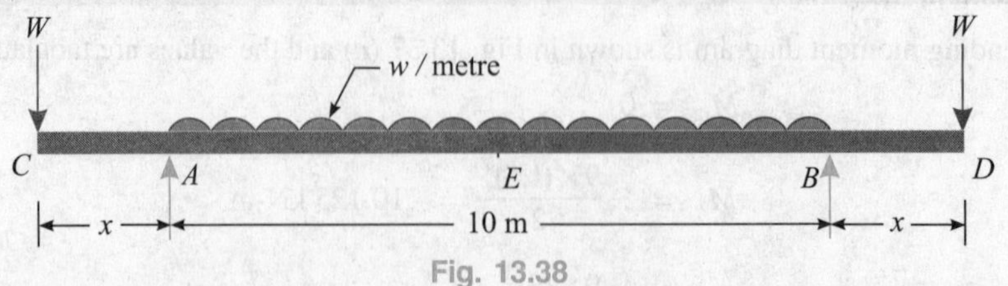


Fig. 13.38

*If W = 10 w, what is the overhanging length on each side, such that the bending moment at the middle of the beam, is zero? Sketch the shear force and bending moment diagrams.*

**SOLUTION.** Given : Span ($l$) = 10 m ; Point loads at $C$ and $D = W$ and uniformly distributed load between $A$ and $B = w$/metre.

Since the beam is symmetrically loaded, therefore, the two reactions (*i.e.*, $R_A$ and $R_B$) will be equal. From the geometry of the figure, we find that the reaction at $A$,

$$R_A = R_B = \frac{1}{2}(W + 10\,w + W) = W + 5\,w$$

$$= 10\,w + 5\,w = 15\,w \qquad (\because W = 10\,w)$$

***Overhanging length of the beam on each side***

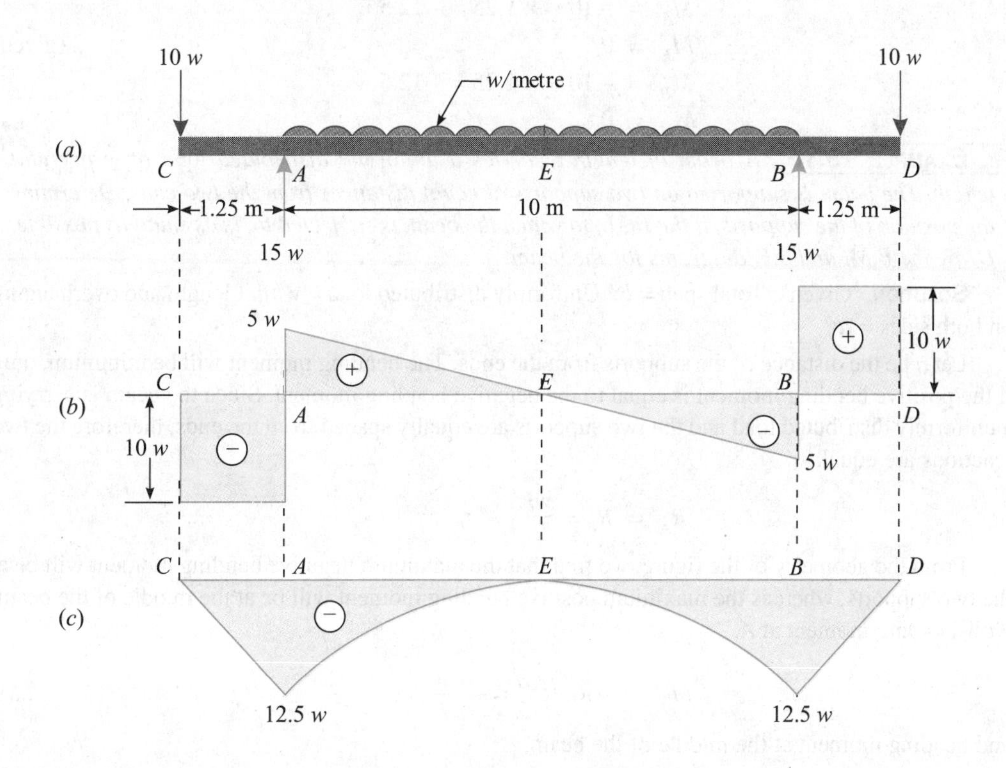


Fig. 13.39

We know that the bending moment at the middle of the beam $A$,

$$M_E = W(5 + x) + w \times 5 \times \frac{5}{2} - 15\,w \times 5$$

$$= 10\,w\,(5 + x) + 12.5\,w - 75\,w \qquad (\because W = 10\,w)$$

$$= 50\,w + 10\,wx - 62.5\,w$$

$$= 10\,wx - 12.5\,w \qquad ...(i)$$

Since the bending moment at the middle of the beam is zero, therefore equating the above equation to zero,

$$10\,wx - 12.5\,w = 0$$

$$\therefore \qquad x = \frac{12.5}{10} = 1.25 \text{ m} \qquad \textbf{Ans.}$$

***Shear force***

The shear force diagram is shown in Fig. 13.39 (*b*), and the values are tabulated here:

$$F_C = -10\,w$$
$$F_A = -10\,w + 15\,w = +5\,w$$
$$F_B = +5\,w - 10\,w + 15\,w = +10\,w$$
$$F_D = +10\,w$$

***Bending moment***

The bending moment diagram is shown in Fig. 13.39 (*c*) and the values are tabulated here:

$$M_C = 0$$
$$M_A = -10\,w \times 1.25 = -12.5\,w$$
$$*M_E = 0 \quad \text{...(given)}$$
$$M_B = -10\,w \times 1.25 = -12.5\,w$$
$$M_D = 0$$

**EXAMPLE 13.18.** *A beam of length l carries a uniformly distributed load of w per unit length. The beam is supported on two supports at equal distances from the two ends. Determine the position of the supports, if the B.M., to which the beam is subjected to, is as small as possible. Draw the B.M. and S.F. diagrams for the beam.*

**SOLUTION.** Given : Total span = $l$ ; Uniformly distributed load = $w$/unit length and overhanging on both sides = $a$

Let $a$ be the distance of the supports from the ends. The bending moment will be minimum, only if the positive bending moment is equal to the negative bending moment. Since the beam is carrying a uniformly distributed load and the two supports are equally spaced from the ends, therefore the two reactions are equal.

or
$$R_A = R_B = \frac{wl}{2}$$

From the geometry of the figure, we find that the maximum negative bending moment will be at the two supports, whereas the maximum positive bending moment will be at the middle of the beam. Now bending moment at $A$,

$$M_A = -wa \times \frac{a}{2} = -\frac{wa^2}{2} \quad \text{...}(i)$$

and bending moment at the middle of the beam,

$$M_M = R_A\left(\frac{l}{2} - a\right) - \left(\frac{wl}{2} \times \frac{l}{4}\right)$$
$$= \frac{wl}{2}\left(\frac{l}{2} - a\right) - \frac{wl^2}{8} \quad \text{...}(ii)$$

Equating (*i*) and (*ii*) and ignoring the nature of $M_A$,

$$\frac{wa^2}{2} = \frac{wl}{2}\left(\frac{l}{2} - a\right) - \frac{wl^2}{8}$$
$$a^2 = \frac{l^2}{2} - la - \frac{l^2}{4} = \frac{l^2}{4} - la$$

* The moment at $E$ (*i.e.*, $M_E$) may also be found out as discussed below:
$M_E = (10\,w \times 6.25) + (5\,w \times 2.5) - (15\,w \times 5) = 0$

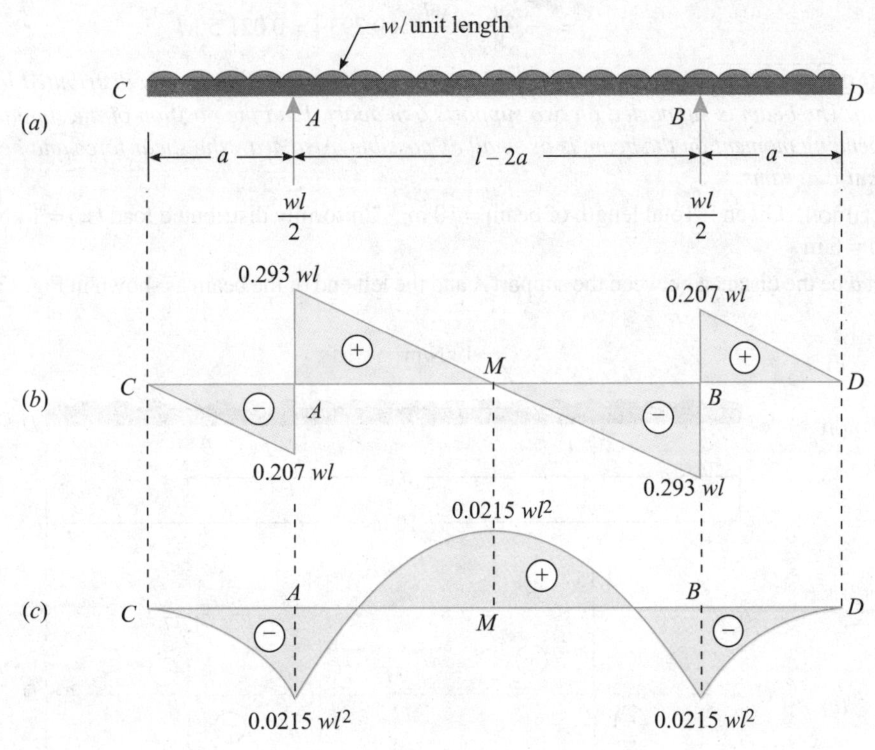


Fig. 13.40

or $\qquad a^2 + la - \frac{l^2}{4} = 0$

Solving it as a quadratic equation for $a$,

$$a = \frac{-l \pm \sqrt{l^2 + \frac{4 \times l^2}{4}}}{2} = \frac{-l \pm \sqrt{2\,l^2}}{2}$$

$$= 0.5\,l + 0.707\,l = 0.207\,l \qquad \text{(Taking + sign)}$$

***Shear force***

The shear force diagram is shown in Fig. 13.40 ($b$), and values are tabulated here:

$$F_C = 0$$
$$F_A = 0 - w \times 0.207\,l + 0.5\,wl = +\,0.293\,wl$$
$$F_M = +\,0.293\,wl - w \times 0.293\,l = 0$$
$$F_B = 0 - w \times 0.293\,l + 0.5\,wl = +\,0.207\,wl$$
$$F_D = +\,0.207\,wl - w \times 0.207\,l = 0$$

***Bending moment***

The bending moment diagram is shown in Fig. 13.40 ($c$) and the values are tabulated here:

$$M_C = 0$$

$$M_A = M_B = -\frac{wa^2}{2} = -\frac{w}{2}\,(0.207\,l)^2 = -\,0.0215\,wl^2$$

$$M_M = -\frac{wl}{2} \times \frac{l}{4} + \frac{wl}{2}\left(\frac{l}{2} - a\right) = -\frac{wl^2}{8} + \frac{wl}{2}\,(0.5\,l - 0.207\,l)$$

$$= -\frac{wl^2}{8} + \frac{wl}{2} \times 0.293\, l = 0.021\,5\, wl^2$$

**EXAMPLE 13.19.** *A horizontal beam 10 m long is carrying a uniformly distributed load of 1 kN/m. The beam is supported on two supports 6 m apart. Find the position of the supports, so that bending moment on the beam is as small as possible. Also draw the shear force and bending moment diagrams.*

**SOLUTION.** Given : Total length of beam = 10 m ; Uniformly distributed load ($w$) = 1 kN/m and span ($l$) = 6 m

Let $a$ be the distance between the support $A$ and the left end of the beam as shown in Fig. 13.41 ($a$).

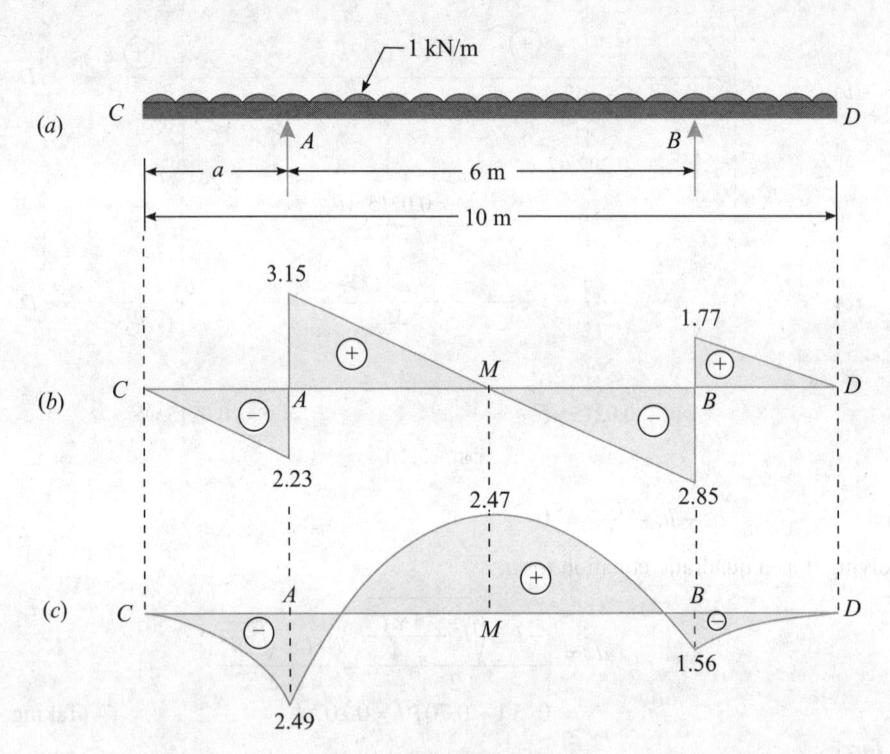


**Fig. 13.41**

Taking moments about ,

$$R_B \times 6 = 1 \times 10\,(5 - a) = 10\,(5 - a)$$

$$\therefore \qquad R_B = \frac{10\,(5-a)}{6} = \frac{5}{3}\,(5 - a)$$

and

$$R_A = 10 - \frac{5}{3}\,(5 - a) = \frac{5}{3}\,(1 + a)$$

From the geometry of the figure, we find that the maximum negative bending moment will be at either of the two supports and the maximum positive bending moment will be in the span $AB$. Let the maximum positive bending moment be at $M$ at a distance of $x$ from $C$.

Since the shear force at $M$ is zero, therefore

$$1 \times x - R_A = 0$$

$$\therefore \qquad x = R_A = \frac{5}{3}\,(1 + a)$$

We know that the bending moment at $A$,

$$M_A = -1 \times a \times \frac{a}{2} = -\frac{x^2}{2} \qquad ...(i)$$

and bending moment, where shear force is zero (*i.e.*, at a distance of $x$ from $C$),

$$M_M = 1 \times x \times \frac{x}{2} + R_A (x - a) = R_A (x - a) - \frac{x^2}{2}$$

$$= \frac{5}{3}(1+a)\left[\frac{5}{3}(1+a) - a\right] - \frac{1}{2}\left[\frac{5}{3}(1+a)\right]^2$$

...($\because$ Substituting the values of $R_A$ and $x$)

$$= \frac{5}{3}(1+a)\left[\frac{5}{3} + \frac{5a}{3} - a\right] - \frac{25}{18}(1+a)^2$$

$$= \frac{5}{3}(1+a)\frac{5}{3}\left[1 + a - \frac{3a}{5}\right] - \frac{25}{18}(1+a)^2$$

$$= \frac{25}{9}(1+a)\left[1 + \frac{2a}{5}\right] - \frac{25}{18}(1+a)^2$$

$$= \frac{25}{9}(1+a)\left[\left(\frac{5+2a}{5}\right) - \frac{1}{2}(1+a)\right]$$

$$= \frac{25}{9}(1+a)\left[\frac{10 + 4a - 5 - 5a}{10}\right]$$

$$= \frac{25}{9}(1+a)\left(\frac{5-a}{10}\right)$$

$$= \frac{5}{9}(1+a)\left[\frac{5-a}{2}\right] = \frac{5}{18}(1+a)(5-a)$$

$$= \frac{5}{18}(5 - a + 5a - a^2)$$

$$= \frac{5}{18}(5 + 4a - a^2) \qquad ...(ii)$$

Equating (*i*) and (*ii*) and ignoring the nature of $M_A$,

$$\frac{a^2}{2} = \frac{5}{18}(5 + 4a - a^2) = \frac{25}{18} + \frac{20a}{18} - \frac{5a^2}{18}$$

$$\therefore \quad a^2 = \frac{25}{9} + \frac{20a}{9} - \frac{5a^2}{9}$$

or $\quad 14a^2 - 20a - 25 = 0$

Solving it as a quadratic equation for $a$,

$$a = \frac{20 \pm \sqrt{(20)^2 + (4 \times 14 \times 25)}}{2 \times 14} = 2.23 \text{ m}$$

$$\therefore \quad x = \frac{5}{3}(1+a) = \frac{5}{3}(1 + 2.23) = 5.38 \text{ m}$$

Now reaction at $B$,

$$R_B = \frac{5}{3}(5-a) = \frac{5}{3}(5 - 2.23) = 4.62 \text{ kN}$$

and

$$R_A = \frac{5}{3}(1+a) = \frac{5}{3}(1 + 2.23) = 5.38 \text{ kN}$$

*Shear force diagram*

The shear force diagram is shown in Fig. 13.41 (*b*), and the values are tabulated here:

$$F_C = 0$$
$$F_A = 0 - 1 \times 2.23 + 5.38 = +3.15 \text{ kN}$$
$$F_B = +3.15 - 1 \times 6 + 4.62 = +1.77 \text{ kN}$$
$$F_D = +1.77 - 1.77 = 0$$

*Bending moment diagram*

The bending moment diagram is drawn in Fig. 13.41 (*c*), and the values are tabulated here:

$$M_C = 0$$
$$M_D = 0$$
$$M_A = -1 \times 2.23 \times \frac{2.23}{2} = -2.49 \text{ kN-m}$$
$$M_M = -1 \times 5.38 \times \frac{5.38}{2} + 5.38 \times 3.15 = 2.47 \text{ kN-m}$$
$$M_B = 1 \times 1.77 \times \frac{1.77}{2} = 1.56 \text{ kN-m}$$

## 13.17. Load and Bending Moment Diagrams from a Shear Force Diagram

Sometimes, instead of load diagram, a shear force diagram for a beam is given. In such cases, we first draw the actual load diagram and then the bending moment diagram. The load diagram for the beam may be easily drawn by keeping the following points in view:

1. If there is a sudden increase or decrease (*i.e.*, vertical line of the shear force diagram), it indicates that there is either a point load or reaction (*i.e.*, support) at that point.
2. If there is no increase or decrease in shear force diagram between any two points (*i.e.*, the shear force line is horizontal and consists of rectangle), it indicates that there is no loading between the two points.
3. If the shear force line is an inclined straight line between any two points, it indicates that there is a uniformly distributed load between the two points.
4. If the shear force line is a parabolic curve between any two points, it indicates that there is a uniformly varying load between the two points.

After drawing the load diagram, for the beam the bending moment diagram may be drawn as usual.

**EXAMPLE 13.20.** *The diagram shown in Fig. 13.42 is the shear force diagram in metric units, for a beam, which rests on two supports, one being at the left hand end.*

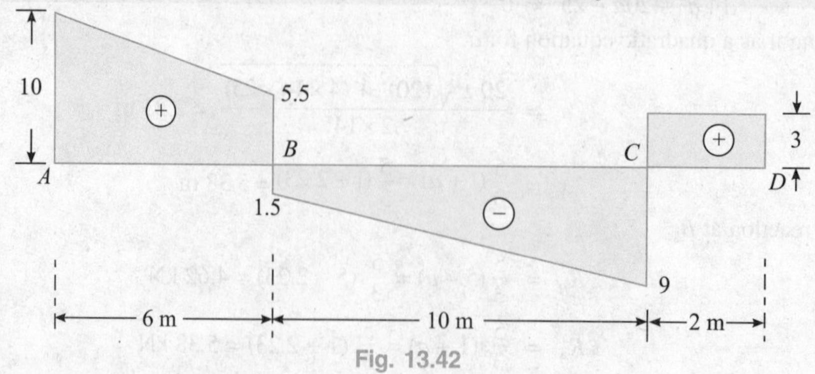


Fig. 13.42

*Deduce directly from the shear force diagram, (a) loading on the beam, (b) bending moment at 2 m intervals along the beam and (c) position of the second support. Also draw bending moment diagram for the beam and indicate the position and magnitude of the maximum value on it.*

**SOLUTION.** Given : Total length ($l$) = 18 m ; Shear force at $A$ = 10 kN and shear force at $D$ = 3 kN.

First of all, let us analyse the shear force diagram as discussed below:

1. ***At A***

We see that the shear force increases suddenly from 0 to 10 kN. Therefore there is a support reaction of 10 kN at *A*.

2. ***Between A and B***

We see that the shear force diagram has an inclined straight line between *A* and *B*. Therefore the beam is carrying a uniformly distributed load between *A* and *B*. We also see that there is a decrease of 10 – 5.5 = 4.5 kN shear force in 6 m length of beam. Therefore the beam carries a uniformly distributed load of 4.5/6 = 3/4 kN/m.

3. ***At B***

We see that the shear force has a sudden decrease of 5.5 + 1.5 = 7 kN. Thus there is a point load of 7 kN at *B*.

4. ***Between B and C***

We see that the shear force diagram has an inclined straight line between *B* and *C*. Therefore the beam is carrying a uniformly distributed load between *B* and *C*. We also see that there is a decrease of 9 – 1.5 = 7.5 kN shear force in 10 m length of beam. Therefore the beam carries a uniformly distributed load of 7.5/10 = 3/4 kN/m.

5. ***At C***

We see that the shear force has a sudden increase of 9 + 3 = 12 kN. Thus there is a support reaction of 12 kN at *C*.

6. ***Between C and D***

We see that the shear force diagram has a straight horizontal line between *C* and *D*. Therefore there is no load between *C* and *D*.

7. ***At D***

We see that the shear force decreases suddenly from + 3 kN to 0. Therefore there is a point load of 3 kN at *D*.

The load diagram is shown in Fig. 13.43 (*b*).

***Bending Moment***

Let us calculate bending moments at 2 meters interval along the beam.

$$M_0 = 0$$

$$M_2 = 10 \times 2 - \frac{3}{4} \times 2 \times 1 = 18.5 \text{ kN-m}$$

$$M_4 = 10 \times 4 - \frac{3}{4} \times 4 \times 2 = 34 \text{ kN-m}$$

$$M_6 = 10 \times 6 - \frac{3}{4} \times 6 \times 3 = 46.5 \text{ kN-m}$$

$$M_8 = 10 \times 8 - \frac{3}{4} \times 8 \times 4 - 7 \times 2 = 42 \text{ kN-m}$$

$$M_{10} = 10 \times 10 - \frac{3}{4} \times 10 \times 5 - 7 \times 4 = 34.5 \text{ kN-m}$$

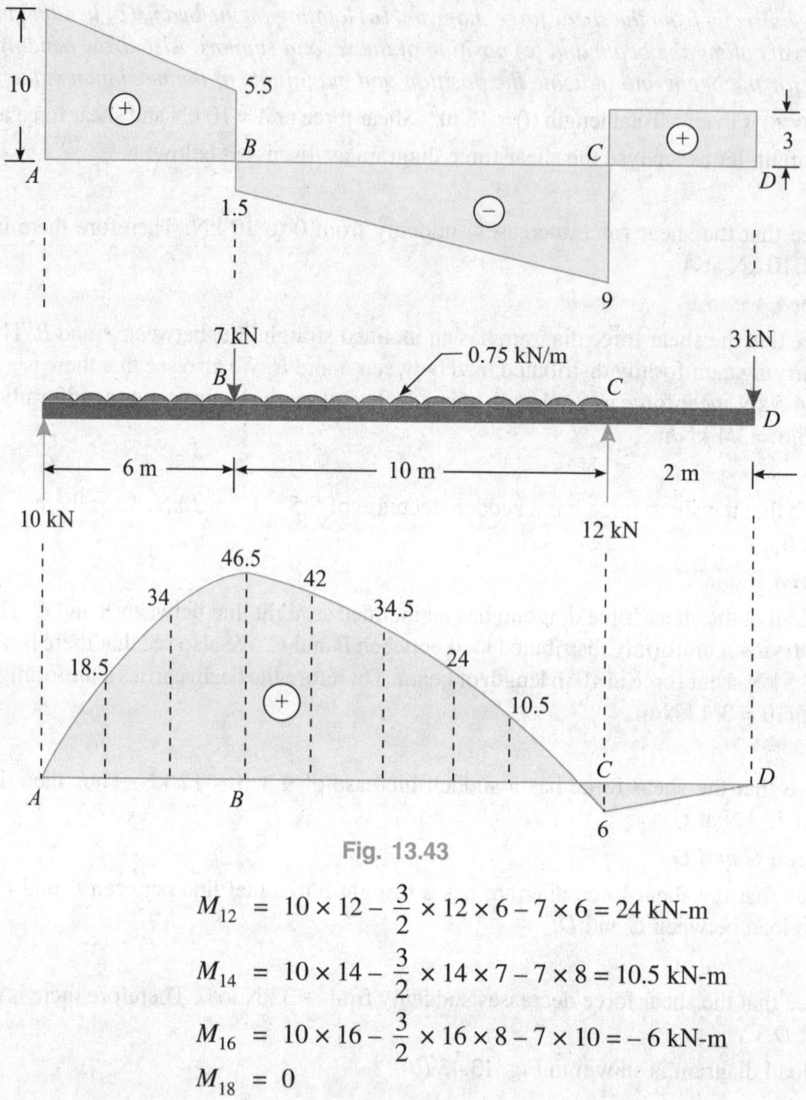


Fig. 13.43

$$M_{12} = 10 \times 12 - \frac{3}{2} \times 12 \times 6 - 7 \times 6 = 24 \text{ kN-m}$$

$$M_{14} = 10 \times 14 - \frac{3}{2} \times 14 \times 7 - 7 \times 8 = 10.5 \text{ kN-m}$$

$$M_{16} = 10 \times 16 - \frac{3}{2} \times 16 \times 8 - 7 \times 10 = -6 \text{ kN-m}$$

$$M_{18} = 0$$

***Maximum bending moment***

The maximum bending moment, positive or negative will occur at $B$ (*i.e.*, 6 m from $A$) and $C$ (*i.e.*, 16 m from $A$) because the shear force changes sign at both the points. But from the bending moment diagram, we see that maximum positive bending moment occurs at $B$ and the maximum negative bending moment at $C$. Now complete the diagram as shown in Fig. 13.43 (*c*).

**EXAMPLE 13.21.** *Figure 13.44 shows the shear force diagram of a loaded beam.*

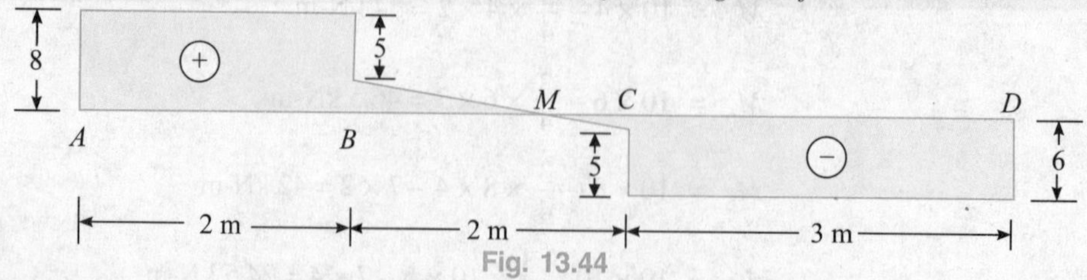


Fig. 13.44

*Find the loading on the beam and draw the bending moment diagram.*

**Solution.** Given : Total length ($L$) = 7 m ; Shear force at $A$ = 8 kN and shear force at $D$ = 6 kN

First of all, let us analyse the shear force diagram as discussed below:

1. ***At A***

We see that the shear force increase suddenly from 0 to 8 kN. Therefore there is a support reaction of 8 kN at $A$.

2. ***Between A and B***

We see that shear force diagram has a straight horizontal line between $A$ and $B$. Therefore there is no load between $A$ and $B$.

3. ***At B***

We see that the shear force has a sudden decrease of $8 - 3 = 5$ kN. Therefore there is a point load of 5 kN at $B$.

4. ***Between B and C***

We see that the shear force diagram has an inclined straight line between $B$ and $C$. Therefore the beam is carrying a uniformly distributed load between $B$ and $C$. We also see that there is a decrease of $3 + 1 = 4$ kN in 2 m length of the beam. Therefore the beam is carrying a uniformly distributed load of $4/2 = 2$ kN/m.

5. ***At C***

We see that the shear force has sudden decrease of $6 - 1 = 5$ kN. Therefore there is a point load of 5 kN at $C$.

6. ***Between C and D***

We see that the shear force has a straight horizontal line between $C$ and $D$. Therefore there is no load between $C$ and $D$.

7. ***At D***

We see that the shear force suddenly decreases from – 6 kN to 0. Therefore there is a section of 6 kN at $D$.

The load diagram is shown in Fig. 13.45.

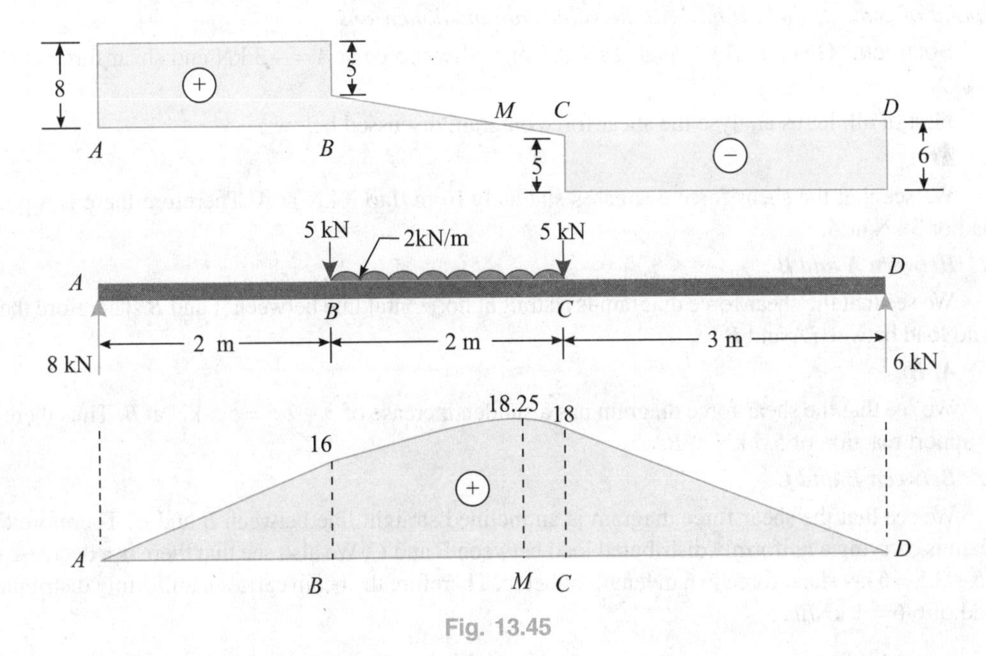


Fig. 13.45

***Bending moment diagram***

The bending moment diagram is shown in Fig. 13.45 and the values are tabulated here:

$$M_A = 0$$
$$M_B = 8 \times 2 = 16 \text{ kN-m}$$
$$M_C = 6 \times 3 = 18 \text{ kN-m}$$
$$M_D = 0$$

We know that the maximum bending moment will occur at $M$, where the shear force changes sign. Let $x$ be the distance between $B$ and $M$. From the geometry of the figure between $B$ and $M$,

$$\frac{x}{3} = \frac{2-x}{1} \quad \text{or} \quad x = 6 - 3x$$
$$4x = 6 \quad \text{or} \quad x = 1.5 \text{ m}$$

$$\therefore \quad M_M = (8 \times 3.5) - (5 \times 1.5) - (2 \times 1.5 \times \frac{1.5}{2})$$
$$= 18.25 \text{ kN-m}$$

**EXAMPLE 13.22.** *Shear force diagram for a loaded beam is shown in Fig. 13.46.*

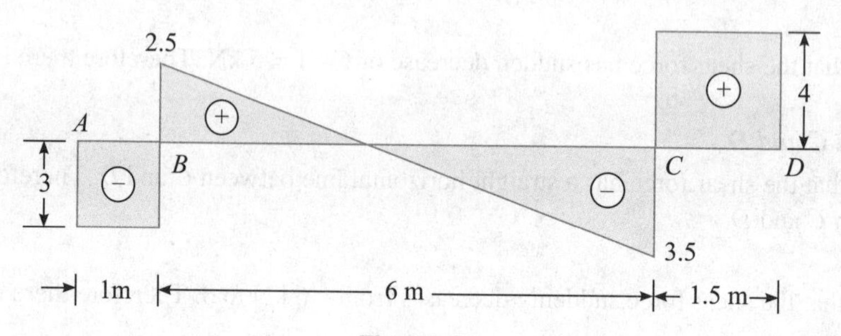


Fig. 13.46

*Determine the loading on the beam and bence draw bending moment diagram. Locate the point of contraflexure, if any. All the values are in kilonewtons.*

**SOLUTION.** Given : Total span ($L$) = 8.5 m ; Shear force at $A$ = – 3 kN and shear force at $D$ = + 4 kN

First of all, let us analyse the shear force diagram discussed below:

1. ***At A***

We see that the shear force decreases suddenly from 0 to 3 kN at $A$. Therefore there is a point load of 3 kN at $A$.

2. ***Between A and B***

We see that the shear force diagram is a straight horizontal line between $A$ and $B$. Therefore there is no load between $A$ and $B$.

3. ***At B***

We see that the shear force diagram has a sudden increase of 3 + 2.5 = 5.5 kN at $B$. Thus there is a support reaction of 5.5 kN at $B$.

4. ***Between B and C***

We see that the shear force diagram is an inclined straight line between $B$ and $C$. Therefore the beam is carrying a uniformly distributed load between $B$ and $C$. We also see that there is a decrease of 2.5 + 3.5 = 6 kN shear force in 6 m length of beam. Therefore the beam carries a uniformly distributed load of 6/6 = 1 kN/m.

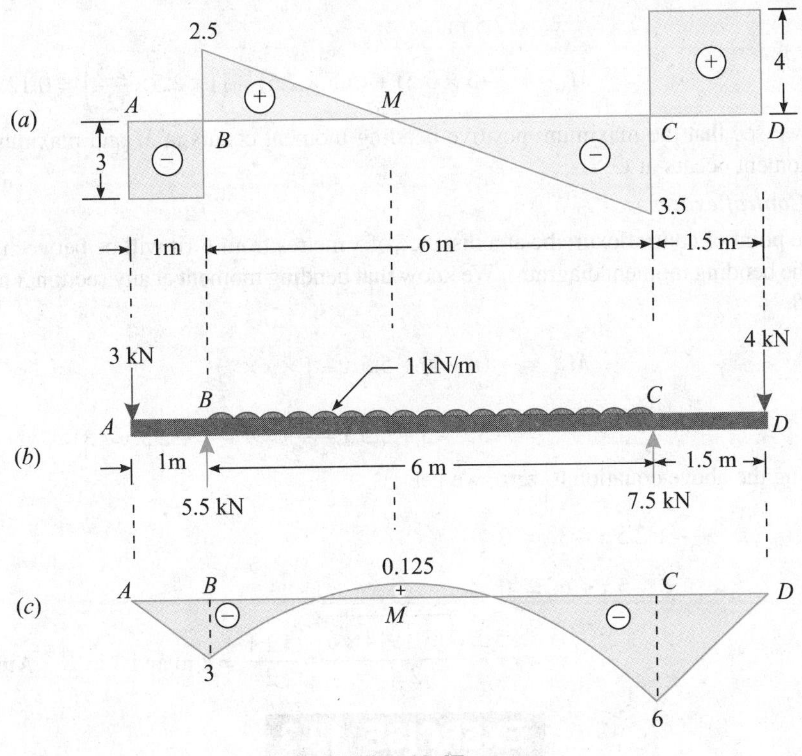


Fig. 13.47

**5.** ***At C***

We see that the shear force diagram has a sudden increase of 3.5 + 4 = 7.5 kN. Thus there is a support reaction of 7.5 kN at *C*.

**6.** ***Between C and D***

We see that the shear force diagram is a straight horizontal line between *C* and *D*. Therefore there is no load between *C* and *D*.

**7.** ***At D***

We see that the shear force decreases suddenly from + 4 kN to 0. Therefore there is a point load of 4 kN at *D*.

The load diagram is shown in Fig. 13.47 (*b*).

***Bending moment diagram***

The bending moment diagram is shown in Fig. 13.47 (*c*) and the values are tabulated here:

$$M_A = 0$$
$$M_B = -3 \times 1 = -3 \text{ kN-m}$$
$$M_C = -4 \times 1.5 = -6 \text{ kN-m}$$
$$M_D = 0$$

***Maximum bending moment***

The maximum bending moment, positive or negative will occur at *B*, *M* or *C* because shear force changes sign at all three points. Let $x$ be the distance between *B* and *M*. From the geometry of the figure between *B* and *C*,

$$\frac{x}{2.5} = \frac{6-x}{3.5}$$

$$3.5\,x = 15 - 2.5\,x$$

or $$x = 2.5 \text{ m}$$

$\therefore$ $$M_M = -(3 \times 3.5) + (5.5 \times 2.5) - \left(1 \times 2.5 \times \frac{2.5}{2}\right) = 0.125 \text{ kN-m}$$

Thus we see that the maximum positive bending moment occurs at $M$ and maximum negative bending moment occurs at $C$.

***Points of Contraflexures***

Let the point of contraflexure be at a distance of $x$ metres from $B$ (it will be between $B$ and $C$ as is seen in the bending moment diagram). We know that bending moment at any section $X$ at a distance of $x$ from $B$,

$$M_X = -(x+1) + 5.5\,x - 1 \times x \times \frac{x}{2}$$

$$= -3\,x - 3 + 5.5\,x - \frac{x^2}{2} = -\frac{x^2}{2} + 2.5\,x - 3$$

Equating the above equation to zero, we get

$$-\frac{x^2}{2} + 2.5\,x - 3 = 0$$

or $$x^2 - 5\,x + 6 = 0$$

$$x = \frac{5 \pm \sqrt{(5)^2 - 4 \times 6}}{2} = \frac{5 \pm 1}{2} = 2 \text{ m and } 3 \text{ m} \qquad \textbf{Ans.}$$

## EXERCISE 13.3

1. A beam 6 m long rests on two supports 5 m apart. The right end is overhanging by 1 m. The beam carries a uniformly distributed load of 1.5 kN/m over the entire length of the beam. Draw S.F. and B.M. diagram and find the amount and position of maximum bending moment.
[**Ans.** 4.32 kN-m at 2.4 m from left end]

2. Draw the shear force and bending moment diagrams, for the overhanging beam carrying loads as shown in Fig. 13.48.

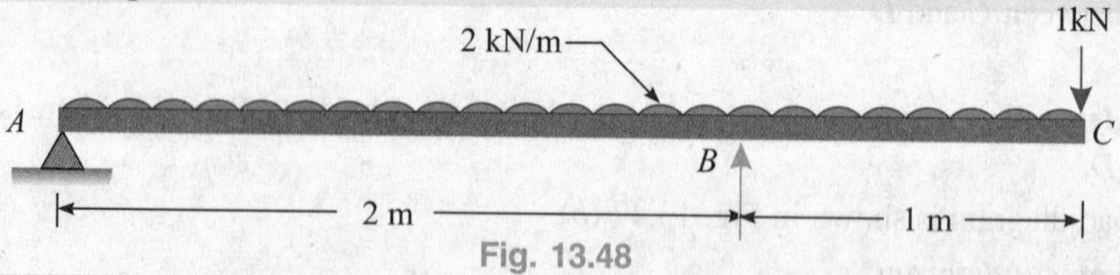


Fig. 13.48

Mark the values of the principal ordinates and locate the point of contraflexure, if any.
[**Ans.** 1 m from $A$]

3. A beam 10 m long carries load as shown in Fig. 13.49.

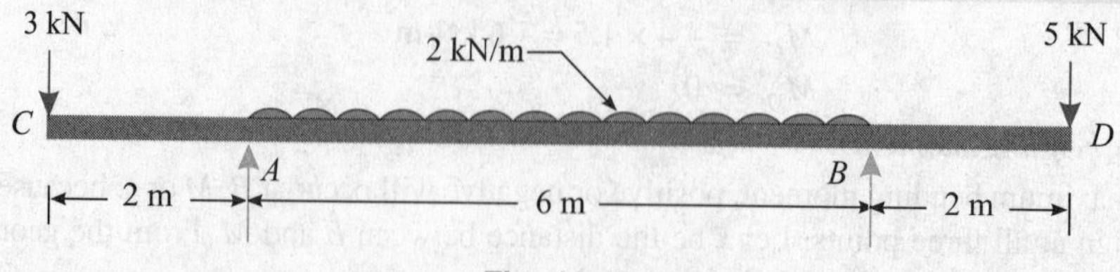


Fig. 13.49

Draw shear force and bending moment diagrams for the beam and determine the points of contraflexures, if any.
[**Ans.** 3.62 m and 5.72 m from $C$]

**4.** A beam $AB$ 20 metres long, carries a uniformly distributed load 0.6 kN/m together with concentrated loads of 3 kN at left hand end $A$ and 5 kN at right hand-end $B$ as shown in Fig. 13.50.

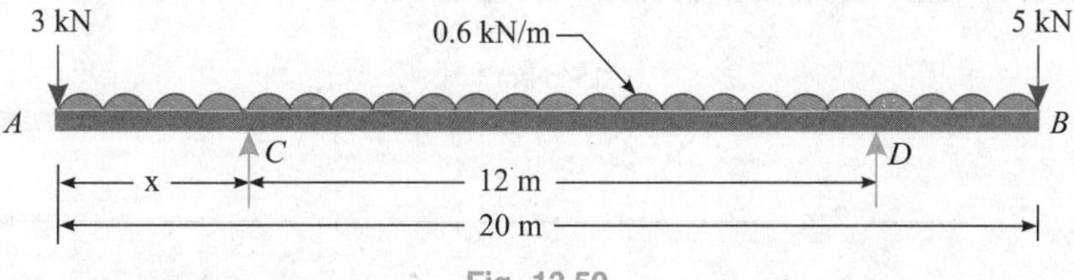


Fig. 13.50

The props are so located that the reaction is the same at each support. Determine the position of the props and draw bending moment and shear force diagrams. Mark the values of the maximum bending moment and maximum shear force. **[Ans.** 5 m ; 17 m]

## 13.18. Beams Subjected to a Moment

Sometimes, a beam is subjected to a clockwise or anticlockwise moment (or couple) at a section. In such a case, the magnitude of the moment is considered while calculating the reactions. The bending moment at the section of the couple changes suddenly in magnitude equal to that of the couple. This may also be found out by calculating the bending moment separately with the help of both the reactions. Since the bending moment does not involve any load, therefore the shear force does not change at the section of couple.

**Notes:** 1. A clockwise moment (called positive moment) causes negative shear force over the beam and positive bending moment at the section. Similarly, an anticlockwise moment (called negative moment) causes positive shear force over the beam and negative bending moment at the section.

2. The bending moment will suddenly increase due to clockwise moment and decrease due to anticlockwise moment at the point of its application when we move from left to right along the beam.

**Example 13.23.** *A simply supported beam of 5 m span is subjected to a clockwise moment of 15 kN-m at a distance of 2 m from the left end as shown in Fig. 13.51.*

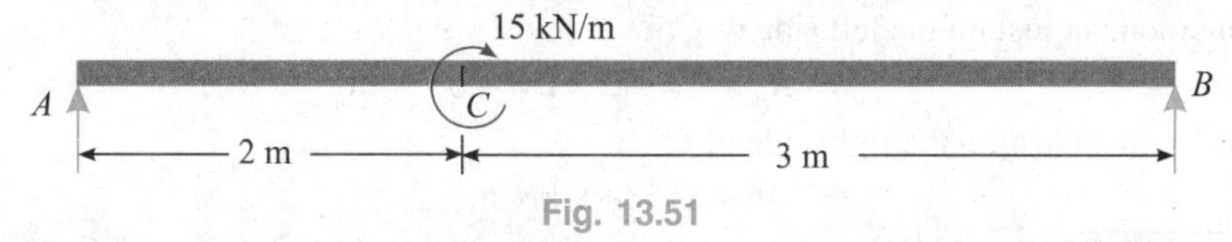


Fig. 13.51

*Draw the shear force and bending moment diagrams for the beam.*

**Solution.** Given : Span ($l$) = 5 m and couple at $C$ ($\mu$) = 15 kN-m

We know that the tendency of the moment is to uplift the beam from its support $A$ and to depress it at its support $B$. It is thus obvious that the reaction at $A$ will be downwards and that at $B$ will be upwards as shown in Fig. 13.52 (*a*).

Taking moments about ,

$$R_B \times 5 = 15 \qquad \text{...(Since the beam is subjected to moment only)}$$

$$\therefore \qquad R_B = \frac{15}{5} = 3 \text{ kN} \quad \text{(upwards)}$$

Since there is no external loading on the beam, therefore the reaction at $A$ will be of the same magnitude but in opposite direction. Therefore reaction at $A$,

$$R_A = 3 \text{ kN} \quad \text{(downwards)}$$

*Shear force diagram*

We know the shear force is constant from $A$ to $B$ and is equal to – 3 kN (because of downward reaction at $A$ or upward reaction at $B$) as shown in Fig. 13.52 ($b$).

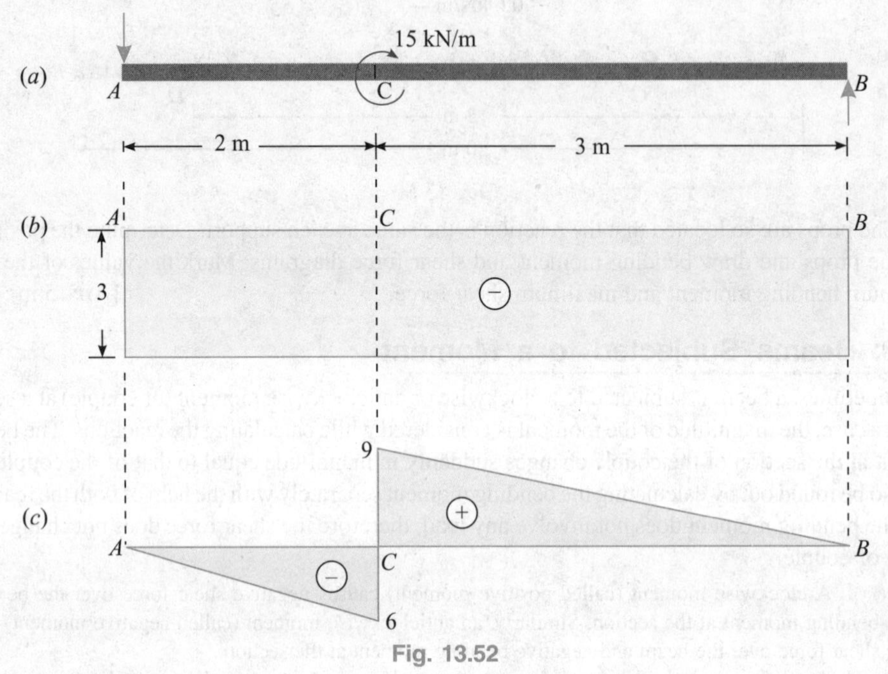


Fig. 13.52

*Bending moment diagram*

The bending moment diagram is shown in Fig. 13.52 ($c$) and the values are tabulated here:

$$M_A = 0$$

$$M_B = 0$$

Bending moment just on the left side of $C$,

$$= R_A \times 2 = -3 \times 2 = -6 \text{ kN-m}$$

and bending moment just on the right side of $C$*

$$= -6 + 15 = +9 \text{ kN-m}$$

**EXAMPLE 13.24.** *A simply supported beam of span 2.5 m is subjected to a uniformly distributed load and a clockwise couple as shown in Fig. 13.53.*

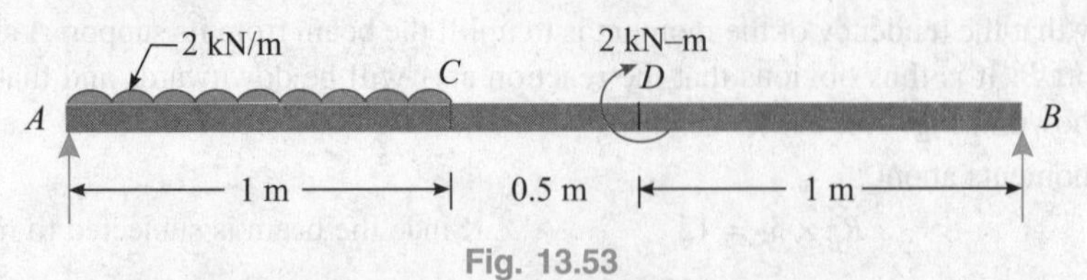


Fig. 13.53

*Draw the shear force and bending moment diagrams for the beam.*

**SOLUTION.** Given : Span ($l$) = 2.5 m ; Uniformly distributed load between $A$ and $C$ ($w$)=2 kN/m and couple at $D$ ($\mu$) = 2 kN-m

* At $C$, the bending moment will suddenly increase due to clockwise moment at $C$. The bending moment just on the right side $C$ may also be found out from the reaction $R_B$, *i.e.*, $= R_B \times 3 = +3.0 \times 3 = +9.0$ kN-m .

First of all, let us find out the reactions $R_A$ and $R_B$. Taking moments about $A$ and equating the same,

$$R_B \times 2.5 = \left(2 \times 1 \times \frac{1}{2}\right) + 2 = 3 \quad \text{...(+ 3 due to clockwise moment)}$$

$\therefore$ $R_B = 3/2.5 = 1.2$ kN

and RA $= (2 \times 1) - 1.2 = 0.8$ kN

***Shear force diagram***

The shear force diagram is shown in Fig. 13.54 (*b*) and the values are tabulated here:

$$F_A = +R_A = +8 \text{ kN}$$
$$F_C = +0.8 - (2 \times 1) = -1.2 \text{ kN}$$
$$F_B = -1.2 \text{ kN}$$

***Bending moment diagram***

The bending moment diagram is shown in Fig. 13.54 (*c*) and the values are tabulated here:

$$M_A = 0$$
$$M_C = (0.8 \times 1) - (2 \times 1 \times 0.5) = -0.2 \text{ kN-m}$$
$$M_D = (0.8 \times 1.5) - (2 \times 1 \times 1) = -0.8 \text{ kN-m}$$

...(With the help of $R_A$)

$= 1.2 \times 1 = 1.2$ kN-m ...(With the help of $R_B$)

We know that maximum bending moment will occur either at $E$ where shear force changes sign or at $D$ due to couple. Let $x$ be the distance between $A$ and $E$. From the geometry of the figure between $A$ and $C$, we find that

$$\frac{x}{0.8} = \frac{1-x}{1.2}$$

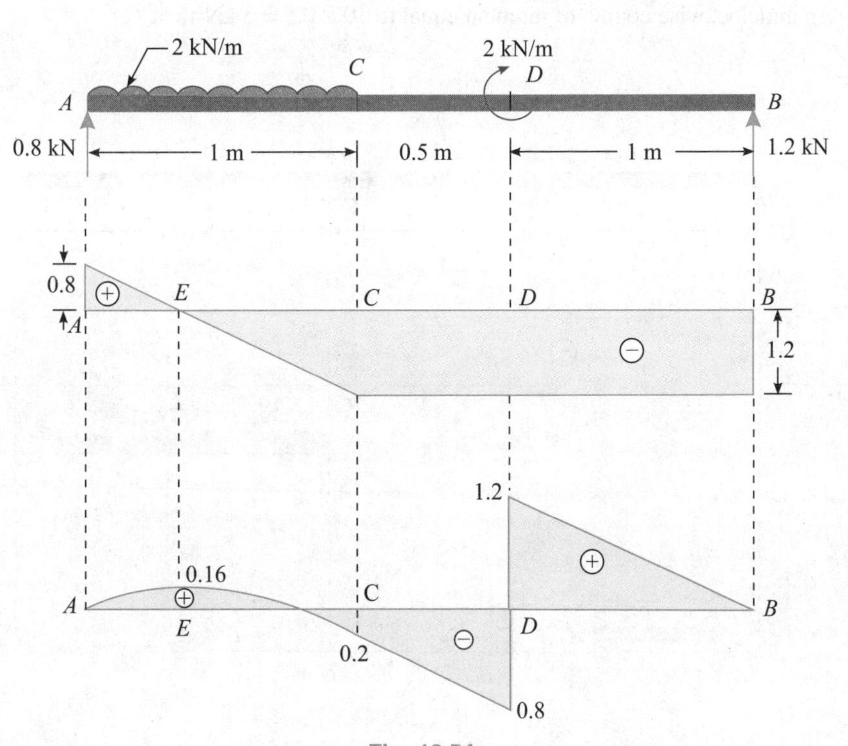


Fig. 13.54

or $$1.2\,x = 0.8 - 0.8\,x$$

$$2\,x = 0.8 \quad \text{or} \quad x = \frac{0.8}{2} = 0.4 \text{ m}$$

$$\therefore \quad M_E = (0.8 \times 0.4) - \left(2 \times 0.4 \times \frac{0.4}{2}\right) = +\,0.16 \text{ kN-m}$$

From the above two values of $M_D$, we find that it will suddenly increase from – 0.8 kN-m to + 1.2 kN-m due to the clockwise moment of 2 kN-m,

$$M_B = 0$$

**EXAMPLE 13.25.** *A simply supported beam 5 metres long carries a load of 10 kN on a bracket welded to the beam as shown in Fig. 13.55.*

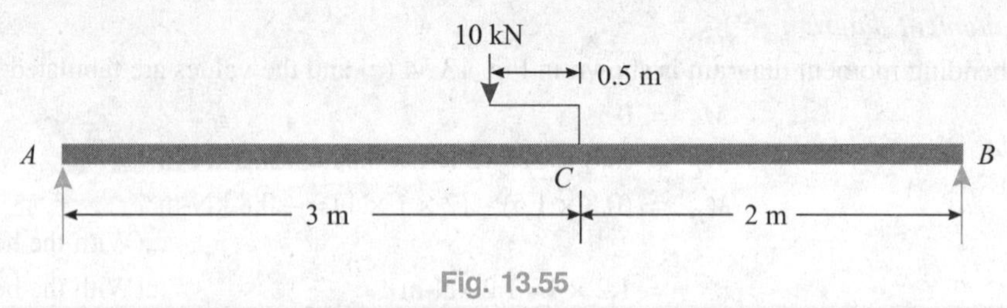


Fig. 13.55

*Draw the shear force and bending moment diagrams for the beam.*

**SOLUTION.** Given : Span ($l$) = 5 m and load on the bracket at $C$ = 10 kN.

It will be interesting to know that the 10 kN load, applied on the bracket will have the following two effects:

1. Vertical load of 10 kN at $C$,
2. An anticlockwise couple of moment equal to 10 × 0.5 = 5 kN-m at $C$.

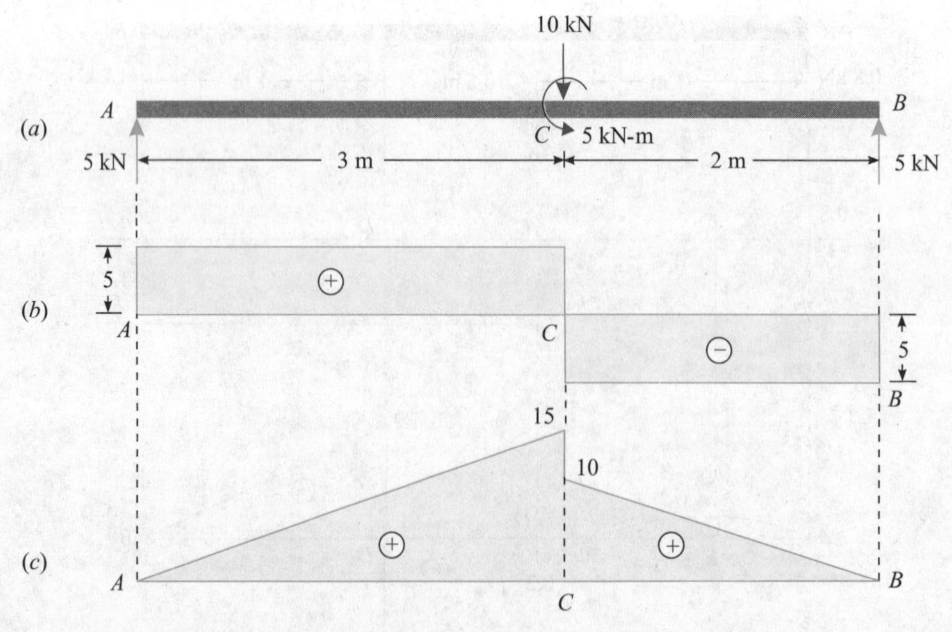


Fig. 13.56

Now the shear force and bending moment diagrams should be drawn by combining the above two mentioned effects as shown in Fig. 13.56 (*a*). First of all, let us find out the reactions $R_A$ and $R_B$. Taking moments about *A* and equating the same,

$$R_B \times 5 = (10 \times 3) - 5 = 25 \qquad \text{...(– 5 due to anticlockwise moment)}$$

$$\therefore \qquad R_B = 25/5 = 5 \text{ kN}$$

and $$R_A = 10 - 5 = 5 \text{ kN}$$

***Shear force diagram***

The shear force diagram is shown in Fig. 13.56 (*b*) and the values are tabulated here:

$$F_A = +R_A = +5 \text{ kN}$$

$$F_C = +5 - 10 = -5 \text{ kN}$$

$$F_B = -5 \text{ kN}$$

***Bending moment diagram***

The bending moment diagram is shown in Fig. 13.56 (*c*) and the values are tabulated here:

$$M_A = 0$$

$$M_C = 5 \times 3 = 15 \text{ kN-m} \qquad \text{...(With the help of } R_A)$$

$$= 5 \times 2 = 10 \text{ kN-m} \qquad \text{...(With the help of } R_B)$$

$$M_B = 0$$

From the above two values of $M_C$ we find that it will suddenly decrease from 15 kN-m to 10 kN-m due to the anticlockwise moment of 5 kN-m.

**EXAMPLE 13.26.** *A beam is loaded as shown in Fig. 13.57.*

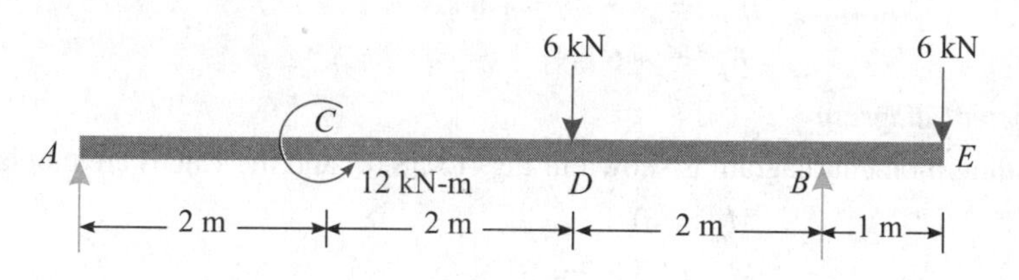


Fig. 13.57

*Construct the shear force and bending moment diagrams for the beam and mark the values of the important ordinates.*

**SOLUTION.** Given : Span ($l$) = 7 m ; Couple at *C* ($\mu$) = 12 kN-m ; Point load at *D* ($W_1$) = 6 kN and point load at *E* ($W_2$) = 6 kN

Taking moments about *A*,

$$R_B \times 6 = (6 \times 4) + (6 \times 7) - 12 = 54$$

...(– 12 due to anticlockwise moment)

$$R_B = \frac{54}{6} = 9 \text{ kN}$$

$$\therefore \qquad R_A = (6 + 6) - 9 = 3 \text{ kN}$$

***Shear force diagram***

The shear force diagram is shown in Fig. 13.58 (*b*) and the values are tabulated here:

$$F_A = +3 \text{ kN}$$

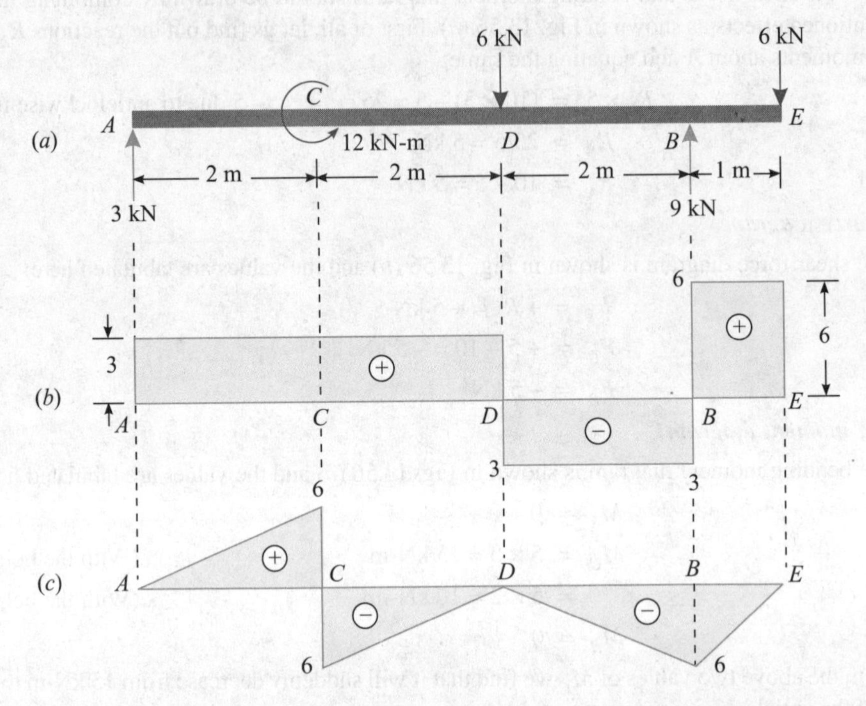


Fig. 13.58

$$F_D = +3 - 6 = -3 \text{ kN} \quad \text{...(With the help of } R_A)$$

$$F_B = -3 + 9 = +6 \text{ kN}$$

$$F_E = +6 \text{ kN}$$

***Bending moment diagram***

The bending moment diagram is shown in Fig. 13.58 (*c*) and the values are tabulated here:

$$M_A = 0$$

$$M_C = 3 \times 2 = 6 \text{ kN-m}$$

$$M_D = 3 \times 4 - 12 = 0$$

$$M_B = -6 \times 1 = -6 \text{ kN-m}$$

$$M_E = 0$$

At *C*, the bending moment will suddenly decrease from 6 kN-m to 6 – 12 = – 6 kN-m because of anticlockwise couple as shown in Fig. 13.58 (*c*).

## 13.19. Beams Subjected to Inclined Loads

In the previous articles, we have been discussing the cases, when the load used to act at right angles to the axis of the beam. But in actual practice, there may be cases when a beam is subjected to inclined loads. These inclined loads are resolved at right angles and along the axis of the beam. A little consideration will show that the transverse components (*i.e.*, components, which are resolved at right angles to the axis of the beam) will cause shear force and bending moments. The axial components (*i.e.*, components, which are resolved along the axis of the beam) will cause thrust *i.e.*, pulls or pushes in the beam, depending upon its end position.

In such cases, one end of the beam is hinged, whereas the other is simply supported or supported on rollers. The hinged end will be subjected to horizontal thrust equal to the unbalanced horizontal force of the axial components of the inclined loads. In such cases, like shear force and bending moment diagrams, an axial force diagram is drawn, which represents the horizontal thrust. The general practice, to draw the axial force diagram is that the tensile force is taken as positive, whereas the compressive force as negative.

**EXAMPLE 13.27.** *Analyse the beam shown in Fig. 13.59 and draw the bending moment and shear force diagrams.*

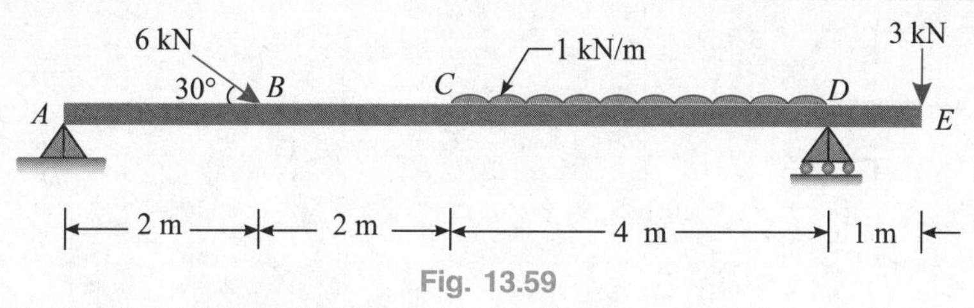


Fig. 13.59

*Locate the points of contraflexure, if any.*

**SOLUTION.** Given : Span $l = 9$ m ; Inclined load at $B = 6$ N ; Uniformly distributed load between $C$ and $D$ ($w$) = 1 kN/m and point load at $E$ = 3 kN.

Resolving vertically the force of 6 kN at $B$

$$= 6 \sin 30° = 6 \times 0.5 = 3 \text{ kN}$$

and now resolving horizontally the force of 6 kN at $B$

$$= 6 \cos 30° = 6 \times 0.866 = 5.196 \text{ kN}$$

*Taking moments about $A$, $R_D \times 8 = (3 \times 9) + (1 \times 4 \times 6) + (3 \times 2) = 57$

$$\therefore \quad R_D = \frac{57}{8} = 7.125 \text{ kN}$$

and $$R_A = (3 + 4 + 3) - 7.125 = 2.875 \text{ kN}$$

The load diagram and reactions are shown in Fig. 13.60 (*a*).

***Shear force diagram***

The shear force diagram is shown in Fig. 13.60 (*b*) and the values are tabulated here:

$$F_A = +\,2.875 \text{ kN}$$
$$F_B = +\,2.875 - 3 = -\,0.125 \text{ kN}$$
$$F_C = -\,0.125 \text{ kN}$$
$$F_D = -\,0.125 - (1 \times 4) + 7.125 = +\,3 \text{ kN}$$
$$F_E = +\,3 \text{ kN}$$

***Bending moment diagram***

The bending moment diagram is shown in Fig. 13.60 (*c*) and the values are tabulated here:

$$M_A = 0$$
$$M_B = 2.875 \times 2 = 5.75 \text{ kN-m}$$
$$M_C = (2.875 \times 4) - (3 \times 2) = 5.5 \text{ kN-m}$$
$$M_D = -\,3 \times 1 = -\,3 \text{ kN-m}$$

* The moment of axial component *i.e.*, horizontal component of the 6 kN force will have no moment about $A$.

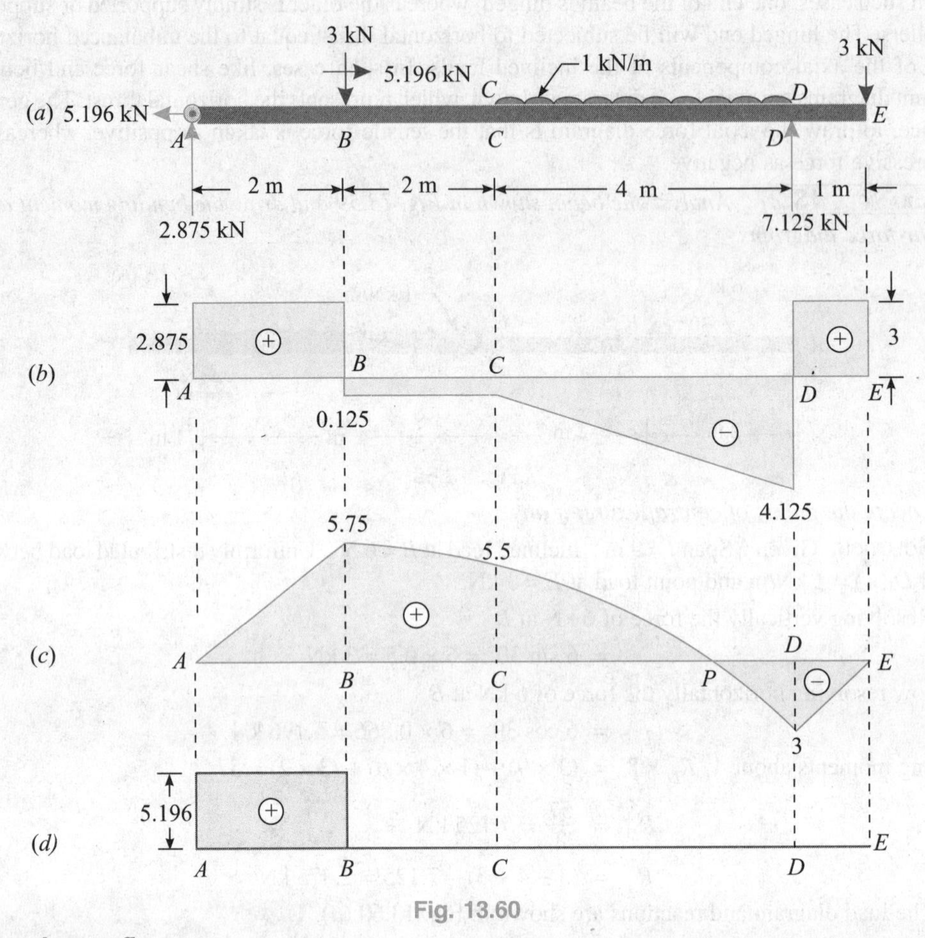


Fig. 13.60

***Point of contraflexure***

Let the point of contraflexure ($P$) be at a distance of $x$ from $D$ (It will be between $C$ and $D$ as is seen in the bending moment diagram). We know that the bending moment at any section $X$ in $CD$ at a distance $x$ from $D$,

$$M_X = 3(x+1) + \left(1 \times x \times \frac{x}{2}\right) - 7.125\,x$$

Equating the above equation to zero,

$$3(x+1) + \frac{x^2}{2} - 7.125\,x = 0$$

$$3x + 3 + \frac{x^2}{2} - 7.125\,x = 0$$

$$\frac{x^2}{2} - 4.125\,x + 3 = 0$$

$$x^2 - 8.25\,x + 6 = 0$$

Solving it as a quadratic equation for $x$,

$$x = \frac{8.25 \pm \sqrt{(8.25)^2 - (4 \times 6)}}{2} = 0.8 \text{ m} \qquad \textbf{Ans.}$$

*Axial force diagram*

From the load diagram, we see that horizontal reaction at $A$ (being a hinged end) is equal to 5.196 kN (←). Therefore the section $AB$ of the beam is subjected to an axial tensile force ($A_{AB}$) of 5.196 kN. The beam from $B$ to $E$ is not subjected to any axial force. The axial force diagram is drawn in Fig. 13.60 (*d*).

**EXAMPLE 13.28.** *A horizontal beam AB 6 m long is hinged at A and freely supported at B. The beam is loaded as shown in Fig. 13.61.*

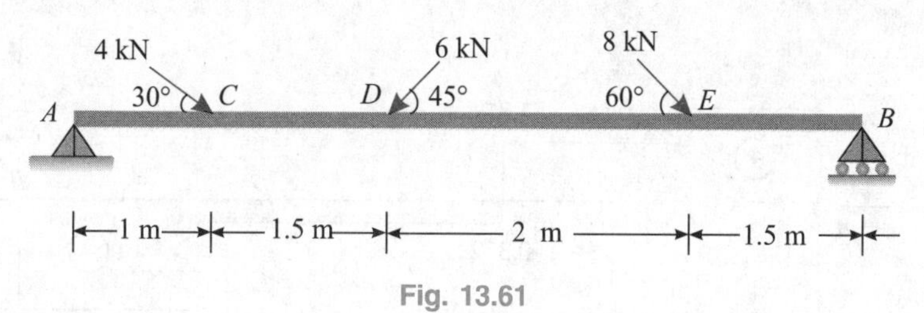


Fig. 13.61

*Draw the shear force, bending moment and thrust diagrams for the beam.*

**SOLUTION.** Given : Span ($l$) = 6 m ; Inclined load of $C$ = 4 kN ; Inclined load of $D$ = 6 kN and inclined load of $E$ = 8 kN.

Resolving vertically the force of 4 kN at $C$

$$= 4 \sin 30° = 4 \times 0.5 = 2 \text{ kN}$$

and now resolving horizontally the force of 4 kN at $C$

$$= 4 \cos 30° = 4 \times 0.866 = 3.464 \text{ kN}$$

Similarly, resolving vertically the force of 6 kN at $D$

$$= 6 \sin 45° = 6 \times 0.707 = 4.242 \text{ kN}$$

and now resolving horizontally the force of 6 kN at $D$

$$= 6 \cos 45° = 6 \times 0.707 = 4.242 \text{ kN}$$

Similarly, resolving vertically the force of 8 kN at $E$

$$= 8 \sin 60° = 8 \times 0.866 = 6.928 \text{ kN}$$

and now resolving horizontally the force of 8 kN at $E$

$$= 8 \cos 60° = 8 \times 0.5 = 4 \text{ kN } (\rightarrow)$$

Taking moments about $A$,

$$R_B \times 6 = (2 \times 1) + (4.242 \times 2.5) + 6.928 \times 4.5 = 43.78$$

$$\therefore \quad R_B = \frac{43.78}{6} = 7.3 \text{ kN}$$

and

$$R_A = 2 + 4.242 + 6.928 - 7.3 = 5.87 \text{ kN}$$

The load diagram and reactions are shown in Fig. 13.62 (*a*).

***Shear force diagram***

The shear force diagram is shown in Fig. 13.62 (*b*) and the values are tabulated here:

$$F_A = +5.87 \text{ kN}$$

$$F_C = +5.87 - 2 = +3.87 \text{ kN}$$

$$F_D = +3.87 - 4.242 = -0.372 \text{ kN}$$

$$F_E = -0.372 - 6.928 = -7.3 \text{ kN}$$

$$F_B = -7.3 + 7.3 = 0$$

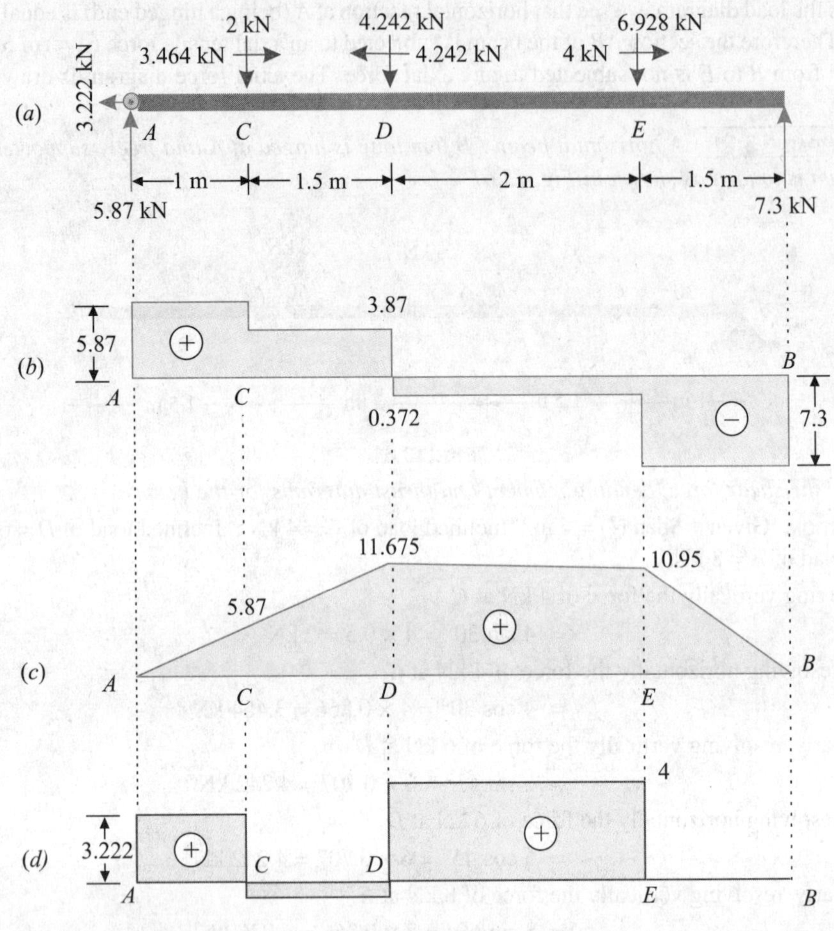


Fig. 13.62

***Bending moment diagram***

The bending moment diagram is shown in Fig. 13.62 (*c*) and the values are tabulated here:

$$M_A = 0$$
$$M_C = 5.87 \times 1 = 5.87 \text{ kN-m}$$
$$M_D = 5.87 \times 2.5 - 2 \times 1.5 = 11.675 \text{ kN-m}$$
$$M_E = 7.3 \times 1.5 = 10.95 \text{ kN-m}$$
$$M_B = 0$$

***Maximum bending moment***

It will occur at *D*, where shear force changes sign. Thus we see that maximum bending moment occurs at *D*.

***Axial force diagram***

From the load diagram, we see that the horizontal reaction at *A* (being a hinged end) is $\overleftarrow{3.464} + \overleftarrow{4.0} - \overrightarrow{4.242} = 3.222$ kN (←) The axial force diagram is shown in Fig. 13.62 (*d*) and the values are tabulated here:

$$A_{AC} = 3.222 \text{ kN (Tensile)}$$

$$A_{CD} = 3.464 - 3.222 = 0.242 \text{ kN (Compressive)}$$
$$A_{DE} = 4.242 - 0.242 = 4 \text{ kN (Tensile)}$$
$$A_{EB} = 0$$

## 13.20. Shear Force and Bending Moment Diagrams for Inclined Beams

In the previous articles, we have discussed the cases of horizontal beams, subjected to various types of loadings. But sometimes, we come across inclined beams or members (such as ladders etc.) and carrying vertical loads. In such cases, the given loads are resolved at right angles and along the axis of the beam. The beam is further analysed in the same manner as a beam is subjected to inclined loads. The horizontal and vertical reactions at the two supports of the inclined beam are found out from the simple laws of statics.

**EXAMPLE 13.29.** *A ladder AB 5 m long, weighing 500 N/m, rests against a smooth wall and on a rough floor as shown in Fig. 13.63.*

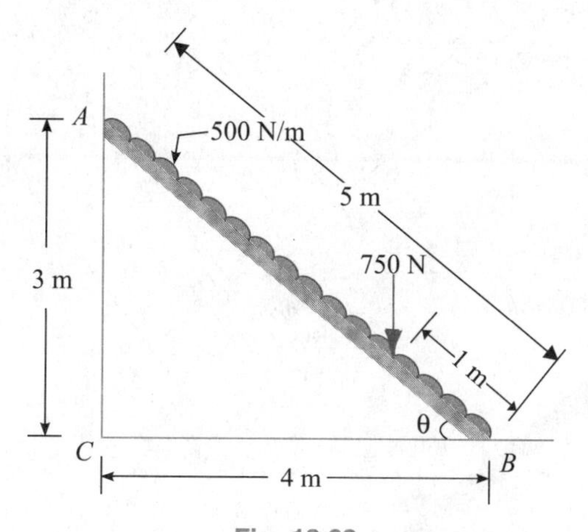


Fig. 13.63

*Find the reactions at A and B and construct the shear force, bending moment and axial thrust diagrams for the ladder.*

**SOLUTION.** Given : Span ($l$) = 5 m ; Uniformly distributed load ($w$) = 500 N/m and point load at $D$ = 750 N.

From the geometry of the figure, we find that

$$\tan \theta = \frac{3}{4} = 0.75$$

$$\therefore \quad \sin \theta = \frac{3}{5} = 0.6 \quad \text{and} \quad \cos \theta = \frac{4}{5} = 0.8$$

$R_A$ and $R_B$ = Normal reactions at the wall and floor,

$R_f$ = *Frictional resistance at the floor.

Equating the vertical and horizontal forces,

$$R_B = (500 \times 5) + 750 = 3250 \text{ N}$$

and

$$R_A = R_f$$

Taking moments about $B$,

$$R_A \times 3 = (500 \times 5 \times 2) + (750 \times 0.8) = 5600$$

* Since the wall is smooth, therefore there is no frictional resistance at the wall.

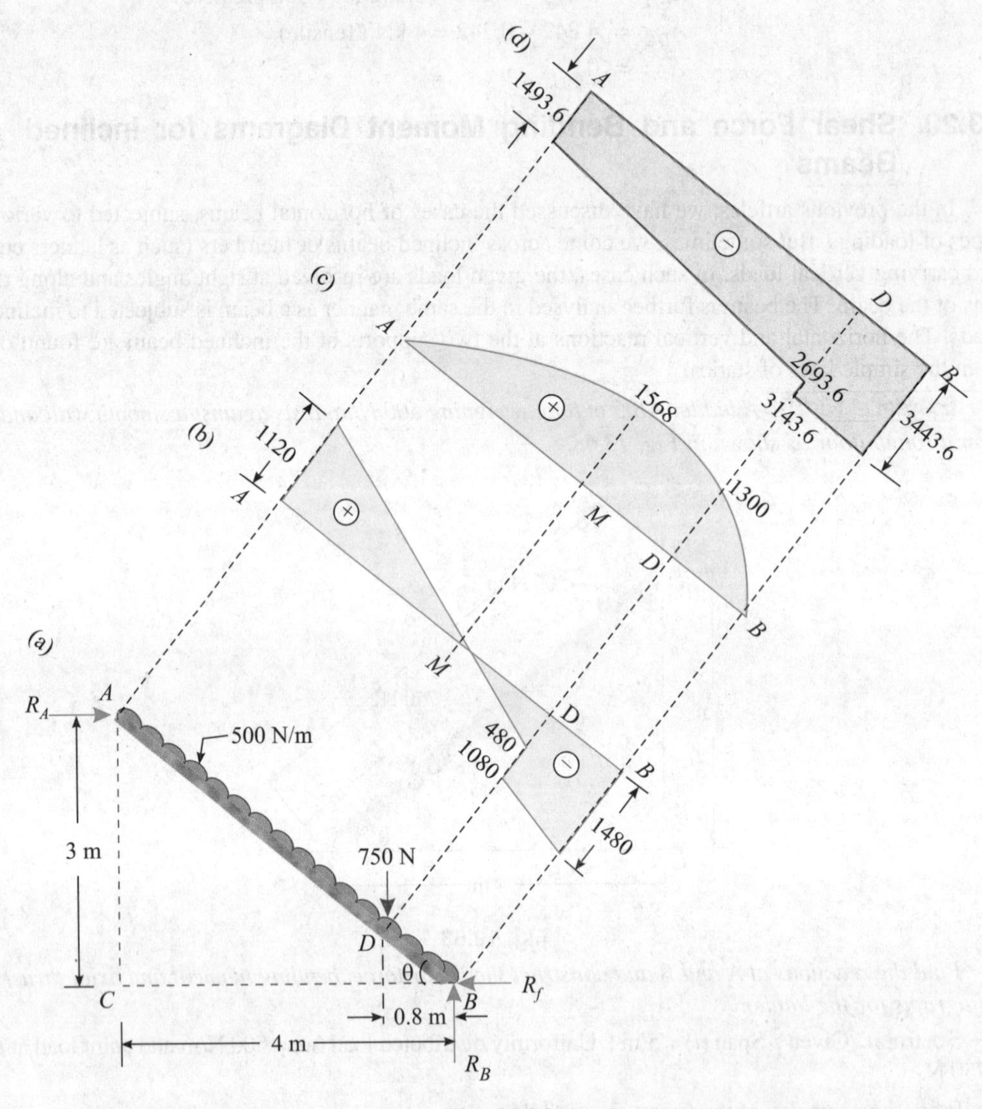


**Fig. 13.64**

$$\therefore \qquad R_A = R_f = \frac{5600}{3} = 1867 \text{ N}$$

Resolving the reaction $R_A$ at $A$ along the beam

$$= R_A \cos\theta = 1867 \times 0.8 = 1493.6 \text{ N}$$

and now resolving the reaction $R_A$ at right angles to the beam

$$= R_A \sin\theta = 1867 \times 0.6 = 1120 \text{ N}$$

Similarly, resolving the reactions $R_B$ and $R_f$ at $B$ along the beam

$$= R_B \sin\theta + R_f \cos\theta$$

$$= 3250 \times 0.6 + 1867 \times 0.8 = 3443.6 \text{ N}$$

and now resolving the reactions $R_B$ and $R_f$ at right angles to the beam

$$= R_B \cos\theta - R_f \sin\theta$$

$$= 3250 \times 0.8 - 1867 \times 0.6 = 1480 \text{ N}$$

Resolving the force 750 N at $D$ along the beam

$$= 750 \sin \theta = 750 \times 0.6 = 450 \text{ N}$$

and now resolving this force 750 N at right angle to the beam

$$= 750 \cos \theta = 750 \times 0.8 = 600 \text{ N}$$

Resolving the weight of ladder 500 N/m along the beam

$$= 500 \sin \theta = 500 \times 0.6 = 300 \text{ N/m}$$

and now resolving this weight of 500 N/m at right angles to the beam

$$= 500 \cos \theta = 500 \times 0.8 = 400 \text{ N/m}$$

***Shear force***

The shear force diagram is shown in Fig. 13.64 (*b*) and the values are tabulated here:

$$F_A = +1120 \text{ N}$$

$$F_D = +1120 - (400 \times 4) - 600 = -1080 \text{ N}$$

$$F_B = -1080 - (400 \times 1) + 1480 = 0$$

***Bending moment***

The bending moment diagram is shown in Fig. 13.64 (*c*) and the values are tabulated here:

$$M_A = 0$$

$$M_D = 3250 \times 0.8 - 1867 \times 0.6 - 400 \times 1 \times 0.5 \text{ N}$$

$$= 1279.8 \text{ N}$$

$$M_B = 0$$

***Maximum bending moment***

It will occur at $M$, where shear force changes sign. Let $x$ be the distance between $D$ and $M$. From the geometry of the figure, distance between $A$ and $D$, we find that

$$\frac{x}{1120} = \frac{4-x}{480}$$

or $$480\,x = 4480 - 1120\,x$$

$$x = 2.8$$

$\therefore$ $$M_M = 1120 \times 2.8 - 400 \times 2.8 \times \frac{2.8}{2} = 1568 \text{ N}$$

***Axial force diagram***

The axial force diagram as shown in Fig. 13.64 (*d*) and the values are tabulated here:

$$P_A = -1493.6 \text{ N}$$

$$P_D = -1493.6 - (300 \times 4) - 450 \text{ N}$$

$$= -3143.6 \text{ N}$$

$$P_B = -3143.6 - (300 \times 1) = -3443.6 \text{ N}$$

## EXERCISE 13.4

1. A simply supported beam *AB* of 4 m span is subjected to a clockwise moment of 20 kN-m at its centre. Draw the S.F. and B.M. diagrams. [**Ans.** $R_A = R_B = 5$ kN ; M = 10 kN-m]
2. A simply supported beam 7.5 m long is subjected to a couple of 30 kN-m in an anticlockwise direction at a distance of 5.5 m from the left support. Draw the S.F. and B.M. diagrams for the beam. [**Ans.** $R_A = R_B = 4$ kN ; M = – 22 kN-m ; + 8 kN-m]
3. Analyse the beam subjected to the moment and uniformly distributed load as shown in Fig. 13.65.

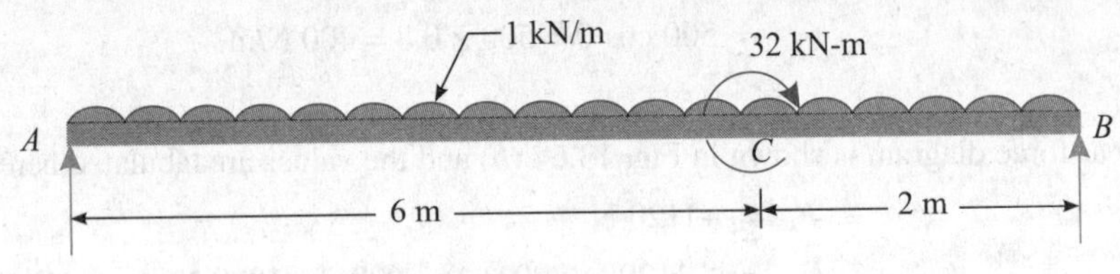


Fig. 13.65

Draw the moment and bending diagrams. [**Ans.** $M_{max}$ = – 18.0 kN.m at *C*]

4. Calculate the reactions at *A* and *B* for the beam shown in Fig. 13.66 and draw the bending moment diagram and shear force diagram. $\left[\textbf{Ans.}\ \frac{4W}{3} ; \frac{2W}{2}\right]$

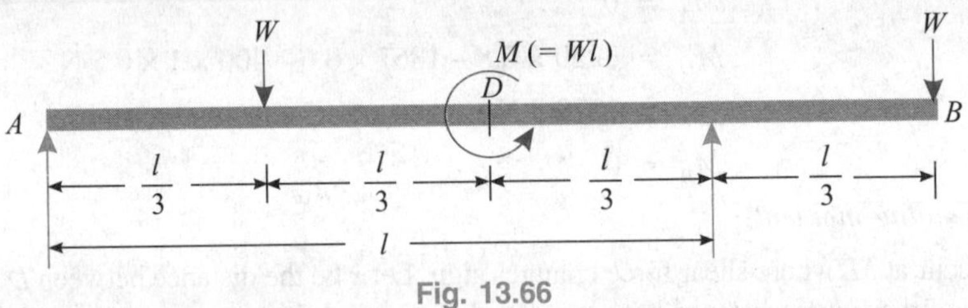


Fig. 13.66

5. Analyse the beam shown in Fig. 13.67.

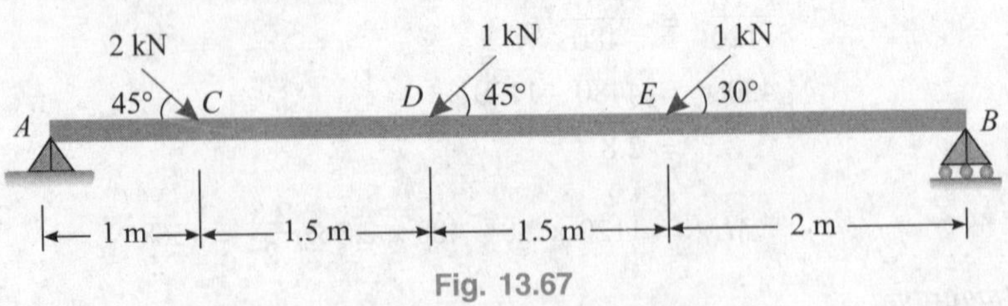


Fig. 13.67

Draw the shear force, bending moment and thrust diagrams.
[**Ans.** = 2.09 kN ; $R_B = 1.53$ kN ; $M_C = 2.09$ kN-m ; $M_D = 3.11$ kN-m ; $M_A = 3.06$ kN-m ; $P_A = -1.893$ kN ; $P_C = -3.307$ kN ; $P_D = 2.6$ kN ; $P_E = -2.6$ kN]

## QUESTIONS

1. Define the terms shear force and bending moment.
2. Discuss the utility of shear force and bending moment diagrams.
3. Explain briefly the relationship between shear force and bending moment at a section.

4. How will you determine the maximum bending moment in a simply supported beam?
5. What do you understand by the term, 'point of contraflexture'?
6. Describe the effect of a couple on the S.F. and B.M. diagram of a beam.
7. Explain the procedure adopted for analysing simply supported beam subjected to inclined loads.

## OBJECTIVE TYPE QUESTIONS

1. If a cantilever beam is subjected to a point load at its free end, then the shear force under the point load is
   (*a*) zero (*b*) less than the load
   (*c*) equal to the load (*d*) more than the load.
2. The bending moment at the free end of a cantilever beam carrying any type of load is
   (*a*) zero (*b*) minimum (*c*) maximum (*d*) equal to the load.
3. The B.M. at the centre of a simply supported beam carrying a uniformly distributed load is
   (*a*) $w \cdot l$ (*b*) $\frac{wl}{2}$ (*c*) $\frac{wl^2}{4}$ (*d*) $\frac{wl^2}{8}$
   When $w$ = Uniformly distributed load and
   $l$ = Span of the beam.
4. When shear force at a point is zero, then bending moment at that point will be
   (*a*) zero (*b*) minimum (*c*) maximum (*d*) infinity.
5. The point of contraflexure is a point where
   (*a*) shear force changes sign (*b*) bending moment changes sign
   (*c*) shear force is maximum (*d*) bending moment is maximum.

## ANSWERS

1. (*c*) 2. (*a*) 3. (*d*) 4. (*c*) 5. (*b*)

Chapter **14**

# Bending Stresses in Simple Beams

## Contents



## 14.1. Introduction

We have already discussed in Chapter 13 that the bending moments and shearing forces are set up at all sections of a beam, when it is loaded with some external loads. We have also discussed the methods of estimating the bending moments and shear forces at various sections of the beams and cantilevers.

As a matter of fact, the bending moment at a section tends to bend or deflect the beam and the internal stresses resist its bending. The process of bending stops, when every cross-section sets up full resistance to the bending moment. The *resistance, offered by the internal stresses, to the

* The resistance offered by the internal stresses to the shear force is called shearing stresses. It will be discussed in the next chapter.

bending, is called bending stress, and the relevant theory is called the theory of simple bending.

## 14.2. Assumptions in the Theory of Simple Bending

The following assumptions are made in the theory of simple bending:

1. The material of the beam is perfectly homogeneous (*i.e.*, of the same kind throughout) and isotropic (*i.e.*, of equal elastic properties in all directions).
2. The beam material is stressed within its elastic limit and thus, obeys Hooke's law.
3. The transverse sections, which were plane before bending, remains plane after bending also.
4. Each layer of the beam is free to expand or contract, independently, of the layer above or below it.
5. The value of $E$ (Young's modulus of elasticity) is the same in tension and compression.
6. The beam is in equilibrium *i.e.*, there is no resultant pull or push in the beam section.

## 14.3. Theory of Simple Bending

Consider a small length of a simply supported beam subjected to a bending moment as shown in Fig. 14.1 (*a*). Now consider two sections *AB* and *CD*, which are normal to the axis of the beam *RS*. Due to action of the bending moment, the beam as a whole will bend as shown in Fig. 14.1 (*b*).

Since we are considering a small length of $dx$ of the beam, therefore the curvature of the beam in this length, is taken to be circular. A little consideration will show that all the layers of the beam, which were originally of the same length do not remain of the same length any more. The top layer of the beam has suffered compression and reduced to $A'C'$. As we proceed towards the lower layers of the beam, we find that the layers have no doubt suffered compression, but to lesser degree; until we come across the layer *RS*, which has suffered no change in its length, though bent into $R'S'$. If we further proceed towards the lower layers, we find the layers have suffered tension, as a result of which the layers are stretched. The amount of extension increases as we proceed lower, until we come across the lowermost layer *BD* which has been stretched to $B'D'$.

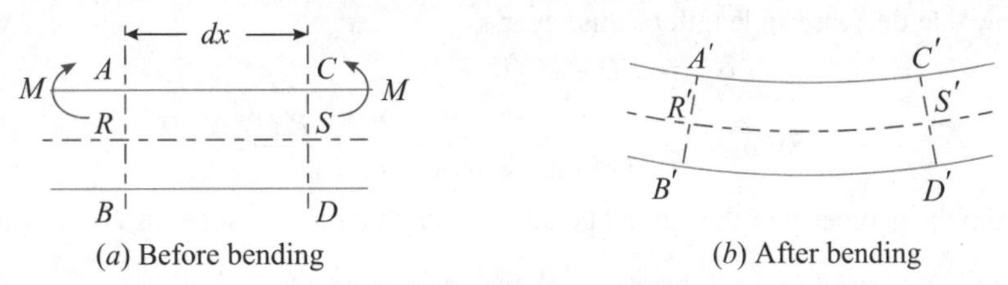

(*a*) Before bending (*b*) After bending

Fig. 14.1. Simple bending

Now we see that the layers above have been compressed and those below *RS* have been stretched. The amount, by which layer is compressed or stretched, depends upon the position of the layer with reference to *RS*. This layer *RS*, which is neither compressed nor stretched, is known as neutral plane or neutral layer. This theory of bending is called theory of simple bending.

## 14.4. Bending Stress

Consider a small length $dx$ of a beam subjected to a bending moment as shown in Fig. 14.2 (*a*). As a result of this moment, let this small length of beam bend into an arc of a circle with $O$ as centre as shown in Fig. 14.2 (*b*).

Let $M$ = Moment acting at the beam,

$\theta$ = Angle subtended at the centre by the arc and

$R$ = Radius of curvature of the beam.

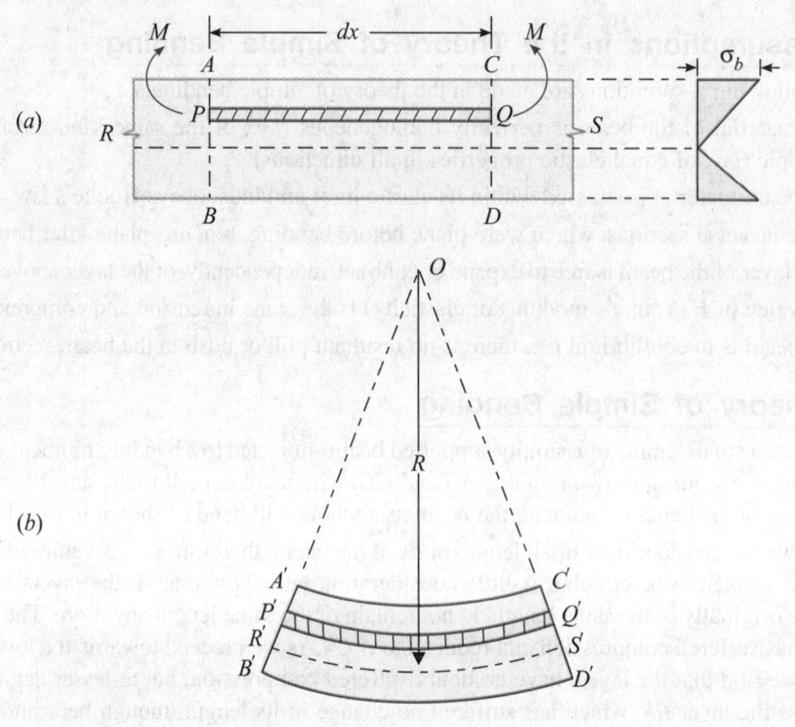


Fig. 14.2. Bending stress

Now consider a layer $PQ$ at a distance $y$ from $RS$ the neutral axis of the beam. Let this layer be compressed to $P'Q'$ after bending as shown in Fig. 14.2 (*b*).

We know that decrease in length of this layer,

$$\delta l = PQ - P'Q'$$

$$\therefore \quad \text{Strain } \varepsilon = \frac{\delta l}{\text{Original length}} = \frac{PQ - P'Q'}{PQ}$$

Now from the geometry of the curved beam, we find that the two sections $OP'Q'$ and $OR'S'$ are similar.

$$\therefore \quad \frac{P'Q'}{R'S'} = \frac{R-y}{R}$$

$$\text{or} \quad 1 - \frac{P'Q'}{R'S'} = 1 - \frac{R-y}{R}$$

$$\text{or} \quad \frac{R'S' - P'Q'}{PQ} = \frac{y}{R}$$

$$\frac{PQ - P'Q'}{PQ} = \frac{y}{R} \quad ...(PQ = R'S' = \text{Neutral axis})$$

$$\varepsilon = \frac{y}{R} \quad ...\left(\because \varepsilon = \frac{PQ - P'Q'}{PQ}\right)$$

It is thus obvious, that the strain (ε) of a layer is proportional to its distance from the neutral axis. We also know that the bending stress,

$$\sigma_b = \text{Strain} \times \text{Elasticity} = \varepsilon \times E$$

$$= \frac{y}{R} \times E = y \times \frac{E}{R} \qquad \ldots \left( \because \varepsilon = \frac{y}{R} \right)$$

Since $E$ and $R$ are constants in this expression, therefore the stress at any point is directly proportional to $y$, *i.e.*, the distance of the point from the neutral axis. The above expression may also be written as,

$$\frac{\sigma_b}{y} = \frac{E}{R} \quad \text{or} \quad \sigma_b = \frac{E}{R} \times y$$

**Note.** Since the bending stress is inversely proportional to the radius ($R$), therefore for maximum stress the radius should be minimum and vice versa.

**Example 14.1.** *A steel wire of 5 mm diameter is bent into a circular shape of 5 m radius. Determine the maximum stress induced in the wire. Take E = 200 GPa.*

**Solution.** Given : Diameter of steel wire ($d$) = 5 mm ; Radius of circular shape ($R$) = 5 m = $5 \times 10^3$ mm and modulus of elasticity ($E$) = 200 GPa = $200 \times 10^3$ N/mm$^2$.

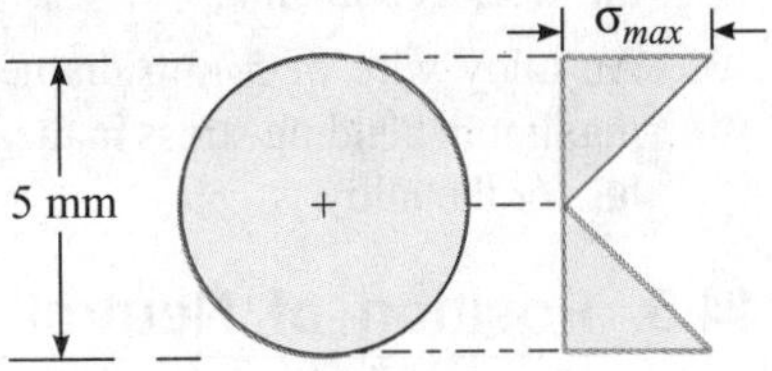

$\sigma_{max}$

5 mm

Fig. 14.3

We know that distance between the neutral axis of the wire and its extreme fibre,

$$y = \frac{d}{2} = \frac{5}{2} = 2.5 \text{ mm}$$

and maximum bending stress induced in the wire,

$$\sigma_{b\,(max)} = \frac{E}{R} \times y = \frac{200 \times 10^3}{5 \times 10^3} \times 2.5 = 100 \text{ N/mm}^2 = 100 \text{ MPa} \qquad \textbf{Ans.}$$

**Example 14.2.** *A copper wire of 2 mm diameter is required to be wound around a drum. Find the minimum radius of the drum, if the stress in the wire is not to exceed 80 MPa. Take modulus of elasticity for the copper as 100 GPa.*

**Solution.** Given : Diameter of wire ($d$) = 2 mm ; Maximum bending stress $\sigma_{b\,(max)}$ = 80 MPa = 80 N/mm$^2$ and modulus of elasticity ($E$) = 100 GPa = $100 \times 10^3$ N/mm$^2$.

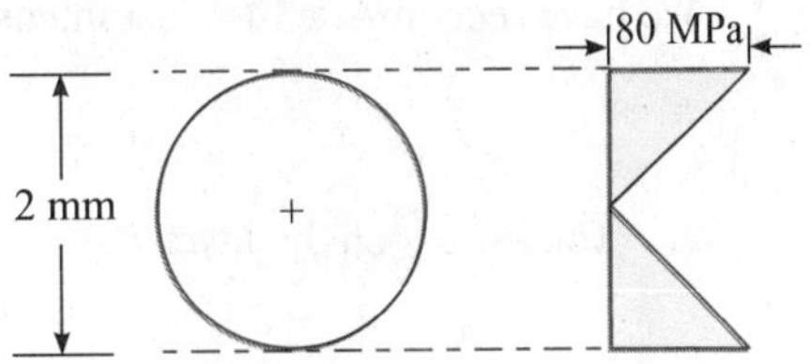

80 MPa

2 mm

Fig. 14.4

We know that distance between the neutral axis of the wire and its extreme fibre

$$y = \frac{2}{2} = 1 \text{ mm}$$

∴ Minimum radius of the drum

$$R = \frac{y}{\sigma_{b(\max)}} \times E = \frac{1}{80} \times 100 \times 10^3 \qquad \ldots \left( \because \frac{\sigma_b}{y} = \frac{E}{R} \right)$$

$$= 1.25 \times 10^3 \text{ mm} = 1.25 \text{ m} \qquad \textbf{Ans.}$$

**Example 14.3.** *A metallic rod of 10 mm diameter is bent into a circular form of radius 6 m. If the maximum bending stress developed in the rod is 125 MPa, find the value of Young's modulus for the rod material.*

**Solution.** Given : Diameter of rod ($d$) = 10 mm ; Radius ($R$) = 6 m = $6 \times 10^3$ mm and maximum bending stress $\sigma_{b\,(max)}$ = 125 MPa = 125 N/mm$^2$.

We know that distance between the neutral axis of the rod and its extreme fibre,

$$y = \frac{10}{2} = 5$$

∴ Value of Young's modulus for the rod material,

$$E = \frac{\sigma_{b(max)}}{y} \times R = \frac{125}{5} \times (6 \times 10^3) \text{ N/mm}^2 \quad \ldots \left( \because \frac{\sigma_b}{y} = \frac{E}{R} \right)$$

$$= 150 \times 10^3 \text{ N/mm}^2 = 150 \text{ GPa} \quad \textbf{Ans.}$$

## EXERCISE 14.1

1. A copper rod 20 mm diameter is bent into a circular arc of 8 m radius. Determine the intensity of maximum bending stress induced in the metal. Take $E$ = 100 GPa. [**Ans.** 125 MPa]
2. A steel wire of 3 mm diameter is to be wound around a circular component. If the bending stress in the wire is limited to 80 MPa, find the radius of the component. Take Young's modulus for the steel as 200 GPa. [**Ans.** 3.75 m]
3. An alloy wire of 5 mm diameter is wound around a circular drum of 3 m diameter. If the maximum bending stress in the wire is not to exceed 200 MPa, find the value of Young's modulus for the alloy. [**Ans.** 120 GPa]

## 14.5. Position of Neutral Axis

The line of intersection of the neutral layer, with any normal cross-section of a beam, is known as neutral axis of that section. We have seen in Art. 14.2 that on one side of the neutral axis there are compressive stresses, whereas on the other there are tensile stresses. At the neutral axis, there is no stress of any kind.

Consider a section of the beam as shown in Fig. 14.5. Let be the neutral axis of the section. Consider a small layer $PQ$ of the beam section at a distance from the neutral axis as shown in Fig. 14.5.

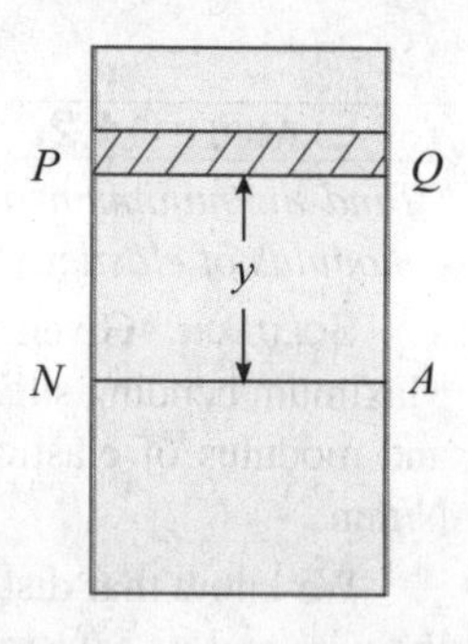

Fig. 14.5. Neutral axis

Let $\delta a$ = Area of the layer $PQ$.

We have seen in Art. 14.4 that intensity of stress in the layer $PQ$,

$$\sigma = y \times \frac{E}{R}$$

∴ Total stress on the layer $PQ$

$$= \text{Intensity of stress} \times \text{Area}$$

$$= y \times \frac{E}{R} \times \delta a$$

and total stress of the section.

$$= \Sigma y \times \frac{E}{R} \times \delta a = \frac{E}{R} \Sigma y . \delta a$$

Since the section is in equilibrium, therefore total stress, from top to bottom, must be equal to zero.

$$\therefore \quad \frac{E}{R} \Sigma y . \delta a = 0$$

or

$$\Sigma y . \delta a = 0 \quad \ldots \left( \because \frac{E}{R} \text{ cannot be equal to zero} \right)$$

A little consideration will show that $y \times \delta a$ is the moment of the area about the neutral axis and $\Sigma y \times \delta a$ is the moment of the entire area of the cross-section about the neutral axis. It is thus obvious that the neutral axis of the section will be so located that moment of the entire area about the axis is zero. We know that the moment of any area about an axis passing through its central axis of a section always passes through its centroid. Thus to locate the neutral axis of a section, first find out the centroid of the section and then draw a line passing through this centroid and normal to the plane of bending. This line will be the neutral axis of the section.

## 14.6. Moment of Resistance

We have already seen in Art. 14.2 that on one side of the neutral axis there are compressive stresses and on the other there are tensile stresses. These stresses form a couple, whose moment must be equal to the external moment (*M*). The moment of this couple, which resists the external bending moment, is known as moment of resistance.

Consider a section of the beam as shown in Fig. 14.6. Let *NA* be the neutral axis of the section. Now consider a small layer *PQ* of the beam section at a distance *y* from the neutral axis as shown in Fig. 14.6.

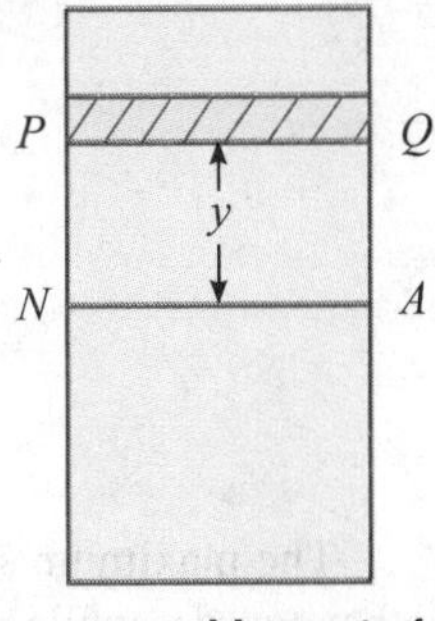


Fig. 14.6. Moment of resistance

Let $\delta a$ = Area of the layer *PQ*.

We have seen in Art. 14.4 that the intensity of stress in the layer *PQ*,

$$\sigma = y \times \frac{E}{R}$$

∴ Total stress in the layer *PQ*

$$= y \times \frac{E}{R} \times \delta a$$

and moment of this total stress about the neutral axis

$$= y \times \frac{E}{R} \times \delta a \times y = \frac{E}{R} y^2 . \delta a \qquad ...(i)$$

The algebraic sum of all such moments about the neutral axis must be equal to *M*. Therefore

$$M = \Sigma \frac{E}{R} y^2 . \delta a = \frac{E}{R} \Sigma y^2 . \delta a$$

The expression $\Sigma y^2 . \delta a$ represents the moment of inertia of the area of the whole section about the neutral axis. Therefore

$$M = \frac{E}{R} \times I \qquad ...(\text{where } I = \text{moment of inertia})$$

or

$$\frac{M}{I} = \frac{E}{R}$$

We have already seen in Art 14.4 that,

$$\frac{\sigma}{y} = \frac{E}{R}$$

$$\therefore \qquad \frac{M}{I} = \frac{\sigma}{y} = \frac{E}{R}$$

It is the most important equation in the theory of simple bending, which gives us relation between various characteristics of a beam.

## 14.7. Distribution of Bending Stress across the Section

We have seen in the previous articles that there is no stress at the neutral axis. In a *simply supported beam, there is a compressive stress above the neutral axis and a tensile stress below it.

* In a cantilever, there is a tensile stress above the neutral axis and compressive stress below it.

We have also discussed that the stress at a point is directly proportional to its distance from the neutral axis. If we plot the stresses in a simply supported beam section, we shall get a figure as shown in Fig. 14.7.

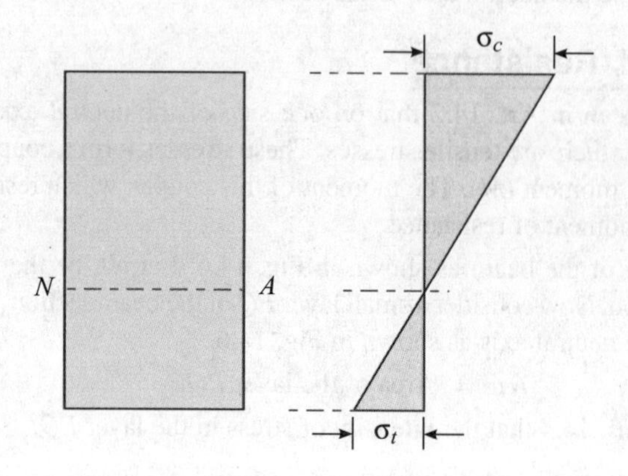


Fig. 14.7. Distribution of Bending Stress

The maximum stress (either compressive or tensile) takes place at the outermost layer. Or in other words, while obtaining maximum bending stress at a section, the value of $y$ is taken as maximum.

## 14.8. Modulus of Section

We have already discussed in the previous article, the relation for finding out the bending stress on the extreme fibre of a section, *i.e.*,

$$\frac{M}{I} = \frac{\sigma}{y} \qquad \text{or} \qquad M = \sigma \times \frac{I}{y}$$

From this relation, we find that the stress in a fibre is proportional to its distance from the *c.g.* If $y_{max}$ is the distance between the *c.g.* of the section and the extreme fibre of the stress, then

$$M = \sigma_{max} \times \frac{I}{y_{max}} = \sigma_{max} \times Z$$

where $Z = \frac{I}{y_{max}}$. The term '$Z$' is known as modulus of section or section modulus. The general practice of writing the above equation is $M = \sigma \times Z$, where $\sigma$ denotes the maximum stress, tensile or compressive in nature.

We know that if the section of a beam to, is symmetrical, its centre of gravity and hence the neutral axis will lie at the middle of its depth. We shall now consider the modulus of section of the following sections:

**1.** Rectangular section. **2.** Circular section.

**1.** ***Rectangular section***

We know that moment of inertia of a rectangular section about an axis through its centre of gravity.

$$I = \frac{bd^3}{12}$$

$\therefore$ Modulus of section $\quad Z = \frac{I}{y} = \frac{bd^3}{12} \times \frac{2}{d} = \frac{bd^2}{6} \qquad \ldots \left(\because y = \frac{d}{2}\right)$

2. *Circular section*

We know that moment of inertia of a circular section about an axis through its *c.g.*,

$$I = \frac{\pi}{64}(d)^4$$

∴ Modulus of section $$Z = \frac{I}{y} = \frac{\pi}{64}(d)^4 \times \frac{2}{d} = \frac{\pi}{32}(d)^2 \qquad ...\left(\because y = \frac{d}{2}\right)$$

**Note :** If the given section is hollow, then the corresponding values for external and internal dimensions should be taken.

## 14.9. Strength of a Section

It is also termed as flexural strength of a section, which means the moment of resistance offered by it. We have already discussed the relations :

$$\frac{M}{I} = \frac{\sigma}{y} \quad \text{or} \quad M = \frac{\sigma}{y} \times I \quad \text{and} \quad M = \sigma Z$$

It is thus obvious that the moment of resistance depends upon moment of inertia (or section modulus) of the section. A little consideration will show that the moment of inertia of beam section does not depend upon its cross-section area, but its disposition in relation to the neutral axis.

We know that in the case of a beam, subjected to transverse loading, the bending stress at a point is directly proportional to its distance from the neutral axis. It is thus obvious that a larger area near the neutral axis of a beam is uneconomical. This idea is put into practice, by providing beams of section, where the flanges alone withstand almost all the bending stress.

**Example 14.4.** *For a given stress, compare the moments of resistance of a beam of a square section, when placed (i) with its two sides horizontal and (ii) with its diagonal horizontal.*

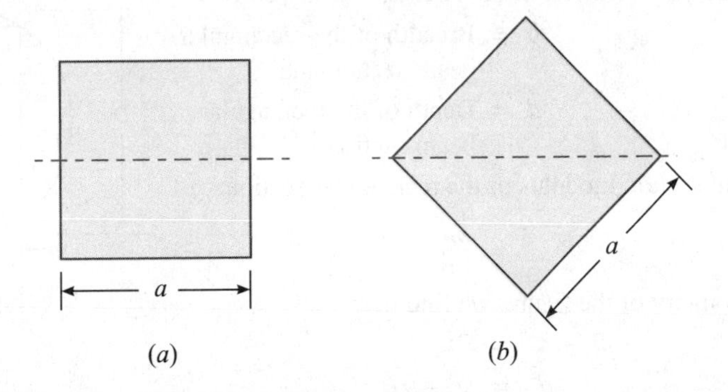


**Fig. 14.8**

**Solution.** Given: The square section with its two horizontal sides and with its diagonal horizontal are shown in Fig. 14.8 (*a*) and (*b*).

Let $a$ = Side of the square beam,

$M_1$ = Moment of resistance of section 1 and

$M_2$ = Moment of resistance of section 2.

We know that the section modulus of the beam section with its two sides horizontal,

$$Z_1 = \frac{bd^2}{6} = \frac{a \times a^2}{6} = \frac{a^3}{6} \qquad ...(i)$$

and moment of inertia of the beam section with its diagonal horizontal may be found out by splitting up the section into two triangles and then adding the moments of inertia of the two triangles about their base.

$$\therefore \qquad I_2 = 2 \times \frac{bh^3}{12} = 2 \times \frac{a\sqrt{2}\left(\frac{a}{\sqrt{2}}\right)^3}{12} = \frac{a^4}{12}$$

and
$$y_{max} = \frac{a}{\sqrt{2}}$$

$$\therefore \qquad Z_2 = \frac{I}{y_{max}} = \frac{\frac{a^4}{12}}{\frac{a}{\sqrt{2}}} = \frac{a^3}{6\sqrt{2}} \qquad ...(ii)$$

Sine the moment of resistance of a section is directly proportional to their moduli of section, therefore

$$\frac{M_1}{M_2} = \frac{Z_1}{Z_2} = \frac{\frac{a^3}{6}}{\frac{a^3}{6\sqrt{2}}} = \sqrt{2} = 1.414 \qquad \textbf{Ans.}$$

**EXAMPLE 14.5.** *A rectangular beam is to be cut from a circular log of wood of diameter D. Find the ratio of dimensions for the strongest section in bending.*

**SOLUTION.** Given : Diameter of the circular log of wood = $D$.

Let $b$ = Breadth of the rectangular beam section and

$d$ = Depth of the rectangular beam section.

D d b

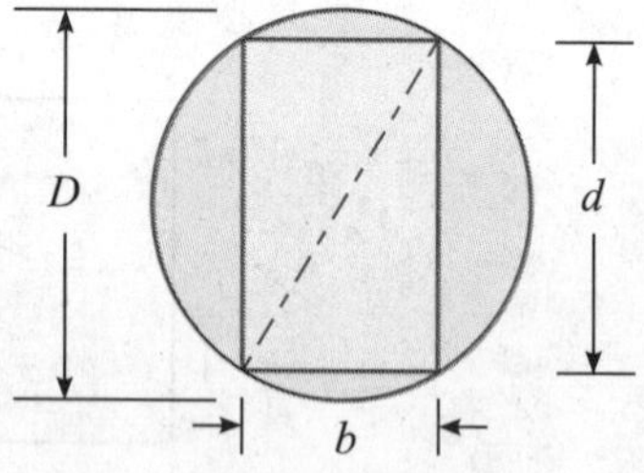

Fig. 14.9

We know that section modulus of the rectangular section.

$$Z = \frac{bd^2}{6}$$

From the geometry of the figure, we find that

$$b^2 + d^2 = D^2$$

or
$$d^2 = D^2 - b^2$$

Substituting the value of $d^2$ in equation ($i$),

$$Z = \frac{b \times (D^2 - b^2)}{6} = \frac{bD^2 - b^3}{6}$$

We also know that for the strongest section, let us differentiate the above equation and equate it to zero. *i.e.*,

$$\frac{dZ}{db} = \frac{d}{db}\left[\frac{bD^2 - b^3}{6}\right] = \frac{D^2 - 3b^2}{6}$$

or
$$\frac{D^2 - 3b^2}{6} = 0 \qquad \text{or} \qquad D^2 - 3b^2 = 0 \qquad \text{or} \qquad b = \frac{D}{\sqrt{3}}$$

Substituting this value of b in equation (*ii*),

$$d^2 = D^2 - \frac{D^2}{3} = \frac{2D^2}{3} \quad \text{or} \quad d = D\sqrt{\frac{2}{3}} \quad \textbf{Ans.}$$

**EXAMPLE 14.6.** *Two beams are simply supported over the same span and have the same flexural strength. Compare the weights of these two beams, if one of them is solid and the other is hollow circular with internal diameter half of the external diameter.*

**SOLUTION.** Given : Span of the solid beam = Span of the hollow beam and flexural strength of solid beam = Flexural strength of the hollow section.

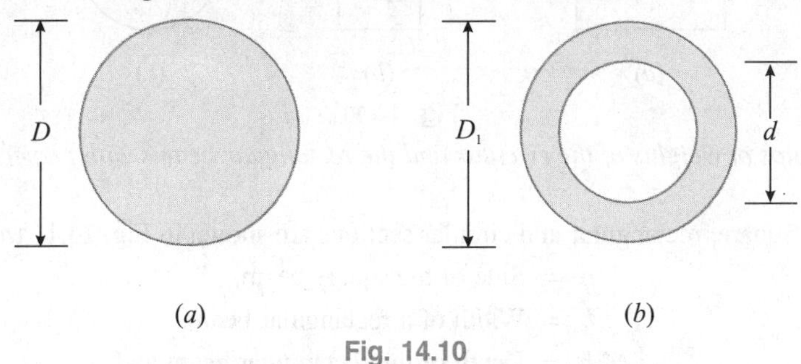


Fig. 14.10

Let $D$ = Diameter of the solid beam and

$D_1$ = Diameter of the hollow beam.

First of all consider the solid beam as shown in Fig. 14.10 (*a*). We know that section modulus of the solid section,

$$Z_1 = \frac{\pi}{32} \times (D)^3 = \frac{\pi}{32} \times D^3 \qquad ...(i)$$

and now consider the hollow beam as shown in Fig. 14.10 (*b*). We also know that section modulus of the hollow section,

$$Z_2 = \frac{\pi}{32\,D_1} \times \left[D_1^4 - d^4\right] = \frac{\pi}{32\,D_1} \times \left[D_1^4 - (0.5\,D_1)^4\right]$$

$$= \frac{\pi}{32} \times 0.9375\,D_1^3 \qquad ...(ii)$$

Since both the beams are supported over the same span ($l$) and have the same flexural strength, therefore section modulus of both the beams must be equal. Now equating equations (*i*) and (*ii*),

$$\frac{\pi}{32} \times D^3 = \frac{\pi}{32} \times 0.9375\,D_1^3 \quad \text{or} \quad D^3 = 0.9375\,(D_1)^3$$

$$\therefore \quad D = (0.9375)^{1/3}\,D_1 = 0.98\,D_1$$

We also know that wights of two beams are proportional to their respective cross-sectional areas. Therefore

$$\frac{\text{Weight of solid beam}}{\text{Weight of hollow beam}} = \frac{\text{Area of solid beam}}{\text{Area of hollow beam}}$$

or

$$= \frac{\frac{\pi}{4} \times D^2}{\frac{\pi}{4} \times \left[(D_1)^2 - d^2\right]} = \frac{D^2}{(D_1)^2 - (0.5\,D_1)^2}$$

$$= \frac{D^2}{0.75(D_1)^2} = \frac{D^2}{(D_1)^2} \times \frac{1}{0.75} = (0.98)^2 \times \frac{1}{0.75} = 1.28 \quad \textbf{Ans.}$$

**EXAMPLE 14.7.** *Three beams have the same length, the same allowable stress and the same bending moment. The cross-section of the beams are a square, a rectangle with depth twice the width and a circle as shown in Fig. 14.11.*

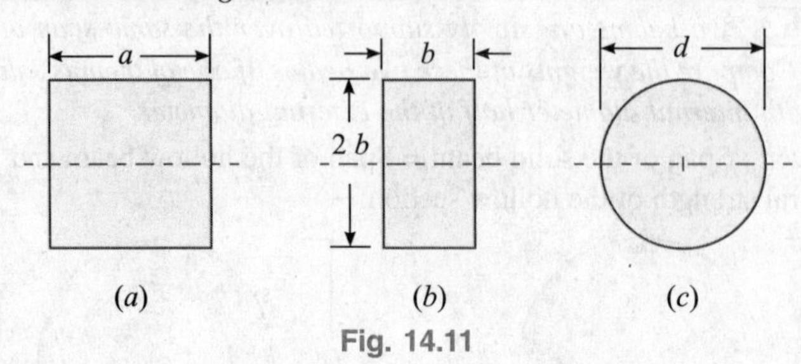


Fig. 14.11

*Find the ratios of weights of the circular and the rectanguar beams with respect to the square beam.*

**SOLUTION.** Square, rectangular and circular sections are shown in Fig. 14.11 (*a*), (*b*) and (*c*).

Let $a$ = Side of the square beam,

$b$ = Width of a rectangular beam,

$\therefore$ $2b$ = Depth of the rectangular beam and

$d$ = Diameter of a circular section.

Since all the three beams have the same allowable stress (σ) and bending moment (*M*), therefore the modulus of section of the three beams must be equal.

We know that the section modulus for a square beam,

$$Z_1 = \frac{bd^2}{6} = \frac{a \times a^2}{6} = \frac{a^3}{6} \qquad ...(i)$$

Similarly, modulus of section for rectangular beam,

$$Z_2 = \frac{bd^2}{6} = \frac{b(2b)^2}{6} = \frac{2b^3}{3} \qquad ...(ii)$$

and modulus of section for a circular beam,

$$Z_3 = \frac{\pi}{32} \times d^3 \qquad ...(iii)$$

Equating equations (*i*) and (*ii*),

$$\frac{a^3}{6} = \frac{2b^3}{3} \quad \text{or} \quad a^3 = 6 \times \frac{2b^3}{3} = 4b^3$$

$$\therefore \quad b = 0.63\, a \qquad ...(iv)$$

Now equating equations (*i*) and (*iii*),

$$\frac{a^3}{6} = \frac{\pi}{32} \times d^3$$

$$\therefore \quad a^3 = 6 \times \frac{\pi}{32} \times d^3 = \frac{3\pi}{16} \times d^3$$

$$\text{or} \quad d = 1.19\, a \qquad ...(v)$$

We know that weights of all the beams are proportional to the cross sectional areas of their sections. Therefore

$$\frac{\text{Weight of square beam}}{\text{Weight of rectangular beam}} = \frac{\text{Area of square beam}}{\text{Area of rectangular beam}}$$

$$= \frac{a^2}{2b^2} = \frac{a^2}{2 \times (0.63a)^2} = \frac{1}{0.79} \quad \textbf{Ans.}$$

and
$$\frac{\text{Weight of square beam}}{\text{Weight of circular beam}} = \frac{\text{Area of square beam}}{\text{Area of circular beam}}$$

$$= \frac{a^2}{\frac{\pi}{4} \times d^2} = \frac{a^2}{\frac{\pi}{4} \times (1.19a)^2} = \frac{1}{1.12} \quad \textbf{Ans.}$$

**EXAMPLE 14.8.** *Prove that moment of resistance of a beam of square section, with its diagonal in the plane of bending is increased by flatting top and bottom corners as shown in Fig. 14.12. Also prove that the moment of resistance is a maximum when y = 8Y/9.*

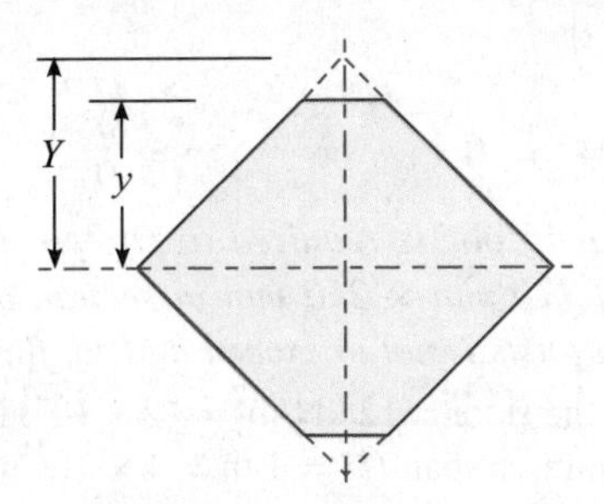


Fig. 14.12

**SOLUTION.** First of all, let us divide the section into a square with diagonal $2y$ and a rectangle with sides as $2y$ and $2(Y - y)$ as shown in Fig. 14.13 (*a*) and (*b*).

The moment of inertia of the square section with its diagonal in the plane of bending may be found out by splitting up the section into two triangles, and then adding the moments of inertia of the two triangles about its base.

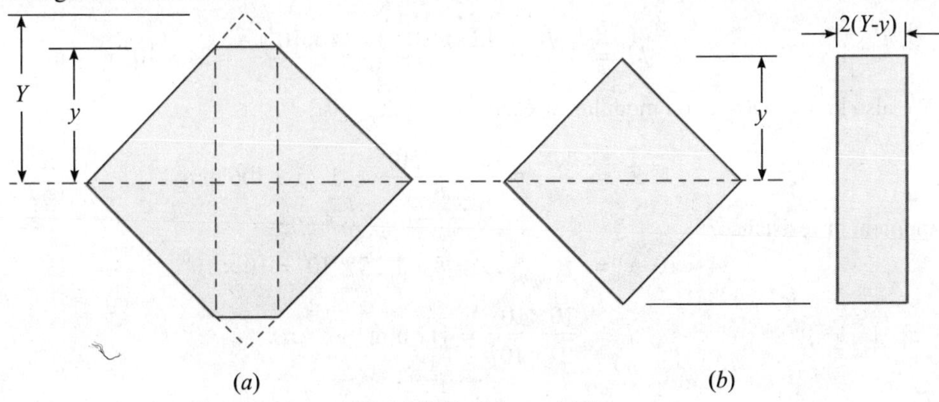


Fig. 14.13. (*a*) and (*b*)

We know that moment of inertia for the square section,

$$I_1 = 2 \times \frac{bh^3}{12} = 2 \times \frac{2y \times y^3}{12} = \frac{y^4}{3}$$

and moment of inertia for the rectangular section,

$$I_2 = \frac{2(Y - y) \times (2y)^3}{12} = \frac{4}{3}(Yy^3 - y^4)$$

∴ Total moment of inertia of the section,

$$I = I_1 + I_2 = \frac{y^4}{3} + \frac{4}{3}(Yy^3 - y^4) = \frac{4}{3}Yy^3 - y^4$$

We also know that the bending stress at a distance $x$ from the neutral axis,

$$\sigma_b = \frac{M}{I} \times y = \frac{M}{\frac{4}{3}Yy^3 - y^4} \times y = \frac{M}{\frac{4}{3}Yy^2 - y^3}$$

Now for maximum bending stress, differentiating the above equation and equating the same to zero,

$$\frac{d}{dy}\left(\frac{M}{\frac{4}{3}Y \cdot y^2 - y^3}\right) = 0 \quad \text{or} \quad \frac{4}{3}Y \times 2y - 3y^2 = 0$$

$$\frac{8Y}{3} - 3y = 0 \quad \text{or} \quad y = \frac{8Y}{9} \quad \textbf{Ans.}$$

**EXAMPLE 14.9.** *A wooden floor is required to carry a load of 12 kN/m² and is to be supported by wooden joists of 120 mm × 250 mm in section over a span of 4 metres. If the bending stress in these wooden joists is not to exceed 8 MPa, find the spacing of the joists.*

**SOLUTION.** Given : Load on the floor = 12 kN/m² = $12 \times 10^{-3}$ N/mm² ; Width of joist ($b$) = 120 mm ; Depth of joist ($d$) = 250 mm ; Span ($l$) = 4 m = $4 \times 10^3$ mm and maximum bending stress $\sigma_{b\,(max)}$ = 8 MPa = 8 N/mm².

Let $x$ = Spacing of the joists in mm.

We know that rate of loading on the joist,

$$w = 12 \times 10^{-3} \times x \times 1 = 12 \times 10^{-3}\, x \text{ N/mm}$$

and maximum bending moment at the centre of a simply supported beam subjected to a uniformly distributed load,

$$M = \frac{wl^2}{8} = \frac{(12x \times 10^{-3}) \times (4 \times 10^3)^2}{8} = 24 \times 10^3\, x \text{ N-m} \quad ...(i)$$

We also know that section modulus of each rectangular joist,

$$Z = \frac{bd^2}{6} = \frac{120 \times (250)^2}{6} = 1.25 \times 106 \text{ mm}^3$$

and moment of resistance,

$$24 \times 10^3\, x = \sigma_{b\,(max)} \,.\, Z = 8 \times 1.25 \times 10^6 = 10 \times 10^6$$

$$\therefore \quad x = \frac{10 \times 10^6}{24 \times 10^3} = 417 \text{ mm} \quad \textbf{Ans.}$$

## 14.10. Bending Stresses in Symmetrical Sections

Fig. 14.14. Symmetrical sections.

We know that in a symmetrical section (*i.e.*, circular, square or rectangular), the centre of gravity of the section lies at the geometrical centre of the section as shown in Fig. 14.14. Since the neutral axis of a section passes through its centre of gravity, therefore neutral axis of a symmetrical section passes through its geometrical centre. In such cases, the outermost layer or extreme fibre is at a distance of *d*/2 from its geometrical centre, where d is the diameter (in a circular section) or depth (in square or rectangular sections).

NOTE : In most or the cases, we are required to find the maximum bending stress in the section. We know that the bending stress at a point, in a section is directly proportional to its distance from the neutral axis. Therefore, maximum bending stress in a section will occur in the extreme fibre of the section.

**EXAMPLE 14.10.** *A rectangular beam 60 mm wide and 150 mm deep is simply supported over a span of 6 m. If the beam is subjected to central point load of 12 kN, find the maximum bending stress induced in the beam section.*

**SOLUTION.** Given : Width ($b$) = 60 mm ; Depth ($d$) = 150 mm ; Span ($l$) = $6 \times 10^3$ mm and load ($W$) = 12 kN = $12 \times 10^3$ N.

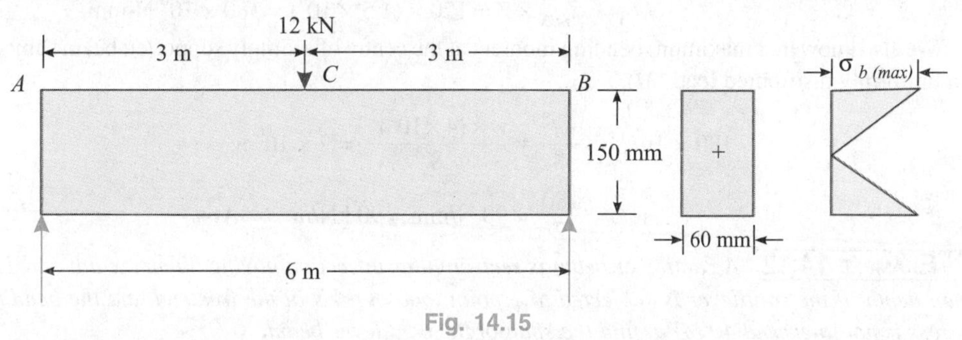


Fig. 14.15

We know that maximum bendint moment at the centre of a simply supported beam subjected to a central point load,

$$M = \frac{Wl}{4} = \frac{(12 \times 10^3) \times (6 \times 10^3)}{4} = 18 \times 10^6 \text{ N-mm}$$

and section modulus of the rectangular section,

$$Z = \frac{bd^2}{6} = \frac{60 \times (150)^2}{6} = 225 \times 10^3 \text{ mm}^3$$

∴ Maximum bending stress,

$$\sigma_{max} = \frac{M}{Z} = \frac{18 \times 10^6}{225 \times 10^3} = 80 \text{ N/mm}^2 = 80 \text{ MPa} \qquad \textbf{Ans.}$$

**EXAMPLE 14.11.** *A rectangular beam 300 mm deep is simply supported over a span of 4 metres. What uniformly distributed load the beam may carry, if the bending stress is not to exceed 120 MPa. Take I = 225 × 10⁶ mm⁴.*

**SOLUTION.** Given : Depth ($d$) = 300 mm ; Span ($l$) = 4 m = $4 \times 10^3$ mm ; Maximum bending stress ($\sigma_{max}$) = 120 MPa = 120 N/mm$^2$ and moment of inertia of the beam section ($I$) = $225 \times 10^6$ mm$^4$.

Let $w$ = Uniformly distributed load the beam can carry.

We know that distance between the neutral axis of the section and extreme fibre,

$$y = \frac{d}{2} = \frac{300}{2} = 150 \text{ mm}$$

and section modulus of the rectangular section,

$$Z = \frac{I}{y} = \frac{225 \times 10^6}{150} = 1.5 \times 10^6 \text{ mm}^3$$

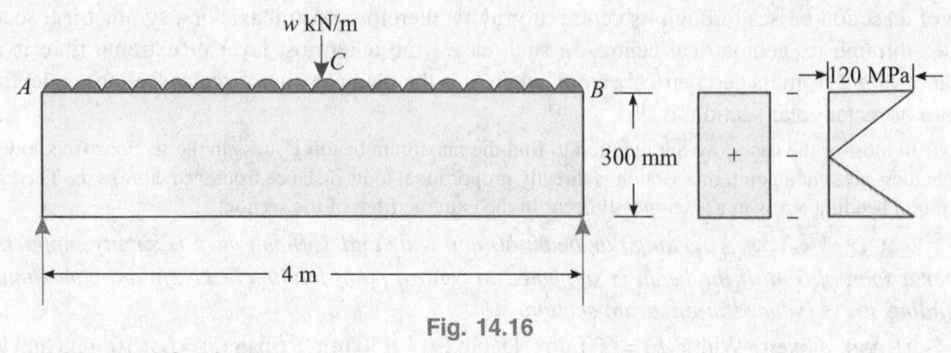


Fig. 14.16

∴ Moment of resistance,

$$M = \sigma_{max} \times Z = 120 \times (1.5 \times 10^6) = 180 \times 10^6 \text{ N-mm.}$$

We also know that maximum bending moment at the centre of a simply supported beam subjected **to** a uniformly distributed load ($M$),

$$180 \times 10^6 = \frac{wl^2}{8} = \frac{w \times (4 \times 10^3)^2}{8} = 2 \times 10^6\, w$$

$$\therefore \quad w = \frac{180}{2} = 90 \text{ N/mm} = 90 \text{ kN/m} \qquad \textbf{Ans.}$$

**EXAMPLE 14.12.** *A cantilever beam is rectrangular in section having 80 mm width and 120 mm depth. If the cantilever is subjected to a point load of 6 kN at the free end and the bending stress is not to exceed 40 MPa, find the span of the cantilever beam.*

**SOLUTION.** Given : Width ($b$) = 80 mm ; Depth ($d$) = 120 mm ; Point load ($W$) = 6 kN = $6 \times 10^3$ N and maximum bending stress ($\sigma_{max}$) = 40 MPa = 40 N/mm$^2$.

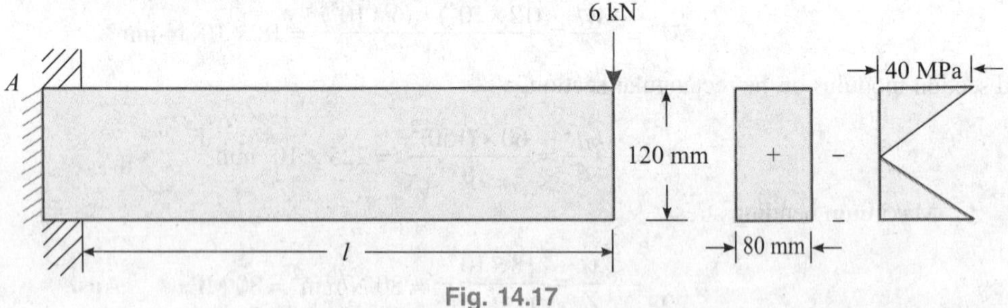


Fig. 14.17

Let $\quad l$ = Span of the cantilever beam.

We know that section modulus of the rectangular section,

$$Z = \frac{bd^2}{6} = \frac{80 \times (120)^2}{6} = 192 \times 10^3 \text{ mm}^3$$

and maximum bending moment at the fixed end of the cantilever subjected to a point load at the free end,

$$M = Wl = (6 \times 10^3) \times l$$

∴ Maximum bending stress [$\sigma_{b\,(max)}$]

$$40 = \frac{M}{Z} = \frac{6 \times 10^3 \times l}{192 \times 10^3} = \frac{l}{32}$$

or

$$l = 40 \times 32 = 1280 \text{ mm} = 1.28 \text{ m} \qquad \textbf{Ans.}$$

**EXAMPLE 14.13.** *A rectangular beam 60 mm wide and 150 mm deep is simply supported over a span of 4 metres. If the bneam is subjected to a uniformly distributed load of 4.5 kN/m, find the maximum bending stress induced in the beam.*

**SOLUTION.** Given : Width ($b$) = 60 mm ; Depth ($d$) = 150 mm ; Span ($l$) = 4 m = $4 \times 10^3$ mm and uniformly distributed load ($w$) = 4.5 kN/m = 4.5 N/mm.

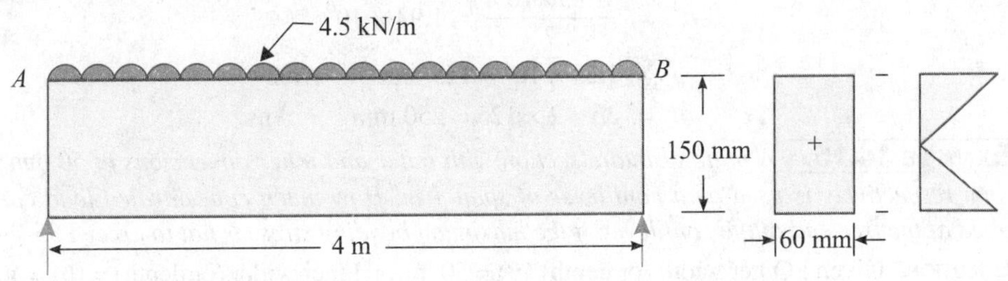


Fig. 14.18

We know that section modulus of the rectangular section,

$$Z = \frac{bd^2}{6} = \frac{60 \times (150)^2}{6} = 225 \times 10^3 \text{ mm}^3$$

and maxmum bending moment at the centre of a simply supported beam subjected to a uniformly distributed load,

$$M = \frac{wl^2}{8} = \frac{4.5 \times (4 \times 10^3)^2}{8} = 9 \times 10^6 \text{ N-mm}$$

∴ Maximum bending stress,

$$\sigma_{max} = \frac{M}{Z} = \frac{9 \times 10^6}{225 \times 10^3} = 40 \text{ N/mm}^2 = 40 \text{ MPa} \qquad \textbf{Ans.}$$

**EXAMPLE 14.14.** *A timber beam of rectangular section supports a load of 20 kN uniformly distributed over a span of 3.6 m. If depth of the beam section is twice the width and maximum stress is not to exceed 7 MPa, find the dimensions of the beam section.*

**SOLUTION.** Given : Total load ($W$) = 20 kN = $20 \times 10^3$ N ; Span ($l$) = $3.6 \times 10^3$ mm ; Depth of beam section ($d$) = $2b$ and ($\sigma_{max}$) = 7 MPa = 7 N/mm$^2$.

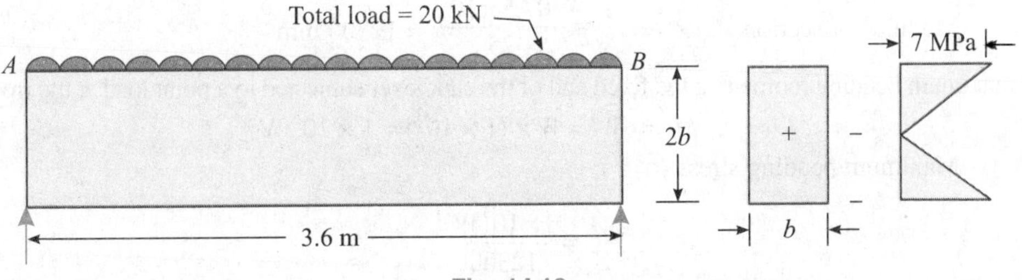


Fig. 14.19

We know that section modulus of the rectangular section,

$$Z = \frac{bd^2}{6} = \frac{b \times (2b)^2}{6} = \frac{2b^3}{3}$$

and maximum bending moment at the centre of a smiply supported beam subject to a uniformly distributed load,

$$M = \frac{wl^2}{8} = \frac{Wl}{8} = \frac{(20 \times 10^3) \times (3.6 \times 10^3)}{8} = 9 \times 10^6 \text{ N-mm}$$

∴ Maximum bending stress ($\sigma_{max}$),

$$7 = \frac{M}{Z} = \frac{9 \times 10^6}{\frac{2b^2}{3}} = \frac{13.5 \times 10^6}{b^3}$$

or
$$b^3 = \frac{(13.5 \times 10^6)}{7} = 1.93 \times 10^6$$

∴
$$b = 1.25 \times 10^2 = 125 \text{ mm} \quad \textbf{Ans.}$$

and
$$d = 2b = 2 \times 125 = 250 \text{ mm} \quad \textbf{Ans.}$$

**EXAMPLE 14.15.** *A hollow square section with outer and inner dimensions of 50 mm and 40 mm respectively is used as a cantilever of span 1 m. How much concentrated load can be applied at the free end of the cantilever, if the maximum bending stress is not to exceed 35 MPa?*

**SOLUTION.** Given : Outer width (or depth) ($B$) = 50 mm ; Inner width (or depth) = ($b$) = 40 mm; Span ($l$) = $1 \times 10^3$ mm and maximum bending stress $\sigma_{b\,(max)}$ = 35 MPa = 35 N/mm².

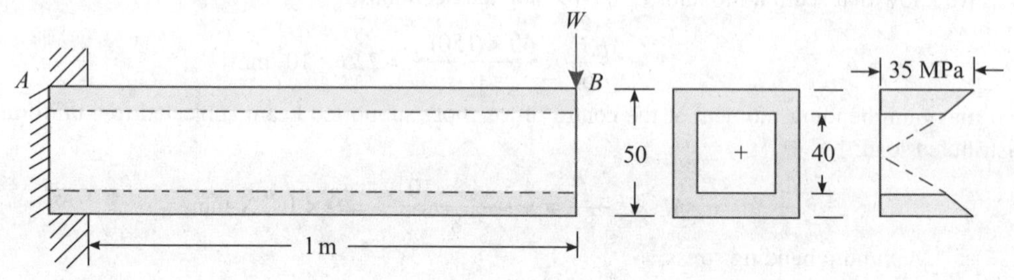


Fig. 14.20

Let $W$ = Concentreated load that be applied at the free end of the cantilever.

We know that moment of inertia of the hollow square section,

$$I = \frac{BD^3}{12} - \frac{bd^3}{12} = \frac{BB^3}{12} - \frac{bb^3}{12} = \frac{B^4}{12} - \frac{b^4}{12} = \frac{(50)^4}{12} - \frac{(40)^4}{12} \text{ mm}^4$$
$$= 307.5 \times 10^3 \text{ mm}^4$$

∴ Modulus of section,
$$Z = \frac{I}{y} = \frac{307.5 \times 10^3}{25} = 12300 \text{ mm}^3$$

and maximum bending moment at the fixed end of the cantilever subjected to a point load at the free end,

$$M = Wl = W \times (1 \times 10^3) = 1 \times 10^3\ W$$

∴ Maximum bending stress ($\sigma_{max}$),

$$35 = \frac{M}{Z} = \frac{1 \times 10^3 W}{12300}$$

or
$$W = \frac{35 \times 12300}{1 \times 10^3} = 430.5 \text{ N} \quad \textbf{Ans.}$$

**EXAMPLE 14.16.** *A hollow steel tube having external and internal diameter of 100 mm and 75 mm respectively is simply supported over a span of 5 m. The tube carries a concentrated load of W at a distance of 2 m from one of the supports. What is the value of W, if the maximum bending stress is not to exceed 100 MPa.*

**SOLUTION.** Given : External diameter ($D$) = 100 mm ; Internal diameter ($d$) = 75 mm ; Span ($l$) = 5 m = 5 × $10^3$ mm ; Distance $AC$ ($a$) = 2m = 2 × $10^3$ mm or Distance $BC$ ($b$) = 5 – 2 = 3 m = 3 × $10^3$ mm and maximum bending stress ($\sigma_{max}$) = 100 MPa = 100 N/mm$^2$.

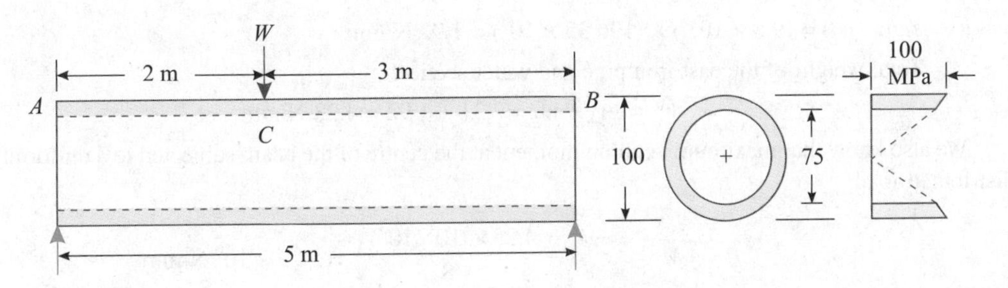


Fig. 14.21

We know that maximum bending moment over a simply supported beam subjected to an eccentric load,

$$M = \frac{Wab}{l} = \frac{W \times (2 \times 10^3) \times (3 \times 10^3)}{5 \times 10^3} = 1.2 \times 10^3\ W$$

and section modulus of a hollow circular section,

$$Z = \frac{\pi}{32 \times D} \times \left[D^4 - d^4\right] = \frac{\pi}{32 \times 100} \times \left[(100)^4 - (75)^4\right] \text{mm}^3$$

$$= 67.1 \times 10^3\ \text{mm}^3$$

We also know that maximum bending stress [$\sigma_{b\,(max)}$],

$$100 = \frac{M}{Z} = \frac{1.2 \times 10^3\ W}{67.1 \times 10^3} = 0.018\ W$$

$$\therefore \quad W = \frac{100}{0.018} = 5.6 \times 10^3\ \text{N} = 5.6\ \text{kN} \qquad \textbf{Ans.}$$

**EXAMPLE 14.17.** *A cast iron water pipe of 500 mm inside diameter and 20 mm thick is supported over a span of 10 meters. Find the maximum stress in the pipe metal, when the pipe is running full. Take density of cast iron as 70.6 kN/m$^3$ and that of water as 9.8 kN/m$^3$.*

**SOLUTION.** Given : Inside diameter ($d$) = 500 mm ; Thickness ($t$) = 20 mm or outside diameter ($D$) = $d + 2t$ = 500 + (2 × 20) = 540 mm ; Span ($l$) = 10 m = 10 × $10^3$ mm ; density of cast iron = 70.6 kN/m$^3$ = 70.6 × $10^{-6}$ N/mm$^2$ and density of water = 9.8 kN/m$^3$ = 9.8 × $10^{-6}$ N/mm$^2$.

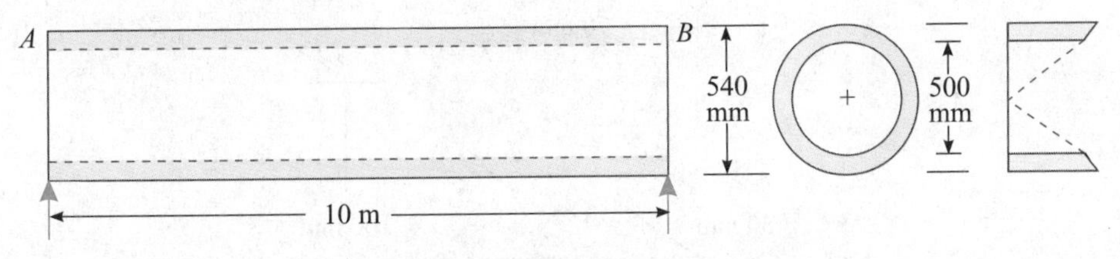


Fig. 14.22

We know that cross-sectional area of the cast iron pipe,

$$= \frac{\pi}{4} \times \left[D^2 - d^2\right] = \frac{\pi}{4} \times \left[(540)^2 - (500)^2\right] = 32.67 \times 10^3\ \text{mm}^2$$

and its weight ($w_1$) = (70.6 × $10^{-6}$) × (32.67 × $10^3$) = 2.31 N/mm

We also know that cross-sectional area of the water section

$$= \frac{\pi}{4} \times (d)^2 = \frac{\pi}{4} \times (500)^2 = 196.35 \times 10^3 \text{ mm}^2$$

and its weight $(w_2) = (9.8 \times 10^{-6}) \times (196.35 \times 10^3) = 1.92$ N/mm

∴ Total weight of the cast iron pipe and water section

$$w = w_1 + w_2 = 2.31 + 1.92 = 4.23 \text{ N/mm}$$

We also know that maximum bending moment at the centre of the beam subjected to a uniformly distributed load,

$$M = \frac{wl^2}{8} = \frac{4.23 \times (10 \times 10^3)^2}{8} = 52.9 \times 10^6 \text{ N-mm}$$

and section modulus of a hollow circular section,

$$Z = \frac{\pi}{32D} \times \left[D^2 - d^4\right] = \frac{\pi}{32 \times 540} \times \left[(540)^4 - (500)^4\right] \text{mm}^3$$

$$= 4.096 \times 10^6 \text{ mm}^3$$

∴ Maximum bending stress,

$$\sigma_{b\,(max)} = \frac{M}{Z} = \frac{52.9 \times 10^6}{4.096 \times 10^6} = 12.9 \text{ N/mm}^2 = 12.9 \text{ MPa} \qquad \textbf{Ans.}$$

## EXERCISE 14.2

1. A beam 3 m long has rectangular section of 80 mm width and 120 mm depth. If the beam is carrying a uniformly distributed load of 10 kN/m, find the maximum bending stress developed in the beam. [**Ans.** 58.6 MPa]
2. A rectangular beam 200 mm deep is simply supported over a beam of span 2 m. Find the uniformly distributed load, the beam can carry if the bending stress is not to exceed 30 MPa. Take I for the beam as $8 \times 10^6$ mm$^4$. [**Ans.** 4.8 N/mm]
3. A rectangular beam, simply supported over a span of 4 m, is carrying a uniformly distributed load of 50 kN/m. Find the dimensions of the beam, if depth of the beam section is 2.5 times its width. Take maximum bending stress in the beam section as 60 MPa. [**Ans.** 125 mm; 300 mm]
4. Calculate the cross-sectional dimensions of the strongest rectangular beam, that can be cut out of a cylindrical log of wood whose diameter is 500 mm. [**Ans.** 288.5 mm × 408.5 mm]

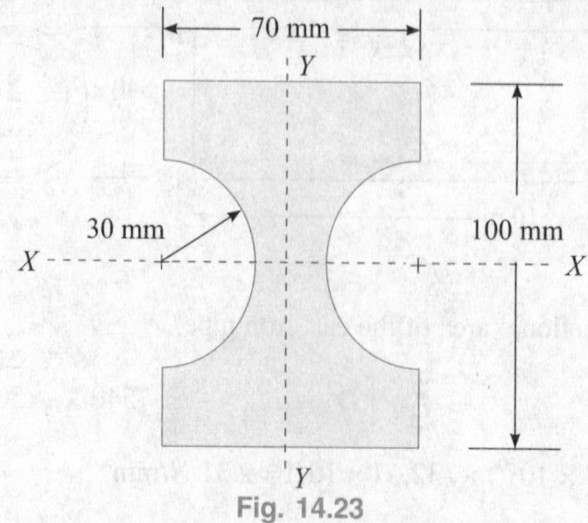

Fig. 14.23

5. Fig. 14.23 shows the section of a beam. What is the ratio of its moment of resistance to bending in the plane $Y$-$Y$ to that for bending in the plane $X$-$X$, if the maximum stress due to bending **is** same in both the cases.

For a semi-circle of radius $r$, the centroid is at a distance of $4r/3\pi$ from the centre. [**Ans.** 2.85]

## QUESTIONS

1. Define the term 'bending stress' and explain clearly the theory of simply bending.
2. State the assumptions made in the theory of simple bending.
3. Prove the relations,

$$\frac{M}{I} = \frac{\sigma}{y} = \frac{E}{R}$$

where $M$ = Bending moment,
$I$ = Moment of inertia,
$\sigma$ = Bending stress in a fibre, at a distance $y$ from the neutral axis,
$E$ = Young's modulus, and
$R$ = Radius of curvature.

4. Discuss the procedure in finding out the bending stress in a symmetrical section.
5. How will you find the bending stress in a hollow circular section?

## OBJECTIVE TYPE QUESTIONS

1. The neutral axis of a section is an axis, at which the bending stress is
   (*a*) minimum (*b*) zero (*c*) maximum (*d*) infinity
2. In the theory of simply bending, the bending stress in the beam section varies
   (*a*) linearly (*b*) parabolically (*c*) elliptically (*d*) none of them
3. When a cantilever is loaded at its free end, maximum compressive stress shall develop at
   (*a*) bottom fibre (*b*) top fibre (*c*) neutral axis (*d*) centre of gravity
4. The section modulus of a rectangular section having width ($b$) and depth ($d$) is
   (*a*) $\frac{bd}{6}$ (*b*) $\frac{bd^2}{6}$ (*c*) $\frac{bd^3}{6}$ (*d*) $\frac{b^2d}{6}$
5. The section modulus of a circular section of diameter ($d$) is
   (*a*) $\frac{\pi}{32}(d)^2$ (*b*) $\frac{\pi}{32}(d)^3$ (*c*) $\frac{\pi}{64}(d)^3$ (*d*) $\frac{\pi}{64}(d)^4$

## ANSWERS

**1.** (*b*) **2.** (*a*) **3.** (*a*) **4.** (*b*) **5.** (*b*)

Chapter **15**

# Bending Stresses in Composite Beams

## Contents

## 15.1. Introduction

In the last chapter, we have discussed the bending stresses in simple beams, and the pattern in which these stresses vary along the symmetrical sections. But sometimes we come across beams of composite sections. And we are required to study the pattern in which these stresses vary along such sections.

## 15.2. Types of Composite Beams

Though there are many types of composite beams that we come across, yet the following are important from the subject point of view:

1. Beams of unsymmetrical sections
2. Beams of uniform strength
3. Flitched beams.

## 15.3. Beams of Unsymmetrical Sections

We have already discussed in the last chapter that in a symmetrical section, the distance of extreme fibre from the c.g. of the section $y = d/2$. But this is not the case, in an unsymmetrical section ($L$, $I$, $T$, etc.), since the neutral axis of such a section does not pass through the geometrical centre of the section. In such cases, first the centre of gravity of the section is obtained as discussed in Chapter 6 and then the values of $y$, in the tension and compression sides, is studied. For obtaining the bending stress in a beam, the bigger value of $y$ (in tension or compression) is used in the equation. This will be illustrated by the following examples.

**EXAMPLE 15.1.** *Two wooden planks 150 mm × 50 mm each are connected to form a T-section of a beam. If a moment of 6.4 kN-m is applied around the horizontal neutral axis, inducing tension below the neutral axis, find the bending stresses at both the extreme fibres of the cross- section.*

**SOLUTION.** Given: Size of wooden planks = 150 mm × 50 mm and moment ($M$) = 6.4 kN-m = 6.4 × $10^6$ N-mm.

Two planks forming the $T$-section are shown in Fig. 15.1. First of all, let us find out the centre of gravity of the beam section. We know that distance between the centre of gravity of the section and its bottom face,

$$\bar{y} = \frac{(150 \times 50)\,175 + (150 \times 50)\,75}{(150 \times 50) + (150 \times 50)} = \frac{1875000}{15000} = 125 \text{ mm}$$

∴ Distance between the centre of gravity of the section and the upper extreme fibre,

$$y_t = 20 - 125 = 75 \text{ mm}$$

and distance between the centre of gravity of the section and the lower extreme fibre,

$$y_c = 125 \text{ mm}$$

We also know that Moment of inertia of the $T$ section about an axis passing through its c.g. and parallel to the botom face,

$$I = \left[\frac{150 \times (50)^3}{12} + (150 \times 50)\,(175 - 125)^2\right] + \left[\frac{50 \times (150)^3}{12} + (150 \times 50\,(125 - 75)^2\right] \text{mm}^4$$

$$= (20.3125 \times 10^6) + (32.8125 \times 10^6) \text{ mm}^4$$

$$= 53.125 \times 10^6 \text{ mm}^4$$

∴ Bending stress in the upper extreme fibre,

$$\sigma_1 = \frac{M}{I} \times y_t = \frac{6.4 \times 10^6}{53.125 \times 10^6} \times 125 \text{ N/mm}^2$$

$$= 15.06 \text{ N/mm}^2 = 15.06 \text{ MPa (compression)} \quad \textbf{Ans.}$$

and bending stress in the lower extreme fibre,

$$\sigma_2 = \frac{M}{I} \times y_c = \frac{6.4 \times 10^6}{53.125 \times 10^6} \times 75 \text{ N/mm}^2$$

$$= 9.04 \text{ N/mm}^2 = 9.04 \text{ MPa (tension)} \quad \textbf{Ans.}$$

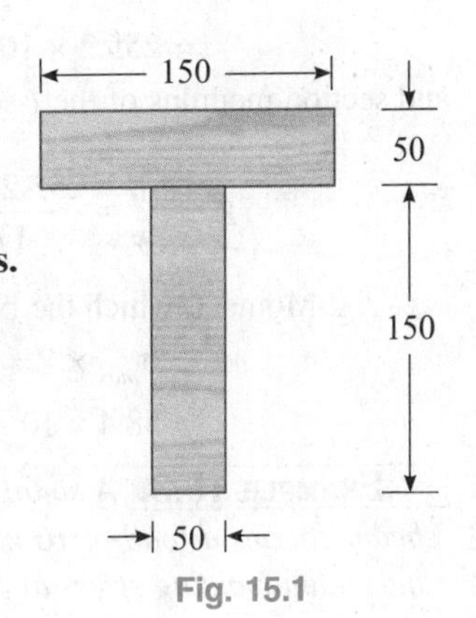

Fig. 15.1

**EXAMPLE 15.2.** *Figure 15.2 shows a rolled steel beam of an unsymmetrical I-section.*

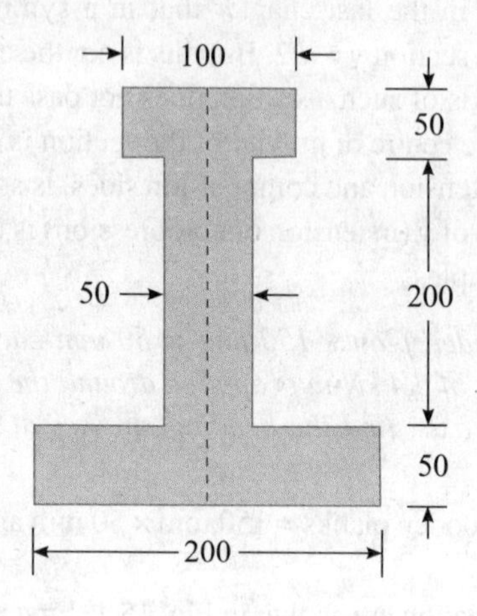


**Fig. 15.2**

*If the maximum bending stress in the beam section is not to exceed 40 MPa, find the moment, which the beam can resist.*

**SOLUTION.** Given: Maximum bending stress $(\sigma_{max}) = 40 \text{ MPa} = 40 \text{ N/mm}^2$.

We know that distance between the centre of gravity of the section and bottom face,

$$\bar{y} = \frac{(100 \times 50)\, 275 + (200 \times 50)\, 150 + (200 \times 50)\, 25}{(100 \times 50) + (200 \times 50) + (200 \times 50)} = 125 \text{ mm}$$

$$\therefore \quad y_1 = 300 - 125 = 175 \text{ mm} \quad \text{and} \quad y_2 = 125 \text{ mm}$$

Thus we shall take the value of $y = 175$ mm (*i.e.*, greater of the two values between $y_1$ and $y_2$).

We also know that moment of inertia of the *I*-section about an axis passing through its centre of gravity and parallel to the bottom face,

$$I = \left[\frac{100 \times (50)^3}{12} + (100 \times 50)\,(275 - 125)^2\right] + \left[\frac{50 \times (200)^3}{12} + (50 \times 200)\,(150 - 125)^2\right]$$

$$+ \left[\frac{200 \times (50)^3}{12} + (200 \times 50)\,(125 - 25)^2\right] \text{mm}^4$$

$$= 255.2 \times 10^6 \text{ mm}^4$$

and section modulus of the *I*-section,

$$Z = \frac{I}{y} = \frac{255.2 \times 10^6}{175} = 1.46 \times 10^6 \text{ mm}^3$$

$\therefore$ Moment, which the beam can resist,

$$M = \sigma_{max} \times Z = 40 \times (1.46 \times 10^6) \text{ N-mm}$$

$$= 58.4 \times 10^6 \text{ N-mm} = 58.4 \text{ kN-m} \quad \textbf{Ans.}$$

**EXAMPLE 15.3.** *A simply supported beam and its cross-section are shown in Fig. 15.3. The beam carries a load of 10 kN as shown in the figure. Its self weight is 3.5 kN/m. Calculate the maximum bending stress at X-X.*

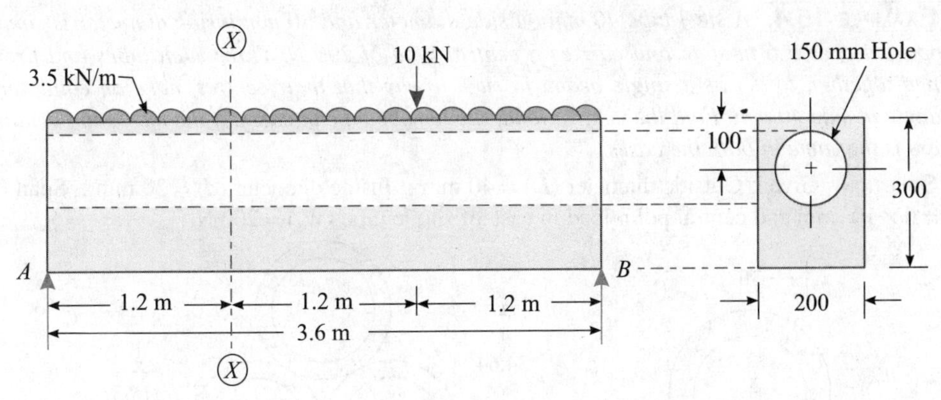


Fig. 15.3

**Solution.** Given: Point load ($W$) = 10 kN = $10 \times 10^3$ N and self weight of the beam ($w$) = 3.5 kN/m = 3.5 N/mm.

First of all, let us find out the centre of gravity of the beam section. We know that distance between the centre of gravity of the section and its bottom face,

$$\bar{y} = \frac{[(200 \times 300)\,150] - \left[\frac{\pi}{4}(150)^2 \times 200\right]}{[200 \times 300] - \left[\frac{\pi}{4}(150)^2\right]} = 129 \text{ mm}$$

∴ Distance between centre of gravity of the section and the upper extreme fibre,

$$y_t = 30 - 129 = 171 \text{ mm}$$

and distance between the centre of gravity of the section and the lower extreme fibre,

$$y_c = 129 \text{ mm}$$

Therefore for maximum bending stress, we shal use the value of y equal to 171 mm (*i.e.*, greater of the two values of $y_t$ and $y_c$). We know that moment of inertia of the section passing through its centre of gravity and parallel to *x-x* axis,

$$I = \left[\frac{200\,(300)^3}{12} + (200 \times 300) \times (150 - 129)^2\right] - \left[\frac{\pi}{64}(150)^4 + \frac{\pi}{4} \times (150)^2 \times (200 - 129)^2\right] \text{mm}^4$$

$$= (476.5 \times 10^6) - (113.9 \times 10^6) = 362.6 \times 10^6 \text{ mm}^4$$

Now let us find out the bending moment at *x-x*. Taking moments about *A* and equating the same,

$$R_B \times 3.6 = (3.5 \times 3.6 \times 1.8) + (10 \times 2.4) = 46.68$$

$$\therefore \quad R_B = \frac{46.68}{3.6} = 13.0 \text{ kN}$$

$$\text{or} \quad R_A = [(3.5 \times 3.6) + 10] - 13.0 = 9.6 \text{ kN}$$

and bending moment at *X*,

$$M = (9.6 \times 1.2) - (3.5 \times 1.2 \times 0.6) = 9 \text{ kN-m} = 9 \times 10^6 \text{ N-mm}$$

∴ Maximum bending stress at *X*,

$$\sigma_b = \frac{M}{I} \times y = \frac{9 \times 10^6}{362.6 \times 10^6} \times 171 = 4.24 \text{ N/mm}^2$$

$$= 4.24 \text{ MPa} \qquad \textbf{Ans.}$$

**EXAMPLE 15.4.** *A steel tube 40 mm outside diameter and 30 mm inside diameter is simply supported over a 6 m span and carries a central load of 200 N. Three such tubes and firmly joined together, to act as a single beam, in such a way that their centres make an equilateral triangle of side 40 mm. Find the central load, the new beam can carry, if the maximum bending stress is the same in both the cases.*

**SOLUTION.** Given: Outside diameter ($D$) = 40 mm ; Inside diameter ($d$) = 30 mm ; Span ($l$) = 6 m = $6 \times 10^3$ mm and central point load in case of single tube ($W_1$) = 200 N.

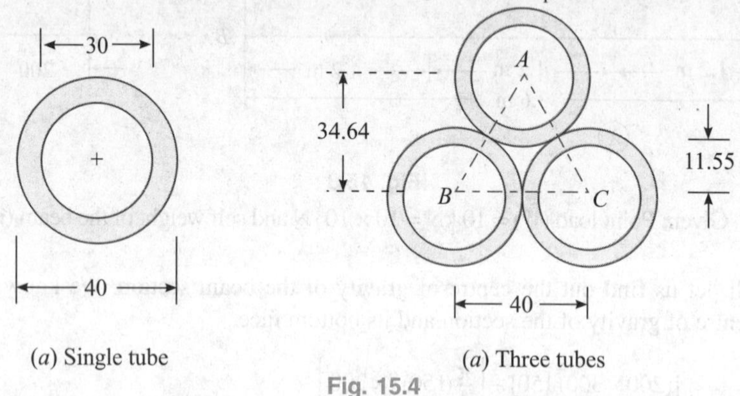


(*a*) Single tube (*a*) Three tubes

**Fig. 15.4**

First of all, consider a single tube as shown in Fig. 15.4 (*a*). We know that maximum bending moment over simply supported load subjected to a central point load

$$M = \frac{Wl}{4} = \frac{200 \times (6 \times 10^3)}{4} = 300 \times 10^3 \text{ N-mm}$$

and section modulus of a hollow circular section

$$Z = \frac{\pi}{32D} \times \left[D^4 - d^4\right] = \frac{\pi}{32 \times 40} \times \left[(40)^4 - (30)^4\right] \text{ mm}^3$$

$$= 4.295 \times 10^3 \text{ mm}^3$$

∴ Maximum bending stress,

$$\sigma_{max} = \frac{M}{Z} = \frac{300 \times 10^3}{4.295 \times 10^3} = 69.85 \text{ N/mm}^2$$

Now consider these tubes firmly joined together as shown in Fig. 15. 4 (*b*). We know that vertical height of the equilateral triangle,

$$= AB \sin 60° = 40 \times 0.866 = 34.64 \text{ mm}$$

∴ Centre of gravity of the section will lie at a height of 34.64/3 = 11.5 mm from the base *BC*. Thus distance between the centre of gravity of the section and upper extreme fibre,

$$y_c = (34.64 - 11.55) + 20 = 43.09 \text{ mm}$$

and distance between the centre of gravity of the section and the lower extreme fibre,

$$y_t = 11.55 + 20 = 31.55 \text{ mm}$$

Therefore for maximum bending stress, we shall use the value of $y$ equal to 43.09 mm (*i.e.*, greater of two values of $y_c$ and $y_t$). We know that cross-sectional area of one tube,

$$A = \frac{\pi}{4} \times \left[D^2 - d^2\right] = \frac{\pi}{4} \times \left[(40)^2 - (30)^2\right] = 549.8 \text{ mm}^2$$

and moment of inertia of one hollow tube

$$= \frac{\pi}{64} \times \left[D^4 - d^4\right] = \frac{\pi}{64} \times \left[(40)^4 - (30)^4\right] = 85.9 \times 10^3 \text{ mm}^4$$

∴ Moment of inertia of whole section passing through its centre of gravity and parallel to $X$-$X$ axis,

$$I = 2\left[85.9\times10^3 + 549.8\,(11.55)^2\right] + \left[85.9\times10^3 + 549.8\,(34.64-11.55)^2\right]$$

$$= (318.5 \times 10^3) + (379.0 \times 10^3) = 697.5 \times 10^3 \text{ mm}^4$$

and maximum bending moment at the centre of beam due to the central load $W_2$,

$$M = \frac{W_2 l}{4} = \frac{W_2 \times (6\times10^3)}{4} = 1.5 \times 10^3\, W_2 \text{ N-mm}$$

We know that maximum bending stress ($\sigma_{max}$)

$$69.85 = \frac{M}{I}\times y = \frac{1.5\times10^3\, W_2}{697.5\times10^3} \times 43.09 = 0.093\, W_2$$

$$\therefore \quad W_2 = \frac{69.85}{0.093} = 751 \text{ N} \qquad \textbf{Ans.}$$

**EXAMPLE 15.5.** *Figure 15.5 shows a rolled steel beam of an unsymmetrical I-section.*

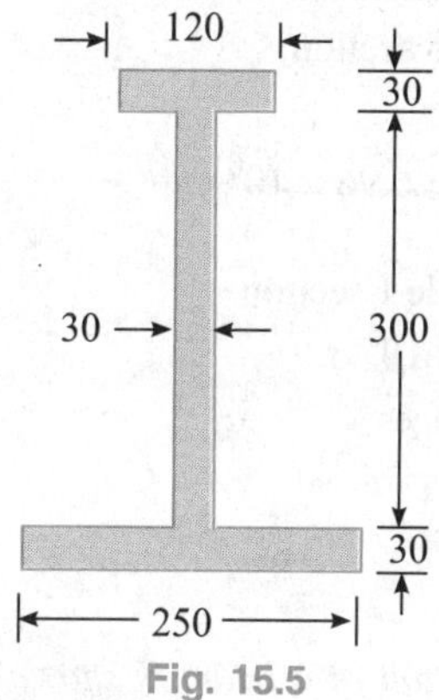


Fig. 15.5

*If a similar I-section is welded on the top of it to form a symmetrical section, determine the ratio of the moment of resistance of the new section to that of the single section. Assume the permissible bending stress in tension and compression to be the same.*

**SOLUTION.** Given: Permissible bending stess in tension = Permissible bending stress in compression.

First of all, let us find the centre of gravity of the section. We know that distance between the centre of gravity of the section and bottom face,

$$\bar{y} = \frac{(120\times30)\,345 + (300\times30)\,180 + (250\times30)\,15}{(120\times30) + (300\times30) + (250\times30)} \text{ mm}$$

$$= \frac{2974500}{20100} = 148 \text{ mm}$$

$$\therefore \quad y_1 = 360 - 148 = 212 \text{ mm} \qquad \text{and} \qquad y_2 = 148 \text{ mm}$$

Thus for the prupose of calculating moment of resistance of the section, we shall take the value of $y$ equal to 212 mm (*i.e.*, greater of the two values between $y_1$ and $y_2$). We also know that moment of inertia of the $I$-section about an axis through its centre of gravity and parallel to its $x$-$x$ axis,

$$I_1 = \left[\frac{120\times(30)^3}{12} + (120\times30)\,(345-148)^2\right] + \left[\frac{30\times(300)^3}{12} + (30\times300)\,(180-148)^2\right]$$

$$+ \left[\frac{250\times(30)^3}{12} + (250\times30)\,(148-15)^2\right] \text{ mm}^4$$

$= 350 \times 10^6 \text{ mm}^4$

$\therefore$ Section modulus of the *I*-section,

$$Z_1 = \frac{1}{y} = \frac{350 \times 10^6}{212} = 1.65 \times 10^6 \text{ mm}^3$$

and moment of resistance of the *I*-section

$$M_1 = \sigma \times Z_1 = \sigma \times 1.65 \times 10^6 = 1.65 \times 10^6 \sigma \quad ...(i)$$

Now, let us consider the double section as shown in Fig. 15.6. We know that in this case, centre of gravity of the section will lie at the junction of the two sections.

Therefore moment of inertia of the double section about its axis through its c.g. and parallel to *x-x* axis,

$$I_2 = 2\,[(350 \times 10^6) + 20100 \times (212)^2] \text{ mm}^4$$
$$= 2\,[(350 \times 10^6) + (903.4 \times 10^6)] = 2506.8 \times 10^6 \text{ mm}^4$$

$\therefore$ Section moulus of the double section,

$$Z_2 = \frac{I}{y} = \frac{2506.8 \times 10^6}{360} = 6.96 \times 10^6 \text{ mm}^3$$

and moment of resistance of the double I-section

$$M_2 = \sigma \times Z_2 = \sigma \times 6.96 \times 10^6 \sigma \quad ...(ii)$$

$\therefore$ Ratio of moments of resistances

$$\frac{M_2}{M_1} = \frac{6.96 \times 10^6 \sigma}{1.65 \times 10^6 \sigma} = 4.22 \quad \textbf{Ans.}$$

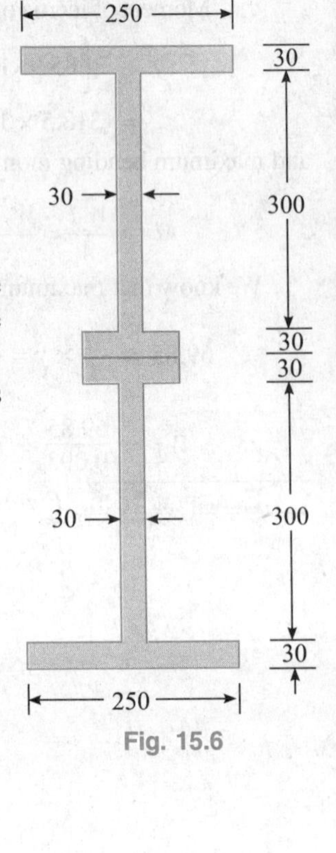


Fig. 15.6

**EXAMPLE 15.6.** *The cross-section of a beam is shown in Fig. 15.7. The beam is made of material with permissible stress in compression and tension equal to 100 MPa and 140 MPa respectively.*

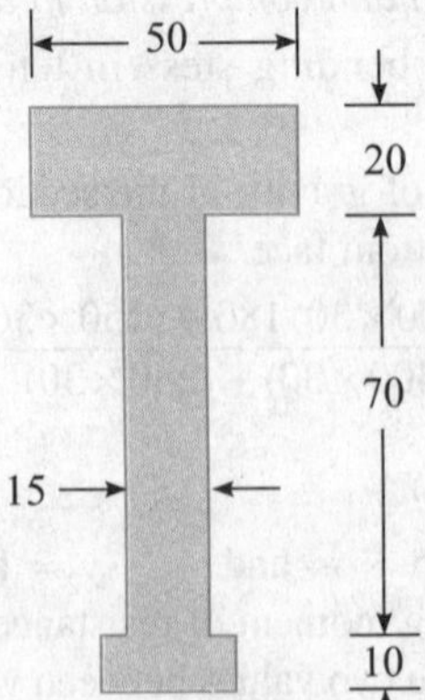

Fig. 15.7

*Calculate the moment of resistance of the cross-section, when subjected to a moment causing compression at the top and tension at the bottom.*

**SOLUTION.** Given: Permissible stress in compression ($\sigma_c$) = 100 MPa = 100 N/mm$^2$ and permissible stress in tension ($\sigma_t$) = 140 MPa = 140 N/mm$^2$.

***Moment of resistance of the cross-section***

First of all, let us find the centre of gravity of the section. We know that the distance between the centre of gravity of the section and its bottom face,

$$\bar{y} = \frac{(50 \times 20)\,90 + (70 \times 15)\,45 + (25 \times 10)5}{(50 \times 20) + (70 \times 15) + (25 \times 10)} = 60.2 \text{ mm}$$

$$\therefore \qquad y_1 = 100 - 60.2 = 39.8 \text{ mm} \qquad \text{and} \qquad y_2 = 60.2 \text{ mm}$$

Thus for the purpose of calculating moment of resistance of the section, we shall take the value of $y$ equal to 60.2 mm (*i.e.*, greater of the two values between $y_1$ and $y_2$). We also know that moment of inertia of the section about an axis through its c.g. and parallel to $x$-$x$ axis,

$$I = \left[\frac{50 \times (20)^3}{12} + (50 \times 20)\,(90 - 60.2)^2\right] + \left[\frac{15 \times (70)^3}{12} + (70 \times 15)\,(60.2 - 45)^2\right]$$

$$+ \left[\frac{25 \times (10)^3}{12} + (25 \times 10)\,(60.2 - 5)^2\right] \text{mm}^4$$

$$= 2356.6 \times 10^3 \text{ mm}^4$$

$\therefore$ Section modulus of the section (in compression zone),

$$Z_1 = \frac{I}{y_1} = \frac{2356.6 \times 10^3}{39.8} = 59.2 \times 10^3 \text{ mm}^3$$

and moment of resistance of the compression zone,

$$M_1 = \sigma_c \times Z_1 = 100 \times 59.2 \times 10^3 = 5920 \times 10^3 \text{ N-mm}$$

Similarly, section modulus of the section (in tension zone),

$$Z_2 = \frac{M}{y_2} = \frac{2356.6 \times 10^3}{60.2} = 39.1 \times 10^3 \text{ mm}^3$$

and moment of resistance of the tension zone,

$$M_2 = \sigma_t \times Z_2 = 140 \times 39.1 \times 10^3 = 5474 \times 10^3 \text{ N-mm}$$

$\therefore$ Moment of resistance of the cross-section is the least of the two values *i.e.*,

$5474 \times 10^3$ N-mm **Ans.**

## EXERCISE 15.1

1. Cantilever beam of span 2.5 m has a *T*-section as shown in Fig. 15.8. Find the point load, which the cantilever beam can carry at its free end, if the bending stress is not to exceed 50 MPa. (**Ans.** 1.6 kN)
2. An *I*-section shown in Fig. 15.9 is simply supported over a span of 5 metres. If the tensile stress is not to exceed 20 MPa, find the safe uniformly distributed load, the beam can carry. (**Ans.** 6.82 kN/m)

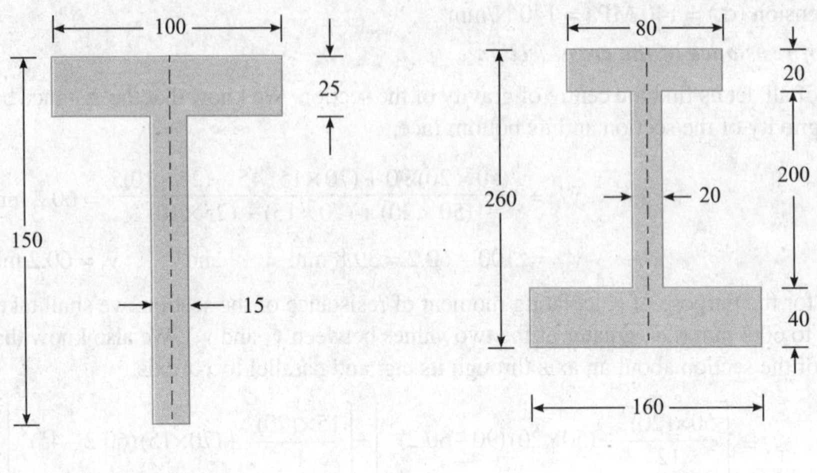


Fig. 15.8 Fig. 15.9

3. Two beams are simply supported over the same span and have the same flexural strength. Compare the weights of these two beams, if one of them is solid circular and the other hollow circular with internal diameter half of the external diameter. **(Ans.** 1.28)

## 15.4. Beams of Uniform Strength

We have already discussed that in a simply supported beam, carrying a *uniformly distributed load, the maximum bending moment will occur at its centre. It is thus obvious that the bending stress is also maximum at the centre of the beam. As we proceed, from the centre of the beam towards the supports, the bending moment decreases and hence the maximum stress developed is below the permissible limit. It results in the wastage of material. This wastage is negligible in case of small spans, but considerable in case of large spans.

The beams of large spans are designed in such a way that their cross-sectional area is decreased towards the supports so that the maximum bending stress developed is equal to the allowable stress (as is done at the centre of the beam). Such a beam, in which bending stress developed is constant and is equal to the allowable stress at every section is called a beam of uniform strength. The section of a beam of uniform strength may be varied in the following ways:

1. By keeping the width uniform and varying the depth.
2. By keeping the depth uniform and varying the width.
3. By varying both width and depth.

The most common way of keeping the beam of uniform strength is by keeping the width uniform and varying the depth.

**EXAMPLE 15.7.** *A simply supported beam of 2.4 meters span has a constant width of 100 mm throughout its length with varying depth of 150 mm at the centre to minimum at the ends as shown in Fig. 15.10. The beam is carrying a point load W at its mid-point.*

* This is the most practical case. However, if a beam is carrying some other type loading, the maximum bending moment will occur, at a point, near its centre.

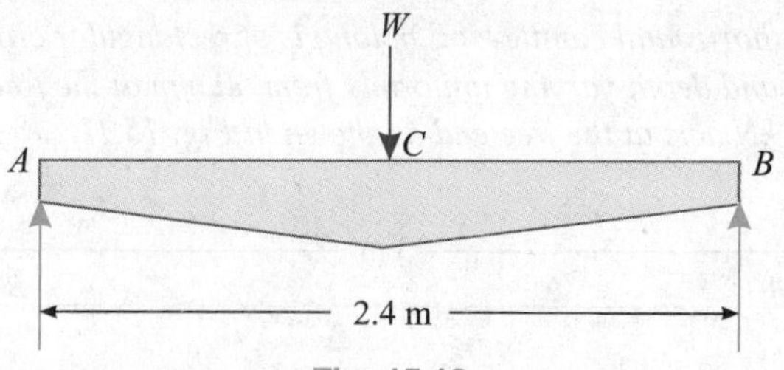


Fig. 15.10

*Find the minimum depth of the beam at a section 0.6 m from the left hand support, such that the maximum bending stress at this section is equal to that at the mid-span of the beam.*

**Solution.** Given: Span $(l) = 2.4 \text{ m} = 2.4 \times 10^3 \text{ mm}$ ; Width $(b) = 100$ mm and depth at the centre $(d_C) = 150$ mm.

Let $d_X$ = Depth at the section $X$ *i.e.*, 0.6 m (*i.e.*, 600 mm) from the left end.

$f_X$ = Bending stress at $X$ and

$f_C$ = Bending stress at $C$.

Since the beam is carrying a central point load, therefore the reaction at $A$,

$$R_A = R_B = \frac{W}{2}$$

Bending moment at $C$,

$$M_C = \frac{W}{2} \times 1200 = 600 \text{ W}$$

Similarly,

$$M_X = \frac{W}{2} \times 600 = 300 \text{ W}$$

We know that section modulus at the centre of beam,

$$Z_X = \frac{b.d_X^2}{6} = \frac{100 d_X^2}{6} = 50 \frac{d_X^2}{3} \text{ mm}^3$$

and

$$Z_C = \frac{b.d_C^2}{6} = \frac{100 \times (150)^2}{6} = 375\,000 \text{ mm}^3$$

We also know that bending moment at $C$ $(M_C)$,

$$600\, W = \sigma_C \times Z_C = \sigma_C \times 375\,000$$

$$\therefore \quad \sigma_C = \frac{600\, W}{375\,000} \quad ...(i)$$

Similarly bending moment at $X$ $(M_X)$

$$300\, W = \sigma_X \times Z_X = \sigma_X \times \frac{50\, d_X^2}{3}$$

$$\therefore \quad \sigma_X = 300\, W \times \frac{3}{50\, d_X^2} = \frac{18\, W}{d_X^2} \quad ...(ii)$$

Since $\sigma_C$ is equal to $\sigma_X$, therefore equating (*i*) and (*ii*),

$$\frac{600W}{375000} = \frac{18W}{d_X^2}$$

$$\therefore \quad d_X^2 = \frac{18 \times 375000}{600} = 11250 \text{ mm}^2$$

or

$$d_X = 106.01 \text{ mm} \quad \textbf{Ans.}$$

**Example 15.8.** *A horizontal cantilever 3 m long is of rectangular cross-section 60 mm wide throughout its length, and depth varying uniformly from 60 mm at the free end to 180 mm at the fixed end. A load of 4 kN acts at the free end as shown in Fig. 15.11.*

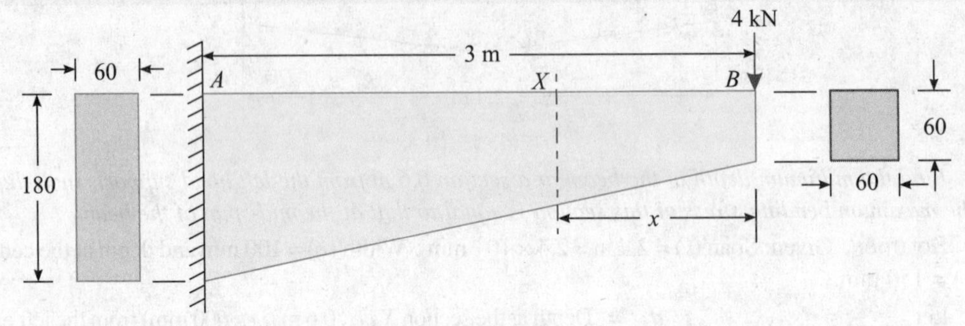


Fig. 15.11

*Find the position of the highest stressed section and the value of the maximum bending stress induced. Neglect the weight of the cantilever itself.*

**Solution.** Given: Span $(l) = 3 \text{ m} = 3 \times 10^3$ mm and point load at the free end $(W) = 4 \text{ kN} = 4 \times 10^3$ N.

***Position of the highest stressed section***

Let $x$ = Distance in metres of the section from $B$, which is highest stressed.

We know that the moment at $X$,

$$M_X = (4 \times 10^3)(x \times 10^3) = 4 \times 10^6 \, x \text{ N-mm} \quad ...(i)$$

and depth of the cantilever at $X$,

$$d = 60 + \frac{180 - 60}{3} x = 60 + 40\,x \text{ mm} \quad ...(ii)$$

∴ Section modulus at $X$,

$$Z_X = \frac{bd^2}{6} = \frac{60}{6}(60 + 40\,x)^2 \text{ mm}^3$$

$$= 10\,[20\,(3 + 2\,x)]^2 = 4000\,(3 + 2\,x)^2 \text{ mm}^3 \quad ...(iii)$$

We also know that bending stress at $X$,

$$\sigma = \frac{M_X}{Z_X} = \frac{4 \times 10^6\,x}{4000\,(3 + 2\,x)^2} = \frac{10^3\,x}{(3 + 2\,x)^2} \text{ N/mm}^2 \quad ...(iv)$$

Now for σ to be maximum, differentiate the above equation and equate it to zero, *i.e.*,

$$\frac{d\sigma}{dx} = \frac{d}{dx}\left(\frac{10^3\,x}{(3 + 2\,x)^2}\right) = 0 \quad \text{or} \quad 2\,(3 + 2\,x) = 0$$

$$\therefore \quad x = 1.5 \text{ m} \quad \textbf{Ans.}$$

***Value of the maximum bending stress***

Now substituting the value of $x$ in equation ($iv$),

$$\sigma_{max} = \frac{10^3 \times 1.5}{(3 + 2 \times 1.5)^2} = 41.7 \text{ N/mm}^2 = 41.7 \text{ MPa} \quad \textbf{Ans.}$$

## 15.5 Beams of Composite Section (Flitched Beams)

A composite section may be defined as a section made up of two or more different materials, joined together in such a manner that they behave like a single piece and, each material bends to the same radius of curvature. Such beams are used when a beam of one material, if used alone, requires quite a large cross-sectional area; which does not suit the space available. A material is then reinforced with some other material, of higher strength, in order to reduce the cross-sectional area of the beam and to suit the space available (as is done in the case of reinforced cement concrete beams).

In such cases, the total moment of resistance will be equal to the sum of the moments of individual sections.

Consider a beam of a composite section made up of two different materials as shown in Fig. 15.12.

Let $E_1$ = Modulus of elasticity of part 1,

$I_1$ = Moment of inertia of the part 1,

$M_1$ = Moment of resistance for part 1,

$\sigma_1$ = Stress in part 1,

$Z_1$ = Modulus of section for part 1,

$E_2, I_2, M_2, \sigma_2, Z_2$ = Corresponding values for part 2 and

$R$ = Radius of the bend up beam.

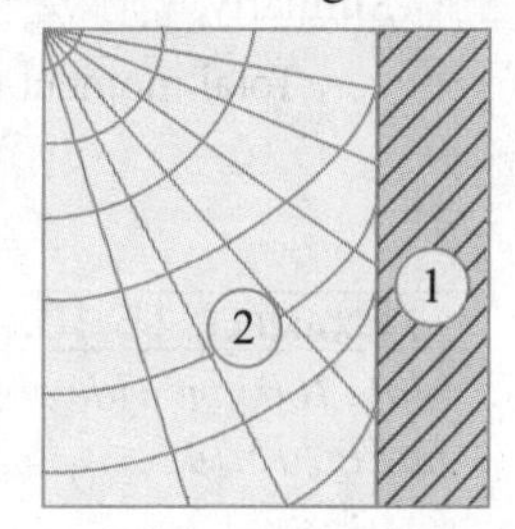


Fig. 15.12

We know that the moment of resistance for beam 1,

$$M_1 = \sigma_1 \times Z_1 \qquad (\because M = \sigma \times Z)$$

Similarly,

$$M_2 = \sigma_2 \times Z_2$$

∴ Total moment of resistance of the composite section,

$$M = M_1 + M_2 = (\sigma_1 \times Z_1) + (\sigma_2 \times Z_2) \qquad ...(i)$$

We also know that at any distance from the neutral axis, the strain in both the materials will be the same.

$$\frac{\sigma_1}{E_1} = \frac{\sigma_2}{E_1} \quad \text{or} \quad \sigma_1 = \frac{E_1}{E_2} \times \sigma_2 = m \times \sigma_2$$

where $m = \frac{F_1}{E_2}$ *i.e.*, Modulus ratio.

From the above two relations, we can find out the total moment of resistance of a composite beam or stresses in the two materials. But, if the sections of both the materials are not symmetrical, then one area of the components is converted into an equivalent area of the other.

**EXAMPLE 15.9.** *A flitched timber beam made up of steel and timber has a section as shown in Fig. 15.13.*

*Determine the moment of resistance of the beam. Take $\sigma_S$ = 100 MPa and $\sigma_T$ = 5 MPa.*

**SOLUTION.** Width of each timber section $(b_T)$ = 60 mm ; Depth of each timber section $(d_T)$ = 200 mm ; Stress in timber $(\sigma_T)$ = 5 MPa = 5 $N/mm^2$ ; Width of steel section $(b_S)$ = 15 mm ; Depth of steel section $(d_S)$ = 20 mm and stress in steel $(\sigma_S)$ = 100 MPa = 100 $N/mm^2$.

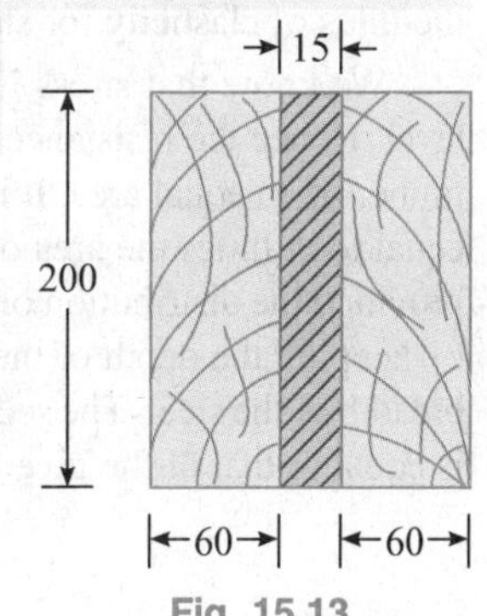


Fig. 15.13

We know that the section modulus of a rectangular body,

$$Z = \frac{bd^2}{6}$$

∴ Modulus of section for both the timber sections,

$$Z_T = 2\left[\frac{60 \times (200)^2}{2}\right] = 800 \times 10^3 \text{ mm}^3 \qquad ...(\because \text{ of two sections})$$

Similarly, modulus of section for the steel section

$$Z_S = \frac{15 \times (200)^2}{6} = 100 \times 10^3 \text{ mm}^3$$

We also know that moment of resistance for timber,

$$M_T = \sigma_T \times Z_T = 5 \times (800 \times 10^3) = 4 \times 10^6 \text{ N-mm}$$

Similarly,

$$M_S = \sigma_S \times Z_S = 100 \times (100 \times 10^3) = 10 \times 10^6 \text{ N-mm}$$

∴ Total moment of resistance of the beam,

$$M = M_T + M_S = (4 \times 10^6) + (10 \times 10^6) = 14 \times 10^6 \text{ N-mm}$$

$$= 14 \text{ kN-m} \qquad \textbf{Ans.}$$

**EXAMPLE 15.10.** *A timber beam 100 mm wide and 200 mm deep is strengthened by a steel plate 100 mm wide and 100 mm thick, screwed at the bottom surface of the timber beam as shown in Fig. 15.14.*

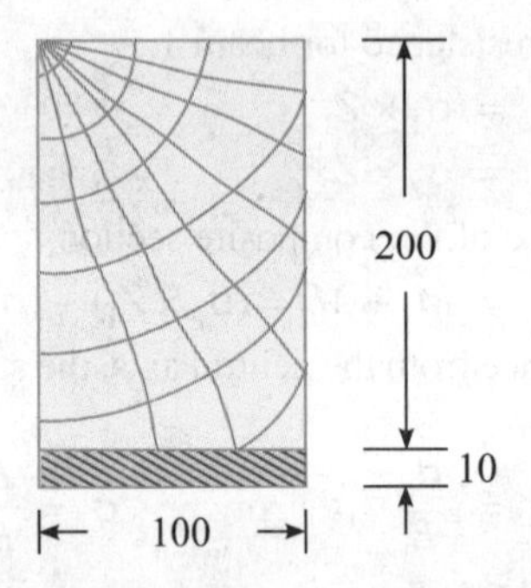


Fig. 15.14

*Calculate the moment of resistance of the beam, if the safe stresses in timber and steel are 10 MPa and 150 MPa respectively. Take $E_S = 20\ E_T$.*

**SOLUTION.** Given : Width of timber section ($b_T$) = 100 mm ; Depth of timber section ($d_T$) = 200 mm ; Safe stress in timber ($\sigma_T$) = 10 MPa = 10 N/mm$^2$ ; Width of steel section ($b_S$) = 100 mm ; Depth of steel section ($d_S$) = 10 mm ; Safe stress in steel ($\sigma_S$) = 150 MPa = 150 N/mm$^2$ and modulus of elasticity for steel ($E_S$) = 20 $E_T$.

We know that stress in steel is *m* times (20 times in this case) the stress in timber at the same level. Hence the resistance offered by the steel is also equal to *m* times the resistance offered by the timber of an equal area. It is thus obvious that if we replace steel by timber (or *vice versa*) of an area equal to *m* times the area of the steel, the total resistance to bending offered will remain unchanged; provided the distribution of the area about the neutral axis also remains unchanged. This can be done, by keeping the depth of the area unchanged and by increasing the breadth of the timber *m* times the breadth of the steel. The section thus obtained is called equivalent section and its moment of resistance is equal to that of the given section.

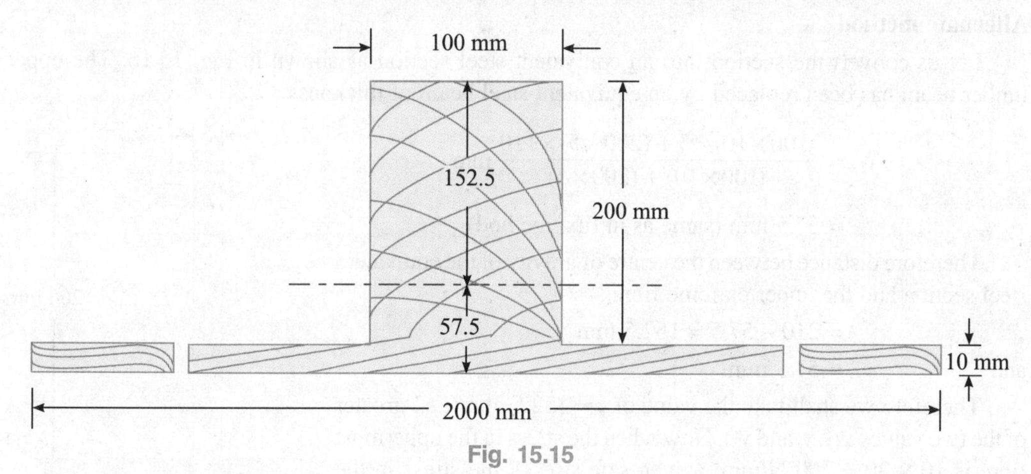


Fig. 15.15

In this case the equivalent section (of wood) is shown in Fig. 15.15. The bottom steel plate has been replaced by an equivalent timber of 100 × 20 = 2000 mm.

We know that distance between the centre of gravity of the equivalent timber section and its bottom face,

$$\bar{y} = \frac{(2000 \times 10) \times 5 + (100 \times 200) \times 110}{(2000 \times 10) + (100 \times 200)} = 57.5 \text{ mm}$$

Therefore distance between the centre of gravity of the equivalent timber section and the upper extreme fibre,

$$y_C = 210 - 57.5 = 152.5 \text{ mm}$$

and $y_T = 57.5$ mm

Therefore we shall take the value of $y = 152.5$ mm (*i.e.*, greater of the two values among, $y_T$ and $y_C$). Now when the stress in uppermost fibre is 10 N/mm$^2$ (given safe stress), then the stress in the lowermost fibre,

$$= \frac{10 \times 57.5}{152.5} = 3.77 \text{ N/mm}^2$$

∴ Actual stress in steel at this fibre = 3.77 × 20 = 75.4 N/mm$^2$

It is below the given safe stress (*i.e.*, 150 N/mm$^2$). We also know that moment of inertia of the equivalent timber section about an axis passing through its centre of gravity and parallel to *x-x* axis,

$$I = \left[\frac{2000 \times (10)^3}{12} + (2000 \times 10)(57.5 - 5)^2\right] + \left[\frac{100 \times (200)^3}{12} + (100 \times 200)(110 - 57.5)^2\right] \text{mm}^4$$

$$= (55.3 \times 10^6) + (121.8 \times 10^6) = 177.1 \times 10^6 \text{ mm}^4$$

and section modulus of the equivalent section,

$$Z = \frac{I}{y} = \frac{177.1 \times 10^6}{152.5} = 1.16 \times 10^6 \text{ mm}^3$$

∴ Moment of resistance of the equivalent section,

$$M = \sigma_1 \times Z = 10\,(1.16 \times 10^6) = 11.6 \times 10^6 \text{ N-mm}$$

$$= 11.6 \text{ kN-m}$$ **Ans.**

**Alternate method**

Let us convert the section into an equivalent steel section as shown in Fig. 15.16. The upper timber beam has been replaced by an equivalent steel beam of thickness

$$\bar{y} = \frac{(100 \times 10) \times 5 + (200 \times 5) \times 110}{(100 \times 10) + (200 \times 5)} \text{ mm}$$

$$= 57.5 \text{ mm (same as in first method)}$$

Therefore distance between the centre of gravity of the equivalent steel section and the upper extreme fibre,

$$y_c = 210 - 57.5 = 152.5 \text{ mm}$$

and $\quad y_t = 57.5$ mm

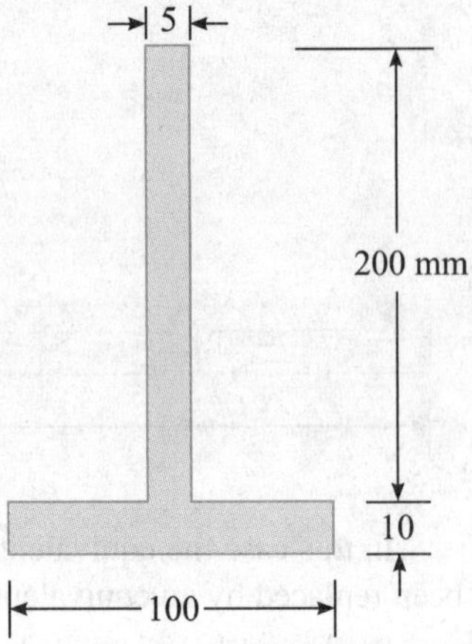

Therefore we shall take the value of $y = 152.5$ mm (*i.e.*, greater of the two values *i.e.*, $y_t$ and $y_c$). Now when the stress in the uppermost fibre is 10 × 20 = 200 N/mm² (given safe stress), the stress in the lowermost fibre

$$= \frac{200 \times 57.5}{152.5} = 75.4 \text{ N/mm}^2$$

Fig. 15.16

It is below the given safe stress (*i.e.*, 150 N/mm²). We also know that moment of inertia of the equivalent steel section, about an axis passing through its c.g. and parallel to *x-x* axis,

$$I = \left[\frac{100 \times (10)^3}{12} + (100 \times 10)(57.5 - 5)^2\right] + \left[\frac{5 \times (200)^3}{12} + (5 \times 200)(110 - 57.5)^2\right] \text{mm}^4$$

$$= (2.76 \times 10^6) + (6.09 \times 10^6) = 8.85 \times 10^6 \text{ mm}^4$$

and section modulus of the equivalent section,

$$Z = \frac{I}{y} = \frac{8.85 \times 10^6}{152.5} = 0.058 \times 10^6 \text{ mm}^3$$

∴ Moment of resistance of the equivalent section,

$$M = \sigma_2 \times Z = (20 \times 10) \times (0.058 \times 10^6) = 11.6 \times 10^6 \text{ N-mm}$$

$$= 11.6 \text{ kN-m} \quad \textbf{Ans.}$$

**EXAMPLE 15.11.** *A compound beam is formed by joining two bars, one of brass and the other of steel, each 40 mm wide and 10 mm deep. This beam is supported over a span of 1 mm with the brass bar placed over the steel bar as shown in Fig. 15.17.*

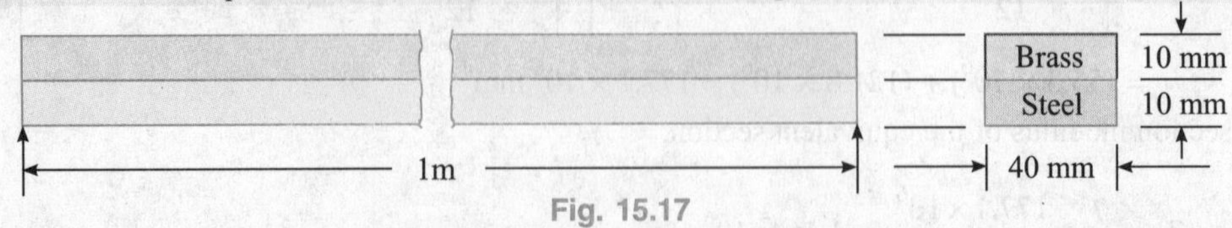


Fig. 15.17

*Determine the maximum load, which can be applied at the centre of the beam, when the bars are:*

*(a) separate and can beand independently,*

*(b) firmly secured to each other, throughout their length.*

*Take* $E_S = 200$ *GPa ;* $E_B = 80$ *GPa and* $\sigma_S = 112.5$ *MPa ;* $\sigma_B = 75$ *MPa*

**SOLUTION.** Given: Width ($b$) = 40 mm ; Depth of brass bar ($d_B$) = $d_S$ = 10 mm ; Span ($l$) = 1 m = 1 × $10^3$ mm ; Modulus of elasticity for steel ($E_S$) = 200 GPa = 200 × $10^3$ N/mm$^2$ ; Modulus of elasticity for brass ($E_B$) = 80 GPa = 80 × 10 N/mm$^2$ ; Allowable stress in steel ($\sigma_S$) = 112.5 MPa = 112.5 N/mm$^2$ and allowable stress in brass $\sigma_B$ = 75 MPa = 75 N/mm$^2$.

***When the bars are separate and can bend independently***

Let $W$ = Maximum load, which can be applied at the centre of the beam.

We know that section modulus for steel,

$$Z_S = Z_B = \frac{bd^2}{6} = \frac{40 \times (10)^2}{6} = \frac{2000}{3} \text{ mm}^3$$

A little consideration will show that each bar will bend about its own axis independently. But for the sake of simplicity, let us assume that each bar has the same radius of curvature. We know that

$$\frac{M}{I} = \frac{E}{R} \quad \text{or} \quad R = \frac{EI}{M}$$

$$\therefore \quad \frac{E_S \cdot I_S}{M_S} = \frac{E_B \cdot I_B}{M_B}$$

$$\frac{M_S}{M_B} = \frac{E_S}{E_B} = \frac{200 \times 10^3}{80 \times 10^3} = 2.5 \qquad ...(\because I_S = I_B)$$

or $$M_S = 2.5\, M_B$$

$$\therefore \quad \sigma_S \cdot Z_S = 2.5\, \sigma_B \cdot Z_B$$

$$\sigma_S = 2.5\, \sigma_B \qquad ...\left(\because Z_S = Z_B = \frac{2000}{3}\right)$$

Thus stress in brass when the *stress in steel is 112.5 N/mm$^2$,

$$\sigma_B = \frac{\sigma_S}{2.5} = \frac{112.5}{2.5} = 45 \text{ N/mm}^2$$

It is below the permissible stress (*i.e.*, 75 N/mm$^2$). Therefore moment of resistance of the steel beam,

$$M_S = \sigma_S \times Z_S = 112.5 \times \frac{2000}{3} = 75\,000 \text{ N-mm}$$

and $$M_B = \sigma_B \times Z_B = 45 \times \frac{2000}{3} = 30\,000 \text{ N-mm}$$

Therefore total moment of resistance,

$$M = M_S + M_B = 75\,000 + 30\,000 = 105\,000 \text{ N-mm} \qquad ...(i)$$

We know that maximum bending moment at the centre, when it is to support a load $W$ at the centre,

$$M = \frac{Wl}{4} = \frac{W \times (1 \times 10^3)}{4} = 250\, W \qquad ...(ii)$$

* If the maximum stress in brass is considered to be 75 N/mm$^2$, then the stress in steel

$$\sigma_S = 2.5\, \sigma_B = 2.5 \times 75 = 187.5 \text{ N/mm}^2$$

But it is more than the permissible limit. Therefore we shall consider stress in steel as 112.5 N/mm$^2$.

Equating equations (*i*) and (*ii*),

$$105\,000 = 250\,W$$

$$\therefore \qquad W = \frac{105000}{250} = 420 \text{ N} \qquad \textbf{Ans.}$$

***When the bars are firmly secured to each other throughout their length***

Now let us convert the whole section into an equivalent *brass section as shown in Fig. 15.18.

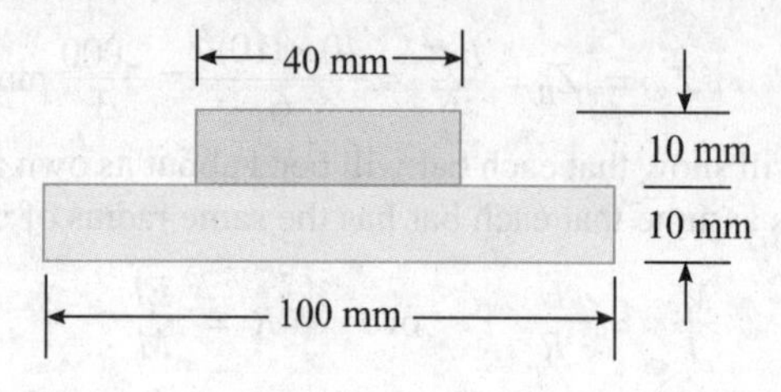


Fig. 15.18

The bottom steel plate has been replaced by an equivalent brass plate of thickness

$$= 40 \times \frac{200 \times 10^3}{80 \times 10^3} = 100 \text{ mm}$$

We know that distance between the centre of gravity of the section and bottom face of the equivalent brass section,

$$\bar{y} = \frac{(100 \times 10)\,5 + (40 \times 10)\,15}{(100 \times 10) + (40 \times 10)} = 7.86 \text{ mm}$$

$\therefore$ Distance of centre of gravity from the upper extreme fibre,

$$y_1 = 20 - 7.86 = 12.14 \text{ mm} \qquad \text{and} \qquad y_2 = 7.86 \text{ mm}$$

Therefore we shall take the value of $y = 12.14$ mm (*i.e.*, greater of the two values among $y_T$ and $y_C$).

Now when the stress in the uppermost fibre is 75 N/mm$^2$ (given stress) then the stress in the lowermost fibre is

$$= \frac{75 \times 7.86}{12.14} = 48.6 \text{ N/mm}^2$$

Therefore actual stress in steel in the lowermost fibre

$$= 48.6 \times 2.5 = 121.5 \text{ N/mm}^2$$

It is more than the given safe stress in steel (*i.e.*, 112.5 N/mm$^2$). It is thus obvious that the brass cannot be fully stressed. Now taking maximum stress in steel at the bottom to be 112.5 N/mm$^2$, we find that the stress in brass at the bottom fibre,

$$\sigma_B = \frac{\sigma_S}{2.5} = \frac{112.5}{2.5} = 45 \text{ N/mm}^2$$

We also know that moment of inertia of the equivalent section about an axis passsing through its centre of gravity and parallel to *x-x* axis,

$$I = \left[\frac{100 \times (10)^3}{12} + (100 \times 10)(7.86 - 5.0)^2\right] + \left[\frac{40 \times (10)^3}{12} + (40 \times 10)(15 - 7.86)^2\right] \text{mm}^4$$

* We may also convert the whole section into an equivalent steel section.

$$= 40.24 \times 10^3 \text{ mm}^4$$

and section modulus of the equivalent section,

$$Z = \frac{I}{y} = \frac{40.24 \times 10^3}{12.14} = 3.31 \times 10^3 \text{ mm}^3$$

∴ Moment of resistance of the equivalent section,

$$M = \sigma \times Z = 45 \times (3.31 \times 10^3) = 149 \times 10^3 \text{ N-mm} \quad ...(iii)$$

We know that the maximum bending moment at the centre, when it is to support a load $W$ at the centre,

$$M = \frac{Wl}{4} = \frac{W \times (1 \times 10^3)}{4} = 250\ W \quad ...(iv)$$

Equating equations (*iii*) and (*iv*)

$$149 \times 10^3 = 250\ W$$

$$\therefore \quad W = \frac{149 \times 10^3}{250} = 596 \text{ N} \quad \textbf{Ans.}$$

## EXERCISE 15.2

1. A cantilever beam 2.5 m long has 50 mm width throughout its length and depth varying uniformly from 50 mm at the free end to 150 mm at the fixed end. If a load of 3 kN acts at the free end, find the position of highest stressed section and value of maximum bending stress induced. Neglect the weight of the beam itself. (**Ans.** 1.25 m ; 45 MPa)
2. A timber beam 150 mm deep and 150 mm wide is reinforced by a steel plate 100 mm wide and 10 mm deep attached at the lower face of the timber beam. Calculate the moment of resistance of the beam, if allowable stresses in timber and steel are 6 MPa and 60 MPa respectively. Take $E_s = 166\ E_t$. (**Ans.** 9.45 kN-m)
3. A timber joist 100 mm wide and 150 mm deep is reinforced by fixing two steel plates each 100 mm wide and 10 mm thick attached symmetrically at the top and the bottom. Find the moment of resistance of the beam, if allowable stresses in timber and steel are 7 MPa and 100 MPa respectively. Take $E_s = 16\ E_t$. (**Ans.** 17.15 kN-m)

## QUESTIONS

1. Discuss the difference of procedure in finding out the bending stress in (*a*) symmetrical section, and (*b*) an unsymmetrical section.
2. Explain the term 'strength of a section'.
3. Illustrate the term 'beam of uniform strength'. Explain its necessity.
4. What do you understand by the term flitched beam? How would you find out the bending stresses in such a beam when it is of (*a*) a symmetrical section and (*b*) an unsymmetrical section?
5. Define the term 'equivalent section' used in a flitched beam.

## OBJECTIVE TYPE QUESTIONS

1. Which of the following is a composite section?
   (*a*) hollow circular section (*b*) *T*-section
   (*c*) *Z*-section (*d*) both '*b*' and '*c*'
2. A beam of uniform strength has constant
   (*a*) shear force (*b*) bending moment
   (*c*) cross-sectional area (*d*) deflection
3. In a flitched beam, one section is reinforced with another section. The purpose of such a beam is to improve
   (*a*) shear force over the section (*b*) moment of resistance over the section
   (*c*) appearance of the section (*d*) all of these

## ANSWERS

1. (*d*) 2. (*b*) 3. (*b*)

# Chapter 16

# Shearing Stresses in Beams

**Contents**



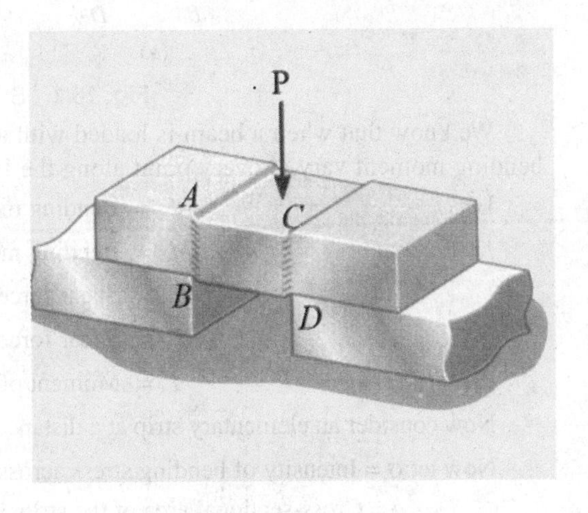


## 16.1. Introduction

In the previous chapter, we discussed the theory of simple bending. In this theory, we assumed that no shear force is acting on the section. But in actual practice when a beam is loaded, the shear force at a section always comes into play, alongwith the bending moment. It has been observed that the effect of shearing stress, as compared to the bending stress, is quite negligible, and is not of much importance. But, sometimes, the shearing stress at a section assumes much importance in the design criterion. In this chapter, we shall discuss the shearing stress for its own importance.

## 16.2. Shearing Stress at a Section in a Loaded Beam

Consider a small portion *ABDC* of length *dx* of a beam loaded with uniformly distributed load as shown in Fig. 16.1 (*a*).

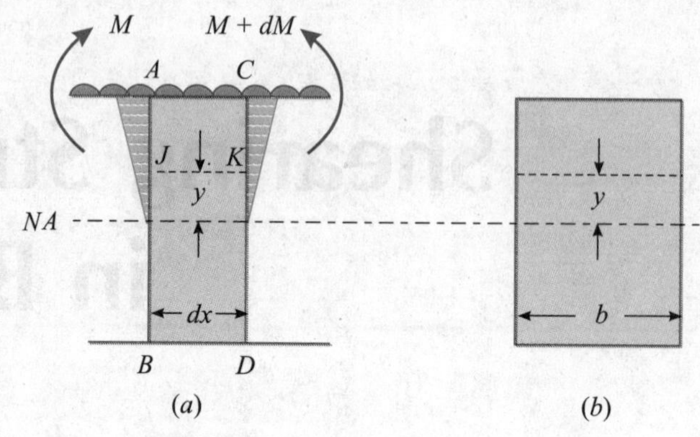


Fig. 16.1. Shearing stress

We know that when a beam is loaded with a uniformly distributed load, the shear force and bending moment vary at every point along the length of the beam.

Let $M$ = Bending moment at *AB*,

$M + dM$ = Bending moment at *CD*,

$F$ = Shear force at *AB*,

$F + dF$ = Shear force at *CD*, and

$I$ = Moment of inertia of the section about its neutral axis.

Now consider an elementary strip at a distance *y* from the neutral axis as shown in Fig. 16.1 (*b*).

Now let $\sigma$ = Intensity of bending stress across *AB* at distance *y* from the neutral axis and

$a$ = Cross-sectional area of the strip.

We have already discussed that

$$\frac{M}{I} = \frac{\sigma}{y} \quad \text{or} \quad \sigma = \frac{M}{I} \times y \qquad \text{... (See Art. 14.6)}$$

Similarly,
$$\sigma + d\sigma = \frac{M + dM}{I} \times y$$

where $\sigma + d\sigma$ = Intensity of bending stress across *CD*.

We know that the force acting across *AB*

$$= \text{Stress} \times \text{Area} = \sigma \times a = \frac{M}{I} \times y \times a \qquad \text{...}(i)$$

Similarly, force acting across *CD*

$$= (\sigma + d\sigma) \times a = \frac{M + dM}{I} \times y \times a \qquad \text{...}(ii)$$

$\therefore$ Net unbalanced force on the strip

$$= \frac{M + dM}{I} \times y \times a - \frac{M}{I} \times y \times a = \frac{dM}{I} \times y \times a$$

The total *unbalanced force ($F$) above the neutral axis may be found out by integrating the above equation between 0 and $d/2$.

$$\text{or} \qquad = \int_0^{\frac{d}{2}} \frac{dM}{I} a \cdot y \cdot dy = \frac{dM}{I} \int_0^{\frac{d}{2}} a \cdot y \cdot dy = \frac{dM}{I} A\bar{y} \qquad ...(iii)$$

where $A$ = Area of the beam above neutral axis, and $\bar{y}$ = Distance between the centre of gravity of the area and the neutral axis.

We know that the intensity of the shear stress,

$$\tau = \frac{\text{Total force}}{\text{Area}} = \frac{\frac{dM}{I} \cdot A\bar{y}}{dx \cdot b} \qquad \text{...(Where } b \text{ is the width of beam)}$$

$$= \frac{dM}{dx} \times \frac{A \cdot \bar{y}}{Ib}$$

$$= F \times \frac{A\bar{y}}{Ib} \qquad \left(\text{Substituting } \frac{dM}{dx} = F = \text{Shear force}\right)$$

## 16.3. Distribution of Shearing Stress

In the previous article, we have obtained a relation, which helps us in determining the value of shear stress at any section on a beam. Now in the succeeding articles, we shall study the distribution of the shear stress along the depth of a beam. For doing so, we shall calculate the intensity of shear stress at important sections of a beam and then sketch a shear stress diagram. Such a diagram helps us in obtaining the value of shear stress at any section along the depth of the beam. In the following pages, we shall discuss the distribution of shear stress over the following sections:

1. Rectangular sections,
2. Triangular sections,
3. Circular sections,
4. *I*-sections,
5. *T*-sections and
6. Miscellaneous sections.

## 16.4. Distribution of Shearing Stress over a Rectangular Section

Consider a beam of rectangular section $ABCD$ of width and depth as shown in Fig. 16.2 ($a$). We know that the shear stress on a layer $JK$ of beam, at a distance $y$ from the neutral axis,

$$\tau = F \times \frac{A\bar{y}}{Ib} \qquad ...(i)$$

* This may also be found out by splitting up the beam into number of strips at distance of from the neutral axis.

We know that unbalanced force on strip 1 $= \frac{dM}{I} \times a_1 \cdot y_1$

Similarly, unbalanced force on strip 2 $= \frac{dM}{I} \times a_2 \cdot y_2$

and unbalanced force on strip 3 $= \frac{dM}{I} \times a_3 \cdot y_3$ and so on

$\therefore$ Total force, $F = \frac{dM}{I} \times a_1 \cdot y_1 + \frac{dM}{I} \times a_2 \cdot y_2 + \frac{dM}{I} \times a_3 \cdot y_3 + ....$

$$= \frac{dM}{I} (a_1 . y_1 + a_2 . y_2 + a_3 . y_3 + ...) = \frac{dM}{I} A\bar{y}$$

where $F$ = Shear force at the section,

$A$ = Area of section above $y$ (*i.e.*, shaded area *AJKD* ),

$\bar{y}$ = Distance of the shaded area from the neutral axis,

∴ $A\bar{y}$ = Moment of the shaded area about the neutral axis,

$I$ = Moment of inertia of the whole section about its neutral axis, and

$b$ = Width of the section.

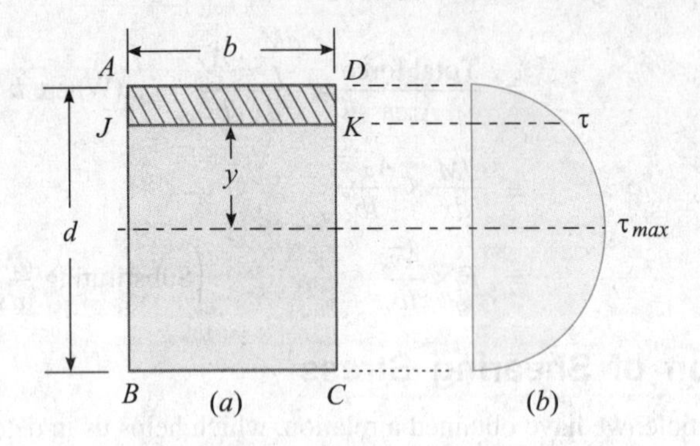


Fig. 16.2. Rectangular section

We know that area of the shaded portion *AJKD*,

$$A = b\left(\frac{d}{2} - y\right) \quad ...(ii)$$

$$\therefore \quad \bar{y} = y + \frac{1}{2}\left(\frac{d}{2} - y\right) = y + \frac{d}{4} - \frac{y}{2}$$

$$= \frac{y}{2} + \frac{d}{4} = \frac{1}{2}\left(y + \frac{d}{2}\right) \quad ...(iii)$$

Substituting the above values of $A$ and $\bar{y}$ in equation (*i*),

$$\tau = F \times \frac{A\bar{y}}{Ib} = F \times \frac{b\left(\frac{d}{2} - y\right) \times \frac{1}{2}\left(y + \frac{d}{2}\right)}{Ib}$$

$$= \frac{F}{2I}\left(\frac{d^2}{4} - y^2\right) \quad ...(iv)$$

We see, from the above equation, that $\tau$ increase as $y$ decreases. At a point, where $y = d/2$, $\tau = 0$; and where $y$ is zero, $\tau$ is maximum. We also see that the variation of $\tau$ with respect to $y$ is a parabola.

At neutral axis, the value of $\tau$ is maximum. Thus substituting $y = 0$ and $I = \frac{bd^3}{12}$ in the above equation,

$$\tau_{max} = \frac{F}{2 \times \frac{ba^3}{12}}\left(\frac{d^2}{4}\right) = \frac{3F}{2bd} = 1.5\,\tau_{av} \quad ...\left(\because \tau_{av} = \frac{F}{\text{Area}} = \frac{F}{bd}\right)$$

Now draw the shear stress distribution diagram as shown in Fig. 16.2 (*b*).

**EXAMPLE 16.1.** *A wooden beam 100 mm wide, 250 mm deep and 3 m long is carrying a uniformly distributed load of 40 kN/m. Determine the maximum shear stress and sketch the variation of shear stress along the depth of the beam.*

**SOLUTION.** Given: Width ($b$) = 100 mm ; Depth ($d$) = 250 mm ; Span ($l$) = 3 m = $10^3$ mm and uniformly distributed load ($w$) = 40 kN/m = 40 N/mm.

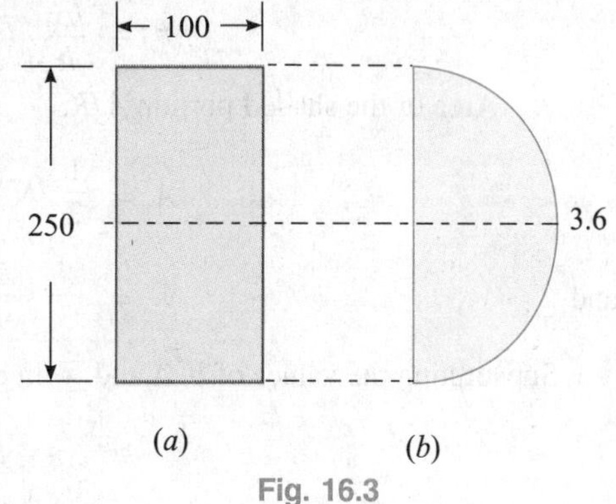

Fig. 16.3

We know that shear force at one end of the beam,

$$F = \frac{wl}{2} = \frac{40 \times (3 \times 10^3)}{2} \text{ N}$$

$$= 60 \times 10^3 \text{ N}$$

and area of beam section,

$$A = b \cdot d = 100 \times 250 = 25\,000 \text{ mm}^2$$

∴ Average shear stress across the section,

$$\tau_{av} = \frac{F}{A} = \frac{60 \times 10^3}{25000} = 2.4 \text{ N/mm}^2 = 2.4 \text{ MPa}$$

and maximum shear stress,

$$\tau_{max} = 1.5 \times \tau = 1.5 \times 2.4 = 3.6 \text{ MPa} \qquad \textbf{Ans.}$$

The diagram showing the variation of shear along the depth of the beam is shown in Fig. 16.3 (*b*).

## 16.5. Distribution of Shearing Stress over a Triangular Section

Consider a beam of triangular cross-section *ABC* of base *b* and height *h* as shown in Fig. 16.4 (*a*).

We know that the shear stress on a layer *JK* at a distance *y* from the neutral axis,

$$\tau = F \times \frac{A\bar{y}}{Ib} \qquad ...(i)$$

where $F$ = Shear force at the section,

$A\bar{y}$ = Moment of the shaded area about the neutral axis and

$I$ = Moment of inertia of the triangular section about its neutral axis.

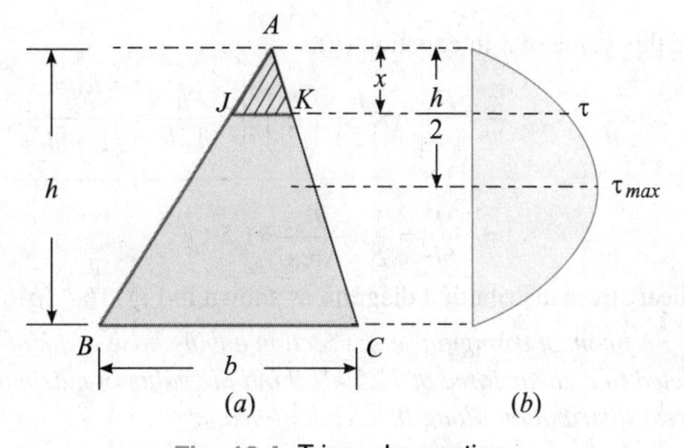

Fig. 16.4. Triangular section.

We know that width of the strip *JK*,

$$b = \frac{bx}{h}$$

$\therefore$ Area of the shaded portion *AJK*,

$$A = \frac{1}{2} JK \times x = \frac{1}{2}\left(\frac{bx}{h} \times x\right) = \frac{bx^2}{2h}$$

and

$$\bar{y} = \frac{2h}{3} - \frac{2x}{3} = \frac{2}{3}(h - x)$$

Substituting the values of *b*, *A* and $\bar{y}$ in equation (*i*),

$$\tau = F \times \frac{\left(\frac{bx^2}{2h}\right) \times \frac{2}{3}(h - x)}{I \times \frac{bx}{h}} = \frac{F}{3I} \times [x(h - x)]$$

$$= \frac{F}{3I} \times \left[hx - x^2\right] \qquad \text{...(ii)}$$

Thus we see that the variation of $\tau$ with respect to *x* is parabola. We also see that as a point where $x = 0$ or $x = h$, $\tau = 0$. At neutral axis, where $x = \frac{2h}{3}$,

$$\tau = \frac{F}{3I}\left[h \times \frac{2h}{3} - \left(\frac{2h}{3}\right)^2\right] = \frac{F}{3I} \times \frac{2h^2}{9} = \frac{2Fh^2}{27I}$$

$$= \frac{2Fh^2}{27 \times \frac{bh^3}{36}} = \frac{8F}{3bh} \qquad \text{...}\left(\because I = \frac{bh^3}{36}\right)$$

$$= \frac{4}{3} \times \frac{F}{\text{Area}} = 1.33\,\tau_{av} \qquad \text{...}\left(\because \text{Area} = \frac{bh}{2}\right)$$

Now for maximum intensity, differentiating the equation (*ii*) and equating to zero,

$$\frac{d\tau}{dx}\left[\frac{F}{3I}(hx - x^2)\right] = 0$$

$$\therefore \qquad h - 2x = 0 \quad \text{or} \quad x = \frac{h}{2}$$

Now substituting this value of *x* in equation (*ii*),

$$\tau_{max} = \frac{F}{3I}\left[h \times \frac{h}{2} - \left(\frac{h}{2}\right)^2\right] = \frac{Fh^2}{12I} = \frac{Fh^2}{12 \times \frac{bh^3}{36}} \qquad \text{...}\left(\because I = \frac{bh^3}{36}\right)$$

$$= \frac{3F}{bh} = \frac{3}{2} \times \frac{F}{\text{Area}} = 1.5\,\tau_{av}$$

Now draw the shear stress distribution diagram as shown in Fig. 16.4 (*b*).

**EXAMPLE 16.2.** *A beam of triangular cross section having base width of 100 mm and height of 150 mm is subjected to a shear force of 13.5 kN. Find the value of maximum shear stress and sketch the shear stress distribution along the depth of beam.*

**SOLUTION.** Given: Base width (*b*) = 100 mm ; Height (*h*) = 150 mm and shear force (*F*) = 13.5 kN = $13.5 \times 10^3$ N

We know that area of beam section,

$$A = \frac{b \cdot h}{2} = \frac{100 \times 150}{2} \text{ mm}^2$$
$$= 7500 \text{ mm}^2$$

∴ Average shear stress across the section,

$$\tau_{av} = \frac{F}{A} = \frac{13.5 \times 10^3}{7500} \text{ N/mm}^2$$
$$= 1.8 \text{ N/mm}^2 = 1.8 \text{ MPa}$$

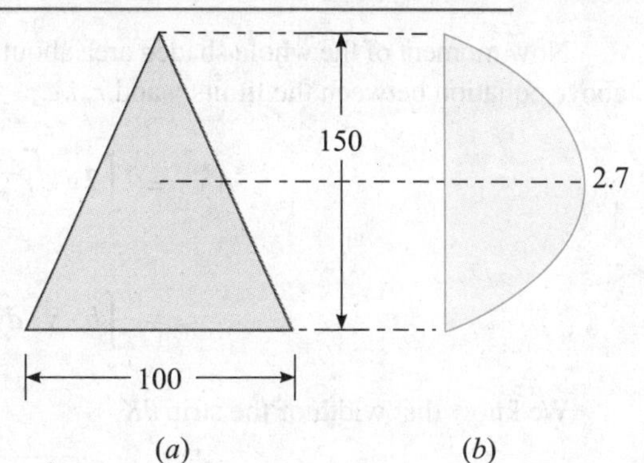

Fig. 16.5

and maximum shear stress,

$$\tau_{av} = 1.5 \times \tau_{av} = 1.5 \times 1.8 = \mathbf{2.7 \text{ MPa}} \quad \textbf{Ans.}$$

The diagram showing the variation of shear stress along the depth of the beam is shown in Fig. 16.5(*b*).

## 16.6. Distribution of Shearing Stress over a Circular Section

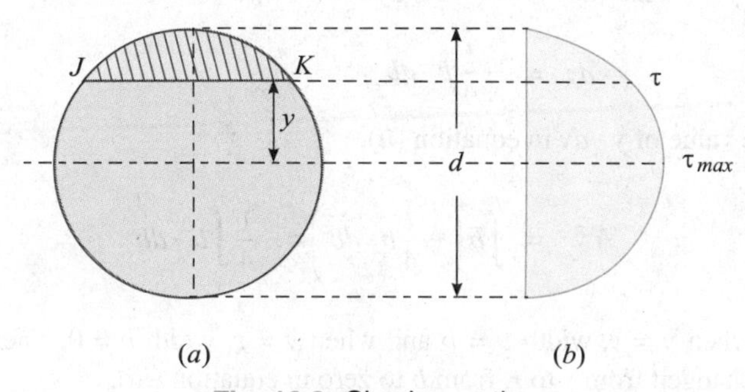

Fig. 16.6. Circular section.

Consider a circular section of diameter *d* as shown in Fig. 16.6 (*a*). We know that the shear stress on a layer *JK* at a distance *y* from the neutral axis,

$$\tau = F \times \frac{A\bar{y}}{Ib}$$

where $F$ = Shear force at the section,

$A\bar{y}$ = Moment of the shaded area about the neutral axis,

$r$ = Radius of the circular section,

$I$ = Moment of inertia of the circular section and

$b$ = Width of the strip *JK*.

We know that in a circular section,

width of the strip *JK*, $$b = 2\sqrt{r^2 - y^2}$$

and area of the shaded strip,

$$A = 2\sqrt{r^2 - y^2} \cdot dy$$

∴ Moment of this area about the neutral axis

$$= 2y\sqrt{r^2 - y^2} \cdot dy \qquad ...(i)$$

Now moment of the whole shaded area about the neutral axis may be found out by integrating the above equation between the limits *y* and *r*, *i.e.*,

$$A\bar{y} = \int_y^r 2y\sqrt{r^2 - y^2} \cdot dy$$

$$= \int_y^r b \cdot y \cdot dy \qquad \ldots (\because b = 2\sqrt{r^2 - y^2}) \quad \ldots (ii)$$

We know that width of the strip *JK*,

$$b = 2\sqrt{r^2 - y^2}$$

or

$$b^2 = 4\sqrt{r^2 - y^2} \qquad \ldots \text{(Squaring both sides)}$$

Differentiating both sides of the above equation,

$$2b \cdot db = 4(-2y)\, dy = -8y \cdot dy$$

or

$$y \cdot dy = -\frac{1}{4} b \cdot db$$

Substituting the value of $y \cdot dy$ in equation (*ii*),

$$A\bar{y} = \int_y^r b\left(-\frac{1}{4} b \cdot db\right) = -\frac{1}{4}\int_y^r b^2 \cdot db \qquad \ldots (iii)$$

We know that when *y* = *y*, width *b* = *b* and when *y* = *r*, width *b* = 0. Therefore, the limits of integration may be changed from *y* to *r*, from *b* to zero in equation (*iii*),

or

$$A\bar{y} = -\frac{1}{4}\int_b^0 b^2 \cdot db$$

$$= \frac{1}{4}\int_0^b b^2 \cdot db \qquad \ldots \text{(Eliminating –ve sign)}$$

$$= -\frac{1}{4}\left[\frac{b^3}{3}\right]_0^b = \frac{b^3}{12}$$

Now substituting this value of $A\bar{y}$ in our original formula for the shear stress, *i.e.*,

$$\tau = F \times \frac{A\bar{y}}{Ib} = F \times \frac{\frac{b^3}{12}}{Ib} = F \times \frac{b^2}{12I}$$

$$= F \times \left[\frac{(2\sqrt{r^2 - y^2})^2}{12I}\right] \qquad \ldots (\because b = 2\sqrt{r^2 - y^2})$$

$$= F \times \frac{r^2 - y^2}{3I}$$

Thus we again see that $\tau$ increases as $y$ decreases. At a point, where $y = r$, $\tau = 0$, $= 0$ and where $y$ is zero, $\tau$ is maximum. We also see that the variation of $\tau$ with respect to $y$ is a parabolic curve. We see that at neutral axis $\tau$ is maximum.

Substituting $y = 0$ and $I = \frac{\pi}{64} \times d^4$ in the above equation,

$$\tau_{max} = F \times \frac{r^2}{3 \times I} = F \times \frac{\left(\frac{d}{2}\right)^2}{3 \times \frac{\pi}{64} \times d^4} = \frac{4F}{3 \times \frac{\pi}{4} \times d^2} = 1.33\,\tau_{av}$$

Now draw the shear stress distribution diagram as shown in Fig. 16.6 (*b*).

**EXAMPLE 16.3.** *A circular beam of 100 mm diameter is subjected to a shear force of 30 kN. Calculate the value of maximum shear stress and sketch the variation of shear stress along the depth of the beam.*

**SOLUTION.** Given: Diameter ($d$) = 100 mm and shear force ($F$) = 30 kN = $30 \times 10^3$ N

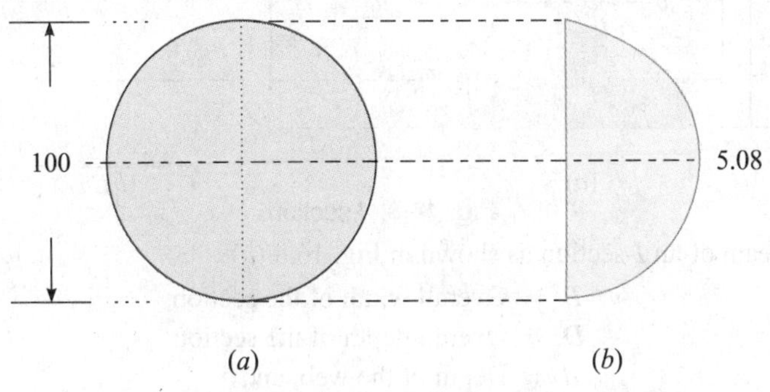


Fig. 16.7

We know that area of the beam section,

$$A = \frac{\pi}{4}(d)^2 = \frac{\pi}{4}(100)^2 \text{ mm}^2 = 7854 \text{ mm}^2$$

$\therefore$ Average shear stress across the section

$$\tau_{av} = \frac{F}{A} = \frac{30 \times 10^3}{7854} = 3.82 \text{ N/mm}^2 = 3.82 \text{ MPa}$$

and maximum shear stress,

$$\tau_{max} = 1.33 \times \tau_{av} = 1.33 \times 3.82 = \mathbf{5.08\ MPa} \quad \textbf{Ans.}$$

The diagram showing the variation of shear stress along the depth of the beam is shown in Fig. 16.7.

## EXERCISE 16.1

1. A rectangular beam 80 mm wide and 150 mm deep is subjected to a shearing force of 30 kN. Calculate the maximum shear stress and draw the distribution diagram for the shear stress.]

   [**Ans.** 3.75 MPa]

2. A rectangular beam 100 mm wide is subjected to a maximum shear force of 50 kN. Find the depth of the beam, if the maximum shear stress is 3 MPa. [**Ans.** 250 mm]

3. A triangular beam of base width 80 mm and height 100 mm is subjected to a shear force of 12 kN. What is the value of maximum shear stress? Also draw the shear stress distribution diagram over the beam section. **[Ans.** 4.5 MPa]
4. A circular beam of diameter 150 mm is subjected to a shear force of 70 kN. Find the value of maximum shear stress and sketch the shear stress distribution diagram over the beam section. **[Ans.** 5.27 MPa]

## 16.7. Distribution of Shearing Stress over an *I*-Section

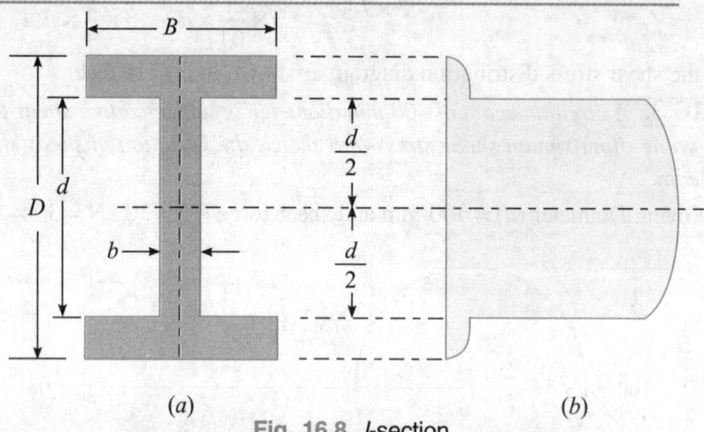

Fig. 16.8. *I*-section.

Consider a beam of an *I*-section as shown in Fig. 16.8 (*a*)

Let $B$ = Overall width of the section,
$D$ = Overall depth of the section,
$d$ = Depth of the web, and
$b$ = Thickness of the web.

We know that the shear stress on a layer *JK* at a distance *y* from the neutral axis,

$$\tau = F \times \frac{A\overline{y}}{Ib} \quad \text{... (i)}$$

Now we shall discuss two important cases

(*i*) when *y* is greater than $\frac{d}{2}$

(*ii*) when *y* is less than $\frac{d}{2}$.

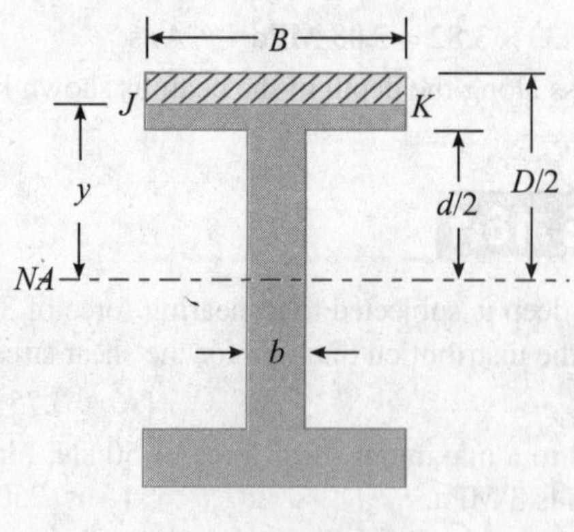

(*a*) *y* greater than *d*/2

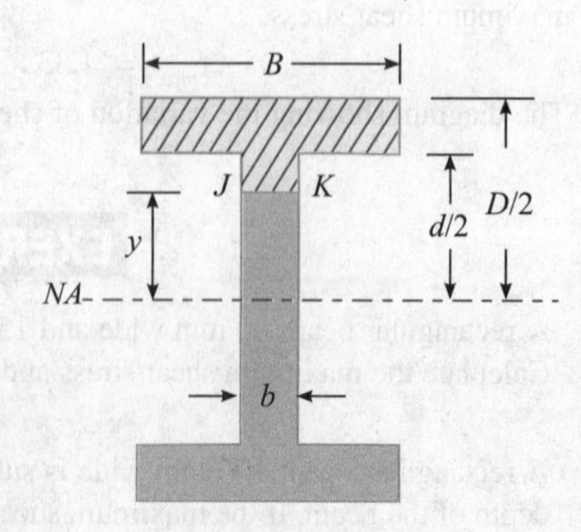

(*a*) *y* less than *d*/2

Fig. 16.9

***(i) When y is greater than $\frac{d}{2}$***

It means that $y$ lies in the flange as shown in Fig. 16.9 (*a*). In this case, shaded area of the flange,

$$A = B\left(\frac{D}{2} - y\right)$$

and
$$\bar{y} = y + \frac{1}{2}\left(\frac{D}{2} - y\right)$$

Now substituting these values of $A$ and $\bar{y}$ from the above equations, in our original equation (*i*) of shear force, *i.e.*,

$$\tau = F \times \frac{A\bar{y}}{Ib} = F \times \frac{B\left(\frac{D}{2} - y\right) \times \left[y + \frac{1}{2}\left(\frac{D}{2} - y\right)\right]}{IB}$$

$$= \frac{F}{2I}\left(\frac{D^2}{4} - y^2\right)$$

Thus we see that $\tau$ increases as $y$ decreases. We also see that the variation of $\tau$ with respect to $y$ is a parabolic curve. At the upper edge of the flange, where $y = \frac{D}{2}$, shear stress is zero and at the lower edge where $y = \frac{d}{2}$, the shear stress,

$$\tau = \frac{F}{2I}\left[\frac{D^2}{4} - \left(\frac{d}{2}\right)^2\right] = \frac{F}{8I}(D^2 - d^2)$$

***(ii) When y is less than $\frac{d}{2}$***

It means that $y$ lies in the web as shown in Fig. 10.9 (*b*). In this case, the value of $A\bar{y}$ for the flange

$$= B\left(\frac{D}{2} - \frac{d}{2}\right) \times \left[\frac{d}{2} + \frac{1}{2}\left(\frac{D}{2} - \frac{d}{2}\right)\right]$$

$$= B\left(\frac{D-d}{2}\right)\left[\frac{1}{2}\left(\frac{D+d}{2}\right)\right] = B\,\frac{(D^2 - d^2)}{8} \qquad ...(i)$$

and the value of $A\bar{y}$ for the web above $AB$

$$= b\left(\frac{d}{2} - y\right) \times \left[y + \frac{1}{2}\left(\frac{d}{2} - y\right)\right]$$

$$= b\left(\frac{d}{2} - y\right) \times \left[\frac{1}{2}\left(\frac{d}{2} + y\right)\right] = \frac{b}{2}\left(\frac{d^2}{4} - y^2\right) \qquad ...(ii)$$

$$\therefore \quad \text{Total } A\bar{y} = \frac{B(D^2 - d^2)}{8} + \frac{b}{2}\left(\frac{d^2}{4} - y^2\right)$$

Now substituting the value of $A\bar{y}$ from the above equation, in our original equation of shear stress on a layer at a distance $y$ from the neutral axis, *i.e.*,

$$\tau = F \times \frac{A\bar{y}}{Ib} = F \times \frac{\frac{B(D^2 - d^2)}{8} + \frac{b}{2}\left(\frac{d^2}{4} - y^2\right)}{Ib}$$

$$= \frac{F}{Ib}\left[\frac{B(D^2 - d^2)}{8} + \frac{b}{2}\left(\frac{d^2}{4} - y^2\right)\right]$$

Thus we see that in the web also $\tau$ increases as $y$ decreases. We also see that the variation of $\tau$ with respect to $y$ in the web also is a parabolic curve. At neutral axis where $y = 0$, the shear stress is maximum.

∴ Maximum shear stress,

$$\therefore \quad \tau_{max} = \frac{F}{Ib}\left[\frac{B}{8}(D^2 - d^2) + \frac{bd^2}{8}\right] \quad \text{...(Substituting } y = 0\text{)}$$

Now, shear stress at the junction of the top of the web and bottom of the flange

$$= \frac{F}{Ib}\left[\frac{B}{8}(D^2 - d^2)\right] \quad \text{...}\left(\text{Substituting } y = \frac{d}{2}\right)$$

$$= \frac{F}{8I} \times \frac{B}{b}(D^2 - d^2)$$

**NOTES: 1.** We see that the shear stress at the junction of the top of web and bottom of the flange is different from both the above expressions $\left(i.e., \text{ when } y > \frac{d}{2} \text{ and } y < \frac{d}{2}\right)$.

We also see that the shear stress changes, abruptly from $\frac{F}{8I}(D^2 - d^2)$ to $\frac{F}{8I} \times \frac{B}{b}(D^2 - d^2)$.

Thus the shear stress at this junction, suddenly increases by $B/b$ times as shown in Fig. 16.8(*b*).

**2.** If the I-section is symmetrical, the shear stress distribution diagram will also be symmetrical.

**3.** From the shear stress distribution diagram, we see that most of the shear stress is taken up by the web. It is an important factor in the design of various important structures.

**EXAMPLE 16.4.** *An I-sections, with rectangular ends, has the following dimensions:*

*Flanges =150 mm × 20 mm, Web = 300 mm 10 mm.*

*Find the maximum shearing stress developed in the beam for a shear force of 50 kN.*

**SOLUTION.** Given: Flange width ($B$) = 150 mm ; Flange thickness = 20 mm ; Depth of web ($d$) = 300 mm; Width of web = 10 mm; Overall depth of the section ($D$) = 340 mm and shearing force ($F$) = 50 kN = $50 \times 10^3$ N.

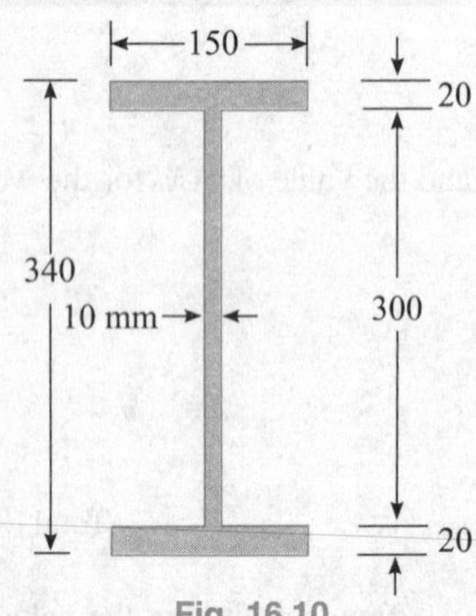


Fig. 16.10

We know that moment of inertia of the *I*-section about its centre of gravity and parallel to *x-x* axis,

$$I_{XX} = \frac{150 \times (340)^3}{12} - \frac{140 \times (300)^3}{12} \text{ mm}^4$$

$$= 176.3 \times 10^6 \text{ mm}^4$$

and maximum shearing stress,

$$\tau_{max} = \frac{F}{Ib}\left[\frac{B}{8}(D^2 - d^2) + \frac{bd^2}{8}\right]$$

$$= \frac{50 \times 10^3}{(176.3 \times 10^6) \times 10}\left[\frac{150}{8}[(340)^2 - (300)^2] + \frac{10 \times (300)^2}{8}\right] \text{ N/mm}^2$$

$= 16.8 \text{ N/mm}^2 =$ **16.8 MPa** **Ans.**

**EXAMPLE 16.5.** *An I-section beam 350 mm × 200 mm has a web thickness of 12.5 mm and a flange thickness of 25 mm. It carries a shearing force of 200 kN at a section. Sketch the shear stress distribution across the section.*

**SOLUTION.** Given: Overall depth ($D$) = 350 mm ; Flange width ($B$) = 200 mm ; Width of Web = 12.5 mm ; Flange thickness = 25 mm and the shearing force ($F$) = 200 kN = $200 \times 10^3$ N.

We know that moment of inertia of the $I$-section about it centre of gravity and parallel to $x$-$x$ axis,

$$I_{XX} = \frac{200 \times (350)^3}{12} - \frac{187.5 \times (300)^3}{12} = 292.7 \times 10^6 \text{ mm}^4$$

We also know that shear stress at the upper edge of the upper flange is zero. And shear stress at the joint of the upper flange and web

$$= \frac{F}{8I}[D^2 - d^2] = \frac{200 \times 10^3}{8 \times (292.7 \times 10^6)}[(350)^2 - (300)^2] \text{ N/mm}^2$$

$$= 2.78 \text{ N/mm}^2 = 2.78 \text{ MPa}$$

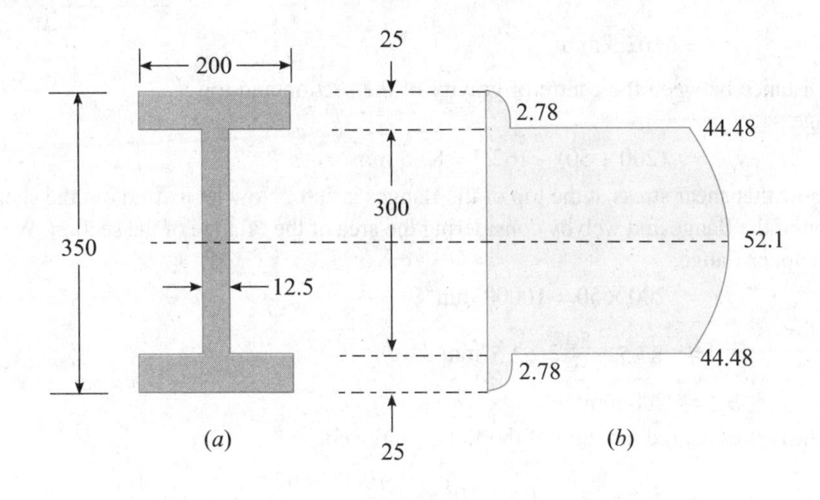


Fig. 16.11

The shear stress at the junction suddenly increases from 2.78 MPa to $2.78 \times \frac{200}{12.5} = 44.48$ MPa.

We also know that the maximum shear stress,

$$\tau_{max} = \frac{F}{I \cdot b}\left[\frac{B}{8}(D^2 - d^2) + \frac{bd^2}{8}\right]$$

$$= \frac{200 \times 10^3}{(292.7 \times 10^6) \times 12.5}\left[\frac{200}{8}(350)^2 - (300)^2 + \frac{12.5 \times (300)^2}{8}\right]$$

$$= 52.1 \text{ N/mm}^2 = 52.1 \text{ MPa}$$

Now complete the shear stress distribution diagram across the section as shown in Fig 16.11 ($b$).

## 16.8. Distribution of Shearing Stress over a T-section

The procedure for determining the distribution of stress over a *T*-section is the same as discussed in Art. 16.7. In this case, since the section is not symmetrical about *x-x* axis, therefore, the shear stress distribution diagram will also not be symmetrical.

**EXAMPLE 16.6.** *A T-shaped cross-section of a beam shown in Fig. 16.12 is subjected to a vertical shear force of 100 kN. Calculate the shear stress at important points and draw shear stress distribution diagram. Moment of inertia about the horizontal neutral axis is* $mm^4$.

**SOLUTION.** Given: Shear stress $(F) = 100$ kN $= 100 \times 10^3$ N and moment of inertia $(I) = 113.4 \times 10^6$ mm$^4$.

First of all let us find out the position of the neutral axis. We know that distance between the centre of gravity of the section and bottom of the web,

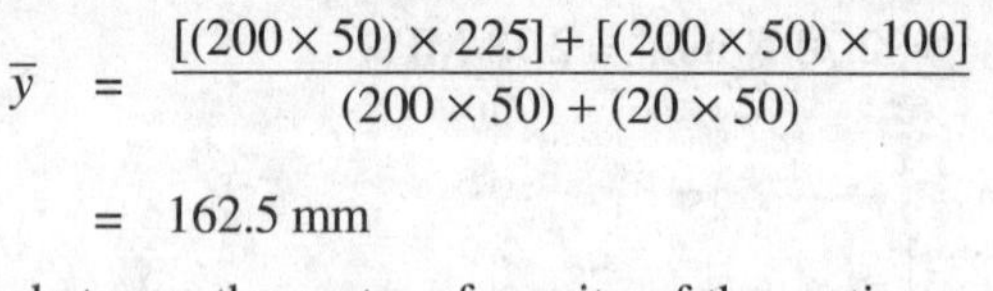

$$\bar{y} = \frac{[(200 \times 50) \times 225] + [(200 \times 50) \times 100]}{(200 \times 50) + (20 \times 50)}$$

$$= 162.5 \text{ mm}$$

Fig. 16.12

$\therefore$ Distance between the centre of gravity of the section and top of the flange,

$$y_C = (200 + 50) - 162.5 = 87.5 \text{ mm}$$

We know that shear stress at the top of the flanges is zero. Now let us find out the shear stress at the junction of the flange and web by considering the area of the *flange of the section. We know that area of the upper flange,

$$A = 200 \times 50 = 10000 \text{ mm}^2$$

$$\bar{y} = 87.5 - \frac{50}{2} = 62.5 \text{ mm}$$

$$B = 200 \text{ mm}$$

$\therefore$ Shear stress at the junction of the flange and web,

$$\tau = F \times \frac{A \cdot \bar{y}}{I \cdot B} = 100 \times 10^3 \times \frac{10000 \times 62.5}{(113.4 \times 10^6) \times 200} \text{ N/mm}^2$$

$$= 2.76 \text{ N/mm}^2 = 2.76 \text{ MPa}$$

* It may also be found out by considering the area of web of the section as discussed below. We know that area of the web,

$$A = 200 \times 50 = 10000 \text{ mm}^2$$

$$\bar{y} = 162.5 - 200/2 = 62.5 \text{ mm}, \; b = 50 \text{ mm}$$

$\therefore$ Shear stress at the junction of the flange and web,

$$\tau = F \times \frac{A \cdot \bar{y}}{I \cdot b} = 100 \times 10^3 \times \frac{10000 \times 62.5}{(113.4 \times 10^6) \times 50}$$

$$= 11.04 \text{ N/mm}^2 = 11.04 \text{ MPa}$$

In this case, the shear stress at the junction suddenly decreases from 11.04 MPa to $11.04 \times \dfrac{50}{200} = 2.76$ MPa.

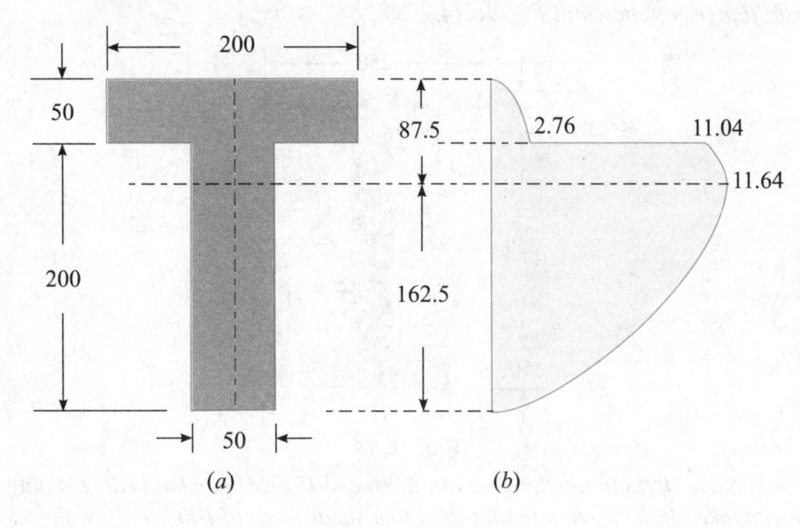


Fig. 16.13

The shear stress at the junction suddenly increases from 2.76 MPa to $2.76 \times \frac{200}{50} = 11.04$ MPa.

Now let us find out the shear stress at the neutral axis, where the shear stress is maximum.

Considering the area of the *T*-section above the neutral axis of the section, we know that

$$^*A\bar{y} = [(200 \times 50) \times 62.5] + \left[(37.5 \times 50) \times \frac{37.5}{2}\right] \text{mm}^3$$

$$= 660.2 \times 10^3 \text{ mm}^3$$

and $$b = 50 \text{ mm}$$

∴ Maximum shear stress,

$$\tau_{max} = F \times \frac{A \cdot \bar{y}}{I \cdot b} = 100 \times 10^3 \times \frac{660.2 \times 10^3}{(113.4 \times 10^6) \times 50} \text{ N/mm}^2$$

$$= 11.64 \text{ N/mm}^2 = 11.64 \text{ MPa}$$

Now draw the shear stress distribution diagram across the section as shown in Fig. 16.13(*b*).

## 16.9. Distribution of Shearing Stress over a Miscellaneous Section

The procedure for determining the distribution of shear stress over a miscellaneous section, is the same as discussed in the previous articles. The shear stress at all the important points should be calculated and then shear stress distribution diagram should be drawn as usual.

---

* It may also be found out by considering the area below neutral axis as discussed below. We know that

$$A\bar{y} = (162.5 \times 50) \times \frac{162.5}{2} = 660.2 \times 10^3 \text{ mm}^3$$

**EXAMPLE 16.7.** *A cast-iron bracket subjected to bending, has a cross-section of I-shape with unequal flanges as shown in Fig. 16.14.*

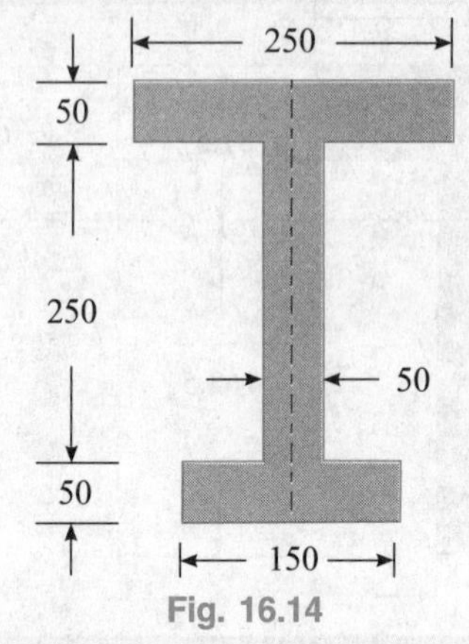


Fig. 16.14

*If the compressive stress in top flange is not to exceed 17.5 MPa, what is the bending moment, the section can take? If the section is subjected to a shear force of 100 kN, draw the shear stress distribution over the depth of the section.*

**SOLUTION.** Given: Compressive stress ($\sigma_c$) = 17.5 MPa = 17.5 N/mm$^2$ and shear force ($F$) = 100 kN = $100 \times 10^3$ N

***Bending moment the section can take***

First of all, let us find out the position of the neutral axis. We know that distance between centre of gravity of the section and bottom face,

$$\bar{y} = \frac{(250 \times 50)\,325 + (250 \times 50)\,175 + (150 \times 50)\,25}{(250 \times 50) + (250 \times 50) + (150 \times 50)}$$

$$= \frac{6\,437\,500}{32\,500} = 198 \text{ mm}$$

$\therefore$ Distance of centre of gravity from the upper extreme fibre,

$$y_c = 350 - 198 = 152 \text{ mm}$$

and moment of inertia of the section about an axis passing through its centre of gravity and parallel to $x$-$x$ axis,

$$I = \left[\frac{250 \times (50)^3}{12} + (250 \times 50)\,(325 - 198)^2\right]$$

$$+ \left[\frac{50 \times (250)^3}{12} + (50 \times 250)\,(198 - 175)^2\right]$$

$$+ \left[\frac{150 \times (50)^3}{12} + (150 \times 50)\,(198 - 25)^2\right] \text{mm}^4$$

$$= 502 \times 10^6 \text{ mm}^4$$

$\therefore$ Bending moment the section can take

$$= \frac{\sigma_c}{y_c \times I} = \frac{17.5}{152} \times 502 \times 10^6 = 57.8 \times 10^6 \text{ N-mm}$$

$= 57.8$ kN-m **Ans.**

*Shear stress distribution diagram*

We know that the shear stress at the extreme edges of both the flanges is zero. Now let us find out the shear stress at the junction of the upper flange and web by considering the area of the upper flange. We know that area of the upper flange,

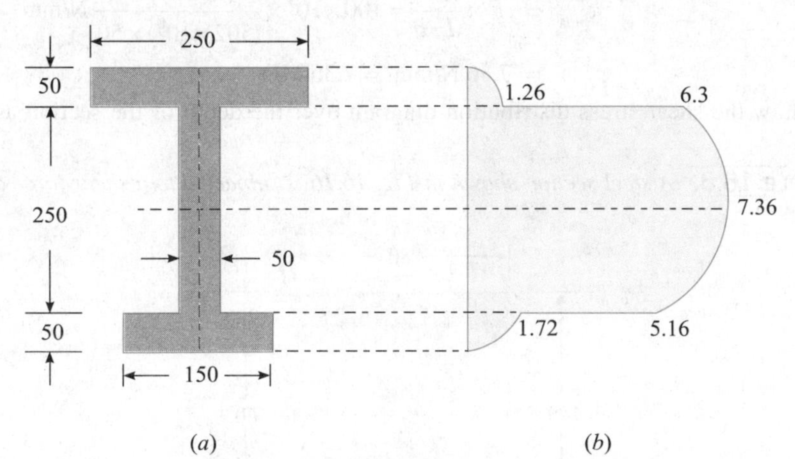


Fig. 16.15

$$A = 250 \times 50 = 12500 \text{ mm}$$

$$\bar{y} = 152 - \frac{50}{2} = 127 \text{ mm}$$

and $$B = 250 \text{ mm}$$

∴ Shear stress at the junction of the upper flange and web,

$$\tau = F \times \frac{A \cdot \bar{y}}{I \cdot B} = 100 \times 10^3 \frac{12500 \times 127}{(502 \times 10^6) \times 250} \text{ N/mm}^2$$

$$= 1.26 \text{ N/mm}^2 = 1.26 \text{ MPa}$$

The shear stress at the junction suddenly increases from 1.26 MPa to $1.26 \times \frac{250}{50} = 6.3$ MPa. Now let us find out the shear stress at the junction of the lower flange and web by considering the area of the lower flange. We know that area of the lower flange,

$$A = 150 \times 50 = 7500 \text{ mm}^2$$

$$\bar{y} = 198 - \frac{50}{2} = 173 \text{ mm}$$

and $$B = 150 \text{ mm}$$

∴ Shear stress at the junction of the lower flange and web,

$$\tau = F \times \frac{A \cdot \bar{y}}{I \cdot B} = 100 \times 10^3 \times \frac{7500 \times 173}{(502 \times 10^6) \times 150}$$

$$= 1.72 \text{ N/mm}^2 = 1.72 \text{ MPa}$$

The shear stress at the function suddenly increases from 1.72 MPa to $1.72 \times \frac{150}{50} = 5.16$ MPa.

Now let us find out the shear stress at the neutral axis, where the shear stress is maximum. Considering the area of the *I*-section above neutral axis, we know that

$$A\bar{y} = [(250 \times 50) \times 127] + \left[(102 \times 50) \times \frac{102}{2}\right] \text{mm}^3$$

$$= 1.848 \times 106 \text{ mm}^3$$

and $b = 50$ mm

∴ Maximum shear stress,

$$\tau_{max} = F \times \frac{A \cdot \bar{y}}{I \cdot b} = 100 \times 10^3 \times \frac{1.848 \times 10^6}{(502 \times 10^6) \times 50} \text{ N/mm}^2$$

$$= 7.36 \text{ N/mm}^2 = 7.36 \text{ MPa}$$

Now draw the shear stress distribution diagram over the depth of the section as shown in Fig. 16.15.

**EXAMPLE 16.8.** *A steel section shown in Fig. 16.16 is subjected to a shear force of 20 kN.*

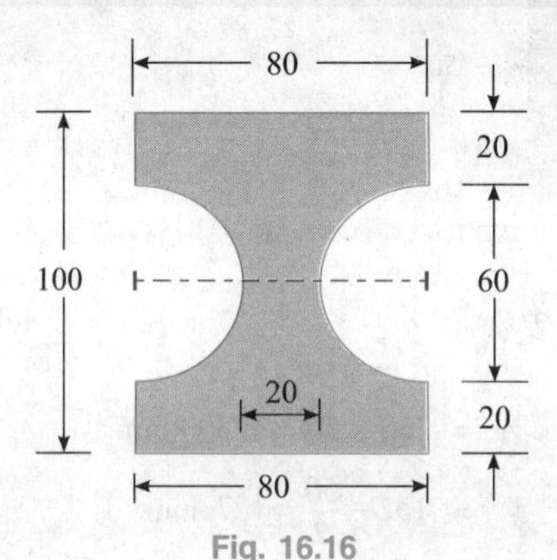


Fig. 16.16

*Determine the shear stress at the important points and sketch the shear distribution diagram.*

**SOLUTION.** Given: Shear force $(F) = 20 \text{ kN} = 20 \times 10^3$ N

Since the section is symmetrical about $x$-$x$ and $y$-$y$ axes therefore, centre of the section will lie on the geometrical centroid of the section. For the purpose of moment of inertia and shear stress, the two semi-circular grooves may be assumed to be together and considered as one circular hole of 60 mm diameter. Therefore moment of inertia of the section about an axis passing through its centre of gravity and parallel to $x$-$x$ axis,

$$I = \left[\frac{80 \times (100)^3}{12}\right] - \left[\frac{\pi}{64}(60)^4\right] = 6.03 \times 10^6 \text{ mm}^4$$

We know that shear stress at the extreme edges of $A$ and $E$ of the section is zero. Now let us find out the shear stress at $B$ by considering the area between $A$ and $B$.

We know that area of the upper portion between $A$ and $B$

$$A = 80 \times 20 = 1600 \text{ mm}^2$$

$$\bar{y} = 30 + \frac{20}{2} = 40 \text{ mm}$$

and $B = 80$ mm

∴ Shear stress at $B$,

$$\tau = F \times \frac{A \cdot \bar{y}}{I \cdot B} = 20 \times 10^3 \times \frac{1600 \times 40}{(6.03 \times 10^6) \times 80} \text{ N/mm}^2$$

$$= 2.65 \text{ N/mm}^2 = 2.65 \text{ MPa}$$

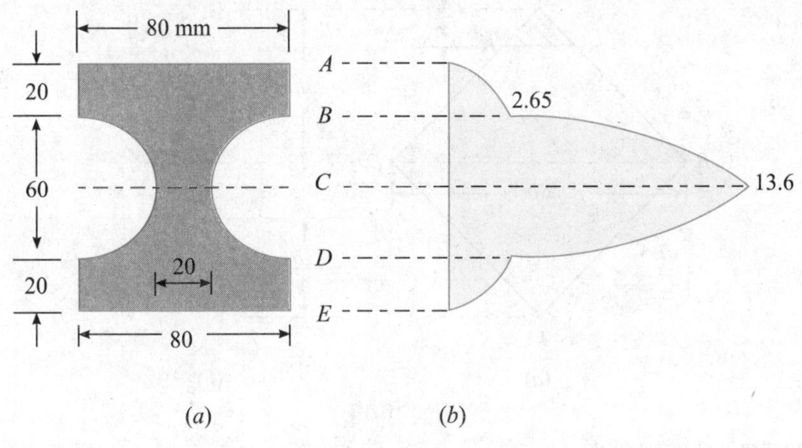


Fig. 16.17

Now let us find out the shear stress at the neutral axis, where the shear stress is maximum. Considering the area above the neutral axis, we know that

$$A\bar{y} = [(80 \times 50) \times 25] - \left[\frac{\pi}{2}(30)^2 \times \frac{4 \times 30}{3\pi}\right] \text{mm}^3$$

$$= 100\ 000 - 18\ 000 = 82\ 000 \text{ mm}^3$$

and $b = 20$ mm

∴ Maximum shear stress,

$$\tau_{max} = F \times \frac{A \cdot \bar{y}}{I \cdot b} = 20 \times 10^3 \times \frac{82000}{(6.03 \times 10^6) \times 20} \text{ N/mm}^2$$

$$= 13.6 \text{ N/mm}^2 = 13.6 \text{ MPa}$$

Now draw the shear stress distribution diagram over the section as shown in Fig. 16.17 (*b*).

**EXAMPLE 16.9.** *A beam of square section is used as a beam with one diagonal horizontal. Find the maximum shear stress in the cross section of the beam. Also sketch the shear stress distribution across the depth of the section.*

**SOLUTION.** Given: A square section with its diagonal horizontal.

The beam with horizontal diagonal is shown in Fig. 16.18 (*a*).

Let $2b$ = Diagonal of the square, and

$F$ = Shear force at the section.

Now consider the shaded strip *AJK* at a distance *x* from the corner *A*. From the geometry of the figure, we find that length $JK = 2x$

∴ Area of *AJK*, $A = \frac{1}{2} \times 2x \cdot x = x^2$

and $\bar{y} = b - \frac{2x}{3}$

We know that moment of inertia of the section *ABCD* about the neutral axis,

$$I = 2 \times \frac{2b \times b^3}{12} = \frac{b^4}{3}$$

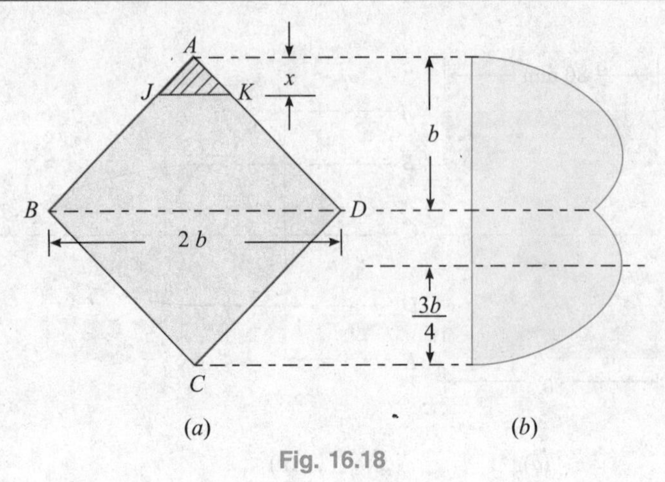


Fig. 16.18

and shearing stress at any point,

$$\tau = F \times \frac{A\bar{y}}{Ib} = F \times \frac{x^2\left(b - \frac{2x}{3}\right)}{\frac{b^4}{3} \times 2x} \qquad \text{(Here } b = JK = 2x\text{)}$$

$$= \frac{F}{2b^4}(3bx - 2x^2) \qquad \text{... }(i)$$

We also know that when $x = 0$, $\tau = 0$ and when $x = b$, then

$$\tau = \frac{F}{2b^2} = \frac{F}{\text{Area}} = \tau_{mean}$$

Now for maximum shear stress, differentiating the equation (*i*) and equating it to zero.

$$\frac{d\tau}{dx} = \frac{d}{dx}\left[\frac{F}{2b^4}(3bx - 2x^2)\right] = 0$$

$$\therefore \qquad 3b - 4x = 0 \qquad \text{or} \qquad x = \frac{3b}{4}$$

Substituting this value of $x$ in equation (*i*),

$$\tau_{max} = \frac{F}{2b^4}\left[3b \times \frac{3b}{4} - 2\left(\frac{3b}{4}\right)^2\right] = \frac{F}{2b^4} \times \frac{9b^2}{8}$$

$$= \frac{9}{8} \times \frac{F}{2b^2} = \frac{9}{8} \times \frac{F}{\text{Area}} = \frac{9}{8} \times \tau_{mean}$$

Now complete the shear stress distribution diagram as shown in Fig. 16.18 (*b*).

**EXAMPLE 16.10.** *A rolled steel joist 200 mm × 160 mm wide has flange 22 mm thick and web 12 mm thick. Find the proportion, in which the flanges and web resist shear force.*

**SOLUTION.** Given : Overall depth ($D$) = 200 mm ; Flange width ($B$) = 160 mm ; Flange thickness ($t_f$) = 22 mm ; Web thickness ($b$) = 12 mm and web depth ($d$) = 156 mm.

Let $\qquad F$ = Shear force resisted by the section.

From the geometry of the figure, we find that the moment of inertia of the section through its c.g. and parallel to $x$-$x$ axis,

$$I = \frac{1}{12}[(160) \times (200)^3 - (148)(156)^3] \text{ mm}^4$$

$$= 59.84 \times 10^6 \text{ mm}^4$$

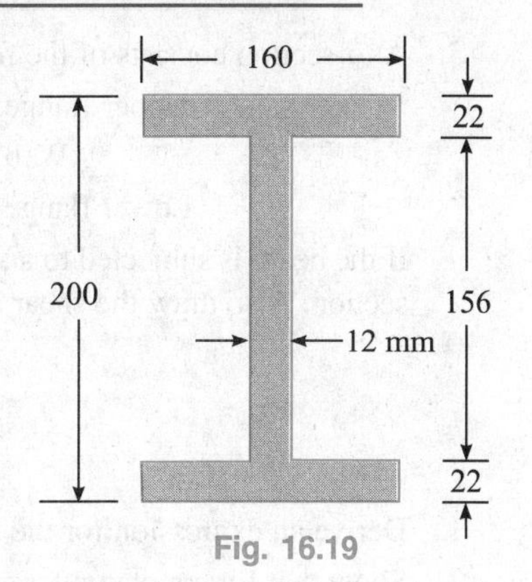

Fig. 16.19

Now consider an elementary strip of thickness $dy$ of the flange at a distance $y$ from the neutral axis. Therefore area of the elementary strip,

$$dA = 160\, dy$$

We know that the intensity of shear stress at the strip,

$$\tau = \frac{F}{2I}\left(\frac{D^2}{4} - y^2\right) = \frac{F}{2I}\left(\frac{(200)^2}{4} - y^2\right)$$

$$= \frac{F}{2I}(10000 - y^2)$$

∴ Resistance offered to shear by this strip

$$= \tau \cdot dA = \frac{F}{2I}(10\,000 - y^2) \times 160\, dy$$

$$= 160\, dy \times \frac{F}{2I}(10\,000 - y^2) = \frac{80F}{I}(10\,000 - y^2)\, dy$$

Now total resistance offered to shear by the flange

$$= \int_{78}^{100} \frac{80F}{I}(10\,000 - y^2)\, dy$$

$$= \frac{80F}{I}\left[10\,000\, y - \frac{y^3}{3}\right]_{78}^{100}$$

$$= \frac{80F}{I}\left[\frac{2 \times 10^6}{3} - \frac{1.865 \times 10^6}{3}\right]$$

$$= \frac{80F}{I} \times \frac{0.135 \times 10^6}{3}$$

$$= \frac{80F}{59.84 \times 10^6} \times 0.045 \times 10^6 = 0.06\, F$$

∴ Total resistance offered to shear by both the flanges

$$= 0.06\, F \times 2 = 0.12\, F$$

and total resistance offered to shear by the web

$$= F - 0.12\, F = 0.88\, F$$

It is obvious that the resistance offered by flanges is 12% and by web is **88%** **Ans.**

## EXERCISE 16.2

1. An *I*-section beam consists of two flanges 150 mm × 20 mm and a web of 310 mm × 10 mm. Find the magnitude of maximum shear stress when it is subjected to a shear force of 40 kN and draw the shear stress distribution diagram over the depth of the section. [**Ans.** 13.1 MPa]
2. A *T*-section beam with 100 mm × 15 mm flange and 150 × 15 mm web is subjected to a shear force of 10 kN at a section. Draw the variation of shear stress across the depth of the beam and obtain the value of maximum shear stress at the section. [**Ans.** 6.3 MPa]

3. An *I*-section consists of the following sections:

Upper flange = 130 mm × 50 mm

Web = 200 mm × 50 mm

Lower flange = 200 mm × 50 mm

If the beam is subjected to a shearing force of 50 kN, find the maximum shear stress across the section. Also draw the shear stress distribution diagram. Take *I* as $284.9 \times 10^6$ mm$^4$.

[**Ans.** 4.42 MPa]

## QUESTIONS

1. Derive an expression for the shear stress at any point in the cross-section of a beam.
2. Show that for a rectangular section, the distribution of shearing stress is parabolic.
3. The cross-section of a beam is a circle with the diameter *D*. If *F* is the total shear force at the cross-section, show that the shear stress at a distance *y* from the neutral axis.

$$= \frac{16F}{3\pi D^2}\left[1-\left(\frac{2v}{D}\right)^2\right]$$

4. Explain by mathematical expression, that the shear stress abruptly changes at the junction of the flange and web of an *I*-section and a *T*-section.
5. Describe the procedure for drawing the shear stress distribution diagram for composite sections.

## OBJECTIVE TYPE QUESTIONS

1. When a rectangular section of a beam is subjected to a shearing force, the ratio of maximum shear stress to the average shear stress is
(*a*) 2.0 (*b*) 1.75 (*c*) 1.5 (*d*) 1.25
2. In a triangular section, the maximum shear stress occurs at
(*a*) apex of the triangle (*b*) mid of the height
(*c*) 1/3 of the height (*d*) base of the triangle
3. A square with side *x* of a beam is subjected to a shearing force of *F*. The value of shear stress at the top edge of the section is
(*a*) zero (*b*) 0.5 $F/a^2$ (*c*) $F/a^2$ (*d*) 1.5 $F/a^2$
4. An inverted *T*-section is subjected to a shear force *F*. The maximum shear stress will occur at
(*a*) top of the section (*b*) neutral axis of the section
(*c*) junction of web and flange (*d*) none of these

## ANSWERS

**1.** (*c*) **2.** (*b*) **3.** (*a*) **4.** (*b*)

Chapter

# 17

# Direct and Bending Stresses

## Contents

## 17.1. Introduction

We have already discussed in Chapter 2, that whenever a body is subjected to an axial tension or compression, a direct stress comes into play at every section of the body. We also know that whenever a body is subjected to a bending moment a bending stress comes into play. It is thus obvious that if a member is subjected to an axial loading, along with a transverse bending, a direct stress as well as a bending stress comes into play. The magnitude and nature of these stresses may be easily found out from the magnitude and nature of the load and the moment. A little consideration will show that since both these stresses act normal to a cross-section, therefore the two stresses may be algebraically added into a single resultant stress.

## 17.2. Eccentric Loading

A load, whose line of action does not coincide with the axis of a column or a strut, is known as an eccentric load. A bucket full of water, carried by a person in his hand, is an excellent example of an eccentric load. A little consideration will show that the man will feel this load as more severe than the same load, if he had carried the same bucket over his head. The simple reason for the same is that if he carries the bucket in his hand, then in addition to his carrying bucket, he has also to lean or bend on the other side of the bucket, so as to counteract any possibility of his falling towards the bucket. Thus we say that he is subjected to :

1. Direct load, due to the weight of bucket (including water) and
2. Moment due to eccentricity of the load.

## 17.3. Columns with Eccentric Loading

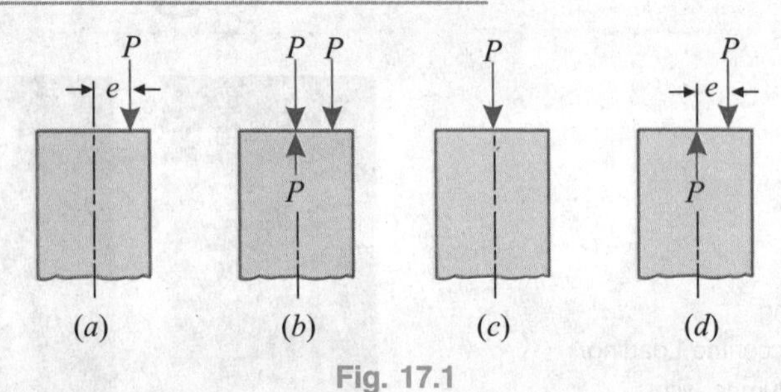


Fig. 17.1

Consider a column subjected to an eccentric loading. The eccentric load may be easily analysed as shown in Fig. 17.1 and as discussed below :

1. The given load $P$, acting at an eccentricity of $e$, is shown in Fig. 17.1 (*a*).
2. Let us introduce, along the axis of the strut, two equal and opposite forces $P$ as shown in Fig. 17.1 (*b*).
3. The forces thus acting, may be split up into three forces.
4. One of these forces will be acting along the axis of the strut. This force will cause a direct stress as shown in Fig. 17.1 (*c*).
5. The other two forces will form a couple as shown in Fig. 17.1 (*d*). The moment of this couple will be equal to $P \times e$ (This couple will cause a bending stress).

**Note :** A column may be of symmetrical or unsymmetrical section and subjected to an eccentric load, with eccentricity about one of the axis or both the axes. In the succeeding pages, we shall discuss these cases one by one.

## 17.4. Symmetrical Columns with Eccentric Loading about One Axis

Consider a column *ABCD* subjected to an eccentric load about one axis (*i.e.*, about *y-y* axis) as shown in Fig. 17.2

Let $P$ = Load acting on the column,

$e$ = Eccentricity of the load,

$b$ = Width of the column section and

$d$ = Thickness of the column.

∴ Area of column section,

$$A = b \cdot d$$

and moment of inertia of the column section about an axis through its centre of gravity and parallel to the axis about which the load is eccentric (*i.e.*, *y-y* axis in this case),

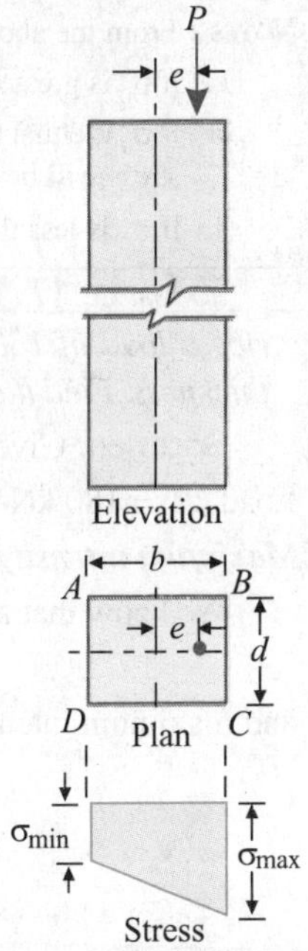

$$I = \frac{d \cdot b^2}{12}$$

and modulus of section, $$Z = \frac{I}{y} = \frac{db^2/12}{b/2} = \frac{db^2}{6}$$

We know that direct stress on the column due to the load,

$$\sigma_0 = \frac{P}{A}$$

and moment due to load, $$M = P \cdot e$$

$\therefore$ Bending stress at any point of the column section at a distance *y* from *y-y* axis,

$$\sigma_b = \frac{M \cdot y}{I} = \frac{M}{Z} \qquad \ldots\left(\because Z = \frac{I}{y}\right)$$

Now for the bending stress at the extreme, let us substitute $y = \frac{b}{2}$ in the above equation,

$$\sigma_b = \frac{M \cdot \frac{b}{2}}{I} = \frac{M \cdot \frac{b}{2}}{\frac{db^3}{12}} \qquad \ldots\left(\because I = \frac{db^3}{12}\right)$$

$$= \frac{6M}{db^3} = \frac{6P \cdot e}{db^2} \qquad \ldots(\because M = P \cdot e)$$

$$= \frac{6P \cdot e}{A \cdot b} \qquad \ldots(\text{Substituting } db = A)$$

P

e

Elevation

b

A B

e

d

D C

Plan

$\sigma_{min}$

$\sigma_{max}$

Stress

Fig. 17.2

We have already discussed in the previous article, that an eccentric load causes a direct stress as well as bending stress. It is thus obvious that the total stress at the extreme fibre,

$$= \sigma_0 \pm \sigma_b = \frac{P}{A} \pm \frac{6P \cdot e}{A \cdot b} \qquad \ldots(\text{In terms of eccentricity})$$

$$= \frac{P}{A} \pm \frac{M}{Z} \qquad \ldots(\text{In terms of modulus of section})$$

The +ve or –ve sign will depend upon the position of the fibre with respect to the eccentric load. A little consideration will show that the stress will be maximum at the corners *B* and *C* (because these corners are near the load), whereas the stress will be minimum at the corners *A* and *D* (because these corners are away from the load). The total stress along the width of the column will vary by a straight line law. The maximum stress,

$$\sigma_{max} = \frac{P}{A} + \frac{6P \cdot e}{Ab} = \frac{P}{A}\left(1 + \frac{6e}{b}\right) \qquad \ldots(\text{In terms of eccentricity})$$

$$= \frac{P}{A} + \frac{M}{Z} \qquad \ldots(\text{In terms of section modulus})$$

and $$\sigma_{min} = \frac{P}{A} - \frac{6P \cdot e}{Ab} = \frac{P}{A}\left(1 - \frac{6e}{b}\right) \qquad \ldots(\text{In terms of eccentricity})$$

$$= \frac{P}{A} - \frac{M}{Z} \qquad \ldots(\text{In terms of section modulus})$$

**Notes :** From the above equations, we find that

1. If $\sigma_0$ is greater than $\sigma_b$, the stress throughout the section, will be of the same nature (*i.e.*, compressive).
2. If $\sigma_0$ is equal to $\sigma_b$, even then the stress throughout the section will be of the same nature. The minimum stress will be equal to zero, whereas the maximum stress will be equal to $2 \times \sigma_0$.
3. If $\sigma_0$ is less than $\sigma_b$, then the stress will change its sign (partly compressive and partly tensile).

**Example 17.1.** *A rectangular strut is 150 mm and 120 mm thick. It carries a load of 180 kN at an eccentricity of 10 mm in a plane bisecting the thickness. Find the maximum and minimum intensities of stress in the section.*

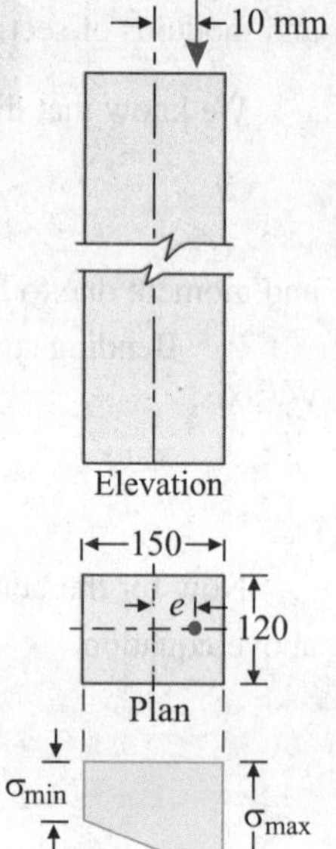

Fig. 17.3

**Solution.** Given: Width ($b$) = 150 mm ; Thickness ($d$) = 120 mm ;

Load ($P$) = 180 kN = $180 \times 10^3$ N and eccentricity ($e$) = 10 mm.

***Maximum intensity of stress in the section***

We know that area of the strut,

$$A = b \times d = 150 \times 120 = 18\,000 \text{ mm}^2$$

and maximum intensity of stress in the section,

$$\sigma_{max} = \frac{P}{A}\left(1 + \frac{6e}{b}\right) = \frac{1800 \times 10^3}{18\,000}\left(1 + \frac{6 \times 10}{150}\right) \text{ N/mm}^2$$

$$= 10\,(1 + 0.4) = 14 \text{ N/mm}^2 = \textbf{14 MPa} \quad \textbf{Ans.}$$

***Minimum intensity of stress in the section***

We also know that minimum intensity of stress in the section,

$$\sigma_{min} = \frac{P}{A}\left(1 - \frac{6e}{b}\right) = \frac{1800 \times 10^3}{18\,000}\left(1 - \frac{6 \times 10}{150}\right) \text{ N/mm}^2$$

$$= 10\,(1 - 0.4) = 6 \text{ N/mm}^2 = \textbf{6 MPa} \quad \textbf{Ans.}$$

**Example 17.2.** *A rectangular column 200 mm wide and 150 mm thick is carrying a vertical load of 120 kN at an eccentricity of 50 mm in a plane bisecting the thickness. Determine the maximum and minimum intensities of stress in the section.*

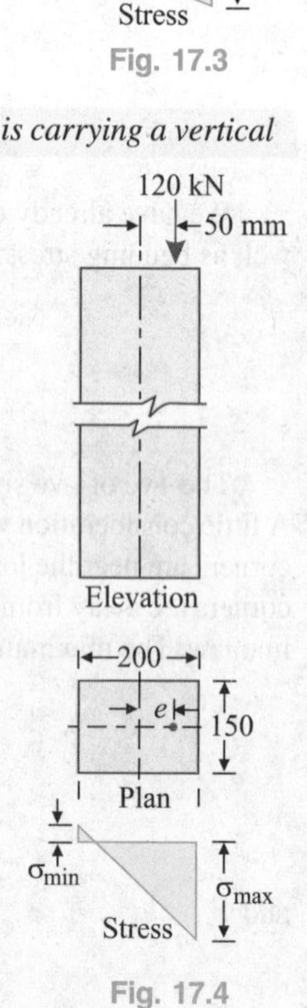

Fig. 17.4

**Solution.** Given: Width ($b$) = 200 mm; Thickness ($d$) = 150 mm ; Load ($P$) = 120 kN = $120 \times 10^3$ N and eccentricity ($e$) = 50 mm.

***Maximum intensity of stress in the section***

We know that area of the column,

$$A = b \times d = 200 \times 150 = 30\,000 \text{ mm}^2$$

and maximum intensity of stress in the section,

$$\sigma_{max} = \frac{P}{A}\left(1 + \frac{6e}{b}\right) = \frac{120 \times 10^3}{30\,000}\left(1 + \frac{6 \times 50}{200}\right) \text{ N/mm}^2$$

$$= 4\,(1 + 1.5) = 10 \text{ N/mm}^2 = \textbf{10 MPa} \quad \textbf{Ans.}$$

***Minimum intensity of stress in the section***

We also know that minimum intensity of stress in the section,

$$\sigma_{min} = \frac{P}{A}\left(1 - \frac{6e}{b}\right) = \frac{120 \times 10^3}{30\,000}\left(1 - \frac{6 \times 50}{200}\right) \text{ N/mm}^2$$

$$= 4\,(1 - 1.5) = 4\,(-0.5) = -2 \text{ N/mm}^2$$

$$= 2 \text{ N/mm}^2 \text{ (tension)} = \textbf{2 MPa (tension)} \quad \textbf{Ans.}$$

**EXAMPLE 17.3.** *In a tension specimen 13 mm in diameter the line of pull is parallel to the axis of the specimen but is displaced from it. Determine the distance of the line of pull from the axis, when the maximum stress is 15 per cent greater than the mean stress on a section normal to the axis.*

**SOLUTION.** Given: Diameter ($d$) = 13 mm and maximum stress ($\sigma_{max}$) = 1.15 $\sigma_{mean}$

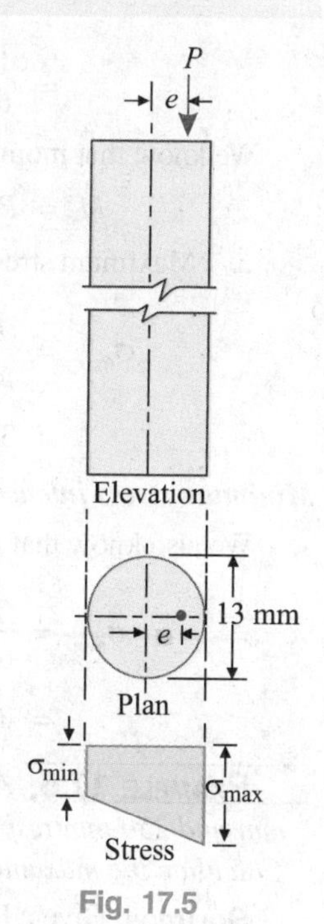

Fig. 17.5

We know that area of the specimen,

$$A = \frac{\pi}{4}(d)^2 = \frac{\pi}{4}(13)^2 = 132.7 \text{ mm}^2$$

and its section modulus,

$$Z = \frac{\pi}{32}(d)^3 = \frac{\pi}{32}(13)^3 = 215.7 \text{ mm}^3$$

Let $P$ = Pull on the specimen in N, and

$e$ = Distance of the line of pull from the axis in mm.

∴ Moment due to load,

$$M = P \cdot e$$

We also know that the mean stress,

$$\sigma_{mean} = \frac{P}{A} = \frac{P}{132.7} \text{ N/mm}^2 \quad ...(i)$$

and maximum stress,

$$\sigma_{max} = \sigma_{mean} + \frac{M}{Z} = \frac{P}{132.7} + \frac{P \cdot e}{215.7}$$

Since $\sigma_{max}$ is 15% greater than $\sigma_{mean}$, therefore

$$\frac{P}{132.7} + \frac{P \cdot e}{215.7} = \frac{P}{132.7} \times \frac{115}{100}$$

or

$$\frac{1}{132.7} + \frac{e}{215.7} = \frac{115}{13270}$$

∴

$$e = \left(\frac{115}{13270} - \frac{1}{132.7}\right) \times 215.7 = \mathbf{0.25 \text{ mm}} \quad \textbf{Ans.}$$

**EXAMPLE 17.4.** *A hollow rectangular masonry pier is 1.2 m × 0.8 m wide and 150 mm thick. A vertical load of 2 MN is transmitted in the vertical plane bisecting 1.2 m side and at an eccentricity of 100 mm from the geometric axis of the section.*

*Calculate the maximum and minimum stress intensities in the section.*

**SOLUTION.** Given: Outer width ($B$) = 1.2 m = 1.2 × $10^3$ mm ; Load ($P$) = 2 MN = 2 × $10^6$ N ; Outer thickness ($D$) = 0.8 m = 0.8 × $10^3$ mm ; Thickness ($t$) = 150 mm and eccentricity ($e$) = 100 mm.

***Maximum stress intensity in the section***

We know that area of the pier,

$$A = (BD - bd)$$
$$= [(1.2 \times 10^3) \times (0.8 \times 10^3)] - [(0.9 \times 10^3) \times (0.5 \times 10^3)]$$
$$= (0.96 \times 10^6) - (0.45 \times 10^6) = 0.51 \times 10^6 \text{ mm}^2$$

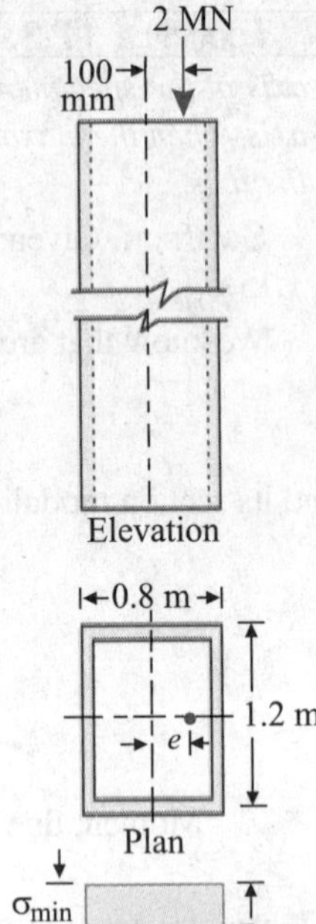

Fig. 17.6

and its section modulus,

$$Z = \frac{1}{6}[BD^2 - bd^2] = \frac{1}{6}[(1.2 \times 10^3) \times (0.8 \times 10^3)^2]$$
$$- [(0.9 \times 10^3) \times (0.5 \times 10^3)^2] \text{ mm}^3$$
$$= \frac{1}{6}[(768 \times 10^6) - (225 \times 10^6) = 90.5 \times 10^6 \text{ mm}^3$$

We know that moment due to eccentricity of load,

$$M = P \cdot e = (2 \times 10^6) \times 100 = 200 \times 10^6 \text{ N-mm}$$

∴ Maximum stress intensity in the section,

$$\sigma_{max} = \frac{P}{A} + \frac{M}{Z} = \frac{2 \times 10^6}{0.51 \times 10^6} + \frac{200 \times 10^6}{90.5 \times 10^6} \text{ N/mm}^2$$
$$= 3.92 + 2.21 = 6.13 \text{ N/mm}^2 = \mathbf{6.13\ MPa} \quad \textbf{Ans.}$$

***Minimum stress intensity in the section***

We also know that minimum stress intensity in the section,

$$\sigma_{min} = \frac{P}{A} - \frac{M}{Z} = \frac{2 \times 10^6}{0.51 \times 10^6} - \frac{200 \times 10^6}{90.5 \times 10^6} \text{ N/mm}^2$$
$$= 3.92 - 2.21 = 1.71 \text{ N/mm}^2 = \mathbf{1.71\ MPa} \quad \textbf{Ans.}$$

**EXAMPLE 17.5.** *A hollow circular column having external and internal diameters of 300 mm and 250 mm respectively carries a vertical load of 100 kN at the outer edge of the column. Calculate the maximum and minimum intensities of stress in the section.*

**SOLUTION.** Given: External diameter $(D) = 300$ mm ; Internal diameter $(d) = 250$ mm and load $(P) = 100$ kN $= 100 \times 10^3$ N

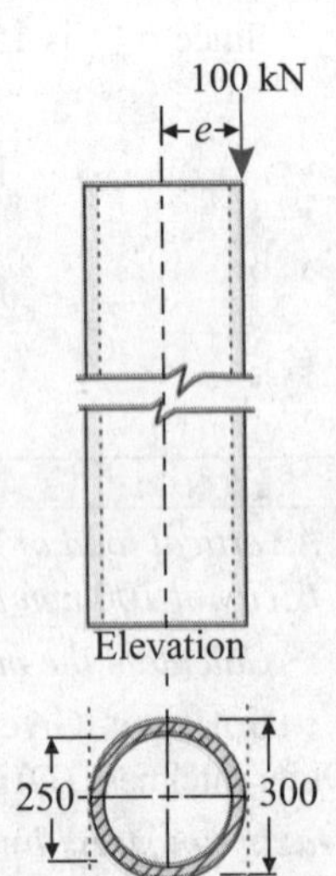

Fig. 17.7

***Maximum intensity of stress in the section***

We know that area of the column,

$$A = \frac{\pi}{4}(D^2 - d^2) = \frac{\pi}{4}[(300)^2 - (250)^2] \text{ mm}^2$$
$$= 21.6 \times 10^3 \text{ mm}^2$$

and its section modulus,

$$Z = \frac{\pi}{32} \times \left[\frac{D^4 - d^4}{D}\right] = \frac{\pi}{32} \times \left[\frac{(300)^4 - (250)^4}{300}\right] \text{ mm}^3$$
$$= 1372 \times 10^3 \text{ mm}^3$$

Since the column carries the vertical load at its outer edge, therefore eccentricity,

$$e = 150 \text{ mm}$$

and moment due to eccentricity of load,

$$M = P \cdot e = (100 \times 10^3) \times 150 = 15 \times 10^6 \text{ N-mm}$$

∴ Maximum intensity of stress in the section,

$$\sigma_{max} = \frac{P}{A} + \frac{M}{Z} = \frac{100 \times 10^3}{21.6 \times 10^3} + \frac{15 \times 10^6}{1372 \times 10^3} \text{ N/mm}^2$$
$$= 4.63 \times 10.93 = 15.56 \text{ N/mm}^2 = \mathbf{15.56\ MPa} \quad \textbf{Ans.}$$

***Minimum intensity of stress in the section***

We also know that minimum intensity of stress in the section,

$$\sigma_{min} = \frac{P}{A} - \frac{M}{Z} = \frac{100 \times 10^3}{21.6 \times 10^3} - \frac{15 \times 10^6}{1372 \times 10^3} \text{ N/mm}^2$$

$$= 4.63 - 10.93 = -6.3 \text{ N/mm}^2$$

$$= 6.3 \text{ N/mm}^2 \text{ (tension)} = \mathbf{6.3\ MPa\ (tension)} \quad \textbf{Ans.}$$

## EXERCISE 17.1

1. A rectangular strut 200 mm wide and 150 mm thick carries a load of 60 kN at an eccentricity of 20 mm in a plane bisecting the thickness. Find the maximum and minimum intensities of stresses in the section. (**Ans.** 3200 kPa ; 800 kPa)
2. A circular column of 200 mm diameter is subjected to a load of 300 kN, which is acting 5 mm away from the geometric centre of the column. Find the maximum and minimum stress intensities in the section. (**Ans.** 11.94 MPa ; 7.16 MPa)
3. A rectangular hollow masonry pier of 1200 mm × 800 mm with wall thickness of 150 mm carries a vertical load of 100 kN at an eccentricity of 100 mm in the plane bisecting to 1200 mm side. Calculate the maximum and minimum stress intensities in the section (**Ans.** 291.6 kPa ; 100.4 kPa)
4. A hollow circular column of 200 mm external diameter and 180 mm internal diameter is subjected to a vertical load of 75 kN at an eccentricity of 35 mm. What are the maximum and minimum stress intensities ? (**Ans.** 22.28 MPa ; 2.84 MPa)

## 17.5. Symmetrical Columns with Eccentric Loading about Two Axes

In the previous articles, we have discussed the cases of eccentric loading about one axis only. But, sometimes the load is acting eccentrically about two axes as shown in Fig. 17.8. Now consider a column *ABCD* subjected to a load with eccentricity about two axes as shown in Fig. 17.8.

Let $P$ = Load acting on the column ,

$A$ = Cross-sectional area of the column,

$e_X$ = Eccentricity of the load about *X-X* axis,

Moment of the load about *X-X* axis,

$$M_X = P \cdot e_X$$

Let $I_{XX}$ = Moment of inertia of the column section about *X-X* axis and

$e_Y, M_Y, I_{YY}$ = Corresponding values of *Y-Y* axis.

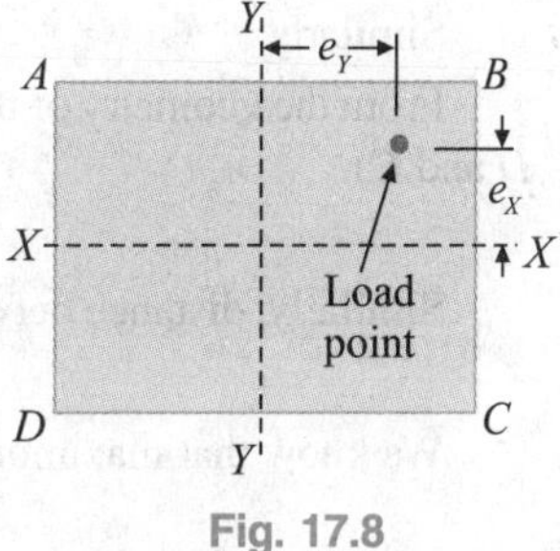

Fig. 17.8

The effect of such a load may be split up into the following three parts :

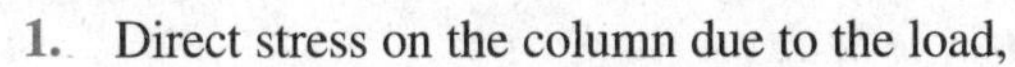

1. Direct stress on the column due to the load,

$$\sigma_0 = \frac{P}{A} \quad ...(i)$$

2. Bending stress due to eccentricity $e_X$,

$$\sigma_{bX} = \frac{M_X \cdot y}{I_{XX}} = \frac{P \cdot e_X \cdot y}{I_{XX}} \quad ...(ii)$$

3. Bending stress due to eccentricity $e_Y$,

$$\sigma_{bY} = \frac{M_Y \cdot x}{I_{YY}} = \frac{P \cdot e_Y \cdot x}{I_{YY}} \quad ...(iii)$$

∴ Total stress at the extreme fibre

$$= \sigma_0 \pm \sigma_{bX} \pm \sigma_{bY} = \frac{P}{A} \pm \frac{M_X \cdot y}{I_{XX}} \pm \frac{M_Y \cdot x}{I_{YY}}$$

The +ve or –ve sign depends upon the position of the fibre with respect to the load. A little consideration will show that the stress will be maximum at *B*, where both the +ve signs are to be adopted. The stress will be minimum at *D*, where both the –ve signs are to be adopted. While calculating the stress at *A*, the value of $M_X$ is to be taken as +ve, whereas the value of $M_Y$ as –ve. Similarly for the stress at *C*, the value of $M_Y$ is to be taken as +ve, whereas the value of $M_X$ as –ve.

**EXAMPLE 17.6.** *A column 800 mm × 600 mm is subjected to an eccentric load of 60 kN as shown in Fig. 17.9.*

*What are the maximum and minimum intensities of stresses in the column ?*

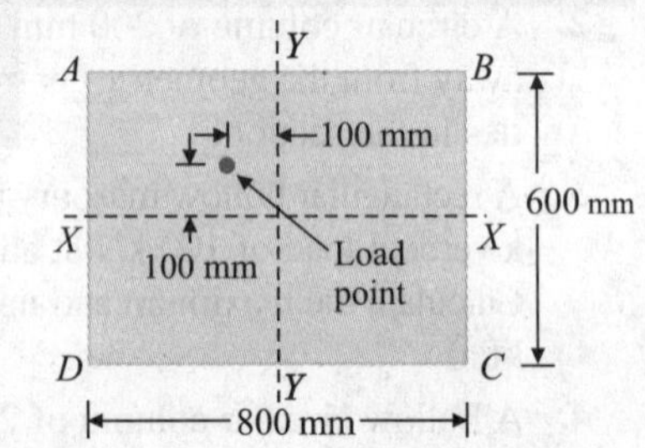

Fig. 17.9

**SOLUTION.** Given: Width (*b*) = 800 mm ; Thickness (*d*) = 600 mm ; Load (*P*) = 60 kN = $60 \times 10^3$ N ; Eccentricity along *X-X* axis ($e_X$) = 100 mm and eccentricity along *Y-Y* axis ($e_Y$) = 100 mm.

***Maximum intensity of stress in the column***

We know that area of the column,

$$A = b \times d = 800 \times 600 = 480 \times 10^3 \text{ mm}^2$$

and moment of inertia of the column about *X-X* axis,

$$I_{XX} = \frac{bd^3}{12} = \frac{800 \times (600)^3}{12} = 14.4 \times 10^9 \text{ mm}^4$$

Similarly,

$$I_{YY} = \frac{bd^3}{12} = \frac{600 \times (800)^3}{12} = 25.6 \times 10^9 \text{ mm}^4$$

We also know that moment due to eccentricity of load along *X-X* axis,

$$M_X = P \cdot e_X = (60 \times 10^3) \times 100 = 6 \times 10^6 \text{ N-mm}$$

Similarly,

$$M_Y = P \cdot e_Y = (60 \times 10^3) \times 100 = 6 \times 10^6 \text{ N-mm}$$

From the geometry of the loading, we find that distance between *Y-Y* axis and corners *A* and *B* (or *D* and *C*).

$$x = 400 \text{ mm}$$

Similarly, distance between *X-X* axis and corners *A* and *D* (or *B* and *C*).

$$y = 300 \text{ mm}$$

We know that maximum intensity of stress at *A*,

$$\sigma_A = \frac{P}{A} + \frac{M_X y}{I_{XX}} + \frac{M_Y x}{I_{YY}}$$

$$= \frac{60 \times 10^3}{480 \times 10^3} + \frac{(6 \times 10^6) \times 300}{14.4 \times 10^9} + \frac{(6 \times 10^6) \times 400}{25.6 \times 10^9} \text{ N/mm}^2$$

$$= 0.125 + 0.125 + 0.094 = 0.344 \text{ N/mm}^2 = \mathbf{0.344 \text{ MPa}} \quad \textbf{Ans.}$$

***Minimum intensity of stress in the column***

We also know that minimum intensity of stress in the column,

$$\sigma_C = \frac{P}{A} - \frac{M_X \cdot y}{I_{XX}} - \frac{M_Y \cdot x}{I_{YY}}$$

$$= \frac{60 \times 10^3}{480 \times 10^3} - \frac{(6 \times 10^6) \times 300}{14.4 \times 10^9} - \frac{(6 \times 10^6) \times 400}{25.6 \times 10^9} \text{ N/mm}^2$$

$$= 0.125 - 0.125 - 0.094 = -0.094 \text{ N/mm}^2$$

$$= 0.094 \text{ N/mm}^2 \text{ (tension)} = \mathbf{0.094 \text{ MPa (tension)}} \quad \textbf{Ans.}$$

**EXAMPLE 17.7.** *A masonry pier of 3 m × 4 m supports a vertical load of 80 kN as shown in Fig. 17.10.*

*(a) Find the stresses developed at each corner of the pier.*

*(b) What additional load should be placed at the centre of the pier, so that there is no tension anywhere in the pier section ?*

*(c) What are the stresses at the corners with the additional load in the centre.*

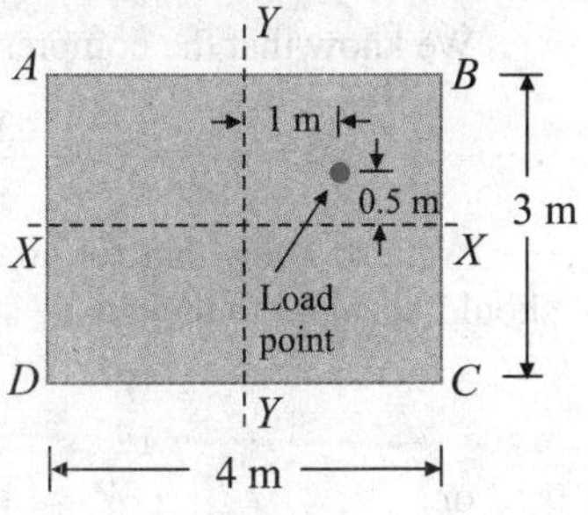

Fig. 17.10

**SOLUTION.** Given: Width ($b$) = 4 m ; Thickness ($d$) = 3 m ; Load ($P$) = 80 kN ; Eccentricity along $X$-$X$ axis ($e_X$) = 0.5 m and eccentricity along $Y$-$Y$ axis ($e_Y$) = 1 m.

***(a) Stresses developed at each corner***

We know that area of the pier,

$$A = b \times d = 4 \times 3 = 12 \text{ m}^2$$

and moment of inertia of the pier about $X$-$X$ axis,

$$I_{XX} = \frac{bd^3}{12} = \frac{4 \times (3)^3}{12} = 9 \text{ m}^4$$

Similarly,
$$I_{YY} = \frac{bd^3}{12} = \frac{3 \times (4)^3}{12} = 16 \text{ m}^4$$

We also know that moment due to eccentricity of load along $X$-$X$ axis,

$$M_X = P \cdot e_X = 80 \times 0.5 = 40 \text{ kN-m}$$

Similarly,
$$M_Y = P \cdot e_Y = 80 \times 1.0 = 80 \text{ kN-m}$$

From the geometry of the loading, we find that distance between $Y$-$Y$ axis and the corners $A$ and $B$,

$$x = 2 \text{ m}$$

Similarly distance between $X$-$X$ axis and the corners $A$ and $D$,

$$y = 1.5 \text{ m}$$

We know that stress at $A$,

$$\sigma_A = \frac{P}{A} + \frac{M_X \cdot y}{I_{XX}} - \frac{M_Y \cdot x}{I_{YY}} = \frac{80}{12} + \frac{40 \times 1.5}{9} - \frac{80 \times 2}{16} \text{ kN/m}^2$$

$$= 6.67 + 6.67 - 10 = 3.34 \text{ kN/m}^2 = \mathbf{3.34 \text{ kPa}} \quad \textbf{Ans.}$$

Similarly,
$$\sigma_B = \frac{P}{A} + \frac{M_X \cdot y}{I_{XX}} + \frac{M_Y \cdot x}{I_{YY}} = \frac{80}{12} + \frac{40 \times 1.5}{9} + \frac{80 \times 2}{16} \text{ kN/m}^2$$

$$= 6.67 + 6.67 + 10.0 = 23.34 \text{ kN/m}^2 = \mathbf{23.34 \text{ kPa}} \quad \textbf{Ans.}$$

$$\sigma_C = \frac{P}{A} - \frac{M_X \cdot y}{I_{XX}} + \frac{M_Y \cdot x}{I_{YY}} = \frac{80}{12} - \frac{40 \times 1.5}{9} + \frac{80 \times 2}{16} \text{ kN/m}^2$$

$$= 6.67 - 6.67 + 10.0 = 10.0 \text{ kN/m}^2 = \mathbf{10.0 \text{ kPa}} \qquad \textbf{Ans.}$$

and
$$\sigma_D = \frac{P}{A} - \frac{M_X \cdot y}{I_{XX}} - \frac{M_Y \cdot x}{I_{YY}} = \frac{80}{12} - \frac{40 \times 1.5}{9} - \frac{80 \times 2}{16} \text{ kN/m}^2$$

$$= 6.67 - 6.67 - 10.0 = -10.0 \text{ kN/m}^2 = \mathbf{10 \text{ kPa (tension)}} \qquad \textbf{Ans.}$$

***(b) Additional load at the centre for no tension in the pier section***

Let $W$ = Additional load (in kN) that should be placed at the centre for no tension in the pier section.

We know that the compressive stress due to the load

$$= \frac{W}{A} = \frac{W}{12} \text{ kN/m}^2$$

We also know that for no tension, in the pier section the compressive stress due to the load. $W$ should be equal to the tensile stress at $D$, *i.e.*, 10.0 kN/m$^2$.

$$\therefore \quad \frac{W}{12} = 10.0$$

or
$$W = 10.0 \times 12 = \mathbf{120 \text{ kN}} \qquad \textbf{Ans.}$$

***(c) Stresses at the corners with the additional load in the centre***

We find that the stress due to the additional load

$$= \frac{W}{A} = \frac{120}{12} = 10.0 \text{ kN/m}^2$$

$\therefore$ Stress at $A$, $\sigma_A = 3.34 + 10.0 = \mathbf{13.34 \text{ kPa}}$ **Ans.**

Similarly, $\sigma_B = 23.34 + 10.0 = \mathbf{33.34 \text{ kPa}}$ **Ans.**

$\sigma_C = 10.0 + 10.0 = \mathbf{20.0 \text{ kPa}}$ **Ans.**

and $\sigma_D = 10.0 + 10.0 = \mathbf{0}$ **Ans.**

## 17.6. Unsymmetrical Columns with Eccentric Loading

In the previous articles, we have discussed the symmetrical column sections subjected to eccentric loading. But in an unsymmetrical column, first *c.g.* and then moment of inertia of the section is found out. After that the distances between the *c.g.* of the section and its corners are calculated. The stresses on the corners are then found out as usual, by using the respective values of moment of inertia and distance of the corner from the *c.g.* of the section.

**EXAMPLE 17.8.** *A hollow cylindrical shaft of 200 mm external diameter has got eccentric bore of 140 mm diameter, such that the thickness varies from 20 mm at one end to 40 mm at the other. Calculate the extreme stress intensities, if the shaft is subjected to a load of 400 kN along the axis of the bore.*

**SOLUTION.** Given: External diameter ($D$) = 200 mm ; Internal diameter ($d$) = 140 mm and load ($P$) = 400 kN = 400 × 10$^3$ N.

We know that net area of the shaft,

$$A = \frac{\pi}{4}[(200)^2 - (140)^2] = 5\,100\,\pi \text{ mm}^2$$

First of all, let find out the centre of gravity of the section. Let the left end $A$ be the point of reference.

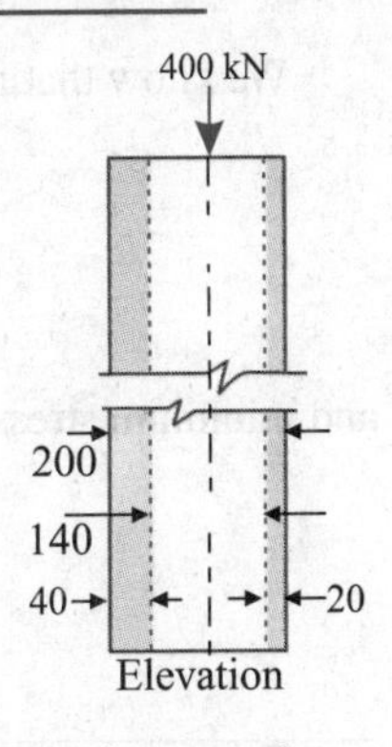


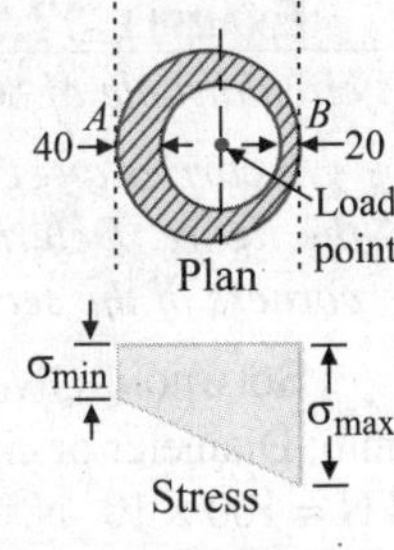


Fig. 17.11

*(i) Main circle*

$$a_1 = \frac{\pi}{4} \times D^2 = \frac{\pi}{4} \times (200)^2 = 10\,000\ \pi \text{ mm}^2$$

$$x_1 = \frac{200}{2} = 100 \text{ mm}$$

*(ii) Bore*

$$a_2 = \frac{\pi}{4} \times d^2 = \frac{\pi}{4} \times (140)^2 = 4\,900\ \pi \text{ mm}^2$$

$$x_2 = 40 + \frac{140}{2} = 110 \text{ mm}$$

We know that distance between the centre of gravity of the section and the left end $A$,

$$\bar{x} = \frac{a_1 x_1 - a_2 x_2}{a_1 - a_2} = \frac{(10000\ \pi \times 100) - (4900\ \pi \times 100)}{10000\ \pi \times 4900\ \pi}$$

$$= 90.4 \text{ mm}$$

From the geomety of the figure, we find that the eccentricity of the load,

$$e = 110 - 90.4 = 19.6 \text{ mm}$$

∴ Moment due to eccenticity of load,

$$M = P \cdot e = (400 \times 10^3) \times 19.6$$

$$= 7.84 \times 10^6 \text{ N-mm}$$

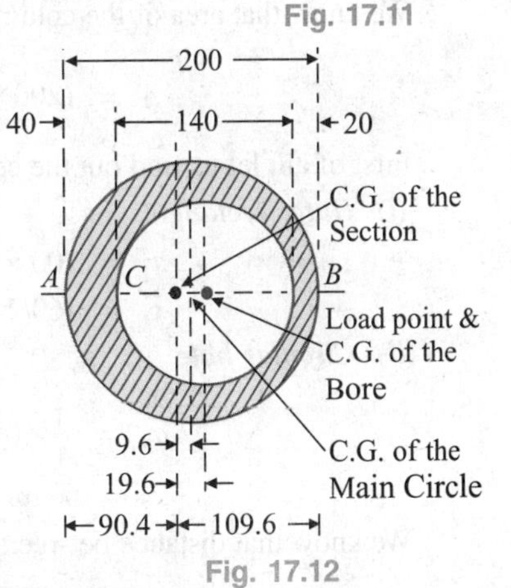


Fig. 17.12

Distance of corner $A$ from the centre of gravity of the section,

$$y_A = 90.4 \text{ mm}$$

Similarly, $y_B = 200 - 90.4 = 109.6$ mm

We know that the moment of inertia of the main circle about its centre of gravity,

$$I_{G1} = \frac{\pi}{64} \times (200)^4 = 25 \times 10^6\ \pi \text{ mm}^4$$

and distance between the centre of gravity of the main circle and centre of gravity of the section,

$$h_1 = 100 - 90.4 = 9.6 \text{ mm}$$

∴ Moment of inertia of the main circle about the centre of gravity of the section

$$= I_{G1} + a_1 h_1^2 = (25 \times 10^6\ \pi) + (10\,000\ \pi)\,(9.6)^2 \text{ mm}^4$$

$$= 25.92 \times 10^6\ \pi \text{ mm}^4$$

Similarly, moment of inertia of the bore about its centre of gravity

$$I_{G2} = \frac{\pi}{64} \times (140)^4 = 6.0 \times 10^6\ \pi \text{ mm}^4$$

and distance between the centre of gravity of the bore and the centre of gravity of the section,

$$h_2 = 110 - 90.4 = 19.6 \text{ mm}$$

∴ Moment of inertia of the bore about the centre of gravity of the section

$$= I_{G2} + a_2 h_2^2 = (6.0 \times 10^6\ \pi) + (4\,900\ \pi)\,(19.6)^2 \text{ mm}^4$$

$$= 7.88 \times 10^6\ \pi \text{ mm}^4$$

and net moment of inertia of the section about its centre of gravity,

$$I = (25.92 \times 10^6\ \pi) - (7.88 \times 10^6\ \pi) = 18.04 \times 10^6\ \pi \text{ mm}^4$$

We know that maximum stress intensity,

$$\sigma_{max} = \frac{P}{A} + \frac{M \cdot y_B}{I} = \frac{400 \times 10^3}{5100\,\pi} + \frac{(7.84 \times 10^6) \times 109.6}{18.04 \times 10^6\,\pi} \text{ N/mm}^2$$

$$= 24.97 + 15.16 = 40.13 \text{ N/mm}^2 = \mathbf{0.13\ MPa} \quad \textbf{Ans.}$$

and minimum stress intensity,

$$\sigma_{min} = \frac{P}{A} - \frac{M \cdot y_A}{I} = \frac{400 \times 10^3}{5100\,\pi} - \frac{(7.84 \times 10^6) \times 90.4}{18.04 \times 10^6\,\pi} \text{ N/mm}^2$$

$$= 24.97 - 12.51 = 12.46 \text{ N/mm}^2 = \mathbf{12.46\ MPa} \quad \textbf{Ans.}$$

**EXAMPLE 17.9.** *A short C.I. column has a rectangular section 160 mm  200 mm with a circular hole of 80 mm diameter as shown in Fig. 17.13.*

*It carries an eccentric load of 100 kN, located as shown in the figure. Determine the values of the stresses at the four corners of the section.*

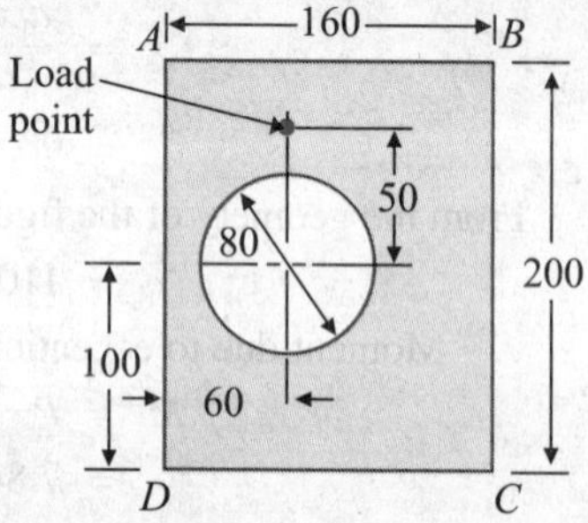

**SOLUTION.** Given: Width ($B$) = 160 mm ; Depth ($D$) = 200 mm; Diameter of circular hole ($d$) = 80 mm and load ($P$) = 100 kN = $100 \times 10^3$ N.

Fig. 17.13

We know that area of the column section,

$$A = (200 \times 160) - \left(\frac{\pi}{4} \times (80)^2\right) = 26\,970 \text{ mm}^2$$

First of all, let us find out the centre of gravity of the section. Let $AD$ be the line of reference.

***(i) Outer rectangle***

$$a_1 = 200 \times 160 = 32\,000 \text{ mm}^2$$

$$x_1 = 160/2 = 80 \text{ mm}$$

***(ii) Circular hole***

$$a_2 = \frac{\pi}{4} \times (80)^2 = 5\,027 \text{ mm}^2$$

$$x_2 = 60 \text{ mm}$$

We know that distance between the centre of gravity of the section and $AD$,

$$\bar{x} = \frac{a_1 x_1 - a_2 x_2}{a_1 - a_2} = \frac{(32000 \times 80) - (5027 \times 60)}{(32000 - 5027)} = 83.7 \text{ mm}$$

From the geometry of the figure, we find that eccentricity of load about $X$-$X$ axis

$$e_X = 50 \text{ mm} \quad \text{and} \quad e_Y = 83.7 - 60 = 23.7 \text{ mm}$$

∴ Moment due to eccentricity of load along $X$-$X$ axis,

$$M_X = P \cdot e_X = (100 \times 10^3) \times 50 = 5 \times 10^6 \text{ N-mm}$$

Similarly $\quad M_Y = P \cdot e_Y = (100 \times 10^3) \times 23.7 = 2.37 \times 10^6$ N-mm

and distance of the corner $A$ from $X$-$X$ axis passing through centre of gravity of the section,

$$y_A = y_B = y_C = y_D = 100 \text{ mm}$$

Similarly, distance of corner $A$ from $Y$-$Y$ axis passing through centre of gravity of the section,

$$x_A = x_D = 83.7 \text{ mm}$$

and $\quad x_B = x_C = 160 - 83.7 = 76.3$ mm

We know that the moment of inertia of the main rectangle *ABCD*, passing through its centre of gravity and parallel to *X-X* axis,

$$I_{G1} = \frac{160 \times (200)^2}{12} = 106.7 \times 10^6 \text{ mm}^4$$

and moment of inertia of the circular hole, passing through its centre of gravity and parallel to *X-X* axis,

$$I_{G2} = \frac{\pi}{4} \times (80)^2 = 2.01 \times 10^6 \text{ mm}^4$$

Since the centre of gravity of the rectangle and the circular hole coincides with the *X-X* axis, therefore moment of inertia of the section about *X-X* axis,

$$I_{XX} = (106.7 \times 10^6) - (2.01 \times 10^6) = 104.69 \times 10^6 \text{ mm}^4 \quad ...(i)$$

We also know that the moment of inertia of the main rectangle *ABCD*, passing through its centre of gravity and parallel to *Y-Y* axis,

$$I_{G3} = \frac{200 \times (160)^3}{12} = 68.26 \times 10^6 \text{ mm}^4$$

and distance between the centre of gravity of the rectangle from *Y-Y* axis,

$$h_1 = 83.7 - 80 = 3.7 \text{ mm}$$

∴ Moment of inertia of the rectangle through centre of gravity of the section and about *Y-Y* axis

$$= I_{G3} + a_1 h_1^2 = (68.26 \times 10^6) + 32\,000 \times (3.7)^2 \text{ mm}^4$$

$$= 68.7 \times 10^6 \text{ mm}^4$$

Similarly, moment of inertia of the circular hole through its centre of gravity and parallel to *Y-Y* axis,

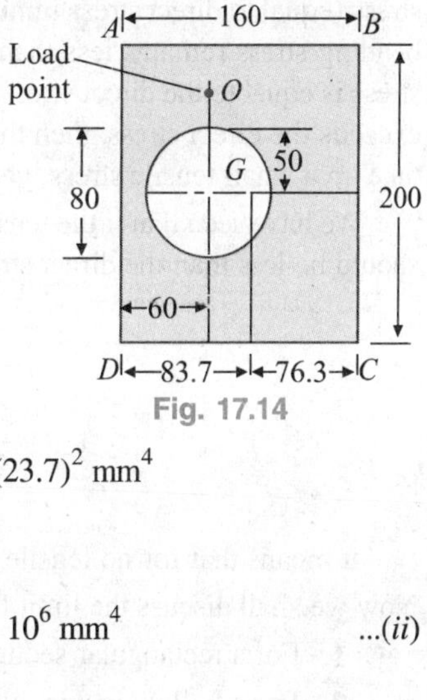

$$I_{G4} = \frac{\pi}{64} \times (80)^4 = 2.01 \times 10^6 \text{ mm}^4$$

and distance between the centre of gravity of the circular section from *Y-Y* axis,

$$h_2 = 83.7 - 60 = 23.7 \text{ mm}$$

∴ Moment of inertia of the circular hole through centre of gravity of the section and about *Y-Y* axis

$$= I_{G4} + a_2 h_2^2 = (2.01 \times 10^6) + 5\,027 \times (23.7)^2 \text{ mm}^4$$

$$= 4.84 \times 10^6 \text{ mm}^4$$

Fig. 17.14

and net moment of inertia of the section about *Y-Y* axis,

$$I_{YY} = (68.7 \times 10^6) - (4.84 \times 10^6) = 63.86 \times 10^6 \text{ mm}^4 \quad ...(ii)$$

Now from the geometry of the figure, we find that stress at *A*,

$$\sigma_A = \frac{P}{A} + \frac{M_X \cdot y_A}{I_{XX}} + \frac{M_Y \cdot x_A}{I_{YY}}$$

$$= \frac{100 \times 10^3}{26970} + \frac{(5 \times 10^6) \times 100}{104.69 \times 10^6} + \frac{(2.3 \times 10^6) \times 83.7}{63.86 \times 10^6} \text{ N/mm}^2$$

$$= 11.5 \text{ N/mm}^2 = \mathbf{11.5 \text{ MPa}} \quad \textbf{Ans.}$$

Similarly,

$$\sigma_B = \frac{P}{A} + \frac{M_X \cdot y_B}{I_{XX}} - \frac{M_Y \cdot x_B}{I_{YY}}$$

$$= \frac{100 \times 10^3}{26970} + \frac{(5 \times 10^6) \times 100}{104.69 \times 10^6} - \frac{(2.3 \times 10^6) \times 76.3}{63.86 \times 10^6} \text{ N/mm}^2$$

$$= 5.74 \text{ N/mm}^2 = \mathbf{5.74 \text{ MPa}} \quad \textbf{Ans.}$$

$$\sigma_C = \frac{P}{A} - \frac{M_X \cdot y_C}{I_{XX}} - \frac{M_Y \cdot x_C}{I_{YY}}$$

$$= \frac{100 \times 10^3}{26970} - \frac{(5 \times 10^6) \times 100}{104.69 \times 10^6} - \frac{(2.3 \times 10^6) \times 76.3}{63.86 \times 10^6} \text{ N/mm}^2$$

$$= -3.82 \text{ N/mm}^2 = \mathbf{3.82 \text{ MPa (tensile)}} \quad \textbf{Ans.}$$

and

$$\sigma_D = \frac{P}{A} - \frac{M_X \cdot y_D}{I_{XX}} + \frac{M_Y \cdot x_D}{I_{YY}}$$

$$= \frac{100 \times 10^3}{26970} - \frac{(5 \times 10^6) \times 100}{104.69 \times 10^6} + \frac{(2.3 \times 10^6) \times 83.7}{63.86 \times 10^6} \text{ N/mm}^2$$

$$= 1.95 \text{ N/mm}^2 = \mathbf{1.95 \text{ MPa}} \quad \textbf{Ans.}$$

## 17.7. Limit of Eccentricity

We have seen in Art. 17.2 and 17.3, that when an eccentric load is acting on a column, it produces direct stress as well as bending stress. On one side of the neutral axis there is a maximum stress (equal to the sum of direct and bending stress) and on the other side of the neutral axis there is a minimum stress (equal to direct stress minus bending stress). A little consideration will show that so long as the bending stress remains less than the direct stress, the resultant stress is compressive. If the bending stress is equal to the direct stress, then there will be a zero stress on one side. But if the bending stress exceeds the direct stress, then there will be a tensile stress on one side. Though cement concrete can take up a small tensile stress, yet it is desirable that no tensile stress should come into play.

We have seen that if the tensile stress is not to be permitted to come into play, then bending stress should be less than the direct stress, or maximum, it may be equal to the direct stress, *i.e.*,

$$\sigma_b \le p_0$$

$$\frac{P \cdot e}{Z} \le \frac{P}{A} \quad \text{...}(\because M = P \cdot e)$$

or

$$e \le \frac{Z}{A}$$

It means that for no tensile condition, the eccentricity $e$ should be less than $\frac{Z}{A}$ or equal to $\frac{Z}{A}$. Now we shall discuss the limit for eccentricity in the following cases :

1. For a rectangular section,
2. For a hollow rectangular section,
3. For a circular section and
4. For a hollow circular section.

***(a) Limit of eccentricity for a rectangular section***

Consider a rectangular section of width ($b$) and thickness ($d$) as shown in Fig. 17.15. We know that the modulus of section,

$$Z = \frac{1}{6}bd^2 \quad \text{...}(i)$$

and area of the section,

$$A = bd \quad \text{...}(ii)$$

We also know that for no tension condition,

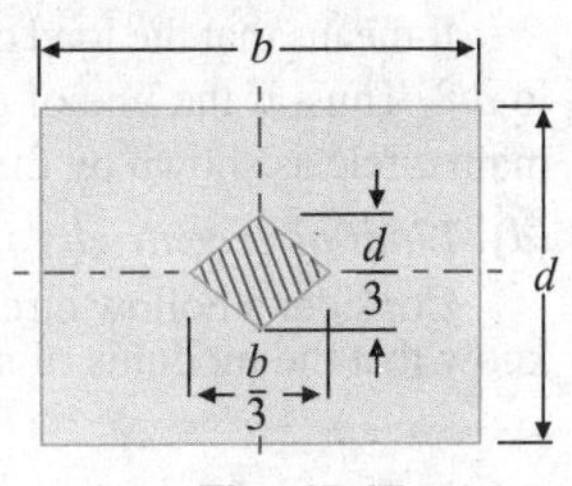

Fig. 17.15

$$e \le \frac{Z}{A}$$

$$\le \frac{\frac{1}{6}bd^2}{bd}$$

$$\le \frac{1}{6}d$$

It means that the load can be eccentric, on either side of the geometrical axes, by an amount equal to $d/6$. Thus if the line of action of the load is within the middle third, as shown by the dotted area in Fig. 17.15, then the stress will be compressive throughout.

***(b) Limit of eccentricity for a hollow rectangular section***

Consider a hollow rectangular section with $B$ and $D$ as outer width and thickness and $b$ and $d$ internal dimensions respectively. We know that the modulus of section,

$$Z = \frac{(BD^3 - bd^3)}{6D} \qquad ...(i)$$

and area of the hollow rectangular section,

$$A = BD - bd \qquad ...(ii)$$

We also know that for no tension condition,

$$e \le \frac{Z}{A}$$

$$\le \frac{\frac{(BD^3 - bd^3)}{6D}}{BD - bd}$$

$$\le \frac{(BD^3 - bd^3)}{6D\,(BD - bd)}$$

It means that the load can be eccentric, on either side of the geometrical axis, by an amount equal to $\dfrac{(BD^3 - bd^3)}{6D\,(BD - bd)}$.

***(c) Limit of eccentricity of a circular section***

Consider a circular section of diameter $d$ as shown in Fig. 17.16. We know that the modulus of section,

$$Z = \frac{\pi}{32} \times d^3 \qquad ...(i)$$

and area of circular section, $\quad A = \dfrac{\pi}{4} \times d^2 \qquad ...(ii)$

We also know that for no tension condition,

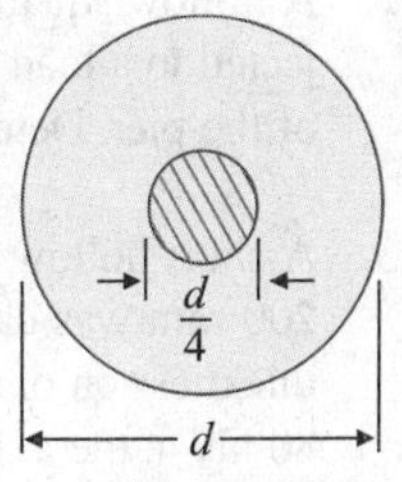

Fig. 17.16

$$e \le \frac{Z}{A}$$

$$\le \frac{\frac{\pi}{32} \times d^3}{\frac{\pi}{4} \times d^2}$$

$$\le \frac{d}{8}$$

It means that the load can be eccentric, on any side of the geometrical centre, by an amount equal to $d/8$. Thus, if the line of action of the load is within a circle of diameter equal to one-fourth of the main circle as shown by the dotted area in Fig. 17.16, then the stress will be compressive throughout.

***(d) Limit of eccentricity for hollow circular section***

Consider a hollow circular section of external and internal diameters as $D$ and $d$ respectively. We know that the modulus of section,

$$Z = \frac{\pi}{32} \times \frac{(D^4 - d^4)}{D} \qquad ...(i)$$

and area of hollow circular section,

$$A = \frac{\pi}{4} \times (D^2 - d^2) \qquad ...(ii)$$

We also know that for no tension condition,

$$e \le \frac{Z}{A}$$

$$\le \frac{\frac{\pi}{32} \times \frac{(D^4 - d^4)}{D}}{\frac{\pi}{4} \times (D^2 - d^2)}$$

$$\le \frac{(D^2 - d^2)}{8D} \qquad ...[\because (D^4 - d^4) = (D^2 + d^2)(D^2 - d^2)]$$

It means that the load can be eccentric, on any side of the geometrical centre, by an amount equal to $\frac{(D^2 - d^2)}{8D}$.

## EXERCISE 17.2

1. A rectangular pier is 1500 mm × 1000 mm is subjected to a compressive load of 450 kN as shown in Fig. 17.17.

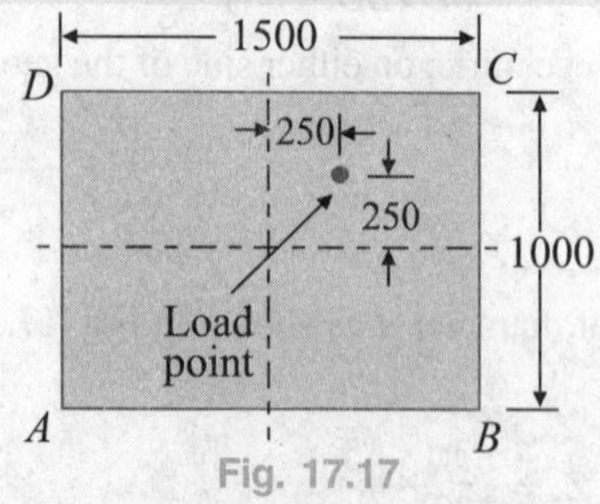


Fig. 17.17

Find the stress intensities on all the four corners of the pier.

[**Ans.** $\sigma_A = -4.5$ kPa ; $\sigma_B = +1.5$ kPa ; $\sigma_C = 10.5$ kPa ; $\sigma_D = 4.5$ kPa]

2. A hollow square column has 1.5 m outside length and 1 m inside length. The column is subjected to a load of 7 kN located on a diagonal and at a distance of 0.8 m from the vertical axis of the pier. Determine the stress intensities on the outside corners of the column.

[**Ans.** 23.15 kPa ; 5.6 kPa ; 11.95 kPa ; 5.6 kPa]

3. A short hollow cylindrical cast iron column of outside diameter 300 mm and inside diameter 200 mm was casted. On inspection, it was found the bore is eccentric in such a way that the thickness on one side is 70 mm and 30 mm on the other. If the column is subjected to a load of 80 kN at the axis of the bore, find the extreme intensities of stresses in the base.

[**Ans.** 3.66 kPa ; 0.73 MPa]

## QUESTIONS

1. Distinguish clearly between direct stress and bending stress.
2. What is meant by eccentric loading? Explain its effects on a short column.
3. Derive the relation for the maximum and minimum stress intensities due to eccentric loading.
4. Obtain a relation for the maximum and minimum stresses at the base of a symmetrical column. When it is subjected to
   (*a*) an eccentric load about one axis and (*b*) an eccentric load about two axes.
5. Show that for no tension in the base of a short column, the line of action of the load should be within the middle third.
6. Define the term limit of eccentricity. How will you find out this limit in case of a hollow circular section?

## OBJECTIVE TYPE QUESTIONS

1. The maximum stress intensity at the base of a square column of area $A$ and side $b$ subjected to a load $W$ at an eccentricity e is equal to

   (*a*) $\frac{W}{A}\left(1+\frac{2e}{b}\right)$ (*b*) $\frac{W}{A}\left(1-\frac{4e}{b}\right)$ (*c*) $\frac{W}{A}\left(1+\frac{6e}{b}\right)$ (*d*) $\frac{W}{A}\left(1+\frac{8e}{b}\right)$

2. The minimum stress intensity in the above case is

   (*a*) $\frac{W}{A}\left(1-\frac{e}{b}\right)$ (*b*) $\frac{W}{A}\left(1-\frac{2e}{b}\right)$ (*c*) $\frac{W}{A}\left(1-\frac{3e}{b}\right)$ (*d*) $\frac{W}{A}\left(1-\frac{6e}{b}\right)$

3. The maximum eccentricity of a load on a circular section to have same type of stress is
   (*a*) one-eighth of diameter (*b*) one-sixth of diameter
   (*c*) one-fourth of diameter (*d*) one-third of diameter

## ANSWERS

**1.** (*c*) **2.** (*d*) **3.** (*c*)

Chapter 18

# Dams and Retaining Walls

## Contents



## 18.1. Introduction

A dam* is constructed to store large quantity of water, which is used for the purposes of irrigation and power generation. A dam may be of any cross-section, but the dams of trapezoidal cross-section are very popular these days. A retaining wall is generally constructed to retain earth in hilly areas. Though there are many types of dams, yet the following are important from the subject point of view :

1. Rectangular dams.
2. Trapezoidal dams having water face vertical,
3. Trapezoidal dams having water face inclined.

* A dam constructed with earth is called an earthen dam; whereas a dam constructed with cement concrete is called a concrete dam or gravity dam.

We shall discuss the above three types of dams one by one.

## 18.2. Rectangular Dams

Consider a unit length of a rectangular dam, retaining water on one of its vertical sides as shown in Fig. 18.1.

Let $b$ = Width of the dam,

$H$ = Height of he dam,

$\rho$ = Specific weight of the dam masonry

$h$ = Height of water reatined by the dam, and

$w$ = *Specific weight of the water

∴ Weight of dam per unit length,

$$W = \rho \cdot b \cdot H$$

This weight will act through centre of gravity of the dam section.

We know that the intensity of water pressure will be zero at the water surface and will **increase by a straight line law to $wh$ at the bottom. Thus the average intensity of water pressure on the face of the dam

$$= \frac{wh}{2} \quad ...(i)$$

∴ Total pressure per unit length of the dam,

$$P = h \times \frac{wh}{2} = \frac{wh^2}{2} \quad ...(ii)$$

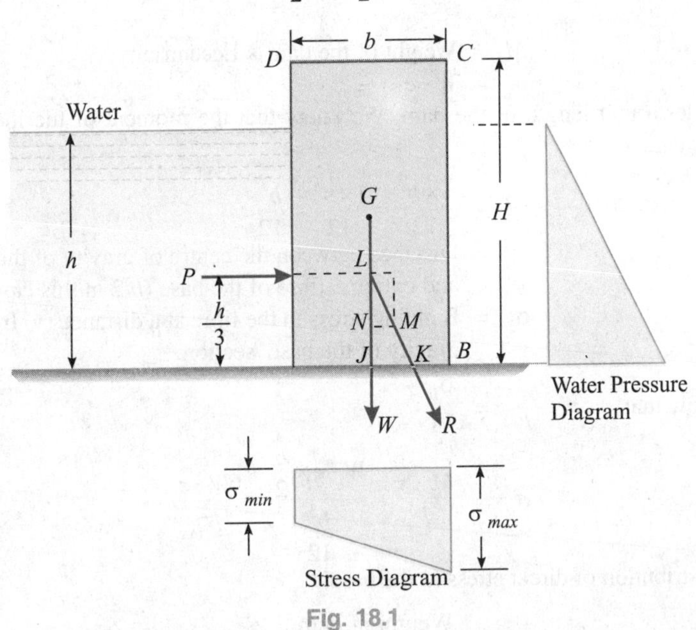


Fig. 18.1

* Unless mentioned, otherwise, the specific weight of the water is taken as 9.8 kN/m$^3$.

** Sometimes, the dam is subjected to two kinds of liquids (*e.g.*, some insoluble oil over water). In such a case, the pressure diagram will be zero at the top and will increase by a straight line law to $(w_1h_1)$ up to the depth of oil. It will further increase by a straight line law to $(w_1h_1 + w_2h_2)$ at the bottom of the water; where $w_1$, $w_2$ and $h_1$, $h_2$ are the specific weights and heights of the oil and water respectively. The total pressure on the dam may be found out as usual.

This water pressure acts at a height of $h/3$ from the bottom of the dam as shown in Fig. 18.1. Now with $P$ and $W$ as adjacent sides complete the parallelogram. The resultant of water pressure ($P$) and weight of the dam ($W$) will be given by the relation,

$$R = \sqrt{P^2 + W^2} \quad ...(iii)$$

Let $x$ be the horizontal distance between the centre of gravity of the dam and the point through which the resultant $R$ cuts the base (*i.e.*, the distance $JK$ in Fig. 18.1). From similar triangles $LNM$ and $LJK$, we see that

$$\frac{JK}{LJ} = \frac{NM}{LN}$$

$$\therefore \quad \frac{x}{h/3} = \frac{P}{W}$$

or

$$x = \frac{P}{W} \times \frac{h}{3} \quad ...(iv)$$

Let $d$* be the distance between the toe of the dam $A$ and the point, where the resultant $R$ cuts the base (*i.e.*, the distance $AK$ in Fig. 18.1)

$$\therefore \quad d = AJ + JK = \frac{b}{2} + x = \frac{b}{2} + \left(\frac{P}{W} \times \frac{h}{3}\right)$$

and the eccentricity of the resultant,

$$e = d - \frac{b}{2} \quad ...(x \text{ in the case})$$

A little consideration will show that as a result of the eccentricity, some moment will come into play, which will cause some bending stress at the base section of the dam. The magnitude of this moment,

$$M = \text{Weight of the dam} \times \text{Eccentricity}$$
$$= W \cdot e$$

Now consider a unit length of the dam. We know that the moment of the inertia of the base section about its c.g.,

$$I = \frac{l \times b^3}{12} = \frac{1 \times b^3}{12} = \frac{b^3}{12}$$

Now let $y$ = Distance between the centre of gravity of the base section and extreme fibre of the base ($b/2$ in this case)

and $\sigma_b$ = Bending stress in the fibre at a distance ($y$) from the centre of gravity of the base section.

We also know that

$$\frac{M}{I} = \frac{\sigma_b}{y}$$

$$\therefore \quad \sigma_b = \frac{M \cdot y}{I} = \frac{W \cdot e \times \frac{b}{2}}{\frac{b^3}{12}} = \frac{6W \cdot e}{b^2}$$

Now the distribution of direct stress at the base,

$$\sigma_0 = \frac{\text{Weight of dam}}{\text{Width of dam}} = \frac{W}{b}$$

* The distance $d$ may also be found out by taking moments of (*i*) water pressure, (*ii*) weight of dam and (*iii*) resultant force about $A$ and equating the same, *i.e.*,

$$Wd = P \cdot \frac{h}{3} + W \cdot \frac{b}{2}$$

($\because$ Vertical component of the resultant force is $W$ and is acting at a distance $d$ from $A$ and its horizontal component is acting through $A$.)

Now a little consideration will show that the stress across the base at $B$ will be maximum, whereas the stress across the base at $A$ will be minimum.

$$\therefore \quad \sigma_{max} = \sigma_0 + \sigma_b = \frac{W}{b} + \frac{6W \cdot e}{b^2} = \frac{W}{b}\left(1 + \frac{6e}{b}\right)$$

and

$$\sigma_{min} = \sigma_0 - \sigma_b = \frac{W}{b} - \frac{6W \cdot e}{b^2} = \frac{W}{b}\left(1 - \frac{6e}{b}\right)$$

**Notes. 1.** When the reservoir is empty, there will be no water pressure on the dam. In this case, there will be no eccentricity and thus the weight of the dam $W$ will act through the c.g. of the base section, which will cause direct stress only.

**2.** Sometimes, the value of $\sigma_{min}$ comes out to be negative. In such a case, there will be a tensile stress at the base of the dam.

**EXAMPLE 18.1.** *A water tank contains 1.3 m deep water. Find the pressure exerted by the water per metre length of the tank. Take specific weight of water as 9.8 kN/m³.*

**SOLUTION.** Given: Height of water $(h) = 1.3$ m and $w = 9.8$ kN/m³.

We know that pressure exerted by the water per metre length of the tank,

$$P = \frac{wh^2}{2} = \frac{9.8 \times (1.3)^2}{2} = \mathbf{8.28\ kN} \quad \textbf{Ans.}$$

**EXAMPLE 18.2.** *Find the magnitude and line of action of the pressure exerted on the side of a tank, which is 1.5 m square and 1 metre deep. The tank is filled half full with a liquid having specific gravity of 2, while the remainder is filled with a liquid having a specific gravity of 1. Take specific weight of water as 10 kN/m³.*

**SOLUTION.** Given: Side of the square tank = 1.5 m; Depth of the tank = 1 m; Depth of liquid of specific gravity 2 $(h_2) = 0.5$ m; Depth of liquid of specific gravity 1 $(h_1) = 0.5$ m and specific weight of water $(w) = 10$ kN/m³.

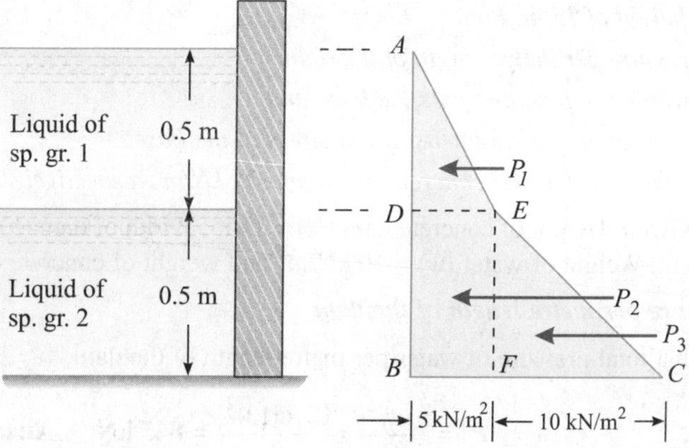


Fig. 18.2

***Magnitude of the pressure***

We know that intensity of pressure at $D$ (or $B$) due to liquid of sp. gr. 1.

$$= DE = BF = w_1 h_1$$
$$= (1 \times 10) \times 0.5 = 5 \text{ kN/m}^2$$

$\therefore$ Total pressure at $D$, due to liquid of sp. gr. 1,

$$P_1 = \text{Area of triangle } ADE \times \text{Length of the tank wall}$$
$$= \left(\frac{1}{2} \times 5.0 \times 0.5\right) \times 1.5 = 1.875 \text{ kN} \qquad ...(i)$$

and total pressure at $B$ due to liquid of sp. gr. 1,

$$P_2 = \text{Area of rectangle } BDFE \times \text{Length of the tank wall}$$
$$= (5.0 \times 0.5) \times 1.5 = 3.75 \text{ kN} \quad \text{...(ii)}$$

Similarly, intensity of pressure at $B$ due to liquid of sp. gr. 2,

$$FC = w_2 h_2 = (2 \times 10) \times 0.5 = 10 \text{ kN/m}^2$$

and total pressure from $E$ to $F$ or $D$ to $F$ (or $B$) due to liquid of sp. gr. 2,

$$P_3 = \text{Area of triangle } EFC \times \text{Length of the tank wall}$$
$$= \left(\frac{1}{2} \times 10 \times 0.5\right) = 3.75 \text{ kN} \quad \text{...(iii)}$$

∴ Magnitude of the pressure exerted on the side of the tank,

$$P = P_1 + P_2 + P_3 = 1.875 + 3.75 + 3.75 = \mathbf{9.375\ kN} \quad \textbf{Ans.}$$

***Line of action of the resultant force (i.e., pressure)***

Let $\bar{h}$ = Depth of the line of action of the resultant pressure from $A$.

Taking moments of all the pressures about A and equating the same,

$$P \times \bar{h} = \left[P_1 \times \frac{2 \times 0.5}{3}\right] + \left[P_2 \times \left(0.5 + \frac{0.5}{2}\right)\right] + \left[P_3 \times \left(0.5 + \frac{2 \times 0.5}{3}\right)\right]$$

$$9.375 \times \bar{h} = \left[1.875 \times \frac{1}{3}\right] + \left[3.75 \times \frac{3}{4}\right] + \left[3.75 \times \frac{5}{6}\right]$$
$$= 0.625 + 2.81 + 3.125 = 6.56$$

$$\therefore \quad \bar{h} = \frac{6.56}{9.375} = \mathbf{0.7\ m} \quad \textbf{Ans.}$$

**EXAMPLE 18.3.** *A concrete dam of rectangular section 15 m high and 6 m wide contains water up to a height of 13 m. Find*

*(a) total pressure per metre length of the dam,*

*(b) point, where the resultant cuts the base and*

*(c) maximum and minimum intensities of stress at the base.*

*Assume weight of water and concrete as 10 and 25 kN/m$^3$ respectively.*

**SOLUTION.** Given: Height of concrete dam ($H$) = 15 m; Width of dam ($b$) = 6 m; Height of water in dam ($h$) = 13 m; Weight of water ($w$) = 10 kN/m$^3$ and weight of concrete ($\rho$) = 25 kN/m$^3$.

***(a) Total pressure per metre length of the dam***

We know that total pressure of water per metre length of the dam,

$$P = \frac{wh^2}{2} = \frac{10 \times (13)^2}{2} = \mathbf{845\ kN} \quad \textbf{Ans.}$$

***(b) Point where the resultant cuts the base***

Let the resultant ($R$) cut the base at $K$ as shown in Fig.18.3.

We know that weight of the concrete per metre length,

$$W = \rho \times b \times H = 25 \times 6 \times 15 = 2250 \text{ kN}$$

and horizontal distance between the centre of gravity of the dam section and the point where the resultant cuts the base (*i.e.*, distance $JK$),

$$x = \frac{P}{W} \times \frac{h}{3} = \frac{845}{2250} \times \frac{13}{3} = \mathbf{1.63\ m} \quad \textbf{Ans.}$$

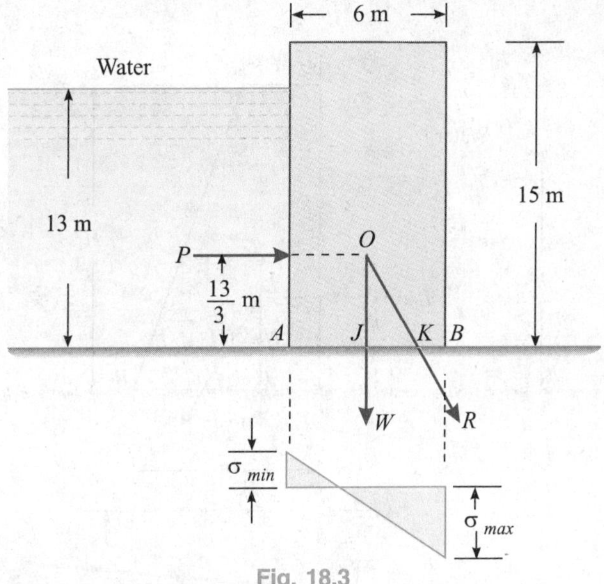


Fig. 18.3

**(c) *Maximum and minimum intensities of stress at the base***

We know that

*eccentricity of the resultant,

$$e = x = 1.63 \text{ m}$$

∴ Maximum intensity of stress at the base,

$$\sigma_{max} = \frac{W}{b}\left(1+\frac{6e}{b}\right) = \frac{2250}{6}\left(1+\frac{6\times 1.63}{6}\right) \text{ kN/m}^2$$

$$= 986.25 \text{ kN/m}^2 = \textbf{986.25 kPa (Compression)} \quad \textbf{Ans.}$$

and minimum intensity of stress at the base,

$$\sigma_{min} = \frac{W}{b}\left(1-\frac{6e}{b}\right) = \frac{2250}{6}\left(1-\frac{6\times 1.63}{6}\right) \text{ kN/m}^2$$

$$= -236.25 \text{ kN/m}^2 = \textbf{236.25 kPa (Tension)} \quad \textbf{Ans.}$$

## 18.3. Trapezoidal Dams with Water Face Vertical

Consider a unit length of a trapezoidal dam having its water face vertical as shown in Fig. 18.4.

Let $a$ = Top width of the dam,

$b$ = Bottom width of the dam,

$H$ = Height of the dam,

$\rho$ = Specific weight of the dam masonry,

$h$ = Height of water retained by the dam, and

$w$ = Specific weight of the water.

---

* The Eccentricity ($e$) may also be found out by taking moments about $A$. Let $d$ be the distance $AK$. Therefore

$$W\cdot d = P\times\frac{h}{3}+W\times\frac{b}{2}$$

and $$d = \frac{b}{2}+\left(\frac{P}{W}\times\frac{h}{3}\right) = \frac{6}{2}+\frac{845}{2250}\times\frac{13}{3} = 4.63 \text{ m}$$

∴ Eccentricity, $$e = d-\frac{b}{2} = 4.63 - 3.0 = 1.63 \text{ m}$$

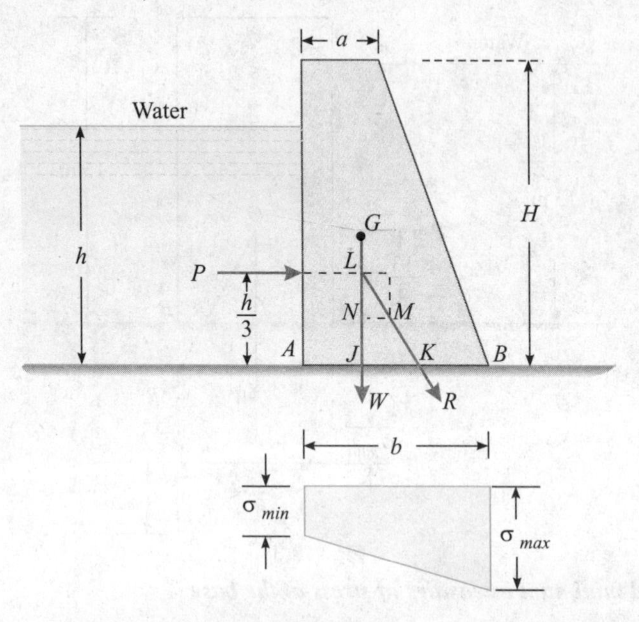


Fig. 18.4

We know that the weight of the dam per unit length, ...(i)

$$W = \rho \times \frac{(a+b)}{2} \times H \quad \text{...(ii)}$$

Like a rectangular dam, the total pressure on a unit length of the trapezoidal dam,

$$P = \frac{wh^2}{2} \quad \text{...(iii)}$$

and the horizontal distance between the centre of gravity of the dam and the point, where the resultant $R$ cuts the base,

$$x = \frac{P}{W} \times \frac{h}{3} \quad \text{...(iv)}$$

The distance between the toe of the dam $A$ and the point where the resultant $R$ cuts the base (*i.e.*, distance $AK$ in Fig. 18.4),

$$d = AJ + JK = AJ + \left(\frac{P}{W} \times \frac{h}{3}\right) \quad \text{...(v)}$$

Now the distance $AJ$* may be found out by splitting the dam section into a rectangle and a triangle. Now taking their moments about $A$ and equating the same with the moment of the dam section about $A$.

$\therefore$ Eccentricity, $e = d - AJ$

* The distance $AJ$ may also be found out from the relation,

$$AJ = \frac{a^2 + ab + b^2}{3(a+b)}$$

The stress across the base at will be maximum, whereas the stress across the base at $A$ will be minimum, such that

$$\sigma_{max} = \frac{W}{b}\left(1+\frac{6e}{b}\right)$$

and
$$\sigma_{min} = \frac{W}{b}\left(1-\frac{6e}{b}\right)$$

**Note.** When the reservoir is empty, there will be no water pressure on the dam. In this case, the eccentricity of the weight of the dam,

$$e = \frac{b}{2} - AJ$$

Since the eccentricity in this case will be minus, therefore the total stress across the base at $B$ will be minimum, whereas the stress across the base at $A$, *will be maximum, such that*

$$\sigma_{min} = \frac{W}{b}\left(1-\frac{6e}{b}\right)$$

and
$$\sigma_{max} = \frac{W}{b}\left(1+\frac{6e}{b}\right)$$

Dam

**EXAMPLE 18.4.** *A concrete dam of trapezoidal section having water on vertical face is 16 m high. The base of the dam is 8 m wide and top 3 m wide. Find*

(*a*) *resultant thrust on the base per metre length of the dam,*

(*b*) *point, where the resultant thrust cuts the base and*

(*c*) *intensities of maximum and minimum stresses across the base.*

*Take weight of the concrete as 25 kN/m³ and the water level coinciding with the top of the dam.*

**SOLUTION.** Given: Height of the dam ($H$) = 16 m ; Height of water retained by the dam ($h$) = 16 m; Bottom width of the dam ($b$) = 8 m ; Top width of the dam ($a$) = 3 m and weight of concrete ($\rho$) = 25 kN/m$^3$.

**(a) *Resultant thrust on the base per metre length***

We know that total water pressure per metre length of the dam,

$$P = \frac{wh^2}{2} = \frac{9.81 \times (16)^2}{2} \text{ kN}$$

$$= 1255.7 \text{ kN} \quad \text{...}(i)$$

and weight of concrete per metre, length,

$$W = \rho \times \left(\frac{a+b}{2}\right) \times H$$

$$= 25 \times \left(\frac{3+8}{2}\right) \times 16 \text{ kN}$$

$$= 2200 \text{ kN} \quad \text{...}(ii)$$

∴ Resultant thrust per metre length,

$$R = \sqrt{P^2 + W^2} = \sqrt{(1255.7)^2 + (2200)^2}$$

$$= \mathbf{2533 \text{ kN}} \quad \textbf{Ans.}$$

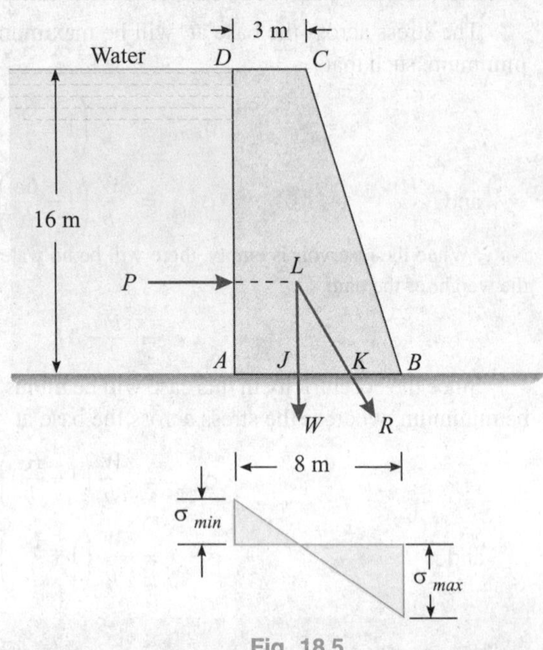


Fig. 18.5

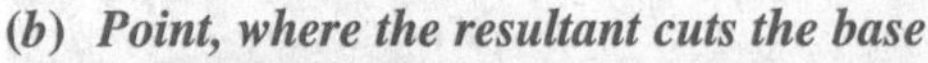

**(b) *Point, where the resultant cuts the base***

Let the resultant ($R$) cut the base at $K$ as shown in Fig. 18.5. First of all, let us find out the position* of the centre of gravity of the dam section. Now taking moment of the area about $A$ and equating the same,

$$\left[(16 \times 3) + \left(\frac{16 \times 5}{2}\right)\right] AJ = \left(16 \times 3 \times \frac{3}{2}\right) + \left[16 \times \frac{5}{2}\left(3 + \frac{5}{3}\right)\right]$$

$$88 \times AJ = 72 + 186.7 = 258.7$$

or

$$^*AJ = \frac{258.7}{88} = 2.94 \text{ m}$$

We know that horizontal distance between the centre of gravity of dam section and the point, where the resultant cuts the base (*i.e.*, distance $JK$),

$$x = \frac{P}{W} \times \frac{h}{3} = \frac{1255.7}{2200} \times \frac{16}{3} = 3.04 \text{ m}$$

∴ Horizontal distance $AK$,

$$^{**} \quad d = AJ + x = 2.94 + 3.04 = \mathbf{5.98 \text{ m}} \quad \textbf{Ans.}$$

---

* The horizontal distance may also be found out from the following relation.

$$AJ = \frac{a^2 + ab + b^2}{3(a+b)} = \frac{(3)^2 + (3 \times 8) + (8)^2}{3(3+8)}$$

$$= \frac{97}{33} = 2.94 \text{ m}$$

** The horizontal distance $d$ may also be found out by taking moment about $A$ and equating the same, *i.e.*,

$$W \cdot d = \left(P \times \frac{h}{3}\right) + (W \times AJ)$$

or

$$d = AJ + \frac{P}{W} \times \frac{h}{3} = 2.94 + \left(\frac{1255}{2200} \times \frac{16}{3}\right) = 5.98 \text{ m}$$

***(c) Intensities of maximum and minimum stresses across the base***

We know that eccentricity of the resultant,

$$e = d - \frac{b}{2} = 5.98 - \frac{8}{2} = 1.98 \text{ m}$$

∴ Intensity of maximum stress across the base,

$$\sigma_{max} = \frac{W}{b}\left(1 + \frac{6e}{b}\right) = \frac{2200}{8}\left(1 + \frac{6 \times 1.98}{8}\right) \text{kN/m}^2$$

$$= 683.3 \text{ kN/m}^2 = \mathbf{683.3 \text{ kPa}} \quad \textbf{Ans.}$$

and intensity of minimum stress across the base,

$$\sigma_{min} = \frac{W}{b}\left(1 + \frac{6e}{b}\right) = \frac{2200}{8}\left(1 + \frac{6 \times 1.98}{8}\right) \text{kN/m}^2$$

$$= -133.4 \text{ kN/m}^2 = \mathbf{133.4 \text{ kPa (tension)}} \quad \textbf{Ans.}$$

**EXAMPLE 18.5.** *A masonry trapezoidal dam 4 m high, 1 m wide at its top and 3 m wide at its bottom retains water on its vertical face. Determine the maximum and minimum stresses at the base, (i) when the reservoir is full and (ii) when the reservoir is empty. Take weight of water as 10 kN/m³ and that of masonry as 24 kN/m³.*

**SOLUTION.** Given: Height of the dam ($H$) = 4 m ; Top width of the dam ($a$) = 1 m ; Bottom width of the dam ($b$) = 3 m ; Weight of water ($w$) = 10 kN/m³ and weight of masonry ($\rho$) = 24 kN/m³.

***(i) Maximum and minimum stresses at the base when the reservoir is full***

Let the resultant ($R$) cut the base at $K$ as shown in Fig. 18.6($a$).

We know that the total pressure of water per metre length of the dam,

$$P = \frac{wh^2}{2} = \frac{10 \times (4)^2}{2} = 80 \text{ kN} \quad \text{... } (i)$$

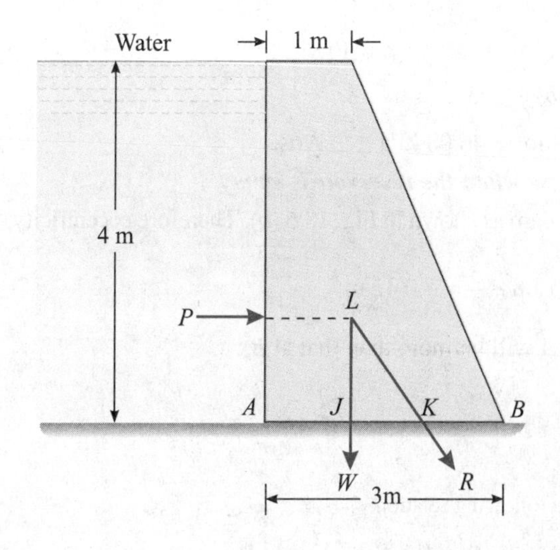


($a$) When the reservoir is full

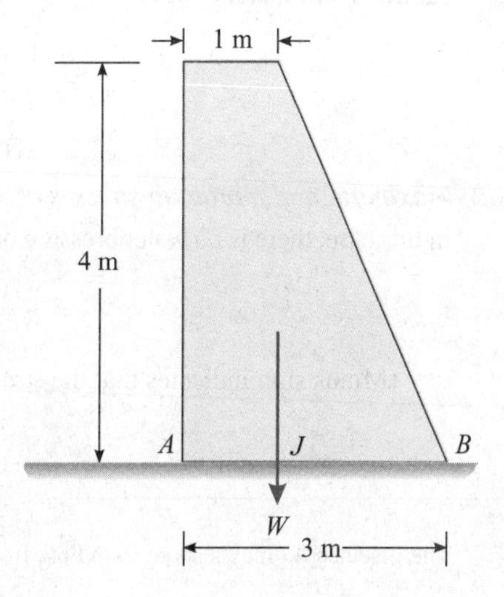


($b$) When the reservoir is empty

Fig. 18.6

and weight of concrete per metre length,

$$W = \rho\left(\frac{a+b}{2}\right)\times H = 24\times\left(\frac{1+3}{2}\right)\times 4 \text{ kN} = 192 \text{ kN} \quad \text{... (ii)}$$

Now let us find out the position* of the centre of gravity of the dam section. Taking moments of the area about $A$ and equating the same,

$$\left(4\times1+\frac{4\times2}{2}\right)AJ = \left(4\times1\times\frac{1}{2}\right)+\left[\frac{4\times2}{2}\left(1+\frac{2}{3}\right)\right]AJ$$

$$8AJ = 2+6.67 = 8.67$$

or $$*\,AJ = \frac{8.67}{8} = 1.08 \text{ m}$$

We know that horizontal distance between the centre of gravity of the dam section and the point, where the resultant cuts the base (*i.e.*, distance $JK$),

$$x = \frac{P}{W}\times\frac{h}{3} = \frac{80}{192}\times\frac{4}{3} = 0.56 \text{ m}$$

∴ Horizontal distance $AK$,

$$d = AJ + x = 1.08 + 0.56 = 1.64 \text{ m}$$

and eccentricity, $$e = d - \frac{b}{2} = 1.64 - \frac{3}{2} = 0.14 \text{ m}$$

We also know that maximum stress at the base

$$\sigma_{max} = \frac{W}{b}\left(1+\frac{6e}{b}\right) = \frac{192}{3}\left(1+\frac{6\times0.14}{3}\right)\text{kN/m}^2$$

$$= 81.92 \text{ kN/m}^2 = \mathbf{81.92\ kPa} \qquad \textbf{Ans.}$$

and mininum stress at the base,

$$\sigma_{min} = \frac{W}{b}\left(1-\frac{6e}{b}\right) = \frac{192}{3}\left(1-\frac{6\times0.14}{3}\right)\text{kN/m}^2$$

$$= 46.08 \text{ kN/m}^2 = 46.08 \text{ kPa} \qquad \textbf{Ans.}$$

***(ii) Maximum and minimum stresses at the base when the reservoir is empty***

In this case, there is no water pressure on the dam as shown in Fig. 18.6 (*b*). Therefore eccentricity,

$$e = d - \frac{b}{2} = 1.08 - \frac{3}{2} = -\,0.42 \text{ m}$$

∴ (Minus sign indicates that the stress at $A$ will be more than that at $B$).

---

* The distance $AJ$ may also be found out from the following relation:

$$AJ = \frac{a^2+ab+b^2}{3\,(a+b)} = \frac{(1)^2+(1\times3)+(3)^2}{3\,(1+3)} = \frac{13}{12} = 1.08 \text{ m}$$

We know that maximum stress at the base ($A$),

$$\sigma_{max} = \frac{W}{b}\left(1+\frac{6e}{b}\right) = \frac{192}{3}\left(1+\frac{6\times 0.42}{3}\right)\text{kN/m}^2$$

$$= 117.76 \text{ kN/m}^2 = \mathbf{117.76 \text{ kPa}} \quad \textbf{Ans.}$$

and minimum stress at the base, ($B$)

$$\sigma_{min} = \frac{W}{b}\left(1-\frac{6e}{b}\right) = \frac{192}{3}\left(1-\frac{6\times 0.42}{3}\right)\text{kN/m}^2$$

$$= 10.24 \text{ kN/m}^2 = \mathbf{10.24 \text{ kPa}} \quad \textbf{Ans.}$$

**EXAMPLE 18.6.** *A masonry dam as shown in Fig. 18.7 has a total height of 20 m with a top width of 5 m and a free board of 2 m. Its upstream face is vertical while the downstream face has a batter of 0.66 horizontal to 1.0 vertical. The specific gravity of masonry may be taken as 2.4.*

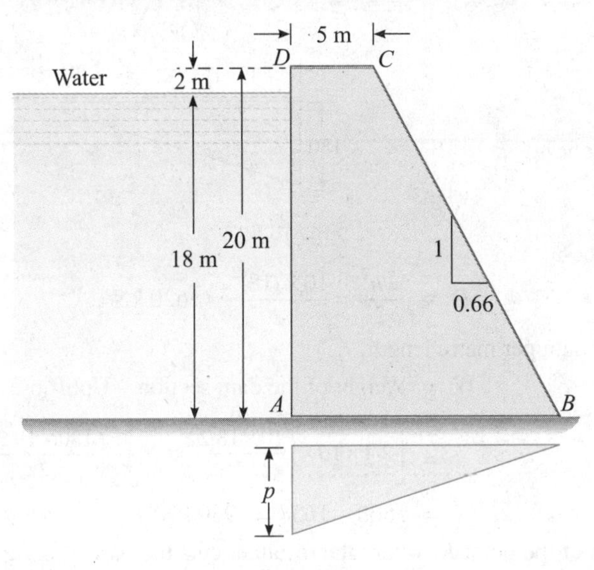


Fig. 18.7

*In addition to the hydrostatic pressure on the upstream face, there is an uplift pressure at the foundation, which may be taken to vary linearly from a value equal to the hydrostatic pressure at the upstream end, to zero at the downstream end.*

*Calculate the extreme values of the normal stresses on the foundation, when the reservoir is full. Take specific weight of water as 10 kN/m³.*

**SOLUTION.** Given: Height of the dam ($H$) = 20 m; Top width of the dam ($a$) = 5 m ; Free board = 2 m; Slope of downward face = 0.66 horizontal to 1.0 vertical ; Specific gravity of masonry = 2.4 ; Uplift pressure at the downstream point = 0 and specific weight of water ($w$) = 10 kN/m³.

We know that height of water, ($h$) = 20 – 2 = 18 m

and bottom width $\quad (b) = 5 + 20 \times \dfrac{0.66}{1.0} = 18.2 \text{ m}$

From the geometry of the uplift pressure, we know that pressure at the upstream ($A$),

$$p = wh = 10 \times 18 = 180 \text{ kN/m}^2$$

Let the resultant ($R$) cut the base at $K$ as shown in Fig.18.8 Let $d$ be the horizontal distance $AK$.

We know that total water pressure per metre length of the dam,

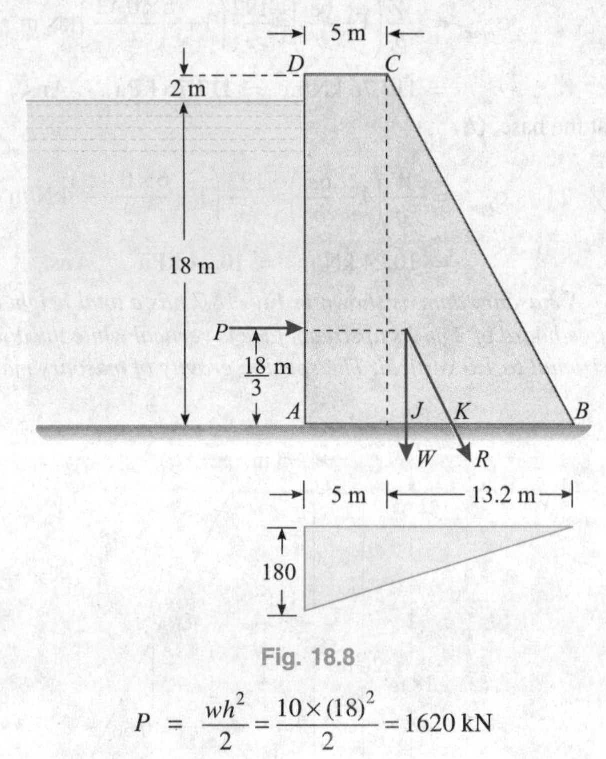


Fig. 18.8

$$P = \frac{wh^2}{2} = \frac{10 \times (18)^2}{2} = 1620 \text{ kN}$$

and net weight of the dam per metre length,

$$W = \text{Weight of the dam section} - \text{Uplift pressure}$$

$$= \left[2.4 \times 10 \times \left(\frac{10 \times 18.2}{2}\right) \times 20\right] - \left(\frac{180 \times 18.2}{2}\right) \text{kN}$$

$$= 5568 - 1638 = 3930 \text{ kN}$$

Now let us find out the point $K$, where the resultant cuts the base. Taking moments of the dam section about $A$ and equating the same,

$$W \times d = [(1620 \times 6)] + [(2.4 \times 10) \times (20 \times 5 \times 2.5)]$$

$$+ \left[(2.4 \times 10) \times \frac{20 \times 13.2}{2}\left(5 + \frac{13.2}{3}\right)\right] - \left[\frac{180 \times 18.2}{2} \times \frac{18.2}{3}\right]$$

$$3930\, d = 9720 + 6000 + 29780 - 9937 = 35563$$

$$d = \frac{35563}{3930} = 9.05 \text{ m}$$

and eccentricity,

$$e = d - \frac{b}{2} = 9.05 - \frac{18.2}{2} = -0.05 \text{ m}$$

(Minus sign indicates that the stress at $A$ will be more than that at $B$).

∴ Maximum stress at the base point ($A$),

$$\sigma_{max} = \frac{W}{b}\left(1 + \frac{6e}{b}\right) = \frac{3930}{18.2}\left(1 + \frac{6 \times 0.05}{18.2}\right) \text{kN/m}^2$$

$$= 219.5 \text{ kN/m}^2 = \mathbf{219.5 \text{ kPa}} \quad \textbf{Ans.}$$

and minimum stress at the base point (*B*),

$$\sigma_{min} = \frac{W}{b}\left(1 - \frac{6e}{b}\right) = \frac{3930}{18.2}\left(1 - \frac{6 \times 0.05}{18.2}\right) \text{kN/m}^2$$

$$= 212.4 \text{ kN/m}^2 = \mathbf{212.4 \text{ kPa}} \quad \textbf{Ans.}$$

## 18.4. Trapezoidal Dams with Water Face Inclined

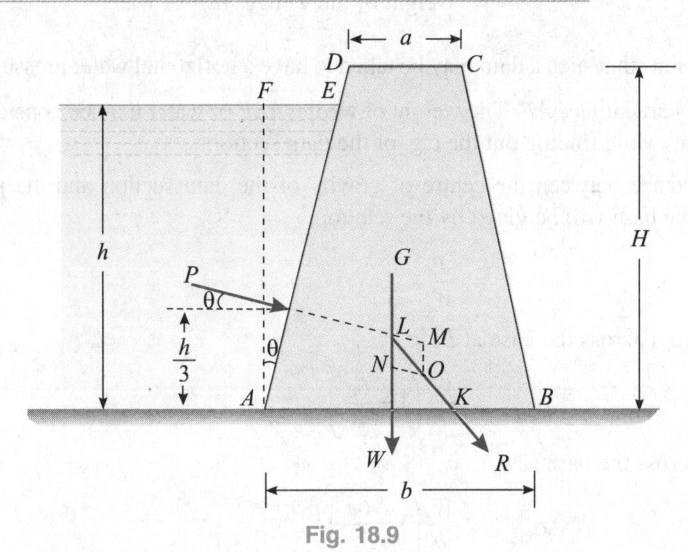


Fig. 18.9

Consider a unit length of a trapezoidal dam, having its water surface inclined as shown in Fig. 18.9.

Let
$a$ = Top width of the dam,
$b$ = Bottom width of the dam,
$H$ = Height of the dam,
$\rho$ = Specific weight of the dam masonry,
$h$ = Height of water retained by the dam,
$w$ = Specific weight of the water, and
$\theta$ = Inclination of the water face with the vertical.

∴ Length of the sloping side *AE*, which is subjected to water pressure,

$$l = \frac{h}{\cos \theta}$$

Now we see that the weight of the dam per unit length,

$$W = \rho \times \frac{(a+b)}{2} \times H \quad ...(i)$$

The intensity of water pressure will be zero at the water surface and will increase by a straight line law to *wh* at the bottom. Therefore the total pressure on a unit length of the dam,

$$P = \frac{wh}{2} \times l = \frac{whl}{2} \quad ...(ii)$$

This water pressure *P* will act at a height of *h*/3 from the bottom of the dam as shown in Fig. 18.9.

∴ Horizontal component of this water pressure,

$$P_H = P \cos \theta = \frac{whl}{2} \times \frac{h}{l} = \frac{wh^2}{2} \qquad ...(iii)$$

and vertical component of this water pressure,

$$P_V = P \sin \theta = \frac{whl}{2} \times \frac{EF}{l} = \frac{w}{2} \times EF \times h$$

$$= \text{Weight of the wedge } AEF \text{ of water}$$

It is thus obvious that such a dam may be taken to have a horizontal water pressure equal to $\frac{wh^2}{2}$ on the imaginary vertical face $AF$. The weight of wedge $AEF$ of water may be considered as a part of the weight of dam, while finding out the c.g. of the dam section.

Now the distance between the centre of gravity of the dam section and the point, where the resultant $R$ cuts the base will be given by the relation.

$$x = \frac{P}{W} \times \frac{h}{3} \qquad \text{... (As usual)}$$

∴ Total stress across the base at $B$,

$$\sigma_{max} = \frac{W}{b}\left(1 + \frac{6e}{b}\right) \qquad \text{... (As usual)}$$

and total stress across the base at $A$,

$$\sigma_{min} = \frac{W}{b}\left(1 - \frac{6e}{b}\right) \qquad \text{... (As usual)}$$

**Note:** When the reservoir is empty, there will be neither water pressure on the dam nor there will be the weight of wedge $AEF$ of water. In this case the eccentricity of the weight of the dam,

$$e = \frac{b}{2} - AJ$$

Since the eccentricity will be minus, therefore total stress across the base at $B$,

$$\sigma_{min} = \frac{w}{b}\left(1 - \frac{6e}{b}\right)$$

and total stress across the base at $A$,

$$\sigma_{max} = \frac{W}{b}\left(1 + \frac{6e}{b}\right)$$

**EXAMPLE 18.7.** *An earthen dam of trapezoidal section is 10 m high. It has top width of 1 m and bottom width 7 m. The face exposed to water has a slope of 1 horizontal to 10 vertical as shown in Fig. 18.10.*

*Calculate the maximum and minimum stresses on the base, when the water level coincides with the top of the dam. Take weight of the masonry as 20 kN/m³ and that of water as 10 kN/m³.*

**SOLUTION.** Given: Height of the dam $(H) = 10$ m ; Top width of the dam $(a) = 1$ m ; Bottom width of the dam $(b) = 7$ m ; Height of water retained by the dam $(h) = 10$ m ; Weight of masonry $(\rho) = 20$ kN/m³ and weight of water $(w) = 10$ kN/m³.

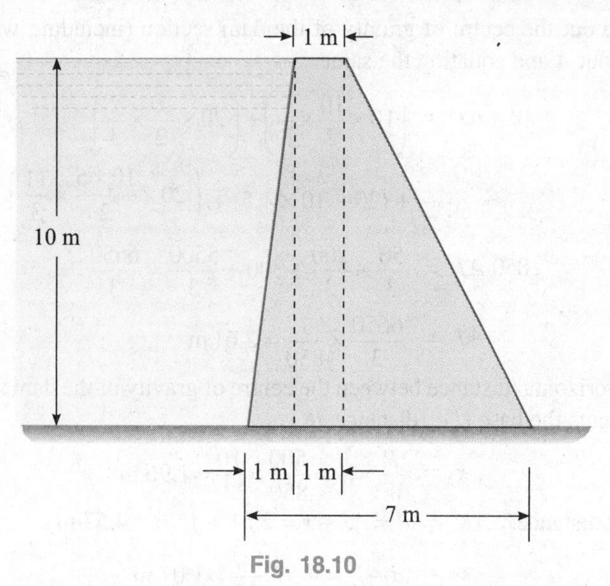


Fig. 18.10

Let the resultant ($R$) cut the base at $K$ as shown in Fig. 18.11. We know that total water pressure per metre length of the dam,

$$*P = \frac{wh^2}{2} = \frac{10 \times (10)^2}{2} = 500 \text{ kN}$$

and weight of the dam per metre length (including wedge $AED$ of water)

$$W = \left(w \times \frac{h}{2}\right) + \left(\rho \times \frac{a+b}{2} \times H\right) = \left(10 \times \frac{10}{2}\right) + \left(20 \times \frac{1+7}{2} \times 10\right) \text{kN}$$

$$= 50 + 800 = 850 \text{ kN}$$

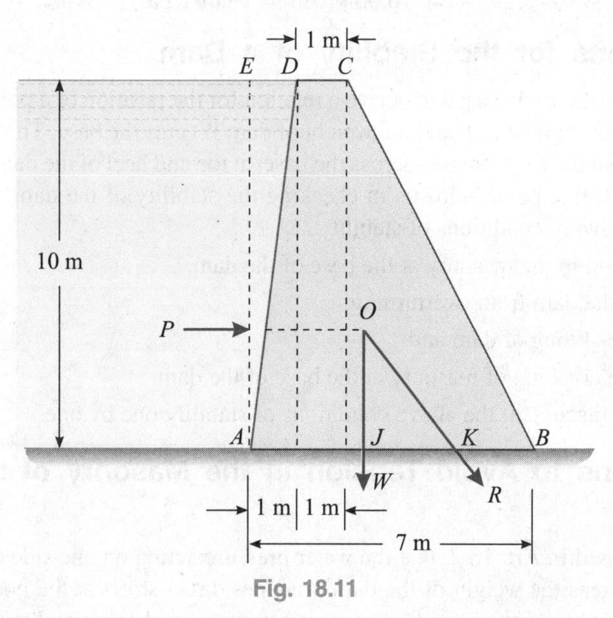


Fig. 18.11

* Strictly speaking, the total pressure is acting normally to the face $AD$ of the dam. But here we shall assume the pressure to act normally to the imaginary vertical plane $AE$ as discussed in Art. 18.4.

Now let us find out the centre of gravity of the dam section (including wedge *AED* of water). Taking moments about *A* and equating the same,

$$W \times AJ = \left(10 \times \frac{10}{2} \times \frac{1}{3}\right) + \left(20 \times \frac{10}{2} \times \frac{2}{3}\right) + (20 \times 10 \times 1.5) + \left(20 \times \frac{10 \times 5}{2} \times \frac{11}{3}\right).$$

$$850\ AJ = \frac{50}{3} + \frac{200}{3} + 300 + \frac{5500}{3} = \frac{6650}{3}$$

$$\therefore \quad AJ = \frac{6650}{3} \times \frac{1}{850} = 2.61 \text{ m}$$

We know that horizontal distance between the centre of gravity of the dam section and the points where the resultant cuts the base (*i.e.*, distance *JK*),

$$x = \frac{P}{W} \times \frac{h}{3} = \frac{500}{850} \times \frac{10}{3} = 1.96 \text{ m}$$

$\therefore$ Horizontal distance $AK = d = AJ + x = 2.61 + 1.96 = 4.57$ m

and eccentricity, $e = d - \frac{b}{2} = 4.57 - \frac{7}{2} = 1.07$ m

We also know that maximum stress at the base,

$$\sigma_{max} = \frac{W}{b}\left(1 + \frac{6e}{b}\right) = \frac{850}{7}\left(1 + \frac{6 \times 1.07}{7}\right) \text{kN/m}^2$$

$$= 232.8 \text{ kN/m}^2 = \mathbf{232.8\ kPa} \qquad \textbf{Ans.}$$

and minimum stress at the base,

$$\sigma_{min} = \frac{W}{b}\left(1 - \frac{6e}{b}\right) = \frac{850}{7}\left(1 - \frac{6 \times 1.07}{7}\right) \text{kN/m}^2$$

$$= 10.06 \text{ kN/m}^2 = \mathbf{10.06\ kPa} \qquad \textbf{Ans.}$$

## 18.5. Conditions for the Stability of a Dam

In the previous articles, we used to derive a relation for the position of a point through which the resultant *R* (of the water pressure *P* and the weight of dam *W*) cuts the base. The position of this point helps us in finding out the total stresses across the base, at toe and heel of the dam. Apart from finding out the total stresses, this point helps us in checking the stability of the dam. In general, a dam is checked for the following conditions of stability:

1. To avoid tension in the masonry at the base of the dam,
2. To safeguard the dam from overturning,
3. To prevent the sliding of dam and
4. To prevent the crushing of masonry at the base of the dam.

Now we shall discuss all the above conditions of stability one by one.

## 18.6. Conditions to Avoid Tension in the Masonry of the Dam at its Base

We have discussed in Art. 18.2. that the water pressure acting on one side of the dam, produces bending stress; whereas the weight of the dam produces direct stress at the bottom of the dam. We have also seen that on one side of the dam, there is a maximum stress (equal to sum of the direct and bending stress); whereas on other side of the dam, there is a minimum stress (equal to direct stress minus bending stress). A little consideration will show that so long as the bending stress remains less than the direct stress, the resultant stress is compressive. But when the bending stress is equal to the direct stress, there will be zero stress on one side. But when the bending stress exceeds the direct stress, there will be a tensile stress on one side. Though cement concrete can take up a small amount of tensile stress, yet it is desirable to avoid tension in the masonry of the dam at its base.

It is thus obvious, that in order to avoid the tension in the masonry of the dam at its base, the bending stress should be less than the direct stress or it may be equal to the direct stress, *i.e.*,

$$\sigma_b \leq \sigma_0$$

$$\frac{6W \cdot e}{b^2} \leq \frac{W}{b}$$

or
$$e \leq \frac{b}{6}$$

It means that the eccentricity of the resultant can be equal to $b/6$ on either side of geometrical axis of base section. Thus the resultant must lie within the middle third of the base width in order to avoid tension.

## 18.7. Condition to Prevent Overturning of the Dam

We have already discussed that when a dam is retaining water, it is subjected to some water pressure. We can easily find out the resultant $R$ of the water pressure $P$ and the weight of dam $W$. Since the dam is in equilibrium, therefore the resultant $R$ must be balanced by equal and opposite reaction acting at $K$. This reaction may be split up into two components *viz.*, horizontal and vertical. The horizontal component must be equal to the water pressure $P$, whereas the vertical component must be equal to the weight $W$. Thus the following four forces acting on the dam, keep it in equilibrium:

1. Water pressure $P$,
2. Horizontal component of the reaction,
3. Weight of the dam $W$ and
4. Vertical component of the reaction.

These four forces may be grouped into two sets or couples. The moment of a couple consisting of the first two forces,

$$M_1 = \text{Force} \times \text{Arm} = P \times \frac{h}{3} \qquad ...(i)$$

Similarly, moment of a couple consisting of the last two forces,

$$M_2 = W \times JK \qquad ...(ii)$$

A little consideration will show that the moment of the first two forces will tend to overturn the dam about $B$; whereas the moment of the last two forces will tend to restore the dam. Since the dam is in equilibrium and a couple can only be balanced by a couple, therefore overturning moment must be equal to the restoring moment, *i.e.*,

$$P \times \frac{h}{3} = W \times JK$$

or
$$JK = \frac{P}{W} \times \frac{h}{3}$$

Incidentally, this equation is the same which we derived in Art. 18.6 and gives the position of the point $K$, where the resultant cuts the base. Since the dam will tend to overturn about $B$, therefore balancing moment about $B$,

$$M_3 = W \times JB$$

Now, we see that the dam is safe against overturning, so long as the balancing moment is more than the overturning moment (or restoring moment, which is equal to overturning moment), *i.e.*,

$$W \times JB > W \times JK$$

or
$$JB > JK$$

It is thus obvious that the condition to prevent the dam from overturning, is that the point $K$ should be between $J$ and $B$ or more precisely between $A$ and $B$.

As a matter of fact, this is a superfluous condition. We know that to avoid tension in the masonry of a dam at its bottom, the resultant must lie within the middle third of the base width. Since we have to check the stability of a dam for tension in the base masonry, therefore the stability of the dam for overturning is automatically checked.

## 18.8. Condition to Prevent the Sliding of Dam

We have already discussed in Art. 18.7 that there are four forces which act on a dam and keep it in equilibrium. Out of these four forces, two are vertical and the following two are horizontal:

(*a*) Water pressure $P$ and

(*b*) Horizontal component of the reaction.

A little consideration will show that the horizontal component of the reaction will be given by the frictional force at the base of the dam.

Let $\mu$ = Coefficient of friction between the base of dam and the soil.

We know that the maximum available force of friction,

$$F_{max} = \mu W$$

It is thus obvious that so long as $F_{max}$ is more* than the water pressure $P$, the dam is safe against sliding.

## 18.9. Condition to Prevent Crushing of Masonry at the Base of the Dam

We have already discussed in Art. 18.2 that whenever a dam is retaining water, the masonry of dam at its bottom is subjected to some stress. This stress varies from $\sigma_{max}$ to $\sigma_{min}$ by a straight line law. A little consideration will show that the condition to prevent the crushing of masonry at the base of the dam, is that the maximum stress $\sigma_{max}$ should be less than the permissible stress in the masonry.

**EXAMPLE 18.8.** *A masonry wall 5 metres high and 1.8 metre wide is containing water up to a height of 4 metres. If the coefficient of friction between the wall and the soil is 0.6, check the stability of the wall. Take weight of the masonry and water as 22 kN/m³ and 9.81 kN/m³.*

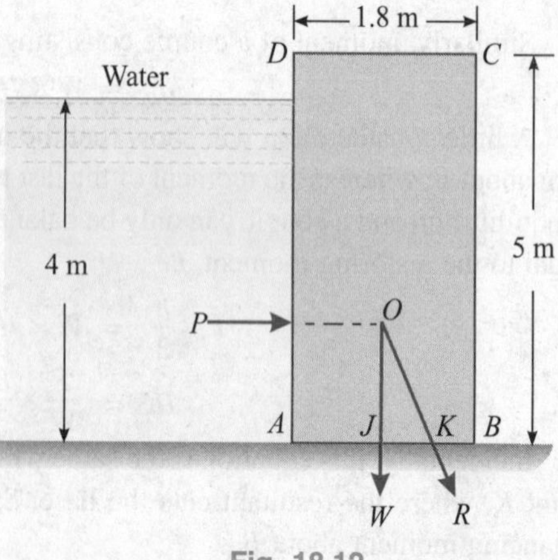


Fig. 18.12

**SOLUTION.** Given: Height of the wall ($H$) = 5 m ; Width of the wall ($b$) = 1.8 m; Height of the water ($h$) = 4 m ; Coefficient of friction between the wall and the soil ($\mu$) = 0.6 and weight of masonry ($\rho$) = 22 kN/m³.

Let the resultant ($R$) cut the base at $K$ as shown in Fig.18.12 We know that total water pressure per metre length of the wall,

$$P = \frac{wh^2}{2} = \frac{9.81 \times (4)^2}{2} = 78.48 \text{ kN}$$

* Some authorities feel that the dam will be safe, when the force of friction is at least 1.5 times the total water pressure per metre length. *i.e.*, $\frac{\mu W}{P} = 1.5$

and weight of the wall per metre length,

$$W = 22 \times 5 \times 1.8 = 198 \text{ kN}$$

1. ***Check for tension in the masonary at the base***

We know that horizontal distance between the centre of gravity of the wall and point, where the resultant thrust $(R)$ cuts the base,

$$x = \frac{P}{W} \times \frac{h}{3} = \frac{78.48}{198} \times \frac{4}{3} = 0.53 \text{ m}$$

$$\therefore \qquad AK = AJ + x = 0.9 + 0.53 = 1.43 \text{ m}$$

Since the resultant thrust lies beyond the middle third of the base width (*i.e.*, from 0.6 to 1.2 m), therefore the wall shall fail due to tension in its base. **Ans.**

2. ***Check for overturning.***

Since the resultant thrust is passing within the base as obtained above, therefore the wall is safe against overturning. **Ans.**

3. ***Check for sliding the wall.***

We know that horizontal pressure due to water, $(P) = 78.48$ kN.

And the frictional force $= \mu W = 0.6 \times 198 = 118.8$ kN

Since the frictional force (118.8 kN) is *more than the horizontal pressure (78.48 kN), therefore the wall is safe against sliding. **Ans.**

**EXAMPLE 18.9.** *A trapezoidal masonry dam having 3 m top width, 8 m bottom width and 12 m high is retaining water as shown in Fig. 18.13.*

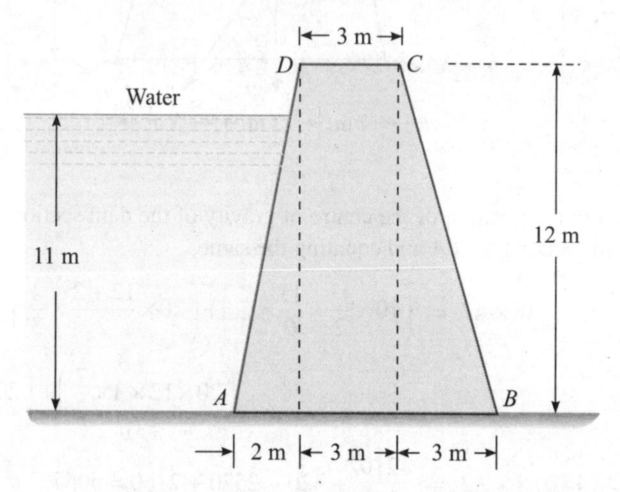


Fig. 18.13

*Check the stability of the dam, when it is retaining water to a height of 11 m. The masonry weighs 20 kN/m3 and coefficient of friction between the dam masonry and soil is 0.6. Take the allowable compressive stress as 400 kN/m².*

**SOLUTION.** Given: Top width of the dam $(a) = 3$ m ; Bottom width of the dam $(b) = 8$ m ; Height of the dam $(H) = 12$ m ; Height of water retained by the dam $(h) = 11$ m ; Weight of masonry $(\rho) = 20$ kN/m$^3$ ; Coefficient of friction between the dam masonry and soil $(\mu) = 0.6$ and allowable compressive stress $(\sigma_{max}) = 400$ kN/m$^3$.

* Certain authorities on the subject are of the opinion that magnitude of the weight should preferably be 1.5 times the horizontal pressure due to water.

1. ***Check for tension in the masonry at its base***

Let the resultant ($R$) cut the base at $K$ as shown in Fig. 18.14.

We know that water pressure per metre length of the dam,

$$P = \frac{wh^2}{2} = 10 \times \frac{(11)^2}{2} = 605 \text{ kN}$$

and weight of the dam per metre length (including the wedge $AED$ of water),

$$W = \left(10 \times \frac{1}{2} \times 11 \times \frac{11}{6}\right) + \left(20 \times \frac{3+8}{2} \times 12\right) \text{kN}$$

$$= 100.8 + 1320 = 1420.8 \text{ kN}$$

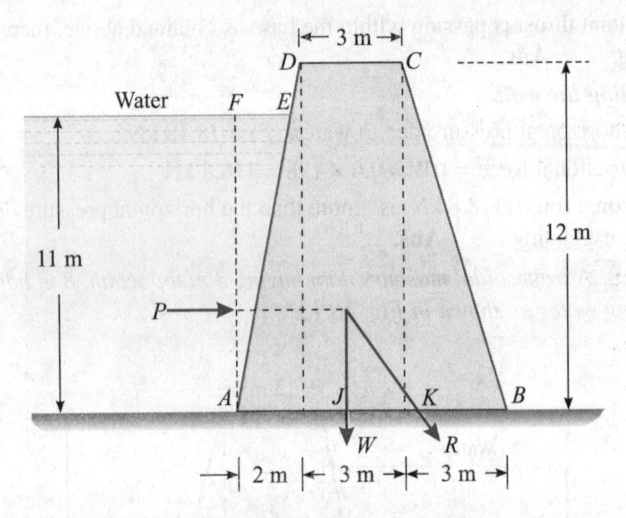


Fig. 18.14

Now let us find out the position of the centre of gravity of the dam section. Taking moments of the weight of the dam section about $A$ and equating the same,

$$W \times AJ = \left(10 \times \frac{11}{2} \times \frac{11}{6} \times \frac{2}{3}\right) + \left(20 \times \frac{12+2}{2} \times \frac{4}{3}\right)$$

$$+ \left(20 \times 12 \times 3 \times \frac{7}{2}\right) + \left(20 \times \frac{12 \times 3}{2} \times 6\right)$$

$$1420.8 \times AJ = \frac{1210}{18} + 320 + 2520 + 2160 = 5067$$

$$\therefore \quad AJ = \frac{5067}{1420.8} = 3.57 \text{ m}$$

We know that horizontal distance between the centre of gravity of the dam section and the point, where the resultant cuts the base (*i.e.*, distance $JK$),

$$x = \frac{P}{W} \times \frac{h}{3} = \frac{605}{1420.8} \times \frac{11}{3} = 1.56 \text{ m}$$

Horizontal distance $AK$, $\quad d = AJ + x = 3.57 + 1.56 = 5.13 \text{ m}$

Since the resultant force lies within the middle third of the base width (*i.e.*, from 2.67 m to 5.33 m), therefore the dam is safe against the tension in its masonry at the base. **Ans.**

2. ***Check for overturning***

Since the resultant force lies within the base *AB* and obtained above, therefore the dam is safe against overturning. **Ans.**

3. ***Check for sliding of the dam***

We know that the frictional force at the base

$$= \mu W = 0.6 \times 1420.8 = 852.5 \text{ kN}$$

Since the frictional force (852.5 kN) is more than the horizontal pressure (605 kN), therefore the dam is safe against sliding. **Ans.**

4. ***Check for crushing of the masonry at the base of the dam***

We know that eccentricity,

$$e = d - \frac{b}{2} = 5.13 - \frac{8}{2} = 1.13 \text{ m}$$

and maximum stress,

$$\sigma_{max} = \frac{W}{b}\left(1 + \frac{6e}{b}\right) = \frac{1420.8}{8}\left(1 + \frac{6 \times 1.13}{8}\right) \text{kN/m}^2$$

$$= 328.1 \text{ kN/m}^2$$

Since the maximum stress (328.1 kN/m$^2$) is less than the allowable stress (400 kN/m$^2$), therefore masonry of the dam is safe against crushing. **Ans.**

## 18.10. Minimum Base Width of a Dam

We have already discussed in Arts. 18.6 to 18.9, the general conditions for the stability of a dam, when the section is given. But sometimes, while designing a dam, we have to calculate its necessary base width. This can be easily found out by studying the conditions for the stability of a dam. Thus the base width (*b*) of a dam may be obtained from the following three conditions:

1. To avoid tension in the masonry at the base of the dam, the eccentricity $(e) = \dfrac{b}{6}$.

   In this case, the maximum stress $\sigma_{max} = \dfrac{2W}{b} = 5$ and the minimum stress $\sigma_{min} = 0$.

   The stress diagram at the base will be a triangle.

2. To avoid the sliding of dam, the force of friction between the dam and soil, is at least 1.5 times the total water pressure per metre length, *i.e.*,

$$\frac{\mu W}{P} = 1.5$$

3. To prevent the crushing of masonry at the base of the dam, the maximum stress should be less than the permissible stress of the soil.

**Note.** If complete data of a dam is given, then the base width for all the above three conditions should be found out separately. The maximum value of the base width from the above three conditions will give the necessary base width of the dam. But sometimes, sufficient data is not given to find out the values of base width for all the above mentioned three conditions. In such a case, the value of minimum base width may be found out, for any one of the above three conditions.

**EXAMPLE 18.10.** *A mass concrete dam shown in Fig. 18.15 has a trapezoidal cross-section. The height above the foundation is 64 m and its water face is vertical. The width at the top is 4.5 m.*

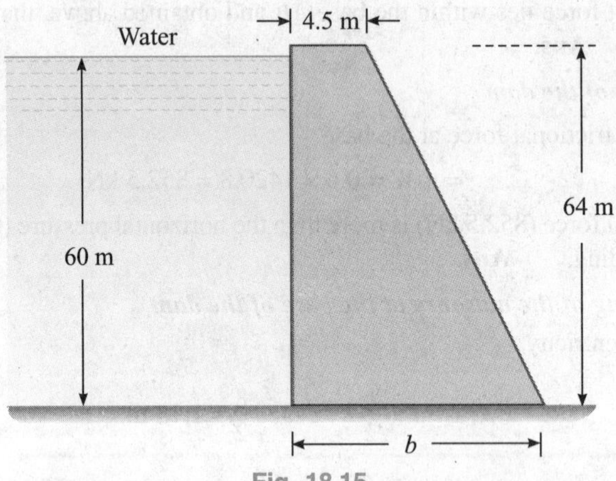


Fig. 18.15

*Calculate the necessary minimum width of the dam at its bottom, to ensure that no tension should be developed when water is stored up to 60 metres. Draw the pressure diagram at the base of the dam,* for this condition and indicate the maximum pressure developed.

*Take density of concrete as 22.6 kN/m³ and density of water as 9.81 kN/m³.*

**SOLUTION.** Given: Height of dam ($H$) = 64 m ; Top width of dam ($a$) = 4.5 m ; Height of water restored by the dam ($h$) = 60 m; Density of concrete ($\rho$) = 22.6 kN/m³ and density of water ($w$) = 9.81 kN/m³.

***Minimum width of the dam at its bottom***

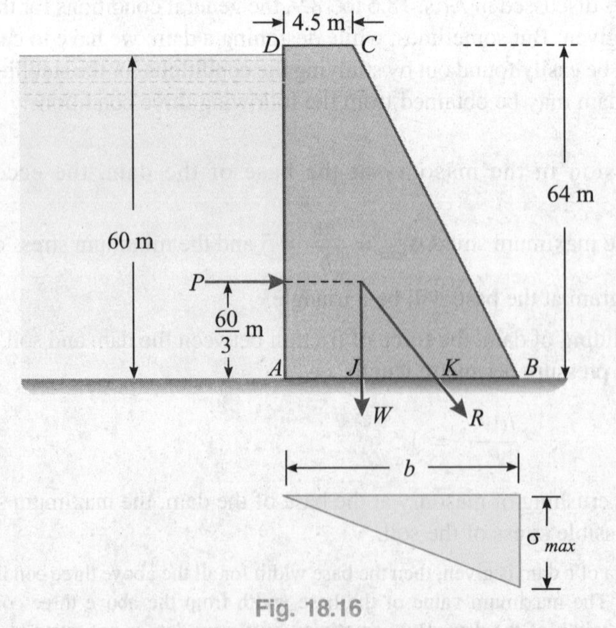


Fig. 18.16

Let $\quad b$ = Minimum width of the dam at its bottom in metres.

We see that the minimum width of the dam at its bottom is to be found out only for one condition *i.e.*, no tension shall be developed at the base.

Let the resultant ($R$) cut the base at $K$ as shown in Fig. 18.16.

We know that water pressure per metre length of the dam,

$$P = \frac{wh^2}{2} = \frac{9.81 \times (60)^2}{2} = 17660 \text{ kN} \qquad ...(i)$$

and weight of the dam per metre length,

$$W = \rho \times \frac{(a+b)}{2} \times H = 22.6 \times \frac{(4.5+b)}{2} \times 64 \text{ kN}$$

$$= 723.2\,(4.5+b) \text{ kN} \qquad ...(ii)$$

Now let us find out the position of the centre of gravity of the dam section. We know that the distance $AJ$,

$$AJ = \frac{a^2+ab+b^2}{3\,(a+b)} = \frac{(4.5)^2+4.5b+b^2}{3\,(4.5+b)} = \frac{20.25+4.5b+b^2}{3\,(4.5+b)}$$

We know that horizontal distance between the centre of gravity of the dam section and the point, where the resultant cuts the base (*i.e.*, distance $JK$),

$$x = \frac{P}{W} \times \frac{h}{3} = \frac{17660}{723.2\,(4.5+b)} \times \frac{60}{3} = \frac{488}{(4.5+b)}$$

∴ Horizontal distance $AK$,

$$d = AJ + x = \frac{20.25+4.5b+b^2}{3\,(4.5+b)} + \frac{488}{4.5+b}$$

$$= \frac{20.25+4.5b+b^2+1464}{3\,(4.5+b)} = \frac{1484.25+4.5b+b^2}{3\,(4.5+b)}$$

∴ Eccentricity of the resultant,

$$e = d - \frac{b}{2} = \frac{1484.25+4.5b+b^2}{3\,(4.5+b)} - \frac{b}{2}$$

We know that in order to avoid tension in the masonry at the base of the dam, the eccentricity,

$$e = \frac{b}{6}$$

or

$$\frac{1484.25+4.5b+b^2}{3\,(4.5+b)} - \frac{b}{2} = \frac{b}{6}$$

$$\frac{1484.25+4.5b+b^2}{3\,(4.5+b)} = \frac{b}{6} + \frac{b}{2} = \frac{2b}{3}$$

$$1484.25 + 4.5b + b^2 = 2b\,(4.5+b) = 9b + 2b^2$$

or

$$b^2 + 4.5b - 1484.25 = 0$$

Solving this equation as a quadratic equation for $b$, we get

$$b = \frac{-4.5 \pm \sqrt{(4.5)^2 + (4 \times 1484.25)}}{2} = \mathbf{36.35\ m} \qquad \textbf{Ans.}$$

***Pressure diagram***

Substituting the value of $b$ in equation (*ii*)

$$W = 723.2\,(4.5 + 36.35) = 29\,540 \text{ kN}$$

Since no tension should be developed at the base, therefore pressure diagram will be a triangle with zero pressure stress at $B$. Therefore pressure at the base

$$\sigma_{max} = \frac{2W}{b} = \frac{2 \times 29540}{36.35} = \textbf{1625 kPa} \quad \textbf{Ans.}$$

$$= \textbf{1625 kPa} \quad \textbf{Ans.}$$

$$\sigma_{min} = 0$$

The pressure diagram at the base of the dam is shown in Fig. 18.16 (*b*). **Ans.**

**EXAMPLE 18.11.** *A concrete dam has its upstream face vertical and a top width of 3 m. Its downstream face has a uniform batter. It stores water to a depth of 15 m with a free board of 2 m as shown in Fig. 18.17.*

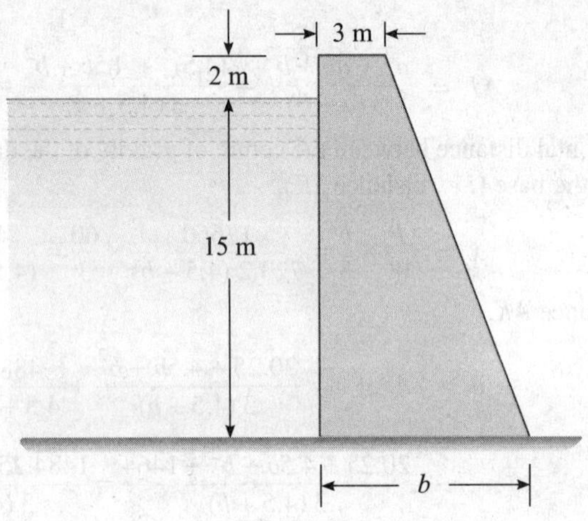


Fig. 18.17

*The weights of water and concrete may be taken as 10 kN/m$^3$ and 25 kN/m$^3$. Calculate*

*(a) the minimum dam width at the bottom for no tension in concrete. Neglect uplift.*

*(b) the extreme intensities of pressure on the foundation, when the reservoir is empty.*

**SOLUTION.** Given: Top width of the dam ($a$) = 3 m ; Height of water retained by the dam ($h$) = 15 m ; Height of the dam ($H$) = 15 + 2 = 17 m ; Weight of water ($w$) = 10 kN/m$^3$ and weight of concrete ($\rho$) = 25 kN/m$^3$.

***Minimum dam width at the bottom***

Let $b$ = Minimum dam width at bottom

We see that the minimum dam width at the bottom is to be found out only for one condition *i.e.*, no tension should be developed at the base.

We also know that total pressure on the dam per metre length,

$$P = \frac{wh^2}{2} = \frac{10 \times (15)^2}{2} = 1125 \text{ kN} \quad ...(i)$$

and weight of concrete of the dam per metre length

$$W = \rho \times \frac{(a+b)}{2} \times H = 25 \times \frac{(3+b)}{2} \times 17$$

$$= 212.5\ (3 + b) \text{ kN}$$

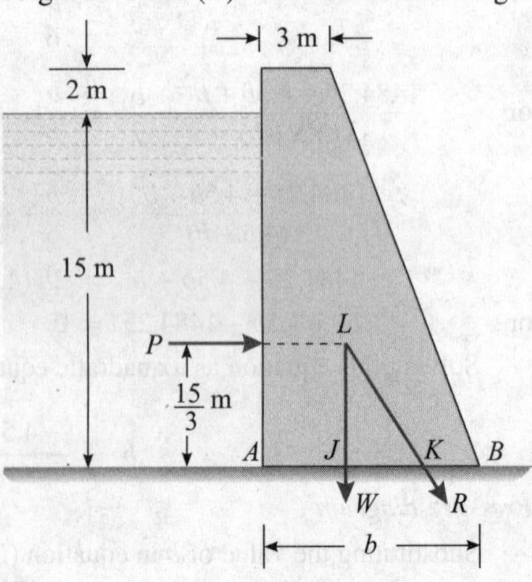


Fig. 18.18

Now, let us find out the position of the centre of gravity of the dam section. We know that the distance $AJ$

$$= \frac{a^2 + ab + b^2}{3\,(a+b)} = \frac{(3)^2 + 3b + b^2}{3\,(3+b)} = \frac{9 + 3b + b^2}{3\,(3+b)}$$

We know that horizontal distance between the centre of gravity of the dam section and the point, where the resultant cuts the base (*i.e.*, distance $JK$),

$$x = \frac{P}{W} \times \frac{h}{3} = \frac{1125}{212.5\,(3+b)} \times \frac{15}{3} = \frac{26.5}{(3+b)}$$

∴ Horizontal distance $AK$,

$$d = AJ + x = \frac{9 + 3b + b^2}{3\,(3+b)} + \frac{26.5}{(3+b)}$$

$$= \frac{9 + 3b + b^2 + 79.5}{3\,(3+b)} = \frac{88.5 + 3b + b^2}{3\,(3+b)}$$

and eccentricity of the resultant,

$$e = d - \frac{b}{2} = \frac{88.5 + 3b + b^2}{3\,(3+b)} - \frac{b}{2}$$

We know that in order to avoid tension in the concrete at the dam base, the eccentricity,

$$e = \frac{b}{6}$$

or
$$\frac{88.5 + 3b + b^2}{3\,(3+b)} - \frac{b}{2} = \frac{b}{6}$$

∴
$$\frac{88.5 + 3b + b^2}{3\,(3+b)} = \frac{b}{6} + \frac{b}{2} = \frac{2b}{3}$$

$$88.5 + 3b + b^2 = 2b\,(3 + b) = 6b + 2b^2$$

$$b^2 + 3b - 88.5 = 0$$

Solving this equation, as a quadratic equation for $b$, we get

$$b = \frac{-3 \pm \sqrt{(3)^2 + (4 \times 88.5)}}{2} = \frac{-3 \pm 19}{2}$$

$= \mathbf{8\ m}$ **Ans.**

***Extreme intensities of pressure on the foundation when the reservoir is empty***

We know that the weight of dam per metre length,

$$W = 25 \times \frac{(3+8)}{2} \times 17 \text{ kN}$$

$$= 2337.5 \text{ kN}$$

We also know that distance $AJ$,

$$d = \frac{a^2 + ab + b^2}{3\,(a+b)}$$

$$= \frac{(3)^2 + (3 \times 8) + (8)^2}{3\,(3+8)}$$

$$= 2.94 \text{ m}$$

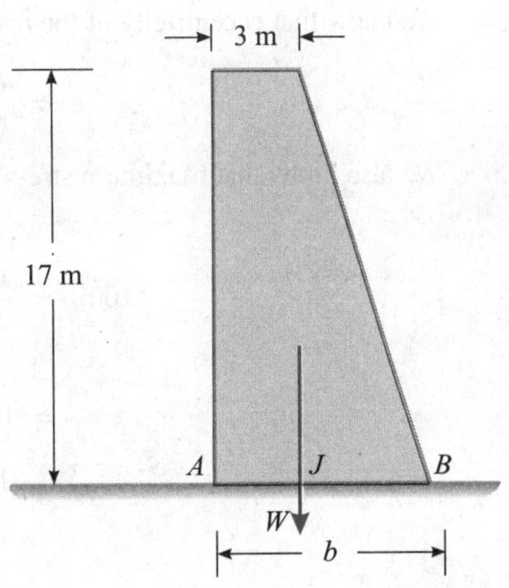


Fig. 18.19

and eccentricity,

$$e = d - \frac{b}{2} = 2.94 - \frac{8}{2} = -1.06 \text{ m}$$

(Minus sign indicates that the pressure at point will be more than that at point *B*).

We also know that maximum intensity of pressure at point *A*,

$$\sigma_{max} = \frac{W}{b}\left(1 + \frac{6e}{b}\right) = \frac{2337.5}{8}\left(1 + \frac{6 \times 1.06}{8}\right) \text{kN/m}^2$$

$$= 524.5 \text{ kN/m}^2 = \mathbf{524.5\ kPa} \quad \textbf{Ans.}$$

and minimum intensity of pressure at point *B*,

$$\sigma_{min} = \frac{W}{b}\left(1 - \frac{6e}{b}\right) = \frac{2337.5}{8}\left(1 - \frac{6 \times 1.06}{8}\right) \text{kN/m}^2$$

$$= 59.9 \text{ kN/m}^2 = \mathbf{59.9\ kPa} \quad \textbf{Ans.}$$

## 18.11. Maximum Height of a Dam

We have already discussed in Art. 18.10, the various conditions for the minimum base width of a dam. The same conditions also hold good for the maximum height of a dam.

**EXAMPLE 18.12.** *Assuming uniformly varying stress across the base, find the limit of height of a triangular masonry dam, with water upto the top of the vertical face, in order that the vertical compressive stress across the base shall not exceed 1 MPa. the masonry weighs 20 kN/m³.*

**SOLUTION.** Given: Maximum compressive stress ($\sigma_{max}$) = 1 MPa = 1000 kPa = 1000 kN/m² and Weight of masonry ($\rho$) = 20 kN/m³.

Let $e$ = Eccentricity of the resultant,

$H$ = Height of the dam in metres, and

$b$ = Bottom width of the dam in metres.

and weight of dam per metre length,

$$W = 20 \times \frac{bH}{2} = 10\,bH$$

We know that eccentricity of the resultant for maximum stress,

$$e = \frac{b}{6}$$

We also know that maximum stress across the base ($\sigma_{max}$),

$$1000 = \frac{W}{b}\left(1 + \frac{6e}{b}\right) = \frac{10\,bH}{b}\left(1 + \frac{6 \times \frac{b}{6}}{b}\right)$$

$$= 10\,H\,(1 + 1) = 20\,H$$

$$H = \frac{1000}{20} = \mathbf{50\ m} \quad \textbf{Ans.}$$

## EXERCISE 18.1

1. A wall 5 m long contains 3 m deep water. What is the total pressure on the wall? Take specific weight of water as 10 kN/m$^3$. [**Ans.** 225 kN]
2. A rectangular masonry dam 6 m high and 3 m wide has water level with its top. Find (*i*) total pressure per metre length of the dam, (*ii*) Point at which the resultant cuts the base and (*iii*) maximum and minimum intensities of stresses at the bottom of the dam. Assume the weight of water and masonry as 10 kN/m$^3$ and 20 kN/m$^3$ respectively. [**Ans.** 180 kN ; 1.0 m ; 360 kPa ; – 120 kPa]
3. A masonry trapezoidal dam 1 m wide at top, 4 m at its base and 6 m high is retaining water on its vertical face to a height equal to the top of the dam. Determine the maximum and minimum intensities of stress. Take density of masonry as 22.5 kN/m$^3$. [**Ans.** 143.9 kPa ; 24.9 kPa]
4. A concrete trapezoidal dam 2.5 m wide at the top and 10 m wide at the bottom is 25 m high. It contains water on its vertical side. Check the stability of the dam, when it contains water for a depth of 20 m. Take coefficients of friction between the wall and soil as 0.6 and weight of the concrete as 24 kN/m$^3$.

   [**Ans. 1.** The dam shall fail due to tension. **2.** Safe against overturning. **3.** Safe against sliding]
5. A masonry dam 12 metres high trapezoidal in section has top width 1 metre and bottom width 7.2 metres. The face exposed to water has a slope of 1 horizontal to 10 vertical. Check the stability of the dam, when the water level rises 10 m high. The coefficient of friction between the bottom of the dam and the soil as 0.6. Take the weight of the masonry as 22 kN/m$^3$.

   [**Ans.** Safe against tension; Safe against overturning; Safe against sliding]
6. A trapezoidal dam 4 m high has top width of 1 m with vertical face exposed to water. Find minimum bottom width of the dam, if no tension is to develop at the base. [**Ans.** 2.55 m]

## 18.12. Retaining Walls

We have already discussed in Art. 18.1 that a retaining wall is generally, constructed to retain earth in hilly areas. The analysis of a retaining wall is, somewhat like a dam. The retaining wall is subjected to pressure, produced by the retained earth in a similar manner, as the dam is subjected to water pressure.

## 18.13. Earth Pressure on a Retaining Wall

It has been established since long that the earth particles lack in cohesion and hence have a definite *angle of repose. These earth particles always exert some lateral pressure on the walls, which retain or support them. The magnitude of this lateral pressure depends upon type of earth particles and the manner, in which they have been deposited on the back of the retaining wall. It has been experimentally found that the lateral pressure is minimum, when the earth particles have been loosely dumped, whereas the pressure is relatively high, when the same particles have been compacted by tamping or rolling. The earth pressures may be classified into the following two types:

1. Active earth pressure and
2. Passive earth pressure.

## 18.14. Active Earth Pressure

The pressure, exerted by the retained material called backfill, on the retaining wall is known as active earth pressure. As a result of the active pressure, the retaining wall tends to slide away from the retained earth. It has been observed that the active pressure of the retained earth, acts on the retaining wall, in the same way as the pressure of the stored water on the dam.

* It may be defined as the maximum natural slope, at which the soil particles will rest due to their internal friction, if left unsupported for a sufficient length of time.

## 18.15. Passive Earth Pressure

Sometimes, the retaining wall moves laterally against the retained earth, which gets compressed. As a result of the movement of the retaining wall, the compressed earth is subjected to a pressure (which is in the opposite direction of the active pressure) known as passive earth pressure.

It may be noted that the active pressure is the practical pressure, which acts on the retaining walls; whereas the passive earth pressure is a theoretical pressure, which rarely comes into play.

## 18.16. Theories for Active Earth Pressure

There are many theories and hypothesis for the active earth pressure, on the retaining walls. But none of these gives the exact value of the active pressure. The following two theories are considered to give a fairly reliable values:

**1.** Rankine's theory and **2.** Coulomb's wedge theory.

## 18.17. *Rankine's Theory for Active Earth Pressure

It is one of the most acceptable theories, for the determination of active earth pressure on the retaining wall. This theory is based on the following assumptions:

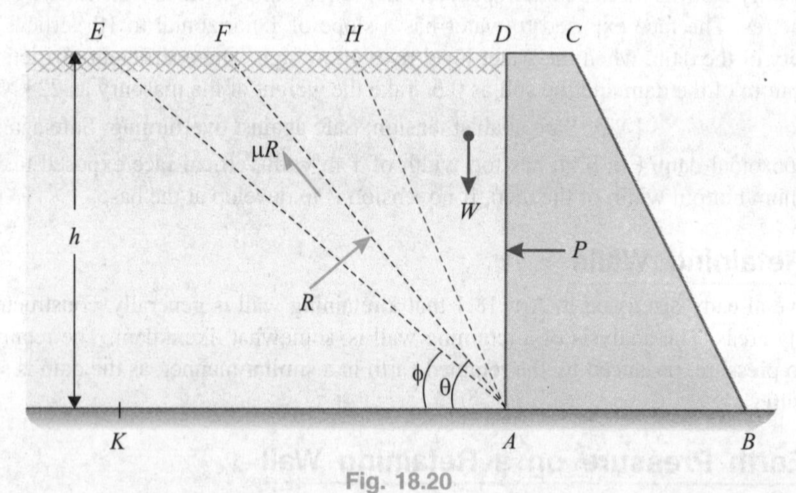


Fig. 18.20

**1.** The retained material is homogeneous and cohesionless.

**2.** The back of the wall is smooth, *i.e.*, the frictional resistance between the retaining wall and the retained material is neglected.

**3.** The failure of the retained material takes place along a plane, called rupture plane.

Consider a trapezoidal retaining wall *ABCD*, retaining earth up to a height of $h$ on its vertical face *AD*. Let the retained earth be levelled with the top of the wall *CD*. Draw *AE* at an angle $\phi$ with *AK* (where $\phi$ is the angle of repose of the retained earth). A little consideration will show that if retaining wall is removed, the retained earth will be subjected to tension and will slide down along certain plane, whose inclination will be more than that of the angle of repose with *AK*. Let such a plane *AF* be inclined at an angle $\theta$ with *AK* as shown in Fig. 18.20. Now consider a horizontal force $P$ offered by the retaining wall, which will keep the wedge *AFD* of the retained earth in equilibrium.

We see that, the wedge *AFD* of the retained earth is in equilibrium, under the action of the following forces.**

* This theory was given by Prof. W.J. Rankine, a British Engineer in 1857.

** The frictional force, along the face *AD* of the retaining wall, is neglected.

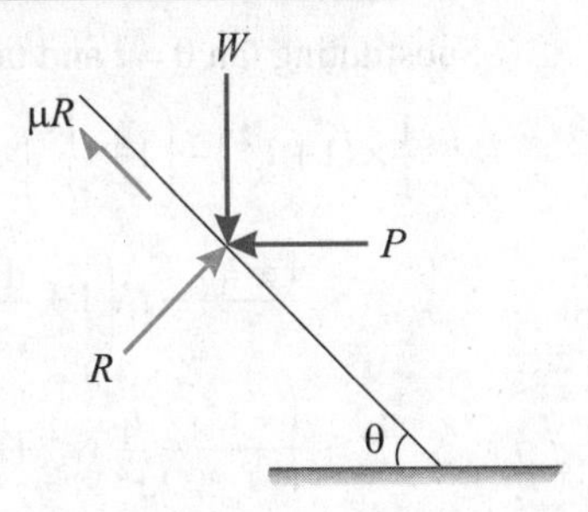

Fig. 18.21

1. Weight of the wedge *AFD*,

$$W = \frac{1}{2} w \times AD \times DF = \frac{wh^2}{2} \cot \theta$$

where $w$ = Specific weight of the material.

2. Horizontal thrust $P$ offered by the retaining wall on the retained material.
3. Normal reaction $R$ acting at right angle to the plane *AF*.
4. The frictional force, $F = \mu R$ acting on the opposite direction of the motion of the retained earth (where $\mu$ is the coefficient of friction of the retained material).

The above condition is similar to the equilibrium of a body of a weight $W$ on a rough inclined plane, when it is subjected to a horizontal force $P$ as shown in Fig. 18.21. From the geometry of the figure, we find that

$$R = W \cos \theta + P \sin \theta$$

and

$$\begin{aligned} P \cos \theta &= W \sin \theta - \mu R \\ &= W \sin \theta - \mu (W \cos \theta + P \sin \theta) \\ &= W \sin \theta - \mu W \cos \theta - \mu P \sin \theta \end{aligned}$$

or

$$\begin{aligned} P \cos \theta + \mu P \sin \theta &= W \sin \theta - \mu W \cos \theta \\ P (\cos \theta + \mu \sin \theta) &= W (\sin \theta - \mu \cos \theta) \end{aligned}$$

$$\therefore \quad P = \frac{W (\sin \theta - \mu \cos \theta)}{(\cos \theta + \mu \sin \theta)}$$

Substituting the value of $\mu = \tan \phi$ in the above equation,

$$P = \frac{W (\sin \theta - \tan \phi \cos \theta)}{(\cos \theta + \tan \phi \sin \theta)}$$

Multiplyying the numerator and denominator by $\cos \phi$,

$$P = \frac{W (\sin \theta \cos \phi - \sin \phi \cos \theta)}{(\cos \theta \cos \phi + \sin \phi \sin \theta)}$$

$$= W \frac{\sin (\theta - \phi)}{\cos (\theta - \phi)} = W \tan (\theta - \phi)$$

Substituting the value of $W$ in the above equation,

$$P = \frac{wh^2}{2} \cot \theta \cdot \tan (\theta - \phi)$$

A little consideration will show that if the retaining wall is removed, the retained earth will immediately slide down across a plane, where the tendency for the material to slide down is greatest. Let such a plane be *AH*. Therefore maximum value of the force $P$ is required to retain the wedge *AHD* of the earth. In order to locate the plane *AH* (*i.e.*, the plane of rupture), differentiate the equation for $P$ and equal to zero *i.e.*,

$$\frac{dP}{d\theta} \left[ \frac{wh^2}{2} [\cot \theta \tan (\theta - \phi] \right] = 0$$

or

$$\frac{wh^2}{2} [\cot \theta \sec^2 (\theta - \phi - \operatorname{cos ec}^2 \theta \tan (\theta - \phi)] = 0$$

$$\therefore \quad \cot \theta \sec^2 (\theta - \phi) - (\operatorname{cosec}^2 \theta \tan (\theta - \phi) = 0$$

Substituting $\tan\theta = t$ and $\tan(\theta - \phi) = t_1$ in the above equation,

$$\frac{1}{t}\times\left(1+t_1^2\right)-\left(1+\frac{1}{t^2}\right)\times t_1 = 0$$

$$\frac{1+t_1^2}{t}-t_1\left(1+\frac{1}{t^2}\right) = 0$$

$$\frac{1+t_1^2}{t}-\frac{t_1}{t^2}(r^2+1) = 0$$

$$t(1+t_1^2)-t_1(1+t^2) = 0$$

$$t+tt_1^2-t_1-t_1t^2 = 0$$

$$t-t_1+tt_1^2-t_1t^2 = 0$$

$$(t-t_1)-tt_1(t-t_1) = 0$$

$$(t-t_1)(1-tt_1) = 0$$

Therefore either $t = t_1$ or $1 - tt_1 = 0$. Since $\tan\theta$ cannot be equal to $\tan(\theta - \phi)$, therefore $1 - tt_1 = 0$

or $$1 - \tan\theta\cdot\tan(\theta-\phi) = 0$$

This statement is possible, only if

$$\theta + (\theta - \phi) = \frac{\pi}{2} \quad \text{or} \quad \theta = \frac{\pi}{4}+\frac{\phi}{2}$$

Thus the plane of rupture is inclined at $\frac{\pi}{4}+\frac{\phi}{2}$ with the horizontal. We also see that

$$\angle HAE = \angle HAK - \angle KAE$$

$$= \left(\frac{\pi}{4}+\frac{\phi}{2}\right)-\phi = \left(\frac{\pi}{4}+\frac{\phi}{2}\right) = \frac{1}{2}\left(\frac{\pi}{2}-\phi\right)$$

$$= \frac{1}{2}\angle DAE$$

Now substituting the values in the equation for $P$,

$$P = \frac{wh^2}{2}\cot\theta\tan(\theta-\phi) = \frac{wh^2}{2}\cot\left(\frac{\pi}{4}+\frac{\phi}{2}\right)\tan\left(\frac{\pi}{4}-\frac{\phi}{2}\right) = \frac{wh^2}{2}\times\frac{\tan\left(\frac{\pi}{4}-\frac{\phi}{2}\right)}{\tan\left(\frac{\pi}{4}+\frac{\phi}{2}\right)}$$

or $$P = \frac{wh^2}{2}\times\frac{1-\sin\phi}{1+\sin\phi}$$

**Notes: 1.** Similarly, it can be proved that if the retained material is surcharged (*i.e.*, the angle of surcharge is $\alpha$ with the horizontal), the total pressure on the retaining wall per unit length,

$$P = \frac{wh^2}{2}\cos\alpha\cdot\frac{\cos\alpha-\sqrt{\cos^2\alpha-\cos^2\phi}}{\cos\alpha+\sqrt{\cos^2\alpha-\cos^2\phi}}$$

This pressure may now be resolved into horizontal and vertical components. The horizontal component $P_H = P\cos\alpha$ will act at a height $h/3$ from the base and vertical component $P_V = P\sin\alpha$. It will act along $DA$.

**2.** If the retained material is subjected to some superimposed or surcharged load (*i.e.*, the pressure due to traffic *etc.*) it will cause a constant pressure on the retaining wall from top to bottom. The total horizontal pressure due to surcharged load,

$$P = p\times\frac{1-\sin\phi}{1+\sin\phi}$$

where $p$ is the intensity of the surcharged load.

**EXAMPLE 18.13.** *Find the resultant lateral pressure and the distance of the point of application from the bottom in the case of retaining wall as shown in Fig. 18.22.*

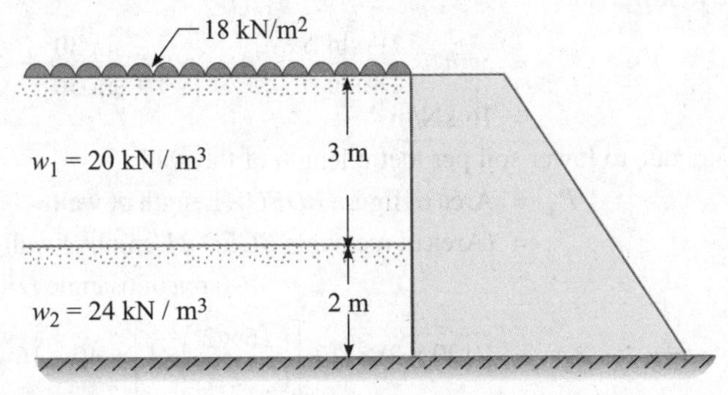


Fig. 18.22

*Take weight of upper soil as 20 kN/m³ and φ = 30° and weight of lower soil as 24 kN/m³ and φ = 30°*

**SOLUTION.** Given: Surcharge = 18 kN/m$^2$ ; Weight of upper soil ($w_1$) = 20 kN/m$^3$ ; Depth of upper soil ($h_1$)= 3 m ; Weight of lower soil ($w_2$) = 24 kN/m$^3$; Depth of lower soil ($h_2$) = 2 m and angle of repose for both the soils $\phi$ = 30°.

***Resultant lateral pressure per metre length of the wall***

The pressure diagram on the retaining wall is shown in Fig. 18.23. In this figure, the pressure *HA* or *GC* is due to surcharge. Pressure *BD* is due to upper soil and pressure *EF* is due to lower soil.

We know that pressure *HA* or *GC* due to surcharge

$$= 18 \times \frac{1 - \sin\phi}{1 + \sin\phi} = 18 \times \frac{1 - \sin 30°}{1 + \sin 30°} = 18 \times \frac{1 - 0.5}{1 + 0.5} \text{ kN/m}^2$$

$$= 6 \text{ kN/m}^2$$

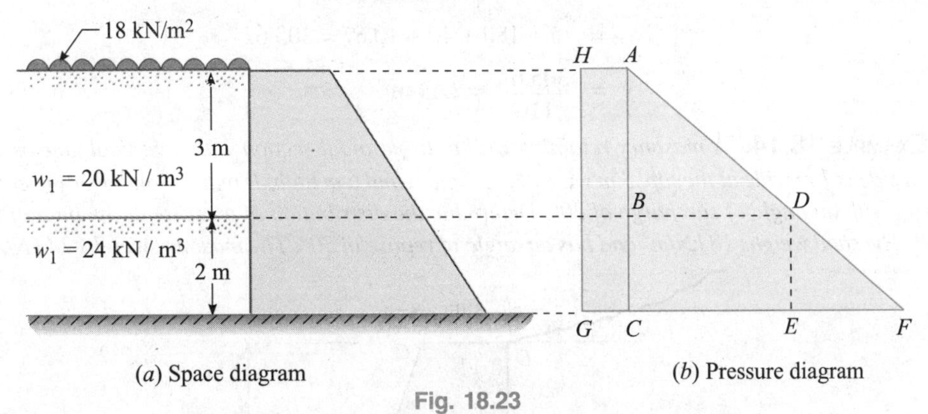


(*a*) Space diagram (*b*) Pressure diagram

Fig. 18.23

$\therefore$ Total pressure due to surcharge per metre length of the wall,

$$P_1 = \text{Area of rectangle } HGCA \times \text{length of the wall}$$

$$= (6 \times 5) \times 1 = 30 \text{ kN} \quad ...(i)$$

Similarly, pressure *BD* due to upper soil

$$= w_1 h_1 \times \frac{1 - \sin\phi}{1 + \sin\phi} = 20 \times 3 \times \frac{1 - \sin 30°}{1 + \sin 30°} = 60 \times \frac{1 - 0.5}{1 + 0.5}$$

$$= 20 \text{ kN/m}^2$$

$\therefore$ Total pressure due to upper soil per metre length of the wall

$$P_2 = \text{Area of tiangle } ABD \times \text{Length of wall}$$

$$= \left(\frac{20 \times 3}{2}\right) = 30 \text{ kN} \qquad \ldots(ii)$$

and pressure *EF* due to lower soil

$$= w_2 h_2 \times \frac{1 - \sin \phi}{1 + \sin \phi} = 24 \times 2 \times \frac{1 - \sin 30°}{1 + \sin 30°} = 48 \times \frac{1 - 0.5}{1 + 0.5}$$
$$= 16 \text{ kN/m}^2$$

∴ Total pressure due to lower soil per metre length of the wall,

$$P_3 = \text{Area of figure } BDFC \times \text{Length of wall}$$
$$= (\text{Area of rectangle } BCED \times \text{Length of wall}) + (\text{area of triangle } DEF \times \text{Length of wall})$$
$$= [(120 \times 2) \times 1] + \left[\left(\frac{16 \times 2}{2}\right) \times 1\right] = 40 + 16 = 56 \text{ kN} \qquad \ldots(iii)$$

and total pressure per metre length of the wall,

$$P = P_1 + P_2 + P_3 = 30 + 30 + 56 = 116 \text{ kN} \quad \textbf{Ans.}$$

***Point of application of the resultant pressure***

Let $y$ = Height of the point of application of the resultant pressure from the bottom of the wall.

Taking moments of all pressures about *G* and equating the same,

$$P \times y = \left[\text{Pressure } ACGH \times \frac{5}{2}\right] + \left[\text{Pressure } ABD \times \left(2 + \frac{3}{3}\right)\right] + \left[\text{Pressure } BCDE \times \frac{2}{2}\right] + \left[\text{Pressure } DEF \times \frac{2}{3}\right]$$

$$116 \times y = \left(30 \times \frac{5}{2}\right) + (60 \times 3) + (40 \times 1) + \left(16 \times \frac{2}{3}\right)$$
$$= 75 + 180 + 40 + 10.67 = 305.67$$

$$y = \frac{305.67}{116} = \mathbf{2.64 \text{ m}} \quad \textbf{Ans.}$$

**EXAMPLE 18.14.** *A masonry retaining wall of trapezoidal section with a vertical face on the earth side is 1 m wide at the top, 3 m wide at the bottom and 6 m high. It retains sand over the entire height with an angle of surcharge of 20°. Determine the distribution of pressure at the base of the wall. The sand weighs 18 kN/m³ and has an angle of repose of 30°. The masonry weighs 24 kN/m³.*

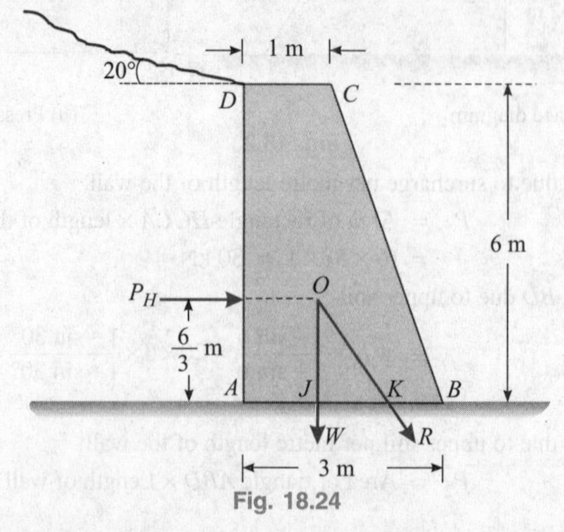


Fig. 18.24

**SOLUTION.** Given: Top width $(a) = 1$ m ; Bottom width $(b) = 3$ m ; Height of the wall $(h) = 6$ m; Angle of surcharge $(\alpha) = 20°$ ; Specific weight of sand $(w) = 18$ kN/m$^3$ ; Angle of repose $(\phi) = 30°$ and specific weight of masonry $(\rho) = 24$ kN/m$^3$.

Let the resultant $(R)$ cut the base at $K$ as shown in Fig. 18.24. We know that total pressure per metre length of the wall,

$$P = \frac{wh^2}{2}\cos\alpha \times \frac{\cos\alpha - \sqrt{\cos^2\alpha - \cos^2\alpha}}{\cos\alpha + \sqrt{\cos^2\alpha - \cos^2\alpha}}$$

$$= \frac{18 \times (6)^2}{2}\cos 20° \times \frac{\cos 20° - \sqrt{\cos^2 20° - \cos^2 30°}}{\cos 20° + \sqrt{\cos^2 20° - \cos^2 30°}} \text{ kN}$$

$$= 324 \times 0.9397 \times \frac{0.9397 - \sqrt{(0.9397)^2 - (0.866)2}}{0.9397 + \sqrt{(0.9397)^2 - (0.866)^2}} \text{ kN}$$

$$= 304.5 \times \frac{0.575}{1.3044} = 134.2 \text{ kN}$$

∴ Horizontal component of the pressure,

$$P_H = 134.2 \cos 20° = 134.2 \times 0.9397 = 126.1 \text{ kN}$$

and vertical component of the pressure,

$$P_V = 134.2 \sin 20° = 134.2 \times 0.3420 = 45.9 \text{ kN}$$

We also know that weight of the retaining wall

$$= 24 \times \frac{(1+3)}{2} \times 6 = 288 \text{ kN}$$

∴ Total weight acting vertically down,

$$W = 45.9 + 288 = 333.9 \text{ kN}$$

Now let us find out the position of the c.g. of the vertical load. Taking moments of the vertical loads about $A$ and equating the same,

$$W \times AJ = (P_V \times 0) + (24 \times 1 \times 6 \times 0.5) + \left(24 \times \frac{(6 \times 2)}{2} \times 2\right)$$

$$333.9\ AJ = 72 + 288 = 360$$

$$\therefore \quad AJ = \frac{360}{333.9} = 1.08 \text{ m}$$

We know that the horizontal distance between the centre of gravity of wall section and the point where the resultant cuts the base (*i.e.*, distance $JK$),

$$x = \frac{P_H}{W} \times \frac{h}{3} = \frac{126.1}{333.9} \times \frac{6}{3} = 0.75 \text{ m}$$

∴ *Horizontal distance $AK$,

$$d = AJ + JK = 1.08 + 0.75 = 1.83 \text{ m}$$

* The horizontal distance $d$ may also be found out by taking moments about A and equating the same,

$$W \cdot d = \left(P_H \times \frac{h}{3}\right) + (24 \times 1 \times 6 \times 0.5) + \left(24 \times \frac{6 \times 2}{2} \times 2\right)$$

$$333.9 \times d = \left(126.1 \times \frac{6}{3}\right) + 72 + 288 = 612.2 \quad \text{or} \quad d = \frac{612.2}{339.9} = 1.8 \text{ m}$$

and eccentricity, $e = d - \frac{b}{2} = 1.83 - \frac{3}{2} = 0.33 \text{ m}$

We also know that maximum intensity of pressure at the base,

$$\sigma_{max} = \frac{W}{b}\left(1 + \frac{6e}{b}\right) = \frac{333.9}{3}\left(1 + \frac{6 \times 0.33}{3}\right) \text{kN/m}^2$$

$$= 184.8 \text{ kN/m}^2 = \mathbf{184.8 \text{ kPa}} \quad \textbf{Ans.}$$

and minimum intensity of pressure at the base,

$$\sigma_{min} = \frac{W}{b}\left(1 - \frac{6e}{b}\right) = \frac{333.9}{3}\left(1 - \frac{6 \times 0.33}{3}\right) \text{kN/m}^2$$

$$= 37.84 \text{ kN/m}^2 = \mathbf{37.84 \text{ kPa}} \quad \textbf{Ans.}$$

## 18.18. *Coulomb's Wedge Theory for Active Earth Pressure

In Rankine's theory for active earth pressure, we considered the equilibrium of an element within the mass of the retained material. But in this theory, the equilibrium of the whole material supported by the retaining wall is considered, when the wall is at the point of slipping away from the retained material.

This theory is based on the concept of sliding wedge, which is torn off from the backfill on the movement of the wall and is based on the following assumptions:

1. The retained material is homogeneous and cohesionless.
2. The sliding wedge itself acts as a rigid body and the earth pressure is obtained by considering the limiting equilibrium of the sliding wedge as a whole.
3. The position and direction of the earth pressure is known *i.e.*, the pressure acts on the back of the wall and at a height of one-third of the wall height from the base. The pressure is inclined at an angle $\delta$ (called the angle of wall friction) to the normal to the back.

Consider a trapezoidal retaining wall *ABCD* retaining surcharged earth up to a height of $h$ on the inclined face *AD* as shown in Fig. 18.25.

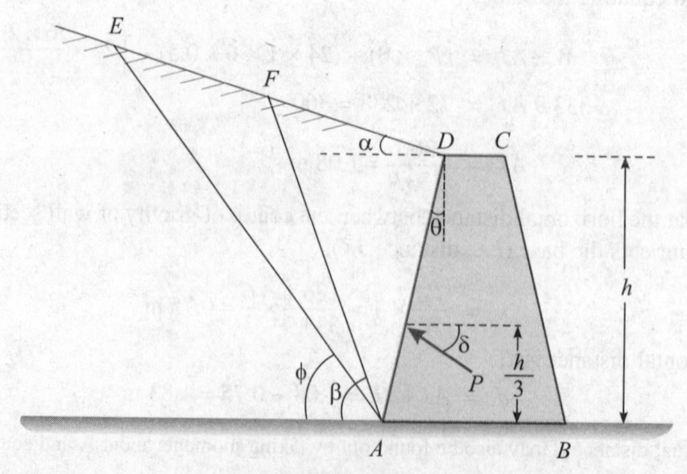


Fig. 18.25

Let $h$ = Height of the wall,

$w$ = Specific weight of the retained earth,

* This theory was given by Prof. C.A. Coulomb a French scientist in 1876.

$\phi$ = Angle of repose of the retained earth,

$\alpha$ = Angle of surcharge,

$\theta$ = Angle, which the inclined face $AD$ makes with the vertical and

$\delta$ = Angle of friction between the retaining wall and the retained earth.

In this case, the earth pressure is given by the relation,

$$P = \frac{wh^2}{2} \times \frac{\cos^2(\phi - \theta)}{\cos^2\theta \cos(\delta + \theta)\left[1 + \sqrt{\frac{\sin(\delta + \phi)\sin(\phi - \alpha)}{\cos(\delta + \theta)\cos(\theta - \alpha)}}\right]^2}$$

## 18.19. Conditions for the Stability of a Retaining Wall

The conditions, for the stability of a retaining wall are the same as those for the stability of a dam. In general, a retaining wall is checked for the following conditions of stability:

1. To avoid tension in the masonry at the base of the wall.
2. To safeguard the wall from overturning.
3. To prevent the sliding of wall.
4. To prevent the crushing of masonry at the base of the wall.

**EXAMPLE 18.15.** *Find the stability of the retaining wall shown in Fig. 18.26.*

*Also find the extreme stresses at the base of the wall, taking the densities of soil retained and masonry of the wall as 16 kN/m³ and 22 kN/m³ respectively. Assume angle of internal friction of the soil as 30°.*

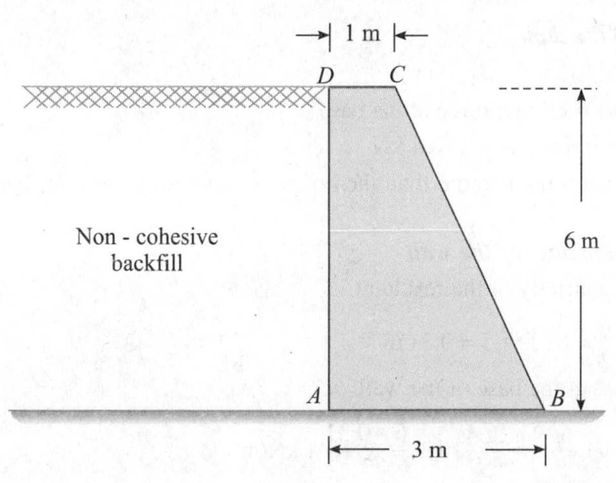


Fig. 18.26

**SOLUTION.** Given: Top width ($a$) = 1 m ; Bottom width ($b$) = 3 m ; Height of wall ($h$) = 6 m ; Density of soil ($w$) = 16 kN/m$^3$ ; Density of masonry ($\rho$) = 22 kN/m$^3$ ; and angle of internal friction ($\phi$) = 30°

***Check for tension in the masonry***

Let the resultant ($R$) cut the base at $K$ as shown in Fig. 18.27. We know that earth pressure per metre length of the wall,

$$P = \frac{wh^2}{2} \times \frac{1 - \sin\phi}{1 + \sin\phi} = \frac{16 \times (6)^2}{2} \times \frac{1 - \sin 30°}{1 + \sin 30°} \text{ kN}$$

$$= 288 \times \frac{1-0.5}{1+0.5} = 288 \times \frac{0.5}{1.5} = 96 \text{ kN}$$

and weight of the wall per metre length,

$$W = \rho \times \frac{(a+b)}{2} \times h = 22 \times \frac{(1+3)}{2} \times 6 = 264 \text{ kN}$$

Now let us find out the centre of gravity of the wall section. Taking moments of the wall section about $A$ and equating the same, *i.e.*,

$$\left(6 \times 1 + \frac{6 \times 2}{2}\right) AJ = \left(6 \times 1 \times \frac{1}{2}\right) + \left[6 \times \frac{2}{2}\left(1 + \frac{2}{3}\right)\right]$$

$$12\ AJ = 3 + 10 = 13$$

$$\therefore \quad {}^{*}AJ = \frac{13}{12} = 1.08 \text{ m}$$

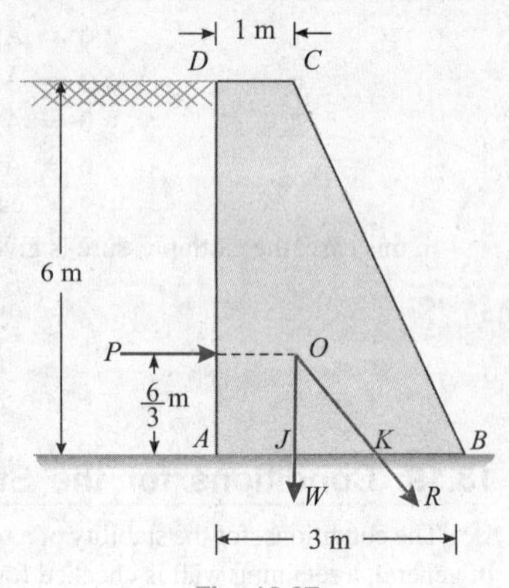

Fig. 18.27

We know that horizontal distance between the centre of gravity of the wall section and the point, where the resultant cuts the base (*i.e.*, distance $JK$),

$$x = \frac{P}{W} \times \frac{h}{3} = \frac{96}{264} \times \frac{6}{3} = 0.73 \text{ m}$$

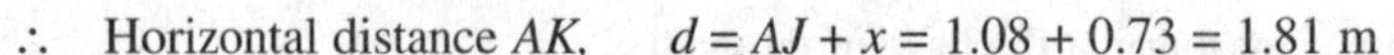

$\therefore$ Horizontal distance $AK$, $\quad d = AJ + x = 1.08 + 0.73 = 1.81$ m

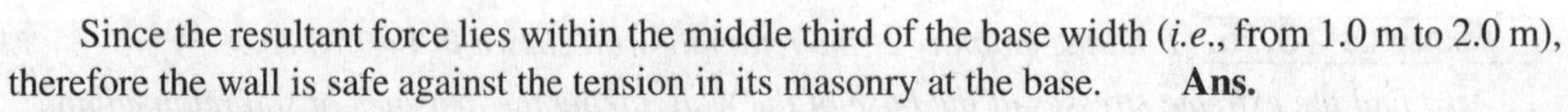

Since the resultant force lies within the middle third of the base width (*i.e.*, from 1.0 m to 2.0 m), therefore the wall is safe against the tension in its masonry at the base. **Ans.**

***Check for overturning***

Since the resultant force lies within the base $AB$ as obtained above, therefore the wall is safe against overturning also. **Ans.**

***Check for sliding of the dam***

Let coefficient of friction, $(\mu) = 0.6$

We know that the frictional force at the base

$$= \mu W = 0.6 \times 264 = 158.4 \text{ kN}$$

Since the frictional force is more than the horizontal pressure, therefore the wall is safe against sliding. **Ans.**

***Extreme stresses at the base of the wall***

We know that the eccentricity of the resultant,

$$e = d - \frac{b}{2} = 1.81 - 1.5 = 0.31 \text{ m}$$

$\therefore$ Maximum stress at the base of the wall,

$$\sigma_{max} = \frac{W}{b}\left(1 + \frac{6e}{b}\right) = \frac{264}{3}\left(1 + \frac{6 \times 0.31}{3}\right) \text{ kN/m}^2$$

$$= 142.6 \text{ kN/m}^2 = 142.6 \text{ kPa} \quad \textbf{Ans.}$$

and minimum stress at the base of the wall,

$$\sigma_{min} = \frac{W}{b}\left(1 - \frac{6e}{b}\right) = \frac{264}{3}\left(1 - \frac{6 \times 0.31}{3}\right) \text{ kN/m}^2$$

$$= 33.4 \text{ kN/m}^2 = 33.4 \text{ kPa} \quad \textbf{Ans.}$$

---

* The distance $AJ$ may also be found out from the following relation :

$$AJ = \frac{a^2 + ab + b^2}{3(a+b)} = \frac{(1)^2 + (1 \times 3) + (3)^2}{3(1+3)} = \frac{13}{12} = 1.08 \text{ m}$$

**EXAMPLE 18.16.** *A masonry retaining wall 4 m high above ground level as shown in Fig. 18.28 sustains earth with a positive surcharge of 10°. The width of the wall at top is 0.75 m and at the base 2.5 m. The earth face of the wall makes an angle of 20° with the vertical.*

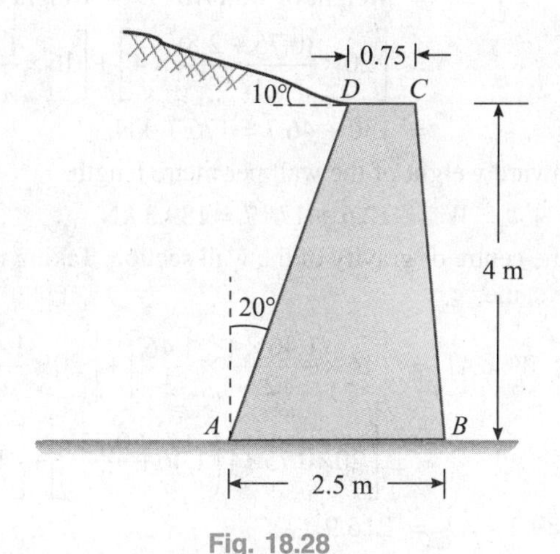


Fig. 18.28

*Determine the thrust on the wall and examine the safety of the wall for no tension, overturning and sliding.*

*Given the weight of earth = 16 kN/m³, masonry = 20 kN/m³. Maximum pressure allowable on soil 120 kPa; angle of repose of the soil = 30°; angle of friction between the soil and wall = 20° and angle of surcharge = 10°.*

**SOLUTION.** Given: Height of wall $(h) = 4$ m ; Angle of surcharge $(\alpha) = 10°$ ; Top width $(a) = 0.75$ m; Bottom width $(b) = 2.5$ m ; Angle of $AD$ with vertical $(\theta) = 20°$ ; Weight of earth $(w) = 16$ kN/m³; Weight of masonry $(\rho) = 20$ kN/m³ ; Maximum allowable pressure $(\sigma_{max}) = 120$ kPa = 120 kN/m² ; Angle of repose $(\phi) = 30°$ and angle of friction $(\delta) = 20°$.

***Check for tension***

We know that active earth pressure on the wall per metre length,

$$P = \frac{wh^2}{2} \times \frac{\cos^2 (\phi - \theta)}{\cos^2 \theta \cos (\delta + \theta) \left[1 + \sqrt{\dfrac{\sin (\delta + \phi) \sin (\phi - \alpha)}{\cos (\delta + \theta) \cos (\theta - \alpha)}}\right]^2}$$

$$= \frac{16 \times (4)^2}{2} \times \frac{\cos^2 (30° - 20°)}{\cos^2 20° \cos 40° \left[1 + \sqrt{\dfrac{\sin 50° \times \sin 20°}{\cos 40° \times \cos 10°}}\right]^2}$$

$$= 128 \times \frac{\cos^2 10°}{(0.9397)^{2 \times 0.766} \left[1 + \sqrt{\dfrac{0.766 \times 0.342}{0.766 \times 0.9848}}\right]^2}$$

$$= 128 \times \frac{(0.9848)^2}{(0.9397)^2 \times 0.766 \times (1.5893)^2} = 72.7 \text{ kN}$$

∴ Horizontal component of the pressure,

$$P_H = P \cos \theta = 72.7 \cos 10° = 72.7 \times 0.9848 = 71.6 \text{ kN}$$

and vertical component of the pressure,

$$P_V = P\sin\theta = 72.7\sin 10° = 772.7 \times 0.1736 = 12.6 \text{ kN}$$

We also know that the weight of the wall per metre length

$$= \text{Weight of wall } ABCD + \text{ Weight of wedge } ADE \text{ of earth}$$

$$= \left[20 \times \frac{(0.75 + 2.5)}{2} \times 4\right] + \left[16 \times \frac{1.46}{2} \times 4\right]$$

$$= 130 + 46.7 = 176.7 \text{ kN}$$

Therefore total downward weight of the wall per metre length,

$$W = 12.6 + 176.7 = 189.3 \text{ kN}$$

Now let us find out the centre of gravity of the wall section. Taking moments of the wall section about $A$, and equating the same,

$$W \times AJ = \left[16 \times \frac{1.46 \times 4}{2} \times \frac{1.46}{3}\right] + \left[20 \times \frac{1.46 \times 4}{2} \times \frac{1.46 \times 2}{3}\right]$$

$$+ \left[20 \times 0.75 \times 4\left(1.46 + \frac{0.75}{2}\right)\right] + \left[20 \times \frac{0.29 \times 4}{2}\left(2.21 + \frac{0.29}{2}\right)\right]$$

$$189.3 \times AJ = 216.9$$

$$AJ = \frac{216.9}{189.3} = 1.14 \text{ m}$$

We know that horizontal distance between the centre of gravity of the wall section and the point, where the resultant cut the base (*i.e.*, distance $JK$),

$$x = \frac{P_H}{W} \times \frac{h}{3} = \frac{71.6}{189.3} = 0.5 \text{ m}$$

Horizontal distance $AK$, $\quad d = AJ + x = 1.14 + 0.5 = 1.64 \text{ m}$

Since the resultant force lies at a point, which is at a distance of 2/3 from $A$, therefore the wall is safe against the tension in its masonry at the base. **Ans.**

***Check for overturning***

Since the resultant force lies within the base $AB$ as obtained above, therefore the wall is safe against overturning also. **Ans.**

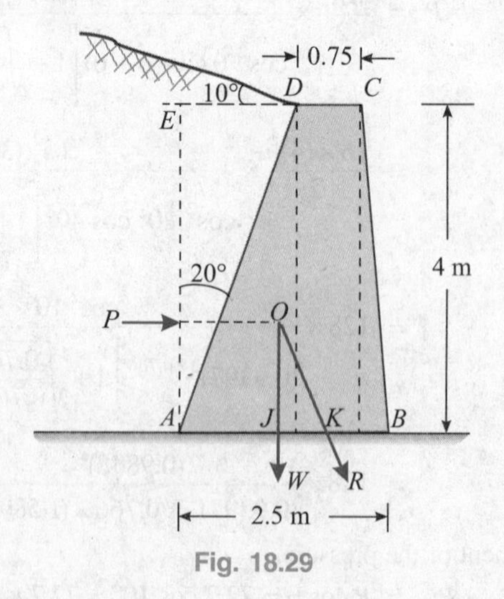


Fig. 18.29

*Check for sliding of the wall*

Let coefficient of friction, $\mu$ = 0.6

We know that the frictional force at the base

$$= \mu W = 0.6 \times 189.3 = 113.6 \text{ kN}$$

Since the frictional force is more than the horizontal component of the pressure, therefore the wall is safe against sliding. **Ans.**

*Check for maximum stress at the base*

We know that eccentricity of the resultant,

$$e = d - \frac{b}{2} = 1.64 - \frac{2.5}{2} = 0.39 \text{ m}$$

∴ Maximum stress at the base

$$\sigma_{max} = \frac{W}{b}\left(1 + \frac{6e}{b}\right) = \frac{189.3}{2.5}\left(1 + \frac{6 \times 0.39}{2.5}\right) \text{kN/m}^2$$

$$= 146.6 \text{ kN/m}^2 = 146.6 \text{ kPa}$$

Since the maximum stress is more than the permissible stress (120 kN/m2), therefore the wall is not safe against maximum stress at the base. **Ans.**

## EXERCISE 18.2

1. A trapezoidal masonry retaining wall 1 m wide at top, 3 m wide at its bottom is 8 m high. It is retaining earth having level with the top of the wall on its vertical face. Find the maximum and minimum intensities of stress at the base of the wall, if the weight of masonry and earth is 24 $\text{kN/m}^3$ and 18 $\text{kN/m}^3$. Angle of repose of the earth is 40° **[Ans.** 244 kPa ; 12 kPa]
2. A trapezoidal masonry retaining wall 1 m wide at top, 3 m wide at its bottom is 6 m high. The vertical face is retaining earth with angle of repose 30° at surcharge of 20° with the horizontal. Determine the maximum and minimum intensities of stress at the base of the dam. Take the densities of earth and masonry as 20 $\text{kN/m}^3$ and 24 $\text{kN/m}^3$. **[Ans.** 169.5 kPa ; 56.9 kPa]
3. A masonry wall 8 m high and 3 m wide contains water for a height of 7 m. Check the stability of the wall, if the coefficient of friction between the wall and the soil is 0.55. Take weight of masonry as 22.2 $\text{kN/m}^3$.

[**Ans. 1.** The wall shall fail due to tension. **2.** Safe for overturning. **3.** Safe against sliding]

## QUESTIONS

1. What do you understand by the term dam? Name the various types of dams commonly used these days.
2. Derive an equation for the maximum and minimum intensities of stress at the base of a trapezoidal dam.
3. Name the various conditions for the stability of a dam. Describe any two of them.
4. How will you find out the (*i*) minimum base width and (*ii*) maximum height of a dam?
5. What is a retaining wall? Discuss its uses.
6. Explain what do you understand by active and passive earth pressures of soil?
7. What are the assumptions made in Rankine's theory for calculating the magnitude of earth pressure behind retaining walls.
8. State and explain Rankine's theory of earth pressure.

## OBJECTIVE TYPE QUESTIONS

**1.** The water pressure per metre length on a vertical wall is

(*a*) $wh$ (*b*) $\frac{wh}{2}$ (*c*) $\frac{wh^2}{2}$ (*d*) $\frac{wh^2}{4}$

where $w$ = Specific weight of water and $h$ = Height of the water

**2.** The maximum and minimum stress intensities at the base of a dam containing water are

(*a*) $\frac{W}{b}\left(1+\frac{6e}{b}\right)$ and $\frac{W}{b}\left(1-\frac{6e}{b}\right)$ (*b*) $\frac{W}{2b}\left(1+\frac{6e}{b}\right)$ and $\frac{W}{2b}\left(1-\frac{6e}{b}\right)$

(*c*) $\frac{2W}{b}\left(1+\frac{6e}{b}\right)$ and $\frac{2W}{b}\left(1-\frac{6e}{b}\right)$ (*d*) $\frac{3W}{2b}\left(1+\frac{6e}{b}\right)$ and $\frac{3W}{2b}\left(1-\frac{6e}{b}\right)$

**3.** The stability of a dam is checked for

(*a*) tension at the base (*b*) overturning of the dam

(*c*) sliding of the dam (*d*) all of these

**4.** Total pressure per unit length of a retaining wall is given by

(*a*) $\frac{wh}{2}\times\frac{1-\sin\phi}{1+\sin\phi}$ (*b*) $\frac{wh^2}{2}\times\frac{1-\sin\phi}{1+\sin\phi}$ (*c*) $\frac{wh}{2}\times\frac{1+\sin\phi}{1-\sin\phi}$ (*d*) $\frac{wh^2}{2}\times\frac{1+\sin\phi}{1-\sin\phi}$

## ANSWERS

**1.** (*c*) **2.** (*a*) **3.** (*d*) **4.** (*b*)

Chapter **19**

# Deflection of Beams

## Contents

## 19.1. Introduction

We see that whenever a cantilever or a beam is loaded, it deflects from its original position. The amount, by which a beam deflects, depends upon its cross-section and the bending moment. In modern design offices, following are the two design criteria for the deflection of a cantilever or a beam:

**1.** Strength **2.** Stiffness.

As per the strength criterion of the beam design, it should be strong enough to resist bending moment and shear force. Or in other words, the beam should be strong enough to resist the bending stresses and shear stresses. And as per the stiffness criterion of the beam design, which is equally important, it should be stiff enough to resist the deflection of the beam. Or in other words, the beam should be stiff enough not to deflect more than the permissible limit* under the action of the loading. In actual practice, some specifications are always laid to limit the maximum deflection of a cantilever or a beam to a small fraction of its span.

In this chapter, we shall discuss the slope and deflection of the centre line of beams under the different types of loadings.

## 19.2. Curvature of the Bending Beam

Consider a beam *AB* subjected to a bending moment. As a result of loading, let the beam deflect from *ABC* to *ADB* into a circular arc as shown in Fig. 19.1.

Let $l$ = Length of the beam *AB*,

$M$ = Bending moment,

$R$ = Radius of curvature of the bent up beam,

$I$ = Moment of inertia of the beam section,

$E$ = Modulus of elasticity of beam material,

$y$ = Deflection of the beam (*i.e.*, *CD*) and

$i$ = Slope of the beam (*i.e* angle which the tangent at *A* makes with *AB*).

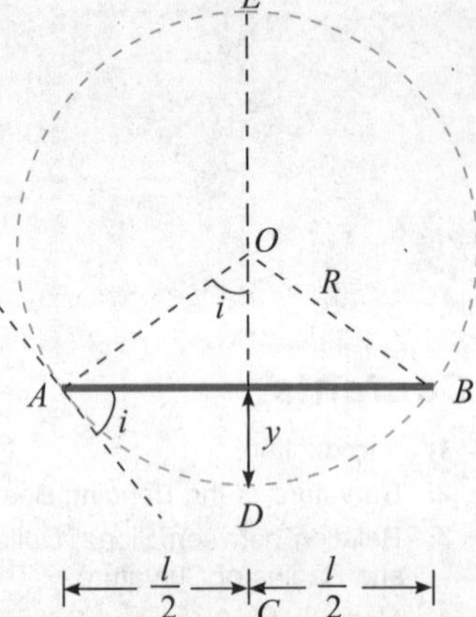

Fig. 19.1. Curvature of the beam.

From the geometry of a circle, we know that

$$AC \times CB = EC \times CD$$

or
$$\frac{1}{2} \times \frac{1}{2} = (2R - y) \times y$$

$\therefore$
$$\frac{l^2}{4} = 2Ry - y^2 = 2Ry \qquad \text{...(Neglecting } y^2)$$

or
$$y = \frac{l^2}{8R} \qquad \text{...}(i)$$

We have already discussed in Art. 14.6 that for a loaded beam,

$$\frac{M}{I} = \frac{E}{R} \quad \text{or} \quad R = \frac{EI}{M}$$

Now substituting this value of *R* in equation (*i*),

$$y = \frac{l^2}{8 \times \frac{EI}{M}} = \frac{El^2}{8\,EI}$$

From the geometry of the figure, we find that the slope of the beam *i* at *A* or *B* is also equal to angle *AOC*.

$\therefore$
$$\sin i = \frac{AC}{OA} = \frac{l}{2R}$$

Since the angle *i* is very small, therefore, sin *i* may be taken equal to *i* (in radians).

$\therefore$
$$i = \frac{l}{2R} \text{ radians} \qquad \text{...}(ii)$$

Again substituting the value of *R* in equation (*ii*),

$$i = \frac{l}{2R} = \frac{l}{2 \times \frac{EI}{M}} = \frac{Ml}{2El} \text{ radians} \qquad \text{...}(iii)$$

---

* As per Indian Standard Specifications, this limit is Span/325.

**NOTES:** 1. The above equations for deflection ($y$) and slope ($i$) have been derived from the bending moment only *i.e.*, the effect of shear force has been neglected. This is due to the reason that the effect of shear force is extremely small as compared to the effect of bending moment.

2. In actual practice the beams bend into an arc of a circle only in a few cases. A little consideration will show that a beam will bend to an arc of a circle only if (*i*) the beam is of uniform section and (*ii*) the beam is subjected to a constant moment throughout its length or the beam is of uniform strength.

## 19.3. Relation between Slope, Deflection and Radius of Curvature

Consider a small portion $PQ$ of a beam, bent into an arc as shown in Fig. 19.2.

Let $ds$ = Length of the beam $PQ$,

$R$ = Radius of the arc, into which the beam has been bent,

$C$ = Centre of the arc,

$\Psi$ = Angle, which the tangent at $P$ makes with $x$-$x$ axis and

$\Psi + d\Psi$ = Angle which the tangent at $Q$ makes with $x$-$x$ axis.

From the geometry of the figure, we find that

$$\angle PCQ = d\Psi$$

and

$$ds = R \, . \, d\Psi$$

$$\therefore \quad R = \frac{ds}{d\Psi} = \frac{dx}{d\Psi} \qquad \text{... (Considering } ds = dx\text{)}$$

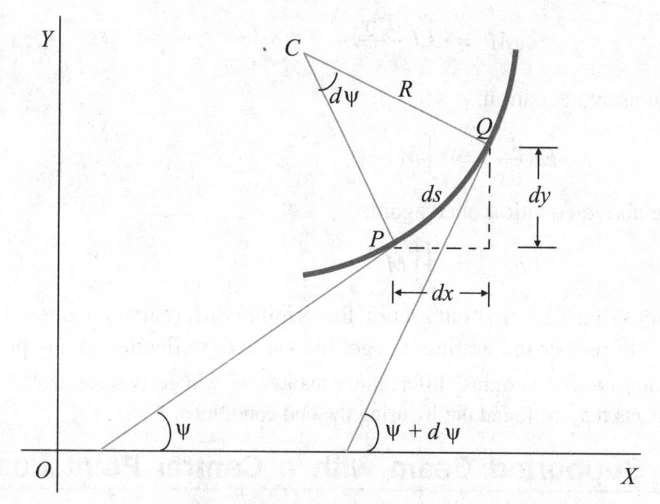


Fig. 19.2. Beam bent into an arc.

or

$$\frac{1}{R} = \frac{d\Psi}{dx} \qquad \text{...(}i\text{)}$$

We know that if $x$ and $y$ be the co-ordinates of point P, then

$$\tan \Psi = \frac{dy}{dx}$$

Since $\Psi$ is a very small angle, therefore taking $\tan \Psi = \Psi$,

$$\therefore \quad \frac{d\Psi}{dx} = \frac{d^2 y}{dx^2} \qquad \text{...} \left( \because \frac{1}{R} = \frac{d\Psi}{dx} \right)$$

We also know that

$$\frac{M}{I} = \frac{E}{R} \quad \text{or} \quad M = EI \times \frac{I}{R}$$

$$\therefore \qquad M = EI \times \frac{d^2y}{dx^2} \qquad \text{... } \left(\text{Substituting value of } \frac{1}{R}\right)$$

**NOTE.** The above equation is also based only on the bending moment. The effect of shear force, being very small as compared to the bending moment, is neglected.

## 19.4. Methods for Slope and Deflection at a Section

Though there are many methods to find out the slope and deflection at a section in a loaded beam, yet the following two methods are important from the subject point of view:

1. Double integration method.
2. Macaulay's method.

It will be interesting to know that the first method is suitable for a single load, whereas the second method is suitable for several loads.

## 19.5. Double Integration Method for Slope and Deflection

We have already discussed in Art. 19.3 that the bending moment at a point,

$$M = EI \frac{d^2y}{dx^2}$$

Integrating the above equation,

$$EI \frac{dy}{dx} = \int M \qquad \text{... (i)}$$

and integrating the above equation once again,

$$EI \cdot y = \iint M \qquad \text{... (ii)}$$

It is thus obvious that after first integration the original differential equation, we get the value of slope at any point. On further integrating, we get the value of deflection at any point.

**NOTE.** While integrating twice the original differential equation, we will get two constants $C_1$ and $C_2$. The values of these constants may be found out by using the end conditions.

## 19.6. Simply Supported Beam with a Central Point Load

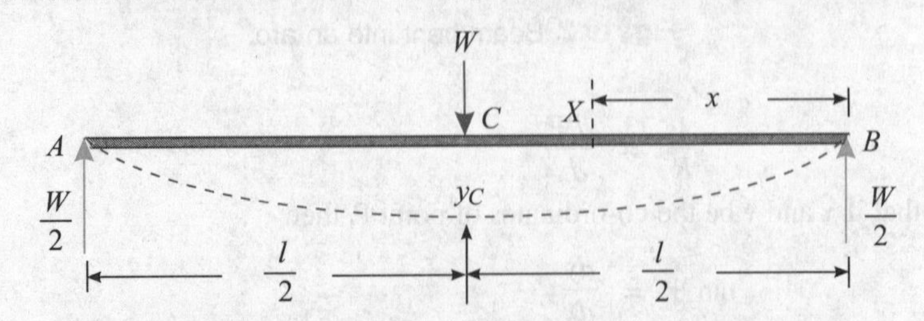


Fig. 19.3. Simply supported beam with a central point load.

Consider a simply supported beam $AB$ of length $l$ and carrying a point load $W$ at the centre of beam $C$ as shown in Fig. 19.3. From the geometry of the figure, we find that the reaction at $A$,

$$R_A = R_B = \frac{W}{2}$$

Consider a section $X$ at a distance $x$ from $B$. We know that the bending moment at this section,

$$M_X = R_B \cdot x = \frac{W}{2} x = \frac{Wx}{2} \quad \text{... (Plus sign due to sagging)}$$

$$\therefore \quad EI \frac{d^2y}{dx^2} = \frac{Wx}{2} \quad \text{...(i)}$$

Integrating the above equation,

$$EI \frac{dy}{dx} = \frac{Wx^2}{4} + C_1 \quad \text{...(ii)}$$

where $C_1$ is the first constant of integration. We know that when $x = \frac{l}{2}$, then $\frac{dy}{dx} = 0$. Substituting these values in equation (*ii*),

$$0 = \frac{Wl^2}{16} + C_1 \quad \text{or} \quad C_1 = -\frac{Wl^2}{16}$$

Substituting this value of $C_1$ in equation (*ii*),

$$EI \frac{dy}{dx} = \frac{Wx^2}{4} - \frac{Wl^2}{16} \quad \text{...(iii)}$$

This is the required equation for the slope, at any section. It will be interesting to know that the maximum slope occurs at $A$ and $B$. Thus for maximum slope at $B$, substituting $x = 0$ in equation (*iii*),

$$EI \cdot i_B = -\frac{Wl^2}{16}$$

$$\therefore \quad i_B = -\frac{Wl^2}{16EI}$$

...(Minus sign means that the tangent at $B$ makes an angle with $AB$ in the negative or anticlockwise direction)

or

$$i_B = \frac{Wl^2}{16EI} \text{ radians}$$

By symmetry,

$$i_A = \frac{Wl^2}{16EI} \text{ radians}$$

Integrating the equation (*iii*) once again,

$$\therefore \quad EI.y = \frac{Wx^3}{12} - \frac{Wl^2x}{16} + C_2 \quad \text{...(iv)}$$

where $C_2$ is the second constant of integration. We know that when $x = 0$, then $y = 0$, Substituting these values in equation (*iv*), we get $C_2 = 0$.

$$\therefore \quad EI.y = \frac{Wx^3}{12} - \frac{Wl^2x}{16} \quad \text{...(v)}$$

This is the required equation for the deflection, at any section. A little consideration will show that maximum deflection occurs at the mid-point $C$. Thus for maximum deflection, substituting $x = \frac{l}{2}$ in equation (*v*),

$$EIy_C = \frac{W}{12}\left(\frac{l}{2}\right)^3 - \frac{Wl^2}{16}\left(\frac{l}{2}\right)$$

$$= \frac{Wl^3}{96} - \frac{Wl^3}{32} = -\frac{Wl^3}{48}$$

or

$$y_C = -\frac{Wl^3}{48EI}$$

... (Minus sign means that the deflection is downwards)

$$= \frac{Wl^3}{48EI}$$

**Example 19.1.** *A simply supported beam of span 3 m is subjected to a central load of 10 kN. Find the maximum slope and deflection of the beam. Take I = 12 × 10⁶ mm⁴ and E = 200 GPa.*

**Solution.** Given: Span ($l$) = 3 m = $3 \times 10^3$ mm ; Central load ($W$) = 10 kN = $10 \times 10^3$ N ; Moment of inertia ($I$) = $12 \times 10^6$ mm$^4$ and modulus of elasticity ($E$) = 200 GPa = $200 \times 10^3$ N/mm$^2$.

***Maximum slope of the beam***

We know that maximum slope of the beam,

$$i_A = \frac{Wl^2}{16EI} = \frac{(10 \times 10^3) \times (3 \times 10^3)^2}{16 \times (200 \times 10^3) \times (12 \times 10^6)} = \mathbf{0.0023 \text{ rad}} \quad \textbf{Ans.}$$

***Maximum deflection of the beam***

We also know that maximum deflection of the beam,

$$y_C = \frac{Wl^3}{48EI} = \frac{(10 \times 10^3) \times (3 \times 10^3)^3}{48 \times (200 \times 10^3) \times (12 \times 10^6)} = \mathbf{2.3 \text{ mm}} \quad \textbf{Ans.}$$

**Example 19.2.** *A wooden beam 140 mm wide and 240 mm deep has a span of 4 m. Determine the load, that can be placed at its centre to cause the beam a deflection of 10 mm. Take E as 6 GPa.*

**Solution.** Given: Width ($b$) = 140 mm ; Depth ($d$) = 240 mm ; Span ($l$) = 4 m = $4 \times 10^3$ mm ; Central deflection ($y_C$) = 10 mm and modulus of elasticity ($E$) = 6 GPa = $6 \times 10^3$ N/mm$^2$.

Let $W$ = Magnitude of the load,

We know that moment of inertia of the beam section,

$$I = \frac{bd^3}{12} = \frac{140 \times (240)^3}{12} = 161.3 \times 10^6 \text{ mm}^4$$

and deflection of the beam at its centre ($y_C$),

$$10 = \frac{Wl^3}{48EI} = \frac{W \times (4 \times 10^3)^3}{48 \times (6 \times 10^3) \times (161.3 \times 10^6)}$$

$$\therefore \quad W = \frac{10}{1.38 \times 10^{-3}} = 7.25 \times 10^3 \text{ N} = \mathbf{7.25 \text{ kN}} \quad \textbf{Ans.}$$

## 19.7. Simply Supported Beam with an Eccentric Point Load

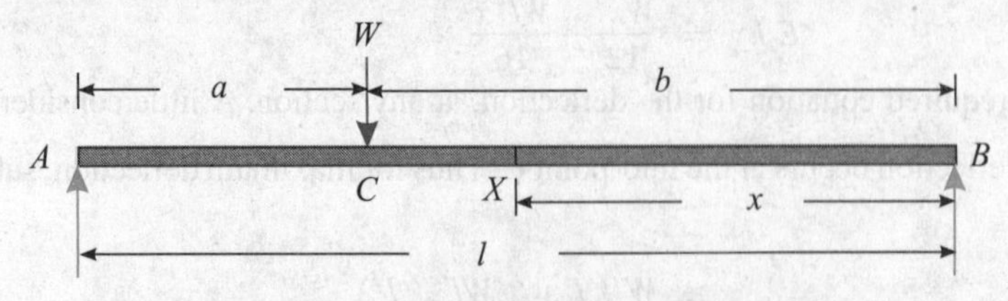


Fig. 19.4. Section $X$ in $CB$.

Consider a simply supported beam $AB$ of length $l$ and carrying an eccentric point load $W$ at $C$ as shown in Fig. 19.4. From the geometry of the figure, we find that the reaction at $A$,

$$R_A = \frac{Wb}{l} \quad \text{and} \quad R_B = \frac{Wa}{l}$$

Now consider a section $X$ in $CB$ at a distance $x$ from $B$, such that $x$ is less than $b$ (*i.e.*, $x < b$). We know that the bending moment at this section,

$$M_X = R_B \cdot x = \frac{Wax}{l} \qquad \text{...(Plus sign due to saging)}$$

$$\therefore \qquad E\,l.\frac{d^2y}{dx^2} = \frac{Wax}{l}$$

Integrating the above equation,

$$E\,l.\frac{dy}{dx} = \frac{Wax^2}{2l} + C_1 \qquad \text{...(}i\text{)}$$

where $C_1$ is the first constant of integration. We know that at $C$, $x = b$ and $\frac{dy}{dx} = i_C$.

$$\therefore \qquad E\,li_C = \frac{Wab^2}{2l} + C_1$$

or
$$C_1 = (El \cdot i_C) - \frac{Wab^2}{2l}$$

Substituting this value of $C_1$ in equation (*i*),

$$El\frac{dy}{dx} = \frac{Wax^2}{2l} + (El \cdot i_C) - \frac{Wab^2}{2l} \qquad \text{...(}ii\text{)}$$

Integrating the above equation once again,

$$El\,.\,y = \frac{Wax^3}{6l} + (El \cdot i_C \cdot x) - \frac{Wab^2x}{2l} + C_2 \qquad \text{...(}iii\text{)}$$

where $C_2$is the second constant of integration. We know that when $x = 0$, then $y = 0$. Substituting these values in equation (*iii*), we get $C_2 = 0$.

$$\therefore \qquad El\,.\,y = \frac{Wax^3}{6l} + (El \cdot i_C x) - \frac{Wab^2x}{2l} \qquad \text{...(}iv\text{)}$$

The equations (*ii*) and (*iv*) are the required equations for slope and deflection at any point in the section $AC$. A little consideration will show that these equations are useful, only if the value of $i_C$ is known.

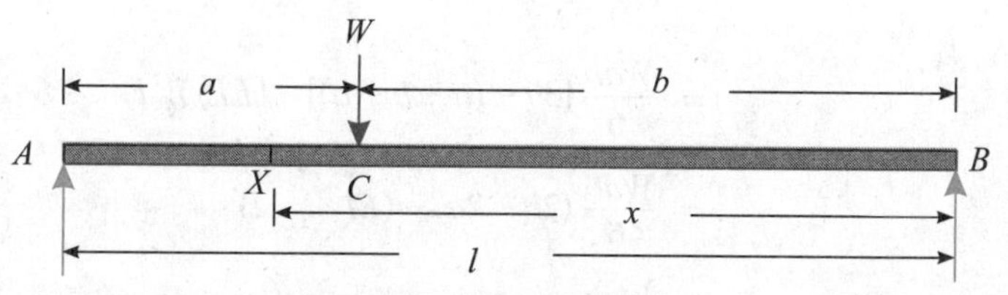


Fig. 19.5. Section $X$ in $AC$.

Now consider a section $X$ in $AC$, at a distance $x$ from $B$ such that $x$ is greater than $b$ (*i.e.*, $x > b$) as shown in Fig. 19.5. We know that bending moment at this section,

$$M_X = \frac{Wax}{l} - W\,(x - b)$$

$$\therefore \qquad El\frac{d^2y}{dx^2} = \frac{Wax}{l} - W\,(x - b) \qquad \text{...(}v\text{)}$$

Integrating the above equation,

$$El \cdot \frac{dy}{dx} = \frac{Wax^2}{2l} - \frac{W\,(x - b)^2}{2} + C_3 \qquad \text{...(}vi\text{)}$$

where $C_3$ is the third constant of integration. We know that at $C$, $x = b$ and $\frac{dy}{dx} = i_C$.

$$\therefore \quad EI \cdot i_C = \frac{Wab^2}{2l} + C_3$$

or

$$C_3 = (EI \cdot i_C) - \frac{Wab^2}{2l}$$

Substituting this value of $C_3$ in equation (*vi*),

$$EI \cdot \frac{dy}{dx} = \frac{Wax^2}{2l} - \frac{W(x-b)^2}{2} + (EI \cdot i_C) - \frac{Wab^2}{2l} \qquad \text{...(vii)}$$

Integrating the above equation once again,

$$EI \cdot y = \frac{Wax^3}{6l} - \frac{W(x-b)^3}{6} + (EI \cdot i_C)x - \frac{Wab^2}{2l}x + C_4 \qquad \text{...(viii)}$$

where $C_4$ is the fourth constant of integration. We know that when $x = l$, then $y = 0$. Substituting these values in the above equation,

$$0 = \frac{Wal^2}{6} - \frac{Wa^3}{6} - \frac{Wab^2}{2} + (EI \cdot i_C \cdot l) + C_4$$

$$C_4 = \frac{Wab^2}{2} + \frac{Wa^3}{6} - \frac{Wal^2}{6} - (EI \cdot i_C) \cdot l \qquad \text{... } [\because \ (x-b) = a]$$

$$\therefore \quad = \frac{Wab^2}{2} + \frac{Wa}{6}(a^2 - l^2) - (EI \cdot i_C \cdot l)$$

$$= \frac{Wab^2}{2} + \frac{Wa}{6}[(l+a)(l-a)] - (EI \cdot i_C \cdot l)$$

$$\text{... } [\because \ l^2 - a^2 = (l+a)(l-ay)]$$

$$= \frac{Wab^2}{2} - \frac{Wab}{6}(l+a) - (EI \cdot i_C \cdot l) \qquad [\because \ (l-a) = b]$$

$$= \frac{Wab}{6}[3b - (l+a)] - (EI \cdot i_C \cdot l)$$

$$= \frac{Wab}{6}[3b - (a+b+a)] - (EI \cdot i_C \, l) \qquad (\because \ l = a+b)$$

$$= \frac{Wab}{6}(2b - 2a) - (EI \cdot i_C \cdot l)$$

$$= \frac{Wab}{3}(b-a) - (EI \cdot i_C \cdot l)$$

Substituting this value of $C_4$ in equation (*viii*),

$$EI \cdot y = \frac{Wax^3}{6l} - \frac{W(x-b)^3}{6} + (EI \cdot i_C \cdot x) - \frac{Wab^2 x}{2l}$$

$$+ \frac{Wab}{3}(b-a) - (EI \cdot i_C \cdot l) \qquad \text{...(ix)}$$

The equations (*vii*) and (*ix*) are the required equations for the slope and deflection at any point in the section $AC$. A little consideration will show that these equations are useful, only if the value of $i_C$

is known. Now to obtain the value of $i_C$, let us first find out the deflection at $C$ from the equations for sections $AC$ and $CB$.

Now substituting $x = b$ in equation (*iv*) and equating the same with equation (*ix*),

$$\frac{Wab^3}{6l} + (EI \cdot i_C \cdot b) - \frac{Wab^3}{2l} = \frac{Wab^3}{6l} - \frac{W(b-b)^3}{6} + (EI \cdot i_C \cdot b) - \frac{Wab^3}{2l} + \frac{Wab}{3}(b-a) - (EI \cdot i_C \cdot l)$$

$$\therefore \quad EI \,.\, i_C = \frac{Wab}{3l}(b-a)$$

Substituting the value of $EI \,.\, i_C$ in equation (*ii*)

$$EI\frac{dy}{dx} = \frac{Wax^2}{2l} + \frac{Wab}{3l}(b-a) - \frac{Wab^2}{2l}$$

$$= \frac{Wa}{6l}[3x^2 + 2b(b-a) - 3b^2]$$

$$= \frac{Wa}{6l}(3x^2 + b^2 - 2ab)$$

This is required equation for slope at any section in $BC$. We know that the slope is maximum at $B$. Thus for maximum slope, substituting $x = 0$ in equation (*x*),

$$EI \,.\, i_B = \frac{Wa}{6l}(-b^2 - 2ab)$$

$$= -\frac{Wa}{6l}(b^2 + 2ab) \qquad \text{...(Taking minus sign outside)}$$

$$= -\frac{Wab}{6l}(b + 2a)$$

$$= -\frac{Wa}{6l}(l-a)(l+a) \qquad \text{...}(\because\ a = l - b \text{ and } a + b = l)$$

$$= -\frac{Wa}{6l}(l^2 - a^2)$$

or

$$i_B = -\frac{Wa}{6EIl}(l^2 - a^2)$$

[Minus sign means that the tangent at $B$, makes an angle with $AB$ in the negative or anticlockwise direction.]

$$= \frac{Wa}{6EIl}(l^2 - a^2)$$

Similarly,

$$i_A = \frac{Wb}{6EIl}(l^2 - b^2) \qquad \text{...(Substituting } b \text{ for } a)$$

For deflection at any point in $AC$, substituting the value of $EI \,.\, i_C$ in equation (*iv*),

$$EI \cdot y = \frac{Wax^3}{6l} + \frac{Wab}{3l}(b-a)\,x - \frac{Wab^2 x}{2l}$$

$$= \frac{Wax}{6l}[x^2 + 2b(b-a) - 3b^2]$$

$$= \frac{Wax}{6l}(x^2 + 2b^2 - 2ab - 3b^2)$$

$$= \frac{Wax}{6l}(x^2 - b^2 - 2ab)$$

$$= -\frac{Wax}{6l}[b(b + 2a) - x^2]$$

$$= -\frac{Wax}{6l}[(l - a) - x^2] \qquad ...(\because\ b = l - a \text{ and } a + b = l)$$

$$= -\frac{Wax}{6l}[l^2 - a^2 - x^2]$$

or
$$y = -\frac{Wax}{6lEI}[l^2 - a^2 - x^2] \qquad \text{... (Minus sign means that the deflection is downwards)}$$

$$= -\frac{Wax}{6EIl}(l^2 - a^2 - x^2) \qquad ...(xi)$$

For deflection at $C$ (*i.e.*, under the load) substituting $x = b$ in the above equation,

$$y_C = -\frac{Wab}{6EIl}(l^2 - a^2 - b^2) \qquad ...(xii)$$

We know that maximum deflection will occur in $CB$ since $b > a$. Now for maximum deflection, let us substitute $\frac{dy}{dx} = 0$. Therefore equating the equation $(x)$ to zero,

$$\frac{Wa}{6l}(3x^2 - b^2 - 2ab) = 0$$

or
$$3x^2 - b(b + 2a) = 0$$

$$3x^2 - (l - a)(l + a) = 0 \qquad ...(\because\ b = l - a \text{ and } a + b = l)$$

$$3x^2 - (l^2 - a^2) = 0$$

$$3x^2 = l^2 - a^2$$

$$\therefore \qquad x = \sqrt{\frac{l^2 - a^2}{3}}$$

For maximum deflection, substituting this value of $x$ in equation $(xi)$,

$$y_{max} = \frac{Wa}{6EIl}\sqrt{\frac{l^2 - a^2}{3}} \times \left[l^2 - a^2 - \left(\frac{l^2 - a^2}{3}\right)\right] = \frac{Wa}{6EIl}\sqrt{\frac{l^2 - a^2}{3}} \times \left[\frac{2}{3}(l^2 - a^2)\right]$$

$$y_{max} = \frac{Wa}{9\sqrt{3} \cdot EIl}(l^2 - a^2)^{3/2}$$

**EXAMPLE 19.3.** *A beam of uniform section of span l is simply supported at its ends. It is carrying a point load of W at a distance of l/3 from one end. Find the deflection of the beam under the load.*

**SOLUTION.** Given: Span $= l$; Point load $= W$ and distance between the point load and left end $(a)$ $= l/3$ or distance between point load and right end $(b) = l - l/3 = 2l/3$.

We know that deflection under the load

$$= \frac{Wab}{6EIl}(l^2 - a^2 - b^2) = \frac{W \times \frac{l}{3} \times \frac{2l}{3}}{6EIl} \times \left[l^2 - \left(\frac{l}{3}\right)^2 - \left(\frac{2l}{3}\right)^2\right]$$

$$= 0.0165\frac{Wl^3}{EI} \qquad \textbf{Ans.}$$

**EXAMPLE 19.4.** *A steel joist, simply supported over a span of 6 m carries a point load of 50 kN at 1.2 m from the left hand support. Find the position and magnitude of the maximum deflection. Take EI = 14 × $10^{12}$ N-mm².*

**SOLUTION.** Given: Span ($l$) = 6 m = 6 × $10^3$ mm ; Point load = ($W$) = 50 kN = 50 × $10^3$ N ; Distance between point load and left end ($a$) = 1.2 m = 1.2 × $10^3$ mm and flexural rigidity ($EI$) = 14 × $10^{12}$ N-mm².

***Position of the maximum deflection***

We know that position of the maximum deflection (or distance between the point of maximum deflection and left hand support),

$$x = \sqrt{\frac{l^2 - a^2}{3}} = \sqrt{\frac{(6 \times 10^3) - (1.2 \times 10^3)^2}{3}} \text{ mm}$$

$$= 3.39 \times 10^3 \text{ mm} = \mathbf{3.39\ m} \qquad \textbf{Ans.}$$

***Magnitude of the maximum deflection***

We also know that magnitude of the maximum deflection,

$$y_{max} = \frac{Wa}{9\sqrt{3} \cdot EIl}(l^2 - a^2)^{3/2}$$

$$= \frac{(50 \times 10^3) \times (1.2 \times 10^3)}{9\sqrt{3} \times (14 \times 10^{12}) \times (6 \times 10^3)} \times [(6 \times 10^3)^2 - (1.2 \times 10^3)^2]^{3/2} \text{ mm}$$

$$= (0.0458 \times 10^{-9}) \times (2.052 \times 10^{11}) = \mathbf{9.4\ mm} \qquad \textbf{Ans.}$$

**EXAMPLE 19.5.** *A simply supported beam AB of span 5 metres is carrying a point load of 30 kN at a distance 3.75 m from the left end A. Calculate the slopes at A and B and deflection under the load. Take EI = 26 × $10^{12}$ N-mm².*

**SOLUTION.** Given: Span ($l$) = 5 m = 5 × $10^3$ mm ; Point load ($W$) = 30 kN = 30 × $10^3$ N ; Distance between point load and left end ($a$) = 3.75 m = 3.75 × $10^3$ mm ; Distance between point load and right end ($b$) = 5 – 3.75 = 1.25 m = 1.25 × $10^3$ mm and flexural rigidity ($EI$) = 26 × $10^{12}$ N-mm².

***Slope at A***

We know that slope at $A$,

$$i_A = \frac{Wb}{6EIl}(l^2 - b^2)$$

$$= \frac{(30 \times 10^3) \times (1.25 \times 10^3)}{6\,(26 \times 10^{12}) \times (5 \times 10^3)} \times [(5 \times 10^3)^2 - (1.25 \times 10^3)^2 \text{ rad}$$

$$= (0.0481 \times 10^{-9}) \times (23.4375 \times 10^6) = \mathbf{0.00113\ rad} \qquad \textbf{Ans.}$$

***Slope at B***

We also know that slope at $B$,

$$i_B = \frac{Wa}{6EIl}(l^2 - a^2)$$

$$= \frac{(30 \times 10^3) \times (3.75 \times 10^3)}{6\,(26 \times 10^{12}) \times (5 \times 10^3)} \times [(5 \times 10^3)^2 - (3.75 \times 10^3)^2]$$

$$= (0.1442 \times 10^{-9}) \times (10.9375 \times 10^6) = \mathbf{0.00158\ rad} \qquad \textbf{Ans.}$$

***Deflection under the load***

We also know that deflection under the load,

$$y_C = \frac{Wb}{6EIl}(l^2 - a^2 - b^2)$$

$$= \frac{(30\times10^3)\times(3.75\times10^3)\times(1.25\times10^3)}{6\,(26\times10^{12})\times(5\times10^3)}$$

$$\times\,[(15\times10^3)^2-(3.75\times10^3)^2-(1.25\times10^3)^2]\text{ mm}$$

$$= (0.18\times10^{-6})\times(9.375\times10^6) = \mathbf{1.69\ mm} \qquad \textbf{Ans.}$$

## 19.8. Simply Supported Beam with a Uniformly Distributed Load

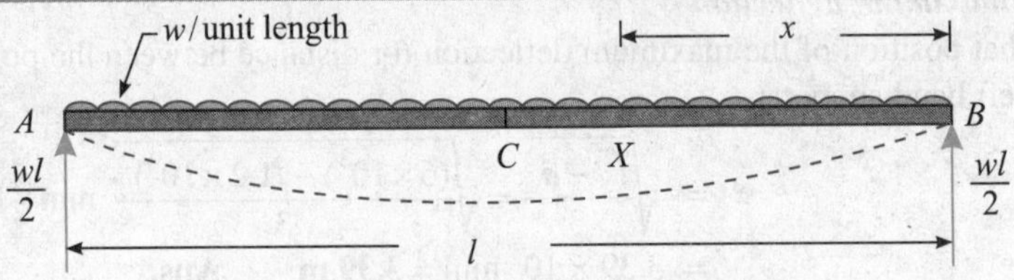


Fig. 19.6. Uniformly distributed load.

Consider a simply supported beam of length and carrying a uniformly distributed load of per unit length as shown in Fig. 19.6. From the geometry of the figure, we know that the reaction at $A$,

$$R_A = R_B = \frac{wl}{2}$$

Consider a section $X$ at a distance $x$ from $B$. We know that the bending moment at this section,

$$M_X = \frac{wlx}{2} - \frac{wx^2}{2} \qquad \text{...(Plus sign due to sagging)}$$

$$\therefore \quad EI\,\frac{d^2y}{dx^2} = \frac{wlx}{2} - \frac{wx^2}{2} \qquad ...(i)$$

Integrating the above equation,

$$EI\,\frac{dy}{dx} = \frac{wlx^2}{4} - \frac{wx^3}{6} + C_1 \qquad ...(ii)$$

where $C_1$ is the first constant of integration. We know when $x = \frac{l}{2}$, then $\frac{dy}{dx} = 0$

Substituting these values in the above equation,

$$0 = \frac{wl}{4}\left(\frac{l}{2}\right)^2 - \frac{w}{6}\left(\frac{l}{2}\right)^3 + C_1 = \frac{wl^3}{16} - \frac{wl^3}{48} + C_1$$

or

$$C_1 = -\frac{wl^3}{24}$$

Substituting this value of $C_1$ in equation (*ii*),

$$\therefore \quad EI\,\frac{dy}{dx} = \frac{wlx^2}{4} - \frac{wx^3}{6} - \frac{wl^3}{24} + C_1 \qquad ...(iii)$$

This is the required equation for the slope at any section. We know that maximum slope occurs at $A$ and $B$. Thus for maximum slope, substituting $x = 0$ in equation (*iii*),

$$EI \cdot i_B = -\frac{wl^3}{24}$$

... (Minus sign means that the tangent at $A$ makes an angle with $AB$ in the negative or anticlockwise direction)

$$\therefore \quad i_B = -\frac{wl^3}{24EI}$$

or
$$i_B = \frac{wl^3}{24EI}$$

By symmetry,
$$i_A = \frac{wl^3}{24EI}$$

Integrating the equation (*iii*) once again,

$$EI \cdot y = \frac{wlx^3}{12} - \frac{wx^4}{24} - \frac{wl^3 x}{24} + C_2 \quad \text{...(iv)}$$

where $C_2$ is the second constant of integration. We know when $x = 0$, then $y = 0$. Substituting these values in equation (*iv*), we get $C_2 = 0$

$$\therefore \quad EI \cdot y = \frac{wlx^3}{12} - \frac{wx^4}{24} - \frac{wl^3 x}{24} \quad \text{...(v)}$$

This is the required equation for the deflection at any section. We know that maximum deflection occurs at the mid-point *C*. Thus maximum deflection, substituting $x = l/2$ in equation (*v*),

$$EI \cdot y_C = \frac{wl}{12}\left(\frac{l}{2}\right)^3 - \frac{w}{24}\left(\frac{l}{2}\right)^4 - \frac{wl^3}{24}\left(\frac{l}{2}\right) = \frac{wl^4}{96} - \frac{wl^4}{384} - \frac{wl^4}{48} = -\frac{5wl^4}{384}$$

or
$$y_C = -\frac{5wl^4}{384EI}$$
...(Minus sign means that the deflection is downwards)

$$= \frac{5wl^4}{384EI}$$

**Note.** The above expression for slope and deflection may also be expressed in terms of total load. Such that $W = wl$.

$$\therefore \quad i_B = i_A = \frac{wl^3}{24EI} = \frac{wl^2}{24EI}$$

and
$$y_C = \frac{5wl^3}{384EI}$$

**Example 19.6.** *A simply supported beam of span 4 m is carrying a uniformly distributed load of 2 kN/m over the entire span. Find the maximum slope and deflection of the beam. Take EI for the beam as 80 × 10⁹ N-mm².*

**Solution.** Given: Span $(l) = 4\text{ m} = 4 \times 10^3\text{ mm}$ ; Uniformly distributed load $(w) = 2\text{ kN/m} = 2\text{ N/mm}$ and flexural rigidity $(E) = 80 \times 10^9\text{ N-mm}^2$.

***Maximum slope of the beam***

We know that maximum slope of the beam,

$$i_A = \frac{wl^3}{24EI} = \frac{2 \times (4 \times 10^3)^3}{34 \times (80 \times 10^9)} = \mathbf{0.067\ rad} \quad \textbf{Ans.}$$

***Maximum deflection of the beam***

We also know that maximum deflection of the beam,

$$y_C = \frac{5wl^4}{384EI} = \frac{5 \times 2 \times (4 \times 10^3)^4}{384 \times (80 \times 10^9)} = \mathbf{83.3\ mm} \quad \textbf{Ans.}$$

**EXAMPLE 19.7.** *A simply supported beam of span 6 m is subjected to a uniformly distributed load over the entire span. If the deflection at the centre of the beam is not to exceed 4 mm, find the value of the load. Take E = 200 GPa and I = 300 × 10⁶ mm⁴.*

**SOLUTION.** Given: Span (l) = 6 m = $6 \times 10^3$ mm ; Deflection at the centre ($y_C$) = 4 mm ; modulus of elasticity ($E$) = 200 GPa = $200 \times 10^3$ N/mm² and moment of inertia ($I$) = $300 \times 10^6$ mm⁴.

Let $w$ = Value of uniformly distributed load in N/mm or kN/m.

We know that deflection at the centre of the beam ($y_C$),

$$4 = \frac{5wl^4}{384EI} = \frac{5 \times w \times (6 \times 10^3)^4}{384 \times (200 \times 10^3) \times (300 \times 10^6)} = 0.281\, w$$

$$\therefore \quad w = \frac{4}{0.281} = \mathbf{14.2\ kN/m} \qquad \textbf{Ans.}$$

**EXAMPLE 19.8.** *A timber beam of rectangular section has a span of 4.8 metres and is simply supported at its ends. It is required to carry a total load of 45 kN uniformly distributed over the whole span. Find the values of the breadth (b) and depth (d) of the beam, if maximum bending stress is not to exceed 7 MPa and maximum deflection is limited to 9.5 mm. Take E for timber as 10.5 GPa.*

**SOLUTION.** Given: Span ($l$) = 4.8 m = $4.8 \times 10^3$ mm ; Total load ($W$) = ($wl$) = 45 kN = $45 \times 10^3$ N; Maximum bending stress $\sigma_{b\,(max)}$ = 7 MPa = 7 N/mm² ; Maximum deflection ($y_C$) = 9.5 mm and modulus of elasticity ($E$) = 10.5 GPa = $10.5 \times 10^3$ N/mm².

Let $b$ = Breadth of the beam and

$d$ = Depth of the beam.

We know that in a simply supported beam, carrying a uniformly distributed load, the maximum bending moment,

$$M = \frac{wl^2}{8} = \frac{wl \times l}{8} = \frac{W \times l}{8} = \frac{45 \times 4.8}{8}$$

$$= 27 \text{ kN-m} = 27 \times 10^6 \text{ N-mm}$$

and moment of inertia of a rectangular section,

$$I = \frac{bd^3}{12}$$

We also know that distance between the neutral axis of the section and extreme fibre,

$$y = \frac{d}{2}$$

∴ Maximum bending stress [$\sigma_{b\,(max)}$],

$$7 = \frac{M}{I} \times y = \frac{27 \times 10^6}{\frac{bd^3}{12}} \times \frac{d}{2} = \frac{162 \times 10^6}{bd^2}$$

or

$$bd^2 = \frac{162 \times 10^6}{7} = 23.14 \times 10^6$$

We know that maximum deflection ($y_C$),

$$9.5 = \frac{5wl^4}{384EI} = \frac{5\,(wl)\,l^3}{384EI} = \frac{5\,(45 \times 10^3) \times (4.8 \times 10^3)^3}{384 \times (10.5 \times 10^3) \times \frac{bd^3}{12}} = \frac{74.1 \times 10^9}{bd^3}$$

$$\therefore \qquad bd^3 = \frac{74.1 \times 10^9}{9.5} = 7.8 \times 10^9$$

Dividing equation (*ii*) by equation (*i*),

$$d = \frac{7.8 \times 10^9}{23.14 \times 10^6} = \mathbf{337\ mm} \qquad \textbf{Ans.}$$

Substituting this value of *d* in equation (*i*),

$$b \times (337)^2 = 23.14 \times 106$$

$$\therefore \qquad b = \frac{23.14 \times 10^6}{(337)^2} = \mathbf{204\ mm} \qquad \textbf{Ans.}$$

## 19.9. Simply Supported Beam with a Gradually Varying Load

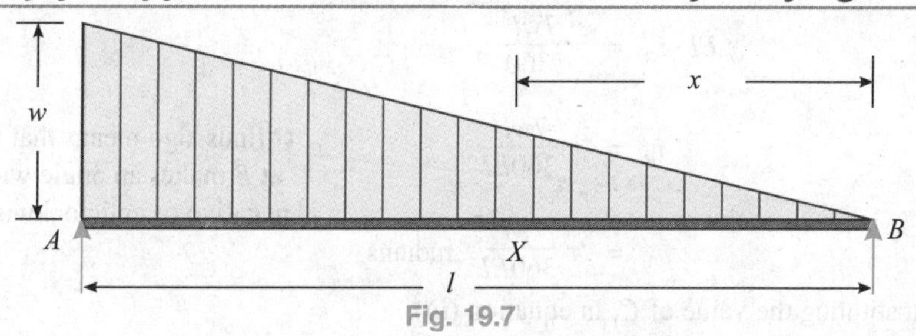


Fig. 19.7

Consider a simply supported beam *AB* of length *l* and carrying a gradually varying load from zero at *B* to *w* per unit length at *A* as shown in Fig. 19.7. From the geometry of the figure, we find that the reaction at *A*,

$$R_A = \frac{wl}{3} \qquad \text{and} \qquad R_B = \frac{wl}{6}$$

Now consider a section *X* at a distance *x* from *B*. We know that the bending moment at this section,

$$M_X = R_B \cdot x - \left( \frac{wx}{l} \times \frac{x}{2} \times \frac{x}{3} \right) = \frac{wlx}{6} - \frac{wx^3}{6l}$$

$$\therefore \qquad EI \frac{d^2y}{dx^2} = \frac{wlx}{6} - \frac{wx^3}{6l} \qquad ...(i)$$

Integrating the above equation,

$$EI \frac{dy}{dx} = \frac{wlx^2}{12} - \frac{wx^4}{24l} + C_1 \qquad ...(ii)$$

where $C_1$ is the first constant of integration. Integrating the equation (*ii*) once again,

$$EI \cdot y = \frac{wlx^3}{36} - \frac{wx^5}{120l} + C_1 x + C_2 \qquad ...(iii)$$

where $C_2$ is the second constant of integration. We know that when $x = 0$, then $y = 0$. Therefore $C_2 = 0$. We also know that when $x = l$, then $y = 0$. Substituting these values in equation (*iii*),

$$0 = \frac{wl}{36} \times l^3 - \frac{w}{120l} \times l^5 + C_1 l = \frac{wl^4}{36} - \frac{wl^4}{120} + C_1 l$$

$$\therefore \qquad C_1 = \frac{wl^3}{36} + \frac{wl^3}{120} = -\frac{7wl^3}{360}$$

Now substituting this value of $C_1$ in equation (*ii*),

$$EI\frac{dy}{dx} = \frac{wlx^2}{12} - \frac{wx^4}{23l} - \frac{7wl^3}{360} \qquad ...(iv)$$

This is the required equation for slope at any section, by which we can get the slope at any section on the beam. A little consideration will show that the maximum slope will be either at the support *A* or *B*. Thus for slope at A, substituting $x = 1$ in equation (*iv*),

$$EI \cdot i_A = \frac{wl}{12} \times l^2 - \frac{w}{24l} \times l^4 - \frac{7wl^3}{360} = \frac{wl^3}{45}$$

$$\therefore \quad i_A = \frac{wl^3}{45EI}$$

Now for slope at *B*, substituting $x = 0$ in equation (*iv*),

$$EI \cdot i_B = -\frac{7wl^3}{360}$$

$$\therefore \quad i_B = -\frac{7wl^3}{360EI}$$

... (Minus sign means that the tangent at *B* makes an angle with *AB* in the negative or anticlockwise direction)

$$= -\frac{7wl^3}{360EI} \text{ radians}$$

Now substituting the value of $C_1$ in equation (*iii*),

$$EI \cdot y = -\frac{wlx^3}{36} - \frac{wx^5}{120l} - \frac{7wl^3x}{360}$$

$$\therefore \quad y = \frac{1}{EI}\left(\frac{wlx^3}{36} - \frac{wx^5}{120l} - \frac{7wl^3x}{360}\right)$$

This is the required equation for the deflection at any section, by which we can get deflection at any section on the beam. For deflection at the centre of the beam, substituting $x = l/2$ in equation (*v*),

$$y_C = \frac{1}{EI}\left[\frac{wl}{36}\left(\frac{l}{2}\right)^3 - \frac{w}{120l}\left(\frac{l}{2}\right) - \frac{7wl^3}{360}\left(\frac{l}{2}\right)\right]$$

$$= -\frac{0.00651wl^4}{EI}$$

... (Minus sign means that the deflection is downwards)

$$= \frac{0.00651wl^4}{EI}$$

We know that the maximum deflection will occur, where slope of the beam is zero. Therefore equating the equation (*iv*) to zero,

$$\frac{wlx^2}{12} - \frac{wx^4}{24l} - \frac{7wl^3}{360} = 0$$

$$\therefore \quad x = 0.519\,l$$

Now substituting this value of *x* in equation (*v*),

$$y_{max} = \frac{1}{EI}\left[\frac{wl}{36}(0.519l)^3 - \frac{w}{120l}(0.519l)^5 - \frac{7wl^3}{360}(0.519l)\right]$$

$$= -\frac{0.006\,52wl^4}{EI}$$

... (Minus sign means that the deflection is downwards)

**EXAMPLE 19.9.** *A simply supported beam AB of span 4 metres is carrying a triangular load varying from zero at A to 5 kN/m at B. Determine the maximum deflection of the beam. Take rigidity of the beam as $1.25 \times 10^{12}$ N-mm$^2$.*

**SOLUTION.** Given: Span $(l) = 4 \text{ m} = 4 \times 10^3$ mm ; Load at $A = (w) = 5$ kN/m $= 5$ N/mm and flexural rigidity $(EI) = 1.25 \times 10^{12}$ N-mm$^2$.

We know that maximum deflection of the beam

$$y_{max} = \frac{0.006\,52wl^4}{EI} = \frac{0.006\,52 \times 5 \times (4 \times 10^3)^4}{1.25 \times 10^{12}} = \mathbf{6.68\ mm} \qquad \textbf{Ans.}$$

## EXERCISE 19.1

1. A simply supported beam of span 2.4 m is subjected to a central point load of 15 kN. What is the maximum slope and deflection at the centre of the beam? Take $EI$ for the beam as $6 \times 10^{10}$ N-mm$^2$. **[Ans.** 0.09 rad ; 72 mm]
2. A beam 3 m long, simply supported at its ends, is carrying a point load at its centre. If the slope at the ends of the beam is not to exceed 1°, find the deflection at the centre of the beam. **[Ans.** 17.5 mm]

**Hint:** $$y_C = \frac{wl^3}{48EI} = \frac{wl^2}{16EI} \times \frac{l}{3} = i_A \times \frac{l}{3} = \frac{1 \times \pi}{180} \times \frac{3 \times 10^3}{3} = 0.0175 \times 10^3 = 17.5 \text{ mm}$$

3. A rolled steel beam simply supported over a span of 6 m carries a point load of 40 kN at a distance of 4 m from left end supports. What is the position of the maximum deflection of the beam. Take $E$ as 200 GPa and $I = 70 \times 10^6$. **[Ans.** 2.58 m from the left end]
4. A simply supported beam of 3 m span is subjected to a point load of 40 kN at a distance of 1 m from the left end. Determine the deflection of the beam under the load. Take $EI$ for the beam as $12 \times 10^9$ N-mm$^2$. **[Ans.** 1.5 mm]
5. A simply supported beam $AB$ of span 4 m is subjected to a point load of 40 kN at a distance of 1 m from $A$. Determine the slopes at both the ends $A$ and $B$. Take $EI = 500 \times 10^{12}$ N-mm$^2$. **[Ans.** 0.07 rad ; 0.05 rad]
6. A beam simply supported at its both ends carries a uniformly distributed load of 16 kN/m. If the deflection of the beam at its centre is limited to 2.5 mm, find the span of the beam. Take $EI$ for the beam as $9 \times 10^{12}$ N-mm$^2$. **[Ans.** 3.22 m]

## 19.10. Macaulay's Method* for Slope and Deflection

We have seen in the previous articles and examples that the problems of deflections in beams are bit tedious and laburious, specially when the beam is carrying some point loads. Mr. W.H. Macaulay devised a method, a continuous expression, for bending moment and integrating it in such a way, that the constants of integration are valid for all sections of the beam ; even though the law of bending moment varies from section to section. Now we shall discuss the application of Macaulay's method for finding out the slopes and deflection of a few types of beams:

**NOTES.** The following rules are observed while using Macaulay's method:

1. Always take origin on the extreme left of the beam.
2. Take left clockwise moment as negative and left anticlockwise moment as positive.
3. While calculating the slopes and deflections, it is convenient to use the values first in terms of kN and metres.

* This method was original proposed by Mr. A. Clebsch, which was further developed by Mr. W.H. Macaulay.

***(i) Simply supported beam with a central point load.***

Consider a simply supported beam *AB* of length *l* and carrying a point load *W* at the centre of the beam *C* as shown in Fig. 19.8.

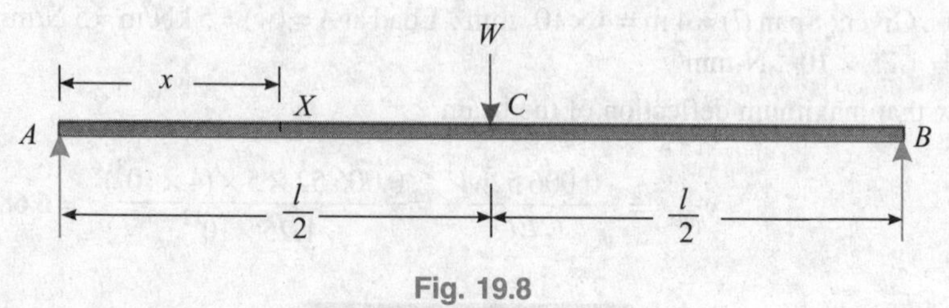


**Fig. 19.8**

Take *A* as the origin. We know that bending moment at any point, in section *AC* at a distance *x* from *A*,

$$M_X = -\frac{W}{2}x \qquad \text{....(Minus sign due to left clockwise)}$$

and the bending moment at any point in section *CB* and at a distance *x* from *A*,

$$M_X = -\frac{W}{2}x + W\left(x - \frac{1}{2}\right) \qquad ...(i)$$

Thus we can express the bending moment, for all the sections of the beam in a single equation, *i.e.*,

$$M_X = -\frac{W}{2}x \;\vdots\; + W\left(x - \frac{1}{2}\right)$$

For any point in section *AC*, stop at the dotted line, and for any point in section *CB* add the expression beyond the dotted line also.

Now re-writing the above equation,

$$EI\frac{d^2y}{dx^2} = -\frac{Wx}{2} \;\vdots\; + W\left(x - \frac{1}{2}\right) \qquad ...(ii)$$

Integrating the above equation,

$$EI\frac{dy}{dx} = -\frac{Wx^2}{4} + C_1 \;\vdots\; + \frac{W}{2}\left(x - \frac{1}{2}\right) \qquad ...(iii)$$

It may be noted that the integration of $\left(x - \frac{l}{2}\right)$ has been made as a whole and not for individual terms for the expression. This is only due to this simple integration that the Macaulay's method is more effective. This type of integration is also justified as the constant of integration $C_1$ is not only valid for the section *AC*, but also for section *CB*.

Integrating the equation (*iii*) once again,

$$EI \cdot y = -\frac{Wx^3}{12} + C_1 x + C_2 \;\vdots\; + \frac{W}{6}\left(x - \frac{l}{2}\right)^3 \qquad ...(iv)$$

It may again be noted that the integration of $\left(x-\frac{l}{2}\right)^2$ has again been made as a whole and not for individual terms. We know that when $x = 0$, then $y = 0$. Substituting these values in equation (*iv*), we find $C_2 = 0$. We also know that when $x = l$, then $y = 0$. Substituting these values of $x$ and $y$ and $C_2 = 0$ in equation (*iv*),

$$0 = -\frac{Wl^3}{12} + C_1 l + \frac{W}{6}\left(\frac{l}{2}\right)^3$$

$$\therefore \qquad C_1 l = -\frac{Wl^3}{12} - \frac{Wl^3}{48} = \frac{3Wl^3}{48} = \frac{Wl^3}{16}$$

or

$$C_1 = \frac{Wl^2}{16}$$

Now substituting this value of $C_1$ in equation (*iii*),

$$\therefore \qquad EI\frac{dy}{dx} = \frac{Wx^2}{4} + \frac{Wl^2}{16} \;\vdots\; + \frac{W}{2}\left(x-\frac{l}{2}\right)^2$$

This is the required equation for slope at any section. We know that maximum slope occurs at $A$ and $B$. Thus for maximum slope at $A$, substituting $x = 0$ in equation ($v$) upto the dotted line only,

$$EI \cdot i_A = \frac{WL^2}{16}$$

$$\therefore \qquad i_A = \frac{Wl^2}{16EI} \qquad \text{...(As before)}$$

By symmetry,

$$i_B = \frac{Wl^2}{16EI} \qquad \text{...(As before)}$$

Substituting the value of $C_1$ again in equations (*iv*) and $C_2 = 0$,

$$EI \cdot y = -\frac{Wx^3}{12} + \frac{Wl^2 x}{16} \;\vdots\; + \frac{W}{6} + \left(x-\frac{l}{2}\right)^3 \qquad \text{...(vi)}$$

This is required equation for deflection at any section. We know that maximum deflection occurs at $C$. Thus for maximum deflection, substituting $x = l/2$ in equation (*vi*) for the portion $AC$ only (remembering that $C$ lies in $AC$),

$$EI \cdot y_C = -\frac{W}{12}\left(\frac{l}{2}\right)^3 \;\vdots\; + \frac{Wl^2}{16}\left(\frac{l}{2}\right) = \frac{Wl^3}{48}$$

or

$$y_C = \frac{Wl^3}{48EI} \qquad \text{...(As before)}$$

***(ii) Simply supported beam with aneccentric point load.***

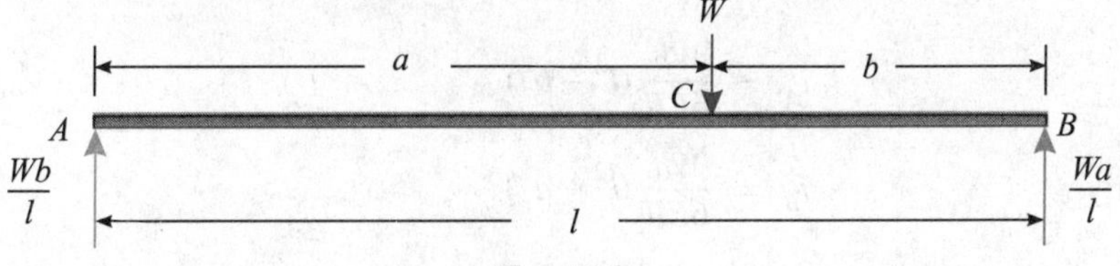


Fig. 19.9

Consider a simply supported beam $AB$ of length $l$ and carrying an eccentric point load $W$ at $C$ such that $AC = a$ and $CB = b$ as shown in Fig. 19.9. Take $A$ as the origin. The bending moment at any point in section $AC$ at a distance $x$ from $A$,

$$M_X = -\frac{Wb}{l}x \quad \text{(Minus sign due to left clockwise)}$$

Bending moment at any point in section $CB$ at a distance $x$ from $A$,

$$M_X = -\frac{Wbx}{l} \;\vdots\; + W(x-a) \quad ...(i)$$

or

$$EI\frac{d^2y}{dx^2} = \frac{Wbx}{l} \;\vdots\; + W(x-a) \quad ...(ii)$$

Integrating the above equation,

$$EI\frac{dy}{dx} = -\frac{Wbx^2}{2l} + C_1 \;\vdots\; + \frac{W(x-a)^2}{2} \quad ...(iii)$$

Integrating the above equation once again,

$$EIy = -\frac{Wbx^3}{6l} + C_1x + C_2 \;\vdots\; + \frac{W(x-a)^3}{6} \quad ....(iv)$$

We know that when $x = 0$, then $y = 0$. Substituting these values in equation (*iv*) upto the dotted line only. Therefore $C_2 = 0$. We also know that when $x = l$, then $y = 0$. Substituting these values again in equation (*iv*) and $C_2 = 0$.

$$EI \cdot y = -\frac{Wb}{6l}(l)^3 + C_1l \;\vdots\; + \frac{W(l-a)^3}{6}$$

$$= -\frac{Wbl^2}{6} + C_1l \;\vdots\; + \frac{Wb^3}{6} \quad ...[\because (l-a) = b]$$

$$\therefore \quad C_1l = \frac{Wbl^2}{6} + \frac{Wb^3}{6} = \frac{Wb}{6}(l^2 - b^2)$$

or

$$C_1 = \frac{Wb}{6l}(l^2 - b^2)$$

Now substituting this value of $C_1$ in equation (*iii*),

$$EI\frac{dy}{dx} = -\frac{Wbx^2}{2l} + \frac{Wb}{6l}(l^2 - b^2) \;\vdots\; + \frac{W(x-a)^2}{2} \quad ...(v)$$

This is the required equation for slope at any point. We know that slope is maximum at $A$ or $B$. Substituting $x = 0$ upto dotted line only (remembering that $C$ lies in $AC$),

$$EI \cdot i_A = \frac{Wb}{6l}(l^2 - b^2)$$

or

$$i_A = \frac{Wb}{6EIl}(l^2 - b^2) \quad ...\text{(As before)}$$

Similarly,

$$i_B = \frac{Wa}{6EIl}(l^2 - a^2)$$

Substituting the value of $C_1$ again in equation (*iv*) and $C_2 = 0$,

$$EI \cdot y = -\frac{Wbx^3}{6l} + \frac{Wbx}{6l}(l^2 - b^2) \;\vdots\; + \frac{W(x-a)^3}{6}$$

This is the required equation for deflection at any point. For deflection in *AC*, consider the equation up to the dotted line only,

$$EI \cdot y = -\frac{Wbx^3}{6l} + \frac{Wbx}{6l}(l^2 - b^2) = \frac{Wbx}{6l}(l^2 - b^2 - x^2)$$

$$\therefore \quad y = \frac{Wbx}{6EIl}(l^2 - b^2 - x^2) \quad \text{... (As before)}$$

**Note.** The Macaulay's method may also be used for cantilever beams or for beams subjected to some moment.

**Example 19.10.** *A horizontal steel girder having uniform cross-section is 14 m long and is simply supported at its ends. It carries two concentrated loads as shown in Fig. 19.10.*

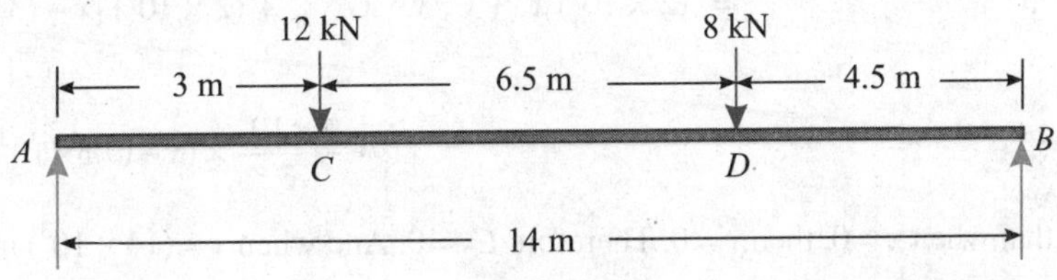


Fig. 19.10

*Calculate the deflections of the beam under the loads C and D. Take E = 200 GPa and I = 160 × $10^6$ mm$^4$.*

**Solution.** Given: Span (l) = 14 m = 14 × $10^3$ mm ; Load at *C* $(W_1)$ = 12 kN = 12 × $10^3$ N ; Load at *D* $(W_2)$ = 8 kN = 8 × $10^3$ N ; Modulus of elasticity (*E*) = 200 GPa = 200 × $10^3$ N/mm$^2$ and moment of inertia (*I*) = 160 × $10^6$ mm$^4$.

Taking moments about *A* and equating the same,

$$R_B \times 14 = (12 \times 3) + (8 \times 9.5) = 112$$

$$\therefore \quad R_B = \frac{112}{14} = 8 \text{ kN} = 8 \times 10^3 \text{ N}$$

and $\quad R_A = (12 + 8) - 8 = 12 \text{ kN} = 12 \times 10^3 \text{ N}$

Now taking *A* as the origin and using Macaulay's method, the bending moment at any section *X* at a distance *x* from *A*,

$$EI\frac{d^2y}{dx^2} = -(12 \times 10^3)\,x + \;\vdots\; (12 \times 10^3) \times [x - (3 \times 10^3)]$$
$$+ (8 \times 10^3) \times [x - (9.5 \times 10^3)]$$

Integrating the above equation,

$$EI\frac{dy}{dx} = -(12\, t\, 10^3)\frac{x^2}{2} + \;\vdots\; C_1 + (12 \times 10^3) \times \frac{[x - (3 \times 10^3)]^2}{2}$$
$$+ (8 \times 10^3) \times \frac{[x - (9.5 \times 10^3)]^2}{2}$$

$$= -(6 \times 10^3)\, x^2 + C_1 \quad \vdots \quad + (6 \times 10^3) \times [x - (3 \times 10^3)]^2$$

$$\vdots \quad + (4 \times 10^3) \times [x - (9.5 \times 10^3)] \qquad ...(i)$$

Integrating the above equation once again,

$$EI \cdot y = -(6 \times 10^3) \times \frac{x^3}{3} + C_1\, x + C_2 + \quad \vdots \quad (6 \times 10^3) \times \frac{[x - (3 \times 10^3)]^3}{3}$$

$$\vdots \quad + (4 \times 10^3) \times \frac{[x - (9.5 \times 10^3)]^3}{3}$$

$$= (2 \times 10^3)\, x^3 + C_1\, x + C^2 \quad \vdots \quad + (2 \times 10^3)\, [x - (3 \times 10^3)]^3$$

$$\vdots \quad + \frac{4 \times 10^3}{3} \times (x - (9.5 \times 10^3)]^3 \qquad ...(ii)$$

We know that when $x = 0$, then $y = 0$. Therefore $C_2 = 0$. And when $x = (14 \times 10^3)$ mm, then $y = 0$. Therefore

$$0 = -(2 \times 10^3) \times (14 \times 10^3)^3 + C_1 \times (14 \times 10^3)$$

$$+ (2 \times 10^3) \times [(14 \times 10^3) - (3 \times 10^3)]^3$$

$$+ \frac{4 \times 10^3}{3} \times [(14 \times 10^3)\ (9.5 \times 10^3)]^3$$

$$= -(5488 \times 10^{12}) + (14 \times 10^3)\, C_1 + (2662 \times 10^{12}) + 121.5 \times 10^{12}$$

$$= -(2704.5 \times 10^{12}) + (14 \times 10^3)\, C_1$$

$$\therefore \quad C_1 = \frac{2704.5 \times 10^{12}}{14 \times 10^3} = 193.2 \times 10^9$$

Substituting the value of $C_1$ equal to $193.2 \times 10^9$ and $C_2 = 0$ in equation (*ii*),

$$EIy = -2 \times 10^3\, x^3 + 193.2 \times 10^9\, x \quad \vdots \quad + 2 \times 10^3\, [x - (3 \times 10^3)]3$$

$$\vdots \quad + \frac{4 \times 10^3}{3} \times [x - (9.5 \times 10^3)]^3 \qquad ...(iii)$$

Now for deflection under the 12 kN load, substituting $x = 3$ m ( or $3 \times 10^3$ mm) in equation (*iii*) up to the first dotted line only,

$$EIy_C = -2 \times 10^3 \times (3 \times 10^3)^3 + 193.2 \times 10^9 \times (3 \times 10^3)$$

$$= -(54 \times 10^{12}) + (579.6 \times 10^{12}) = 525.6 \times 10^{12}$$

$$\therefore \quad y_C = \frac{525.6 \times 10^{12}}{EI} = \frac{525.6 \times 10^{12}}{(200 \times 10^3) \times (160 \times 10^6)} = \mathbf{16.4\ mm} \qquad \textbf{Ans.}$$

Similarly, for deflection under the 8 kN load, substituting $x = 9.5$ m (or $9.5 \times 10^3$ mm) in equation (*iii*) up to the second dotted line only,

$$EI\, y_D = -2 \times 10^3 \times (9.5 \times 10^3)^3 + 193.2 \times 10^9 \times (9.5 \times 10^3) + 2 \times 10^3 \times [(9.5 \times 10^3) - (3 \times 10^3)]^3$$

$$= -(1714.75 \times 10^{12}) + (1835.4 \times 10^{12}) + (549.25 \times 10^{12})$$

$$= 669.9 \times 10^{12}$$

$$\therefore \quad y_D = \frac{669.9 \times 10^{12}}{EI} = \frac{669.9 \times 10^{12}}{(200 \times 10^3) \times (160 \times 10^6)} = \mathbf{20.9\ mm} \quad \textbf{Ans.}$$

**EXAMPLE 19.11.** *A horizontal beam AB is freely supported at A and B, 8 m apart and carries a uniformly distributed load of 15 kN/m run (including its own weight). A clockwise moment of 160 kN-m is applied to the beam at a point C, 3 m from the left hand support A. Calculate the slope of the beam at C, if EI = 40 × 10³ kN-m².*

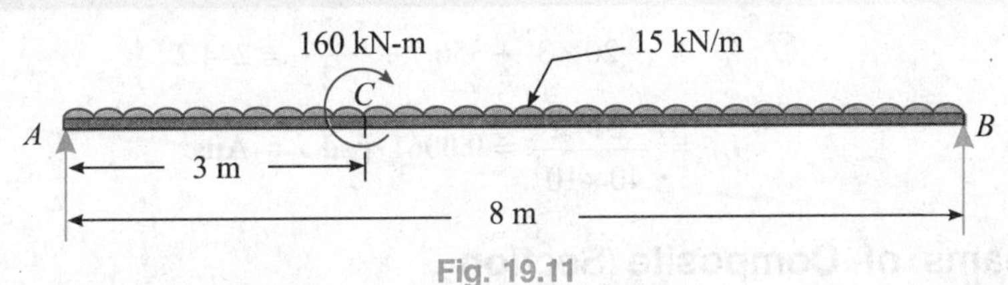


Fig. 19.11

**SOLUTION.** Given: Span ($l$) = 8 m ; Uniformly distributed load ($w$) = 15 kN/m ; Moment at $C$ ($\mu$) = 160 kN-m (clockwise) and flexural rigidity ($EI$) = $40 \times 10^3$ kN-m².

Taking moments about $A$ and equating the same,

$$R_B \times 8 = (15 \times 8 \times 4) + 160 = 640 \text{ kN-m}$$

$$\therefore \quad R_B = \frac{640}{8} = 80 \text{ kN}$$

and

$$R_A = (15 \times 8) - 80 = 40 \text{ kN}$$

Let $\quad i_C$ = Slope at $C$.

Taking $A$ as origin and using Macaulay's method, the bending moment at any section $X$ at a distance $x$ from $A$,

$$EI \frac{d^2 y}{dx^2} = -40x \; \vdots \; + 15x \times \frac{x}{2} - 160\,(x-3)$$

$$= -40x \; \vdots \; \frac{15x^2}{2} - 160\,(x-3)$$

Integrating the above equation,

$$EI \frac{dy}{dx} = -40\frac{x^2}{2} + C_1 \; \vdots \; + \frac{15x^3}{6} \; \vdots \; -160\,(x-3)$$

$$= -20x^2 + C_1 \; \vdots \; + \frac{5x^3}{2} \; \vdots \; -160\,(x-3) \qquad \text{...}(i)$$

Integrating the above equation once again,

$$EI \cdot y = -\frac{20x^3}{3} + C_1 x + C_2 \; \vdots \; + \frac{5x^4}{8} - \frac{160\,(x-3)^2}{2} \qquad \text{...}(ii)$$

We know that when $x = 0$, then $y = 0$. Therefore $C_2 = 0$ and when $x = 8$, then $y = 0$. Therefore

$$0 = \frac{-20 \times (8)^3}{3} + (C_1 \times 8) + \frac{5 \times (8)^4}{8} - \frac{160 \times (5)^2}{2}$$

$$= 8C_1 - 2853.3$$

$$\therefore \quad C_1 = \frac{2853.3}{8} = 356.7$$

Substituting the values of $C_1 = 356.7$ and $C_2 = 0$ in equation (*i*),

$$EI\frac{dy}{dx} = -20x^2 + 356.7 \;\vdots\; + \frac{5x^3}{2} \;\vdots\; - 160\,(x-3)$$

Now for the slope at *C*, substituting $x = 3$ m in the above equation up to *C i.e.*, neglecting the *last term.

$$EI \cdot i_C = -20 \times 3^2 + 356.7 + \frac{5 \times 3^3}{2} = 244.2$$

$$\therefore \quad i_C = \frac{244.2}{40 \times 10^3} = \mathbf{0.0061\ rad} \qquad \textbf{Ans.}$$

## 19.11. Beams of Composite Section

It is a beam made up of two or more different materials, joined together in such a manner, that they behave like a single piece, and the deflection of each piece is equal.

The slope and deflection of such a beam, is found out by algebraically adding the flexural rigidities of the two or more different materials, in the application of the respective relation. Mathematically,

$$\Sigma EI = E_1 I_1 + E_2 I_2$$

**Note.** The moment of inertia of the composite section is to be found out about the c.g. of the section.

**Example 19.12.** *A composite beam of span 8 m consists of a timber section 180 mm wide and 300 mm deep. Two steel plates 180 mm long and 20 mm thick are fixed at the top and bottom of the timber section. The composite beam in subjected to a point load of 100 kN at middle of the beam. Determine the deflection of the beam under the load. Take E for steel and timber as 200 GPa and 10 GPa respectively.*

**Solution.** Given: Span $(l) = 8\text{ m} = 8 \times 10^3$ mm ; Timber section = 180 mm wide and 300 mm deep ; Steel plates = 180 mm × 20 mm; Point load $(W) = 100\text{ kN} = 100 \times 10^3$ N ; *E* for steel $(E_s) = 200$ GPa $= 200 \times 10^3$ N-mm$^2$ and *E* for timber $(E_t) = 10$ GPa $= 10 \times 10^3$ N-mm$^2$.

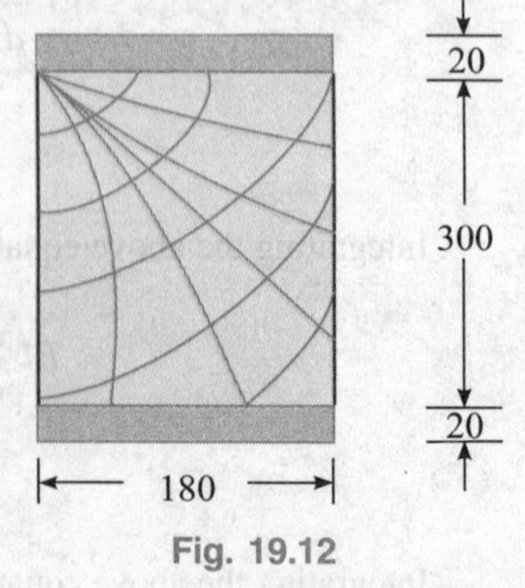

Fig. 19.12

From the geometry of the composite beam, we find that the centre of gravity of the composite section coincides with the centre of gravity of the timber section. Therefor flexural rigidity for the timber section about its centre of gravity,

$$EI_{(timber)} = (10 \times 10^3) \times \left[\frac{180 \times (300)^3}{12}\right] \text{N-mm}^2$$

$$= 4050 \times 10^9 \text{ N-mm}^2$$

* If, however, it is included by mistake, its value will be zero.

Similarly, $$E_{(steel)} = (200 \times 10^3) \times \left[ 2 \times \frac{180 \times (20)^3}{12} + 12\,(180 \times 20) \times (160)^2 \right] \text{N-mm}^2$$
$$= (200 \times 10^3) \times [(0.24 \times 10^6) + (184.3) \times 10^6] \text{ N-mm}^2$$
$$= 36910 \times 10^9 \text{ N-mm}^2$$

$\therefore$ Total flexural rigidity for the composite section about its centre of gravity
$$\Sigma\, EI = (4050 \times 10^9) + (36910 \times 10^9) = 40962 \times 10^9 \text{ N-mm}^2$$

We know that deflection at the centre of the beam,
$$y_C = \frac{Wl^3}{48\,\Sigma\, EI} = \frac{(100 \times 10^3) \times (8 \times 10^3)^3}{48 \times (40960 \times 10^9)} = \mathbf{26\ mm} \qquad \textbf{Ans.}$$

**EXAMPLE 19.13.** *A flitched beam consists of two timber joists 120 mm wide and 300 mm deep with a steel plate 250 mm deep and 12 mm thick fixed symmetrically between the timber joists. The beam carries a uniformly distributed load of 5 kN per metre and is simply supported over a span of 6 metres. If $E_s$ and $E_t$ are 200 GPa and 10 GPa respectively, find the slopes at the supports and deflection of the beam at its centre.*

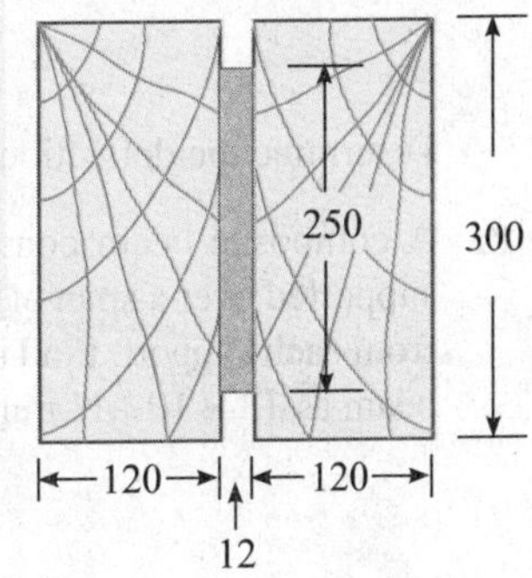

**SOLUTION.** Given: Timber joists = 120 mm wide and 300 mm deep ; Steel plate = 250 mm deep and 12 mm thick ; Uniformly distributed load ($w$) = 5 kN/m = 5 N/mm ; Span ($l$) = 6 m = $6 \times 10^3$ mm $E$ for steel ($Es$) = 200 GPa = $200 \times 10^3$ N/mm$^2$ and $E$ for timber ($E_t$) = 10 GPa = $10 \times 10^3$ N/mm$^2$.

Fig. 19.13

From the geometry of the flitched beam, we find that the centre of gravity of the beam section coincides with the centre of gravity of the steel plate. Therefore flexural rigidity for the timber joists,
$$EI_{(timber)} = (10 \times 10^3) \left( 2 \times \frac{120 \times (300)^3}{12} \right)$$
$$= 5400 \times 10^9 \text{ N-mm}^2$$

Similarly,
$$EI_{(steel)} = 200 \times 10^3 \left( \frac{12 \times (250)^3}{12} \right)$$
$$= 3125 \times 10^9 \text{ N-mm}^2$$

$\therefore$ Total flexural rigidity of the flitched beam about its centre of gravity,
$$\Sigma\, EI = 5400 \times 10^9 + 3125 \times 10^9 = 8525 \times 10^9 \text{ N-mm}^2$$

***Slope at the supports***

We know that slope at the support,
$$i_A = \frac{wl^3}{24\,\Sigma\, EI} = \frac{5 \times (6 \times 10^3)^3}{24 \times (8525 \times 10^9)} = \mathbf{0.0053\ rad} \qquad \textbf{Ans.}$$

***Deflection of the beam at its centre***

We also know that deflection of the beam at its centre,
$$y_C = \frac{5wl^4}{384\,\Sigma\, EI} = \frac{5 \times 5 \times (6 \times 10^3)^4}{384 \times (8525 \times 10^9)} = \mathbf{9.9\ mm} \qquad \textbf{Ans.}$$

## EXERCISE 19.2

1. A horizontal beam of uniform section and 6 m long is simply supported at its ends. Two vertical concentrated loads of 48 kN and 40 kN act at 1 m and 3 m respectively from the left hand support. Determine the position and magnitude of the maximum deflection, if $E = 200$ GPa and $EI = 85 \times 10^6$ mm$^4$. **[Ans.** 16.75 mm]
2. An overhanging beam *ABC* is loaded as shown in Fig. 19.14.

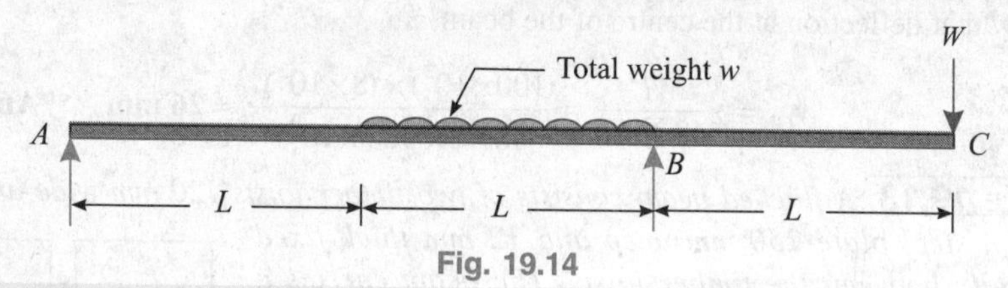


**Fig. 19.14**

Determine the deflection of the beam at point *C* in terms of *E, I, W* and *L*. $\left[\textbf{Ans. } \dfrac{39Wl^3}{48.EI}\right]$

3. A composite beam consisting of two timber sections and centrally embedded steel plate, is supported over a span of 4 metres. It carries two concentrated loads of 20 kN each at points 1 m from each support. Find the deflection of the beam under each load. Take flexural rigidity of the beam as $13 \times 10^{12}$ N-mm$^2$. **[Ans.** 2.04 mm]

## QUESTIONS

1. What is the relation between slope, deflection and radius of curvature of a simply supported beam?
2. A simply supported beam *AB* of span *l* and stiffness *EI* carries a concentrated load *P* at its centre. Find the expression for slope of the beam at the support *A* and deflection of the beam at its centre.
3. Derive a relation for the slope and deflection of a simply supported beam subjected to a uniformly distributed load of w/m length.
4. What is Macaulay's method for finding the slope and deflection of a beam? Discuss the cases, where it is of a particular use.
5. Explain the procedure for finding out the deflection of a beam of composite section.

## OBJECTIVE TYPE QUESTIONS

1. A simply supported beam carriers a point load at its centre. The slope at its supports is

(*a*) $\dfrac{Wl^2}{16EI}$ (*b*) $\dfrac{Wl^3}{16EI}$ (*c*) $\dfrac{Wl^2}{48EI}$ (*d*) $\dfrac{Wl^3}{48EI}$

where $W$ = Magnitude of the point load,
$l$ = Span of the beam and
$EI$ = Rigidity of the beam.

2. A simply supported beam *AB* of span (*l*) carries a point load (*W*) at *C* at a distance (*a*) from the left end *A*, such that $a < b$. The maximum deflection will be

(*a*) at *C* (*b*) between *A* and *C*
(*c*) between *C* and *B* (*d*) any where between *A* and *B*

**3.** A simply supported beam of span ($l$) is subjected to a uniformly distributed load of ($w$) per unit length over the whole span. The maximum deflections at the centre of the beam is

(*a*) $\frac{5wl^5}{48EI}$ (*b*) $\frac{5wl^4}{96EI}$ (*c*) $\frac{5wl^4}{192EI}$ (*d*) $\frac{5wl^3}{384EI}$

**4.** Two simply supported beams of the same span carry the same load. If the first beam carries the total load as a point load at its centre and the other uniformly distributed over the whole span, then ratio of maximum slopes of first beam to the second will be

(*a*) 1 : 1 (*b*) 1 : 1.5 (*c*) 1.5 : 1 (*d*) 2 : 1

## ANSWERS

**1.** (*a*) **2.** (*c*) **3.** (*d*) **4.** (*c*)

Chapter

# 20

# Deflection of Cantilevers

## Contents

## 20.1. Introduction

In the previous chapter, we have discussed the slope and deflection of beams, subjected to various types of loadings. The same formulae may also be used for finding out the slope and deflection of cantilevers.

## 20.2. Methods for Slope and Deflection at a Section

Though there are many methods for the slope and deflection at a section in a loaded cantilever, yet the following are important from the subject point of view:

1. Double integration method and
2. Moment area method.

## 20.3. Double Integration Method for Slope and Deflection

We have already discussed in the previous chapter, the double integration method for finding out the slope and deflection at any section of a beam. We shall use the same method for finding out the slope and deflection in cantilever also.

## 20.4. Cantilever with a Point Load at its Free End

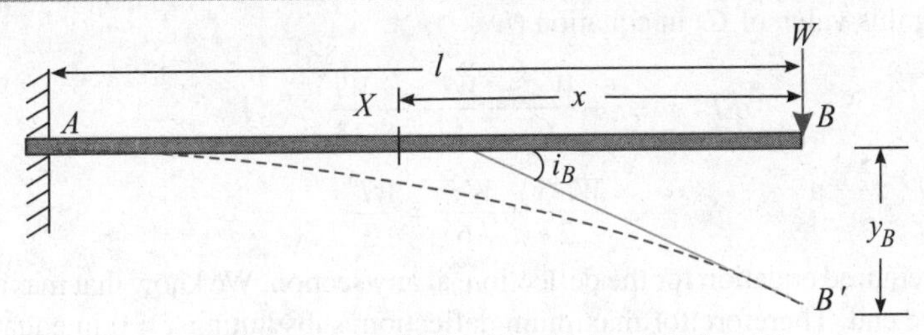


Fig. 20.1. Point load at the free end.

Consider a cantilever *AB* of length *l* and carrying a point load *W* at the free end as shown in Fig. 20.1. Consider a section *X*, at a distance *x* from the free end *B*.

We know that bending moment at this section,

$$M_X = -W \cdot x \quad \text{...(Minus sign due to hogging)}$$

$$\therefore \quad EI\frac{d^2y}{dx^2} = -W \cdot x \quad \text{...}(i)$$

Integrating the above equation,

$$EI\frac{dy}{dx} = -\frac{Wx^2}{2} + C_1 \quad \text{...}(ii)$$

where $C_1$ is thė first constant of integration. We know that when $x = l$, $\frac{dy}{dx} = 0$. Substituting these values in the above equation,

$$0 = -\frac{Wl^2}{2} + C_1 \quad \text{or} \quad C_1 = -\frac{Wl^2}{2}$$

Now substituting this value of $C_1$ in equation (*ii*),

$$EI\frac{dy}{dx} = -\frac{Wx^2}{2} + \frac{Wl^2}{2} \quad \text{...}(iii)$$

This is the required equation for the slope, at any section by which we can get the slope at any point on the cantilever. We know that maximum slope occurs at the free end. Now let us see the abbreviation *i* for the angle of inclination (in radian) and considering $i = \tan i$, for very small angles. Thus for maximum slope, substituting $x = 0$ in equation (*iii*),

$$EI \cdot i_B = \frac{Wl^2}{2}$$

$$\therefore \quad I_B = \frac{Wl^2}{2EI} \text{ radians}$$

Plus sign means that the tangent at *B* makes an angle with *AB* in the positive or clockwise direction. Integrating the equation (*iii*) once again,

$$EI \cdot y = \frac{Wx^3}{6} + \frac{Wl^2x}{2} + C_2 \quad \text{...}(iv)$$

where $C_2$ is the second constant of integration. We know that when $x = l$, $y = 0$. Substituting these values in the above equation,

$$0 = -\frac{Wl^3}{6} + \frac{Wl^3}{2} + C_2 = \frac{Wl^3}{3} + C_2$$

or $$C_2 = -\frac{Wl^3}{3}$$ ...(Minus sign means that the deflection is downwards)

Substituting this value of $C_2$ in equation (*iv*),

$$EI \cdot y = -\frac{Wx^3}{6} + \frac{Wl^2 x}{2} - \frac{Wl^3}{3}$$

$$= \frac{Wl^2 x}{2} - \frac{Wx^3}{6} = \frac{Wl^3}{3} \quad ...(v)$$

This is the required equation for the deflection, at any section. We know that maximum deflection occurs at the free end. Therefore for maximum deflection, substituting $x = 0$ in equation (*vi*),

$$EI \cdot y_B = -\frac{Wl^3}{3}$$

or $$y_B = -\frac{Wl^3}{3EI}$$

$$= \frac{Wl^3}{3EI}$$

**EXAMPLE 20.1.** *A cantilever beam 120 mm wide and 150 mm deep is 1.8 m long. Determine the slope and deflection at the free end of the beam, when it carries a point load of 20 kN at its free end. Take E for the cantilever beam as 200 GPa.*

**SOLUTION.** Given: Width ($b$) = 120 mm; Depth ($d$) = 150 mm ; Span ($l$) = 1.8 m = $1.8 \times 10^3$ mm ; Point load ($W$) = 20 kN = $20 \times 10^3$ N and modulus of elasticity ($E$) = 200 GPa = $200 \times 10^3$ N/mm$^2$.

***Slope at the free end***

We know that moment of inertia of the beam section,

$$I = \frac{bd^3}{12} = \frac{120 \times (150)^3}{12} = 33.75 \times 10^6 \text{ mm}^4$$

and slope at the free end, $$i_B = \frac{Wl^2}{2EI} = \frac{(20 \times 10^3) \times (1.8 \times 10^3)^2}{2 \times (200 \times 10^3) \times (33.75 \times 10^6)} = \mathbf{0.0048 \text{ rad}}$$ **Ans.**

***Deflection at the free end***

We also know that deflection at the free end,

$$y_B = \frac{Wl^3}{3EI} = \frac{(20 \times 10^3) \times (1.8 \times 10^3)^3}{3 \times (200 \times 10^3) \times (33.75 \times 10^6)} = \mathbf{5.76 \text{ mm}}$$ **Ans.**

**EXAMPLE 20.2.** *A cantilever beam of 160 mm width and 240 mm depth is 1.75 m long. What load can be placed at the free end of the cantilever, if its deflection under the load is not to exceed 4.5 mm. Take E for the beam material as 180 GPa.*

**SOLUTION.** Given: Width ($b$) = 160 mm; Depth ($d$) = 240 mm ; Span ($l$) = 1.75 m = $1.75 \times 10^3$ mm ; Deflection under the load ($y_B$) = 4.5 mm and modulus of elasticity ($E$) = 180 GPa = $180 \times 10^3$ N/mm$^2$.

Let $W$ = Load, which can be placed at the free end of the cantilever.

We know that moment of inertia of the beam section,

$$I = \frac{bd^3}{12} = \frac{160 \times (240)^3}{12} = 184.32 \times 10^6 \text{ mm}^4$$

and deflection of the cantilever under the load $(y_B)$,

$$4.5 = \frac{Wl^3}{3EI} = \frac{W \times (1.75 \times 10^3)^3}{3 \times (180 \times 10^3) \times (184.32 \times 10^6)} = \frac{W}{18571.72}$$

$$\therefore \quad W = 4.5 \times 18571.72 = 83572.74 \text{ N} = \mathbf{83.57 \text{ kN}} \qquad \textbf{Ans.}$$

## 20.5. Cantilever with a Point Load not at the Free End

Consider a cantilever $AB$ of length $l$ and carrying a point load $W$ at at a distance $l_1$ from the fixed end as shown in Fig. 20.2.

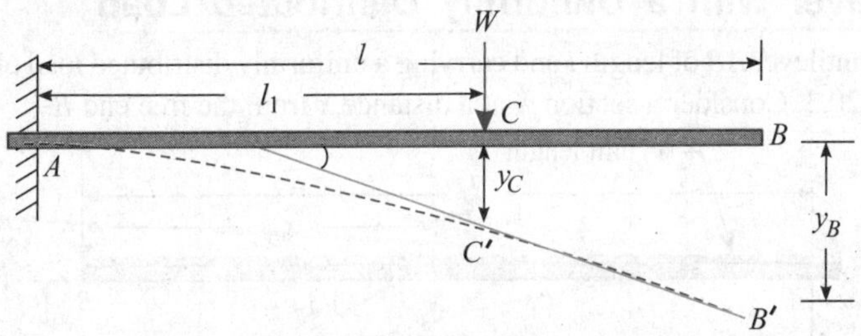


Fig. 20.2. Point load not at the free end.

A little consideration will show that the portion $AC$ of the cantilever will bend into $AC'$, while the portion $CB$ will remain straight and displaced to $C' B'$, as shown in Fig. 20.2. The portion $AC$ of the cantilever may be taken as similar to a cantilever in Art. 20.4 (*i.e.*, load at the free end).

$$i_C = \frac{Wl_1^2}{2EI}$$

Since the portion of the cantilever is straight, therefore

$$i_B = i_C = \frac{Wl_1^2}{2EI}$$

and

$$y_C = \frac{Wl_1^3}{3EI}$$

From the geometry of the figure, we find that

$$y_B = y_C + i_C (l - l_1) = \frac{Wl_1^3}{3EI} + \frac{Wl_1^2}{2EI}(l - l_1)$$

**Cor.** If

$$l_1 = \frac{l}{2}, \; y_B = \frac{W}{3EI}\left(\frac{l}{2}\right)^3 + \frac{W}{2EI}\left(\frac{l}{2}\right)^2 \times \frac{l}{2} = \frac{5Wl^3}{48EI}$$

**EXAMPLE 20.3.** *A cantilever beam 3 m long carries a point load of 20 kN at a distance of 2 m from the fixed end. Determine the slope and deflection at the free end of the cantilever. Take EI = 8 × $10^{12}$ N-mm$^2$.*

**SOLUTION.** Given: Span $(l) = 3 \text{ m} = 3 \times 10^3$ mm ; Point load $(W) = 20 \text{ kN} = 20 \times 10^3$ N ; Distance of the load from the fixed end $(l_1) = 2 \text{ m} = 2 \times 10^3$ mm and flexural rigidity $(EI) = 8 \times 10^{12}$ N-mm$^2$.

***Slope at the free end of the cantilever***

We know that slope at the free end of the contilever

$$i_B = \frac{Wl_1^2}{2EI} = \frac{(20 \times 10^3) \times (2 \times 10^3)^2}{2 \times (8 \times 10^{12})} = \mathbf{0.005 \text{ rad}} \qquad \textbf{Ans.}$$

***Deflection at the free end of the cantilever***

We also know that deflection at the free end of the cantilever,

$$y_B = \frac{Wl_1^3}{3EI} + \frac{Wl_1^2}{2EI}(l - l_1)$$

$$= \left[\frac{(20\times10^3)\times(2\times10^3)^3}{3\times(8\times10^{12})}\right]\left[(3\times10^3)-(2\times10^3)\right] \text{ mm}$$

$$+\frac{(20\times10^3)\times(2\times10^3)^2}{2\times(8\times10^{12})}[(3\times10^3)-(2\times10^3)) \text{ mm}$$

$$= 6.7 + 5.0 = \mathbf{11.7\ mm} \quad \textbf{Ans.}$$

## 20.6. Cantilever with a Uniformly Distributed Load

Consider a cantilever $AB$ of length $l$ and carrying a uniformly distributed load of $w$ per unit length as shown in Fig. 20.3. Consider a section $X$ at a distance $x$ from the free end $B$.

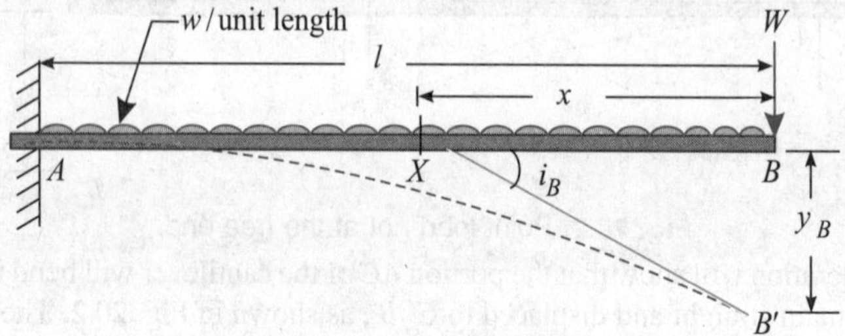


Fig. 20.3. Uniformly distributed load.

We know that bending moment at the section,

$$M_X = \frac{wx^2}{2} \qquad \text{...(Minus sign due to hogging)}$$

$$\therefore \quad EI\frac{d^2y}{dx^2} = \frac{wx^2}{2} \qquad \text{...}(i)$$

Integrating the above equation,

$$EI\frac{dy}{dx} = \frac{wx^3}{6} + C_1 \qquad \text{...}(ii)$$

where $C_1$ is the first constant of integration. We know that when $x = l$, then $\frac{dy}{dx} = 0$. Substituting these values in equation (*ii*),

$$0 = -\frac{wl^3}{6} + C_1 \quad \text{or} \quad C_1 = \frac{wl^3}{6}$$

Substituting this value of $C_1$ in equation (*ii*),

$$EI\frac{dy}{dx} = -\frac{wx^3}{6} + \frac{wl^3}{6} \qquad \text{...}(iii)$$

This is the required equation for the slope at any section. We know that maximum slope occurs at the free end $B$. Therefore for maximum slope, substituting $x = 0$ in equation (*iii*),

$$El \cdot i_B = \frac{wl^3}{6}$$

or
$$i_B = \frac{wl^3}{6EI} \text{ radians}$$
...(Plus sign means that the tangent at $B$ makes an angle with $AB$ in the positive or clockwise direction)

Integrating the equation (*iii*) once again,

$$EI \cdot y = -\frac{wx^4}{24} + \frac{wl^3 x}{6} + C_2$$

where $C_2$ is the second constant of integration. We know that when $x = l$, then $y = 0$. Substituting these values in the above equation,

$$0 = -\frac{wl^4}{24} + \frac{wl^4}{6} + C_2 \quad \text{or} \quad C_2 = -\frac{wl^4}{8}$$

Substituting this value of $C_2$ is equation (*iv*),

$$EI \cdot y = -\frac{wx^4}{24} + \frac{wl^3 x}{6} - \frac{wl^4}{8} = \frac{wl^3 x}{6} - \frac{wx^4}{24} - \frac{wl^4}{8} \quad ...(v)$$

This is the required equation for the deflection at any section. We know that maximum deflection occurs at the free end.

Therefore for maximum slope, substituting $x = 0$ in equation (*v*),

$$EI \cdot y_B = \frac{wl^4}{8}$$

or
$$y_B = -\frac{wl^4}{8EI}$$
...(Minus sign means that the deflection is downwards)

$$= \frac{wl^4}{8EI}$$

**Note.** The above expression for slope and deflecion may also be expressed in terms of total load. Such that $W = wl$.

$$i_B = i_A = \frac{wl^3}{6EI} = \frac{wl^2}{6EI} \quad \text{and} \quad y_B = \frac{wl^4}{8EI} = \frac{wl^3}{8EI}$$

**Example 20.4.** *A cantilever beam 2 m long is subjected to a uniformly distributed load of 5 kN/m over its entire length. Find the slope and deflection of the cantilever beam at its free end. Take $EI = 2.5 \times 10^{12}$ $mm^2$.*

**Solution.** Given: Span ($l$) = 2 m = $2 \times 10^3$ mm ; Uniformly distributed load ($w$) = 5 kN/m = 5N/mm and flexural rigidity ($EI$) = $2.5 \times 10^{12}$ N-mm$^2$.

***Slope of the cantilever beam at its free end***

We know that slope of the cantilever at its free end,

$$i_B = \frac{wl^3}{6EI} = \frac{5 \times (2 \times 10^3)^3}{6 \times (2.5 \times 10^{12})} = \mathbf{0.0027 \text{ rad}} \quad \textbf{Ans.}$$

***Deflection of the cantilever beam at its free end***

We also know that deflection of the cantilever at its free end,

$$y = \frac{wl^4}{8EI} = \frac{5 \times (2 \times 10^3)^4}{8 \times (2.5 \times 10^{12})} = \mathbf{4.0 \text{ mm}} \quad \textbf{Ans.}$$

**Example 20.5.** *A cantilever beam 100 mm wide and 180 mm deep is projecting 2 m from a wall. Calculate the uniformly distributed load, which the beam should carry, if the deflection of the free end should not exceed 3.5 mm. Take E as 200 GPa.*

**Solution.** Given: Width ($b$) = 100 mm ; Depth ($d$) = 180 mm ; Span ($l$) = 2 m = $2 \times 10^3$ mm ; Deflection at the free end ($y_B$) = 3.5 mm and modulus of elasticity ($E$) = 200 GPa = $200 \times 10^3$ N/mm$^2$.

Let $w$ = Uniformly distributed load, which the beam should carry.

We know that moment of inertia of the beam reaction,

$$I = \frac{bd^3}{12} = \frac{100 \times (180)^3}{12} = 48.6 \times 10^6 \text{ mm}^4$$

and deflection of the free end of the beam ($y_B$),

$$3.5 = \frac{wl^4}{8EI} = \frac{w \times (2 \times 10^3)^4}{8 \times (200 \times 10^3) \times (48.6 \times 10^6)} = 0.206\, w$$

$$\therefore \quad w = 3.5/0.206 = 17 \text{ N/mm} = \mathbf{17\ kN/m} \quad \textbf{Ans.}$$

**Example 20.6.** *A cantilever beam of length 3 m is carrying a uniformly distributed load of w kN/m. Assuming rectangular cross-section with depth (d) equal to twice the width (b), determine the dimensions of the beam, so that vertical deflection at the free end does not exceed 8 mm. Take maximum bending stress = 100 MPa and E = 200 GPa.*

**Solution.** Given: Span ($l$) = 3 m = $3 \times 10^3$ mm ; Uniformly distributed load = $w$ kN/m = $w$ N/mm; Depth ($d$) = 2 $b$ ; Deflection at the free end ($y_B$) = 8 mm ; Maximum bending stress ($\sigma_{max}$) = 100 MPa = 100 N/mm$^2$ and modulus of elasticity ($E$) = 200 GPa = $200 \times 10^3$ N/mm$^2$.

We know that moment of inertia of the beam section,

$$I = \frac{bd^3}{12} = \frac{b\,(2b)^3}{12} = \frac{2b^4}{3} \text{ mm}^4$$

and deflection at the free end of the cantilever ($y_B$),

$$8 = \frac{wl^4}{8EI} = \frac{w \times (3 \times 10^3)^4}{8 \times (200 \times 10^3) \times 2b^4/3} = \frac{75.9 \times 10^6 \times w}{b^4}$$

$$\therefore \quad b^4 = \frac{(75.9 \times 10^6)\, w}{8} = 9.5 \times 10^6\, w \qquad \ldots(i)$$

We also know that moment at the fixed end of the cantilever,

$$M = \frac{wl^2}{2} = \frac{w \times (3 \times 10^3)^2}{2} = 4.5 \times 10^6\, w \text{ N-mm}$$

and from the bending stress equation,

$$\frac{M}{I} = \frac{\sigma_b}{y} \quad \text{or} \quad \frac{4.5 \times 10^6\, w}{2b^4/3} = \frac{100}{b}$$

$$\text{or} \quad 13.5 \times 10^6\, w = 100 \times 2b^3 \qquad \ldots\left(\because\ y = \frac{d}{2} = \frac{2b}{2} = b\right)$$

$$b^3 = \frac{13.5 \times 10^6 \times w}{200} = 67.5 \times 10^3\, w \qquad \ldots(ii)$$

Dividing equation ($i$) by ($ii$),

$$b = \frac{(9.5 \times 10^6\, w)}{(67.5 \times 10^3\, w)} = \mathbf{141\ mm} \quad \textbf{Ans.}$$

$$\text{and} \quad d = 2b = 2 \times 141 = \mathbf{282\ mm} \quad \textbf{Ans.}$$

## 20.7. Cantilever Partially Loaded with a Uniformly Distributed Load

Consider a cantilever *AB* of length $l$ and carrying a uniformly distributed load $w$ per unit length for a length of $l$ from the fixed end as shown in Fig. 20.4.

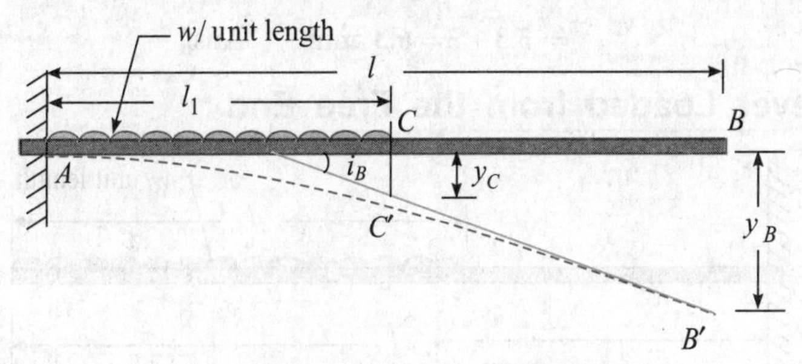


Fig. 20.4. Partially uniformly distributed load.

A little consideration will show that the portion *AC* of the cantilever will bend into *AC′*, while the portion *CB* will remain straight, but will displace to *C′ B′* as shown in the figure. The portion *AC* of the cantilever may be taken as similar to a cantilever in Art. 20.6 (*i.e.*, Cantilever with uniformly distributed load).

$$\therefore \qquad I_C = \frac{wl_1^3}{6EI} \text{ rad}$$

Since the portion *CB* of the cantilever is straight, therefore

$$I_B = I_C = \frac{wl_1^3}{6EI}$$

and

$$y_C = \frac{wl_1^4}{8EI}$$

From the geometry of the figure, we find that

$$y_B = y_C + i_C(l - l_1) = \frac{wl_1^4}{8EI} + \frac{wl_1^3}{6EI}[l - l_1]$$

**Cor.** If $l_1 = \frac{l}{2}$, then $y_B = \frac{w}{8EI}\left(\frac{l}{2}\right)^4 + \frac{w}{6EI}\left(\frac{l}{2}\right)^3 \times \frac{l}{2} = \frac{7wl^4}{384EI}$

**EXAMPLE 20.7.** *A cantilever 2.5 m long is loaded with a uniformly distributed load of 10 kN/m over a length of 1.5 m from the fixed end. Determine the slope and deflection at the free end of the cantilever. Take flexural rigidity of the beam as $1.9 \times 10^{12}$ N-mm$^2$.*

**SOLUTION.** Given: Span ($l$) = 2.5 m = $2.5 \times 10^3$ mm ; Uniformly distributed load ($w$) = 10 kN/m = 10 N/mm ; Loaded length ($l_1$) = 1.5 m = $1.5 \times 10^3$ mm and flexural rigidity ($EI$) = $1.9 \times 10^{12}$ N-mm$^2$.

***Slope at the free end***

We know that slope at the free end.

$$i_B = \frac{wl_1^3}{6EI} = \frac{10 \times (1.5 \times 10^3)^3}{6 \times (1.9 \times 10^{12})} = \mathbf{0.003 \text{ rad}} \qquad \textbf{Ans.}$$

***Deflection at the free end***

We also know that deflection at the free end,

$$y_B = \frac{wl_1^4}{8EI} + \frac{wl_1^3}{6EI}[l - l_1]$$

$$= \frac{10 \times (1.5 \times 10^3)^4}{8 \times (1.9 \times 10^{12})} + \frac{10 \times (1.5 \times 10^3)^3}{6 \times (1.9 \times 10^{12})} \times \left[ (2.5 \times 10^3) - (1.5 \times 10^3) \right]$$

$$= 3.3 + 3 = \mathbf{6.3 \text{ mm}} \quad \textbf{Ans.}$$

## 20.8. Cantilever Loaded from the Free End

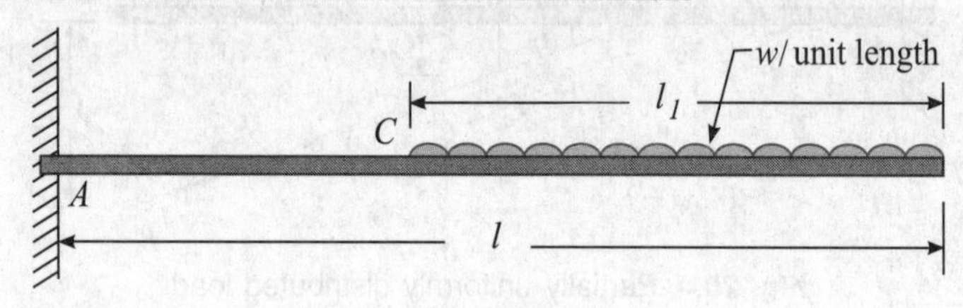


Fig. 20.5

Consider a cantilever *AB* of length *l* and carrying a uniformly distributed load *w* per unit length for a length of $l_1$, from the free end as shown in Fig. 20.5.

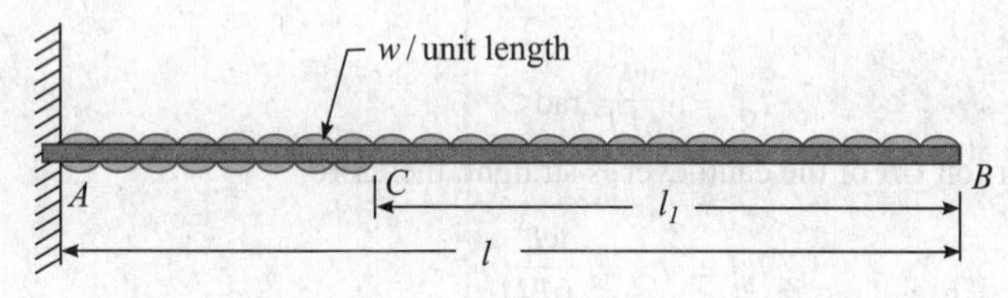


Fig. 20.6

The slope and deflection of the cantilever, in this case may be obtained as discussed below:

1. First of all, consider the whole cantilever from *A* to *B* to be loaded with a uniformly distributed load of *w* per unit lenght as shown in Fig. 20.6.
2. Then superimpose an upward uniformly distributed load of *w* per unit length from *A* to *C* as shown in Fig. 20.6.
3. Then obtain the slopes and deflections due to the above mentioned loading as per Art. 20.6 and 20.7.
4. Then the slope at *B* is equal to the slope due to the total load *minus* the slope due to the superimposed load.
5. Similarly, the deflection at *B* is equal to the deflection due to the total load *minus* the deflection due to the superimposed load,

$$\therefore \qquad i_B = \left[ \frac{wl^3}{6EI} \right] - \left[ \frac{w(l - l_1)^3}{6EI} \right]$$

Similarly,

$$y_B = \left[ \frac{wl^4}{8EI} \right] - \left[ \frac{w(l - l_1)^4}{8EI} + \frac{w(l - l_1)^3 \, l_1}{6EI} \right]$$

**Note.** The slope and deflection at *A* due to superimposed uniformly distributed load from *A* to *C* is obtained by substituting $(l - l_1)$ for *l*, and vice versa in Art. 20.7.

**EXAMPLE 20.8.** *A cantilever 75 mm wide and 200 mm deep is loaded as shown in Fig. 20.7. Find the slope and deflection at B. Take E = 200 GPa.*

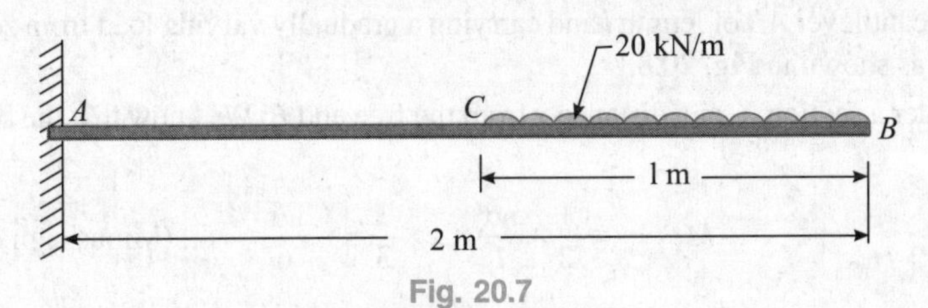


Fig. 20.7

**Solution.** Given: Width ($b$) = 75 mm ; Depth ($d$) = 200 mm ; Uniformly distributed load ($w$) = 20 kN/m = 20 N/mm ; Span ($l$) = 2 m = $2 \times 10^3$ mm ; Loaded length ($l_1$) = 1 m = $1 \times 10^3$ mm and Young's modulus ($E$) = 200 GPa = $200 \times 10^3$ N/mm$^2$

***Slope at B***

We know that moment of inertia of the cantilever section,

$$I = \frac{bd^3}{12} = \frac{75 \times (200)^3}{12} = 50 \times 10^6 \text{ mm}^4$$

and slope at $B$

$$i_B = \left[\frac{wl^3}{6EI}\right] - \left[\frac{w(l-l_1)^3}{6EI}\right]$$

$$= \left[\frac{20 \times (2 \times 10^3)^3}{6 \times (200 \times 10^3) \times (50 \times 10^6)}\right] \left[\frac{20\,[(2 \times 10^3) - (1 \times 10^3)]^3}{6 \times (200 \times 10^3) \times (50 \times 10^6)}\right]$$

$$= 0.00267 - 0.000333 = \mathbf{0.00234 \text{ rad}} \quad \textbf{Ans.}$$

***Deflection at B***

We also know that deflection at $B$,

$$y_B = \left[\frac{wl^4}{8EI}\right] - \left[\frac{w(l-l_1)^4}{8EI} + \frac{w(l-l_1)^3\, l_1}{6EI}\right]$$

$$= \left[\frac{20 \times (2 \times 10^3)^4}{8\,(200 \times 10^3) \times (50 \times 10^6)}\right]$$

$$- \left[\frac{20\,[(2 \times 10^3) - (1 \times 10^3)]^4}{8\,(200 \times 10^3) \times (50 \times 10^6)} + \frac{20\,[(2 \times 10^3) - (1 \times 10^3)]^3\,(1 \times 10^3)}{6\,(200 \times 10^3) \times (50 \times 10^6)}\right]$$

$$= 4.0 - 0.58 = \mathbf{3.42 \text{ mm}} \quad \textbf{Ans.}$$

## 20.9. Cantilever with a Gradually Varying Load

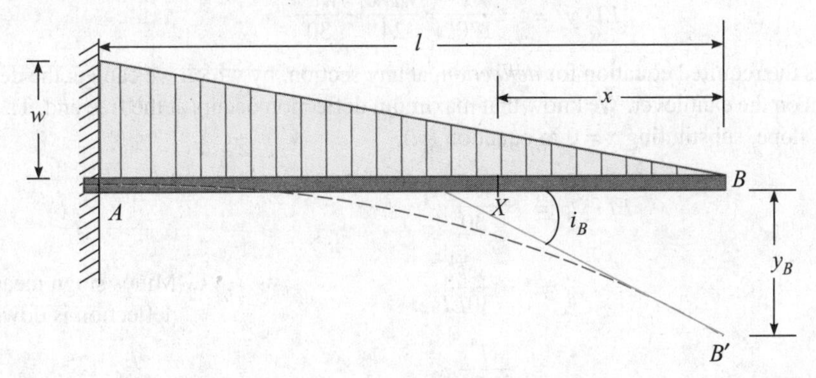


Fig. 20.8

Consider a cantilever $AB$ of length $l$ and carrying a gradually varying load from zero at $B$ to $w$ per unit length at $A$ as shown in Fig. 20.8.

Now consider a section $X$, at a distance $x$ from the free end $B$. We know that the bending moment at the section,

$$M_X = -\frac{1}{2} \times \frac{wx}{l} \times x \times \frac{x}{3} = -\frac{wx^3}{6l} \quad \text{...(Minus sign due to hogging)}$$

$$\therefore \quad EI \frac{d^2 y}{dx^2} = -\frac{wx^3}{6l} \quad ...(i)$$

Integrating the above equation,

$$EI \cdot \frac{dy}{dx} = -\frac{wx^4}{24l} + C_1 \quad ...(ii)$$

where $C_1$ is the constant of integration. We know that when $x = l$, then $\frac{dy}{dx} = 0$. Substituting these values in equation (*ii*),

$$0 = -\frac{wl^4}{24l} + C_1 \quad \text{or} \quad C_1 = -\frac{wl^3}{24}$$

$$\therefore \quad EI \cdot \frac{dy}{dx} = -\frac{wx^4}{24l} + \frac{wl^3}{24} \quad ...(iii)$$

This is required equation for the slope at any section, by which we can get the slope at any section on the cantilever. We know that the maximum slope occurs at the free end. Therefore for maximum slope, substituting $x = 0$ in equation (*iii*),

$$EI \cdot i_B = \frac{wl^3}{24}$$

$$\text{or} \quad i_B = \frac{wl^3}{24EI} \text{ radians} \quad ...(iv)$$

Integrating the equation (*iii*) once again,

$$EI \cdot y = -\frac{wx^5}{120l} \times \frac{wl^3}{24} + C_2 \quad ...(v)$$

where $C_2$ is the constant of integration. We know that when $x = l$, then $y = 0$. Substituting these values in the above equation,

$$0 = -\frac{wl^4}{120} + \frac{wl^4}{24} + C_2 \quad \text{or} \quad C_2 = -\frac{wl^4}{30}$$

$$\therefore \quad EI \cdot y = -\frac{wx^5}{120l} + \frac{wl^3 x}{24} - \frac{wl^4}{30} \quad ...(vi)$$

This is the required equation for *deflection*, at any section, by which we can get the deflection at any section on the cantilever. We know that maximum deflection occurs at the free end. Therefore for maximum slope, substituting $x = 0$ in equation (*vi*),

$$EI \cdot y_B = -\frac{wl^4}{30}$$

$$\therefore \quad y_B = -\frac{wl^4}{30EI} \quad \text{...(Minus singn means that the deflection is downwards)}$$

$$= \frac{wl^4}{30EI}$$

**EXAMPLE 20.9.** *A cantilever of 2 m span carries a triangular load of zero intensity at the free end and 100 kN/m at the fixed end. Determine the slope and deflection at the free end. Take I = 100 × 10⁶ mm⁴ and E = 200 GPa.*

**SOLUTION.** Given: Span ($l$) = 2 m = $2 \times 10^3$ mm ; Load at the fixed end ($w$) = 100 kN/m = 100 N/mm; Moment of inertia ($I$) = $100 \times 10^6$ mm$^4$ and modulus of elasticity ($E$) = 200 GPa = $200 \times 10^3$ N/mm$^2$.

***Slope at the free end***

We know that slope at the free end,

$$i_B = \frac{wl^3}{24EI} = \frac{1000 \times (2 \times 10^3)}{24 \times (200 \times 10^3) \times (100 \times 10^6)} = \mathbf{0.00167\ rad} \qquad \textbf{Ans.}$$

***Deflection at the free end***

We also know that deflection at the free end,

$$y_B = \frac{wl^4}{30EI} = \frac{100 \times (2 \times 10^3)^4}{30 \times (200 \times 10^3) \times (100 \times 10^6)} = \mathbf{2.67\ mm} \qquad \textbf{Ans.}$$

## EXERCISE 20.1

1. A cantilever 2.4 m long carries a point load of 30 kN at its free end. Find the slope and deflection of the cantilever under the load. Take flexural rigidity for the cantilever beam as $25 \times 10^{12}$ N-mm$^2$.
   [**Ans.** 0.0035 rad ; 5.5 mm]
2. A cantilever 150 mm wide and 200 mm deep projects 1.5 m out of a wall. Find the slope and deflection of the cantilever at the free end, when it carries a point load of 50 kN at its free end. Take $E$ = 200 GPa. [**Ans.** 0.0028 rad ; 2.8 mm]
3. A cantilever beam 120 mm wide and 180 mm deep is 2 m long. Find the maximum load, which can be placed at the free end, the deflection of the cantilever at its free end should not exceed 5 mm. Take $E$ as 200 GPa. [**Ans.** 21.87 kN]
4. A cantilever beam of length 1.8 m is carrying a uniformly distributed load of 10 kN/m on its entire length. What is the slope and deflection of the beam at its free end? Take flexural rigidity of the beam as $3.2 \times 10^{12}$ N-mm$^2$.
   [**Ans.** 0.003 rad ; 4.1 mm]
5. A cantilever beam 120 mm wide and 200 mm deep is 2.5 m long. Find the uniformly distributed load, the beam should carry to produce a deflection of 5 mm at its free end. Take $E$ = 200 GPa.
   [**Ans.** 16.4 kN/m]
6. A cantilever beam of 2.5 m span carries a load which is gradually varying from zero at the free end to 200 kN/m over the fixed end. Find the deflection of the free end. Take flexural rigidity of the section as $160 \times 10^{12}$ N-mm$^2$.
   [**Ans.** 1.63 mm]

## 20.10. Cantilever with Several Loads

If a cantilever is loaded with several point or uniformly distributed loads, the slope as well as the deflection at any point on the cantilever, is equal to the *algebraic sum* of the slopes and deflections at that point due to various loads acting individually.

**EXAMPLE 20.10.** *A cantilever AB 2 m long is carrying a load of 20 kN at free end and 30 kN at a distance 1 m from the free end. Find the slope and deflection at the free end. Take E = 200 GPa and I = 150 × $10^6$ $mm^4$.*

**SOLUTION.** Given: Span *AB* ($l$) = 2 m = 2 × $10^3$ mm ; Load at the free end ($W_1$) = 20 kN = 20 × $10^3$ N; Load at *C* ($W_2$) = 30 kN = 30 × $10^3$ N ; Length *AC* ($l_1$) = 1 m = 1 × $10^3$ mm ; Modulus of elasticity (*E*) = 200 GPa = 200 × $10^3$ N/mm² and moment of inertia (*I*) = 150 × $10^6$ mm⁴.

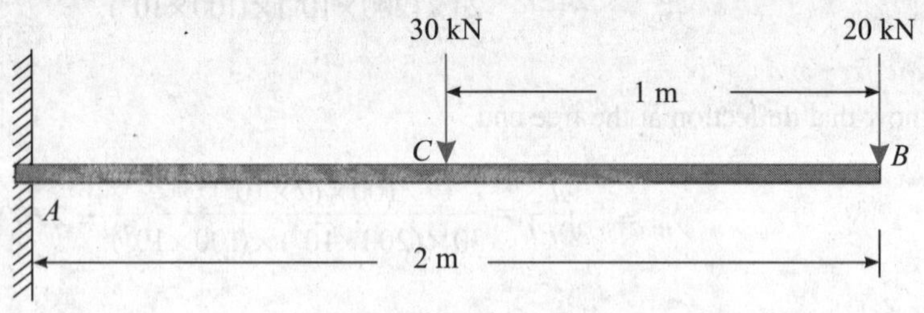


Fig. 20.9

***Slope at the free end***

We know that slope at the free end

$$i_B = \left[\frac{W_1 l^2}{2EI}\right] + \left[\frac{W_2\, l_1^2}{2EI}\right]$$

$$= \left[\frac{(20\times10^3)\times(2\times10^3)^2}{2\times(200\times10^3)\times(150\times10^6)}\right] + \left[\frac{(30\times10^3)\times(1\times10^3)^2}{2\times(200\times10^3)\times(150\times10^6)}\right] \text{ rad}$$

$$= 0.00133 + 0.0005 = \mathbf{0.00183\ rad} \quad \textbf{Ans.}$$

***Deflection at the free end***

We also know that deflection at the free end,

$$y_B = \left[\frac{W_1\, l^3}{3EI}\right] + \left[\frac{W_2\, l_1^3}{3EI}\right] + \left[\frac{W_2\, l_1^2}{2EI}(l - l_1)\right]$$

$$= \left[\frac{(20\times10^3)\times(2\times10^3)^3}{3\times(200\times10^3)\times(150\times10^6)}\right] + \left[\frac{(30\times10^3)\times(1\times10^3)^3}{3\,(200\times10^3)\times(150\times10^6)}\right] + \left[\frac{(30\times10^3)\times(1\times10^3)^2}{2\,(200\times10^3)\times(150\times10^6)}\times(2\times10^3)-(1\times10^3)\right]$$

$$= 1.78 + 0.33 + 0.5 = \mathbf{2.61\ mm} \quad \textbf{Ans.}$$

**EXAMPLE 20.11.** *A cantilever 2 m long carries a point load 20 kN at its free end and a uniformly distributed load of 8 kN/m over the whole length. Determine the slope and deflection of the cantilever at its free end. Take E = 200 GPa and I = 50 × 10⁶ mm⁴.*

**SOLUTION.** Given: Span $(l)$ = 2 m = $2 \times 10^3$ mm ; Point load $(W)$ = 20 kN = $20 \times 20^3$ N ; Uniformly distributed load $(w)$ = 8 kN/m = 8 N/mm ; Modulus of elasticity $(E)$ = 200 GPa = $200 \times 10^3$ N/mm$^2$ and moment of inertia $(I)$ = $50 \times 10^6$ mm$^4$.

***Slope of the cantilever at its free end***

We know that slope of the cantilever at its free end,

$$y_B = \left[\frac{Wl^3}{3EI}\right] + \left[\frac{wl^4}{8EI}\right]$$

$$= \left[\frac{(20\times10^3)\times(2\times10^3)^2}{2\times(200\times10^3)\times(50\times10^6)}\right] + \left[\frac{8\times(2\times10^3)^3}{8\times(200\times10^3)\times(50\times10^6)}\right] \text{rad}$$

$$= 0.004 + 0.0011 = \mathbf{0.0051\ rad} \qquad \textbf{Ans.}$$

***Deflection of the cantilever at its free end***

We also know that deflection of the cantilever at its free end,

$$y_B = \left[\frac{Wl^3}{3EI}\right] + \left[\frac{wl^4}{8EI}\right]$$

$$= \left[\frac{(20\times10^3)\times(2\times10^3)^3}{3\times(200\times10^3)\times(50\times10^6)}\right] + \left[\frac{8\times(2\times10^3)^4}{8\times(200\times10^3)\times(50\times10^6)}\right]$$

$$= 5.3 + 1.6 = \mathbf{6.9\ mm} \qquad \textbf{Ans.}$$

**EXAMPLE 20.12.** *A cantilever 100 mm wide and 180 mm deep projects 2.0 m from a wall into which it is cast. The cantilever carries a uniformly distributed load of 20 kN/m over a length of 1 m from the free end, and point load of 10 kN at the free end as shown in Fig. 20.10.*

*Find the slope and deflection at the free end. Take E = 200 GPa.*

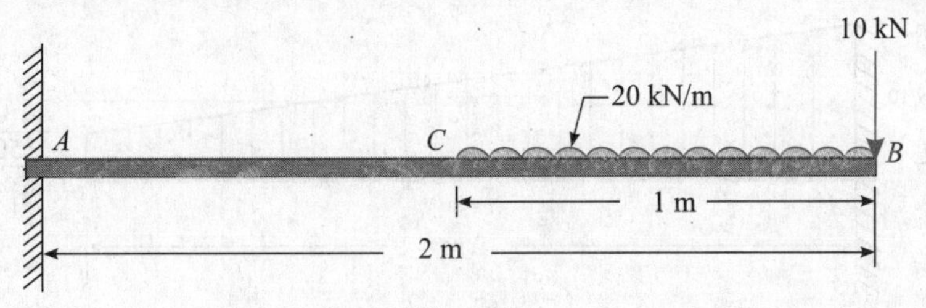


Fig. 20.10

**SOLUTION.** Given: Width $(b)$ = 100 mm ; Depth $(d)$ = 180 mm ; Load at the free end $(W)$ = 10 kN = $10 \times 10^3$ N ; Modulus of elasticity $(E)$ = 200 GPa = $200 \times 10^3$ N/mm$^2$ ; Length $AB$ $(l)$ = 2 m = $2 \times 10^3$ mm ; Load in $CB$ $(w)$ = 20 kN/m = 20 N/mm and length $CB$ $(l_1)$ = 1 m = $1 \times 10^3$ mm.

***Slope at the free end***

We know that moment of inertia of the cantilever section,

$$I = \frac{bd^3}{12} = \frac{100 \times (180)^3}{12} = 48.6 \times 10^6 \text{ mm}^4$$

and slope at the free end,

$$i_B = \left[\frac{Wl^2}{2EI}\right] + \left[\left(\frac{Wl^3}{6EI}\right) - \left(\frac{w\,(l - l_1)^3}{6EI}\right)\right]$$

$$= \left[\frac{(10 \times 10^3) \times (2 \times 10^3)^2}{2\,(200 \times 10^3)\,(48.6 \times 10^6)}\right]$$

$$+ \left[\left(\frac{20\,(2 \times 10^3)^3}{6\,(200 \times 10^3)\,(48.6 \times 10^6)}\right) - \left(\frac{20 \times [(2 \times 10^3) - (1 \times 10^3)]^3}{6\,(200 \times 10^3)\,(48.6 \times 10^6)}\right)\right] \text{ rad}$$

$$= 0.00206 + (0.00274 - 0.00034) = \mathbf{0.00446 \text{ rad}} \quad \textbf{Ans.}$$

***Deflection at the free end***

We also know that deflection at the free end,

$$y_B = \left[\frac{Wl^3}{3EI}\right] + \left[\frac{wl^4}{8EI}\right] - \left[\left(\frac{w\,(l - l_1)^4}{8EI}\right) + \left(\frac{w\,(l - l_1)^3\, l}{6EI}\right)\right]$$

$$= \left[\frac{(10 \times 10^3) \times (2 \times 10^3)^3}{3\,(200 \times 10^3)\,(48.6 \times 10^6)}\right] + \left[\frac{20 \times (2 \times 10^3)^4}{8\,(200 \times 10^3)\,(48.6 \times 10^6)}\right]$$

$$- \left[\left(\frac{20\,[(2 \times 10^3) - (1 \times 10^3)^4}{8\,(200 \times 10^3)\,(48.6 \times 10^6)}\right)\right.$$

$$\left. + \left(\frac{20\,[(2 \times 10^3) - (1 \times 10^3)]^3 \times (2 \times 10^3)}{6\,(200 \times 10^3)\,(48.6 \times 10^6)}\right)\right]$$

$$= 2.74 + 2.06 - (0.13 + 0.69) = \mathbf{3.98 \text{ mm}} \quad \textbf{Ans.}$$

**EXAMPLE 20.13.** *A metallic cantilever 150 mm wide, 200 mm deep and of 2 m span carries a uniformly varying load of 50 kN/m at the free end to 150 kN/m at the fixed end as shown in Fig. 20.11. Find the slope of the cantilever at the free end. Take E = 100 GPa.*

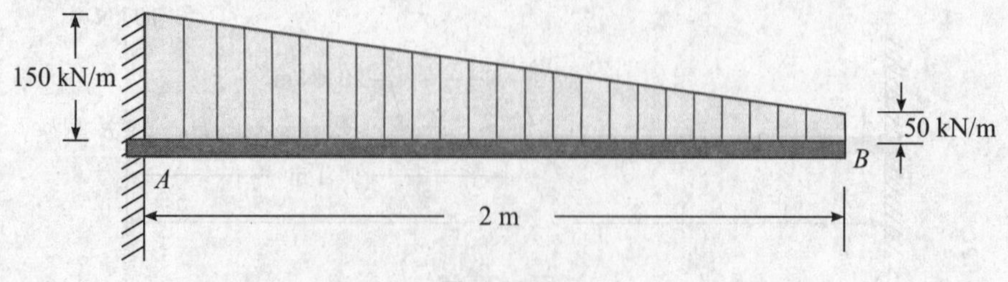


Fig. 20.11

**Solution.** Given: Width ($b$) = 150 mm ; Depth ($d$) = 200 mm ; Span ($l$) = 2 m = $2 \times 10^3$ mm ; Load at $A$ = 150 kN/m = 150 N/mm ; Load at $B$ = 50 kN/m = 50 N/mm and modulus of elasticity ($E$) = 100 GPa = $100 \times 10^3$ N/mm$^3$.

***Slope at the free end***

Let us split up the trapezoidal load into a uniformly distributed load ($w_1$) of 50 N/mm and a triangular load ($w_2$) of 100 N/mm at $A$ to zero at $B$.

We know that moment of inertia of the cantilever section,

$$I = \frac{bd^3}{12} = \frac{150 \times (200)^3}{12} = 100 \times 10^6 \text{ mm}^4.$$

$\therefore$ Slope at the free end $B$,

$$i_B = \left[\frac{w_1 l^3}{6EI}\right] + \left[\frac{w_2 l^3}{24EI}\right]$$

$$= \left[\frac{50 \times (2 \times 10^3)^3}{6 \times (100 \times 10^3) \times 100 \times 10^6}\right] + \left[\frac{100 \times (2 \times 10^3)^3}{24 \times (100 \times 10^3) \times (100 \times 10^6)}\right] \text{ rad}$$

$$= 0.0067 + 0.0033 = \mathbf{0.01 \text{ rad}} \quad \textbf{Ans.}$$

***Deflection at the free end***

We also know that deflection at the free end,

$$y_B = \left[\frac{w_1 l^4}{8EI}\right] + \left[\frac{w_2 l^4}{30EI}\right]$$

$$= \left[\frac{50 \times (2 \times 10^3)^4}{8 \times (100 \times 10^3) \times (100 \times 10^6)}\right] + \left[\frac{100 \times (2 \times 10^3)^4}{30 \times (100 \times 10^3) \times (100 \times 10^6)}\right] \text{ mm}$$

$$= 10 + 5.3 = \mathbf{15.3 \text{ mm}} \quad \textbf{Ans.}$$

## 20.11. Cantilever of Composite Sections

We have already discussed in the previous chapter, the slope and deflection at any section of composite section of a beam. We shall use the same method for finding out the slope and deflection in cantilevers.

**Example 20.14.** *A composite cantilever beam 2 m long consists of a rectangular timber joist 150 mm × 240 mm deep. Two steel plates 150 mm × 10 mm thick are fixed at the top and bottom faces of the timber joist as shown in fig. 20.12.*

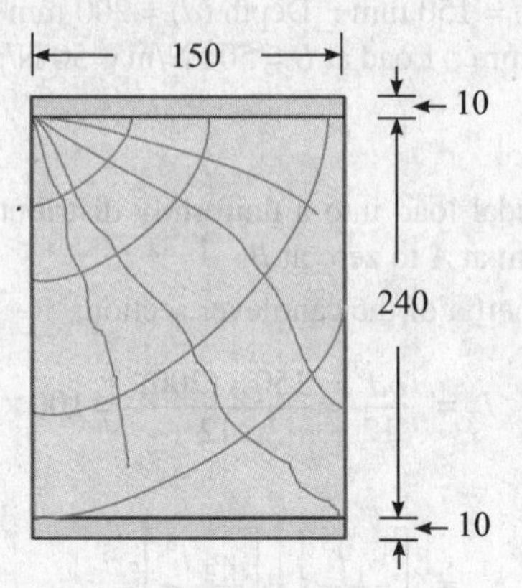


Fig. 20.12

*Find the slope and deflection of the cantilever at its free end, when it is carrying a uniformly distributed load of 10 kN/m. Take $E_s$ = 200 GPa and $E_t$ = 10 GPa.*

**SOLUTION.** Given: Span ($l$) = 2 m = $2 \times 10^3$ mm ; Uniformly distributed load ($w$) = 10 kN/m = 10 N/mm; Modulus of elasticity for steel ($E_s$) = 200 GPa = $200 \times 10^3$ N/mm$^2$ and modulus of elasticity for timber ($E_t$) = 10 GPa = $10 \times 10^3$ N/mm$^2$.

***Slope at the free end***

From the geometry of the composite beam, we find that the centre of gravity of the composite section coincides with the centre of gravity of the timber section. Therefore flexural rigidity for the timber section about its centre of gravity,

$$EI_{(timber)} = (10 \times 10^3)\left[\frac{150 \times (240)^3}{12}\right] \text{ N-mm}^2$$

$$= 1728 \times 10^9 \text{ N-mm}^2$$

Similarly,

$$EI_{(steel)} = (200 \times 10^3)\left[2\left(\frac{150 \times (10)^3}{12}\right) + 2\,(150 \times 10) \times (125)^2\right] \text{ N-mm}^2$$

$$= (200 \times 10^3) \times [(0.025 \times 10^6) + (46.875 \times 10^6)] \text{ N-mm}^2$$

$$= 9380 \times 10^9 \text{ N-mm}^2$$

∴ Total flexural rigidity for the composite section about its centre of gravity,

$$\Sigma EI = (1728 \times 10^9) + (9380 \times 10^9) = 11108 \times 10^9 \text{ N-mm}^2$$

We know that slope at the free end,

$$i_B = \frac{wl^3}{6\,\Sigma EI} = \frac{10 \times (2 \times 10^3)^3}{6 \times (11108 \times 10^9)} = \mathbf{0.0012 \text{ rad}} \quad \textbf{Ans.}$$

***Deflection at the free end***

We also know that deflection at the free end,

$$y_B = \frac{wl^4}{8\,\Sigma EI} = \frac{10 \times (2 \times 10^3)^4}{8 \times (11108 \times 10^9)} = \mathbf{1.8 \text{ mm}} \quad \textbf{Ans.}$$

## EXERCISE 20.2

1. A cantilever beam $AB$ having length $L$ and stiffness $EI$ is fixed at the end $A$. A uniformly distributed load of intensity $w$/unit length acts over half of the beam from the fixed end. Obtain the expressions for slope and deflection at the end $B$.

$$\left[\textbf{Ans. } \frac{wL^3}{48EI};\ \frac{7wL^4}{384EI}\right]$$

2. A cantilever 2 m long carries a point load of 1 kN at the free end, and a uniformly distributed load of 2 kN/m over a length of 1.25 m from the fixed end. Find the deflection at the free end, if $E = 200$ GPa. Take $I = 138.24 \times 10^6\ \text{mm}^4$.

[**Ans.** 1.46 m]

3. A horizontal cantilever of uniform section and length $L$ carries a load $W$ at a distance $L/4$ from the free end. Derive from the first principles the deflection at the free end in terms of $W$, $L$, $E$ and $I$.

$$\left[\textbf{Ans. } \frac{27WL^3}{128EI}\right]$$

4. A horizontal cantilever of length $3a$ carries two concentrated loads, $W$ at a distance $a$ from the fixed end and $W'$ at the free end. Obtain the formula for the maximum deflection due to the loading.

$$\left[\textbf{Ans. } \frac{3a^2}{3EI}(2W + 7W'\right]$$

## QUESTIONS

1. Derive an expression for the slope and deflection at the free end of a cantilever $AB$ of span $l$ and flexural rigidity $EI$, when it is subjected to a point load at the free end.
2. Obtain an expression for the slope and deflection at the free end of a cantilever $AB$ of span $l$ and stiffness $EI$ when it is carrying a point load at a distance $l_1$ from the fixed end.
3. Show that the deflection of a cantilever at its free end $B$ is given by the relation:

$$y_B = \frac{wl^4}{8EI}$$

where $w$ = Uniformly distributed load per unit length of the cantilever,
$l$ = Span of the cantilever and
$EI$ = Flexural rigidity of the cantilever.

4. Derive an expression for the slope and deflection of a cantilever subjected to a triangular load uniformly varying from zero at the free end to $w$ at the fixed end.

## OBJECTIVE TYPE QUESTIONS

1. Maximum deflection of a cantilever beam of span $l$ carrying a point load $W$ at its free end is

(a) $\frac{Wl^3}{2EI}$ (b) $\frac{Wl^3}{3EI}$ (c) $\frac{Wl^3}{8EI}$ (d) $\frac{Wl^3}{16EI}$

where $EI$ = Rigidity of the cantilever beam.

2. The maximum slope of a cantilever carrying a point load at its free end is at the
   (*a*) fixed end (*b*) centre of span
   (*c*) free end (*d*) none of these
3. A cantilever beam of span $l$ caries a 1 uniformly distributed load $w$ over the entire span. The maximum slope of the cantilever is

   (*a*) $\dfrac{wl^2}{3EI}$ (*b*) $\dfrac{wl^2}{4EI}$ (*c*) $\dfrac{wl^3}{6EI}$ (*d*) $\dfrac{wl^3}{8EI}$

   where $EI$ = Rigidity of the beam.
4. Maximum deflection of a cantilever is equal to

   (*a*) $\dfrac{wl^4}{2EI}$ (*b*) $\dfrac{wl^4}{3EI}$ (*c*) $\dfrac{wl^4}{8EI}$ (*d*) $\dfrac{wl^4}{16EI}$

   where $w$ = Uniformly distributed load per unit length over the entire span,
   $l$ = Span of the cantilever beam and
   $EI$ = Rigidity of the cantilever beam,

## ANSWERS

| 1. (*b*) | 2. (*c*) | 3. (*c*) | 4. (*c*) |
|---|---|---|---|

Chapter 21

# Deflection by Moment Area Method

## Contents

## 21.1. Introduction

In the last chapters, we have discussed the slope and deflection of various types of beams and cantilevers. But the derivations of the relations are difficult and lengthy. But in this chapter, we shall discuss a graphical method for the slope and deflection of beams and cantilevers. This method is simple and enables us quicker solutions. It is popularly known as moment area method and is based on Mohr's theorems which are stated below:

## 21.2. Mohr's Theorems

The deflection of beams and cantilevers by moment area method is based on the following two theorems, which were stated by Mohr.

**Mohr's Theorem – I**

It states, "*The change of slope between any two points, on an elastic curve is equal to the net area of B.M. diagram between these points divided by EI.*"

**Mohr's Theorem – II**

It states, "*The intercept taken on a vertical reference line of tangents at any two points on an elastic curve, is equal to the moment of the B.M. diagram between these points about the reference line divided by EI.*"

## 21.3. Area and Position of the Centre of Gravity of Parabolas

A parabola is defined as a figure having at least one of its sides a parabolic curve. In Fig. 21.1, the side is a parabolic curve, whereas and are straight lines.

A parabolic curve is generally, expressed as , where is the degree of parabolic curve. In this chapter, we have to find the areas and positions of the centre of gravity of various parabolas. The following table gives these two values for various degrees of concave parabolic curves:

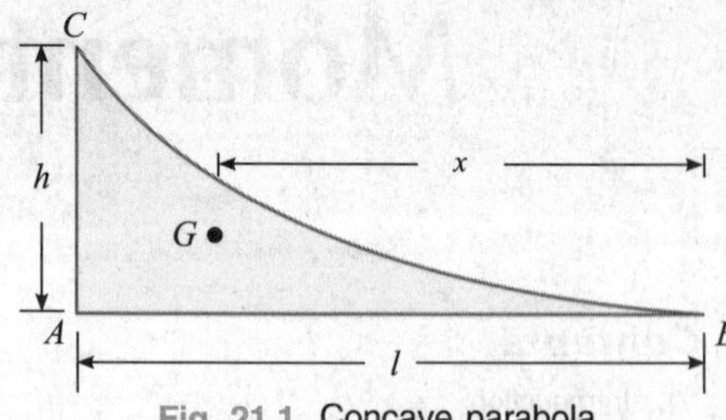


Fig. 21.1. Concave parabola

Table 21.1.

| *S. No.* | *Value of n* | *Area (A)* | *Distance between B and G (x)* |
|---|---|---|---|
| 1 | 2 | $(l \times h) \times \frac{1}{3}$ | $l \times \frac{3}{4}$ |
| 2 | 3 | $(l \times h) \times \frac{1}{4}$ | $l \times \frac{4}{5}$ |
| 3 | 4 | $(l \times h) \times \frac{1}{5}$ | $l \times \frac{5}{6}$ |

The above values of area (*A*) and distance (*x*) may also be expressed as given below:

$$\text{Area } (A) = (l \times h) \times \frac{1}{n+1}$$

and distance

$$(x) = l \times \frac{n+1}{n+2}$$

## 21.4. Simply Supported Beam with a Central Point Load

Consider a simply supported beam *AB* of length *l* and carrying a point load *W* at *i.e.*, the centre of the beam as shown in Fig. 21.2 (*a*). We know that the reaction at *A*,

$$R_A = R_B = \frac{W}{2}$$

∴ Bending moment at A due to reaction $R_B$,

$$M_1 = +\frac{Wl}{2}$$

Similarly, bending moment at *A* due to the load *W*,

$$M_2 = -W \times \frac{1}{2} = -\frac{Wl}{2}$$

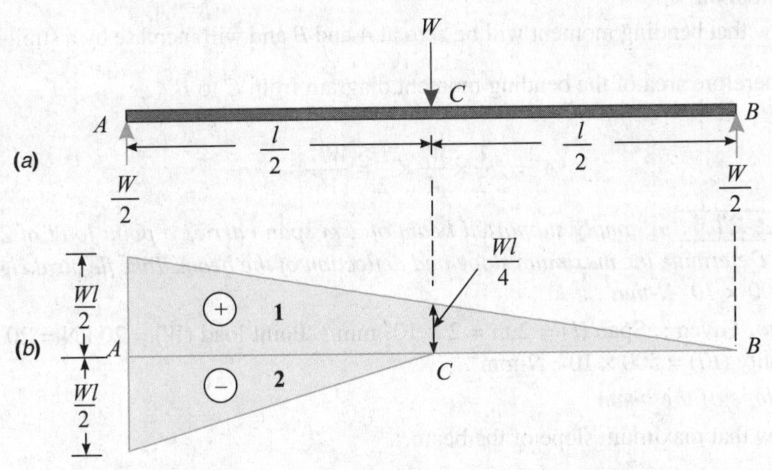


Fig. 21.2. Simply supported beam with a central point load

Now draw the bending moment diagram with the above two moments. The positive bending moment is drawn above the reference line, where negative is below it as shown in Fig. 21.2 (*b*). Such a bending moment diagram is called *component bending moment diagram.*

Now in order to find the slope at *B*, let us consider the bending moment diagram from *C* to *B*.

∴ Area of bending moment diagram from *C* to *B*,

$$*A = \frac{1}{2} \times \frac{Wl}{2} \times \frac{l}{2} = \frac{Wl^2}{16}$$

and distance of centre of gravity of the bending moment diagram from *B*,

$$\bar{x} = \frac{2}{3} \times \frac{l}{2} = \frac{l}{3}$$

$$\therefore \quad i_B = \frac{A}{EI} = \frac{Wl^2}{16EI} \quad \text{...(As before)}$$

By symmetry,

$$i_A = \frac{Wl^2}{16EI} \quad \text{...(As before)}$$

and

$$y_C = \frac{A\bar{x}}{EI} = \frac{\frac{Wl^2}{16} \times \frac{l}{3}}{EI} = \frac{Wl^3}{48EI} \quad \text{...(As before)}$$

* It may also be found out by studying the component bending moment diagram *A* to *C*. Area of bending moment digram from *A* to *C*.

$$A = \left[\frac{1}{2}\left(\frac{Wl}{2} + \frac{Wl}{4}\right) \times \frac{l}{2}\right] - \left[\frac{1}{2} \times \frac{Wl}{2} \times \frac{l}{2}\right] = \frac{Wl^2}{16}$$

and

$$A\bar{x} = A_1\bar{x}_1 - A_2\bar{x}_2 = \left(\frac{3Wl^2}{16} \times \frac{2l}{9}\right) - \left(\frac{Wl^2}{8} \times \frac{l}{6}\right)$$

$$= \frac{Wl}{24} - \frac{Wl^3}{48} = \frac{Wl^3}{48}$$

***Alternative method***

We know that bending moment will be zero at $A$ and $B$ and will increase by a straight line law to $\frac{Wl}{4}$ at $C$. Therefore area of the bending moment diagram from $C$ to $B$,

$$A = \frac{1}{2} \times \frac{Wl}{4} \times \frac{l}{2} = \frac{Wl^2}{16}$$

**EXAMPLE 21.1.** *A simply supported beam of 2 m span carries a point load of 20 kN at its mid-point. Determine the maximum slope and deflection of the beam. Take flexural rigidity of the beam as 500 × $10^9$ N-mm².*

**SOLUTION.** Given : Span ($l$) = 2 m = 2 × $10^3$ mm ; Point load ($W$) = 20 kN = 20 × $10^3$ N and flexural rigidity ($EI$) = 500 × $10^9$ N-mm².

***Maximum slope of the beam***

We know that maximum slope of the beam,

$$i_B = \frac{Wl^2}{16\,EI} = \frac{(20 \times 10^3) \times (2 \times 10^3)^2}{16 \times (500 \times 10^9)} = 0.01 \text{ rad} \quad \textbf{Ans.}$$

***Maximum deflection of the beam***

We also know that maximum defection of the beam at its centre,

$$y_C = \frac{Wl^3}{48\,EI} = \frac{(20 \times 10^3) \times (2 \times 10^3)^3}{48 \times (500 \times 10^9)} = 6.67 \text{ mm} \quad \textbf{Ans.}$$

## 21.5. Simply supported Beam with an Eccentric Point Load

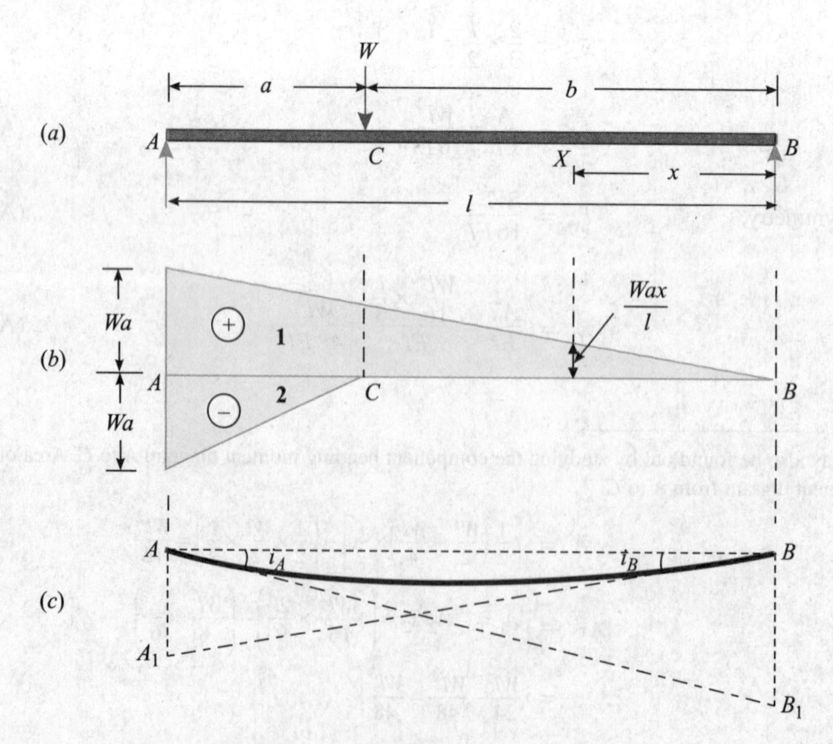


Fig. 21.3. Eccentric point load.

Consider a simply supported beam $AB$ of length $l$ and carrying a point load $W$ at $C$, such that $AC = a$ and $CB = b$ as shown in Fig. 21.3 (*a*). We know that the reaction at $A$,

$$R_A = \frac{Wb}{l} \quad \text{and} \quad R_B = \frac{Wa}{l}$$

and bending moment at $A$ due to reaction $R_B$,

$$M_1 = +\frac{Wa}{l} \times l = +Wa$$

Similarly, bending moment at A due to the load $W$,

$$M_2 = -Wa$$

Now draw the compound bending moment Diagram as shown in Fig. 21.3 (*b*). We know that area of the positive bending moment Diagram,

$$A_1 = \frac{1}{2} \times Wa \times l = \frac{Wal}{2}$$

and area of negative bending moment diagram.

$$A_2 = \frac{1}{2} \times Wa \times a = \frac{Wa^2}{2}$$

From the geometry of the loading, we see that the slope at any section is not known. It is thus obvious that the slope and deflection cannot be found out directly. Now draw vertical lines through $A$ and $B$. Let $AA_1$ and $BB_1$ be equal to intercepts of the tangents at $A$ and $B$ as shown in Fig. 21.3 (*c*). We see that,

$$AA_1 = i_B \times l$$

But $$AA_1 = \frac{A_1 x_1 - A_2 x_2}{EI} = \frac{I}{EI}\left[\left(\frac{Wal}{2} \times \frac{l}{3}\right) - \left(\frac{Wa^2}{2} \times \frac{a}{3}\right)\right] = \frac{Wa}{6EI}(l^2 - a^2)$$

$$\therefore \quad i_B = \frac{AA_1}{l} = \frac{Wa}{6EIl}(l^2 - a^2) \qquad \text{...(As before)}$$

Similarly, $$i_A = \frac{Wb}{6EIl}(l^2 - a^2) \qquad \text{...(As before)}$$

Now consider any section $X$ at a distance $x$ from $B$. We find that the area of bending moment diagram between $X$ and $B$,

$$A = \frac{1}{2} \times \frac{Wax}{l} \times x = \frac{Wax^2}{2l} \qquad \text{...(Plus sign due to anti-clockwise on the right)} \quad ...(ii)$$

$\therefore$ Change of slope between $X$ and $B$

$$= \frac{A}{WI} = \frac{Wax^2}{2lEI} \qquad \text{...(Minus sign due to clockwise on the right)} \quad ...(iii)$$

Now for maximum deflection, the slope at $X$ should be equal to zero. Or in other words, the change of slope between $B$ and $X$ should be equal to the slope at $B$. *i.e.*,

$$\frac{Wa}{6lEI}(l^2 - a^2) = \frac{Wax^2}{2lEI}$$

$$\therefore \quad x^2 = \frac{l^2 - a^2}{3} \quad \text{or} \quad x = \sqrt{\frac{l^2 - a^2}{3}}$$

We have seen in equation (*ii*) that the area of bending moment diagram between *X* and *B*,

$$A = \frac{Wax^2}{2l}$$

and distance of centre of gravity of bending moment diagram from *B*,

$$\bar{x} = \frac{2x}{3}$$

$$\therefore \qquad y_x = \frac{A\bar{x}}{EI} = \frac{\frac{Wax^2}{2l} \times \frac{2x}{3}}{EI} = \frac{Wax^3}{3\,EIl} \qquad ...(iv)$$

Now for maximum deflection, substituting the value of $x = \sqrt{\frac{l^2 - a^2}{3}}$ in equation (*iv*),

$$y_{max} = \frac{Wa}{3lEI}\left(\sqrt{\frac{l^2 - a^2}{3}}\right)^3 = \frac{Wa}{9\sqrt{3}\,lEI}(l^2 - a^2)^{\frac{3}{2}} \qquad ...(\text{As before})$$

**EXAMPLE 21.2.** *A simply supported beam AB of 2.8 m span carries a point load of 60 kN at a distance of 1 m from the left hand support A. What is the position of the maximum deflection of the beam? Also find the magnitude of the deflection under the load. Take EI for the beam section as $4 \times 10^{12}$ N-mm².*

**SOLUTION.** Given: Span ($l$) = 2.8 m = $2.8 \times 10^3$ mm ; Point load ($W$) = 60 kN = $60 \times 10^3$ N ; Distance between the point load and the left hand support ($a$) = 1 m = $1 \times 10^3$ mm and flexural rigidity of the beam section ($EI$) = $4 \times 10^{12}$ N-mm².

***Position of the maximum deflection***

We know that position of the maximum deflection (or distance between the point of maximum deflection and left hand support *A*).

$$x = \sqrt{\frac{l^2 - a^2}{3}} = \sqrt{\frac{(2.8 \times 10^3)^2 - (1 \times 10^3)^2}{3}} \text{ mm}$$

$$= 1.51 \times 10^3 \text{ mm} = 1.51 \text{ m} \qquad \textbf{Ans.}$$

***Magnitude of deflection under the load***

We know that distance between the point load and right hand support *B*,

$$b = 1 - a = (2.8 \times 10^3) - (1 \times 10^3) = 1.8 \times 10^3 \text{ m}$$

and magnitude of deflection under the load

$$= \frac{Wab}{6\,EIl}(l^2 - a^2 - b^2)$$

$$= \frac{(60 \times 10^3) \times (1 \times 10^3) \times (1.8 \times 10^3)}{6 \times (4 \times 10^{12}) \times (2.8 \times 10^3)} \times [(2.8 \times 10^3)^2 - (1 \times 10^3) - (1.8 \times 10^3)^2]$$

$$= (1.61 \times 10^{-6}) \times (3.6 \times 10^6) = 5.8 \text{ mm} \qquad \textbf{Ans.}$$

## 21.6. Simply Supported Beam with a Uniformly Distributed Load

Consider a simply supported beam *AB* of length *l* and carrying a uniformly distributed load of *w* per unit length as shown in Fig. 21.4 (*a*). We know that the reaction at *A*,

$$R_A = R_B = \frac{wl}{2}$$

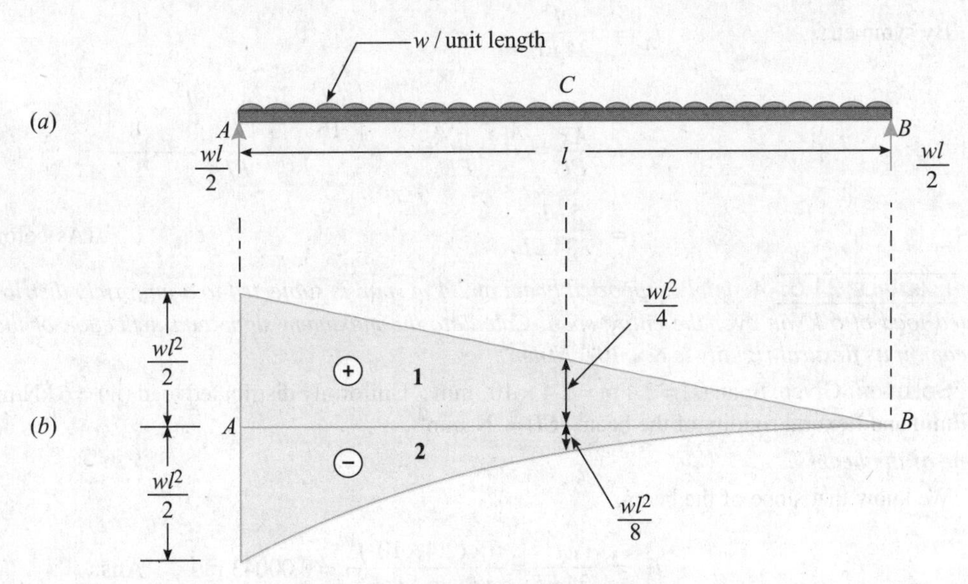


Fig. 21.4. Uniformly distributed load.

and bending moment at $A$ due to reaction $R_B$,

$$M_1 = \frac{wl}{2} \times l = \frac{wl^2}{2}$$

...(Plus sign due to anti-clock wise on the right)

Similarly, bending moment at A due to load $w$,

$$M_2 = -wl \times \frac{1}{2} = \frac{wl^2}{2}$$

Now draw the compound bending moment diagram as shown in Fig. 21.4 ($b$). We know that area of the positive bending moment diagram between $C$ and $B$,

$$A_1 = +\frac{1}{2} \times \frac{wl^2}{4} \times \frac{l}{2} = \frac{wl^3}{16}$$

...(Minus sign due to clock wise on the right)

and area of the negative bending moment diagram between $C$ and $B$,

$$A_2 = \frac{1}{3} \times \frac{wl^2}{8} \times \frac{l}{2} = \frac{wl^3}{48}$$

∴ Net area bending moment diagram from $C$ to $B$,

$$A = A_1 - A_2 = \frac{wl^3}{16} - \frac{wl^3}{48} = \frac{wl^3}{24}$$

and distance of centre gravity of the positive bending moment diaram on $CB$ from $B$,

$$x_1 = \frac{2}{3} \times \frac{l}{2} = \frac{l}{3}$$

Similarly, distance of the negative bending moment diagram on $CB$ from $B$,

$$\bar{x} = \frac{3}{4} \times \frac{1}{2} = \frac{3l}{8}$$

$$\therefore \quad i_B = \frac{A}{EI} = \frac{wl^3}{24\,EI}$$

...(As before)

By symmetry,
$$i_A = \frac{wl^3}{24\,EI}$$

and
$$y_C = \frac{A\bar{x}}{EI} = \frac{A_1\,\bar{x}_1 - A_2\,\bar{x}_2}{EI} = \frac{\left(\frac{wl^3}{16} \times \frac{l}{3}\right) - \left(\frac{wl^3}{48} \times \frac{3l}{8}\right)}{EI}$$

$$= \frac{5\,wl^4}{384\,EI} \qquad \text{...(As before)}$$

**EXAMPLE 21.3.** *A simply supported beam of 2.4 m span is subjected to a uniformly distributed load of 6 kN/m over the entire span. Calculate the maximum slope and deflection of the beam, if its flexural rigidity is $8 \times 10^{12}$ N-mm$^2$.*

**SOLUTION.** Given: Span ($l$) = 2.4 m = $2.4 \times 10^3$ mm ; Uniformly distributed load ($w$) = 6 kN/m = 6 N/mm and flexural rigidity of the beam ($EI$) = N-mm$^2$.

***Slope of the beam***

We know that slope of the beam,
$$i_A = \frac{wl^3}{24\,EI} = \frac{6 \times (2.4 \times 10^3)^3}{24 \times (8 \times 10^{12})} = 0.00043 \text{ rad} \qquad \textbf{Ans.}$$

***Deflection of the beam***

We also know that maximum deflection of the beam,
$$y_C = \frac{5\,wl^4}{384\,EI} = \frac{5 \times 6 \times (2.4 \times 10^3)^4}{384 \times (8 \times 10^{12})} = 0.324 \text{ mm} \qquad \textbf{Ans.}$$

**EXAMPLE 21.4** *A beam AB of length l is loaded with a uniformly distributed load as shown in Fig. 21.5.*

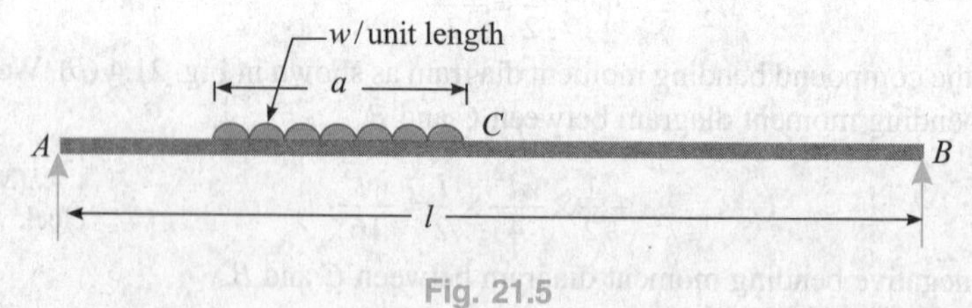


Fig. 21.5

*Determine by moment area method, the central deflection of the beam.*

**SOLUTION.** Given: Span = $l$ ; Uniformly distributed load = $w$ / unit length and loaded portion of the beam = $a$.

For the sake of convenience, let us assume another load of $w$ / unit length, to act for a length of $a$ in $CB$ as shown in Fig. 21.6 ($a$). We know that the reaction at $A$,
$$R_A = R_B = wa$$

∴ Bending moment at $A$ due to reaction $R_B$,
$$M_1 = wal$$

Similarly, bending moment at $A$ due to load,
$$M_2 = -wa \times \frac{l}{2} = -\,wal$$

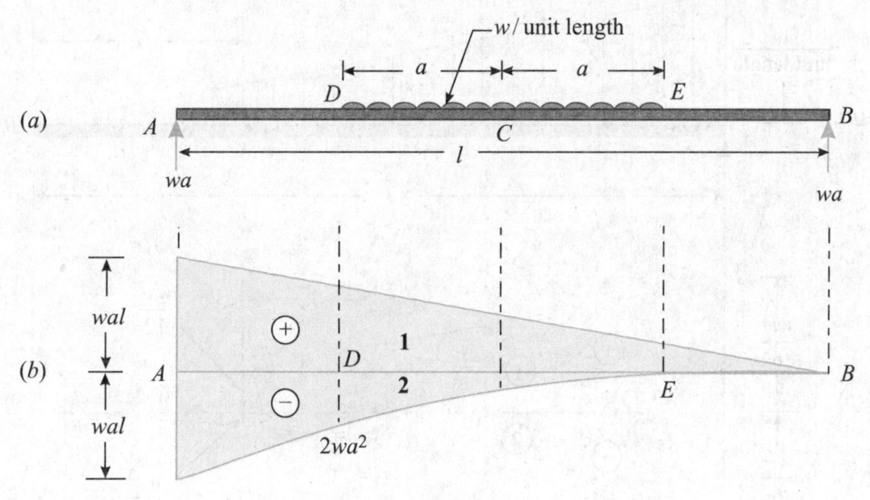


Fig. 21.6

Now draw the compound bending moment diagram as shown in Fig. 21.6 (*b*).

∴ Area of the positive bending moment diagram from *C* to *B*,

$$A_1 = \frac{1}{2} \times \frac{wal}{2} \times \frac{l}{2} = \frac{wal^2}{8}$$

and area of the negative bending moment as diagram from C to B,

$$A_2 = -\frac{1}{3} \times \frac{wa^2}{2} \times a = -\frac{wa^3}{6}$$

We know that the distance of centre of gravity of the bending moment diagram on CB from B,

$$\bar{x}_1 = \frac{2}{3} \times \frac{l}{2} = \frac{l}{3}$$

Similarly,

$$\bar{x}_2 = \frac{l}{2} - a + \frac{3a}{4} = \left(\frac{l}{2} - \frac{a}{4}\right)$$

∴

$$2y_C = \frac{A\bar{x}}{EI} = \frac{A_1\bar{x}_1 + A_2\bar{x}_2}{EI} = \frac{\left[\frac{wal^2}{8} \times \frac{l}{3}\right] + \left[-\frac{wa^3}{6} \times \left(\frac{l}{2} - \frac{a}{4}\right)\right]}{EI}$$

$$= \frac{\frac{wal^3}{24} - \frac{wa^3}{24}(2l - a)}{EI} = \frac{wa}{24\,EI}(l^3 - 2la^2 + a^3)$$

or

$$y_C = \frac{wa}{48\,EI}(l^3 - 2la^2 + a^3) \quad \textbf{Ans.}$$

## 21.7. Simply Supported Beam with a Gradually Varying Load

Consider a simply supported beam *AB* of length *l* and carrying a gradually varying load from zero at *B* to *w* per unit length at *A* as shown in Fig. 21.7 (*a*). We know that the reaction at *A*,

$$R_A = \frac{wl}{3} \quad \text{and} \quad R_B = \frac{wl}{6}$$

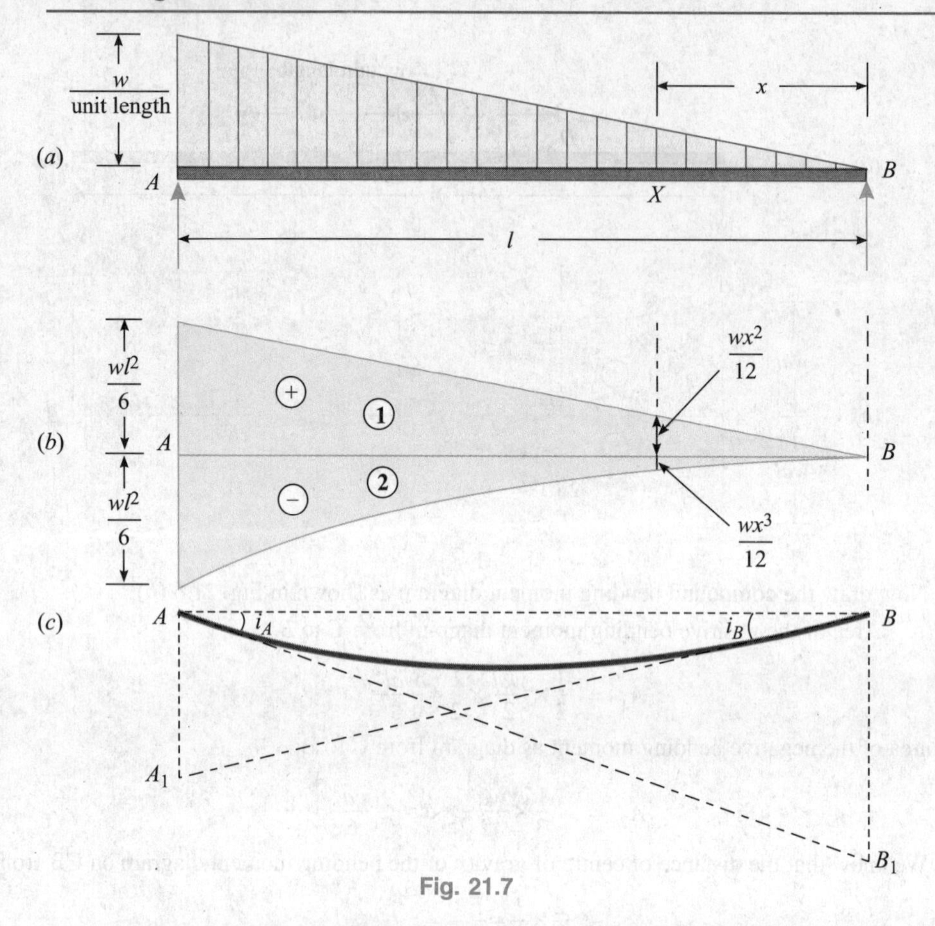


Fig. 21.7

∴ Bending moment at A due to reaction $R_B$,

$$M_1 = \frac{wl}{6} \times l = +\frac{wl^2}{6}$$

Similarly, bending moment at A due to the load,

$$M_2 = -\frac{wl}{2} \times \frac{l}{3} = -\frac{wl^2}{6}$$

Now draw the compound bending moment diagram as shown in Fig. 21.7 (*b*),

∴ Area of the positive bending moment diagram,

$$A_1 = \frac{1}{2} \times \frac{wl^2}{6} \times l = \frac{wl^3}{12}$$

and area of the negative bending moment diagram,

$$A_2 = \frac{1}{4} \times \frac{wl^2}{6} \times l = \frac{wl^3}{24}$$

From the geometry of the loading, we see that the slope at any section is not known. It is thus obvious that the slope and deflection cannot be found out directly. Now draw vertical lines through $A$ and $B$. Let $AA_1$ and $BB_1$ be equal to the intercepts of the tangents at $A$ and $B$ as shown in Fig. 14.7 (*c*).

We see that $AA_1 = i_B \times l$ and $BB_1 = i_A \times l$

But
$$AA_1 = \frac{A_1\bar{x}_1 - A_2\bar{x}_2}{EI} = \frac{I}{EI}\left[\left(\frac{wl^3}{12}\times\frac{l}{3}\right)-\left(\frac{wl^3}{24}\times\frac{l}{5}\right)\right]=\frac{7\,wl^4}{360\,EI}$$

$$\therefore \quad i_B = \frac{7\,wl^4}{360\,EI} \text{ radians} \qquad \text{...(As before)}$$

Similarly
$$BB_1 = \frac{A_1 x_1 - A_2 x_2}{EI} = \frac{1}{EI}\left[\left(\frac{wl^3}{12}\times\frac{2l}{3}\right)-\left(\frac{wl^3}{24}\times\frac{4l}{5}\right)\right]=\frac{wl^4}{45\,EI}$$

$$\therefore \quad i_A = \frac{wl^4}{45\,EI} \text{ radians} \qquad \text{...(As before)}$$

Now consider any section $X$, at a distance $x$ from $B$. We find that the area of bending moment diagram between $X$ and $B$,

$$A = \left(\frac{1}{2}\times\frac{wlx}{6}\times x\right)-\left(\frac{1}{4}\times\frac{wx^3}{6l}\times x\right)=\frac{wlx^2}{12}-\frac{wx^4}{24l}$$

$$\therefore \quad \text{Slope at } X, \quad i_X = \frac{A}{EI}=\frac{I}{EI}\left(\frac{wlx^2}{12}-\frac{wx^4}{24l}\right)$$

Now for maximum deflection, the slope at $X$ should be equal to zero, or in other words the change of slope between $B$ and $x$ should be equal to the slope at $X$,

$$\therefore \quad \frac{7\,xl^4}{360\,EI} = \frac{1}{EI}\left(\frac{wlx^2}{12}-\frac{wx^4}{24l}\right)$$

$$7\,l^4 = 30\,lx^2-\frac{15x^4}{l}$$

or
$$x = 0.519\,1$$

We know that deflection of the beam at $X$ (considering the portion $XB$ of the beam),

$$y_X = \frac{A_1 x_1 - A_2 x_2}{EI} = \frac{1}{EI}\left[\left(\frac{wlx^2}{12}\times\frac{2x}{3}\right)-\left(\frac{wx^4}{24l}\times\frac{4x}{5}\right)\right]$$

$$= \frac{1}{EI}\left[\frac{wlx^3}{18}-\frac{wx^5}{30l}\right]$$

Now for the deflection at the centre substituting $x = l/2$ in the above equation,

$$y_C = \frac{0.00651\,wl^4}{EI} \qquad \text{...(As before)}$$

For maximum deflection, substituting the value of $x = 0.519\,1$ and the above equation,

$$y_{max} = \frac{0.00652\,wl^4}{EI} \qquad \text{...(As before)}$$

**EXAMPLE 21.5.** *A beam of span 3.6 m is simply supported over its both ends. If the beam is subjected to a triangular load of 3 kN/m at A to zero at B, find the values of slopes at A and B. Take flexural rigidity for the beam section as 6 × $10^{12}$ N-mm$^2$.*

**Solution.** Given: Span ($l$) = 3.6 m = 3.6 × $10^3$ mm ; Load at $A$ ($w$) = 3 kN/m = 3 N/mm and flexural rigidity ($EI$) = 6 × $10^{12}$ N-mm².

We know that slope at $A$,

$$i_A = \frac{wl^3}{45\,EI} = \frac{3 \times (3.6 \times 10^3)^3}{45 \times (6 \times 10^{12})} = 0.00052 \text{ rad} \quad \textbf{Ans.}$$

and slope at $B$,

$$i_B = \frac{7\,wl^3}{360\,EI} = \frac{7 \times 3 \times (3.6 \times 10^3)^3}{360 \times (6 \times 10^{12})} = 0.00045 \text{ rad} \quad \textbf{Ans.}$$

## EXERCISE 21.1

1. A simply supported beam 2.4 m span is subjected to a central point load of 15 kN. Determine the maximum slope and deflection of the beam at its centre. Take $EI$ for the beam section as 6 × $10^{10}$ N-mm². [**Ans.** 0.09 rad ; 7.2 mm]
2. A simply supported beam of span 6 meters is subjected a point load of 40 kN at a distance 4 m from the left hand support. Calculate the position of maximum deflection of the beam. [**Ans.** 2.58 m from the left end support]
3. A simply supported beam of span 3 m is carrying a uniformly distributed load of 10 kN/m. Find the values of maximum slope and deflection of the beam. Take modulus of rigidity for the beam section as 10 × $10^9$ N-mm². [**Ans.** 0.0113 rad ; 10.5 mm]
4. A simply supported beam of span 2.5 m carries a gradually varying load from zero to 10 kN/m. What is the maximum deflection of the beam? Take $EI$ for the beam section as 1.2 × $10^{12}$ N-mm². [**Ans.** 2.1 mm]

## 21.8. Cantilever with a Point Load at the Free End

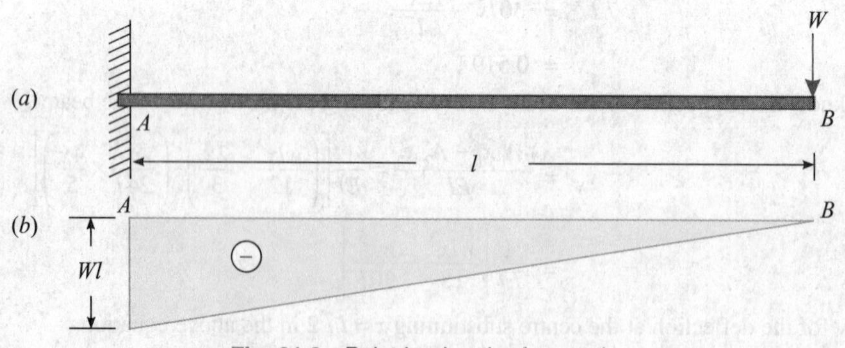


Fig. 21.8. Point load at the free end.

Consider a cantilever $AB$ of length $l$ and carrying a point load $W$ at the free end as shown in Fig. 21.8 ($a$). We know that the bending moment will be zero at $B$ and will increase by a straight line law to $Wl$ at $A$ as shown in Fig. 21.8 ($b$).

∴ Area of bending moment diagram,

$$A = \frac{1}{2} \times Wl \cdot l = \frac{Wl^2}{2}$$

and distance between the centre of gravity of bending moment diagram and $B$,

$$\bar{x} = \frac{2l}{3}$$

$$\therefore \qquad i_B = \frac{A}{EI} = \frac{Wl^2}{2EI} \text{ radians} \qquad \text{...(As before)}$$

and

$$y_B = \frac{A\bar{x}}{EI} = \frac{\frac{Wl^2}{2} \times \frac{2l}{3}}{EI} = \frac{Wl^3}{3EI} \qquad \text{...(As before)}$$

**EXAMPLE 21.6.** *A cantilever beam of 2.0 m span is subjected to a point load of 30 kN at its free end. Find the slope and deflection of the free end. Take EI for the beam as $8 \times 10^{12}$ N-mm$^2$.*

**SOLUTION.** Given : Span $(l) = 2 \text{ m} = 2 \times 10^3$ mm ; Point load $(W) = 30 \text{ kN} = 30 \times 10^3$ N and flexural rigidity $(EI) = 8 \times 10^{12}$ N-mm$^2$.

***Slope at the free end***

We know that slope of the free end,

$$i_B = \frac{Wl^2}{2EI} = \frac{(30 \times 10^3) \times (2 \times 10^3)^2}{2 \times (8 \times 10^{12})} = 0.0075 \text{ rad} \qquad \textbf{Ans.}$$

***Deflection of the free end***

We also know that deflection of the free end,

$$y_B = \frac{Wl^3}{3EI} = \frac{(30 \times 10^3) \times (2 \times 10^3)^3}{3 \times (8 \times 10^{12})} = 10 \text{ mm} \qquad \textbf{Ans.}$$

## 21.9. Cantilever with a Point Load at Any Point

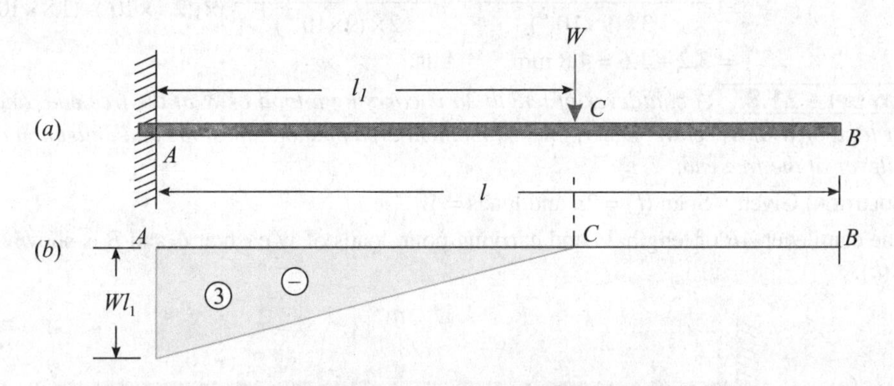


Fig. 21.9. Point load not at the face end.

Consider a cantilever $AB$ of length $l$ and carrying a point load $W$ at a distance $l_1$ from the fixed end as shown in Fig. 21.9 (*a*). We know that the bending moment will be zero at $B$ and $C$, and will increase by a straight line law to $Wl_1$ and $A$ as shown in Fig. 21.9 (*b*).

Therefore area of bending moment diagram,

$$A = \frac{1}{2} \times Wl_1 \times l_1 = \frac{Wl_1^2}{2}$$

and distance between the centre of gravity of bending moment diagram and $B$,

$$\bar{x} = \frac{2l_1}{3} + (l - l_1)$$

$$\therefore \qquad i_B = \frac{A}{EI} = \frac{Wl_1^2}{2\,EI} \text{ radians} \qquad \text{...(As before)}$$

and

$$y_B = \frac{A\bar{x}}{EI} = \frac{Wl_1^2 \times \left[\frac{2l_1}{3} + (l - l_1)\right]}{EI}$$

$$= \frac{Wl_1^3}{3\,EI} + \frac{Wl_1^2}{2\,EI}(l - l_1) \qquad \text{....(As before)}$$

**EXAMPLE 21.7.** *A cantilever beam of span 2.4 m carries a point load of 15 kN at a distance of 1.8 m from the fixed end. What are the values of slope and deflection at the free end of the cantilever, if the flexural rigidity for the beam section is $9 \times 10^{12}$ N-mm$^2$.*

**SOLUTION.** Given : Span $(l) = 2.4 \text{ m} = 2.4 \times 10^3$ mm ; Point load $(W) = 15 \text{ kN} = 15 \times 10^3$ N ; Distance of the load from the fixed end $(l_1) = 1.8 \text{ m} = 1.8 \times 10^3$ mm and flexural rigidity $(EI) = 9 \times 10^{12}$ N-mm$^2$.

***Value of slope at the free end***

We know that value of slope at the free end,

$$i_B = \frac{Wl_1^2}{2\,EI} = \frac{(15 \times 10^3) \times 1.8 \times 10^3)^3}{2 \times (9 \times 10^{12})} = 0.0027 \text{ rad} \qquad \textbf{Ans.}$$

***Value of deflection at the free end***

We also know that value of deflection at the free end,

$$y_B = \frac{Wl_1^3}{3\,EI} + \frac{Wl_1^2}{2\,EI}(l - l_1)$$

$$= \frac{(15 \times 10^3) \times (1.8 \times 10^3)^3}{3 \times (9 \times 10^{12})} + \frac{(15 \times 10^3) \times (1.8 \times 10^3)^2}{2 \times (9 \times 10^{12})} \times [(2.4 \times 10^3) - (1.8 \times 10^3)] \text{ mm}$$

$$= 3.2 + 1.6 = 4.8 \text{ mm} \qquad \textbf{Ans.}$$

**EXAMPLE 21.8.** *A cantilever of length 2a is carrying a load of W at the free end, and another load of W at its centre. Determine, by moment area method, the slope and deflection of the cantilever at the free end.*

**SOLUTION.** Given : Span $(l) = 2a$ and loads $= W$

The cantilever $AB$ of length $2a$ and carrying point loads of $W$ each at $C$ and $B$ is shown in Fig. 21.10 (*a*).

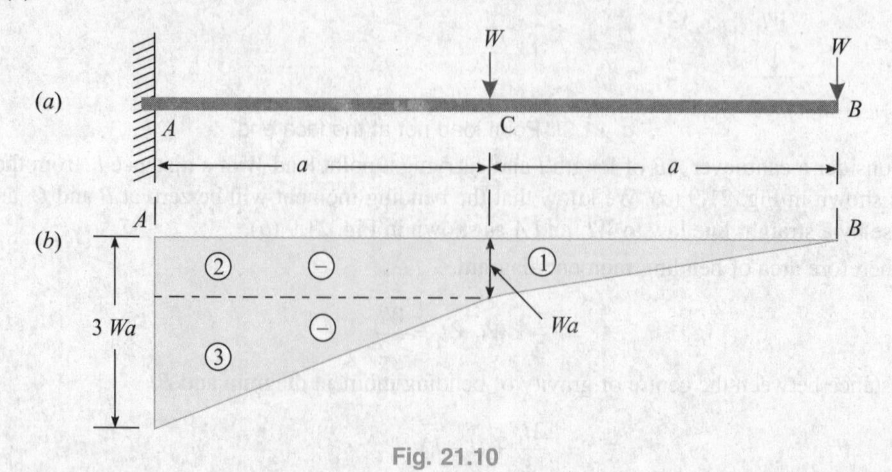


Fig. 21.10

*Slope at the free end*

We know that the bending moment at *B*,

$$M_B = 0$$
$$M_C = -Wa$$
$$M_A = -(W \times 2a) - (Wa) = -3Wa$$

Now draw the bending moment diagram as shown in Fig. 21.10 (*b*).

We know that area of bending moment diagram 1,

$$A_1 = \frac{1}{2} \times Wa \times a = \frac{Wa^2}{2}$$

Similarly, area of the bending moment diagram 2,

$$A_2 = Wa \times a = Wa^2$$

and area of bending moment 3,

$$A_3 = \frac{1}{2} \times 2Wa \times a = Wa^2$$

∴ Total area of bending moment diagram,

$$A = A_1 + A_2 + A_3 = \frac{Wa^2}{2} + Wa^2 + Wa^2 = \frac{5Wa^2}{2}$$

and slope of the cantilever at the free end *B*,

$$i_B = \frac{A}{EI} = \frac{5Wa^2}{2EI} \quad \textbf{Ans.}$$

*Deflection at the free end*

We also know that total moment of the bending moment diagram about *B*,

$$A\bar{x} = A_1 \bar{x}_1 + A_2 \bar{x}_2 + A_3 \bar{x}_3$$
$$= \left(\frac{Wa^2}{2} \times \frac{2a}{3}\right) + \left(Wa^2 \times \frac{3a}{2}\right) + \left(Wa^2 \times \frac{5a}{3}\right) = \frac{7Wa^3}{2}$$

∴ Deflection of the cantilever at the free end *B*,

$$y_B = \frac{A\bar{x}}{EI} = \frac{7Wa^3}{2EI}$$

## 21.10. Cantilever with a Uniformly Distributed Load

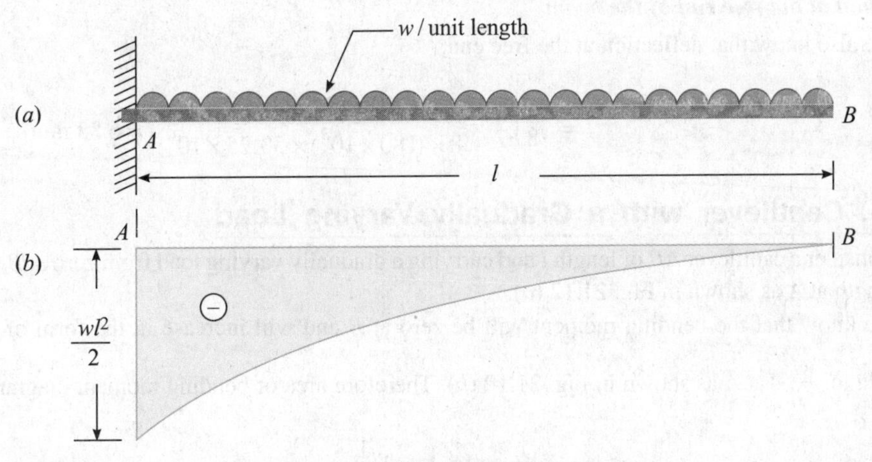


Fig. 21.11. Uniformly distributed load.

Consider a cantilever *AB* of length *l*, and carrying a uniformly distributed load of *w* per unit length as shown in Fig. 21.11 (*a*).

We know that the bending moment will be zero at and will increase in the form of a parabola to $\frac{wl^2}{2}$ at *A* as shown in Fig. 21.11 (*b*). Therefore area of bending moment diagram,

$$A = \frac{wl^2}{2} \times l \times \frac{1}{3} = \frac{Wl^3}{6}$$

and distance between the centre of gravity of bending moment diagram and *B*,

$$\bar{x} = \frac{3l}{4}$$

$$\therefore \qquad i_B = \frac{A}{EI} = \frac{wl^3}{6\,EI} \text{ radians} \qquad \text{...(As before)}$$

and

$$y_B = \frac{A\bar{x}}{EI} = \frac{\frac{wl^3}{6} \times \frac{3l}{4}}{EI} = \frac{wl^4}{8\,EI} \qquad \text{...(As before)}$$

**EXAMPLE 21.9.** *A cantilever beam 120 mm wide and 150 mm deep carries a uniformly distributed load of 10 kN/m over its entire length of 2.4 meters. Find the slope and deflection of the beam at its free end. Take E = 180 GPa.*

**SOLUTION.** Given : Width (*b*) = 120 mm ; Depth (*d*) = 150 mm ; Uniformly distributed load (*w*) = 10 kN/m = 10 N/mm ; Length (*l*) = 2.4 m = $2.4 \times 10^3$ mm and modulus of elasticity (*E*) = 180 GPa = $180 \times 10^3$ N-mm$^2$.

***Slope at the free end of the beam***

We know that moment of inertia of the cantilever beam section,

$$I = \frac{bd^3}{12} = \frac{120 \times (150)^3}{12} = 33.75 \times 10^6 \text{ mm}^4$$

and slope at the free end,

$$i_B = \frac{wl^3}{6\,EI} = \frac{10 \times (2.4 \times 10^3)^3}{6 \times (180 \times 10^3) \times (33.75 \times 10^6)} = 0.0038 \text{ rad} \qquad \textbf{Ans.}$$

***Deflection at the free end of the beam***

We also know that deflection at the free end,

$$y_B = \frac{wl^4}{8\,EI} = \frac{10 \times (2.4 \times 10^3)^4}{8 \times (180 \times 10^3) \times 33.75 \times 10^6)} = 6.83 \text{ mm} \qquad \textbf{Ans.}$$

## 21.11. Cantilever with a Gradually Varying Load

Consider a cantilever *AB* of length *l* and carrying a gradually varying load from zero at *B* to *w* per unit length at *A* as shown in Fig. 21.12 (*a*).

We know that the bending moment will be zero at *B* and will increase in the form of a cubic parabola to $\frac{wl^2}{6}$ at *A* as shown in Fig. 21.12 (*b*). Therefore area of bending moment diagram,

$$A = \frac{wl^2}{6} \times l \times \frac{1}{4} = \frac{wl^3}{24}$$

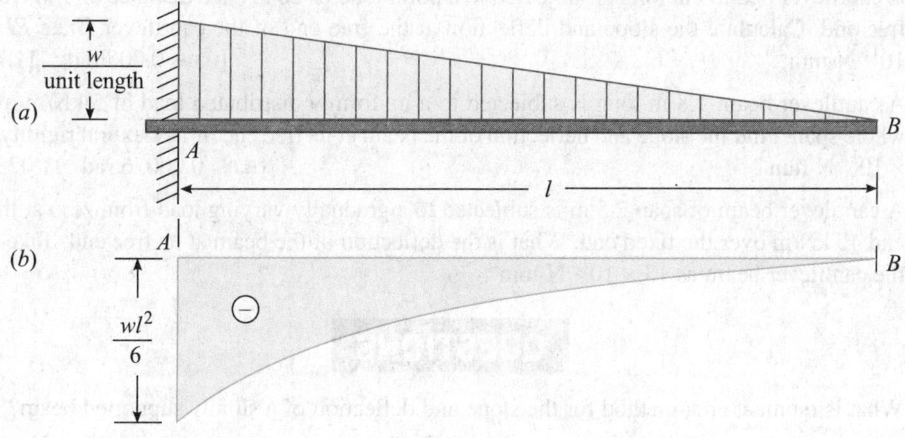


Fig. 21.12

and distance between centre of gravity of the bending moment diagram and $B$,

$$\bar{x} = l \times \frac{4}{5}$$

$$\therefore \quad i_B = \frac{A}{EI} = \frac{wl^3}{24\,EI} \quad \text{...(As before)}$$

and

$$y_B = \frac{A\bar{x}}{EI} = \frac{\frac{wl^3}{24} \times l \times \frac{4}{5}}{EI} = \frac{wl^4}{30\,EI} \quad \text{...(As before)}$$

**EXAMPLE 21.10.** *A cantilever beam of span 2.8 m metres carries a gradually varying load from zero at free end to 20 kN/m at the fixed end. Find the value of slope and deflection at the free end, if the flexural rigidity for the beam section is.*

**SOLUTION.** Given : Span ($l$) = 2.8 m = 2.8 × $10^3$ mm ; Load at the fixed end ($w$) = 20 kN/m = 20 N/mm and flexural rigidity ($EI$) = 8 × $10^{12}$ N-mm$^2$.

***Value of slope at the free end***

We know that value of slope at the free end,

$$i_B = \frac{wl^3}{24\,EI} = \frac{20 \times (2.8 \times 10^3)^3}{24 \times (8 \times 10^{12})} = 0.0023 \text{ rad} \qquad \textbf{Ans.}$$

***Value of deflection at the free end***

We also know that value of deflection at the free end,

$$y_B = \frac{wl^4}{30\,EI} = \frac{20 \times (2.8 \times 10^3)^4}{30 \times (8 \times 10^{12})} = 5.1 \text{ mm} \qquad \textbf{Ans.}$$

## EXERCISE 21.2

1. A cantilever 2.4 m long carries a point load of 37.5 kN at its free end. Find the slope and deflection under the load. Take flexural rigidity for the beam section as 20 × $10^{12}$ N-mm$^2$.
[**Ans.** 0.0054 rad ; 8.64 mm]

2. A cantilever beam 3 m long is subjected to a point load of 20 kN at a distance of 1 m from the free end. Calculate the slope and deflection at the free end of the cantilever. Take $EI = 8 \times 10^{12}$ N-mm$^2$. [**Ans.** 0.005 rad ; 11.7 mm]

3. A cantilever beam 1.8 m long is subjected to a uniformly distributed load of 5 kN/m over its whole span. Find the slope and deflection of the beam at its free end, if its flexural rigidity is $6.4 \times 10^{12}$ N-mm$^2$. [**Ans.** 0.00076 rad ; 1.025 mm]

4. A cantilever beam of span 2.5 m is subjected to a gradually varying load from zero at the free end 40 kN/m over the fixed end. What is the deflection of the beam at its free end. Take *EI* for the cantilever beam as $13 \times 10^{12}$ N-mm$^2$. [**Ans.** 4 mm]

## QUESTIONS

1. What is moment area method for the slope and deflection of a simply supported beam?
2. What are the uses of moment area method in finding out the slope and deflection of beams?
3. Derive with the help of moment area method a relation for the deflection of a simply supported beam carrying a gradually varying load of zero intensity from one end to w per. metre on the other?
4. With the help of moment area method obtain a relation for the slope of a cantilever of span $l$ subjected to a uniformly distributed load of w per unit length?

## OBJECTIVE TYPE QUESTIONS

1. A simply supported beam of span l is carrying a point load W at its centre. The deflection of the beam at its centre is

   (*a*) $\frac{Wl^2}{12EI}$ (*b*) $\frac{Wl^2}{16EI}$ (*c*) $\frac{Wl^2}{24EI}$ (*d*) $\frac{Wl^2}{48EI}$

2. A beam of length $l$ is simply supported over its both ends. If it is carrying a uniformly distributed load of w per unit length, then its slope at the ends is

   (*a*) $\frac{Wl^3}{24EI}$ (*b*) $\frac{Wl^4}{24EI}$ (*c*) $\frac{5wl^2}{24EI}$ (*d*) $\frac{5wl^3}{24EI}$

3. A cantilever beam of span $l$ carries a uniformly distributed load of $w$ per unit length over its entire span. If its span is halved, then its slope will become

   (*a*) half (*b*) one-fourth (*c*) one-eight (*d*) one-sixteenth

4. A cantilever beam of span $l$ is carrying a triangular load of zero intensity at its free end to $w$ per unit length at its fixed end. The deflection at its free end will be

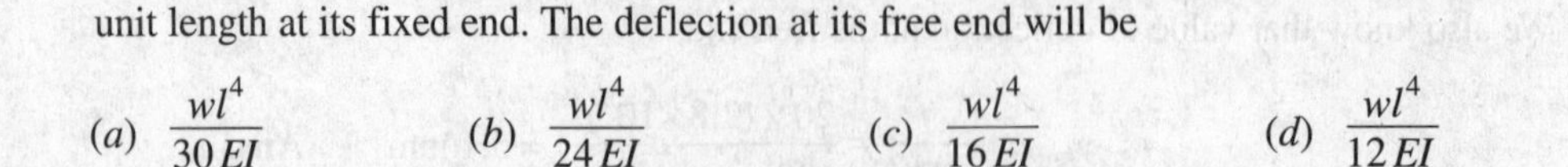

   (*a*) $\frac{wl^4}{30EI}$ (*b*) $\frac{wl^4}{24EI}$ (*c*) $\frac{wl^4}{16EI}$ (*d*) $\frac{wl^4}{12EI}$

## ANSWERS

1. (*b*) 2. (*a*) 3. (*c*) 4. (*a*)

Chapter 22

# Deflection by Conjugate Beam Method

## Contents

## 22.1. Introduction

The conjugate beam method* is a modified form of moment area method and may be conveniently used for finding out the slope and deflection of cantilevers and simply supported beams. This method is specially useful for beams and cantilevers with varying flexural rigidities.

## 22.2. Conjugate Beam

It is an *imaginary* beam of length equal to that of the original beam, width equal to $1/EI$ and

---

* This method was proposed by Prof. O. Mohr in 1868 and later on systematically developed by Prof. H.F.B. Muller-Breslau in 1885.

loaded with the usual bending moment diagram.* The slope and deflection is then found out by following two theorems known as Mohr's theorems.

**1. Mohr's first theorem**

It states, "The shear force at any section, of the conjugate beam is equal to the slope of the elastic curve at the corresponding section of the actual beam."

**2. Mohr's second theorem**

It states, "The bending moment at any section of the conjugate beam is equal to the deflection of the elastic curve at the corresponding section of actual beam."

## 22.3. Relation between an Actual Beam and the Conjugate Beam

The relations between an actual beam and the conjugate beam are given in table 22.1.

**Table 22.1**

| *S. No.* | *Actual beam* | *Conjugate beam* | *Remarks* |
|---|---|---|---|
| **1.** | Fixed end | Free end | Slope and deflection at fixed end of the actual beam is zero. S.F. and B.M. at the free end of the conjugate beam is also zero. |
| **2.** | Free end | Fixed end | Slope and deflection at the free end of the ac tual beam exist. S.F. and B.M. at the fixed end of the conjugate beam also exist. |
| **3.** | Simply supported or roller supported end | Simply supported end | Slope at the free end of the actual beam exists. But deflection is zero. S.F. at the simply supported end of the conjugate beam exists. But the B.M. is zero. |

Now we shall discuss the applications of the conjugate beam with various types of loadings on the cantilevers and beams.

## 22.4. Cantilever with a Point Load at the Free End

Consider a cantilever $AB$ of length $l$ and carrying a point load $W$ at the free end $B$ as shown in Fig. 22.1 (*a*). We know that the bending moment will be zero at $B$ and will increase by a straight line law to $(Wl)$ at $A$.

Now draw the conjugate beam $AB$ (free at $A$, fixed at $B$ and loaded with the bending moment diagram) as shown in Fig. 22.1 (*b*).

From the geometry of the figure, we find that the total load of the conjugate beam,

$$P = \frac{1}{2}(-Wl) \times l \times \frac{l}{EI} = -\frac{Wl^2}{2EI}$$

Now shear force at $B$ (on the conjugate beam $AB$),

$$F_B = -P = -\left(-\frac{Wl^2}{2EI}\right) \qquad \text{...(Minus sign due to left downwards)}$$

$$= \frac{Wl^2}{2EI}$$

* Sometimes, for the sake of convenience, the beam is assumed to be of uniform thickness and the moment diagram is adjusted according to the relation $\frac{M}{I}$.

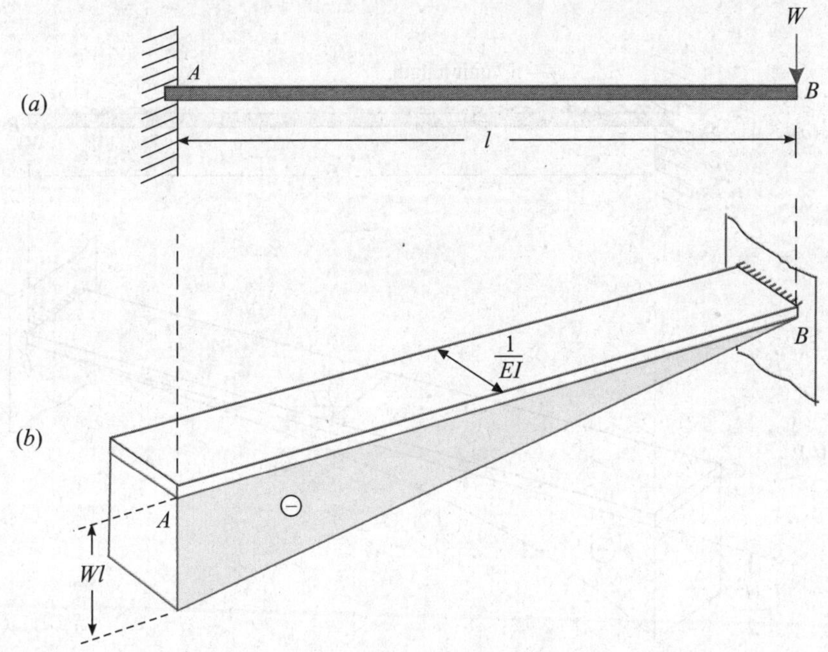


Fig. 22.1

$\therefore$ Slope at $B$, $i_B = F_B = \dfrac{Wl^2}{2EI}$ radians ...(As before)

and bending moment at $B$ (on the conjugate beam $AB$),

$$M_B = -P \times \frac{2l}{3} = -\left(-\frac{Wl^2}{2EI} \times \frac{2l}{3}\right) \quad \text{...(Minus sign due to hogging)}$$

$$= \frac{Wl^3}{3EI}$$

$\therefore$ Deflection at $B$, $y_B = M_B = \dfrac{Wl^3}{3EI}$ ...(As before)

## 22.5. Cantilever with a Uniformly Distributed Load

Consider a cantilever $AB$ of length ($l$) and carrying a uniformly distributed load $w$ per unit length as shown in Fig. 22.2 ($a$). We know that the bending moment will be zero at $B$, and will increase in the form of a parabola to $\left(-\dfrac{wl^2}{2}\right)$ at $A$.

Now draw the conjugate beam $AB$ (free at $A$, fixed at $B$ and loaded with the bending moment diagram) as shown in Fig. 22.2 ($b$).

From the geometry of the figure, we find that the total load of the conjugate beam,

$$P = \frac{1}{3}\left(-\frac{wl^2}{2}\right) \times l \times \frac{1}{EI} = -\frac{wl^3}{6EI}$$

Now shear force at $B$ (on the conjugate beam),

$$F_B = -P = \left(-\frac{wl^3}{6EI}\right) \quad \text{...(Minus sign due to left downwards)}$$

$$= \frac{wl^3}{6EI}$$

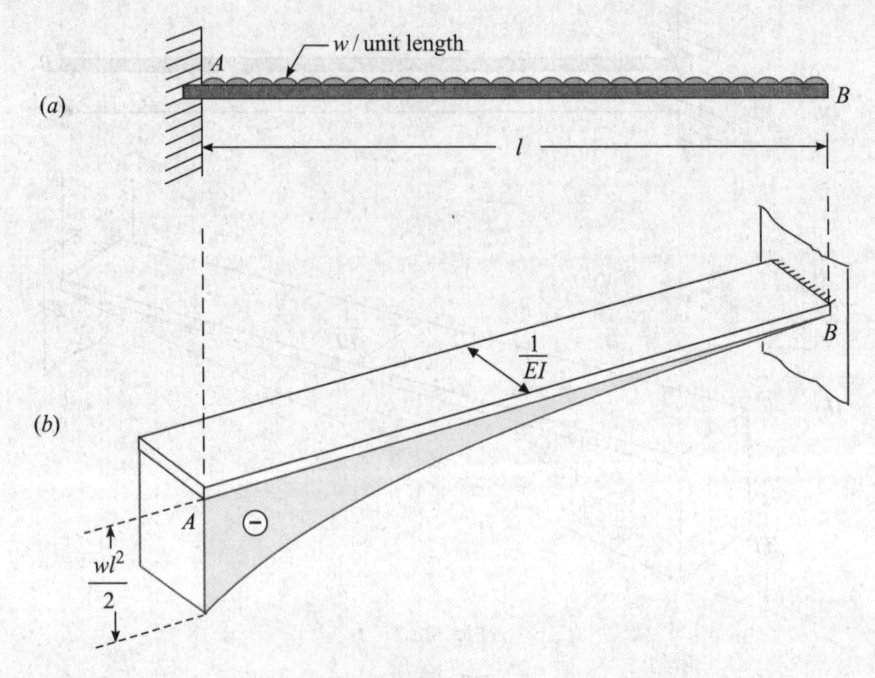


Fig. 22.2

$\therefore$ Slope at $B$, $$i_B = F_B = \frac{wl^3}{6\,EI}$$ ...(As before)

and bending moment at B (on the conjugate beam),

$$M_B = -P \times \frac{3l}{4} = -\left(-\frac{wl^3}{6\,EI} \times \frac{3l}{4}\right)$$ ...(Minus sign due to hogging)

$$= \frac{wl^4}{8\,EI}$$

$\therefore$ Deflection at $B$, $$y_B = M_B = \frac{wl^4}{8\,EI}$$ ...(As before)

## 22.6. Cantilever with a Gradually Varying Load

Consider a cantilever $AB$ of length l and carrying a gradually varying load from zero at $B$ to w per unit length at $A$ as shown in Fig. 22.3 ($a$). We know that the bending moment will be zero at $A$ and will increase in the form of cubic parabola to $\left(-\frac{wl^2}{6}\right)$ at $A$.

Now draw the conjugate beam (free at $A$, fixed at $B$ and loaded with bending moment diagram) as shown in Fig. 22.3 ($b$).

From the geometry of the figure, we find that the total load of the conjugate beam,

$$P = \frac{1}{4}\left(-\frac{wl^2}{6}\right) \times l \times \frac{1}{EI} = -\frac{wl^3}{24\,EI}$$

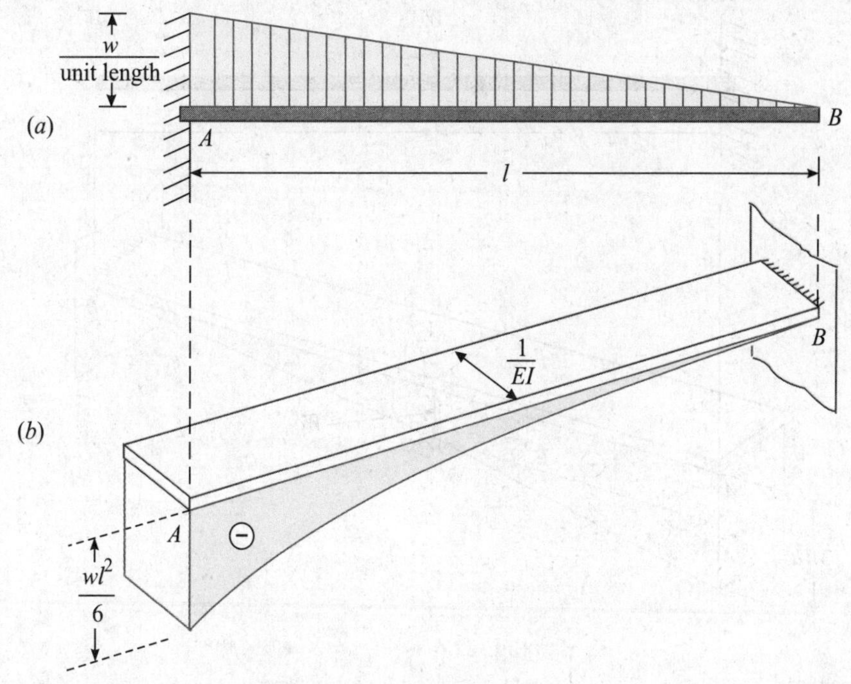


Fig. 22.3

Now shear force at $B$ (on the conjugate beam),

$$F_B = -P = -\left(-\frac{wl^3}{24\,EI}\right) \qquad \text{...(Minus sign due to left downward)}$$

$$= \frac{wl^3}{24\,EI}$$

$\therefore$ Slope at $B$, $$i_B = F_B = \frac{wl^3}{24\,EI} \qquad \text{...(As before)}$$

and bending moment at $B$ (on the conjugate beam),

$$M_B = -P \times \frac{4l}{5} = -\left(\frac{wl^3}{24\,EI} \times \frac{4l}{5}\right) \qquad \text{...(Minus sign due to hogging)}$$

$$= \frac{wl^4}{30\,EI}$$

$\therefore$ Deflection at $B$, $$y_B = M_B = \frac{wl^4}{30\,EI} \qquad \text{...(As before)}$$

**EXAMPLE 22.1.** *A cantilever of length 2a is carrying a load of W at the free end and another load of W at its centre. Determine the slope and deflection of the cantilever at the free end.*

**SOLUTION.** Given : Span = $2a$ and load = $W$

We know that the bending moment will be zero at $B$, and will increase by a straight line law to $(-Wa)$ at $C$ and $[-(Wa) - (W \cdot 2a)] = [-3\,Wa]$ at $A$. Now draw the conjugate beam $AB$ (free at $A$, fixed at $B$ and loaded with bending moment diagram) as shown in Fig.22.4 ($b$).

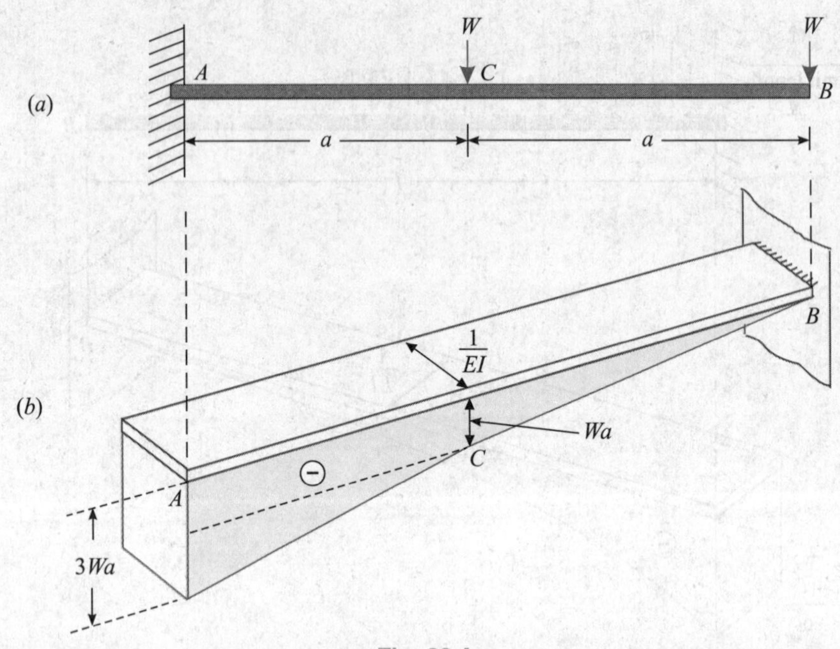


Fig. 22.4

***Slope at the free end***

From the geometry of the figure, we find that the total load of the conjugate beam,

$$P = \left[\frac{1}{2}(-Wa)\times a\times\frac{1}{EI}\right]+\left[\frac{1}{2}(-Wa-3Wa)\times a\times\frac{1}{EI}\right]$$

$$= -\frac{5Wa^2}{2\,EI}$$

Now shear force at $B$ (on the conjugate beam),

$$F_B = -P = -\left(-\frac{5Wa^2}{2\,EI}\right) \quad \text{...(Minus sign due to left downward)}$$

$$= \frac{5Wa^2}{2\,EI}$$

$\therefore$ Slope at $B$, $\quad i_B = F_B = \dfrac{5Wa^2}{2\,EI}$ radians **Ans.**

***Deflection at the free end***

The bending moment at $B$ (in the cantilever beam),

$$M_B = -\left[\left(-\frac{Wa}{EI}\times\frac{a}{2}\times\frac{2a}{3}\right)+\left(-\frac{Wa}{EI}\times\frac{a}{2}\times\frac{4a}{3}\right)+\left(-\frac{3Wa}{EI}\times\frac{a}{2}\times\frac{5a}{3}\right)\right]$$

...(Minus sing due to hogging)

$$= \frac{7Wa^3}{2\,EI}$$

$\therefore$ Deflection at $B$, $\quad y_B = M_B = \dfrac{7Wa^3}{2\,EI}$ **Ans.**

**EXAMPLE 22.2.** *A cantilever beam AB of span l is carrying a point load W at B. The moment of inertia for the left half is 2I, whereas that for the right half is I. Find the slope and deflection at B in terms of EI, W and L.*

**SOLUTION.** Given : Span = $l$ ; Point load at free end = $W$ ; Moment of inertia for left half = $2I$ and moment of inertia for right half = $I$.

We know that the bending moment will be zero at $B$ and will increase by a straight line law to $\left(-\dfrac{Wl}{2}\right)$ at $C$ and $(-Wl)$ at $A$.

Now draw the conjugate beam $AB$ (free at $A$, fixed at $B$ and loaded with the bending moment diagram) as shown in Fig. 22.5 (*b*).

***Slope at the free end***

From the geometry of the figure, we find that the total load of the conjugate beam,.

$$P = \left[\frac{1}{2}\left(-\frac{Wl}{2}\right)\times\frac{l}{2}\times\frac{1}{EI}\right]+\left[\frac{1}{2}\left(-\frac{Wl}{2}-Wl\right)\times\frac{l}{2}\times\frac{1}{EI}\right]$$

$$= -\frac{Wl^2}{2\,EI}$$

Now shear force at $B$ (on the conjugate beam),

$$F_B = -P = -\left(-\frac{Wl^2}{2\,EI}\right) \quad \text{...(Minus sign due to left downward)}$$

$$= \frac{Wl^2}{2\,EI}$$

∴ Slope at $B$, $\qquad i_B = F_B = \dfrac{Wl^2}{2\,EI}$ **Ans.**

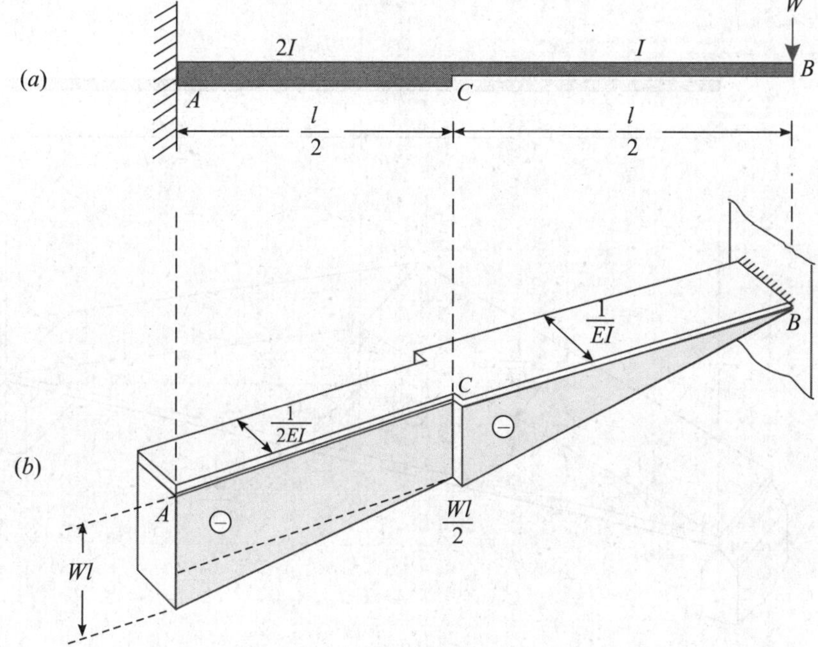


Fig. 22.5

*Deflection at the free end*

The bending moment at $B$ (on the cantilever beam),

$$M_B = -\left[\left(-\frac{Wl}{2\times EI}\times\frac{l}{2}\times\frac{1}{2}\times\frac{l}{3}\right)+\left(-\frac{Wl}{2\times 2\,EI}\times\frac{l}{2}\times\frac{1}{2}\times\frac{2l}{3}\right)+\left(-\frac{Wl}{2\,EI}\times\frac{l}{2}\times\frac{1}{2}\times\frac{5l}{6}\right)\right]=\frac{9Wl^3}{48\,EI}$$

$\therefore$ Deflection at $B$, $\quad y_B = M_B = \dfrac{9Wl^3}{48\,EI}$ **Ans.**

## EXERCISE 22.1

1. A simply supported beam of 2m span carries a point load of 20 kN at its mid-point. Determine the maximum slope and deflection of the beam. Take flexural rigidity of the beam as 500 × $10^9$ N-mm$^2$. [**Ans.** 0.01 rad ; 6.67 mm]
2. A simply supported beam of span 3 m is carrying a uniformly distributed load of 10 kN/m. Find the values of maximum slope and deflection of the beam. Take modulus of rigidity for the beam section as 10 × $10^9$ N-mm$^2$. [**Ans.** 0.0113 rad ; 10.5 mm]
3. A simply supported beam of span 2.5 m carries a gradually varying load from zero to 10 kN/m. What is the maximum deflection of the beam? Take $EI$ for the beam section as 1.2 × $10^{12}$ N-mm$^2$. [**Ans.** 2.1 mm]

## 22.7. Simply Supported Beam with a Central Point Load

Consider a simply supported beam $AB$ of length $l$ and carrying a point load $W$ at $C$ (*i.e.*, centre of the beam) as shown in Fig. 22.6 (*a*). We know that the bending moment will be zero at $A$ and $B$ and will increase by a straight line law to $\left(+\dfrac{Wl}{4}\right)$ at $C$. Now draw the conjugate beam $AB$ (simply supported at $A$ and $B$ and loaded with the bending moment diagram) as shown in Fig. 22.6 (*b*).

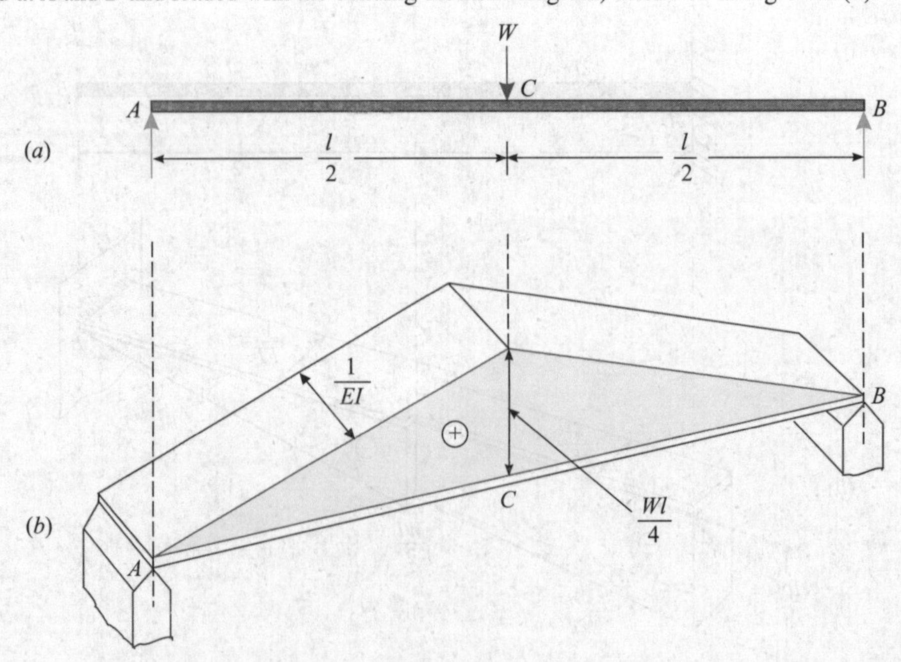


Fig. 22.6

From the geometry of figure, we find that the total load of the conjugate beam,

$$P = \frac{1}{2}\left(+\frac{Wl}{4}\right)\times l\times\frac{1}{EI} = \frac{Wl^2}{8\,EI}$$

Now shear force at $AB$ on the conjugate beam,

$$F_A = +R_A = +\frac{P}{2} = +\frac{1}{2}\times\frac{Wl^2}{8\,EI} \quad \text{...(Plus sign due to left upward)}$$

$$= \frac{Wl^2}{16\,EI}$$

∴ Slope at $A$,

$$i_A = F_A = \frac{Wl^2}{16\,EI} \quad \text{...(As before)}$$

Similarly,

$$i_B = \frac{Wl^2}{16\,EI}$$

and bending moment at $C$ (on the conjugate beam $AB$),

$$M_C = +\left[\left(\frac{Wl^2}{16\,EI}\times\frac{l}{2}\right)-\left(\frac{Wl^2}{16\,EI}\times\frac{l}{6}\right)\right] \quad \text{...(Plus sign due to sagging)}$$

$$= \frac{Wl^3}{48\,EI}$$

∴ Deflection at $C$,

$$y_C = M_C = \frac{Wl^3}{48\,EI} \quad \text{...(As before)}$$

## 22.8. Simply Supported Beam with an Eccentric Point Load

Consider a simply supported beam $AB$ of length $l$ and carrying a point load $W$ at $C$, such that $AC = a$ and $CB = b$ as shown in Fig. 22.7 ($a$). We know that the bending moment will be zero at $A$ and $B$ and will increase by a straight line law to $\left(+\frac{Wab}{l}\right)$ at $C$.

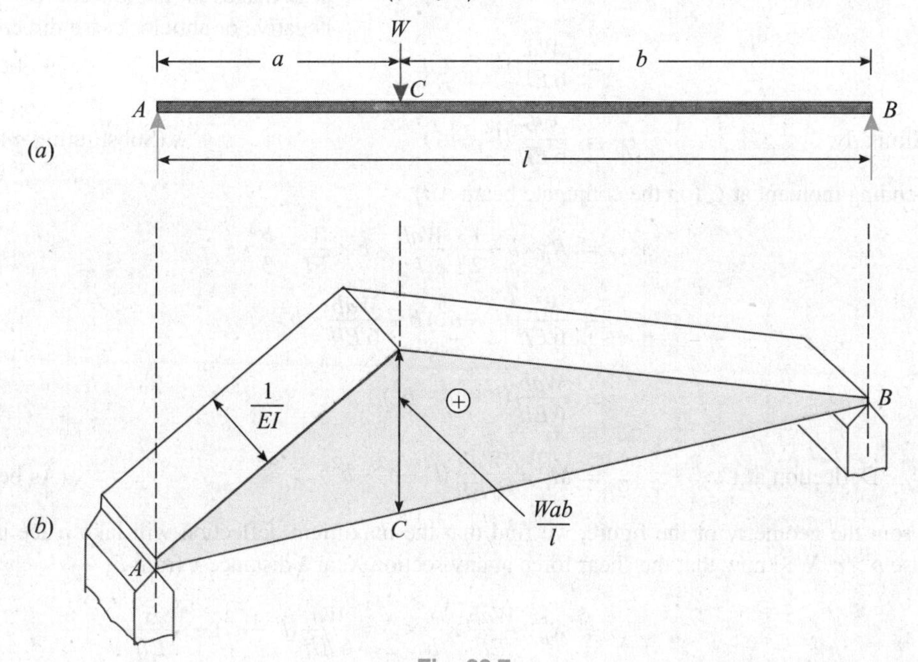


Fig. 22.7

Now draw the conjugate beam *AB* (simply supported at *A* and *B*, and loaded with bending moment diagram) as shown in Fig. 22.7 (*b*).

The reaction at *B* of the conjugate beam may be found out by taking moments about *A*,

$$R_B \times l = \left[\frac{1}{2} \times \frac{Wab}{l} \times a \times \frac{1}{EI} \times \frac{2a}{3}\right] + \left[\frac{1}{2} \times \frac{Wab}{l} \times b \times \frac{1}{EI}\left(a + \frac{b}{3}\right)\right]$$

$$= \frac{Wab}{3\,EIl} \times a^2 + \frac{Wab}{2\,EIl}\left(ab + \frac{b^2}{3}\right)$$

$$= \frac{Wab}{6\,EIl}[2a^2 + 3ab + b^2]$$

$$= \frac{Wab}{6\,EIl}[a^2 + b^2 + 2ab + a^2 + ab]$$

$$= \frac{Wab}{6\,EIl}[(a+b)^2 + a(a+b)]$$

$$= \frac{Wa\,(l-a)}{6\,EIl}[l^2 + al] \qquad ...(\because\ a + b = l)$$

$$R_B \times l = \frac{Wal\,(l-a)}{6\,EIl}(l + a)$$

or

$$R_B = \frac{Wa}{6\,EIl}(l^2 - a^2)$$

Now shear force at *B* (on the conjugate beam *AB*),

$$F_B = -R_B \qquad ...\text{(Minus sign due to right upward)}$$

$\therefore$ Slope at *B*,

$$i_B = F_B = -\frac{Wa}{6\,EIl}(l^2 - a^2)$$

...(minus sign means that the tangent at *B* makes an angle with *AB* in the negative or anticlockwise direction)

$$= \frac{Wa}{6\,EIl}(l^2 - a^2) \qquad ...\text{(As before)}$$

Similarly,

$$i_A = \frac{Wb}{6\,EIl}(l^2 - b^2) \qquad ...\text{(Substituting } a \text{ for } b\text{)}$$

and bending moment at *C* (on the conjugate beam *AB*),

$$M_C = R_B \times b - \frac{1}{2} \times \frac{Wab}{l} \times b \times \frac{1}{EI} \times \frac{b}{3}$$

$$= \frac{Wa}{6\,EIl}(l^2 - a^2)\,b - \frac{Wab}{6\,EIl} \times b^2$$

$$= \frac{Wab}{6\,EIl}(l^2 - a^2 - b^2)$$

$\therefore$ Deflection at *C*,

$$y_C = M_C = \frac{Wab}{6\,EIl}(l^2 - a^2 - b^2) \qquad ...\text{(As before)}$$

From the geometry of the figure, we find that the maximum deflection will take place in *CB* because $b > a$. We know that the shear force at any section *X*, at a distance *x* from *B*,

$$F_X = R_B - \frac{Wab}{EIl} \times \frac{x}{b} \times \frac{x}{2} = \frac{Wa}{6\,EIl}(l^2 - a^2) - \frac{Wax^2}{2\,EIl}$$

$$= \frac{Wa}{6\,EIl}(l^2 - a^2 - 3x^2)$$

We know that the maximum deflection will take place at a section, where the slope of the elastic curve is zero (or in other words maximum bending moment takes place at a section, where shear force is zero), therefore equating the above equation to zero,

$$\frac{Wa}{6\,EIl}(l^2 - a^2 - 3x^2) = 0$$

$$\therefore \quad x = \sqrt{\frac{l^2 - a^2}{3}}$$

We also know that the bending moment at any section $X$ at a distance $x$ from $B$,

$$M_X = R_B \cdot x - \frac{Wab}{EIl} \times \frac{x}{b} \times \frac{x}{2} \times \frac{x}{3} = \frac{Wa}{6\,EIl}(l^2 - a^2)\,x - \frac{Wax^3}{6\,EIl}$$

$$= \frac{Wax}{6EIl}(l^2 - a^2 - x^2)$$

Now for maximum bending moment, substituting the value of $x$ in the above equation,

$$M_{max} = \frac{Wax}{6EIl}\sqrt{\frac{l^2 - a^2}{3}} \times \left[ l^2 - a^2 - \left( \frac{l^2 - a^2}{3} \right) \right]$$

$$M_{max} = \frac{Wax}{6\,EIl}\sqrt{\frac{l^2 - a^2}{3}} \times \left[ \frac{2}{3}(l^2 - a^2) \right] = \frac{Wa}{9\sqrt{3} \cdot EIl}(l^2 - a^2)^{\frac{3}{2}}$$

$\therefore$ Maximum deflection,

$$y_{max} = M_{max} = \frac{Wa}{9\sqrt{3} \cdot EIl}(l^2 - a^2)^{\frac{3}{2}} \qquad \text{...(As before)}$$

**EXAMPLE 22.3.** *A rolled steel joist ISMB 250 × 125 mm shown in Fig. 22.8 carries a single concentrated load of 20 kN at the right third point over a simply supported span of 9 m.*

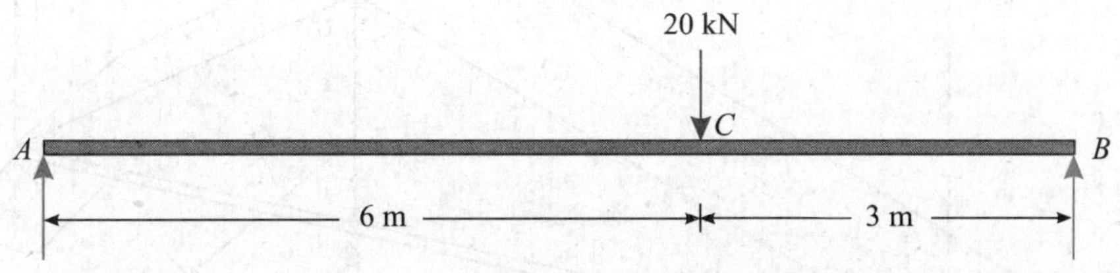


Fig. 22.8

*If the value of $I_{XX}$ for the beam is 51.316 × $10^6$ mm$^4$ and the value of E for the material is 200 GPa, calculate by the use of conjugate beam method (i) deflection under the load and (ii) maximum deflection on the span.*

**SOLUTION.** Given : Load ($W$) = 20 kN ; Span ($l$) = 9 m ; Distance $AC$ ($a$) = 6 m ; Distance $BC$ ($b$) = 3 m ; Moment of inertia ($I_{XX}$) = 51.316 × $10^6$ mm$^4$ = 51.316 × $10^{-6}$ m$^4$ and modulus of elasticity ($E$) = 200 GPa = 200 × $10^9$ N/m$^2$ = 200 × $10^6$ kN/m$^2$.

***Deflection under the load***

We know that the bending moment will be zero at $A$ and $B$, and will increase by a straight line law to

$$= \left(\frac{Wab}{l}\right) = \frac{20 \times 6 \times 3}{9} = 40 \text{ kN-m}$$

Now draw the conjugate beam as shown in Fig. 22.9 (*b*). Taking moments about *A*,

$$R_B \times 9 = \frac{1}{EI}\left(\frac{1}{2} \times 40 \times 6 \times 4\right) + \frac{1}{EI}\left(\frac{1}{2} \times 40 \times 3 \times 7\right)$$

$$= \frac{480}{EI} + \frac{420}{EI} = \frac{900}{EI}$$

$$\therefore \quad R_B = \frac{900}{EI \times 9} = \frac{100}{EI}$$

and
$$R_A = \left(\frac{1}{EI} \times \frac{1}{2} \times 40 \times 9\right) - \frac{100}{EI} = \frac{180}{EI} - \frac{100}{EI} = \frac{80}{EI}$$

$\therefore$ Bending moment at *C* on the conjugate beam *AB*,

$$M_C = \left(\frac{100}{EI} \times 3\right) - \left(\frac{1}{EI} \times \frac{1}{2} \times 3 \times 40 \times 1\right) = \frac{300}{EI} - \frac{60}{EI} = \frac{240}{EI}$$

and deflection at *C*

$$y_C = M_C = \frac{240}{(200 \times 10^6) \times (51.316 \times 10^{-6})} \text{ m}$$

$$= 0.0234 \text{ m} = 23.4 \text{ mm} \qquad \textbf{Ans.}$$

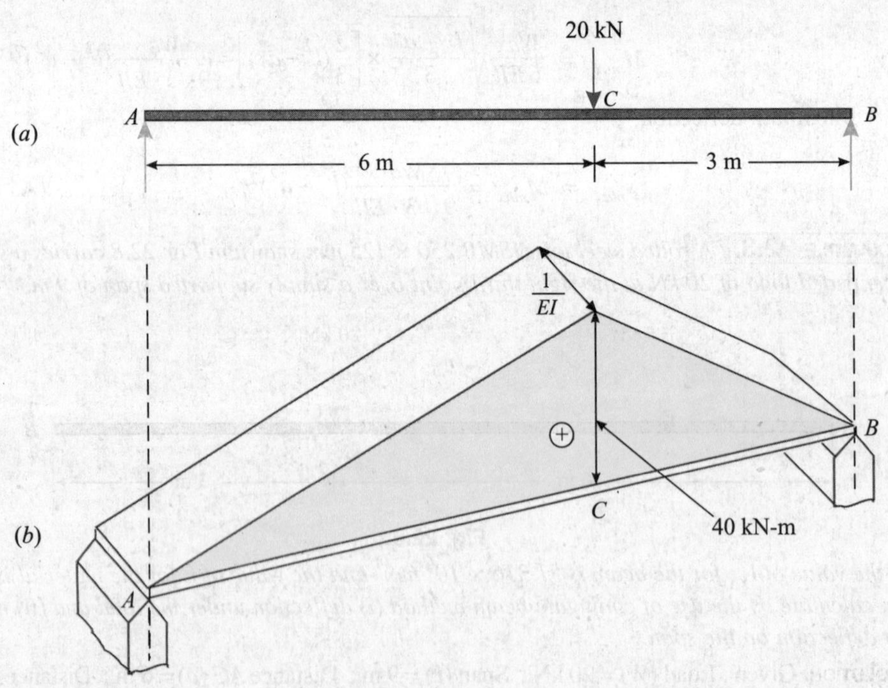


Fig. 22.9

***Maximum deflection***

From the geometry of the figure, we find that the maximum deflection will take place in *AC* (because *AC* > *CB*). We know that the shear force at any section *X* at a distance *x* from *A*.

$$F_X = \left(\frac{80}{EI}\right) - \left(\frac{1}{EI} \times \frac{1}{2} \times \frac{40}{6} \times x\right) = \frac{80}{EI} - \frac{10\,x^2}{2\,EI}$$

We know that the maximum deflection will take place at a section, where slope of the elastic curve or shear force is zero. Therefore equating the above equation to zero,

$$\frac{80}{EI} - \frac{10x^2}{3EI} = 0 \quad \text{or} \quad 10x^2 = 240$$

$$\therefore \quad x = \sqrt{24} = 4.9 \text{ m}$$

We also know that the bending moment at any section $X$ is $AC$ at a distance $x$ from $A$,

$$M_X = \left(\frac{80}{EI}x\right) - \left(\frac{40}{EI} \times \frac{x}{6} \times \frac{x}{2} \times \frac{x}{3}\right) = \frac{80}{EI}x - \frac{10x^3}{9EI}$$

∴ Maximum bending moment,

$$M_{max} = \frac{80}{EI} \times 4.9 - \frac{10 \times (4.9)^3}{9EI} = \frac{261.3}{EI}$$

and maximum deflection,
$$y_{max} = M_{max} = \frac{261.3}{(200 \times 10^6) \times (51.316 \times 10^{-6})} \text{ m}$$

$$= 0.0255 \text{ m} = 25.5 \text{ mm} \quad \textbf{Ans.}$$

## 22.9. Simply Supported Beam with a Uniformly Distributed Load

Consider a simply supported beam $AB$ of length $l$ and carrying a uniformly distributed load of $w$ per unit length as shown in Fig. 22.10 (*a*). We know that the bending moment at $A$ and $B$ will be zero and will increase in the form of *parabola* to $\left(+\frac{wl^2}{8}\right)$ at $C$ (*i.e.*, the centre of the beam).

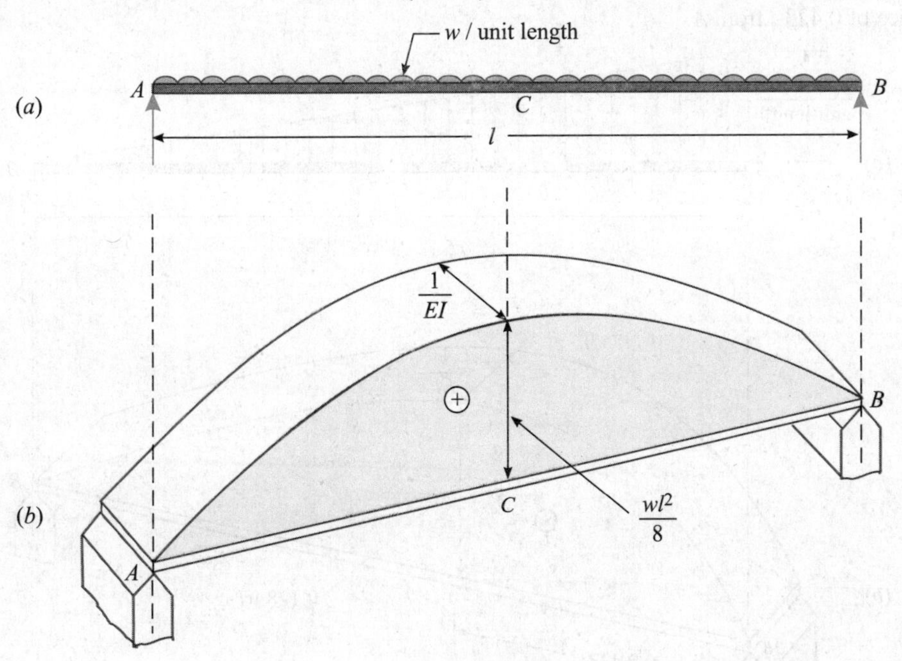


Fig. 22.10

Now draw the conjugate beam $AB$ (simply supported at $A$ and $B$ and loaded with the bending moment diagram) as shown in Fig. 22.10 (*b*).

From the geometry of the figure, we find that the total load of the conjugate beam,

$$P = \frac{2}{3}\left(+\frac{wl^2}{8}\right)\times l \times \frac{1}{EI} = \frac{wl^3}{12\,EI}$$

Now shear force at $A$ (on the conjugate beam $A$ ore $B$),

$$F_A = +R_A = +\frac{P}{2} = \frac{1}{2}\times\frac{wl^3}{12\,EI} \qquad \text{...(Plus sign due to left upward)}$$

$$= \frac{wl^3}{24\,EI}$$

∴ Slope at $A$,

$$i_A = F_A = \frac{wl^3}{24\,EI} \qquad \text{...(As before)}$$

Similarly,

$$i_B = \frac{wl^3}{24\,EI}$$

and bending moment at $C$ (on the conjugate beam $AB$),

$$M_C = \left(\frac{wl^3}{24\,EI}\times\frac{l}{2}\right) - \left(\frac{wl^3}{24\,EI}\times\frac{3l}{16}\right) = \frac{5\,wl^4}{384\,EI}$$

∴ Deflection at $C$,

$$y_C = M_C = \frac{5\,wl^4}{384\,EI} \qquad \text{...(As before)}$$

## 22.10. Simply Supported Beam with a Gradually Varying Load

Consider a simply supported beam $AB$ of length $l$ and carrying a gradually varying load from zero at $B$ to $w$ per unit length at $A$ is shown in Fig. 22.11 ($a$). We know that the bending moment will be zero at $A$ and $B$ and will increase in the form of parabola to maximum equal to 0.128 $wl$ at a distance of 0.423 $l$ from $A$.

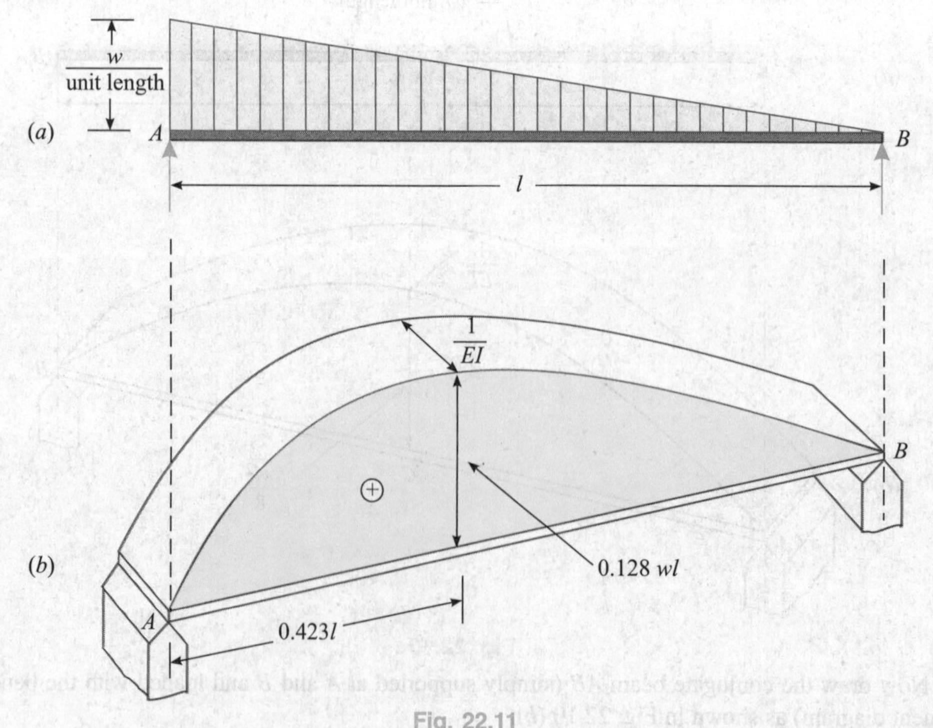


Fig. 22.11

Now draw the conjugate beam *AB* (simply supported at *A* and *B* and loaded with the bending moment diagram) as shown in Fig. 22.11 (*b*).

The reaction at *B* of the conjugate beam may be found out by taking moments about *A*.,

$$R_B \times l = \frac{wl^3}{24EI} \times \frac{8l}{15} = \frac{wl^4}{45EI}$$

$$\therefore \quad R_B = \frac{wl^4}{45EI}$$

Now shear force at *B* (on the conjugate beam *AB*),

$$F_B = -R_B$$ ...(Minus sign due to right upward)

∴ Slope at *B*,

$$i_B = F_B = -\frac{wl^3}{45EI}$$ ...(Minus sign means that the tangent at *B* makes an angle with *AB* in the negative or anticlockwise direction)

$$= \frac{wl^3}{45EI}$$ ...(As before)

Similarly,

$$R_A = \frac{wl^3}{24EI} - \frac{wl^3}{45EI} = \frac{7wl^3}{360EI}$$

∴ Slope at *A*,

$$i_A = F_A = R_A = \frac{7wl^3}{360EI}$$ ...(As before)

**EXAMPLE 22.4.** *A simply supported beam AB of span 4 m, carrying a load of 100 kN at its mid span C has cross-sectional moment of inertia 24 × 10⁶ mm⁴ over the left half of the span and 48 × 10⁶ mm⁴ over the right half. Find the slopes at the two supports and the deflection under the load. Take E = 200 GPa.*

**SOLUTION.** Given : Span *AB* ($l$) = 4 m ; Load ($W$) = 100 kN ; Moment of inertia of *AC* ($I_{AC}$) = $24 \times 10^6$ mm$^4$ = $24 \times 10^{-6}$ m$^4$ ; Moment of inertia of *BC* ($I_{BC}$) = $48 \times 10^6$ mm$^4$ = $48 \times 10^{-6}$ m$^4$ and modulus of elasticity ($E$) = 200 GPa = $200 \times 10^9$ N/m$^2$ = $200 \times 10^6$ kN/m$^2$.

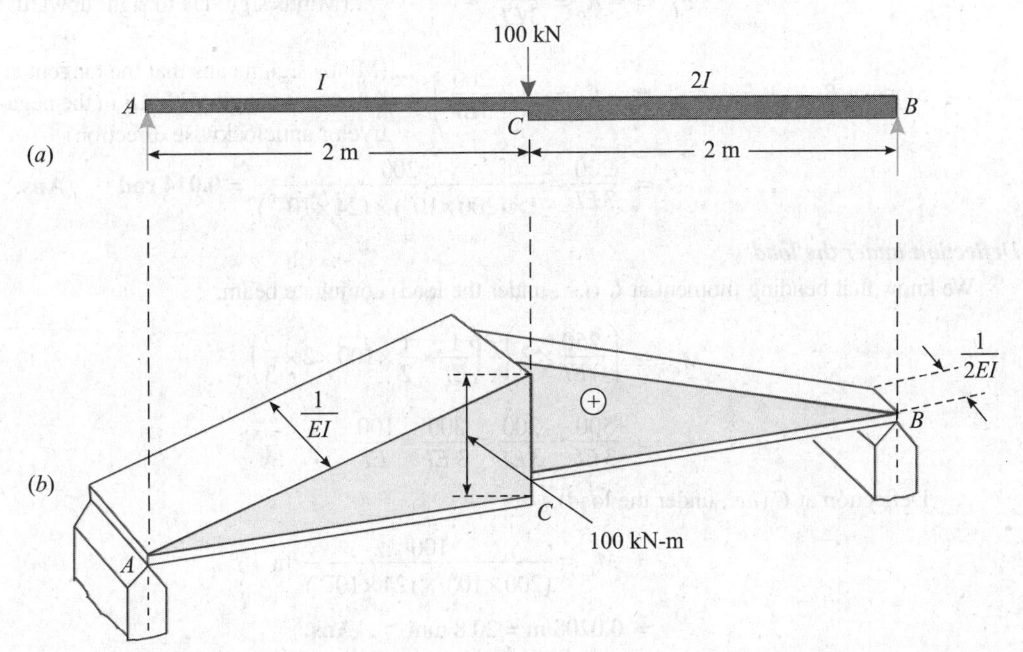


Fig. 22.12

*Slope at the two supports*

We know that the bending moment will be zero at $A$ and $B$ and increase by a *straight line law* to

$$\left(+\frac{Wl}{4}\right) = \frac{100\times 4}{4} = 100 \text{ kN-m}$$

Now draw the conjugate beam as shown in Fig. 22.12 (*b*). Taking moments about $A$,

$$R_B \times A = \frac{1}{EI}\left(\frac{1}{2}\times 100\times 2\times\frac{4}{3}\right)+\frac{1}{2\,EI}\left(\frac{1}{2}\times 100\times 2\times\frac{8}{3}\right)$$

$$= \frac{400}{3\,EI}+\frac{400}{3\,EI}=\frac{800}{3\,EI}$$

$$\therefore \quad R_B = \frac{800}{3\,EI\times 4}=\frac{200}{3\,EI}$$

and

$$R_A = \frac{1}{EI}\left(\frac{1}{2}\times 100\times 2\right)+\frac{1}{2\,EI}\left(\frac{1}{2}\times 100\times 2\right)-\frac{200}{3\,EI}$$

$$= \frac{100}{EI}+\frac{50}{EI}-\frac{200}{3\,EI}=\frac{250}{3\,EI}$$

We know that shear force at $A$,

$$F_A = +R_A = +\frac{250}{3\,EI} \qquad \text{...(Plus sign due to left upwards)}$$

$\therefore$ Slope at $A$,

$$i_A = \frac{250}{3\,EI} = \frac{250}{3\times(200\times 10^6)\times(24\times 10^{-6})} \text{ rad}$$

$$= \mathbf{0.017 \text{ rad}} \qquad \textbf{Ans.}$$

Similarly, shear force at $B$,

$$F_B = -R_B = \frac{200}{3\,EI} \qquad \text{...(Minus sign due to right upwards)}$$

$\therefore$ Slope at $B$,

$$i_B = -F_B = -\left(-\frac{200}{3\,EI}\right)$$

...(Minus sign means that the tangent at $B$ makes an angle with $AB$ in the negative or anticlockwise direction)

$$= \frac{200}{3\,EI} = \frac{200}{3\times(200\times 10^6)\times(24\times 10^{-6})} = \mathbf{0.014 \text{ rad}} \qquad \textbf{Ans.}$$

*Deflection under the load*

We know that bending moment at $C$ (*i.e.*, under the load) conjugate beam,

$$M_C = \left(\frac{250}{3\,EI}\times 2\right)-\left(\frac{1}{EI}\times\frac{1}{2}\times 100\times 2\times\frac{2}{3}\right)$$

$$= \frac{500}{3\,EI}-\frac{200}{3\,EI}=\frac{300}{3\,EI}=\frac{100}{EI}$$

$\therefore$ Deflection at $C$ (*i.e.*, under the load),

$$y_C = M_C = \frac{100}{(200\times 10^6)\times(24\times 10^{-6})} \text{ m}$$

$$= 0.0208 \text{ m} = 20.8 \text{ mm} \qquad \textbf{Ans.}$$

***Alternate Solution***

Let us draw the conjugate beam by keeping its width constant and equal to $\frac{1}{EI}$ as shown in Fig. 22.13 (*b*).

***Slope at the two supports***

Therefore moment *CD*

$$= \frac{Wl}{4} = \frac{100 \times 4}{4} = 100 \text{ kN-m}$$

and moment *CE* $= \frac{100}{2} = 50$ kN-m ...( because of $2I$)

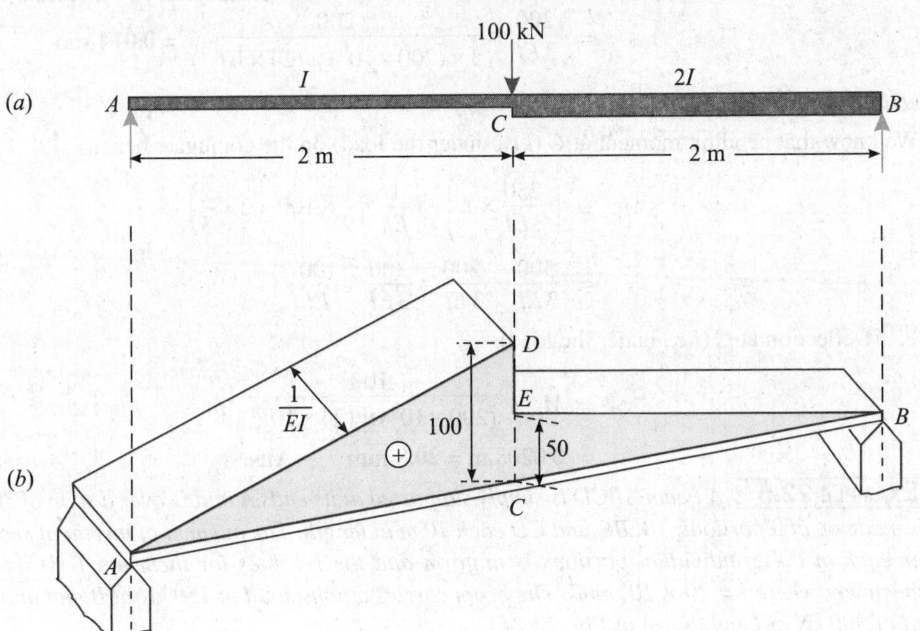


Fig. 22.13

Taking moments about *A*,

$$R_B \times 4 = \frac{1}{EI}\left[\left(\frac{1}{2} \times 100 \times 2 \times \frac{4}{3}\right) + \left(\frac{1}{2} \times 50 \times 2 \times \frac{8}{3}\right)\right]$$

$$= \frac{1}{EI}\left(\frac{400}{3} + \frac{400}{3}\right) = \frac{800}{3\,EI}$$

$$\therefore \quad R_B = \frac{800}{3\,EI \times 4} = \frac{200}{3\,EI}$$

and

$$R_A = \frac{1}{EI}\left[\left(\frac{1}{2} \times 100 \times 2\right) + \left(\frac{1}{2} \times 50 \times 2\right)\right] - \frac{200}{3\,EI}$$

$$= \frac{1}{EI}(100 + 50) - \frac{200}{3\,EI} = \frac{250}{3\,EI}$$

We know that shear force at *A*,

$$F_A = +R_A = +\frac{250}{3\,EI}$$ ...(Plus sign due to left upward)

$\therefore$ Slope at $A$,
$$i_A = F_A = \frac{250}{3\,EI} = \frac{250}{3 \times (200 \times 10^6) \times (24 \times 10^{-6})} \text{ rad}$$
$$= \mathbf{0.017 \text{ rad}} \quad \textbf{Ans.}$$

Similarly, shear force at $B$,
$$F_B = -R_B = -\frac{200}{3\,EI} \quad \text{...(Minus sign due to left upward)}$$

$\therefore$ Slope at $B$,
$$i_B = R_B = -\left(-\frac{200}{3\,EI}\right)$$
...(Minus sign means that the tangent at $B$ makes an angle with $AB$ in the negative or anticlockwise direction)
$$= \frac{200}{3\,EI} = \frac{200}{3 \times (200 \times 10^6) \times (24 \times 10^{-6})} = \mathbf{0.014 \text{ rad}} \quad \textbf{Ans.}$$

***Deflection under the load***

We know that bending moment at $C$ (*i.e.*, under the load) on the conjugate beam,
$$M_C = \left(\frac{250}{3\,EI} \times 2\right) - \left(\frac{1}{EI} \times \frac{1}{2} \times 100 \times 2 \times \frac{2}{3}\right)$$
$$= \frac{500}{3\,EI} - \frac{200}{3\,EI} = \frac{300}{3\,EI} = \frac{100}{EI}$$

$\therefore$ Deflection at $C$ (*i.e.*, under the load)
$$y_C = M_C = \frac{100}{(200 \times 10^6) \times (24 \times 10^{-6})} \text{ m}$$
$$= 0.0208 \text{ m} = \mathbf{20.8 \text{ mm}} \quad \textbf{Ans.}$$

**EXAMPLE 22.5.** *A beam ABCD is simply supported at its ends A and D over a span of 30 m. It is made up of 3 portions AB, BC and CD each 10 m in length. The moment of inertia of section over each of these individual portions is uniform and the I values for them are I, 3I and 2I respectively, where $I = 20 \times 10^9$ mm$^4$. The beam carries a point load of 150 kN at B and another load of 300 kN at C as shown in Fig. 22.14.*

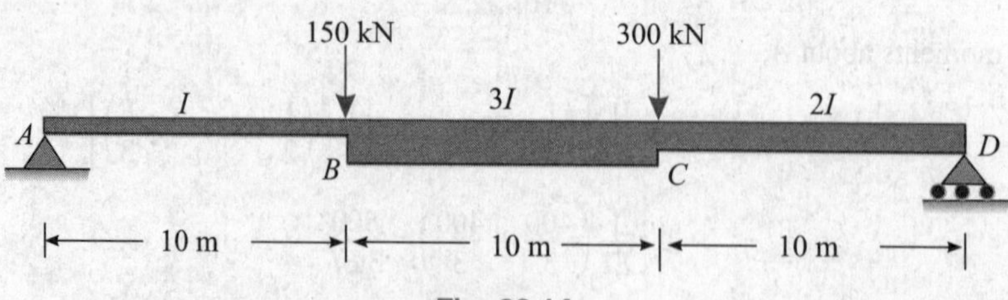


**Fig. 22.14**

*Neglecting the self load of the beam, calculate the deflection at B and the slope at C. Take the values of E, the modulus of material of the beam 200 GPa uniform throughout the length.*

**SOLUTION.** Given : Length $AB$ ($l_{AB}$) = $l_{BC}$ = $l_{CD}$ = 10 m ; Moment of inertia of $AB$ ($I_{AB}$) = $I$ = 20 × 10$^9$ mm$^4$ = 20 × 10$^{-3}$ m$^4$ ; Moment of inertia of $BC$ ($I_{BC}$) = 3 $I$ = 3 × (20 × 10$^{-3}$) = 60 × 10$^{-3}$m$^4$ ; Moment of inertia of $CD$ ($I_{CD}$) = 2 $I$ = 2 × (20 × 10$^{-3}$) = 40 × 10$^{-3}$ m$^4$ ; Load at $B$ = 150 kN ; Load at $D$ = 300 kN and modulus of elasticity ($E$) = 200 GPa = 200 × 10$^9$ N/m$^2$ = 200 × 10$^6$ kN/m$^2$.

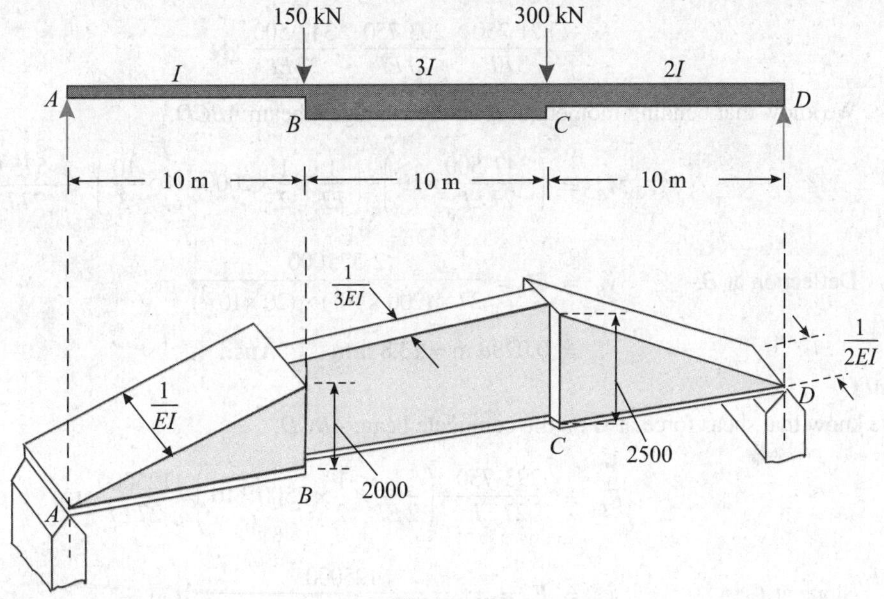


Fig. 22.15

*Deflection at B*

Taking moments about $A$,

$$V_D \times 30 = (150 \times 10) + (300 \times 20) = 7500$$

$$\therefore \quad V_D = \frac{7500}{30} = 250 \text{ kN}$$

and $$V_A = (150 + 300) - 250 = 200 \text{ kN}$$

$$\therefore \quad M_A = \text{VA} \times 10 = 200 \times 10 = 2000 \text{ kN-m}$$

and $$M_C = V_D \times 10 = 250 \times 10 = 2500 \text{ kN-m}$$

We know that the bending moment will be zero at $A$ and $D$ and will increase by a straight line law to 2000 kN-m at $B$ and 2500 kN-m at $C$.

Now draw the conjugate beam as shown in Fig. 22.15. Taking moments about $A$ of the conjugate beam,

$$R_D \times 30 = \frac{1}{EI}\left(\frac{1}{2} \times 2000 \times 10 \times \frac{20}{3}\right) + \frac{1}{3\,EI}\left(\frac{1}{2} \times 2000 \times 10 \times \frac{40}{3}\right)$$

$$+ \frac{1}{2\,EI}\left(\frac{1}{2} \times 2500 \times 10 \times \frac{50}{3}\right) + \frac{1}{2\,EI}\left(\frac{1}{2} \times 2500 \times 10 \times \frac{70}{3}\right)$$

$$= \frac{200000}{3\,EI} + \frac{400000}{9\,EI} + \frac{625000}{9\,EI} + \frac{437500}{9\,EI} = \frac{2937500}{9\,EI}$$

$$\therefore \quad R_D = \frac{2937500}{9\,EI \times 30} = \frac{293750}{27\,EI} \text{ kN}$$

and $$R_A = \left[\frac{1}{EI}\left(\frac{1}{2} \times 2000 \times 10\right) + \frac{1}{3\,EI}\left(\frac{2000 + 2500}{2} \times 10\right)\right.$$

$$\left. + \frac{1}{2\,EI}\left(\frac{1}{2} \times 2500 \times 10\right) - \left(\frac{293750}{27\,EI}\right) \text{kN}\right]$$

$$= \frac{71\ 250}{3EI} - \frac{293\ 750}{27EI} = \frac{347\ 500}{27EI} \text{ kN}$$

∴ We know that bending moment at *B* on the conjugate beam *ABCD*.

$$M_B = \left(\frac{347\ 500}{27\ EI} \times 10\right) - \left(\frac{1}{EI} \times \frac{1}{2} \times 2000 \times 10 \times \frac{10}{3}\right) = \frac{2\ 575\ 000}{27\ EI}$$

∴ Deflection at *B*,

$$y_B = M_B = \frac{2\ 575000}{27 \times (200 \times 10^6) \times (20 \times 10^{-3})} \text{ m}$$

$$= 0.0238 \text{ m} = \mathbf{23.8\ mm} \quad \textbf{Ans.}$$

***Slope at C***

We know that shear force at *C* on the conjugate beam *ABCD*,

$$F_C = \frac{293\ 750}{27\ EI} - \left(\frac{1}{2\ EI} \times \frac{1}{2} \times 2500 \times 10\right) = \frac{125000}{27\ EI} \text{ kN}$$

∴ Slope at *C*,

$$i_C = F_C = \frac{125000}{27 \times (200 \times 10^6) \times (20 \times 10^{-3})} \text{ rad}$$

$$= \mathbf{0.00116\ rad} \quad \textbf{Ans.}$$

## EXERCISE 22.2

1. A cantilever 2.4 m long carries a point load of 10 kN at its free end. Find the slope and deflection of the cantilever under the load. Take flexural rigidity for the cantilever beam as $25 \times 10^{12}$ N-mm$^2$.

   [**Ans.** 0.0012 rad ; 1.84 mm]

2. A cantilever beam 2 m long is subjected to a uniformly distributed load of 5 kN/m over its entire length. Find the slope and deflection of the cantilever at the free end.

   [**Ans.** 0.0027 rad ; 4.0 mm]

3. A cantilever of length *l* is loaded with a uniformly distributed load of *w* per unit length for the middle half of its length. Using conjugate beam method, determine the ratio of slopes at the centre and free end of the cantilever.

   $\left[\textbf{Ans. } \frac{25}{26}\right]$

4. In the above example, determine the ratio of deflection at the centre and free end of the cantilever.

   $\left[\textbf{Ans. } \frac{43}{112}\right]$

## QUESTIONS

1. What is a conjugate beam? Discuss its utilities.
2. State Mohr's theorems.
3. Give the relation between an actual beam and a conjugate beam when the former has a fixed end.
4. How will you find the slope and deflection by conjugate method of a cantilever carrying a uniformly distributed load.

5. With the help of conjugate beam, show that the maximum deflection of a simply supported beam of span $l$ carrying an eccentric load ($W$) at a distance $a$ from one of the supports is given by:

$$y_{max} = \frac{Wa}{9\sqrt{3} \cdot lEI}(l^2 - a^2)^2$$

where $EI$ is the flexural rigidity of the beam.

6. Derive an expression for the slopes at the two ends of a simply supported beam, when it is subjected to a triangular load of zero intensity at one end and $w$ per unit length at the other.

## OBJECTIVE TYPE QUESTIONS

1. If the actual beam has both ends fixed, then the ends of the conjugate beam will be
   (*a*) fixed at both ends (*b*) free at both ends
   (*c*) fixed at one end and free at the other (*d*) either '*a*' or '*b*'.
2. The width of a conjugate beam is
   (*a*) $1/EI$ (*b*) $EI$ (*c*) $2\,EI$ (*d*) $4\,EI$
3. With the help of conjugate beam method, we can find out the slope in case of
   (*a*) cantilevers (*b*) fixed beams
   (*c*) simply supported beams (*d*) both '*a*' and '*c*'.
4. We can find the deflection of a beam carrying
   (*a*) uniformly distributed load (*b*) central point load
   (*c*) gradually varying load (*d*) all of these

## ANSWERS

**1.** (*b*) **2.** (*a*) **3.** (*c*) **4.** (*d*)

Chapter **23**

# Propped Cantilevers and Beams

Contents

## 23.1. Introduction

We have already discussed in chapters 19 and 20 that whenever a cantilever or a beam is loaded, it gets deflected. As a matter of fact, the amount by which a cantilever or a beam may deflect, is so small that it is hardly detected by the residents. But sometimes, due to inaccurate design or bad workmanship, the deflection of the free end of a cantilever (or centre of the beam) is so much that the residents are always afraid of its falling down and it effects their health. In order to set right the deflected cantilever or a beam or more precisely to avoid the deflection to some extent, it is propped up (*i.e.*, supported by some vertical pole at the original level before deflection) at some suitable point. Such an arrangement of providing a support is known as propping and the cantilever or beam is known as *propped cantilever* or *propped beam.*

## 23.2. Reaction* of a Prop

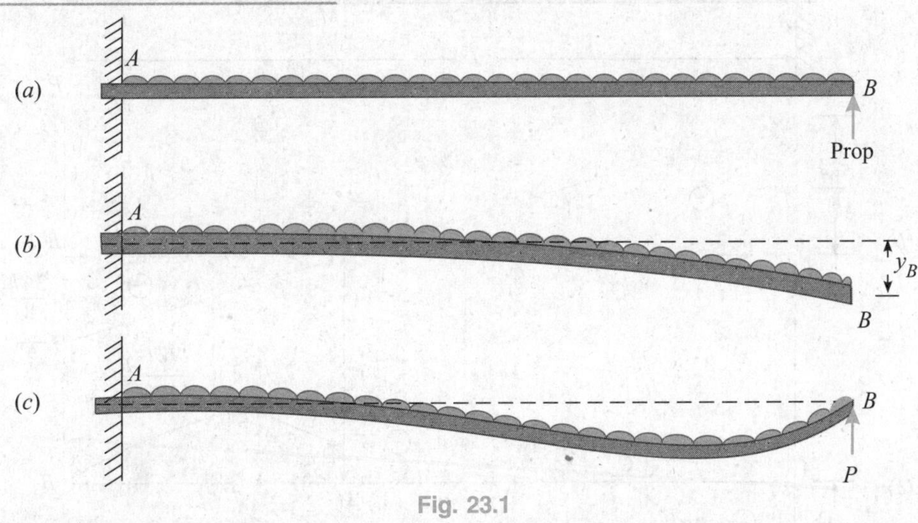


Fig. 23.1

Consider a cantilever beam *AB* fixed at *A* and propped at *B* as shown in Fig. 23.1 (*a*). Let the cantilever be subjected to some loading (say uniformly distributed load) as shown in the figure.

It has been experimentally found that this prop will be subjected to some reaction. This reaction can be obtained as discussed below:

1. Imagine the prop to be removed and calculate the deflection of the free end *B* as shown in Fig. 23.1 (*b*).
2. Now imagine a prop to be introduced at *B*, which will exert an upward force *P* equal to the reaction of the prop. It will cause an upward deflection of *B* due to the prop reaction as shown in Fig. 23.1 (*c*).
3. Now by equating the downward deflection due to the load and the upward deflection due to the prop reaction, the reaction of the prop may be found out.

## 23.3. Propped Cantilever with a Uniformly Distributed Load

Consider a cantilever *AB* fixed at and propped at *B* and carrying a uniformly distributed load over its entire span as shown in Fig. 23.2 (*a*).

Let $l$ = Span of the cantilever *AB*,

$w$ = Uniformly distributed load per unit length over the entire span and

$P$ = Reaction at the prop.

We know that the downward deflection of *B* due to uniformly distributed load (neglecting prop reaction),

$$y_B = \frac{wl^4}{8EI} \quad ...(i)$$

* Very often the students commit the mistake of finding out the prop reactions by equating the clockwise moments (due to load on cantilever) to the anticlockwise moment (due to the prop reaction) about the fixed end; as they would do in the case of a simply supported beam. This practice does not hold good in this case, as the net moment at the fixed end is not zero. There exists a fixing moment, which can not be determined unless the prop reaction is known.

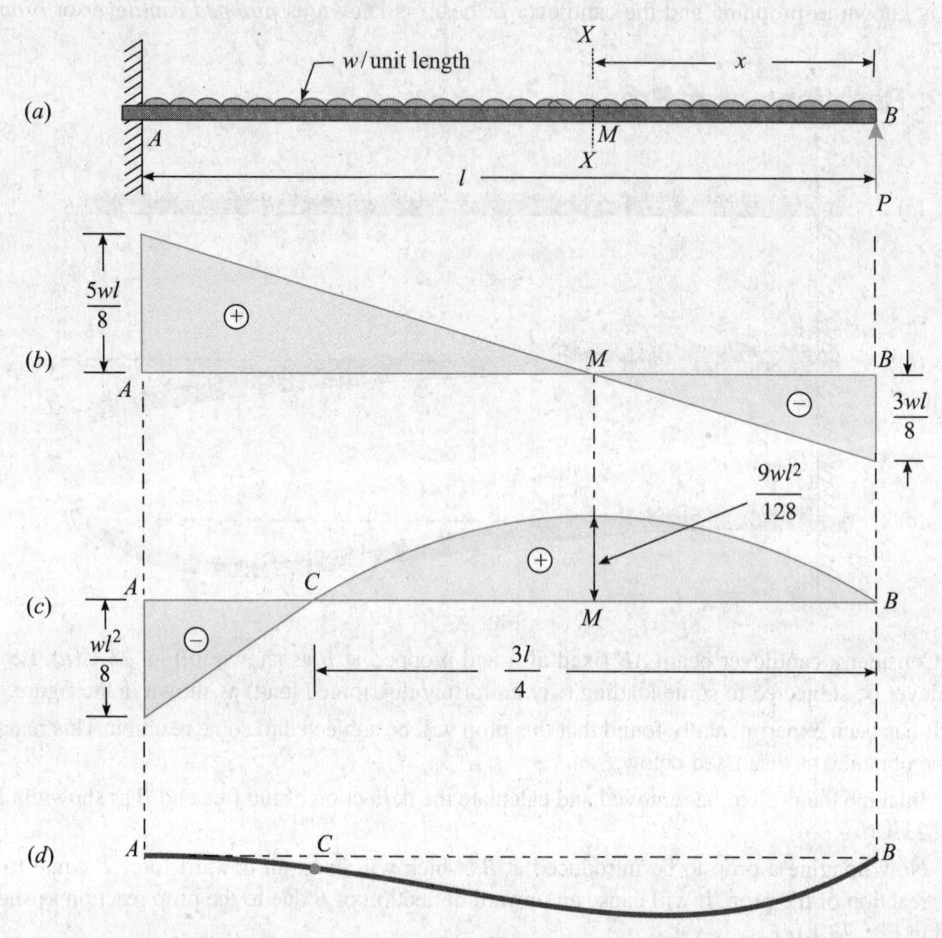


Fig. 23.2

and the upward deflection of the cantilever due to the force $P$ (neglecting uniformly distributed load),

$$y_B = \frac{Pl^3}{3EI} \qquad \text{...}(ii)$$

Since both the deflections are equal, therefore, equating equations (*i*) and (*ii*),

$$\frac{Pl^3}{3EI} = \frac{wl^4}{8EI}$$

$$\therefore \qquad P = \frac{3wl}{8} = \frac{3W}{8} \qquad \text{...(where } W = wl = \text{total load)}$$

Now we shall analyse the propped cantilever for shear force, bending moment, slope and deflection at important sections of the cantilever.

***(i) Shear force diagram***

We know that the shear force at $B$,

$$F_B = -\frac{3wl}{8} \qquad \text{... (Minus sign due to right upwards)}$$

and

$$F_A = \frac{5wl}{8} \qquad \text{... (Plus sign due to left upwards)}$$

Let $M$ be the point at a distance $x$ from $B$, where shear force changes sign,

$$\therefore \quad \frac{x}{l-x} = \frac{3}{5} \quad \text{or} \quad 5x = 3l - 3x$$

or
$$x = \frac{3l}{8}$$

Thus the shear force is zero at a distance $\frac{3l}{8}$ from $B$. The shear force diagram is shown in Fig. 23.2($b$).

***(ii) Bending moment diagram***

We know that the bending moment at the propped end $B$,

$$M_B = 0$$

and
$$M_A = \frac{3wl}{8} \cdot l - \frac{wl^2}{2} = -\frac{wl^2}{8} \qquad ...(iii)$$

We also know that the bending moment will be maximum at $M$, where shear force changes sign.

$$\therefore \quad M_M = \frac{3wl}{8}\left(\frac{3l}{8}\right) - \frac{w}{2}\left(\frac{3l}{8}\right)^2 = \frac{9wl^2}{128} \qquad ...(iv)$$

and bending moment at any section $X$, at a distance $x$ from the propped end $B$,

$$M_M = \frac{3wl}{8} \cdot x - \frac{wx^2}{2}$$

Now in order to find out the point of contraflexure, let us equate this bending moment to zero. Therefore

$$\frac{3wl}{8} x - \frac{wx^2}{2} = 0 \quad \text{or} \quad x = \frac{3l}{4}$$

The bending moment diagram is shown in Fig. 23.2($c$).

***(iii) Slope at the propped end***

We know that the bending moment at any section $X$ at a distance $x$ from $B$,

$$M_M = \frac{3wl}{8} x - \frac{wx^2}{2}$$

$$\therefore \quad EI \frac{d^2y}{dx^2} = \frac{3wlx}{8} - \frac{wx^2}{2}$$

Integrating the above equation,

$$EI \frac{dy}{dx} = \frac{3wlx^2}{16} - \frac{wx^3}{6} + C_1$$

where $C_1$ is the constant of integration. We know that when $x = l$, then $\frac{dy}{dx} = 0$. Therefore substituting these values in the above equation,

$$0 = \frac{3wl \cdot l^2}{16} - \frac{wl^3}{6} + C_1$$

or
$$C_1 = -\frac{wl^3}{48}$$

$$\therefore \quad EI \frac{dy}{dx} = \frac{3wlx^2}{16} - \frac{wx^3}{6} - \frac{wl^3}{48} \qquad ...(v)$$

This is the required equation for slope at any section of the cantilever. Now for the slope at $B$, substituting $x = 0$ in the above equation,

$$EI \cdot i_B = -\frac{wl^3}{48}$$

$$\therefore \quad i_B = -\frac{wl^3}{48EI}$$

... (Minus sign means the tangent at $B$ makes an angle with $AB$ in the negative or anti-clockwise direction)

$$= \frac{wl^3}{48EI} \text{ radians} \qquad ...(vi)$$

**(iv) *Deflection at the centre of the beam***

Integrating the equation (*vi*) once again,

$$EI \cdot y = \frac{3wlx^3}{48} - \frac{wx^4}{24} - \frac{wl^3x}{48} + C_2$$

$$= \frac{wlx^3}{16} - \frac{wx^4}{24} - \frac{wl^3x}{48} + C_2$$

where $C_2$ is the constant of integration. We know that when $x = l$ then $y = 0$. Therefore substituting the values in the above equation, we get $C_2 = 0$. Therefore

$$EI \cdot y = \frac{wlx^3}{16} - \frac{wx^4}{24} - \frac{wl^3x}{48} \qquad ...(vii)$$

This is the required equation, for deflection at any section of the cantilever. Now for the deflection at the centre of the cantilever, substituting $x = l/2$,

$$EI \cdot y_C = \frac{wl}{16}\left(\frac{l}{2}\right)^3 - \frac{w}{24}\left(\frac{l}{2}\right)^4 - \frac{wl^3}{48}\left(\frac{l}{2}\right) = -\frac{wl^4}{192}$$

$$\therefore \quad y_C = -\frac{wl^4}{192EI}$$

... (Minus sign means that the deflection is downwards)

$$= \frac{wl^4}{192EI} \qquad ...(viii)$$

**(v) *Maximum deflection of the beam***

We know that the maximum deflection takes place at a point, where slope is zero. Therefore, equating the equation (*v*) to zero,

$$\frac{3wlx^2}{16} - \frac{wx^3}{6} - \frac{wl^3}{48} = 0$$

$$\therefore \quad 9lx^2 - 8x^3 - l^3 = 0$$

Solving this equation by trial and error, we get $x = 0.422l$.

$$\therefore \quad EI \cdot y_{max} = \frac{wl}{16}(0.422l)^3 - \frac{W}{24}(0.442l)^4 - \frac{wl^3}{48}(0.422l)$$

$$= -0.005\,415\text{wl4}$$

or

$$y_{max} = -\frac{0.005\,415wl^4}{EI}$$

... (Minus sign means that the deflection is downwards)

$$= \frac{0.005\,415wl^4}{EI} \quad ...(ix)$$

**(vi) *Elastic curve***

It may be noted that the *elastic curve between *A* and *C* will be convex upwards (due to negative bending moment) and between *C* and *B* it will be convex downwards (due to positive bending moment). The elastic curve at *C* will be straight line (due to zero bending moment).

Now draw the elastic curve of the cantilever as shown in Fig. 23.2(*d*).

**EXAMPLE 23.1.** *A beam AB of span 3 m is fixed at A and propped at B. Find the reaction at the prop, when it is loaded with a uniformly distributed load of 20 kN/m over its entire span.*

**SOLUTION.** Given: Span = ($l$) = 3 m and uniformly distributed load ($w$) = 20 kN/m

We know that prop reaction,

$$P = \frac{3wl}{8} = \frac{3 \times 20 \times 3}{8} = \mathbf{22.5\ kN} \quad \textbf{Ans.}$$

**EXAMPLE 23.2.** *A propped cantilever beam 3 m long has 100 mm wide and 150 mm deep cross-section. If the allowable bending stress and the deflection at the centre is 45 MPa and 2.5mm respectively, determine the safe uniformly distributed load the cantilever can carry. Take E = 120 GPa.*

**SOLUTION.** Given: Length ($l$) = 3 m = $3 \times 10^3$ mm ; Width ($b$) = 100 mm ; Depth ($d$) = 150 mm; Allowable bending stress ($\sigma_{b\,(max)}$) = 45 MPa = 45 N/mm$^2$ ; Deflection at the centre ($y_C$) = 2.5 mm and modulus of elasticity ($E$) = 120 GPa = $120 \times 10^3$ N/mm$^2$.

Let $w$ = Uniformly distributed load over the cantilever.

Now we shall solve the value of $w$ from bending stress and deflection one by one.

We know that moment of inertia of the beam section,

$$I = \frac{bd^3}{12} = \frac{100 \times (150)^3}{12} = 28.125 \times 10^6 \text{ mm}^4$$

∴ Section modulus of the beam section,

$$Z = \frac{I}{b/2} = \frac{28.125 \times 10^6}{150/2} = 375 \times 10^3 \text{ mm}^3$$

We also know that maximum bending moment on a propped cantilever,

$$M = \frac{wl^2}{8} = \frac{w \times (3 \times 10^3)^2}{8} = 1.125 \times 10^6\ w$$

∴ Maximum bending stress ($\sigma_{b\,max}$),

$$45 = \frac{M}{Z} = \frac{1.125 \times 10^6\ w}{375 \times 10^3} = 3w$$

or

$$w = \frac{45}{3} = 15 \text{ N/mm} = 15 \text{ kN} \quad ...(i)$$

We also know that deflection at the centre of the propped cantilever ($y_C$),

$$2.5 = \frac{wl^4}{192EI} = \frac{w \times (3 \times 10^3)^4}{192 \times (120 \times 10^3) \times (28.125 \times 10^6)}$$

---

* It is the curved shape of the centre line of the propped cantilever, into which the cantilever will bend due to its elasticity.

$$= \frac{81\times10^{12}\ w}{648\times10^{12}} = 0.125\ w$$

$$\therefore \quad w = \frac{2.5}{0.125} = 20 \text{ N/mm} = 20 \text{ kN/m} \quad ...(ii)$$

Thus the safe load over the propped cantilever is the minimum of the values obtained from equations (*i*) and (*ii*), *i.e.*, 15 kN/m. **Ans.**

**EXAMPLE 23.3.** *A beam AB 2 m long and carrying a uniformly distributed load of 15 kN/m is resting over a similar beam CD 1 m long as shown in Fig. 23.3.*

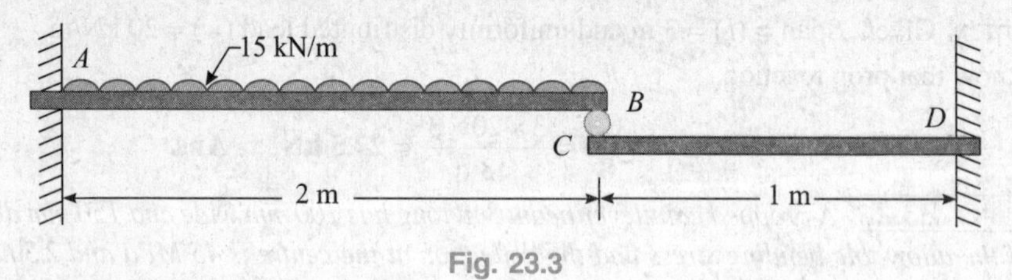


Fig. 23.3

*Find the reaction at C.*

**SOLUTION.** Given: Length of cantilever *AB* ($l_{AB}$) = 2 m ; Uniformly distributed load on *AB* (*w*) = 15 kN/m and length of cantilever *CD* ($l_{CD}$) = 1 m.

Let $R_C$ = Reaction at *C*.

A little consideration will show that the deflection of *B* (of beam *AB*) will be the resultant of (*a*) downward due to load on *AB* and (*b*) upward due to reaction at *B*. But the deflection of *C* (of beam *CD*) will be downward due to reaction at *C*.

We know that downward deflection of the cantilever beam *AB* at *B* due to the load of 15 kN/m (neglecting the reaction at *B*),

$$y = \frac{wl^4}{8EI} = \frac{15\times(2)^4}{8EI} = \frac{30}{EI}$$

and upward deflection of the cantilever *AB* at *B* due to the reaction at *C* (neglecting the load on the beam *AB*),

$$= \frac{R_C\cdot l^3}{3EI} = \frac{R_C\times(2)^3}{3EI} = \frac{8\ R_C}{3EI}$$

∴ Net downward deflection of the cantilever *AB* at *B*,

$$y_B = \frac{30}{EI} - \frac{8\ R_C}{3EI} \quad ...(i)$$

We also know that the downward deflection of the beam *CD* at *C* due to the reaction $R_C$

$$= \frac{R_C\cdot l^3}{3EI} = \frac{R_C\times(1)^3}{3EI} = \frac{R_C}{3EI} \quad ...(ii)$$

Since both the deflections of *B* and *C* are equal, therefore equating (*i*) and (*ii*),

$$\frac{30}{EI} - \frac{8R_C}{3EI} = \frac{R_C}{3EI} \quad \text{or} \quad 30 = 3\ R_C$$

$$\therefore \quad R_C = \mathbf{10\ kN} \quad \textbf{Ans.}$$

**EXAMPLE 23.4.** *A cantilever ABC is fixed at A and propped at C is loaded as shown in Fig. 23.4.*

*Find the reaction at C.*

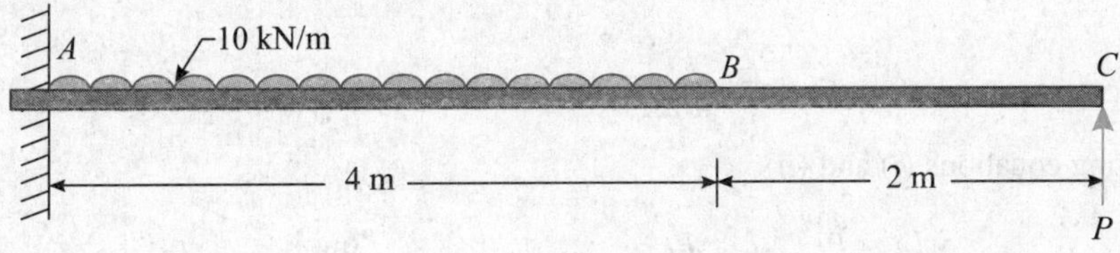


Fig. 23.4

**SOLUTION.** Given: Span ($l$) = 6 m ; Uniformly distributed load ($w$) = 10 kN/m and length of the loaded portion ($l_1$) = 4 m.

Let $\quad P$ = Reaction at the end $C$.

First of all, let us find out the deflection of cantilever at $C$ due to load on $AB$ (neglecting the prop reaction),

We know that deflection at $C$ due to load on $AB$,

$$y_C = \frac{wl_1^4}{8EI} + \frac{wl_1^3}{6EI}(l - l_1) = \frac{10 \times (4)^4}{8EI} + \frac{10 \times (4)^3}{6EI}(6-4)$$

$$= \frac{320}{EI} + \frac{640}{3EI} = \frac{1600}{3EI} \qquad ...(i)$$

Now let us find out the deflection of the cantilever at $C$ due to the reaction on the prop (neglecting the load on $AB$),

We know that deflection due to reaction at the prop,

$$y_C = \frac{Wl^3}{3EI} = \frac{P \times (6)^3}{3EI} = \frac{72P}{EI} \qquad ...(ii)$$

Since both the deflections are equal, therefore equating (*i*) and (*ii*),

$$\frac{1600}{3EI} = \frac{72P}{EI} \quad \text{or} \quad 1600 = 216\,P$$

$\therefore \quad P = 1600/216 =$ **7.41 kN** **Ans.**

## EXERCISE 23.1

1. A horizontal cantilever of length $l$ supports a uniformly distributed load of $w$ per unit length. If the cantilever is propped at a distance of $l/4$ from the free end, find the reaction of the prop.
$$\left[\textbf{Ans. } \frac{19wl}{32}\right]$$
2. A cantilever $ABC$ of uniform section is fixed at $A$ and propped at $B$. A point load $W$ is applied at the free end $C$. Find the ratio of $AB$ to $BC$, so that the reaction at $B$ is 1.5 $W$. [**Ans.** 3 : 1]
3. The free end of a cantilever of length $l$ rests on the middle of a simply supported beam of the same span, and having the same section. Determine the reaction of the cantilever at its free end, if it is carrying a uniformly distributed load of $w$ per unit length.

[**Hint:** Net deflection of free end of the cantilever. $\left[\textbf{Ans. } \frac{6wl}{17l}\right]$

$$= \frac{wl^2}{8EI} - \frac{Pl^3}{3EI}$$

and deflection of the centre of the beam

$$= \frac{Pl^3}{48EI} \quad ...(i)$$

Eqauting equations (*i*) and (*ii*),

$$\frac{wl^2}{8EI} - \frac{Pl^3}{3EI} = \frac{Pl^3}{48EI} \quad \text{or} \quad P = \frac{6w}{17l} \quad ...(ii)$$

4. A propped cantilever of span $l$ propped at the free end, is subjected to a load $W$ at mid of the span. Find the prop reaction.

$$\left[\textbf{Ans. } \frac{5W}{16}\right]$$

## 23.4. Cantilever Propped at an Intermediate Point

Sometimes, a cantilever is subjected to a point load or uniformly distributed load and is propped at an intermediate point. In such a case, the reaction of the prop is found out first by calculating the deflection of the cantilever at the point of prop and then following the usual procedure, as already discussed.

**EXAMPLE 23.5.** *A cantilever of span l carries a point load W at the free end as shown in Fig. 23.5. It is propped at a distance l/4 from the free end.*

*Find the prop reaction.*

**SOLUTION.** Given: Span = $l$; Point load = $W$ and distance between the free end and the prop $(x) = l/4 = 0.25l$.

Let $P$ = Prop reaction.

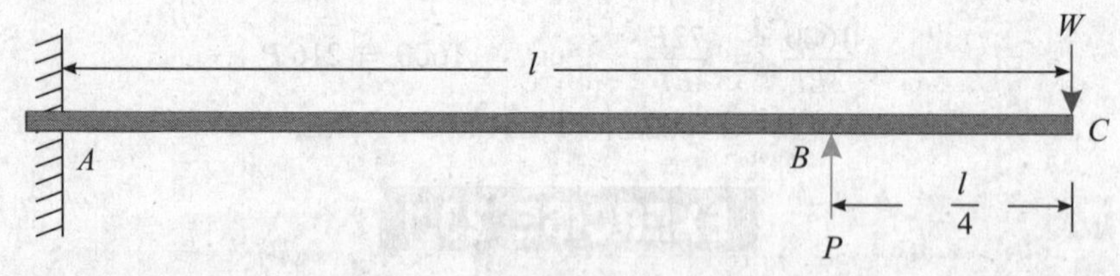


Fig. 23.5

First of all, let us find out the deflection of the cantilever at $B$ due to the load $W$ at $C$ (neglecting the prop),

We know that deflection at $B$ due to load

$$y_B = \frac{Wl^2 x}{2EI} - \frac{Wx^3}{6EI} - \frac{Wl^3}{3EI} = \frac{Wl^2\,(0.25\,l)}{2} - \frac{W\,(0.25\,l)^3}{6} - \frac{Wl^3}{3}$$

$$= -\frac{27\,Wl^3}{128EI}$$

...(Minus sign means that the deflection is downwards)

$$= \frac{27\,Wl^3}{128EI} \quad ...(i)$$

Now, let us find out the deflection of the cantilever at $B$ due to the prop reaction $P$.

We know that deflection of the cantilever at $B$ due to prop reaction $P$,

$$y_B = \frac{Pl_1^3}{3EI} = \frac{P\left(\dfrac{3l}{4}\right)^3}{3EI} = \frac{9Pl^3}{64EI} \qquad ...(ii)$$

Since both the deflections are equal, therefore equating (*i*) and (*ii*),

$$\frac{27Wl^3}{128\,EI} = \frac{9Pl^3}{64\,EI} \quad \text{or} \quad P = \frac{27W \times 64}{128 \times 9} = \frac{3W}{2} \quad \textbf{Ans.}$$

**EXAMPLE 23.6.** *A rigid beam ABC is pinned to a wall to O and is supported by two springs at A and B as shown in Fig. 23.6.*

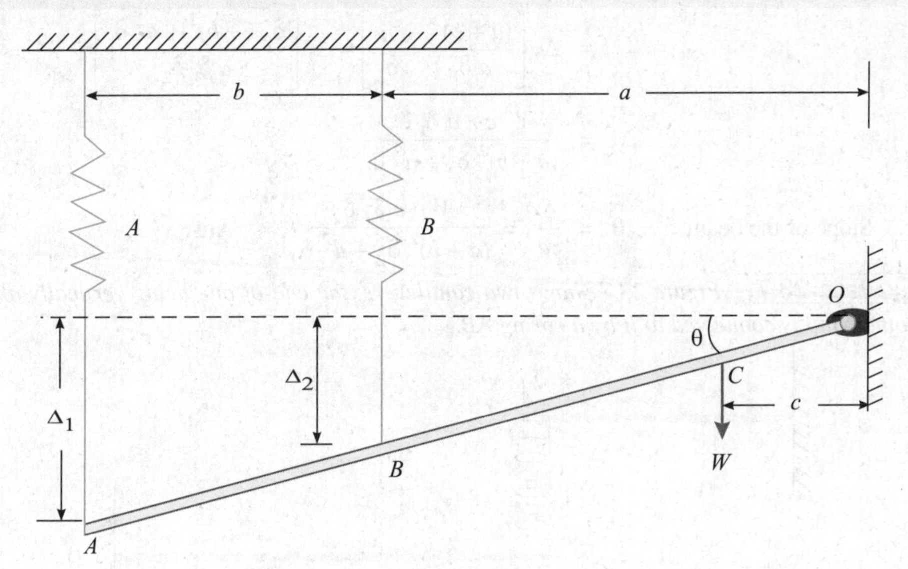


Fig. 23.6

*It carries a load W at C at a distance c from the pinned end. The deformations under the unit load of the springs are $\delta_1$ and $\delta_2$ respectively. Show that the slope of the beam,*

$$\theta = \frac{W_C \cdot \delta_1\,\delta_2}{(a+b)^2\,\delta_2 + a^2\,\delta_1}$$

**SOLUTION.** Given: Deformation of the spring $A$ under the unit load $= \delta_1$ and deformation of the spring $B$ under the unit load $= \delta_2$.

Let $F_1$ = Force in spring $A$, and

$F_2$ = Force in spring $B$.

$\therefore$ Deformation of spring $A$ due to force $F_1$,

$$D_1 = F_1 \cdot \delta_1$$

or

$$F_1 = \frac{\Delta_1}{\delta_1} \qquad ...(i)$$

Similarly,

$$F_2 = \frac{\Delta_2}{\delta_2} \qquad ...(ii)$$

We know that the rigid beam $ABC$ is hinged at $O$. Therefore equating the anticlockwise moments and the clockwise moments about $O$,

$$W \cdot c = F_1\,(a+b) + F_2 \cdot a = \frac{\Delta_1}{\delta_1}(a+b) + \frac{\Delta_2}{\delta_2}\,a \qquad ...(iii)$$

Now from the geometry of the rigid beam, we find that

$$\frac{\Delta_1}{\Delta_2} = \frac{a+b}{a}$$

$$\therefore \qquad \Delta_1 = \frac{(a+b)\,\Delta_2}{a}$$

Substituting is value of $\Delta_1$ in equation (*iii*),

$$W \cdot c = \frac{(a+b)^2}{a} \times \frac{\Delta_2}{\delta_1} + \frac{\Delta_2}{\delta_2}\,a$$

$$= \Delta_2 \left[\frac{(a+b)^2}{a\,\delta_1} + \frac{a}{\delta_2}\right] = \Delta_2 \left[\frac{\delta_2\,(+b)^2 + a^2\,\delta_1}{a\,\delta_1\,\delta_2}\right]$$

or $$\Delta_2 = \frac{W \cdot c \times a\,\delta_1\,\delta_2}{(a+b)^2\,\delta_2 + a^2\,\delta_1} \qquad ...(iv)$$

$\therefore$ Slope of the beam $$\theta = \frac{\Delta_2}{a} = \frac{W \cdot c\,\delta_1\,\delta_2}{(a+b)^2\,\delta_2 + a^2\,\delta_1} \qquad \textbf{Ans.}$$

**EXAMPLE 23.7.** *Figure 23.7 shows two cantilevers, the end of one being vertically above the other, and is connected to it by a spring AB.*

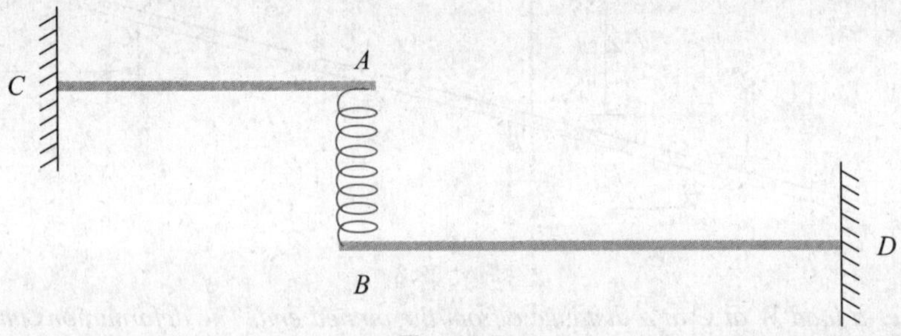


Fig. 23.7

*Initially, the system is unstrained. A weight W placed at A causes a vertical deflection at A of $\delta_1$ and a vertical deflection at B of $\delta_2$. When the spring is removed, the weight W at A causes a deflection at A of $\delta_3$. Find the extension in the spring, when it is replaced and the weight W is transferred to B.*

**SOLUTION.** Given: Weight at $A = W$; Deflection of $A$ (with spring) $= \delta_1$; Deflection of $B$ (with spring) $= \delta_2$ and deflection of $A$ without spring $= \delta_3$.

Let
$l_1$ = Length of cantilever $AC$,
$l_2$ = Length of cantilever $BD$,
$P$ = Force in the spring when the load is at $A$,
$T$ = Force in the spring when the load is at $B$,
$\delta_4$ = Deflection of $A$ with load at $B$, and
$\delta_5$ = Deflection of $B$ with load at $B$.

We know that when the cantilever $AC$ is loaded with $W$ at $A$, the deflection of $A$,

$$\delta_1 = \frac{(W-P)\,l_1^3}{3EI} \qquad ...(i)$$

Similarly,
$$\delta_2 = \frac{Pl_2^3}{3EI} \quad ...(ii)$$

or
$$\frac{l_2^3}{3EI} = \frac{\delta_2}{P} \quad ...(iii)$$

Now compression of the spring $AB$
$$= \delta_1 - \delta_2$$

and stiffness of the spring,
$$s = \frac{\text{Force}}{\text{Compression}} = \frac{P}{\delta_1 - \delta_2} \quad ...(iv)$$

We also know that when the spring is removed and the cantilever $AC$ is loaded with $W$ at $A$, then deflection of $A$,
$$\delta_3 = \frac{Wl_1^3}{3EI} \quad ...(v)$$

or
$$\frac{l_1^3}{3EI} = \frac{\delta_3}{W} \quad ...(vi)$$

Dividing equation (*i*) by (*ii*),
$$\frac{\delta_1}{\delta_2} = \frac{W-P}{W} = 1 - \frac{P}{W}$$

$$\therefore \quad \frac{P}{W} = 1 - \frac{\delta_1}{\delta_3} = \frac{\delta_3 - \delta_1}{\delta_3} \quad ...(vii)$$

or
$$P = \frac{W(\delta_3 - \delta_1)}{\delta_3}$$

Substituting this value of $P$ in equation (*iv*),
$$\delta = \frac{\frac{W(\delta_3 - \delta_1)}{\delta_3}}{(\delta_1 - \delta_2)} = \frac{W(\delta_3 - \delta_1)}{\delta_3(\delta_1 - \delta_2)} \quad ...(viii)$$

We know that when the cantilever $BD$ is loaded with $W$ at $B$, the deflection of $B$,
$$\delta_5 = \frac{(W-T)\,l_2^3}{3EI} = \frac{(W-T)\,\delta_2}{P} \quad ...(ix)$$

$$...\left[\text{Substituting } \frac{l_2^3}{3EI} = \frac{\delta_2}{P} \text{ from equation } (iii)\right]$$

$$= \frac{(W-T)\,\delta_2 \times \delta_3}{W(\delta_3 - \delta_1)}$$

...[Substituting the value of $P$ from equation (*vii*)]

Similarly,
$$\delta_4 = \frac{T \cdot l_1^3}{3EI} = \frac{T \cdot \delta_3}{W} \quad ...(x)$$

$$...\left[\text{Substituting } \frac{l_1^3}{3EI} = \frac{\delta_3}{W} \text{ from equation } (iv)\right]$$

When the cantilever is loaded with $W$ at $B$, the stiffness of the spring,

$$s = \frac{\text{Force}}{\text{Extention}} = \frac{T}{\delta_5 - \delta_4}$$

$$\delta_5 - \delta_4 = \frac{T}{s} = \frac{T \times \delta_3 (\delta_1 - \delta_2)}{W (\delta_3 - \delta_1)} \quad ...(xi)$$

... (Substituting value of $s$ from equation ($viii$)

We also know that extension of the spring,

$$\delta_5 - \delta_4 = \frac{(W - T)\, \delta_2 \times \delta_3}{W (\delta_3 \times \delta_1)} - \frac{T \cdot \delta_3}{W} \quad \text{... [From equation } (ix) \text{ and } (x)]$$

$$= \frac{\delta_3}{W}\left[\frac{(W - T)\, \delta_2}{(\delta_3 - \delta_1)} - T\right] \quad ...(xii)$$

Equating both the values of $(\delta_5 - \delta_4)$,

$$\frac{T \times \delta_3 (\delta_1 - \delta_2)}{W (\delta_3 - \delta_1)} = \frac{\delta_3}{W}\left[\frac{(W - T)\, \delta_2}{\delta_3 - \delta_1} - T\right]$$

$$\frac{T (\delta_1 - \delta_2)}{(\delta_3 - \delta_1)} = \frac{(W - T)\, \delta_2 - T (\delta_3 - \delta_1)}{(\delta_3 - \delta_1)}$$

or $$T \cdot \delta_1 - T \cdot \delta_2 = W \cdot \delta_3 - T \cdot \delta_2 - T \cdot \delta_3 + T\, \delta_1$$

$\therefore$ $$T \cdot \delta_3 = W \cdot \delta_2$$

or $$T = \frac{W \cdot \delta_2}{\delta_3}$$

Substituting this value of $T$ in equation ($xi$),

$$\delta_5 - \delta_4 = \frac{\frac{W \cdot \delta_2}{\delta_3} \times (\delta_1 - \delta_2)}{W (\delta_3 - \delta_1)} = \frac{\delta_2 (\delta_1 - \delta_2)}{(\delta_3 - \delta_1)} \quad \textbf{Ans.}$$

**EXAMPLE 23.8.** *A horizontal cantilever beam of length l and of uniform cross-section carries a uniformly distributed load of w per unit length for the full span. The cantilever is supported by a rigid prop at a distance kl from the fixed end, the level of the beam at the prop being the same as that of the fixed end as shown in Fig. 23.8.*

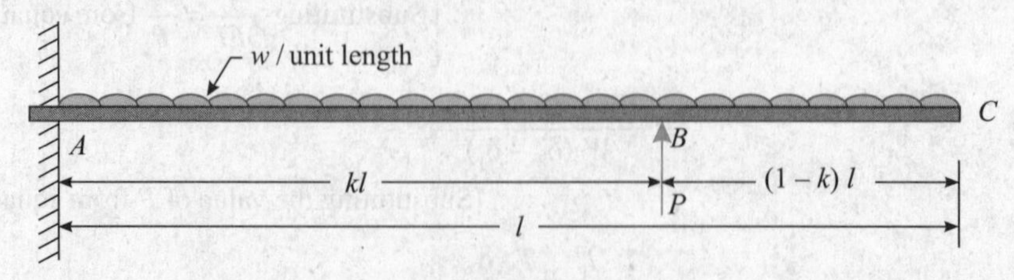


**Fig. 23.8**

*Evaluate k in terms of l for the condition, that the bending moment at the prop is equal to the bending moment at the fixed end. Also determine the reaction at the prop and draw the shear force and bending moment diagrams.*

**Solution.** Given: Span = $l$ ; Load on the beam = $w$ per unit length and distance of prop from the fixed end = $kl$.

Let $P$ = Prop reaction.

From the geometry of the cantilever, we find that the bending moment at the prop

$$= -\frac{w(1-k)^2 l^2}{2}$$

and bending moment at the fixed end $A$

$$= P \cdot kl - \frac{wl^2}{2}$$

Since both the bending moments are equal (as given), therefore equating the same,

$$P \cdot kl - \frac{wl^2}{2} = -\frac{w(1-k)^2 l^2}{2}$$

$$\therefore \quad P \cdot k = \frac{wl}{2} - \frac{wl(1-k)^2}{2} = \frac{wl}{2} - \frac{wl(1+k^2-2k)}{2}$$

$$= \frac{wl}{2}(1 - 1 - k^2 + 2k) = \frac{wlk}{2}(2-k)$$

or $$P = \frac{wl}{2}(2-k) \quad \text{...(i)}$$

First of all, let us find out the deflection of the cantilever at $B$ due to load, but neglecting the prop. We know that the bending moment at any section $X$, at a distance $x$ from the fixed end,

$$M_X = \frac{w(l-x)^2}{2}$$

$$\therefore \quad EI\frac{d^2y}{dx^2} = -\frac{w(l-x)^2}{2} = -\frac{w}{2}(l^2 + x^2 - 2lx)$$

Integrating the above equation,

$$EI\frac{dy}{dx} = -\frac{w}{2}\left(l^2x + \frac{x^3}{3} - \frac{2lx^2}{2}\right) + C_1$$

$$= -\frac{w}{2}\left(l^2x + \frac{x^3}{3} - lx^2\right) + C_1$$

where $C_1$ is the constant of integration. We know that when $x = 0$, then $\frac{dy}{dx} = 0$. Substituting these values of $x$ and $\frac{dy}{dx}$ in the above equation, we get $C_1 = 0$.

$$\therefore \quad EI \cdot \frac{dy}{dx} = -\frac{w}{2}\left(l^2x + \frac{x^3}{3} - lx^2\right)$$

Integrating the above equation once again,

$$EI \cdot y = -\frac{w}{2}\left(\frac{l^2x^2}{2} + \frac{x^4}{12} - \frac{lx^3}{3}\right) + C_2$$

where $C_2$ is the constant of integration. We know that when $x = 0$, then $y = 0$. Substituting these values of $x$ and $y$ in the above equation, we get $C_2 = 0$.

$$\therefore \quad EI \cdot y = -\frac{w}{2}\left(\frac{l^2 x^2}{2} + \frac{x^4}{12} - \frac{lx^3}{3}\right)$$

Now for deflection at $B$, substituting $x = kl$ in the above equation,

$$EI \cdot y_B = -\frac{w}{2}\left(\frac{l^2 \cdot k^2 l^2}{2} + \frac{k^4 l^4}{12} - \frac{l \cdot k^3 l^3}{3}\right)$$

$$= -\frac{wl^4 k^2}{24}(6 + k^2 - 4k)$$

or
$$y_B = -\frac{wl^4 k^2}{24EI}(k^2 - 4k + 6)$$

(Minus sign means that the deflection is downwards)

$$= \frac{wl^4 k^2}{24EI}(k^2 - 4k + 6) \quad ...(ii)$$

Similarly, upward deflection of the cantilever due to the prop reaction,

$$y_B = \frac{P(kl)^3}{3EI} = \frac{Pk^3 l^3}{3EI}$$

Substituting the value of $P$ from equation (i),

$$y_B = \frac{wl}{2}(2-k) \times \frac{k^3 l^3}{3EI}$$

$$= \frac{wl^4 (2-k) k^3}{6EI} \quad ...(iii)$$

Since the level of the beam, at the prop, is the same as that of the fixed end, therefore the net deflection at $B$ is zero. Now equating (*ii*) and (*iii*),

$$\frac{wl^4 k^2}{24EI}(k^2 - 4k + 6) = \frac{wl^4 (2-k) k^3}{6EI}$$

$$\therefore \quad k^2 - 4k + 6 = 4k(2-k) = 8k - 4k^2$$

or
$$5k^2 - 12k + 6 = 0$$

Solving the above equation as a quadratic equation for $k$,

$$k = \frac{12 \pm \sqrt{144 - 4 \times 5 \times 6}}{2 \times 5} = \frac{12 - \sqrt{24}}{10} = \mathbf{0.71} \quad \textbf{Ans.}$$

***Reaction at the prop***

Substituting the value of $k$ in equation (*i*), we get the reaction at the prop,

$$P = \frac{wl}{2}(2-k) = \frac{wl}{2}(2 - 0.71) = \mathbf{0.645\ wl} \quad \textbf{Ans.}$$

***Shear force and bending moment diagrams***

From the geometry of the cantilever, we find that the shear force at $C$,

$$F_C = 0$$

$$F_B = +0.29\ wl - 0.645\ wl = -0.355\ wl$$

$$F_A = -0.355\ wl + 0.71\ wl = +0.355\ wl$$

Now draw the shear force diagram as shown in Fig. 23.9 (*b*). From the geometry of the shear force diagram, we find that the shear force changes sign at *M i.e.*, at the middle of *AB i.e.*, at a distance of 0.355 *l* from *A*.

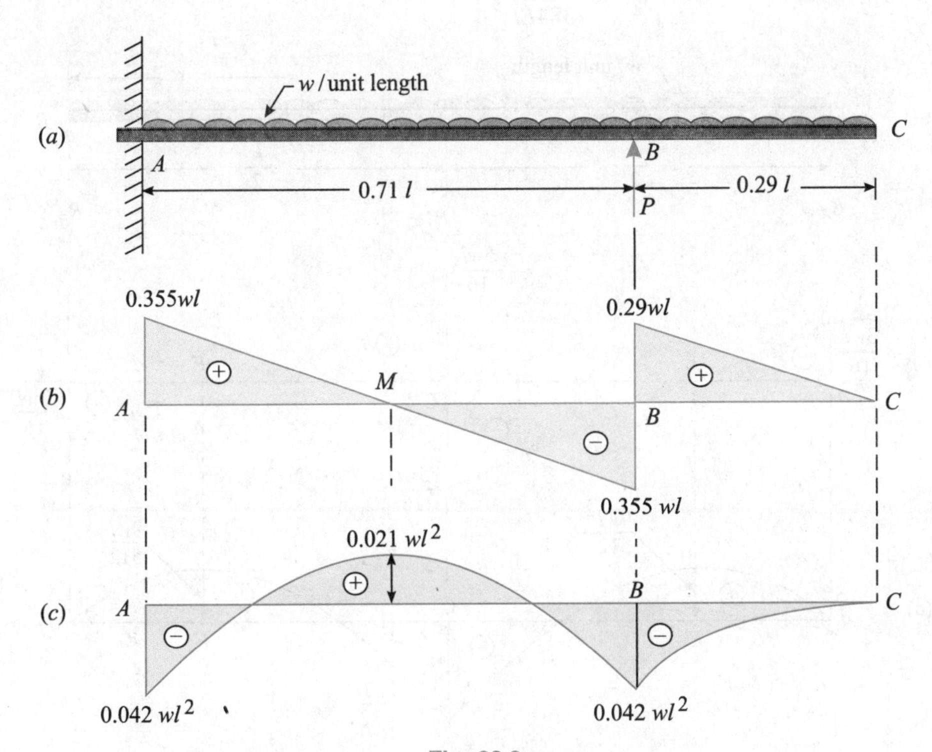


Fig. 23.9

From the geometry of the cantilever, we also find that the bending moment at *C*,

$$M_C = 0$$

$$M_B = -\frac{w\,(0.29\,l)^2}{2} = -\,0.042\,wl^2$$

$$M_A = -\frac{wl^2}{2} + 0.642\,wl \times 0.71\,l$$
$$= -\,0.042\,wl^2$$

$$M_M = -\frac{w\,(0.645\,l)^2}{2} + 0.645\,wl \times 0.355\,l$$
$$= +\,0.021\,wl^2$$

Now draw the bending moment diagram as shown in Fig. 23.9 (c).

## 23.5. Simply Supported Beam with a Uniformly Distributed Load and Propped at the Centre

Consider a simply supported beam *AB* propped at its centre *C* and carrying a uniformly distributed load over its entire span as shown in Fig. 23.10 (*a*).

Let $l$ = Span of the beam *AB*,

$w$ = Uniformly distributed load per unit length over the entire span, and

$P$ = Reaction at the prop.

We know that the downward deflection of $C$ due to uniformly distributed load (neglecting the prop reaction),

$$y_C = \frac{5wl^4}{384EI} \quad \text{...(i)}$$

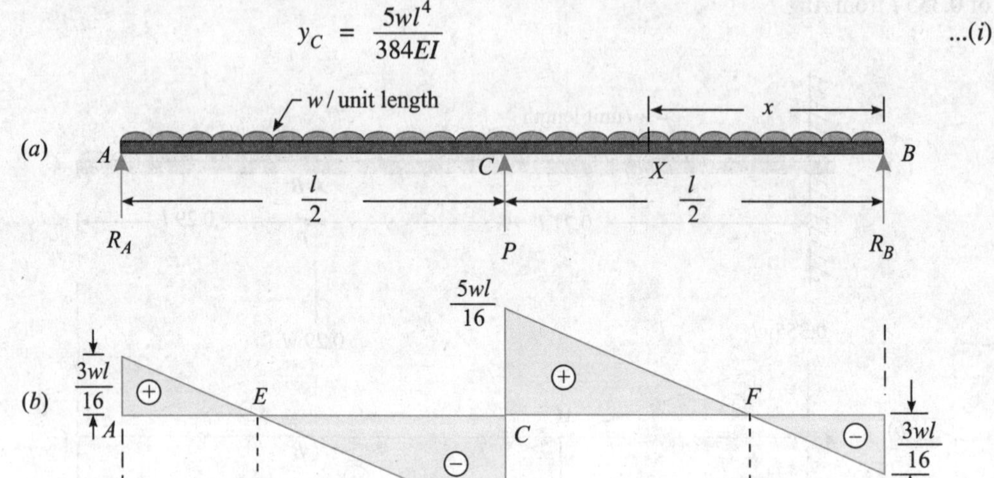


Fig. 23.10

and the upward deflection of the beam at $C$ due to the prop reaction $P$ (neglecting uniformly distributed load),

$$y_C = \frac{Pl^3}{48EI} \quad \text{...(ii)}$$

Since both the deflections are equal, therefore equating equations (*i*) and (*ii*),

$$\frac{Pl^3}{48EI} = \frac{5wl^4}{384EI}$$

or

$$P = \frac{5wl}{8} = \frac{5W}{8} \quad \text{...(where } W = wl\text{)}$$

$\therefore$ Reaction of $A$,

$$R_A = R_B = \frac{1}{2}\left(wl - \frac{5wl}{8}\right) = \frac{3wl}{16}$$

$$= \frac{3W}{16} \quad \text{...(where } W = wl\text{)}$$

Now let us analyse the propped beam for shear force, bending moment, slope and deflection at important sections of the beam.

(*i*) ***Shear force diagram***

We know that the shear force at $A$,

$$F_A = \frac{3wl}{16} \quad \text{...(Plus sign due to left upwards)}$$

$$F_C = \frac{3wl}{16} - \frac{wl}{2} + \frac{5wl}{8} = \frac{5wl}{16}$$

$$F_B = \frac{5wl}{16} - \frac{wl}{2} = -\frac{3wl}{16}$$

Let $F$ be the point, where the shear force changes sign in $CB$ at the distance $x$ from $B$. Therefore

$$\therefore \quad \frac{x}{\frac{l}{2} - x} = \frac{3}{5}$$

or $$x = \frac{3l}{16} \qquad ...(iii)$$

Thus, the shear force is zero at a distance of $\frac{3l}{16}$ from $B$. Similarly, the shear force, is zero at a distace of $\frac{3l}{16}$ from $A$. The shear force diagram is shown in Fig. 23.10($b$).

***(ii) Bending moment diagram***

We know that the bending moment at $A$,

$$M_A = 0$$

$$M_C = \frac{3wl}{16} \times \frac{1}{2} - w \times \frac{1}{2} \times \frac{1}{4}$$

$$= -\frac{wl^2}{32} \qquad \text{...(Max. negative bending moment)}$$

We also know that the bending moment will be maximum at $F$ and $E$, where shear force changes sign.

$$\therefore \quad M_F = M_E = \frac{3wl}{16} \times \frac{3l}{16} - w \times \frac{3l}{16} \times \frac{3l}{32}$$

$$= \frac{9wl^2}{512} \qquad \text{...(Max. positive bending moment)}$$

Now, in order to find out the point of contraflexure, let us equate the bending moment at a distance $x$ from $A$, to zero.

$$\frac{3wl}{16} \cdot x - \frac{wx^2}{2} = 0$$

or $$x = \frac{3l}{8}$$

The bending moment diagram is shown in Fig. 23.10 ($b$).

***(iii) Slope at the ends***

We know that the bending moment at any section $X$, at a distance $x$ from $B$,

$$M_X = \frac{3wlx}{16} - \frac{wx^2}{2}$$

$$\therefore \quad EI \frac{d^2y}{dx^2} = \frac{3wlx}{16} - \frac{wx^2}{2}$$

Integrating the above equation,

$$EI \frac{dy}{dx} = \frac{3wlx^2}{32} - \frac{wx^3}{6} + C_1$$

where $C_1$ is the constant of integration. We know that when $x = \frac{l}{2}$, then $\frac{dy}{dx} = 0$. Substituting these values in the above equation,

$$0 = \frac{3wl}{32}\left(\frac{l}{2}\right)^3 - \frac{w}{6}\left(\frac{l}{2}\right)^3 + C_1$$

or

$$C_1 = -\frac{wl^3}{384}$$

$$EI\frac{dy}{dx} = \frac{3wlx^2}{32} - \frac{wx^3}{6} - \frac{wl^3}{384} \qquad \text{...(iv)}$$

Now for the slope at $B$, substituting $x = 0$ in the above equation,

$$EI \cdot i_B = -\frac{wl^3}{384}$$

$$\therefore \quad i_B = -\frac{wl^3}{384EI}$$

...(Minus sign means that the tangent at $B$ makes an angle with $AB$ in the negative or anticlockwise direction)

$$= \frac{wl^3}{384EI} \text{ radians} \qquad \text{...(v)}$$

By symmetry,

$$i_A = \frac{wl^3}{384EI} \text{ radians}$$

***(iv) Deflection of the beam***

Integeating the equation (*iv*) once again,

$$EI \cdot y = \frac{3wlx^3}{96} - \frac{wx^4}{24} - \frac{wl^3x}{384} + C_2$$

where $C_2$ is the constant of integration. We know that when $x = 0$, then $y = 0$. Therefore substituting these values in the above equations, we get $C_2 = 0$.

$$\therefore \quad EI \cdot y = \frac{3wlx^3}{96} - \frac{wx^4}{24} - \frac{wl^3x}{384} \qquad \text{...(vi)}$$

This is the required equation for *deflection* at any section of the beam.

***(v) Maximum deflection***

We know that the maximum deflection takes place at a point, where slope is zero. Therefore equating the equation (*iv*) to zero,

$$\frac{3wlx^2}{32} - \frac{wx^3}{6} - \frac{wl^3}{384} = 0$$

$$\therefore \quad 64x^4 - 36x^2 + l^3 = 0$$

Solving the equation by trial and error, we get $x = 0.27\ l$

$$\therefore \quad EI \cdot y_{max} = \frac{3wl}{96}(0.27\ l)^3 - \frac{w}{24}(0.27\ l)^4 - \frac{wl^3}{384}(0.27\ l)$$

$$= -0.000\ 306\ 2\ wl^4$$

$$y_{max} = -\frac{0.00\ 306\ 2\ wl^4}{EI}$$

...(Minus sign means that the deflection is downwards)

$$= \frac{0.00\ 306\ 2\ wl^4}{EI}$$

**EXAMPLE 23.9.** *A uniform girder of length 8 m is subjected to a total load of 20 kN, uniformly distributed over the entire length. The girder is freely supported at the ends. Calculate the deflection and bending moment at the mid-span.*

*If a prop is introduced at the centre of the beam, so as to nullify the deflection already worked out, what would be the net bending moment at mid-point?*

**SOLUTION.** Given: Length ($l$) = 8 m and total uniformly distributed load ($W$) = 20 kN.

***Deflection at the mid-span of the beam without prop***

Let $EI$ = Stiffness of the beam.

We know that the uniformly distributed load,

$$w = \frac{W}{l} = \frac{20}{8} = 2.5 \text{ kN/m}$$

∴ Deflection at mid-span without prop,

$$y_C = \frac{5wl^4}{384EI} = \frac{5 \times 2.5 \times (8)^4}{384EI} = \frac{400}{3EI} \quad \textbf{Ans.}$$

***Bending moment at the mid-span of the beam without prop***

We know that bending moment at the mid-span of the beam without prop,

$$M_1 = \frac{wl^2}{8} = \frac{2.5 \times (8)^2}{8} = \textbf{20 kN-m} \quad \textbf{Ans.}$$

***Bending moment at the mid-span of the beam with prop***

We also know that bending moment at the mid-span of the beam with prop,

$$M_2 = \frac{wl^2}{32} = \frac{2.5 \times (8)^2}{32} = \textbf{5 kN-m} \quad \textbf{Ans.}$$

## 23.6. Sinking of the Prop

In the previous articles, we have assumed that the prop in a cantilever or beam behaves like a rigid one *i.e.*, it does not yield down due to the load acting on the beam. But sometimes, the prop sinks down, due to its elastic property and the reaction. A sinking prop is called an *elastic prop* or *yielding prop*.

Let $\delta$ = Distance through which the prop has sunk down, due to load.

$y_1$ = Downward deflection of the beam, at the point of prop and

$y_2$ = Upward deflection of the beam, due to the prop reaction,

A little consideration will show that if the prop would not have sunk down, then

$$y_1 = y_2$$

But due to sinking of the prop,

$$y_1 = y_2 + \delta$$

Now the prop reaction may be found out as usual.

**EXAMPLE 23.10.** *A cantilever of length l is subjected to a point load W at its free end. The cantilever is also propped with an elastic prop at its free end. The prop sinks down in proportion to the load applied on it. Determine the value of proportionality k for sinking, when the reaction on the prop is half of the load W.*

**SOLUTION.** Given: Span = $l$; Load at the free end = $W$; Prop reaction ($P$) = $\frac{W}{2}$ and constant of proportionality of sinking to the load = $k$.

From the given data, we find that sinking of the prop,

$$\delta = \frac{k \cdot W}{2}$$

We know that the downward deflection of the cantilever due to load $W$ at its free end (neglecting prop reaction),

$$y_1 = \frac{Wl^3}{3EI} \quad ...(i)$$

and upward deflection of the cantilever due to prop reaction (neglecting load),

$$y_2 = \frac{Pl^3}{3EI} = \frac{\frac{W}{2} \times l^3}{3EI} = \frac{Wl^3}{6EI} \quad ...(ii)$$

∴ Sinking of the prop,

$$\delta = y_1 - y_2$$

$$\therefore \quad \frac{k \cdot W}{2} = \frac{Wl^3}{3EI} - \frac{Wl^3}{6EI} = \frac{Wl^3}{6EI}$$

or $$k = \frac{l^3}{3EI} \quad \textbf{Ans.}$$

**EXAMPLE 23.11.** *A simply supported beam of span l carries a uniformly distributed load of w per unit length. The beam was propped at the middle of the span. Find the amount, by which the prop should yield, in order to make all the three reactions equal.*

**SOLUTION.** Given: Span = $l$; Uniformly distributed load = $w$ per unit length and each reaction $(P) = \frac{wl}{3}$.

We know that the downward deflection of the beam, due to uniformly distributed load (neglecting prop reaction),

$$y_1 = \frac{5wl^4}{384EI} \quad ...(i)$$

and upward deflection due to the prop reaction (neglecting load),

$$y_2 = \frac{Pl^3}{48EI} = \frac{\frac{wl}{3} \times l^3}{48EI} = \frac{wl^4}{144EI} \quad ...(ii)$$

∴ Yield of the prop,

$$\delta = y_1 - y_2 = \frac{5wl^4}{384EI} - \frac{wl^4}{144EI} = \frac{7wl^4}{1152EI} \quad \textbf{Ans.}$$

## EXERCISE 23.2

1. A cantilever $AB$, 9 m long is fixed at $A$ and propped at $C$ at a distance 1 m from $B$. The cantilever carries a load, which varies gradually from zero at the free end to 6 kN/m at the fixed end. Calculate the prop reaction. **[Ans. 3.11 kN]**
2. A simply supported beam of length $l$ is carrying a uniformly distributed load of $w$ per unit length over its entire span. What upward load should be applied at the centre of the beam in order to neutralise the deflection? $\left[\textbf{Ans.}\ \frac{5wl}{8}\right]$

3. A cantilever of length $l$ is propped at its free end. The cantilever carries a uniformly distributed load of $w$ per unit length over entire span. If the prop sinks by $\delta$, find the prop reaction.

$$\left[\textbf{Ans. } \frac{3EI}{l^3}\left(\frac{wl^4}{8EI} - \delta\right)\right]$$

4. A uniform beam of cross-section 200 mm wide and 300 mm deep is simply supported on a span of 8 m and carries a load of 5 kN/m. If the centre of the beam is propped at the level of the supports, find the prop reaction.

If the prop sinks down by 20 mm, find the new prop reaction. Take $E$ as 120 GPa.

[**Ans.** 25 kN ; 16.9 kN]

## QUESTIONS

1. What do you understand by the term "prop"? Discuss its importance.
2. Describe the procedure for finding out the prop reaction of a cantilever.
3. Derive an equation for the prop reaction in (*a*) a cantilever carrying a u.d.l. over the entire span and propped at the free end and (*b*) a simply supported beam carrying a u.d.l. over the entire span and propped at the mid-span.
4. From first principles, derive a relation for the maximum deflection of a cantilever carrying a uniformly distributed load and propped at the free end.
5. Define 'sinking of a prop'. How does it differ from a rigid prop?
6. Explain the procedure for finding out the reaction on an elastic prop.

## OBJECTIVE TYPE QUESTIONS

1. A cantilever of span $l$ is fixed at $A$ and propped at the other end $B$. If it is carrying a uniformly distributed load of $w$ per unit length, then the prop reaction will be

   (*a*) $\frac{3wl}{8}$  (*b*) $\frac{5wl}{8}$  (*c*) $\frac{3wl}{16}$  (*d*) $\frac{5wl}{16}$

2. The deflection at the centre of a propped cantilever of span $l$ carrying a uniformly distributed load $w$ per unit length is

   (*a*) $\frac{wl^4}{48EI}$  (*b*) $\frac{wl^4}{96EI}$  (*c*) $\frac{wl^4}{128EI}$  (*d*) $\frac{wl^4}{192EI}$

3. The maximum deflection of a propped cantilever of span $l$ subjected to a uniformly distributed load of $w$ per unit length will occur at a distance of

   (*a*) 0.25 $l$ from the propped end  (*b*) 0.33 $l$ from the propped end
   (*c*) 0.422 $l$ from the propped end  (*d*) 0.615 $l$ from the propped end.

4. A simply supported beam of span $l$ is carrying a uniformly distributed load of $w$ per unit length. If the beam is propped at its mid-point, then the prop reaction is equal to

   (*a*) $\frac{3wl}{8}$  (*b*) $\frac{5wl}{8}$  (*c*) $\frac{3wl^2}{8}$  (*d*) $\frac{5wl^2}{8}$

## ANSWERS

1. (*a*)  2. (*d*)  3. (*c*)  4. (*b*)

Chapter

# 24

# Fixed Beams

## Contents



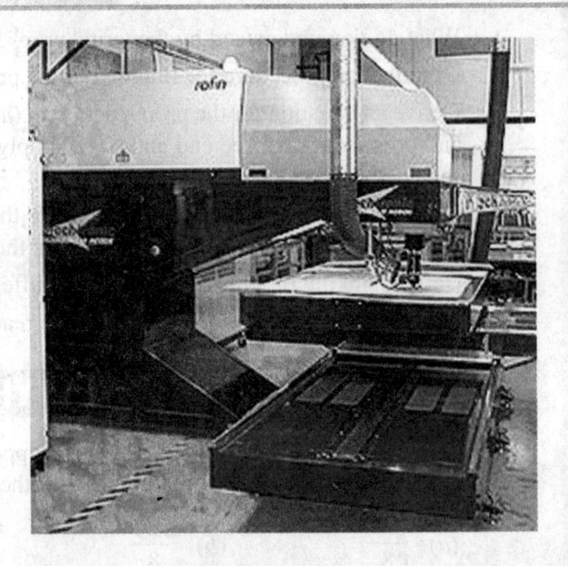

## 24.1. Introduction

A beam, which is built-in at its two supports, is called a constrained beam or a fixed beam. Since the beam is fixed at its two supports, therefore the slope of the elastic curve of the beam at its two ends, even after loading will be zero. Thus, a fixed beam *AB* may be looked upon as a simply supported beam, subjected to end moments $M_A$ and $M_B$, such that the slopes at two supports are zero. A little consideration will show that this is only possible, if the magnitude and directions of the restraining moments $M_A$ and $M_B$ are equal and opposite to that of the bending moments under a given system of loading.

## 24.2. Advantages of Fixed Beams

A fixed beam has the following advantages over a simply supported beam:

1. The beam is stiffer, stronger and more stable.
2. The slope at both the ends is zero.
3. The fixing moments are developed at the two ends, whose effect is to reduce the maximum bending moment at the centre of the beam.
4. The deflection of a beam, at its centre is very much reduced.

## 24.3. Bending Moment Diagram for Fixed Beams

Consider a fixed beam *AB*, of span *l* subjected to various types of loading as shown in Fig. 24.1 (*a*). Now we shall analyse the beam into the following two categories:

1. A simply supported beam *AB* subjected to vertical loads and reactions.
2. A simply supported beam *AB* subjected to end moments.

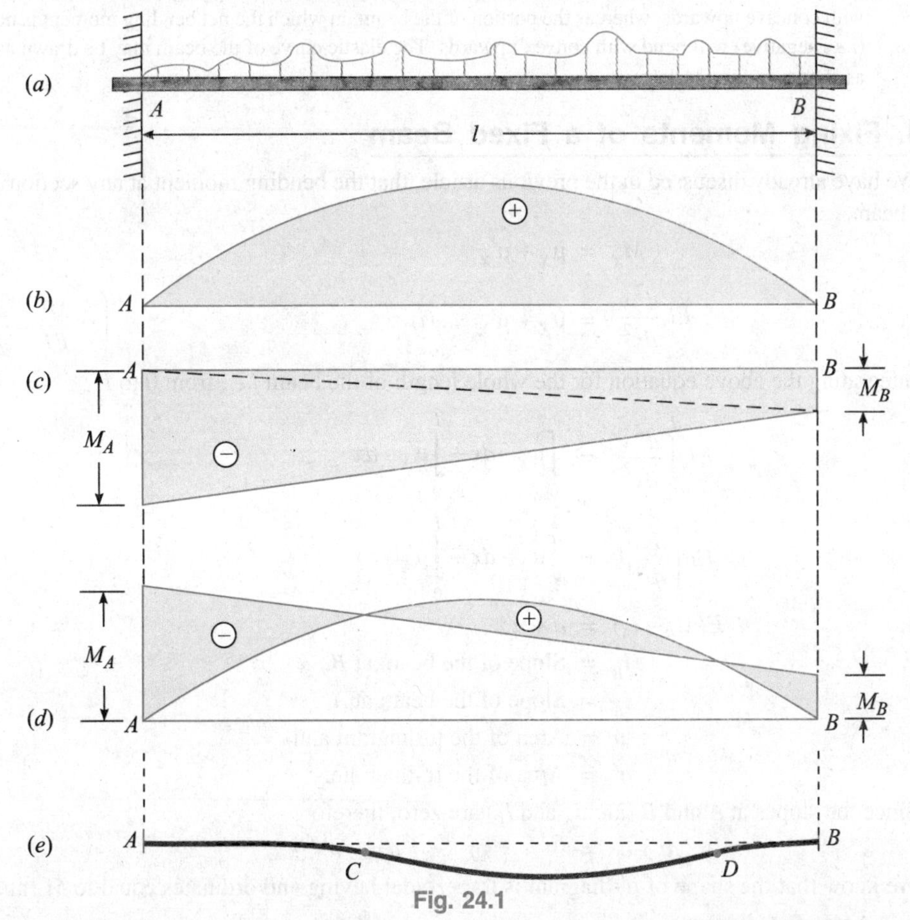


Fig. 24.1

The complete bending moment diagram may be drawn by superimposing the bending moment diagrams for the above two cases. We know that the beam *AB*, when treated as a simply supported beam carrying vertical loads and reactions will be subjected to positive bending moment (*i.e.*, sagging) as shown in Fig. 24.1 (*b*). But the beam *AB*, when treated as a simply supported beam, having fixing moments $M_A$ and $M_B$ will be subjected to negative bending moment (*i.e.*, hogging) as shown in

Fig. 24.1 (*c*). Since the directions of the above two moments are opposite to each other, therefore their resultant effect may be seen by drawing the two moments, on the same side of the base *AB* as shown in Fig. 24.1(*d*).

Now consider any section *X*, at a distance *x* from *A*. Let the bending moment due to vertical loading be $\mu_X$. The bending moment due to fixing moments $M_A$ and $M_B$ at *X*,

$$\mu_X = -\left[M_A + (M_B - M_A)\frac{x}{l}\right]$$

∴ Total bending moment at *X*,

$$M_X = \mu_X + \mu'_X = \mu_X - \left[M_A + (M_B - M_A)\frac{x}{l}\right]$$

**Notes: 1.** The total bending moment at any section may be found out from the above relation, if the values of $M_A$ and $M_B$ are known.

**2.** The shear force diagram for the beam may now be drawn as usual.

**3.** The portion of the beam *AB*, in which the net bending moment is sagging (*i.e.*, positive) will bend with concave upwards, whereas the portion of the beam, in which the net bending moment is hogging (*i.e.*, negative) will bend with convex upwards. The elastic curve of the beam may be drawn as usual as shown in Fig. 24.1 (*e*).

## 24.4. Fixing Moments of a Fixed Beam

We have already discussed in the previous article, that the bending moment at any section *X* of a fixed beam,

$$M_X = \mu_X + \mu'_X$$

∴ $$EI\frac{d^2y}{dx^2} = \mu_X + \mu'_X \quad ...(i) \qquad \left(\because \frac{M}{EI} = \frac{d^2y}{dx^2}\right)$$

Integrating the above equation for the whole length of the beam *i.e.*, from 0 to *l*,

$$EI\int_0^l \frac{d^2y}{dx^2} = \int_0^l \mu_X \cdot dx + \int_0^l \mu'_X \cdot dx$$

$$EI\left[\frac{dy}{dx}\right]_0^l = \int_0^l \mu_X \cdot dx + \int_0^l \mu'_X \cdot dx$$

or $$EI\,(i_B - i_A) = a + a' \quad ...(ii)$$

where $i_B$ = Slope of the beam at *B*,

$i_A$ = Slope of the beam at *A*,

$a$ = Area of the μ-diagram and

$a'$ = Area of the μ′-diagram.

Since the slopes at *A* and *B* (*i.e.*, $i_A$ and $i_B$) are zero, therefore

$$a + a' = 0 \quad \text{or} \quad a = -a'$$

We know that the shape of μ′-diagram is trapezoidal having end ordinates equal to $M_A$ and $M_B$.

∴ Area of μ′-diagram,

$$a' = \frac{l}{2}(M_A + M_B)$$

or $$\frac{l}{2}(M_A + M_B) = -a \qquad (\because a = -a')$$

$$\therefore \qquad M_A + M_B = -\frac{2a}{l} \qquad ...(iii)$$

From equation (*i*) we know that

$$EI\frac{d^2y}{dx^2} = \mu_X + \mu'_X$$

Multiplying the above equation by $x$ and integrating the same for the whole length of the beam *i.e.*, from 0 to $l$,

$$EI\int_0^l x \cdot \frac{d^2y}{dx^2} = \int_0^l x \cdot \mu_X \, dx + \int_0^l x \cdot \mu'_X \, dx$$

$$EI\left[x\frac{dy}{dx} - y\right]_0^l = a\bar{x} + a'\bar{x}'$$

$$\text{or} \quad EI\,[l\,(i_B - y_B) - 0\,(I_A - y_A) = a\bar{x} + a\bar{x}'$$

Since $i_B$ and $y_B$ are equal to zero, therefore

$$a\bar{x} + a'\bar{x}' = 0 \qquad \text{or} \qquad a\bar{x} = -a'\bar{x}'$$

where $\bar{x}$ = Distance of centre of gravity of μ-diagram from $A$ and

$\bar{x}'$ = Distance of centre of gravity of μ′-diagram from $A$.

We know that the shape of the μ′-diagram is trapezoidal with end ordinates equal to $M_A$ and $M_B$. Therefore splitting up the μ′-diagram into two triangles as shown in Fig. 24.1(*c*).

$$a'\bar{x}' = \left(M_A \times \frac{l}{2} \times \frac{l}{3}\right) + \left(M_B \times \frac{l}{2} \times \frac{2l}{3}\right)$$

$$= (M_A + 2M_B)\frac{l^2}{6}$$

$$\text{or} \qquad (M_A + 2M_B)\frac{l^2}{6} = -a\bar{x}$$

$$\therefore \qquad M_A + 2M_B = -\frac{6a\bar{x}}{l^2} \qquad ...(iv)$$

Now subtracting equation (*iii*) and (*iv*),

$$M_B = -\frac{6a\bar{x}}{l^2} + \frac{2a}{l^2} = \frac{2a}{l^2}(-3\bar{x} + l)$$

and substituting the value of $M_B$ in equation (*iii*),

$$M_A + \frac{2a}{l^2}(-3x + l) = -\frac{2a}{l}$$

$$M_A = -\frac{2a}{l} - \frac{2a}{l^2}(-3\bar{x} + l)$$

$$= -\frac{2a}{l^2}[l + (-3x + l)]$$

$$= -\frac{2a}{l^2}(2l - 3\bar{x})$$

These are the required equations for the fixing moments $M_A$ and $M_B$ of a fixed beam *AB*. Here we shall discuss the following standard cases for the fixing moments.

1. A fixed beam carrying a central point load.
2. A fixed beam carrying an eccentric point load.
3. A fixed beam carrying a uniformly distributed load.
4. A fixed beam carrying a gradually varying load from zero at one end to *w* per unit length at the other end.

## 24.5. Fixing Moments of a Fixed Beam Carrying a Central Point Load

Consider a beam *AB* of length *l* fixed at *A* and *B* and carrying a central point load *W* as shown in Fig. 24.2 (*a*).

(*i*) ***Bending moment diagram***

Let $M_A$ = Fixing moment at *A* and

$M_B$ = Fixing moment at *B*.

Since the beam is symmetrical, therefore $M_A$ and $M_B$ will also be equal. Moreover, the μ′-diagram (*i.e.*, bending moment diagram due to fixing moments $M_A$ and $M_B$) will be a rectangle as shown i n Fig. 24.2 (*b*). We know that μ-diagram *i.e.*, bending moment diagram due to central point lead will be a triangle with the central ordinate equal to $\frac{wl}{4}$ as shown in Fig. 24.2 (*b*).

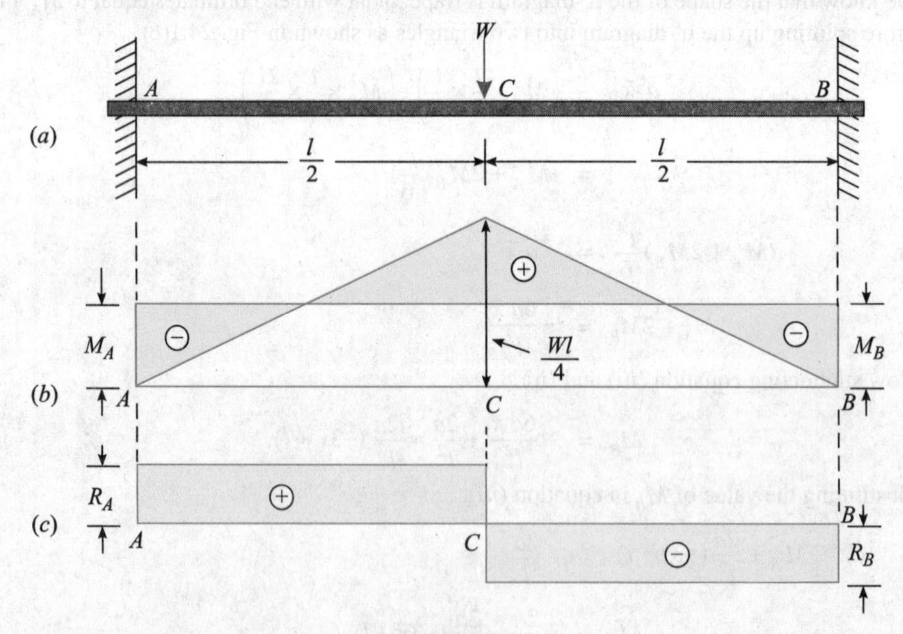


Fig. 24.2

Now equating the areas of the two diagrams,

$$M_A \cdot l = -\frac{1}{2} \cdot l \cdot \frac{Wl}{4} = -\frac{Wl^2}{8}$$

$$M_A = -\frac{Wl}{8}$$

Similarly, $M_B = -\dfrac{Wl}{8}$ ...[By symmetry]

Now complete the bending moment diagrams as shown in Fig. 24.2 (*b*).

***(ii) Shear force diagram***

Let $R_A$ = Reaction at *A* and

$R_B$ = Reaction at *B*.

Equating clockwise moments and anticlockwise moments about *A*,

$$R_B \times l + M_A = M_B + W \times \frac{l}{2}$$

$$\therefore \quad R_B = \frac{W}{2} \qquad ...(\because M_A = M_B)$$

Similarly, $R_A = \dfrac{W}{2}$ ...(By symmetry)

Now complete the S.F. diagram as shown in Fig. 24.2 (*c*).

***(iii) Deflection of the beam***

From the geometry of the figure, we find that the points of contraflexure will be at a distance of $l/4$ from both the ends of the beam.

We know that bending moment at any section *X*, at a distance *x* from *A*,

$$M_X = \mu_X - \mu'_X$$

or $$EI\frac{d^2y}{dx^2} = \frac{Wx}{2} - \frac{Wl}{8} \qquad ...(i)$$

Integrating the above equation,

$$EI\frac{dy}{dx} = \frac{Wx^2}{4} - \frac{Wlx}{8} + C_1$$

where $C_1$ is the first constant of integration. We know that when $x = 0$, then $\dfrac{dy}{dx} = 0$. Therefore $C_1 = 0$.

or $$EI\frac{dy}{dx} = \frac{Wx^2}{4} - \frac{Wlx}{8} \qquad ...(ii)$$

This is the required equation for the slope of the beam at any section.

Now integrating the equation (*ii*) once again,

$$EI \cdot y = \frac{Wx^3}{12} - \frac{Wlx^2}{16} + C_2$$

where $C_2$ is the second constant of integration. We know that when $x = 0$, then $y = 0$. Therefore $C_2 = 0$.

or $$EI \cdot y = \frac{Wx^3}{12} - \frac{Wlx^2}{16} \qquad ...(iii)$$

This is the required equation for the *deflection* of the beam at any section. We know that the maximum deflection occurs at the centre of the beam. Therefore substituting $x = l/2$ in the above equation,

$$EI \cdot y_C = \frac{W}{12}\left(\frac{l}{2}\right)^3 - \frac{Wl}{16}\left(\frac{l}{2}\right)^2 = \frac{Wl^3}{96} - \frac{Wl^2}{64} = -\frac{Wl^3}{192}$$

or $$y_C = -\frac{Wl^3}{192\,EI}$$ ...(Minus sign means that the deflection is downwares)

$$= \frac{Wl^3}{192\,EI}$$

NOTE: The term $EI$ is known as flexural rigidity.

**EXAMPLE 24.1.** *A fixed beam AB, 4 metres long, is carrying a central point load of 3 tonnes. Determine the fixing moments and deflection of the beam under the load. Take flexural rigidity of the beam as $5 \times 10^3$ kN-m$^2$.*

**SOLUTION.** Given: Length ($l$) = 4 m ; Central point load ($W$) = 3 kN and flexural rigidity ($EI$) = $5 \times 10^3$ kN-m$^2$.

*Fixing moments*

We know that fixing moment at $A$,

$$M_A = -\frac{Wl}{8} = -\frac{3 \times 4}{8} = -\mathbf{1.5\ kN\text{-}m} \qquad \textbf{Ans.}$$

Similarly, fixing moment at $B$,

$$M_B = -\frac{Wl}{8} = \frac{3 \times 4}{8} = -\mathbf{1.5\ kN\text{-}m} \qquad \textbf{Ans.}$$

*Deflection of the beam under the load*

We also know that deflection of the beam under the load,

$$y_C = \frac{Wl^3}{192\,EI} = \frac{3 \times (4)^3}{192 \times (5 \times 10^3)} = 0.2 \times 10^{-3}\ \text{m} = \mathbf{0.2\ mm} \qquad \textbf{Ans.}$$

## 24.6. Fixing Moments of a Fixed Beam Carrying an Eccentric Point Load

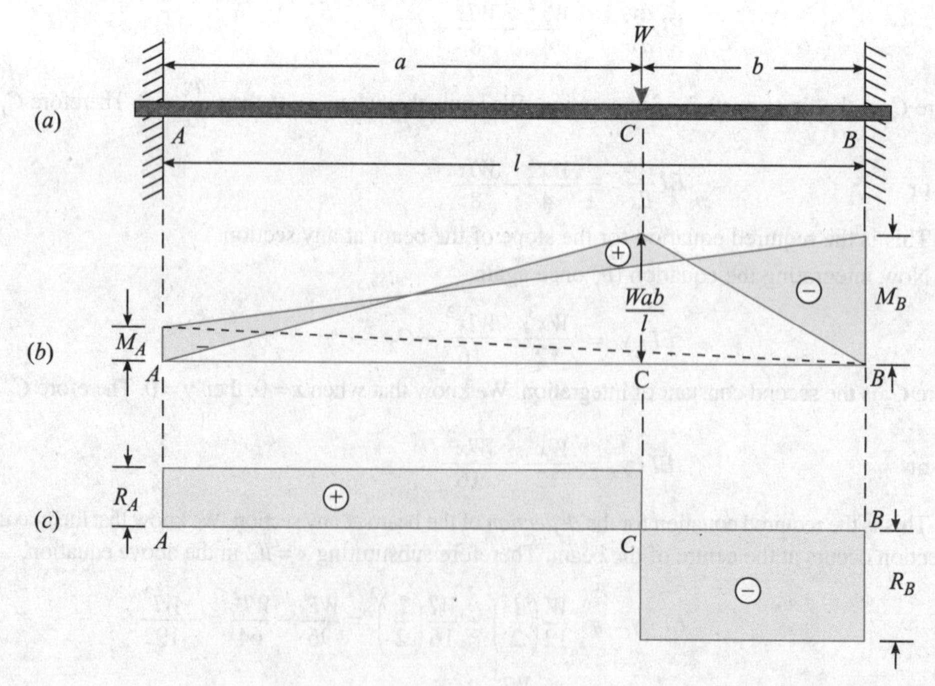


Fig. 24.3

Consider a beam $AB$ fixed at $A$ and $B$ and carrying an eccentric point load as shown in Fig. 24.3($a$).

Let
$l$ = Span of the beam,
$W$ = Load on the beam,
$a$ = Distance between the fixed end $A$ and the axis of the load,
$b$ = Distance between the fixed end $B$ and the axis of the load,
$M_A$ = Fixing moment at $A$ and
$M_B$ = Fixing moment at $B$.

***(i) Bending moment diagram***

Since the beam is not symmetrical, therefore $M_A$ and $M_B$ will also not be equal. Moreover, the μ′-diagram will be a trapezium as shown in Fig. 24.3 ($b$).

We know that the μ-diagram will be triangle with ordinate equal to $\frac{Wab}{l}$ as shown in Fig. 24.3 ($b$). Now equating the areas of the two diagrams,

$$(M_A + M_B)\frac{l}{2} = -\frac{Wab}{l} \times \frac{l}{2}$$

$$\therefore \quad M_A + M_B = -\frac{Wab}{l} \quad \text{...(i)}$$

The moment of μ-diagram area about $A$ (by splitting up the triangle into two right angled triangles)

$$= -\left[\left(\frac{Wab}{l} \times \frac{a}{2} \times \frac{2a}{3}\right) + \frac{Wab}{l} \times \frac{b}{2}\left(a + \frac{b}{3}\right)\right]$$

$$= -\left[\left(\frac{Wab}{6l} \times 2a^2\right) + \frac{Wab}{2l}\left(ab + \frac{b^2}{3}\right)\right] \quad \text{...(ii)}$$

and moment of μ′-diagram about $A$ (by splitting up the trapezium into two triangles) as shown in Fig. 24.3 ($b$)

$$= \left(M_A \times \frac{l}{2} \times \frac{l}{3}\right) + \left(M_B \times \frac{l}{2} \times \frac{2l}{3}\right) = (M_A + 2M_B)\frac{l^2}{6} \quad \text{...(iii)}$$

Now equating these two moments,

$$(M_A + 2M_B)\frac{l^2}{6} = -\left[\left(\frac{Wab}{6l} \times 2a^2\right) + \frac{Wab}{2l}\left(ab + \frac{b^2}{3}\right)\right]$$

$$= -\left[\left(\frac{Wab}{6l} \times 2a^2\right) + \frac{Wab}{2l}\left(\frac{3ab + b^2}{3}\right)\right]$$

$$= -\left[\left(\frac{Wab}{6l} \times 2a^2\right) + \frac{Wab}{6l}\left(3ab + b^2\right)\right]$$

$$= -\frac{Wab}{6l}(2a^2 + 3ab + b^2)$$

$$= -\frac{Wab}{6l}[2(l-b)^2 + 3(l-b)b + b^2] \quad \text{...}(\because a + b = l)$$

$$= -\frac{Wab}{6l}[2(l^2 + b^2 - 2lb) + 3(lb - b^2) + b^2]$$

$$= -\frac{Wab}{6l}[2l^2 + 2b^2 - 4lb + 3lb - 3b^2 + b^2]$$

$$= -\frac{Wab}{6l}(2l^2 - lb) = -\frac{Wab}{6}(2l - b)$$

$$= -\frac{Wab}{6}[2(a + b) - b] \quad ...(\because a + b = l)$$

$$= -\frac{Wab\,(2a + b)}{6}$$

$$\therefore \quad M_A + 2M_B = -\frac{Wab\,(2a + b)}{l^2} \quad ...(iv)$$

Subtracting equation (*i*) from (*iv*),

$$M_B = -\frac{Wab\,(2a + b)}{l^2} + \frac{Wab}{l} = -\frac{Wab}{l}\left(\frac{2a + b}{l} - 1\right)$$

$$= -\frac{Wab}{l}\left(\frac{2a + b - l}{l}\right) = -\frac{Wab}{l^2}[2a + (l - a) - l]$$

$$= -\frac{Wa^2b}{l^2}$$

Subtracting this value of $M_B$ in equation (*i*),

$$M_A - \frac{Wa^2b}{l^2} = -\frac{Wab}{l}$$

$$\therefore \quad M_A = -\frac{Wab}{l} + \frac{Wa^2b}{l^2} = -\frac{Wab}{l}\left(1 - \frac{a}{l}\right)$$

$$= -\frac{Wab}{l}\left(\frac{l - a}{l}\right) = -\frac{Wab^2}{l^2} \quad (\because l - a = b)$$

Now complete the bending moment diagram as shown in Fig. 24.3(*b*).

***(ii) Shear force diagram***

Let $R_A$ = Reaction at *A* and

$R_B$ = Reaction at *B*.

Equating clockwise moments and anticlockwise moments about *A*,

$$R_B \times l + M_A = M_B + W \cdot a$$

$$\therefore \quad R_B = \frac{(M_B - M_A) + W \cdot a}{l}$$

Similarly,

$$R_A = \frac{(M_A - M_B) + W \cdot b}{l}$$

Now, complete the shear force diagram as shown in Fig. 24.3(*b*).

***(iii) Deflection of the beam***

We know that the bending moment at any section *X* at a distance *x* from *A*.

$$M_X = \mu_X - \mu'_X$$

or
$$EI\frac{d^2y}{dx^2} = \frac{Wb}{l}x - \left[M_A + (M_B - M_A)\frac{x}{l}\right]$$

$$= \frac{Wbx}{l} - \left[\frac{Wab^2}{l^2} + \left(\frac{Wa^2b}{l^2} - \frac{Wab^2}{l^2}\right)\frac{x}{l}\right]$$

$$= \frac{Wbx}{l} - \left[\frac{Wab^2}{l^2} + \frac{Wab\,(a-b)x}{l^3}\right]$$

$$= \frac{Wbx}{l} - \frac{Wab^2}{l^2} - \frac{Wab\,(a-b)x}{l^3}$$

Integrating the above equation,

$$EI\frac{dy}{dx} = \frac{Wbx^2}{2l} - \frac{Wab^2x}{l^2} - \frac{Wab\,(a-b)x^2}{2l^3} + C_1$$

where $C_1$ is the first constant of integration. We know that when $x = 0$, then $\frac{dy}{dx} = 0$. Therefore $C_1 = 0$.

or
$$EI\frac{dy}{dx} = \frac{Wbx^2}{2l} - \frac{Wab^2x}{l^2} - \frac{Wab\,(a-b)x^2}{2l^3}$$

$$= \frac{Wbx^2}{2l}\left(1 - \frac{a\,(a-b)}{l^2}\right) - \frac{Wab^2x}{l^2}$$

$$= \frac{Wbx^2}{2l}\left(\frac{l^2 - a^2 + ab}{l^2}\right) - \frac{Wab^2x}{l^2}$$

$$= \frac{Wbx^2}{2l^3}((a+b)^2 - a^2 + ab) - \frac{Wab^2x}{l^2} \qquad (\because l = a + b)$$

$$EI\frac{dy}{dx} = \frac{Wbx^2}{2l^3}(a^2 + b^2 + 2ab - a^2 + ab) - \frac{Wab^2x}{l^2}$$

$$= \frac{Wbx^2}{2l^3}(3ab + b^2) - \frac{Wab^2x}{l^2}$$

$$= \frac{Wb^2x^2\,(3a+b)}{2l^3} - \frac{Wab^2x}{l^2} \qquad ...(v)$$

Integrating the above equation once again,

$$EI \cdot y = \frac{Wb^2x^3\,(3a+b)}{6l^3} - \frac{Wab^2x^2}{2l^2} + C_2$$

where $C_2$ is the second constant of integration. We know that when $x = 0$, then $y = 0$. Therefore $C_2 = 0$.

or
$$EI \cdot y = \frac{Wb^2x^3\,(3a+b)}{6l^3} - \frac{Wab^2x^3}{2l^2}$$

$$= \frac{Wb^2x^2}{6l^3}[x\,(3a+b) - 3al] \qquad ...(vi)$$

We know that for maximum deflection, $\frac{dy}{dx}$ should be equal to zero. Therefore, equating the equation (*v*) to zero.

$$\frac{Wb^2 x^2 (3a+b)}{2l^3} - \frac{Wab^2 x}{l^2} = 0$$

$$\therefore \qquad x = \frac{2al}{(3a+b)}$$

Substituting this value of $x$ in equation (*vi*),

$$EI \cdot y_{max} = \frac{Wb^2}{6l^3}\left(\frac{2al}{3a+b}\right)^2 \left[\frac{2al}{(3a+b)}(3a+b) - 3al\right]$$

$$= \frac{Wb^2}{6l^3} \times \frac{4a^2l^2}{(3a+b)^2}(2al - 3al)$$

$$= -\frac{2}{3} \times \frac{Wa^3b^2}{(3a+b)^2}$$

$$\therefore \qquad y_{max} = -\frac{2}{3} \times \frac{Wa^3b^2}{(3a+b)^2\,EI}$$ ... (Minus sign indicated that deflection is downwards)

The deflection under the load may be found out by substituting $x = a$ in equation (*vi*),

$$EI \cdot y = \frac{Wb^2a^2}{6l^3}[a\,(3a+b) - 3al]$$

$$= \frac{Wb^2a^2}{6l^3}[a\,(3a+b) - 3a\,(a+b)] \qquad [\because l = a+b]$$

$$= \frac{Wb^2a^2}{6l^3}[3a^2 + ab - 3a^2 - 3ab] = -\frac{Wa^3b^3}{3l^3}$$

$$\therefore \qquad y = -\frac{Wa^3b^3}{3l^3\,EI}$$ ... (Minus sign indicated that deflection is downwards)

$$= \frac{Wa^3b^3}{3l^3\,EI}$$

**EXAMPLE 24.2.** *A fixed beam AB of 5 m span carries a point load of 20 kN at a distance of 2 m from A. Determine the values of fixing moments and the deflection under the load, if flexural rigidity of the beam is $10 \times 10^3$ kN-m$^2$.*

**SOLUTION.** Given: Span ($l$) = 5 m ; Point load ($W$) = 20 kN ; Distance between load and $A$ ($a$) = 2 m or distance between load and $B$ ($b$) = 5 – 2 = 3 m and flexural rigidity ($EI$) = $10 \times 10^3$ kN-m$^2$.

***Fixing moments***

We know that fixing moment at $A$,

$$M_A = -\frac{Wab^2}{l^2} = -\frac{20 \times 2 \times (3)^2}{(5)^2} = -14.4 \text{ kN-m}$$ **Ans.**

and fixing moment at $B$,

$$M_B = -\frac{Wa^2b}{l^2} = -\frac{20 \times (2)^2 \times 3}{(5)^2} = -9.6 \text{ kN-m}$$ **Ans.**

***Deflection under the load***

We also know that deflection under the load,

$$y = \frac{Wa^3b^3}{3l^2\,EI} = \frac{20 \times (2)^3 \times (3)^3}{3 \times (5)^3 \times (10 \times 10^3)} \text{ m}$$

$$= 1.15 \times 10^{-3} \text{ m} = \mathbf{1.15 \text{ mm}} \quad \textbf{Ans.}$$

**EXAMPLE 24.3.** *A beam of span l is fixed at its both ends. It carries two concentrated loads of W each at a distance of l/3 from both the ends. Find the fixing moments and draw the bending moment diagram.*

**SOLUTION.** Given: Span = $l$ and concentrated load = $W$.

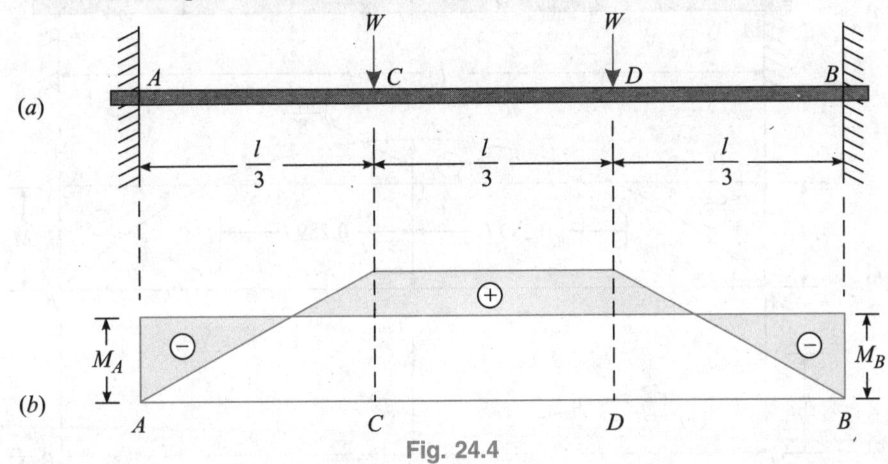


Fig. 24.4

For the sake of convenience, let us first find out the fixing moments, separately due to loads at $C$ and $D$ and then add up the moments. Since the beam and loading is symmetrical, therefore both the fixing moments must be equal. Now consider the load $W$ at $C$. From the geometry of the figure, we find that $a = l/3$ and $b = 2l/3$.

We know that fixing moment at $A$ due to the load $W$ at $C$,

$$M_{A_1} = -\frac{Wab^2}{l^2} = -\frac{W \times \frac{l}{3} \times \left(\frac{2l}{3}\right)^2}{l^2} = -\frac{4Wl}{27} \quad \text{...}(i)$$

Now consider the load $W$ at $D$. From the geometry of the figure, we find that $a = 2l/3$ and $b = l/3$. We know that fixing moment at $A$ due to load $W$ at $D$,

$$M_{A_2} = -\frac{Wa^2b}{l^2} = -\frac{W \times \frac{2l}{3} \times \left(\frac{l}{3}\right)^2}{l^2} = -\frac{2Wl}{27} \quad \text{...}(ii)$$

∴ Total fixing moment at $A$,

$$M_A = M_B = M_{A1} + M_{A2} = -\left(\frac{4Wl}{27} + \frac{2Wl}{27}\right)$$

$$= -\frac{6Wl}{27} = -\frac{\mathbf{2Wl}}{\mathbf{9}} \quad \textbf{Ans.}$$

We know that when the beam is considered as a simply supported, the reaction at $A$,

$$R_A = W$$

∴ Bending moment at $C$, $$M_C = R_A \times \frac{l}{3} = W \times \frac{l}{3} = \frac{Wl}{3}$$

Now complete the bending moment diagram as shown in Fig. 24.4.

## 24.7. Fixing Moments of a Fixed Beam Carrying a Uniformly Distributed Load

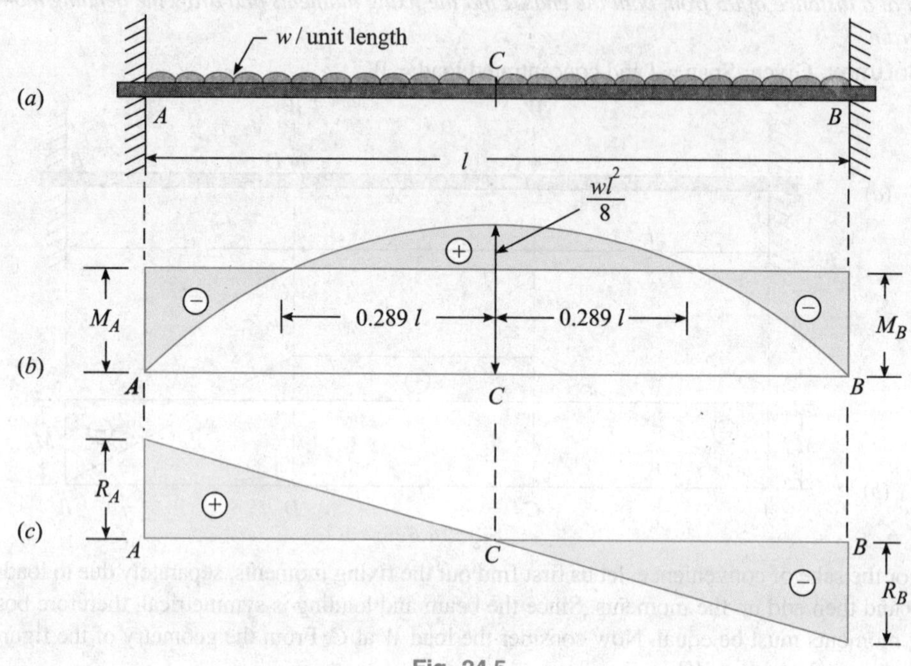


Fig. 24.5

Consider a beam *AB* of length *l* fixed at *A* and *B* and carrying a uniformly distributed load *w* per unit length over the entire span as shown in Fig. 24.5 (*a*).

*(i)* ***Bending moment diagram***

Let $M_A$ = Fixing moment at *A*, and

$M_B$ = Fixing moment at *B*.

Since the beam is symmetrical, therefore $M_A$ and $M_B$ will also be equal. Moreover, the μ′-diagram will be a rectangle, as shown in Fig. 24.5 (*b*). We know that the μ-diagram will be a parabola with the central ordinate equal to $\frac{wl^2}{8}$ as shown in Fig. 24.5 (*b*).

Now equating the areas of the two diagrams,

$$M_A \cdot l = -\frac{2}{3} \cdot l \cdot \frac{wl^2}{8} = -\frac{wl^3}{12}$$

$$\therefore \quad M_A = -\frac{wl^2}{12}$$

Similarly,

$$M_B = -\frac{wl^2}{12} \quad \text{...(By symmetry)}$$

We know that maximum positive bending moment at the centre of the beam (neglecting fixing moments)

$$= \frac{wl^2}{8}$$

∴ Net positive bending moment at the centre of the beam

$$= \frac{wl^2}{8} - \frac{wl^2}{12} = \frac{wl^2}{24}$$

Now complete the bending moment diagram as shown in Fig. 24.5 (*b*)

***(ii) Shear force diagram***

Let $R_A$ = Reaction at *A*, and

$R_B$ = Reaction at *B*.

Equating the clockwise moments and anticlockwise moments about *A*,

$$R_B \times l + M_A = M_B + w \times l \times \frac{l}{2}$$

$$\therefore \quad R_B = \frac{wl}{2} \qquad ...(\because M_A = M_B)$$

Similarly, $$R_A = \frac{wl}{2} \qquad ...\text{(By symmetry)}$$

Now complete the shear force diagram as shown in Fig. 24.5 (*c*).

***(iii) Deflection of the beam***

We know that bending moment at any section *X*, at a distance *x* from *A*,

$$M_X = \mu_X = \mu'_X = \left(\frac{wl}{2} \cdot x - \frac{wx^2}{2}\right) - \frac{wl^2}{12}$$

$$\therefore \quad EI\frac{d^2y}{dx^2} = \left(\frac{wlx}{2} - \frac{wx^2}{2}\right) - \frac{wl^2}{12} \qquad ...(i)$$

Integrating the above equation,

$$EI\frac{dy}{dx} = \frac{wlx^2}{4} - \frac{wx^3}{6} - \frac{wl^2x}{12} + C_1$$

where $C_1$ is the first constant of integration. We know that when $x = 0$, then $\frac{dy}{dx} = 0$. Therefore $C_1 = 0$.

or $$EI\frac{dy}{dx} = \frac{wlx^2}{4} - \frac{wx^3}{6} - \frac{wl^2x}{12} \qquad ...(ii)$$

Integrating the equation (*ii*) once again,

$$EI \cdot y = \frac{wlx^3}{12} - \frac{wx^4}{24} - \frac{wl^2x^2}{24} + C_2$$

where $C_2$ is the second constant of integration. We know that when $x = 0$, then $y = 0$. Therefore $C_2 = 0$.

or $$EI \cdot y = \frac{wlx^3}{12} - \frac{wx^4}{24} - \frac{wl^2x^2}{24} \qquad ...(iii)$$

We know that the maximum deflection occurs at the centre of the beam. Therefore substituting $x = l/2$ in the above equation,

$$EI \cdot y_C = \frac{wl}{12}\left(\frac{l}{2}\right)^3 - \frac{w}{24}\left(\frac{l}{2}\right)^4 - \frac{wl^2}{24}\left(\frac{l}{2}\right)^2 = \frac{wl^4}{96} - \frac{wl^4}{384} - \frac{wl^4}{96} = -\frac{wl^4}{384}$$

or $$y_C = -\frac{wl^4}{384\,EI} \qquad ...\text{(Minus sign means that deflection is downwards}$$

$$= \frac{wl^4}{384\,EI}$$

***(iv) Points of contraflexures***

The points of contraflexures may be found out by equating (*i*) to zero,

$$\left(\frac{wlx}{2}-\frac{wx^2}{2}\right)-\frac{wl^2}{12} = 0$$

or
$$lx - x^2 - \frac{l^2}{6} = 0$$

$\therefore$
$$x^2 - lx + \frac{l^2}{6} = 0$$

Solving this quadratic equation for *x*,

$$x = \frac{l \pm \sqrt{l^2 - \frac{4l^2}{6}}}{2} = \frac{l}{2} \pm \frac{l}{2\sqrt{3}}$$

$$= 0.5\,l \pm 0.289\,l = 0.789\,l \qquad \text{and} \qquad 0.211\,l$$

**EXAMPLE 24.4.** *An encastre beam AB 4 m long is subjected to uniformly distributed load of 3 kN/m over the entire length. Determine the values of maximum negative and positive bending moments. Also calculate the maximum deflection of the beam. Take flexural rigidity of the beam as 10 MN-m².*

**SOLUTION.** Given: Length ($l$) = 4 m ; Uniformly distributed load ($w$) = 3 kN/m and flexural rigidity ($EI$) = 10 MN-m = $10 \times 10^3$ kN-m$^2$.

***Maximum negative bending moment***

We know that maximum negative bending moment,

$$= M_A = M_B = -\frac{wl^2}{12} = -\frac{3\times(4)^2}{12} = \mathbf{-4\ kN\text{-}m} \qquad \textbf{Ans.}$$

***Maximum positive bending moment***

We know that maximum positive bending moment,

$$M_C = \frac{wl^2}{24} = -\frac{3\times(4)^2}{24} = \mathbf{2\ kN\text{-}m} \qquad \textbf{Ans.}$$

***Maximum deflection of the beam***

We also know that maximum deflection of the beam,

$$y_C = \frac{wl^4}{384\,EI} = \frac{3\times(4)^4}{384\times(10\times10^3)} = 0.2\times10^{-3}\ \text{m}$$

$$= \mathbf{0.2\ mm} \qquad \textbf{Ans.}$$

**EXAMPLE 24.5.** *A fixed beam AB of span 6 m is carrying a uniformly distributed load of 4kN/m over the left half of the span. Find the fixing moments and support reactions.*

**SOLUTION.** Given: Span ($l$) = 6 m ; Uniformly distributed load ($w$) = 4 kN/m and loaded portion ($l_1$) = 3 m.

***Fixing moments***

Let
$M_A$ = Fixing moment at *A* and,
$M_B$ = Fixing moment at *B*.

First of all, consider the beam *AB* on a simply supported. Taking moments about *A*,

$$R_B \times 6 = 4 \times 3 \times 1.5 = 18$$

$$\therefore \qquad R_B = \frac{18}{6} = 3 \text{ kN}$$

and $$R_A = 3 \times 4 - 3 = 9 \text{ kN}$$

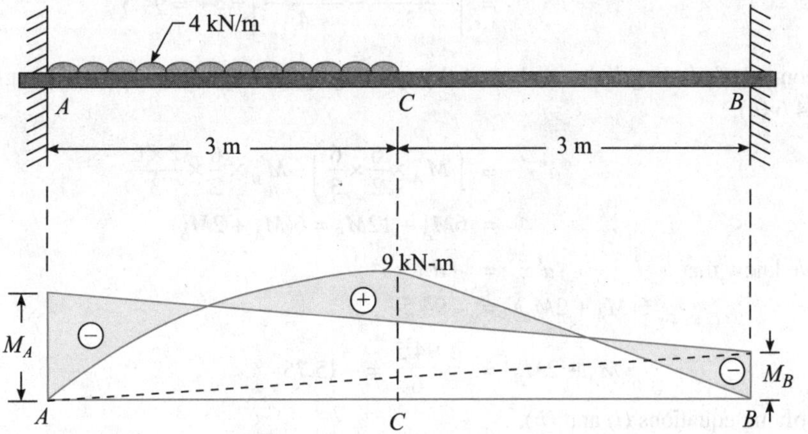

(a)

(b)

Fig. 24.6

We know that μ-diagram will be parabolic from $A$ to $C$ and triangular from $C$ to $B$ as shown in Fig. 24.6 (*b*). The bending moment at $C$ (treating the beam as a simply supported),

$$M_C = R_B \times 3 = 3 \times 3 = 9 \text{ kN-m}$$

The bending moment at any section $X$ in $AC$, at a distance $x$ from $A$ (treating the beam as a simply supported),

$$M_X = 9x - 4x \cdot \frac{x}{2} = 9x - 2x^2$$

∴ Area μ-diagram from $A$ to $B$,

$$a = \int_0^3 (9x - 2x^2)\, dx + \frac{1}{2} \times 9.0 \times 3$$

$$= \left[\frac{9x^2}{2} - \frac{2x^3}{3}\right]_0^3 + 13.5$$

$$= \frac{9 \times (3)^2}{2} - \frac{2 \times (3)^3}{3} + 13.5 = 36$$

and area of μ′-diagram, $$a' = (M_A + M_B) \times \frac{6}{2} = 3\,(M_A + M_B)$$

We know that $$a' = -a$$

$$\therefore \qquad 3\,(M_A + M_B) = -36$$

or $$M_A + M_B = -\frac{36}{3} = -12 \qquad ...(i)$$

Moment of μ-diagram area about $A$ (by splitting up the diagram into $AC$ and $CB$),

$$a\bar{x} = \int_0^3 (9x^2 - 2x^3)\, dx + \frac{1}{2} \times 9 \times 3 \times 4$$

$$a\bar{x} = \left[\frac{9x^3}{3} - \frac{2x^4}{4}\right]_0^3 + 54$$

$$= \left[\frac{9\times(3)^3}{3} - \frac{2\times(3)^4}{4}\right] + 54 = 94.5$$

and moment of $\bar{x}'$ -diagram area about $A$ (by splitting up the trapezium into two triangles) as shown in Fig. 24.6 (*a*),

$$a'\bar{x}' = \left(M_A \times \frac{6}{2} \times \frac{6}{3}\right) + M_B \times \frac{6}{2} \times \frac{2\times 6}{3}$$

$$= 6M_A + 12M_B = 6(M_A + 2M_B)$$

We know that

$$a'\bar{x} = -a\bar{x}$$

$$6(M_A + 2M_B) = -94.5$$

$$\therefore \quad M_A + 2M_B = -\frac{94.5}{6} = -15.75 \quad \text{...(ii)}$$

Solving equations (*i*) and (*ii*),

$M_A$ = **– 8.25 kN-m** **Ans.**

$M_B$ = **–3.75 kN-m** **Ans.**

Now complete the bending moment diagram as shown in Fig. 24.6 (*b*).

***Support reactions***

Let $R_A$ = Reaction at $A$, and

$R_B$ = Reaction at $B$.

Equating the clockwise moments and anticlockwise moments about $A$,

$$R_B \times 6 + 8.25 = (4 \times 3 \times 1.5) + 3.75 = 21.75$$

$$\therefore \quad R_B = \frac{21.75 - 8.25}{6} = \mathbf{2.25\ kN} \quad \textbf{Ans.}$$

and $R_A = 4 \times 3 - 2.25 =$ **9.75 kN** **Ans.**

**EXAMPLE 24.6.** *A beam AB of uniform section and 6 m span is built-in at the ends. A uniformly distributed load of 3 kN/m runs over the left half of the span and there is in addition a concentrated load of 4 kN at right quarter as shown in Fig. 24.7.*

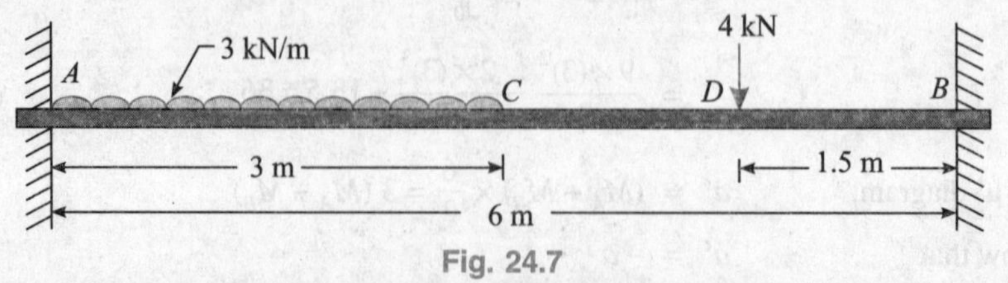


Fig. 24.7

*Determine the fixing moments at the ends, and the reactions. Sketch neatly the bending moment and shearing force diagram marking thereon salient values.*

**SOLUTION:** Given: Span ($l$) = 6 m ; Uniformly distributed load on $AC$ ($w$) = 3 kN/m ; Loaded portion ($l_1$) = 3 m and concentrated load at $D$ ($W$) = 4 kN.

*Fixing moments at the ends*

Let $M_A$ = Fixing moment at $A$ and

$M_B$ = Fixing moment at $B$.

First of all, consider the beam $AB$ as a simply supported. Taking moments about $A$,

$$R_B \times 6 = (3 \times 3 \times 1.5) + (4 \times 4.5) = 31.5$$

$$\therefore \quad R_B = \frac{31.5}{6} = 5.25 \text{ kN}$$

and

$$R_A = (3 \times 3 + 4) - 5.25 = 7.75 \text{ kN}$$

We know that the μ-diagram will be parabolic from $A$ to $C$, trapezoidal from $C$ to $D$ and triangular from $D$ to $B$ as shown in Fig. 24.8(*b*). The bending moment at $D$ (treating the beam at a simply supported),

$$M_D = 5.25 \times 1.5 = 7.875 \text{ kN-m}$$

and

$$M_C = 5.25 \times 3 - 4 \times 1.5 = 9.75 \text{ kN-m}$$

The bending moment at any section $X$ in $AC$, at a distance $x$ from $A$ (treating the beam as a simply supported),

$$M_X = 7.75x - 3x \frac{x}{2} = 7.75x - 1.5x^2$$

∴ Area of μ-diagram from $A$ to $B$,

$$\therefore \quad a = \int_0^3 (7.75x - 1.5x^2)\, dx + \left(\frac{1}{2}(9.75 + 7.875) \times 1.5\right) + \left(\frac{1}{2} \times 7.875 \times 1.5\right)$$

$$= \left[\frac{7.75x^2}{2} - \frac{1.5x^3}{3}\right]_0^3 + 19.125$$

$$= \frac{7.75 \times (3)^2}{2} - \frac{1.5 \times (3)^3}{3} + 19.125 = 40.5$$

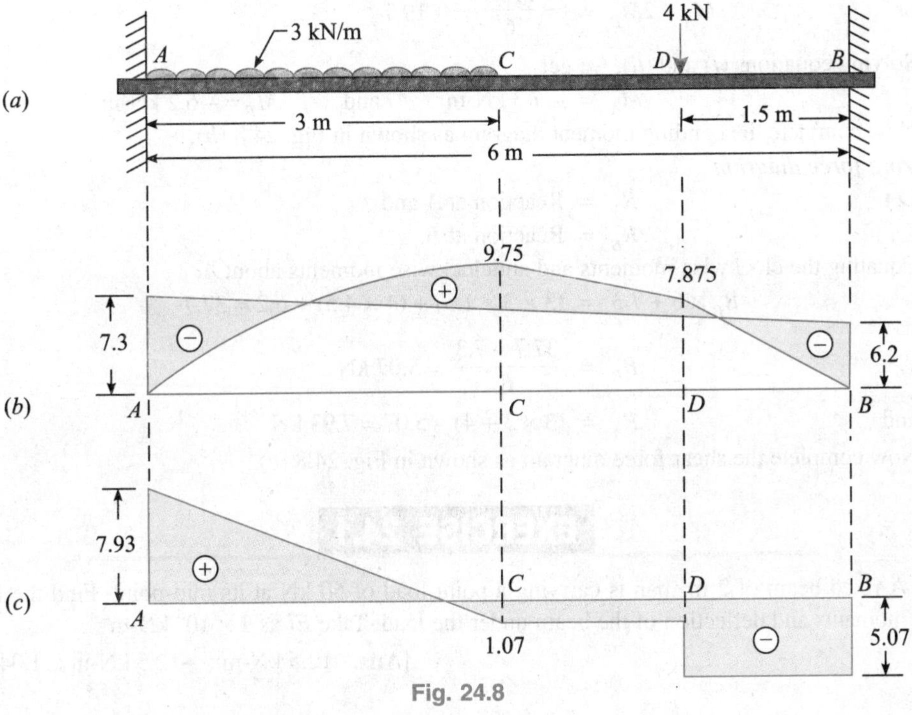

(*a*) 3 kN/m, 4 kN, A, C, D, B, 3 m, 1.5 m, 6 m

(*b*) 9.75, 7.875, 7.3, 6.2, A, C, D, B

(*c*) 7.93, 1.07, 5.07, A, C, D, B

Fig. 24.8

and area of μ-diagram, $a' = (M_A + M_B) \times \frac{6}{2} = 3\,(M_A + M_B)$

We know that $a' = -a$

$\therefore \quad 3\,(M_A + M_B) = -40.5 \quad ...(\because a = 40.5)$

or $M_A + M_B = -13.5 \quad ...(i)$

Moment of μ-diagram area about $A$ (by splitting up the diagram into $AC$, $CD$ and $DB$),

$$a\bar{x} = \int_0^3 (7.75x^2 - 1.5x^3)\,dx + \left(\frac{1}{2} \times 9.75 \times 1.5 \times 3.5\right)$$

$$+ \left(\frac{1}{2} \times 7.875 \times 1.5 \times 4\right) + \left(\frac{1}{2} \times 7.875 \times 1.5 \times 5\right)$$

$$= \left[\frac{7.75x^3}{3} - \frac{1.5x^4}{4}\right]_0^3 + 78.75$$

$$= \left[\frac{7.75 \times (3)^3}{3} - \frac{1.5 \times (3)^4}{4}\right] + 78.75 = 118.1$$

and moment of μ′-diagram area about $A$ (by splitting up the trapezium into two triangles),

$$a'\bar{x}' = \left(M_A \times \frac{6}{2} \times \frac{6}{3}\right) + \left(M_B \times \frac{6}{2} \times \frac{2 \times 6}{3}\right)$$

$$= 6M_A + 12M_B = 6\,(M_A + 2M_B)$$

We know that $a'\bar{x}' = -a\bar{x}$

$\therefore \quad 6\,(M_A + 2M_B) = -118.1$

or $M_A + 2M_B = -\frac{118.1}{6} = -19.7 \quad ...(ii)$

Solving equations (*i*) and (*ii*), we get

$M_A = -7.3$ kN-m and $M_B = -6.2$ kN-m

Now complete the bending moment diagram as shown in Fig. 24.8 (*b*).

***Shearing force diagram***

Let $R_A$ = Reaction at $A$ and

$R_B$ = Reaction at $B$.

Equating the clockwise moments and anticlockwise moments about $A$,

$$R_B \times 6 + 7.3 = (3 \times 3 \times 1.5) + (4 \times 4.5) + 6.2 = 37.7$$

$$\therefore \quad R_B = \frac{37.7 - 7.3}{6} = 5.07 \text{ kN}$$

and $R_A = (3 \times 3 + 4) - 5.07 = 7.93$ kN

Now complete the shear force diagram as shown in Fig. 24.8 (*c*).

## EXERCISE 24.1

1. A fixed beam of 2 m span is carrying a point load of 50 kN at its mid-point. Find the fixing moments and deflection of the beam under the load. Take $EI$ as $2 \times 10^3$ kN-m$^2$.

**[Ans.** –12.5 kN-m ; –12.5 kN-m ; 1.04 mm]

2. A fixed beam $AB$ of span 3 m is subjected to a point load of 15 kN at a distance of 1 m from $A$. Determine the fixing moments of $A$ and $B$. **[Ans.** –6.67 kN-m ; –3.33 kN-m]
3. A built-in beam of span 3.6 m is carrying a uniformly distributed load of 15 kN/m. Find the fixing moments at the supports. Also find the maximum positive bending moment. **[Ans.** –16.2 kN-m ; 8.1 kN-m]
4. A fixed beam $AB$ of span 6 m is subjected to two point loads of 20 kN and 15 kN at distances of 2 m and 4 m from $A$. Calculate the fixing moments at $A$ and $B$.**[Ans.** –24.4 kN-m ; –22.2 kN-m]

## 24.8. Fixing Moments of a Fixed Beam Carrying a Gradually Varying Load from Zero at One End to *w* per unit length at the Other

Consider a beam $AB$ fixed at $A$ and $B$ and carrying a gradually varying load from zero at $A$ to $w$ per unit length at $B$ as shown in Fig. 24.9 ($a$).

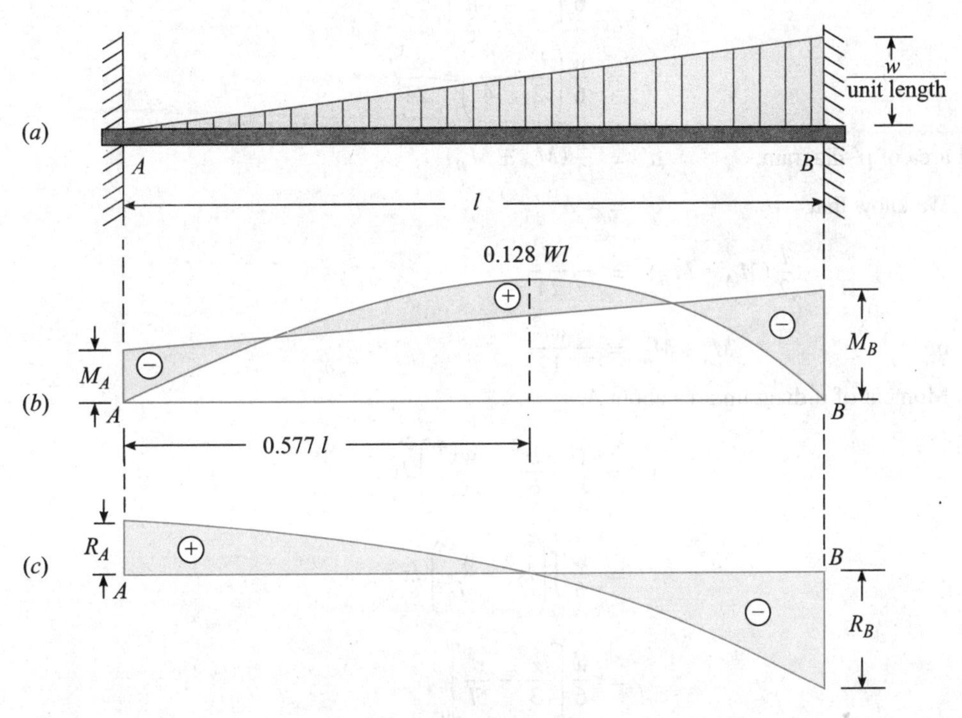


Fig. 24.9

Let
$$l = \text{Span of the beam,}$$
$$M_A = \text{Fixing moment at } A \text{ and}$$
$$M_B = \text{Fixing moment at } B.$$

First of all, consider the beam $AB$ as a simply supported and taking moments about $A$,

$$R_B \times l = w \times \frac{l}{2} \times \frac{2l}{3} = \frac{wl^2}{3}$$

$$\therefore \quad R_B = \frac{wl}{3}$$

and $$R_A = \frac{wl}{2} - \frac{wl}{3} = \frac{wl}{6}$$

We know that the μ-diagram will be parabolic from $A$ to $B$. The bending moment at any section $X$, at a distance $x$ from $A$ (treating the beam as a simply supported),

$$M_X = \frac{wl}{6} \times x - \frac{wx}{l} \times \frac{x}{2} \times \frac{x}{3} = \frac{wlx}{6} - \frac{wx^3}{6l}$$

∴ Area of μ-diagram, $$a = \int_0^l \left( \frac{wlx}{6} - \frac{wx^3}{6l} \right) dx$$

$$= \frac{w}{6} \int_0^l \left( lx - \frac{x^3}{l} \right) dx$$

$$= \frac{w}{6} \left[ \frac{lx^2}{2} - \frac{x^4}{4l} \right]_0^l$$

$$= \frac{w}{6} \left( \frac{l^3}{2} - \frac{l^3}{4} \right) = \frac{wl^3}{24}$$

and area of μ′-diagram, $$a' = \frac{l}{2}(M_A + M_B)$$

We know that $$a' = -a$$

∴ $$\frac{l}{2}(M_A + M_B) = -\frac{wl^3}{24}$$

or $$M_A + M_B = -\frac{wl^2}{12} \quad ...(i)$$

Moment of μ-diagram area about $A$,

$$a\bar{x} = \int_0^l \left( \frac{wlx^2}{6} - \frac{wx^4}{6l} \right) dx$$

$$= \frac{w}{6} \int_0^l \left( lx^2 - \frac{x^4}{l} \right) dx$$

$$= \frac{w}{6} \left[ \frac{lx^3}{3} - \frac{x^5}{5l} \right]_0^l$$

$$= \frac{w}{6} \left( \frac{l^4}{3} - \frac{l^4}{5} \right) = \frac{wl^4}{45}$$

and moment of μ′-diagram about $A$ (by splitting up the trapezium into two triangles),

$$a'\bar{x}' = M_A \times \frac{l}{2} \times \frac{l}{3} + M_B \times \frac{l}{2} \times \frac{l}{3}$$

$$= \frac{l^2}{6}(M_A + 2M_B)$$

We know that $$a'\bar{x}' = -a\bar{x}$$

∴ $$\frac{l^2}{6}(M_A + 2M_B) = -\frac{wl^4}{45}$$

or $$M_A + 2M_B = -\frac{2wl^2}{15} \qquad ...(ii)$$

Solving equations (*i*) and (*ii*),

$$M_A = -\frac{wl^2}{30} = -\frac{Wl}{15} \qquad ...\left(\because W = \frac{wl}{2}\right)$$

and $$M_B = -\frac{wl^2}{20} = -\frac{Wl}{10} \qquad ...\left(\because W = \frac{wl}{2}\right)$$

***Alternative method***

Consider a strip of width $dx$ at a distance $x$ from the support $A$. We see that the load at this section

$$= \frac{wx}{l}$$

$\therefore$ Weight of the strip, $$W = \frac{wx}{l} \cdot dx$$

We know that the fixing moment at $A$, due to the strip

$$= -\frac{Wab^2}{l^2} = \frac{\frac{wx}{l}\, dx \cdot x\,(l - x)^2}{l^2}$$

$$= -\frac{wx^2\,(l - x)^2\, dx}{l^3}$$

Total fixed end moment at $A$ will be given by integrating the above equation from 0 to $l$,

$$M_A = \int_0^l -\frac{wx^2\,(l - x)^2\, dx}{l^3}$$

$$= -\frac{w}{l^3}\int_0^l x^2(l^2 + x^2 - 2lx)\, dx$$

$$= -\frac{w}{l^3}\int_0^l (l^2x^2 + x^4 - 2lx^3)\, dx$$

$$= -\frac{w}{l^3}\left[\frac{l^2\,x^3}{3} + \frac{x^5}{5} - \frac{2lx^4}{4}\right]_0^l$$

$$= -\frac{w}{l^3}\left[\frac{l^5}{3} + \frac{l^5}{5} - \frac{l^5}{2}\right]$$

$$= -\frac{wl^2}{30} = -\frac{Wl}{15} \qquad ...\left(\because W = \frac{wl}{2}\right)$$

Similarly, $$M_B = \int_0^l -\frac{\frac{wx}{l}\, dx \cdot x^2\,(l - x)}{l^2}$$

$$= -\frac{w}{l^3}\int_0^l x^3\,(l - x)\, dx$$

$$= -\frac{w}{l^3}\int_0^l (lx^3 - x^4)\,dx$$

$$= -\frac{w}{l^3}\left[\frac{lx^4}{4} - \frac{x^5}{5}\right]_0^l$$

$$= -\frac{w}{l^3}\left(\frac{l^5}{4} - \frac{l^5}{5}\right)$$

$$= -\frac{wl^2}{20} = -\frac{Wl}{10} \qquad \ldots\left(\because W = \frac{wl}{2}\right)$$

We know that the maximum* bending moment equal to 0.128 $Wl$ will take place at a distance of 0.577 $l$ from $A$, treating the beam as a simply supported. Now complete the bending moment diagram as shown in Fig. 24.9 (*b*).

***Shear force diagram***

Let $R_A$ = Reaction at $A$ and

$R_B$ = Reaction at $B$.

Equating the clockwise moments and anticlockwise moments about $A$,

$$R_B \times l + M_A = M_B + w \times \frac{l}{2} \times \frac{2l}{3} = M_B + \frac{wl^2}{3}$$

$$\therefore \quad R_B = \frac{M_B - M_A}{l} + \frac{wl}{3}$$

Similarly, equating clockwise moments and anticlockwise moments about $B$,

$$R_A \times l + M_B = M_A + w \times \frac{l}{2} \times \frac{l}{3} = M_A + \frac{wl^2}{6}$$

$$\therefore \quad R_A = \frac{M_A - M_B}{l} + \frac{wl}{6}$$

Now complete the shear force diagram as shown in Fig. 24.9 (*c*).

**EXAMPLE 24.7.** *A beam AB of span 5 m is built-in at its both ends. It carries a gradually varying load from zero at A to 4 kN/m at B. Determine the fixed end moments and reactions at both ends of the beam.*

**SOLUTION.** Given: Span ($l$) = 5 m and load at $B$ ($w$) = 4 kN/m

***Fixed end moments***

We know that fixed end moment at $A$,

$$M_A = -\frac{wl^2}{30} = -\frac{4 \times (5)^2}{30} = \mathbf{-3.33\ kN\text{-}m} \qquad \textbf{Ans.}$$

and fixed end moment at $B$,

$$M_B = -\frac{wl^2}{20} = -\frac{4 \times (5)^2}{20} = \mathbf{-5.0\ kN\text{-}m} \qquad \textbf{Ans.}$$

***Reactions at both ends of the beam***

We also know that reaction at $A$,

$$R_A = \frac{M_A - M_B}{l} + \frac{wl}{6} = \frac{-3.33 - (-5.0)}{5} + \frac{4 \times 5}{6}\ \text{kN}$$

* For details, please refer to Art. 13.14

$$= 0.33 + 3.33 = \mathbf{3.66\ kN} \qquad \textbf{Ans.}$$

and reaction at $B$,

$$R_B = \frac{M_B - M_A}{l} + \frac{wl}{3} = \frac{-5.0 - (-3.33)}{5} + \frac{4 \times 5}{3} \text{ kNm}$$

$$= -0.33 + 6.67 = \mathbf{6.34\ kN} \qquad \textbf{Ans.}$$

## 24.9. Fixing Moments of a Fixed Beam due to Sinking of a Support

Sometimes, one of the two supports of a fixed beam sinks down, whereas the other remains at the same level. The effect of sinking of a support is to produce some additional moment at the two supports. Consider a beam fixed at $A$ and $B$ without any load. Let the support $B$ sink down from its original level as shown in Fig. 24.10 (*a*).

Let $\quad l$ = Span of the fixed beam $AB$ and

$\delta$ = Amount, by which the support $B$ has sunk down.

Since the beam is not loaded, therefore

$$EI \frac{d^4 y}{dx^4} = 0$$

Integrating the above equation,

$$EI \frac{d^3 y}{dx^3} = -F_A \qquad \text{...}(\because F_A \text{ is negative})$$

where $F_A$ is the shear force at $A$.

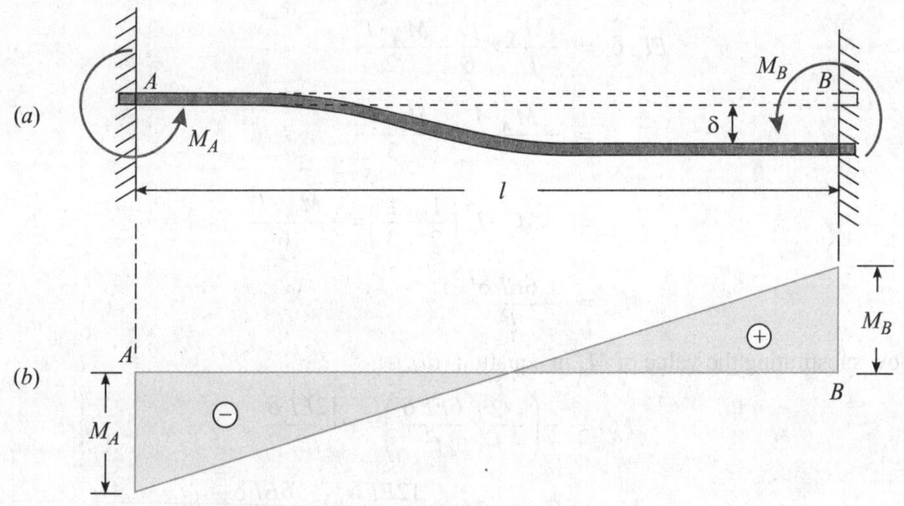


Fig. 24.10

Integrating the above equation again,

$$EI \frac{d^2 y}{dx^2} = -F_A \cdot x - M_A \qquad \text{...}(\because M_A \text{ is negative})$$

where $M_A$ is the bending moment at $A$,

Integrating the above equation once again,

$$EI \frac{dy}{dx} = -\frac{F_A \cdot x^2}{2} - M_A \cdot x + C_1$$

where $C_1$ is the first constant of integration. We know that when $x = 0$, then $\frac{dy}{dx} = 0$. Therefore $C_1 = 0$.

$$\therefore \quad EI\frac{dy}{dx} = -\frac{F_A \cdot x^2}{2} - M_A \cdot x \quad ...(i)$$

Integrating the above equation once again,

$$EI \cdot y = -\frac{F_A \cdot x^3}{6} - \frac{M_A \cdot x^2}{2} + C_2$$

where $C_2$ is the second constant of integration. We know that when $x = 0$, then $y = 0$. Therefore $C_2 = 0$.

$$\therefore \quad EI \cdot y = -\frac{F_A \cdot x^3}{6} - \frac{M_A \cdot x^2}{2} \quad ...(ii)$$

We know that when $x = l$, then $\frac{dy}{dx} = 0$,

$$\therefore \quad 0 = -\frac{F_A \cdot l^2}{2} - M_A \cdot l$$

or

$$F_A = -\frac{2M_A}{l} \quad ...(iii)$$

We also know that when $x = l$, then $y = \delta$

$$\therefore \quad EI \cdot \delta = -\frac{F_A \cdot l^3}{6} - \frac{M_A \cdot l^2}{2}$$

Substituting the value of $F_A$ in the above equation,

$$EI \cdot \delta = \frac{2M_A}{l} \times \frac{l^3}{6} - \frac{M_A \cdot l^2}{2}$$

$$= -\frac{M_A \cdot l^2}{2} + \frac{M_A \cdot l^2}{3}$$

$$= -M_A \cdot l^2\left(\frac{1}{2} - \frac{1}{3}\right) = -\frac{M_A \cdot l^2}{6}$$

$$\therefore \quad M_A = -\frac{6EI\,\delta}{l^2}$$

Now substituting the value of $M_A$ in equation (*iii*),

$$F_A = -\left(-\frac{2}{l} \times \frac{6EI\,\delta}{l^2}\right) = +\frac{12EI\,\delta}{l^3}$$

$$\therefore \quad M_B = F_A \cdot l + M_A = \frac{12EI\,\delta}{l^3} \cdot l - \frac{6EI\,\delta}{l^2}$$

$$= +\frac{6EI\,\delta}{l^2}$$

The bending moment diagram is shown in Fig. 24.10 (*b*).

**Note.** If the support *A* sinks down by $\delta$ from its original level, then

$$M_A = \frac{6EI\,\delta}{l^2} \quad \text{and} \quad M_B = -\frac{6EI\,\delta}{l^2}$$

**Example 24.8.** *A steel fixed beam AB of span 6 m is 60 mm wide and 100 mm deep. The support B sinks down by 6 mm. Find the fixing moments at A and B. Take E = 200 GPa.*

**Solution.** Given: Span ($l$) = 6 m = mm ; Width ($b$) = 60 mm ; Depth ($d$) = 100 mm ; Sinking of the support $B$ ($\delta$) = 6 mm and modulus of elasticity ($E$) = 200 GPa = $200 \times 10^3$ N/mm$^2$.

We know that moment of inertia of the beam section,

$$I = \frac{bd^3}{12} = \frac{60 \times (100)^3}{12} = 5 \times 10^6 \text{ mm}^4$$

∴ Fixing moment at $A$,

$$M_A = -\frac{6EI\,\delta}{l^2} = -\frac{6 \times (200 \times 10^3) \times (5 \times 10^6) \times 6}{(6 \times 10^3)^2} \text{ N-mm}$$

$$= -1 \times 10^6 \text{ N-mm} = \mathbf{-1 \text{ kN-m}} \quad \textbf{Ans.}$$

and fixing moment at $B$,

$$M_B = +\frac{6EI\,\delta}{l^2} = \frac{6 \times (200 \times 10^3) \times (5 \times 10^6) \times 6}{(6 \times 10^3)^2} \text{ N-mm}$$

$$= +1 \times 10^6 \text{ N-mm} = \mathbf{1 \text{ kN-m}} \quad \textbf{Ans.}$$

**Example 24.9.** *A beam AB of 8 m span is fixed at its both ends. When a uniformly distributed load of 20 kN/m is placed on the beam, its support B sinks 12 mm below the support A. What are the support moments, if I for the section is 98.75 × 10$^6$ mm$^4$ and E = 200 GPa.*

**Solution.** Given: Span ($l$) = 8 m ; Uniformly distributed load ($w$) = 20 kN/m ; Sinking of support $B$ ($\delta$) = 12 mm ; Moment of inertia of the beam section ($I$) = $98.75 \times 10^6$ mm$^4$ and modulus of elasticity ($E$) = 200 GPa = $200 \times 10^3$ N/mm$^2$.

First of all, let us find out the support moments due to uniformly distributed load. We know that support moments at $A$,

$$M_{A_1} = -\frac{wl^2}{12} = -\frac{20 \times (8)^2}{12} = -106.67 \text{ kN-m}$$

Similarly, $M_{B_1} = -106.7$ kN-m

Now let us find out the support moments due to sinking of the support $B$. We know that support moment at $A$,

$$M_{A_2} = -\frac{6EI\,\delta}{l^2} = -\frac{6 \times (200 \times 10^3) \times (98.25 \times 10^6) \times 12}{(8 \times 10^3)^2} \text{ N-mm}$$

$$= -22.2 \times 10^6 \text{ N-mm} = -22.2 \text{ kN-m}$$

∴ Support moment at $A$,

$$M_A = M_{A_1} + M_{A_2} = -106.7 + (-22.2) = \mathbf{-128.9 \text{ kN-m}} \quad \textbf{Ans.}$$

We also know that support moment at $B$,

$$M_{B_2} = +\frac{6EI\,\delta}{l^2} = +\frac{6 \times (200 \times 10^3) \times (98.75 \times 10^6) \times 12}{(8 \times 10^3)^2}$$

$$= +22.2 \times 10^6 \text{ N-mm} = 22.2 \text{ kN-m}$$

∴ Support moment at $B$,

$$M_B = M_{B_1} + M_{B_2} = -106.7 + 22.2 = \mathbf{-84.5\ kN\text{-}m} \quad \textbf{Ans.}$$

## EXERCISE 24.2

1. A fixed beam $AB$ of span 3 m carries a gradually varying load from zero at $A$ to 5 kN/m at $B$. Find the fixed end moments at $A$ and $B$. **[Ans.** –1.5 kN-m ; 2.25 kN-m]
2. A fixed beam $AB$ is of span 4 m. The support $B$ sinks down by 10 mm. Find the fixing moments at $A$ and $B$ if the flexural rigidity of the beam is $4 \times 10^3$ kN-m$^2$. **[Ans.** –15 kN-m ; +15 kN-m]
3. A built-in beam $AB$ of span 4 m is carrying a uniformly distributed load of 15 kN/m. The support $B$ sinks down by 10 mm. Determine the fixing moments at $A$ and $B$. Take $E = 200$ Gpa and $I = 8 \times 10^6$ mm$^4$. **[Ans.** –26 kN-m ; –14 kN-m]

**Hints:**
$$M_{A_1} = -\frac{wl^2}{12} = -\frac{15 \times (4)^2}{12} = -20 \text{ kN-m}$$

and
$$M_{A_2} = -\frac{6EI\,\delta}{l^2} = -\frac{6 \times (200 \times 10^3) \times (8 \times 10^6) \times 10}{(4 \times 10^3)^2} \text{ N-mm}$$
$$= -6 \times 10^6 \text{ N-mm} = -6 \text{ kN-m}$$

∴
$$M_A = M_{A_1} + M_{A_2} = -20 - 6 = -26 \text{ kN-m}$$

Similarly,
$$M_B = M_{B_1} + M_{B_2} = -20 + 6 = -14 \text{ kN-m}$$

## QUESTIONS

1. What is your idea of an encastre beam and the arrangement of fixing it at its two supports ?
2. What is meant by an encastre beam? Is there any advantage in using it ?
3. If a fixed beam $AB$ carries a central load $W$, find out the value of maximum deflection.
4. Derive an expression for the maximum deflection of an encastre beam, carrying a uniformly distributed load of $w$ per unit length.
5. A beam, built-in at its both ends, has a uniform flexural rigidity $EI$ throughout its length $l$. It carries a single point load $W$, which is placed at a distance $a$ from the left end. Calculate, from first principles, the fixed end moments developed at two ends.
6. Derive an expression for the fixing moments, when one of the supports of a fixed beam sinks down by δ from its original position.

## OBJECTIVE TYPE QUESTIONS

1. When a fixed beam of span $l$ is subjected to a central point load $W$, then the fixing moments are

   (a) $-\frac{Wl}{4}$ (b) $-\frac{Wl}{8}$ (c) $-\frac{Wl}{12}$ (d) $-\frac{Wl}{16}$

2. A fixed beam $AB$ of span $l$ is carrying a point load $W$ at a distance $a$ from the support $A$ and at a distance of $b$ from the support $B$. The fixing moment at $A$ will be

   (a) $\frac{Wab^2}{l}$ (b) $\frac{Wa^2b}{l}$ (c) $\frac{Wab^2}{l^2}$ (d) $\frac{Wa^2b}{l^2}$

3. A fixed beam $AB$ of length $l$ is loaded with a uniformly distributed load of $w$ per unit length. The support moments are

(a) $-\frac{wl^2}{8}$ (b) $-\frac{wl^2}{12}$ (c) $-\frac{wl^2}{24}$ (d) $-\frac{wl^2}{48}$

4. A built-in beam $AB$ of span $l$ is loaded with a gradually varying load from zero at $A$ to $w$ per unit length at $B$. The fixing moment $A$ will be

(a) $\frac{wl^2}{8}$ (b) $\frac{wl^2}{12}$ (c) $\frac{wl^2}{20}$ (d) $\frac{wl^2}{30}$

## ANSWERS

1. (b) 2. (c) 3. (b) 4. (d)

Chapter **25**

# Theorem of Three Moments

## Contents



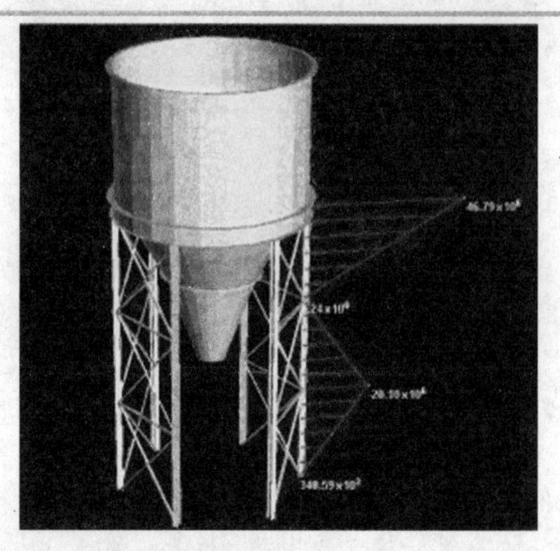

## 25.1. Introduction

A beam, which is supported on more than two supports, is called a *continuous beam*. Such a beam, when loaded will deflect with convexity upwards, over the intermediate supports and with concavity upwards over the mid of the spans. The intermediate supports of a continuous beam are always subjected to some bending moment. The end supports, if simply supported will not be subjected to any bending moment. But the end supports, if fixed, will be subjected to fixing moments and the slope of the beam, at the fixed ends will be zero.

## 25.2. Bending Moment Diagrams for Continuous Beams

The analysis of a continuous beam is similar

to that of a fixed beam. The bending moment diagram for a continuous beam under any system of loading may be drawn in the following two stages:

1. By considering the beam as a series of discontinuous beams, from support to support and drawing the usual μ-diagram due to vertical loads.
2. By superimposing the usual μ′-diagram, due to end moments over μ-diagram.

## 25.3. Claypeyron's Theorem of Three Moments

It states, "*If a beam has n supports, the end ones being fixed, then the same number of equations required to determine the support moments may be obtained from the consecutive pairs of spans i.e., AB-BC, BC-CD, CD-DE and so on.*"

**Proof:**

Consider a continuous beam *ABC*, fixed at *A* and *C* and supported at *B* as shown in Fig. 25.1 (*a*).

Let $l_1$ = Span of the beam,

$I_1$ = Moment of inertia of the beam in span *AB*,

$l_2, I_2$ = Corresponding values for the span *BC*,

$M_A$ = Support moment at *A*,

$M_B$ = Support moment at *B*,

$M_C$ = Support moment at *C*,

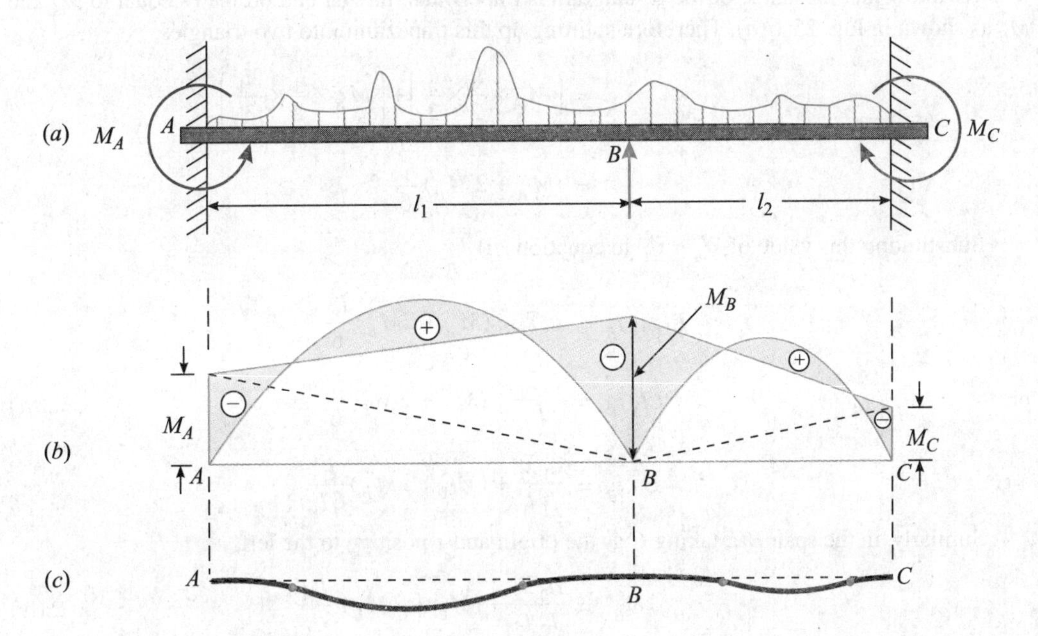


Fig. 25.1

$\mu_X$ = Bending moment at any section *X*, considering the beam between two supports as simply supported and

$\mu_X'$ = Fixing moment at any section *X*, of the beam,

We know that in the span *AB*, the bending moment at any section *X* at a distance *x* from *A*,

$$M_X = \mu_X + \mu_X'$$

$$\therefore \quad EI_1 \frac{d^2y}{dx^2} = \mu_X + \mu_X' \quad \quad ...\left(\because \frac{M}{EI} = \frac{d^2y}{dx^2}\right)$$

Multiplying the above equation by $x$ and integrating the same for the whole span $AB$ *i.e.*, from 0 to $l_1$.

$$EI_1 \int_0^{l_1} \frac{x \cdot d^2 y}{dx^2} = \int_0^{l_1} x \cdot \mu_X \cdot dx + \int_0^{l_1} x \cdot \mu_X' \cdot dx$$

$$EI_1 \left[ x \cdot \frac{dy}{dx} - y \right]_0^{l_1} = a_1 \bar{x}_1 + a'_1 \bar{x}'_1$$

or
$$EI_1 [l_1 i_B - y_B) - 0 (i_A - y_A)] = a_1 \bar{x}_1 + a'_1 \bar{x}_1$$

$$EI_1 [l_1 i_B - y_B] = a_1 \bar{x}_1 + a'_1 \bar{x}'_1 \quad ...(i)$$

Since $y_B$ is equal to zero, therefore

$$EI_1 l_1 i_B = a_1 \bar{x}_1 + a'_1 \bar{x}'_1 \quad ...(ii)$$

where $a_1$ = Area of the μ-diagram in the span $AB$,

$\bar{x}_1$ = Distance of centre of gravity of μ-diagram from A in the span $AB$,

$a'_1 \bar{x}'_1$ = Corresponding values for the μ′-diagram and

$i_B$ = Slope of the beam $AB$ at $B$,

We know that the shape of the μ′-diagram is trapezoidal, having end ordinates equal to $M_A$ and $M_B$ as shown in Fig. 25.1 (*a*). Therefore splitting up this trapezium into two triangles,

$$a'_1 \cdot \bar{x}'_1 = \left( M_A \times \frac{l_1}{2} \times \frac{l_1}{3} \right) + \left( M_B \times \frac{l_1}{2} \times \frac{2l_1}{3} \right)$$

$$= (M_A + 2M_B) \frac{l_1^2}{6}$$

Substituting this value of $a'_1 \cdot \bar{x}'_1$ in equation (*ii*)

$$EI_1 l_1 i_B = a_1 \bar{x}_1 + (M_A + 2M_B) \frac{l_1^2}{6}$$

or
$$EI_1 i_B = \frac{a_1 \bar{x}_1}{l_1} + (M_A + 2M_B) \frac{l_1}{6} \quad ...(iii)$$

∴
$$E \cdot i_B = \frac{a_1 \bar{x}_1}{I_1 l_1} + (M_A + 2M_B) \frac{l_1}{6 I_1}$$

Similarly, in the span $BC$, taking $C$ as the origin and $x$ positive to the left,

$$E_{iB}' = \frac{a_2 \bar{x}_2}{I_2 l_2} + (M_C + 2M_B) \frac{l_2}{6 I_2} \quad ...(iv)$$

where, $a_2$ = Area of the μ-diagram in the span $BC$,

$\bar{x}'_2$ = Distance of centre of gravity of μ-diagram from $C$ in the span $BC$,

$a'_2$ = Area of the μ′-diagram in the span $BC$,

$\bar{x}'_2$ = Distance of the centre of gravity of the μ′-diagram from $C$ in the span $BC$ and

$i'_B$ = Slope of the beam $BC$ at $B$,

Since $i_B$ is equal to $-i'_B$, therefore $E \cdot i_B$ is equal to $-E \cdot i'_B$.

or
$$\frac{a_1 \bar{x}_1}{I_1 l_1} + (M_A + 2M_B)\frac{l_1}{6I_1} = -\left[\frac{a_2 \bar{x}_2}{I_2 l_2} + (M_C + 2M_B)\frac{l_2}{6I_2}\right]$$

$$(M_A + 2M_B)\frac{l_1}{I_1} + (M_C + 2M_B)\frac{l_2}{I_2} = -\frac{6a_1 \bar{x}_1}{I_1 l_1} - \frac{6a_2 \bar{x}_2}{I_2 l_2}$$

$$M_A \frac{l_1}{I_1} + 2M_B \frac{l_1}{I_1} + M_C \frac{l_2}{I_2} + 2M_B \frac{l_2}{I_2} = -\left(\frac{6a_1 \bar{x}_1}{I_1 l_1} + \frac{6a_2 \bar{x}_2}{I_2 l_2}\right)$$

$$\therefore\ M_A\left(\frac{l_1}{I_2}\right) + 2M_B\left(\frac{l_1}{I_1} + \frac{l_2}{I_2}\right) + M_C\left(\frac{l_2}{I_2}\right) = -\left(\frac{6a_1 \bar{x}_1}{I_1 l_1} + \frac{6a_2 \bar{x}_2}{I_2 l_2}\right)$$

**Notes:** 1. For the sake of simplicity, we have considered a continuous beam with two spans only. But this equation can be extended for any number of spans.

2. If moment of inertia of the beam is constant, then

$$M_A l_1 + 2M_B (l_1 + l_2) + M_C l_2 = -\left(\frac{6a_1 \bar{x}_1}{l_1} + \frac{6a_2 \bar{x}_2}{l_2}\right)$$

3. The shear force diagram for the beam may be drawn as usual.
4. The elastic curve of the beam may be drawn as usual as shown in Fig. 25.1 (*c*).

## 25.4. Application of Clapeyron's Theorem of Three Moments to Various Types of Continuous Beams

We have already studied in Art. 25.3 the Clapeyron's theorem of three moments. Now we shall discuss its application to the following types of continuous beams:

1. Continuous beams with simply supported ends,
2. Continuous beams with fixed end supports,
3. Continuous beams with the end span overhanging and
4. Continuous beams with a sinking support.

## 25.5. Continuous Beams with Simply Supported Ends

Sometimes, a continuous beam is simply supported on its one or both the end supports. In such a case, the fixing moment on the simply supported end is zero.

**Example 25.1.** *A continuous beam ABC 10 m long rests on three supports A, B and C at the same level and is loaded as shown in Fig. 25.2.*

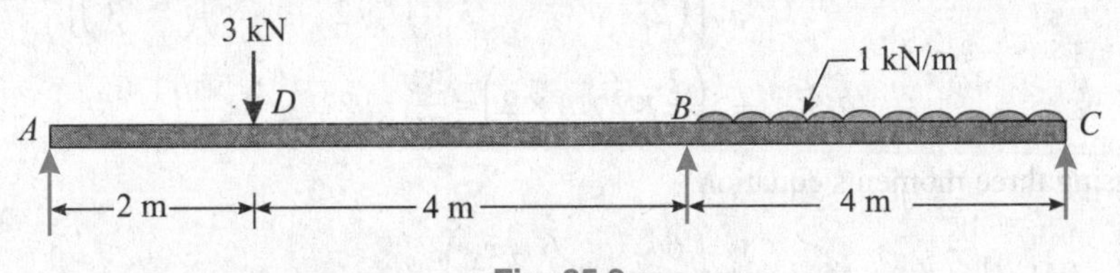


Fig. 25.2

*Determine the moments over the beam and draw the bending moment diagram. Also calculate the reactions at the supports and draw the shear force diagram.*

**Solution.** Given : Length *AB* ($l_1$) = 6 m ; Length *BC* ($l_2$) = 4 m ; Point load at *D* = (*W*) = 3 kN; Distance *AD* (*a*) = 2 m ; Distance *DB* (*b*) = 4 m and uniformly distributed load in *BC* = *w* = 1 kN/m.

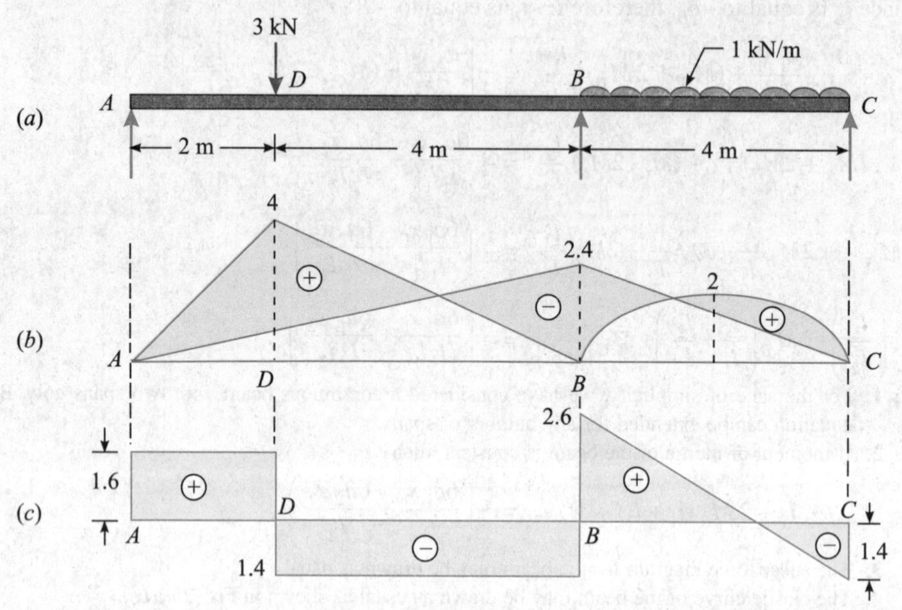


Fig. 25.3

***Moments over the beam***

Let

$$^{*}M_A = \text{Fixing moment at } A,$$

$$M_B = \text{Fixing moment at } B \text{ and}$$

$$^{*}M_C = \text{Fixing moment at } C.$$

First of all, let us consider the beam *AB* as a simply supported. Therefore, bending moment at *D*,

$$M_D = \frac{Wab}{l_1} = \frac{3 \times 2 \times 4}{6} = 4 \text{ kN-m}$$

Similarly, bending moment at the mid of the span *BC*

$$= \frac{wl_2^2}{8} = \frac{1 \times (4)^2}{8} = 2 \text{ kN-m}$$

Now draw the μ-diagrams with the help of above bending moments as shown in Fig. 25.3 (*a*). From the geometry of the above bending moment diagrams, we find that

$$a_1 \bar{x}_1 = \left[\left(\frac{1}{2} \times 2 \times 4 \times \frac{2 \times 2}{3}\right) + \left(\frac{1}{2} \times 4 \times 4\right)\left(2 + \frac{4}{3}\right)\right] = 32$$

and

$$a_2 \bar{x}_2 = \left(\frac{2}{3} \times 2 \times 4 \times 2\right) = \frac{32}{3}$$

Now using three moments equation,

$$M_A l_1 + 2M_B (l_1 + l_2) + M_C l_2 = \left(\frac{6a_1 \bar{x}_1}{l_1} + \frac{6a_2 \bar{x}_2}{l_2}\right)$$

$$0 + 2 M_B (6 + 4) + 0 = \left(\frac{6 \times 32}{6} + \frac{6 \times \frac{32}{3}}{4}\right)$$

* Since the beam is simply supported at *A* and *C*, therefore fixing moments $M_A$ and $M_C$ will be zero.

$\therefore \quad 20\ M_B = -(32+16) = -48$

or $\quad M_B = -\dfrac{48}{20} = -2.4$ kN-m **Ans.**

Now complete the bending moment diagram as shown in Fig. 25.3 (*b*).

Shear force diagram

Let $\quad R_A$ = Reaction at *A*,

$R_B$ = Reaction at *B* and

$R_C$ = Reaction at *C*.

Taking moments about B,

$$R_A \times 6 - (3 \times 4) = -2.4 \qquad ...(\because M_B = -2.4 \text{ kN-m})$$

$$\therefore \quad R_A = \frac{-2.4+12.0}{6} = \frac{9.6}{6} = 1.6 \text{ kN} \quad \textbf{Ans.}$$

Similarly,
$$R_C \times 4 - (4 \times 2) = -2.4 \qquad ...(\because M_B = -2.4 \text{ kN-m})$$

$$\therefore \quad R_C = \frac{-2.4+8.0}{4} = \frac{5.6}{4} = 1.4 \text{ kN} \quad \textbf{Ans.}$$

and
$$R_B = (3 + 1 \times 4) - (1.6 + 1.4) = 4.0 \text{ kN} \quad \textbf{Ans.}$$

Now complete the shear force diagram as shown in Fig. 25.3 (*c*)

**EXAMPLE 25.2.** *A continuous beam ABCD, simply supported at A, B, C and D, is loaded as shown in Fig. 25.4.*

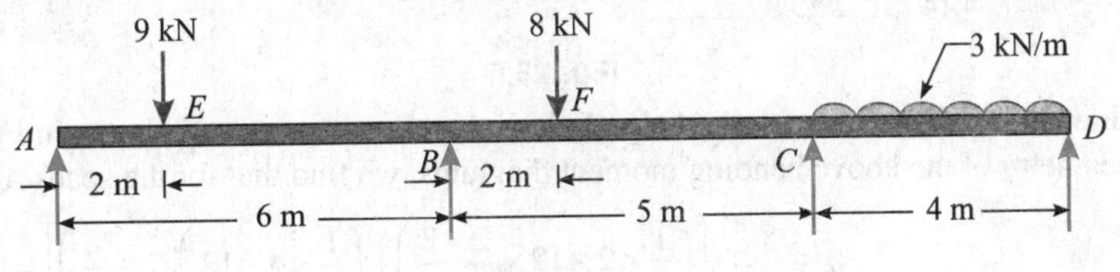


Fig. 25.4

*Find the moments over the beam and draw bending moment and shear force diagrams.*

**SOLUTION.** Given : Length *AB* ($l_1$) = 6 m ; Length *BC* ($l_2$) = 5 m ; Length *CD* ($l_3$) = 4 m ; Load at *E* ($W_1$) = 9 kN ; distance *AE* ($a_1$) = 2 m ; Distance *EB* ($b_1$) = 4 m ; Load at *F* ($W_2$) = 8 kN ; Distance *BF* ($a_2$) = 2 m ; Distance *FC* ($b_2$) = 3 m and uniformly distributed load in *CD* ($w$) = 3 kN/m.

***Moments over the beam***

Let $\quad$ * $M_A$ = Fixing moment at *A*,

$M_B$ = Fixing moment at *B*,

$M_C$ = Fixing moment at *C* and

* $M_D$ = Fixing moment at *D*.

First of all, let us consider the beam *AB* as a simply supported beam. Therefore bending moment at *E*,

$$M_E = \frac{W_1 a_1 b_1}{l_1} = \frac{9 \times 2 \times 4}{6} = 12 \text{ kN-m}$$

Similarly,
$$M_F = \frac{W_2 a_2 b_2}{l_2} = \frac{8 \times 2 \times 3}{5} = 9.6 \text{ kN-m}$$

* Since the beam is simply supported at *A* and *D*, therefore fixing moments $M_A$ and $M_D$ will be zero.

and bending moment at the mid of the span *CD*

$$= \frac{wl_3^2}{8} = \frac{3 \times (4)^2}{8} = 6 \text{ kN-m}$$

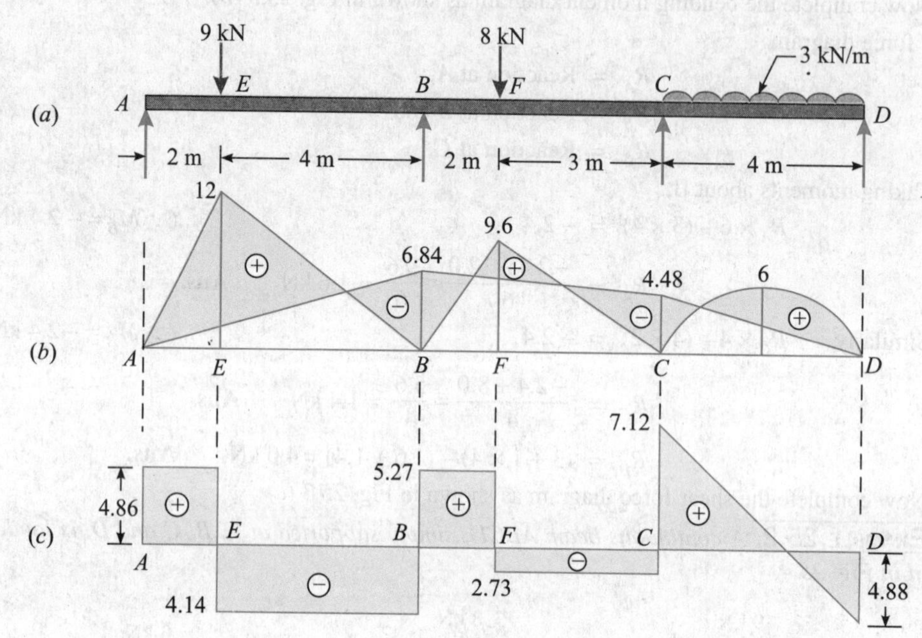


Fig. 25.5

Now draw the m-diagrams with the help of above bending moments as shown in Fig. 25.5 (*b*). From the geometry of the above bending moment diagrams, we find that for the spans *AB* and *BC*,

$$a_1 \bar{x}_1 = \left[\left(\frac{1}{2} \times 2 \times 12 \times \frac{2 \times 2}{3}\right) + \left(\frac{1}{2} \times 4 \times 12\right)\left(2 + \frac{4}{3}\right)\right] = 96$$

and
$$a_2 \bar{x}_2 = \left[\left(\frac{1}{2} \times 3 \times 9.6 \times \frac{2 \times 3}{3}\right) + \left(\frac{1}{2} \times 2 \times 9.6\right)\left(3 + \frac{2}{3}\right)\right] = 64$$

Similarly, for the spans *BC* and *CD*,

$$^* a_2 \bar{x}_2 = \left[\left(\frac{1}{2} \times 3 \times 9.6 \times \frac{2 \times 3}{3}\right) + \left(\frac{1}{2} \times 2 \times 9.6\right)\left(2 + \frac{3}{3}\right)\right] = 56$$

and
$$a_3 \bar{x}_3 = \left(\frac{2}{3} \times 6 \times 4 \times 2\right) = 32$$

Now using three moments equation for the spans *AB* and *BC*,

$$M_A l_1 + 2M_B (l_1 + l_2) + M_C l_2 = -\left(\frac{6a_1 x_1}{l_1} + \frac{6a_2 \bar{x}_2}{l_2}\right)$$

$$0 + 2 M_B (6 + 5) + M_C \times 5 = -\left(\frac{6 \times 96}{6} + \frac{6 \times 64}{5}\right)$$

$$22 M_B + 5 M_C = -172.8 \quad \text{...(i)}$$

* The previous value of $a_2 \bar{x}_2$ is with reference to the support *C* (being the end support of spans *AB* and *BC*). this value of $a_2 \bar{x}_2$ is with reference to the support *B* (being the end support of spans *BC* and *CD*).

Again using three moments equation for the spans $BC$ and $CD$,

$$M_B\, l_2 + 2\, M_C\, (l_2 + l_3) + M_D\, l_3 = -\left(\frac{6a_2\, \overline{x}_2}{l_2} + \frac{6a_3\, \overline{x}_3}{l_3}\right)$$

$$M_B \times 5 + 2\, M_C\, (5 + 4) + 0 = \left(\frac{6 \times 56}{5} + \frac{6 \times 32}{4}\right)$$

$$5\, M_B + 18\, M_C = -\,115.2 \qquad \text{...}(ii)$$

Solving equations ($i$) and ($ii$),

$$M_B = \mathbf{-\,6.84\ kN\text{-}m} \quad \text{and} \quad M_C = \mathbf{-\,4.48\ kN\text{-}m} \qquad \textbf{Ans.}$$

Now complete the bending moment diagram as shown in Fig. 25.5 ($b$).

Reactions at the supports

Let $R_A$ = Reaction at $A$,

$R_C$ = Reaction at $B$,

$R_C$ = Reaction at $C$ and

$R_D$ = Reaction at $D$.

Taking moments about $B$,

$$R_A \times 6 - (9 \times 4) = -\,6.84 \qquad \text{...}(\because M_B = -\,6.84 \text{ kN-m})$$

$$\therefore \quad R_A = \frac{-6.84 + 36}{6} = \frac{29.16}{6} = \mathbf{4.86\ kN} \qquad \textbf{Ans.}$$

Now taking moments about C,

$$R_D \times 4 - (12 \times 2) = -\,4.48 \qquad \text{...}(\because M_C = -\,4.48 \text{ kN-m})$$

$$\therefore \quad R_D = \frac{-4.48 + 24}{4} = \frac{19.52}{4} = \mathbf{4.88\ kN} \qquad \textbf{Ans.}$$

Again taking moments about $C$,

$$R_A \times 11 - (9 \times 9) + R_B \times 5 - (8 \times 3) = -\,4.48 \qquad \text{...}(\because M_C = -\,4.48 \text{ kN-m})$$

$$4.86 \times 11 - 81 + 5\, R_B - 24 = -\,4.48$$

$$R_B = \frac{-4.48 - 53.46 + 81 + 24}{5} = \frac{47.06}{5} = \mathbf{9.41\ kN} \qquad \textbf{Ans.}$$

and $R_C = (9 + 8 + 12) - (4.86 + 4.88 + 9.41) = \mathbf{9.85\ kN}$ **Ans.**

Now draw the shear force diagram as shown in Fig. 25.5 ($c$)

## 25.6. Continuous Beams with Fixed End Supports

Sometimes, a continuous beam is fixed at its one or both ends. If the beam is fixed at the left end $A$, then an imaginary zero span is taken to the left of $A$ and the three moments theorem is applied as usual. Similarly, if the beam is fixed at the right end, then an imaginary zero span is taken after the right end support and the three moments theorem is applied as usual.

**Notes. 1.** The fixing moment, at the imaginary support of the zero span *i.e.*, $M_0$ is always equal to zero.

**2.** Propped cantilevers and beams may also be analyses by Clapeyron's theorem of three moments.

**Example 25.3.** *A continuous beam ABC of uniform section, with span AB as 8 m and BC as 6 m, is fixed at A and simply supported at B and C. The beam is carrying a uniformly distributed load of 1 kN/m throughout its length. Find the moments along the beam and the reactions at the supports. Also draw the bending moment and shear force diagrams.*

**SOLUTION.** Given : Length $AB$ $(l_1) = 8$ m ; Length $BC$ $(l_2) = 6$ m and uniformly distributed load $(w) = 1$ kN/m.

***Moments along the beam***

Since the beam is fixed at $A$, therefore assume a zero span to the left of $A$.

Let

$$*M_0 = \text{Fixing moment at the left hand support of zero span,}$$
$$M_A = \text{Fixing moment at } A,$$
$$M_B = \text{Fixing moment at } B \text{ and}$$
$$*M_C = \text{Fixing moment at } C.$$

First of all, consider the beam $AB$ as a simply supported beam. Therefore bending moment at the mid of the span $AB$.

$$= \frac{wl_1^2}{8} = \frac{10 + (8)^2}{8} = 8.0 \text{ kN-m}$$

Similarly, bending moment at the mid of the span $BC$

$$= \frac{wl_2^2}{8} = \frac{1 \times (6)^2}{8} = 4.5 \text{ kN-m}$$

Now draw the μ-diagram with the help of above bending moments as shown in Fig. 25.6 (*b*). From the geometry of the above bending moment diagrams, we find that for the span $0A$ and $AB$,

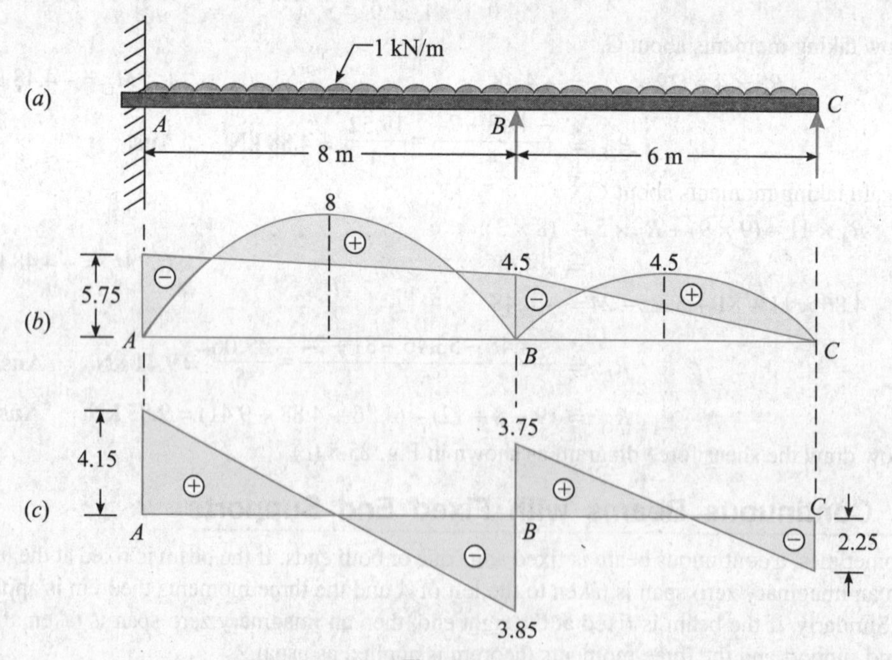


Fig. 25.6

$$a_0 \bar{x}_0 = 0$$

and

$$a_1 \bar{x}_1 = \left(\frac{2}{3} \times 8 \times 8 \times 4\right) = \frac{512}{3} = 170.67$$

* Since there is a zero span on the left of $A$, therefore the fixing moment $M_0$ will be zero. Moreover, as the beam is simply supported at $C$, therefore fixing moment $M_C$ will also be zero.

Similarly, for the span $AB$ and $BC$,

$$a_1 \bar{x}_1 = \left(\frac{2}{3}\times 8\times 8\times 4\right) = \frac{512}{3} = 170.67$$

and
$$a_2 \bar{x}_2 = \left(\frac{2}{3}\times 4.5\times 6\times 3\right) = 54$$

Now using three moments equation for the spans $0A$ and $AB$,

$$M_0\, l_0 + 2\, M_A\, (0 + l_1) + M_B\, l_1 = -\left(\frac{6\, a_0\, \bar{x}_0}{l_0} + \frac{6\, a_1\, \bar{x}_1}{l_1}\right)$$

$$0 + 2\, M_A\, (0 + 8) + M_B \times 8 = -\left(0 + \frac{6\times 170.67}{8}\right)$$

$$16\, M_A + 8\, M_B = -128 \quad \text{...(i)}$$

Again using three moments equation for the spans $AB$ and $BC$,

$$M_A\, l_1 + 2\, M_B\, (l_1 + l_2) + M_C\, l_2 = -\left(\frac{6 a_1\, \bar{x}_1}{l_1} + \frac{6 a_2\, \bar{x}_2}{l_2}\right)$$

$$M_A \times 8 + 2\, M_B\, (8 + 6) + 0 = -\left(\frac{6\times 170.67}{8} + \frac{6\times 54}{6}\right)$$

$$8\, M_A + 28\, M_B = -182 \quad \text{...(ii)}$$

Solving the equations (*i*) and (*ii*)

$$M_A = -5.75 \text{ kN-m} \quad \text{and} \quad M_B = -4.5 \text{ kN-m} \quad \textbf{Ans.}$$

Now complete the bending moment diagram as shown in Fig. 25.6 (*b*).

Reactions at supports

Let
$R_A$ = Reaction at $A$,
$R_B$ = Reaction at $B$, and
$R_C$ = Reaction at $C$,

Taking moments about $B$,

$$R_C \times 6 - (6 \times 3) = -4.5 \quad (\because M_B = -4.5 \text{ kN-m})$$

$$\therefore \quad R_C = \frac{-4.5 + 18}{6} = \frac{13.5}{6} = 2.2 \text{ kN} \quad \textbf{Ans.}$$

Now taking moments about $A$,

$$R_C \times 14 + R_B \times 8 - (1 \times 14 \times 7) = -5.75 \quad (\because M_A = -5.57 \text{ kN-m})$$

$$(2.25 \times 14) - 8 \cdot R_B - 98 = -5.75$$

$$\therefore \quad R_B = \frac{-5.75 + 98 - 31.5}{8} = 7.6 \text{ kN} \quad \textbf{Ans.}$$

and
$$*R_A = 14.0 - (2.25 + 7.6) = 4.15 \text{ kn} \quad \textbf{Ans.}$$

Now completed the shear force diagram as shown in Fig. 25.7 (*c*).

---

* The reaction at $R_A$ may also be found out by taking mokments about $B$, *i.e.*,

$$R_A \times 8 - (8 \times 4) - M_A = -4.5 \quad ...(\because M_B = -4.5 \text{ kN-m})$$

$$8\, R_A - 32 - 5.75 = -4.5$$

$$\therefore \quad R_A = \frac{-4.5 + 32 + 5.75}{8} = \frac{33.25}{8} = 4.15 \text{ kN} \quad \textbf{Ans.}$$

**EXAMPLE 25.4.** *Evaluate the bending moment and shear force diagrams of the beam shown in Fig. 25.7.*

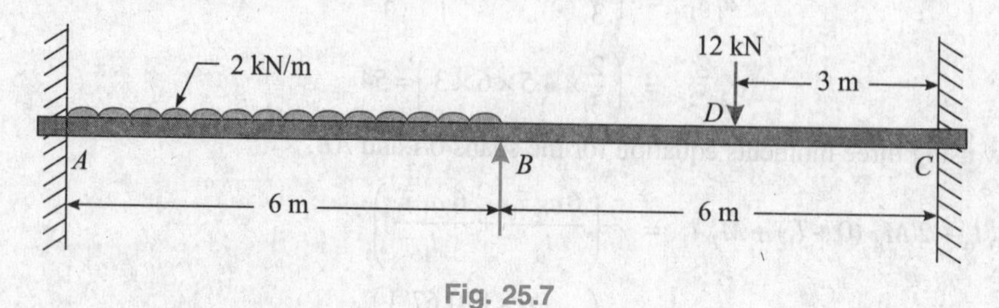


Fig. 25.7

*What are the reactions at the supports?*

**SOLUTION.** Given : Length $AB$ $(l_1) = 6$ m ; Length $BC$ $(l_2) = 6$ m ; Uniformly distributed load in $AB$ $(w) = 2$ kN/m and point load at $D$ $(W) = 12$ kN.

***Bending moment at A, B and C***

Since the beam is fixed at $A$ and $C$, therefore assume a zero span to the left of $A$ and right of $C$.

Let $M_0$ = Fixing moments at the imaginary supports of zero span (on the left of A and right of $C$).

$M_A$ = Fixing moment at $A$,

$M_B$ = Fixing moment at $B$ and

$M_C$ = Fixing moment at C.

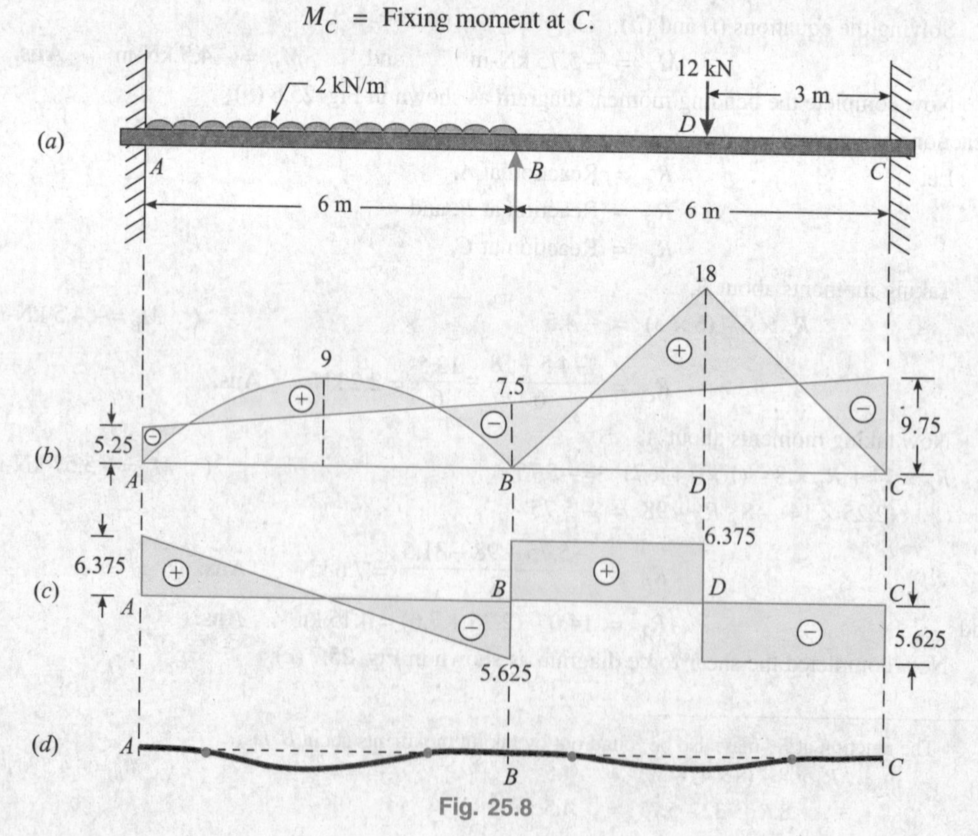


Fig. 25.8

First of all, consider the beam $AB$ as a simply supported beam. Therefore bending moment at the mid of span $AB$

$$= \frac{wl_1^2}{8} = \frac{2 \times (6)^2}{8} = 9 \text{ kN-m}$$

Similarly, bending moment at the mid of the span BC

$$= \frac{Wl}{4} = \frac{12 \times 6}{4} = 18 \text{ kN-m}$$

Now draw the μ-diagram with the help of above bending moments as shown in Fig. 25.8 (*b*). From the geometry of bending moment diagrams, we find that for the spans $0A$ and $AB$,

$$a_0 \bar{x}_0 = 0$$

$$a_1 \bar{x}_1 = \left(\frac{1}{2} \times 9.0 \times 6 \times 3\right) = 108$$

Similarly, for the span $AB$ and $BC$,

$$a_1 \bar{x}_1 = \left(\frac{2}{3} \times 9.0 \times 6 \times 3\right) = 108$$

$$a_2 \bar{x}_2 = \left(\frac{1}{2} \times 18.0 \times 6 \times 3\right) = 162$$

and for span $BC$ and $CD$,

$$a_3 \bar{x}_3 = \left(\frac{1}{2} \times 18.0 \times 6 \times 3\right) = 162$$

$$a_0 \bar{x}_0 = 0$$

Now using three moments equation for the spans $0A$ and $AB$,

$$M_0 \, l_0 + 2 \, M_A \, (0 + l_1) + M_0 \, l_1 = -\left(\frac{6 \, a_0 \, \bar{x}_0}{l_0} + \frac{6 \, a_1 \, \bar{x}_1}{l_1}\right)$$

$$0 + 2 \, M_A \, (0 + 6) + M_B \times 6 = \left(0 + \frac{6 \times 108}{6}\right)$$

$$12 \, M_A + 6 \, M_B = -108$$

or $$2 \, M_A + M_B = -18 \quad \text{...}(i)$$

Again using three moments equation for the spans $AB$ and $BC$,

$$M_A \, l_1 + 2 \, M_B \, (l_1 + l_2) + M_C \, l_2 = \left(\frac{6 a_1 \, x_1}{l_1} + \frac{6 a_2 \, x_2}{l_2}\right)$$

$$M_A \times 6 + 2 \, M_B \, (6 + 6) + M_C \times 6 = -\left(\frac{6 \times 108}{6} + \frac{6 \times 162}{6}\right)$$

$$6 \, M_A + 24 \, M_B + 6 \, M_C = -270$$

or $$M_A + 4 \, M_B + M_C = -45 \quad \text{...}(ii)$$

Again using three moments equation for the spans $BC$ and $C\,0$,

$$M_B \, l_2 + 2 \, M_C \, (l_2 + l_0) + M_C \, l_0 = -\left(\frac{6 \, a_2 \, \bar{x}_2}{l_2} + \frac{6 \, a_0 \, \bar{x}_0}{l_0}\right)$$

$$M_B \times 6 + 2 \, M_C \, (6 + 0) + 0 = -\left(\frac{6 \times 162}{6} + 0\right)$$

$$6 \, M_B + 12 \, M_C = -162$$

or $$M_B + 2 \, M_C = -27 \quad \text{...}(iii)$$

Solving equations (*i*), (*ii*) and (*iii*)

$$M_A = -5.25 \text{ kN-m},$$

$$M_B = -7.5 \text{ kN-m}$$

and $$M_C = -9.75 \text{ kN-m}$$

Now complete the bending moment diagram as shown in Fig. 25.8 (*b*).

Reactions at the supports

Let $R_A$ = Reaction at $A$,

$R_B$ = Reaction at $B$ and

$R_C$ = Reaction at $C$.

Taking moments about $B$ and equating the same,

$$R_A \times 6 + M_B = M_A + (2 \times 6 \times 3)$$

$$R_A \times 6 - 7.5 = -5.25 + 36 = 30.75$$

$$\therefore \quad R_A = \frac{30.75 + 7.5}{6} = 6.375 \text{ kN} \quad \textbf{Ans.}$$

Again taking moments about $B$ and equating the same,

$$R_C \times 6 + M_B = M_C + 12 \times 3$$

$$R_C \times 6 - 7.5 = -9.75 + 36 = 26.25$$

$$\therefore \quad R_C = \frac{26.25 + 7.5}{6} = 5.625 \text{ kN} \quad \textbf{Ans.}$$

and $$R_B = (2 \times 6 + 12) - (6.375 + 5.625) = 12 \text{ kN} \quad \textbf{Ans.}$$

Now draw the shear force diagram and elastic curve as shown in Fig. 25.8 (*c*) and 25.8 (*d*).

## EXERCISE 25.1

1. A continuous beam is simply supported over three spans, such that $AB$ = 8 m, $BC$ = 12 m and $CD$ = 5 m. It carries uniformly distributed load of 4 kN/m in span $AB$, 3 kN/m in span $BC$ and 6 kN/m in span $CD$. Find the moments over the supports $B$ and $C$.
[**Ans.** – 35.9 kN-m ; – 31.0 kN-m]

2. A simply supported beam $ABC$ is continuous over two spans $AB$ and $BC$ of 6 m and 5 m respectively. The span $AB$ is carrying a uniformly distributed load of 2 kN/m and the span $BC$ is carrying a point load of 5 kN at a distance of 2 m from $B$. Find the support moment and the reactions. [**Ans.** – 7.1 kN-m ; 4.82 kN ; 11.6 kN ; 0.58 kN]

3. A continuous beam $ABC$ is fixed at $A$ and is simply supported at $B$ and $C$. The span $AB$ is 6 m and carries a uniformly distributed load of 1 kN/m. The span $AC$ is 4 m and carries a uniformly distributed load of 3 kN/m. Determine the fixed end moments.
[**Ans.** $M_A$ = 2.143 kN-m ; $M_B$ = 1.714 kN-m ; $M_C$ = 0]

4. A continuous beam $ABCD$ is simply supported over three spans of 5 m, 5 m and 4 m respectively. The first two spans are carrying a uniformly distributed load of 4 kN/m, whereas the last span is carrying a uniformly distributed load of 5 kN/m. Find the support moments at $B$ and $C$.
[**Ans.** – 10.37 kN-m ; – 8.52 kN-m]

5. A continuous beam $ABCD$ is simply supported over three spans of 6 m, 5 m and 4 m respectively. The beam carries point loads of 90 kN and 80 kN at 2 m and 8 m from the support $A$ and a uniformly distributed load 30 kN/m over the span $CD$. Find the moments and reactions at the supports. [**Ans.** 68.4 kN-m ; 44.8 kN-m ; 48.6 kN-m ; 90.1 kN ; 98.5 kN]

## 25.7. Continuous Beams with End Span Overhanging

Sometimes, a continuous beam is overhanging, at its one or both ends. In such a case, the overhanging part of the beam behaves like a cantilever. The fixing moments on the end support may be found out by the cantilever action of the overhanging part of the beam.

**EXAMPLE 25.5.** *A beam ABCD 9 m long is simply supported at A, B and C, such that the length AB is 3 m, length BC is 4.5 m and the overhung CD is 1.5 m. It carries a uniformly distributed load of 1.5 kN/m in span AB and a point load of 1 kN at the free end D. The moments of inertia of the beam in span AB and CD is I and that in the span BC is 2I. Draw the bending moment and shear force diagrams for the beam.*

**SOLUTION.** Given : Length *AB* ($l_1$) = 3 m ; Length *BC* ($l_2$) = 4.5 m ; Length *CD* ($l_3$) = 1.5 m ; Uniformly distributed load in *AB* (*w*) = 1.5 kN/m ; Point load at *D* = 1 kN ; Moment of inertia of lengths *AB* and *CD* ($I_{AB}$) = *I* and moment of inertia of length *BC* ($I_{BC}$) = 2 *I*.

***Bending moment diagram***

Let *$M_A$ = Fixing moment at *A*,

$M_B$ = Fixing moment at *B* and

$M_C$ = Fixing moment at *C*.

First of all, consider the beam *AB* as a simply supported beam. Therefore bending moment at the mid of span *AB*

$$= \frac{wl_1^2}{8} = \frac{1.5 \times (3)^2}{8} = 1.69 \text{ kN-m}$$

From the geometry of the figure, we find that the dixing moment at *C*,

$$M_C = -1.0 \times 1.5 = -1.5 \text{ kN-m}$$

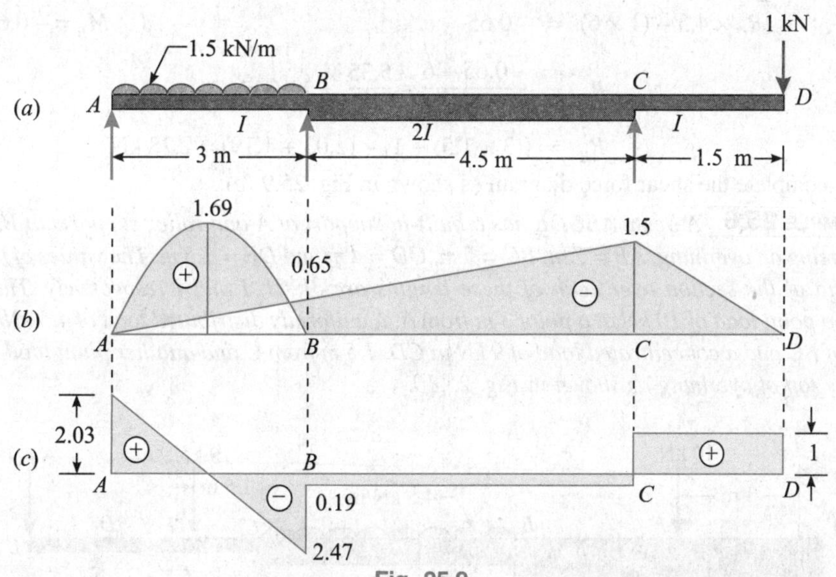


Fig. 25.9

Now draw the μ-diagram, with the help of above bending moments as shown in Fig. 25.9 (*b*). From the geometry of the above bending moment diagram, we find that for the spans *AB* and *BC*,

$$a_1 \bar{x}_1 = \frac{2}{3} \times 1.69 \times 3 \times 1.5 = 5.07$$

* Since the beam is simply supported at A, therefore fixing moment $M_A$ will be zero.

Now using three moments equation for the spans *AB* and *BC*,

$$M_A \frac{l_1}{I_1} + 2M_B\left(\frac{l_1}{I_1} + \frac{l_2}{I_2}\right) + M_C \frac{l_2}{I_2} = -\left[\frac{6a_1\bar{x}_1}{I_1 l_1} + \frac{6a_2\bar{x}_2}{I_2 l_2}\right]$$

$$0 + 2M_B\left(\frac{3}{I} + \frac{4.5}{2I}\right) + 1.5 \times \frac{4.5}{2I} = -\frac{6 \times 5.07}{I \times 3}$$

$$\frac{10.5\,M_B}{I} - \frac{6.75}{2I} = \frac{10.14}{I}$$

or $$10.5\,M_B - 3.375 = -10.14$$

$$\therefore \quad M_B = \frac{-10.14 + 3.375}{10.5} = -0.65 \text{ kN-m}$$

Now complete the final bending moment diagram as shown in Fig. 25.9 (*b*).

***Shear force diagram***

Let
$R_A$ = Reaction at *A*,
$R_B$ = Reaction at *B* and
$R_C$ = Reaction at *C*,

Taking moments at *B*,

$$R_A \times 3 - (1.5 \times 3 \times 1.5) = -0.65 \qquad ...(\because M_B = -0.65 \text{ kN-m})$$

$$\therefore \quad R_A = \frac{-0.65 + 6.75}{3} = \frac{6.1}{3} = 2.03 \text{ kN}$$

Again taking moments about *B*,

$$R_C \times 4.5 - (1 \times 6) = -0.65 \qquad ...(\because M_B = -0.65 \text{ kN-m})$$

$$\therefore \quad R_C = \frac{-0.65 + 6}{4.5} = \frac{5.35}{4.5} = 1.19 \text{ kN}$$

and $$R_B = [(3 \times 1.5) + 1] - (2.03 + 1.19) = 2.28 \text{ kN}$$

Now complete the shear force diagram as shown in Fig. 25.9 (*c*).

**EXAMPLE 25.6.** *A beam ABCDE has a built-in support at A and roller supports at B, C and D, DE being an overhung. AB = 7 m, BC = 5 m, CD = 4 m and DE = 1.5 m. The values of moment of inertia of the section over each of these lengths are 3I, 2I, I and I respectively. The beam carries a point load of 10 kN at a point 3 m from A, a uniformly distributed load of 4.5 kN/m over whole of BC and a concentrated load of 9 kN in CD, 1.5 m from C and another point load of 3 kN at E, the top of overhung as shown in Fig. 25.10.*

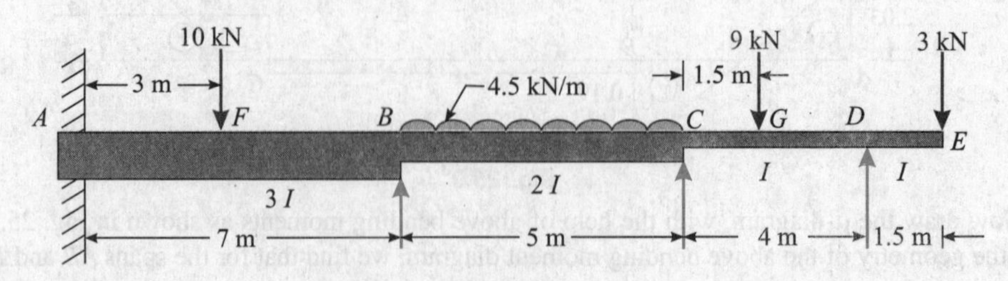


Fig. 25.10

*Determine (i) moments developed over each support and (ii) draw diagram for the entire beam, stating values at salient points.*

**SOLUTION.** Given : Length $AB$ ($l_1$) = 7 m ; Length $BC$ ($l_2$) = 5 m ; Length $CD$ ($l_3$) = 4 m ; Length $DE$ ($l_4$) = 1.5 m ; Moment of inertia of length $AB$ ($I_{AB}$) = 3 $I$ ; Moment of inertia for length $BC$ ($I_{BC}$) = 2 $I$ ; Moment of inertia of length $CD$ ($I_{CD}$) = $I_{DE}$ = $I$ ; Point load at $F$ ($W_1$) = 10 kN ; Uniformly distributed load between $B$ and $C$ ($w_2$) = 4.5 kN/m ; Point load at $G$ ($W_2$) = 9 kN and point load at $E$ = 3 kN.

(i) ***Moments developed over each support***

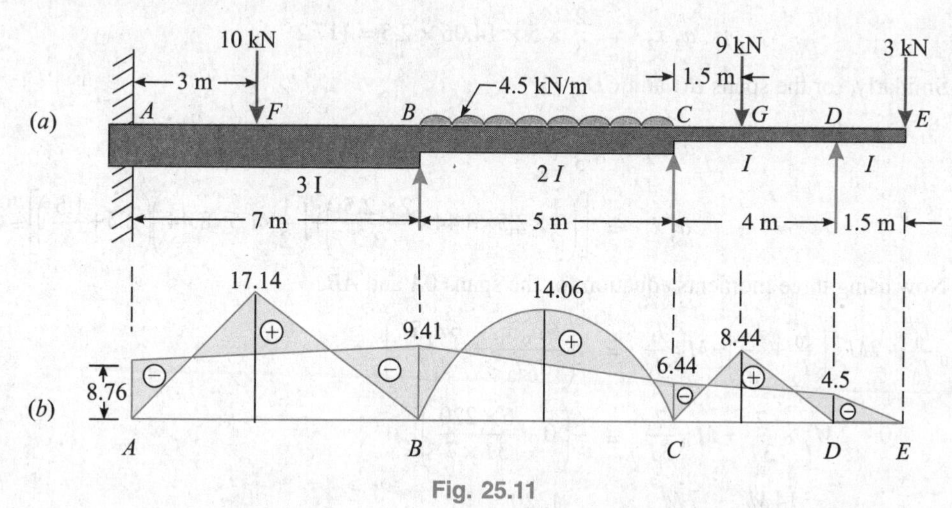


Fig. 25.11

Since the beam is fixed at $A$, therefore, assume a zero span to the left of $A$.

Let *$M_0$ = Fixing moment at left hand support of zero span,

$M_A$ = Fixing moment at $A$,

$M_B$ = Fixing moment at $B$,

$M_C$ = Fixing moment at $C$ and

$M_D$ = Fixing moment at $D$.

First of all, let us consider the beam $AB$ as a simply supported beam. Therefore bending moment under the 10 kN load

$$= \frac{W_1 a_1 b_1}{l_1} = \frac{10 \times 3 \times 4}{7} = 17.14 \text{ kN-m}$$

Similarly, bending moment under the 9 kN load in span $CD$

$$= \frac{W_3 a_3 b_3}{l_3} = \frac{9 \times 1.5 \times 2.5}{4} = 8.44 \text{ kN-m}$$

and bending moment at the mid of the span $BC$

$$= \frac{w_2 l_2^2}{8} = \frac{4.5 \times (5)^2}{8} = 14.06 \text{ kN-m}$$

From the geometry of the figure, we find that the fixing moment at $D$, due to load at $E$,

$$M_D = -3 \times 1.5 = -4.5 \text{ kN-m}$$

Now draw μ-diagram with the help of above bending moments as shown in Fig. 25.11 (*b*). From the geometry of the above bending moment diagram, we find that for the span $0A$ and $AB$,

$$a_0 \bar{x}_0 = 0$$

* Since there is a zero span on the left to $A$, therefore the fixing moment $M_0$ will be zero.

and $$a_1 \bar{x}_1 = \left[\left(\frac{1}{2}\times 4\times 17.14\times\frac{2\times 4}{3}\right)+\left(\frac{1}{2}\times 3\times 17.14\right)\left(4+\frac{3}{3}\right)\right]=220$$

Similarly, for the spans *AB* and *BC*,

$$a_1 \bar{x}_1 = \left[\left(\frac{1}{2}\times 3\times 17.14\times\frac{3\times 2}{3}\right)+\left(\frac{1}{2}\times 4\times 17.14\right)\left(3+\frac{4}{3}\right)\right]=200$$

and $$a_2 \bar{x}_2 = \frac{2}{3}\times 5\times 14.06\times 2.5 = 117.2$$

Similarly, for the spans *BC* and *CD*,

$$a_2 \bar{x}_2 = \frac{2}{3}\times 5\times 14.06\times 2.5 = 117.2$$

and $$a_3 \bar{x}_3 = \left[\left(\frac{1}{2}\times 2.5\times 8.44\times\frac{2\times 2.5}{3}\right)+\left(\frac{1}{2}\times 1.5\times 8.44\right)\left(2.5+\frac{1.5}{3}\right)\right]=36.6$$

Now using three moments equation for the spans 0*A* and *AB*,

$$M_0\frac{l_0}{I_0}+2M_A\left(\frac{l_0}{I_0}+\frac{l_1}{I_1}\right)+M_B\frac{l_1}{I_1} = -\left(\frac{6a_0\bar{x}_0}{I_0 l_0}+\frac{6a_1\bar{x}_1}{I_1 l_1}\right)$$

$$0+2M_A\times\frac{7}{3I}+M_B\frac{7}{3I} = -\left(0+\frac{6\times 220}{3I\times 7}\right)$$

$$\frac{14M_A}{3I}+\frac{7M_B}{3I} = -\frac{1320}{3I\times 7}$$

or $$2M_A + M_B = -26.94 \quad ...(i)$$

Again using three moments equation for the span *AB* and *BC*,

$$M_A\frac{l_1}{I_1}+2M_B\left(\frac{l_1}{I_1}+\frac{l_2}{I_2}\right)+M_C\frac{l_2}{I_2} = -\left(\frac{6a_1\bar{x}_1}{I_1 l_1}+\frac{6a_2\bar{x}_2}{I_2 l_2}\right)$$

$$M_A\frac{7}{3I}+2M_B\left(\frac{7}{3I}+\frac{5}{2I}\right)+M_C\frac{5}{2I} = -\left(\frac{6\times 200}{3I\times 7}+\frac{6\times 117.2}{2I\times 5}\right)$$

$$\frac{7M_A}{3I}+\frac{29M_B}{3I}+\frac{5M_C}{2I} = -\left(\frac{171.4}{3I}+\frac{140.6}{2I}\right)$$

$$14M_A + 58M_B + 15M_C = -764.6 \quad ...(ii)$$

Again using three moments equation for the spans *BC* and *CD*,

$$M_B\frac{l_2}{I_2}+2M_C\left(\frac{l_2}{I_2}+\frac{l_3}{I_3}\right)+M_D\frac{l_3}{I_3} = \left(\frac{6a_2\bar{x}_2}{I_2 l_2}\right)+\frac{6a_3\bar{x}_3}{I_3 l_3}$$

$$M_B\frac{5}{2I}+2M_C\left(\frac{5}{2I}+\frac{4}{I}\right)-4.5\frac{4}{I} = -\left(\frac{6\times 117.2}{2I\times 5}+\frac{6\times 36.6}{I\times 4}\right)$$

$$\frac{5M_B}{2I}+\frac{26M_C}{2I}-\frac{18}{I} = -\left(\frac{140.6}{2I}+\frac{54.9}{I}\right)$$

$$5M_B + 26M_C = -214.4 \quad ...(iii)$$

Now solving equations (*i*), (*ii*) and (*iii*), we get

$$M_A = -8.76 \text{ kN-m}$$ **Ans.**

$$M_B = -9.41 \text{ kN-m} \quad \textbf{Ans.}$$
$$M_C = -6.44 \text{ kN-m} \quad \textbf{Ans.}$$

Now complete the bending moment diagram as shown in Fig. 25.11 (*b*)

## 25.8. Continuous Beams with a Sinking Support

Sometimes, one of the supports of a continuous beam sinks down due to loading with respect to the other supports which remain at the same level. The sinking of a support effects the moments at the supports.

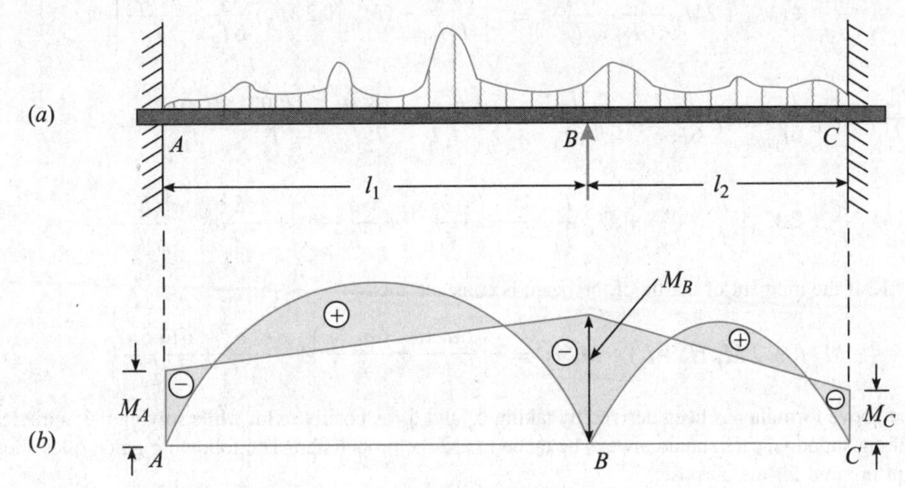


Fig. 25.12

Now consider a continuous beam *ABC* fixed at *A* and *C* and supported at *B* as shown in Fig. 25.12. Let the support *B** sink down through a more distance than the support *C* sinks down from its original position (or in other words from the support *A*),

Let $\delta_A$ = Height of support *A* from B

and $\delta_C$ = Height of support *C* from *B*.

We have already discussed in Art. 25.3 that in the span *AB*

$$dEI_1\,[l_1\,i_B - y_B] = a_1\,\bar{x}_1 + a'_1\,\bar{x}'_1$$

or

$$EI_1\,[l_1\,i_B - d_A] = a_1\,\bar{x}_1 + a'_1\,\bar{x}'_1$$

We have also discussed that

$$a'_1 + \bar{x}'_1 = (M_A + 2\,M_B)\,\frac{l_1^2}{6}$$

$$\therefore \quad EI_1\,[l_1\,i_B + \delta_A) = a_1\,\bar{x}_1 + (M_A + 2M_B)\,\frac{l_1^2}{6}$$

$$EI_1\,l_1\,i_B + EI_1\,\delta_A = a_1\,\bar{x}_1 + (M_A + 2M_B)\,\frac{l_1^2}{6}$$

$$Ei_B + \frac{E\,\delta_A}{l_1} = \frac{a_1\,\bar{x}_1}{I_1\,l_1} + (M_A + 2M_B)\,\frac{l_1}{6I_1}$$

* It is also possible if all three supports sink down. But the support *B* sinks down more than *A* and *C*.

or $$Ei_B = \frac{a_1 \bar{x}_1}{I_1 l_1} + (M_A + 2M_B)\frac{l_1}{6I_1} - \frac{E\delta_A}{l_1}$$

Similarly, in the span *BC* taking *C* as the origin and *x* positive to the left,

$$Ei_E' = \frac{a_2 \bar{x}_2}{I_2 l_2} + (M_C + 2M_B)\frac{l_2}{6I_2} - \frac{E\delta_C}{l_2}$$

Since $i_B$ is equal to $-i_B'$, therefore $Ei_B$ is equal to $-Ei_B'$.

or $$\frac{a_1\bar{x}_1}{I_1 l_1} + (M_A + 2M_B)\frac{l_1}{6I_1} - \frac{E\delta_A}{l_1} = -\left[\frac{a_2 \bar{x}_2}{I_2 l_2} + (M_C + 2M_B)\frac{l_2}{6I_2} - \frac{E\delta_C}{l_2}\right]$$

$$M_A \frac{l_1}{6I_1} + 2M_B \frac{l_2}{6I_1} + M_C \frac{l_2}{6I_2} + 2M_B \frac{l_2}{6I_2} = -\left(\frac{a_1 \bar{x}_1}{I_1 l_1} + \frac{a_2 \bar{x}_2}{I_2 l_2}\right) + \frac{E\delta_A}{l_2} + \frac{E\delta_C}{l_2}$$

$$M_A \frac{l_1}{I_1} + 2M_B\left(\frac{l_1}{I_1} + \frac{l_2}{I_2}\right) + M_C \frac{l_2}{I_2} = -\left(\frac{6a_1 \bar{x}_1}{I_1 l_1} + \frac{6a_2 \bar{x}_2}{I_1 l_2}\right) + \frac{6E\delta_A}{l_1} + \frac{6E\delta_C}{l_2}$$

**Notes: 1.** If the moment of inertia of the beam is constant, then

$$M_A l_1 + 2M_B(l_1 + l_2) + M_C l_2 = -\left(\frac{6a_1 \bar{x}_1}{l_1} + \frac{6a_2 \bar{x}_2}{l_2}\right) + \frac{6EI\delta_A}{l_1} + \frac{6EI\delta_C}{l_2}$$

**2.** The above formula has been derived by taking $\delta_A$ and $\delta_C$ as positive. But while solving the numericals on sinking supports, care should always be taken to use the proper sign. The following guide rules should be kept in mind for the purpose:

(*i*) The three moments equation always refer to two adjacent spans. The reference level should always be taken as that of the common support.

(*ii*) The sign of δ for the left and right support, should then be used by comparing its height with the central support (positive for higher and negative for lower), *e.g.*, Consider a continuous beam *ABCD* in which let the support *B* sink down by an amount equal to δ. The three supports, namely *A*, *C* and *D* will remain at the same level. Now, while using three moments equation for the spans *AB* and *BC*, the values of $\delta_A$ and $\delta_C$ will be positive (because both the supports *A* and *C* are at higher level than that of *B*). But while using three moments equation for the spans *BC* and *CD*, the value of $\delta_B$ will be negative (because the support *B* is at a lower level than that of *C*) and the value of $\delta_D$ will be zero (because the support *D* is at the same level as that of *C*).

**Example 25.7.** *A continuous beam ABC, shown in Fig. 25.13 carries a uniformly distributed load of 50 kN on AB and BC. The support B sinks by 5 mm below A and C and the values of EI is constant throughout the beam.*

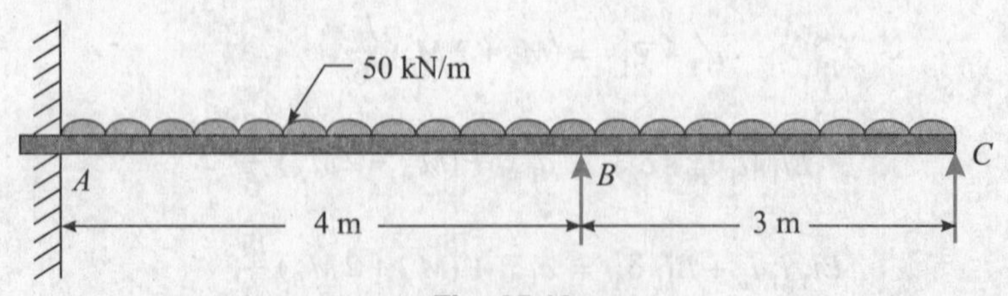


**Fig. 25.13**

*Taking E = 200 GPa and I = 332 × $10^6$ mm$^4$, find the bending moment at supports A and B and draw the bending moment diagram.*

**Solution.** Given : Length *AB* ($l_1$) = 4 m ; Length *BC* ($l_2$) = 3 m ; Uniformly distributed load over *AC* (*w*) = 50 kN-m ; Sinking of support of ($\delta_B$) = – 5 mm = – 0.005 m or $\delta_A = \delta_C$ = + 0.005 m ; Modulus of elasticity (*E*) 200 GPa = 200 × $10^6$ kN/m$^2$ and moment of inertia (*I*) = 332 × $10^6$ mm$^4$ = 332 × $10^{-6}$ m$^4$.

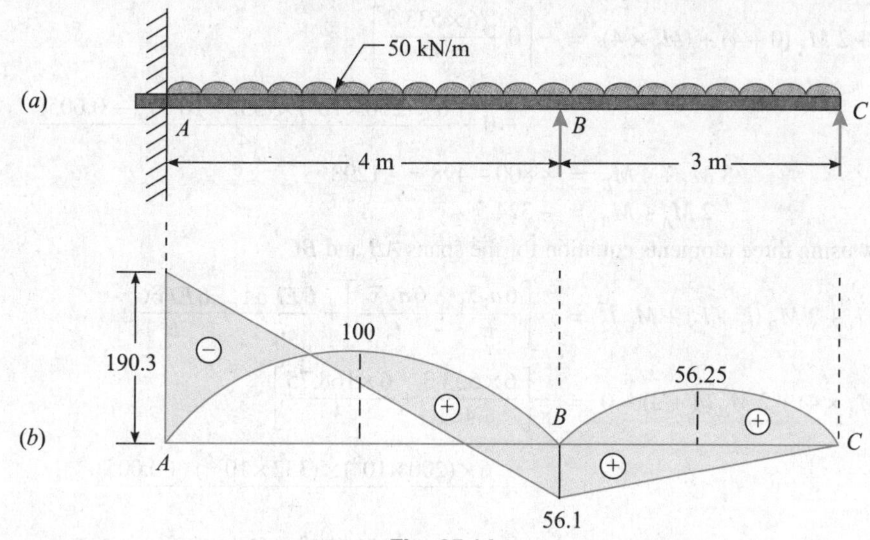


**Fig. 25.14**

Since the beam is fixed at *A*, therefore let us assume a zero span to the left of *A*,

$$M_A = \text{Fixing moment at } A,$$

$$M_B = \text{Fixing moment at } B, \text{ and}$$

$$^*M_C = \text{Fixing moment at } C,$$

First of all, consider the beam *AB* as a simply supported. Therefor bending moment at the mid of the span *AB*

$$= \frac{wl_1^2}{8} = \frac{50 \times (4)^2}{8} = 100 \text{ kN-m}$$

Similarly, bending moment at the mid of the span BC

$$= \frac{wl_2^2}{8} = \frac{50 \times (3)^2}{8} = 56.25 \text{ kN-m}$$

Now draw μ-diagram, with the help of above bending moments as shown in Fig. 25.14 (*b*). From the geometry of the above bending moment diagram, we find that for the spans 0*A* and *AB*,

$$a_0 \bar{x}_0 = 0$$

and $$a_1 \bar{x}_1 = \frac{2}{3} \times 100 \times 4 \times 2 = \frac{1600}{3} = 533.3$$

Similarly, for the spans *AB* and *BC*,

$$a_1 \bar{x}_1 = \frac{2}{3} \times 100 \times 4 \times 2 = \frac{1600}{3} = 533.3$$

and $$a_2 \bar{x}_2 = \frac{2}{3} \times 56.25 \times 3 \times 1.5 = 168.75$$

Now using three moments equation for the spans 0*A* and *AB*,

$$M_0\, l_0 + 2\, M_A\, (0 + l_1) + M_B\, l_1 = -\left[\frac{6\, a_0\, \bar{x}_0}{l_0} + \frac{6\, a_1\, \bar{x}_1}{l_1}\right] + \frac{6\, EI\, \delta_0}{l_1} + \frac{6\, EI\, \delta_B}{l_2}$$

* Since the beam is simply supported at *C*, therefore fixing moment *C* will be zero.

$$0 + 2\,M_A\,(0 + 4) + (M_B \times 4) = -\left[0 + \frac{6 \times 533.3}{4}\right]$$

$$+ 0 + \frac{6 \times (200 \times 10^6) \times (332 \times 10^{-6})\ (-\ 0.005)}{4}$$

$$8\,M_A + 4\,M_B = -\ 800 - 498 = -\ 1298$$

$$\therefore \quad 2\,M_A + M_B = -\ 324.5 \qquad ...(i)$$

Now using three moments equation for the spans *AB* and *BC*,

$$M_A\,l_1 + 2\,M_B\,(l_1 + l_2) + M_C\,l_2 = -\left[\frac{6\,a_1\,\bar{x}_1}{l_1} + \frac{6\,a_2\,\bar{x}_2}{l_2}\right] + \frac{6\,EI\,\delta A}{l_1} + \frac{6\,EI\,\delta C}{l_2}$$

$$(M_A \times 4) + 2\,M_B\,(4 + 3) + 0 = -\left[\frac{6 \times 533.3}{4} + \frac{6 \times 168.75}{3}\right]$$

$$+ \frac{6 \times (200 \times 10^6) \times (332 \times 10^{-6})\ (+0.005)}{4}$$

$$+ \frac{6 \times 200 \times 10^6 \times 332 \times 10^{-6}\ (+0.005)}{3}$$

$$4\,M_A + 14\,M_B = -\ (800 + 337.5) + 498 + 664 = 24.5$$

$$\therefore \quad 2\,M_A + 7\,M_B = 12.25 \qquad ...(ii)$$

Solving equations (*i*) and (*ii*),

$$M_A = -\ 190.3 \text{ kN-m} \quad \textbf{Ans.}$$

and $\quad M_B = 56.1$ kN-m **Ans.**

Now complete the bending moment diagram as shown in Fig. 25.14 (*b*).

NOTES: 1. While considering the spans *0A* and *AB*, the assumed support 0 and fixed support *A* are at the same level. But the support *B* has sunk down. Therefore the value of $\delta_0$ is taken as zero and that for $\delta_B$ as negative.

2. While considering the spans *AB* and *BC*, the supports *A* and *C* are at the same level. But the support *B* has sunk down. Therefore the values $\delta_A$ and $\delta_C$ are taken as positive.

**EXAMPLE 25.8.** *A continuous beam is built-in at A and is carried over rollers at B and C as shown in Fig. 25.15. AB = BC = 12 m.*

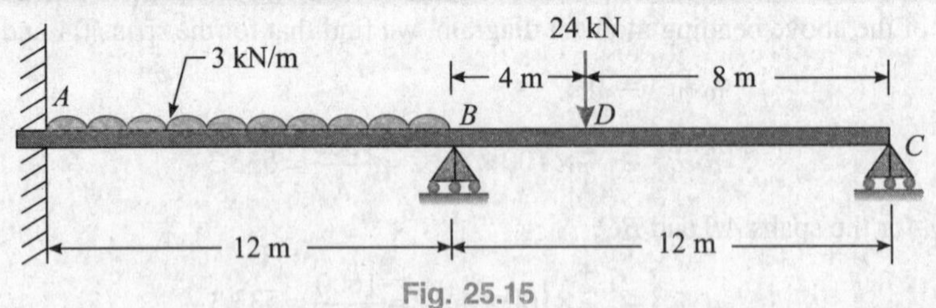


Fig. 25.15

*It carries a uniformly distributed load of 3 kN/m over AB and a point load of 24 kN in BC, 4 m from the support B, which sinks 30 mm. The values of E and I are 200 GPa and 0.2 × 10⁹ respectively and uniform throughout. Calculate the support moments and draw bending moment diagram and shear force diagram, giving critical values. Also draw the deflected shape of the centre line of the beam.*

**SOLUTION.** Given : Length *AB* ($l_1$) = 12 m ; Length *BC* ($l_2$) = 12 m ; Uniformly distributed loas in *AB* ($w$) = 3 kN/m ; Point loas at *D* ($W$) = 24 kN ; Sinking of support *B* ($\delta_B$) = – 30 mm = – 0.03 m or $\delta A = \delta C = +\ 0.03$ m ; modulus of elasticity ($E$) = 200 GPa = $200 \times 10^6$ kN-m$^2$ and sinking of support *B* ($I$) = $0.2 \times 10^9$ mm$^4$ = $0.2 \times 10^{-3}$ m$^4$.

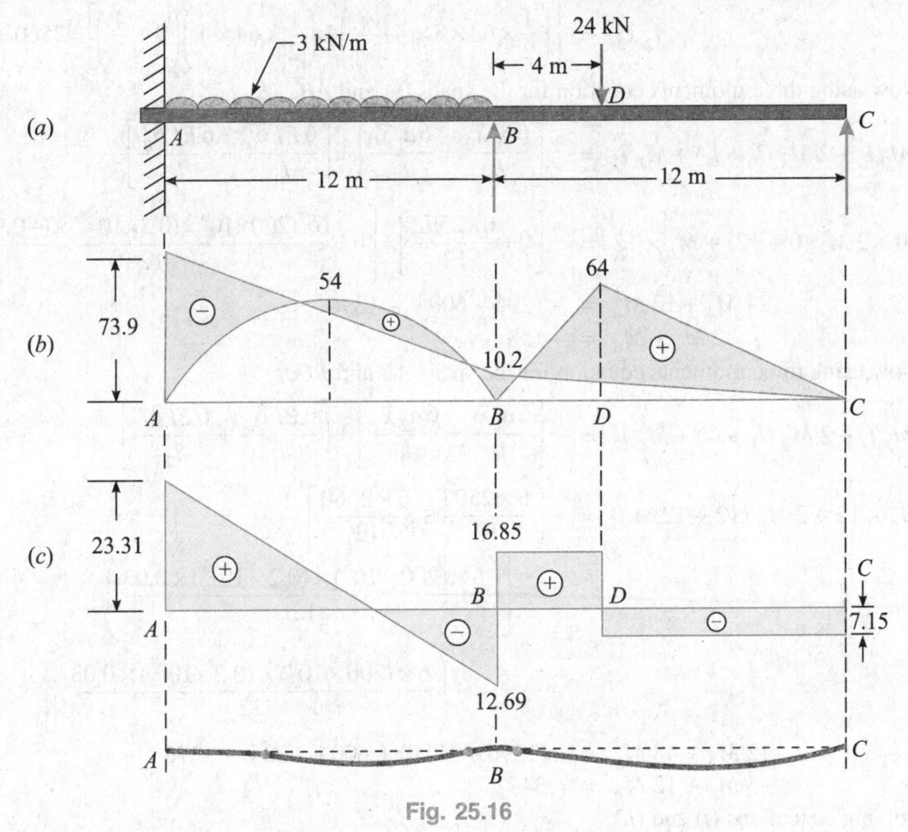


Fig. 25.16

Since the beam is fixed at $A$, therefore let us assume a zero span to the left of $A$,

*Support moments at A, B and C*

Let $M_A$ = Support moment at $A$,

$M_B$ = Support moment at $B$, and

$M_C$ = Support moment at $C$,

First of all, consider the beam $AB$ as a simply supported beam. Therefore bending moment at the mid of span $AB$

$$= \frac{wl_1^2}{8} = \frac{3 \times (12)^2}{8} = 54.0 \text{ kN-m}$$

Similarly, bending moment under the 24 kN load

$$= \frac{Wab}{l_2} = \frac{24 \times 4 \times 8}{12} = 64.0 \text{ kN-m}$$

Now draw μ-diagram with the help of above bending moments as shown in Fig. 25.16 (*b*).

From the geometry of the above bending moment diagram, we find that for the spans $0A$ and $AB$,

$$a_0 \,\bar{x}_0 = 0$$

and

$$a_1 \,\bar{x}_1 = \frac{2}{3} \times 54 \times 12 \times 6 = 2592$$

Similarly, for the spans $AB$ and $BC$,

$$a_1 \bar{x}_1 = \frac{2}{3} \times 54 \times 12 \times 6 = 2592$$

and $$a_2\,\bar{x}_2 = \left[\left(\frac{1}{2}\times 64\times 8\times\frac{2\times 8}{3}\right)+\left(\frac{1}{2}\times 64\times 4\right)\left(8+\frac{4}{3}\right)\right]=2560$$

Now using three moments equation for the spans $0A$ and $AB$,

$$M_0\,l_0 + 2\,M_A\,(l_0 + l_1) + M_B\,l_1 = -\left[\frac{6\,a_0\,\bar{x}_0}{l_0}+\frac{6\,a_1\,\bar{x}_1}{l_1}\right]+\left[\frac{6\,EI\,\delta_0}{l_0}+\frac{6\,EI\,\delta_B}{l_1}\right]$$

$$0 + 2\,M_A\,(0 + 12) + M_B \times 12 = -\left[0+\frac{6\times 2592}{12}\right]+\left[0+\frac{6\times(200\times 10^6\times(0.2\times 10^{-3})\times(-0.03)}{12}\right]$$

$$24\,M_A + 12\,M_B = -1296 - 600 = -1896$$

$$\therefore \quad 2\,M_A + M_B = -158 \qquad \text{...}(i)$$

Now using three moments equation for the spans $AB$ and $BC$,

$$M_A\,l_1 + 2\,M_B\,(l_1 + l_2) + M_C\,l_2 = -\left[\frac{6\,a_1\,\bar{x}_1}{l_1}+\frac{6\,a_2\,\bar{x}_2}{l_1}\right]+\left[\frac{6\,EI\,\delta_A}{l_1}+\frac{6\,EI\,\delta_C}{l_2}\right]$$

$$M_A \times 12 + 2\,M_B\,(12 + 12) + 0 = -\left[\frac{6\times 2592}{12}+\frac{6\times 2560}{12}\right] - \left[\left(\frac{6\times(200\times 10^6)\times(0.2\times 10^{-3})\times 0.03}{12}\right) + \left(\frac{6\times(200\times 10^6)\times(0.2\times 10^{-3})\times 0.03}{12}\right)\right]$$

$$12\,M_A + 48\,M_B = -(1296 + 1280) + 600 + 600 = -1376$$

$$\therefore \quad 3\,M_A + 12\,M_B = -344 \qquad \text{...}(ii)$$

Solving eqauations (*i*) and (*ii*)

$M_A = -73.9$ kN-m **Ans.**

$M_B = 10.2$ kN-m **Ans.**

$M_C = 0$ **Ans.**

***Bending moment diagram***

Now complete the bending moment diagram as shown in Fig. 25.16 (*b*).

***Shear force diagram***

Let $R_A$ = Reaction at $A$,

$R_B$ = Reaction at $B$, and

$R_C$ = Reaction at $C$.

Taking moments about $B$,

$$-10.2 = R_C \times 12 - (24 \times 4) \qquad \text{...}(\because M_B = -10.2 \text{ kN-m})$$

$$\therefore \quad R_C = \frac{-10.2 + 96}{12} = 7.15 \text{ kN}$$

Now taking moments about $A$,

$$-73.9 = 7.15 \times 24 + R_B \times 12 - (24 \times 16) - (3 \times 12 \times 6) \qquad \text{...}(\because M_A = -73.9 \text{ kN-m})$$

$$= 12\,R_B - 428.4$$

$$\therefore \quad R_B = \frac{-73.9 + 428.4}{12} = 29.54 \text{ kN}$$

and $$R_A = (3 \times 12 + 24) - (7.15 + 29.54) = 23.31 \text{ kN}$$

Now complete the Shear Force diagram as shown in Fig. 25.16 (*c*).

The elastic curve *i.e.*, deflected shape of the centre line of the beam is shown in Fig. 25.16 (*d*).

## 25.9. Continuous Beams Subjected to a Couple

Sometimes, a continuous beam is subjected to a couple in one (or more) of the spans. Such a beam is also analyses in the similar manner. The couple will cause negative moments in one part and positive moments in the other part of the span. While taking moment of the bending moment diagram, due care should be taken for the positive and negative bending moments.

**EXAMPLE 25.9.** *A continuous beam ABC of constant moment of inertia is simply supported at A, B and C. The beam carries a central point load of 4 kN in span AB and a central clockwise couple of moment 30 kN-m in span BC as shown in Fig. 25.17.*

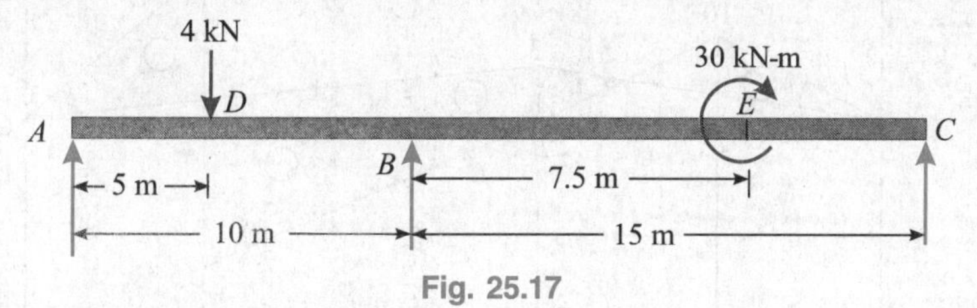


Fig. 25.17

*Find the support moments and plot the shear force and bending moment diagrams.*

**SOLUTION.** Given : Length $AB$ ($l_1$) = 10 m ; Length $BC$ ($l_2$) = 15 m ; Load at $D$ ($w$) = 4 kN and couple at $E$ ($\mu$) = 30 kN-m.

***Support moments***

Let

$$*M_A = \text{Support moment at } A,$$

$$M_B = \text{Support moment at } B \text{ and}$$

$$*M_C = \text{Support moment at } C.$$

First of all, consider the beam $AB$ and $BC$ as a simply supported beam. Therefore bending moment at $D$,

$$M_D = \frac{Wl_1}{4} = \frac{4 \times 10}{4} = 10 \text{ kN-m}$$

We know that the **moment just on the right side of E

$$= \frac{\mu}{2} = \frac{30}{2} = 15.0 \text{ kN-m}$$

and **moment just on the left side of $E$

$$= -\frac{\mu}{2} = -\frac{30}{2} = 15 \text{ kN-m}$$

Now draw the -diagram with the help of above bending moments as shown in Fig. 25.18 (*b*).

From the geometry of the bending moment diagrams, we find that for spans $AB$ and $BC$,

$$a_1 \bar{x}_1 = \frac{1}{2} \times 10 \times 10 \times 5 = 250$$

$$a_2 \bar{x}_2 = 0 \qquad (\because + \text{B.M.} = - \text{B.M.})$$

Now using three moments equation for the spans $AB$ and $BC$,

$$M_A \, l_1 + 2 \, M_B \, (l_1 + l_2) + M_C \, l_2 = -\left(\frac{6 a_1 \bar{x}_1}{l_1} + \frac{6 a_2 \bar{x}_2}{l_2}\right)$$

* Since the beam is simply supported at A and C, therefore fixing moment $M_A$ and $M_C$ will be zero.

** For details, please refer to Art. 13.18.

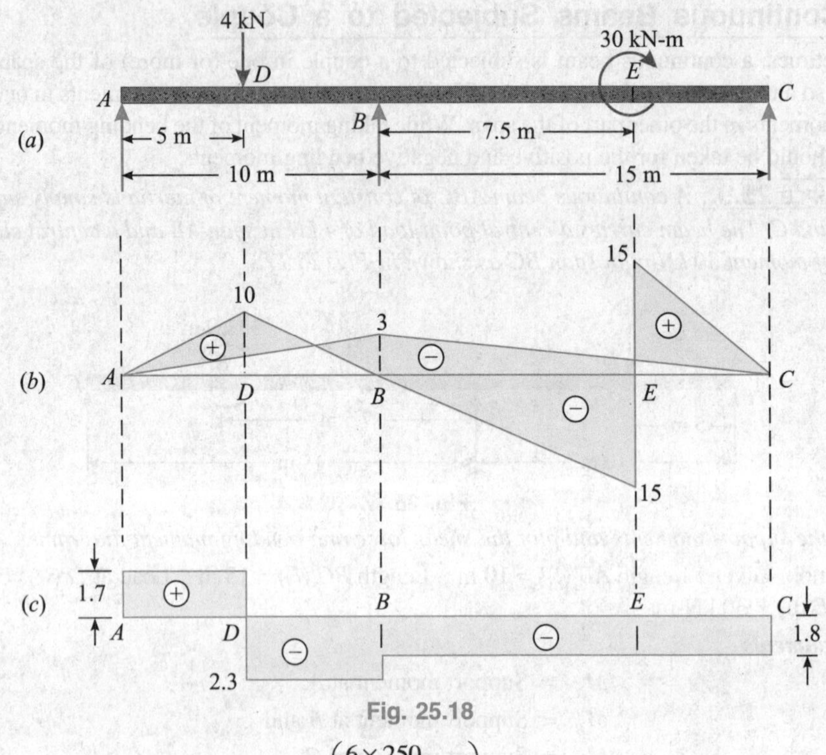


Fig. 25.18

$$0 + 2\,M_B\,(10 + 15) + 0 = -\left(\frac{6 \times 250}{10} + 0\right)$$

$$\therefore \quad 50\,M_B = -150$$

or $\quad M_B = -3$ kN-m

Now complete the bending moment diagram as shown in Fig. 25.18 (*b*).

***Shear force diagram***

Let $\quad R_A$ = Reaction at *A*,

$R_B$ = Reaction at *B*, and

$R_C$ = Reaction at *C*.

Taking moments about *B*,

$$R_A \times 10 - 4 \times 5 = -3 \qquad ...(\because M_B = -3 \text{ kN-m})$$

$$\therefore \quad R_A = \frac{-3 + 20}{10} = 1.7 \text{ kN}$$

Similarly, $\quad R_C \times 15 - 30 = -3 \qquad ...(\because M_B = -3$ kN-m$)$

$$\therefore \quad R_C = \frac{-3 + 30}{15} = 1.8 \text{ kN}$$

and $\quad R_B = 4 - (1.7 + 1.8) = 0.5$ kN

Now draw the shear force diagram as shown in Fig. 25.18 (*c*)

## EXERCISE 25.2

1. A continuous beam *ABCD* is fixed at *A* and simply supported at *B* and *C*, the beam *CD* is overhanging. The spans *AB* = 6 m, *BC* = 5 m and overhanging *CD* = 2.5 m. The moment of inertia of the span *BC* is 2 *l* and that of *AB* and *CD* is *l*. The beam is carrying a uniformly distributed load of 2 kN/m over the span *AB*, a point load of 5 kN in *BC* at a distance of 3 m

from $B$ and a point load of 8 kN at the free end.

Determine the fixing moments at $A$, $B$ and $C$ and draw the bending moment diagram.

[**Ans.** 8.11 kN-m ; 1.79 kN-m ; 20.0 kN-m]

2. A beam $ABCD$ is continuous over three spans $AB = 8$ m, $BD = 4$ m and $CD = 8$ m. The beam $AB$ and $BC$ is subjected to a uniformly distributed load of 1.5 kN/m, whereas there is a central point load of 4 kN in $CD$. The moment of inertia of $AB$ and $CD$ is 2 $l$ and that of $BC$ is $I$. The ends $A$ and $D$ are fixed. During loading, the support $B$ sinks down by 10 mm. Find the fixed end moments. Take $E = 200$ GPa and $l = 1600 \times 10^6 \text{mm}^4$.

[**Ans.** – 16.53 kN-m ; – 30.66 kN-m ; – 77.33 kN-m ; – 21.33 kN-m]

3. A continuous beam $ABCD$ 20 m long is supported at $B$ and $C$ and fixed at $A$ and $D$. The spans $AB$, $BC$ and $CD$ are 6 m, 8 m and 6 m respectively. The span $AB$ carries a uniformly distributed load of 1 kN/m, the span $BC$ carries a central point load of 10 kN and the span $CD$ carries a point load of 5 kN/m at a distance of 3 m from $C$. During loading, the support $B$ sinks by 10 mm. Find the fixed end moments and draw the bending moment diagram. Take $E = 200$ GPa and $I = 3 \times 10^9 \text{mm}^4$. The moment of inertia of the spans $AB$ and $CD$ is $I$ and that of $BC$ is $2I$.

[**Ans.** – 5.74 kN-m ; – 2.57 kN-m ; – 9.2 kN-m ; – 2.07 kN-m]

## QUESTIONS

1. What a continuous beam is subjected to a couple in a span, it will cause negative moment in one part and positive moment in other part of the span.
2. Explain the theorem of three moments.
3. Prove the Clapeyron's theorem of three moments.
4. How will you apply the theorem of three moments to a fixed beam?
5. Explain the effect on a continuous beam, when one of the intermediate supports sinks down.
6. Describe the effect of a couple acting in one of the spans of a continuous beam.

## OBJECTIVE TYPE QUESTIONS

1. Fixing moment over a simply supported end is

   (*a*) zero (*b*) negative (*c*) positive (*d*) infinity

2. When a continuous beam is fixed at the left end, then an imaginary span is taken to the left of the beam. The support moment at the imaginary support is

   (*a*) negligible (*b*) considerable (*c*) zero (*d*) calculated

3. If a continuous beam is fixed at its both ends, then imaginary support is

   (*a*) not taken (*b*) taken on left side only
   (*c*) taken on both the ends (*d*) taken on right side only.

4. If one of the span of a continuous beam is subjected to a clockwise couple, then

   (*a*) the span will be subjected to positive moment.
   (*b*) the span will be subjected to negative moment.
   (*c*) entire beam will be subjected to positive moment.
   (*d*) one part of the span is subjected to positive moment and the other part to negative moment.

## ANSWERS

1. (*a*) 2. (*c*) 3. (*d*) 4. (*d*)

# Chapter 26

# Moment Distribution Method

## Contents

## 26.1. Introduction

The moment distribution method, which was first introduced by Prof. Hardy Cross in 1930, is widely used for the analysis of all types of indeterminate structures. In this method, all the members of a structure are first assumed to be fixed in position and direction and fixed end moments due to external loads are obtained. Now all the hinged joints are released, by applying an equal and opposite moment and their effects are evaluated on the opposite joints. The unbalanced moment at a joint, is distributed in the two spans in the ratio of their distribution factors. This process is continued, till we reach the required degree of precision.

## 26.2. Sign Conventions

Though different types of sign conventions are adopted by different authors in their books, yet the following sign conventions, which are widely used and internationally recognised will be used in this book.

1. All the clockwise moments at the ends are taken as *positive*.
2. All the anticlockwise moments at the ends are taken as *negative*.

## 26.3. Carry Over Factor

We have already discussed in Art. 26.1 that the moments are applied on all the end joints of a structure, whose effects are evaluated on the other joints. The ratio of moment produced at a joint to the moment applied at the other joint, without displacing it, is called *carry over factor*. Now, we shall find out the value of carry over factor in the following two cases of beam:

1. When the beam is fixed at one end and simply supported at the other.
2. When the beam is simply supported at both the ends.

## 26.4. Carry Over Factor for a Beam Fixed at One End and Simply Supported at the Other

Consider a beam $AB$ fixed at $A$ and simply supported at $B$. Let a clockwise moment be applied at the support $B$ of the beam as shown in Fig. 26.1.

Let $l$ = Span of the beam,

$\mu$ = Clockwise moment applied at $B$ (*i.e.*, $M_B$) and

$M_A$ = Fixing moment at $A$.

Since the beam is not subjected to any external loading, therefore the two reactions ($R$) must be equal and opposite as shown in Fig. 26.1.

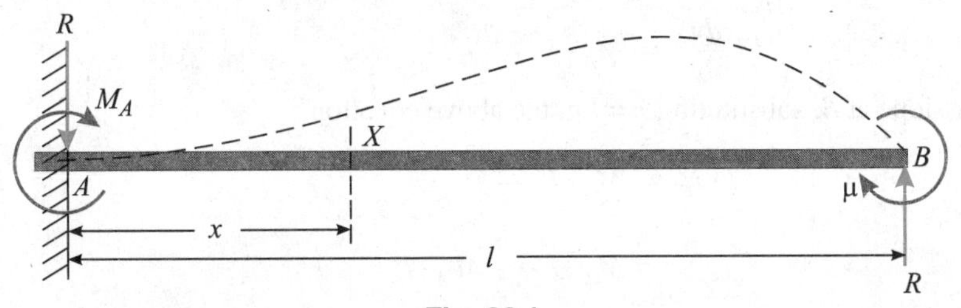


Fig. 26.1

Taking moments about $A$ and equating the same,

$$R \cdot l = M_A + \mu \qquad ...(i)$$

Now consider any section $X$, at a distance $x$ from $A$. We know that the moment at $X$,

$$M_X = M_A - R \cdot x$$

or

$$EI \frac{d^2y}{dx^2} = M_A - R \cdot x \qquad ...\left(\because M = EI \frac{d^2y}{dx^2}\right)$$

Integrating the above equation,

$$EI \frac{dy}{dx} = M_A \cdot x - \frac{Rx^2}{2} + C_1$$

where $C_1$ is the first constant of integration. We know that when $x = 0$, then $\frac{dy}{dx} = 0$. Therefore $C_1 = 0$.

or $$EI\frac{dy}{dx} = MA \cdot x - \frac{Rx^2}{2} \quad ...(ii)$$

Integrating the above equation once again,

$$EI \cdot y = \frac{M_A \cdot x^2}{2} - \frac{Rx^3}{6} + C_2$$

where $C_2$ is the second constant of integration. We know that when $x = 0$, then $y = 0$. Therefore $C_2 = 0$.

or $$EI \cdot y = \frac{M_A \cdot x^2}{2} - \frac{Rx^3}{6} \quad ...(iii)$$

We also know that when $x = l$, then $y = 0$. Therefore substituting these values in equation (*iii*),

$$0 = \frac{M_A \cdot l^2}{2} - \frac{R \cdot l^3}{6}$$

$$\therefore \quad \frac{Rl^3}{6} = \frac{M_A \cdot l^2}{2}$$

or $$R \cdot l = 3M_A$$

Substituting the value of in equation (*i*),

$$3M_A = M_A + \mu$$

or $$M_A = \frac{\mu}{2} = \frac{M_B}{2} \quad ...(iv)$$

$$\therefore \quad \frac{M_A}{M_B} = \frac{1}{2}$$

It is thus obvious, that carry over factor is *one-half* in this case,

We see from equation (*ii*) that

$$EI\frac{dy}{dx} = M_A \cdot x - \frac{Rl^2}{2}$$

Now for slope at *B*, substituting $x = l$ in the above equation,

$$EI \cdot i_B = M_A \cdot l - \frac{Rl^2}{2} \quad ...(\because \; R \cdot l = 3M_A)$$

$$= M_A \cdot l - \frac{3}{2} M_A \cdot l$$

$$= -\frac{M_A \cdot l}{2} = -\frac{\mu l}{4} \quad ...\left(\because \; M_A = \frac{\mu}{2}\right)$$

$$\therefore \quad i_B = -\frac{\mu l}{4EI}$$

$$= \frac{\mu l}{4EI}$$

...(Minus sign means that the tangent at *B* makes an angle with *AB* in the negative or anticlockwise direction)

or $$\mu = \frac{4EI \cdot i_B}{l}$$

## 26.5. Carry Over Factor for a Beam Simply Supported at Both Ends

Consider a beam *AB* simply supported at *A* and *B*. Let a clockwise moment be applied at the support *B* of the beam as shown in Fig. 26.2.

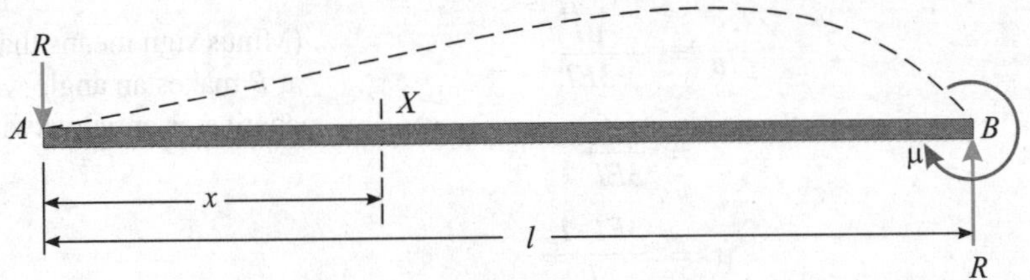


Fig. 26.2

Let $l$ = Span of the beam, and

$\mu$ = Clockwise moment at $B$.

Since the beam is simply supported at $A$, therefore there will be no fixing moment at $A$. Moreover, as the beam is not subjected to any external loading, therefore the two reactions must be equal and opposite as shown in Fig. 26.2.

Taking moments about $A$,

$$R \cdot l = \mu \qquad ...(i)$$

Now consider any section $X$, at a distance $x$ from $A$. We know that the moment at $X$,

$$M_X = -R \cdot x$$

or

$$EI \frac{d^2y}{dx^2} = -R \cdot x \qquad ... \left( \because M = EI \frac{d^2y}{dx^2} \right)$$

Integration the above equation,

$$EI \frac{dy}{dx} = -\frac{Rx^2}{2} + C_1 \qquad ...(ii)$$

where $C_1$ is the first constant of integration. Integrating the above equation once again,

$$EI \cdot y = -\frac{Rx^3}{6} + C_1 x + C_2$$

where $C_2$ is the second constant of integration. We know that when $x = 0$, then $y = 0$. Therefore $C_2 = 0$.

or

$$EI \cdot y = -\frac{Rx^3}{6} + C_1 x \qquad ...(iii)$$

We also know that when $x = l$, then $y = 0$. Therefore substituting these values in the above equation,

$$0 = -\frac{Rl^3}{6} + C_1 l$$

$$\therefore \quad C_1 = \frac{Rl^2}{6} = \frac{\mu l}{6} \qquad ...(\because R \cdot l = \mu)$$

Substituting this value of $C_1$ in equation (*ii*),

$$EI \frac{dy}{dx} = -\frac{Rx^2}{2} + \frac{\mu l}{6} = -\frac{Rlx^2}{2l} + \frac{\mu l}{6}$$

$$= -\frac{\mu x^2}{2l} + \frac{\mu l}{6} \qquad ...(\because R \cdot l = \mu)$$

Now for slope at $B$, substituting $x = l$ in the above equation,

$$EI \cdot i_B = -\frac{\mu l^2}{2l} + \frac{\mu l}{6} = -\frac{\mu l}{2} + \frac{\mu l}{6} = -\frac{\mu l}{3}$$

$\therefore \quad i_B = -\dfrac{\mu l}{3EI}$ ... (Minus sign means that the tangent at $B$ makes an angle with $AB$ in the negative or anticlockwise direction)

$$= \frac{\mu l}{3EI}$$

$$\therefore \quad \mu = \frac{3EI \cdot i_B}{l}$$

## 26.6. Stiffness Factor

It is the moment required to rotate the end, while acting on it, through a unit angle without translation of the far end. We have seen in Art. 26.4 that the moment on a beam having one end fixed and the other freely supported,

$$\mu = \frac{4EI \cdot i_B}{l}$$

$\therefore$ Stiffness factor for such a beam (substituting $i_B = 1$),

$$k_1 = \frac{4EI}{l}$$

Similarly, we have seen in Art. 26.5 that the moment on a beam having simply supported ends,

$$\mu = \frac{3EI \cdot i_B}{l}$$

$\therefore$ Stiffness factor for such a beam (substituting $i_B = 1$),

$$k_2 = \frac{3EI}{l}$$

## 26.7. Distribution Factors

Sometimes, a moment is applied on a structural joint to produce rotation, without the translation of its members. This moment is distributed among all the connecting members of the joint in the proportion of their stiffness.

Consider four members $OA$, $OB$, $OC$ and $OD$ meeting at $A$. Let the members $OA$ and $OC$ be fixed at $A$ and $C$, whereas the members $OB$ and $OD$ be hinged at $B$ and $D$. Let the joint $O$ be subjected to a moment $\mu$ as shown in Fig. 26.3.

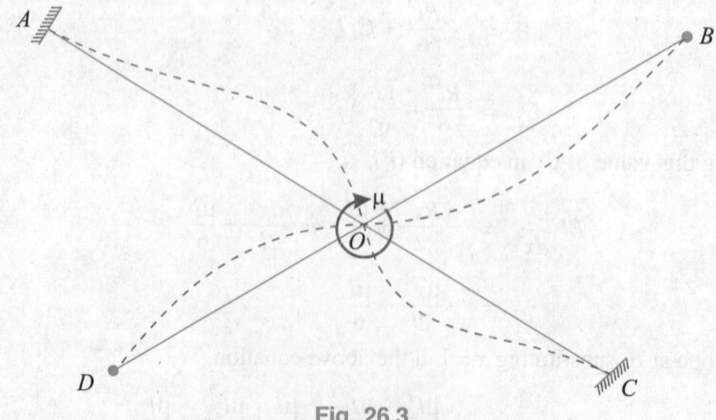


Fig. 26.3

Let $\quad l_1$ = Length of the member $OA$,

$I_1$ = Moment of inertia of the member $OA$,

$E_1$ = Modulus of elasticity of the member $OA$,

$l_1, I_2, E_2$ = Corresponding values for the member $OB$,

$l_3, I_3, E_3$ = Corresponding values for the member $OC$,

and $l_4, I_4, E_4$ = Corresponding values for the member $OD$.

A little consideration will show that as a result of the moment μ, each member gets rotated through some equal angle. Let this angle through which each member is rotated be θ.

We know that the stiffness of member $OA$,

$$k_1 = \frac{4E_1I_1}{l_1} \qquad ...(\because \text{ End } A \text{ is fixed})$$

Similarly,
$$k_2 = \frac{3E_2\,I_2}{l_2} \qquad ...(\because \text{ End } B \text{ is hinged})$$

and
$$k_3 = \frac{4E_3\,I_3}{l_3} \qquad ...(\because \text{ End } C \text{ is fixed})$$

and
$$k_4 = \frac{3E_4\,I_4}{l_4} \qquad ...(\because \text{ End } D \text{ is hinged})$$

Now total stiffness of all the members,

$$k = k_1 + k_2 + k_3 + k_4$$

and total moment applied at the joint,

$$\mu = k\theta$$

∴ Moment on the member $OA$,

$$\mu_1 = k_1\theta$$

Similarly,
$$\mu_2 = k_2\theta : \quad \mu_3 = k_3\theta \quad \text{and} \quad \mu_4 = k_4\theta$$

∴
$$\frac{\mu_1}{\mu} = \frac{k_1\theta}{k\theta} = \frac{k_1}{k}$$

Similarly,
$$\frac{\mu_2}{\mu} = \frac{k_2}{k}\ ; \quad \frac{\mu_3}{\mu} = \frac{k_3}{k} \quad \text{and} \quad \frac{\mu_4}{\mu} = \frac{k_4}{k}$$

∴
$$\mu_1 = k_1\frac{\mu}{k}$$

Similarly,
$$\mu_2 = k_2\frac{\mu}{k}\ ; \ \mu_3 = k_3\frac{\mu}{k} \ \text{and}\ \mu_4 = k_4\frac{\mu}{k}$$

The quantities $\frac{k_1}{k}, \frac{k_2}{k}, \frac{k_3}{k}$ and $\frac{k_4}{k}$ are known as distribution factors for the members $OA$, $OB$, $OC$ and $OD$ respectively. The moments $\mu_1$, $\mu_2$, $\mu_3$ and $\mu_4$ are known as distributed moments.

**EXAMPLE 26.1.** *Five members OA, OB, OC, OD and OE meeting at O, are hinged at A and C and fixed at B, D and E. The lengths of OA, OB, OC, OD and OE are 3 m, 4 m, 2 m, 3 m and 5 m and their moments of inertia are 400 mm$^4$, 300 mm$^4$, 200 mm$^4$, 300 mm$^4$ and 250 mm$^4$ respectively. Determine the distribution factors for the members and the distributed moments, when a moment of 4000 kN-m is applied at O.*

**SOLUTION:** Given: Length $OA$ = 3 m ; Length $OB$ = 4 m ; Length $OC$ = 2 m ; Length $OD$ = 3 m; Length $OE$ = 5 m ; Moment of inertia of $OA$ = 400 mm$^4$ ; Moment of inertia of $OB$ = 300 mm$^4$ ; Moment of inertia of $OC$ = 200 mm$^4$ ; Moment of inertia of $OD$ = 300 mm$^4$ ; Moment of inertia of $OE$ = 250 mm$^4$ and moment on $D$ = 4000 kN-m.

We know that stiffness factor for *OA*,

$$k_A = \frac{3EI}{I} = \frac{3 \times E \times 400}{4} = 400\,E \quad ...(\because \text{ Member is hinged at } A)$$

Similarly,
$$k_B = \frac{4EI}{I} = \frac{4 \times E \times 300}{4} \quad ...(\because \text{ Member is fixed at } B)$$

Now complete the column for stiffness for all the members, keeping in mind whether the member is hinged or fixed at the end. Now find out the distribution factor and distribution moments for each member as shown in the above chart.

Now from the above chart, we find that the distribution factors for *OA, OB, OC, OD* and *OE* are $\frac{1}{4}, \frac{3}{16}, \frac{3}{16}, \frac{1}{4}$ and $\frac{1}{8}$ respectively **Ans.**

The moment of 4000 kN-m applied at the joint *O* will be distributed among the members as obtained from the above in the following table:

| *Member* | *Length (m)* | *M.I. (mm⁴)* | *Stiffness (k)* | *Distribution factor* | *Distributed moments N-m* |
|---|---|---|---|---|---|
| *OA* | 3 | 400 | $\frac{3E \times 400}{3} = 400$ | $\frac{400E}{1600E} = \frac{1}{4}$ | $\frac{1}{4} \times 4000 = 1000$ |
| *OB* | 4 | 300 | $\frac{4E \times 300}{4} = 300$ | $\frac{300E}{1600E} = \frac{3}{16}$ | $\frac{3}{16} \times 4000 = 750$ |
| *OC* | 2 | 200 | $\frac{3E \times 200}{2} = 300$ | $\frac{300E}{1600E} = \frac{3}{16}$ | $\frac{3}{16} \times 4000 = 750$ |
| *OD* | 3 | 300 | $\frac{4E \times 300}{3} = 400$ | $\frac{400E}{1600E} = \frac{1}{4}$ | $\frac{1}{4} \times 4000 = 1000$ |
| *OE* | 5 | 250 | $\frac{4E \times 250}{5} = 200$ | $\frac{200E}{1600E} = \frac{1}{8}$ | $\frac{1}{8} \times 4000 = 500$ |
| | **Total** | | **Σ = 1600E** | | **Σ = 400** |

Thus distributed moments for *OA, OB, OC, OD* and *OE* are 1000, 750, 750, 1000 and 500 N-m respectively. **Ans.**

## 26.8. Application of Moment Distribution Method to Various Types of Continuous Beams

In the previous articles, we have studied the principles of the moment distribution method. First of all, all the supports are assumed to be clamped and fixing moments are found out for all the spans. The unbalanced moment at a support is distributed among the two spans in the ratio of their stiffness factors and their effects are evaluated on the opposite joints. This process is continued till we reach the desired degree of precision. Though the moment distribution method has very wide applications yet in the proceeding articles, we shall discuss its application to the following types of beams only:

1. Beams with fixed end supports,
2. Beams with simply supported ends,
3. Beams with end span overhanging and
4. Beams with a sinking support.

## 26.9. Beams with Fixed End Supports

Sometimes, a continuous beam is fixed at its one or both ends. In such a case, the fixed end moments are obtained and the unbalanced moment is distributed in two spans in the ratio of their distribution factors.

**EXAMPLE 26.2.** *Evaluate the bending moment and shear force diagrams of the beam shown in Fig. 26.4.*

*What are the reactions at the supports?*

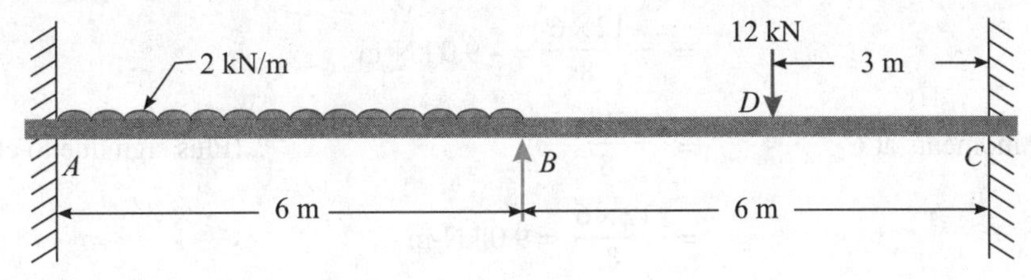


**Fig. 26.4**

* **SOLUTION.** Given: Length $AB$ $(l_1) = 6$ m ; Length $BC$ $(l_2) = 6$ m ; Uniformly distributed load in $AB$ $(w) = 2$ kN/m and point load at $D$ $(W) = 12$ kN.

First of all, let us consider the continuous beam $ABC$ to be split into two fixed beams $AB$ and $BC$ as shown in Fig. 26.5 (*b*) and (*c*).

We know that in span $AB$ the fixing moment at $A$

$$= -\frac{wl^2}{12}$$ ...(Minus sign due to anticlockwise)

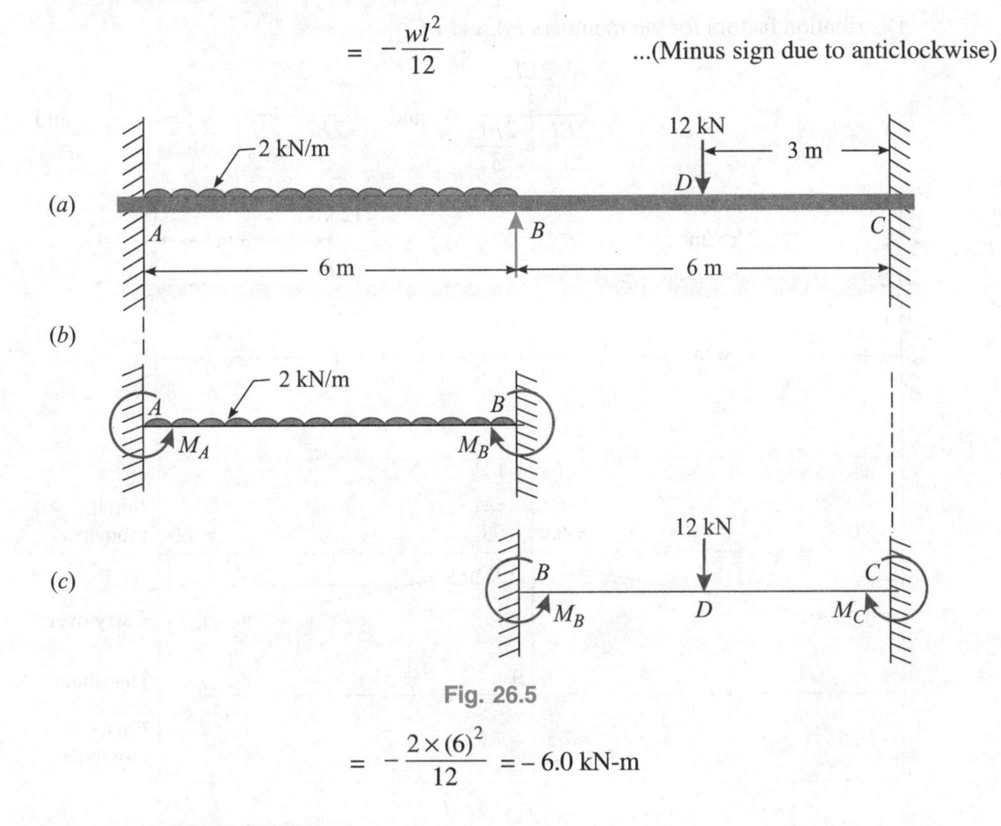


**Fig. 26.5**

$$= -\frac{2 \times (6)^2}{12} = -6.0 \text{ kN-m}$$

* We have already solved this question by Theorem of three moments in Example 25.4.

and fixing moment at $B = +\frac{wl^2}{12}$ ...(Plus sign due to clockwise)

$$= \frac{2\times(6)^2}{12} = 6.0 \text{ kN-m}$$

Similarly, in span *BC* the fixing moment at *B*

$$= -\frac{Wl}{8} \quad \text{...(Minus sign due to anticlockwise)}$$

$$= -\frac{12\times 6}{8} = -9.0 \text{ kN-m}$$

and fixing moments at *C* $= +\frac{Wl}{8}$ ...(Plus sign due to clockwise)

$$= \frac{12\times 6}{8} = 9.0 \text{ kN-m}$$

Now let us find out the distribution factors at *B*. From the geometry of the figure, we find that the stiffness factor for *BA*,

$$k_B = \frac{4EI}{l} = \frac{4EI}{6} = \frac{2EI}{3} \quad \text{...}(\because \text{ The beam is fixed at } A)$$

Similarly, $k_B = \frac{4EI}{l} = \frac{4EI}{6} = \frac{2EI}{3}$ ...($\because$ The beam is fixed at *C*)

$\therefore$ Distribution factors for the members *BA* and *BC*

$$= \frac{\frac{2EI}{3}}{\frac{2EI}{3}+\frac{2EI}{3}} \text{ and } \frac{\frac{2EI}{3}}{\frac{2EI}{3}+\frac{2EI}{3}} = \frac{1}{2} \text{ and } \frac{1}{2}$$

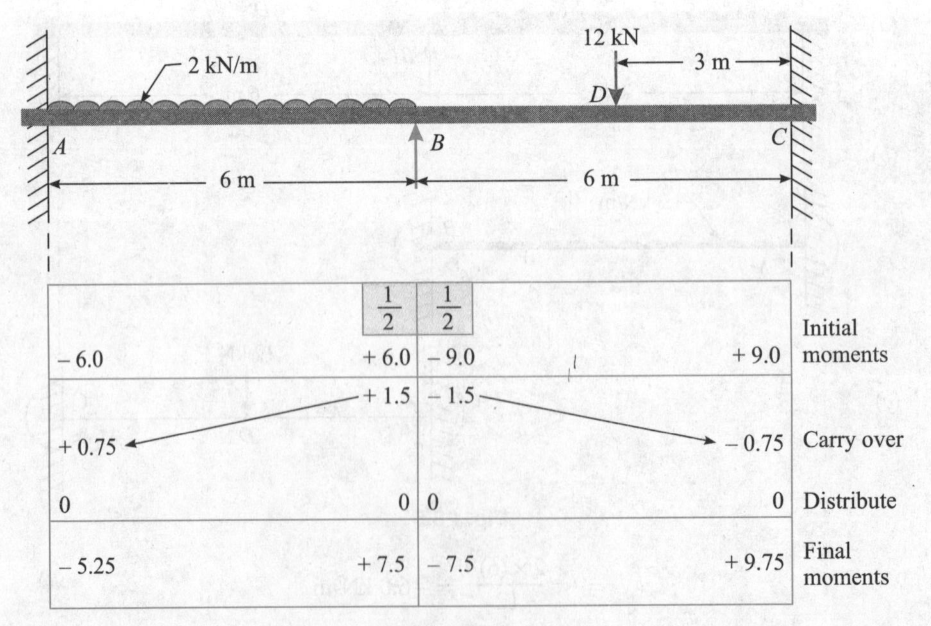


Fig. 26.6

Now draw the beam and fill up the distribution factors and initial moments as shown in Fig. 26.6.

We find that the fixing moment at $B$ in span $AB$, is + 6.0 kN-m and that in span $BC$, is – 9.0 kN-m. Thus there is an unbalanced moment equal to (– 9.0 + 6.0) = – 3.0 kN-m. Now, distribute this unbalanced moment (equal to – 3.0 kN-m) into the spans $AB$ and $BC$ in the ratio of their distribution factors *i.e.*, + 1.5 kN and + 1.5 kN. Now *carry over the effects of these distributed moments at $A$ and $D$ equal to $\frac{1}{2} \times 1.5 = +0.75$ kN. Then distribute the unbalanced moment at $B$ (In this case, there is no carry over moment from $A$ or $D$ at $B$. So the distribution of moment at $B$ is zero). Now, find out the final moments at $A$, $B$ and $C$ in the spans $AB$ and $BC$, by algebraically adding the respective values as shown in the last column of the table.

We know that the bending moment at the centre of the span $AB$, treating it as a simply supported beam

$$= \frac{wl^2}{8} = \frac{2 \times (6)^2}{8} = 9.0 \text{ kN-m}$$

Similarly, bending moment under the load in the span $BC$

$$= \frac{Wl}{4} = \frac{12 \times 6}{4} = 18.0 \text{ kN-m}$$

Now complete the final bending moment **diagram as shown in Fig. 26.7 (*b*).

Let $R_A$ = Reaction at $A$,

$R_B$ = Reaction at $B$ and

$R_C$ = Reaction at $C$.

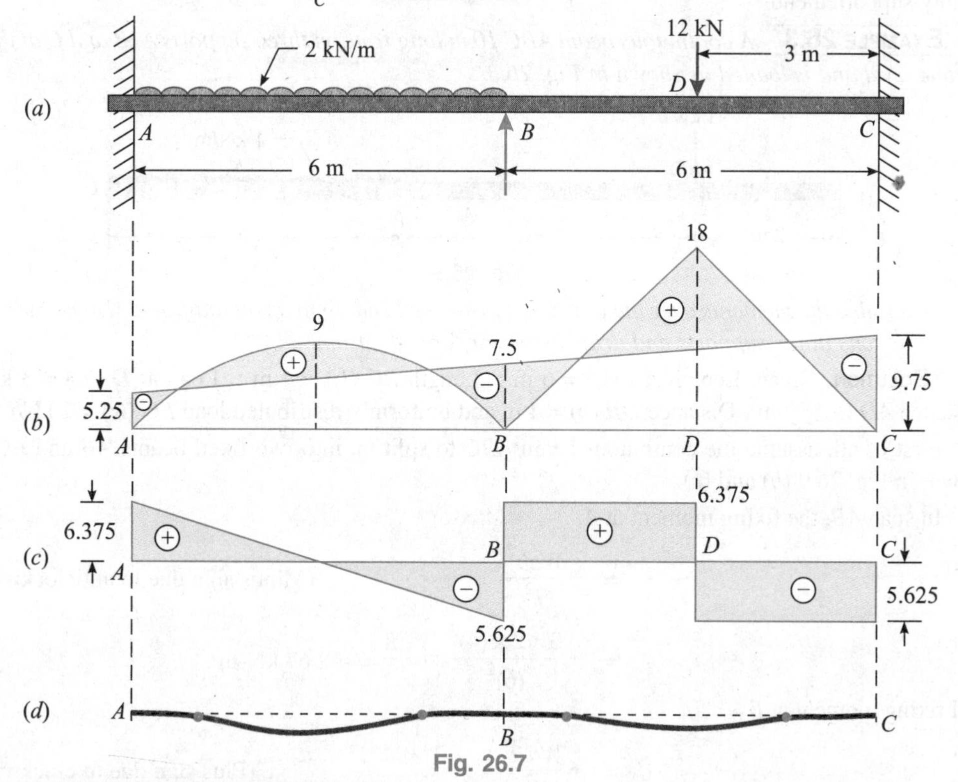


Fig. 26.7

* As per At. 26.4 the carry over factor is $\frac{1}{2}$.

** Though the moment at $B$ in the span $AB$, is *positive* and moment at $C$ in the span $BC$ is *positive* in the table, yet these moments are taken as *negative* in the bending moment diagram. The reason for the same is that these points the moments tend to bend the beam with convexity upwards.

Taking moments at $B$ and equating the same,

$$R_A \times 6 + M_B = M_A + 2 \times 6 \times 3$$

$$R_A \times 6 - 7.5 = -5.25 + 36 = 30.75$$

$$\therefore \quad R_A = \frac{30.75 + 7.5}{6} = \mathbf{6.375\ kN} \quad \textbf{Ans.}$$

Again taking moments about $B$ and equating the same,

$$R_C \times 6 + M_B = M_C + 12 \times 3$$

$$R_C \times 6 - 7.5 = -9.75 + 36 = 26.25$$

$$\therefore \quad R_C = \frac{26.25 + 7.5}{6} = \mathbf{5.625\ kN} \quad \textbf{Ans.}$$

and $$R_B = (2 \times 6 + 12) - (6.375 + 5.625) = \mathbf{12.0\ kN} \quad \textbf{Ans.}$$

Now complete the shear force diagram and elastic curve as shown in Fig. 26.7 (*c*) and (*d*).

## 26.10. Beams with Simply Supported Ends

Sometimes, a continuous beam is simply supported curve on one or both of its ends. We know that the fixing moment on a simply supported end is *zero*. Therefore in such a case, the simply supported ends are released by applying equal and opposite moments and their effects are carried over on the opposite joints. It may also be noted that no moment is carried over from the opposite joint to the simply supported end.

**EXAMPLE 26.3** *A continuous beam ABC 10 m long rests on three supports A, B and C at the same level and is loaded as shown in Fig. 26.8.*

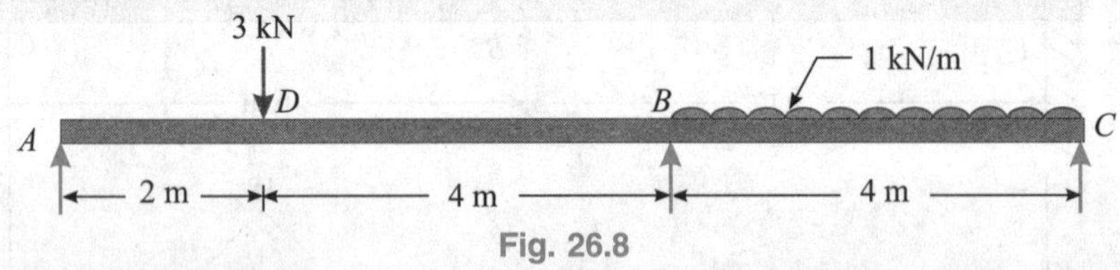


Fig. 26.8

*Determine the moments over the beam and draw the bending moment diagram. Also calculate the reactions at the supports and draw the shear force diagram.*

***SOLUTION.** Given: Length $AB$ $(l_1) = 6$ m ; Length $BC$ $(l_2) = 4$ m ; Load at $D$ $(W) = 3$ kN ; Distance $AD$ $(a) = 2$ m ; Distance $DB$ $(b) = 4$ m and uniformly distributed load $BC$ $(w) = 1$ kN/m.

First of all, assume the continuous beam $ABC$ to split up into two fixed beams $AB$ and $BC$ as shown in Fig. 26.9 (*b*) and (*c*).

In span $AB$, the fixing moment at $A$

$$= -\frac{Wab^2}{l^2} \quad \text{...(Minus sign due to anticlockwise)}$$

$$= -\frac{3 \times 2 \times (4)^2}{(6)^2} = -\frac{8}{3} = -2.67 \text{ kN-m}$$

and fixing moment at $B$

$$= +\frac{Wa^2b}{l^2} \quad \text{...(Plus sign due to clockwise)}$$

$$= +\frac{3 \times (2)^2 \times 4}{(6)^2} = +\frac{4}{3} = +1.33 \text{ kN-m}$$

* We have already solved this question by Theorem of three moments in Example 25.1.

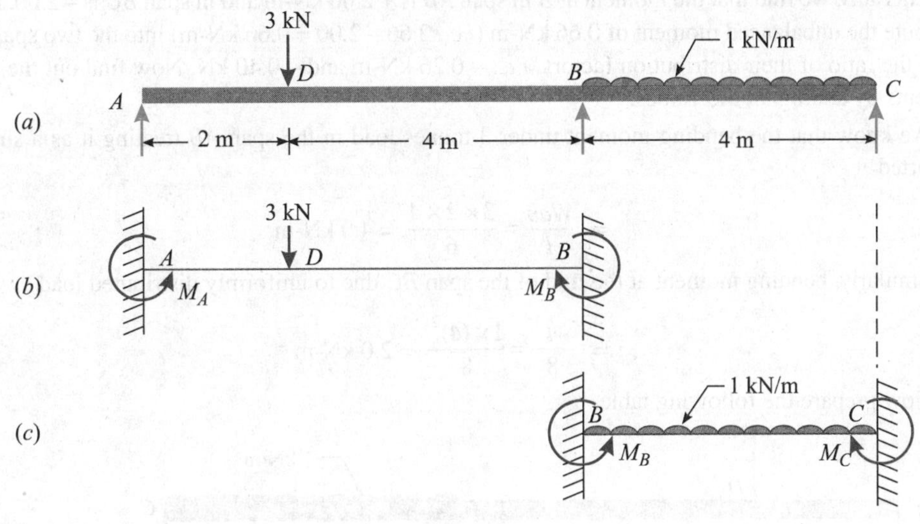


Fig. 26.9

Similarly, in span $BC$, the fixing moment at $B$

$$= -\frac{wl^2}{12} \qquad \text{...(Minus sign due to anticlockwise)}$$

$$= -\frac{1 \times (4)^2}{12} = -\frac{4}{3} = -1.33 \text{ kN-m}$$

and fixing moment at $C$

$$= +\frac{wl^2}{12} \qquad \text{...(Plus sign due to clockwise)}$$

$$= +\frac{1 \times (4)^2}{12} = +\frac{4}{3} = +1.33 \text{ kN-m}$$

Now let us find out the distribution factors at $B$. From the geometry of the figure, we find that the stiffness factor for $BA$,

$$k_{BA} = \frac{3EI}{l} = \frac{3EI}{6} = \frac{EI}{2} \qquad \text{...}(\because \text{ The beam is hinged at } A)$$

Similarly, stiffness factor for $BC$,

$$k_{BC} = \frac{3EI}{l} = \frac{3EI}{4} = \frac{3EI}{4} \qquad \text{...}(\because \text{ The beam is hinged at } C)$$

Distribution factors for $BA$ and $BC$

$$= \frac{\frac{EI}{2}}{\frac{EI}{2} + \frac{3EI}{4}} \text{ and } \frac{\frac{3EI}{4}}{\frac{EI}{2} + \frac{3EI}{4}} = \frac{2}{5} \text{ and } \frac{3}{5}$$

Now draw the beam and fill up the distribution factors and fixing moments as shown in Fig. 26.10 (*a*). Now release the ends $A$ and $C$ (because of simply supported ends) by applying equal and opposite moments. Now *carry over the moments from $A$ to $B$ and $C$ to $B$. After adding the carry over

* As per article 26.5, the carry over factor is $\frac{1}{2}$

moments at $B$, we find that the moment at $B$ in span $AB$ is + 2.66 kN-m and in span $BC$ is – 2.00. Now distribute the unbalanced moment of 0.66 kN-m (*i.e.*, 2.66 – 2.00 = 0.66 kN-m) into the two spans of $AC$ in the ratio of their distribution factors, *i.e.*, – 0.26 kN-m and – 0.40 kN. Now find out the final moments by adding all the values.

We know that the bending moment under 3 tonnes load in the span $AB$ treating it as a simply supported

$$= \frac{Wab}{l} = \frac{3 \times 2 \times 4}{6} = 4.0 \text{ kN-m}$$

Similarly, bending moment at the mid of the span $BC$ due to uniformly distributed load

$$= \frac{wl^2}{8} = \frac{1 \times (4)^2}{8} = 2.0 \text{ kN-m}$$

Now prepare the following table.

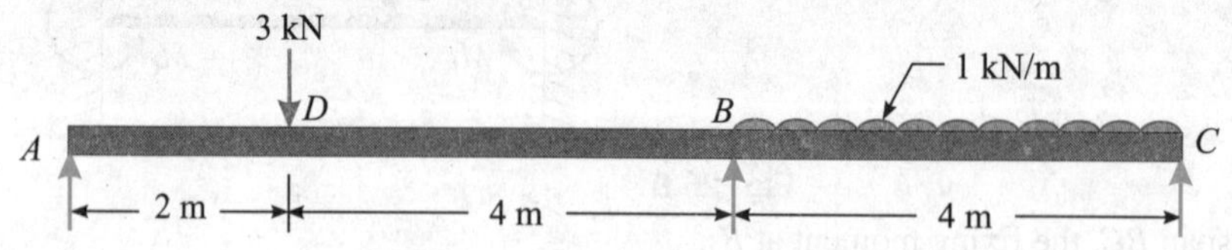


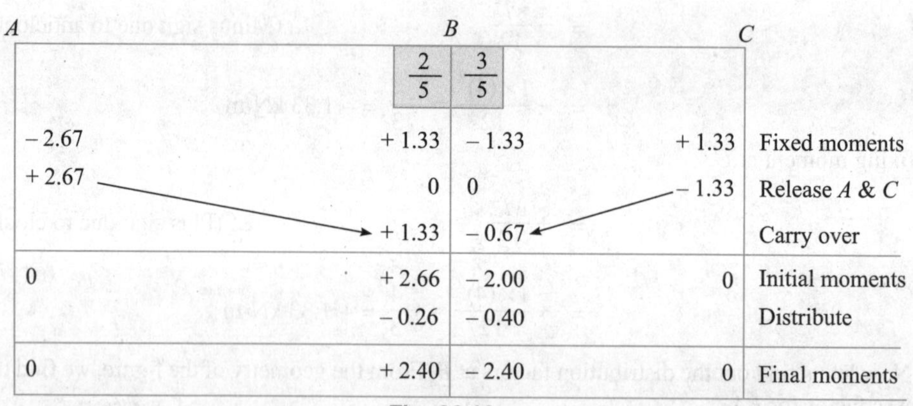

| A | B | | C | |
|---|---|---|---|---|
| | 2/5 | 3/5 | | |
| – 2.67 | + 1.33 | – 1.33 | + 1.33 | Fixed moments |
| + 2.67 | 0 | 0 | – 1.33 | Release $A$ & $C$ |
| | + 1.33 | – 0.67 | | Carry over |
| 0 | + 2.66 | – 2.00 | 0 | Initial moments |
| | – 0.26 | – 0.40 | | Distribute |
| 0 | + 2.40 | – 2.40 | 0 | Final moments |

Fig. 26.10

Now complete the final bending moment diagram as shown in Fig. 26.11(*b*).

***Shear force diagram***

Let
$R_A$ = Reaction at $A$,
$R_B$ = Reaction at $B$ and
$R_C$ = Reaction at $C$.

Taking moments about $B$,

$$R_A \times 6 - 3 \times 4 = -2.4 \qquad ...(\because M_B = 2.4 \text{ kN-m})$$

$$\therefore \quad R_A = \frac{-2.4 + 12.0}{6} = \frac{9.6}{6} = \mathbf{1.6 \text{ kN}} \quad \textbf{Ans.}$$

Similarly,

$$R_C \times 4 - 4 \times 2 = -2.4 \qquad ...(\because M_B = -2.4 \text{ kN-m})$$

$$\therefore \quad R_C = \frac{-2.4 + 8.0}{4} = \frac{5.6}{4} = \mathbf{1.4 \text{ kN}} \quad \textbf{Ans.}$$

and $R_B = (3 + 4) - (1.6 + 1.4) = \mathbf{4.0\ kN}$ **Ans.**

Now draw shear force diagram as shown in Fig. 26.11 (*c*).

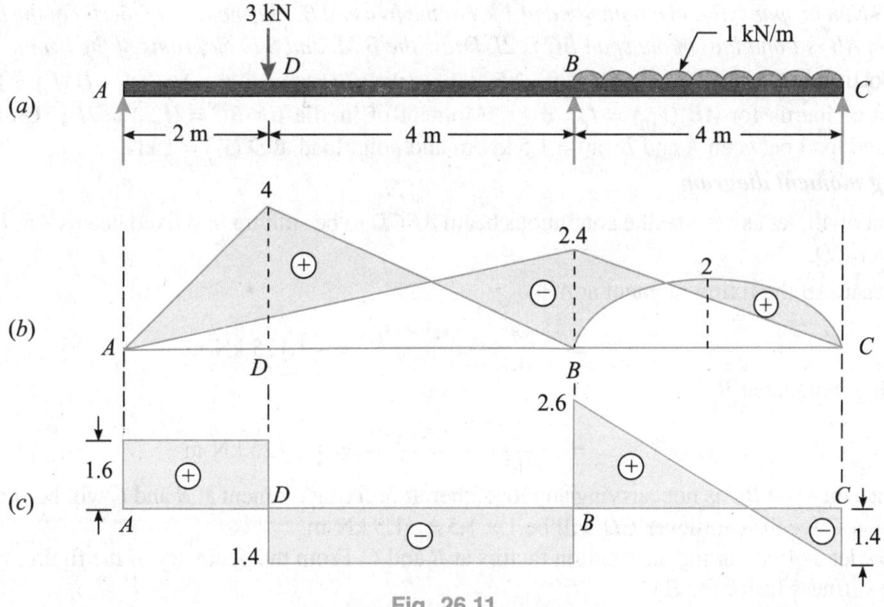


Fig. 26.11

## EXERCISE 26.1

1. A continuous beam *ABC* is fixed at *A* and is simply supported at *B* and *C*. The span *AB* is 6 m and carries a uniformly distributed load of 1 kN/m. The span *AC* is 4 m and carries a uniformly distributed load of 3 kN/m. Determine the fixed end moments.
[**Ans.** ( = 2.143 kN-m ; $M_B$ = 4.714 kN-m ; $M_C$ = 0)]
2. A simply supported beam *ABC* is continuous over two spans *AB* and *BC* of 6 m and 5 m respectively. The span *AB* is carrying a uniformly distributed load of 2 kN-m and the span *BC* is carrying a point load of 5 kN at a distance of 2 m from *B*. Find the support moment at *B*.
[**Ans.** 7.1 kN-m]
3. A continuous beam *ABCD* is simply supported over three spans, such that *AB* = 8 m *BC* = 12 m and *CD* = 5 m. It carries uniformly distributed load of 4 kN/m in span *AB*, 3 kN/m in span *BC* and 6 kN/m in span *CD*. Find the moments over the supports *B* and *C*.
[**Ans.** 35.9 kN-m ; 31.0 kN-m]
4. A continuous beam *ABCD* is simply supported over three spans of 6 m, 5 m and 4 m respectively. The beam carries point loads of 90 kN and 80 kN at 2 m and 8 m from the support *A* and a uniformly distributed load of 30 kN-m over the span *CD*. Find the moments and reactions at the supports. [**Ans.** 68.4 kN-m ; 44.8 kN ; 48.6 kN ; 94.1 kN ; 98.5 kN]

## 26.11. Beams with End Span Overhanging

Sometimes, a beam is overhanging at its one or both the end supports. In such a case, the bending moment at the support near the overhanging end will be due to the load over the overhanging portion and will remain constant, irrespective of the moments on the other supports. It is thus obvious that the distribution factors over the support having one span overhanging will be 1 and 0. Moreover, this support is considered as a simply supported for the purpose of calculating distribution factors in the span, adjoining the overhanging span.

**EXAMPLE 26.4.** *A beam ABCD 9 m long is simply supported at A, B and C such that the span AB is 3 m, span BC is 4.5 m and the overhanging CD is 1.5 m. It carries a uniformly distributed load of 1.5 kN/m in span AB and a point load of 1 kN at the free end B. The moment of inertia of the beam in span AB is I and that in the span BC is 2I. Draw the B.M. and S.F. diagrams of the beam.*

*SOLUTION. Given : Length $AB$ ($l_1$) = 3 m ; Length $BC$ ($l_2$) = 4.5 m ; Length $CD$ ($l_3$) = 1.5 m ; Moment of inertia for $AB$ ($I_{AB}$) = $I_{CD}$ = $I$ ; Moment of inertia for $BC$ = ($I_{BC}$) = $2I$ ; Uniformly distributed load between $A$ and $B$ ($w$) = 1.5 kN/m and point load at $D$ ($W$) = 1 kN.

***Bending moment diagram***

First of all, let us assume the continuous beam $ABCD$ to be split up into fixed beams $AB$, $BC$ and cantilever $CD$.

In span $AB$ the fixing moment at $AP$

$$= -\frac{wl^2}{12} = -\frac{1.5 \times (3)^2}{12} = -1.125 \text{ kN-m}$$

and fixing moment at $B$

$$= +\frac{wl^2}{12} = +\frac{1.5 \times (3)^2}{12} = +1.125 \text{ kN-m}$$

Since the span $BC$ is not carrying any load, therefore fixing moment at $B$ and $C$ will be zero. The moment at $C$ for the cantilever $CD$ will be 1 × 1.5 = –1.5 kN-m.

Now let us find out the distribution factors at $B$ and $C$. From the geometry of the figure, we find that the stiffness factor for $BA$,

$$k_{BA} = \frac{3EI}{l} = \frac{3EI}{3} = 1EI \qquad \text{...(∵ The end } A \text{ is simply supported at } A)$$

Similarly,

$$k_{BC} = \frac{3EI}{l} = \frac{3E \times 2I}{4.5} = \frac{6EI}{4.5} = \frac{4EI}{3} \qquad \text{...(∵ Beam is over handing beyond } C)$$

Distribution factor for $BA$ and $BC$

$$= \frac{2EI}{1EI + \frac{4EI}{3}} \text{ and } \frac{\frac{4EI}{3}}{1EI + \frac{4EI}{3}} = \frac{3}{7} \text{ and } \frac{4}{7}$$

We know that distribution factors for $CA$ and $CD$ will be 1 and 0, because the beam is overhanging at $C$.

Now prepare the following table.

| A | B | | C | | D |
|---|---|---|---|---|---|
| | $\frac{3}{7}$ | $\frac{4}{7}$ | 1 | 0 | |
| – 1.125 | + 1.125 | 0 | 0 | – 1.5 | Fixed end moments |
| + 1.125 | 0 | 0 | + 1.5 | 0 | Release $A$ and balance $C$ |
| | + 0.562 | + 0.75 | 0 | 0 | Carry over |
| 0 | + 1.687 | + 0.75 | + 1.5 | – 1.5 | Initial moments |
| | – 1.037 | – 1.40 | – | – | Distribute |
| 0 | + 0.65 | – 0.65 | + 1.5 | – 1.5 | Final moments |

* We have already solved, this question by Theorem of three momens in Example 25.5

We know that the bending moment in the mid of span $AB$ due to uniformly distributed load, by considering it as a simply supported beam

$$= \frac{wl^2}{8} = \frac{1.5 \times (3)^2}{8} = 1.69 \text{ kN-m}$$

Now complete the final bending moment diagram as shown in Fig. 26.12($b$).

***Shear force diagram***

Let
$R_A$ = Reaction at $A$,
$R_B$ = Reaction at $B$ and
$R_C$ = Reaction at $C$.

Taking moments at $B$

$$R_A \times 3 - (1.5 \times 3 \times 1.5) = -0.65 \quad ...(\because M_B = -0.65 \text{ kN-m})$$

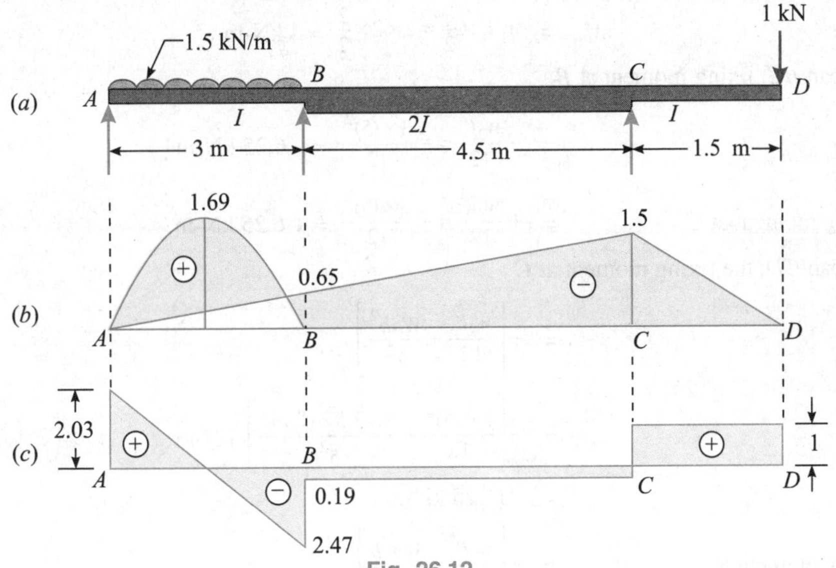


Fig. 26.12

or
$$R_A = \frac{-0.65 + 6.75}{3} = 2.03 \text{ kN}$$

Again taking moments about $B$,

$$R_C \times 4.5 - (1 \times 6) = -0.65 \quad ...(\because M_B = -0.65 \text{ kN-m})$$

or
$$R_C = \frac{-0.65 + 6}{4.5} = 1.19 \text{ kN}$$

$\therefore$
$$R_B = (3 \times 1.5 + 1) - (2.03 + 1.19) = 2.28 \text{ kN}$$

Now complete the shear force diagram as shown in Fig. 26.12 ($c$).

**EXAMPLE 26.5.** *A continuous beam ABCDE, with uniform flexural rigidity throughout has roller supports at B, C and D, a built-in support E and an overhang AB as shown in Fig. 26.13.*

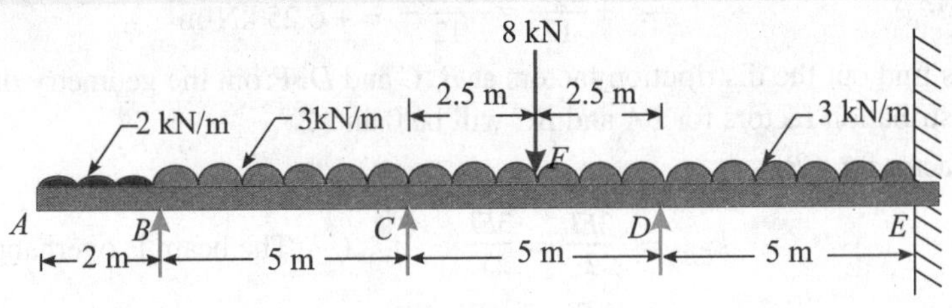


Fig. 26.13

*It carries a uniformly distributed load of intensity of 2 kN/m on AB and another of intensity of 3 kN/m over BCDE. In addition to it, a point load of 8 kN is placed midway between C and D. The span lengths are AB = 2 m, BC = CD = DE = 5 m. Obtain the support moments by the moment distribution method and sketch the B.M. diagram giving values at salient points.*

**SOLUTION.** Given: Length *AB* $(l_1) = 2$ m ; Length *BC* $(l_2) =$ Length *CD* $(l_3) =$ Length *DE* $(l_4) = 5$ m ; Uniformly distributed load on *AB* $(w_1) = 2$ kN/m ; Uniformly distributed load in *BC, CD, DE* $(w_2) = 3$ kN/m and point load at *F* $(W) = 8$ kN.

*Support moments*

First of all, let us assume the continuous beam *ABCDE* to be split up into cantilever *AB* and fixed beams *BC, CD* and *DE*.

We know that the bending moment at *B* for the cantilever *AB*,

$$M_B = w_1 l_1 \times \frac{l_1}{2} = 2 \times 2 \times \frac{2}{2} = 4 \text{ kN-m}$$

In span *BC*, fixing moment at *B*

$$= -\frac{w_2 l^2}{12} = -\frac{3 \times (5)^2}{12} = -6.25 \text{ kN-m}$$

and fixing moment at *C*

$$= +\frac{w_2 l^2}{12} = +\frac{3 \times (5)^2}{12} = +6.25 \text{ kN-m}$$

In span *CD*, the fixing moment at *C*

$$= -\left[\frac{w_2 l^2}{12} + \frac{Wab^2}{l^2}\right]$$

$$= -\left[\frac{3 \times (5)^2}{12} + \frac{8 \times 2.5 \times (2.5)^2}{5^2}\right] \text{ kN-m}$$

$$= -11.25 \text{ kN-m}$$

and fixing moment at *D*

$$= +\left[\frac{w_2 l^2}{12} + \frac{Wa^2 b}{l^2}\right]$$

$$= +\left[\frac{3 \times (5)^2}{12} + \frac{8 \times 2.5 \times (2.5)^3}{5^2}\right] \text{ kN-m}$$

$$= +11.25 \text{ kN-m}$$

In span *DE*, the fixing moment at *D*

$$= -\frac{w_2 l^2}{12} = -\frac{3 \times (5)^2}{12} = -6.25 \text{ kN-m}$$

and fixing moment at *E*

$$= +\frac{w_2 l^2}{12} = \frac{3 \times (5)^2}{12} = +6.25 \text{ kN-m}$$

Now let us find out the distribution factors at *B, C* and *D*. From the geometry of the figure, we find that the distribution factors for *BA* and *BC* will be 0 and 1.

Stiffness factor for *CB*,

$$k_{CB} = \frac{3EI}{l} = \frac{3EI}{5} \quad \text{...(∵ The beam is overhanging beyond } C)$$

and

$$k_{CD} = \frac{4EI}{l} = \frac{4EI}{5} \quad \text{...(∵ The beam is continuous at } D)$$

Distribution factors for *CB* and *CD*

$$= \frac{\frac{3EI}{5}}{\frac{3EI}{5}+\frac{4EI}{5}} \quad \text{and} \quad \frac{\frac{4EI}{5}}{\frac{3EI}{5}+\frac{4EI}{5}} = \frac{3}{7} \quad \text{and} \quad \frac{4}{7}$$

Similarly, stiffness factor for *DC*,

$$k_{CD} = \frac{4EI}{l} = \frac{4EI}{5} \qquad \text{...}(\because \text{ The beam is continuous at } C)$$

and

$$k_{DE} = \frac{4EI}{l} = \frac{4EI}{5} \qquad \text{...}(\because \text{ The beam is fixed at } E)$$

Distribution factors for *DC* and *DE*

$$= \frac{\frac{4EI}{5}}{\frac{4EI}{5}+\frac{4EI}{5}} \quad \text{and} \quad \frac{\frac{4EI}{5}}{\frac{4EI}{5}+\frac{4EI}{5}} = \frac{1}{2} \quad \text{and} \quad \frac{1}{2}$$

Now prepare the following table:

| *A* | *B* | *C* | | *D* | | *E* | |
|---|---|---|---|---|---|---|---|
| 0 | 1 | $\frac{3}{7}$ | $\frac{4}{7}$ | $\frac{1}{2}$ | $\frac{1}{2}$ | | |
| + 4.00 | – 6.25 | + 6.25 | – 11.25 | + 11.25 | – 6.25 | + 6.25 | Fixed end moments |
| | + 2.25 | | | | | | Balance *B* |
| | | + 1.13 | | | | | Cary over |
| + 4.00 | – 4.00 | + 7.38 | – 11.25 | + 11.25 | – 6.25 | + 6.25 | Initial moments |
| | | + 1.66 | + 2.21 | – 2.50 | – 2.50 | | Distribute |
| | | | – 1.25 | + 1.10 | | – 1.25 | Carry over |
| | | + 0.54 | + 0.71 | – 0.55 | – 0.55 | | Distribute |
| | | | – 0.28 | + 0.36 | | – 0.28 | Cary over |
| | | + 0.12 | + 0.16 | – 0.18 | – 0.18 | | Distribute |
| | | | – 0.09 | + 0.08 | | – 0.09 | Carry over |
| | | + 0.04 | + 0.05 | – 0.04 | – 0.04 | | Distribute |
| + 4.00 | – 4.00 | + 9.74 | – 9.74 | + 9.52 | – 9.52 | + 4.63 | Final moments |

We know that the bending moment in the middle of span *BC*, by considering it as a simply supported beam

$$= \frac{w_2 l^2}{8} = \frac{3 \times (5)^2}{8} = 9.38 \text{ kN-m}$$

Similarly, bending in the middle of span *CD*, by considering it as a simply supported beam

$$= \frac{w_2 l^2}{8} + \frac{Wab}{1} = \frac{3 \times (5)^2}{8} + \frac{8 \times 2.5 \times 2.5}{5} = 19.38 \text{ kN-m}$$

and bending moment in the middle of span *DE*, by considering it as a simply supported beam

$$= \frac{w_2 l^2}{8} = \frac{3 \times (5)^2}{8} = 9.38 \text{ kN-m}$$

Now complete the final bending moment diagram as shown in Fig. 26.14 (*b*).

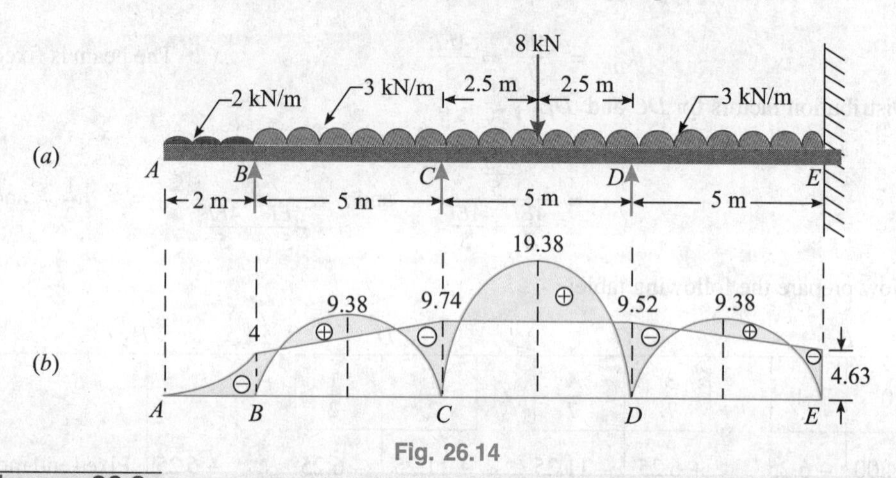


Fig. 26.14

**EXAMPLE 26.6.** *A beam ABCDE has a built-in support A and roller supports at B, C and D, DE being an overhung. AB = 7 m, BC = 3 m, CD = 4m and DE = 1.5 m. The values of I, the moment of inertia of the section, over each of these lengths are 3I, 2I, I and I respectively. The beam carries a point load of 10 kN at a point 3 m from A, a uniformly distributed load of 4.5 kN/m over whole of BC and a concentrated load of 9 kN in CD 1.5 m from C and another point load of 3 kN at E the top of overhang as shown in Fig. 26.15.*

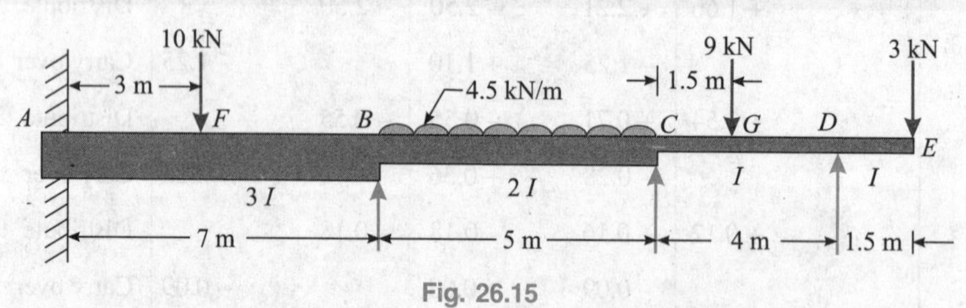


Fig. 26.15

*Determine (i) moments developed over each of the supports A, B, C and D and (ii) draw bending moment diagram for the entire beam, stating values at salient points.*

***SOLUTION.** Given : Length *AB* ($l_1$) = 7 m ; Length *BC* ($l_2$) = 5 m ; Length *CD* ($l_3$) = 4 m ; Length *DE* ($l_4$) = 1.5 m ; Moment of inertia for *AB* ($I_{AB}$) = 3*I* ; Moment of inertia for *BC* ($I_{BC}$) = 2*I* ; Moment of inertia for *CD* ($I_{CD}$) = $I_{DE}$ = *I* ; Point load at *F* ($W_1$) = 10 kN ; Uniformly distributed load on *BC* ($w_1$) = 4.5 kN/m ; Point load at *G* ($W_2$) = 9 kN and point load at *E* = 3 kN.

***Moments developed over each of the support***

First of all, let us assume the continuous beam *ABCDE* to be split up into fixed beams *AB*, *BC*, *CD* and cantilever *DE*.

* This question is also solved in Example 25.6.

In span $AB$, fixing moment at $A$

$$= -\frac{W_1 ab^2}{l^2} = -\frac{10 \times 3 \times (4)^2}{(7^2)} = -9.80 \text{ kN-m}$$

and fixing moment at $B$

$$= +\frac{W_1 a^2 b}{l^2} = \frac{10 \times (3)^2 \times 4}{(7)^2} = 7.35 \text{ kN-m}$$

In span $BC$, fixing moment at $A$

$$= -\frac{wl^2}{12} = -\frac{4.5 \times (5)^2}{12} = -9.38 \text{ kN-m}$$

and fixing moment at $C$

$$= +\frac{wl^2}{12} = -\frac{4.5 \times (5)^2}{12} = 9.38 \text{ kN-m}$$

In span $CD$, fixing moment at $C$

$$= -\frac{W_2 ab^2}{l^2} = -\frac{9 \times 1.5 \times (2.5)^2}{(4)^2} = -5.27 \text{ kN-m}$$

and fixing moment at $D$

$$= +\frac{W_2 a^2 b}{l^2} = -\frac{9 \times (1.5)^2 \times 2.5}{(4)^2} = 3.16 \text{ kN-m}$$

and the moment at $D$, for the cantilever at $DE$

$$= 3 \times 1.5 = -4.5 \text{ kN-m}$$

Now let us find out the distribution factors at $B$, $C$ and $D$. From the geometry of the figure, we find that the stiffness factor for $BA$,

$$k_{BA} = \frac{4EI}{l} = \frac{4E \times 3I}{7} = \frac{12EI}{7} \quad \text{...}(\because \text{ The beam is fixed at } A)$$

and

$$k_{BC} = \frac{4EI}{l} = \frac{4E \times 2I}{5} = \frac{8EI}{5} \quad \text{...}(\because \text{ The beam is continuous at } C)$$

Distribution factors for $BA$ and $BC$

$$= \frac{\frac{12EI}{7}}{\frac{12EI}{7} + \frac{8EI}{5}} \text{ and } \frac{\frac{8EI}{5}}{\frac{12EI}{7} + \frac{8EI}{5}} = \frac{15}{29} \text{ and } \frac{14}{29}$$

Similarly, stiffness factor for $CB$,

$$k_{CB} = \frac{4EI}{l} = \frac{4E \times 2I}{5} = \frac{8EI}{5} \quad \text{...}(\because \text{ The beam is continuous at } B)$$

and

$$k_{CD} = \frac{3EI}{l} = \frac{3EI}{4} \quad \text{...}(\because \text{ The beam is overhanging beyond } D)$$

Distribution factors for $CB$ and $CD$

$$= \frac{\frac{8EI}{5}}{\frac{8EI}{5} + \frac{3EI}{4}} \text{ and } \frac{\frac{3EI}{4}}{\frac{8EI}{5} + \frac{3EI}{4}} = \frac{32}{47} \text{ and } \frac{15}{47}$$

Distribution factors for $DC$ and $DE$ will be 1 and 0, because the beam is overhanging at $D$.

Now prepare the following table:

| A | B | | C | | D | | E |
|---|---|---|---|---|---|---|---|
| | $\frac{15}{29}$ | $\frac{14}{29}$ | $\frac{32}{47}$ | $\frac{15}{47}$ | 1 | 0 | |
| – 9.80 | + 7.35 | – 9.38 | + 9.38 | – 5.27 | + 3.16 | – 4.5 | Fixed end moments |
| | | | | | + 1.34 | | Balance *D* |
| | | | | + 0.67 | | | Carry over |
| – 9.80 | + 7.35 | – 9.38 | + 9.38 | – 4.60 | + 4.5 | – 4.5 | Initial moments |
| | + 1.05 | + 0.98 | – 3.25 | – 1.53 | | | Distribute |
| + 0.53 | | – 1.63 | + 0.49 | | 0 | | Carry over |
| | + 0.84 | + 0.79 | – 0.33 | – 0.16 | | | Distribute |
| + 0.42 | | – 0.17 | + 0.40 | | 0 | | Carry over |
| | + 0.09 | + 0.08 | – 0.27 | – 0.13 | | | Distribute |
| + 0.05 | | – 0.13 | + 0.04 | | 0 | | Carry over |
| | 0.07 | + 0.06 | – 0.03 | – 0.01 | | | Distribute |
| + 0.04 | | – 0.02 | + 0.03 | | 0 | | Carry over |
| | 0.01 | 0.01 | – 0.02 | – 0.01 | | | Distribute |
| – 8.76 | 9.41 | – 9.41 | + 6.44 | – 6.44 | + 4.5 | – 4.5 | Final moments |

The moments developed at each of the support are given in the above table. **Ans.**

***Bending moment diagram***

We know that the bending moment under the 10 kN load in span *AB*, treating it as a simply supported

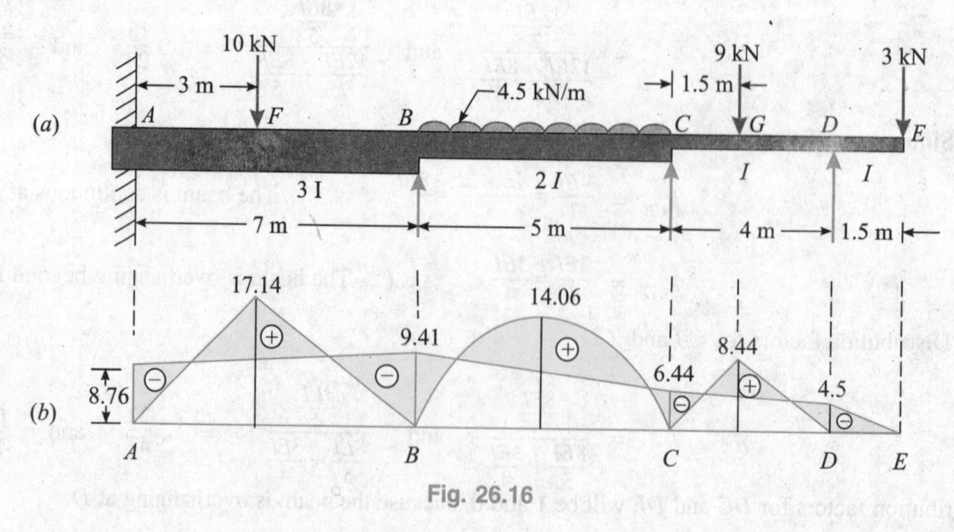


Fig. 26.16

$$= \frac{W_1 ab}{l} = \frac{10 \times 3 \times 4}{7} = 17.14 \text{ kN-m}$$

Similarly, bending moment at the mid of the span *BC*

$$= \frac{wl^2}{8} = \frac{4.5 \times (5)^2}{8} = 14.06 \text{ kN-m}$$

and bending moment under the 9 kN load in span *CD*

$$= \frac{W_2 ab}{l} = \frac{9 \times 1.5 \times 2.5}{4} = 8.44 \text{ kN-m}$$

Now complete the final bending moment diagram as shown in Fig. 26.16 (*b*).

## 26.12. Beams with a Sinking Support

Sometimes, one of the supports of a continuous beam sinks down with respect to others, as a result of loading. As a result of sinking, some moments are caused on the two supports, in addition to the moments due to loading.

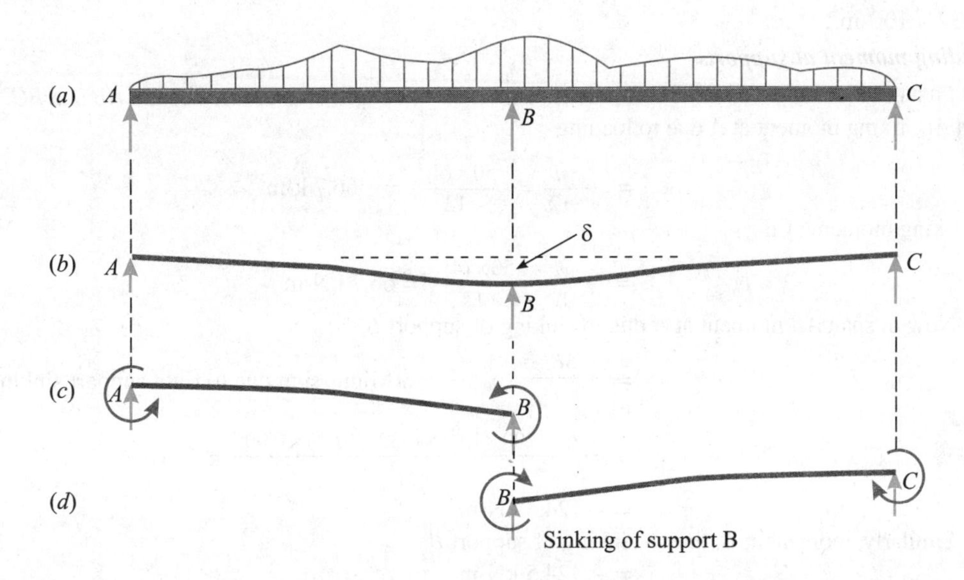


Fig. 26.17

Consider a beam *ABC* simply supported at *A*, *B* and *C* subjected to any loading. As a result of this loading, let the support *B* sink down by an amount equal to δ as shown in Fig. 26.17 (*b*). Now assume the beam *ABC* to be split up into two beams *AB* and *BC* as shown in Fig. 26.17 (*c*) and (*d*).

Now from the geometry of the beam *AB*, we find that the moment caused at *A* and *B* due to sinking of the support *B*

$$= -\frac{6EI\,\delta}{l^2} \quad \text{...(Minus sign due to right support sinking down)}$$

Similarly, moment caused at *B* and *C* of the beam *BC* due to sinking of the support *B*

$$= +\frac{6EI\,\delta}{l^2} \quad \text{...(Plus sign due to left support sinking down)}$$

**EXAMPLE 26.7.** *A continuous beam ABC shown in Fig. 26.18 carries a uniformly distributed load of 50 kN/m on AB and BC. The support B sinks by 5 mm below A and C and the value of EI is constant throughout the beam.*

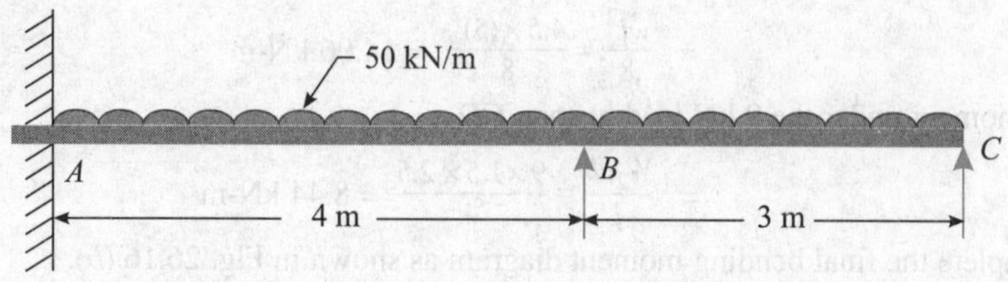


Fig. 26.18

*Taking E = 200 GPa and I = 332 × 10⁶ mm⁴, find the bending moment at supports A and B and draw the bending moment diagram.*

***Solution.** Given : Length $AB$ ($l_1$) = 4 m ; Length $BC$ ($l_2$) = 3 m ; Uniformly distributed load ($w$) = 50 kN/m ; Sinking of support $B$ ($\delta_B$) = –5 mm = – 0.005 m or $\delta_A = \delta_C$ = + 0.005 m ; Modulus of elasticity ($E$) = 200 GPa = 200 × $10^6$ kN/m$^2$ and moment of inertia ($I$) = 332 × $10^6$ mm$^4$ = 332 × $10^{-6}$ m$^4$.

***Bending moment at supports***

First of all, let us assume the continuous beam $ABC$ to be split up into fixed beams $AB$ and $BC$. In span $AB$, fixing moment at $A$ due to loading

$$= -\frac{wl^2}{12} = -\frac{50 \times (4)^2}{12} = -66.7 \text{ k-m}$$

and fixing moment at $B$

$$= +\frac{wl^2}{12} = \frac{50 \times (4)^2}{12} = 66.7 \text{ kN-m}$$

Now in span $AB$ moment at $A$ due to sinking of support $B$

$$= -\frac{6EI\,\delta}{l^2} \quad \text{...(Minus sign due to right support sinking)}$$

$$= -\frac{6 \times (200 \times 10^6) \times (332 \times 10^{-6}) \times 0.005}{(4)^2} \text{ kN-m}$$

$$= -124.5 \text{ kN-m}$$

Similarly, moment at $B$, due to sinking of support $B$

$$= -124.5 \text{ kN-m}$$

∴ Total moment at $A$ in span $AB$

$$= -66.7 - 124.5 = -191.2 \text{ kN-m}$$

and total moment at $B$ in span $AB$

$$= +66.7 - 124.5 = -57.8 \text{ kN-m}$$

In span $BC$, fixing moment at $B$ due to loading

$$= -\frac{wl^2}{12} = -\frac{50 \times (3)^2}{12} = -37.5 \text{ kN-m}$$

and fixing moment at $C$

$$= +\frac{wl^2}{12} = \frac{50 \times (3)^2}{12} = +37.5 \text{ kN-m}$$

Now in span $BC$, moment at $B$ due to sinking of support $B$

$$= +\frac{6EI\,\delta}{l^2} \quad \text{...(Plus sign due to left support sinking)}$$

* We have already solved this question by Theorem of there moments in Example 25.7.

$$= +\frac{6\times(200\times10^6)\times(332\times10^{-6})\times0.005}{(3)^2} \text{ kN-m}$$

$$= 221.3 \text{ kN-m}$$

Similarly, moment at $C$, due to sinking of support $B$

$$= +\,221.3 \text{ kN-m}$$

∴ Total moment at $B$, in span $BC$

$$= -\,37.5 - 221.3 = 183.8 \text{ kN-m}$$

and total moment at $C$, in span $BC$

$$= +\,37.5 - 221.3 = 258.8 \text{ kN-m}$$

Now, let us find out the distribution factors at $B$. From the geometry of the figure, we find that the stiffness factor for $BA$,

$$k_{BA} = \frac{4EI}{l} = \frac{4EI}{4} = 1\,EI \qquad ...(\because \text{ The beam is fixed at } A)$$

and

$$k_{BC} = \frac{3EI}{l} = \frac{3EI}{3} \qquad ...(\because \text{ The beam is simply suppoted at } C)$$

Distribution factors for $BA$ and $BC$ will be $\frac{1}{2}$ and $\frac{1}{2}$.

Now prepare the following table:

| A | B | B | C | |
|---|---|---|---|---|
| | $\frac{1}{2}$ | $\frac{1}{2}$ | | |
| – 191.2 | – 57.8 | + 183.8 | + 258.8 | Fixed end moment |
| | | | – 258.8 | Release $C$ |
| | | – 129.4 | | Carry over |
| – 191.2 | – 57.8 | + 54.4 | 0 | Initial moments |
| | + 1.7 | + 1.7 | | Distribute |
| + 0.9 | | | | Carry over |
| 0 | 0 | 0 | 0 | Distribute |
| – 190.3 | – 56.1 | + 56.1 | 0 | Final moments |

The bending moments at the supports are given in the above table. **Ans.**

***Bending moment diagram***

We know that the bending moment at the mid of the span $AB$, treating it as a simply supported beam

$$= \frac{wl^2}{8} = \frac{50\times(4)^2}{8} = 100 \text{ kN-m}$$

Similarly, bending moment at the mid of the span $BC$

$$= \frac{wl^2}{8} = \frac{50\times(3)^2}{8} = 56.25 \text{ kN-m}$$

Now complete the final bending moment diagram as shown in Fig. 26.19 (*b*).

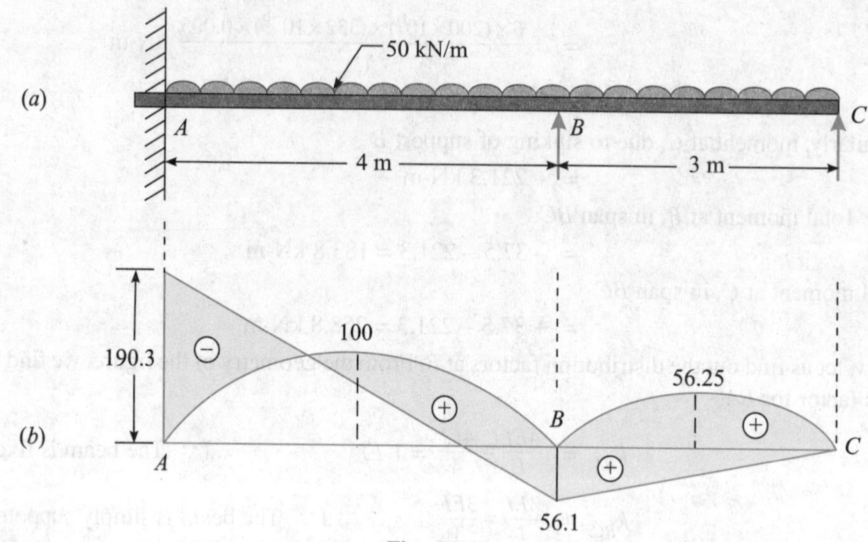


Fig. 26.19

**EXAMPLE 26.8.** *A continuous beam is built-in at A and is carried over rollers at B and C as shown in Fig. 26.20. AB = BC = 12 m.*

*It carries a uniformly distributed load of 3 kN/m over AB and a point load of 24 kN over BC, 4 m from the support B, which sinks 30 mm. Values of E and I are 200 GPa and 0.2 × $10^9$ $m^4$ respectively uniform throughout.*

*Calculate the support moments and draw bending moment diagram and shear force diagram, giving critical values. Also draw the deflected shape of the centre line of the beam.*

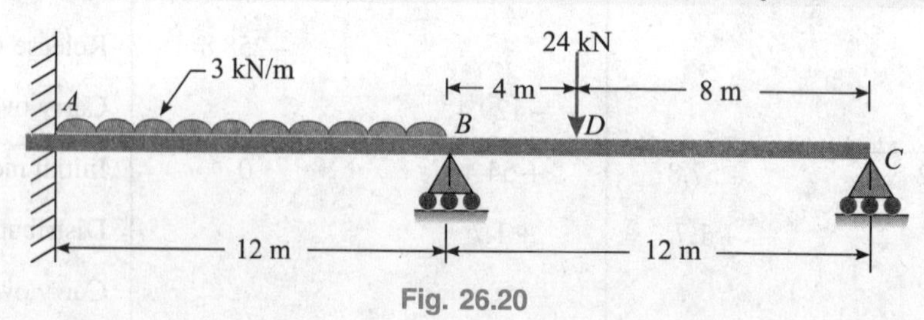


Fig. 26.20

***SOLUTION.** Given : Length *AB* ($l_1$) = 12 m ; Length *BC* ($l_2$) = 12 m ; Uniformly distributed load on *AB* (*w*) = 3 kN/m ; Point load at *D* (*W*) = 24 kN ; Sinking of support *B* ($\delta_B$) = – 30 mm = – 0.03 m or $\delta_A = (\delta_C) = + 0.03$ m ; Modulus of elasticity (*E*) = 200 GPa = 200 × $10^6$ kN-$m^2$ and moment of inertia (*I*) = 0.2 × $10^9$ $mm^4$ = 0.2 × $10^{-3}$ $m^4$.

***Support moments at A, B and C***

First of all, let us assume the continuous beam *ABC* to be split up into fixed beams *AB* and *BC*. In span *AB*, the fixing moment at *A* due to loading

$$= -\frac{wl^2}{12} = -\frac{3 \times (12)^2}{12} = -36.0 \text{ kN-m}$$

and fixing moment at *B*

$$= +\frac{wl^2}{12} = +\frac{3 \times (12)^2}{12} = +36.0 \text{ kN-m}$$

Now in span *AB*, the moment at *A* due to sinking of support

* We have already solved this question by Three moments theorem in Example 25.8.

$$= -\frac{6EI\,\delta}{l^2} \quad \text{... (Minus sign due to right support sinking)}$$

$$= -\frac{6\times(200\times10^6)\times(0.2\times10^{-3})\times0.03}{(12)^2} = -50.0 \text{ kN-m}$$

Similarly, moment at $B$ due to sinking of support $B$

$$= -50.0 \text{ kN-m}$$

∴ Total moment at $A$ in span $AB$

$$= -36.0 - 5.00 = -86.0 \text{ kN-m}$$

and total moment at $B$ in span $AB$

$$M = +36.0 - 50.0 = -14.0 \text{ kN-m}$$

Now in span $BC$, the fixing moment at $B$ due to loading

$$= -\frac{Wab^2}{l^2} = -\frac{24\times4\times(8)^2}{(12)^2} = -42.67 \text{ kN-m}$$

and fixing moment at $B$

$$= +\frac{Wa^2b}{l^2} = \frac{24\times(4)^2\times8}{(12)^2} = +21.33 \text{ kN-m}$$

Now in span $BC$, the moment at $B$, due to sinking of support $B$

$$= +\frac{6EI\,\delta}{l^2} \quad \text{... (Plus sign due to left support sinking)}$$

$$= +\frac{6\times(200\times10^6)\times(0.2\times10^{-3})\times0.03}{(12)^2} = +50.0 \text{ kN-m}$$

Similarly, moment at $C$ due to sinking of support $B$ = + 50.0 kN-m

∴ Total moment at $B$ in span $BC$

$$= -42.67 + 50.0 = +7.33 \text{ kN-m}$$

and total moment at $C$ in span $BC$

$$= +21.33 + 50.0 = +71.33 \text{ kN-m}$$

Now, let us find out the distribution factors at $B$. From the geometry of the figure, we find that the stiffness factor for $BA$,

$$k_{BA} = \frac{4EI}{l} = \frac{4EI}{12} = \frac{EI}{3} \quad \text{...(∵ The beam is fixed at } A)$$

and

$$k_{BC} = \frac{3EI}{l} = \frac{3EI}{12} = \frac{EI}{4} \quad \text{...(∵ The beam is simply supported at } C)$$

Distribution factors for $BA$ and $BC$

$$= \frac{\frac{EI}{3}}{\frac{EI}{3}+\frac{EI}{4}} \text{ and } \frac{\frac{EI}{4}}{\frac{EI}{3}+\frac{EI}{4}} = \frac{4}{7} \text{ and } \frac{3}{7}$$

Now prepare the following table:

| A | B | | C | |
|---|---|---|---|---|
| | $\frac{4}{7}$ | $\frac{3}{7}$ | | |
| – 86.00 | – 14.00 | + 7.33 | + 71.33 | Fixed end moment |
| | | | – 71.33 | Release C |
| | | – 35.67 | | Carry over |
| – 86.00 | – 14.00 | – 28.34 | 0 | Initial moments |
| | + 24.20 | + 18.14 | | Distribute |
| + 12.10 | | | | Carry over |
| 0 | 0 | 0 | 0 | Distribute |
| – 73.9 | + 10.2 | – 10.2 | 0 | Final moments |

The bending moments at the supports are given in the above table. **Ans.**

***Bending moment diagram***

From the geometry of the figure, we find that the bending moment at the mid of the span *AB* treating it as a simply supported beam

$$= \frac{wl^2}{8} = \frac{3 \times (12)^2}{8} = 54.0 \text{ kN-m}$$

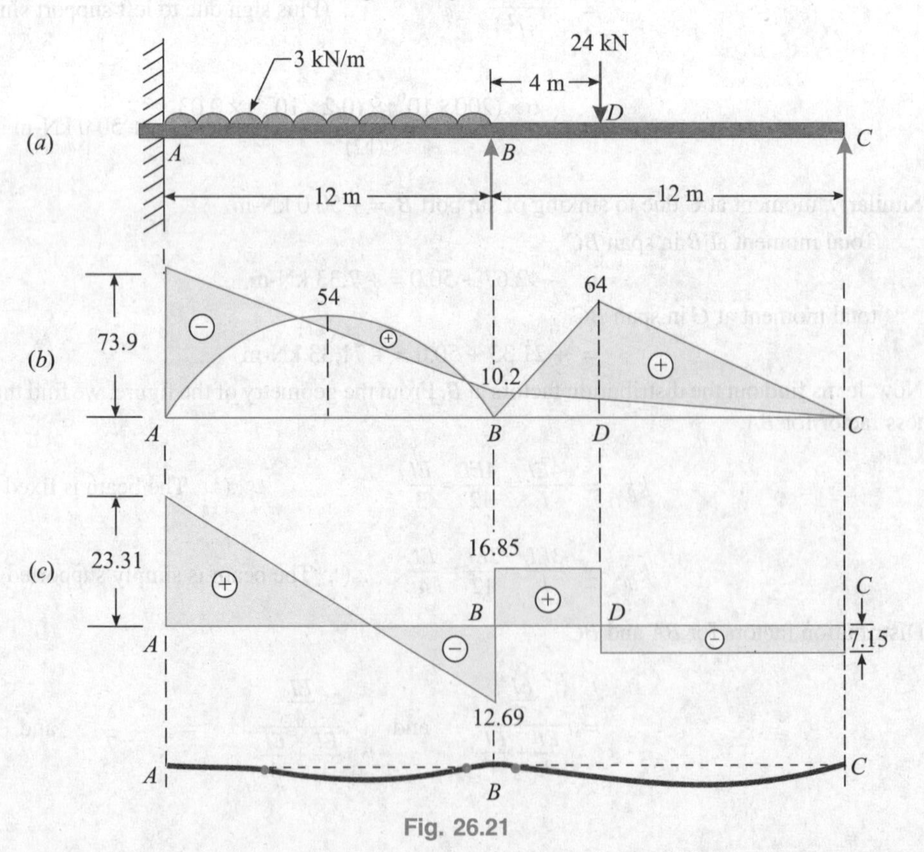


Fig. 26.21

Similarly, bending moment under the 24 kN load

$$= \frac{Wab}{l} = \frac{24 \times 4 \times 8}{12} = 64.0 \text{ kN-m}$$

Now complete the bending moment as shown in Fig. 26.21 (*b*).

Shear force diagram

Let $R_A$ = Reaction at *A*,

$R_B$ = Reaction at *B* and

$R_C$ = Reaction at *C*.

Taking moments about *B*, and equating the same

$$R_C \times 12 - (24 \times 4) = -10.2 \qquad ...(\because M_B = -10.2 \text{ kN-m})$$

$$\therefore \quad R_C = \frac{-10.2 + 96.0}{12} = 7.15 \text{ kN}$$

Now taking moments about *A*, and equating the same

$$-73.9 = (7.15 \times 24) + R_B \times 12 - (24 \times 16) - (3 \times 12 \times 6)$$

$$...(\because \text{ Bending moment } A = -73.9 \text{ kN-m})$$

$$= 12R_B - 428.4$$

$$\therefore \quad R_B = \frac{-73.9 + 428.4}{12} = 29.54 \text{ kN}$$

and $R_A = (3 \times 12 + 24) - (7.15 + 29.54) = 23.31$ kN

Now complete the shear force diagram as shown in Fig. 26.21 (*c*).

The elastic curve *i.e.*, deflected shape of the centre line of the beam is shown in Fig. 26.21 (*d*).

## EXERCISE 26.2

1. A continuous beam *ABCD* is fixed at *A* and simply supported at *B* and *C*, the beam *CD* is overhanging. The spans *AB* = 6 m, *BC* = 5 m and overhanging *CD* = 2.5 m. The moment of inertia of the span *BC* is 2*l* and that of *AB* and *CD* is *l*. The beam is carrying a uniformly distributed load 2 kN/m over the span *AB*, a point load of 5 kN in *BC* at a distance of 3 m from *B*, and a point load of 8 kN at the free end.

   Determine the fixing moments at *A*, *B* and *C* and draw the bending moment diagram.

   [**Ans.** 8.11 kN-m ; 1.79 kN-m ; 20.0 kN-m]

2. A beam *ABCD* is continuous over three spans *AB* = 8 m, *BC* = 4 m and *CD* = 8 m. The beam *AB* and *BC* is subjected to a uniformly distributed load of 1.5 kN/m, whereas there is a central point load of 4 kN in *CD*. The moment of inertia of *AB* and *CD* is 2*I* and that of *BC* is *I*. The end *A* and *D* are fixed. During loading the support *A* sinks down by 10 mm. Find the fixed end moments. Take $E$ = 200 Gpa and $I = 16 \times 10^6$ mm$^4$.

   [**Ans.** –16.53 kN-m ; – 30.66 kN-m ; – 77.33 kN-m ; – 21.33 kN-m]

3. A continuous beam *ABCD* 20 m long is supported at *B* and *C* and fixed at *A* and *D*. The spans *AB*, *BC* and *CD* are 6 m, 8 m and 6 m respectively. The span *AB* carries a uniformly distributed load of 1 kN/m, the span *BC* carries a central point load of 10 kN and the span *CD* carries a point load of 5 kN at a distance of 3 m from *C*. During loading, the support *B* sinks by 10 mm. Find the fixed end moments and draw the bending moment diagram. Take $E$ = 10 mm and $I = 3 \times 10^9$ mm$^4$. The moment of inertia of the spans *AB* and *CD* is *I* and that of *BC* = 2*I*.

   [**Ans.** – 5.74 kN-m ; – 2.57 kN-m ; – 9.2 kN-m ; – 2.07 kN-m]

## QUESTIONS

1. Define the term 'carry over factor'. Derive a relation for the stiffness factor for a beam simply supported at its both ends.
2. What do you understand by the term 'distribution factor'? Discuss its importance in the method of moment distribution.
3. Explain the procedure for finding out the fixed end moments in:
   (*a*) beams with fixed end supports. (*b*) beams with simply supported ends.
   (*c*) beams with end span overhanging. (*d*) beams with a sinking support.

## OBJECTIVE TYPE QUESTIONS

1. Stiffness factor for a beam fixed at one end and freely supported at the other is
   (*a*) $\frac{3EI}{l}$ (*b*) $\frac{4EI}{l}$ (*c*) $\frac{6EI}{l}$ (*d*) $\frac{8EI}{l}$
2. Stiffness factor for beam simply supported at both the ends is"
   (*a*) $\frac{3EI}{l}$ (*b*) $\frac{4EI}{l}$ (*c*) $\frac{6EI}{l}$ (*d*) $\frac{8EI}{l}$
3. If the end *B* of a continuous beam *ABC* sinks down, then the moment at *A* will be
   (*a*) zero (*b*) negative (*c*) positive (*d*) infinity

## ANSWERS

1. (*b*) 2. (*a*) 3. (*b*)

Chapter 27

# Torsion of Circular Shafts

## Contents

## 27.1. Introduction

In workshops and factories, a turning force is always applied to transmit energy by rotation. This turning force is applied either to the rim of a pulley, keyed to the shaft or at any other suitable point at some distance from the axis of the shaft. The product of this turning force and the distance between the point of application of the force and the axis of the shaft is known as torque, turning moment or twisting moment. And the shaft is said to be subjected to torsion. Due to this torque, every cross-section of the shaft is subjected to some shear stress.

## 27.2. Assumptions for Shear Stress in a Circular Shaft Subjected to Torsion

Following assumptions are made, while

finding out shear stress in a circular shaft subjected to torsion:

1. The material of the shaft is uniform throughout.
2. The twist along the shaft is uniform.
3. Normal cross-sections of the shaft, which were plane and circular before the twist, remain plane and circular even after the twist.
4. All diameters of the normal cross-section, which were straight before the twist, remain straight with their magnitude unchanged, after the twist.

A little consideration will show that the above assumptions are justified, if the torque applied is small and the angle of twist is also small.

## 27.3. Torsional Stresses and Strains

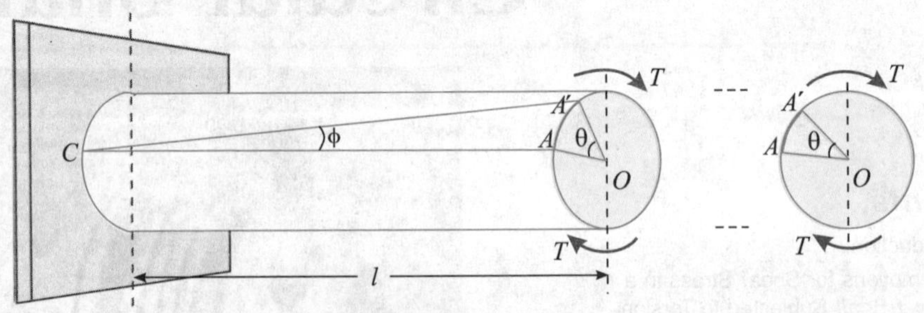


Fig. 27.1

Consider a circular shaft fixed at one end and subjected to a torque at the other end as shown in Fig. 27.1.

Let $T$ = Torque in N-mm,

$l$ = Length of the shaft in mm and

$R$ = Radius of the circular shaft in mm.

As a result of this torque, every cross-section of the shaft will be subjected to shear stresses. Let the line $CA$ on the surface of the shaft be deformed to $CA'$ and $OA$ to $OA'$ as shown in Fig. 27.1.

Let $\angle ACA'$ = $\phi$ in degrees

$\angle AOA'$ = $\theta$ in radians

$\tau$ = Shear stress induced at the surface and

$C$ = Modulus of rigidity, also known as torsional rigidity of the shaft material.

We know that shear strain = Deformation per unit length

$$= \frac{AA'}{l} = \tan\theta$$

$$= \phi \qquad \text{...(}\phi\text{ being very small, } \tan\phi = \phi\text{)}$$

We also know that the arc $AA' = R \cdot \theta$

$$\therefore \qquad \phi = \frac{AA'}{l} = \frac{R \cdot \theta}{l} \qquad \text{...(i)}$$

If $\tau$ is the intensity of shear stress on the outermost layer and $C$ the modulus of rigidity of the shaft, then

$$\phi = \frac{\tau}{C} \qquad \text{...(ii)}$$

From equations (*i*) and (*ii*), we find that

$$\frac{\tau}{C} = \frac{R \cdot \theta}{l} \quad \text{or} \quad \frac{\tau}{R} = \frac{C \cdot \theta}{l}$$

If $\tau_X$ be the intensity of shear stress, on any layer at a distance $x$ from the centre of the shaft, then

$$\frac{\tau_X}{x} = \frac{\tau}{R} = \frac{C \cdot \theta}{l} \quad ...(iii)$$

## 27.4. Strength of a Solid Shaft

The term, strength of a shaft means the maximum torque or power, it can transmit. As a matter of fact, we are always interested in calculating the torque, a shaft can withstand or transmit.

Let $R$ = Radius of the shaft in mm and

$\tau$ = Shear stress developed in the outermost layer of the shaft in N/mm$^2$

Consider a shaft subjected to a torque $T$ as shown in Fig. 27.2. Now let us consider an element of area $da$ of thickness $dx$ at a distance $x$ from the centre of the shaft as shown in Fig. 27.2.

$$\therefore \quad da = 2\pi x \cdot dx \quad ...(i)$$

and shear stress at this section,

$$\therefore \quad \tau_X = \tau \times \frac{x}{R} \quad ...(ii)$$

where $\tau$ = Maximum shear stress.

$$\therefore \quad \text{Turning force} = \text{Shear Stress} \times \text{Area}$$

$$= \tau_x \cdot dx$$

$$= \tau \times \frac{x}{R} \times da$$

$$= \tau \frac{x}{R} \times 2\pi x \cdot dx$$

$$= \frac{2\pi x}{R} \cdot x^2\, dx$$

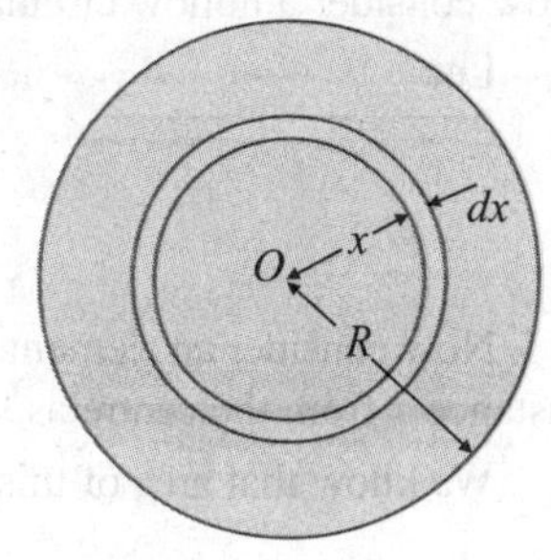


Fig. 27.2

We know that turning moment of this element,

$$dT = \text{Turning force} \times \text{Distance of element from axis of the shaft}$$

$$= \frac{2\pi\tau}{R} x^2\, dx \cdot x = \frac{2\pi\tau}{R} x^3 \cdot dx \quad ...(iii)$$

The total torque, which the shaft can withstand, may be found out by integrating the above equation between 0 and $R$ *i.e.*,

$$T = \int_0^R \frac{2\pi\tau}{R} x^3 \cdot dx = \frac{2\pi\tau}{R} \int_0^R x^2 \cdot dx$$

$$= \frac{2\pi\tau}{R} \left[\frac{x^4}{4}\right]_0^R = \frac{\pi}{2} \tau \cdot R^3 = \frac{\pi}{16} \times \tau \times D^3 \quad \text{N-mm}$$

where $D$ is the diameter of the shaft and is equal to $2R$.

**EXAMPLE 27.1.** *A circular shaft of 50 mm diameter is required to transmit torque from one sh~~ft to another. Find the safe torque, which the shaft can transmit, if the shear stress is not to exceed 40 MPa.*

**SOLUTION.** Given: Diameter of shaft ($D$) = 50 mm and maximum shear stress ($\tau$) = 40 MPa = 40 N/mm$^2$.

We know that the safe torque, which the shaft can transmit,

$$T = \frac{\pi}{16} \times \tau \times D^3 = \frac{\pi}{16} \times 40 \times (50)^3 \text{ N-mm}$$

$$= 0.982 \times 10^6 \text{ N-mm} = \mathbf{0.982\ kN\text{-}m} \quad \textbf{Ans.}$$

**EXAMPLE 27.2.** *A solid steel shaft is to transmit a torque of 10 kN-m. If the shearing stress is not to exceed 45 MPa, find the minimum diameter of the shaft.*

**SOLUTION.** Given: Torque ($T$) = 10 kN-m = $10 \times 10^6$ N-mm and maximum shearing stress ($\tau$) = 45 MPa = 45 N/mm$^2$.

Let $D$ = Minimum diameter of the shaft in mm.

We know that torque transmitted by the shaft ($T$),

$$10 \times 10^6 = \frac{\pi}{16} \times \tau \times D^3 = \frac{\pi}{16} \times 45 \times D^3 = 8.836\ D^3$$

$$\therefore \quad D^3 = \frac{10 \times 10^6}{8.836} = 1.132 \times 10^6$$

or $D = 1.04 \times 10^2 = \mathbf{104\ mm}$ **Ans.**

## 27.5. Strength of a Hollow Shaft

It means the maximum torque or power a hollow shaft can transmit from one pulley to another. Now consider a hollow circular shaft subjected to some torque.

Let $R$ = Outer radius of the shaft in mm,

$r$ = Inner radius of the shaft in mm, and

$\tau$ = Maximum shear stress developed in the outer most layer of the shaft material.

Now consider an elementary ring of thickness $dx$ at a distance $x$ from the centre as shown in Fig. 27.3.

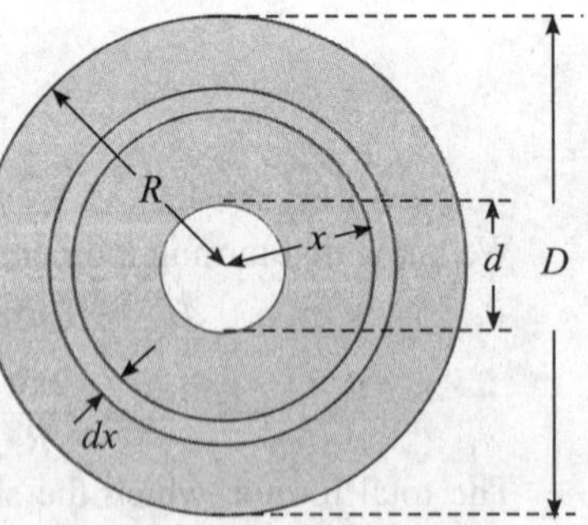

We know that area of this ring,

$$da = 2\pi x \cdot dx \qquad ...(i)$$

and shear stress at this section,

$$\tau_X = \tau \times \frac{x}{R}$$

$\therefore$ Turning force = Stress × Area

$$= \tau_X \cdot dx \qquad ...\left(\because \tau_X = \tau \times \frac{x}{R}\right)$$

$$= \tau \times \frac{x}{R} \times 2\pi\, xdx \qquad ...(\because da = 2\pi\, xdx)$$

$$= \frac{2\pi\tau}{R} x^2 \cdot dx \qquad ...(ii)$$

Fig. 27.3

We know that turning moment of this element,

$dT$ = Turning force × Distance of element from axis of the shaft

$$= \frac{2\pi\tau}{R} x^2 \cdot dx \cdot x = \frac{2\pi\tau}{R} x^3 \cdot dx \qquad ...(iii)$$

The total torque, which the shaft can transmit, may be found out by integrating the above equation between $r$ and $R$.

$$\therefore \qquad T = \int_r^R \frac{2\pi\tau}{R} x^3 \cdot dx = \frac{2\pi\tau}{R} \int_r^R x^3 \cdot dx$$

$$= \frac{2\pi\tau}{R} \left[ \frac{x^4}{4} \right]_r^R = \frac{2\pi\tau}{R} \left( \frac{R^4 - r^4}{4} \right) = \frac{\pi}{16} \times \tau \times \left( \frac{D^4 - d^4}{D} \right) \text{N-mm}$$

where $D$ is the external diameter of the shaft and is equal to $2R$ and $2d$ is the internal diameter of the shaft and is equal to $2r$.

**Note:** We have already discussed in Art. 27.3 that the shear stress developed at a point is proportional to its distance from the centre of the shaft. It is thus obvious that in the central portion of a shaft, the shear stress induced is very small. In order to utilize the material to the fuller extent, hollow shafts are used.

**Example 27.3.** *A hollow shaft of external and internal diameter of 80 mm and 50 mm is required to transmit torque from one end to the other. What is the safe torque it can transmit, if the allowable shear stress is 45 MPa ?*

**Solution.** Given: External diameter ($D$) = 80 mm; Internal diameter ($d$) = 50 mm and allowable shear stress ($\tau$) = 45 MPa = 45 N/mm$^2$.

We know that torque transmitted by the shaft,

$$T = \frac{\pi}{16} \times \tau \times \left[ \frac{D^4 - d^4}{D} \right] = \frac{\pi}{16} \times 45 \times \left[ \frac{(80)^4 - (50)^4}{80} \right] \text{N-mm}$$

$$= 3.83 \times 10^6 \text{ N-mm} = \mathbf{3.83 \text{ kN-m}} \qquad \textbf{Ans.}$$

## 27.6. Power Transmitted by a Shaft

We have already discussed that the main purpose of a shaft is to transmit power from one shaft to another in factories and workshops. Now consider a rotating shaft, which transmits power from one of its ends to another.

Let $N$ = No. of revolutions per minute and

$T$ = Average torque in kN-m.

Work done per minute = Force × Distance = $T \times 2\pi N = 2\pi NT$

Work done per second = $\dfrac{2\pi NT}{60}$ kN-m

Power transmitted = Work done in kN-m per second

$$= \frac{2\pi NT}{60} \text{ kW}$$

**Note:** If the torque is in the N-m, then work done will also be in N-m and power will be in watt ($W$).

**Example 27.4.** *A circular shaft of 60 mm diameter is running at 150 r.p.m. If the shear stress is not to exceed 50 MPa, find the power which can be transmitted by the shaft.*

**Solution.** Given: Diameter of the shaft ($D$) = 60 mm ; Speed of the shaft ($N$) = 150 r.p.m. and maximum shear stress ($\tau$) = 50 MPa = 50 N/mm$^2$.

We know that torque transmitted by the shaft,

$$T = \frac{\pi}{16} \times \tau \times D^3 = \frac{\pi}{16} \times 50 \times (60)^3 \text{ N-mm}$$

$$= 2.12 \times 10^6 \text{ N-mm} = 2.12 \text{ kN-m}$$

and power which can be transmitted by the shaft,

$$P = \frac{2\pi NT}{60} = \frac{2\pi \times 150 \times 2.12}{60} = \mathbf{33.3\ kW} \qquad \textbf{Ans.}$$

**EXAMPLE 27.5.** *A hollow shaft of external and internal diameters as 100 mm and 40 mm is transmitting power at 120 r.p.m. Find the power the shaft can transmit, if the shearing stress is not to exceed 50 MPa.*

**SOLUTION.** Given: External diameter ($D$) = 100 mm ; Internal diameter ($d$) = 40 mm ; Speed of the shaft ($N$) = 120 r.p.m. and allowable shear stress ($\tau$) = 50 MPa = 50 N/mm$^2$.

We know that torque the shaft can transmit,

$$T = \frac{\pi}{16} \times \tau \times \left[\frac{D^4 - d^4}{D}\right] = \frac{\pi}{16} \times 50 \times \left[\frac{(100)^4 - (40)^4}{100}\right] \text{ N-mm}$$

$$= 9.56 \times 10^6 \text{ N-mm} = 9.56 \text{ kN-m}$$

and power the shaft can transmit,

$$P = \frac{2\pi NT}{60} = \frac{2\pi \times 120 \times 9.56}{60} = \mathbf{120\ kW} \qquad \textbf{Ans.}$$

**EXAMPLE 27.6.** *A solid circular shaft of 100 mm diameter is transmitting 120 kW at 150 r.p.m. Find the intensity of shear stress in the shaft.*

**SOLUTION.** Given : Diameter of the shaft ($D$) = 100 mm ; Power transmitted ($P$) = 120 kW and speed of the shaft ($N$) = 150 r.p.m.

Let $T$ = Torque transmitted by the shaft, and

$\tau$ = Intensity of shear stress in the shaft.

We know that power transmitted by the shaft ($P$),

$$120 = \frac{2\pi NT}{60} = \frac{2\pi \times 150 \times T}{60} = 15.7\, T$$

$$\therefore \quad T = \frac{120}{15.7} = 7.64 \text{ kN-m} = 7.64 \times 10^6 \text{ N-mm}$$

We also know that torque transmitted by the shaft ($T$),

$$7.64 \times 10^6 = \frac{\pi}{16} \times \tau \times D^3 = \frac{\pi}{16} \times \tau \times (100)^3 = 0.196 \times 10^6\, \tau$$

$$\tau = \frac{7.64}{0.196} = 39 \text{ N/mm}^2 = \mathbf{39\ MPa} \qquad \textbf{Ans.}$$

**EXAMPLE 27.7.** *A hollow shaft is to transmit 200 kW at 80 r.p.m. If the shear stress is not to exceed 60 MPa and internal diameter is 0.6 of the external diameter, find the diameters of the shaft.*

**SOLUTION.** Given : Power ($P$) = 200 kW ; Speed of shaft ($N$) = 80 r.p.m. ; Maximum shear stress ($\tau$) = 60 MPa = 60 N/mm$^2$ and internal diameter of the shaft ($d$) = 0.6$D$ (where $D$ is the external diameter in mm).

We know that torque transmitted by the shaft,

$$T = \frac{\pi}{16} \times \tau \times \left[\frac{D^4 - d^4}{D}\right] = \frac{\pi}{16} \times 60 \times \left[\frac{D^4 - (0.6D)^4}{D}\right] \text{ N-mm}$$

$$= 10.3\ D^3 \text{ N-mm} = 10.3 \times 10^{-6}\ D^3 \text{ kN-m} \qquad ...(i)$$

We also know that power transmitted by the shaft ($P$),

$$200 = \frac{2\pi NT}{60} = \frac{2\pi \times 80 \times (10.3 \times 10^{-6}\ D^3)}{60} = 86.3 \times 10^{-6}\ D^3$$

$$\therefore \quad D^3 = \frac{200}{(86.3 \times 10^{-6})} = 2.32 \times 10^6 \text{ mm}^3$$

or $D = 1.32 \times 10^2 =$ **132 mm** **Ans.**

and $d = 0.6\ D = 0.6 \times 132 =$ **79.2 mm** **Ans.**

**Example 27.8.** *A solid steel shaft has to transmit 100 kW at 160 r.p.m. Taking allowable shear stress as 70 MPa, find the suitable diameter of the shaft. The maximum torque transmitted in each revolution exceeds the mean by 20%.*

**Solution.** Given: Power ($P$) = 100 kW ; Speed of the shaft ($N$) = 160 r.p.m. ; Allowable shear stress ($\tau$) = 70 MPa = 70 N/mm² and maximum torque ($T_{max}$) = 1.2 $T$ (where $T$ is the mean torque).

Let $D$ = Diameter of the shaft in mm.

We know that power transmitted by shaft ($P$),

$$100 = \frac{2\pi NT}{60} = \frac{2\pi \times 160 \times T}{60} = 16.8\ T$$

$$\therefore \quad T = \frac{100}{16.8} = 5.95 \text{ kN-m} = 5.95 \times 10^6 \text{ N-mm}$$

and maximum torque, $T_{max} = 1.2T = 1.2 \times (5.95 \times 10^6) = 7.14 \times 10^6$ N-mm

We also know that maximum torque ($T_{max}$),

$$7.14 \times 10^6 = \frac{\pi}{16} \times \tau \times D^3 = \frac{\pi}{16} \times 70 \times D^3 = 13.7\ D^3$$

$$\therefore \quad D^3 = \frac{7.14 \times 10^6}{13.7} = 0.521 \times 10^6$$

or $D = 0.8 \times 10^2 =$ **80 mm** **Ans.**

## EXERCISE 27.1

1. A circular shaft of 80 mm diameter is required to transmit torque in a factory. Find the torque, which the shaft can transmit, if the allowable shear stress is 50 MPa. (**Ans.** 5.03 kN-mm)
2. A solid steel shaft is required to transmit a torque of 6.5 kN-m. What should be the minimum diameter of the shaft, if the maximum shear stress is 40 MPa? (**Ans.** 94 mm)
3. A solid shaft of 40 mm diameter is subjected to a torque of 0.8 kN-m. Find the maximum shear stress induced in the shaft. (**Ans.** 63.7 MPa)
4. A hollow shaft of external and internal diameters of 60 mm and 40 mm is transmitting torque. Find the torque it can transmit, if the shear stress is not to exceed 40 MPa. (**Ans.** 1.36 kN-m)
5. A circular shaft of 80 mm diameter is required to transmit power at 120 r.p.m. If the shear stress is not to exceed 40 MPa, find the power transmitted by the shaft. (**Ans.** 50.5 kW)

6. A hollow shaft of external and internal diameters as 80 mm and 50 mm respectively is transmitting power at 150 r.p.m. Determine the power, which the shaft can transmit, if the shearing stress is not to exceed 40 MPa. **(Ans.** 53.6 kW)
7. A hollow shaft has to transmit 53 kW at 160 r.p.m. If the maximum shear stress is 50 MPa and internal diameter is half of the external diameter, find the diameters of the shaft. **(Ans.** 70 mm; 35 mm)

## 27.7. Polar Moment of Inertia

The moment of inertia of a plane area, with respect to an axis perpendicular to the plane of the figure, is called polar moment of inertia with respect to the point, where the axis intersects the plane. In a circular plane, this point is always the centre of the circle. We know that

$$\frac{\tau}{R} = \frac{C \cdot \theta}{l} \qquad ...(i) \qquad ... \text{(from Art. 27.3)}$$

and

$$T = \frac{\pi}{16} \times \tau \times D^3 \qquad ...(ii) \qquad ... \text{(from Art. 27.3)}$$

or

$$\tau = \frac{16T}{\pi D^3}$$

Substituting the value of $\tau$ in equation (*i*),

$$\frac{16T}{\pi D^3 \times R} = \frac{C \cdot \theta}{l}$$

or

$$\frac{T}{\frac{\pi}{16} \times D^3 \times R} = \frac{C \cdot \theta}{l}$$

$$\frac{T}{\frac{\pi}{32} \times D^4} = \frac{C \cdot \theta}{l} \qquad ...\left(\text{Radius, } R = \frac{D}{2}\right)$$

$$\frac{T}{J} = \frac{C \cdot \theta}{l} \qquad ...(iii)$$

where $J = \frac{\pi}{32} \times D^4$. It is known as polar moment of inertia.

The above equation (*iii*) may also be written as :

$$\frac{\tau}{R} = \frac{T}{J} = \frac{C \cdot \theta}{l} \qquad ...\left(\because \frac{\tau}{R} = \frac{C \cdot \theta}{l}\right)$$

NOTES. 1. In a hollow circular shaft the polar moment of inertia,

$$J = \frac{\pi}{32}(D^4 - d^4)$$

where $d$ is the internal diameter of the shaft.

2. The term $\frac{J}{R}$ is known as torsional section modulus or polar modulus. It is similar to section modulus which is equal to $\frac{I}{y}$.

3. Thus polar modulus for a solid shaft,

$$Zp = \frac{2\pi}{32D} \times D^4 = \frac{\pi}{16} D^3$$

and the polar modulus for a hollow shaft,

$$Zp = \frac{2\pi}{32D}(D^4 - d^4) = \frac{\pi}{16D}(D^4 - d^4)$$

**EXAMPLE 27.9.** *Calculate the maximum torque that a shaft of 125 mm diameter can transmit, if the maximum angle of twist is 1° in a length of 1.5 m. Take C = 70 GPa.*

**SOLUTION.** Given: Diameter of shaft $(D) = 125$ mm ; Angle of twist $(\theta) = 1° = \frac{\pi}{180}$ rad ; Length of the shaft $(l) = 1.5 \text{ m} = 1.5 \times 10^3$ mm and modulus of rigidity $(C) = 70 \text{ GPa} = 70 \times 10^3 \text{ N/mm}^2$.

Let $T$ = Maximum torque the shaft can transmit.

We know that polar moment of inertia of a solid circular shaft,

$$J = \frac{\pi}{32} \times (D)^4 = \frac{\pi}{32}(125)^4 = 24.0 \times 10^6 \text{ mm}^4$$

and relation for torque transmitted by the shaft,

$$\frac{T}{J} = \frac{C \cdot \theta}{l}$$

$$\frac{T}{24.0 \times 10^6} = \frac{(70 \times 10^3)\pi/180}{1.5 \times 10^3} = 0.814$$

$$\therefore \quad T = 0.814 \times (24.0 \times 10^6) = 19.5 \times 10^6 \text{ N-mm}$$

$$= \mathbf{19.5 \text{ kN-m}} \quad \textbf{Ans.}$$

**EXAMPLE 27.10.** *Find the angle of twist per metre length of a hollow shaft of 100 mm external and 60 mm internal diameter, if the shear stress is not to exceed 35 MPa. Take C = 85 GPa.*

**SOLUTION.** Given: Length of the shaft $(l) = 1 \text{ m} = 1 \times 10^3$ mm ; External diameter $(D) = 100$ mm; Internal diameter $(d) = 60$ mm ; Maximum shear stress $(\tau) = 35 \text{ MPa} = 35 \text{ N/mm}^2$ and modulus of rigidity $(C) = 85 \text{ GPa} = 85 \times 10^3 \text{ N/mm}^2$.

Let $\theta$ = Angle of twist in the shaft.

We know that torque transmitted by the shaft,

$$T = \frac{\pi}{16} \times \tau \times \left[\frac{D^4 - d^4}{D}\right] = \frac{\pi}{16} \times 35 \times \left[\frac{(100)^4 - (60)^4}{100}\right] \text{ N-mm}$$

$$= 5.98 \times 10^6 \text{ N-mm}$$

We also know that polar moment of inertia of a hollow circular shaft,

$$J = \frac{\pi}{32}[D^4 - d^4] = \frac{\pi}{32}[(100)^4 - (60)^4] = 8.55 \times 10^6 \text{ mm}^4$$

and relation for the angle of twist,

$$\frac{T}{J} = \frac{C \cdot \theta}{l} \quad \text{or} \quad \frac{5.98 \times 10^6}{8.55 \times 10^6} = \frac{(85 \times 10^3)\,\theta}{1 \times 10^3} = 85.\theta$$

$$\therefore \quad \theta = \frac{5.98 \times 10^6}{(8.55 \times 10^6) \times 85} = 0.008 \text{ rad} = \mathbf{0.5°} \quad \textbf{Ans.}$$

**EXAMPLE 27.11.** *A solid shaft of 120 mm diameter is required to transmit 200 kW at 100 r.p.m. If the angle of twists not to exceed 2°, find the length of the shaft. Take modulus of rigidity for the shaft material as 90 GPa.*

**SOLUTION.** Given : Diameter of shaft $(D) = 120$ mm; Power $(P) = 200$ kW ; Speed of shaft $(N) = 100$ r.p.m. ; Angle of twist $(\theta) = 2° = \frac{2\pi}{180}$ rad. and modulus of rigidity $(C) = 90 \text{ GPa} = 90 \times 10^3 \text{ N/mm}^2$.

Let $T$ = Torque transmitted by the shaft, and

$l$ = Length of the shaft.

We know that power transmitted by the shaft ($P$),

$$200 = \frac{2\pi NT}{60} = \frac{2\pi \times 100 \times T}{60} = 10.5T$$

$$\therefore \quad T = \frac{200}{10.5} = 19 \text{ kN-m} = 19 \times 10^6 \text{ N-mm}$$

We also know that polar moment of inertia of a solid shaft,

$$J = \frac{\pi}{32} \times (D)^4 = \frac{\pi}{32} \times (120)^4 = 0.4 \times 10^6 \text{ mm}^4$$

and relation for the length of the shaft,

$$\frac{T}{J} = \frac{C \cdot \theta}{l} \quad \text{or} \quad \frac{19 \times 10^6}{20.4 \times 10^5} = \frac{(90 \times 10^3) \times (2\pi/180)}{l}$$

$$0.931 = \frac{3.14 \times 10^3}{l}$$

$$\therefore \quad l = \frac{(3.14 \times 10^3)}{0.931} = 3.37 \times 10^3 = \mathbf{3.37 \text{ m}} \quad \textbf{Ans.}$$

**EXAMPLE 27.12.** *Find the maximum torque, that can be safely applied to a shaft of 80 mm diameter. The permissible angle of twist is 1.5 degree in a length of 5 m and shear stress not to exceed 42 MPa. Take C = 84 GPa.*

**SOLUTION.** Given: Diameter of shaft ($D$) = 80 mm ; Angle of twist ($\theta$) = $1.5° = \frac{1.5\pi}{180}$ rad ; Length of shaft ($l$) = 5 m = $5 \times 10^3$ mm ; Maximum shear stress ($\tau$) = 42 MPa = 42 N/mm$^2$ and Modulus of rigidity ($C$) = 84 GPa = $84 \times 10^3$ N/mm$^2$.

First of all, let us find out the values of torques based on shear stress and angle of twist.

1. ***Torque based on shear stress***

We know that the torque which can be applied to the shaft,

$$T_1 = \frac{\pi}{16} \times \tau \times D^3 = \frac{\pi}{16} \times 42 \times (80)^3 = 4.22 \times 10^6 \text{ N-mm} \quad ...(i)$$

2. ***Torque based on angle of twist***

We also know that polar moment of inertia of a solid circular shaft,

$$J = \frac{\pi}{32}(D)^4 = \frac{\pi}{32} \times (80)^4 = 4.02 \times 10^6 \text{ mm}^4$$

and relation for the torque that can be applied:

$$\frac{T_2}{J} = \frac{C \cdot \theta}{l} \quad \text{or} \quad \frac{T_2}{4.02 \times 10^6} = \frac{(84 \times 10^3) \times (1.5\pi/180)}{5 \times 10^3} = 0.44$$

$$\therefore \quad T_2 = 0.44 \times (4.02 \times 10^6) = 1.77 \times 10^6 \text{ N-mm} \quad ...(ii)$$

We shall apply a torque of $1.77 \times 10^6$ N-mm (*i.e.*, lesser of the two values). **Ans.**

**EXAMPLE 27.13.** *A solid shaft is subjected to a torque of 1.6 kN-m. Find the necessary diameter of the shaft, if the allowable shear stress is 60 MPa. The allowable twist is 1° for every 20 diameters length of the shaft. Take C = 80 GPa.*

**SOLUTION.** Given: Torque $(T) = 1.6$ kN-m $= 1.6 \times 10^6$ N-mm; Allowable shear stress $(\tau) = 60$ MPa $= 60$ N/mm$^2$; Angle of twist $(\theta) = 1° = \frac{\pi}{180}$ rad ; Length of shaft $(l) = 20\,D$ and modulus of rigidity $(C) = 80$ GPa $= 80 \times 10^3$ N/mm$^2$.

First of all, let us find out the value of diameter of the shaft for its strength and stiffness.

**1.** ***Diameter for strength***

We know that torque transmitted by the shaft $(T)$,

$$1.6 \times 10^6 = \frac{\pi}{16} \times \tau \times D_1^3 = \frac{\pi}{16} \times 60 \times D_1^3 = 11.78\,D_1^{\,3}$$

$$\therefore \quad D_1^{\,3} = \frac{1.6 \times 10^6}{11.78} = 0.136 \times 10^6 \text{ mm}^3$$

$$\text{or} \quad D_1 = 0.514 \times 10^2 = 51.4 \text{ mm} \qquad ...(i)$$

**2.** ***Diameter for stiffness***

We know that polar moment of inertia of a solid circular shaft,

$$J = \frac{\pi}{32} \times (D_2)^4 = 0.098\,D_2^4$$

and relation for the diameter,

$$\frac{T}{J} = \frac{C \cdot \theta}{l} \quad \text{or} \quad \frac{1.6 \times 10^6}{0.098\,D_2^4} = \frac{(80 \times 10^3) \times (\pi/180)}{20 D_2}$$

$$\therefore \quad D_2^3 = \frac{(1.6 \times 10^6) \times 20}{0.098 \times (80 \times 10^3) \times (\pi/180)} = 234 \times 10^3 \text{ mm}^3$$

$$\text{or} \quad D_2 = 6.16 \times 10^1 = 61.6 \text{ mm} \qquad ...(ii)$$

We shall provide a shaft of diameter of 61.6 mm (*i.e.*, greater of the two values). **Ans.**

**EXAMPLE 27.14.** *A solid shaft of 200 mm diameter has the same cross-sectional area as a hollow shaft of the same material with inside diameter of 150 mm. Find the ratio of*

(*a*) *powers transmitted by both the shafts at the same angular velocity.*

(*b*) *angles of twist in equal lengths of these shafts, when stressed to the same intensity.*

**SOLUTION.** Given: Diameter of solid shaft $(D_1) = 200$ mm and inside diameter of hollow shaft $(d) = 150$ mm.

(*a*) ***Ratio of powers transmitted by both the shafts***

We know that cross-sectional area of the solid shaft,

$$A_1 = \frac{\pi}{4} \times D_1^2 = \frac{\pi}{4} \times (200)^2 = 10\,000\,\pi \text{ mm}^2$$

and cross-sectional area of hollow shaft,

$$A_2 = \frac{\pi}{4} \times (D^2 - d^2) = \frac{\pi}{4} \times [D^2 - (150)^2] = \frac{\pi}{4}(D^2 - 22\,500)$$

Since the cross-sectional areas of both the shafts are same, therefore equating $A_1$ and $A_2$,

$$\frac{\pi}{4}(200)^2 = \frac{\pi}{4}(D^2 - 22\,500)$$

$$\therefore \quad 40\,000 = D^2 - 22\,500$$

$$D^2 = 40\,000 + 22\,500 = 62\,500 \text{ mm}^2$$

$$\text{or} \quad D = 250 \text{ mm}$$

We also know that torque transmitted by the solid shaft,

$$T_1 = \frac{\pi}{16} \times \tau \times D_1^3 = \frac{\pi}{16} \times \tau \times (200)^3 = 500 \times 10^3 \, \pi \, \tau \text{ N-mm} \quad ...(i)$$

Similarly, torque transmitted by the hollow shaft,

$$T_2 = \frac{\pi}{16} \times \tau \times \left[\frac{D^4 - d^4}{D}\right] = \frac{\pi}{16} \times \tau \times \left[\frac{(250)^4 - (150)^4}{250}\right] \text{ N-mm}$$

$$= 850 \times 10^3 \, \pi \, \tau \text{ N-mm}$$

$$\therefore \quad \frac{\text{Power transmitted by hollow shaft}}{\text{Power transmitted by solid shaft}}$$

$$= \frac{T_2}{T_1} = \frac{50 \times 10^3 \, \pi \, \tau}{500 \times 10^3 \, \pi \, \tau} = \mathbf{1.7} \quad \textbf{Ans.}$$

**(b)** ***Ratio of angles of twist in both the shafts***

We know that relation for angle of twist for a shaft,

$$\frac{\tau}{R} = \frac{C \cdot \theta}{l} \quad \text{or} \quad \theta = \frac{\tau l}{RC}$$

∴ Angle of twist for the solid shaft,

$$\theta_1 = \frac{\tau l}{RC} = \frac{\tau l}{100C} \quad ... \left(\text{where } R = \frac{D_1}{2} = \frac{200}{2} = 100 \text{ mm}\right)$$

Similarly angle of twist for the hollow shaft,

$$\theta_2 = \frac{\tau l}{RC} = \frac{\tau l}{125C} \quad ... \left(\text{where } R = \frac{D_1}{2} = \frac{250}{2} = 125 \text{ mm}\right)$$

$$\therefore \quad \frac{\text{Angle of twist of hollow shaft}}{\text{Angle of twist of solid shaft}} = \frac{\theta_2}{\theta_1} = \frac{\frac{\tau l}{125C}}{\frac{\tau l}{100C}} = \frac{100}{125} = \mathbf{0.8} \quad \textbf{Ans.}$$

## 27.8. Replacing a Shaft

Sometimes, we are required to replace a solid shaft by a hollow one, or vice versa. In such cases, the torque transmitted by the new shaft should be equal to that by the replaced shaft. But sometimes, there are certain other conditions which have also to be considered while designing the new shaft.

**EXAMPLE 27.15.** *A solid steel shaft of 60 mm diameter is to be replaced by a hollow steel shaft of the same material with internal diameter equal to half of the external diameter. Find the diametres of the hollow shaft and saving in material, if the maximum allowable shear stress is same for both shafts.*

**SOLUTION.** Given: Diameter of solid shaft ($D$) = 60 mm.

***Diameter of the hollow shaft***

Let $D$ = External diameter of the hollow shaft,

$d$ = Internal diameter of the hollow shaft (equal to $D_1/2$) and

$\tau$ = Shear stress developed in both the shafts.

We know that torque transmitted by the solid shaft,

$$T = \frac{\pi}{16} \times \tau \times D^3 = \frac{\pi}{16} \times \tau \times (60)^3 \quad ...(i)$$

and torque transmitted by the hollow shaft,

$$T_1 = \frac{\pi}{16} \times \tau \times \left[\frac{D^4 - d^4}{D_1}\right] = \frac{\pi}{16} \times \tau \times \left[\frac{D_1^4 - (0.5D_1)^4}{D_1}\right]$$

$$= \frac{\pi}{16} \times \tau \times 0.9375\, D_1^3 \qquad ...(ii)$$

Since the torque transmitted and allowable shear stress in both the cases are same, therefore equating the equations (*i*) and (*ii*),

$$\frac{\pi}{16} \times \tau \times (60)^3 = \frac{\pi}{16} \times \tau \times 0.9375\, D_1^3$$

$$\therefore \qquad D_1^3 = \frac{(60)^3}{0.9375} = 230400 \text{ mm}^3$$

or $\qquad D_1 = \mathbf{61.3\ mm}$ **Ans.**

and $\qquad d = \frac{61.3}{2} = \mathbf{30.65\ mm}$ **Ans.**

***Saving in material***

We know that saving in material

$$= \frac{\left[\frac{\pi}{4}(60)^2\right] - \left[\frac{\pi}{4}((61.3)^2 - (30.65)^2)\right]}{\frac{\pi}{4}(60)^2} = \frac{3600 - 2819}{3600}$$

$$= 0.217 = \mathbf{21.7\%} \qquad \textbf{Ans.}$$

**EXAMPLE 27.16.** *A solid shaft of 80 mm diameter is to be replaced by a hollow shaft of external diameter 100 mm. Determine the internal diameter of the hollow shaft if the same power is to be transmitted by both the shafts at the same angular velocity and shear stress.*

**SOLUTION.** Given: Diameter of solid shaft ($D$) = 80 mm and external diameter of hollow shaft ($D_1$) = 100 mm.

Let $\qquad d$ = Internal diameter of the hollow shaft, and

$\qquad \tau$ = Shear stress developed in both the shafts.

We know that torque transmitted by the solid shaft,

$$T = \frac{\pi}{16} \times \tau \times D^3 = \frac{\pi}{16} \times \tau \times (80)^3 \qquad ...(i)$$

and torque transmitted by the hollow shaft,

$$T_1 = \frac{\pi}{16} \times \tau \times \left[\frac{D^4 - d^4}{D}\right] = \frac{\pi}{16} \times \tau \times \left[\frac{(100)^4 - d^4}{100}\right] \qquad ...(ii)$$

Since both the torques are equal, therefore equating the equations (*i*) and (*ii*),

$$\frac{\pi}{16} \times \tau \times (80)^2 = \frac{\pi}{16} \times \tau \times \left[\frac{(100)^4 - d^4}{100}\right]$$

$$(80)^3 = \frac{(100)^4 - d^4}{100} = (100)^3 - \frac{d^4}{100}$$

$$\frac{d^4}{100} = (100)^3 - (80)^3 = 488 \times 10^3$$

$$d^4 = (488 \times 10^3) \times 100 = 488 \times 10^5 = 4880 \times 10^4$$

$$\therefore \qquad d = 8.36 \times 10 = \mathbf{83.6\ mm} \qquad \textbf{Ans.}$$

**EXAMPLE 27.17.** *A solid aluminium shaft 1 m long and of 50 mm diameter is to be replaced by a hollow shaft of the same length and same outside diameter, so that the hollow shaft could carry the same torque and has the same angle of twist. What must be the inner diameter of the hollow shaft ?*

*Take modulus of rigidity for the aluminium as 28 GPa and that for steel as 85 GPa.*

**SOLUTION.** Given: Length of aluminium shaft ($l_A$) = 1 m = $1 \times 10^3$ mm ; Diameter of aluminium shaft ($D_A$) = 50 mm ; Length of steel shaft ($l_S$) = 1 m = $1 \times 10^3$ mm ; Outside diameter of steel shaft ($D_S$) = 50 mm; Modulus of rigidity for aluminium ($C_A$) = 28 GPa = $28 \times 10^3$ N/mm$^2$ and modulus of rigidity for steel = 85 GPa = $85 \times 10^3$ N/mm$^2$.

Let $d_S$ = Inner diameter of steel shaft in mm.

We know that polar moment of inertia of the solid aluminium shaft,

$$J_A = \frac{\pi}{32} \times D^4 = \frac{\pi}{32} \times (50)^4 \text{ mm}^4$$

We also know that relation for angle of twist

$$\frac{T}{J} = \frac{C \cdot \theta}{l} \quad \text{or} \quad \theta = \frac{T \cdot l}{JC}$$

$$\therefore \quad \theta_A = \frac{T_A \cdot l_A}{\frac{\pi}{32} \times (50)^4 \times 28 \times 10^3} \text{ rad.}$$

and

$$\theta_S = \frac{T_S \cdot l_S}{\frac{\pi}{32} \times [(50)^4 - (d)^4] \times 85 \times 10^3} \text{ rad.}$$

Since both the angles of twists (*i.e.*, $\theta_A$ and $\theta_B$) are same, therefore equating these values,

$$\frac{T_A \cdot l_A}{\frac{\pi}{32} \times (50)^4 \times 28 \times 10^3} = \frac{T_S \cdot l_S}{\frac{\pi}{32} \times [(50)^4 - (d)^4] \times 85 \times 10^3}$$

Substituting $T_A = T_S$ and $l_A = l_S$ in the above equation,

$$(50)^4 \times 28 = [(50)^4 - d^4] \times 85$$

$$175 \times 10^6 = (531.25 \times 10^6) - 85\, d^4$$

$$85\, d^4 = (531.25 \times 10^6) - (175 \times 10^6) = 356.25 \times 10^6$$

$$d^4 = \frac{356.25 \times 10^6}{85} = 4.191 \times 10^6 \text{ mm}^4$$

$$\therefore \quad d = \mathbf{45.25 \text{ mm}} \quad \textbf{Ans.}$$

**EXAMPLE 27.18.** *A hollow steel shaft of 300 mm external diameter and 200 mm internal diameter has to be replaced by a solid alloy shaft. Assuming the same values of polar modulus for both, calculate the diameter of the latter and work out the ratio of their torsional rigidities. Take C for steel as 2.4 C for alloy.*

**SOLUTION.** Given: External diameter of steel shaft ($D$) = 300 mm ; Internal diameter of steel shaft ($d$) = 200 mm and modulus of rigidity for steel ($C_S$) = 2.4 (where $C_A$ is the modulus of rigidity for the alloy).

***Diameter of the solid alloy shaft***

Let $D_1$ = Diameter of the solid alloy shaft.

We know that polar modulus of hollow steel shaft,

$$Z_S = \frac{\pi}{16D}(D^4 - d^4) = \frac{\pi}{16 \times 300}[(300)^4 - (200)^4] \text{ mm}^3$$

$$= \frac{8.125 \times 10^6 \pi}{6} \text{ mm}^3 \quad ...(i)$$

Similarly, polar modulus of solid alloy shaft,

$$Z_A = \frac{\pi}{16} D_1^3 \text{ mm}^2 \quad ...(ii)$$

Since the polar modulus for both the shafts are the same, therefore equating (*i*) and (*ii*),

$$\frac{8.125 \times 10^6 \pi}{6} = \frac{\pi}{16} D_1^3$$

or

$$D_1^3 = \frac{8.125 \times 10^6 \times 16}{6} = 21.67 \times 10^6$$

$$\therefore \quad D_1 = \mathbf{278.8\ mm} \quad \textbf{Ans.}$$

***Ratio of torsional rigidities***

We know that the torsional rigidity of hollow steel shaft

$$= C_S \times J_S = 2.4\, C_A \times \frac{\pi}{32} [(300)^4 - (200)^4] \quad ...(iii)$$

Similarly, torsional rigidity for solid alloy shaft

$$= C_A \times J_A = C_A \times \frac{\pi}{32} \times D^4 = C_A \times \frac{\pi}{32} \times (278.8)^4 \quad ...(iv)$$

$$\therefore \quad \frac{\text{Torsional rigidity of hollow steel shaft}}{\text{Torsional rigidity of solid alloy shaft}}$$

$$= \frac{2.4\, C_A \times \frac{\pi}{32} [(300)^4 - (200)^4]}{C_A \times \frac{\pi}{32} (278.8)^4} = \mathbf{2.58} \quad \textbf{Ans.}$$

## EXERCISE 27.2

1. Find the torque a solid shaft of 100 mm diameter can transmit, if the maximum angle of twist is 1.5° in a length of 2 m. Take $C = 70$ GPa. (**Ans.** 9.0 kN-m)
2. A hollow shaft of external and internal diameters as 80 mm and 40 mm is required to transmit torque from one pulley to another. What is the value of torque transmitted, if angle of twist is not to exceed 1° in a length of 2 m. Take $C = 80$ GPa (**Ans.** 2.63 kN-m)
3. A solid shaft and a hollow circular shaft, whose inside diameter is 3/4 of the outside diameter are of equal lengths and are required to transmit a given torque. Compare the weights of these two shafts, if maximum shear stress developed in both the shaft is also equal. (**Ans.** 1.76)
4. A solid shaft of 150 mm diameter is to be replaced by a hollow shaft of the same material with internal diameter equal to 60% of the external diameter. Find the saving in material, if maximum allowable shear stress is the same for both the shafts. (**Ans.** 30.9%)
5. A shaft is transmitting 100 kW at 180 r.p.m. If the allowable shear stress in the shaft material is 60 MPa, determine the suitable diameter for the shaft. The shaft is not to twist more than 1° in a length of 3 metres. Take $C = 80$ GPa. (**Ans.** 103.8 mm)

## 27.9. Shaft of Varying Sections

Sometimes a shaft, made up of different lengths having different cross-sectional areas, is required to transmit some torque (or horse power) from one pulley to another.

A little consideration will show that for such a shaft, the torque transmitted by individual sections have to be calculated first and the minimum value of these torques will be the strength of such a shaft. The angle of twist for such a shaft may be found out as usual.

**EXAMPLE 27.19.** *The stepped steel shaft shown in Fig. 27.4 is subjected to a torque (T) at the free end, and a torque (2T) in the opposite direction at the junction of the two sizes.*

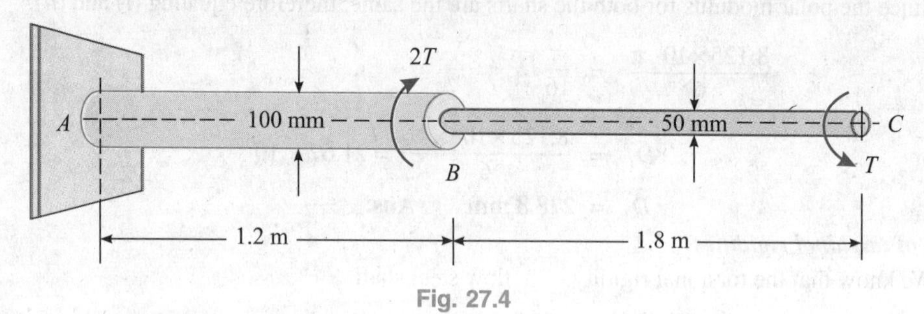


Fig. 27.4

*What is the total angle of twist at the free end, if maximum shear stress in the shaft is limited to 70 MPa? Assume the modulus of rigidity to be 84 GPa.*

**SOLUTION.** Given: Torque at $C = T$ (anticlockwise); Torque at $B = 2T$ (clockwise) ; Diameter of shaft $AB$ ($D_{AB}$) = 100 mm ; Diameter of shaft $BC$ ($D_{BC}$) = 50 mm ; Maximum shear stress ($\tau$) = 70 MPa = 70 N/mm$^2$ and modulus of rigidity ($C$) = 84 GPa = $84 \times 10^3$ N/mm$^2$.

Since the torques at $B$ and $C$ are in opposite directions, therefore the effect of these two torques will be studied first independently, sum of the two twists (one in clockwise direction and the other in anticlockwise direction).

First of all, let us first find out the value of torque $T$ at $C$. It may be noted that if the value of torque is obtained for the portion $AB$, it will induce more stress in the portion $BC$ (because the portion $BC$ is of less diameter). Therefore we shall calculate the torque for the portion $BC$ (because it will not induce stress more than the permissible in the portion $AB$).

We know that the torque at $C$,

$$T = \frac{\pi}{16} \times \tau \times (D_{BC})^3 = \frac{\pi}{16} \times 70 \times (50)^3 = 1.718 \times 10^6 \text{ N-mm}$$

We also know that polar moment of inertia of the solid circular shaft $AB$,

$$J_{AB} = \frac{\pi}{32} \times (D_{AB})^4 = \frac{\pi}{32} \times (100)^4 = 9.82 \times 10^6 \text{ mm}^4$$

Similarly,

$$J_{BC} = \frac{\pi}{32} \times (D_{BC})^4 = \frac{\pi}{32} \times (50)^4 = 0.614 \times 10^6 \text{ mm}^4$$

$\therefore$ Angle of twist at $C$ due to torque ($T$) at $C$,

$$\theta = \frac{T \cdot l}{J \cdot C} = \frac{T}{C}\left(\frac{l_{AB}}{J_{AB}} + \frac{l_{BC}}{J_{BC}}\right)$$

$$= \frac{1.718 \times 10^6}{84 \times 10^3}\left(\frac{1.2 \times 10^3}{9.82 \times 10^6} + \frac{1.8 \times 10^3}{0.614 \times 10^6}\right) \text{rad}$$

$$= 20.45 \times (30.54 \times 10^{-4}) = 0.0624 \text{ rad} \quad ...(i)$$

Similarly, angle of twist at $C$ due to torque $(2T)$ at $B$,

$$\theta = \frac{2T}{C} \times \frac{l_{AB}}{J_{AB}} = \frac{2 \times (1.718 \times 10^6)}{84 \times 10^3} \times \frac{1.2 \times 10^3}{9.82 \times 10^6} \text{ rad}$$

$$= 40.9 \times (1.222 \times 10^{-4}) = 0.005 \text{ rad} \quad ...(ii)$$

From the geometry of the shaft, we find that the twist at $B$ (due to torque of $2T$ at $B$) will continue at $C$ also. Since the directions of both the twists are opposite to each other, therefore net angle of twist at $C$

$$= 0.0624 - 0.005 = 0.0574 \text{ rad} = \mathbf{3.29°} \quad \textbf{Ans.}$$

**EXAMPLE 27.20.** *A shaft ABC of 500 mm length and 40 mm external diameter is bored, for a part of its length AB to a 20 mm diameter and for the remaining length BC to a 30 mm diameter bore as shown in Fig. 27.5. If the shear stress is not to exceed 80 MPa, find the maximum power, the shaft can transmit at a speed of 200 r.p.m.*

*If the angle of twist in the length of 20 mm diameter bore is equal to that in the 30 mm diameter bore, find the length of the shaft that has been bored to 20 mm and 30 mm diameter.*

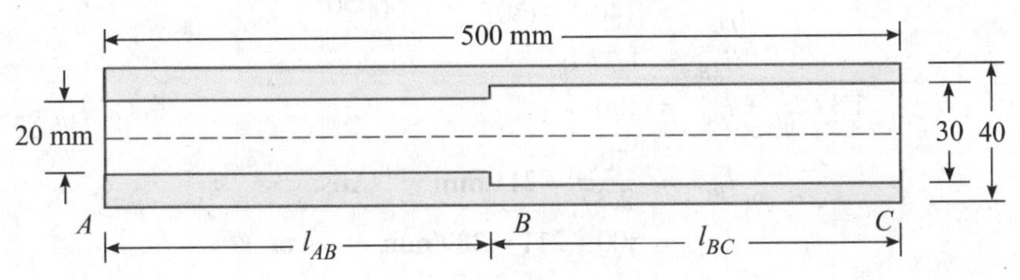


Fig. 27.5

**SOLUTION.** Given: Total length of the shaft $(l)$ = 500 mm; External diameter of the shaft $(D)$ = 40 mm ; Internal diameter of shaft $AB$ $(d_{AB})$ = 20 mm ; Internal diameter of shaft $BC$ $(d_{BC})$ = 30 mm ; Maximum shear stress $(\tau)$ = 80 MPa = 80 N/mm$^2$ and speed of the shaft $(N)$ = 200 r.p.m.

***Maximum power the shaft can transmit***

We know that torque transmitted by the shaft $AB$,

$$T_{AB} = \frac{\pi}{16} \times \tau \times \left(\frac{D^4 - d_{AB}^4}{D}\right) = \frac{\pi}{16} \times 80 \times \left[\frac{(40)^4 - (20)^4}{40}\right] \text{ N-mm}$$

$$= 942.5 \times 10^3 \text{ N-mm} \quad ...(i)$$

Similarly,

$$T_{BC} = \frac{\pi}{16} \times \tau \times \left(\frac{D^4 - d_{BC}^4}{D}\right) = \frac{\pi}{16} \times 80 \times \left[\frac{(40)^4 - (30)^4}{40}\right] \text{ N-mm}$$

$$= 687.3 \times 10^3 \text{ N-mm} \quad ...(ii)$$

From the above two values, we see that the safe torque transmitted by the shaft is minimum of the two, *i.e.*, 687.3 × 10$^3$ N-mm = 687.3 N-m. Therefore maximum power the shaft can transmit,

$$P = \frac{2\pi NT}{60} = \frac{2 \times \pi \times 200 \times (687.3)}{60} = 14\,394 \text{ W}$$

$$= \mathbf{14.39 \text{ kW}} \quad \textbf{Ans.}$$

***Length of the shaft, that has been bored to 20 mm diameter***

Let $l_{AB}$ = Length of the shaft $AB$ (*i.e.*, 20 mm diameter bore) and

$l_{BC}$ = Length of the shaft $BC$ (*i.e.*, 30 mm diameter bore) equal to $(500 - l_{AB})$ mm.

We know that polar moment of inertia for the shaft $AB$,

$$J_{AB} = \frac{\pi}{32} \times (D^4 - d_{AB}^4) = \frac{\pi}{32} \times [(40)^4 - (20)^4] \text{ mm}^4$$

Similarly,

$$J_{BC} = \frac{\pi}{32} \times (D^4 - d_{BC}^4) = \frac{\pi}{32} \times [(40)^4 - (30)^4] \text{ mm}^4$$

We know that relation for the angle of twist:

$$\frac{T}{J} = \frac{C\,\theta}{l} \quad \text{or} \quad \theta = \frac{T \cdot l}{JC}$$

$$\therefore \quad \theta_{AB} = \frac{T \cdot l_{AB}}{J_{AB} \cdot C} \quad \text{and} \quad \theta_{BC} = \frac{T \cdot l_{BC}}{J_{BC} \cdot C}$$

Since $\theta_{AB} = \theta_{AC}$ and $T$ as well as $C$ is equal in both these cases, therefore

$$\frac{l_{AB}}{J_{AB}} = \frac{l_{BC}}{J_{BC}} \quad \text{or} \quad \frac{l_{AB}}{\frac{\pi}{32} \times [(40)^4 - (20)^4]} = \frac{l_{BC}}{\frac{\pi}{32} \times [(40)^4 - (30)^4]}$$

or

$$\frac{l_{AB}}{l_{BC}} = \frac{(40)^4 - (20)^4}{(40)^4 - (30)^4} = \frac{2400000}{1750000} = 1.37$$

$$\therefore \quad l_{AB} = 1.37\, l_{BC}$$

$$1.37\, l_{BC} + l_{BC} = 500 \quad \quad ...(\because l_{AB} + l_{BC} = 500)$$

$$\therefore \quad l_{BC} = \frac{500}{2.37} = \mathbf{211\ mm} \quad \textbf{Ans.}$$

and

$$l_{AB} = 500 - 211 = \mathbf{289\ mm} \quad \textbf{Ans.}$$

## 27.10. Composite Shaft

Sometimes, a shaft is made up of composite section *i.e.*, one type of shaft rigidly sleeved over another type of shaft. At the time of sleeving, the two shafts are joined together in such a way, that the composite shaft behaves like a single shaft. The total torque transmitted by the composite shaft is shared by the two shafts, depending upon their diameters and elastic properties.

**EXAMPLE 27.21.** *A composite shaft consists of copper rod of 30 mm diameter enclosed in a steel tube of external diameter 40 mm and 5 mm thick. The shaft is required to transmit a torque of 0.5 kN-m. Determine the shearing stresses developed in the copper and steel, if both the shafts have equal lengths and welded to a plate at each end, so that their twists are equal. Take $C_C$ = 40 GPa and $C_S$ = 80 GPa.*

**SOLUTION.** Given: Diameter of copper rod ($D_C$) = 30 mm ; External diameter of steel tube ($D_S$) = 40 mm ; Thickness of steel tube = 5 mm ; Therefore internal diameter of steel tube ($d_S$) = 40 – (2 × 5) = 30 mm; Total torque to be transmitted ($T$) = 0.5 kN-m = 0.5 × $10^6$ N-mm ; Modulus of rigidity for copper ($C_C$) = 40 GPa = 40 × $10^3$ N/mm$^2$ and modulus of rigidity for steel ($C_S$) = 80 GPa = 80 × $10^3$ N/mm$^2$.

Let $T_C$ = Torque shared by copper rod,

$\tau_C$ = Shear stress developed in the copper rod and

$T_S, \tau_S$ = Corresponding values for steel tube.

∴ Total torque ($T$)

$$T_C + T_S = 0.5 \times 10^6 \text{ N-mm} \quad \quad ...(i)$$

We know that polar moment of inertia of copper rod,

$$J_C = \frac{\pi}{32} \times (D_C^4) = \frac{\pi}{32} \times (30)^4 = \frac{0.81 \times 10^6\, \pi}{32} \text{ mm}^4$$

and polar moment of inertia of steel tube,

$$J_S = \frac{\pi}{32} \times (D_S^4 - d_S^4) = \frac{\pi}{32} [(40)^4 - (30)^4] = \frac{1.75 \times 10^6 \, \pi}{32} \text{ mm}^4$$

We also know that relation for angle of twist:

$$\frac{T}{J} = \frac{C\theta}{l} \quad \text{or} \quad \theta = \frac{T \cdot l}{JC}$$

$$\therefore \quad \theta_C = \frac{T_C \cdot l}{J_C \cdot C_C} = \frac{T_C \cdot l}{\frac{0.81 \times 10^6 \, \pi}{32} \times (40 \times 10^3)} = \frac{T_C \cdot l}{1012.5 \times 10^6 \, \pi} \text{ rad.}$$

Similarly,

$$\theta_S = \frac{T_S \cdot l}{J_S \cdot C_S} = \frac{T_S \cdot l}{\frac{1.75 \times 10^6 \, \pi}{32} \times (80 \times 10^3)} = \frac{T_S \cdot l}{4375 \times 10^6 \, \pi} \text{ rad.}$$

Since $\theta_C$ is equal to $\theta_S$, therefore equating these values,

$$\frac{T_C \cdot l}{1012.5 \times 10^6 \, \pi} = \frac{T_S \cdot l}{4375 \times 10^6 \, \pi} \quad \text{or} \quad T_C = \frac{81\, T_S}{350}$$

Substituting this value of $T_C$ in equation (*i*),

$$\frac{81\, T_S}{350} + T_S = 0.5 \times 10^6 \quad \text{or} \quad \frac{431\, T_S}{350} = 0.5 \times 10^6$$

$$\therefore \quad T_S = \frac{(0.5 \times 10^6) \times 350}{431} = 0.406 \times 10^6 \text{ N-mm}$$

and

$$T_C = \frac{81\, T_S}{350} = \frac{81 \times (0.406 \times 10^6)}{350}$$

$$= \mathbf{0.094 \times 10^6 \text{ N-mm}} \quad \textbf{Ans.}$$

We know that torque transmitted by copper rod ($T_C$),

$$0.094 \times 10^6 = \frac{\pi}{16} \times \tau_C \times D_C^3 = \frac{\pi}{16} \times \tau_C \times (30)^3 = 5301\, \tau_C$$

$$\therefore \quad \tau_C = \frac{0.094 \times 10^6}{5301} = 17.7 \text{ N/mm}^2 = \mathbf{17.7 \text{ MPa}} \quad \textbf{Ans.}$$

Similarly, torque transmitted by steel tube ($T_S$),

$$0.406 \times 10^6 = \frac{\pi}{16} \times \tau_S \times \left( \frac{D_S^4 - d_S^4}{D_S} \right) = \frac{\pi}{16} \times \tau_S \times \left( \frac{(40)^4 - (30)^4}{40} \right) = 8590\, \tau_S$$

$$\therefore \quad \tau_S = \frac{0.406 \times 10^6}{8590} = 47.3 \text{ N/mm}^2 = \mathbf{47.3 \text{ MPa}} \quad \textbf{Ans.}$$

**EXAMPLE 27.22.** *A composite shaft consists of a steel rod of 60 mm diameter surrounded by a closely fitting tube of brass. Find the outside diameter of the brass tube, when a torque of 1 kN-m is applied on the composite shaft and shared equally by the two materials. Take C for steel as 84 GPa and C for brass as 42 GPa.*

*Also determine the common angle of twist in a length of 4 metres.*

**SOLUTION.** Given: Diameter of steel rod ($D_S$) = 60 mm ; Inner diameter of brass tube ($d_S$) = 60 mm ; Total torque ($T$) = 1 kN-m = $1 \times 10^6$ N-mm ; Torque shared by steel rod ($T_S$) = $\frac{1}{2} \times (1 \times 10^6) = 0.5 \times 10^6$ N-mm ; Torque shared by brass tube ($T_B$) = $(1 \times 10^6) - (0.5 \times 10^6) = 0.5 \times 10^6$ N-mm ; Modulus of rigidity for steel ($C_S$) = 84 GPa = $84 \times 10^3$ N/mm$^2$ ; Modulus of rigidity for brass ($\sigma_B$) = 42 GPa = $42 \times 10^3$ N/mm$^2$ and length of shaft ($l$) = 4 m = $4 \times 10^3$ mm.

***Outside diameter of the brass tube***

Let $D_B$ = Outside diameter of the brass tube in mm.

We know that polar moment of inertia of steel rod,

$$J_S = \frac{\pi}{32} \times (D_S)^4 = \frac{\pi}{32} \times (60)^4 = \frac{12.96 \times 10^6 \, \pi}{32} \text{ mm}^4$$

and polar moment of inertia of brass tube,

$$J_B = \frac{\pi}{32} \times (D_B^4 - d_B^4) = \frac{\pi}{32} \times [D_B^4 - (60)^4] \text{ mm}^4$$

We also know that the relation for angle of twist:

$$\frac{T}{J} = \frac{C\theta}{l} \quad \text{or} \quad \theta = \frac{T \cdot l}{JC}$$

$$\therefore \quad \theta_S = \frac{T \cdot l}{J_S \cdot C_S} = \frac{T \cdot l}{\frac{\pi}{32} \times (60)^4 \times (84 \times 10^3)} \quad ...(i)$$

Similarly,

$$\theta_B = \frac{T \cdot l}{J_B \cdot C_B} = \frac{T \cdot l}{\frac{\pi}{32} \times [D_B^4 - (60)^4] \times 42 \times 10^3} \quad ...(ii)$$

Since $\theta_S$ is equal to $\theta_B$, therefore equating these two values,

$$\frac{T \cdot l}{\frac{\pi}{32} \times (60)^4 \times (84 \times 10^3)} = \frac{T \cdot l}{\frac{\pi}{32} \times [D_B^4 - (60)^4] \times 42 \times 10^3}$$

$$D_B^4 - 12.96 \times 10^6 = 2 \times (12.96 \times 10^6) = 25.92 \times 10^6$$

or

$$D_B^4 = (25.92 \times 10^6) + (12.96 \times 10^6) = 38.88 \times 10^6 \text{ mm}^4$$

$$\therefore \quad D_B = \mathbf{79 \text{ mm}} \quad \textbf{Ans.}$$

***Common angle of twist***

Let $\theta$ = Common angle of twist.

Substituting the values of $T$ (equal to $0.5 \times 10^6$ N-mm) and $l$ (equal to $4 \times 10^3$ mm) in equation ($i$),

$$\theta = \frac{(0.5 \times 10^6) \times (4 \times 10^3)}{\frac{\pi}{32} \times (60)^4 \times (84 \times 10^3)} = 0.0187 \text{ rad} = \mathbf{1.07°} \quad \textbf{Ans.}$$

## 27.11. Strain Energy due to Torsion

We have already discussed in Chapter of Strain Energy that when a body is subjected to a shear stress, the strain energy stored is,

$$U = \frac{\tau^2}{2C} \times V \quad \text{...(See Art. 8.10)}$$

where $\tau$ = Shear stress,

$C$ = Modulus of rigidity or shear modulus and

$V$ = Volume of the body.

But in the case of a solid circular shaft, the torsional stress varies from zero at the central axis to a maximum τ at the surface. Now, consider a circular shaft of diameter $D$, subjected to shear stress (τ) at its surface. Now, let us consider an elementary ring of thickness $dx$ and a distance $x$ from the axis of the shaft.

∴ Area of the ring, $da = 2\pi x\, dx$

and its volume $V = l \cdot 2\pi x\, dx$

We know that shear stress at this section,

$$\theta = \tau \times \frac{x}{R}$$

∴ Strain energy stored in this ring

$$= \frac{q^2}{2C} \times l \cdot 2\pi x\, dx = \frac{\tau^2}{2C} \times \frac{x^2}{R^2} \cdot l \cdot 2\pi x\, dx = \frac{\tau^2}{2CR^2} \cdot 2\pi l \cdot x^3\, dx$$

The total strain energy stored in the shaft may be found out by integrating the above equation from zero to $R$.

$$\therefore \quad U = \int_0^R \frac{\tau^2}{2\,CR^2} \cdot 2\pi l \cdot x^3\, dx$$

$$= \frac{\tau^2\, \pi l}{CR^2} \int_0^R x^3\, dx = \frac{\tau^2\, \pi l}{CR^2} \left[\frac{x^4}{4}\right]_0^R$$

$$= \frac{\tau^2\, \pi l\, R^2}{4C} = \frac{\tau^2}{4C} \times \pi l \times \left(\frac{D}{2}\right)^2 \qquad \ldots\left(\because R = \frac{D}{2}\right)$$

$$= \frac{\tau^2}{4C} \times V \qquad \ldots\left(\because V = \frac{\pi}{4} \times D^2 \times l\right)$$

$$\therefore \text{ For Solid Circular Shaft, } U = \frac{\tau^2}{4C} \times V$$

**Note.** If the shaft is a hollow one, then by integrating the equation between $r$ and $R$ instead of integrating it from zero to $R$, we get the strain energy stored,

$$\therefore \text{ For Hollow Circular Shaft, } U = \frac{\tau^2}{4C}\left(\frac{D^2 + d^2}{D^2}\right) \times V$$

**Example 27.23.** *A solid steel shaft 120 mm diameter and 1.5 m long is used to transmit power from one pulley to another. Determine the maximum strain energy that can be stored in the shaft, if maximum allowable shear stress is 50 MPa. Take shear modulus as 80 GPa.*

**Solution.** Given: Diameter of shaft ($D$) = 120 mm ; Length of shaft ($l$) = 1.5 m = $1.5 \times 10^3$ mm; Allowable shear stress (τ) = 50 MPa = 50 N/mm$^2$ and shear modulus ($C$) = 80 GPa = $80 \times 10^3$ N/mm$^2$.

We know that volume of the shaft,

$$V = \frac{\pi}{4} \times (120)^2 \times (1.5 \times 10^3) = 16.96 \times 10^6 \text{ mm}^3$$

and strain energy stored in the shaft,

$$U = \frac{\tau^2}{4C} \times V = \frac{(50)^2}{4 \times (80 \times 10^3)} \times 16.96 \times 10^6 \text{ N-mm}$$

$$= \mathbf{132.5 \times 10^3 \text{ N-mm}} \qquad \textbf{Ans.}$$

## 27.12. Shaft Couplings

Sometimes, due to the non-availability of a single shaft of the required length, it becomes necessary to connect two shafts together. This is usually done by means of flanged couplings as shown in Fig. 27.6 (*a*) and (*b*).

The flanges of the two shafts are joined together by bolts and nuts or rivets and the torque is then transferred from one shaft to another through the couplings. A little consideration will show that as the torque is transferred through the bolts, it is thus obvious that the bolts are subjected to shear stress. As the diameter of bolts is small, as compared to the diameter of the flanges, therefore shear stress is as assumed to be uniform in the bolts. The design of a shaft coupling means (*a*) design of bolts and (*b*) design of keys.

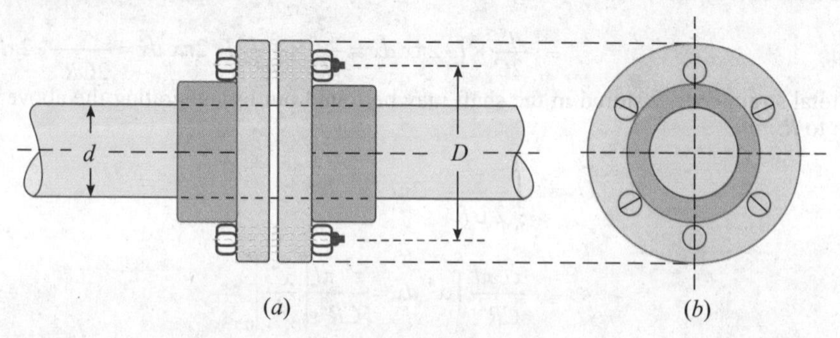


Fig. 27.6

Now we shall discuss the above two designs one by one.

## 27.13. Design of Bolts

Consider a shaft coupling, transmitting torque from one shaft to another.

Let $\tau_S$ = Shear stress in the shaft,

$d$ = Diameter of the shaft,

$D$ = Diameter of the bolt pitch circle,

(*i.e.*, the circle on which the bolts are arranged)

$d_b$ = Diameter of the bolts,

$n$ = No. of bolts and

$\tau_b$ = Shear stress in the bolts.

We know that the torque transmitted by the shaft,

$$T = \frac{\pi}{16} \times \tau_S \times d^3 \quad \text{...}(i)$$

and torque resisted by one bolt

$$= \text{Area} \times \text{Stress} \times \text{Radius of bolt circle}$$

$$= \frac{\pi}{4} \times d_b^2 \times \tau_b \times R = \frac{\pi}{4} \times d_b^2 \times \tau_b \times \frac{D}{2} = \frac{\pi\, d_b^2 \times \tau_b \cdot D}{8}$$

∴ Total torque resisted by the bolts

$$= n \times \frac{\pi \times d_b^2 \times \tau_b\, D}{8} \quad \text{...}(ii)$$

Since the torque resisted by the bolts should be equal to the torque transmitted by the shaft, therefore equating (*i*) and (*ii*),

$$\frac{\pi}{16}\tau_S\, d^3 = \frac{n \times \pi\, d_b^2 \cdot \tau_b\, D}{8}$$

This is the required equation for the number of bolts or the diameter of bolts.

## 27.14. Design of Keys

As a matter of fact, a flange is attached to a shaft by means of a key. A rectangular notch is cut on the circumference of the shaft and a similar notch is cut on inner side of the flange. The flange is then placed over the shaft in such a way, that the two notches form a rectangular hole. A rectangular key is then inserted into the hole, and the flange is said to be keyed to the shaft is shown in Fig. 27.7.

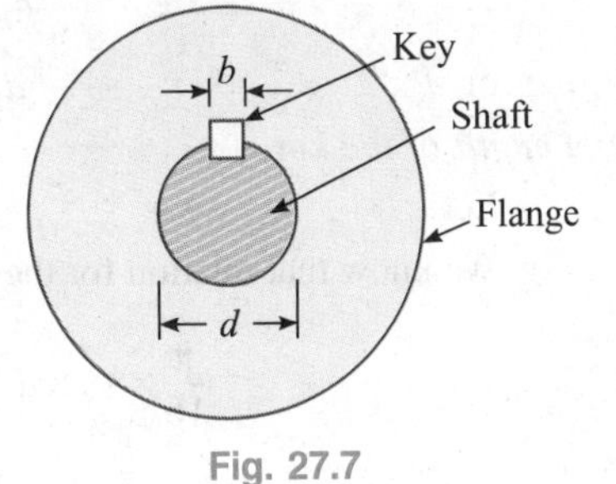


Fig. 27.7

A little consideration will show that the torque is transmitted by the shaft to the flange through the key. It is thus obvious that the key is also subjected to the shear stress.

Let $l$ = Length of the key,

$b$ = Width of the key and

$\tau_K$ = Shear stress in the key.

We know that the torque resisted by the key,

$$T = \text{Area} \times \text{Stress} \times \text{Radius}$$

$$= l \cdot b \times \tau_K \times r = \frac{l \cdot b \cdot \tau_K \cdot d}{2} \quad \text{...(i)}$$

We also know that the torque transmitted by the shaft,

$$T = \frac{\pi}{16} \times \tau_S \times d^3 \quad \text{...(ii)}$$

Since the torque resisted by the key should be equal to the torque transmitted by the shaft, therefore equating (*i*) and (*ii*),

$$\frac{\pi}{16} \times \tau_S \times d^3 = \frac{l \cdot b \cdot \tau_K \cdot d}{2}$$

This is the required equation for the length or width of the key.

**Note.** Sometimes the torque is not transmitted by flange coupling and key. But it is transmitted through gears. In such a case, the gear ratio should also be taken into account, for calculating the revolutions of the shaft.

**Example 27.24.** *A 80 mm shaft transmits power at maximum shear stress of 60 MPa when the stress in key and coupling bolts is 50 MPa and 40 MPa respectively. The coupling has 4 bolts arranged symmetrically along a circle of 200 mm diameter. Determine the diameter of bolts. If the key is 20 mm wide, determine its length.*

**Solution.** Given: Diameter of shaft ($d$) = 80 mm ; Shear stress in shaft ($\tau_S$) = 60 MPa = 60 N/mm$^2$; Stress in key ($\tau_K$) = 50 MPa = 50 N/mm$^2$ ; Stress in bolts ($\tau_b$) = 40 MPa = 40 N/mm$^2$ ; No. of bolts ($n$) = 4 ; Bolt circle diameter ($D$) = 200 mm and width of the key ($b$) = 20 mm.

***Diameter of bolts***

Let $d_b$ = Diameter of the bolts.

We know that relation for the diameter of bolts:

$$\frac{\pi}{16} \times \tau_S \times d^3 = n \times \frac{\pi\, d_b^2 \cdot \tau_b \cdot D}{8}$$

$$\frac{\pi}{16} \times 60 \times (80)^3 = 4 \times \frac{\pi\, d_b^2 \times 40 \times 200}{8}$$

$$1\,920\,000 = 4000\, d_b^2$$

$$\therefore \quad d_b^2 = \frac{1\,920\,000}{4000} = 480$$

or $\quad d_b =$ **21.9 mm say 22 mm Ans.**

*Length of the key*

Let $\quad l =$ Length of the key in mm.

We know that relation for the length of the key:

$$\frac{\pi}{16} \times \tau_S \times d^3 = \frac{l \cdot b \cdot \tau_K \cdot d}{2}$$

$$\frac{\pi}{16} \times 60 \times (80)^3 = \frac{l \times 20 \times 50 \times 80}{2}$$

$$1\,920\,000\,\pi = 40\,000\, l$$

$$\therefore \quad l = \frac{1\,920\,000\,\pi}{40\,000} = \mathbf{150.8\ mm} \quad \textbf{Ans.}$$

**EXAMPLE 27.25.** *A motor shaft consists of a steel tube 30 mm external diameter and 3 mm thick. The engine develops 10 kW at 2000 r.p.m. What will be the maximum stress in the tube, when the power is transmitted through 4 : 1 gearing?*

**SOLUTION.** Given: External diameter of shaft ($D$) = 30 mm ; Thickness = 3 mm or internal diameter ($d$) = 30 – (2 × 3) = 24 mm ; Power ($P$) = 10 kW; Engine speed = 2000 r.p.m. and gearing = 4 : 1.

Let $\quad \tau =$ Torque transmitted by the shaft and

$\quad \tau =$ Maximum shear stress in the shaft.

Since the power is transmitted through 4 : 1 gearing, therefore speed of the shaft,

$$N = \frac{2000}{4} = 500 \text{ r.p.m.}$$

We know that power transmitted by the shaft ($P$)

$$10 = \frac{2\pi NT}{60} = \frac{2\pi \times 500 \times T}{60} = 52.36\, T$$

$$\therefore \quad T = \frac{10}{52.36} = 0.19 \text{ kN-m} = 0.19 \times 10^6 \text{ N-mm}$$

We also know that torque transmitted by the shaft ($T$)

$$0.19 \times 10^6 = \frac{\pi}{16} \times \tau \times \left[\frac{D^4 - d^4}{D}\right] = \frac{\pi}{16} \times \tau \times \left[\frac{(30)^4 - (24)^4}{30}\right] = 3130\,\tau \text{ N-mm}$$

$$= 3130\,\tau$$

$$\therefore \quad \tau = \frac{0.19 \times 10^6}{3130} = 60.7 \text{ N/mm}^2 = \mathbf{60.7\ MPa} \quad \textbf{Ans.}$$

## EXERCISE 27.3

1. A steel shaft *ABC* is subjected to two equal and opposite torques as shown in Fig. 27.8.

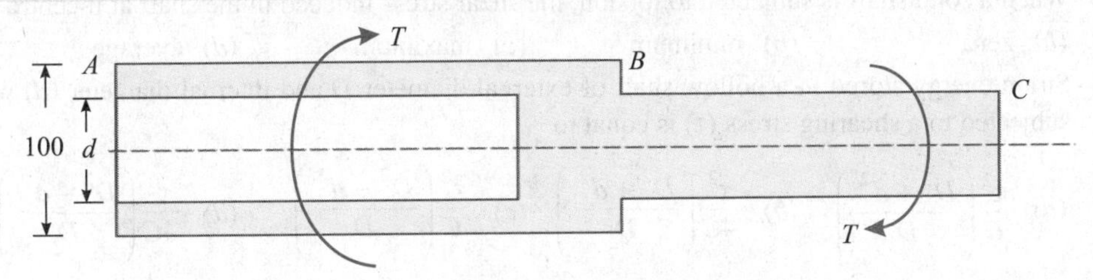


Fig. 27.8

If the shearing stresses developed in *AB* and *BC* are equal, then find the value of internal diameter of the hollow shaft. **(Ans.** 72 mm)

2. A solid steel shaft *AB* of 30 mm diameter has enlarged ends at *A* and *B*. On this enlarged portion is held a steel tube of internal diameter 6 cm and 2 mm thick. If a torque of is applied on the composite shaft, determine torque shared by the shaft and sleeve. **(Ans.** 66 kN-m ; 14 kN-m)
3. A solid shaft 1 m long and 30 mm diameter is transmitting power from one pulley to another. Find the strain energy that can be stored in the shaft, when the shaft is subjected to a shear stress of 40 MPa. Take *C* as 80 GPa. **(Ans.** $3.53 \times 10^3$ N-mm)
4. Two 100 mm diameter shafts are to be connected by means of flanges with 20 mm diameter bolts equally spaced in a circle of diameter 200 mm. If the maximum shear stress in the shaft is not to exceed 75 MPa and the average shear stress in the bolts is not to exceed 60 MPa, determine the number of rivets. **(Ans.** 8)

## QUESTIONS

1. Define the term 'torque'.
2. Write the assumptions for finding out the shear stress in a circular shaft, subjected to torsion.
3. Prove $\frac{\tau}{R} = \frac{C\theta}{l}$ in case of torsion of a circular shaft.
4. Obtain a relation for the torque and power, a solid shaft can transmit.
5. Explain the term 'polar modulus'.
6. Derive an expression for the angle of twist in the case of a member of circular cross-section subjected to torsional moment.

## OBJECTIVE TYPE QUESTIONS

1. Torque transmitted by a solid shaft of diameter (*D*), when subjected to a shear stress ($\tau$) is equal to

   (*a*) $\frac{\pi}{16} \times \tau \times D^2$ (*b*) $\frac{\pi}{16} \times \tau \times D^3$ (*c*) $\frac{\pi}{32} \times \tau \times D^2$ (*d*) $\frac{\pi}{32} \times \tau \times D^3$

2. A shaft revolving at r.p.m. transmits torque (*T*) in kg-m. The power developed is

   (*a*) $2\pi NT$ kW (*b*) $\frac{2\pi NT}{30}$ kW (*c*) $\frac{2\pi NT}{60}$ kW (*d*) $\frac{2\pi NT}{120}$ kW

3. Polar moment of inertia of a solid shaft of diameter (D) is

(*a*) $\frac{\pi}{16} \times D^3$ (*b*) $\frac{\pi}{16} \times D^4$ (*c*) $\frac{\pi}{32} \times D^3$ (*d*) $\frac{\pi}{32} \times D^4$

4. When a solid shaft is subjected to torsion, the shear stress induced in the shaft at it centre is

(*a*) zero (*b*) minimum (*c*) maximum (*d*) average

5. Strain energy stored in a hollow shaft of external diameter *D* and internal diameter (*d*) when subjected to a shearing stress (τ) is equal to

(*a*) $\frac{\tau^2}{C}\left(\frac{D^2 + d^2}{D}\right)$ (*b*) $\frac{\tau^2}{4C}\left(\frac{D^2 + d^2}{D}\right)$ (*c*) $\frac{\tau^2}{C}\left(\frac{D^2 - d^2}{D}\right)$ (*d*) $\frac{\tau^2}{4C}\left(\frac{D^2 - d^2}{D^2}\right)$

## ANSWERS

1. (*b*) 2. (*c*) 3. (*d*) 4. (*a*) 5. (*b*)

Chapter **28**

# Springs

## Contents

## 28.1. Introduction

A spring is a device, in which the material is arranged in such a way that it can undergo a considerable change, without getting permanently distorted. A spring is used to absorb energy due to resilience, which may be restored as and when required. The quality of a spring is judged from the energy it can absorb e.g., in a watch the spring is wound to absorb strain energy. This energy is released to run the watch, when the spring regains its original shape. A carriage spring is used to absorb shocks. It is thus obvious that a spring, which is capable of absorbing the greatest amount of energy for the given stress is known to be the best one.

## 28.2. Stiffness of a Spring

The load required to produce a unit deflection in a spring is called spring stiffness or *stiffness of a spring*.

## 28.3. Types of Springs

We have already discussed that a spring is used for absorbing energy due to resilience. Thus in general, the springs are of the following two types depending upon the type of resilience.

1. Bending spring and 2. Torsion spring.

## 28.4. Bending Springs

A spring, which is subjected to bending only and the resilience is also due to it, is known as *bending spring*. Laminated springs or leaf springs are also called bending springs.

## 28.5. Torsion Springs

A spring, which is subjected to torsion or twisting moment only and the resilience is also due to it, is known as a *torsion spring*. Helical springs are also called torsion springs. Some springs are subjected to bending as well as torsion.

## 28.6. Forms of Springs

Though there are many forms of springs, which are made by the manufacturers, yet the following types of springs are commonly used in Engineering practice.

1. Carriage springs or leaf springs 2. Helical springs.

## 28.7. Carriage Springs or Semi-elliptical Type Leaf Springs

These are also called laminated springs and are of two types: (1) semi-elliptical types (*i.e.*, simply supported at its ends subjected to central load) and (2) quarter-elliptical (*i.e.*, cantilever) types.

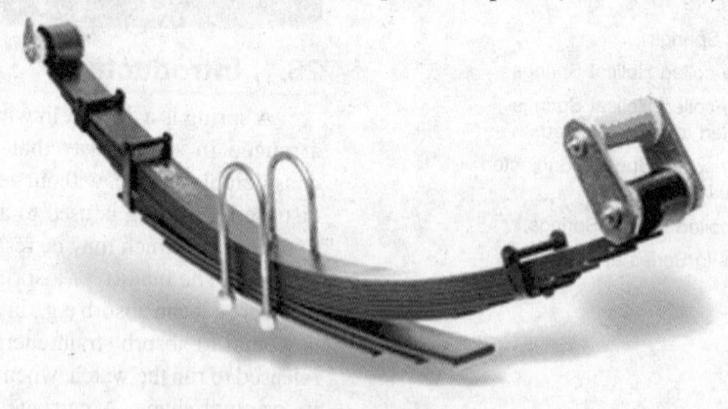

Carriage Springs or Leaf Springs

The carriage springs are widely used in railway wagons, coaches and road vehicles these days. These are used to absorb shocks, which give an unpleasant feeling to the passengers. The energy absorbed by a laminated spring, during a shock, is released immediately without doing any useful work.

A laminated spring, in its simplest form, consists of a number of parallel strips of a metal having different lengths but same width and placed one over the other in laminations as shown in Fig. 28.1.

All the plates are initially bent to the same radius and are free to slide one over the other. When the spring is loaded to the designed load, all the plates become flat and the central deflection disappears. The purpose of this type of arrangement of plates is to make the spring of uniform strength throughout. This is achieved by tapering the ends of the laminations. The semi-elliptical type spring rests on the axis of the vehicle and its top plate is pinned at the ends to the chassis of the vehicle.

Now consider a carriage spring pinned at its both ends, and carrying an upward load at its centre as shown in Fig. 28.1.

Let $l$ = Span of the spring,

$t$ = Thickness of plates,

$b$ = Width of the plates,

$n$ = Number of plates,

$W$ = Load acting on the spring,

$\sigma$ = Maximum bending stress developed in the plates,

$\delta$ = Original deflection of the top spring and

$R$ = Radius of the spring.

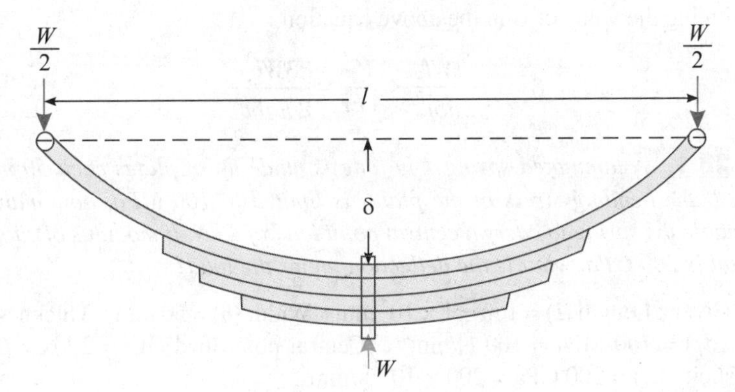


Fig. 28.1. Carriage spring

A little consideration will show, that the load will be acting on the spring on the lowermost plate and it will be shared equally on the two ends of the top plate as shown in Fig. 28.1. We know that the bending moment, at the centre of the span due to this load,

$$M = \frac{Wl}{4} \qquad ...(i)$$

and moment resisted by one plate

$$= \frac{\sigma \cdot I}{y} \qquad ...\left(\because \frac{M}{I} = \frac{\sigma}{y}\right)$$

$$= \frac{\sigma \times \frac{bt^3}{12}}{\frac{t}{2}} = \frac{\sigma \cdot bt^2}{6} \qquad ...\left(\because I = \frac{bt^3}{12} \text{ and } y = \frac{t}{2}\right)$$

∴ Total moment resisted by $n$ plates,

$$M = \frac{n\sigma bt^2}{6} \qquad ...(ii)$$

* For details please refer to Art. 19.2.

Since the maximum bending moment due to load is equal to the total resisting moment, therefore equating (*i*) and (*ii*),

$$\frac{Wl}{4} = \frac{n\sigma bt^2}{6} \quad \text{or} \quad \sigma = \frac{3Wl}{2nbt^2}$$

From the geometry of the spring figure, we know that the central *deflection,

$$\delta = \frac{l^2}{8R} \qquad ...(iii)$$

We also know that in the case of a bending beam,

$$\frac{\sigma}{y} = \frac{E}{R} \quad \text{or} \quad R = \frac{E \cdot y}{\sigma} = \frac{Et}{2\sigma} \qquad ...\left(\because y = \frac{t}{2}\right)$$

Substituting this value of $R$ in equation (*iii*)

$$\delta = \frac{l^2}{8 \times \frac{E \cdot t}{2\sigma}} = \frac{\sigma l^2}{4Et}$$

Now substituting the value of $\sigma$ in the above equation,

$$\delta = \frac{3Wl}{2nbt^2} \times \frac{l^2}{4Et} = \frac{3Wl^3}{8Enbt^3}$$

**EXAMPLE 28.1.** *A laminated spring 1 m long is made up of plates each 50 mm wide and 10 mm thick. If the bending stress in the plates is limited to 100 MPa, how many plates are required to enable the spring to carry a central point load of 2 kN. If modulus of elasticity for the spring material is 200 GPa, what is the deflection under the load?*

**SOLUTION.** Given : Length ($l$) = 1 m = $1 \times 10^3$ mm ; Width ($b$) = 50 mm ; Thickness ($t$) = 10 mm; Bending stress ($\sigma_b$) = 100 MPa = 100 N/mm$^2$ ; Central point load ($W$) = 2 kN = $2 \times 10^3$ N and modulus of elasticity ($E$) = 200 GPa = $200 \times 10^3$ N/mm$^2$.

***No. of plates required in the spring***

Let $n$ = No. of plates required in the spring.

We know that bending stress ($\sigma_b$)

$$100 = \frac{3Wl}{2nbt^2} = \frac{3 \times (2 \times 10^3) \times (1 \times 10^3)}{2n \times 50 \times (10)^2} = \frac{600}{n}$$

$$\therefore \qquad n = \frac{600}{100} = 6 \qquad \textbf{Ans.}$$

***Deflection under the load***

We also know that deflection under the load

$$\delta = \frac{\sigma_b l^2}{4Et} = \frac{100 \times (1 \times 10^3)^2}{4 \times (200 \times 10^3) \times 10} = 12.5 \text{ mm} \qquad \textbf{Ans.}$$

**EXAMPLE 28.2.** *A leaf spring is to be made of seven steel plates 65 mm wide and 6.5 mm thick. Calculate the length of the spring, so that it may carry a central load of 2.75 kN, the bending stress being limited to 160 MPa. Also calculate the deflection at the centre of the spring. Take E for the spring material as 200 GPa.*

**SOLUTION.** Given : No. of plates ($n$) = 7 ; Width ($b$) = 65 mm ; Thickness ($t$) = 6.5 mm ; Central load ($W$) = 2.75 kN = $2.75 \times 10^3$ N ; Maximum bending stress ($\sigma_b$) = 160 MPa = 160 N/mm$^2$ and modulus of elasticity for the spring material ($E$) = 200 GPa = $200 \times 10^3$ N/mm$^2$.

***Length of the spring***

Let $l$ = Length of the spring in mm.

We know that bending stress ($\sigma_b$),

$$160 = \frac{3Wl}{2nbt^2} = \frac{3 \times (2.75 \times 10^3) \times l}{2 \times 7 \times 65 \times (6.5)^2} = 0.215\, l$$

$$\therefore \quad l = \frac{160}{0.215} = 744.2 \text{ mm} \quad \textbf{Ans.}$$

***Deflection at the centre of the spring***

We also know that deflection at the centre of the spring,

$$\delta = \frac{\sigma l^2}{4\,Et} = \frac{160 \times (744.2)^2}{4 \times (200 \times 10^3) \times 6.5} = 17.0 \text{ mm} \quad \textbf{Ans.}$$

**EXAMPLE 28.3.** *A leaf spring 750 mm long is required to carry a central point load of 8 kN. If the central deflection is not to exceed 20 mm and the bending stress is not greater than 200 MPa, determine the thickness, width and number of plates.*

*Also compute the radius, to which the plates should be curved. Assume width of the plate equal to 12 times its thickness and E equal to 200 GPa.*

**SOLUTION.** Given : Length ($l$) = 750 mm ; Point load ($W$) = 8 kN = $8 \times 10^3$ N ; Central deflection ($\delta$) = 20 mm ; Bending stress ($\sigma_b$) = 200 MPa = 200 N/mm$^2$ ; Width of plates ($b$) = 12 $t$ (wheree $t$ is the thickness of the plates) and modulus of elasticity ($E$) = 200 GPa = $200 \times 10^3$ N/mm$^2$.

***Thickness of the plates***

We know that central deflection ($\delta$)

$$20 = \frac{\sigma_b l^2}{4\,Et} = \frac{200 \times (750)^2}{4 \times (200 \times 10^3) \times t} = \frac{140.6}{t}$$

$$\therefore \quad t = \frac{140.6}{20} = 7.0 \text{ mm} \quad \textbf{Ans.}$$

***Width of the plates***

We know that width of the plates,

$$b = 12\,t = 12 \times 7 = 84 \text{ mm} \quad \textbf{Ans.}$$

***Number of plates***

Let $n$ = Number of plates

We also know that bending stress ($\sigma_b$)

$$200 = \frac{3Wl}{2nbt^2} = \frac{3 \times (8 \times 10^3) \times 750}{2n \times 84 \times (7)^2} = \frac{2187}{n}$$

$$\therefore \quad n = \frac{2187}{200} = 10.9 \text{ say } 11 \quad \textbf{Ans.}$$

## 28.8 Quarter-Elliptical Type Leaf Springs

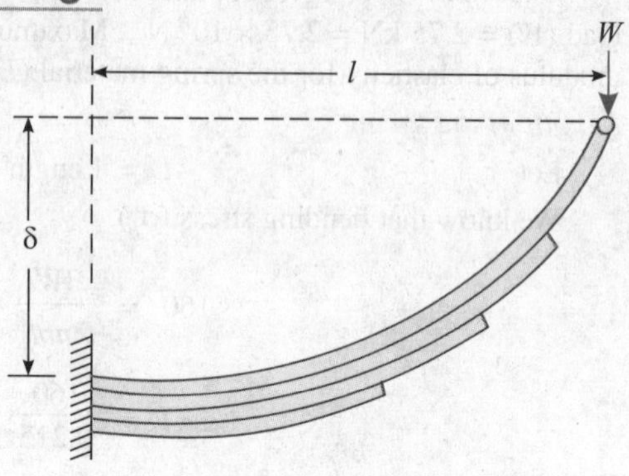

Fig. 28.2. Quarter-elliptical spring.

The quarter-elliptical type leaf springs are rarely used, except as certain parts in some machines. Like a carriage spring, a quarter-elliptical type leaf spring consists of a number of parallel strips of a metal having different lengths but same width and placed one over the other in laminations as shown in Fig. 28.2. All the plates are initially bent to the same radius and are free to slide one over the other.

Now consider a quarter-elliptical type leaf spring subjected to a load at its free end as shown in Fig. 28.2.

Let $l$ = Length of the spring,
$t$ = Thickness of the plates,
$b$ = Width of the plates,
$n$ = Number of plates,
$W$ = Load acting at the free end of the spring, and
$\delta$ = Original deflection of the spring.

We know that the bending moment at the fixed end of the leaf,

$$M = Wl \qquad ...(i)$$

and moment resisted by one plate

$$= \frac{\sigma \cdot I}{y}$$

$$= \frac{\sigma \times \frac{bt^3}{12}}{\frac{t}{2}} = \frac{\sigma bt^2}{6} \qquad ...\left(\because I = \frac{bt^3}{12} \text{ and } y = \frac{t}{2}\right)$$

∴ Total moment resisted by $n$ plates,

$$M = \frac{n\sigma bt^2}{6} \qquad ...(ii)$$

Since the maximum bending moment due to load is equal to the total resisting moment, therefore, equating (*i*) and (*ii*),

$$Wl = \frac{n\sigma bt^2}{6} \quad \text{or} \quad \sigma = \frac{6Wl}{nbt^2}$$

From the geometry of the spring figure, we know that

$$\delta(2R - \delta) = l \cdot l = l^2$$

$$\therefore \quad \delta = \frac{l^2}{2R} \qquad ...(\text{Neglecting } \delta^2) \qquad ...(iii)$$

We know that in the case of a bending cantilever,

$$= \frac{E}{R}$$

or $$R = \frac{E \cdot y}{\sigma} = \frac{Et}{2\sigma} \qquad \ldots\left(\because\ y = \frac{t}{2}\right)$$

Substituting this value of $R$ in equation (*iii*),

$$\delta = \frac{l^2}{2 \times \frac{Et}{2\sigma}} = \frac{\sigma l^2}{Et}$$

Now substituting the value of $\sigma$ in the above equation,

$$\delta = \frac{6Wl}{nbt^2} \times \frac{l^2}{Et} = \frac{6Wl^3}{Enbt^3}$$

**EXAMPLE 28.4.** *A quarter-elliptic leaf spring 800 mm long is subjected to a point load of 10 kN. If the bending stress and deflection is not to exceed 320 MPa and 80 mm respectively, find the suitable size and number of plates required by taking the width as 8 times the thickness. Take E as 200 GPa.*

**SOLUTION.** Given : Length ($l$) = 800 mm ; Point load ($W$) = 10 kN = $10 \times 10^3$ N ; Bending stress ($\sigma_b$) = 320 MPa = 320 N/mm$^2$ ; Deflection ($\delta$) = 80 mm ; Plate width $b = 8\,t$ (where $t$ is the thickness of the plates) and modulus of elasticity ($E$) = 200 GPa = $200 \times 10^3$ N/mm$^2$.

***Thickness of the plates***

Let $t$ = Thickness of plates in mm, and

$n$ = Number of the plates

We know that bending stress ($\sigma_b$),

$$320 = \frac{6Wl}{nbt^2} = \frac{6 \times (10 \times 10^3) \times 800}{nbt^2} = \frac{48 \times 10^6}{nbt^2} \qquad \ldots(i)$$

and deflection ($\delta$),

$$80 = \frac{6Wl^3}{Enbt^3} = \frac{6 \times (10 \times 10^3) \times (800)^3}{(200 \times 10^3) \times nbt^2} = \frac{153.6 \times 10^6}{nbt^3} \qquad \ldots(ii)$$

Dividing equation (*ii*) by (*i*),

$$\frac{80}{320} = \frac{\frac{153.6 \times 10^6}{nbt^3}}{\frac{48 \times 10^6}{nbt^2}} = \frac{3.2}{t}$$

$$\therefore \quad t = \frac{3.2 \times 320}{80} = 12.8 \text{ say } 13 \text{ mm} \qquad \textbf{Ans.}$$

***Width of plates***

We know that width of the plates

$$b = 8\,t = 8 \times 13 = 104 \text{ mm} \qquad \textbf{Ans.}$$

***Number of plates required***

Substituting the values of $t$ and $b$ in equation (*i*)

$$320 = \frac{48 \times 10^6}{n \times 104 \times (13)^2} = \frac{2.731 \times 10^3}{n}$$

$$\therefore \quad n = \frac{2.731 \times 10^3}{320} = 8.5 \text{ say } 9 \qquad \textbf{Ans.}$$

## EXERCISE 28.1

1. A laminated spring 1 m long is built in 100 mm × 10 mm plates. If the spring is to carry a load of 10 kN at its centre, determine the number of plates required for the spring. Take allowable bending stress as 150 MPa. **[Ans.** $n$ = 10]
2. A leaf spring 1 m long is made up with steel plates with width equal to 6 times its thickness. Design the spring for a load of 15 kN when the maximum stress is 100 MPa and deflection is not to exceed 16 mm. **[Ans.** $t$ = 12.5 mm ; $b$ = 75 mm ; $n$ = 12]
3. A carriage spring 800 mm long is made of 12 plates of 40 mm width. Determine the thickness of the plates, if bending stress is not to exceed 200 MPa and spring is to carry a load of 6 kN at its centre. Also determine the central deflection of the spring. Take $E$ as 200 GPa. **[Ans.** 9 mm ; 16.5 mm)
4. A laminated spring of the quarter elliptic type 600 mm long is to provide a deflection of 75 mm under an end load of 1960 N. If the leaf material is 60 mm wide and 6 mm thick, find the number of leaves required and the maximum stress. **[Ans.** 13 ; 252 N/mm$^2$]

## 28.9. Helical Springs

It is a torsion springs and made up of a wire coiled into a helix. Though, there are many types of helical springs, yet the following are important from the subject point of view:

1. Closely-coiled helical springs and
2. Open-coiled helical springs.

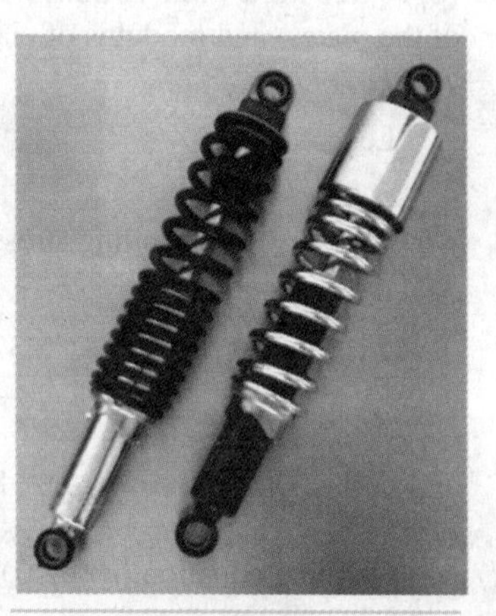

Helical springs

## 28.10. Closely-coiled Helical Springs

In a closely coiled helical spring, the spring wire is coiled so close that the each turn is practically a plane at right angles to the axis of the helix and the wire is subjected to torsion. The bending stress is negligible as compared to the torsional stress. A closely-coiled helical spring may be subjected to (1) axial loading and (2) axial twist. In this chapter, we shall discuss both the cases one by one.

## 28.11. Closely-coiled Helical Springs Subjected to an Axial Load

Consider a closely-coiled helical spring subjected to an axial load as shown in Fig. 28.3.

Let $d$ = Diameter of the spring wire,

$R$ = Mean radius of the spring coil,

$n$ = No. of turns of coils,

$C$ = Modulus of rigidity for the spring material,

$W$ = Axial load on the spring,

$\tau$ = Maximum shear stress induced in the wire due to twisting,

$\theta$ = Angle of twist in the spring wire and

$\delta$ = Deflection of the spring, as a result of axial load.

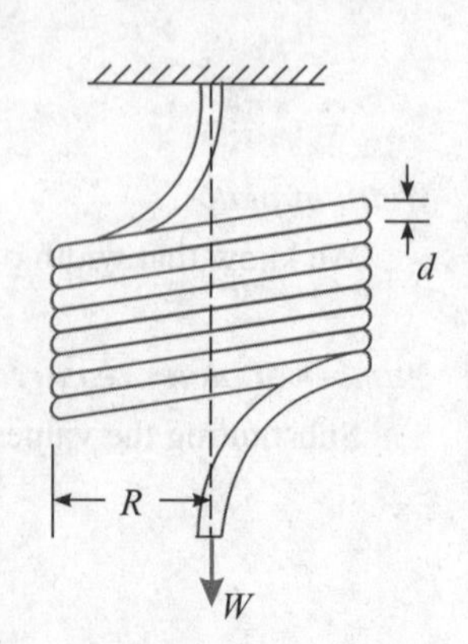


Fig. 28.3

A little consideration will show that the load $W$ will cause a twisting moment,

$$T = W \cdot R \qquad ...(i)$$

We know that the twisting moment,

$$T = \frac{\pi}{16} \times \tau \times d^3 \qquad ...(ii)$$

$$\therefore \quad W \cdot R = \frac{\pi}{16} \times \tau \times d^3$$

We also know that the length of the wire,

$$l = \text{Length of one coil} \times \text{No. of coils}$$

$$= 2\pi R \cdot n$$

We have discussed in Chapter of Torsion of Circular Shafts that

$$\frac{T}{J} = \frac{C.\theta}{l} \qquad \text{... (See Art. 27.7)}$$

or

$$\theta = \frac{T \cdot l}{J \cdot C} = \frac{WR \cdot 2\pi R n}{\frac{\pi}{32} \times d^4 C} \qquad ...(\because T = WR)$$

$$= \frac{64 WR^2 n}{Cd^4}$$

$\therefore$ Deflection of the spring,

$$\delta = R \cdot \theta = R \times \frac{64 WR^2 n}{Cd^4} = \frac{64 WR^3 n}{Cd^4}$$

We know that the energy stored in the spring,

$$U = \frac{1}{2} W \cdot \delta$$

and stiffness of the spring,

$$s = \frac{W}{\delta} = \frac{Cd^4}{64 R^3 n}$$

**EXAMPLE 28.5.** *A close-coiled helical spring is required to carry a load of 150 N. If the mean coil diameter is to be 8 times that of the wire, calculate these diameters. Take maximum shear stress as 100 MPa.*

**SOLUTION.** Given : Load $(W)$ = 150 N ; Diameter of coil $(D) = 8d$ (where $d$ is the diameter of the wire) or radius $(R) = 4\,d$ and maximum shear stress $(\tau)$ = 100 MPa = 100 N/mm$^2$.

We know that relation for the twisting moment,

$$WR = \frac{\pi}{16} \times \tau \times d^3$$

or

$$150 \times 4\,d = \frac{\pi}{16} \times 100 \times d^3$$

or

$$d^2 = \frac{150 \times 4 \times 16}{\pi \times 100} = 30.6$$

or

$$d = 5.53 \text{ say } 6 \text{ mm} \qquad \textbf{Ans.}$$

and

$$D = 8\,d = 8 \times 6 = 48 \text{ mm} \qquad \textbf{Ans.}$$

**EXAMPLE 28.6.** *A closely-coiled helical spring of round steel wire 5 mm in diameter having 12 complete coils of 50 mm mean diameter is subjected to an axial load of 100 N. Find the deflection of the spring and the maximum shearing stress in the material. Modulus of rigidity (C) = 80 GPa.*

**SOLUTION.** Given : Diameter of spring wire ($d$) = 5 mm ; No. of coils ($n$) = 12 ; Diameter of spring ($D$) = 50 mm or radius ($R$) = 25 mm ; Axial load ($W$) = 100 N and modulus of rigidity ($C$) = 80 GPa = $80 \times 10^3$ N/mm$^2$.

*Deflection of the spring*

We know that deflection of the spring,

$$\delta = \frac{64\, WR^3\, n}{Cd^4} = \frac{64 \times 100 \times (25)^3 \times 12}{(80 \times 10^3) \times (5)^4} = 24 \text{ mm} \quad \textbf{Ans.}$$

*Maximum shearing stress in the material*

Let $\tau$ = Maximum shearing stress in the material.

We also know that relation for the torque,

$$WR = \frac{\pi}{16} \times \tau \times d^3$$

or

$$100 \times 25 = \frac{\pi}{16} \times \tau \times (5)^3$$

$$2500 = 24.54\, \tau$$

$$\therefore \quad \tau = \frac{2500}{24.54} = 101.9 \text{ N/mm}^2 = 101.9 \text{ MPa} \quad \textbf{Ans.}$$

## 28.12. Closely-coiled Helical Springs Subjected to an Axial Twist

Consider a closely-coiled helical spring subjected to an axial twist as shown in Fig. 28.4.

Let $d$ = Diameter of the spring wire,

$R$ = Mean radius of the spring coil,

$n$ = No. of turns of coils,

$C$ = Modulus of rigidity for the spring material and

$M$ = Moment or axial twist applied on the spring.

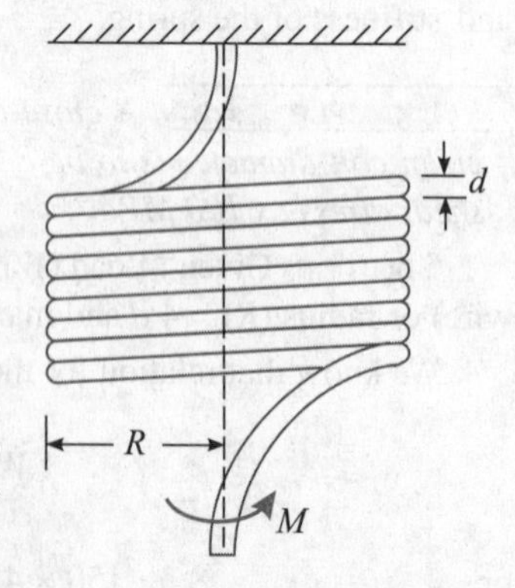

Fig. 28.4. Closely-coiled helical spring

A little consideration will show that the number of spring coils will tend to increase or decrease depending upon the sense of the moment. Moreover, if the number of turns tend to increase then the mean radius of the spring coil will decrease. Now let us consider that the number of turns increase from $n$ to $n'$ and the mean radius decreases from $R$ to $R'$.

Now length of the spring,

$$l = 2\,\pi\, Rn = 2\pi\, R'\, n' \quad \text{...}(i)$$

$$\therefore \quad \frac{1}{R} = \frac{2\,\pi n}{l} \quad \text{and} \quad \frac{1}{R'} = \frac{2\pi n'}{l}$$

We know that $\frac{M}{I} = E \times \text{Change of curvature}$

$$= E\left(\frac{1}{R'} - \frac{1}{R}\right) = E\left(\frac{2\pi n'}{l} - \frac{2\pi n}{l}\right) = \frac{2\pi E}{l}(n' - n)$$

or $$2\pi\,(n' - n) = \frac{Ml}{EI} \qquad ...(ii)$$

We also know that the total angle of bend,

$$\phi = 2\pi\,(n' - n)$$

Substituting the value of $2\pi\,(n' - n)$ from equation (*ii*),

$$\phi = \frac{Ml}{EI}$$

Differentiating the above equation with respect to $l$,

$$\frac{d\phi}{dl} = \frac{M}{EI}$$

It is thus obvious that the change in curvature or angle of bend per unit length, is constant throughout the spring.

We know that the energy stored in the spring,

$$U = \frac{1}{2} M \cdot \phi$$

**EXAMPLE 28.7.** *A closely-coiled helical spring is made up of 10 mm diameter steel wire having 10 coils with 80 mm mean diameter. If the spring is subjected to an axial twist of 10 kN-mm, determine the bending stress and increase in the number of turns. Take E as 200 GPa.*

**SOLUTION.** Given : Diameter of spring wire ($d$) = 10 mm ; No. of coils ($n$) = 10 ; Diameter of coil ($D$) = 80 mm or radius ($R$) = 40 mm ; Axial twist ($M$) = 10 kN-mm = $10 \times 10^3$ N-mm and modulus of elasticity ($E$) = 200 GPa = $200 \times 10^3$ N/mm$^2$.

***Bending stress in the wire***

We know that moment of inertia of the spring wire section,

$$I = \frac{\pi}{64} \times d^4 = \frac{\pi}{64} \times (10)^4 = 490.9 \text{ mm}^4$$

∴ Bending stress in the wire

$$\sigma = \frac{M}{I} \times y = \frac{10 \times 10^3}{490.9} \times 5 = 101.9 \text{ N/mm}^2 = 101.9 \text{ MPa} \qquad \textbf{Ans.}$$

***Increase in the number of turns***

We know that length of the coil,

$$l = 2\pi\, Rn = 2\pi \times 40 \times 10 = 800\,\pi \text{ mm}$$

and increase in the no. of turns,

$$n' - n = \frac{Ml}{EI} \times \frac{1}{2\pi} = \frac{(10 \times 10^3) \times 800\pi}{(200 \times 10^3) \times 490.9} \times \frac{1}{2\pi} = 0.04 \qquad \textbf{Ans.}$$

## 28.13. Open-coiled Helical Springs

In an open helical spring, the spring wire is coiled in such a way, that there is large gap between the two consecutive turns. As a result of this, the spring can take compressive load also. An open helical spring, like a closed helical spring, may be subjected to (1) axial loading or (2) axial twist. In this chapter, we shall discuss only the first case.

Now consider an open coiled helical spring subjected to an axial load as shown in Fig. 28.5.

Let $d$ = Diameter of the spring wire,

$R$ = Mean radius of the spring coil,

$P$ = Pitch of the spring coils,

$n$ = No. of turns of coils,

$C$ = Modulus of rigidity for the spring materials,

$W$ = Axial load on the spring,

$\tau$ = Maximum shear stress induced in the spring wire due to loading,

$\sigma_b$ = Bending stress induced in the spring wire due to bending,

$\delta$ = Deflection of the spring as a result of axial load and

$\alpha$ = Angle of helix.

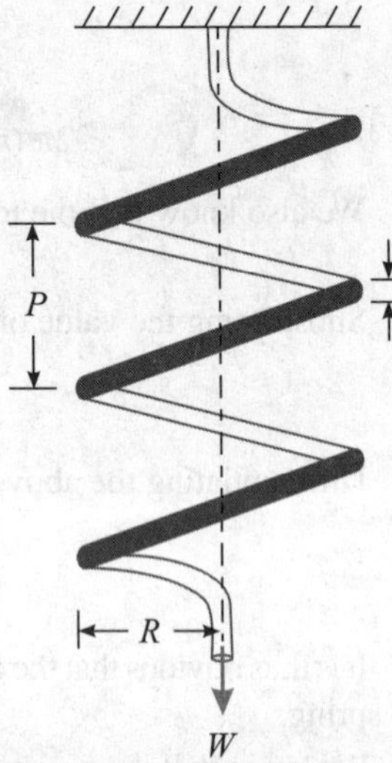

Fig. 28.5. Open coiled helical spring

A little consideration will show that the load $W$ will cause a moment $WR$. This moment may be resolved into the following two components,

$$T = WR \cos \alpha \quad \text{...(It causes twisting of coils)}$$

$$M = WR \sin \alpha \quad \text{...(It causes bending of coils)}$$

Let $\delta$ = Angle of twist, as a result of twisting moment, and

$\phi$ = Angle of bend, as a result of bending moment.

We know that the length of the spring wire,

$$l = 2\pi \, nR \sec \alpha \quad \text{...(i)}$$

and twisting moment,

$$W \cdot R \cos \alpha = \frac{\pi}{16} \times \tau \times d^3 \quad \text{...(ii)}$$

We also know that bending stress,

$$\sigma_b = \frac{M \cdot y}{I} \frac{WR \sin \alpha \cdot \frac{d}{2}}{\frac{\pi}{64} \times d^4} \quad \text{...}\left(\because \frac{M}{I} = \frac{\sigma_b}{y}\right)$$

$$= \frac{32\, WR \sin \alpha}{\pi d^3} \quad \text{...(iii)}$$

and angle of twist

$$\theta = \frac{Tl}{JC} = \frac{WR \cos \alpha \cdot l}{JC} \quad \text{...}\left(\because \frac{T}{J} = \frac{C\theta}{l}\right)$$

We have also seen in the previous article, that angle of bend due to bending moment,

$$\phi = \frac{Ml}{EI} = \frac{WR \sin \alpha \cdot l}{EI}$$

We know that the work done by the load in deflecting the spring, is equal to the stress energy of the spring.

$$\therefore \quad \frac{1}{2} W \cdot \delta = \frac{1}{2} T.\theta + \frac{1}{2} M\phi$$

or

$$W.\delta = T.\theta + M.\phi$$

$$= \left[ WR \cos\alpha \times \frac{WR\cos\alpha \cdot l}{JC} \right] + \left[ WR \sin\alpha \times \frac{WR \sin\alpha \cdot l}{EI} \right]$$

or

$$\delta = WR^2 l \left[ \frac{\cos^2\alpha}{JC} + \frac{\sin^2\alpha}{EI} \right] \quad ...(iv)$$

Now substituting the values of $l = 2\pi nR \sec\alpha$, $J = \frac{\pi}{32}(d)^4$ and $I = \frac{\pi}{64}(d)^4$ in the above equation

$$\delta = WR^2 \times 2\pi nR \sec\alpha \left[ \frac{\cos^2\alpha}{\frac{\pi}{32} d^4 C} + \frac{\sin^2\alpha}{E \times \frac{\pi}{64} d^4} \right]$$

$$= \frac{64\,WR^3 n \sec\alpha}{d^4} \left[ \frac{\cos^2\alpha}{C} + \frac{2\sin^2\alpha}{E} \right] \quad ...(v)$$

**Note.** If we substitute $\alpha = 0$ in the above equation, it gives deflection of a closed coiled spring *i.e.*,

$$\delta = \frac{64\,WR^2\, n}{Cd^4}$$

**Example 28.8.** *An open coil helical spring made up of 10 mm diameter wire and of mean diameter of 100 mm has 12 coils, angle of helix being 15°. Determine the axial deflection and the intensities of bending and shear stresses under an axial load of 500 N. Take C as 80 GPa and E as 200 GPa.*

**Solution.** Given : Diameter of wire $(d)$ = 10 mm ; Mean diameter of spring $(D)$ = 100 mm or radius $(R)$ = 50 mm ; No. of coils $(n)$ = 12 ; Angle of helix $(\alpha)$ = 15° ; Load $(W)$ = 500 N ; Modulus of rigidity $(C) = 80 \text{ GPa} = 80 \times 10^3 \text{ N/mm}^2$ and modulus of elasticity $(E) = 200 \text{ GPa} = 200 \times 10^3 \text{ N/mm}^2$.

***Deflection of the spring***

We know that deflection of the spring,

$$\delta = \frac{64\,WR^3 n \sec\alpha}{d^4} \left[ \frac{\cos^2\alpha}{C} + \frac{2\sin^2\alpha}{E} \right]$$

$$= \frac{64 \times 500 \times (50)^3 \times 12 \sec 15°}{(10)^4} \left[ \frac{\cos^2 15°}{80 \times 10^3} + \frac{2\sin^2 15°}{200 \times 10^3} \right] \text{mm}$$

$$= 4\,800\,000 \times 1.0353 \left[ \frac{(0.9659)^2}{80 \times 10^3} + \frac{2 \times (0.2588)^2}{200 \times 10^3} \right] \text{mm}$$

$$= 4969\,440 \times \frac{2.467}{200 \times 10^3} = 61.3 \text{ mm} \quad \textbf{Ans.}$$

***Bending stress in the section***

We know that bending moment in the coil,

$$M = WR \sin\alpha = 500 \times 50 \sin 15° \text{ N-mm}$$

$$= 25\,000 \times 0.2588 = 6470 \text{ N-mm}$$

and moment of inertia of the spring wire,

$$I = \frac{\pi}{64} \times (d)^4 = \frac{\pi}{64} \times (10)^4 = 490.9 \text{ mm}^4$$

∴ Bending stress in the section

$$\sigma_b = \frac{M}{I} \times y = \frac{6470}{490.9} \times 5 = 65.9 \text{ N/mm}^2 = 65.9 \text{ MPa} \qquad \textbf{Ans.}$$

***Shear stress induced in the wire***

Let $\tau$ = Shear stress induced in the wire in N/mm².

We know that twisting moment (or torque) in the coil,

$$T = WR \cos \alpha = 500 \times 50 \cos 15° \text{ N-mm}$$
$$= 25\,000 \times 0.9659 = 24\,150 \text{ N-mm}$$

We also know that twisting moment ($T$)

$$24\,150 = \frac{\pi}{16} \times \tau \times d^3 = \frac{\pi}{16} \times \tau \times (10)^3 = 196.4\,\tau$$

$$\therefore \quad \tau = \frac{24150}{196.4} = 123 \text{ N/mm}^2 = 123 \text{ MPa} \qquad \textbf{Ans.}$$

## 28.14. Springs in Series and Parallel

In the previous articles, we have been discussing the use of one helical spring only. But sometimes two (or even more) springs are used at one place. Though there are many ways of using these springs, yet the springs in (*i*) series and (*ii*) parallel are important from the subject point of view.

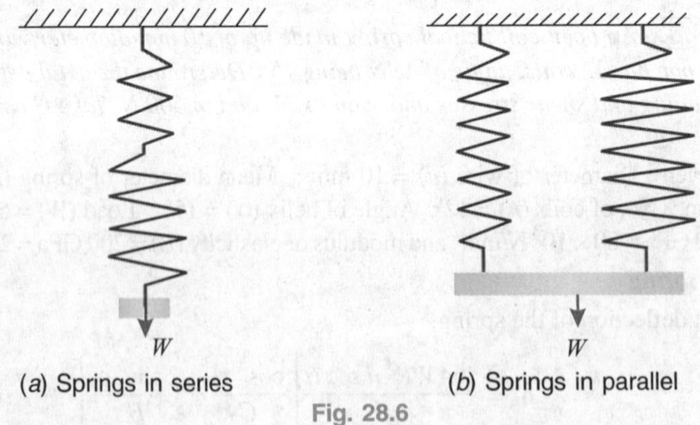


(*a*) Springs in series (*b*) Springs in parallel

Fig. 28.6

1. ***Springs in series***

In this case, the two springs are connected in series as shown in Fig. 28.6 (*a*). Each spring is subjected to the same load applied at the end of one spring. A little consideration will show that the total extension of the assembly is equal to the algebraic sum of the extensions of the two springs.

2. ***Springs in parallel***

In this case, the two springs are connected in parallel as shown in Fig. 28.6 (*b*). The extension of each spring is the same. A little consideration will show that the load applied on the assembly is shared by the two springs.

**EXAMPLE 28.9.** *Two-close coiled helical springs wound from the same wire, but with different core radii having equal no. of coils are compressed between rigid plates at their ends. Calculate the maximum shear stress induced in each spring, if the wire diameter is 10 mm and the load applied between the rigid plates is 500 N. The core radii of the springs 100 mm and 75 mm respectively.*

**SOLUTION.** Given : No. of coils in the outer spring $(n_1) = n_2$ (where $n_2$ is the no. of coils in the inner spring) ; Diameter of spring wire $(d) = 10$ mm ; Load $(W) = 500$ N ; Radius of outer spring $(R_1)$ = 100 mm and radius of inner spring $(R_2) = 75$ mm.

Let $\tau_1$ = Shear stress developed in the outer spring,

$W_1$ = Load shared by the outer spring and

$\tau_2, W_2$ = Corresponding values for the inner spring.

We know that deflection of the outer spring,

$$\delta_1 = \frac{64\,W_1 R_1^3\, n_1}{Cd^4} = \frac{64 \times W_1 \times (100)^3 \times n_1}{C \times (10)^4} = \frac{6400\,W_1\, n_1}{C} \quad ...(i)$$

Similarly,
$$\delta_2 = \frac{64\,W_2 R_2^3\, n_2}{Cd^4} = \frac{64 \times W_2 \times (75)^3 \times n_2}{C \times (10)^4} = \frac{2700\,W_2\, n_2}{C} \quad ...(ii)$$

Since the springs are held between two rigid plates, therefore deflections in both the springs must be equal. Now equating *(i)* and *(ii)*,

$$\frac{6400\,W_1\, n_1}{C} = \frac{2700\,W_2\, n_2}{C}$$

$$6400\,W_1 = 2700\,W_2 \quad ...(\because n_1 = n_2)$$

or
$$W_1 = \frac{27\,W_2}{64}$$

We also know that $W_1 + W_2 = 500$

$$\therefore \quad \frac{27\,W_2}{64} + W_2 = 500$$

or
$$\frac{91\,W_2}{64} = 500$$

$$\therefore \quad W_2 = \frac{500 \times 64}{91} = 351.6 \text{ N}$$

and
$$W_1 = 500 - W_2 = 500 - 351.6 = 148.4 \text{ N}$$

We know that relation for torque,

$$W_1 R_1 = \frac{\pi}{16} \times \tau_1 \times d^3$$

or
$$148.4 \times 100 = \frac{\pi}{16} \times \tau_1 \times (10)^3$$

$$\therefore \quad \tau_1 = \frac{148.4 \times 100 \times 16}{\pi\,(10)^3} = 75.6 \text{ N/mm}^2 = 75.6 \text{ MPa} \qquad \textbf{Ans.}$$

Similarly,
$$\tau_2 = \frac{351.6 \times 75 \times 16}{\pi\,(10)^3} = 134.3 \text{ N/mm}^2 = 134.4 \text{ MPa} \qquad \textbf{Ans.}$$

## EXERCISE 28.2

1. A closely coiled helical spring of mean diameter 140 mm is made up of 12 mm diameter steel wire. Calculate the direct axial load, the spring can carry if the maximum stress is not to exceed 100 MPa. [**Ans.** 484 N]

2. A closely coiled helical spring is made of 6 mm wire. The maximum shear stress and deflection under a 200 N load is not to exceed 80 MPa and 11 mm respectively. Determine the no. of coils and their mean diameter. Take $C$ = 84 MPa. [**Ans.** 20 ; 34 mm]

3. A open coil helical spring made of 10 mm diameter wire has 15 coils of 50 mm radius with a 20° angle of helix. Determine the deflection of the spring, when subjected to an axial load of 300 N. Take $E$ = 200 GPa and = 80 GPa.

[**Ans.** 47.4 mm]

## QUESTIONS

1. What is a spring? Explain its uses.
2. What are various types of springs? Distinguish clearly between bending springs and torsion springs.
3. What is a laminated spring? Where is it used?
4. Derive from first principles, making usual assumptions the formula for the maximum bending stress and for the central deflection of a leaf spring consisting of $n$ leaves and subjected to a central load.
5. What are helical springs? Differentiate between a closely coiled helical spring and an open coiled helical spring.
6. A closely coiled helical spring with $D$ as diameter of the coil and $d$ as diameter of the wire is subjected to an axial load $W$. Prove that the maximum shear stress produced is equal to $\dfrac{8WD}{\pi d^3}$.
7. Derive an equation for the deflection of an open coiled helical spring.

## OBJECTIVE TYPE QUESTIONS

1. In a leaf spring, maximum bending stress developed in the plates is

   (a) $\dfrac{Wl}{nbt^2}$ (b) $\dfrac{2Wl}{nbt^2}$ (c) $\dfrac{3Wl}{nbt^2}$ (d) $\dfrac{3Wl}{2nbt^2}$

2. The maximum deflection at the centre of a leaf spring is

   (a) $\dfrac{\sigma_b l}{Et}$ (b) $\dfrac{\sigma_b l^2}{2Et}$ (c) $\dfrac{\sigma_b l^2}{3Et}$ (d) $\dfrac{\sigma_b l^2}{4Et}$

3. When a closely coiled spring is subjected to an axial load, it is said to be under"

   (a) bending (b) shear (c) torsion (d) all of these

4. The deflection of a closely-coiled helical spring of diameter ($D$) subjected to an axial load ($W$) is

   (a) $\dfrac{64\,WR^3 n}{Cd^4}$ (b) $\dfrac{64\,WR^2 n}{Cd^4}$ (c) $\dfrac{64\,WRn}{Cd^4}$ (d) $\dfrac{64\,WRn^2}{Cd^4}$

## ANSWERS

1. (d) 2. (d) 3. (c) 4. (a)

Chapter **29**

# Riveted Joints

### Contents

## 29.1. Introduction

We see that in a steel structure, such as plate girder, roof truss, boiler shells, ship-building, etc., its various members, which constitute the structure, are connected to one another. These members are connected either by welding or by rivets. In riveting, specially made ductile metal pins, called rivets, are inserted into the holes of the members. The rivets are hammered for permanent fastenings. It may be noted that a riveted joint is used in a structure only when it is a must, as it weakens the section due to hole.

## 29.2. Types of Riveted Joints

Following are two types of riveted joints, depending upon the way in which the members are connected:

**1.** Lap joint  **2.** Butt joint.

## 29.3. Lap Joint

It is a joint in which a part of one member overlaps the other and the two members are then riveted together as shown in Fig. 29.1.

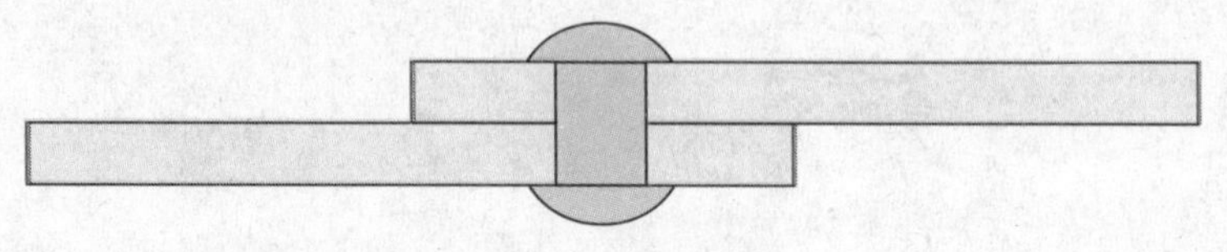

Fig. 29.1. Lap joint.

## 29.4. Butt Joint

It is a joint in which the edges of the two members butt (*i.e.*, touch) against each other and a cover plate is placed either on one side or on both sides of the two members. The cover plate is then riveted together with both the members. Butt joints are of the following two types:

**1.** Single cover butt joint, **2.** Double cover butt joint

## 29.5. Single Cover Butt Joint

It is a joint in which the edges of the two members butt against each other, and only one cover plate is placed on one side of the members and riveted together as shown in Fig. 29.2.

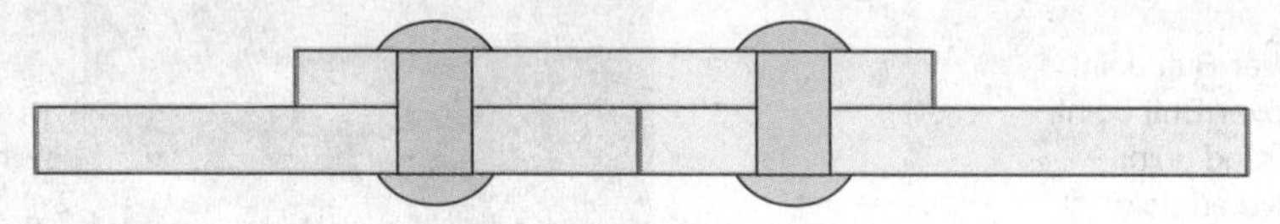

Fig. 29.2. Single cover butt joint.

## 29.6. Double Cover Butt Joint

It is a joint in which the edges of the two members butt against each other, and two cover plates placed on both the sides of the member and riveted together as shown in Fig. 29.3.

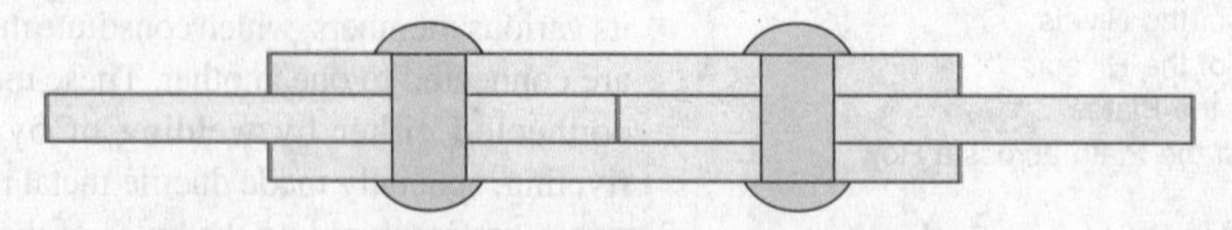

Fig. 29.3. Double cover butt joint.

In addition to the above, following are the types of the riveted joints, depending upon the way in which the rivets are connected:

**1.** Single riveted joint,

**2.** Double riveted joint,

**3.** Multiple riveted joint,

**4.** Chain riveted joint and

**5.** Zig-zag riveted joint.

## 29.7. Single Riveted Joint

It is a joint in which there is a single row of rivets in a lap joint as shown in Fig. 29.4(*a*). Or there is a single row of rivets on each side in a butt joint as shown in Fig. 29.4(*b*).

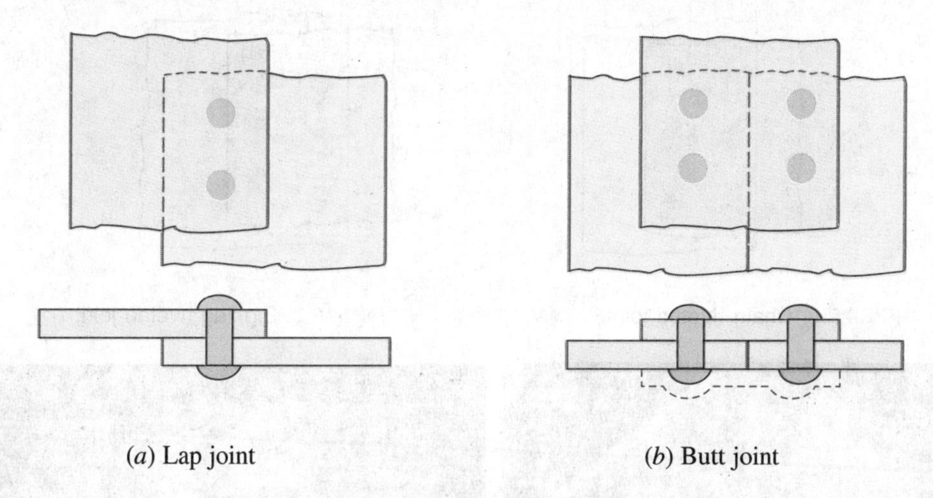

(*a*) Lap joint (*b*) Butt joint

Fig. 29.4. Single riveted joint.

## 29.8. Double Riveted Joint

It is a joint in which there are two rows of rivets in a lap joint as shown in Fig. 29.5 (*a*). Or there are two rows of rivets on each side in a butt joint as shown in Fig. 29.5 (*b*).

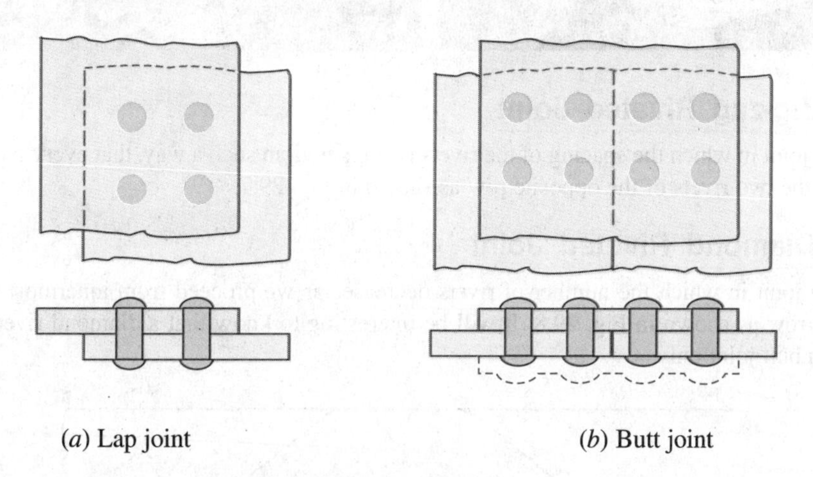

(*a*) Lap joint (*b*) Butt joint

Fig. 29.5. Double riveted joint.

## 29.9. Multiple Riveted Joint

It is a joint in which there are more than two rows of rivets in a lap joint or more than two rows of rivets on each side in a butt joint. A multiple riveted joint may be a triple riveted joint or quadruple riveted joint.

## 29.10. Chain Riveted Joint

It is a joint in which every rivet of a row is opposite to other rivet of the other row as shown in Fig. 29.6.

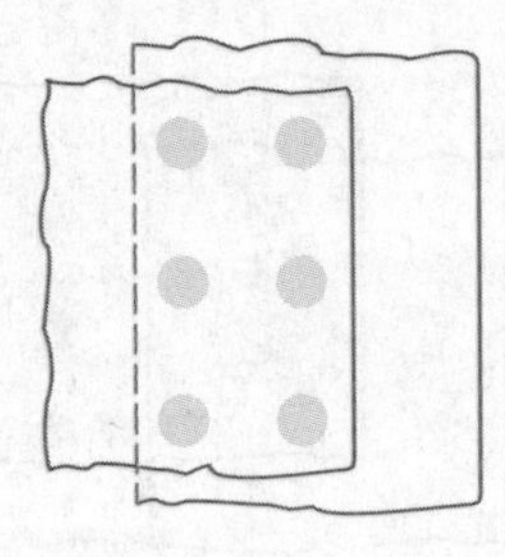

Fig. 29.6. Chain riveted joint.

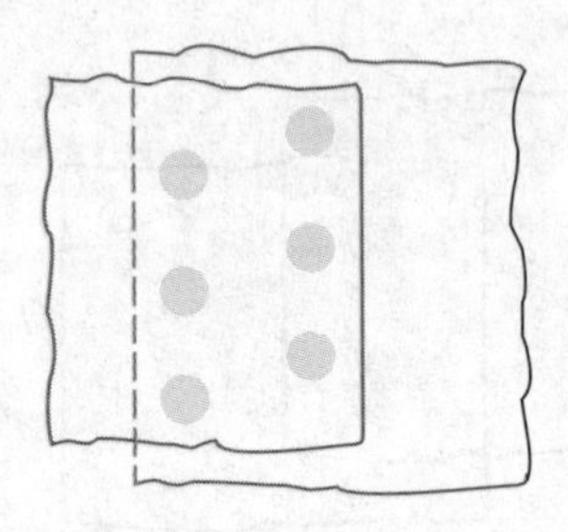

Fig. 29.7. Zig-zag riveted joint

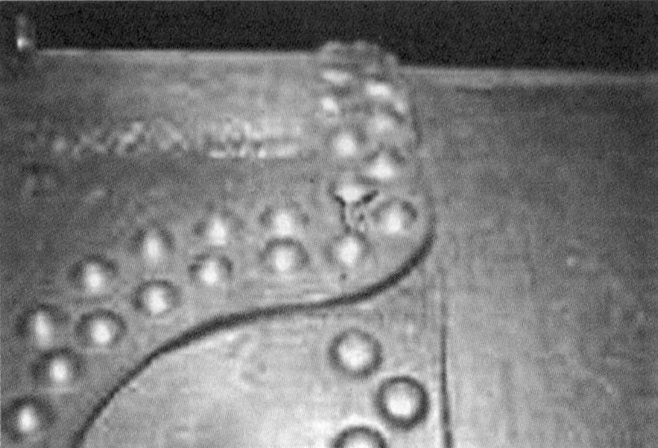

## 29.11. Zig-zag Riveted Joint

It is a joint in which the spacing of the rivets is staggered, in such a way, that every rivet is in the middle of the two rivets of the opposite row as shown in Fig. 29.7.

## 29.12. Diamond Riveted Joint

It is a joint in which the number of rivets decreases as we proceed from innermost row to the outermost row as shown in Fig. 29.8. It will be interesting to know that a diamond riveted joint is adopted in butt joint only.

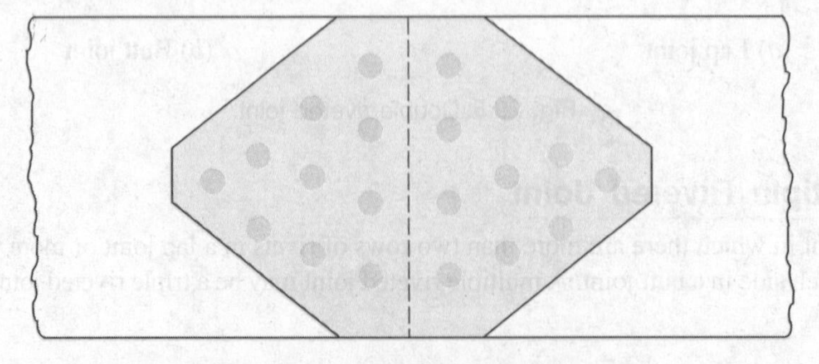

Fig. 29.8. Diamond riveted joint.

## 29.13. Pitch of Rivets

The centre to centre distance between the two consecutive rivets in a row, is called the pitch of the rivets and is denoted by the letter *p* as shown in Fig. 29.9.

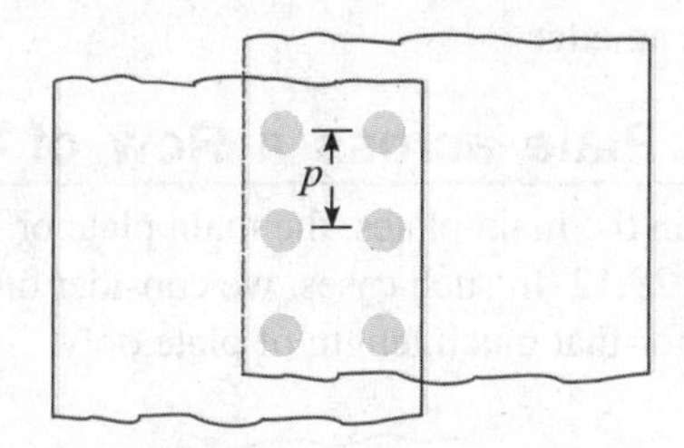


Fig. 29.9. Pitch of rivets

In some cases, the pitch is different for different rows. In such cases, the pitch is the least in the innermost row.

## 29.14. Failure of Riveted Joint

A riveted joint may fail due to many ways. But the following are more important from the subject point of view:

**1.** Failure of the rivets. **2.** Failure of the plates.

We shall discuss both the above mentioned cases of failure, in more details, one by one.

## 29.15. Failure of the Rivets

A rivet may fail due to either of the following two reasons:

**1.** Shearing of the rivet. **2.** Crushing of the rivet.

Now we shall discuss both the cases one by one.

## 29.16. Shearing of the Rivets

The plates, which are connected by the rivets, exert tensile stress on the rivets. If the rivets are unable to resist the stresses, they may be sheared off as shown in Fig. 29.10. It will be interesting to know that

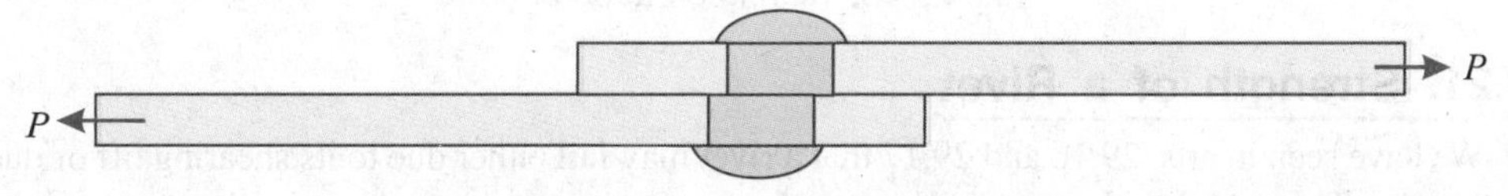


Fig. 29.10. Shearing of rivets

1. The rivets are in single shear in a lap joint and in a single cover butt joint.
2. The rivets are in double shear in a double cover butt joint.

## 29.17. Crushing of the Rivets

Sometimes, the rivets do not actually shear off, under the tensile stress, but are crushed as shown in Fig. 29.11. Such a failure of rivet is called crushing of the rivet.

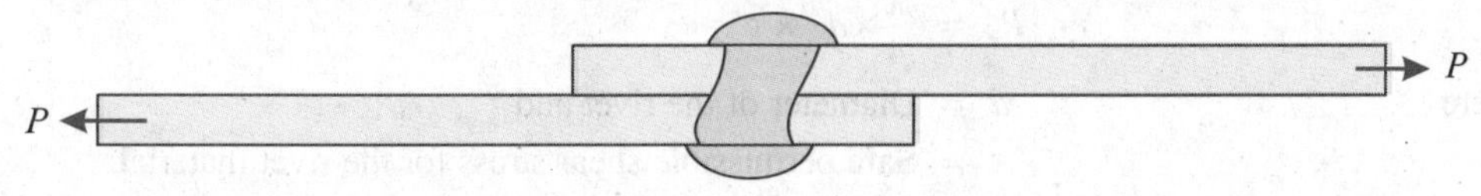


Fig. 29.11. Crushing of rivets

## 29.18. Failure of the plates

A plate may fail in many ways. But the following are important from the subject point of view:

1. Tearing off the plate across a row of rivets,
2. Tearing off the plate at an edge.

## 29.19. Tearing off the Plate across a Row of Rivets

Due to the tensile stresses, in the main plates, the main plate or cover plates may tear off across a row of rivets as shown in Fig. 29.12. In such cases, we consider only one pitch length of the plate, since every rivet is responsible for that much length of plate only.

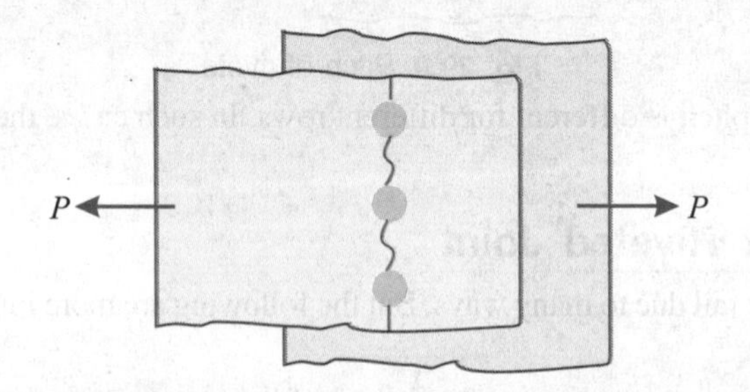


Fig. 29.12. Tearing across a row of rivets

## 29.20. Tearing off the Plate at an Edge

A plate may also fail due to tearing at an edge as shown in Fig. 29.13. This can be avoided by keeping the centre of the nearest rivet, from the edge of the plate, at least two times the diameter of the rivet.

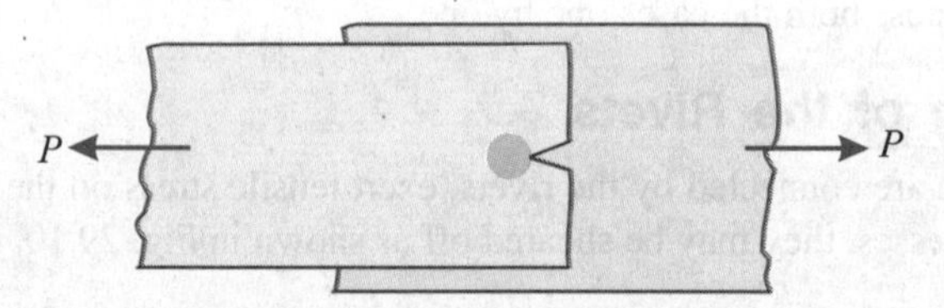


Fig. 29.13. Tearing off at an edge

## 29.21. Strength of a Rivet

We have seen in arts. 29.16 and 29.17 that a rivet may fail either due to its shearing off or due to its crushing. Thus, while calculating the strength of a rivet we see as to how much resistance it can offer. The resistance offered by a rivet is known as its strength or the value of the rivet. Following two values of a rivet are important from the subject point of view:

1. Shearing value. 2. Bearing value.

1. ***Shearing value***

The resistance offered by a rivet, to be sheared off is known as its shearing value. Mathematically, pull required to shear off the rivet,

$$P_s = \frac{\pi}{4} \times d^2 \times \tau$$

where $d$ = Diameter of the rivet and

$\tau$ = Safe permissible shear stress for the rivet material.

If the rivet is in double shear (*i.e.*, in the case of a double cover butt joint), the corresponding equation becomes:

$$P_s = 2 \times \frac{\pi}{4} \times d^2 \times \tau$$

2. *Bearing value*

The resistance offered by a rivet to be crushed is known as the bearing value. Mathematically, pull required to crush the rivet,

$$P_b = \sigma_b \times t \times d$$

where $\sigma_b$ = Safe permissible bearing stress for the rivet material,

$t$ = Thickness of the main plate and

$d$ = Diameter of the rivet.

NOTE: Sometimes, the bearing strength of a rivet is also termed as its crushing strength. And permissible bearing stress is termed as permissible crushing stress ($\sigma_c$).

## 29.22. Strength of a Plate

We have seen in Art. 29.19 that the plate may fail due to its tearing off across the row of rivets. Thus, while calculating the strength of a plate, we see as to how much resistance it can offer. This resistance offered by the plate, against tearing off, is known as the strength of the plate or value of the plate. Mathematically, pull required to tear off the plate,

$$P_t = \sigma_t (p - d)\, t$$

where $\sigma_t$ = Permissible tensile stress for the plate material,

$p$ = Pitch of the rivet and

$t$ = Thickness of the plate.

## 29.23. Strength of a Riveted Joint

The strength of a joint may be definéd as the maximum force, which it can transmit, without causing it to fail. We have seen in Arts. 29.21 and 29.22 that $P_s$, $P_b$ and $P_t$ are the pulls required to shear of the rivet, crushing of the rivet and tearing of the plate. A little consideration will show that if we go on increasing the pull on a riveted joint, it will fail when the least of these three pulls is reached; because a higher value of the other pulls will never reach, since the joint has already failed, either by shearing of the rivet or by crushing of the rivet or by tearing of the plate.

If the joint is continuous (as in the case of boilers) the strength is calculated per pitch length. But if the length of the joint is small, the strength is calculated for the whole length of the plate.

## 29.24. Efficiency of a Riveted Joint

The efficiency of a riveted joint is the ratio of the strength of the joint, to the strength of the unriveted plate. Mathematically

$$\text{Efficiency} \quad \eta = \frac{\text{Least of } P_s, P_b \text{ and } P_t}{P}$$

where $P = \sigma_1 \times p \times t$

$P$ = Pull required to tear off the unriveted plate,

$\sigma_1$ = Permissible tensile stress for the plate material,

$p$ = Pitch of the rivets and

$t$ = Thickness of the plate.

**EXAMPLE 29.1.** *A single riveted lap joint is made in 12 mm thick plates with 22 mm diameter rivets. Determine the strength of the rivet, if the pitch of the rivets is 60 mm. Take allowable stresses in shearing as 60 MPa, in bearing as 150 MPa and in tearing as 80 MPa respectively.*

**SOLUTION.** Given: Thickness of plates ($t$) = 12 mm; Diameter of rivets ($d$) = 22 mm ; Pitch ($p$) = 60 mm ; Allowable shear stress ($\tau$) = 60 MPa = 60 N/mm$^2$ ; Allowable stress in bearing ($\sigma_b$) = 150 MPa = 150 N/mm$^2$ and allowable stress in tearing ($\sigma_t$) = 80 MPa = 80 N/mm$^2$

First of all, let us calculate the pulls required for shearing and crushing of the rivets as well as tearing of the main plates.

1. ***Shearing of rivets.*** We know that in a lap joint, the rivets are in single shear. Thus, in a single riveted lap joint, the strength of one rivet in single shear is taken. Therefore shearing strength of the rivet,

$$P_s = \tau \times \frac{\pi}{4} \times (d)^2 = 60 \times \frac{\pi}{4} \times (22)^2 = 22\,810 \text{ N} = 22.81 \text{ kN} \quad ...(i)$$

2. ***Bearing of rivets.*** We know that in a single riveted joint, the strength of one rivet in bearing is taken. Therefore bearing strength of the rivet,

$$P_b = \sigma_b \times t \times d = 150 \times 12 \times 22 = 39\,600 \text{ N} = 39.6 \text{ kN} \quad ...(ii)$$

3. ***Tearing of the plate.*** We also know that in a continuous joint, the strength of one pitch length is taken. Therefore tearing strength of the main plate,

$$P_t = \sigma_t (p - d)\, t = 80\,(60 - 22)\,12 = 36\,480 \text{ N} = 36.48 \text{ kN} \quad ...(iii)$$

Thus strength of the rivet joint is the least of the above mentioned three values $P_s$, $P_b$ and $P_t$. *i.e.,* 22.81 kN **Ans.**

**EXAMPLE 29.2.** *Two plates 10 mm thick are joined by a double riveted lap joint. The pitch of each row of rivets is 50 mm. The rivets are 20 mm diameter and the permissible stresses are:*

*Shearing of rivets = 70 MPa*
*Bearing of rivets = 160 MPa*
*Tearing of the plate = 100 MPa*

*Determine the maximum tensile force on the joint and efficiency of the joint.*

**SOLUTION.** Given: Thickness of plates ($t$) = 10 mm ; Pitch ($p$) = 50 mm; Diameter of rivets ($d$) = 20 mm ; Permissible stress in shearing ($\tau$) = 70 MPa = 70 N/mm$^2$ ; Permissible stress in bearing ($\sigma_b$) = 160 MPa = 160 N/mm$^2$ and permissible stress in tearing ($\sigma_t$) = 100 MPa = 100 N/mm$^2$.

***Maximum tensile force of the joint***

First of all, let us calculate the pulls required for shearing and crushing of the rivets as well as tearing of the main plates.

1. ***Shearing of rivets.*** We know that in a lap joint, the rivets are in single shear. Thus, in a double riveted lap joint, the strength of two rivets in single shear is taken. Therefore shearing strength of the rivets.

$$P_s = 2 \times \tau \times \frac{\pi}{4} \times (d)^2 = 2 \times 70 \times \frac{\pi}{4} \times (20)^2 \text{ N}$$

$$= 44\,000 \text{ N} = 44 \text{ kN} \quad ...(i)$$

2. ***Bearing of rivets.*** We know that in a double riveted joint, the strength of two rivets is taken. Therefore bearing strength of the rivets,

$$P_b = 2 \times \sigma_b \times t \times d = 2 \times 160 \times 10 \times 20 = 64\,000 \text{ N} = 64 \text{ kN} \quad ...(ii)$$

3. ***Tearing of the plates.*** We know that in a continuous joint, the strength of one pitch length is taken. Therefore tearing strength of the main plates,

$$P_t = \sigma_t (p - d)\, t = 100\,(50 - 20)\,10 \text{ N} = 30\,000 \text{ N} = 30 \text{ kN} \quad ...(iii)$$

Thus minimum tensile force on the joint is the least of the above mentioned three values $P_s$, $P_b$ and $P_t$, *i.e.*, **30 kN.** **Ans.**

*Efficiency of the joint*

We also know that strength of the unriveted plate per pitch length,

$$P = \sigma_t \times p \times t = 100 \times 50 \times 10 \text{ N} = 50\,000 \text{ N} = 50 \text{ kN} \qquad ...(iv)$$

∴ Efficiency of the joint,

$$\eta = \frac{30}{50} = 0.6 = \mathbf{60\%} \qquad \textbf{Ans.}$$

**EXAMPLE 29.3.** *A double riveted double cover butt joint is made in 20 mm thick plates with 25 mm diameter rivets and 100 mm pitch. The permissible stresses are;*

$$\tau = 80 \text{ MPa};\ \sigma_c = 160 \text{ MPa and } \sigma_t = 100 \text{ MPa}$$

*Find the pull per pitch length of the joint and efficiency of the joint.*

**SOLUTION.** Given: Thickness of Plate ($t$) = 20 mm ; Diameter of rivets ($d$) = 25 mm; Pitch of rivets ($p$) = 100 mm ; Permissible shear stress in rivets ($\tau$) = 80 MPa = 80 N/mm$^2$ ; Permissible crushing stress of plates ($\sigma_c$)= 160 MPa = 160 N/mm$^2$ and Permissible tearing stress of plates ($\sigma_t$) = 100 MPa = 100 N/mm$^2$.

*Pull per pitch length of the joint*

First of all, let us calculate the pulls required for shearing and crushing of the rivets as well as tearing of the main plates.

1. ***Shearing of rivets.*** We know that in a double cover butt joints, the rivets are in double shear. Thus, in a double riveted double covered butt joint, the strength of two rivets in double shear is taken. Therefore strength of the rivets,

$$P_s = 2 \times 2 \times \tau \times \frac{\pi}{4} \times (d)^2 = 4 \times 80 \times \frac{\pi}{4} \times (25)^2 = 157\,100 \text{ N}$$

$$= 157.1 \text{ kN} \qquad ...(i)$$

2. ***Crushing of the rivets.*** We know that in a double riveted joint, the strength of two rivets in crushing is taken. Therefore crushing strength of rivets,

$$P_c = 2 \times \sigma_c \times t \times d = 2 \times 160 \times 20 \times 25 = 160\,000 \text{ N}$$

$$= 160\,000 \text{ N} = 160 \text{ kN} \qquad ...(ii)$$

3. ***Tearing of the plates.*** We know that in a continuous joint, the strength of one pitch length is taken. Therefore tearing strength of the main plates,

$$P_t = \sigma_t\,(p - d)\,t = 100\,(100 - 25)\,20 = 150\,000 \text{ N} = 150 \text{ kN} \qquad ...(iii)$$

Thus, pull per pitch length of the joint is the least of the above mentioned three values $P_s$, $P_b$ and $P_t$ *i.e.*, **150 kN.** **Ans.**

*Efficiency of the joint*

We also know strength of unriveted plate per pitch length,

$$P = \sigma_t \times p \times t = 100 \times 100 \times 20 = 200\,000 \text{ N} = 200 \text{ kN} \qquad ...(iv)$$

∴ Efficiency of the joint,

$$\eta = \frac{150}{200} = 0.75 = \mathbf{75\%} \qquad \textbf{Ans.}$$

**EXAMPLE 29.4.** *A 200 × 20 mm steel strap is spliced as shown in Fig. 29.14.*

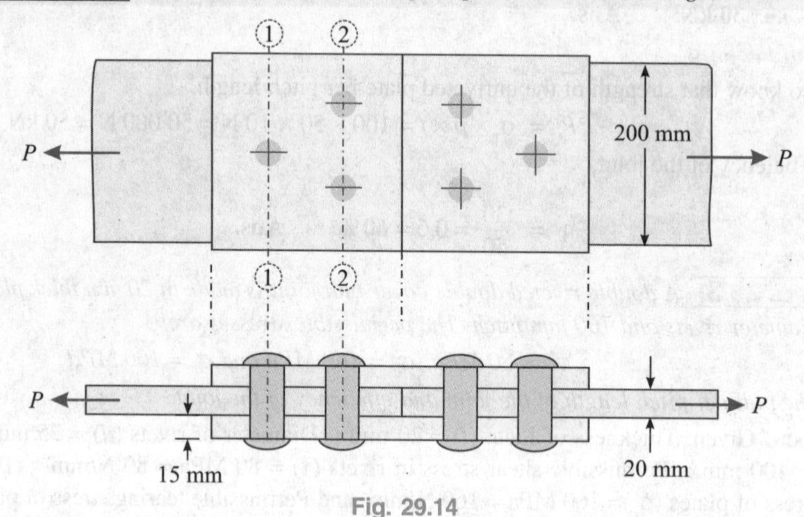


Fig. 29.14

*The rivets are 30 mm in diameter. The allowable working stresses are 70 MPa in shear, 180 MPa in bearing and 90 MPa in tearing. What is the safe load on the spliced strap and efficiency of the joint?*

**SOLUTION.** Given: Width of plates ($b$) = 200 mm; Thickness of main plates ($t$) = 20 mm; diameters of rivets ($d$) = 30 mm; Thickness of cover plates ($t_c$) = 15 mm; Premissible shear stress in rivets ($\tau$) = 70 MPa = 70 N/mm$^2$; Permissible bearing stress in rivets ($\sigma_b$) = 180 MPa = 180 N/mm$^2$ and permissible tearing stress in plates ($\sigma_t$) = 90 MPa = 90 N/mm$^2$.

***Safe load on the spliced strap***

First of all, let us calculate the pulls required for shearing and crushing of the rivets as well as tearing of the main plate and cover plates. Since the joint is not a continuous one, therefore we shall find out the strength of the joint for 3 rivets and full plate width.

1. ***Shearing of rivets.*** We know that in a double covered butt joints, the rivets are in double shear. Therefore strength of 3 rivets in double shear,

$$P_s = 3 \times 2 \times \tau \times \frac{\pi}{4} \times (d)^2 = 6 \times 70 \times \frac{\pi}{4} \times (30)^2 = 297\ 000 \text{ N}$$

$$= 297 \text{ kN} \qquad ...(i)$$

2. ***Bearing of rivets.*** Since there are 3 rivets on either side of the joints, therefore bearing strength of 3 rivets,

$$P_b = 3 \times \sigma_b \times t \times d = 3 \times 180 \times 20 \times 30 = 324\ 000 \text{ N} = 324 \text{ kN} \qquad ...(ii)$$

From the above two values (*i.e.*, $P_s$ and $P_b$), we find that strength of 3 rivets is lesser of these values *i.e.*, 297 kN. Therefore, strength of one rivet is 297/3 = 99 kN.

3. ***Tearing of the main plates.*** We know that as the joint is not continuous, therefore strength of the whole plate is calculated as section (1-1) and (2-2) independently. We also know that at section (1-1), there is only one rivet hole. Therefore tearing strength of the main plate at section (1-1),

$$P_{t1} = \sigma_t\,[b - (1 \times d)]\,t = 90\,[200 - (1 \times 30)]\,20 = 306\ 000 \text{ N}$$

$$= 306 \text{ kN} \qquad ...(iii)$$

We know that at section (2-2), there are two rivet holes. Therefore tearing strength of the main plate at section (2-2),

$$= \sigma_t\,[b - (2 \times d)]\,t = 90\,[200 - (2 \times 30)]\,20 = 252\ 000 \text{ N}$$

$$= 252 \text{ kN}$$

From the above two values of $Pt_1$ and $Pt_2$, we find that tearing strength of the main plate appears to be less at section (2-2). But it will be interesting to know that for tearing of the plate at section (2-2), the rivet in front of it (i.e., at section 1-1) must first give way. Therefore actual strength of the main plate at section (2-2),

$$P_{t2} = 252 + \text{Strength of one rivet} = 252 + 99 = 351 \text{ kN} \quad ...(iv)$$

4. ***Tearing of cover plates.*** We know that the cover plates are weaker at section (2-2), where there are 2 rivet holes. Therefore strength of two cover plates of section (2-2),

$$P_{tc} = \sigma_t [b - (2 \times d)] 2t_c = 90 [200 - (2 \times 30)] 2 \times 15 \text{ N}$$
$$= 378\ 000 \text{ N} = 378 \text{ kN} \quad ...(v)$$

Thus, safe load on the spliced strap is the least of above mentioned five values $P_s$, $P_b$, $P_{t1}$, $^*P_{t2}$ and $P_{tc}$ *i.e.,* **297 kN** **Ans.**

***Efficiency of the Joint***

We also know that strength of the unriveted steel strap,

$$P = \sigma_t \times b \times t = 90 \times 200 \times 20 = 360\ 000 = 360 \text{ kN} \quad ...(vi)$$

∴ Efficiency of the joint,

$$\eta = \frac{297}{360} = 0.825 = \mathbf{82.5\%} \quad \textbf{Ans.}$$

**EXAMPLE 29.5.** *Two 300 mm × 16 mm plates are riveted together with 25 mm diameter rivets and quadruple riveted double cover butt joint as shown in Fig. 29.15.*

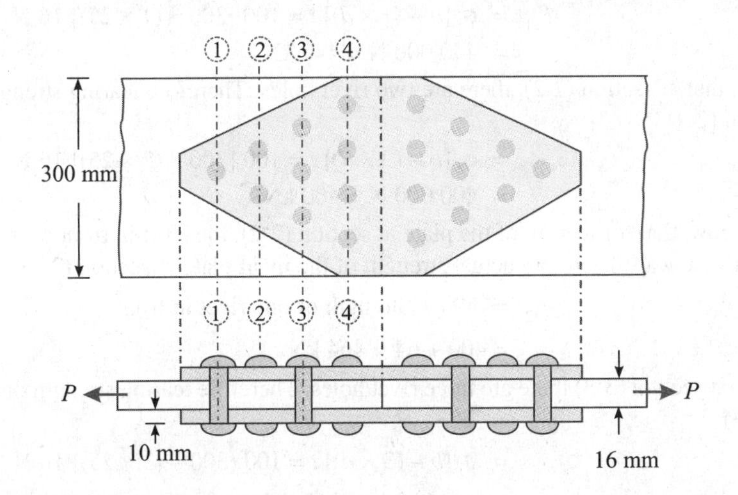


Fig. 29.15

*Find the strength and efficiency of the joint, if the thickness of the cover plates be 10 mm. The permissible stresses are:*

*Shearing ($\tau$) = 80 MPa; Crushing ($\sigma_c$) = 160 MPa; Tearing ($\sigma_t$) = 100 MPa.*

**SOLUTION.** Given : Width of plates ($b$) = 300 mm ; Thickness of plates ($t$) = 16 mm ; Diameter of rivets ($d$) = 25 mm ; Thickness of cover plates ($t_c$) = 10 mm ; Permissible shear stress ($\tau$) = 80 MPa = 80 N/mm$^2$ ; Permissible crushing stress ($\sigma_c$) = 160 MPa = 160 N/mm$^2$ and permissible tearing stress ($\sigma_t$) = 100 MPa = 100 N/mm$^2$.

* The reason for taking the strength of 3 rivets is that, there are only 3 rivets on either side of the joint.

***Strength of the joint***

First of all let, us calculate the pulls required for shearing and crushing of the rivets as well as tearing of the main plate and cover plates. Since the joint is not a continuous one, therefore, we shall find out the strength of the joint for 10 rivets and full plate width (as there are 10 rivets on either side of the joints).

1. ***Shearing of rivets.*** We know that in a double covered butt joint, the rivets are in double shear. Therefore strength of 10 rivets in double shear,

$$P_s = 10 \times 2 \times \tau \frac{\pi}{4} \times (d)^2 = 20 \times 80 \times \frac{\pi}{4} \times (25)^2 \text{ N}$$
$$= 785\,000 \text{ N} = 785.0 \text{ kN} \quad ...(i)$$

2. ***Crushing of rivets.*** Since, there are 10 rivets on either side of the point, therefore crushing strength of 10 rivets,

$$P_c = 10 \times \sigma_c \times t \times d = 10 \times 160 \times 16 \times 25 = 640\,000 \text{ N}$$
$$= 640 \text{ kN} \quad ...(ii)$$

From the above two values (*i.e.*, $P_s$ and $P_c$), we find that strength of 10 rivets is lesser of these values *i.e.*, 640 kN. Therefore strength of one rivet = 640/10 = 64 kN.

3. ***Tearing of the main plate.*** We know that as the joint is not continuous, therefore strength of the whole plate is calculated at sections (1-1), (2-2), (3-3) and (4-4) independently. We also know that at section (1-1), there is only one rivet hole. Therefore tearing strength of the main plate at section (1-1),

$$P_{t1} = \sigma_t\,[b - (1 \times d)]\,t = 100\,[300 - (1 \times 25)]\,16 \text{ N}$$
$$= 440\,000 \text{ N} = 440 \text{ kN} \quad ...(iii)$$

We know, that at section (2-2), there are two rivet holes. Therefore tearing strength of the main plate at section (2-2)

$$= \sigma_t\,[b - (2 \times d)]\,t = 100\,[300 - (2 \times 25)]\,16 \text{ N}$$
$$= 400\,000 \text{ N} = 400 \text{ kN}$$

We also know, that for tearing of the plate at section (2-2), the rivet in front of it (*i.e.*, at section 1-1) must first give way. Therefore actual strength of the main plate at section (2-2),

$$P_{t2} = 400 + \text{Strength of one rivet in front}$$
$$= 400 + 64 = 464 \text{ kN} \quad ...(iv)$$

Similarly, at section (3-3) there are three rivet holes. Therefore tearing strength of the main plate at section (3-3)

$$= \sigma_t\,[b - (3 \times d)]\,t = 100\,[300 - (3 \times 25)]\,16 \text{ N}$$
$$= 360\,000 \text{ N} = 360 \text{ kN}$$

But actual strength of the main plate at section (3-3),

$$P_{t3} = 360 + \text{Strength of three rivets in front}$$
$$= 360 + (3 \times 64) = 552 \text{ kN} \quad ...(v)$$

and at section (4-4) there are four rivet holes. Therefore tearing strength of the main plate at section (4-4)

$$= \sigma_t\,[b - (4 \times d)]\,t = 100\,[300 - (4 \times 25)]\,16 \text{ N}$$
$$= 320\,000 \text{ N} = 320 \text{ kN}$$

But actual strength of the main plate at section (4-4),

$$\sigma_{t4} = 320 + \text{Strength of six rivets in front}$$
$$= 320 + (6 \times 64) = 704 \text{ kN} \quad ...(vi)$$

It will be interesting to know that more and more pull is required for tearing of the main plate, as we move from the outermost section towards the innermost section.

4. ***Tearing of cover plates.*** We know that the cover plates are weaker at section (4-4), where there are 4 rivet holes. Therefore strength of two cover plates at section (4-4),

$$P_{tc} = \sigma_t\,[b - (4 \times d)]\; 2t_c = 100\;[300 - (4 \times 25)]\; 2 \times 10 \text{ N}$$
$$= 400\,000 \text{ N} = 400 \text{ kN}$$

Thus strength of the joint is the least of the above mentioned seven values $P_s$, $P_c$, $P_{t1}$, $P_{t2}$, $P_{t3}$ $P_{t4}$ and $P_{tc}$ *i.e.*, **400 kN.** **Ans.**

*Efficiency of the joint*

We also know that strength of the unriveted plate,

$$P = \sigma_t \times b \times t = 100 \times 300 \times 16 = 480\,000 \text{ N} = 480 \text{ kN}$$

∴ Efficiency of the joint,

$$\eta = \frac{400}{480} = 0.833 = \mathbf{83.3\%} \qquad \textbf{Ans.}$$

## EXERCISE 29.1

1. A single riveted lap joint is made in 10 mm thick plates with 20 mm diameter rivets. Determine the strength and efficiency of the joint, if pitch of the rivets is 50 mm. Take $\tau = 60$ MPa, $\sigma_b = 150$ MPa and $\sigma_t = 80$ MPa. [**Ans.** 18.85 kN ; 47.1%]
2. A double riveted lap joint is provided in 12 mm thick plates with 22 mm diameter rivets with a pitch of 70 mm. What is the strength and efficiency of the joint? Take permissible stresses for shearing of rivets, crushing of rivets and tearing of the plates 60 MPa, 160 MPa and 90 MPa respectively. [**Ans.** 45.6 kN ; 60.3%]
3. Find the efficiencies of the following riveted joints:
   (*i*) Single riveted lap joint in 8 mm thick plates with 18 mm diameter rivets having a pitch of 50 mm.
   (*ii*) Double riveted joint in 10 mm thick plates with 20 mm diameter rivets having a pitch of 60 mm.

   Take allowable stress as $\tau = 70$ MPa, $\sigma_c = 150$ MPa and $\sigma_t = 80$ MPa. [**Ans.** 55.6%, 66.7%]
4. A double riveted double cover butt joint is proved in 12 mm thick plates with 22 mm diameter rivets and 75 mm pitch. Find the strength and efficiency of the joint, if the given stresses are: $\tau = 60$ MPa ; $\sigma_b = 160$ MPa and $\sigma_t = 100$ MPa [**Ans.** 63.6 kN ; 70.7%]

## 29.25. Design of a Riveted Joint

The design of riveted joints is an important job in one modern design office these days. A faulty design can lead to lot many complications. While designing a riveted joint, for structural use, we usually make the following assumptions:

1. The load on joint is equally shared by all the rivets.
2. Initial tensile or shearing stress in the rivets is neglected.
3. Frictional forces between the plates are neglected.
4. Plates are rigid.
5. The shearing stress in all the rivets is uniform.
6. The bearing stress is uniform.
7. Bending of rivets is neglected.

It will be interesting to know that the above mentioned assumptions are very practical in actual operations of the joint. As a matter of fact, we have to calculate a number of details for the given load. But the following are important from the subject point of view:

1. Diameter of the rivets.
2. No. of rivets.
3. Pitch of rivets.
4. Thickness of the cover plates (in case of butt joints).

Following rules are observed, while designing a riveted joint:

**1. *Diameter of rivets***

The diameter of the rivets in mm is obtained by the relation:

$$d = 6\sqrt{t}$$

where $t$ = Thickness of the main plates in mm.

In no way, the diameter of the rivet is provided less than the thickness of the main plate.

**2. *Number of rivets***

The number of rivets are usually calculated, when the length of the joint is small. But when the joint is a continues one (as in the case of boilers), the number of rivets is not calculated. In a small joint, the number of rivets

$$N = \frac{P}{R_v}$$

where $P$ = Thickness of the main plates in mm.

$R_v$ = Rivet value (*i.e.*, least of the rivet value for shearing and bearing).

Sometimes, the number of rivets calculated by the above relation is not an integer. Then the number of rivets *i.e.*, $N$ is taken to the next higher integer (*i.e.*, if by the above relation, we obtain the value of $N = 5.3$, then we shall provide 6 rivets).

**3. *Pitch of rivets***

It is an important factor, while designing a riveted joint. It is calculated on the basis that the plate strength, with one rivet hole, which should not exceed the total value of the rivets in charge of a pitch length. It is done in the following ways:

1. Calculate the shearing strength of the rivet with due consideration, whether the rivet is in single shear or double shear and the number of rivets.
2. Calculate the shearing strength of the rivet with due consideration to the number of rivets.
3. Calculate the tearing strength of the main plate with one rivet hole.
4. The minimum value obtained from (1) and (2) is taken as the rivet value. The strength of the plate obtained from (3) should not exceed the rivet value.

**Note.** In general, the pitch should be between 2.5 to 4 times the diameter of the rivet. From practical considerations, the pitch should not be less than $2d + 12$ mm, where $d$ is the diameter of the rivet in mm.

A little consideration will show, that the strength of the joint in such a case, will be tearing strength of the plate *i.e.*, $P_t$ (because the value of $P_t$ is taken lesser than the rivet values in shearing and bearing).

$\therefore$ Efficiency

$$n = \frac{P_t}{P} = \frac{\sigma_t (p - d)\, t}{\sigma_t \times p \times t} = \frac{(p - d)}{p}$$

**4. *Thickness of cover plates***

The two cover plates are provided each of thickness 0.625 $t$, where $t$ is the thickness of the main plate in mm.

**EXAMPLE 29.6.** *Design a single riveted lap joint to connect two plates of 10 mm thickness. Also find efficiency of the joint. Take $\tau$ = 80 MPa, $\sigma_b$ = 150 MPa and $\sigma_t$ = 100 MPa.*

**SOLUTION.** Given : Thickness of plate ($t$) = 10 mm ; Permissible stress in shear ($\tau$) = 80 MPa = 80 N/mm$^2$ ; Permissible stress in bearing ($\sigma_b$) = 150 MPa = 150 N/mm$^2$ and permissible stress in tearing ($\sigma_t$) = 100 MPa = 100 N/mm$^2$.

***Design of the riveted joint***

We know that as the joint is a continuous lap joint, therefore we have to find out diameter of the rivets ($d$) and pitch of the joint ($p$) only.

1. ***Diameter of the rivets.***

We know that diameter of the rivets,

$$d = 6\sqrt{t} = 6\sqrt{10} = \mathbf{19} \quad \textbf{say} \quad \mathbf{20\ mm} \qquad \textbf{Ans.}$$

2. ***Pitch of rivets.***

First of all, let us calculate the pulls required for shearing and crushing of the rivets as well as tearing of the main plates.

We know that shearing strength of one rivet in simple shear (because of single riveted lap joint),

$$P_s = \tau \times \frac{\pi}{4} \times (d)^2 = 80 \times \frac{\pi}{4} \times (20)^2 \text{ N}$$

$$= 25\ 132 \text{ N} = 25.13 \text{ kN} \qquad \text{... }(i)$$

and bearing strength of one rivet (because of single riveted joint),

$$P_b = \sigma_b \times t \times d = 150 \times 10 \times 20 \text{ N}$$

$$= 30\ 000 \text{ N} = 30 \text{ kN} \qquad \text{... }(ii)$$

We know that strength of one rivet is lesser of these values ($P_s$ and $P_b$) *i.e.*, 25.13 kN. We also know that tearing strength of the main plate in a continuous joint,

$$P_t = \sigma_t (p - d)\, t = 100\,(p - 20)\, 10 \text{ N}$$

$$= 1000\,(p - 20) \text{ N} = (p - 20) \text{ kN} \qquad \text{... }(iii)$$

We also know that tearing strength of the main plate should be greater than or equal to the strength of one rivet. *i.e.*,

$$(p - 20) \geq 25.13$$

$$\therefore \qquad p \geq 25.13 + 20 \geq 45.13 \text{ mm}$$

But the pitch ($p$) should not be less than 2.5 $d$ (*i.e.*, 2.5 × 20 = 50 mm). Moreover, it should not be less than 2$d$ + 12 [*i.e.*, (2 × 20) + 12 = 52 mm]. Therefore let us provide a pitch of 52 mm say **55 mm. Ans.**

***Efficiency of the joint***

We also know that efficiency of the joint,

$$\eta = \frac{p - d}{p} = \frac{55 - 20}{55} = 0.636 = \mathbf{63\%} \qquad \textbf{Ans.}$$

**EXAMPLE 29.7.** *A single riveted double cover butt joint is to be provided in a structure for connecting two plates of 12 mm thickness. Design the joints and find its efficiency. Take permissible stresses in shearing as 75 MPa, in crushing as 175 MPa and in tearing as 100 MPa respectively.*

**SOLUTION.** Given : Thickness of plates ($t$) = 12 mm ; Permissible stress in shearing ($\tau$) = 75 MPa = 75 N/mm$^2$ ; Permissible stress in crushing ($\sigma_c$) = 175 MPa = 175 N/mm$^2$ and permissible stress in tearing ($\sigma_t$) = 100 MPa = 100 N/mm$^2$.

***Design of the riveted joint***

We know that as the joint is a continuous double cover butt joint, therefore we have to find out diameter ($d$) of the rivets, pitch ($p$) of the joint and thickness of cover plates only.

1. ***Diameter of the rivets.*** We know that diameter of the rivets.

$$d = 6\sqrt{t} = 6\sqrt{12} = \mathbf{20.8} \text{ say } \mathbf{22 \text{ mm}} \quad \textbf{Ans.}$$

2. ***Pitch of rivets***. First of all, let us calculate the pulls required for shearing and crushing of the rivets as well as tearing of the main plate. We know that shearing strength of one rivet in double shear (because of single riveted butt joint),

$$P_s = 2 \times \tau \times \frac{\pi}{4} \times (d)^2 = 2 \times 75 \times \frac{\pi}{4} \times (22)^2 = 57\ 027 \text{ n}$$

$$= 57.02 \text{ kN} \qquad ...(i)$$

and crushing strength of one rivet (because of single riveted joint),

$$P_c = \sigma \times t \times d = 175 \times 12 \times 22 = 46\ 200 \text{ N} = 46.2 \text{ kN} \qquad ...(ii)$$

We know that strength of one rivet is lesser of these two values ($P_s$ and $P_c$) *i.e.*, 46.2 kN. We also know that tearing strength of the main plate in a continuous plate,

$$P_t = \sigma_t (p - d)\, t = 100\, (p - d)\, 12 = 1200\, (p - d) \text{ N}$$

$$= 1.2\, (p - d) \text{ kN} \qquad ...(iii)$$

We also know that tearing strength of the main plate should be greater than or equal to the strength of one rivet, *i.e.*,

$$1.2\, (p - d) \geq 46.2$$

or

$$p - 22 \geq 46.2/1.2 = 38.5$$

$$\therefore \quad p \geq 38.5 + 22 = 60.5 \text{ mm}$$

But the pitch ($p$) should not be less than 2.5 $d$ (*i.e.*, 2.5 × 22 = 55 mm). Moreover, it should not be less than $2d + 12$ [*i.e.*, (2 × 22) + 12] = 56 mm. Therefore let us provide a pitch of 60.5 mm say **65 mm. Ans.**

3. ***Thickness of cover plates.***

We know that thickness of cover plates,

$$t_c = 0.625\, t = 0.625 \times 12 = \mathbf{7.5 \text{ mm}} \text{ say } \mathbf{8 \text{ mm}} \quad \textbf{Ans.}$$

***Efficiency of the joint***

We also know that efficiency of the joint,

$$\eta = \frac{p - d}{p} = \frac{65 - 22}{65} = 0.662 = \mathbf{66.2\%} \quad \textbf{Ans.}$$

**EXAMPLE 29.8.** *Design a double cover butt joint to connect two plates 15 mm thick and 240 mm wide. The safe stresses are: $\tau$ = 80 MPa, $\sigma_c$ = 160 MPa and $\sigma_t$ = 100 MPa.*

**SOLUTION.** Given: Thickness of plates ($t$) = 15 mm ; Width of plates ($b$) = 240 mm ; Safe stress in shearing ($\tau$) = 80 MPa = 80 N/mm$^2$; Safe stress in crushing ($\sigma_c$) = 160 MPa = 160 N/mm$^2$ and safe stress in tearing ($\sigma_t$) = 100 MPa = 100 N/mm$^2$.

We know that as width of the plates is given, therefore, the joint is not a continuous one. Thus, it is to be designed for the full width of the plate. Let us design a zig-zag joint with one rivet in the outermost section.

***Diameter of the rivets***

We know that diameter of the rivets,

$$d = 6\sqrt{t} = 6\sqrt{15} = \mathbf{23.2 \text{ mm}} \text{ say } \mathbf{24 \text{ mm}} \quad \textbf{Ans.}$$

*No. of rivets*

First of all, let us calculate the pulls required for shearing and crushing of the rivets as well as tearing of the main plate. We know that shearing strength of one rivet in double shear (because of butt joint),

$$P_s = 2 \times \tau \times \frac{\pi}{4} \times (d)^2 = 2 \times 80 \times \frac{\pi}{4} \times (24)^2$$

$$= 72\ 382 \text{ N} = 72.38 \text{ kN} \quad ...(i)$$

and crushing strength of one rivet,

$$P_c = \sigma_c \times t \times d = 160 \times 15 \times 24 = 57\ 600 \text{ N} = 57.6 \text{ kN} \quad ...(ii)$$

We know that strength of one rivet is lesser of these two values ($P_s$ and $P_c$) *i.e.*, 57.6 kN. In a zig-zag joint, the strength of the plate is minimum at the outermost section with one rivet hole (as assumed by us). We know that tearing strength of the plate at the outermost section,

$$P_t = \sigma_t [b - (1 \times d)]\, t = 100\, [240 - (1 \times 24)]\, 15 \text{ N}$$

$$= 324\ 000 \text{ N} = 324 \text{ kN} \quad ...(iii)$$

∴ No. of rivets required for the joint

$$= \frac{\text{Tearing strength of the plate}}{\text{Strength of one rivet}} = \frac{324}{57.6} = 5.6 \quad \text{say} \quad 6$$

Now let us provide a zig-zag joint with 6 rivets on either side of the joint as shown in Fig. 29.16.

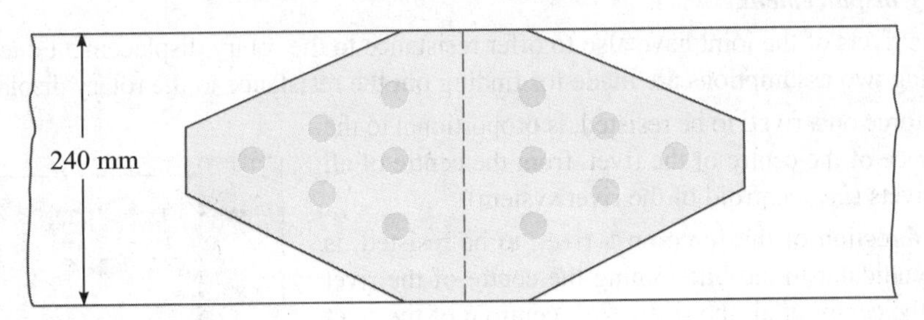


Fig. 29.16

*Thickness of cover plates*

We know that thickness of each cover plate,

$$t_c = 0.625\, t = 0.625 \times 15 = \mathbf{9.4\ mm} \quad \textbf{say} \quad \mathbf{10\ mm} \quad \textbf{Ans.}$$

*Pitch of rivets*

Let us povide a pitch of 2.5 $d$ = 2.5 × 24 = **60 mm** **Ans.**

## EXERCISE 29.2

1. Design a single riveted joint to connect two plates of 8 mm thickness. Take $\tau$ = 70 MPa, $\sigma_c$ =160 MPa and $\sigma_t$ = 90 MPa. [**Ans.** $d$ = 17 say 18 mm; $p$ = 45 mm]
2. Design a double riveted lap joint to connect two plates of 10 mm thickness. Take $\tau$ = 60 MPa, $\sigma_c$ = 150 MPa and $\sigma_t$ = 80 MPa. [**Ans.** $d$ = 19 say 20 mm; $p$ = 45 mm]
3. Design a double riveted double cover built joint to connect two plates 15 mm thick with 23 mm diameter rivets. Take $\tau$ = 95 MPa, $\sigma_t$ = 120 MPa and $\sigma_b$ = 190 MPa. Also find efficiency of the joint. [**Ans.** $p$ = 95 mm; $\eta$ = 75.8%]

## 29.26. Eccentric Riveted Connections

In the previous articles, we have discussed the cases when the rivets have only to resist the linear displacement of the plates. But sometimes, a rivet may have to resist rotary displacement also in addition to linear displacement.

Consider a bracket connection with a girder as shown in Fig. 29.17. It may be noted that all the rivets have to resist the following two types of displacement:

1. Linear displacement and
2. Rotary displacement.

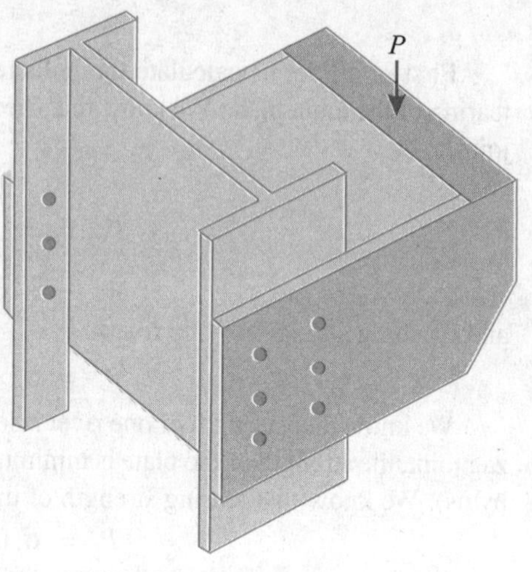


Fig. 29.17. Eccentric riveted connection.

1. ***Linear displacement***

All the rivets have to offer resistance to the linear displacement, due to the load $P$. This resistance *i.e.*, force is assumed to be uniform for all the rivets, and is equal to

$$P_L = \frac{P}{n}$$

where $P$ = Total load on the joint

and $n$ = No. of rivets on the joint.

2. ***Rotary displacement***

All the rivets of the joint have also to offer resistance to the rotary displacement due to the load $P$. Following two assumptions are made for finding out the resistance to the rotary displacement:

1. The force on a rivet, to be resisted, is proportional to the distance of the centre of the rivet, from the centre of all the rivets (*i.e.*, centroid of the rivet system).
2. The direction of the force on a rivet, to be resisted, is perpendicular to the line joining the centre of the rivet and the centre of all the rivets (*i.e.*, centroid of the rivet system).

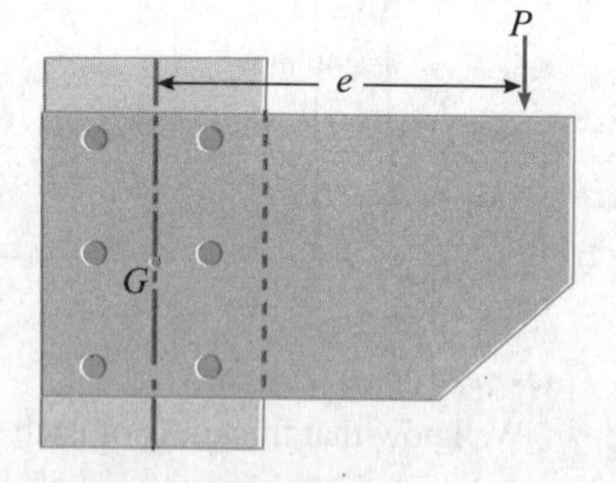


Fig. 29.18

Now consider an eccentric riveted connection as shown in Fig. 29.18.

Let $P$ = Eccentric load on the joint

$e$ = Eccentricity of the load *i.e.*, the distance between the line of action of the load and the centroid of the rivet system *i.e.*, $G$.

Consider a rivet at a distance $r$ from the centroid of the rivet system $G$. The force to be resisted by this rivet due to the moment of $P \times e$ (*i.e.*, load × distance), is directly proportional to the distance between its centre and $G$, *i.e.*,

$$P_R \propto r$$

$$= kr \quad \text{...(i)}$$

where $k$ is a constant.

The moment of this resistance about $G$

$$= P_R \times r = kr \times r = kr^2$$

and total moments of resistance by all the rivets about $G$

$$= \Sigma kr^2 = k\Sigma r^2 \quad \text{...(ii)}$$

This total must be equal to the moment of the load about $G$

$$= P \cdot e \qquad ...(iii)$$

Equating equations (*ii*) and (*iii*),

$$k\Sigma r^2 = P \cdot e \quad \text{or} \quad k = \frac{P \cdot e}{\Sigma r^2}$$

Substituting the value of $k$ in equation (*i*),

$$P_R = kr = \frac{P\,er}{\Sigma r^2} \qquad ...(iv)$$

If $x$ and $y$ are the co-ordinates of rivet (taking $G$ as origin), then

$$r^2 = x^2 + y^2$$

Substituting the value of $r^2$ in equation (*iv*),

$$P_R = \frac{P\,er}{\Sigma (x^2 + y^2)}$$

The direction of this force will be at right angles to the line joining the centre of the rivet and centroid of the rivet system. Since, this force is directly proportional to $r$ (as is clear from the relation) therefore the rivet, which is at the *farthest* distance from the centroid of the rivet system, will be subjected to the *maximum force*. It is thus obvious that while calculating the resistance of a rivet or safety of the connection, a rivet which is at the farthest distance from $G$ is studied, because all the other rivets will be subjected to a lesser force than the farthest one.

The load $P_R$ is resolved horizontally as well as vertically. The resultant load on a rivet will be given by the relation:

$$R = \sqrt{(\Sigma H)^2 + (\Sigma V)^2}$$

where $\Sigma H$ = Horizontal component of $P_R$ and

$\Sigma V = P_L \pm$ Vertical component of $P_R$.

**Note.** Care should always be taken for the +*ve* or –*ve* sign. In general if the rivet, under consideration, is between $G$ and the load +*ve* sign is taken. But if the rivet is away from $G$, then –*ve* sign is taken.

**Example 29.9.** *A bracket is riveted to a column by 6 rivets of equal size as shown in Fig 29.19.*

*It carries a load of 120 kN at a distance of 300 mm from the centre of the column. Calculate the loads carried by two top rivets.*

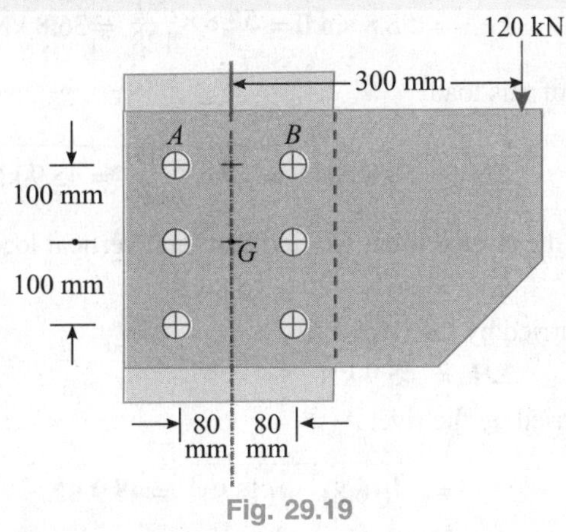


Fig. 29.19

**Solution.** Given: No. of rivets ($n$) = 6; Load ($P$) = 120 kN = $120 \times 10^3$ N and eccentricity ($e$)=300 mm.

***Load carried by the rivet A***

We know that load on each rivet on account of linear displacement,

$$P_L = \frac{P}{n} = \frac{120}{6} = 20 \text{ kN} \quad ...(i)$$

and total moment of resistance by all the rivets about $G$,

$$\Sigma r^2 + \Sigma y^2 = [6 \times (80)^2] + [2 \times (100)^2] + [2 \times (0)2 + [2 \times (100)^2] \text{ mm}^2$$

$$= (38.4 \times 10^3) + (20 \times 10^3) + 0 + (20 \times 10^3) \text{ mm}^2$$

$$= 78.4 \times 10^3 \text{ mm}^2$$

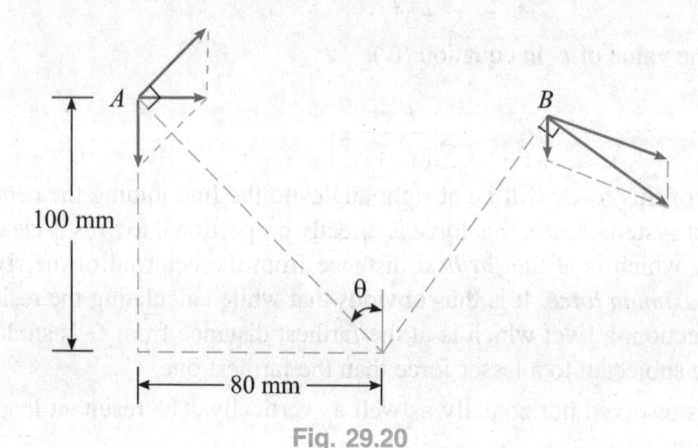


Fig. 29.20

From the geometry of the riveted joint, we find that the distance between the centroid of the joint and the rivet $A$ (or $B$),

$$r = \sqrt{(100)^2 + (80)^2} = 128 \text{ mm}$$

∴ Load on account of rotation,

$$P_R = \frac{P \cdot e \cdot r}{\Sigma r^2} = \frac{(120 \times 10^3) \times 300 \times 128}{78.4 \times 10^3} = 58.8 \times 10^3 \text{ N} = 58.8 \text{ kN}$$

Vertical component of this load

$$= 58.8 \sin \theta = 58.8 \times \frac{80}{128} = 36.8 \text{ kN}$$

and horizontal component of this load

$$\Sigma H = 58.8 \cos \theta = 58.8 \times \frac{100}{128} = 45.9 \text{ kN}$$

From the geometry of the riveted joint, we find that total vertical load carried by rivet $A$,

$$\Sigma V = 36.8 - 20 = 16.8 \text{ kN}$$

and total horizontal load carried by the rivet $A$,

$$\Sigma H = 45.9 \text{ kN} \quad \text{... (As obtained earlier)}$$

∴ Resultant load carried by the rivet $A$

$$= \sqrt{(16.8)^2 + (45.9)^2} = \mathbf{48.9 \text{ kN}} \quad \textbf{Ans.}$$

*Load carried by the rivet B*

From the geometry of the riveted joint, we find that total vertical load carried by rivet *B*,

$$\Sigma V = 36.8 + 20 = 56.8 \text{ kN}$$

and total horizontal load carried by the rivet *B*,

$$\Sigma H = 45.9 \text{ kN} \qquad \text{...(As obtained earlier)}$$

∴ Resultant load carried by the rivet *A*

$$= \sqrt{(56.8)^2 + (45.9)^2} = \mathbf{73\ kN} \qquad \textbf{Ans.}$$

## 29.27. Transmission of Load Through Rods

In the previous articles, we have discussed the transmission of load through plates. While calculating the transmission of load through plate, we used to discuss the strength of a riveted joint. Now we shall discuss the transmission of load through rods.

## 29.28. Types of Rod Joints

Though there are numerous types of joints in the rods, yet the following two types are important from the subject point of view:

1. Knuckle joint and 2. Cotter joint

## 29.29. Knuckle Joint

A knuckle joint or forked joint is used to connect two rods or bars, subjected to tensile load. It may be readily connected or disconnected for adjustments or repairs. The knuckle joints are very common in machines and structures.

In a knuckle joint, one end of one of the rods is made into an eye and the other end of the other rod is made into a fork, with an eye in each of the fork legs. A pin, having a head, is passed through the eyes and fixed by means of a cotter pin as shown in Fig. 29.21.

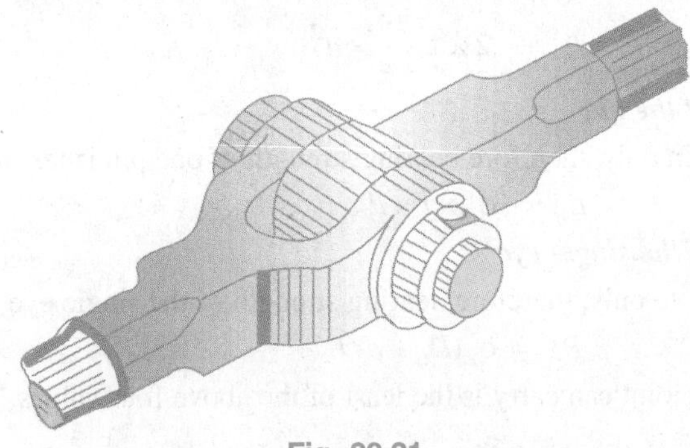

Fig. 29.21

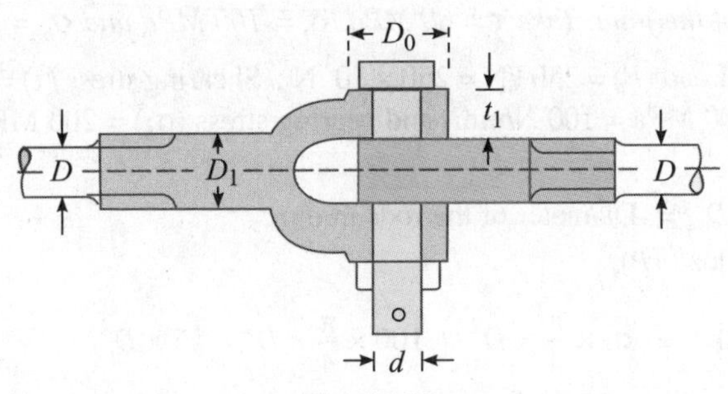


Fig. 29.22

Now consider a knuckle joint connecting two rods as shown in Fig. 29.22.

Let $P$ = Load carried by the rods,

$D$ = Diameter of the rods,

$D_1$ = Diameter of rods near the eyes,

$D_0$ = Outer diameter of the eye,

$t$ = Thickness of the single eye,

$t_1$ = Thickness of each fork leg,

$d$ = Diameter of the pin (or internal diameter of the eye),

$\sigma_t$ = Permissible tensile strength in the rods,

$\sigma_s$ = Permissible shear stress in the pin material and

$\sigma_b$ = Permissible bearing stress for the pin material.

As a matter of facts, the above data is calculated for each joint. But sometimes, a few thumb rules are adopted for the design of the various components of the joint as discussed below:

1. The diameter of the bars near the eyes, is made at least 5 mm more than the bar diameter.
2. The diameter of the pin head is made 1.5 times the diameter of the pin.
3. The outer diameter of the eye is made two times its internal diameter.

Now we shall find out the strength of the knuckle joint, in the same way, as we used to do in the case of the riveted joints.

**1.** ***Strength of the rods***

Since the rod is of uniform diameter, therefore strength of rod is taken for tensile load.

$$\therefore \quad P = \sigma_t \times \frac{\pi}{4} \times D^2 \quad ...(i)$$

**2.** ***Shearing strength of the pin***

Since the pin is in double shear, therefore strength of the pin in double shear is taken.

$$P_s = 2 \times \tau \times \frac{\pi}{4} \times d^2 \quad ...(ii)$$

**3.** ***Bearing strength of the pin***

Since there is one pin only, therefore bearing strength of one pin is taken.

$$\therefore \quad P_b = \sigma_b \times t \times d \quad ...(iii)$$

**4.** ***Tearing strength of the single eye***

Since there is one hole only, therefore, tearing strength of the single eye with one hole is taken.

$$\therefore \quad P_t = \sigma_t (D_0 - d)\, t \quad ...(iv)$$

The load, which the joint can carry is the least of the above four values.

**EXAMPLE 29.10.** *A knuckle joint is required to withstand a tensile load of 250 kN. Design the important parts of the joint. Take $\tau$ = 80 MPa, $\sigma_t$ = 100 MPa and $\sigma_b$ = 200 MPa.*

**SOLUTION.** Given: Load $(P)$ = 250 kN = $250 \times 10^3$ N ; Shearing stress $(\tau)$ = 80 MPa = 80 N/mm$^2$; Tearing stress $(\sigma_t)$ = 100 MPa = 100 N/mm$^2$ and bearing stress $(\sigma_b)$ = 200 MPa = 200 N/mm$^2$.

***Diameter of rods***

Let $D$ = Diameter of the rods in mm.

We know that the load $(P)$,

$$250 \times 10^3 = \sigma_t \times \frac{\pi}{4} \times D^2 = 100 \times \frac{\pi}{4} \times D^2 = 25\,\pi\, D^2$$

$$\therefore \quad D^2 = \frac{250 \times 10^3}{25\,\pi} = 3183$$

or $\quad D = $ **56.4 mm say 60 mm Ans.**

*Diameter of pin*

Let $\quad d$ = Diameter of the pin in mm.

We know that the load ($P$),

$$250 \times 10^3 = 2 \times \tau \times \frac{\pi}{4} \times (d)^2 = 2 \times 80 \times \frac{\pi}{4} \times d^2 = 40\,\pi\,d^2$$

$$\therefore \quad d^2 = \frac{250 \times 10^3}{40\,\pi} = 1989.17$$

or $\quad d = $ **44.6 mm say 45 mm Ans.**

*Thickness of the single eye*

Let $\quad t$ = Thickness of the single eye in mm.

We know that the load ($P$),

$$250 \times 10^3 = \sigma_b \times t \times d = 200 \times t \times 45 = 9000\,t$$

$$\therefore \quad t = \frac{250 \times 10^3}{9000} = \mathbf{27.8\ mm\ \ say\ \ 28\ mm} \quad \textbf{Ans.}$$

*Outer diameter of the eye*

Let $\quad D_0$ = Outer diameter of the eye in mm.

We know that the load ($P$),

$$250 \times 10^3 = \sigma_t\,(D_0 - d)\,t = 100 \times (D_0 - 45)\,28 = 2800\,(D_0 - 45)$$

$$\therefore \quad D_0 = \frac{250 \times 10^3}{2800} + 45 = 89.3 + 45 = \mathbf{134.3\ mm} \quad \textbf{Ans.}$$

## 29.30. Cotter Joint

A cotter joint is used to connect two rods or bars subjected to tensile or compressive loads. Like a knuckle joint, a cotter joint may also be readily connected or disconnected for adjustments or repairs. The cotter joints are widely used in various types of machines.

In a cotter joint, one end of one of the rods is made *spigot and the other end of the other rod is made a socket. A rectangular hole is made in the spigot as well as socket. A cotter which is nothing but a piece of plate with one edge straight and the other tapered from 1 in 24 to 1 in 48, is inserted into the hole and tightened till the socket end comes in contact with the spigot end as shown in Fig. 29.23.

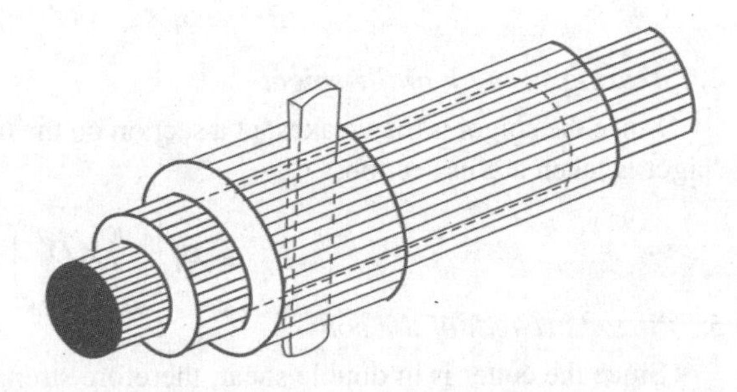

Fig. 29.23

* The end of the rod which goes inside is called a *spigot* and the other end of the other rod which receives the spigot is called a *socket*.

Now consider a cotter joint connecting two rods as shown in Fig. 29.24.

Let $P$ = Load carried by the rods,

$D$ = Diameter of the rods,

$D_1$ = Diameter of the spigot (or internal diameter of the socket),

$D_2$ = External diameter of the socket,

$b$ = Width of the cotter at the axis of the rods,

$t$ = Thickness of the cotter,

$\sigma_t$ = Permissible tensile strength in the rods,

$\tau$ = Permissible shear stress in the cotter material and

$\sigma_b$ = Permissible bearing stress in the cotter material.

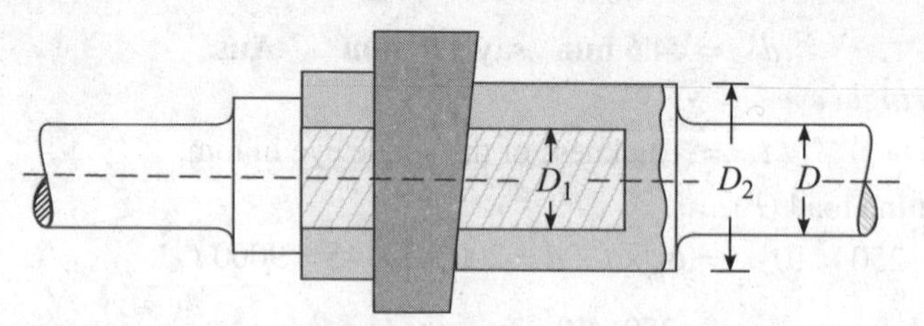


**Fig. 29.24**

As a matter of fact, the above data is calculated for each joint. But sometimes, a few thumb rules are adopted for the design of the various components of the joint as discussed below:

1. The thickness of the cotter is made one-fourth of the diameter of the rods.
2. The diameter of the spigot (or internal diameter of the socket) is made 1.25 times the diameter of the rods.
3. The external diameter of the socket is made 2.5 times the diameter of the rods.

Now we shall find out the strength of the cotter joint, in the same way as we used to do in the case of riveted joints.

1. ***Strength of the rods***

We know that the load on the rod,

$$P = \sigma_t \times \frac{\pi}{4} \times D^2 \qquad ...(i)$$

2. ***Tearing strength of the spigot***

Since the spigot is the weakest at a section on the hole for cotter, therefore tearing strength of the spigot is taken at this section.

$$\therefore \quad P_t = \sigma_t \left[ \left( \frac{\pi}{4} \times D_1^2 \right) - (D_1\, t) \right] \qquad ...(ii)$$

3. ***Shear strength of the cotter***

Since the cotter is in double shear, therefore strength of the cotter in double shear is taken.

$$\therefore \quad P_s = 2 \times t \times b \times t \qquad ...(iii)$$

4. ***Bearing strength of the cotter***

Since there is one cotter only, therefore bearing strength of one cotter is taken.

$$\therefore \quad P_b = \sigma_b \times t \times D_1 \qquad ...(iv)$$

5. ***Tearing strength of the socket***

Since the socket is the weakest at a section on the hole for cotter, therefore tearing strength of the socket is taken at this section.

$$\therefore \quad P_t = \sigma_t \left[ \frac{\pi}{4} (D_2^2 - D_1^2) - t\,(D_2 - D_1) \right] \quad ...(v)$$

The load which the joint can carry, is the least of the above five values.

**EXAMPLE 29.11.** *Two steel rods are to be joined by means of a cotter joint. Design a suitable joint, if the rods have to resist a load of 120 kN. Take $\tau = 70$ MPa, $\sigma_t = 100$ MPa and $\sigma_b = 150$MPa.*

**SOLUTION.** Given: Load $(P)$ = 120 kN = $120 \times 10^3$ N; Permissible shear stress $(\tau)$ = 70 MPa = 70 N/mm$^2$; Permissible tearing stress $(\sigma_t)$ = 100 MPa = 100 N/mm$^2$ and permissible bearing stress $(\sigma_b)$ = 150 MPa = 150 N/mm$^2$.

*Diameter of the rods*

Let $D$ = Diameter of the rods in mm.

We know that the load $(P)$

$$120 \times 10^3 = \sigma_t \times \frac{\pi}{4} \times D^2 = 100 \times \frac{\pi}{4} \times D^2 = 25\,\pi\,D^2$$

$$\therefore \quad D^2 = \frac{120 \times 10^3}{25\,\pi} = 1527.6$$

or $D$ = **39.1 mm say 40 mm Ans.**

*Diameter of spigot and thickness of cotter*

Let $D_1$ = Diameter of the spigot in mm and

$t$ = Thickness of the cotter.

We know that the load $(P)$,

$$120 \times 10^3 = \sigma_b \times t \times D_1 = 150 \times t \times D_1 \quad ...(i)$$

$$\therefore \quad t \times D_1 = \frac{120 \times 10^3}{150} = 800$$

We also know that the load $(P)$,

$$120 \times 10^3 = \sigma_t \left( \frac{\pi}{4} \times D_1^2 - D_1\,t \right) = 100 \left( \frac{\pi}{4} \times D_1^2 - 800 \right) ...(\because D_1\,t = 800)$$

$$120\,000 = 78.54\, D_1^{\,2} - 80\,000$$

$$\therefore \quad D_1^{\,2} = \frac{120\,000 + 80\,000}{78.54} = 2546$$

or $D_1$ = **50 mm Ans.**

and $t = \frac{800}{50}$ = **16 mm say 20 mm Ans.**

*Width of the cotter at the axis of the rod*

Let $b$ = Width of the cotter at the axis of the rod in mm.

We know that the load $(P)$,

$$120 \times 10^3 = 2 \times t \times b \times t = 2 \times 70 \times b \times 20 = 2800\,b$$

$$\therefore \quad b = \frac{120 \times 10^3}{2800}$$ = **42.9 say 50 mm Ans.**

*External diameter of the socket*

Let $D_2$ = External diameter of the socket in mm.

We also know that the load $(P)$

$$120 \times 10^3 = \sigma_t \left[ \frac{\pi}{4} (D_2^2 - D_1^2) - t\,(D_2 - D_1) \right]$$

$$= 100\left[\frac{\pi}{4}[D_2^2 - (50)^2] - 20\,(D_2 - 50)\right]$$

$$1200 = 0.7854\ D_2^{\ 2} - 1964 - 20\ D_2 + 1000$$

$$0.7854\ D_2^{\ 2} - 20\ D_2 - 2164 = 0$$

Solving this equation as a quadratic equation, we get

$$D_2 = \frac{+\,20 \pm \sqrt{(20)^2 + (4 \times 0.7854 \times 2164)}}{2 \times 0.7854}\ \text{mm}$$

$$= \mathbf{66.7\ mm\ \ say\ \ 70\ mm} \qquad \textbf{Ans.}$$

## EXERCISE 29.3

1. A vertical load of 20 kN is applied to a bracket at an eccentricity of 200 mm as shown in Fig. 29.25

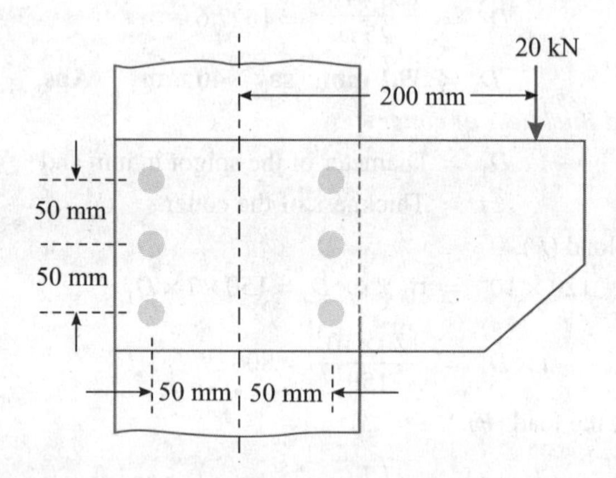


Fig. 29.25

Determine the maximum resistance offered by any rivet. (**Ans.** 13 kN)

2. Two rods are to be joined by knuckle joint, to withstand a tensile load of 25 kN. Design the joint. Take $\tau$ = 40 MPa, $\sigma_t$ = 56 MPa and $\sigma_b$ = 70 MPa

   (**Ans.** $D$ = 24 mm ; $d$ = 20 mm ; $t$ = 18 mm; $D_0$ = 45 mm)

3. Two rods are to be joined by a cotter joint. They are subjected to a tensile load of 300 kN. Design the important points of the joint. Take $\tau$ = 40 MPa, $\sigma_t$ = 60 MPa and $\sigma_b$ = 80 MPa.

   (**Ans.** $D$ = 80 mm ; $D_1$ = 106 mm ; $t$ = 36 mm; $b$ = 104 mm)

## QUESTIONS

1. What do you understand by the term 'riveted joint'? Explain the necessity of such a joint.
2. Describe the common types of riveted joints. Illustrate your answer with neat sketches.
3. Describe briefly the various ways, in which a riveted lap joint or butt joint can fail.
4. Explain briefly:

   (*a*) Efficiency of a riveted joint. (*b*) Failure of riveted joint.
5. Explain the difference between the strength of a riveted joint and efficiency of a riveted joint.

6. What is an eccentric riveted joint? How does it differ from an ordinary riveted joint? Describe the procedure for bringing out the efficiency of such a joint.
7. Describe the procedure for designing a knuckle joint.

## OBJECTIVE TYPE QUESTIONS

1. When one plate overlaps the other and both the plates are riveted with two rows of rivets, the joint is known as
   (*a*) single riveted lap joint (*b*) double riveted lap joint
   (*c*) double riveted single cover butt joint (*d*) double riveted double cover butt joint
2. In a riveted joint, when the rivets in various rows are opposite to each other, the joint is said to be
   (*a*) chain riveted (*b*) zig-zag riveted
   (*c*) diamond riveted (*d*) none of these
3. In a riveted joint, when the number of rivets decreases from the innermost to the outermost row, the joint is said to be
   (*a*) chain (*b*) zig-zag riveted
   (*c*) diamond riveted (*d*) none of these
4. A riveted joint may fail due to
   (*a*) shearing of rivets (*b*) crushing of rivets
   (*c*) tearing of the plates (*d*) all of these.
5. In an eccentric riveted connection, if the magnitude of the load is increased it will effect
   (*a*) linear displacement (*b*) rotary displacement
   (*c*) both '*a*' and '*b*' (*d*) none of these

## ANSWERS

1. (*b*) 2. (*a*) 3. (*c*) 4. (*d*)
5. (*c*)

# Chapter 30

# Welded Joints

## Contents

## 30.1. Introduction

The welding is a process of joining together two or more metal parts. It is done by heating the surfaces, to be connected, to a high temperature and then adding additional molten metal, which fuses with and combines the two surfaces. The molten or fused metal is deposited between the parent metal parts, which are also fused to a specified depth. When the deposited fused metal gets cooled, the parent metal parts are joined by this new metal.

A number of methods are used for the process of fusion. But oxyacetylene gas welding and electric arc welding are most commonly used. The welded joints have proved to be so reliable, that they are replacing the riveted joints in structural and machine joints.

## 30.2. Advantages and Disadvantages of Welded Joints

Following are the advantages and disadvantages of welded joints over riveted joints:

***Advantages***

1. The welded structures are, usually, lighter than riveted structure. This is due to the reason, that in welding gussets or other connecting components are not used.
2. The welding allows the arrangement of the structure components, in such a way, the joint provides maximum efficiency (may be 100%). But it is not possible in case of riveted joints.
3. Additions and alterations can be easily made in the existing structures.
4. As the welded structure is smooth in appearance, therefore it looks pleasing. Moreover, its painting is easier and economical.
5. In welded connections, the tension members are not weakened as in the case of riveted or bolted joint.
6. A welded joint has a great strength. Often a welded joint has the strength of the parent metal itself.
7. Sometimes, the members are of such a shape (*e.g.*, circular steel pipes) that they afford difficulty for riveting. But they can be easily welded.
8. The welding provides very rigid joints. This is in line with the modern trend of providing rigid frames.
9. It is possible to weld any part of a structure at any point. But riveting requires enough clearance.
10. The noise produced in the process of riveting is a source of great nuisance. This is avoided in the process of welding.
11. The process of welding takes less time than the riveting.

***Disadvantages***

1. As there is an uneven heating and cooling, during the fabrication, therefore the members may get distorted or additional stresses may develop.
2. It requires a highly skilled labour and supervision.
3. As the welded joints are rigid joints, therefore they have to be treated as such in their design.
4. Since no provision is kept for expansion and contraction in the frame, therefore there is a possibility of cracks developing in it.
5. The inspection of welding work is more difficult than riveting work.

## 30.3. Types of Welded Joints

Though there are many types of welded joints, yet the following three types are important from the subject point of view:

1. Butt weld point,
2. Fillet weld joint and
3. Plug or slot weld joint.

## 30.4. Butt Weld Joint

It is a joint, in which the edges of the two members butt (*i.e.*, touch) against each other, the two members are joined together by welding. It is used to make butt joint or the joint as shown in Fig. 30.1.

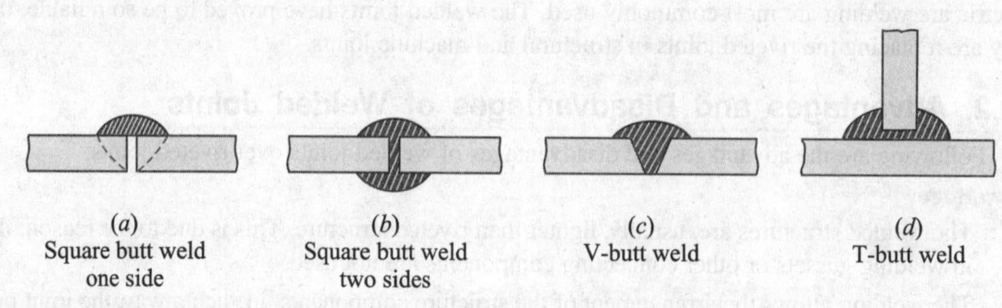


Fig. 30.1

The butt weld is generally made convex on either sides. This extra area (which is about 10%) of the plate is called reinforcement. The butt weld joints are rarely used in engineering structures, except for T-joints.

## 30.5. Fillet Weld Joint

It is a joint, in which the two members either overlap or meet each other at about 90° and the two members are joined together by welding. It is of approximately triangular cross-section and is used for overlap joints and corner joints as shown in Fig. 30.2. The fillet weld joints are widely used in various types of engineering structures.

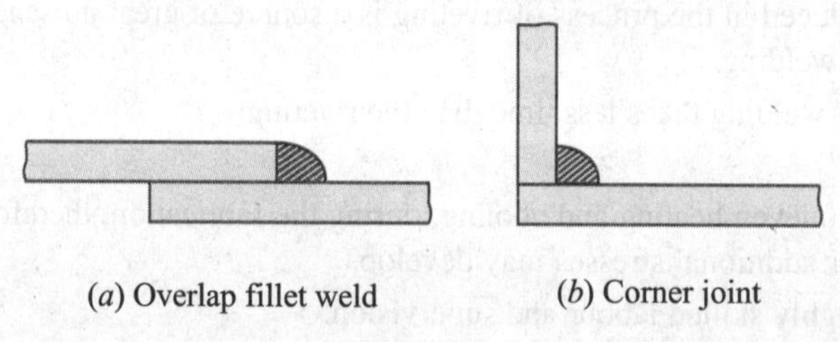


Fig. 30.2

## 30.6. Plug or Slot Weld Joint

Sometimes, sufficient space is not available for providing the necessary length of the fillet weld. In such cases, a circular hole is made, and a fillet weld is provided along the circumference of the hole.

Sometimes, a circular hole or slotted hole is made in one of the members, and weld metal is filled in the hole as shown in Fig. 30.3.

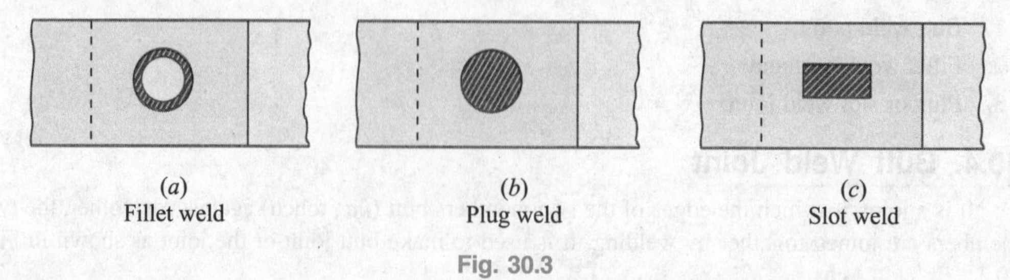


Fig. 30.3

The minimum diameter of the circular hole (or width of the slotted hole) should not be less than (thickness of the member in mm + 8 mm) of the member containing the hole; whereas the maximum diameter should not be more than 2.25 times the thickness of the member.

## 30.7. Technical Terms

Before entering into the details of the welded joints, let us discuss some of the technical terms, which are important from the subject point of view.

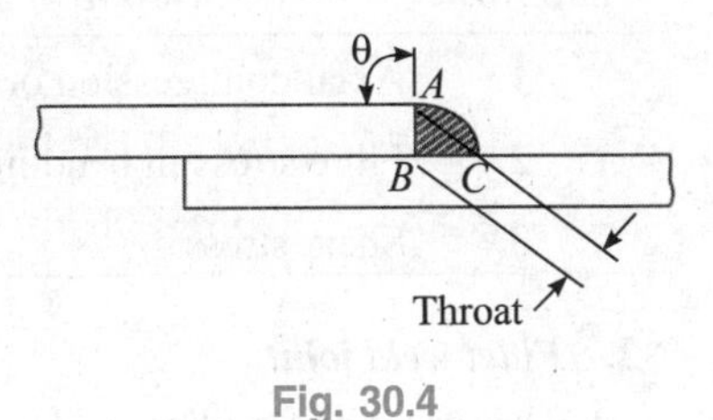

Fig. 30.4

1. *Legs of the weld*

   The sides containing the right angle are called legs of the weld. In Fig. 30.4, the sides *AB* and *BC* are called the legs of the weld.

2. *Size of the fillet weld*

   The minimum length of the leg of a weld is called size of the weld. In Fig. 30.4, the side is called size of the fillet weld.

3. *Throat thickness*

   It is the perpendicular distance between the corner and hypotenuse of the weld cross-section as shown in Fig. 30.4. The thickness of reinforcement is not included, in the throat thickness.

   Effective throat thickness = $k \times$ minimum leg length.

   The value of *k* for different angles between fusion faces is given below:

| Angle | 60° to 90° | 91° to 100° | 101° to 106° | 107° to 113° | 114° to 120° |
|---|---|---|---|---|---|
| *k* | 0.7 | 0.65 | 0.60 | 0.55 | 0.50 |

NOTES: 1. The fillet weld should not be used for connecting parts, whose fusion faces make an angle less than 60° or more than 120°.

2. If no angle between fusion faces is given, it is taken as 90° and the value of *k* is taken as 0.7.

4. *Effective length of the weld*

   The actual length of the weld, which is of specified size and required thickness is called effective length of the weld. For the design purpose, the effective length of a weld is taken as the actual length of the weld *minus* twice the size of the weld.

5. *Side fillet weld*

   The fillet weld, placed parallel to the direction of the force is called a *side fillet weld.* In Fig. 30.5, the welds *A* and *C* are the side fillet welds.

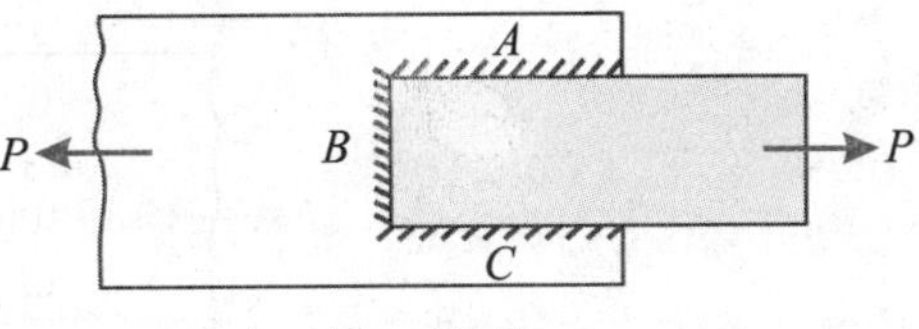

Fig. 30.5

6. *End fillet weld*

   The fillet weld placed at the end of the member, so that it is perpendicular to the force is called *end fillet weld.* In fig. 30.5, the weld *B* is the end fillet weld.

## 30.8. Strength of a Welded Joint

We have already discussed the various types of welded joints. Now we shall discuss their strengths (or in other words, forces of resistance), which are determined as given below:

1. *Butt weld joints*

   The strength of a butt weld is specified by its throat thickness, which is taken as the thickness of the thinner plate. In a butt weld joint, it is assumed that the weld can sustain the full working stresses in tension, compression and shear. The permissible stress in a butt weld, calculated on throat area should not exceed the permissible stresses as given in table 30.1.

**Table 30.1**

| S.No. | Nature of stress | Permissible stress ($\sigma$) in MPa ($N/mm^2$) |
|---|---|---|
| 1. | Axial compression or tension | 140 |
| 2. | Fibre stress in bending | 160 |
| 3. | Shear stress | 100 |

2. ***Fillet weld joint***

The strength of a fillet weld is given by the relation :

$$P = l \cdot t \cdot \sigma = A \cdot \sigma \qquad ...(\because A = l \cdot t)$$

where $l$ = Effective length of the weld,

$t$ = Throat thickness,

$\sigma$ = Permissible stress in the fillet weld as given in Table 30.1

$A$ = Area of the weld.

3. ***Plug or slot weld joint***

The strength of a plug or slot weld joint is given by the relation:

$$P = A \cdot \sigma$$

where $A$ = Cross-sectional area of the plug or slot,

$\sigma$ = Permissible stress in the weld as given in Table 30.1.

**Note:** If a fillet weld is provided, its length is taken to be equal to the average length of the throat of the fillet. This length is, generally, taken to be equal to the length of the line running parallel to the vertical leg of the weld, at a distance of one-fourth the leg from it. In case of circular hole, the length of the fillet is taken as $\pi\left(d - \frac{h}{2}\right)$, where $d$ is the diameter of the hole and $h$ is the height of the vertical leg.

The maximum length of the weld is limited to 10 times of the throat thickness.

**EXAMPLE 30.1.** *A tie bar is welded to a plate as shown in Fig. 30.6.*

*Find the strength of the weld. Take size of the fillet as 6 mm and working stress of the fillet weld as 100 MPa.*

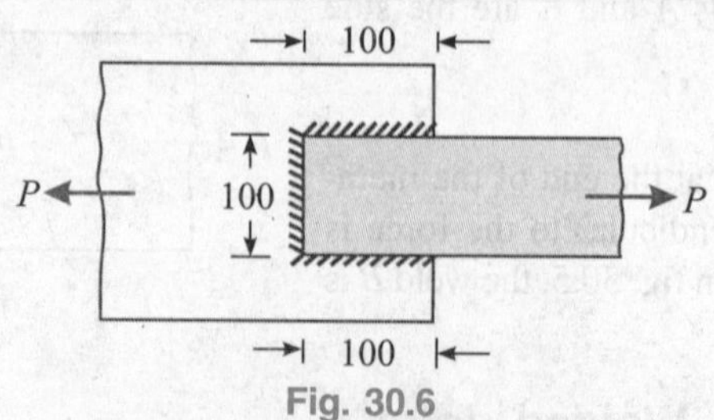


**Fig. 30.6**

**SOLUTION.** Given : Size of the weld ($s$) = 6 mm and stress of the fillet weld ($\sigma$) = 100 MPa = 100 N-mm$^2$.

From the geometry of the weld, we find that length of the fillet weld,

$$l = 100 + 100 + 100 = 300 \text{ mm}$$

and effective throat thickness, $t = 0.7\,s = 0.7 \times 6 = 4.2$ mm

$\therefore$ Strength of the weld, $P = l \cdot t \cdot \sigma = 300 \times 4.2 \times 100 = 126\,000 \text{ N} = 126 \text{ kN}$ **Ans.**

**EXAMPLE 30.2.** *A welded lap joint is provided to connect two tie bars 150 mm × 10 mm as shown in Fig. 30.7.*

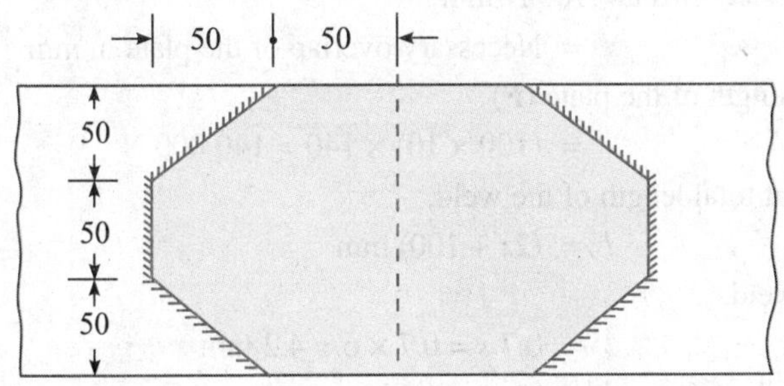


Fig. 30.7

*The working stress in the tie bar is 120 MPa. Investigate the design, if the size of the fillet is 12 mm. Take the working stress in the end fillet as 100 MPa and that in the diagonal fillet as 70 MPa.*

**SOLUTION.** Given : Size of the bar = 150 mm × 10 mm ; Working stress in the bar = 120 MPa = 120 N-mm$^2$ ; Size of the weld ($s$) = 12 mm ; Working stress in the end fillet ($\sigma_e$) = 100 MPa = 100 N-mm$^2$ and working stress in the diagonal fillet ($\sigma_d$) = 70 MPa = 70 N/mm$^2$.

We know that cross-sectional area of the bar,

$$A = 150 \times 10 = 1500 \text{ mm2}$$

∴ Strength of the bar $= 1500 \times 120 = 180\ 000 \text{ N} = 180 \text{ kN}$ ...(*i*)

We also know that effective throat thickness,

$$t = 0.7 \text{ s} = 0.7 \times 12 = 8.4 \text{ mm}$$

From the geometry of the weld, we find that total length of the end fillets

$$l_e = 50 + 50 = 100 \text{ mm}$$

and total length of the diagonal fillets,

$$l_d = 4 \times \sqrt{(50)^2 = (50)^2} = 282.8 \text{ mm}$$

∴ Strength of the end fillet,

$$P_e = l_e \cdot t \cdot \sigma_e = 100 \times 8.4 \times 100 = 84\ 000 \text{ N} = 84 \text{ kN}$$

and strength of the diagonal fillet,

$$P_d = l_d \cdot t \cdot \sigma_d = 282.8 \times 8.4 \times 70 = 166\ 300 \text{ N} = 166.3 \text{ kN}$$

∴ Total strength of the fillet weld,

$$P = P_e + P_d = 84 + 166.3 = 250.3 \text{ kN}$$

Since the strength of the weld is more than the strength of the bar, therefore the joint is safe.

**EXAMPLE 30.3.** *A 100 mm × 10 mm plate is to be welded to another plate 150 mm × 10 mm by fillet weld of 6 mm size on three sides as shown in Fig. 30.8.*

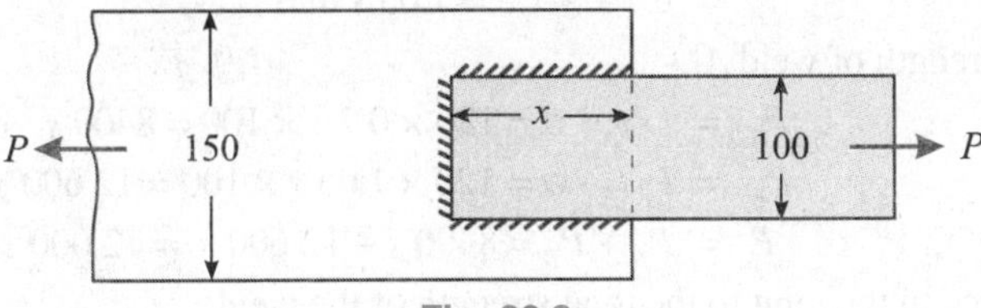


Fig. 30.8

*Determine the necessary overlap of the plate. Take allowable stresses in the plate as 140MPa and allowable stress in the weld as 100 MPa.*

**SOLUTION.** Given : Size of smaller plate = 100 mm × 10 mm ; Size of larger plate = 150 mm × 10 mm ; Size of fillet weld ($s$) = 6 mm ; Allowable stress in plate = 140 MPa = 140 N/mm$^2$ and allowable stress in the weld ($\sigma$) = 100 MPa = 100 N/mm$^2$

Let $x$ = Necessary overlap of the plate in mm.

We know that strength of the plate ($P$)

$$= (100 \times 10) \times 140 = 140\ 000 \text{ N}$$

We also know that total length of the weld,

$$l = (2x + 100) \text{ mm}$$

and thickness of the weld,

$$t = 0.7\,s = 0.7 \times 6 = 4.2 \text{ mm}$$

∴ *Strength of the fillet weld

$$= l \cdot t \cdot \sigma = (2x + 100) \times 4.2 \times 100 = 840x + 42\ 000$$

Equating strength of the plate and strength of the weld,

$$140\ 000 = 840x + 42\ 000$$

or $$x = \frac{140000 - 42000}{840} = 116.7 \text{ mm} \quad \text{say} \quad 120 \text{ mm}$$ **Ans.**

**EXAMPLE 30.4.** *A tie bar (A) = 120 mm × 10 mm, welded to another tie bar (B) = 120 mm × 15 mm, is subjected to a load of 120 kN as shown in Fig. 30.9.*

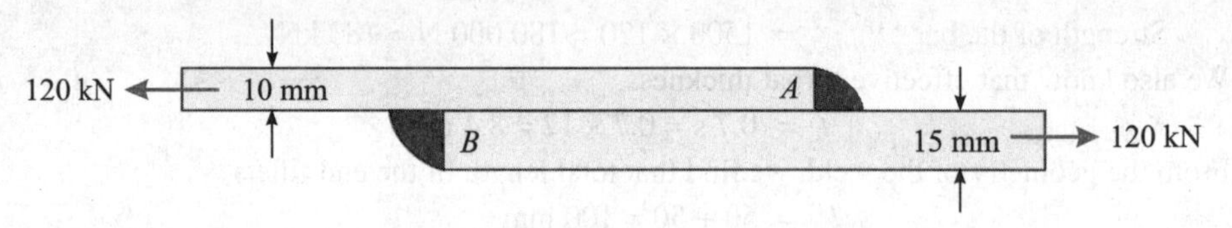


Fig. 30.9

*Determine the sizes of the end fillets, such that the stresses in both the fillets are the same. Take allowable stresses in the weld as 100 MPa.*

**SOLUTION.** Given: Size of tie bar $A$ = 120 mm × 10 mm ; Size of tie bar $B$ = 120 mm × 15 mm ; Load on the plates ($P$) = 120 kN = 120 × 10$^3$ N and allowable stress in the weld ($\sigma$) = 100 MPa = 100 N/mm$^2$.

We know that for equal stresses in the welds $A$ and $B$, the loads shared by fillet welds should be proportional to their effective throat thickness or size of the welds.

Let $s$ = Size of the weld ($A$) in the upper plate (*i.e.*, 10 mm thick plate) in mm

∴ Effective throat thickness of weld $A$,

$$t_A = 0.7\,s \text{ mm}$$

and effective throat thickness of weld $B$,

$$t_B = 0.7\,s \times 1.5 = 1.05\,s \text{ mm}$$

We know that strength of weld $A$,

$$P_A = l \cdot t_A \cdot \sigma = 120 \times 0.7\,s \times 100 = 8400\,s$$

Similarly, $$P_B = l \cdot t_B \cdot \sigma = 120 \times 1.05\,s \times 100 = 12\ 600\,s$$

∴ Total strength, $$P = P_A + P_B = 8400\,s + 12\ 600\,s = 12\ 000\,s$$

Equating the force in the joint to the total strength of the weld,

$$120 \times 10^3 = 21\ 000\,s$$

* Strength of the fillet weld is taken to be equal to the strength of the smaller plate.

$$\therefore \quad s = \frac{120 \times 10^3}{21000} = 5.7 \text{ mm} \quad \text{say} \quad 6 \text{ mm} \qquad \textbf{Ans.}$$

**EXAMPLE 30.5.** *A 100 mm × 12 mm plate is connected to another plate by fillet welds around the end of the bar and also inside a machined slot as shown in Fig. 30.10. All the dimensions are in mm.*

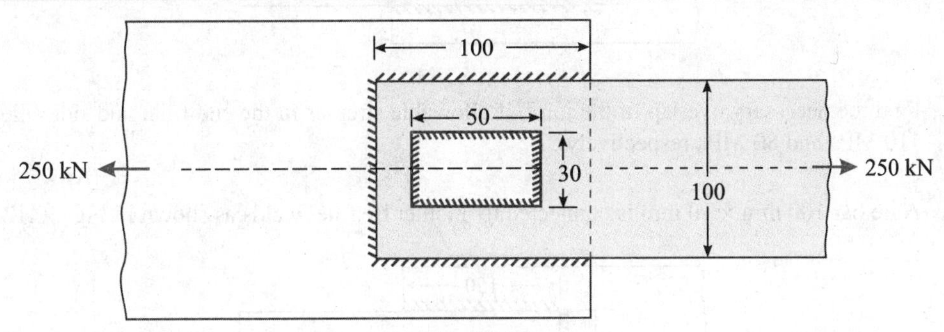


Fig. 30.10

*Determine the size of the weld, if the joint is subjected to a pull of 250 kN. Take working stresses for the transverse welds and longitudinal welds as 100 MPa and 80 MPa respectively.*

**SOLUTION.** Given : Plate size = 100 mm × 12 mm ; Pull on the joint ($P$) = 250 kN = 250 × $10^3$ N; Working stress in transverse weld ($\sigma_t$) = 100 MPa = 100 N/mm$^2$ and working stress in longitudinal welds ($\sigma_l$) = 80 N/mm$^2$.

Let $\quad s$ = Size of the welds in mm.

∴ Effective throat thickness of the welds,

$$t = 0.7\, s \text{ mm}$$

From the geometry of the welds, we find that total length of transverse weld,

$$l_t = (2 \times 200) + (2 \times 50) = 500 \text{ mm}$$

and total length of longitudinal welds

$$l_l = 100 + (2 \times 30) = 160 \text{ mm}$$

We know that strength of the transverse welds,

$$P_t = l_t \cdot t \cdot \sigma_t = 500 \times 0.7\, s \times 100 = 35\,000\, s$$

Similarly,

$$P_l = l_l \cdot t \cdot \sigma_l = 160 \times 0.7\, s \times 80 = 8960\, s$$

∴ Total strength of the welds

$$P = P_t + P_l = 35\,000\, s + 8960\, s = 43\,960\, s$$

Equating the pull in lthe joint to the strength of the welds

$$250 \times 10^3 = 43\,960\, s$$

$$\therefore \quad s = \frac{250 \times 10^3}{43960} = 5.6 \quad \text{say} \quad 6 \text{ mm} \qquad \textbf{Ans.}$$

## EXERCISE 30.1

1. A tie bar is welded to a gusset plate by two side welds of 120 mm length and an end fillet weld of 100 mm length. Find the safe load, which the tie bar can carry, if size of the weld is 7 mm. Take working stress of the weld as 100 MPa. **[Ans.** 166.6 kN]
2. A tie bar 100 mm × 10 mm thick, carrying a load of 144 kN weld is to be connected to another plate with 6 mm fillet weld on all the four sides as shown in Fig. 30.11.

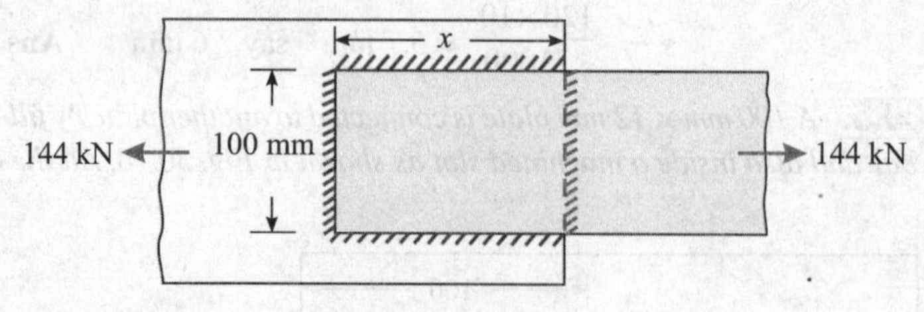


Fig. 30.11

Find the necessary overlap of the joint, if allowable stresses in the end fillet and side fillet are 110 MPa and 80 MPa respectively.

[**Ans.** 75 mm]

3. A tie bar 100 mm × 10 mm is connected to another by fillet welds as shown in Fig. 30.12.

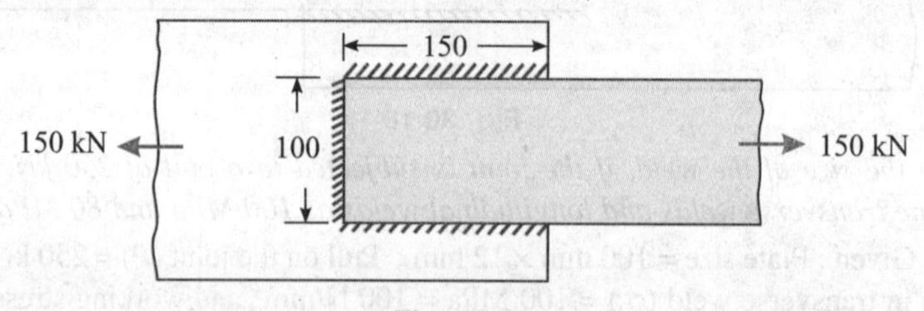


Fig. 30.12

Find the size of the weld, if the permissible stress in the weld is 100 MPa. [**Ans.** 6 mm]

4. A tie bar, carrying a load of 105 kN is connected to a plate with 6 mm fillet welds provided on two parallel sides only. Determine, the minimum lap of the joint, if the permissible stress in the weld is 100 MPa. [**Ans.** 125 mm]

## 30.9. Unsymmetrical Section Subjected to an Axial Load

Sometimes, an unsymmetrical section (say angle section, *T*-section etc.) subjected to an axial load is welded to a plate as shown in Fig. 30.13.

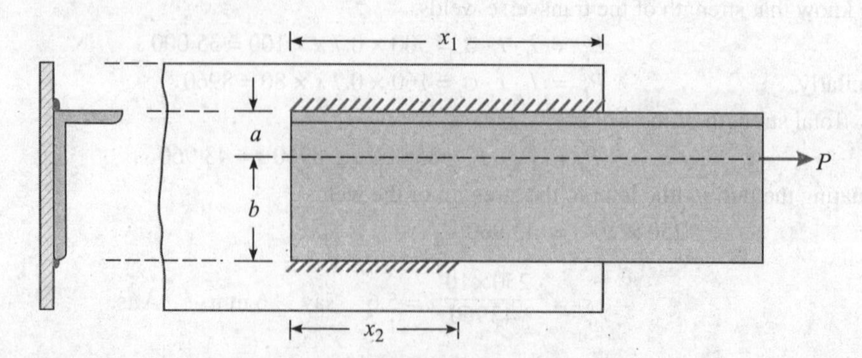


Fig. 30.13

In such cases, the load is applied along the neutral axis of the unsymmetrical section; in order to avoid the effect of eccentricity. A little consideration will show, that whenever an unsymmetrical section is welded to a plate, the fillet weld is applied in such a way that the centre of gravity of the weld lies on the line of action of the load. Or in other words, the centre of gravity of the weld coincides with the neutral axis of the section.

Let $P$ = Load acting on the section,

$\sigma$ = Working stress of the weld,

$x_1, x_2$ = Length of the top and bottom fillet welds respectively,

$a, b$ = Distance between the top and bottom edges of the section respectively from the neutral axis, and

$s$ = Size of the weld.

∴ Effective throat thickness of the weld,

$$t = 0.7\ s$$

We know that the strength of the top weld,

$$P_1 = x_1 \cdot t \cdot \sigma = x_1 \times 0.7\ s \times \sigma = 0.7\ x_1\ s \cdot \sigma$$

Similarly strength of the bottom weld,

$$P_2 = 0.7\ x_2 \cdot s \cdot \sigma$$

Since the load acting on the section is equal to the total strength of the weld, therefore

$$P = 0.7\ x_1 \cdot s \cdot \sigma + 0.7\ x_2 \cdot s \cdot \sigma$$

$$= 0.7\ s \cdot \sigma\ (x_1 + x_2)$$

Now equating the moments of the forces of resistance about the bottom,

$$P_1\ (a + b) = P \cdot b$$

$$\therefore \quad P_1 = \frac{P \cdot b}{(a + b)}$$

Substituting the value of $P_1 = 0.7\ x_1\ s \cdot \sigma$ in the above equation,

$$0.7\ x_1\ s \cdot \sigma = \frac{P \cdot b}{(a + b)}$$

$$\therefore \quad x_1 = \frac{P \cdot b}{0.7\ s \cdot \sigma\ (a + b)} = \frac{P \cdot b}{t \cdot \sigma\ (a + b)}$$

Similarly,

$$x_2 = \frac{P \cdot a}{0.7\ s \cdot \sigma\ (a + b)} = \frac{P \cdot a}{t \cdot \sigma\ (a + b)}$$

where $t$ is the effective throat thickness and equal to 0.7 $s$.

**Note :** Sometimes, the top edge of the angle section is termed as heel, whereas the bottom edge as toe.

**Example 30.6.** *A tie beam of a roof truss consists of an angle 90 mm × 90 mm × 8 mm is subjected to a load of 120 kN. The tie is connected to the gusset plate by welding. Design the joint, if the size of the weld is 6 mm. Take maximum allowable shear stress in the weld as 100 MPa. The distance between the neutral axis and the edges of the angle section are 28.7 mm and 61.3 mm respectively.*

**Solution.** Given : Size of the tie member = 90 mm × 90 mm × 8 mm ; Load ($P$) = 120 kN = 120 × $10^3$ N ; Size of the weld ($s$) = 6 mm ; Allowable shear stress in the weld ($\sigma$) = 100 MPa = 100 N/mm$^2$; Distance between the neutral axis and top edge ($a$) = 28.7 mm and distance between the neutral axis and bottom edge ($b$) = 61.3 mm.

***Length of the fillet weld at the top of the angle iron***

We know that effective throat thickness,

$$t = 0.7\ s = 0.7 \times 6 = 4.2 \text{ mm}$$

∴ Length of the fillet weld at the top of the angle iron,

$$x_1 = \frac{P \cdot b}{t \cdot \sigma\ (a + b)} = \frac{(120 \times 10^3) \times 61.3}{4.2 \times 100 \times (28.7 + 61.3)} = 195 \text{ mm}$$

Let us provide length of the weld equal to 195 + (2 × 6) = 207 mm say 210 mm. **Ans.**

*Length of the fillet weld at the bottom of the angle iron*

We also know that length of the fillet at the bottom of the angle iron,

$$x_2 = \frac{P \cdot b}{t \cdot \sigma\,(a+b)} = \frac{(120 \times 10^3) \times 28.7}{4.2 \times 100 \times (28.7 + 61.3)} = 91 \text{ mm}$$

Let us provide length of the weld equal to 91 + (2 × 6) = 103 mm say 105 mm. **Ans.**

**EXAMPLE 30.7.** *A tie bar in a truss consisting of a double angle section 100 mm × 65 mm × 10 mm is subjected to a load of 250 kN and is welded to a gusset plate as shown in Fig. 30.14.*

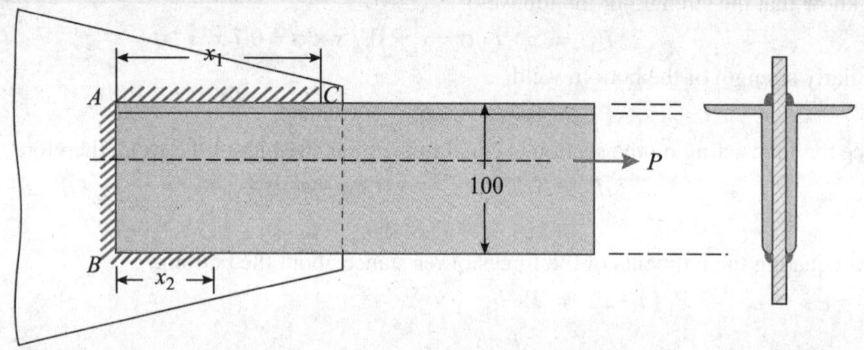


Fig. 30.14

*Design the joint with 8 mm fillet weld, if the permissible stress in the weld is 100 MPa. Take the distances between the neutral axis and the edges of the angle section as 33.7 mm and 66.3 mm respectively.*

**SOLUTION.** Given : Size of angle section = 100 mm × 65 mm × 10 mm ; Total load = 250 kN = $250 \times 10^3$ N ; Size of weld ($s$) = 8 mm ; Permissible stress in the weld ($\sigma$) = 100 MPa = 100 N/mm$^2$; Distance between the neutral axis and top edge ($a$) = 33.7 mm and distance between the neutral axis bottom edge ($b$) = 66.3 mm.

*Length of the top fillet weld*

We know that effective throat thickness,

$$t = 0.7\,s = 0.7 \times 8 = 5.6 \text{ mm}$$

Since there are two angle sections, therefore load on each angle section

$$= \frac{250 \times 10^3}{2} = 125 \times 10^3 = 125\,000 \text{ N}$$

We know that strength of transverse weld $AB$

$$= l \cdot t \cdot \sigma = 100 \times 5.6 \times 100 = 56\,000 \text{ N}$$

∴ Force of rsistance of the top and bottom fillet welds,

$$P = 125\,00 - 56\,000 = 69\,000 \text{ N}$$

and length of the top fillet weld,

$$x_1 = \frac{P \cdot b}{t \cdot \sigma\,(a+b)} = \frac{69000 \times 66.3}{5.6 \times 100 \times (33.7 + 66.3)} = 81.7 \text{ mm}$$

Let us provide length of the weld equal to 81.7 + (2 × 5.6) = 92.9 mm say 95 mm. **Ans.**

*Length of the bottom fillet weld*

We also know that length of the bottom fillet weld,

$$x_2 = \frac{P \cdot a}{t \cdot \sigma\,(a+b)} = \frac{69000 \times 33.7}{5.6 \times 100 \times (33.7 + 66.3)} = 41.5 \text{ mm}$$

Let us provide length of the weld equal to $41.5 + (2 \times 5.6) = 52.7$ say 55 mm. **Ans.**

## 30.10. Eccentric Welded Joints

In the previous articles, we have discussed the cases, where the weld has to resist only the linear displacement of the plate or member connected. But, sometimes, a weld may have to offer resistance to bending or torsion in addition to the linear displacement of the plate. Such welded connections are called eccentric welded connections. Though there are many types of welded connections, yet the following two are important from the subject point of view:

1. Eccentric welded joint subjected to moment, and
2. Eccentric welded joint subjected to torsion.

## 30.11. Eccentric Welded Joint Subjected to Moment

We have already discussed that if load or force acting on a welded joint, does not pass through the centroid of the weld lines, it will cause some bending moment to the weld joint, in addition to the direct load.

Now consider an eccentric welded connection for a bracket carrying a load as shown in Fig. 30.15. In this case, the moment (due to load ) is acting in a plane perpendicular to that of welds. A little consideration will show that the weld has to offer resistance to the following two types of displacements:

(*a*) Linear displacement, and (*b*) Horizontal displacement.

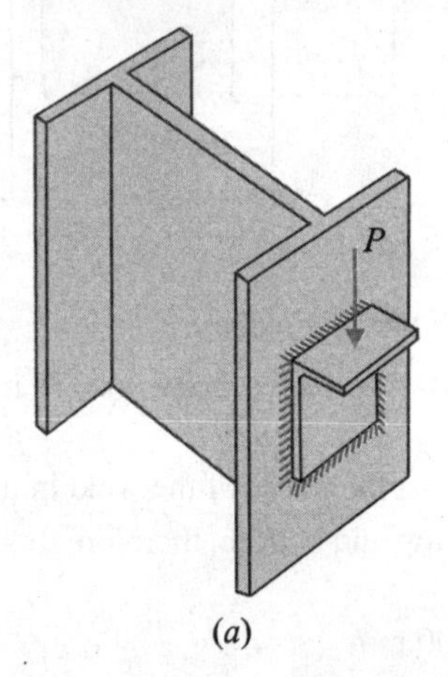


(*a*)

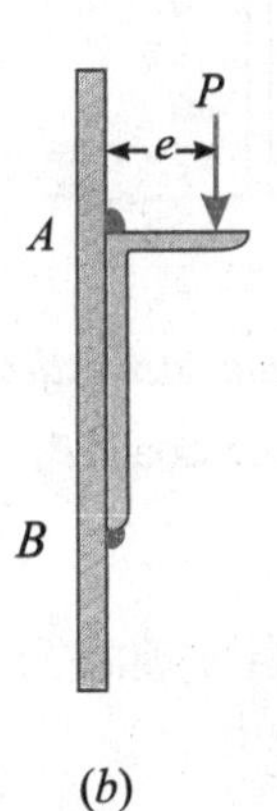


(*b*)

Fig. 30.15

(*a*) ***Resistance against linear displacement***

This resistance (or stress) is assumed to be uniform in all the weld lines and is given by the relation:

$$\mu\sigma_1 = \frac{P}{t \cdot l}$$

where $l$ = Effective throat thickness and

$l$ = Length of the total weld line.

(*b*) ***Resistance against horizontal displacement***

The eccentric load will cause a moment,

$$M = P \cdot e$$

where $e$ = Eccentricity of the load.

This moment will cause tension in the weld at $A$ and compression in the weld at $B$. The stress due to this moment is given by the relation:

$$\sigma_2 = \frac{M}{I} \times y = \frac{P \cdot e \cdot y}{I}$$

where $I$ = Moment of inertia os the weld lines and

$y$ = Distance of the extreme fibre of angle (or in other words weld lines) for neutral axis of the weld.

Now the resultant stress will be given by the relation:

$$\sigma_R = \sqrt{(\sigma_1)^2 + (\sigma_2)^2}$$

**EXAMPLE 30.8.** *A bracket is welded to a column carrying a load of 150 kN at an eccentricity of 75 mm as shown in Fig. 30.16.*

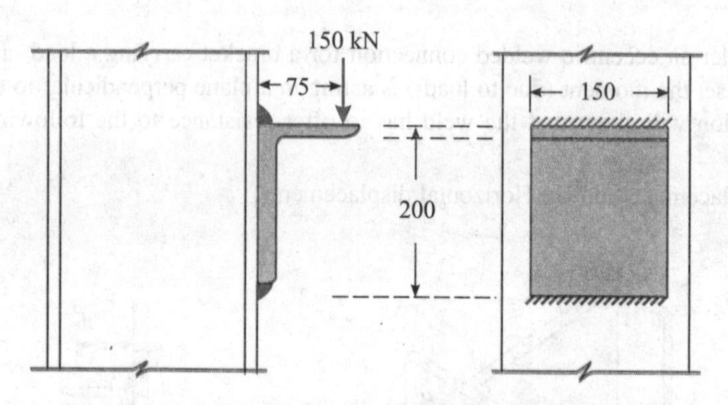


Fig. 30.16

*Find the maximum stress developed in the weld per mm length.*

**SOLUTION.** Given : Load ($P$) = 150 kN = $150 \times 10^3$ N ; Eccentricity ($e$) = 75 mm and depth of the bracket = 200 mm.

Let $t$ = Effective throat thickness of the weld in mm.

Since the bracket is welded to the column at its top and bottom, therefore total length of the fillet weld,

$$l = 2 \times 150 = 300 \text{ mm}$$

and distance of the weld from the neutral axis of the weld,

$$y = \frac{200}{2} = 100 \text{ mm}$$

We know that resistance of the weld against linear motion,

$$R_1 = \frac{P}{t \cdot l} = \frac{150 \times 10^3}{t \times 300} = \frac{500}{t} \text{ N/mm}^2$$

$\therefore$ Linear stress per mm length,

$$\sigma_1 = R_1 \times t = \frac{500}{t} \times t = 500 \text{ N/mm}$$

We know that moment due to eccentricity of the load,

$$M = P \cdot e = (150 \times 10^3) \times 75 = 11\,250 \times 10^3 \text{ N/mm}$$

and moment of inertia of both the welds about *x-x* axis (neglecting moments of inertia of both the welds about their own centres of gravity),

$$I = 2 \times (150 \times t) \times (100)^2 = 3 \times 10^6\ t \text{ mm}^4$$

∴ Resistance of the weld against horizontal motion,

$$*R_2 = \frac{M}{I} = y = \frac{11250 \times 10^3}{3 \times 10^6\ t} \times 100 \text{ N/mm}^2$$

$$= \frac{375}{t} \text{ N/mm}^2$$

and horizontal stress per mm length,

$$\sigma_2 = R_2 \times t = \frac{375}{t} \times t = 375 \text{ N/mm}$$

∴ Resultant stress per mm length,

$$\sigma_R = \sqrt{(\sigma_1)^2 + (\sigma_2)^2} = \sqrt{(500)^2 + (375)^2} = 625 \text{ N/mm} \quad \textbf{Ans.}$$

**EXAMPLE 30.9.** *A bracket consisting of an I-section is connected to the flange of a vertical column as shown in Fig. 30.17.*

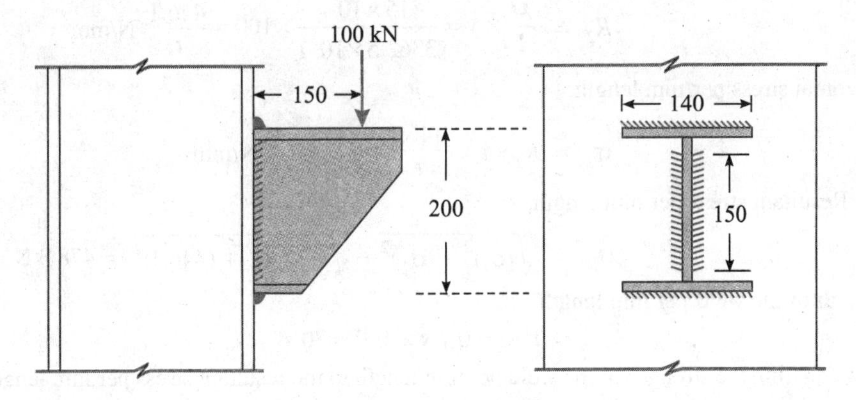


Fig. 30.17

*The bracket carries a load of 100 kN at an eccentricity of 150 mm. Find the size of the weld required for the bracket, if the permissible stress in the weld is 100 MPa.*

**SOLUTION.** given : Load $(P)$ = 100 kN = $100 \times 10^3$ N ; Eccentricity $(e)$ = 150 mm ; Permissible stress in the weld $(\sigma)$ = 100 MPa = 100 MPa = 100 N/mm$^2$ ; Total length of weld lines $(l)$ = 2 (150 + 140) = 580 mm and depth of the bracket = 200 mm.

Let $s$ = Size of the weld in mm.

We know that effective depth of the weld,

$$t = 0.7\ s$$

---

* The value of $R_2$ may also be found out by equating the moment due to load and moment due to resistance about the bottom weld *i.e.*,

$$11\ 250 \times 10^3 = R_2 \times 150 \times 200 \times t$$

$$R_2 = \frac{11250 \times 10^3}{150 \times 200 \times t} = \frac{375}{t}$$

and distance of the horizontal welds from the neutral sxis of the total weld,

$$y = \frac{200}{2} = 100 \text{ mm}$$

∴ Resistance of the weld against linear motion,

$$R_1 = \frac{P}{t \cdot l} = \frac{100 \times 10^3}{t \times 580} = \frac{172.4}{t} \text{ N/mm}^2$$

and linear stress per mm length,

$$\sigma_1 = R_1 \times t = \frac{172.4}{t} \times t = 172.4 \text{ N/mm}$$

We know tha moment due to load,

$$M = P \cdot e = (100 \times 10^3) \times 150 = 15 \times 10^6 \text{ N/mm}$$

and moment of inertia of the welds about $x$-$x$ axis (neglecting moments of inertia of the horizontal welds about their own centres of gravity),

$$I = \frac{2t \times (150)^3}{12} + 2t \times 140 \times (100)^2 = 3362.5 \times 10^3 \, t \text{ mm}^4$$

∴ Resistance of the weld against horizontal motion,

$$R_2 = \frac{M}{I} \times y = \frac{15 \times 10^6}{3362.5 \times 10^3 \, t} \times 100 = \frac{446.1}{t} \text{ N/mm}^2$$

and horizontal stress per mm length,

$$\sigma_2 = R_2 \times t = \frac{446.1}{t} \times t = 446.1 \text{ N/mm}$$

∴ Resultant stress per mm length,

$$\sigma_R = \sqrt{(\sigma_1)^2 + (\sigma_2)^2} = \sqrt{(172.4)^2 + (446.1)^2} = 478.3 \text{ N/mm}$$

and strength of the weld per mm length

$$= t \cdot \sigma = 0.7\, s \times 100 = 70\, s$$

Now equating the strength of the weld per mm length to the resultant stress per mm length of the weld,

$$70\, s = 478.3$$

or

$$s = \frac{478.3}{70} = 6.8 \quad \text{say} \quad 7 \text{ mm} \qquad \textbf{Ans.}$$

## 30.12. Eccentric Welded Joint Subjected to Torsion

In the previous article, we have discussed the eccentric welded joint subjected to bending moment. But in this article, we shall discuss the eccentric welded joint subjected to torsion also. A bracket connection is an example of this type of connection. Now consider an eccentric welded joint carrying a load as shown in Fig. 30.18. In this case, the moment (due to load $P$) is acting in a plane containing the welds. A little consideration will show that the weld has to offer resistance to the following two types of displacements:

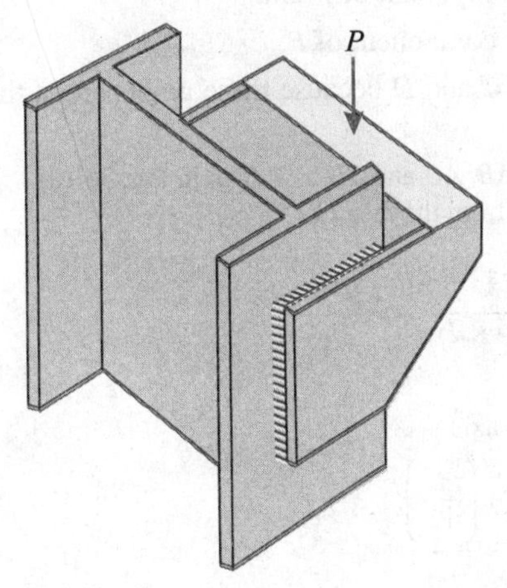


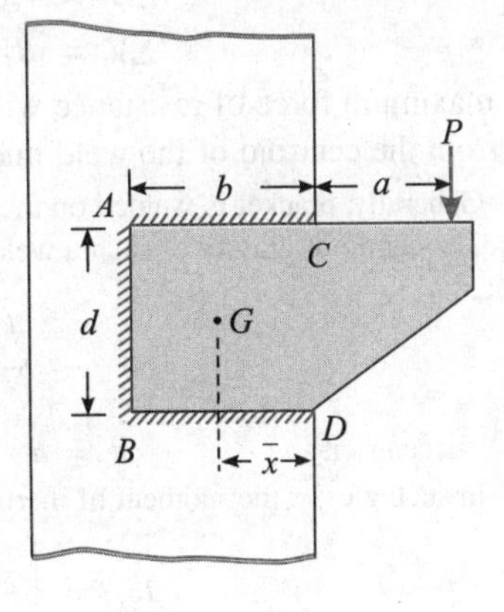


Fig. 30.18

(*a*) Linear displacement and (*b*) Rotary displacement.

(*a*) ***Resistance against linear displacement***

This resistance (or stress) is assumed to be uniform in all the weld lines and is equal to

$$R_l = \frac{P}{t \cdot l}$$

where $t$ = Effective throat thickness and

$l$ = Length of the total weld line.

(*b*) ***Resistance against rotary displacement***

Following two assumptions are made for finding out the resistance against the rotary displacement:

(*i*) The force of resistance at any point in a weld is proportional to its distance from the centroid of the weld lines.

(*ii*) The direction of the force of resistance is perpendicular to the line joining the point and the centroid of the weld lines.

Now consider an eccentric welded connection as shown in Fig. 30.18.

Le $P$ = Load acting on the connection,

$e$ = Eccentricity of the load and

$r$ = Distance between the centroid of the weld lines and the point, where the resistance is required to be found out.

∴ Twisting moment, $T = P \cdot e$

and resistance against twisting or rotary displacement,

or $$R_T = \frac{T}{J} \times r = \frac{P \cdot e \cdot r}{I_{XX} + I_{YY}}$$

where $J$ is the polar moment of inertia and is equal to $I_{XX} + I_{YY}$.

The direction of this force of resistance will be at right angles to the line joining the centroid of weld lines and the point, where this resistance is required to be found out. This force is resolved horizontally as well as vertically. The resultant force against rotary displacement will be given by the relation:

$$R = \sqrt{(\Sigma H)^2 + (\Sigma V)^2}$$

where $\Sigma H$ = Horizontal component of $P$ and

$\Sigma V = R_1$ = Vertical component of $P$.

The maximum force of resistance will occur at $C$ and $D$ because these points are at the farthest distance from the centroid of the weld lines.

**Notes:** Generally, bracket is welded on the sides i.e., $AB$, $AC$ and $BD$ as shown in Fig. 30.18.

1. The centre of gravity of such a welded system from the face $CD$,

$$\bar{x} = \frac{\left(t \times 2b \times \frac{b}{2}\right) + t \times d \times b}{(t \times 2b) + (t \times d)} = \frac{b^2 + bd}{2b + d}$$

∴ Eccentricity, $e = a + \bar{x}$

2. In such a case, the moment of inertia about $x$-$x$ axis,

$$I_{XX} = \left(2 \times bt \times \frac{d^2}{4}\right) + \left(\frac{td^3}{12}\right)$$

and

$$I_{YY} = \frac{2t \times b^3}{12} + 2bt\left(\bar{x} - \frac{b}{2}\right)^2 + t \cdot d\,(b - \bar{x})^2$$

**Example 30.10.** *Figure 30.19 shows an eccentric welded connection with 8 mm fillet welds.*

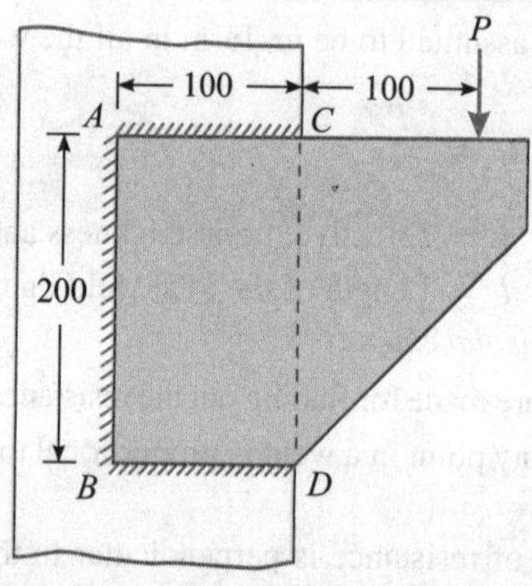


Fig. 30.19

*Calculate the maximum load (P), which the bracket can carry safely. Maximum permissible stress in the weld is 100 MPa.*

**Solution.** Given : Size of weld ($s$) = 8 mm ; Maximum stress in the weld ($s$) = 100 MPa = 100 N/mm$^2$; Width of the weld ($b$) = 100 mm and depth of the weld ($d$) = 200 mm.

We know that effective depth of the weld

$$t = 0.7\,s = 0.7 \times 8 = 5.6 \text{ mm}$$

∴ Resistance of the weld against linear motion,

$$R_l = \frac{P}{t \cdot l} = \frac{P}{400 \times 5.6} = 446.4 \times 10^{-6}\ P \text{ N/mm}^2$$

and linear stress per mm length,

$$\sigma_1 = R_1 \times t = (446.4 \times 10^{-6}\ P) \times 5.6 = 2.5 \times 10^{-3}\ P \text{ N/mm}$$

We know that distance between the centre of gravity of the weld and $D$,

$$\bar{x} = \frac{b^2 + bd}{2b + d} = \frac{(100)^2 + (100 \times 200)}{(2 \times 100) + 200} = 75 \text{ mm}$$

∴ Eccentricity, $e = 100 + 75 = 175$ mm

We also know that moment of inertia of the weld about x-x axis

$$I_{XX} = \left[2bt \times \frac{d^2}{4}\right] + \left[\frac{td^2}{12}\right]$$

$$= \left[2 \times 100 \times 5.6 \times \frac{(200)^2}{4}\right] = \left[\frac{5.6 \times (200)^3}{12}\right] \text{mm}^4$$

$$= (11.2 \times 106) + (3.73 \times 106) = 14.93 \times 10^6 \text{ mm}^4$$

Similarly,

$$I_{YY} = \left[2 \times \frac{tb^3}{12}\right] + \left[2bt\left(\bar{x} - \frac{b}{2}\right)^2 + td\,(b - \bar{x})^2\right] \text{mm}^4$$

$$= \left[2 \times \frac{5.6 \times (100)^3}{12}\right] + \left[2 \times 100 \times 5.6\left(75 - \frac{100}{2}\right)^2 + 5.6 \times 200\,(100 - 75)^2\right] \text{mm}^2$$

$$= (0.93 \times 10^6) + [(0.7 \times 10^6) + (0.7 \times 10^6)] = 2.33 \times 10^6 \text{ mm}^4$$

From the geometry of the weld, we find that maximum stress will occur at $C$ or $D$, because these points are at the farthest distance from the centre of gravity ($G$) of the weld. We also find that the distance $GC$.

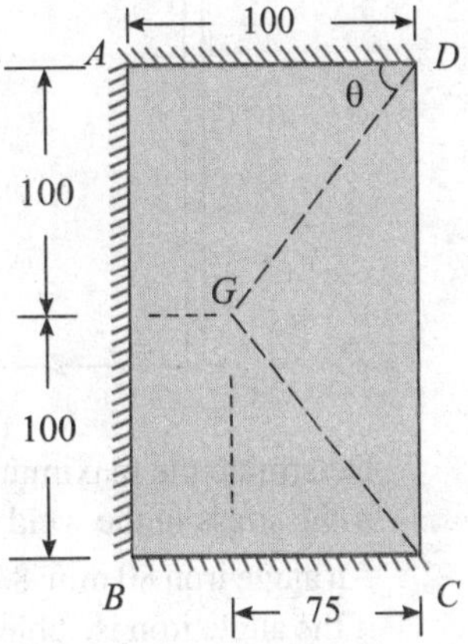

$$r = \sqrt{(100)^2 + (75)^2} \text{ mm}$$

$$= 125 \text{ mm}$$

Fig. 30.20

$$\therefore \quad \cos\theta = \frac{75}{125} = 0.6$$

and

$$\sin\theta = \frac{100}{125} = 0.8$$

∴ Resistance of the weld against rotation,

$$R_T = \frac{P \cdot e \cdot e}{I_{XX} + I_{YY}} = \frac{P \times 175 \times 125}{(14.93 \times 10^6) + (2.33 \times 10^6)}$$

$$= 1267.4 \times 10^{-6}\, P \text{ N/mm}^2$$

and rotational stress per mm length

$$= R_T \times t = 1267.4 \times 10^{-6}\, P \times 5.6 = 7097.4 \times 10^{-6}\, P \text{ N/mm}$$

∴ Horizontal component of this stress,

$$\Sigma H = 7097.4 \times 10^{-6}\, P \sin\theta = (7097.4 \times 10^{-6}\, P) \times 0.6 \text{ N/mm}$$

$$= 4.26 \times 10^{-3}\, P \text{ N/mm}$$

and vertical component of this stress

$$= 7097.4 \times 10^{-6}\, P \sin\theta = (7097.4 \times 10^{-6}\, P) \times 0.8 \text{ N/mm}$$

$$= 5.68 \times 10^{-3}\, P \text{ N/mm}$$

∴ Total vertical stress,

$$\Sigma V = (2.5 \times 10^{-3}\, P) + (5.68 \times 10^{-3}\, P) = 8.18 \times 10^{-3}\, P \text{ N/mm}$$

and resultant stress per mm length,

$$R = \sqrt{\Sigma H^2 + \Sigma V^2}$$

$$= \sqrt{(4.26 \times 10^{-3} P)^2 + (8.18 \times 10^{-3} p)^2} \text{ N/mm}$$

$$= 9.22 \times 10^{-3} \; P \text{ N/mm}$$

We know that strength of the weld per mm length

$$= t \cdot \sigma = 5.6 \times 100 = 560 \text{ N/mm}$$

Now equating strength of the weld per mm length of the resultant force per mm length of the weld,

$$560 = 9.22 \times 10^{-3} \; P$$

$$\therefore \qquad P = \frac{560}{9.22 \times 10^{-3}} = 60.7 \times 10^3 \text{ N} = 60.7 \text{ kN} \qquad \textbf{Ans.}$$

## EXERCISE 30.2

1. An unequal angle bracket is welded to a column by 10 mm fillet welds as shown in Fig. 30.21.

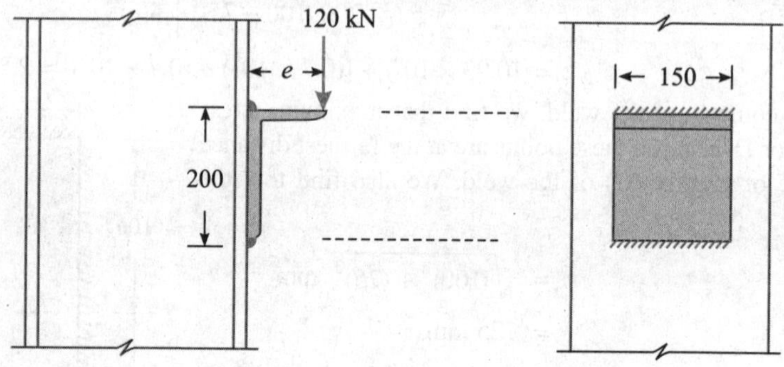


Fig. 30.21

Determine the maximum eccentricity, at which a load of 120 kN can be placed on the bracket, if the stress in the weld is limited to 80 MPa. **[Ans.** 100 mm]

2. An angle iron 80 mm 80 mm 8 mm is to be connected to a plate by 5 mm weld. Design the joint, if the angle iron is subjected to tensile load of 110 kN. The permissible stress in the weld is 100 MPa. The distance of the edges of the angle section from the neutral axis are 22.7 mm and 57.3 mm respectively. **[Ans.** 200 mm ; 90 mm]

3. A bracket carrying a load of 40 kN is welded to a column by four fillet welds as shown in Fig. 30.22.

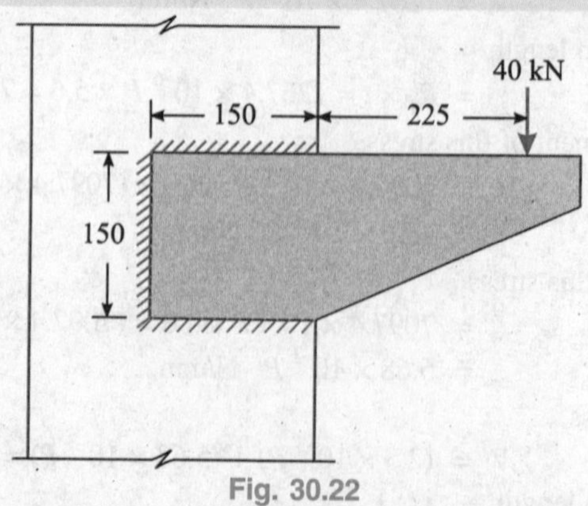


Fig. 30.22

Find the maximum stress per mm length of the weld, so that maximum stress induced in the weld does not exceed 80 MPa. **[Ans.** 333 N/mm]

## QUESTIONS

1. Give the advantages and disadvantages of welded joints.
2. Explain the various types of welded joints.
3. Explain the difference between
   (*a*) Size of the fillet weld and throat thickness
   (*b*) Legs of the weld and length of the weld
   (*c*) Side fillet weld and end fillet weld.
4. What do you understand by the term strength of a welded joint? Give the relation for the same.
5. How will you find out the length of the weld lines, when an unsymmetrical section subjected to an axial load is welded to a plate?
6. What is meant by eccentric welded joint?
7. Derive relation for the resultant stress in the weld lines of an eccentric welded connection, when it is subjected to a
   (*i*) moment and (*ii*) torsion.

## OBJECTIVE TYPE QUESTIONS

1. The effective thickness of a fillet weld is
   (*a*) 0.5 *s* (*b*) 0.6 *s* (*c*) 0.7 *s* (*d*) 0.8 *s*
2. In case of slot weld joint, the shape of the hole is
   (*a*) circular (*b*) square (*c*) rectangular (*d*) any one of these
3. The strength of a welded joint depends upon
   (*a*) length of weld (*b*) size of weld (*c*) stress of the weld (*d*) all of these
4. The strength of a welded joint is equal to
   (*a*) $l\,t\,\sigma$ (*b*) $\frac{lt}{\sigma}$ (*c*) $\frac{l\sigma}{t}$ (*d*) $\frac{t\sigma}{l}$

## ANSWERS

1. (*c*) 2. (*d*) 3. (*d*) 4. (*a*)

Chapter 31

# Thin Cylindrical and Spherical Shells

### Contents



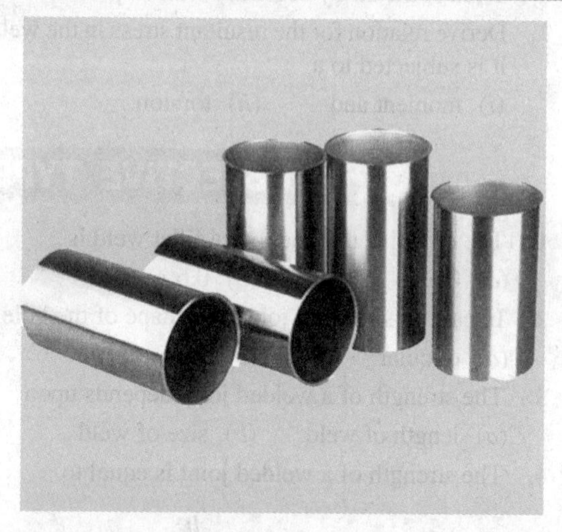

## 31.1. Introduction

In engineering field, we daily come across vessels of cylindrical and spherical shapes containing fluids such as tanks, boilers, compressed air receivers etc. Generally, the walls of such vessels are very thin as compared to their diameters. These vessels, when empty, are subjected to atmospheric pressure internally as well as externally. In such a case, the resultant pressure on the walls of the shell is zero. But whenever a vessel is subjected to internal pressure (due to steam, compressed air etc.) its walls are subjected to tensile stresses.

In general, if the thickness of the wall of a shell is less than 1/10th to 1/15th of its diameter, it is known as a thin shell.

## 31.2. Failure of a Thin Cylindrical Shell due to an Internal Pressure

We have already discussed in the last article that whenever a cylindrical shell is subjected to an internal pressure, its walls are subjected to tensile stresses.

It will be interesting to know that if these stresses exceed the permissible limit, the cylinder is likely to fail in any one of the following two ways as shown in Fig. 31.1 (*a*) and (*b*).

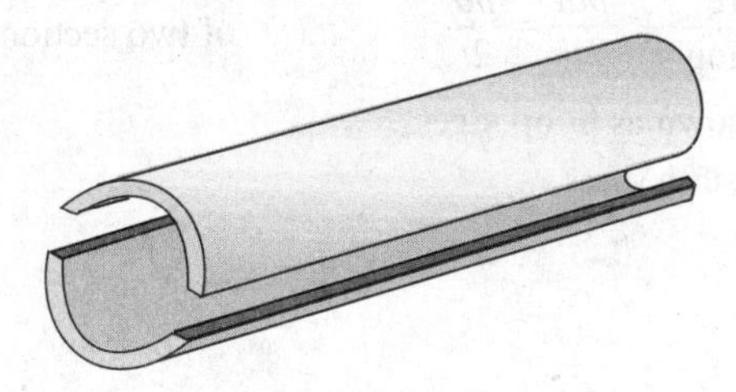

(*a*) Split into two troughs.

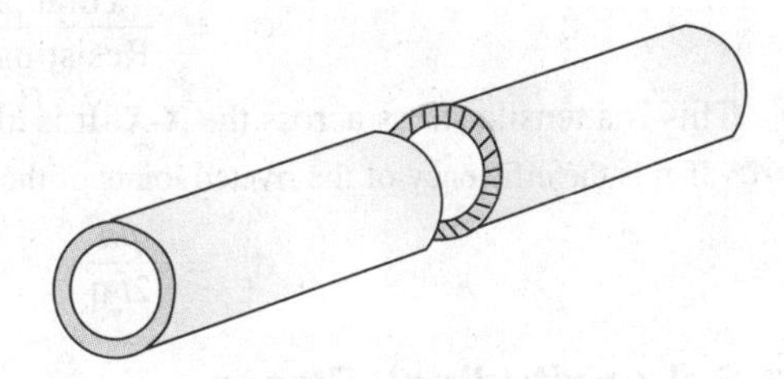

(*b*) Split into two cylinders.

Fig. 31.1

1. It may split up into two troughs and
2. It may split up into two cylinders.

## 31.3. Stresses in a Thin Cylindrical Shell

We have already discussed that whenever a cylindrical shell is subjected to an internal pressure, its walls are subjected to tensile stresses. A little consideration will show that the walls of the cylindrical shell will be subjected to the following two types of tensile stresses:

1. Circumferential stress and
2. Longitudinal stress.

In case of thin shells, the stresses are assumed to be uniformly distributed throughout the wall thickness. However, in case of thick shells, the stresses are no longer uniformly distributed and the problem becomes complex. In this chapter, we shall discuss the stress in thin shells only.

**Note:** The above theory also holds good, when the shell is subjected to compressive stress.

## 31.4. Circumferential Stress

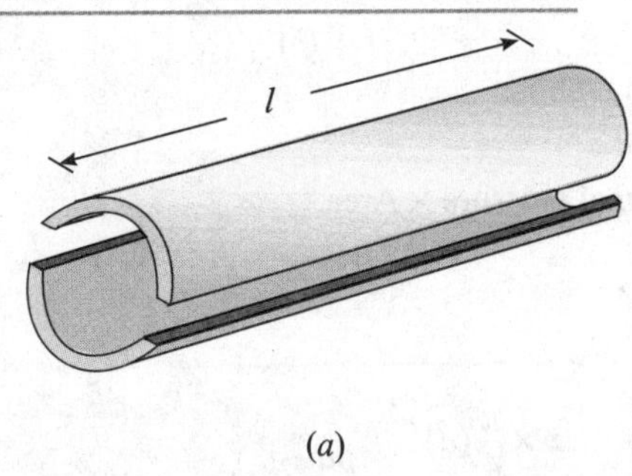


(*a*)

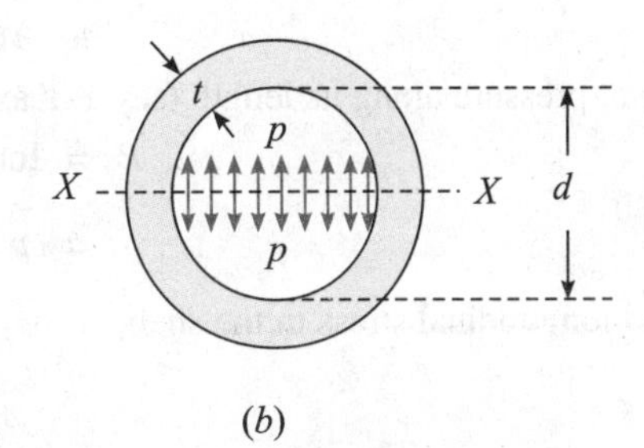


(*b*)

Fig. 31.2

Consider a thin cylindrical shell subjected to an internal pressure as shown in Fig. 31.2(*a*) and (*b*). We know that as a result of the internal pressure, the cylinder has a tendency to split up into two troughs as shown in the figure.

Let $l$ = Length of the shell,

$d$ = Diameter of the shell,

$t$ = Thickness of the shell and

$p$ = Intensity of internal pressure.

Total pressure along the diameter (say $X$-$X$ axis) of the shell,

$$P = \text{Intensity of internal pressure} \times \text{Area} = p \times d \times l$$

and circumferential stress in the shell,

$$\sigma_c = \frac{\text{Total pressure}}{\text{Resisting section}} = \frac{pdl}{2tl} = \frac{pd}{2t} \quad \text{...(}\because \text{ of two sections)}$$

This is a tensile stress across the $X$-$X$. It is also known as **hoop stress.**

NOTE. If $\eta$ is the efficiency of the riveted joints of the shell, then stress,

$$\sigma_c = \frac{pd}{2t\eta}$$

## 31.5. Longitudinal Stress

Consider the same cylindrical shell, subjected to the same internal pressure as shown in Fig. 31.3 (*a*) and (*b*). We know that as a result of the internal pressure, the cylinder also has a tendency to split into two pieces as shown in the figure.

Let

$p$ = Intensity of internal pressure,

$l$ = Length of the shell,

$d$ = Diameter of the shell and

$t$ = Thickness of the shell.

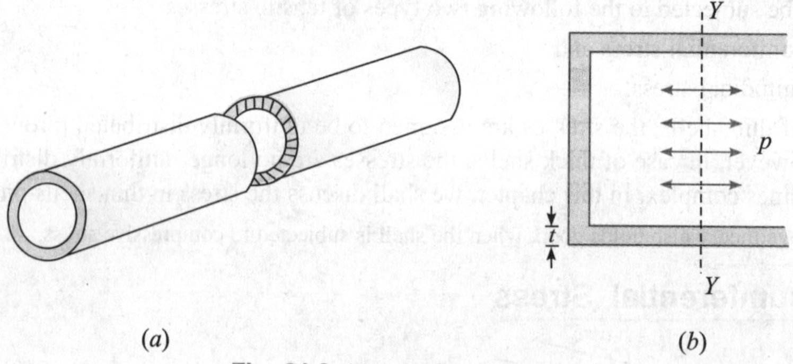


Fig. 31.3. Longitudinal stress.

Total pressure along its length (say $Y$-$Y$ axis) of the shell

$$P = \text{Intensity of internal pressure} \times \text{Area}$$

$$= p \times \frac{\pi}{4}(d)^2$$

and longitudinal stress in the shell,

$$\sigma_l = \frac{\text{Total pressure}}{\text{Resisting section}} = \frac{p \times \frac{\pi}{4}(d)^2}{\pi dt} = \frac{pd}{4t}$$

This is also a tensile stress across the section $Y$-$Y$. It may be noted that the longitudinal stress is half of the circumferential or hoop stress.

NOTE. If $\eta$ is the efficiency of the riveted joints of the shell, then the stress,

$$\sigma_l = \frac{pd}{4t\eta}$$

**EXAMPLE 31.1.** *A stream boiler of 800 mm diameter is made up of 10 mm thick plates. If the boiler is subjected to an internal pressure of 2.5 MPa, find the circumferential and longitudinal stresses induced in the boiler plates.*

**SOLUTION.** Given : Diameter of boiler ($d$) = 800 mm ; Thickness of plates ($t$) = 10 mm and internal pressure ($p$) = 2.5 MPa = 2.5 N/mm$^2$.

***Circumferential stress induced in the boiler plates***

We know that circumferential stress induced in the boiler plates,

$$\sigma_c = \frac{pd}{2t} = \frac{2.5 \times 800}{2 \times 10} = 100 \text{ N/mm}^2 = \mathbf{100 \text{ MPa}} \quad \textbf{Ans.}$$

***Longitudinal stress induced in the boiler plates***

We also know that longitudinal stress induced in the boiler plates,

$$\sigma_l = \frac{pd}{4t} = \frac{2.5 \times 800}{4 \times 10} = 50 \text{ N/mm}^2 = \mathbf{50 \text{ MPa}} \quad \textbf{Ans.}$$

**EXAMPLE 31.2.** *A cylindrical shell of 1.3 m diameter is made up of 18 mm thick plates. Find the circumferential and longitudinal stress in the plates, if the boiler is subjected to an internal pressure of 2.4 MPa. Take efficiency of the joints as 70%.*

**SOLUTION.** Given: Diameter of shell ($d$) = 1.3 m = $1.3 \times 10^3$ mm ; Thickness of plates ($t$) = 18 mm; Internal pressure ($p$) = 2.4 MPa = 2.4 N/mm$^2$ and efficiency ($\eta$) = 70% = 0.7.

***Circumferential stress***

We know that circumferential stress,

$$\sigma_c = \frac{pd}{2t\eta} = \frac{2.4 \times (1.3 \times 10^3)}{2 \times 18 \times 0.7} = 124 \text{ N/mm}^2 = \mathbf{124 \text{ MPa}} \quad \textbf{Ans.}$$

***Longitudinal stress***

We also know that longitudinal stress,

$$\sigma_l = \frac{pd}{4t\eta} = \frac{2.4 \times (1.3 \times 10^3)}{4 \times 18 \times 0.7} = 62 \text{ N/mm}^2 = \mathbf{62 \text{ MPa}} \quad \textbf{Ans.}$$

**EXAMPLE 31.3.** *A gas cylinder of internal diameter 40 mm is 5 mm thick. If the tensile stress in the material is not to exceed 30 MPa, find the maximum pressure which can be allowed in the cylinder.*

**SOLUTION.** Given: Diameter of cylinder ($d$) = 40 mm ; Thickness of plates ($t$) = 5 mm and tensile stress ($\sigma_c$) = 30 MPa = 30 N/mm$^2$.

Let $p$ = Maximum pressure which can be allowed in the cylinder.

We know that circumferential stress ($\sigma_c$),

$$30 = \frac{pd}{2t} = \frac{p \times 40}{2 \times 5} = 4p$$

$$\therefore \quad p = \frac{30}{4} = 7.5 \text{ N/mm}^2 = \mathbf{7.5 \text{ MPa}} \quad \textbf{Ans.}$$

NOTE: 1. Since the circumferential stress ($\sigma_c$) is double the longitudinal stress ($\sigma_l$), therefore in order to find the maximum pressure the given stress should be taken as circumferential stress.

2. If however, we take the given tensile stress of 30 N/mm$^2$ as the longitudinal stress, then

$$30 = \frac{pd}{4t} = \frac{p \times 40}{4 \times 5} = 2p$$

$$\therefore \quad p = \frac{30}{4} = 15 \text{ N/mm}^2 = 15 \text{ MPa}$$

Now we shall provide a pressure of 7.5 MPa *i.e.* (Lesser of the two values) obtained by using the tensile stress as circumferential stress and longitudinal stress.

## 31.6. Design of Thin Cylindrical Shells

Designing of thin cylindrical shell involves calculating the thickness ($t$) of a cylindrical shell for the given length ($l$), diameter ($d$), intensity of maximum internal pressure ($p$) and circumferential stress ($\sigma_c$). The required thickness of the shell is calculated from the relation.

$$t = \frac{pd}{2\sigma_c} \quad \text{... (See Article 31.4)}$$

If the thickness so obtained, is not a round figure, then next higher value is provided.

NOTE: The thickness obtained from the longitudinal stress will be half of the thickness obtained from circumferential stress. Thus, it should not be accepted.

**EXAMPLE 31.4.** *A thin cylindrical shell of 400 mm diameter is to be designed for an internal pressure of 2.4 MPa. Find the suitable thickness of the shell, if the allowable circumferential stress is 50 MPa.*

**SOLUTION.** Given: Diameter of shell ($d$) = 400 mm ; Internal pressure ($p$) = 2.4 MPa = 2.4 N/mm$^2$ and circumferential stress ($\sigma_c$) = 50 MPa = 50 N/mm$^2$.

We know that thickness of the shell,

$$t = \frac{pd}{2\sigma_c} = \frac{2.4 \times 400}{2 \times 50} = \mathbf{9.6 \text{ mm} \quad say \quad 10 \text{ mm}} \quad \textbf{Ans.}$$

**EXAMPLE 31.5.** *A cylindrical shell of 500 mm diameter is required to withstand an internal pressure of 4 MPa. Find the minimum thickness of the shell, if maximum tensile strength in the plate material is 400 MPa and efficiency of the joints is 65%. Take factor of safety as 5.*

**SOLUTION.** Given: Diameter of shell ($d$) = 500 mm ; Internal pressure ($p$) = 4 MPa = 4 N/mm$^2$; Tensile strength = 400 MPa = 400 N/mm$^2$ ; Efficiency ($\eta$) = 65% = 0.65 and factor of safety = 5.

We know that allowable tensile stress (*i.e.*, circumferential stress),

$$\sigma_c = \frac{\text{Tensile strength}}{\text{Factor of safety}} = \frac{400}{5} = 80 \text{ N/mm}^2$$

and minimum thickness of shell,

$$t = \frac{pd}{2\sigma_c \eta} = \frac{4 \times 500}{2 \times 80 \times 0.65} = \mathbf{19.2 \text{ mm} \quad say \quad 20 \text{ mm}} \quad \textbf{Ans.}$$

## 31.7. Change in Dimensions of a Thin Cylindrical Shell due to an Internal Pressure

We have already discussed in the chapter on Elastic Constants that lateral strain is always accompanied by a linear strain. It is thus obvious that in a thin cylindrical shell subjected to an internal pressure, its walls will also be subjected to lateral strain. The effect of the lateral strains is to cause some change in the dimensions (*i.e.*, length and diameter) of the shell. Now consider a thin cylindrical shell subjected to an internal pressure.

Let $l$ = Length of the shell,

$d$ = Diameter of the shell,

$t$ = Thickness of the shell and

$p$ = Intensity of the internal pressure.

We know that the circumferential stress,

$$\sigma_c = \frac{pd}{2t}$$

and longitudinal stress, $$\sigma_l = \frac{pd}{4t}$$

Now let $\delta d$ = Change in diameter of the shell,

$\delta l$ = Change in the length of the shell and

$\frac{1}{m}$ = Poisson's ratio.

Now changes in diameter and length may be found out from the above equations, as usual (*i.e.*, by multiplying the strain and the corresponding linear dimension).

$$\therefore \quad \delta d = \varepsilon_1 \cdot d = \frac{pd}{2tE}\left(1 - \frac{1}{2m}\right) \times d = \frac{pd^2}{2tE}\left(1 - \frac{1}{2m}\right)$$

and $$\delta l = \varepsilon_2 \cdot l = \frac{pd}{2tE}\left(\frac{1}{2} - \frac{1}{m}\right) \times l = \frac{pdl}{2tE}\left(\frac{1}{2} - \frac{1}{m}\right)$$

**EXAMPLE 31.6.** *A cylindrical thin drum 800 mm in diameter and 4 m long is made of 10 mm thick plates. If the drum is subjected to an internal pressure of 2.5 MPa, determine its changes in diameter and length. Take E as 200 GPa and Poisson's ratio as 0.25.*

**SOLUTION.** Given: Diameter of drum ($d$) = 800 mm ; Length of drum ($l$) = 4 m = $4 \times 10^3$ mm ; Thickness of plates ($t$) = 10 mm ; Internal pressure ($p$) = 2.5 MPa = 2.5 N/mm$^2$ ; Modulus of elasticity ($E$) = 200 GPa = $200 \times 10^3$ N/mm$^2$ and poisson's ratio $\left(\frac{1}{m}\right)$ = 0.25.

***Change in diameter***

We know that change in diameter,

$$\delta d = \frac{pd^2}{2tE}\left(1 - \frac{1}{2m}\right) = \frac{2.5 \times (800)^2}{2 \times 10 \times (200 \times 10^3)}\left(1 - \frac{0.25}{2}\right) \text{mm}$$

$$= \mathbf{0.35 \text{ mm}} \quad \textbf{Ans.}$$

***Change in length***

We also know that change in length,

$$\delta l = \frac{pdl}{2tE}\left(\frac{1}{2} - \frac{1}{m}\right) = \frac{2.5 \times 800 \times (4 \times 10^3)}{2 \times 10 \times (200 \times 10^3)}\left(\frac{1}{2} - 0.25\right) \text{mm}$$

$$= \mathbf{0.5 \text{ mm}} \quad \textbf{Ans.}$$

## 31.8. Change in Volume of a Thin Cylindrical Shell due to an Internal Pressure

We have already discussed in the last article, that there is always an increase in the length and diameter of a thin cylindrical shell due to an internal pressure. A little consideration will show that increase in the length and diameter of the shell will also increase its volume. Now consider a thin cylindrical shell subjected to an internal pressure.

Let $l$ = Original length,

$d$ = Original diameter,

$\delta l$ = Change in length due to pressure and

$\delta d$ = Change in diameter due to pressure.

We know that original volume,

$$V = \frac{\pi}{4}\times d^2 \times l = \left[\frac{\pi}{4}(d+\delta d)^2 \times (l \times \delta l)\right] - \frac{\pi}{4}\times d^2 \times l$$

$$= \frac{\pi}{4}(d^2 \cdot \delta l + 2dl \cdot \delta d) \qquad \text{...(Neglecting small quantities)}$$

$$\therefore \quad \frac{\delta V}{V} = \frac{\frac{\pi}{4}(d^2 \cdot \delta l + 2dl \cdot \delta d)}{\frac{\pi}{4}\times d^2 \times l} = \frac{\delta l}{l} + \frac{2\delta d}{d} = \varepsilon_l + 2\varepsilon_c$$

or $\delta V = V(\varepsilon_l + 2\varepsilon_c)$

where $\varepsilon_c$ = Circumferential strain and

$\varepsilon_l$ = Longitudinal strain.

**EXAMPLE 31.7.** *A cylindrical vessel 2 m long and 500 mm in diameter with 10 mm thick plates is subjected to an internal pressure of 3 MPa. Calculate the change in volume of the vessel. Take E = 200 GPa and Poisson's ratio = 0.3 for the vessel material.*

**SOLUTION.** Given: Length of vessel ($l$) = 2 m = $2 \times 10^3$ mm ; Diameter of vessel ($d$) = 500 mm ; Thickness of plates ($t$) = 10 mm ; Internal pressure ($p$) = 3 MPa = 3 N/mm$^2$ ; Modulus of elasticity ($E$) = 200 GPa = $200 \times 10^3$ N/mm$^2$ and poisson's ratio $\left(\frac{1}{m}\right)$ = 0.3.

We know that circumferential strain,

$$\varepsilon_c = \frac{pd}{2tE}\left(1 - \frac{1}{2m}\right) = \frac{3 \times 500}{2 \times 10 \times (200 \times 10^3)}\left(1 - \frac{0.3}{2}\right) = 0.32 \times 10^{-3} \quad ...(i)$$

and logitudinal strain,

$$\varepsilon_l = \frac{pd}{2tE}\left(\frac{1}{2} - \frac{1}{m}\right) = \frac{3 \times 500}{2 \times 10 \times (200 \times 10^3)}\left(\frac{1}{2} - 0.3\right) = 0.075 \times 10^{-3} \quad ...(ii)$$

We also know that original volume of the vessel,

$$V = \frac{\pi}{4}(d)^2 \times l = \frac{\pi}{4}(500)^2 \times (2 \times 10^3) = 392.7 \times 10^6 \text{ mm}^3$$

∴ Change in volume,

$$\delta V = V(\varepsilon_c + 2\varepsilon_l) = 392.7 \times 10^6 \,[0.32 \times 10^{-3} + (2 \times 0.075 \times 10^{-3})] \text{ mm}^3$$

$$= \mathbf{185 \times 10^3 \text{ mm}^3} \quad \textbf{Ans.}$$

## EXERCISE 31.1

1. A cylindrical shell 2 m long and 1 m internal diameter is made up of 20 mm thick plates. Find the circumferential and longitudinal stresses in the shell material, if it is subjected to an internal pressure of 5 MPa. (**Ans.** 125 MPa ; 62.5 MPa)
2. A steam boiler of 1.25 m in diameter is subjected to an internal pressure of 1.6 MPa. If the steam boiler is made up of 20 mm thick plates, calculate the circumferential and longitudinal stresses. Take efficiency of the circumferential and longitudinal joints as 75% and 60% respectively. (**Ans.** 67 MPa ; 42 MPa)
3. A pipe of 100 mm diameter is carrying a fluid under a pressure of 4 MPa. What should be the minimum thickness of the pipe, if maximum circumferential stress in the pipe material is 12.5 MPa. (**Ans.** 16 mm)

4. A cylindrical shell 3 m long has 1 m internal diameter and 15 mm metal thickness. Calculate the circumferential and longitudinal stresses, if the shell is subjected to an internal pressure of 1.5 MPa. Also calculate the changes in dimensions of the shell. Take $E$ = 200 GPa and Poisson's ratio = 0.3. (**Ans.** 50 MPa ; 25 MPa ; $\delta d$ = 0.21 mm ; $\delta l$ = 0.15 mm)
5. A cylindrical vessel 1.8 m long 800 mm in diameter is made up of 8 mm thick plates. Find the hoop and longitudinal stresses in the vessel, when it contains fluid under a pressure of 2.5 MPa. Also find the changes in length, diameter and volume of the vessel. Take $E$ = 200 GPa and $1/m$ = 0.3. (**Ans.** 125 MPa ; 62.5 MPa ; 0.42 mm ; 0.23 mm ; 1074 mm$^3$)

## 31.9. Thin Spherical Shells

Consider a thin spherical shell subjected to an internal pressure as shown in Fig. 31.4.

Let $p$ = Intensity of internal pressure,

$d$ = Diameter of the shell and

$t$ = Thickness of the shell,

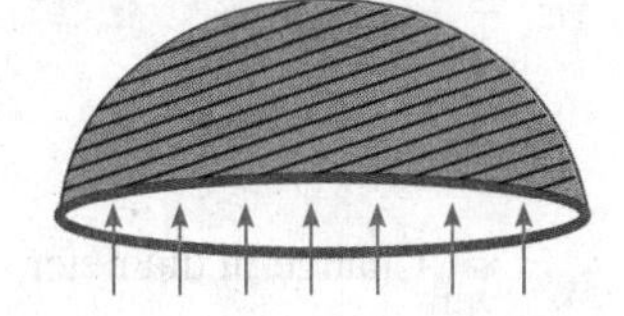

Fig. 31.4. Spherical shell

As a result of this internal pressure, the shell is likely to be torn away along the centre of the sphere. Therefore, total pressure acting along the centre of the sphere,

$$P = \text{Intensity of internal pressure} \times \text{Area}$$

$$= p \times \frac{\pi}{4} \times d^2$$

and stress in the shell material,

$$\sigma = \frac{\text{Total pressure}}{\text{Resisting section}} = \frac{p \times \frac{\pi}{4} \times d^2}{\pi d \times t} = \frac{pd}{4t}$$

**Note.** If $\eta$ is the efficiency of the riveted joints of the spherical shell, then stress,

$$\sigma = \frac{pd}{4t\eta}$$

**EXAMPLE 31.8.** *A spherical gas vessel of 1.2 m diameter is subjected to a pressure of 1.8 MPa. Determine the stress induced in the vessel plate, if its thickness is 5 mm.*

**SOLUTION.** Given: Diameter of vessel ($d$) = 1.2 m = 1.2 × 10$^3$ mm ; Internal pressure ($p$) = 1.8 MPa = 1.8 N/mm$^2$ and thickness of plates ($t$) = 5 mm.

We know that stress in the vessel plates,

$$\sigma = \frac{pd}{4t} = \frac{1.8 \times (1.2 \times 10^3)}{4 \times 5} = 108 \text{ N/mm}^2 = \mathbf{108\ MPa} \qquad \textbf{Ans.}$$

**EXAMPLE 31.9.** *A spherical vessel of 2 m diameter is subjected to an internal pressure of 2 MPa. Find the minimum thickness of the plates required, if the maximum stress is not to exceed 100 MPa. Take efficiency of the joint as 80%.*

**SOLUTION.** Given: Diameter of vessel ($d$) = 2 m = 2 × 10$^3$ mm ; Internal pressure ($p$) = 2 MPa = 2 N/mm$^2$ ; Maximum stress ($\sigma$) = 100 MPa = 100 N/mm$^2$ and efficiency of joint ($\eta$) = 80% = 0.8.

Let $t$ = Minimum thickness of the plates in mm.

We know that stress in the plates ($\sigma$),

$$100 = \frac{pd}{4t\,\eta} = \frac{2 \times (2 \times 10^3)}{4 \times t \times 0.8} = \frac{1250}{t}$$

$$\therefore \quad t = \frac{1250}{100} = \mathbf{12.5\ mm} \qquad \textbf{Ans.}$$

## 31.10. Change in Diameter and Volume of a Thin Spherical Shell due to an Internal Pressure

Consider a thin spherical shell subjected to an internal pressure.

Let $d$ = Diameter of the shell,

$p$ = Intensity of internal pressure and

$t$ = Thickness of the shell.

We have already discussed in the last article that the stress in a spherical shell,

$$\sigma = \frac{pd}{4t}$$

and strain in any one direction,

$$\varepsilon = \frac{\sigma}{E} - \frac{\sigma}{mE} \qquad ...(\because \sigma_1 = \sigma_2 = \sigma)$$

$$= \frac{pd}{4tE} - \frac{pd}{4tEm} = \frac{pd}{4tE}\left(1 - \frac{1}{m}\right)$$

∴ Change in diameter,

$$\delta d = \varepsilon \cdot d = \frac{pd}{4tE}\left(1 - \frac{1}{m}\right) \times d = \frac{pd^2}{4tE}\left(1 - \frac{1}{m}\right)$$

We also know that original volume of the sphere,

$$V = \frac{\pi}{6} \times (d)^3$$

and final volume due to pressure,

$$V + \delta V = \frac{\pi}{6} \times (d + \delta d)^3$$

where $(d + \delta d)$ = Final diameter of the shell.

∴ Volumetric strain,

$$\frac{\delta V}{V} = \frac{(V + \delta V) - V}{V} = \frac{\frac{\pi}{6}(d + \delta d)^3 - \frac{\pi}{6} \times d^3}{\frac{\pi}{6} \times d^3}$$

$$= \frac{d^3 + (3d^2 \cdot \delta d) - d^3}{d^3}$$

...(Ignoring second and higher power of $\delta d$)

$$= \frac{3 \cdot \delta d}{d} = 3\varepsilon$$

and

$$\delta V = V \cdot 3\varepsilon = \frac{\pi}{6}(d)^3 \times 3 \times \frac{pd}{4tE}\left(1 - \frac{1}{m}\right) = \frac{\pi pd^4}{8tE}\left(1 - \frac{1}{m}\right)$$

**EXAMPLE 31.10.** *A spherical shell of 2 m diameter is made up of 10 mm thick plates. Calculate the change in diameter and volume of the shell, when it is subjected to an internal pressure of 1.6 MPa. Take E = 200 GPa and 1/m = 0.3.*

**SOLUTION.** Given: Diameter of shell ($d$) = 2 m = $2 \times 10^3$ mm ; Thickness of plates ($t$) = 10 mm ; Internal pressure ($p$) = 1.6 MPa = 1.6 N/mm$^2$ ; Modulus of elasticity ($E$) = 200 GPa = $200 \times 10^3$ N/mm$^2$ and Poisson's ratio ($1/m$) = 0.3.

***Change in diameter***

We know that change in diameter,

$$\delta d = \frac{pd^2}{4tE}\left(1 - \frac{1}{m}\right) = \frac{1.6 \times (2 \times 10^3)^2}{4 \times 10 \times (200 \times 10^3)}(1 - 0.3)$$

$$= \mathbf{0.56\ mm} \qquad \textbf{Ans.}$$

*Change in volume*

We also know that change in volume,

$$\delta V = \frac{\pi p d^4}{8tE}\left(1-\frac{1}{m}\right) = \frac{\pi \times 1.6 \times (2\times 10^3)^4}{8\times 10 \times (200\times 10^3)}(1-0.3)\ \text{mm}^3$$

$$= \mathbf{3.52\times 10^6\ mm^3} \quad \textbf{Ans.}$$

## 31.11. Riveted Cylindrical Shells

Sometimes, boilers of the desired capacity are made of cylindrical shape by joining different plates usually by rivets. This is generally done : (*i*) by bending the plates to the required diameter and then joining them by a butt joint and (*ii*) by joining individually fabricated shells by a lap joint as shown in Fig. 31.5 (*a*) and (*b*). A little consideration will show that in this case, the plate is weakened by the rivet hole.

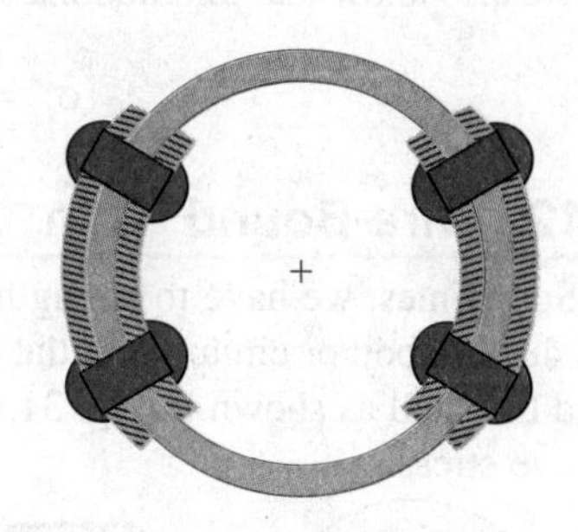

Fig. 31.5 (*a*) Joining by Butt Joint

The circumferential stress in a riveted cylindrical shell,

$$\delta_c = \frac{pd}{2t\eta}$$

Similarly, longitudinal stress,

$$\delta_l = \frac{pd}{4t\eta}$$

where $\eta$ is the efficiency of the riveted joint.

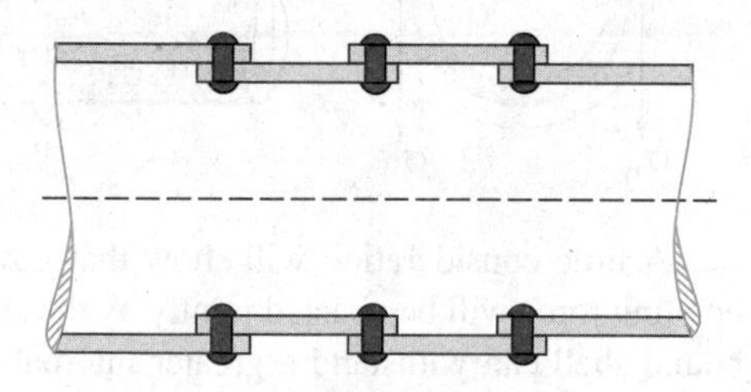

Fig. 31.5 (*b*) Joining by Lap Joint

**Notes:** 1. If the efficiency of the joint is different *i.e.*, the joint has different longitudinal efficiency and circumferential efficiency, then the respective values should be used in the above relation.

2. For designing the shell *i.e.*, determining the thickness of shell, the efficiency of the joint should also be considered.

**Example 31.11.** *A boiler shell of 2 m diameter is made up of mild steel plates of 20 mm thick. The efficiency of the longitudinal and circumferential joints is 70% and 60% respectively. Determine the safe pressure in the boiler, if the permissible tensile stress in the plate section through the rivets is 100 MPa. Also determine the circumferential stress in the plate and longitudinal stress through the rivets.*

**Solution.** Given: Diameter of boiler ($d$) = 2 m = $2 \times 10^3$ mm ; Thickness ($t$) = 20 mm ; Longitudinal efficiency ($\eta_l$) = 70% = 0.7 ; Circumferential efficiency ($\eta_c$) = 60% = 0.6 and permissible stress ($\sigma$) = 100 MPa = 100 N/mm$^2$.

*Safe pressure in boiler*

Let $\quad p$ = Safe pressure in boiler in N/mm$^2$

We know that permissible stress in boiler ($\sigma$),

$$100 = \frac{pd}{2t\eta_l} = \frac{p\times(2\times 10^3)}{2\times 20\times 0.7} = \frac{500p}{7}$$

$$p = \frac{100\times 7}{500} = 1.4\ \text{N/mm}^2 = \mathbf{1.4\ MPa} \quad \textbf{Ans.}$$

*Circumferential stress in the rivets*

We know that circumferential stress in the rivets,

$$\sigma_c = \frac{pd}{2t\eta_l} = \frac{1.4 \times (2 \times 10^3)}{2 \times 20 \times 0.6} = 116.7 \text{ N/mm}^2 = \mathbf{116.7 \text{ MPa}} \quad \textbf{Ans.}$$

*Longitudinal stress in the rivets*

We also know that longitudinal stress in the rivets,

$$\sigma_l = \frac{pd}{4t\eta_l} = \frac{1.4 \times (2 \times 10^3)}{4 \times 20 \times 0.7} = 50 \text{ N/mm}^2 = \mathbf{50 \text{ MPa}} \quad \textbf{Ans.}$$

## 31.12. Wire-Bound Thin Cylindrical Shells

Sometimes, we have to strengthen a cylindrical shell against bursting in longitudinal section (*i.e.*, due to hoop or circumferential stress). This is done by winding a wire under tension, closely round the shell as shown in Fig. 31.6. Its effect will be to put the cylinder wall under initial compressive stress.

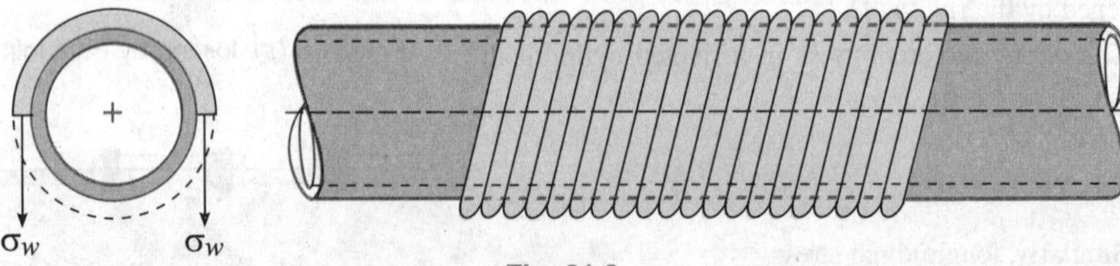


Fig. 31.6

A little consideration will show that now if the internal pressure of the shell is increased, the bursting force will be resisted jointly by the shell section and the wires. It is thus obvious that a wire-bound shell can withstand a greater internal pressure than an ordinary shell. In actual practice the winding wire is made of a high tensile material.

In such cases, the bursting force (due to internal pressure) per cm length is equal to the resisting force (due to pipe section + wire section). The circumferential strain in the pipe is also equal to the strain in the steel wire. The following example will illustrate the above mentioned statements.

**EXAMPLE 31.12.** *A cast iron pipe of 300 mm internal diameter and 12 mm thick is wound closely with a single layer of circular steel wire of 5 mm diameter under a tension of 60 MPa. Find the initial compressive stress in the pipe section. Also find the stresses set up in the pipe and steel wire, when water under a pressure of 4 MPa is admitted into the pipe.*

*Take E for cast iron and steel as 100 GPa and 200 GPa respectively. Poisson's ratio = 0.3.*

**SOLUTION.** Given: Diameter of pipe ($d$) = 300 mm ; Pipe Thickness ($t$) = 12 mm ; Diameter of wire = 5 mm ; Tension in wire = 60 MPa = 60 N/mm$^2$ ; Pressure of water ($p$) = 4 MPa = 4 N/mm$^2$ ; Modulus of elasticity for cast iron ($E_c$) = 100 GPa = 100 × 10$^3$ N/mm$^2$ ; Modulus of elasticity for steel ($E_s$) = 200 GPa = 200 × 10$^3$ N/mm and poisson's ratio $\left(\frac{1}{m}\right)$ = 0.3.

*Initial compressive stress in the pipe section*

A little consideration will show that the pipe will be subjected to compression (due to tension in the wire), before the water under pressure is admitted into the pipe. Now consider one mm length of the pipe. We know that number of wire sections for 1 mm pipe length

$$= \frac{2}{\text{Diameter of wire}} = \frac{2}{5} = 0.4$$

∴ Initial compressive force in the wire (*i.e.*, before the water under pressure is admitted into the pipe)

$$= 0.4 \times \left[\frac{\pi}{4} \times (5)^2\right] \times 60 = 471.3 \text{ N}$$

∴ Initial compressive stress in the pipe section,

$$\sigma_c = \frac{471.3}{2 \times 12} = \mathbf{19.6\ N/mm^2} \quad \textbf{Ans.}$$

***Stresses set up in the pipe and steel wire***

Let $\sigma_p$ = Stress in the pipe section in N.mm$^2$ and

$\sigma_w$ = Stress in the steel wire in N/mm$^2$.

We know that the bursting force per mm length of the pipe, when water under pressure is admitted

$$= p \times d \times l = 4 \times 300 \times 1 = 1200 \text{ N} \quad ...(i)$$

and total resisting force = Resisting force in pipe + Resisting force in wire

$$= [2\,\sigma_p\, t\, l] + \left[\sigma_w \times 0.4 \times \frac{\pi}{4} \times (5)^2\right]$$

$$= [2 \times \sigma_p \times 12 \times 1] + [\sigma_w \times 0.1 \times 3.1416 \times 25]$$

$$= 24\,\sigma_p + 7.854\,\sigma_w \quad ...(ii)$$

Since the bursting force is in the pipe is equal to the total resisting force, therefore equating (*i*) and (*ii*),

$$1200 = 24\,\sigma_p + 7.854\,\sigma_w \quad ...(iii)$$

We know that circumferential strain in the pipe

$$= \frac{\sigma_p}{E_c} - \left[\frac{1}{m} \times \frac{bd}{4t} \times \frac{1}{E_c}\right] = \frac{1}{E_c}\left[\sigma_p - 0.3 \times \frac{4 \times 300}{4 \times 12}\right] = \frac{\sigma_p - 7.5}{E_c} \quad ...(iv)$$

and strain in the steel wire $= \dfrac{\sigma_w}{E_s}$ ...(*v*)

Since the circumferential strain in the pipe is equal to the strain in the steel wire, therefore equating (*iv*) and (*v*),

$$\frac{\sigma_p - 7.5}{E_c} = \frac{\sigma_w}{E_s}$$

$$\sigma_w = \frac{E_s}{E_c} \times (\sigma_p - 7.5) = \frac{2 \times 10^3}{1 \times 10^3} \times (\sigma_p - 7.5) = 2\,(\sigma_p - 7.5) \quad ...(vi)$$

Substituting this value of $\sigma_w$ in equation (*iii*),

$$1200 = (24\,\sigma_p) + [7.854 \times 2\,(\sigma_p - 7.5) = 39.71\,\sigma_p - 117.8$$

or

$$\sigma_p = \frac{1200 + 117.8}{39.71} = 33.2 \text{ N/mm}^2$$

and

$$\sigma_w = (2 \times 33.2) - 15 = 51.4 \text{ N/mm}^2$$

∴ Final stress in the pipe section

$$= 33.2 - 19.6 = 13.6 \text{ N/mm}^2 = \mathbf{13.6\ MPa} \quad \textbf{Ans.}$$

and final stress in steel wire

$$= 60 + 51.4 = 111.4 \text{ N/mm}^2 = \mathbf{111.4\ MPa} \quad \textbf{Ans.}$$

## EXERCISE 31.2

1. A spherical shell of 1 m diameter is subjected to a pressure of 2.4 MPa. What is the stress induced in the vessel plate, if its thickness is 15 mm ? (**Ans.** 40 MPa)
2. A spherical shell of 800 mm diameter is subjected to an internal pressure of 1.5 MPa. What should be the minimum thickness of the shell, if the tensile stress is not to exceed 40 MPa ? (**Ans.** 7.5 mm)
3. A spherical container of 1 m diameter has 15 mm thick plates. Calculate the change in its diameter, if it contains a fluid under a pressure of 2 MPa. Take $E = 200$ GPa and $\mu = 0.28$. (**Ans.** 0.12 mm)
4. A cast iron cylinder of 200 mm internal diameter and 12.5 mm thick is closely wound with a wire of 4 mm, diameter under a stress of 55 MPa. What are the stresses developed in the cylinder and steel wire, when water under a pressure of 3 MPa is admitted into the cylinder. Take $E_s = 200$ GPa, $E_c = 100$ GPa and $1/m = 0.25$. (**Ans.** 3.2 MPa ; 83 MPa)

## QUESTIONS

1. Distinguish between circumferential stress and longitudinal stress in a cylindrical shell, when subjected to an internal pressure.
2. Show that in the case of a thin cylindrical shell subjected to an internal fluid pressure, the tendency to burst lengthwise is twice as great as in a transverse section.
3. Derive a relation for the changes of diameter and length of a thin cylindrical shell, when subjected to an internal pressure.
4. Distinguish between cylindrical shell and spherical shell.
5. Derive a formula for the hoop stress in a thin spherical shell subjected to an internal pressure.

## OBJECTIVE TYPE QUESTIONS

1. A thin cylindrical shell of diameter ($d$), length ($l$) is subjected to an internal pressure ($p$). The circumferential stress in the shell is

   (a) $\frac{pd}{2t}$ (b) $\frac{pd}{4t}$ (c) $\frac{pd}{6t}$ (d) $\frac{pd}{8t}$

2. In a thin shell, the ratio of longitudinal stress to the circumferential stress is

   (a) 1/2 (b) 3/4 (c) 1 (d) 2

3. The design of a thin cylindrical shell is based on

   (a) internal pressure (b) diameter of shell
   (c) longitudinal stress (d) all of these

4. A thin spherical shell of diameter ($d$) and thickness ($t$) is subjected to an internal pressure ($p$). The tensile stress in the shell plates will be

   (a) $\frac{pd}{2t}$ (b) $\frac{pd}{4t}$ (c) $\frac{pt}{2d}$ (d) $\frac{pt}{4d}$

## ANSWERS

1. (a) 2. (a) 3. (d) 4. (b)

Chapter 32

# Thick Cylindrical and Spherical Shells

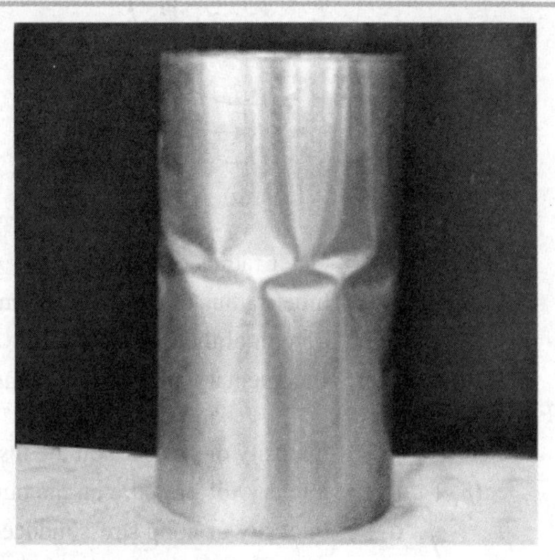

## Contents



## 32.1. Introduction

In the last chapter, we have discussed thin cylindrical and spherical shells. We have studied that in a thin shell, the stresses are assumed to be uniformly distributed throughout the wall thickness. But it is not so in the case of thick shells, and it varies with the thickness.

Thick shells are, generally, used to withstand high pressures. Sometimes, even, compound thick shells are used to withstand very high pressures or to contain chemicals under high pressure.

## 32.2. Lame's Theory

The problem of thick cylinders is somewhat complex and is solved on the following assumptions:

1. The material of the shell is homogeneous and isentropic.
2. The plane sections of the cylinder, perpendicular to the longitudinal axis, remain plane under the effect of pressure *i.e.* the longitudinal strain is constant and is independent of the radius of the shell.

The theory derived after the above mentioned assumptions is popularly known as ***Lame's Theory***.

## 32.3. Stresses in a Thick Cylindrical Shell

Consider a thick cylindrical shell subjected to an internal pressure as shown in Fig. 32.1 (*a*).

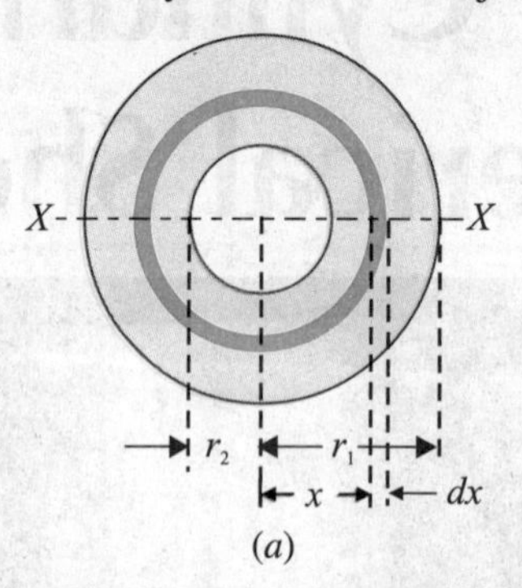


(*a*)

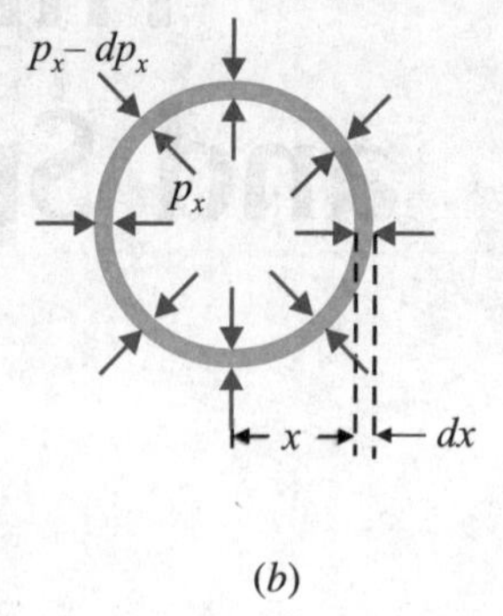


(*b*)

**Fig. 32.1**

Let $r_1$ = Outer radius of the cylinder,

$r_2$ = Inner radius of the cylinder, and

$l$ = Length of the cylinder.

Now consider an elementary ring of the cylinder of thickness *dx,* and of a radius *x* as shown Fig. 32.1 (*a*) and (*b*).

Let $p_x$ = Intensity of pressure on the inside of the ring

$(p_x + dp_x)$ = Intensity of pressure on the outside of the ring and

$\sigma_x$ = Intensity of hoop stress induced in the ring.

We know that the bursting force on any longitudinal section say (*X–X*),

$$P = p_x \cdot 2xl - (p_x + dp_x) \cdot 2(x + dx)l \qquad ...(i)$$

and resisting force

$$= \sigma_x \cdot 2 \cdot dx \cdot l \qquad ...(ii)$$

Since the bursting force is equal to the resisting force, therefore equating (*i*) and (*ii*),

$$\sigma_x \cdot 2 \cdot dx \cdot l = p_x \cdot 2xl - (p_x + dp_x) \cdot 2(x + dx)l$$

$$\therefore \quad \sigma_x \cdot dx = p_x \cdot x - (p_x \cdot x + p_x \cdot dx + dp_x \cdot x + dp_x \cdot dx)$$

$$= -p_x \cdot dx - dp_x \cdot x \qquad \text{(Neglecting } dp_x \,.\, dx\text{)}$$

or

$$\sigma_x = -p_x - x\,\frac{dp_x}{dx} \qquad ...(iii)$$

Now let us obtain another relation between radial pressure ($p_x$) and hoop stress ($\sigma_x$) in order to solve the above equation with the help of the assumption in Lame's theory *i.e.* the longitudinal strain is constant. The equation given in this theory for longitudinal tensile stress,

$$p_0 = \frac{pr_2^2}{r_1^2 - r_2^2}$$

$\therefore$ Longitudinal strain

$$= \frac{p_0}{E} - \frac{\sigma_x}{mE} + \frac{p_x}{mE}$$

Since the longitudinal strain is constant (*i.e.* it is independent of $x$), therefore,

$$\frac{p_0}{E} - \frac{\sigma_x}{mE} + \frac{p_x}{mE} = \text{Constant}$$

$$\therefore \quad \sigma_x - p_x = \text{Constant} \quad \text{(Since } p_0, m \text{ and } E \text{ are constants)}$$

or $$\sigma_x - p_x = 2a \quad \text{(where } a \text{ is constant)}$$

$$\therefore \quad \sigma_x = p_x + 2a \quad ...(iv)$$

Equating the values of $\sigma_x$ from equations (*iii*) and (*iv*),

$$p_x + 2a = -p_x - \frac{x \cdot dp_x}{dx}$$

$$\therefore \quad \frac{dp_x}{dx} = -\frac{2(p_x + a)}{x}$$

or $$\frac{dp_x}{(p_x + a)} = -\frac{2dx}{x}$$

Integrating both sides of the above equation,

$$\log_e (p_x + a) = -2 \log_e x + \log_e b \quad \text{...(where } \log_e b \text{ is constant of integration)}$$

$$\therefore \quad (p_x + a) = \frac{b}{x^2}$$

or $$p_x = \frac{b}{x^2} - a \quad ...(v)$$

Now substituting the value of $p_x$ in equation (*iv*),

$$\sigma_x = \frac{b}{x^2} - a + 2a = \frac{b}{x^2} + a \quad ...(vi)$$

**NOTES:**
1. The values of $a$ and $b$ in the above equation are evaluated from the given conditions.
2. The above two equations (*v*) and (*vi*) are called Lame's equations.
3. It should be remembered that the values $p_x$ is compressive, and that of $\sigma_x$ is tensile in the above equations.
4. Intensity of pressure outside the thick cylinder is taken to be zero, though it has some hoop stress due to pressure inside the thick cylinder.

**EXAMPLE 32.1.** *A cast iron pipe of 400 mm internal diameter and 100 mm thickness carries water under a pressure of 8 N/mm². Determine the maximum and minimum intensities of hoop stress across the section. Also sketch the radial pressure distribution and hoop stress distribution across the section.*

**SOLUTION.** Given : Internal diameter ($d_2$) = 400 mm or internal radius ($r_2$) = 200 mm;

Thickness = 100 mm ; Internal pressure ($p_x$) = 8 N/mm²

We know that External radius = 200 + 100 = 300 mm

Let $\sigma_x$ = Hoop stress at any section.

From the geometry of the pipe, we find that when $x$ = 200 mm then $p_x$ = 8 N/mm². Also when $x$ = 300 mm then $p_x$ = 0.

We know from the Lame's equation that pressure at inner surface

$$8 = \frac{b}{x^2} - a = \frac{b}{(200)^2} - a = \frac{b}{40000} - a$$

or
$$a = \frac{b}{40000} - 8 \quad ...(i)$$

We also know from the Lame's equation that pressure at outer surface

$$0 = \frac{b}{x^2} - a = \frac{b}{(300)^2} - a = \frac{b}{90000} - a$$

or
$$a = \frac{b}{90000} \quad ...(ii)$$

Equating the two values of $a$ from equation ($i$) and ($ii$)

$$\frac{b}{40000} - 8 = \frac{b}{90000}$$

or
$$\frac{b - 320000}{40000} = \frac{b}{90000}$$

or
$$9b - 9 \times 320000 = 4\,b$$

or
$$5\,b = 9 \times 320000$$

∴
$$b = \frac{9 \times 320000}{5} = 576000$$

and
$$a = \frac{b}{90000} = \frac{576000}{90000} = 6.4$$

Now hoop stress at inner surface

$$\sigma_{200} = \frac{b}{x^2} + a = \frac{576000}{(200)^2} + 6.4 = \mathbf{20.8\ N/mm^2} \quad \textbf{Ans.}$$

and hoop stress at outer surface

$$\sigma_{300} = \frac{b}{x^2} + a = \frac{576000}{(300)^2} + 6.4 = \mathbf{12.8\ N/mm^2} \quad \textbf{Ans.}$$

Now draw the radial pressure ($p_x$) and hoop stress ($\sigma_x$) distribution across the section as shown in Fig. 32.2.

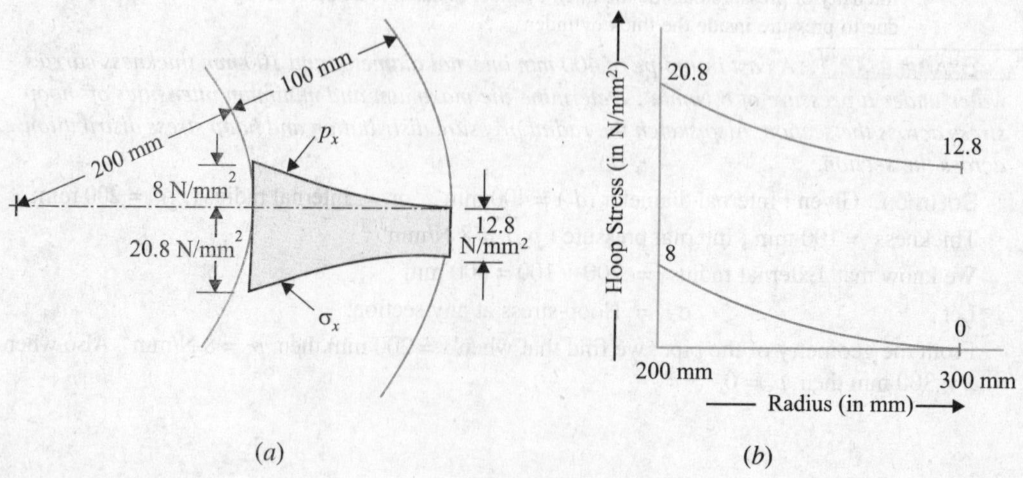


Fig. 32.2

**EXAMPLE 32.2.** *A thick metallic cylindrical shell of 150 mm internal diameter is required to withstand an internal pressure of 8 N/mm². Find the necessary thickness of the shell, if the permissible tensile stress in the section is 20 N/mm².*

**SOLUTION.** Given : Internal diameter of cylindrical shell ($d_2$) = 150 mm or internal radius ($r_2$) = 75 mm ; Internal pressure ($p_x$) = 8 N/mm² and permissible tensile stress ($\sigma_x$) = 20 N/mm².

Let $r_1$ = Outer diameter of the shell.

From the geometry of the shell, we find that when $x$ = 75 mm then $p_x$ = 8 N/mm² and $\sigma_x$ = 20 N/mm².

We know from the Lame's equation that internal pressure ($p_x$)

$$8 = \frac{b}{x^2} - a = \frac{b}{(75)^2} - a = \frac{b}{5625} - a$$

$$\therefore \quad a = \frac{b}{5625} - 8 \qquad ...(i)$$

We also know from the Lame's equation that permissible stress ($\sigma_x$)

$$20 = \frac{b}{x^2} + a = \frac{b}{(75)^2} + a = \frac{b}{5625} + a$$

$$\therefore \quad a = 20 - \frac{b}{5625} \qquad ...(ii)$$

Equating the two values of $a$ from equations (*i*) and (*ii*),

$$\frac{b}{5625} - 8 = 20 - \frac{b}{5625}$$

$$\therefore \quad \frac{2b}{5625} = 20 + 8 = 28$$

$$\text{or} \quad b = \frac{28 \times 5625}{2} = 78750$$

Now substituting the value of $b$ in equation (*i*),

$$a = \frac{78750}{5625} - 8 = 6$$

We know from Lame's equation that internal pressure ($p_x$)

$$0 = \frac{78750}{r_1^2} - 6 \qquad ...(\text{for outer conditions})$$

$$\therefore \quad r_1 = \sqrt{\frac{78750}{6}} = 114.6 \quad \text{say} \quad 115 \text{ mm}$$

Therefore thickness of the shell,

$$t = r_1 - r_2 = 115 - 75 = \mathbf{40\ mm} \quad \textbf{Ans.}$$

## 32.4. Stresses in Compound Thick Cylindrical Shells

We have already discussed in Art. 25.10 that in order to withstand a very high pressure, the thin cylindrical shell is wound with a wire under tension. The effect of the wire is to put the cylinder wall under an initial compressive stress. The same principle is used in thick cylindrical shells. This is done by shrinking one thick shell over another. A little consideration will show, that the outer shell will produce some initial compressive stress in the inner one. The inner shell, in turn, will exert some tensile stress in the outer one. This principle is commonly used in the design of gun tube.

Now when the compound shell is subjected to an internal pressure, both the inner and outer shells will be subjected to hoop tensile stress. The net effect of the initial stresses and those due to internal pressure is to make the resultant stresses more or less uniform.

Now consider a compound thick cylindrical shell made up of two tubes as shown in Fig. 32.3.

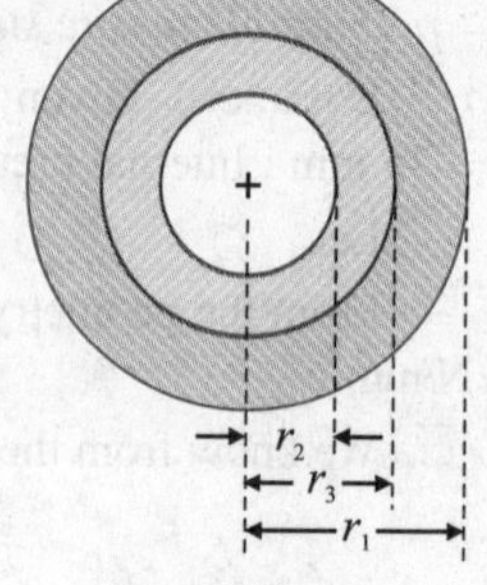

Fig. 32.3

Let $r_1$ = Outer radius of the outer shell,

$r_2$ = Inner radius of the inner shell,

$r_3$ = Outer radius of the inner shell, (also inner radius of the outer shell),

and

$P_1$ = Radial pressure at the junction of the two shells (*i.e.*, at radius $r_3$)

Now the Lame's equations may be applied in this case for the initial conditions (i.e., when the outer tube exerts pressure on the inside shell, or in other words before the fluid under pressure is admitted into the inner shell.

***(i) For inner tube***

$$p_x = \frac{b_1}{x^2} - a_1$$

We know that when $x = r_2$, then $p_x = 0$.

$$\therefore \quad 0 = \frac{b_1}{r_2^2} - a_1 \quad ...(i)$$

Also when $x = r_3$, then $p_x = p_1$.

$$\therefore \quad p_1 = \frac{b_1}{r_3^2} - a_1 \quad ...(ii)$$

***(ii) For outer tube***

$$p_x = \frac{b_2}{x^2} - a_2 \quad \text{and} \quad \sigma_x = \frac{b_2}{x^2} + a_2$$

We know that when $x = r_3$, then $p_x = p_1$.

$$\therefore \quad p_1 = \frac{b_2}{r_3^2} - a_2 \quad ...(iii)$$

Also when $x = r_1$, then $p_x = 0$.

$$\therefore \quad 0 = \frac{b_2}{r_1^2} - a_2 \quad ...(iv)$$

The values of $a_1$, $b_1$, $a_2$ and $b_2$ may be found out from the above four equations, if the radial pressure $p_1$ at the junction of the two shells is known. The hoop stress ($\sigma_x$) may also be obtained with the help of relative expressions.

Now, when the fluid under pressure ($p$) is admitted inside the compound shell it will be resisted jointly by both the shells. The hoop stresses may be calculated by the Lame's formulae as usual. The resultant stresses will be the algebraic sum of the initial stresses an those due to fluid pressure.

**EXAMPLE 32.3.** *A compound cylinder is made by shrinking a tube of 160 mm internal diameter and 20 mm thick over another tube of 160 mm external diameter and 20 mm thick. The radial pressure at the common surface, after shrinking, is 8 N/mm². Find the final stresses set up across the section, when the compound cylinder is subjected to an internal fluid pressure of 60 N/mm².*

**SOLUTION.** Given : Inner diameter of outer cylinder = 160 mm ; Thickness of outer cylinder = 20 mm ; Outer diameter of inner cylinder ($d_3$) = 160 mm or radius $r_3$ = 80 mm ; Thickness of inner cylinder = 20 mm ; Pressure due to shrinkage at the junction of the two cylinders ($p_1$) = 8 N/mm$^2$ and internal fluid pressure in the compound cylinder ($p_x$) = 60 N/mm$^2$.

Outer diameter of outer cylinder = 160 + 20 + 20 = 200 mm

or radius, $r_1$ = 100 mm

Inner diameter of inner cylinder = 160 – 20 – 20 = 120 mm

or radius, $r_2$ = 60 mm

Let $\sigma_x$ = Hoop stress at a radius $x$ in the compound cylinder.

First of all, let us apply all the Lame's four equations for the inner and outer cylinders before the fluid under pressure is admitted.

$$0 = \frac{b_1}{r_2^2} - a_1 = \frac{b_1}{(60)^2} - a_1$$

$$\therefore \quad 0 = \frac{b_1}{3600} - a_1 \quad ...(i)$$

Similarly

$$p_1 = \frac{b_1}{r_3^2} - a_1 = \frac{b_1}{(80)^2} - a_1$$

$$8 = \frac{b_1}{6400} - a_1 \quad ...(ii)$$

and

$$p_1 = \frac{b_2}{r_3^2} - a_2 = \frac{b_2}{(80)^2} - a_2$$

$$8 = \frac{b_2}{6400} - a_2 \quad ...(iii)$$

and

$$0 = \frac{b_2}{r_1^2} - a_2 = \frac{b_2}{(100)^2} - a_2$$

$$0 = \frac{b_2}{10000} - a_2 \quad ...(iv)$$

Solving equations (*i*) and (*ii*) simultaneously, we find that $b_1$ = – 65830 and $a_1$ = – 18.3.

Similarly solving equations (*iii*) and (*iv*) simultaneously, we find that $b_2$ = 142200 and $a_2$ = 14.2.

We know from Lame's equation that permissible stress ($\sigma_x$),

$$\sigma = \frac{b}{x^2} + a$$

$$\sigma_{60} = -\frac{65830}{(60)^2} + (-18.3) \quad \text{(For inner tube)}$$

$$= -\frac{65830}{3600} - 18.3 = -36.6 \text{ N/mm}^2 \quad ...(v)$$

and

$$\sigma_{80} = -\frac{65830}{(80)^2} + (-18.3) \quad \text{(For inner tube)}$$

$$= -\frac{65830}{6400} - 18.3 = -28.6 \text{ N/mm}^2 \quad \text{...(vi)}$$

Similarly

$$\sigma_{80} = \frac{142200}{(80)^2} + 14.2 \quad \text{(For outer tube)}$$

$$= \frac{142200}{6400} + 14.2 = 36.4 \text{ N/mm}^2 \quad \text{...(vii)}$$

and

$$\sigma_{100} = \frac{142200}{(100)^2} + 14.2 \quad \text{(For outer tube)}$$

$$= \frac{142200}{10000} + 14.2 = 28.4 \text{ N/mm}^2 \quad \text{...(viii)}$$

Now let us apply Lame's equation for the inner cylinder only after the fluid under pressure of 60 N/mm$^2$ is admitted. *i.e.*

$$p_x = \frac{b}{x^2} - a$$

$$60 = \frac{b}{(60)^2} - a = \frac{b}{3600} - a \quad \text{...(ix)}$$

and

$$0 = \frac{b}{(100)^2} - a = \frac{b}{10000} - a \quad \text{...(x)}$$

Subtracting equation (*x*) from equation (*ix*)

$$\frac{b}{3600} - \frac{b}{10000} = 60$$

$$\frac{64b}{360000} = 60 \quad \text{or} \quad b = 337500$$

and

$$a = \frac{b}{10000} = \frac{337500}{10000} = 33.75$$

We know from Lame's equation that permissible stress

$$\sigma_x = \frac{b}{x^2} + a$$

$$= \frac{337500}{x^2} + 33.75$$

$\therefore$

$$\sigma_{60} = \frac{337500}{(60)^2} + 33.75 = 127.5 \text{ N/mm}^2$$

$$\sigma_{80} = \frac{337500}{(80)^2} + 33.75 = 86.5 \text{ N/mm}^2$$

$$\sigma_{100} = \frac{337500}{(100)^2} + 33.75 = 67.5 \text{ N/mm}^2$$

Now tabulate the hoop (*i.e.* circumferential) stress at different points as given below:
[(+ve for tension; –ve for compression)]

| Hoop stress | Inner cylinder | | Outer cylinder | |
|---|---|---|---|---|
| N/mm$^2$ | $x$ = 60 mm | $x$ = 80 mm | x = 80 mm | $x$ = 100 mm |
| Initial | −36.6 | −28.6 | +36.4 | +28.4 |
| Due to fluid pressure | +127.5 | +86.5 | +86.5 | +67.5 |
| Final | +90.9 | +57.9 | +122.9 | +95.9 |

## 32.5. Difference of Radii of Shrinkage

We have discussed in the last article, that in the case of compound cylinders the outer shell is shrunk, over the inner one, to produce some initial compressive stress in the inner shell. This is done by making the inner diameter of the outer tube slightly less than the outer diameter of the inner tube. Then the outer cylinder is heated and the inner cylinder is inserted into it. After cooling, the outer cylinder shrinks over the inner one. sometimes, the inner cylinder is also forced into the outer one with the help of machines to obtain the necessary grip or initial stresses. This principle is used in fixing sleeves in automobile engine heads.

Let $r_1$ = Outer radius of the outer shell,

$r_2$ = Inner radius of the inner shell, and

$r_3$ = Common radius of the outer shell at the junction after shrinking.

A little consideration will show, that * before shrinking, the outer radius of the inner shell is slightly more than $r_3$ and inner radius of the outer shell is slightly less than $r_3$.

Now Lame's equations may be applied in this case also. Therefore at the junction, the radial pressure, $(p_1)$ is the same for the shells.

$$\therefore \quad \frac{b_1}{r_3^2} - a_1 = \frac{b_2}{r_3^2} - a_2$$

or
$$\frac{b_2 - b_1}{r_3^2} = (a_2 - a_1)$$

We know that the tensile strain at any point in the shell

$$= \frac{\sigma_x}{E} + \frac{p_x}{mE}$$

$\therefore$ Increase in the inner radius of the outer shell

$$= r_3\left[\frac{\sigma_x}{E} + \frac{p_x}{mE}\right]$$

$$= r_3\left[\frac{1}{E}\left(\frac{b_2}{r_3^2} + a_2\right) + \frac{1}{mE}\left(\frac{b_2}{r_3^2} - a_2\right)\right]$$

Similarly, decrease in the outer radius of the inner shell

$$= -r_3\left[\frac{\sigma_x}{E} + \frac{p}{mE}\right] \qquad \text{(Taking decrease as −ve)}$$

$$= -r_3\left[\frac{1}{E}\left(\frac{b_1}{r_3^2} + a_1\right) + \frac{1}{mE}\left(\frac{b_1}{r_3^2} - a\right)\right]$$

---

* After shrinking, there is some change in the outer radius of the outer shell and inner radius of the inner shell. But this change, being very small, is neglected.

Now we know that the original difference in the outer radius of the inner shell and inner radius of the outer shell

$$= \text{Increase in inner radius of outer shell} + \text{Decrease in outer radius of the inner shell}$$

$$= r_3\left[\frac{1}{E}\left(\frac{b_2}{r_3^2}+a_2\right)+\frac{1}{mE}\left(\frac{b_2}{r_3^2}-a_2\right)\right]$$

$$-r_3\left[\frac{1}{E}\left(\frac{b_1}{r_3^2}+a_1\right)+\frac{1}{mE}\left(\frac{b_1}{r_3^2}-a_1\right)\right]$$

$$= \frac{r_3}{E}\left[\left(\frac{b_2}{r_3^2}+a_2\right)-\left(\frac{b_1}{r_3^2}+a_1\right)\right]+\frac{r_3}{mE}\left[\left(\frac{b_2}{r_3^2}-a_2\right)-\left(\frac{b_1}{r_3^2}-a_1\right)\right]$$

Since $\frac{b_1}{r_3^2}-a_1=\frac{b_2}{r_3^2}-a_2$, therefore second part of the above equation is zero. Therefore original difference of radii at the junction,

$$= \frac{r_3}{E}\left[\left(\frac{b_2}{r_3^2}+a_2\right)-\left(\frac{b_1}{r_3^2}+a_1\right)\right]$$

$$= \frac{r_3}{E}\left[\frac{b_2-b_1}{r_3^2}+a_2-a_1\right]$$

$$= \frac{r_3}{E}\left[(a_2-a_1)+a_2-a_1\right]$$

$$= \frac{2r_3}{E}(a_2-a_1)$$

**Note:** The values of $a_1$ and $a_2$ are evaluated from the given conditions.

**Example 32.4.** *A compound thick cylinder is formed by shrinking a tube of external diameter 300 mm over another tube of internal diameter 150 mm. After shrinking, the diameter at the junction of the tubes is found to be 250 mm and radial compression as 28 N/mm². Find the original difference in radii at the junction. Take E for the cylinder metal as 200 GPa.*

**Solution.** Given : Outer diameter of outer cylinder $(d_1)$ = 300 mm or radius $(r_1)$ = 150 mm ; Inner diameter of inner cylinder $(d_2)$ = 150 mm or radius $(r_2)$ = 75 mm ; Diameter of the junction of the two cylinders $(d_3)$ = 250 mm or radius $(r_3)$ = 125 mm ; Radial compression at the junction $(p_x)$ = 28 N/mm² and modulus of elasticity $(E)$ = 200 GPa = $200 \times 10^3$ N/mm².

Let $dr$ = Original difference in radii.

First of all, consider the inner cylinder. From geometry of the figure, we find that when $x$ = 125 mm, then $p_x$ = 28 N/mm². Also when $x$ = 75 mm, then $p_x = 0$.

We know from the Lame's equation that internal pressure $(p_x)$

$$28 = \frac{b_1}{(125)^2}-a_1=\frac{b_1}{15625}-a_1$$

$$\therefore \quad a_1 = \frac{b_1}{15625}-28 \qquad ...(i)$$

Similarly,
$$0 = \frac{b_1}{(75)^2} - a_1 = \frac{b_1}{5625} - a_1$$

$$\therefore \quad a_1 = \frac{b_1}{5625} \quad ...(ii)$$

Equating the two values of $a_1$ from equations (*i*) and (*ii*),

$$\frac{b_1}{15625} - 28 = \frac{b_1}{5625}$$

$$\frac{b_1 - 437500}{15625} = \frac{b_1}{5625}$$

$$5625\, b_1 - 2.461 \times 10^9 = 15625\, b_1$$

$$\therefore \quad 10000\, b_1 = -2.461 \times 10^9$$

or
$$b_1 = -2.461 \times 10^5$$

and
$$a_1 = -\frac{b_1}{5625} = -\frac{2.461 \times 10^5}{5625} = -43.75$$

Now consider the outer cylinder. From the geometry of the figure, we find that when $x = 125$ mm, then $p_x = 28$ N/mm$^2$. Also when $x = 150$ mm, then $p_x = 0$.

We also know from Lame's equation that internal pressure ($p_x$)

$$28 = \frac{b}{x^2} - a$$

$$28 = \frac{b_2}{(125)^2} - a_2 = \frac{b_2}{15625} - a_2$$

$$\therefore \quad a_2 = \frac{b_2}{15625} - 28 \quad ...(iii)$$

Similarly,

$$0 = \frac{b_2}{(150)^2} - a_2 = \frac{b_2}{22500} - a_2$$

$$\therefore \quad a_2 = \frac{b_2}{22500} \quad ...(iv)$$

Equating the two values of $a_2$ from equations (*iii*) and (*iv*),

$$\frac{b_2}{15625} - 28 = \frac{b_2}{22500}$$

$$\frac{b_2 - 437500}{15625} = \frac{b_2}{22500}$$

$$22500\, b_2 - 9.84 \times 10^9 = 15625\, b_2$$

$$\therefore \quad 6875\, b_2 = 9.84 \times 10^9$$

or
$$b_2 = \frac{9.84 \times 10^9}{6875} = 1.432 \times 10^6$$

and
$$a_2 = \frac{b_2}{22500} = \frac{1.432 \times 10^6}{22500} = 63.64$$

Therefore original difference in radii

$$dr = \frac{2r_3}{E}(a_2 - a_1)$$

$$= \frac{2 \times 125}{200 \times 10^3}[63.64 - (-43.75)] = \mathbf{0.13\ mm} \quad \textbf{Ans.}$$

**EXAMPLE 32.5.** *A steel tube 240 mm external diameter is to be shrunk on another steel tube of 80 mm internal diameter. After shrinking, the diameter at the junction is 160 mm. Before shrinking on, the difference of diameter at the junction was 0.08 mm.*

*Calculate the radial pressure at the junction and hoop stresses developed in the two tubes after shrinking on. Take E as 200 kN/mm².*

**SOLUTION.** Given : External diameter of outer tube $(d_1) = 240$ mm or radius $(r_1) = 120$ mm ; Internal diameter of the tube $(d_2) = 80$ mm or radius $(r_2) = 40$ mm ; Diameter at junction of two tubes $(d_3) = 160$ mm or radius $(r_3) = 80$ mm ; Difference of diameter at junction = 0.08 mm or difference in radius $(dr) = 0.04$ mm ; Modulus of elasticity $(E) = 200$ kN/mm² $= 200 \times 10^3$ N/mm².

***Radial pressure at the junction***

Let $p_1$ = Radial pressure at the junction.

First of all, consider the inner tube. From the geometry of the figure, we find that when $x = 40$ mm, then $p_x = 0$. Also when $x = 80$ mm, then $p_x = p_1$ (where $p_1$ is the radial pressure at the junction).

We know from Lame's equation that internal pressure $(p_x)$

$$0 = \frac{b_1}{(40)^2} - a_1 = \frac{b_1}{1600} - a_1$$

$$\therefore \quad b_1 = 1600\, a_1 \quad \text{...(i)}$$

Similarly,

$$p_1 = \frac{b_1}{(80)^2} - a_1 = \frac{b_1}{6400} - a_1 \quad \text{...(ii)}$$

Now consider the outer tube. From the geometry of the figure, we find that when $x = 120$ mm, then $p_x = 0$. Also when $x = 80$ mm, then $p_x = p_1$ (where $p_1$ is the radial pressure at the junction).

We also know from Lame's equation that internal pressure $(p_x)$

$$0 = \frac{b_2}{(120)^2} - a_2 = \frac{b_2}{14400} - a_2$$

$$\therefore \quad b_2 = 14400\, a_2 \quad \text{...(iii)}$$

Similarly,

$$p_1 = \frac{b_2}{(80)^2} - a_2 = \frac{b_2}{6400} - a_2 \quad \text{...(iv)}$$

Equating the two values of $p_1$ from equations (*ii*) and (*iv*),

$$\frac{b_1}{6400} - a_1 = \frac{b_2}{6400} - a_2$$

Substituting the values of $b_1 = 1600\, a_1$ from equation (*i*) and $b_2 = 14400\, a_2$ from equation (*iii*) in the above equation,

$$\frac{1600\, a_1}{6400\, a_1} - a_1 = \frac{14400\, b_2}{6400} - a_2$$

$$-\frac{3a_1}{4} = \frac{5a_2}{4}$$

$$a_1 = -\frac{5a_2}{3} \quad ...(v)$$

We know that difference in radius

$$dr = \frac{2r_3}{E}(a_2 - a_1)$$

or
$$0.04 = \frac{2 \times 80}{200 \times 10^3}\left[a_2 - \left(-\frac{5a_2}{3}\right)\right] = \frac{4a_2}{1875}$$

∴
$$a_2 = \frac{0.04 \times 1875}{4} = 18.75$$

and
$$b_2 = 14400\, a_2 = 14400 \times 18.75 = 270 \times 10^3$$

Now
$$a_1 = -\frac{5a_2}{3} = -\frac{5 \times 18.75}{3} = -31.25$$

and
$$b_1 = -1600\, a_1 = -1600 \times 31.25 = -50 \times 10^3$$

Now substituting the values of $a_1$ and $b_1$ in equation (*i*),

$$p_1 = \frac{b_1}{6400} - a_1 = -\frac{50 \times 10^3}{6400} - (-31.25)$$

$$= -23.44 \text{ N/mm}^2$$

$$= \mathbf{23.44\ N/mm^2\ (compressive)} \quad \textbf{Ans.}$$

***Hoop stresses developed in the two tubes***

We know that loop stress

$$\sigma_x = \frac{b}{x_2} + a$$

Now
$$\sigma_{40} = \frac{-50 \times 10^3}{(40)^2} - 31.25 \text{ N/mm}^2 \quad \text{(for inner tube)}$$

$$= -62.5 \text{ N/mm}^2$$

$$= \mathbf{62.5\ N/mm^2\ (compressive)} \quad \textbf{Ans.}$$

and
$$\sigma_{80} = \frac{-50 \times 10^3}{(80)^2} - 31.25 \text{ N/mm}^2 \quad \text{(for inner tube)}$$

$$= -39.1 \text{ N/mm}^2$$

$$= \mathbf{39.1\ N/mm^2\ (compressive)} \quad \textbf{Ans.}$$

Similarly,
$$\sigma_{80} = \frac{270 \times 10^3}{(80)^2} + 18.55 \text{ N/mm}^2 \quad \text{(for outer tube)}$$

$$= \mathbf{60.94\ N/mm^2\ (tensile)} \quad \textbf{Ans.}$$

and
$$\sigma_{120} = \frac{270 \times 10^3}{(120)^2} + 18.75 \text{ N/mm}^2 \quad \text{(for outer tube)}$$

$$= \mathbf{37.5\ N/mm^2\ (tensile)} \quad \textbf{Ans.}$$

## 32.6 Thick Spherical Shells

Consider a thick spherical shell subjected to an internal fluid pressure as shown in Fig. 32.4.

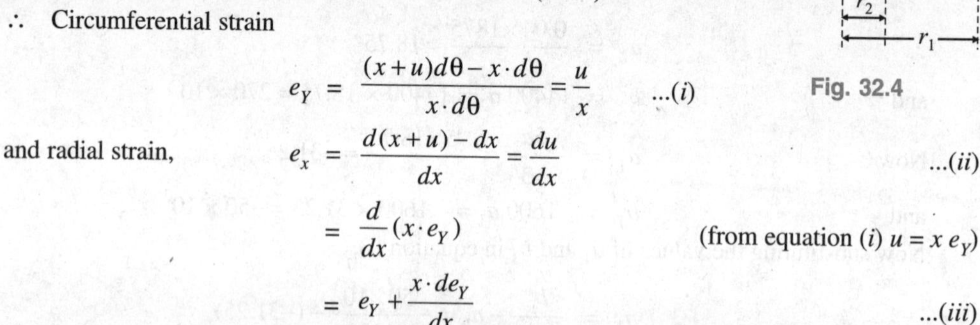

Fig. 32.4

Let $r_1$ = External radius,

$r_2$ = Internal radius, and

$p$ = Internal fluid pressure.

Now consider an elemental disc of the spherical shell of thickness $d_x$ and at a radius $x$. Let this elemental part subtend an angle $d\theta$ at the centre of the shell. As a result of the internal fluid pressure, let the radius of the disc be increased from $x$ to $(x + u)$.

∴ Circumferential strain

$$e_Y = \frac{(x+u)d\theta - x \cdot d\theta}{x \cdot d\theta} = \frac{u}{x} \quad ...(i)$$

and radial strain,

$$e_x = \frac{d(x+u) - dx}{dx} = \frac{du}{dx} \quad ...(ii)$$

$$= \frac{d}{dx}(x \cdot e_Y) \quad \text{(from equation (i) } u = x\, e_Y\text{)}$$

$$= e_Y + \frac{x \cdot de_Y}{dx} \quad ...(iii)$$

Now consider an elemental spherical shell of thickness and at a distance $x$.

Let $p_x$ = Radial pressure at radius $x$,

$(p_x + dp_x)$ = Radial pressure at radius $(x + dx)$, and

$\sigma_x$ = Circumferential stress (which is equal in all directions).

The bursting force on any diameteral plane of the elemental spherical shell

$$= p_x \cdot \pi x^2 - \pi(p_x + dp_x)\,(x + dx)^2 \quad ...(iv)$$

and resisting force

$$= \sigma_x \cdot 2\pi x \cdot dx \quad ...(v)$$

Since the bursting force is equal to the resisting force, therefore equating (*i*) and (*ii*),

$$\sigma_x \cdot 2\pi \cdot dx = p_x \cdot \pi x^2 - \pi(p_x + dp_x)\,(x + dx)^2$$

$$2\sigma_x \cdot xdx = p_x \cdot x^2 - (p_x + dp_x)\,[x^2 + 2x \cdot dx + (dx)^2]$$

$$= p_x x^2 - (p_x \cdot x^2 + 2p_x \cdot x \cdot dx + dp_x \cdot x^2)$$

(Neglecting small quantities)

$$= p_x \cdot x^2 - p_x \cdot x^2 - 2p_x \cdot x \cdot dx - dp_x \cdot x^2$$

$$2\sigma_x \cdot dx = -2p_x \cdot dx - dp_x \cdot x \quad \text{(Dividing both sides by } x\text{)}$$

$$\sigma_x = -p_x \frac{x \cdot dp_x}{2d_x} \quad ...(vi)$$

Differentiating the above equation,

$$\frac{d\sigma_x}{d_x} = -\frac{dp_x}{dx} - \frac{1}{2}\left[x\frac{d^2 p_x}{dx^2} + \frac{dp_x}{dx}\right]$$

Now a little consideration will show, that at any point in the elementary shell at a radius $x$, the three principal stresses are:

(*a*) Radial pressure ($p_x$) compressive,

(*b*) Hoop stress ($\sigma_x$) tensile and

(*c*) Hoop stress ($\sigma_x$) on a plane at right angles also tensile.

$\therefore$ Radial strain,

$$e_x = \frac{p_x}{E} - \frac{2\sigma_x}{mE} \qquad \text{(compressive)} \qquad ...(vii)$$

$$= -\frac{1}{E}\left(p_x + \frac{2\sigma_x}{m}\right) \qquad \text{(tensile)}$$

and circumferential strain.

$$e_Y = \frac{\sigma_x}{E} + \frac{p_x}{mE} - \frac{\sigma_x}{mE} \qquad \text{(tensile)}$$

$$= \frac{1}{E}\left[\frac{\sigma_x(m-1)}{m} + \frac{p_x}{m}\right] \qquad \text{(tensile)} \qquad ...(viii)$$

Now substituting the values of $e_x$ and $e_Y$ from equations (*vii*) and (*viii*) in equation (*iii*),

$$-\frac{1}{E}\left(p_x + \frac{2\sigma_x}{m}\right) = \frac{1}{E}\left[\frac{\sigma_x(m-1)}{m} + \frac{p_x}{m}\right] + \frac{x}{E}\left[\frac{(m-1)}{m}\cdot\frac{d\sigma_x}{dx} + \frac{1}{m}\cdot\frac{dp_x}{dx}\right]$$

Simplifying and rearranging the above equation,

$$(m+1)(p_x + \sigma_x) + (m-1)\, x\frac{d\sigma_x}{dx} + x\frac{dp_x}{dx} = 0$$

Now substituting the value of $\sigma_x$ and $d\sigma_x$ from equation (*vi*) in the above equation and simplifying it,

$$x\frac{d^2 p_x}{dx^2} + \frac{4dp_x}{dx} = 0$$

Substituting $\frac{dp_x}{dx} = a$ (*i.e.*, any constant) in the above equation

$$x\frac{da}{dx} + 4a = 0$$

or

$$\frac{da}{dx} + \frac{4a}{x} = 1 \qquad ...(\text{Dividing by } x)$$

$\therefore$

$$\frac{da}{a} = -\frac{4dx}{x}$$

Integrating the above equation,

$$\log_e a = -4\log_e x + \log_e C_1$$

where $C_1$ is the first constant of integration.

$\therefore$

$$\log_e = \log_e\left(\frac{C_1}{x^4}\right)$$

or

$$a = \frac{C_1}{x^4} \qquad ...(\text{Taking antilog})$$

$\therefore$

$$\frac{dp_x}{dx} = \frac{C_1}{x^4} \qquad \left(\because\ a = \frac{dp_x}{dx}\right)$$

or

$$dp_x = \frac{C_1 \cdot dx}{x^4}$$

Integrating the above equation,

$$p_x = -\frac{C_1}{3x^3} + C \qquad ...(ix)$$

where $C_2$ is the second constant of integration. Substituting this value of $p_x$ in equation (*iv*),

$$\sigma_x = \frac{C_1}{3x^3} - C_2 - \frac{C_1}{2x^3}$$

$$= -\frac{C_1}{6x^3} - C_2 \qquad ...(x)$$

Substituting the values of $C_1 = -6\,b$ and $C_2 = -a$ in equations (*ix*) and (*x*) respectively,

$$p_x = \frac{2b}{x^3} - a$$

and $$\sigma_x = \frac{b}{x^3} + a$$

**Note:** The values of *a* and *b* are evaluated from the given conditions.

**Example 32.6.** *A thick spherical shell of 400 mm inside diameter is subjected to an internal pressure of 1.5 N/mm². Determine the necessary thickness of the shell, if the permissible stress in the shell material is 3 N/mm².*

**Solution.** Given : Inside diameter of the shell ($d_2$) = 400 mm or radius ($r_2$) = 200 mm ; Internal pressure ($p_x$) = 1.5 N/mm² and permissible tensile stress ($\sigma_x$) = 3 N/mm².

Let $r_1$ = Outer radius of the shell.

From the geometry of the shell, we find the that when $x$ = 200 mm then $p_x$ = 1.5 N/mm² and $\sigma_x$ = 3 N/mm².

We know from Lame's equation that pressure at inner surface is

$$p_x = \frac{2b}{x^3} - a$$

or $$1.5 = \frac{2b}{(200)^3} - a = \frac{2b}{8 \times 10^6} - a$$

$$\therefore \quad a = \frac{2b}{8 \times 10^6} - 1.5 \qquad ...(i)$$

We also know from Lame's equation that permissible stress in the shell material ($\sigma_x$)

$$3 = \frac{b}{x^3} + a = \frac{b}{(200)^3} + a = \frac{b}{8 \times 10^6} + a$$

$$\therefore \quad a = 3 - \frac{b}{8 \times 10^6} \qquad ...(ii)$$

Equating two values of *a* from equations (*i*) and (*ii*)

$$\frac{2b}{8 \times 10^6} - 1.5 = 3 - \frac{b}{8 \times 10^6}$$

or $$\frac{3b}{8 \times 10} = 3 + 1.5$$

or $$b = \frac{4.5\left(8 \times 10^6\right)}{3} = 12 \times 10^6$$

Now substituting the value of $b$ in equation $(i)$

$$a = \frac{2\left(12\times10^6\right)}{8\times10^6} - 1.5 = 1.5$$

We know that from Lame's equation that pressure at outer surface is

$$p_x = \frac{2p}{x^3} - a$$

or
$$0 = \frac{2(12\times10^6)}{(r_1)^3} - 1.5$$

or
$$1.5 = \frac{2(12\times10^6)}{(r_1)^3}$$

or
$$r_1 = \sqrt[3]{\frac{2(12\times10^6)}{1.5}} = \sqrt[3]{16\times10^6}$$

$$= 252 \text{ mm}$$

Therefore thickness of the shell

$$t = r_1 - r_2 = 252 - 200 = 52 \text{ mm}$$ **Ans.**

## EXERCISE 32.1

1. A cast iron pipe of 200 mm internal diameter and 50 mm thick carries water under a pressure of 5N/mm$^2$. Find the maximum and minimum intensities of circumferential stress across the section. Also sketch the radial pressure distribution and circumferential stress distribution across the section. [**Ans.** 13 N/mm$^2$; 8 N/mm$^2$]
2. A thick cylindrical shell of 160 mm internal diameter is 45 mm thick. The shell is subjected to an internal pressure of 52.5 N/mm$^2$. Find the maximum and minimum intensities of hoop stress across the section. [**Ans.** 127.5 N/mm$^2$; 75 N/mm$^2$]
3. A thick metallic cylindrical shell of internal diameter 200 mm required to withstand a pressure of 40 N/mm$^2$. Determine the necessary thickness of the shell, if permissible hoop stress in the section is 150 N/mm$^2$. [**Ans.** 31.5 mm]
4. A compound cylinder is formed by shrinking a tube of 200 mm internal diameter and 20 mm thick over another tube of 120 mm diameter and 40 mm thick. If radial pressure at the common surface, after shrinking is 12 N/mm$^2$, then determine the final stresses across the section when a fluid under a pressure of 45 N/mm$^2$ is admitted into the cylinder.

[**Ans.**

| Final stresses (N/mm$^2$) | Inner tube | | Outer tube | |
|---|---|---|---|---|
| | $x$ = 60 mm | $x$ = 100mm | $x$ = 100 mm | $x$ = 120 mm |
| Initial | –37.5 | –25.5 | +66.6 | +54.6 |
| Due to fluid pressure | +75.0 | +36.6 | +36.6 | +30.0 |
| Final | +37.5 | +11.1 | +101.2 | +84.6 |

5. A compound cylinder is made by shrinking a cylinder of external diameter 300 mm over another cylinder of external diameter 250 mm and internal diameter 150 mm. After shrinking, the radial compression at the common junction was 280 N/mm$^2$. Find the original difference in the radii at the junction, if $E$ = 200 GPa. **[Ans.** 0.15 mm]
6. A thick spherical shell of 160 mm internal diameter is subjected to an internal pressure of 40 N/mm$^2$. Find the thickness of the shell, if the permissible tensile stress is 80 N/mm$^2$. **[Ans.** 21 mm]
7. A thick spherical shell 80 mm diameter is required to withstand an internal pressure of 30 N/mm$^2$. Determine the necessary thickness of the shell, if the maximum permissible tensile stress is 80 N/mm$^2$. **[Ans.** 7.8 mm]

## QUESTIONS

1. What is the difference between a thin cylindrical shell and a thick cylindrical shell?
2. What do you understand by the term 'Lame Theory'?
3. Write the assumptions for solving the problems on thick cylindrical shells?
4. Derive a relation for the hoop stress at the junction of a compound thick cylindrical shell.
5. Obtain an expression for the difference of radii for shrinkage of a compound thick cylindrical shell.
6. Write the relations for the hoop stress in a thick spherical shell.

# Chapter 33

# Bending of Curved Bars

## Contents



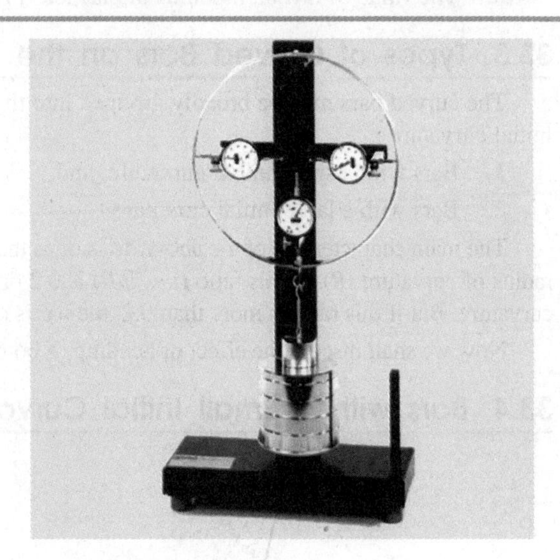

## 33.1. Introduction

In the theory of simple bending, we have discussed :

$$\frac{M}{I} = \frac{\sigma}{y} = \frac{E}{R}$$

where $M$ = Moment of resistance for the bar or bending moment at the bar section,

$I$ = Moment of inertia of the bar section about its centre of gravity.

$\sigma$ = Bending stress in the fibre of the bar section,

$y$ = Distance between the c.g. of the bar section and the fibre,

$E$ = Young's modulus for the bar,

$R$ = Radius of curvature of the bar.

The above mentioned formula is, sometimes, called *straight-beam formula*. The reason for the same is that it can be applied, with sufficient accuracy, to the beams only with small initial curvature.

## 33.2. Assumptions for the Stresses in the Bending of Curved Bars

The stresses in the bending of curved bars are determined on the following assumptions :

**1.** The bar material is stressed within the elastic limit, and thus obeys Hooke's law.

**2.** The transverse sections, which were plane before bending, remain plane after bending also.

**3.** The longitudinal fibres of the bar, parallel to the central axis, exert no pressure on each other.

**4.** The transverse cross-section has at least one axis of symmetry, and the bending moment lies on this plane.

**5.** The value of $E$ (*i.e.*, modulus of elasticity) is the same in tension and compression.

## 33.3. Types of Curved Bars on the Basis of Initial Curvature

The curved bars may be broadly grouped into the following two categories on the basis of their initial curvature :

**1.** Bars with a small initial curvature, and

**2.** Bars with a large initial curvature.

The main characteristic of the above division is the ratio of the depth of bar section ($h$) to the initial radius of curvature ($R$). If this ratio (*i.e.*, $h/R$) is 0.2 or less, the bar is considered to be of small initial curvature. But if this ratio is more than 0.2, the bar is considered to be of large radius of curvature.

Now we shall discuss the effect of bending of both the above types of bars in the following pages.

## 33.4. Bars with a Small Initial Curvature

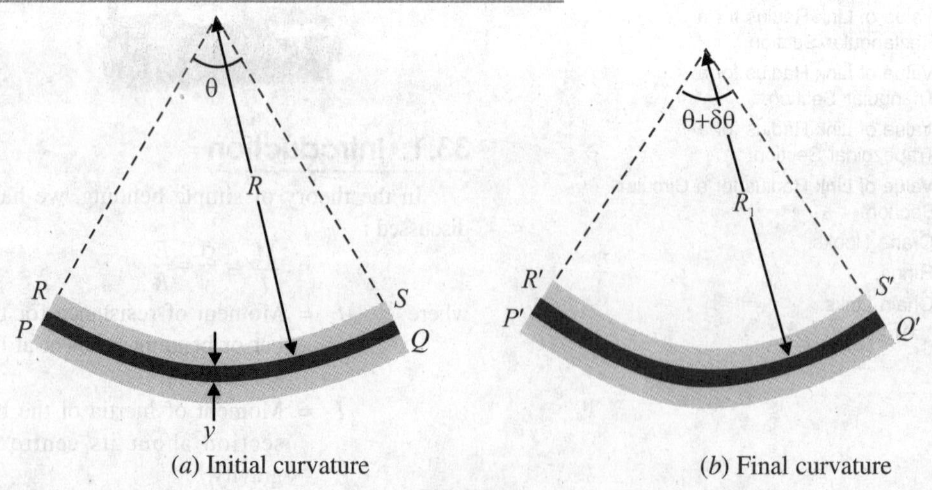


(*a*) Initial curvature (*b*) Final curvature

Fig. 33.1

Consider a curved bar with a small initial curvature as shown in Fig. 33.1(*a*). Let the bar be given more curvature after the application of the end moments as shown in Fig. 33.1 (*b*).

Let $R$ = Initial radius of curvature of the bar,

$R_1$ = Final radius of curvature,

$\theta$ = Initial angle subtended at the centre by the bar, and

$(\theta + \delta\theta)$ = Final angle subtended at the centre of the bar.

Now consider a layer *PQ*, which has been bent up to *P′ Q′* after bending. Let *y* be the distance of the layer *PQ* from *RS*, the neutral axis of the bar. We know that increase in the length of the bar,

$$\delta l = P'Q' - PQ$$

∴ Strain.

$$e = \frac{\delta l}{PQ} = \frac{P'Q' - PQ}{PQ} \quad ...(i)$$

$$= \frac{(R_1 + y)(\theta + \delta\theta) - (R + y)\theta}{(R + y)\theta}$$

$$= \frac{R_1(\theta + \delta\theta) + y\theta + y \cdot \delta\theta - R\theta - y\theta}{(R + y)\theta}$$

$$= \frac{R_1(\theta + \delta\theta) + y \cdot \delta\theta - R\theta}{(R + y)\theta} \quad ...(ii)$$

From the geometry of the bar, we find that

$$RS = R\theta \quad \text{and} \quad R'S' = R_1 (\theta + \delta\theta).$$

Since *RS* is equal to *R′S′* (being neutral axis) therefore,

$$R\theta = R_1 (\theta + \delta\theta) = R_1\theta + R_1 \cdot \delta\theta$$

$$R_1 \cdot \delta\theta = R\theta - R_1\theta = \theta (R - R_1)$$

or

$$\frac{\delta\theta}{\theta} = \frac{(R - R_1)}{R_1} \quad ...(iii)$$

Now substituting $R\theta$ equal to $R_1 (\theta + \delta\theta)$ in equation (*ii*), and by solving it further we get,

$$e = \frac{y \cdot \delta\theta}{(R + y)\theta} - \frac{y}{(R + y)} \times \frac{(R - R_1)}{R}$$

[Substituting the value of $\frac{\delta\theta}{\theta}$ from equation (*iii*)]

Since *y* is small as compared to $R_0$, therefore substituting $(R + y) = R$ in the above equation,

$$e = \frac{y}{R} \times \frac{R - R_1}{R_1} = y\left(\frac{1}{R_1} - \frac{1}{R}\right)$$

$$\frac{\sigma}{E} = y\left(\frac{1}{R_1} - \frac{1}{R}\right) \quad \left(\because \text{Strain} = \frac{\text{Stress}}{E}\right)$$

∴

$$\frac{\sigma}{y} = E\left(\frac{1}{R_1} - \frac{1}{R}\right)$$

or

$$\frac{\sigma}{y} = \frac{M}{I} = E\left(\frac{1}{R_1} - \frac{1}{R}\right) \quad \left(\because \frac{M}{I} = \frac{\sigma}{y}\right)$$

The above expression may also be written as :

$$\sigma = Ey\left(\frac{1}{R_1} - \frac{1}{R}\right)$$

**EXAMPLE 33.1.** *A steel bar 50 mm in diameter, is formed into a circular arc of 4 m radius and supports an angle of 90°. A couple is applied at each end of the bar, which changes the slope to 95° at one end relative to the other. Calculate the maximum bending stress due to the couple. Take E as 200 GPa.*

**SOLUTION.** Given : Diameter of bar (*d*) = 50 mm ; Radius of arc (*R*) = 4 m = 4000 mm ; Initial angle subtended at the centre (θ) = 90° ; Final angle subtended at the centre (θ + δθ) = 95° and modulus of elasticity (*E*) = 200 GPa = $200 \times 10^3$ N/mm²

Distance between centre line of bar and extreme fibre,

$$y = \frac{d}{2} = \frac{50}{2} = 25 \text{ mm}$$

And $\delta\theta = 95^\circ - 90^\circ = 5^\circ$

Let $R_1$ = Final radius of the arc,

Using the relation,

$$\frac{\delta\theta}{\theta} = \frac{(R - R_1)}{R_1}$$

$$\frac{5}{90} = \frac{4000 - R_1}{R_1}$$

or $5\,R_1 = 360000 - 90\,R_1$

$$\therefore \quad R_1 = \frac{360000}{95} = 3789 \text{ mm}$$

Maximum bending stress due to couple

$$\sigma = E.y\left(\frac{1}{R_1} - \frac{1}{R}\right)$$

$$= 200 \times 10^3 \times 25\left(\frac{1}{3789} - \frac{1}{4000}\right)$$

$$= 69.6 \text{ N/mm}^2$$ **Ans.**

## 33.5. Bars with a Large Initial Curvature

In the previous article, we have discussed a case of bars with a small initial curvature (*i.e.* assuming $y$ to be small as compared to $R$). However, there are practical cases of bars, such as hooks, links and rings etc., which have large initial curvature. The analysis of such bars was first dealt with Winkler and later by Andrews and Pearson.

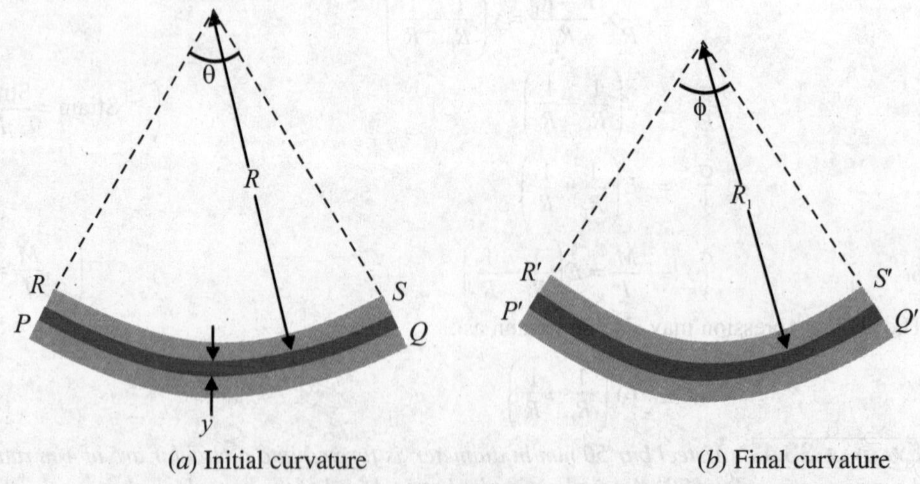


Fig. 33.2

Now consider a curved bar with a large initial curvature as shown in Fig. 33.2 (*a*). Let the bar be given more curvature after the application of the end moments as shown in Fig. 33.2 (*b*).

Let $R$ = Initial radius of curvature,

$R_1$ = Final radius of curvature,

$\theta$ = Initial angle subtended at the centre of the bar,

$\phi$ = Final angle subtended at the centre of the bar,

$\sigma_0$ = Bending stress in the centroidal fibre $R'S'$,

$\sigma$ = Bending stress in the fibre $P'Q'$, and

$dA$ = Area of fibre $P'Q'$.

Now consider a layer $PQ$ which has been bent up to $P'Q'$ after bending. Let $y$ be the distance of the layer $PQ$ from $RS$, the centroidal axis of the bar.

We know that the increase in the length of the bar at the centroidal axis,

$$\delta l = R'S' - RS$$

∴ Strain,
$$e_0 = \frac{\delta l}{l} = \frac{R'S' - RS}{RS} = \frac{R'S'}{RS} - 1$$

or
$$e_0 + 1 = \frac{R'S'}{RS} = \frac{R_1\phi}{R_0\theta} \qquad ...(i)$$

and increase in the length of the bar at a distance $y$ from the centroidal axis

$$\delta l = P'Q' - PQ$$

∴ Strain,
$$e = \frac{\delta l}{l} = \frac{P'Q' - PQ}{PQ} = \frac{P'Q'}{PQ} - 1$$

or
$$e + 1 = \frac{P'Q'}{PQ} = \frac{(R_1 + y)\phi}{(R + y)\theta} \qquad ...(ii)$$

Dividing equation (*ii*) by (*i*),

$$\frac{e+1}{e_0+1} = \frac{\dfrac{(R_1 + y)\phi}{(R + y)\theta}}{\dfrac{R_1\phi}{R\theta}} = \frac{\dfrac{R_1 + y}{R_1}}{\dfrac{R + y}{R}}$$

$$= \frac{1 + \dfrac{y}{R_1}}{1 + \dfrac{y}{R}}$$

∴
$$e + 1 = \frac{(e_0 + 1)\left(1 + \dfrac{y}{R_1}\right)}{1 + \dfrac{y}{R}}$$

$$= \frac{e_0 + e_0\dfrac{y}{R_1} + 1 + \dfrac{y}{R_1}}{1 + \dfrac{y}{R}}$$

∴
$$e = \frac{e_0 + e_0\dfrac{y}{R_1} + 1 + \dfrac{y}{R_1}}{1 + \dfrac{y}{R}} - 1$$

$$= \frac{e_0 + e_0\dfrac{y}{R_1} + 1 + \dfrac{y}{R_1} - 1 - \dfrac{y}{R}}{1 + \dfrac{y}{R}}$$

$$= \frac{e_0 + e_0 \dfrac{y}{R_1} + y\left(\dfrac{1}{R_1} + \dfrac{1}{R}\right)}{1 + \dfrac{y}{R}}$$

$$= \frac{e_0 + e_0 \dfrac{y}{R_1} - \dfrac{e_0 y}{R} + \dfrac{e_0 y}{R} + y\left(\dfrac{1}{R_1} - \dfrac{1}{R}\right)}{1 + \dfrac{y}{R}}$$

$$\left(\because \text{ Adding and subtracting } \frac{e_0 y}{R}\right)$$

$$= \frac{e_0\left(1 + \dfrac{y}{R}\right) + e_0 y\left(\dfrac{1}{R_1} - \dfrac{1}{R}\right) + y\left(\dfrac{1}{R_1} - \dfrac{1}{R}\right)}{1 + \dfrac{y}{R}}$$

$$= e_0 + \frac{e_0\, y + y\left(\dfrac{1}{R_1} - \dfrac{1}{R}\right)}{1 + \dfrac{y}{R}}$$

$$\therefore \qquad e = e_0 + \frac{(e_0 + 1)y\left(\dfrac{1}{R_1} - \dfrac{1}{R}\right)}{1 + \dfrac{y}{R}} \qquad ...(iii)$$

We know that the bending stress in the fibre $P'Q'$,

$$\sigma = E \cdot e = E\left[e_0 + \frac{(e_0 + 1)y\left(\dfrac{1}{R_1} - \dfrac{1}{R}\right)}{1 + \dfrac{y}{R}}\right] \qquad ...(iv)$$

and the force in an element of area $dA$ at a distance $y$ of from the centroidal axis,

$$= \sigma \cdot \delta A = E\left[e_0 + \frac{(e_0 + 1)y\left(\dfrac{1}{R_1} - \dfrac{1}{R}\right)}{1 + \dfrac{y}{R}}\right] dA$$

The total normal force on the cross-section may be found out by integrating the above equation, *i.e.*,

$$P = \int E\left[e_0 + \frac{(e_0 + 1)y\left(\dfrac{1}{R_1} - \dfrac{1}{R}\right)}{1 + \dfrac{y}{R}}\right] dA$$

$$= \int E e_0 \cdot dA + \int \frac{E(e_0 + 1)y\left(\dfrac{1}{R_1} - \dfrac{1}{R}\right)}{1 + \dfrac{y}{R}} dA$$

$$= E\,e_0 A + E(e_0 + 1)\frac{1}{R_1} - \frac{1}{R}\int \frac{y}{1+\frac{y}{R}}\,dA$$

Since the beam is in equilibrium, therefore the total normal force on the cross-section is zero.

$$\text{or} \quad E\,e_0 A + E(e_0 + 1)\left(\frac{1}{R_1} - \frac{1}{R}\right)\int \frac{y}{1+\frac{y}{R}}\,dA = 0 \quad ...(v)$$

Now let us find out the vlaue of $\int \frac{y}{1+\frac{y}{R}}\,dA$ separately.

$$\int \frac{y}{1+\frac{y}{R}}\,dA = \int \frac{yR}{R+y}\,dA$$

$$= \int \frac{yR + y^2 - y^2}{R+y}\,dA \quad (\because \text{ Adding and subtracting } y^2)$$

$$= \int \frac{y(R+y) - y^2}{R+y}\,dA$$

$$= \int y \cdot dA - \int \frac{y^2}{R+y}\,dA$$

Since $\int y \cdot dA$, being the first moment of area, about the central axis is zero, therefore

$$\int \frac{y}{1+\frac{y}{R}}\,dA = -\int \frac{y^2}{R-y}\,dA$$

Now substituting $\int \frac{y^2}{R+y}\,dA = \frac{Ah^2}{R}$, we get

$$\int \frac{y}{1+\frac{y}{R}}\,dA = -\frac{Ah^2}{R} \quad ...(vi)$$

where $h^2$ is the constant of the section; $h$ is called the link radius. It corresponds to the radius of gyration in ordinary case. By using equation (*vi*) the equation (*v*) may be rewritten as

$$E\,e_0 A + E(e_0 + 1)\left(\frac{1}{R_1} - \frac{1}{R}\right)\left(-\frac{Ah^2}{R}\right) = 0$$

$$e_0 - (e_0 + 1)\left(\frac{1}{R_1} - \frac{1}{R}\right) \times \frac{h^2}{R} = 0$$

$$\therefore \quad e_0 = (e_0 + 1)\left(\frac{1}{R_1} - \frac{1}{R}\right) \times \frac{h^2}{R} \quad ...(vii)$$

We know that moment of resistance of the fibre

$$P'Q' = y\,\sigma \cdot dA$$

and the total moment of the section

$$M = \int y\sigma \cdot dA$$

$$= \int yEe \cdot dA$$

Now substituting the value of $e$ from equation (*iii*).

$$M = \int y\,E\left[e_0 + \frac{(e_0+1)\,y\left(\frac{1}{R_1} - \frac{1}{R}\right)}{1+\frac{y}{R}}\right] dA$$

$$= \int E\,y\,e_0 \cdot dA + \int \frac{E(e_0+1)\,y^2\left(\frac{1}{R_1} - \frac{1}{R}\right)}{1+\frac{y}{R}}\, dA$$

$$= E\,e_0 \int y \cdot dA + E(e_0+1)\left(\frac{1}{R_1} - \frac{1}{R}\right)\int \frac{y^2}{1+\frac{y}{R}}\, dA$$

Since $\int y \cdot dA$, being the first moment of area about the central axis, is equal to zero, therefore

$$M = E(e_0+1)\left(\frac{1}{R_1} - \frac{1}{R}\right)\int \frac{y^2}{1+\frac{y}{R}}\, dA$$

$$= E(e_0+1)\left(\frac{1}{R_1} - \frac{1}{R}\right) R \int \frac{y^2}{R+y}\, dA$$

$$= E(e_0+1)\left(\frac{1}{R_1} - \frac{1}{R}\right) Ah^2 \qquad \left(\because \int \frac{y^2}{R+y} = \frac{Ah^2}{R}\right)$$

Substituting the value of $M$ in equation (*vii*),

$$e_0 = \frac{M}{EAR}$$

From equation (*iv*), we know that the bending stress,

$$\sigma = E\left[e_0 + \frac{(e_0+1)\,y\left(\frac{1}{R_1} - \frac{1}{R}\right)}{1+\frac{y}{R}}\right]$$

$$= E\left[e_0 + \frac{(e_0+1)\,y\left(\frac{1}{R_1} - \frac{1}{R}\right)R}{R+y}\right]$$

$$= E\left[\frac{M}{EAR} + \frac{M\,R\,y}{E\,Ah^2(R+y)}\right]$$

$$= \frac{M}{AR} + \frac{M\,R\,y}{Ah^2(R+y)}$$

$$= \frac{M}{AR}\left[1 + \frac{R^2 y}{h^2(R+y)}\right]$$

where $h^2$ is called the link radius, and corresponds to the radius of gyration in ordinary case.

**Notes:** 1. If the answer is positive, the bending stress is tensile.

2. If the answer is negative, the bending stress is compressive.

3. If the fibre $PQ$ is on the other side of $RS$, the value of $y$ is to be taken as negative. In such a case,

$$\sigma = \frac{M}{AR}\left[1 - \frac{R^2 y}{h^2 (R-y)}\right]$$

4. If the bending moment ($M$) tends to decrease the curvature, the bending stress is compressive.

## 33.6. Link radius for Standard Sections

We have discussed in the previous article, the value of link radius, which corresponds to radius of gyration in ordinary case. The equation substituted was :

$$\int \frac{y^2}{R+y} dA = \frac{Ah^2}{R}$$

The above equation may also be written as:

$$h^2 = \frac{R}{A}\int \frac{y^2}{R+y} dA$$

$$= \frac{R}{A}\int \frac{y^2 + Ry - Ry + R^2 - R^2}{R+y} dA$$

($\because$ Adding and subtracting $Ry$ and $R^2$)

$$= \frac{R}{A}\int \frac{y(R+y) - R(y+R) + R^2}{R+y} dA$$

$$= \frac{R}{A}\int \left(y - R + \frac{R^2}{R+y}\right) dA$$

$$= \frac{R}{A}\left[\int y \cdot dA - R\int dA + R^2 \int \frac{dA}{R+y}\right]$$

$$= \frac{R}{A}\left[0 - RA + R^2 \int \frac{dA}{R+y}\right] \qquad \left(\because \int y.dA = 0\right)$$

$$= -R^2 + \frac{R^3}{A}\int \frac{dA}{R+y}$$

$$= \frac{R^3}{A}\int \frac{dA}{R+y} - R^2$$

where $h$ is the link radius.

Now we shall discuss the value of link radius for the following standard sections :

1. Rectangular section.
2. Triangular section.
3. Trapezoidal section.
4. Circular section.

## 33.7. Value of Link Radius for a Rectangular Section

Consider a curved bar of rectangular section as shown in Fig. 33.3.

Let $B$ = Breath of the bar section

$D$ = Depth of the bar section

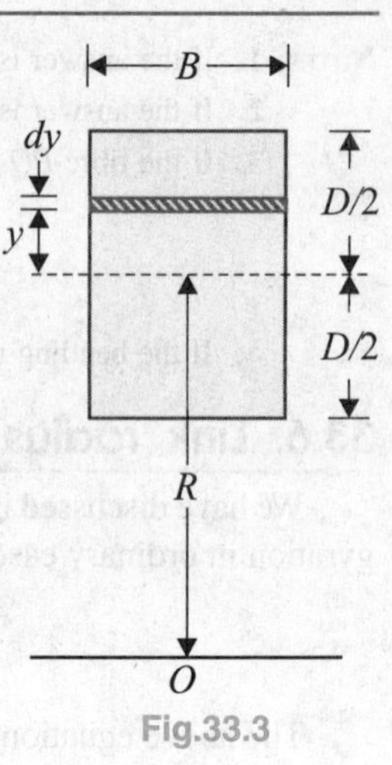

Fig.33.3

$\therefore$ Area, $A = B \times D$

$R$ = Radius of curvature of the bar section (*i.e.*, the distance between the centroidal axis and the axis of curvature.

Now condider a strip of thickness *dy* at a distance *y* from the centroidal axis as shown in Fig. 33.3. Therefore area of the strip.

$$dA = b \cdot dy$$

We know that the general equation for the link radius.

$$h^2 = \frac{R^3}{A}\int \frac{dA}{R+y} - R^2$$

$$= \frac{R^3}{B \times D}\int_{-\frac{D}{2}}^{+\frac{D}{2}} \frac{B \cdot dy}{R+y} - R^2$$

$$= \frac{R^3 B}{B \times D}\log_e\left[R+y\right]_{-\frac{D}{2}}^{+\frac{D}{2}} - R^2$$

$$= \frac{R^3}{D}\log_e\left(\frac{2R+D}{2R-D}\right) - R^2$$

$$= 2{\cdot}3\,\frac{R^3}{D}\log\left(\frac{2R+D}{2R-D}\right) - R^2$$

## 33.8. Value of Link Radius for Triangular Section

Consider curved bar of triangular section as shown in Fig. 33.4.

Let $B$ = Base width of the bar section,

$D$ = Depth of the bar section,

$\therefore$ Area, $A = \dfrac{BD}{2}$

$R$ = Radius of curvature of the bar section (*i.e.* the distance between centroidal axis and the axis of curvature).

$R_1$ = Distance between the base of the bar section and the axis of curvature, and

$R_2$ = Distance between the top of the bar section and the axis of curvature.

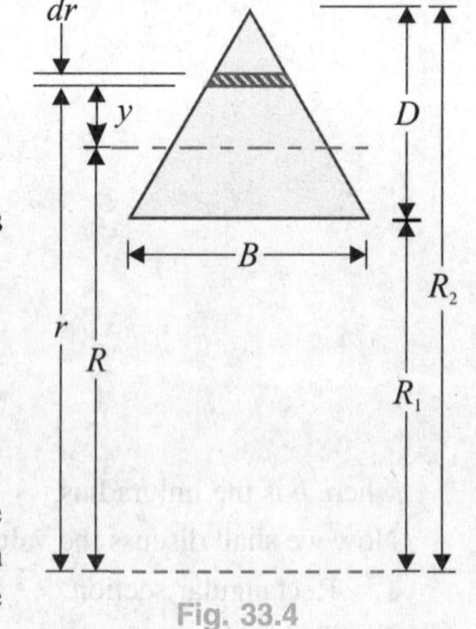

Fig. 33.4

Now consider a strip of thickness *dr* at a distance *y* from the centroidal (*i.e.* at a distance *r* from the axis of curvature) as shown in Fig. 29.4. From the geometry of the figure, we find that the width of the bar,

$$b = \frac{B}{D}(R_2 - r)$$

$\therefore$ Area of strip, $dA = \dfrac{B}{D}(R_2 - r)dr$

We know that the general equation for the link radius,

$$h^2 = \frac{R^3}{A}\int \frac{dA}{R+y} - R^2$$

$$= \frac{R^3}{A}\int_{R_1}^{R_2} \frac{\frac{B}{D}(R_2 - r)\,dr}{R+y} - R^2$$

$$= \frac{R^3}{A}\int_{R_1}^{R_2} \frac{B}{D}(R_2 - r)\frac{dr}{r} - R^2 \qquad [\because (R+y) = r]$$

$$= \frac{R^3}{A}\times\frac{B}{D}\left[\int_{R_1}^{R_2} R_2\frac{dr}{r} - \int_{R_1}^{R_2} dr\right] - R^3$$

$$= \frac{R^3}{A}\times\frac{B}{D}\left\{[R_2 \log_e r\,]_{R_1}^{R_2} - [r\,]_{R_1}^{R_2}\right\} - R^2$$

$$= \frac{R^3}{A}\times\frac{B}{D}\left[R_2 \log_e \frac{R_2}{R_1} - (R_2 - R_1)\right] - R^2$$

$$= \frac{R^3}{A}\times\frac{B}{D}\left[2.3\, R_2 \log \frac{R_2}{R_1} - D\right] - R^2 \qquad (\because R_2 - R = D)$$

## 33.9. Value of Link Radius for a Trapezoidal Section

Consider a curved bar of trapezoidal section as shown in Fig. 33.5.

Let $B_1$ = Base width of the bar section,

$B_2$ = Top width of the bar section,

$D$ = Depth of the bar section,

$\therefore$ Area, $A = \dfrac{D(B_1 + B_2)}{2}$

$R$ = Radius of curvature of the bar section (*i.e.* the distance between the centroidal axis of curvature.

$R_1$ = Distance between the base of the bar section and the axis of curvature, and

$R_2$ = Distance between the top of the bar section and the axis of curvature.

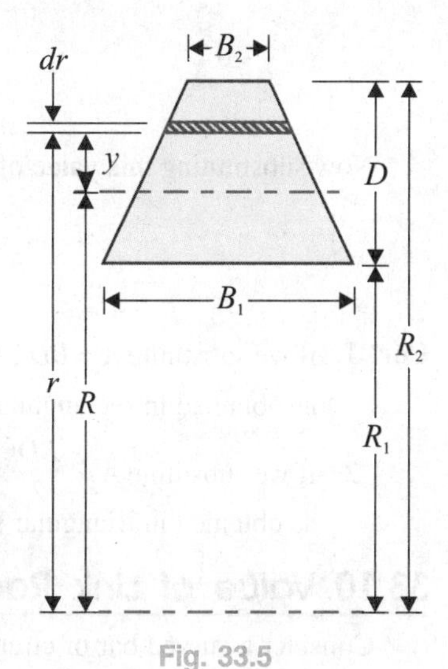

Fig. 33.5

Now consider a strip of thickness *dr* at a distance *y* from the centroidal axis (*i.e.* at a distance *r* from the axis of curvature) as shown in Fig. 29.5. From the geometry of the figure, we find that the width of the bar,

$$b = B_2 + \frac{(B_1 - B_2)}{D}(R_2 - r)$$

$\therefore$ Area of strip,

$$dA = \left[B_2 + \frac{(B_1 - B_2)}{D}(R_2 - r)\right] dr$$

We know that the general equation for the link radius :

$$h^2 = \frac{R^3}{A}\int \frac{dA}{R+y} - R^2 \qquad ...(i)$$

First of all, let us find out the value of $\frac{dA}{R+y}$ separately.

$$\therefore \quad \int \frac{dA}{R+y} = \int_{R_1}^{R_2} \frac{\left[B_2 + \frac{(B_1 - B_2)}{D}(R_2 - r)\right] dr}{R+y}$$

$$= \int_{R_1}^{R_2} \left[B_2 + \frac{(B_1 - B_2)}{D}(R_2 - r)\frac{dr}{r}\right] \qquad (\because R+y=r)$$

$$= \int_{R_1}^{R_2} B_2 \cdot \frac{dr}{r} + \int_{R_1}^{R_2} \frac{(B_1 - B_2)R_2}{D} \frac{dr}{r} - \int_{R_1}^{R_2} \frac{(B_1 - B_2)r}{D} \cdot \frac{dr}{r}$$

$$= B_2[\log_e r]_{R_1}^{R_2} + \frac{(B_1 - B_2)R_2}{D}[\log_e r]_{R_1}^{R_2} - \frac{(B_1 - B_2)}{D}[r]_{R_1}^{R_2}$$

$$= B_2 \log_e \frac{R_2}{R_1} + \frac{(B_1 + B_2)R_2}{D} \log_e \frac{R_2}{R_1} - \frac{(B_1 - B_2)}{D} \times (R_2 - R_1)$$

$$= B_2 \times 2 \cdot 3 \times \log \frac{R_2}{R_1} + \frac{(B_1 - B_2)R_2}{D} \times 2 \cdot 3 \times \log \frac{R_2}{R_1} - (B_1 - B_2)$$

$$(\because R_2 - R_1 = D)$$

$$= 2.3 \times \log \frac{R_2}{R_1}\left[B_2 + \frac{(B_1 - B_2)R_2}{D}\right] - (B_1 - B_2)$$

Now substituting this value of $\int \frac{dA}{R+y}$ in equation (*i*)

$$h^2 = \frac{R^3}{A}\left[2 \cdot 3 \times \log \frac{R_2}{R_1}\left(B_2 + \frac{(B_1 - B_2)R_2}{D}\right) - (B_1 - B_2)\right] - R^2$$

**Cor. 1.** If we substitute $A = BD$ ; $R_2 = R + \frac{D}{2}$ ; $R_1 = R - \frac{D}{2}$ and $B_1 = B_2$ we shall obtain the equation as obtained in rectangular section,

**2.** If we substitute $A = \frac{BD}{2}$; $R_2 = R + \frac{D}{2}$; $R_1 = R - \frac{D}{2}$ and $B_2 = 0$, we shall obtain the equation as obtained in triangular section.

## 33.10. Value of Link Radius for a Circular Section

Consider a curved bar of circular section as shown in Fig. 33.6.

Let $\quad d$ = Diameter of the bar section

$\therefore$ Area, $\quad A = \frac{\pi}{4} \times d^2$

$R$ = Radius of curvature of the bar section (*i.e.*, the distance between the centroidal axis and the axis of curvature).

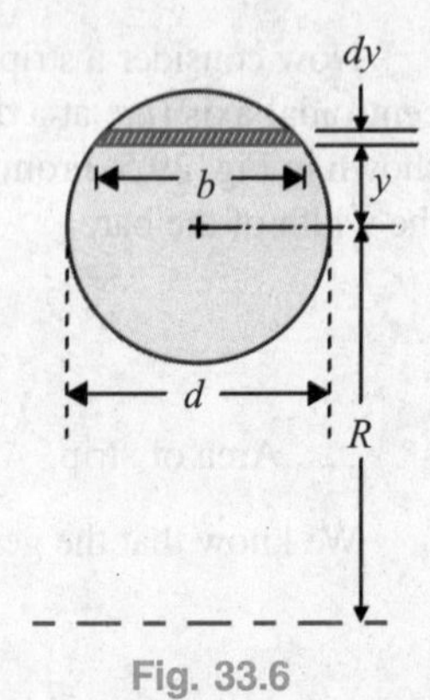

Fig. 33.6

Now consider a strip of thickness $dy$ at a distance $y$ from the centroidal axis as shown in Fig. 33.6. From the geometry of the figure, we find that the width of the bar,

$$b = 2\left[\sqrt{\left(\frac{d}{2}\right)^2 - y^2}\right] = 2\times\sqrt{\left(\frac{d^2}{4} - y^2\right)}$$

∴ Area of strip, $dA = \sqrt{\left(\frac{d^2}{4} - y^2\right)}\,dy$

We know that the general equation for the link radius,

$$h^2 = \frac{R^3}{A}\int\frac{dA}{R+y} - R^2$$

$$= \frac{R^3}{\frac{\pi}{4}d^2}\int_{-\frac{D}{2}}^{+\frac{D}{2}}\frac{2\times\sqrt{\left(\frac{d^4}{4} - y^2\right)}}{R+y}\,dy - R^2$$

$$= \frac{d^2}{16} + \frac{1}{8}\times\frac{d^4}{16\,R^2} \qquad \text{(After simplification)}$$

**EXAMPLE 33.2.** *A beam of rectangular section 20 mm × 40 mm has its centre line curved to a radius of 50 mm. The beam is subjected to a bending moment of 4 × $10^5$ N-mm. Determine the intensity of maximum stresses in the beam. Also plot the bending stress across the section.*

**SOLUTION.** Given : Beam width = 20 mm ; Beam depth ($D$) = 40 mm ; Radius of beam ($R$) = 50 mm and bending moment ($M$) = 4 × $10^5$ N-mm.

We know that beam area, $A = 20 \times 40 = 800 \text{ mm}^2$

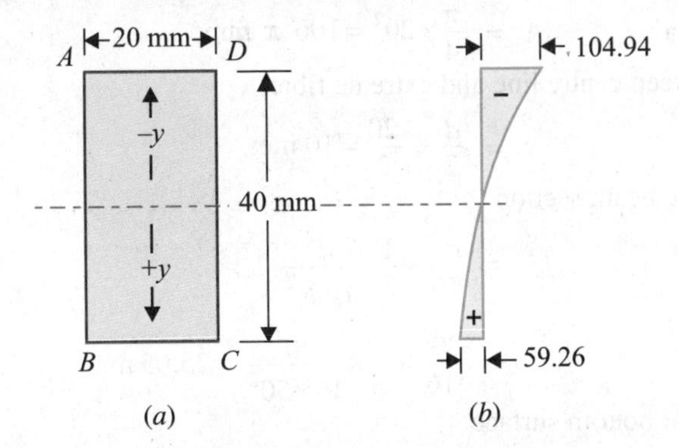


Fig. 33.7

Distance between centre line and extreme fibre,

$$y = \frac{40}{2} = 20 \text{ mm}$$

Link radius for the beam section

$$h^2 = 2.3\frac{R^3}{D}\log\left(\frac{2R+D}{2R-D}\right) - R^2$$

$$= 2.3\times\frac{(50)^3}{40}\log\left(\frac{2\times50+40}{2\times50-40}\right) - (50)^2$$

$$= 7188\log 2.333 - 2500 = 145 \text{ mm}$$

Maximum stress at bottom surface

$$\sigma_1 = \frac{M}{AR}\left(1+\frac{R^2 y}{h^2 (R+y)}\right)$$

$$= \frac{4\times10^5}{800\times50}\left(1+\frac{(50)^2\times20}{145(50+20)}\right)$$

$$= 59.26 \text{ N/mm}^2 \text{ (tensile)} \qquad \textbf{Ans.}$$

Maximum stress at top surface

$$\sigma_2 = \frac{M}{AR}\left(1-\frac{R^2 y}{h^2 (R-y)}\right)$$

$$= \frac{4\times10^5}{800\times50}\left(1-\frac{(50)^2\times20}{145(50-20)}\right)$$

$$= -104.94 \text{ N/mm}^2 \text{ (compressive)} \qquad \textbf{Ans.}$$

The stress distribution is shown in Fig. 33.7 (*b*).

**EXAMPLE 33.3.** *A beam of circular section of diameter 20 mm has its centre line curved to a radius of 50 mm. Find the intensity of maximum stresses in the beam, when subjected to a moment of 5 kN-mm.*

**SOLUTION.** Given : Diameter of section ($d$) = 20 mm ; Radius of curvature ($R$) = 50 mm and moment ($M$) = 5 kN-mm = $5 \times 10^3$ N-mm

We know that area

$$A = \frac{\pi}{4}\times20^2 = 100\,\pi \text{ mm}^2$$

and distance between centre line and extreme fibre,

$$y = \frac{d}{2} = \frac{20}{2} = 10 \text{ mm}$$

Link radius for the beam section

$$h^2 = \frac{d^2}{16}+\frac{1}{8}\times\frac{d^4}{16\,R^2}$$

$$= \frac{20^2}{16}+\frac{1}{8}\times\frac{20^4}{16\times50^2} = 25.05 \text{ mm}$$

Maximum stress at bottom surface

$$\sigma_1 = \frac{M}{AR}\left(1+\frac{R^2 . y}{h^2 (R+y)}\right)$$

$$= \frac{5\times10^3}{100\pi\times50}\left[1+\frac{50^2\times10}{25.05(50+10)}\right] \text{ N/mm}^2$$

$$= 5.61 \text{ N/mm}^2 \text{ (tensile)} \qquad \textbf{Ans.}$$

Maximum stress at top surface

$$\sigma_2 = \frac{M}{AR}\left[1-\frac{R^2 . y}{h^2 (R-y)}\right]$$

$$= \frac{5\times10^3}{100\pi\times50}\left[1-\frac{50^2\times10}{25.05(50-10)}\right] \text{ N/mm}^2$$

$$= -4.98 = 4.98 \text{ N/mm}^2 \text{ (compressive)} \qquad \textbf{Ans.}$$

## 33.11. Crane Hooks

We have already discussed in Art. 33.5 the values of bending stress, due to moment in bars with a large initial curvature. The results of this article may be applied for finding the stresses in the horizontal section ($X_1X_2$) through the centre of curvature ($O$) of a hook, when it is subjected to a load. A little consideration will show, that the horizontal section through the centre of curvature is the most stressed section.

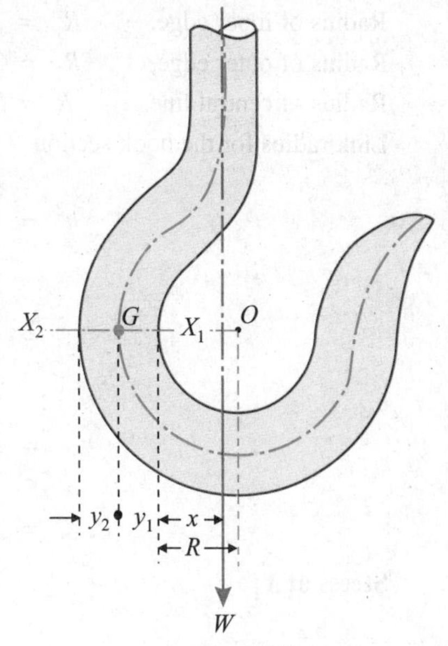

Fig. 33.8

Now the stresses at the inner ($X_1$) and the outer ($X_2$) may be found out by determining the bending stresses at these points, and then adding the direct stresses due to the load $W$. Since the bending moment $M$ tends to increase the curvature, therefore the bending stress at $X_1$ is tensile, and that at $X_2$ is compressive. Thus the total stress at $X_1$.

$$\sigma_1 = \frac{W}{A} + \frac{M}{AR}\left(1 - \frac{R^2 \cdot y_1}{h^2 (R - y_1)}\right)$$

and
$$\sigma_2 = \frac{W}{A} + \frac{M}{AR}\left(1 + \frac{R^2 \cdot y_2}{h^2 (R - y_2)}\right)$$

where
$$M = W.x$$

NOTES : The value of $h^2$ may be found out for the given section.

**EXAMPLE. 33.4.** *A crane hook carries a load of 5 kN the line of load being at a horizontal distance of 32 mm from the inside edge of a horizontal section through the centre of curvature; and the centre of curvature being 38 mm from the same edge. The horizontal section is a trapezium whose parallel sides are 13 mm and 26 mm and height is 32 mm. Determine the greatest tensile and compressive stresses in the hook.*

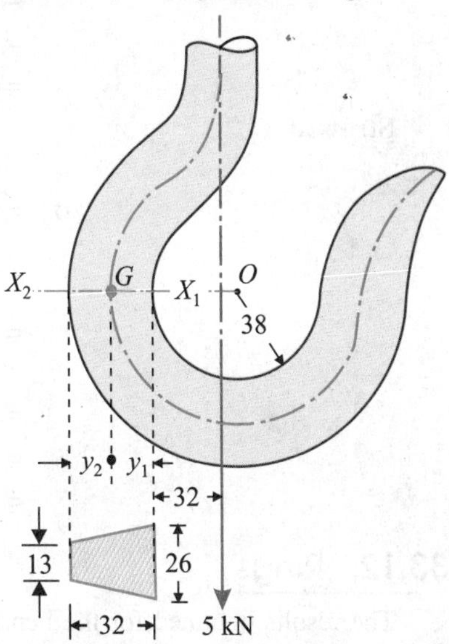

**SOLUTION.**

Given : Load ($W$) = 5 kN = 5 × $10^3$ N ; Distance between the centre line and inner edge ($x$) = 32 mm ; Distance between centre of curvature and inner edge = 38 mm ; Outer width ($B_2$) = 13 mm ; Inner width ($B_1$) = 26 mm and depth ($D$) = 32 mm.

Fig. 33.9

We know that moment,
$$M = W x$$
$$= 5 \times 10^3 \times 32$$
$$= 160 \times 10^3 \text{ N-mm}$$

Area,
$$A = \frac{13 + 26}{2} \times 32 = 624 \text{ mm}^2$$

From the geometry of the hook section, we find that

$$y_1 = \frac{26 + (2 \times 13)}{26 + 13} \times \frac{32}{3} = 14.2 \text{ mm}$$

∴
$$y_2 = 32 - 14.2 = 17.8 \text{ mm}$$

Radius of inner edge. $R_1 = 38$ mm

Radius of outer edge, $R_2 = 38 + 32 = 70$ mm

Radius of central line, $R = 38 + 14.2 = 52.2$

Link radius for the hook section

$$h^2 = \frac{R^3}{A}\left[2.3\log\frac{R_2}{R_1}\left(B_2+\frac{(B_1-B_2)R_2}{D}\right)-(B_1-B_2)\right]-R^2$$

$$= \frac{(52.2)^3}{624}\left[2.3\log\frac{70}{38}\left(13+\frac{(26-13)70}{32}\right)-(26-13)\right]-(52.2)^2$$

$$= \frac{(52.2)^3}{624}\left[(2.3\times0.2653\times41.4)-13\right]-(52.2)^2$$

$$= \frac{142236.6}{624}\times1.225-2725=68$$

**Stress at $X_1$**

$$\sigma_1 = \frac{W}{A}+\frac{M}{AR}\left(1-\frac{R^2 . y_1}{h^2(R-y_1)}\right)$$

$$= \frac{5\times10^3}{624}+\frac{160\times10^3}{624\times52.2}\left(1-\frac{(52.2)^2\times14.2}{68(52.2-14.2)}\right)$$

$$= 8.01 + 4.91\,(1 - 14.97)$$

$$= -60.58 \text{ N/mm}^2$$

$$= 60.58 \text{ N/mm}^2 \text{ (compressive)}$$ **Ans.**

**Stress at $X_2$**

$$\sigma_2 = \frac{W}{A}+\frac{M}{AR}\left(1+\frac{R^2 . y_2}{h^2(R+y_2)}\right)$$

$$= \frac{5\times10^3}{624}+\frac{160\times10^3}{624\times52.2}\left(1+\frac{(52.2)^2\times14.2}{68(52.2+14.2)}\right)$$

$$= 8.01 + 4.91\,(1 + 8.57)$$

$$= 54.99 \text{ N/mm}^2$$

$$= 54.99 \text{ N/mm}^2 \text{ (tensile)}$$ **Ans.**

## 33.12. Rings

The results obtained for the bending stress, due to moment in bars with a large initial curvature, may be applied for rings also.

Consider a ring subjected to a pull (or push) through its centre as shown in Fig. 29.10.

Let $P$ = Pull on the ring,

$R$ = Radius of curvature of the ring, and

$A$ = Cross-sectional area of the ring.

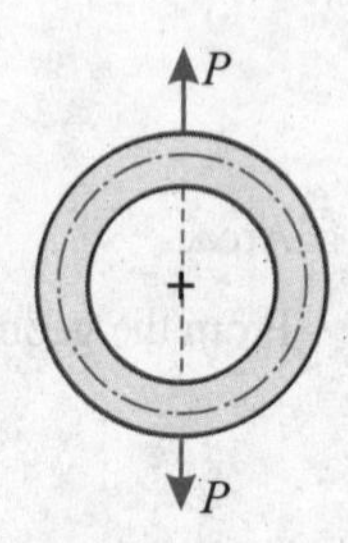


Fig. 33.10

Now consider the ring to be cut into one quadrant $XY$, fixed at $Y$ as shown in Fig. 33.11. We know that the cross-section of the ring at $X$ is subjected to the following :

1. Pull $\left(\frac{P}{2}\right)$ 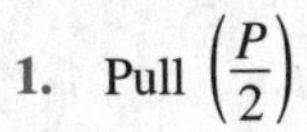

2. Moment $(M_0)$

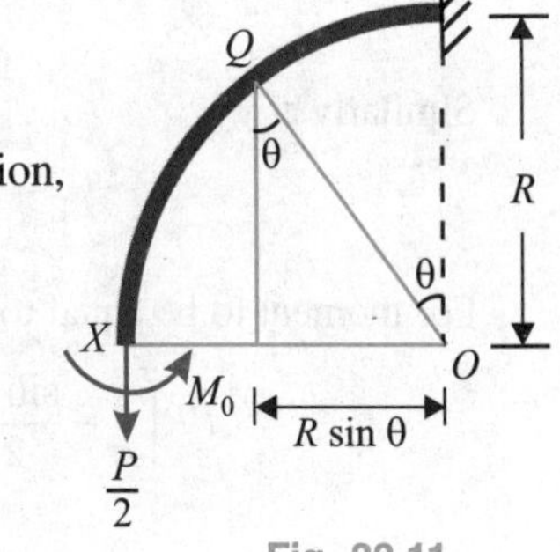

Fig. 33.11

The moment at any point $Q$ of the ring will be given by the relation,

$$M = M_0 + \frac{P}{2}(R - R\sin\theta)$$

$$= M_0 + \frac{PR}{2}(1 - \sin\theta)$$

Now in order to find the value of $M_0$, let us use the principle of minimum strain energy *i.e.*,

$$\frac{dU}{dM_0} = 0$$

Considering a small element of the ring of length $ds$ subtending an angle $d\theta$ at the centre. Therefore change of slope in length $ds$

$$= R \cdot d\theta$$

or $$di = \frac{dy}{ds} = \int \frac{M}{EI} ds$$

Since the change of slope between $X$ and $Y$ is zero, therefore

$$0 = \int_0^{\frac{\pi}{2}} \frac{M}{EI} ds$$

$$= \int_0^{\frac{\pi}{2}} \frac{M}{EI} R \cdot d\theta \qquad ...(\because \quad ds = R \cdot d\theta)$$

$$= \frac{R}{EI} \int_0^{\frac{\pi}{2}} M \cdot d\theta$$

$$= \frac{R}{EI} \int_0^{\frac{\pi}{2}} \left[ M_0 + \frac{PR}{2}(1 - \sin\theta) \right] d\theta \qquad ...\ (i)$$

$\therefore$ $$0 = M_0 - \left(\frac{PR}{\pi} - \frac{\pi}{2}\right)$$

or $$M_0 = \left(\frac{PR}{\pi} - \frac{\pi}{2}\right) \qquad ...(ii)$$

Substituting this value of $M_0$ in equation $(i)$

$$M = \left(\frac{PR}{\pi} - \frac{\pi}{2}\right) + \frac{PR}{2}(1 - \sin\theta)$$

$$= PR\left(\frac{1}{\pi} - \frac{\sin\theta}{2}\right) \qquad ...(iii)$$

We know that at $x$, $\theta = \frac{\pi}{2}$ and $\sin\theta = 1$

$$\therefore \quad M_x = PR\left(\frac{1}{\pi} - \frac{1}{2}\right) = -0.182\,PR$$

Similarly at $y$, $\theta = 0$ and $\sin\theta = 0$

$$\therefore \quad M_y = PR\left(\frac{1}{\pi} - 0\right) = \frac{PR}{\pi} = 0.318\,PR$$

For moment to be equal to zero, equating the equation (*iii*) to zero,

$$PR\left(\frac{1}{\pi} - \frac{\sin\theta}{2}\right) = 0$$

$$\therefore \quad \sin\theta = \frac{2}{\pi}$$

or $\theta = 32.5°$

Now the stress at any point, will be given by the algebraic sum of the direct stress (due to pull $P$) and the bending stress.

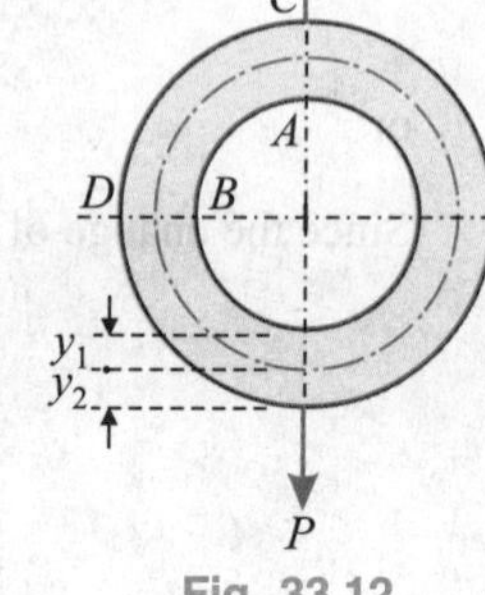

Fig. 33.12

Therefore general expression for the stress at point $Q$ is :

$$\sigma = \sigma_0 + \sigma$$

$$= \frac{P\sin\theta}{2A} + \frac{M}{AR}\left(1 + \frac{R^2}{h^2} \times \frac{y}{R+y}\right)$$

where $M = PR\left(\frac{1}{\pi} - \frac{\sin\theta}{2}\right)$

and $Y$ = Distance between the centre line of the ring and the extreme fibre of the ring.

Now the critical stresses will occur at the points $A$, $B$, $C$ and $D$ of the ring.

***Stress at A***

Substituting the value of $M$, $\theta = 0$ and $y = -y_1$ in the general expression for stress,

$$\sigma_A = 0 + \frac{PR}{\pi AR}\left(1 - \frac{R^2}{h^2} \times \frac{y_1}{R - y_1}\right)$$

$$= \frac{P}{\pi A}\left(1 - \frac{R^2}{h^2} \times \frac{y_1}{R - y_1}\right)$$

Since the value of $\frac{R^2}{h^2} \times \frac{y_1}{R - y_1}$ is always greater than unity, therefore the above expression works out to be *negative*. It means that the net stress at $A$ is compressive.

***Stress at B***

Now substituting the value of $M$, $\theta = \frac{\pi}{2}$ and $y = -y_1$ in the general expression for stress,

$$\sigma_B = \frac{P}{2A} + \frac{PR\left(\frac{1}{\pi} - \frac{1}{2}\right)}{AR}\left(1 - \frac{R^2}{h^2} \times \frac{y_1}{R - y_1}\right)$$

$$= \frac{P}{2A} - \frac{0.182\,P}{A}\left(1 - \frac{R^2}{h^2} \times \frac{y_1}{R - y_1}\right)$$

Since the value of $\frac{R^2}{h^2} \times \frac{y_1}{R - y_1}$ is always greater than unity therefore the above expression works out to be *positive*. It means that the net stress at $B$ is tensile.

***Stress at C***

Now substituting the value of $M$, $\theta = 0$ and $y = +y_2$ in the general expression for stress,

$$\sigma_C = 0 + \frac{PR}{\pi .\, AR}\left(1 + \frac{R^2}{h^2} \times \frac{y_2}{R + y_2}\right)$$

$$= \frac{P}{\pi A}\left(1 + \frac{R^2}{h^2} \times \frac{y_2}{R + y_2}\right)$$

A little consideration will show that the expression works out to be positive. It means that the stress at $C$ is tensile.

***Stress at D***

Now substituting the value of $M$, $\theta = \frac{\pi}{2}$ and $y = +y_2$ in the general expression for stress,

$$\sigma_D = \frac{P}{2A} + \frac{PR\left(\frac{1}{\pi} - \frac{1}{2}\right)}{AR}\left(1 + \frac{R^2}{h^2} \times \frac{y_2}{R + y_2}\right)$$

$$= \frac{P}{2A} - \frac{0.182\,P}{A}\left(1 + \frac{R^2}{h^2} \times \frac{y_2}{R + y_2}\right)$$

A little consideration will show that the above expression works out to be negative. It means that the stress at $D$ is compressive.

**EXAMPLE 33.5.** *A close circular ring made up of 20 mm diameter steel bar is subjected to a pull of 10 kN, whose line of action passes through the centre of the ring. Find the maximum value of tensile and compressive stresses in the ring, if the mean diameter of the ring is 160 mm.*

**SOLUTION.**

Given : Diameter of steel bar ($d$) = 20 mm ; Pull ($P$) = 10 kN = $10^4$ N and diameter of the ring ($D$) = 160 mm or radius of ring ($R$) = 80 mm.

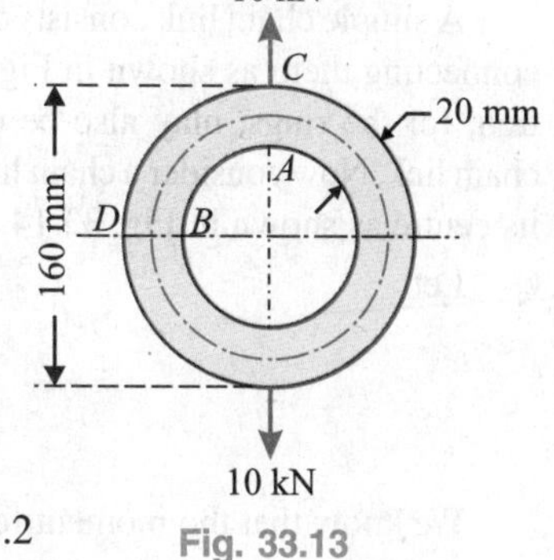

Fig. 33.13

We know that area, $A = \frac{\pi}{4} \times (20)^2 = 100\,\pi \text{ mm}^2$

and distance between centre line of the ring and extreme fibre,

$$y = y_1 = y_2 = 10 \text{ mm}$$

Link radius for the ring section,

$$h^2 = \frac{d^2}{16} + \frac{1}{8} \times \frac{d^4}{16\,R^2}$$

$$= \frac{(20)^2}{16} + \frac{1}{8} \times \frac{(20)^4}{16 \times (80)^2} = 25.2$$

***Stress at A,***

$$\sigma_A = \frac{P}{\pi A}\left(1 - \frac{R^2}{h^2} \times \frac{y_1}{R - y_1}\right)$$

$$= \frac{10^4}{\pi \times 100\pi}\left(1 - \frac{(80)^2}{25.2} \times \frac{10}{(80-10)}\right)$$

$$= -357.83 \text{ N/mm}^2$$

$$= 357.83 \text{ N/mm}^2 \text{ (compressive)} \quad \textbf{Ans.}$$

*Stress at B,*

$$\sigma_B = \frac{P}{2A} - \frac{0.182\,P}{A}\left(1 - \frac{R^2}{h^2} \times \frac{y_1}{R - y_1}\right)$$

$$= \frac{10^4}{2 \times 100\pi} - \frac{0.182 \times 10^4}{100\pi}\left(1 - \frac{(80)^2}{25.2} \times \frac{10}{(80-10)}\right)$$

$$= 220.3 \text{ N/mm}^2 \text{ (tensile)} \quad \textbf{Ans.}$$

*Stress at C*

$$\sigma_C = \frac{P}{\pi A}\left(1 + \frac{R^2}{h^2} \times \frac{y_2}{R + y_2}\right)$$

$$= \frac{10^4}{\pi \times 100\pi}\left(1 + \frac{(80)^2}{25.2} \times \frac{10}{(80+10)}\right)$$

$$= 296 \text{ N/mm}^2 \text{ (tensile)} \quad \textbf{Ans.}$$

*Stress at D*

$$\sigma_D = \frac{P}{2A} - \frac{0.182\,P}{A}\left(1 + \frac{R^2}{h^2} \times \frac{y_2}{R + y_2}\right)$$

$$= \frac{10^4}{2 \times 100\pi} - \frac{0.182 \times 10^4}{100\pi}\left(1 + \frac{(80)^2}{25.2} \times \frac{10}{(80+10)}\right)$$

$$= -153.4 \text{ N/mm}^2$$

$$= 153.4 \text{ N/mm}^2 \text{ (compressive)} \quad \textbf{Ans.}$$

## 33.12. Chain Links

A simple chain link consists of semi-circular ends and straight sides connecting them as shown in Fig. 33.14. The theory of stress distribution, for the rings, may also be extended to determine the stress in a chain link. Now consider a chain link subjected to a pull (or push) through its centre as shown in Fig. 33.14.

Let $P$ = Pull on the link,

$R$ = Radius of curvature of the ring,

$A$ = Cross-sectional area of the ring,

$l$ = Length of the straight portion.

We know that the moment for the circular portion,

$$M = M_0 + \frac{PR}{2}(1 - \sin\theta)$$

and for straight portion,

$$M = M_0 + \frac{PR}{2}$$

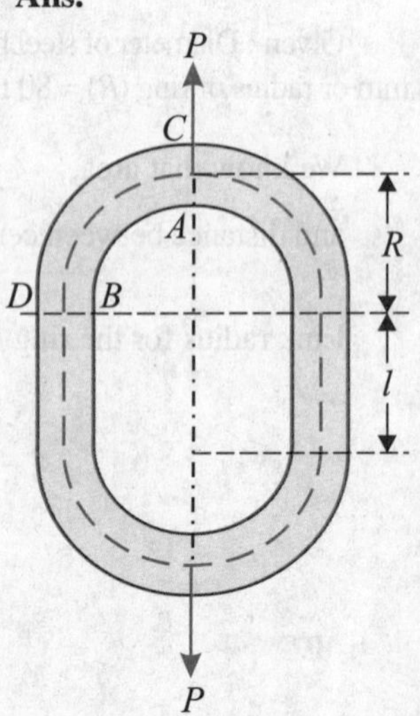

Fig. 33.14

where $$M_0 = \frac{PR}{2}\left(\frac{l+2R}{l+\pi R}\right)$$

Thus the stress at $A$, $$\sigma_A = \frac{P}{2A}\left(\frac{l+2R}{l+\pi R}\right)\left(1-\frac{R^2}{h^2}\times\frac{y_1}{R-y_1}\right)$$

Similarly, $$\sigma_B = \frac{P}{2A}-\frac{PR}{2A}\left(\frac{\pi-2}{l+\pi R}\right)\left(1-\frac{R^2}{h^2}\times\frac{y_1}{R-y_1}\right)$$

$$\sigma_C = \frac{P}{2A}\left(\frac{l+2R}{l+\pi R}\right)\left(1+\frac{R^2}{h^2}\times\frac{y_2}{R-y_2}\right)$$

and $$\sigma_D = \frac{P}{2A}-\frac{PR}{2A}\left(\frac{\pi-2}{l+\pi R}\right)\left(1+\frac{R^2}{h^2}\times\frac{y_2}{R+y}\right)$$

**EXAMPLE 33.6.** *A chain link is made of 20 mm diameter round steel with mean radius of circular ends 25 mm, the length of straight portion being 20 mm. Determine the values of maximum tensile and compressive stresses, when the link is subjected to a pull of 20 kN at its ends.*

**SOLUTION.** Given : Diameter of steel bar $(d)$ = 20 mm ; Radius of link $(R)$ = 25 mm ; Length of straight portion $(l)$ = 20 mm and pull $(P)$ = 20 kN = $2 \times 10^4$ N

We know that area, $$A = \frac{\pi}{4}\times(20)^2 = 100\pi \text{ mm}^2$$

and distance between centre line of the link and the extreme fibre,

$$y = y_1 = y_2 = 10 \text{ mm}$$

Link radius for the link secion,

$$h^2 = \frac{d^2}{16}+\frac{1}{8}\times\frac{d^4}{16R^2}$$

$$= \frac{(20)^2}{16}+\frac{1}{8}\times\frac{(20)^4}{16\times(25)^2} = 27$$

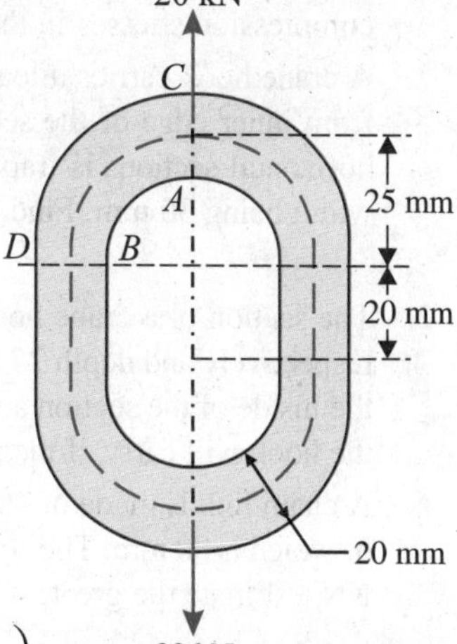

Fig. 33.15

**Stress at *A***

$$\sigma_A = \frac{P}{2A}\left(\frac{l+2R}{l+\pi R}\right)\left(1-\frac{R^2}{h^2}\times\frac{y_1}{R-y_1}\right)$$

$$= \frac{2\times10^4}{2\times100\pi}\left(\frac{20+2\times25}{20+\pi\times25}\right)\left(1-\frac{(25)^2}{27}\times\frac{10}{(25-10)}\right)$$

$$= -\,326.2 \text{ N/mm}^2 = 326.2 \text{ N/mm}^2 \text{ (compressive)}$$

**Stress at *B***

$$\sigma_B = \frac{P}{2A}-\frac{PR}{2A}\left(\frac{\pi-2}{l+\pi R}\right)\left(1-\frac{R^2}{h^2}\times\frac{y_1}{R-y_1}\right)$$

$$= \frac{2\times10^4}{2\times100\pi}-\frac{2\times10^4\times25}{2\times100\pi}\left(\frac{\pi-2}{20+\pi\times25}\right)\left(1-\frac{(25)^2}{27}\times\frac{10}{25-10}\right)$$

$$= 164.8 \text{ N/mm}^2 \text{ (tensile)}$$

**Stress at *C***

$$\sigma_C = \frac{P}{2A}\left(\frac{1+2R}{l+\pi R}\right)\left(1+\frac{R^2}{h^2}\times\frac{y_2}{R+y_2}\right)$$

$$= \frac{2\times10^4}{2\times100\pi}\times\frac{20+2\times25}{20\times\pi\times25}\left(1+\frac{(25)^2}{27}\times\frac{10}{(25+10)}\right)$$

$$= 172.2 \text{ N/mm}^2 \text{ (tensile)}$$

**Stress at *D***

$$\sigma_D = \frac{P}{2A}-\frac{PR}{2A}\left(\frac{\pi-2}{l+\pi R}\right)\left(1+\frac{R^2}{h^2}\times\frac{y_2}{R+y_2}\right)$$

$$= \frac{2\times10^4}{2\times100\pi}-\frac{2\times10^4\times25}{2\times100\pi}\left(\frac{\pi-2}{20+\pi\times25}\right)\left(1+\frac{(25)^2}{27}\times\frac{10}{(25+10)}\right)$$

$$= -38.3 \text{ N/mm}^2 = 38.3 \text{ N/mm}^2 \text{ (compressive)}$$

Thus the maximum tensile stress will occur at *C* equal to 172.2 N/mm$^2$ and maximum compressive stress will occur at *A* equal to 326.3 N/mm$^2$. **Ans.**

## EXERCISE 33.1

1. A beam of rectangular section 30 mm × 40 mm has its central line curved to a radius of 60 mm. The beam is subjected to a bending moment of 120 × 10$^3$ N-mm. Find the greatest tension and compression stresses in the beam. [**Ans.** 13 N/mm$^2$; – 21 N/mm$^2$]
2. A crane hook carries a load of 45 kN, the line of load being at a horizontal distance of 40 mm from inner edge of the section, and the centre of curvature coincides with the load line. The horizontal section is trapezium with 50 mm depth, inner width being 60 mm and the outer width being 30 mm. Find the greatest tensile and compressive stresses in the hook. [**Ans.** 152.8 N/mm$^2$; – 71.6 N/mm$^2$]
3. The section of a crane hook is trapezium, whose inner and outer sides are 20 mm and 10 mm respectively and depth 25 mm. The centre of curvature of the section is at a distance of 25 from the inside of the section and the load line is 20 mm from the same point. Find the greatest load, the hook will carry, if the maximum stress is not to exceed 70 N/mm$^2$. [**Ans.** 3 kN]
4. A chain link is made of 40 mm round steel and is semi-circular at each end, the mean diameter of which is 80 mm. The straight sides of the link are also 80 mm. If the link carries a load of 100 KN, estimate the greatest tensile and compressive stresses in the link. [**Ans.** 45.5 N/mm$^2$; 93.5 N/mm$^2$]

## QUESTIONS

1. Give the assumptions for determining the stresses in the bending of curved bars.
2. Derive an expression for the bending stress on the extreme fibres of a bar (*i*) having a small initial curvature, and (*ii*) having a large initial curvature.
3. What is link radius? Obtain the values of link radius for a (*a*) triangular section and (*b*) trapezoidal section.
4. How will you find out the values of maximum tensile and compressive stresses in a crane hook?
5. Obtain from fundamentals the relation for the maximum compressive and tensile stresses in ring.

Chapter 34

# Columns and Struts

## Contents

## 34.1. Introduction

A structural member, subjected to an axial compressive force, is called a strut. As per definition, a strut may be horizontal, inclined or even vertical. But a vertical strut, used in buildings or frames, is called a *column*.

## 34.2. Failure of a Column or Strut

It has been observed, that when a column or a strut is subjected to some compressive force, then the compressive stress induced,

$$\sigma = \frac{P}{A}$$

where $P$ = Compressive force and

$A$ = Cross-sectional area of the column.

A little consideration will show, that if the force or load is gradually increased the column will reach a stage, when it will be subjected to the ultimate crushing stress. Beyond this stage, the column will fail by crushing. The load corresponding to the crushing stress, is called *crushing load.*

It has also been experienced that sometimes, a compression member does not fail entirely by crushing, but also by bending *i.e.*, buckling. This happens in the case of long columns. It has also been observed that all the short columns fail due to their crushing. But, if a long column is subjected to a compressive load, it is subjected to a compressive stress. If the load is gradually increased, the column will reach a stage, when it will start buckling. The load, at which the column just buckles is called *buckling load, criticial load or crippling load* and the column is said to have developed an elastic instability. A little consideration will show that for a long column, the value of buckling load will be less than the crushing load. Moreover, the value of buckling load is low for long columns and relatively high for short columns.

## 34.3. Euler's Column Theory

The first rational attempt, to study the stability of *long columns, was made by Mr. Euler. He derived an equation, for the buckling load of long columns based on the bending stress. While deriving this equation, the effect of direct stress is neglected. This may be justified with the statement that the direct stress induced in a long column is negligible as compared to the bending stress. It may be noted that the Euler's formula cannot be used in the case of short columns, because the direct stress is considerable and hence cannot be neglected.

## 34.4. Assumptions in the Euler's Column Theory

The following simplifying assumptions are made in the Euler's column theory:

1. Initially the column is perfectly straight and the load applied is truly axial.
2. The cross-section of the column is uniform throughout its length.
3. The column material is perfectly elastic, homogeneous and isotropic and thus obeys Hooke's law.
4. The length of column is very large as compared to its cross-sectional dimensions.
5. The shortening of column, due to direct compression (being very small) is neglected.
6. The failure of column occurs due to buckling alone.

## 34.5. Sign Conventions

Though there are different signs used for the bending of columns in different books, yet we shall follow the following sign conventions which are commonly used and internationally recognised.

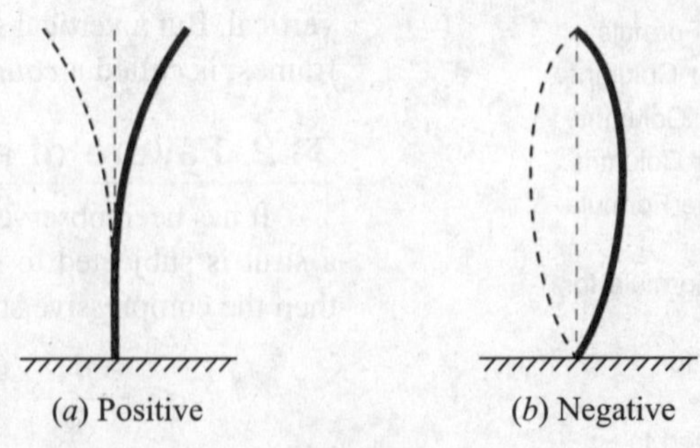

(*a*) Positive (*b*) Negative

Fig. 34.1

* As a matter of fact, mere length is not the only criterion for a column to be called long or short. But it has an important relation with the lateral dimensions of the column.

1. A moment, which tends to bend the column with *convexity* towards its initial central line as shown in Fig. 34.1 (*a*) is taken as *positive*.
2. A moment, which tends to bend the column with its *concavity* towards its initial central line as shown in Fig. 34.1 (*b*) is taken as *negative*.

## 34.6. Types of End Conditions of Columns

In actual practice, there are a number of end conditions, for columns. But, we shall study the Euler's column theory on the following four types of end conditions, which are important from the subject point of view:

1. Both ends hinged,
2. Both ends fixed,
3. One end is fixed and the other hinged, and
4. One end is fixed and the other free.

Now we shall discuss the value of critical load for all the above mentioned type of and conditions of columns one by one.

## 34.7. Columns with Both Ends Hinged

Consider a column *AB* of length *l* hinged at both of its ends *A* and *B* and carrying a critical load at *B*. As a result of loading, let the column deflect into a curved form $AX_1B$ as shown in Fig. 34.2.

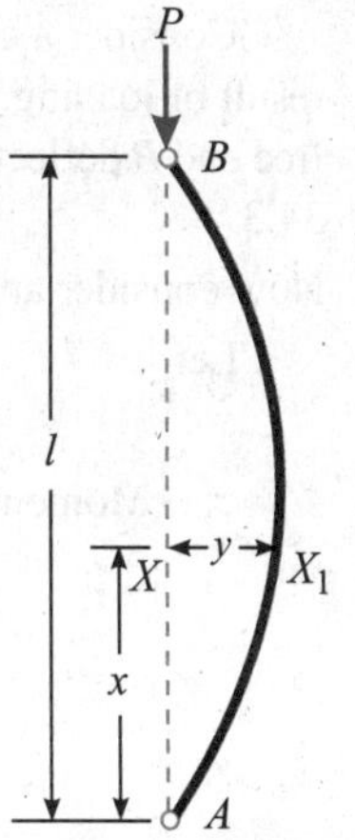

Now consider any section *X*, at a distance *x* from *A*.

Let $P$ = Critical load on the column,

$y$ = Deflection of the column at *X*.

∴ Moment due to the critical load *P*,

$$M = -P \cdot y$$

$$\therefore \quad EI\frac{d^2y}{dx^2} = -P \cdot y \quad \text{... (Minus sign due to concavity towards initial centre line)}$$

$$\therefore \quad EI\frac{d^2y}{dx^2} + P \cdot y = 0$$

or

$$\frac{d^2y}{dx^2} + \frac{P}{EI} \cdot y = 0$$

Fig. 34.2

The general solution of the above differential equation is

$$y = A \cdot \cos\left(x\sqrt{\frac{P}{EI}}\right) + B\sin\left(x\sqrt{\frac{P}{EI}}\right)$$

where *A* and *B* are the constants of integration. We know that when $x = 0$, $y = 0$. Therefore $A = 0$. Similarly when $x = l$, then $y = 0$. Therefore

$$0 = B\sin\left(l\sqrt{\frac{P}{EI}}\right)$$

A little consideration will show that either $B$ is equal to zero or $\sin\left(l\sqrt{\frac{P}{EI}}\right)$ is equal to zero. Now if we consider $B$ to be equal to zero, then it indicates that the column has not bent at all. But if

$$\sin\left(l\sqrt{\frac{P}{EI}}\right) = 0$$

$$\therefore \quad l\sqrt{\frac{P}{EI}} = 0 = \pi = 2\pi = 3\pi = .......$$

Now taking the least significant value,

$$l\sqrt{\frac{P}{EI}} = \pi$$

or

$$P = \frac{\pi^2 EI}{l^2}$$

## 34.8. Columns with One End Fixed and the Other Free

Consider a column $AB$ of length $l$ fixed at $A$ and free at $B$ and carrying a critical load at $B$. As a result of loading, let the beam deflect into a curved form $AX_1B_1$ such that the free end $B$ deflects through $a$ and occupies a new position $B_1$ as shown in Fig. 34.3.

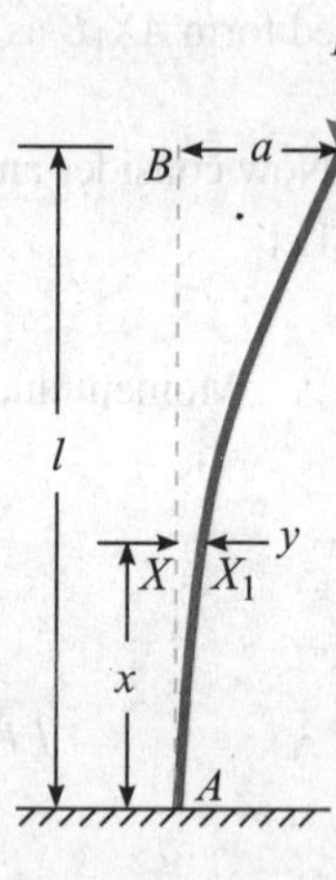

Now consider any section $X$ at a distance $x$ from $A$.

Let $P$ = Critical load on the column and

$y$ = Deflection of the column at $X$.

$\therefore$ Moment due to the critical load $P$,

$$M = +P(a-y)$$

$$= P\,a - P \cdot y \quad \text{... (Plus sign due to convexity towards initial centre line)}$$

$$\therefore \quad EI\frac{d^2y}{dx^2} = P \cdot a - P \cdot y$$

or

$$\frac{d^2y}{dx^2} + \frac{P}{EI} \cdot y = \frac{P \cdot a}{EI}$$


Fig. 34.3

The general solution of the above differential equation is,

$$y = A\cos\left(x\sqrt{\frac{P}{EI}}\right) + B\sin\left(x\sqrt{\frac{P}{EI}}\right) + a \quad ...(i)$$

where $A$ and $B$ are the constants of integration. We know that when $x = 0$, then $y = 0$, therefore $A = -a$. Now differentiating the above equation,

$$\frac{dy}{dx} = -A\sqrt{\frac{P}{EI}}\sin\left(x\sqrt{\frac{P}{EI}}\right) + B\sqrt{\frac{P}{EI}}\cos\left(x\sqrt{\frac{P}{EI}}\right)$$

We also know that when $x = 0$, then $\frac{dy}{dx} = 0$. Therefore

$$0 = B\sqrt{\frac{P}{EI}}$$

A little consideration will show that either $B$ is equal to zero of $\sqrt{\frac{P}{EI}}$ is equal to zero. Since the load $P$ is not equal to zero, it is thus opbvious that $B$ is equal to zero. Now substituting the values $A = -a$ and $B = 0$ is equation (*i*),

$$y = -a\cos\left(x\sqrt{\frac{P}{EI}}\right) + a = a\left[1 - \cos\left(x\sqrt{\frac{P}{EI}}\right)\right]$$

We also know that when $x = l$, then $y = a$. Therefore

$$a = a\left[1 - \cos\left(l\sqrt{\frac{P}{EI}}\right)\right]$$

$$\therefore \quad \cos\left(l\sqrt{\frac{P}{EI}}\right) = 0$$

or

$$l\sqrt{\frac{P}{EI}} = \frac{\pi}{2} = \frac{3\pi}{2} = \frac{5\pi}{2}$$

Now taking the least significant value,

$$l\sqrt{\frac{P}{EI}} = \frac{\pi}{2}$$

$$\therefore \quad P = \frac{\pi^2 EI}{4l^2}$$

## 34.9. Columns with Both Ends Fixed

Consider a column $AB$ of length $l$ fixed at both of its ends $A$ and $B$ and carrying a critical load at $B$. As a result of loading, let the column deflect as shown in Fig. 34.4

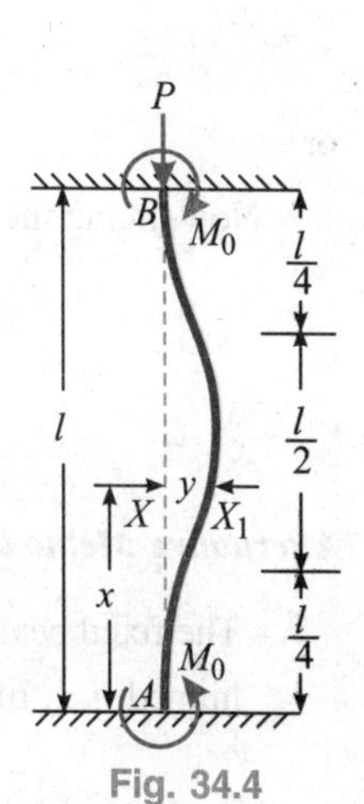


Fig. 34.4

Now consider any section $X$ at a distance $x$ from $A$.

Let $\quad P$ = Critical load on the column and

$\quad y$ = Deflection of the column at $X$.

A little consideration will show that since both the ends of the beam $AB$ are fixed and it is carrying a load, therefore there will be some fixed end moments at $A$ and $B$.

Let $\quad M_0$ = Fixed end moments at $A$ and $B$.

$\therefore$ Moment due to the critical load $P$,

$$M = -P \cdot y$$

$$EI\frac{d^2y}{dx^2} = M_0 - P \cdot y$$

...(Minus sign due to concavity initial centre line)

$$\therefore \quad \frac{d^2y}{dx^2} + \frac{P}{EI} \cdot y = \frac{M_0}{EI}$$

The general solution of the above differential equation is:

$$y = A\cos\left(x\sqrt{\frac{P}{EI}}\right) + B\sin\left(x\sqrt{\frac{P}{EI}}\right) + \frac{M_0}{P} \qquad ...(i)$$

where $A$ and $B$ are the constants of integration. We know that when $x = 0$, then $y = 0$. Therefore $A = -\frac{M_0}{P}$. Now differentiating the above equation,

$$\frac{dy}{dx} = -A\sqrt{\frac{P}{EI}}\sin\left(x\sqrt{\frac{P}{EI}}\right) + B\sqrt{\frac{P}{EI}}\cos\left(x\sqrt{\frac{P}{EI}}\right)$$

We also know that when $x = 0$, then $\frac{dy}{dx} = 0$. Therefore

$$0 = B\sqrt{\frac{P}{EI}}$$

A little consideration will show, that either $B$ is equal to zero, or $\sqrt{\frac{P}{EI}}$ is equal to zero. Since the load $P$ is not equal to zero, it is thus obvious that $B$ is equal to zero. Substituting the values $A = \frac{M_0}{P}$ and $B = 0$ in equation (*i*),

$$y = -\frac{M_0}{P}\cos\left(x\sqrt{\frac{P}{EI}}\right) + \frac{M_0}{P} = \frac{M_0}{P}\left[1 - \cos\left(l\sqrt{\frac{P}{EI}}\right)\right]$$

We also know that when $x = l$, then $y = 0$. Therefore

$$0 = \frac{M_0}{P}\left[1 - \cos\left(l\sqrt{\frac{P}{EI}}\right)\right]$$

$$\therefore \quad \cos\left(l\sqrt{\frac{P}{EI}}\right) = 1$$

or $$l\sqrt{\frac{P}{EI}} = 0 = 2\pi = 4\pi = 6\pi = \ldots\ldots$$

Now taking the least significant value,

$$l\sqrt{\frac{P}{EI}} = 2\pi$$

$$\therefore \quad P = \frac{4\pi^2 EI}{l^2}$$

***Alternative Methods***

1. The fixed beam $AB$ may bne considered as equivalent to a column of length $\frac{l}{2}$ with both ends hinged (*i.e.*, middle portion of the column as shown in Fig. 34.4).

$$\therefore \text{ Critical load,} \quad P = \frac{\pi^2 EI}{\left(\frac{l}{2}\right)^2} = \frac{4\pi^2 EI}{l^2}$$

2. The fixed beam $AB$ may also be considered as equivalent to a column of length $\frac{l}{4}$ with one end fixed and the other free (*i.e.*, lower one-fourth portion of the beam as shown in Fig. 34.4).

$$\therefore \text{ Critical load,} \quad P = \frac{\pi^2 EI}{4\left(\frac{l}{4}\right)^2} = \frac{4\pi^2 EI}{l^2}$$

## 34.10. Columns with One End Fixed and the Other Hinged

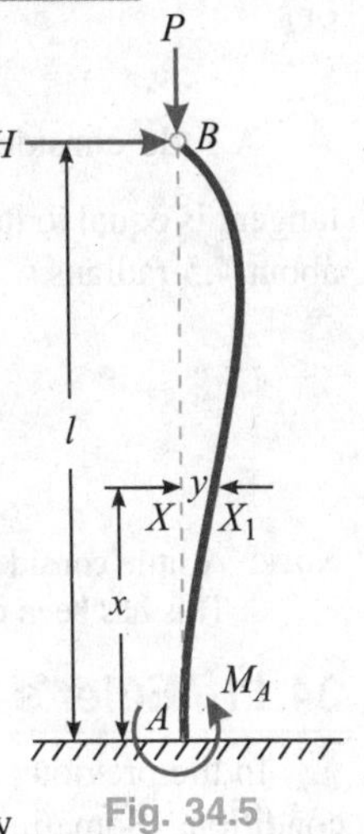

Fig. 34.5

Consider a column $AB$ of length $l$ fixed at $A$ and hinged at $B$ and carrying a critical load at $B$. As a result of loading, let the column deflect as shown in Fig. 34.5.

Now consider any section $X$ at a distance $x$ from $A$.

Let $P$ = Critical load on the column, and

$y$ = Deflection of the beam at $X$,

A little consideration will show, that since the beam $AB$ is fixed at $A$ and it is carrying a load, therefore, there will be some fixed end moment at $A$. In order to balance the fixing moment at $A$, there will be a horizontal reaction at $B$.

Let $M_A$ = Fixed end mement at $A$ and

$H$ = Horizontal reaction at $B$.

∴ Moment due to critical load $P$,

$$M = -P \cdot y \quad \text{...(Minus sign due to conscavity towards initial centre line}$$

or

$$EI \frac{d^2y}{dx^2} = H(l-x) - P \cdot y$$

$$\therefore \quad \frac{d^2y}{dx^2} + \frac{P}{EI} \cdot y = \frac{H(l-x)}{EI}$$

The general solution of the above differential equation is

$$A = y \cos\left(x\sqrt{\frac{P}{EI}}\right) + B \sin\left(x\sqrt{\frac{P}{EI}}\right) + \frac{H(l-x)}{P} \quad \text{...}(i)$$

where $A$ and $B$ are the constants of integration. We know that when $x = 0$, they $y = 0$. Therefore $A = \frac{Hl}{P}$.

Now differentiating the above equation,

$$\frac{dy}{dx} = -A\sqrt{\frac{P}{EI}} \sin\left(x\sqrt{\frac{P}{EI}}\right) + B\sqrt{\frac{P}{EI}} \cos\left(x\sqrt{\frac{P}{EI}}\right) - \frac{H}{P}$$

We know that when $x = 0$, $\frac{dy}{dx} = 0$. Therefore

$$0 = B\sqrt{\frac{P}{EI}} - \frac{H}{P}$$

$$\therefore \quad B = \frac{P}{H} \times \sqrt{\frac{EI}{P}}$$

We also know that when $x = l$, then $y = 0$. Therefore substituting these values of $x$, $A$ and $B$ is equation $(i)$,

$$0 = \frac{Hl}{P} \cos\left(l\sqrt{\frac{P}{EI}}\right) + \frac{H}{P}\sqrt{\frac{EI}{P}} \sin\left(l\sqrt{\frac{P}{EI}}\right)$$

$$\therefore \quad \frac{H}{P}\sqrt{\frac{EI}{P}} \sin\left(l\sqrt{\frac{P}{EI}}\right) = \frac{Hl}{P} \cos\left(l\sqrt{\frac{P}{EI}}\right)$$

or $$\tan\left(l\sqrt{\frac{P}{EI}}\right) = \left(l\sqrt{\frac{P}{EI}}\right)$$

A little consideration will show that the value of $\left(l\sqrt{\frac{P}{EI}}\right)$ in radians, has to be such that its tangent is equal to itself. We know that the only angle, the value of whose tangent is equal to itself, is about 4.5 radians.

$$\therefore \quad l\sqrt{\frac{P}{EI}} = 4.5 \quad \text{or} \quad l^2 \times \frac{P}{EI} = 20.25 \quad \text{or} \quad P = \frac{20.25\ EI}{l^2}$$

$$\therefore \quad P = \frac{2\pi^2\ EI}{l^2}$$

**NOTE:** A little consideration will show that 20.25 is not exactly equal to $2\pi^2$, but approximately equal to $2\pi^2$. This has been done to rationalise the value of $P$, *i.e.*, crippling load in various cases.

## 34.11. Euler's Formula and Equivalent length of a Column

In the previous articles, we have derived the relations for the crippliing load under various end conditions. Sometimes, all these cases are represented by a general equation called Euler's formula,

$$P_E = \frac{\pi^2\ EI}{L_e^{\ 2}}$$

where $L_e$ is the equivalent or effective length of column.

The is another way of representing the equation, for the crippling load by an equivalent length of effective length of a column. The equivalent length of a given column with given end conditions, is the length of an equivalent column of the same material and cross-section with both ends hinged and having the value of the crippling load equal to that of the given column.

The equivalent lengths ($L$) for the given end conditions are given below:

**Table 34.1**

| *S.No.* | *End conditions* | *Relation between equivalent length ($L_e$) and actual length ($l$)* | *Crippling load (P)* |
|---|---|---|---|
| **1.** | Both ends hinged | $L_e = l$ | $P = \frac{\pi^2\ EI}{(l)^2} = \frac{\pi^2\ EI}{l^2}$ |
| **2.** | One end fixed and the other free | $L_e = 2\,l$ | $P = \frac{\pi^2\ EI}{(2l)^2} = \frac{\pi^2\ EI}{4l^2}$ |
| **3.** | Both ends fixed | $L_e = \frac{l}{2}$ | $P = \frac{\pi^2\ EI}{\left(\frac{l}{2}\right)^2} = \frac{4\pi^2\ EI}{l^2}$ |
| **4.** | One end fixed and the other hinged | $L_e = \frac{l}{\sqrt{2}}$ | $P = \frac{\pi^2\ EI}{\left(\frac{l}{\sqrt{2}}\right)^2} = \frac{2\pi^2\ EI}{l^2}$ |

**NOTE.** The vertical column will have two moments of inertia (*viz.*, $I_{XX}$ and $L_{YY}$). Since the column will tend to buckle in the direction of leas moment of inertia, therefore the least value of the two moments of inertia is to be used in tlhe relation.

## 34.12. Slenderness Ratio

We have already discussed in Art. 34.11 that the Euler's formula for the crippling load,

$$P_E = \frac{\pi^2 EI}{L_e^2} \quad ...(i)$$

We know that the buckling of a column under the crippling load will take place about the axis of least resistance. Now substituting $I = Ak^2$ (where $A$ is the area and $k$ is the least radius of gyration of the section) in the above equation,

$$P_E = \frac{\pi^2 E\left(Ak^2\right)}{L_e^2} = \frac{\pi^2 EA}{\left(\frac{L_e}{k}\right)^2} \quad ...(ii)$$

where $\frac{L_e}{k}$ is known as slenderness ratio. Thus slenderness ratio is defined as ratio of equivalent (or unsupported) length of column to the least radius of gyration of the section.

Slenderness ratio does not have any units.

**Note.** It may be noted that the formula for crippling load, in the pervious articles, have been derived on the assumption the the slenderness ratio $\frac{L_e}{k}$ is so large, that the failure of the column occurs only due to bending, the effect of direct stress (*i.e.*, $\frac{P}{A}$) being negligible.

## 34.13. Limitation of Euler's Formula

We have discussed in Art. 32.12 that the Euler's formula for the crippling load,

$$P_E = \frac{\pi^2 EA}{\left(\frac{L_e}{k}\right)^2}$$

∴ Euler's crippling stress,

$$\sigma_E = \frac{P}{A} = \frac{\pi^2 E}{\left(\frac{L_e}{k}\right)^2}$$

A little consideration will show that the crippling stress will be high, when the slenderness ratio is small. We know that the crippling stress for a column cannot be more than the crushing stress of the column material. It is thus obvious that the Euler's formula will give the value of crippling stress of the column (equal to the crushing stress of the column material) corresponding to the slenderness ratio. Now consider a mild steel column. We know that the crushing stress for the mild steel is 320 MPa or 320 N/m$^2$ and Young's modulus for the mild steel is 200 GPa or 200 × 10$^3$ N/mm$^2$.

Now equating the crippling stress to the crushing stress,

$$320 = \frac{\pi^2 E}{\left(\frac{L_e}{k}\right)^2} = \frac{\pi^2 \times (200 \times 10^3)}{\left(\frac{L_e}{k}\right)^2}$$

$$\therefore \quad \left(\frac{L_e}{k}\right)^2 = \frac{\pi^2 \times 200 \times 10^3}{320}$$

or $$\frac{L_e}{k} = 78.5 \text{ say } 80$$

Thus, if the slenderness ratio is less than 80 the Euler's formula for a mild steel column is not valid.

Sometimes, the columns, whose slenderness ratio is *more than* 80 are known as *long columns* and those whose slenderness ratio is *less than* 80 are known as short columns. It is thus obvious that the Euler's formula holds good only for long columns.

NOTE. In the Euler's formula, for crippling load, we have not taken into account the direct stresses induced in the material due to the load, (which increases gradually from zero to its crippling value). As a matter of fact, the combined stress, due to direct load and slight bending reaches its allowable value at a load, lower than that required for buckling ; and therefore this will be the limiting value of the safe load.

**EXAMPLE 34.1.** *A steel rod 5 m long and of 40 mm diameter is used as a column, with on end fixed and the other free. Determine the crippling load by Euler's formula. Take E as 200 GPa.*

**SOLUTION.** Given : Length ($l$) = 5 m = $5 \times 10^3$ mm ; Diameter of column ($d$) = 40 mm and modulus of elasticity ($E$) = 200 GPa = $200 \times 10^3$ N/mm$^2$.

We know that moment of iertia of the column section,

$$I = \frac{\pi}{64} \times (d)^4 = \frac{\pi}{64} \times (40)^4 = 40\,000\,\pi \text{ mm}^4$$

Since the column is fixed at one end and free at the other, therefore equivalent length of the column,

$$L_e = 2\,l = 2 \times (5 \times 10^3) = 10 \times 10^3 \text{ mm}$$

$$\therefore \text{ Euler's crippling load,} \quad P_E = \frac{\pi^2 EI}{L_e^{\,2}} = \frac{\pi^2 \times (200 \times 10^3) \times (40000\,\pi)}{(10 \times 10^3)^2} = 2480 \text{ N}$$

$$= 2.48 \text{ kN} \quad \textbf{Ans.}$$

**EXAMPLE 34.2.** *A hollow alloy tube 4 m long with external and internal diameters of 40 mm and 25 mm respectively was found to extend 4.8 mm under a tensile load of 60 kN. Find the buckling load for the tube with both ends pinned. Also find the safe load on the tube, taking a factor of safety as 5.*

**SOLUTION.** Given : Length $l$ = 4 m ; External diameter of column ($D$) = 40 mm ; Internal diameter of column ($d$) = 25 mm ; Deflection ($\delta l$) = 4.8 mm ; Tensile load = 60 kN = $60 \times 10^3$ N and factor of safety = 5.

***Buckling load for the tube***

We know that area of the tube,

$$A = \frac{\pi}{4} \times [D^2 - d^2] = \frac{\pi}{4}[(40)^2 - (25)^2] = 765.8 \text{ mm}^2$$

and moment of inertia of the tube,

$$I = \frac{\pi}{64}[D^4 = d^4] = \frac{\pi}{64}[(40)^4 - (25)^4] = 106\,500 \text{ mm}^4$$

We also know that strain in the alloy tube,

$$e = \frac{\delta l}{l} = \frac{4.8}{4 \times 10^3} = 0.0012$$

and modulus of elasticity for the alloy,

$$E = \frac{\text{Load}}{\text{Area} \times \text{Strain}} = \frac{60 \times 10^3}{765.8 \times 0.0012} = 65\,290 \text{ N/mm}^2$$

Since the column is pinned at its both ends, therefore equivalent length of the column,

$$L_e = l = 4 \times 10^3 \text{ mm}$$

$$\therefore \quad \text{Euler's buckling load, } P_E = \frac{\pi^2 EI}{L_e^2} = \frac{\pi^2 \times 65\,290 \times 106\,500}{(4 \times 10^3)^2} = 4290 \text{ N}$$

$$= 4.29 \text{ kN} \quad \textbf{Ans.}$$

***Safe load for the tube***

We also know that safe load for the tube

$$= \frac{\text{Buckling load}}{\text{Factor of safety}} = \frac{4.29}{5} = 0.858 \text{ kN} \quad \textbf{Ans.}$$

**EXAMPLE 34.3.** *Compare the ratio of the strength of a solid steel column to that of a hollow of the same cross-sectional area. The internal diameter of the hollow column is 3/4 of the external diameter. Both the columns have the same length and are pinned at both ends.*

**SOLUTION.** Give : Area of solid steel colum $A_S = A_H$ (where $A_H$ = Area of hollow column) ; Internal diameter of hollow column $(d) = 3\,D/4$ (where $D$ = External diameter) and length of solid column $(l_S) = l_H$ (where $l_H$ = Length of hollow column).

Let
$D_1$ = Diameter of the solid column,
$k_H$ = Radius of gyration for hollow column and
$k_S$ = Radius of gyration for solid column.

Since both the columns are pinned at their both ends, therefore equivalent length of the solid column,

$$L_S = l_S = L_H = l_H = L$$

We know that Euler's crippling load for the solid column,

$$P_S = \frac{\pi^2 EI}{L_H^2} = \frac{\pi^2 E \cdot A_S \cdot k_S^2}{L^2} \quad \text{...(i)}$$

Similarly Euler's crippling load for the hollow column

$$P_H = \frac{\pi^2 EI}{L_H^2} = \frac{\pi^2 E \cdot A_H \cdot k_H^2}{L^2} \quad \text{...(ii)}$$

Dividing equation (*ii*) by (*i*),

$$\frac{P_H}{P_S} = \left(\frac{k_H}{k_S}\right)^2 = \frac{\frac{D^2 + d^2}{16}}{\frac{D_1^2}{16}} = \frac{D^2 + d^2}{D_1^2} = \frac{D^2 = \left(\frac{3D}{4}\right)^2}{D_1^2}$$

$$= \frac{25\,D^2}{16\,D_1^2} \quad \text{...(iii)}$$

Since the cross-sectional areas of the both the columns is equal, therefore

$$\frac{\pi}{4} \times D_1^2 = \frac{\pi}{4}(D^2 - d^2) = \frac{\pi}{4}\left[D^2 - \left(\frac{3D}{4}\right)^2\right] = \frac{\pi}{4} \times \frac{7\,D^2}{16}$$

$$\therefore \qquad D_1^2 = \frac{7D^2}{16}$$

Now substituting the value of $D_1^2$ in equation (*iii*),

$$\frac{P_H}{P_S} = \frac{25D^2}{16 \times \frac{7D^2}{16}} = \frac{25}{7} \qquad \textbf{Ans.}$$

**EXAMPLE 34.4.** *An I section joist 400 mm × 200 mm × 20 mm and 6 m long is used as a strut with both ends fixed. What is Euler's cripp;ing load for the column? Take Young's modulus for the joist as 200 GPa.*

**SOLUTION.** Given : Outer depth ($D$) = 400 mm ; Outer width ($B$) = 200 mm ; Length ($l$) = 6 m = $6 \times 10^3$ mm and modulus of elasticity ($E$) = 200 GPa = $200 \times 10^3$ N/mm².

From the geometry of the figure, we find that inner depth,

$$d = 400 - (2 \times 20) = 360 \text{ mm}$$

and inner width,

$$b = 200 - 20 = 180 \text{ mm}$$

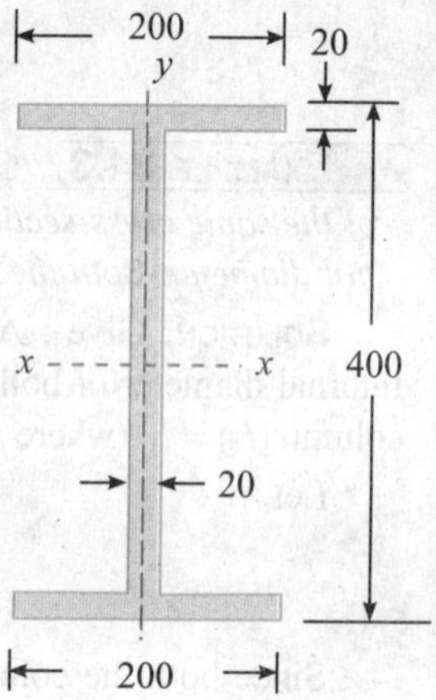

We know that moment of inertia of the joist section about *X-X* axis,

$$I_{XX} = \frac{1}{12}[BD^2 - ba^3]$$

$$= \frac{1}{12}[200 \times (400)^3 - 180 \times (360)^3] \text{ mm}^4$$

$$= 366.8 \times 10^6 \text{ mm}^4 \qquad ...(i)$$

Fig. 34.6

Similarly,

$$I_{YY} = \left[2 \times \frac{2 \times (200)^3}{12}\right] + \frac{360 \times (20)^3}{12} \text{ mm}^4$$

$$= 2.91 \times 10^6 \text{ mm}^4 \qquad ...(ii)$$

Since $I_{YY}$ is less than $I_{XX}$, therefore the joist will tend to buckle in *Y-Y* direction. Thus, we shall take the value of $I$ as $I_{YY} = 2.91 \times 10^6$ mm⁴. Moreover, as the column is fixed at its both ends, therefore equivalent length of the column,

$$L_e = \frac{l}{2} = \frac{(6 \times 10^3)}{2} = 3 \times 10^3 \text{ mm}$$

∴ Euler's crippling load for the column,

$$P_E = \frac{\pi^2 EI}{L_e^2} = \frac{\pi^2 \times (200 \times 10^3) \times (2.91 \times 10^6)}{(3 \times 10^3)^2} = 638.2 \times 10^3 \text{ N}$$

$$= 638.2 \text{ kN} \qquad \textbf{Ans.}$$

**EXAMPLE 34.5.** *A T-section 150 mm × 120 mm × 20 mm is used as a strut of 4 m long with hinged at its both ends. Calculate the crippling load, if Young's modulus for the material be 200 GPa.*

**SOLUTION.** Given : Size of *T*-section = 150 mm × 120 mm × 20 mm ; Length ($l$) = 4 m = $4 \times 10^3$ mm and Young's modulus ($E$) = 200 GPa = $200 \times 10^3$ N/mm².

First of all, let us find the centre of the *T*-section; Let bottom of the web be the axis of reference.

*Web*

$$a_1 = 100 \times 20 = 2000 \text{ mm}^2$$

$$y_1 = \frac{100}{2} = 50 \text{ mm}$$

*Flange*

$$a_2 = 150 \times 20 = 3000 \text{ mm}^2$$

$$y_2 = 120 - \left(\frac{20}{2}\right) = 110 \text{ mm}$$

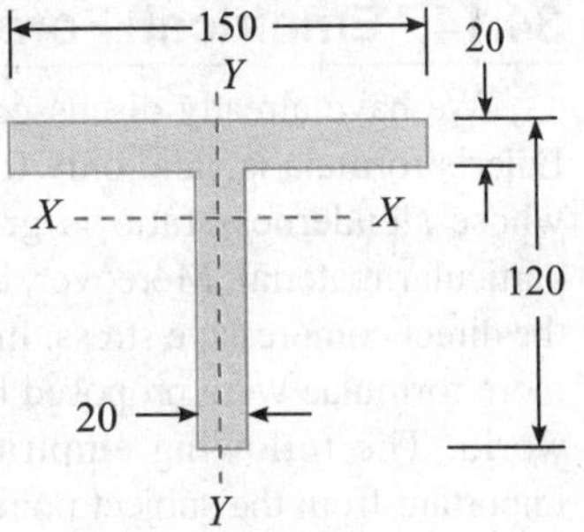

Fig. 34.7

We know that distance between the centre of gravity of the T-section and bottom of the web

$$\bar{y} = \frac{a_1 y_1 + a_2 y_2}{a_1 + a_2} = \frac{(2000 \times 50) + (3000 \times 110)}{200 + 3000} = 86 \text{ mm}$$

We also know that moment of inertia of the *T*-section about *X-X* axis,

$$I_{XX} = \left(\frac{20 \times (100)^3}{12} + 2000 \times (36)^2\right) + \left(\frac{150 \times (20)^3}{12} + 3000 \times (24)^2\right) \text{mm}^4$$

$$= (4.26 \times 10^6) + (1.83 \times 10^6) = 6.09 \times 10^6 \text{ mm}^4$$

Similarly,
$$I_{YY} = \frac{100 \times 9200)^3}{12} + \frac{20 \times (150)^3}{12} = 5.069 \times 10^6 \text{ mm}^4$$

Since $I_{YY}$ is less than $I_{XX}$, therefore the column will tend to buckle in *Y-Y* direction. Thus, we shall take the value of *I* as $I_{YY} = 5.69 \times 10^6 \text{ mm}^4$. Moreover, as the column is hinged at its both ends, therfore length of the column,

$$L_e = l = 4 \times 10^3 \text{ mm}$$

$\therefore$ Euler's crippling load,
$$P_E = \frac{\pi^2 EI}{L_e^2} = \frac{\pi^2 \times (200 \times 10^3) \times (5.69 \times 10^6)}{(4 \times 10^3)^2} = 702 \times 10^3 \text{ N}$$

$$= 702 \text{ kN} \quad \textbf{Ans.}$$

## EXERCISE 34.1

1. A mild steel column of 50 mm diameter is hinged at both of its ends. Find the crippling load for the column, if its length is 2.5 m. Take *E* for the column material as 200 GPa. [**Ans.** 96.9 kN]
2. A hollow cast iron column of 150 mm external diameter and 100 mm internal diameter is 3.5 m long. If one and of the column is rigidity fixed and the other is free, find the critical load on the column. Assume modulus of elasticity for the column material as 120 GPa. [**Ans.** 482 kN
3. A 1.75 m long steel column of rectangular cross-section 120 mm × 100 mm is rigidity fixed at one end and hinged at the other. Determine the buckling load on the column and the corresponding axial stress using Euler's formula. Take *E* for the column material as 200 GPa. [**Ans.** 12.84 MN ; 1070 MPa]
4. An -section 240 mm × 120 mm × 20 mm is used as 6 m long column with both ends fixed. What is the crippling load for the column? Take Young's modulus for the joist as 200 GPa. [**Ans.** 1292.5 kN]

## 34.14. Empirical Formulae for Columns

We have already discussed in the previous articles that the Euler's formula is valid only for long columns *i.e.*, for columns, whose slenderness ratio is greater than a certain value for a particular material. Moreover, it does not take into consideration the direct compressive stress. In order to fill up this lacuna, many more formulae were proposed by different scientists all over the world. The following empirical formulae, out of those, are important from the subject point of view.

1. Rankine's formula,
2. Johnson's formula, and
3. Indian Standard code.

## 34.15. Rankine's Formulae for Columns

We have already discussed that the Euler's formula gives correct results only for very long columns. Though this formula is applicable for columns, ranging from very long to short ones, yet it does not give reliable results. Prof. Rankine, after a number of experiments, gave the following empirical formula for columns.

$$\frac{1}{P_R} = \frac{1}{P_{CS}} + \frac{1}{P_E} \qquad ...(i)$$

where

$P_R$ = Crippling load by Rankine's Formula

$P_{CS} = \sigma_{CS} \cdot A$ = Ultimate crushing load for the column and

$P_E = \frac{\pi^2 EI}{L_e^2}$ = Crippling load obtained by Euler's formula.

A little consideration will show that the value of $P_{CS}$ will remain constant irrespective of the fact whether the column is a long one or short one. Now, we shall study the effect of $P_E$ in short as well as long columns one by one.

1. ***Short columns.*** In case of short columns, the value of $P_E$ will be very high, therefore the value of $\frac{1}{P_E}$ will be quite negligible as compared to $\frac{1}{P_{CS}}$. It is thus obvious that the Rankine's formula will give the value of its crippling load (*i.e.*, *P*) approximately equal to the ultimate crushing load (*i.e.*, ).
2. ***Long columns.*** In case of long columns, the value of $P_E$ will be very small, therefore the value of $\frac{1}{P_E}$ will be quite considerable as compared to $\frac{1}{P_{CS}}$. It is thus obvious that the Rankine's formula will give the value of its crippling load (*i.e.*, *P*) approximately equal to the crippling load by Euler's formula (*i.e.*, $P_E$). Thus, we see that the Rankine's formula gives a fairly correct result for all cases of columns, ranging from short to long columns.

From equation (*i*), we know that

$$\frac{1}{P_R} = \frac{1}{P_{CS}} + \frac{1}{P_E} = \frac{P_E + P_{CS}}{P_{CS} \cdot P_E}$$

$$\therefore \quad P_R = \frac{P_{CS} \cdot P_E}{P_{CS} + P_E} = \frac{P_{CS}}{1 + \frac{P_{CS}}{P_E}}$$

Now substituting the values of $P_{CS}$ and $P_E$ is the above equation

$$P_R = \frac{\sigma_{CS} \cdot A}{1 + \sigma_{CS} \cdot A \times \frac{L_e^2}{\pi^2 E}} = \frac{\sigma_{CS} \cdot A}{1 + \frac{\sigma_{CS}}{\pi^2 E} \times \frac{AL_e^2}{Ak^2}} \qquad ...(\because I = Ak^2)$$

or

$$P_R = \frac{\sigma_{CS} \cdot A}{1 + a\left(\frac{L_e}{k}\right)^2} = \frac{P_{CS}}{1 + a\left(\frac{L_e}{k}\right)^2}$$

where $P_{CS}$ = Crushing load of the column material

$\sigma_{CS}$ = Crushing stress of the column material,

$A$ = Cross-sectional area of the column,

$a$ = Rankine's constant $\left(\text{equal to } \frac{\sigma_C}{\pi^2 E}\right)$

$L_e$ = Equivalent length of the column, and

$k$ = Least radius of gyration.

The folowing table gives the values of crushing stress ($\sigma_C$) and Rankine's constant ($a$) for various materiabls:

**Table 34.2**

| *S.No.* | *Material* | $\sigma_C$ *in MPa* | $a = \frac{\sigma_C}{\pi^2 E}$ |
|---|---|---|---|
| 1. | Mild Steel | 320 | $\frac{1}{7500}$ |
| 2. | Cast Iron | 550 | $\frac{1}{1600}$ |
| 3. | Wrought Iron | 250 | $\frac{1}{9000}$ |
| 4. | Timber | 40 | $\frac{1}{750}$ |

**Note :** The above values are only for a column with both ends hinged. For other end conditions, the equivalent length should be used.

**EXAMPLE 34.6.** *Find the Euler's crippling load for a hollow cylindrical steel column of 38 mm external diameter and 2.5 mm thick. Take length ofthe column as 2.3 m and hinged at its both ends. Take E = 205 GPa.*

*Also determine crippling load by Rankine's formula using constants as 335 MPa and* $\frac{1}{7500}$

**SOLUTION.** Give : External diameter ($D$) = 38 mm ; Thickness = 2.5 mm or inner diameter ($d$) = 38 – (2 × 2.5) = 33 mm ; Length of the column ($l$) = 2.3 m = $2.3 \times 10^3$ mm ; Yield stress ($\sigma_C$) = 335 MPa = 335 N/mm$^2$ and Rankine's constant ($a$) = $\frac{1}{7500}$.

***Euler's crippling load***

We know that moment of inertia of the column section,

$$I = \frac{\pi}{64}(D^4 - d^4) = \frac{\pi}{64}[(38)^4 - (33)^4] = 14.05 \times 10^3 \, \pi \text{ mm}^4$$

Since the column is hinged at its both ends, therefore effective length of the column,

$$L_e = l = 2.3 \times 10^3 \text{ mm}$$

$\therefore$ Euler's crippling load, $P_E = \dfrac{\pi^2 EI}{L_e^{\ 2}} = \dfrac{\pi^2 (205 \times 10^3) \times (14.05 \times 10^3 \ \pi)}{(2.3 \times 10^3)^2} = 16\ 880 \text{ N}$

$= 16.88 \text{ kN}$ **Ans.**

***Rankine's crippling load***

We know that area of the column section,

$$A = \frac{\pi}{4}(D^2 - d^2) = \frac{\pi}{4}[(38)^2 - (33)^2] = 88.75\ \pi \text{ mm}^2$$

and least radius of gyration, $k = \sqrt{\dfrac{I}{A}} = \sqrt{\dfrac{14.05 \times 10^3\ \pi}{88.75\pi}} = 12.6 \text{ mm}$

$\therefore$ Rankine's crippling load,

$$P_R = \frac{\sigma_{CS} \cdot A}{1 + a\left(\dfrac{L_e}{k}\right)^2} = \frac{335 \times 88.75\pi}{1 + \dfrac{1}{7500}\left(\dfrac{2.3 \times 10^3}{12.6}\right)^2} = 17\ 160 \text{ N}$$

$= 17.16 \text{ kN}$ **Ans.**

**EXAMPLE 34.7.** *Figure 34.8 shows a built-up column consisting of 150 mm × 100 mm R.S.J. with 120 mm × 12 mm plate riveted to each flange.*

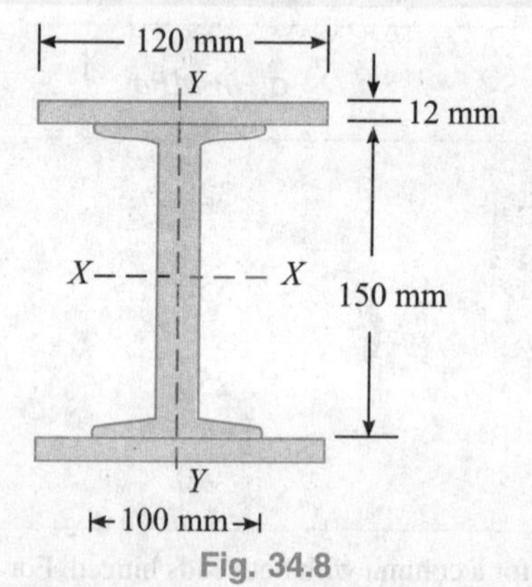


**Fig. 34.8**

*Calculate the safe load, the column can carry, if it is 4 m long having one end fixed and the other hinged with a factor of safety 3.5. Take the properties of the joist as Area = 2167 mm², $I_{XX}$ = 8.391 × 10⁶ mm⁴, $I_{YY}$ = 0.948 × 10⁶ mm⁴. Assume the yield stress as 315 MPa and Rankine's constant (a) =* $\dfrac{1}{7500}$.

**SOLUTION.** Given : Length of the column ($l$) = 4 m = 4 × 10³ mm ; Factor of safety = 3.5 ; Yield stress ($\sigma_C$) = 315 MPa = 315 N/mm² ; Area of joist = 2167 mm² ; Moment of inertia, about $X$-$X$ axis ($I_{XX}$) = 8.391 × 10⁶ mm⁴ ; Moment of inertia about $Y$-$Y$ axis ($I_{YY}$) = 0.948 × 10⁶ mm⁴ and Rankine's constant ($a$) = $\dfrac{1}{7500}$.

From the geometry of the figure, we find that area of the column section,

$$A = 2167 + (2 \times 120 \times 12) = 5047 \text{ mm}^2$$

and moment of inertia of the column section bout $X$-$X$ axis,

$$I_{XX} = (83.91 \times 10^6) + 2\left[\frac{120 \times (12)^3}{12} + 120 \times 12 \times (81)^2\right] \text{mm}^4$$

$$= (8.391) \times 10^6) + (18.93 \times 10^6) = 27.32 \times 10^6 \text{ mm}^4$$

Similarly,
$$I_{YY} = (0.948 \times 10^6) + 2\left[\frac{12 \times (120)^3}{12}\right] \text{mm}^4$$
$$= (0.948 \times 10^6) + (3.456 \times 10^6) = 4.404 \times 10^6 \text{ mm}^4$$

Since $I_{YY}$ is less than $I_{XX}$, therefore the column will tend to buckle in *Y-Y* direction. Thus we shall take *I* equal to $I_{YY} = 4.404 \times 10^6$ mm$^4$ (*i.e.*, least of two). Moreover as the column is fixed at one end and hinged at the other, therefore equivalent length of the column.

$$L_e = \frac{l}{\sqrt{2}} = \frac{4 \times 10^3}{\sqrt{2}} = 2.83 \times 10^3 \text{ mm}$$

We know that least radius of gyration,

$$k = \sqrt{\frac{I}{A}} = \sqrt{\frac{4.404 \times 10^6}{5047}} = 29.5 \text{ mm}$$

∴ Rankine's crippling load on the column

$$P_R = \frac{\sigma_C \cdot A}{1 + a\left(\frac{L_e}{k}\right)^2} = \frac{315 \times 5047}{1 + \frac{1}{7500}\left(\frac{2.83 \times 10^3}{29.5}\right)^2}$$

$$= 714 \times 10^3 \text{ N} = 714 \text{ kN}$$

and safe load on the column

$$= \frac{\text{Crippling load}}{\text{Factor of safety}} = \frac{714}{3.5} = 204 \text{ kN} \qquad \textbf{Ans.}$$

**EXAMPLE 34.8.** *A column is made up of two channels. ISJC 200 and two 250 mm × 10 mm flange plates as shown in Fig. 34.9.*

*Determine by Rankine's formula the safe load, the column of 6 m length, with both ends fixed, can carry with a factor of safety 4. The properties of one channel are Area = 1777 mm$^2$, $I_{XX}$ = 11.612 × 10$^6$ mm$^4$ and $I_{YY}$ = 0.842 × 10$^6$ mm$^4$. Distance of centroid from back to web = 19.7 mm. Take $\sigma_C$ = 320 MPa and Rankine's constant = $\frac{1}{7500}$*

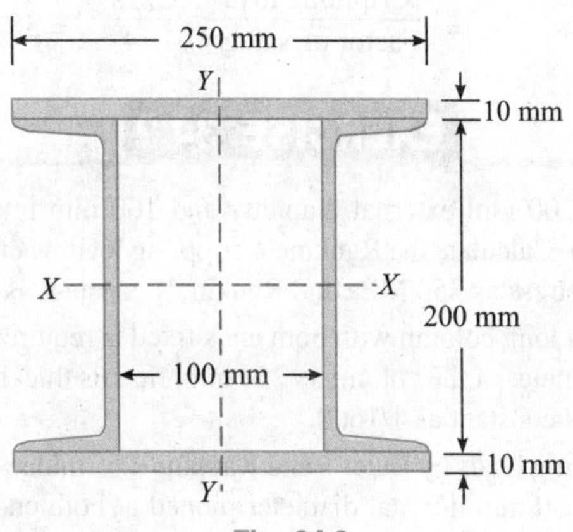


Fig. 34.9

**Solution.** Given : Length of the column ($l$) = 6 m = $6 \times 10^3$ mm ; Factor of safety = 4 ; Area of channel = 1777 mm$^2$ ; Moment of inertia about $X$-$X$ axis ($I_{XX}$) = $11.612 \times 10^6$ mm$^4$ ; Moment of inertia about $y$-$y$ axis ($I_{YY}$) = $0.842 \times 10^6$ mm$^4$ ; Distance of centroid from the back of web = 19.7 mm; Crushing stress ($\sigma_C$) = 320 MPa = 320 N/mm$^2$ and Rankine's constant ($a$) = $\frac{1}{7500}$.

From the geometry of the figure, we find that area of trhe column section,

$$A = 2\,[1777 + (250 \times 10)] = 8554 \text{ mm}^2$$

and moment of inertia of the column section about $X$-$X$ axis,

$$I_{XX} = (2 \times 11.612 \times 10^6) + 2\left[\frac{250 \times (10)^2}{12} + (250 \times 10) \times (105)^2\right] \text{mm}^4$$

$$= (23.224 \times 10^6) + (55.167 \times 10^6) = 78.391 \times 10^5 \text{ mm}^4$$

Similarly,

$$I_{YY} = 2\left[\frac{10 \times (250)^3}{12} = (0.842 \times 10^6) + 1777 \times (50 + 19.7)^2\right] \text{mm}^4$$

$$= 2\,[13.021 \times 10^6) + (9.475 \times 10^6)] = 44.992 \times 10^6 \text{ mm}^4$$

Since $L_{YY}$ is less than $I_{XX}$, therefore the column will tend to buckle in $Y$-$Y$ direction. Thus we shall take $I$ equal to $I_{YY}$ = $44.992 \times 10^6$ mm$^4$ (*i.e.*, least of the two). Moreover as the column is fixed at its both ends, therefore equivalent length of the colum,

$$L_e = \frac{l}{2} = \frac{6 \times 10^3}{2} = 3 \times 10^3 \text{ mm}$$

We know that least radius of gyration,

$$k = \sqrt{\frac{I}{A}} = \sqrt{\frac{44.992 \times 10^6}{8554}} = 72.5 \text{ mm}$$

∴ Rankine's crippling load on the column,

$$P_R = \frac{\sigma_C \cdot A}{1 + a\left(\frac{L_e}{k}\right)^2} = \frac{320 \times 8554}{1 + \frac{1}{7500} \times \left(\frac{3 \times 10^3}{72.5}\right)^2}$$

$$= 2228.5 \times 10^3 \text{ N} = 2228.5 \text{ kN}$$

and Safe load on the column $= \frac{\text{Crippling load}}{\text{Factor of safety}} = \frac{2228.5}{4} = 557.1 \text{ kN}$ **Ans.**

## EXERCISE 34.2

1. A hollow column of 200 mm external diameter and 160 mm internal diameter is used as a column of 4.5m length. Calculate the Rankine's crippling load when the column is fixed at both ends. Take allowable stress as 350 MPa and Rankine's constant as 1/1600. [**Ans.** 2.23 kN]
2. A hollow cast iron 5 m long column with both ends fixed is required for support a load of 1000 kN. If the external diameter of the column is 250 mm, find its thickness. Take working stress as 80 MPa and Rankine's constant as 1/1600. [**Ans.** 29.4 mm]
3. Find the ratio of crippling loads by Euler's and Rankine's formulae for a hollow strut of 40 mm external diameter and 30 mm internal diameter pinned at both ends. Take yield stress as 300 MPa, modulus of elasticity as 200 GPa and Rankine's constant as 1/7500. [**Ans.** 0.945)

4. A steel compound column 4 m long is built up of two steel joints of *I*-section 200 mm × 100 mm × 10 mm joined by two steel plates 300 mm × 10 mm as shown in Fig. 34.10.

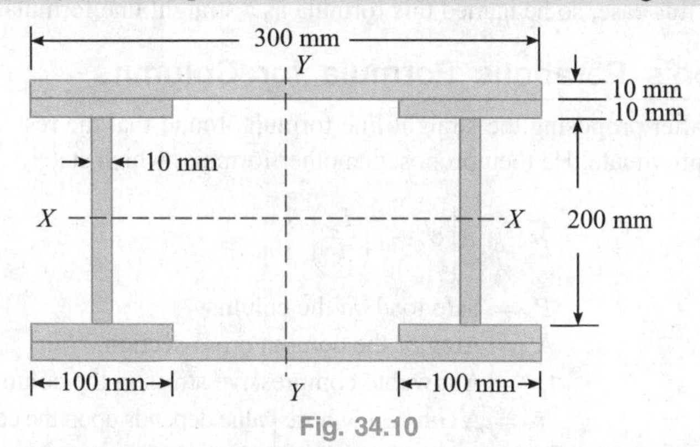


Fig. 34.10

Find the Rankine's crippling load, if both the ends of the column are hinged. Take $P_C$ = 350 MPa and Rankine's constant = 1/ 7500. **[Ans.** 3780 kN]

## 34.16. Johnson's Formula for Columns

Prof. Johnson, after a series of experiments and observations, proposed the following two formulae for columns:

1. Straight line formula and
2. Parabolic formula

## 34.17. Johnson's Straight Line Formula for Columns

This formula was first proposed by Johnson, which states

$$P = A\left[\sigma_C - n\left(\frac{L_e}{k}\right)\right]$$

where
$P$ = Safe load on the column,
$A$ = Area of the column cross-section,
$\sigma_C$ = Allowable compressive stress in the column material,
$n$ = A constant, whose value depends upon the column material and
$\frac{L_e}{k}$ = Slenderness ratio.

The values of $\sigma_C$ and are given in the following table.

**Table 34.3**

| *S. No.* | *Material* | $\sigma_C$ *in MPa* | $n$ |
|---|---|---|---|
| 1. | Mild Steel | 320 | 0.0053 |
| 2. | Wrought Iron | 250 | 0.0053 |
| 3. | Cast Iron | 550 | 0.008 |

A little consideration will show that for short columns that safe load $P = \sigma_C \cdot A$.

But for long columns, there is always a possibility of buckling. It is thus obvious, that the safe load will be small, depending upon the slenderness ratio.

Prof. Johnson, while analysing the safe load, used to plot a curve for $\frac{P}{A}$ and $\frac{L_e}{k}$. Since he used to get a straight line in this case, so he named this formula as a straight line formula.

## 34.18. Johnson's Parabolic Formula for Columns

Prof. Johnson, after proposing the straight line formula, found that the results obtained by this formula are very approximate. He then proposed another formula, which states,

$$P = A\left[\sigma_C - r\left(\frac{L_e}{k}\right)^2\right]$$

where
$P$ = Safe load on the column,
A = Area of the column cross-section,
$\sigma_C$ = Allowable compressive stress in the column material,
$r$ = A constant, whose value depends upon the column material, and
$\frac{L_e}{k}$ = Slenderness ratio with equivalent column length.

The values of $\sigma_C$ and $r$ are given in the following table:

**Table 34.4.**

| S. No. | Material | $\sigma_C$ in MPa | $r$ |
|---|---|---|---|
| 1. | Mild Steel | 320 | 0.000057 |
| 2. | Wrought Iron | 250 | 0.000039 |
| 3. | Cast Iron | 550 | 0.000016 |

Prof. Johnson, while analysing the safe load, used to plot a curve for $\frac{P}{A}$ and $\frac{L_e}{k}$. Since he named this formula as a parabolic formula.

**Note:** Both the Johnson's straight line formula and Johnson's parabolic formula are rarely used these days. They have been mentioned here only for their academic importance only.

## 34.19. Indian Standard Code for Columns

The Bureau of Indian Standards (I.S.I.) has also given a code for the safe stress in I.S. 226–1962, which states

$$\sigma_C = \sigma'_C = \frac{\frac{\sigma_y}{m}}{1 + 0.20 \sec\left[\frac{L_e}{k}\sqrt{\frac{mp'_e}{4E}}\right]} \quad \ldots\left(\text{for } \frac{L_e}{k} = 0 \text{ to } 160\right)$$

and

$$\sigma_C = \sigma'_C\left(1.2 - \frac{L_e}{800k}\right) \quad \ldots\left(\text{for } \frac{L_e}{k} = 160 \text{ and above}\right)$$

where
$\sigma_C$ = Allowable axial compressive stress,
$\sigma'_C$ = A value obtained from the above secant formula,
$\sigma_y$ = The guaranteed minimum yield stress,
$m$ = Factor of safety taken as 1.68,
$\frac{L_e}{k}$ = Slenderness ratio with equivalent column length, and
$E$ = Modulus of elasticity equal to 200 GPa.

The I.S.I. has also given a table in I.S. 800 –1962 which gives the values of $\sigma_C$ for mild steel, for slenderness ratio from 0 to 350. The value of $\sigma_y$ *i.e.*, the guaranteed minimum yield stress for mild steel is taken as 260 MPa. This table is given below:

**Table 34.5.**

| $\frac{L}{k}$ | $\sigma_C$ *in MPa* | $\frac{L_e}{k}$ | $\sigma_C$ *in MPa* | $\frac{L_e}{k}$ | $\sigma_C$ *in MPa* |
|---|---|---|---|---|---|
| 0 | 125 | 90 | 92.8 | 180 | 33.6 |
| 10 | 124.6 | 100 | 84.0 | 190 | 30.0 |
| 20 | 123.9 | 110 | 75.3 | 200 | 27.0 |
| 30 | 122.4 | 120 | 67.1 | 210 | 24.3 |
| 40 | 120.3 | 130 | 59.7 | 220 | 21.9 |
| 50 | 117.2 | 140 | 53.1 | 230 | 19.9 |
| 60 | 113.0 | 150 | 47.4 | 240 | 18.1 |
| 70 | 102.5 | 160 | 42.3 | 300 | 10.9 |
| 80 | 100.7 | 170 | 37.7 | 350 | 3.6 |

**NOTE.** Intermediate values may be obtained by linerar interpolation.

**EXAMPLE 34.9.** *A hollow cylindrical steel tube of 38 mm external diameter and 2.5 mm thick is used as a column of 2.3 m long with both ends hinged. Determine the safe load by I.S. code.*

**SOLUTION.** Given: External diameter ($D$) = 38 mm ; Thickness = 2.5 mm and length of column ($l$) = 3 m = $3 \times 10^3$ mm.

We know that area of the column section,

$$A = \frac{\pi}{4}[D^2 - d^2] = \frac{\pi}{4} \times [(38)^2 - (33)^2] = 278.8 \text{ mm}^2$$

and moment of inertia of column section,

$$I = \frac{\pi}{64}[D^4 - d^4] = \frac{\pi}{64} \times [(38)^4 - (33)^4] = 44.14 \times 10^3 \text{ mm}^4$$

∴ Least radius of gyration,

$$k = \sqrt{\frac{I}{A}} = \sqrt{\frac{44.14 \times 10^3}{278.8}} = 12.6 \text{ mm}$$

Since the column is higed at its both ends, therefore effective length of the column,

$$L_e = l = 2.3 \times 10^3 \text{ mm}$$

and slenderness ratio

$$= \frac{L_e}{k} = \frac{2.3 \times 10^3}{12.6} = 182.5$$

From table, we find that allowable stress for slenderness ratio of 182.5 is 32.7 MPa or 32.7 N/mm$^2$. Therefore safe load on the column,

$$P = A \times \sigma_C = 278.8 \times 32.7 = 9117 \text{ N} = 9.117 \text{ kN} \qquad \textbf{Ans.}$$

## 34.20. Long Columns Subjected to Eccentric Loading

In the previous articles, we have discussed the effect of loading on long columns. We have always referred the cases when the load acts axially on the column (*i.e.*, the line of action of the load coincides with the axis of the column). But in actual practice it is not always possible to have an axial load, on the column, and eccentric loading takes place. Here we shall discuss the effect of eccentric loading on the Rankine's and Euler's formulae for long columns.

***Rankine's formula***

Consider a long column subjected to an eccentric load.

Let

$P$ = Load on the column,

$A$ = Area of cross-section,

$e$ = Eccentricity of the load,

$Z$ = Modulus of section,

$y_e$ = Distance of the extreme fibre (on compressive side) from the axis of the column, and

$k$ = Least radius of gyration.

We have discussed in Chapter of Direct and Bending Stresses, that when a column is subjected to an eccentric load, the maximum intensity of compressive stress is given by the relation,

$$\sigma_{max} = \frac{P}{A} + \frac{M}{Z} = \frac{P}{A} + \frac{P \cdot e}{Z} \qquad \text{... (See Article 17.4)}$$

$$= \frac{P}{A} + \frac{P \cdot e \cdot y_e}{Ak^2} \qquad \text{...} \left( \because Z = \frac{I}{y_e} = \frac{Ak^2}{y_l} \right)$$

$$= \frac{P}{A}\left(1 + \frac{e \cdot y_e}{k^2}\right)$$

If $\sigma_{CS}$ is the permissible crushing stress for the given material, then the safe crushing load for the given column,

$$\sigma_{CS} = \frac{P}{A}\left(1 + \frac{e \cdot y_e}{k^2}\right)$$

or

$$P = \frac{\sigma_C \cdot A}{\left(1 + \dfrac{e \cdot y_e}{k^2}\right)}$$

We have already discussed that the safe load by Rankine's formula for long columns and axial load is given by the relation,

$$P = \frac{\sigma_{CS} \cdot A}{1 + a\left(\dfrac{L_e}{k}\right)^2} \qquad \text{... (See Article 34.15)}$$

It is thus obvious that if the effect of buckling is also to be taken into account, the safe axial load with eccentricity,

$$P = \frac{\sigma_{CS} \cdot A}{\left(1+\frac{e \cdot y_e}{k^2}\right)\left[1+a\left(\frac{L_e}{k}\right)^2\right]}$$

*Euler's formula*

Consider a long column *AB* of length *l* fixed at *A*, free at *B* and carrying an eccentric load as shown in Fig. 34.11.

As a result of loading, let the beam deflect into a curved form $AX_1B_1$ such that the free end *B* deflects through *a* and occupies a new position $B_1$ as shown in Fig. 34.11.

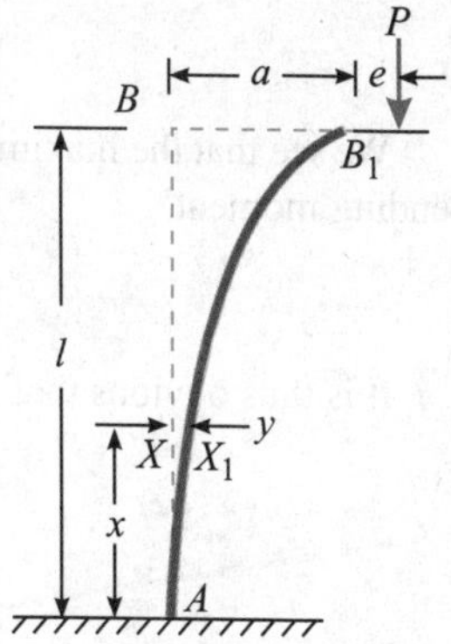

Now consider any section *X*, at a distance *x* from *A*.

Let $P$ = Critical load on the column,

$e$ = Eccentricity of the load,

$y$ = Deflection of the column at *X*.

Thus, the eccentricity of the load *P* at *X*

$$= a + e - y$$

∴ Moment due to load,

$$M = +P(a + e - y)$$

...(Plus sign due to convexity wards initial centre line)

$$= P(a + e) - P \cdot y$$



Fig. 34.11

$$\therefore \quad EI\frac{d^2y}{dx^2} = P(a+e) - P \cdot y$$

or

$$\frac{d^2y}{dx^2} + \frac{P \cdot y}{EI} = \frac{P(a+e)}{EI}$$

The general solution of the above differential equation is

$$y = A\cos\left(x\sqrt{\frac{P}{EI}}\right) + B\sin\left(x\sqrt{\frac{P}{EI}}\right) + (a+e) \qquad ...(i)$$

where *A* and *B* are the constants of integration. We know that when $x = 0$, then $y = 0$, therefore $A = -(a + e)$. Now differentiating the above equation,

$$\frac{dy}{dx} = -A\sqrt{\frac{P}{EI}}\sin\left(x\sqrt{\frac{P}{EI}}\right) + B\sqrt{\frac{P}{EI}}\cos\left(x\sqrt{\frac{P}{EI}}\right)$$

We also know that when $x = 0$, then $\frac{dy}{dx} = 0$, therefore

$$0 = B\sqrt{\frac{P}{EI}}$$

A little consideration will show that either *B* is equal to zero or $\sqrt{\frac{P}{EI}}$ is equal to zero. Since the load *P* is not equal to zero, it is thus obvious that *B* is equal to zero. Substituting the values of $A = -(a + e)$ and $B = 0$ in equation (*i*),

$$y = -(a+e)\cos\left[\left(x\sqrt{\frac{P}{EI}}\right)+(a+e)\right] = a+e\left[1-\cos\left(x\sqrt{\frac{P}{EI}}\right)\right]$$

We also know that when $x = l$, then $y = a$.

Therefore
$$a = a+e\left[1-\cos\left(l\sqrt{\frac{P}{EI}}\right)\right]$$

$$\therefore \quad (a+e)\cos\left(l\sqrt{\frac{P}{EI}}\right) = e$$

or
$$(a+e) = e\cdot\sec\left(l\sqrt{\frac{P}{EI}}\right)$$

We see that the maximum bending moment occurs at $B$ and is equal to $P(a+e)$. Here maximum bending moment,

$$M_{max} = P\cdot e\cdot\sec\left(l\sqrt{\frac{P}{EI}}\right)$$

It is thus obvious that maximum compressive stress will be at such that

$$\sigma_{max} = \frac{P}{A}+\frac{M}{Z} = \frac{P}{A}+\frac{P\cdot e\cdot\sec\left(l\sqrt{\frac{P}{EI}}\right)}{Z}$$

The equation (*ii*) and (*iii*) are valid for the column with one end fixed and other end free, for which $L_e = 2l$ or $l = \frac{L_e}{2}$ (see Table 34.1).

Substituting $l = \frac{L_e}{2}$ in equation (*ii*) and (*iii*), we get the general expression for any end condition. Thus maximum bending moment

$$M_{\max} = P\cdot e\cdot\sec\left(\frac{L_e}{2}\sqrt{\frac{P}{EI}}\right) \quad \ldots (iv)$$

and maximum compressive stress

$$\sigma_{\max} = \frac{P}{A}+\frac{M}{Z} = \frac{P}{A}+\frac{1}{Z}\cdot P\cdot e\cdot\sec\left(\frac{L_e}{2}\sqrt{\frac{P}{EI}}\right)$$

Now for given end conditions, substitute the value of equivalent length ($L_e$) of column from table 34.1.

**Note :** In case of short column (with no buckling), the maximum bending moment is $M_{\max} = P\cdot e$, as compared to long columns for which $M_{\max} = P\cdot e\cdot\sec\left[\frac{L_e}{2}\sqrt{\frac{P}{EI}}\right]$

**EXAMPLE 34.10.** *An alloy hollow circular column of 200 mm external and 160 mm internal diameter is 5 m long and fixed at both of its ends. It is subjected to a load of 120 kN at an eccentricity of 20 mm from the geometrical axis. Determine the maximum stress induced in the column section. Take E as 120 GPa.*

**SOLUTION.** Given: External diameter ($D$) = 200 mm; Internal diameter ($d$) = 160 mm; Length ($l$) = 5 m = $5 \times 10^3$ mm; Load ($P$) = 120 kN = $120 \times 10^3$ N; Eccentricity ($e$) = 20 mm and modulus of elasticity ($E$) = 120 GPa = $120 \times 10^3$ N/mm$^2$.

We know that area of the column section,

$$A = \frac{\pi}{4}[D^2 - d^2] = \frac{\pi}{4}[(200)^2 - (160)^2] = 11.31 \times 10^3 \text{ mm}^2$$

and moment of inertia of column section,

$$I = \frac{\pi}{64}[D^4 - d^4] = \frac{\pi}{64}[(200)^4 - (160)^4] = 46.37 \times 10^6 \text{ mm}^4$$

$$\therefore \text{ Modulus of section, } \quad Z = \frac{I}{D/2} = \frac{46.37 \times 10^6}{200/2} = 463.7 \times 10^3 \text{ mm}^3$$

Since the column is fixed at its both ends, therefore equivalent length of the column,

$$L_e = \frac{l}{2} = \frac{5 \times 10^3}{2} = 2.5 \times 10^3 \text{ mm}$$

and value of

$$\frac{L_e}{2}\sqrt{\frac{P}{EI}} = \frac{2.5 \times 10^3}{2} \times \sqrt{\frac{120 \times 10^3}{(120 \times 10^3) \times (46.37 \times 10^6)}} \text{ rad}$$

$$= 0.1836 \text{ rad} = 10.52°$$

We know that maximum compressive stress induced in the column,

$$\sigma_{max} = \frac{P}{A} + \frac{P \cdot e \cdot \sec\left(\frac{L_e}{2}\sqrt{\frac{P}{EI}}\right)}{Z}$$

$$= \frac{120 \times 10^3}{11.31 \times 10^3} + \frac{(120 \times 10^3) \times 20 \sec 10.52°}{463.7 \times 10^3} \text{ N}$$

$$= 10.61 + 5.26 = 15.87 \text{ N/mm}^2 = \mathbf{15.87 \text{ MPa}} \qquad \textbf{Ans.}$$

## EXERCISE 34.3

1. A 1.5 m long column has a circular cross-section of 50 mm diameter. One end of the column is fixed and the other is free. Using the Indian Standard Code, determine the safe load on the column. **[Ans.** 35.5 kN]
2. A hollow circular column of 200 mm external diameter and 160 mm internal diameter is 4 m long with both ends fixed. If the column carries load of 150 kN at an eccentricity of 25 mm, find the extreme stress in the column. **[Ans.** 21.5 MPa]
3. An alloy tube 60 mm diameter and 2.8 m length is used as a strut with both ends hinged. If the tube is subjected to an eccentric load equal to 60% of the Euler's crippling load. Find the value of eccentricity. Take yield strength as 320 MPa and modulus of elasticity as 210 GPa. **[Ans.** 12.1 mm]

## QUESTIONS

1. What do you understand by the terms 'column' and 'strut'? Distinguish clearly between long columns and short columns.
2. Explain the failure of long columns and short columns.
3. Describe the assumptions in the Euler's column theory.
4. Derive a relation for the Euler's crippling load for a column when (*i*) it has both ends hinged, and (*ii*) both ends fixed.

5. Define the term 'equivalent length'. Discuss its uses.
6. Explain the term 'slenderness ratio' and describe with mathematical expression, how it limits the use of Euler's formula for crippling load.
7. Obtain a relation for the Rankine's crippling load for columns.
8. Give the Johnson's straight line and parabolic formula for columns.
9. What is Indian Standard Code for columns. Are you satisfied with the factor of safety of 1.68. Give explanation to your answer.
10. Explain the effect of eccentric loading on a column. Derive a relation for the maximum stress in an eccentrically loaded column.

## OBJECTIVE TYPE QUESTIONS

1. A column of length is hinged at its both ends. Its equivalent length will be equal to
   (*a*) $2l$ (*b*) $l$ (*c*) $0.5\,l$ (*d*) $0.707\,l$
2. The slenderness ratio of a long column is
   (*a*) 10 –20 (*b*) 20 –30 (*c*) 50 –60 (*d*) above 80
3. The value of Rankine's constant for mild steel is
   (*a*) $\frac{1}{9000}$ (*b*) $\frac{1}{7500}$ (*c*) $\frac{1}{1600}$ (*d*) $\frac{1}{750}$

## ANSWERS

**1.** (*b*) **2.** (*d*) **3.** (*b*)

# Chapter 35

# Introduction to Reinforced Concrete

## Contents

## 35.1. Introduction

We have discussed in Art. 14.3 that whenever a beam is loaded, the upper layers (from the neutral axis) of the beam are in compression; whereas the lower ones are in tension. It has been experimentaly found that the cement concrete is very weak in tension, though it is strong in compression. It has also been observed that the steel is equally strong in tension and compression. But a little consideration will show that a long steel bar can resist tension up to the permissible limit; whereas it cannot resist compression up to the permissible limit because of buckling.

The engineers have taken the full advantages of concrete in compression and steel in tension, by combination, which proved to be an ideal (because the two materials are used to take up the stresses, for which they are most suitable) is called *reinforced cement concrete* (briefly written as R.C.C.).

## 35.2. Advantages of R.C.C. Structures

The following advantages of R.C.C. structures are important from the subject point of view:

1. The concrete, while setting holds a firm grip on the surface of the steel bars.
2. The coefficient of linear expansions for steel and concrete are almost the same. Therefore no internal stresses are set up due to variation in temperature.
3. The coating of cement grout on the surface of steel bars, protects them from corrosion and thus does not produce adverse chemical effects on them.

## 35.3. Assumptions in the Theory of R.C.C.

The following assumptions are made in the theory of reinforcement cement concrete.

1. All the tensile stresses are taken up by the steel reinforcement only.
2. There is a sufficient bond between the steel and concrete.
3. The moduli of elasticity for steel and concrete are constant.
4. The steel and concrete is stressed within its elastic limit and thus obey Hooke's law.

## 35.4. Neutral Axis

It is an imaginary plane, which divides the cross-section of a beam into the tension and compression zones on the opposite sides of the plane.

## 35.5. Types of Neutral Axis

There are two types of neutral axis of reinforced concrete section:

1. Critical neutral axis and
2. Actual neutral axis.

## 35.6. Critical Neutral Axis

The critical neutral axis, of a section is based on the principle that the neutral axis is situated at the centre of gravity of a given section. The depth of critical neutral axis from the top of section may be derived from the first principles and is useful, where the stresses in the concrete and steel are known.

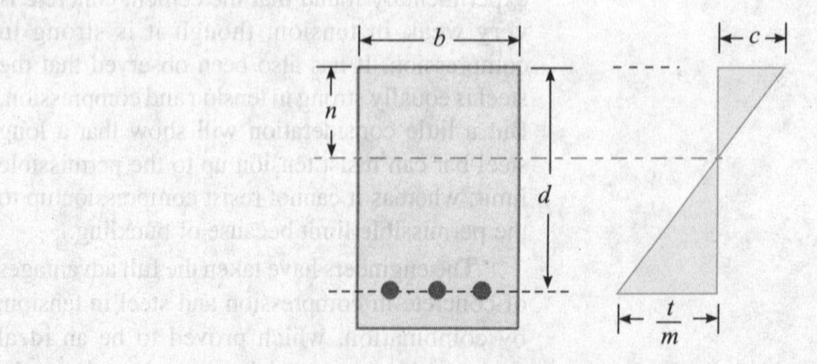


Fig. 35.1

Let us consider a reinforced beam as shown in Fig. 35.1.

Let $d$ = Effective depth of the beam (*i.e.*, depth of the centre of reinforcement from the edge),

$n$ = Depth of neutral axis from the top of the beam,

$t$ = Stress in steel and

$c$ = Stress in concrete.

We have discussed in Art. 14.4 that the strain at a point is proportional to the distance of that point from the neutral axis.

$$\therefore \quad \frac{\text{Strain in the extreme fibre of the concrete}}{\text{Strain in steel}} = \frac{n}{d-n} \qquad \text{... } (i)$$

We know that $\dfrac{\text{Stress}}{\text{Strain}} = E$ or $\text{Strain} = \dfrac{\text{Stress}}{E}$

$$\therefore \quad \text{Strain in steel} = \frac{t}{E_t}$$

and

$$\text{strain in concrete} = \frac{c}{E_c}$$

Substituting these values in equation (*i*),

$$\frac{\dfrac{c}{E_c}}{\dfrac{t}{E_t}} = \frac{n}{d-n} \quad \text{or} \quad \frac{c}{t} \times \frac{E_t}{E_c} = \frac{n}{d-n}$$

$$\therefore \quad \frac{t}{c} = \frac{E_t}{E_c} \times \frac{(d-n)}{n} = \frac{m\,(d-n)}{n} \qquad \text{...} \left(\frac{E_t}{E_c} = m = \text{Modular ratio}\right)$$

**EXAMPLE 35.1.** *Find the position of neutral axis of a beam, having an effective depth of 270 mm. The stresses in concrete and steel may be taken as 5 MPa and 140 MPa respectively. Take m = 18.*

**SOLUTION.** Given : Effective depth ($d$) = 270 mm ; Stress in concrete ($c$) = 5 MPa = 5 N/mm$^2$ ; Stress in steel ($t$) = 140 MPa = 140 N/mm$^2$ and modular ratio ($m$) = 18.

Let $n$ = Depth of the neutral axis from the top of the beam in mm.

We know that relation for neutral axis,

$$\frac{t}{c} = \frac{m\,(d-n)}{n} \quad \text{or} \quad \frac{140}{5} = \frac{18\,(270-n)}{n}$$

$$28\,n = 4860 - 18n \quad \text{or} \quad 46n = 4860$$

$$\therefore \quad n = \frac{4860}{46} = \mathbf{105.6\ mm} \quad \textbf{Ans.}$$

## 35.7. Actual Neutral Axis

The actual neutral axis of a section, is based on the principle that the moment of areas on either side of it (*i.e.*, areas of compression and tension zones) are equal. Since a reinforced section is composed of two dissimilar materials, therefore one of these two materials must be converted into an equivalent section composed of only one material. The general practice is to convert the area of reinforcement into an equivalent area of concrete, by multiplying it by the modular ratio.

Let us consider a reinforced beam as shown in Fig. 35.2.

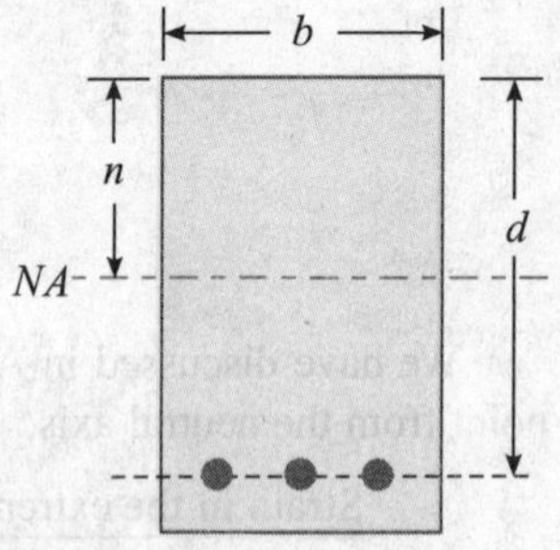

Let $b$ = Breadth of the beam,

$d$ = Effective depth of the beam,

$n$ = Depth of the neutral axis from the top of the beam,

$A_t$ = Area of steel and

$m$ = Modular radio.



Fig. 35.2

∴ Equivalent area of concrete

$$= m \cdot A_t$$

Now moment of concrete area about neutral axis

$$= bn \times \frac{n}{2} = \frac{bn^2}{2} \quad \text{... }(i)$$

Similarly, moment of steel (*i.e.*, equivalent area of concrete)

$$= m \cdot A_t (d - n) \quad \text{... }(ii)$$

Equations (*i*) and (*ii*),

$$\frac{bn^2}{2} = m \cdot A_t (d - n)$$

This will form a quadratic equation for . But only the positive value of $n$ is to be considered.

**Note:** For all sort of calculations, the actual neutral axis is used.

**Example 35.2.** *A reinforced concrete beam is 300 mm wide and has an overall depth of 500 mm. The reinforcement, which consists of 4 number 20 mm dia bars, is placed 50 mm above the bottom of the beam. Find the neutral axis of section, if modular ratio is 18.*

**Solution.** Given: Width of beam ($b$) = 300 mm ; Overall depth of beam = 500 mm ; No. of bars = 4; Diameter of bars = 4 ; Distance between the centre of reinforcement and bottom of beam = 50 mm and modular ratio ($m$) = 18.

Let $n$ = Depth of actual neutral axis from the top of the beam in mm.

We know that area of steel,

$$A_t = 4 \times \frac{\pi}{4} \times (20)^2 = 1257 \text{ mm}^2$$

and effective depth of beam,

$$d = 500 - 50 = 450 \text{ mm}$$

We also know that relation for the depth of actual neutral axis,

$$\frac{bn^2}{2} = n \cdot A_t (d - n)$$

$$\frac{300 \times n^2}{2} = 18 \times 1257 (450 - n)$$

$$150 n^2 = 10\ 181700 - 22\ 626\ n$$

$$n^2 + 150.8n - 67\ 878 = 0$$

This is a quadratic equation for $n$. Therefore

$$n = \frac{-150.8 \pm \sqrt{(150.8)^2 + 4 \times 67878}}{2} \text{ mm}$$

$$= \frac{-150.8 \pm 542.4}{2} = 195.8 \text{ mm} \quad \text{or} \quad -346.6 \text{ mm}$$

Since the value of '$n$' can not be negative, therefore actual value of '$n$' is **195.8 mm** **Ans.**

## 35.8. Moment of Resistance

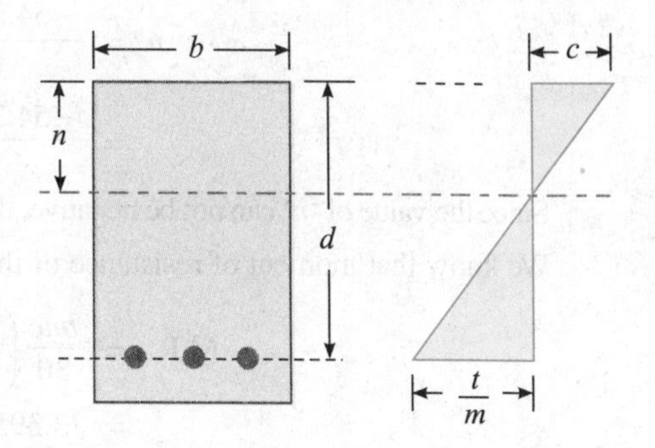


Fig. 35.3

The moment of resistance of a reinforced section is the moment of the couple consisting of the total tensile force in the steel reinforcement (acting at the centre of the reinforcement) and the compressive force in the concrete (acting at the centre of gravity of the compressive forces), as shown in Fig. 35.3.

In order that the section may be stable under the action of the external bending moment, the total moment of resistance should not be less than the external bending moment.

Let $b$ = Width of the beam,

$d$ = Effective depth of the beam and

$n$ = Depth of the neutral axis, from the top of the beam.

$$\therefore \quad \text{Lever arm} = d - \frac{n}{3}$$

Now moment of resistance,

$$\text{M.R.} = \frac{bnc}{2}\left(d - \frac{n}{3}\right) \quad \text{...(where } c \text{ is given)}$$

$$= t \cdot A_t \left(d - \frac{n}{3}\right) \quad \text{...(where } t \text{ is given)}$$

**EXAMPLE 35.3.** *A beam 250 mm wide and 550 mm deep contains 4 no. 12 mm dia bars. The cover of concrete, to the centre of the main reinforcement is 50 mm. Find the moment of resistance of the beam, if maximum allowable stresses in concrete and steel are 4.2 MPa and 125 MPa respectively. Take m = 15.*

**SOLUTION.** Given: Width of beam ($b$) = 250 mm ; Overall depth of beam = 550 mm ; Cover of concrete over reinforcement = 50 mm ; No. of bars = 4 ; Diameter of bars = 12 mm ; Allowable stress in concrete ($c$) = 4.2 MPa = 4.2 N/mm$^2$ ; Allowable stress in steel ($t$) = 125 MPa = 125 N/mm$^2$ and modular ratio ($m$) = 15.

Let $n$ = Depth of actual neutral axis from the top of the beam in mm.

We know that area of steel,

$$A_t = 4 \times \frac{\pi}{4} \times (12)^2 = 452.4 \text{ mm}^2$$

and effective depth of beam,

$$d = 550 - 50 = 500 \text{ mm}$$

We also know that relation for the depth of actual neutral axis,

$$\frac{bn^2}{2} = m \cdot A_t (d - n)$$

$$\frac{250\, n^2}{2} = 15 \times 452.4\, (500 - n)$$

$$125\,n^2 + 6786n - 3\,393000 = 0$$
$$n^2 + 54.3n - 27\,144 = 0$$

This is a quadratic equation for $n$. Therefore

$$n = \frac{-54.3 \pm \sqrt{(543.3)^2 + 4 \times 27\,144}}{2} \text{ mm}$$
$$= \frac{-54.3 \pm 333.9}{2} = 139.8 \text{ mm} \quad \text{or} \quad -194.1 \text{ mm}$$

Since the value of '$n$' can not be negative, therefore we shall take the actual value of '$n$' as 139.8 mm.

We know that moment of resistance of the beam on the basis of stress in concrete,

$$\text{M.R.} = \frac{bnc}{30}\left(d - \frac{n}{3}\right) = \frac{250 \times 139.8 \times 4.2}{2}\left(500 - \frac{139.8}{3}\right) \text{ N-mm}$$
$$= 73\,395 \times 453.4 = 33.27 \times 10^6 \text{ N-mm} \quad \text{... }(i)$$

Similarly, moment of resistance of the beam on the basis of stress in steel,

$$\text{M.R.} = t \cdot A_t\left(d - \frac{n}{3}\right) = 125 \times 452.4\left(500 - \frac{139.8}{3}\right) \text{ N-mm}$$
$$= 56\,550 \times 453.4 = 25.64 \times 10^6 \text{ N-mm} \quad \text{.... }(ii)$$

From the above two values, we find that the moment of resistance of the beam is $25.64 \times 10^6$ N-mm (*i.e.*, lesser of the two values). **Ans.**

**Note:** In this case, the steel develops a maximum stress of 125 N/mm². But the concrete is not stressed to its allowable stress of 4.2 N/mm².

**Example 35.4.** *A reinforced concrete beam of 250 mm width and 600 mm effective depth is subjected to a bending moment of 120 kN-m. The reinforcement consists of 4 bars of 28 mm diameter. Determine the stresses in steel and concrete. Take m = 16.*

**Solution.** Given: Width of beam ($b$) = 250 mm ; Effective depth of beam ($d$) = 600 mm ; Bending moment (M.R.) = 120 kN-m = $120 \times 10^6$ N-mm ; No. of bars = 4 ; Diameter of bars = 28 mm and modular ratio ($n$) = 16

***Stress in steel***

Let $t$ = Stress in steel in N/mm² and

$n$ = Depth of actual neutral axis in mm.

We know that area of reinforcement,

$$A_t = 4 \times \frac{\pi}{4} \times (28)^2 = 2463 \text{ mm}^2$$

and relation for depth of actual neutral axis,

$$\frac{bn^2}{2} = m \cdot A_t\,(d - n)$$

$$\frac{250\,n^2}{2} = 16 \times 2463 \times (600 - n)$$

$$125\,n^2 + 39\,408n - 23\,644\,800 = 0$$
$$n^2 + 315.3 - 189\,160 = 0$$

This is a quadratic equation for $n$. Therefore

$$n = \frac{-315.3 \pm \sqrt{(315.3)^2 + 4 \times 189\,160}}{2} \text{ mm}$$

$$= \frac{-315.3 \pm 925.3}{2} = 305 \text{ mm} \quad \text{or} \quad -620.3 \text{ mm}$$

Since the value of '$n$' can not be negative, therefore we shall take the actual value of '$n$' as 305 mm.

We also know that bending moment (M.R.),

$$120 \times 10^6 = t \cdot A_t \left(d - \frac{n}{2}\right) = t \times 2463 \times \left(600 - \frac{305}{3}\right) = 1.227 \times 10^6\, t$$

$$\therefore \quad t = \frac{120 \times 10^6}{1.227 \times 10^6} = 97.8 \text{ N/mm}^2 = \mathbf{97.8 \text{ MPa}} \quad \textbf{Ans.}$$

***Stress in concrete***

Let $c$ = Stress in concrete in N/mm$^2$.

We know that relation for stress in concrete,

$$\frac{t}{c} = \frac{m(d-n)}{n}$$

$$\frac{97.8}{c} = \frac{16 \times (600 - 305)}{305} = 15.5$$

$$c = \frac{97.8}{15.5} = 6.3 \text{ N/mm}^2 = \mathbf{6.3 \text{ MPa}} \quad \textbf{Ans.}$$

## EXERCISE 35.1

**1.** A reinforced concrete beam 150 mm wide is 250 mm deep to the centre of the reinforcement. Find the position of the neutral axis, if the stress developed in the concrete and steel are 4.5 MPa and 81 MPa respectively. (**Ans.** 113.6 mm)

**2.** A reinforced cement concrete beam is 200 mm deep up to the centre of the reinforcement. Find the position of the neutral axis, if the stresses developed in steel and concrete are 120 MPa and 4 MPa respectively. Take $m$ = 18. (**Ans.** 94.7 mm)

## 35.9. Types of Beam Sections

The beam sections are of the following three types, depending upon the amount of reinforcement:

**1.** Under-reinforced section,

**2.** Balanced section and

**3.** Over-reinforced section.

## 35.10. Under-reinforced Section

It is a section, in which the amount of reinforcement is *less* than the proper requirement. As a result of this, the steel will first attain the maximum permissible stresses and the concrete will not be subjected to full compressive stresses. In this case, the depth of actual neutral axis will be *less* than that of the critical neutral axis.

The moment of resistance of an under-reinforced beam is based on the stresses in steel and will be given by the relation,

$$\text{M.R.} = t \cdot A_t\left(d - \frac{n}{3}\right)$$

## 35.11. Balanced Section

It is a section, in which the amount of reinforcement is equal to the proper requirement. In a balanced beam, the steel as well as concrete will attain maximum permissible stresses simultaneously. In this case, the depth of actual neutral axis will be the *same* as that of the critical neutral axis.

The amount of resistance, of a balanced beam will be based on the stress in steel as well as concrete and will be given by the relations:

$$\text{M.R.} = t \cdot A_t\left(d - \frac{n}{3}\right)$$

$$= \frac{bnc}{2}\left(d - \frac{n}{3}\right)$$

## 35.12. Over-reinforced Section

It is a section, in which the amount of reinforcement is more than the proper requirement. As a result of this, the concrete will first attain the maximum permissible stresses and the steel will not be subjected to full tensile stress. In this case, the depth of actual neutral axis will be *more* than the critical neutral axis.

The moment of resistance of an over-reinforced beam will be based on the stresses in concrete and will be given by the relation :

$$\text{M.R.} = \frac{bnc}{2}\left(d - \frac{n}{3}\right)$$

**EXAMPLE 35.5.** *A rectangular R.C.C. beam 250 mm wide has an effective depth of 330 mm and steel reinforcement 967 mm². Find the moment of resistance, if maximum stresses in concrete and steel are 70 MPa and 140 MPa respectively. Take m = 15.*

**SOLUTION.** Given : Width of beam ($b$) = 250 mm ; Effective depth of beam ($d$) = 330 mm ; Area of reinforcement ($A_t$) = 967 mm$^2$ ; Maximum stress in concrete = 7 MPa = 7 N/mm$^2$ ; Maximum stress in steel ($t$) = 140 MPa = 140 N/mm$^2$ and modular ratio ($m$) = 15.

Let $n$ = Depth of actual neutral axis in mm and

$n_c$ = Depth of critical neutral axis in mm.

We know that relation for the depth of actual neutral axis,

$$\frac{bn^2}{2} = m \cdot A_t (d - n)$$

$$\frac{250\, n^2}{2} = 15 \times 967\,(330 - n)$$

$$125\, n^2 + 14\,500n - 4\,786\,650 = 0$$

$$n^2 + 116n - 38\,293 = 0$$

This is a quadratic equation for $n$. Therefore

$$n = \frac{-116 \pm \sqrt{(116)^2 + 4 \times 38\,293}}{2} \text{ mm}$$

$$= \frac{-116 \pm 408.2}{2} = 146.1 \text{ mm} \quad \text{or} \quad -262.1 \text{ mm} \quad ...(i)$$

Since the value of '*n*' can not be negative, therefore we shall take the actual value of '*n*' as 146.1 mm

We also know that relation for the depth of critical neutral axis,

$$\frac{t}{c} = \frac{m(d - n_c)}{n_c} \quad \text{or} \quad \frac{140}{7} = \frac{15(330 - n_c)}{n_c}$$

$$140\, n_c = 34\,650 - 105\, n_c \quad \text{or} \quad 245\, n_c = 34\,650$$

$$\therefore \quad n_c = \frac{34650}{245} = 141.4 \text{ mm} \quad ...(ii)$$

Since the depth of actual neutral axis (146.1 mm) is more than the critical neutral axis (141.4 mm), therefore the beam is over-reinforced. As a result of this, the concrete will first attain the maximum permissible stress. And the moment of resistance will be based on the stress in concrete. We know that moment of resistance based on stress in concrete,

$$\text{M.R.} = \frac{bnc}{2}\left(d - \frac{n}{3}\right) = \frac{250 \times 146.1 \times 7}{2}\left(330 - \frac{146.1}{3}\right) \text{N-mm}$$

$$= 35.96 \times 10^6 \text{ N-mm} = \mathbf{35.96 \text{ kN-mm}} \quad \textbf{Ans.}$$

**EXAMPLE 35.6.** *A concrete beam having 200 mm width and 500 mm effective depth is reinforced in tension, whose limiting stress is 120 MPa and modulus of elasticity 200 GPa. The concrete has a limiting stress of 6 MPa and modulus of elasticity of 12.5 GPa.*

*Calculate the moment of resistance of the beam, if both the sections are fully stressed. Also find the area of reinforcement.*

**SOLUTION.** Given: Width of beam ($b$) = 200 mm ; Effective depth of beam ($d$) = 500 mm ; Limiting stress in steel ($t$) = 120 MPa = 120 N/mm$^2$ ; Modulus of elasticity for steel ($E_s$) = 200 GPa = $200 \times 10^3$ N/mm$^2$ ; Limiting stress in concrete ($c$) = 6 MPa = 6 N/mm$^2$ and modulus of elasticity for concrete ($E_c$) = 12.5 GPa = $12.5 \times 10^3$ N/mm$^2$.

***Moment of resistance of the beam***

Let $n$ = Depth of neutral axis of the beam in mm.

We know that modular ratio,

$$m = \frac{E_t}{E_c} = \frac{200 \times 10^3}{12.5 \times 10^3} = 16$$

and relation for the depth of neutral axis of the beam,

$$\frac{t}{c} = \frac{m(d - n)}{n} \quad \text{or} \quad \frac{120}{6} = \frac{16(500 - n)}{n}$$

$$120n = 48\,000 - 96n$$

$$216n = 48\,000$$

$$\therefore \quad n = \frac{48000}{216} = 222.2 \text{ mm}$$

We know that momentt of resistance of the beam,

$$\text{M.R.} = \frac{bnc}{2}\left(d - \frac{n}{3}\right) = \frac{200 \times 222.2 \times 6}{2}\left(50 - \frac{222.2}{3}\right)$$

$$= 56.785 \times 10^6 \text{ N-mm} = \mathbf{56.785 \text{ kN-m}} \quad \textbf{Ans.}$$

*Area of reinforcement*

Let $A_t$ = Area of reinforcement in $mm^2$.

We know that moment of resistance of the beam (M.R.)

$$56.785 \times 10^6 = t \cdot A_t \left(d - \frac{n}{3}\right) = 120\, A_t \left(500 - \frac{222.2}{3}\right) = 51110\, A_t$$

$$\therefore \quad A_t = \frac{56.785 \times 10^6}{51110} = \mathbf{1111\ mm^2} \quad \textbf{Ans.}$$

NOTES : 1. This reinforcement is equivalent to 4 bars of 20 mm diameter.

2. Since the steel and concrete is fully stressed, therefore look the neutral axes (i.e., actual and critical) will coincide with each other.

**EXAMPLE 35.7.** *A reinforced concrete beam of 300 mm width and 600 mm overall depth has a 50 mm cover over the bottom reinforcement. The beam is subjected to a load of 24 kN/m including itself weight for a span of 6 m. If the allowable stress in the concrete is 5 MPa, determine the corresponding stress in steel.*

*Also determine the area of reinforcement. Take m = 15.*

**SOLUTION.** Given : Width of beam ($b$) = 300 mm ; Overall depth = 600 mm ; Cover over the bottom reinforcement = 50 mm ; Load over the beam ($w$) = 24 kN/m ; Span ($l$) = 6 m ; Allowable stress in concrete ($c$) = 5 MPa = 5 $N/mm^2$ and modular ratio ($m$) = 15.

*Stress in steel*

Let $t$ = Stress in steel in $N/mm^2$ and

$n$ = Depth of the neutral axis of the beam.

We know that effective depth of the beam,

$$d = 600 - 50 = 550 \text{ mm}$$

and bending moment due to uniformly distributed load over the span,

$$\text{M.R.} = \frac{w \cdot l^2}{8} = \frac{24 \times (6)^2}{8} = 108 \text{ kN-m} = 108 \times 10^6 \text{ N-mm}$$

We also know that bending moment (M.R.)

$$108 \times 10^6 = \frac{bnc}{2}\left(d - \frac{n}{3}\right) = \frac{300 \times n \times 5}{2}\left(550 - \frac{n}{3}\right) = 750n\left(550 - \frac{n}{3}\right)$$

$$144\,000 = 550n - \frac{n^2}{3}$$

$$n^2 - 1650n + 432\,000 = 0$$

This is a quadratic equation for $n$. Therefore

$$n = \frac{+1650 \pm \sqrt{(1650)^2 - 4 \times 432\,000}}{2} \text{ mm}$$

$$= \frac{+1650 \pm 997.2}{2} = 1323.6 \text{ mm} \quad \text{or} \quad 326.4 \text{ mm}$$

We know that relation for stress in steel,

$$\frac{t}{c} = \frac{m\,(d-n)}{n} \quad \text{or} \quad \frac{t}{5} = \frac{15\,(550 - 326.4)}{326.4} = 10.27$$

$$\therefore \quad t = 10.28 \times 5 = 51.4 \text{ N/mm}^2 = \mathbf{51.4\ N/MPa} \quad \textbf{Ans.}$$

***Area of reinforcement***

Let $A_t$ = Area of reinforcement in mm$^2$.

We know that moment of resistance of the beam (M.R.)

$$108 \times 10^6 = t \cdot A_t\left(d - \frac{n}{3}\right) = 51.4 \times A_t\left(550 - \frac{326.4}{3}\right)$$

$$= 22\,655.6\ A_t$$

$$\therefore \quad A_t = \frac{108 \times 10^6}{27\,655.6} = \mathbf{4767\ mm^2} \quad \textbf{Ans.}$$

NOTE : This reinforcement is equivalent to 10 bars of 25 mm diameter.

## 35.13. Design of Beams and Slabs

We have already discussed in the previous articles, that a given section may be under-reinforced, balanced or over-reinforced. We have also discussed that in an under reinforced section, the steel will first attain the maximum permissible stress, whereas the concrete will not be subjected to the full compressive stress. Similarly, in an over-reinforced section, the concrete will first attain the maximum permissible stress, whereas the steel will not be subjected to the full tensile stress. But in a balanced section, the stresses in concrete as well as steel are developed to the maximum permissible limit. It is thus obvious that the beams and slabs should be designed as balanced sections, so that the section is not wasteful. Such designs are also called *economical design.

It has been experimentally found that the maximum permissible stresses in steel and concrete depend upon the type and ratio of cement aggregates, water etc. But in general, the following stresses are adopted for the design of beams and slabs:

Allowable stress in steel, $t$ = 140 N/mm$^2$

Allowable stress in concrete, $c$ = 5 N/mm$^2$

Modular ratio, $m$ = 18

Now consider a beam reinforced at its bottom and subjected to some loading.

Let $b$ = Width of the beam,

$d$ = Effective depth of the beam, and

$n$ = Depth of neutral axis from the top of the beam. (In a design, the depths of actual neutral axis and the critical neutral axis is the same).

We know that

$$\frac{t}{c} = \frac{m(d-n)}{n}$$

$$\frac{140}{5} = \frac{18(d-n)}{n} \quad \text{or} \quad \frac{14}{9} = \frac{d-n}{n}$$

or

$$14n = 9d - 9n \quad \text{or} \quad 23n = 9d$$

$$\therefore \quad n = \frac{9d}{23} = 0.39d \qquad \text{...}(i)$$

We also know that the moment,

$$M = \frac{bnc}{2}\left(d - \frac{n}{3}\right) = \frac{b \times 0.39d \times 5}{2}\left(d - \frac{0.39d}{3}\right)$$

$$= 0.85\ bd^2 \text{ N-mm} \qquad \text{...}(ii)$$

* The term 'economic' refers only to the economy of stresses and not necessarily to the financial economy.

**NOTES: 1.** Sometimes the allowable stresses in steel and concrete as well as modular ratio is different than those assumed above. In such cases, the value of $n$ and $M$ [as obtained in equations (*i*) and (*ii*)] will be different. In such cases, the problem should be solved from fundamentals.

**2.** The bottom reinforcement is given a concrete cover, in order to embed the bars. The general thickness of the cover in beams 40 mm to 60 mm, whereas in slabs it is 2 cm and 3 cm.

**3.** The slabs are designed for 1 metre width and the same design is carried for the whole slab.

**EXAMPLE 35.8.** *Design a reinforced concrete beam of rectangular section to resist a bending moment of 150 kN-m. Take t = 140 MPa, c = 5 MPa and m = 18.*

**SOLUTION.** Given: Moment ($M$) = 150 kN-m = $150 \times 10^6$ N-mm ; Stress in steel ($t$) = 140 MPa = 140 N/mm$^2$ ; Stress in concrete ($c$) = 5 MPa = 5 N/mm$^2$ and modular ratio ($m$) = 18.

***Design of cross-section***

Let $b$ = Width of the beam in mm and

$d$ = Effective depth of the beam in mm.

First of all, let us assume width of the beam ($b$) equal to $0.6d$. We know that moment of resistance (M.R.),

$$150 \times 10^6 = 0.85\, bd^2 = 0.85 \times 0.6d \times d^2 = 0.51\, d^3$$

$$\therefore \quad d^3 = \frac{150 \times 10^6}{0.51} = 294.1 \times 10^6 \quad \text{or} \quad d = 665 \text{ mm}$$

Assuming a cover of 55 mm, the overall depth of the beam,

$= 665 + 55 =$ **720 mm** **Ans.**

and width of the beam, $b = 0.6d = 0.6 \times 665 = 399$ mm say 400 mm. **Ans.**

***Design of reinforcement***

Let $A_t$ = Area of reinforcement in mm$^2$,

Now let us assume depth of the neutral axis to be equal to $0.39d = 0.39 \times 665 = 259$ say 260 mm. We know that moment ($M$),

$$150 \times 10^6 = t \cdot A_t \left( d - \frac{n}{3} \right) = 140 \times A_t \left( 665 - \frac{260}{3} \right) = 80\,970\, A_t$$

Let us provide 4 bars of 25 mm diameter. Therefore area of reinforcement

$$= 4 \times \frac{\pi}{4} \times (25)^2 = \mathbf{1964 \text{ mm}^2} \quad \textbf{Ans.}$$

**EXAMPLE 35.9.** *Design a reinforced concrete slab 160 mm thick over a span of 4 m subjected to a load of 6 kN/m inclusive its own weight. The allowable stress in concrete is 4 MPa and modular ratio is 15. Also determine the stress in steel.*

**SOLUTION.** Given: Slab thickness 160 mm ; Span ($l$) = 4 m = $4 \times 10^3$ mm ; Load ($w$) = 6 kN/m = 6 N/mm ; Stress in concrete ($c$) = 4 MPa = 4 N/mm$^2$ and modular ratio ($m$) = 15.

***Stress in steel***

Let $t$ = Stress in steel in N/mm$^2$ and

$n$ = Depth of neutral axis.

Since width of the slab is not given, therefore we shall design it for 1 m ($1 \times 10^3$ mm) width. And the same reinforcement will be provided for the whole width. Assuming a cover of 30 mm over the reinforcement, we find that effective depth of the beam

$$d = 160 - 30 = 130 \text{ mm}$$

We know that bending moment due to uniformly distributed load,

$$M = \frac{wl^2}{8} = \frac{6 \times (4 \times 10^3)^2}{8} = 12 \times 10^6 \text{ N-mm}$$

We also know that moment of resistance (M.R.)

$$12 \times 10^6 = \frac{bnc}{2}\left(d - \frac{n}{3}\right) = \frac{(1 \times 10^3)\, n \times 4}{2}\left(d - \frac{n}{3}\right) = 2000\, n\left(130 - \frac{n}{3}\right)$$

$$\therefore \quad 6000 = 130n - \frac{n^2}{3}$$

$$n^2 - 390n + 18\,000 = 0$$

This is a quadratic equation for *n*. Therefore

$$n = \frac{+390 \pm \sqrt{(390)^2 - 4 \times 18\,000}}{2} \text{ mm}$$

$$= \frac{+390 \pm 283}{2} = 336.5 \text{ mm} \quad \text{or} \quad 53.5 \text{ mm}$$

Since the value of '*n*' can not be more than the depth of the slab (160 mm), therefore we shall take the value of '*n*' as 53.3 mm.

We know that relation for stress in steel,

$$\frac{t}{c} = \frac{m\,(d-n)}{n} \quad \text{or} \quad \frac{t}{4} = \frac{15\,(130 - 53.3)}{53.3} = 21.6$$

$$\therefore \quad t = 21.6 \times 4 = 86.4 \text{ N/mm}^2 = \mathbf{86.4 \text{ MPa}} \quad \textbf{Ans.}$$

***Design of reinforcement***

Let $A_t$ = Area of reinforcement in mm$^2$

We know that moment of resistance (*M*),

$$12 \times 10^6 = t \cdot A_t\left(d - \frac{n}{3}\right) = 86.4 \times A_t\left(130 - \frac{53.3}{3}\right) = 9697\, A_t$$

$$\therefore \quad A_t = \frac{12 \times 10^6}{9697} = 1237 \text{ mm}^2$$

Let us provide 12 mm diameter bars. We know that area of 1 bar of 12 mm diameter

$$= \frac{\pi}{4} \times (12)^2 = 113.1 \text{ mm}^2$$

∴ Spacing of 12 mm diameter bars

$$= \frac{113.1 \times 1000}{1237} = 91.4 \text{ mm say } 90 \text{ mm}$$

Let us provide 12 mm diameter bars 90 mm centre to centre. **Ans.**

Truck used for carrying premixed concrete

## EXERCISE 35.2

1. A rectangular concrete beam 200 mm wide has effective depth of 400 mm. The limiting stress in steel and concrete are 150 MPa and 50 MPa respectively. Find the moment of resistance of the beam, if both the sections are fully stress. Take m = 18. **[Ans.** 26.25 kN-m]
2. A concrete beam 200 mm wide and effective depth of 350 mm is subjected to a bending moment of 240 kN-m. If the permissible stress in steel and concrete are 140 MPa and 50 MPa respectively, find the amount of reinforcement required. **[Ans.** 77.7 $mm^2$]
3. A reinforced concrete beam of 250 mm width, 550 mm effective depth and 600 mm overall depth, has 3 bars of 22 mm diameter as reinforcement. Determine the moment of resistance for the beam, if the maximum allowable stresses in concrete and steel are 42 MPa and 112 MPa. Take m = 15. If the beam has a span of 5 metres, determine the safe load, which the beam can carry. Take density of concrete as 23 kN/$m^3$. **[Ans.** 53.78 kN-m ; 13.76 kN/m]

## QUESTIONS

1. Discuss the advantages of R.C.C. structures over other structures.
2. Explain clearly the various assumptions made in the theory of R.C.C.
3. Distinguish between the terms 'critical neutral axis' and 'actual neutral axis' and discuss, how the two axes help in finding out the moment of resistance of a beam.
4. Explain the terms under-reinforced beam', 'balanced beam' and 'over-reinforced beam'. How will you find out the types of beam?
5. Explain the procedure for the design of slabs.

## OBJECTIVE TYPE QUESTIONS

1. In a simply reinforced beam, the reinfocement is provided in
   (*a*) tensile zone (*b*) compressive zone
   (*c*) neutral zone (*d*) any where
2. In a single reinforced beam, the depth of lever arm is
   (*a*) $\dfrac{d-n}{3}$ (*b*) $\dfrac{2d-n}{3}$ (*c*) $\dfrac{3d-n}{3}$ (*d*) $\dfrac{4d-n}{3}$
3. If the actual neutral axis of a reinforced section is above the critical section, then the section is
   (*a*) under reinforced (*b*) over reinforced
   (*c*) balanced (*d*) none of these
4. For a balanced reinforced section, the depth of neutral axis (*n*) is given by the relation
   (*a*) $\dfrac{mc}{t}=\dfrac{n}{d-n}$ (*b*) $\dfrac{mc}{t}=\dfrac{d-n}{n}$ (*c*) $\dfrac{mc}{t}=\dfrac{n}{d+n}$ (*d*) $\dfrac{mc}{t}=\dfrac{d+n}{n}$

## ANSWERS

**1.** (*a*) **2.** (*c*) **3.** (*a*) **4.** (*a*)

Chapter

# 36

# Mechanical Properties of Materials

### Contents

## 36.1. Introduction

In modern design offices, a special care is taken at the time of designing a structure that it should be able to withstand the stresses, under the various load conditions, without failure. For doing so, it is very essential to have a complete information about the various properties of the selected material and its dimensions. This information can be obtained by experimental investigations in a well-equipped material testing laboratory.

## 36.2. Classification of Materials

In general all the materials, used by the engineers, may be classified on the basis of their physical properties into the following four types:

1. Elastic materials,
2. Plastic materials,
3. Ductile materials, and
4. Brittle materials

## 36.3. Elastic Materials

We have already discussed that whenever some external system of forces acts on a body, it undergoes some deformation. The deformation increases as the forces increase. If a material regains its original position, on the removal of the external forces, it is called an *elastic material.*

## 36.4. Plastic Materials

If a material does not regain its original position, on the removal of the external forces, it is called a *plastic material.*

## 36.5. Ductile Materials

If a material cannot undergo a considerable deformation, without rupture (*e.g.* if a material can be drawn into wires), it is called a *ductile material.*

## 36.6. Brittle Material

If a material cannot undergo any deformation (like glass, china-ware, etc.) when some external forces act on it and it fails by rupture, it is called a *brittle material.*

## 36.7. Classification of Tests

The various tests carried out, in a material testing laboratory, to know the physical properties of a material, may be grouped under the following three headings :–

**1.** *Types of stress induced*

The tests are carried out by inducing the various types of stresses (*i.e.*, tensile stress, compressive stress, shear stress, torsional stress, etc.)

**2.** *The rate, at which, the stresses are induced*

The tests are also carried by inducing the stresses at different rates. In general, the stresses induced under this heading are gradual and sudden (*i.e.* impact tests).

**3.** *The number of times, the stresses are induced*

The tests are also carried out by inducing the stresses once, and a number of times.

## 36.8. Actual Tests for the Mechanical Properties of Materials

Though there are many tests, which are carried out to know the strength of material, yet the following tests are important from the subject point of view:

1. Tensile test
2. Compressive test
3. Impact test
4. Fatigue test.

## 36.9. Tensile Test of a Mild Steel Specimen

We have studied in Chapter 2, that whenever some external system of forces acts on a body, it undergoes some deformation. If a body is stressed within its elastic limit, the deformation entirely disappears as soon as the forces are removed. It has been also found that beyond the elastic limit, the deformation does not disappear entirely, even after the removal of the forces and there remains some residual deformation. We study this phenomenon, in a greater detail, by referring to a tensile test (or stress-strain diagram) for a mild steel bar.

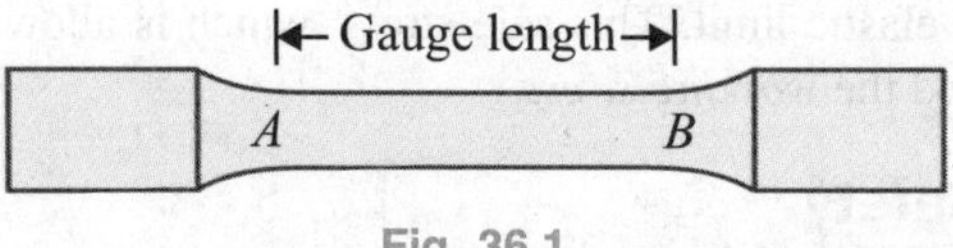


Fig. 36.1

Take a specimen of mild steel bar of uniform section as shown in Fig. 36.1. Let this bar be subjected to a gradually increasing pull (as applied by universal testing machine). If we plot the stresses along the vertical axis, and the corresponding strains along the horizontal axis and draw a curve passing through the vicinity of all such points, we shall obtain a graph as shown in Fig. 36.2.

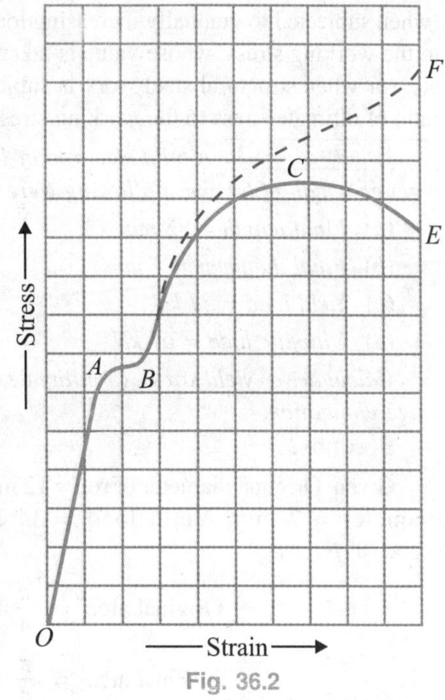


Fig. 36.2

We see, from the graph, that the curve from *O* to *A* is a straight line, which represents that the stress is proportional to strain. Beyond *A*, the curve slightly deviates from the straight. It is thus obvious, that the Hooke's law holds good only up to this limit. Thus the point *A* is the *elastic limit* of the specimen metal. When the specimen is stressed beyond this limit, the strain increases more quickly than the stress (in comparison from *O* to *A*). This happens, because a sudden elongation of the specimen takes place, without an appreciable increase in the stress (or load). The stress, corresponding to the point *B* is called the *yield point*.* It may be noted, that if the load on the specimen is removed, then the elongation from *A* to *B* will not disappear. But will remain as a permanent set. It is thus obvious, that the Hooke's law does not hold good, beyond the point *A* (*i.e.,* elastic limit)

At *B*, the specimen regains some strength and higher values of stresses are required, for higher strains, than those between *A* and *B*. The stress (or load) goes on increasing, till the point *C* is reached. The gradual increase in the length of the specimen is followed with the uniform reduction of its cross-sectional area. The work done, during stretching the specimen, is transformed largely into heat and the specimen becomes hot. At *C*, the stress, which attains its maximum value, is known as *ultimate stress.*

After the specimen has reached the ultimate stress, a neck is formed, which decreases the cross-sectional area of the specimen. A little consideration will show, that the stress (or load) necessary, to break away the specimen, is less than the ultimate stress (or maximum load). The stress is, therefore, reduced until the specimen breaks away at the stress represented by the point *D*. The stress, corresponding to the point *D*, is known as the *breaking stress.***

* At this point, the elongation of a mild steel specimen is about 2%.

** The breaking stress (*i.e.,* stress at *E* which is less than that at *D*, appears to be somewhat misleading. As the formation of a neek takes place at *D*, which reduces the cross-sectional area. It causes the specimen suddenly to fail at *E*. If for each value of the strain between *C* and *E* the tensile load is divided by the reduced cross-sectional area at the narrowest part of the neck, then the true stress-strain curve will follow the dotted line *C.F.* However, it is an established practice, to calculate strains on the basis of original cross-sectional area of the specimen.

## 36.10. Working Stress

The stress-strain curve gives a valuable information about the mechanical properties of a metal. By knowing the elastic limit, yield point and ultimate stress in a material, it is very easy to establish the magnitude of stress, which can be safely undertaken by it. This is, generally, done by keeping the safe stress well below the elastic limit. This safe stress, which is allowed to be undertaken by the material in designs, is called the *working stress.*

## 36.11. Factor of Safety

The ratio of elastic limit to the working stress (or ultimate stress to the working stress) is called the *factor of safety.* In the recent days, the general practice followed is, that for structural steel work (when subjected to gradually increasing loads) the factor of safety is taken as the ratio of elastic limit to the working stress; whose value is taken as 2 to 2.5. But in the case of cast iron, concrete, wood, etc. (or when structural steel work is subjected to sudden loads) the factor of safety is taken as the ratio of ultimate stress to the working stress, whose value is taken as 4 to 6.

**EXAMPLE 36.1.** *A mild steel rod of 12 mm diameter was tested for tensile strength, with the gauge length of 60 mm. Following were the observations :*

(*a*) *Final length = 78 mm*

(*b*) *Final diameter = 7 mm*

(*c*) *Yield load = 34 kN*

(*d*) *Ultimate load = 61 kN*

*Calculate* (*a*) *yield stress,* (*b*) *ultimate tensile stress,* (*c*) *percentage reduction, and* (*d*) *percentage elongation.*

**SOLUTION.**

Given. Original diameter of rod = 12 mm; Original length = 60 mm; Final length = 78 mm; Final diameter = 7 mm; Yield load = 34 kN = $3.4 \times 10^4$ N and ultimate load = 61 kN = $6.1 \times 10^4$ N.

$$\text{Original area} = \frac{\pi}{4} \times (12)^2 = 113 \text{ mm}^2$$

$$\text{Final area} = \frac{\pi}{4} \times (7)^2 = 38.5$$

***Yield stress***

We know that the yield stress

$$= \frac{\text{Yield load}}{\text{Area}} = \frac{3.4 \times 10^4}{113} \text{ N/mm}^2$$

$$= 300.8 \text{ N/mm}^2 \textbf{ Ans.}$$

***Ultimate tensile stress***

We know that the ultimate tensile stress

$$= \frac{\text{Ultimate load}}{\text{Area}} = \frac{6.1 \times 10^4}{113} \text{ N/mm}^2$$

$$= 539.8 \text{ N/mm}^2 \textbf{ Ans.}$$

***Percentage reduction***

We know that the percentage reduction

$$= \frac{\text{Original area} - \text{Final area}}{\text{Original area}} \times 100$$

$$= \frac{113 - 38.5}{113} \times 100 = 65.9\% \textbf{ Ans.}$$

*Percentage elongation*

We also know that the percentage elongation

$$= \frac{\text{Final length} - \text{Original length}}{\text{Final length}} \times 100$$

$$= \frac{78 - 60}{78} \times 100 = 23\% \textbf{ Ans.}$$

## 36.12. Barba's Law and Unwin's Formula

Prof. Barba, after carrying out a series of experiments on tensile testing of a number of specimens, found that the percentage elongation gives a measure of ductility of the metal. Moreover, the amount of local extension depends upon the material and also on the transverse dimensions of the test pieces. He observed that as the test pieces are made from bars, strips, sheets, wires, forgings, castings etc., therefore some standard means of comparison of results are necessary. He enunciated a law which states, "*In tension, similar test pieces deforms similarly.*" He further stated that two test pieces are said to be similar if they have equal value of :

$$\frac{\text{Gauge length}}{\sqrt{\text{Cross-sectional area}}}$$

Prof. Unwin, as a result of the experiments conducted by Barba, gave an expression for total extension,

$$\delta l = bl + c\sqrt{A}$$

or

$$\frac{\delta l}{l} = \left(b + \frac{c\sqrt{A}}{l}\right)$$

$\therefore$

$$\frac{\delta l}{l} \times 100 = \left(b + \frac{cA}{l}\right) \times 100$$

where $\delta l$ = Total extension of the gauge length,

$l$ = Gauge length,

$A$ = Original area of specimen cross-section, and

$b$ and $c$ = Constants depending upon the quality of the specimen material.

**Example 36.2.** *Two specimens A and B of the same material were tested in a laboratory and the results are shown in the table below :*

| *Specimen* | *Gauge length in mm* | *Thickness in mm* | *Width in mm* | *Elongation (%)* |
|---|---|---|---|---|
| *A* | *200* | *8* | *40* | *30* |
| *B* | *250* | *10* | *60* | *32* |

*Estimate the percentage elongations of a third specimen C of the same material having a length of 150 mm and diameter 20 mm.*

**Solution.**

*Specimen A*

Length, $l$ = 200 mm

Area, $A = 8 \times 40 = 320 \text{ mm}^2$.

Elongation $\frac{\delta l}{l} = 30\% = 0.3$

*Specimen B*

Length, $l = 250 \text{ mm}$

Area, $A = 10 \times 60 = 600 \text{ mm}^2$

Elongation $\frac{\delta l}{l} = 32\% = 0.32$

*Specimen C*

Length, $l = 150 \text{ mm}$

Area, $A = \frac{\pi}{4} \times 20^2 = 100\,\pi \text{ mm}^2$

Using Unwin's formula for specimens *A* and *B*

$$\frac{\delta l}{l} = b + \frac{c\sqrt{A}}{l}$$

$$0.3 = b + \frac{c\sqrt{320}}{200} \qquad \text{...(or } A\text{)}$$

$$0.3 = b + 0.089\,c \qquad \text{...}(i)$$

and

$$0.32 = b + \frac{c\sqrt{600}}{250} \qquad \text{...(for } B\text{)}$$

$$0.32 = b + 0.098\,c \qquad \text{...}(ii)$$

Subtracting equation (*i*) from (*ii*).

$$0.02 = 0.009\,c$$

$$\therefore \quad c = \frac{0.02}{0.009} = 2.22$$

Now substituting the value of *c* in equation (*i*),

$$0.3 = b + 0.089 \times 2.22 = b + 0.198$$

$$\therefore \quad b = 0.3 - 0.198 = 0.102$$

Again using Unwin's formula for specimen *C*,

$$\frac{\delta l}{l} = b + \frac{c\sqrt{A}}{l}$$

$$= 0.102 + \frac{2.22\sqrt{100\pi}}{150} = 0.364$$

$$= 36.4\% \quad \textbf{Ans.}$$

## 36.13. Compression Test

It is merely the opposite of the tension test. The compression test is generally performed for testing the brittle materials such as cast iron, concrete, stone etc. The specimens used in this test are usually made of cubical or cylindrical shape. It has been observed that some errors are always introduced in this test due to the following practical difficulties :

1. Since the top and bottom faces of the specimen are seldom absolutely parallel, therefore it is very difficult to ensure axial loading on the specimen.

2. Since the length of the specimen is kept short enough (not more than twice of its diameter) to avoid its buckling, therefore within the elastic limit a small compression takes place, which is difficult to measure accurately.
3. The friction between the ends of the specimen and clutches of the machine, prevents the dimensions of the specimen ends from increasing. This results in the lateral expansion to take place more in the centre, instead of uniform increase in diameter throughout the whole length. Such an effect, which is called barrel effect, is not a case of an ideal compression.

Now we shall discuss the compression test for ductile materials and brittle materials.

1. ***Compression test for ductile materials (i.e., mild steel, copper etc.)***

Consider a ductile material specimen (say a mild steel bar) of uniform section, subjected to a gradually increasing push. If we plot the stresses along the vertical axis, and the corresponding strains along the horizontal axis and draw a curve passing through the vicinity of all such points, we shall obtain a graph as shown in Fig. 36.3.

We see, from the graph, that the curve from *O* to *A* is a straight line, which represents that the stress is proportional to strain. Up to elastic limit, all the metals have approximately the same modulus of elasticity in compression, as that in tension. If the specimen is stressed beyond the elastic limit, the strain increases more quickly than the stress (in comparison from *O* to *A*). This happens, because the sudden contraction of the specimen takes place, without an appreciable increase in the stress. At *B*, the specimen regains some strength and, higher values of stresses are required for higher strains, than those between *A* and *B*. Beyond *B*, the curve continues, almost without any limit, as there is no failure of the material on account of its ductility and the cross-sectional area goes on increasing continuously, with the increase of load. The specimen will shorten and bulge out.

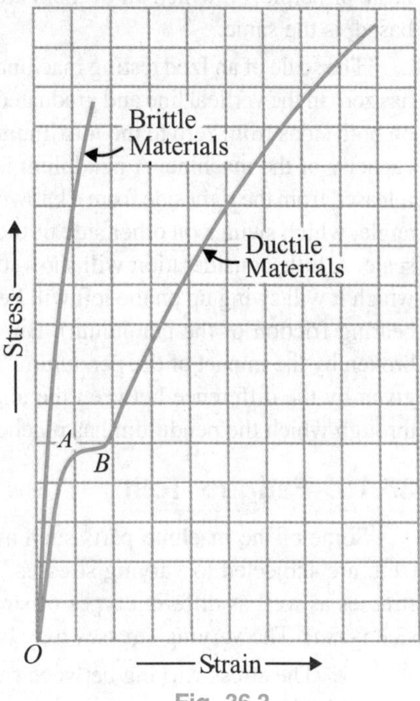

Fig. 36.3

2. ***Compression test for brittle materials (i.e., cast iron, concrete, etc.)***

If we perform the compression test with a specimen of some brittle material (say a cast iron bar) and draw a curve with stresses along vertical axis and the corresponding strains along horizontal axis, we shall obtain a graph as shown is Fig. 36.3. We see, from the graph, that there is a little strain as compared to stress and there is always a point, where the specimen will fail due to shear along a diagonal plane.

The compression test is mainly performed for testing brittle materials only.

## 36.14. Impact Test

Many machines or machine components are subjected to a suddenly applied load, which is called impact blow. For determining the suitability of a material to resist the impacts, Izod and Charpy tests are generally carried out. The Izod test is conducted as discussed below.

The standard test piece, for an Izod test, may be square or round as shown in Fig. 36.4 (*a*) and (*b*).

This test enables us to estimate the property of toughness of a material. The toughness of a material may be defined as the energy absorbed in the failure of its specimen. The energy absorbed by a specimen, during its failure, is the product of its deformation and the average stress. A little consideration will show, that a material, which can be highly stressed and greatly deformed, will be able to with-stand a high blow ; and thus may be said to be more tough.

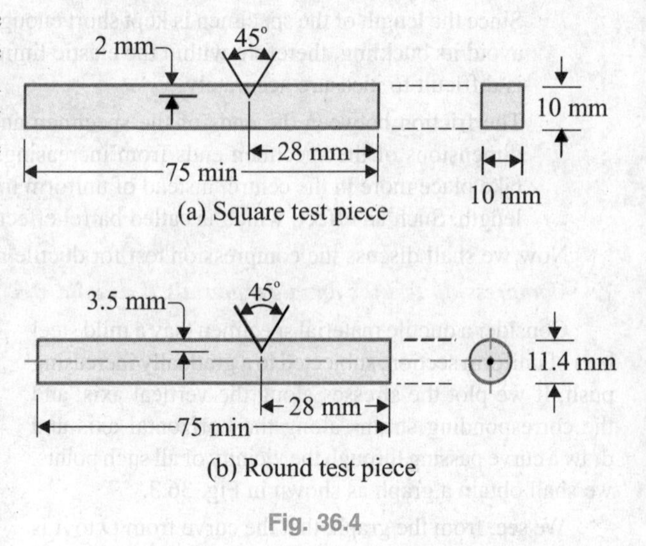

(a) Square test piece

(b) Round test piece

**Fig. 36.4**

There are many types of Izod testing machines available in the market. But the basic principle, on which all of them are based, is the same.

The scale of an Izod testing machine has zero in the vertical line and graduated on both sides from zero to the maximum capacity of the machine. A pendulum is released from the right side from a known angle, which swings on other side of the scale. A little consideration will show, that if the pendulum is not interrupted anywhere, the angle to which it will swing up on the left will be equal to the angle from which it was released (neglecting the bearing friction of the pendulum). But the specimen is kept in the path of the pendulum, which is broken by the impact of the pendulum. The energy absorbed by the specimen, during the breaking, is given by the difference between the angle through which the pendulum was released and the angle through which the pendulum has reached after breaking away the specimen.

## 36.15. Fatigue Test

Some of the machine parts such as axles, shafts, crankshafts, connecting rods, springs, pinion teeth are subjected to varying stresses. It includes the variation in the intensity of the same type of stresses as well as different types of stresses (*i.e.,* change of stress, from tensile to compressive and *vice versa*). The varying stresses may be broadly classified into the following four types :

1. The stress varying between two limits of equal value, but of opposite sign.
2. The stress varying between two limits of unequal values, but of opposite sign.
3. The stress varying between zero and a definite value.
4. The stress varying between two limits of equal values, but of opposite sign.

Though there are numerous ways of fatigue test in a laboratory, yet the basic principle is the same. The most common test is a rotating bar test of cantilever type and beam type. In this test, a number of identical test piece bars (say 8 to 10) are made from a material. One of the piece is fixed to the machine. A load is gradually applied to the test piece. (This load should not be less than that, which can produce a stress equal to 3/4 of the tensile strength of the material up to the elastic limit). The rotation of the electric motor will sometimes push the test piece downwards and sometimes pull upwards, which will reverse the stresses in the upper and lower fibres of the test piece. The speed of the motor will indicate the frequency of the stress reversal.

The speed of the motor shaft is kept constant. It has been experimentally found, that after a sufficient number of stress reversals, a crack in the form of a ring is formed on the outer surface. This crack goes on extending towards the centre of the test piece, till it breaks away. It has also been observed that the speed of the motor (or in other words the frequency of the stress reversal) has no effect on the result. But it is the load or the intensity of the maximum stress, which controls the result.

After the first test piece breaks away, the second piece is tested with a decreased load. The third, fourth, fifth etc. pieces are then tested with decreased loads. It may be noted that the number of reversals, each time, goes on increasing with the decrease in load. After some tests, a limit is reached, when the stress is not sufficient to break the test piece even after $10 \times 10^6$ stress reversals (in some countries, this limit is $20 \times 10^6$). This safe stress, which after reversing for $10 \times 10^6$ time does not cause the test piece to break, is called *endurance limit.*

All the ferrous metals have endurance limit between 0.4 to 0.6 of the tensile strength up to the elastic limit.

## QUESTIONS

1. Describe the necessity of mechanical tests of a material.
2. Give the classifications of materials and distinguish between elastic materials and plastic materials.
3. Describe the tensile test of a mild steel specimen.
4. Write short notes on :
   (*a*) Stress-strain figure.
   (*b*) Yield stress.
   (*c*) Ultimate stress.
   (*d*) Neck of the specimen.
   (*e*) Working stress.
5. Draw the graph for ductile and brittle materials when tested for compression in a laboratory.
6. What is the importance of impact tests. Describe the Izod test.
7. Describe the fatigue test of a material.
8. Write a short note on endurance limit.

# APPENDIX

**Angle of Repose** is the natural slope of the materials which they tend to take up if not acted upon by any external force.

The following table gives the characteristics of most common retained material.

**TABLE 1**

| S.No. | Material | Weight (kN/m$^3$) | Angle of repose ($\phi$) |
|---|---|---|---|
| 1. | Sand (dry) | 14.4 to 16 | 30° |
| | (moist) | 15 to 17.6 | 35° |
| | (wet) | 17.6 to 20 | 25° |
| 2. | Clay (dry) | 19.2 to 22.4 | 30° |
| | (moist) | 12.2 to 25.6 | 45° |
| | (wet) | 19.2 to 25.6 | 15° |
| 3. | Gravel and sand | 16 to 17.6 | 25° to 30° |
| 4. | Gravel | 14.4 | 40° |
| 5. | Ashes | 6.4 | 40° |
| 6. | Mud | 16.8 to 25.6 | 0° |

**NOTE :** These values can be used while Designing of Retaining Walls by the method of Rankine's Theory or Coulomb's Wedge Theory.

**TABLE 2** Slopes and Deflections for Different Loadings on Cantilevers and Beams

| S. No. | Type of Loading | Maximum Bending Moment | Slope | Maximum Deflection |
|---|---|---|---|---|
| 1 | 2 | 3 | 4 | 5 |
| 1. | W, A, B, $i_B$, $y_{max}$, $l$ | $M_A = -Wl$ | $i_B = -\dfrac{Wl^2}{2\,EI}$ | $y_{max} = y_B = -\dfrac{Wl^3}{3\,EI}$ |
| 2. | W, $l_1$, A, B, $i_B$, $y_{max}$, $l$ | $M_A = -Wl_1$ | $i_B = -\dfrac{Wl_1^{\,2}}{2\,EI}$ | $y_{max} = y_B = -\dfrac{Wl_1^{\,2}}{6\,EI}(3l - l_1)$ |
| 3. | w / unit length, A, B, $i_B$, $y_{max}$, $l$ | $M_A = -\dfrac{wl^2}{2}$ | $i_B = -\dfrac{wl^3}{6\,EI}$ | $y_{max} = y_B = -\dfrac{wl^4}{8\,EI}$ |
| 4. | w / unit length, A, C, B, $i_B = i_C$, $l_1$, $y_{max}$, $l$ | $M_A = -\dfrac{wl_1^{\,2}}{2}$ | $i_B = i_C = -\dfrac{wl_1^{\,3}}{6\,EI}$ | $y_{max} = y_B = -\left[\dfrac{wl_1^{\,4}}{8EI} + \dfrac{wl_1^{\,3}}{6EI}(l - l_1)\right]$ |

| S. No. | Type of Loading | Maximum Bending Moment | Slope | Maximum Deflection |
|---|---|---|---|---|
| 1 | 2 | 3 | 4 | 5 |
| 5. | w / unit length; $(l-l_1)$; $l_1$; A; B; $i_B$; $y_{max}$; $l$ | $M_A = -w\,l_1\left[\frac{l_1}{2} + (l-l_1)\right]$ | $i_B = \left[\frac{wl^3}{6EI}\right] - \left[\frac{w(l-l_1)^3}{6EI}\right]$ | $y_{max} = y_B = \left[\frac{wl^4}{8EI}\right] - \left[\frac{w(l-l_1)^4}{8EI} + \frac{w(l-l_1)^3 l_1}{6EI}\right]$ |
| 6. | A; B; M; $i_B$; $y_{max}$; $l$ | $M_A = -M$ | $i_B = -\frac{Ml}{EI}$ | $y_{max} = y_B = -\frac{Ml^2}{2\,EI}$ |
| 7. | w / unit length; A; B; $i_B$; $y_{max}$; $l$ | $M_A = -\frac{wl^2}{6}$ | $i_B = -\frac{wl^3}{24\,EI}$ | $y_{max} = y_B = -\frac{wl^4}{30\,EI}$ |
| 8. | W; $\frac{l}{2}$; $\frac{l}{2}$; A; C; B; $i_A$; $i_B$; $y_{max}$; $l$ | $M_C = \frac{Wl}{4}$ | $i_A = -\frac{Wl^2}{16\,EI}$<br>$i_B = +\frac{Wl^2}{16\,EI}$ | $y_{max} = y_C = -\frac{Wl^3}{48\,EI}$ |

| S. No. | Type of Loading | Maximum Bending Moment | Slope | Maximum Deflection |
|---|---|---|---|---|
| 1 | 2 | 3 | 4 | 5 |
| 9. | W; a; b; C; A; B; $i_A$; $i_B$; $y_{max}$; $y_C$; $l$ | $M_C = \frac{W\,ab}{l}$ | $i_A = -\frac{Wb\,(l^2-b^2)}{6\,EIl}$<br>$i_B = +\frac{Wa\,(l^2-a^2)}{6\,EIl}$ | $y_{max} = -\frac{Wb(l^2-b^2)^{3/2}}{9\sqrt{3}\,EIl}$<br>$\left(at\ x = \sqrt{(l^2-b^2)/3}\right)$<br>$y_C = \frac{Wab(l^2-a^2-b^2)}{6EIl}$<br>$= -\frac{Wa^2b^2}{3\,EIl}$ |
| 10. | w / unit length; C; A; B; $i_A$; $i_B$; $y_{max}$; $l$ | $M_{max} = \frac{wl^2}{8}$ | $i_A = -\frac{wl^3}{24\,EI}$<br>$i_B = +\frac{wl^3}{24\,EI}$ | $y_{max} = y_C = -\frac{5\,wl^4}{384\,EI}$ |
| 11. | w / unit length; C; A; B; $i_A$; $i_B$; $y_{max}$; $l$ | $M_{max} = \frac{wl^3}{9\sqrt{3}}$ | $i_A = -\frac{wl^3}{45\,EI}$<br>$i_B = +\frac{7\,wl^3}{360\,EI}$ | $y_{max} = -\frac{0{\cdot}00652\,wl^4}{EI}$<br>(at $x = 0{\cdot}519\ l$ from $B$)<br>$y_C = -\frac{0{\cdot}00651\,wl^4}{EI}$<br>(at $x = 0{\cdot}5\ l$ from $B$) |

| S. No. | Type of Loading | Maximum Bending Moment | Slope | Maximum Deflection |
|---|---|---|---|---|
| *1* | *2* | *3* | *4* | *5* |
| 12. |  | $M_{max} = \frac{wl^2}{12}$ | $i_A = -\frac{5\,wl^3}{192\,EI}$<br>$i_B = +\frac{5\,wl^3}{192\,EI}$ | $y_{max} = y_C = -\frac{wl^4}{120\,EI}$ |

*Sign conventions used :*

| | | |
|---|---|---|
| **Slope** : | Clockwise | – |
| | Counter-clockwise | + |
| **Deflection** : | Upward | + |
| | Downward | – |

TABLE 3 **Relation between the Actual beam and Conjugate beam**

| | *Actual beam* | *Conjugate beam* |
|---|---|---|
| 1. | (*a*) *Slope* at any section<br>(*b*) *Deflection* at any section<br>(*c*) *Loading* system | (*a*) *Shear force* at the corresponding section<br>(*b*) *Bending moment* at the corresponding section<br>(*c*) $\frac{M}{EI}$ *diagram* is the loading system |
| 2. | A<br>Fixed End |  A′<br>Free End |
| 3. |  A<br>Free End |  A′<br>Fixed End |
| 4. |  A<br>Hinged End | A′<br>Hinged End |
| 5. | A<br>Roller Support | A′<br>Roller Support |
| 6. |  A<br>Internal Hinge | 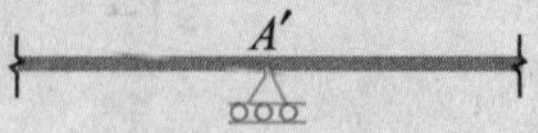 A′<br>Roller Support |
| 7. | 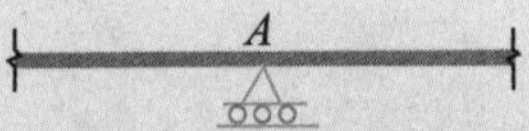 A<br>Roller Support | 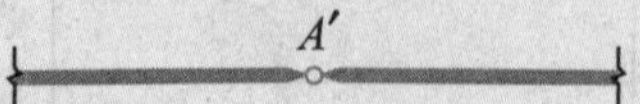 A′<br>Internal Hinge |
| 8. |  A<br>Internal Pin | 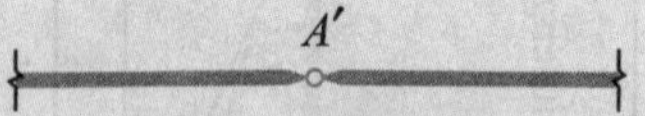 A′<br>Internal Hinge |

NOTE : This table will be used for Evaluating Deflection by Conjugate Beam Method.

## TABLE 4 Fixed Beam Loadings

| *S.No.* | *Type of Loading* | *Fixed End Moments* |
|---|---|---|
| 1. | $W$; $A$, $B$; $\frac{l}{2}$, $\frac{l}{2}$; $l$ | $M_A = M_B = -\frac{Wl}{8}$ |
| 2. | $W$; $A$, $B$; $a$, $b$; $l$ | $M_A = -\frac{Wab^2}{l^2}$ <br> $M_B = -\frac{Wa^2b}{l^2}$ |
| 3. | $M$; $A$, $B$; $a$; $l$ | $M_A = \frac{M}{l^2}(l-a)(l-3a)$ <br> $M_B = \frac{M}{l^2}a(2l-3a)$ |
| 4. | $w$/unit length; $A$, $B$; $l$ | $M_A = M_B = -\frac{wl^2}{12}$ |
| 5. | $w$/unit length; $A$, $B$; $a$; $l$ | $M_A = \frac{wa^2}{12l^2}(6l^2 - 8la + 3a^2)$ <br> $M_B = \frac{wa^3}{12l^2}(4l - 3a)$ |
| 6. | $w$/unit length; $A$, $B$; $l$ | $M_A = -\frac{wl^2}{30}$ <br> $M_B = -\frac{wl^2}{20}$ |
| 7. | $w$/unit length; $A$, $B$; $a$; $l$ | $M_A = -\frac{wa^2}{30l^2}(10l^2 + 6a^2 - 15la)$ <br> $M_B = -\frac{wa^3}{20l^2}(5l - 4a)$ |

| S.No. | Type of Loading | Fixed End Moments |
|---|---|---|
| 8. | w/ unit length; A; B; l/2; l/2 | $M_A = M_B = -\frac{5wl^2}{96}$ |
| 9. | A; B; δ; l | $M_A = -\frac{6EI\delta}{l^2}$<br>$M_B = \frac{6EI\delta}{l^2}$ |

# INDEX

## A

## B

## C



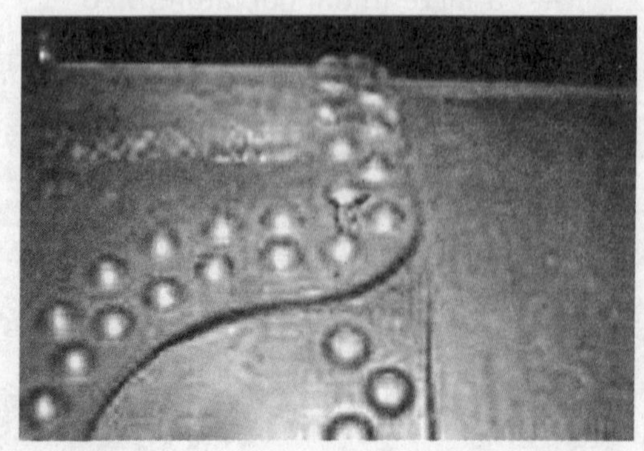

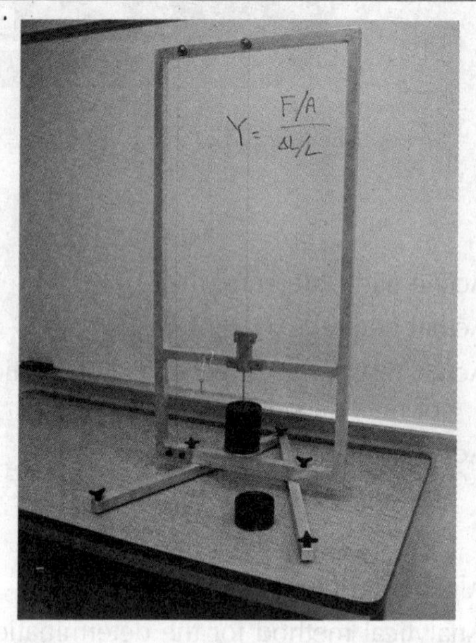

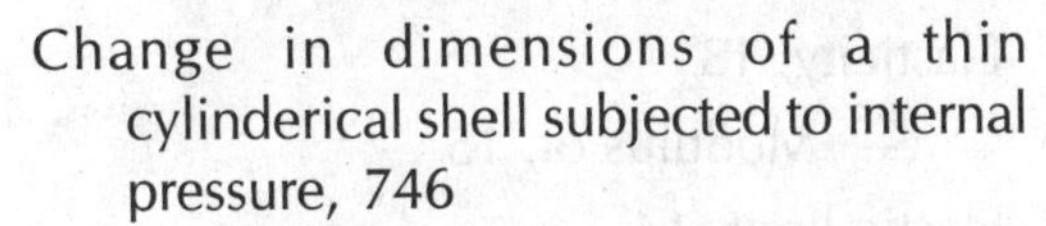

## D

## E



## F



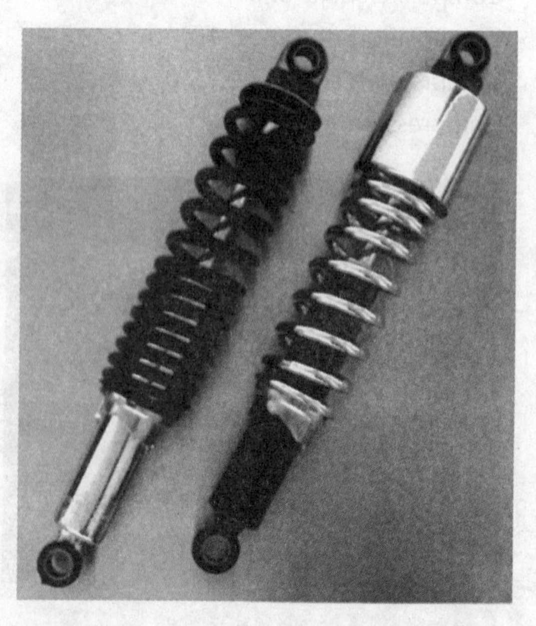

## G



## H



## I



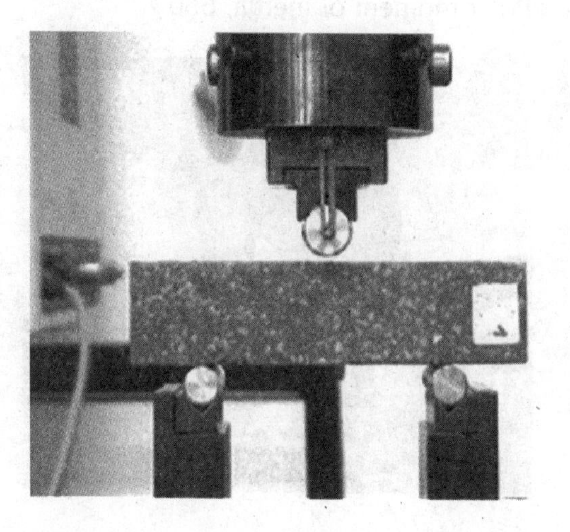

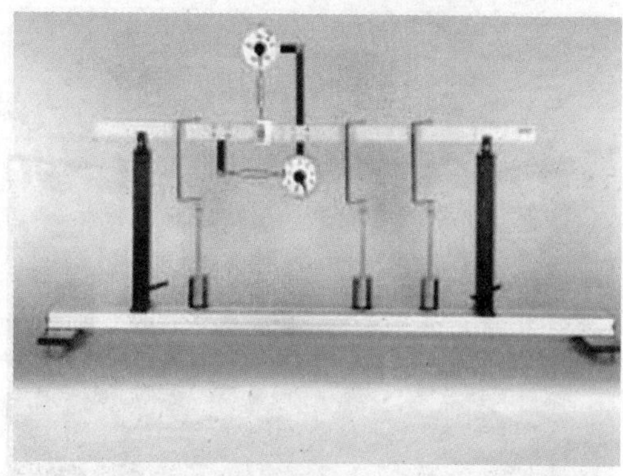



## J



## K



## L



## M

**N**



**O**



**P**

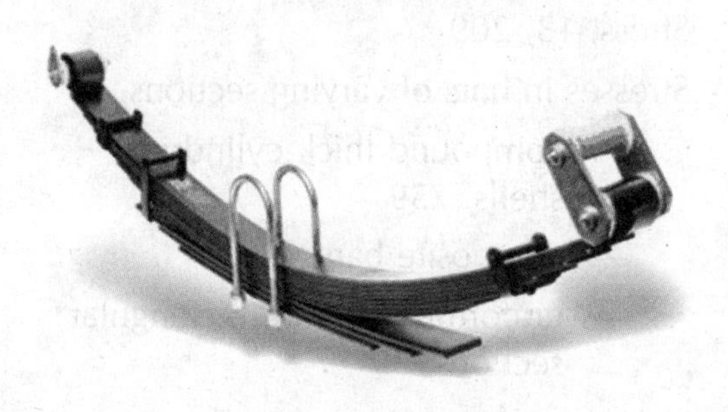



## R



## S

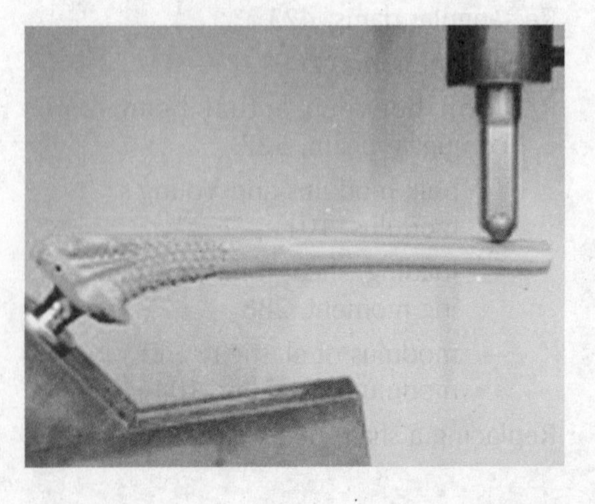

## T



## U



## V



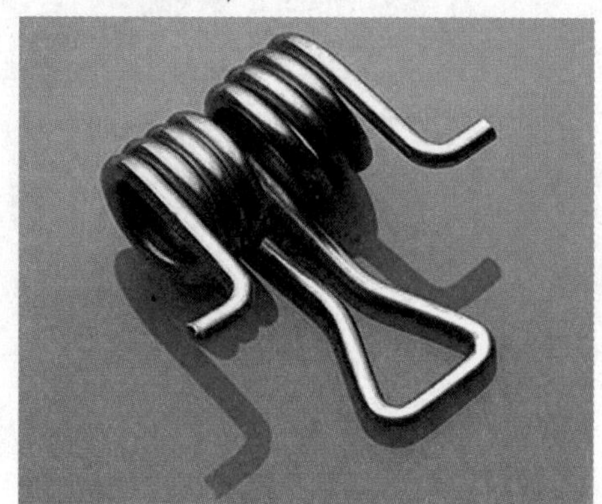

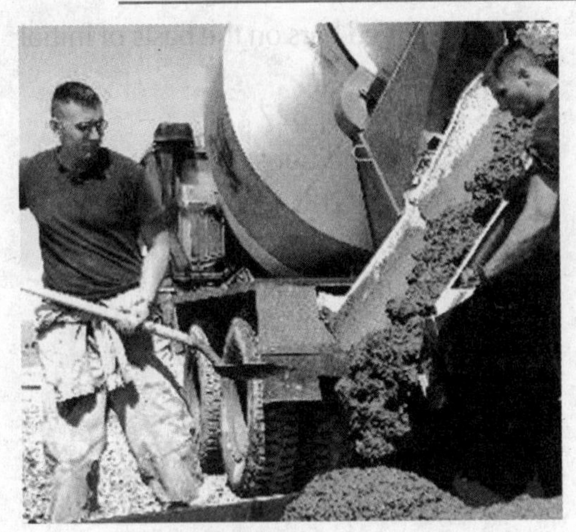



## W



## Y



## Z

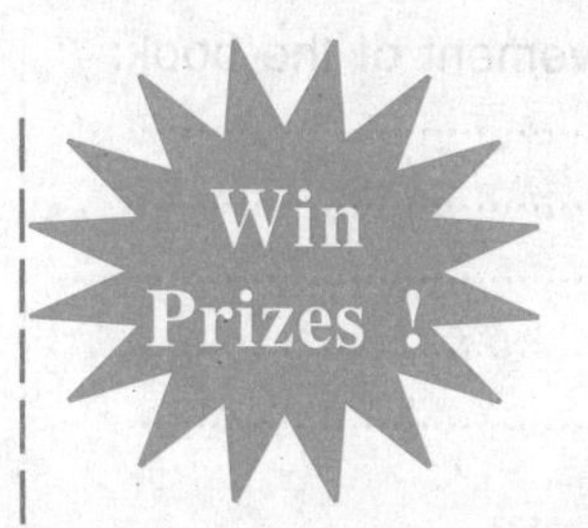


# Attention: Students

**We request you, for your frank assessment, regarding some of the aspects of the book, given as under:**

**10 320 Strength of Materials**
**R.S. Khurmi**

**Reprint 2008**

*Please fill up the given space in neat capital letters. Add additional sheet(s) if the space provided is not sufficient, and if so required.*

(i) What topic(s) of your syllabus that are important from your examination point of view are not covered in the book ?

.........................................................................................................

.........................................................................................................

.........................................................................................................

.........................................................................................................

(ii) What are the chapters and/or topics, wherein the treatment of the subject-matter is not systematic or organised or updated?

.........................................................................................................

.........................................................................................................

.........................................................................................................

.........................................................................................................

.........................................................................................................

(iii) Have you come across misprints/mistakes/factual inaccuracies in the book? Please specify the chapters, topics and the page numbers.

.........................................................................................................

.........................................................................................................

.........................................................................................................

.........................................................................................................

.........................................................................................................

(iv) Name top three books on the same subject (in order of your preference - 1, 2, 3) that you have found/heard better than the present book? Please specify in terms of quality (in all aspects).

1 .........................................................................................................

.........................................................................................................

2 .........................................................................................................

.........................................................................................................

3 .........................................................................................................

.........................................................................................................

(v) Further suggestions and comments for the improvement of the book:

..............................................................................................

..............................................................................................

..............................................................................................

..............................................................................................

..............................................................................................

## Other Details:

(i) Who recommended you the book? (Please tick in the box near the option relevant to you.)

☐ Teacher ☐ Friends ☐ Bookseller

(ii) Name of the recommending teacher, his designation and address:

..............................................................................................

..............................................................................................

..............................................................................................

(iii) Name and address of the bookseller you purchased the book from:

..............................................................................................

..............................................................................................

..............................................................................................

(iv) Name and address of your institution (Please mention the University or Board, as the case may be)

..............................................................................................

..............................................................................................

..............................................................................................

(v) Your name and complete postal address:

..............................................................................................

..............................................................................................

..............................................................................................

(vi) Write your preferences of our publications (1, 2, 3) you would like to have

..............................................................................................

..............................................................................................

**The best assessment will be awarded half-yearly. The award will be in the form of our publications, as decided by the Editorial Board, amounting to Rs. 300 (total).**

*Please mail the filled up coupon at your earliest to*:

Editorial Department
**S. CHAND & COMPANY LTD.,**
Post Box No. 5733, Ram Nagar,
New Delhi 110 055